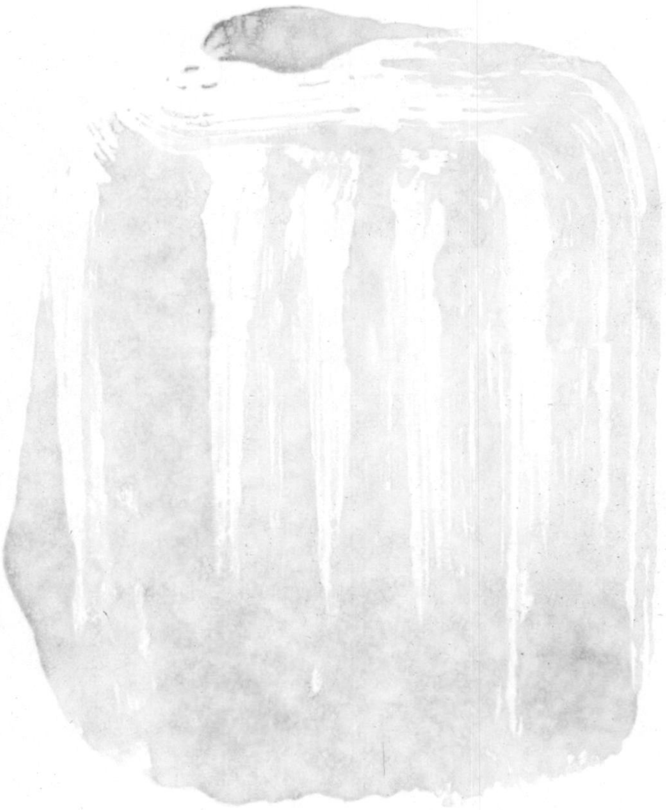

PROCEEDINGS OF THE SIXTH BERKELEY SYMPOSIUM

VOLUME III

PROCEEDINGS *of the* SIXTH BERKELEY SYMPOSIUM ON MATHEMATICAL STATISTICS AND PROBABILITY

Held at the Statistical Laboratory
University of California
June 21–July 18, 1970

with the support of
University of California
National Science Foundation
Air Force Office of Scientific Research
Army Research Office
Office of Naval Research

VOLUME III

PROBABILITY THEORY

EDITED BY LUCIEN M. LE CAM,
JERZY NEYMAN, AND ELIZABETH L. SCOTT

UNIVERSITY OF CALIFORNIA PRESS
BERKELEY AND LOS ANGELES
1972

UNIVERSITY OF CALIFORNIA PRESS
BERKELEY AND LOS ANGELES
CALIFORNIA

CAMBRIDGE UNIVERSITY PRESS
LONDON, ENGLAND

ISBN: 0-520-02185-1

LIBRARY OF CONGRESS CATALOG CARD NUMBER: 49-8189

PRINTED IN THE UNITED STATES OF AMERICA

CONTENTS OF PROCEEDINGS VOLUMES I, II, AND III

Volume I—Theory of Statistics

General Theory

Sequential Analysis

Asymptotic Theory

Nonparametric Procedures

Regression Analysis

Multivariate Analysis

Volume II—Probability Theory

Introduction

Measure Theory

Inequalities

Combinatorial Analysis

Ergodic Theory

Gaussian Processes

Central Limit Theorem

Volume III—Probability Theory

Passage Problems

Markov Processes—Potential Theory

Markov Processes—Trajectories—Functionals

Point Processes, Branching Processes

Information and Control

PREFACE

Berkeley Symposia on Mathematical Statistics and Probability have been held at five year intervals since 1945. The Sixth Berkeley Symposium was divided into four sessions. The first took place from June 21 to July 18, 1970. It covered mostly topics in statistical theory and in theoretical and applied probability. The second session was held from April 9 to April 12, 1971 on the special subject of evolution with emphasis on studies of evolution conducted at the molecular level. The third session held in June 1971 was devoted to problems of biology and health. A fourth session on pollution was held in July 1971.

The first three volumes of the Proceedings cover papers presented in June and July, 1970, as well as papers which were sent to us at that time, but could not be presented in person by their authors. The first volume is entirely devoted to statistics. The second and third are devoted to contributions in probability. Allocation of the papers to the three volumes was made in a manner which we hope is fairly rational, but with an unavoidable amount of arbitrariness and randomness. In the event of doubt, a general index should help the prospective reader locate the desired contribution.

The Berkeley Symposia differ substantially from most other scientific meetings in that they are intended to provide an extended period of contact between participants from all countries in the world. In addition, an effort is made to promote cross contacts between scholars whose fields of specialization cover a broad spectrum from pure probability to applied statistics. However, these fields have expanded so rapidly in the past decades that it is no longer possible to touch upon every domain in a few weeks only. Since time limits the number of invited lectures, the selection of speakers is becoming rapidly an impossible task. We could only sample the abundance of available talent. For this selection, as well as for several other important matters, we were privileged to have the assistance of an advisory committee consisting of Professors Z. W. Birnbaum and L. Schmetterer, representatives of the Institute of Mathematical Statistics, and of Professor Steven Orey, delegate of the American Mathematical Society. The visible success of our gathering is in no small measure attributable to the help we have received from this committee and other scientific friends.

A conference which extends over six weeks with participants from various parts of the world entails expenses. In this respect we feel fortunate that in spite of the general shortage of funds, the University of California and the Federal Agencies found it possible to support our enterprise. We are grateful for the allocation of funds from the Russell S. Springer Memorial Foundation, the National Science Foundation, the Office of Naval Research, the Army Research Office, the Air Force Office of Scientific Research, and the National Institutes of Health, which contributed particularly to the sessions on evolution and on problems of biology and health. In addition the pollution session received support from the Atomic Energy Commission.

The organization of the meetings fell under the responsibility of the undersigned with the able help of the staff of the Statistical Laboratory and of the Department of Statistics. For assistance with travel arrangements and various organizational matters, special thanks are due to Mrs. Barbara Gaugl.

The end of the actual meeting signals the end of a very exciting period, but not the end of our task, since the editing and publishing of over 3,000 pages of typewritten material still requires an expenditure of time and effort.

In this respect we are indebted to Dr. Morris Friedman for translations of Russian manuscripts. We are particularly grateful to Dr. Amiel Feinstein and Mrs. Margaret Stein who not only translated such manuscripts but acted as editors, checking the references and even verifying the accuracy of mathematical results.

The actual editing and marking of manuscripts was not easy since we attempted to follow a uniform style. We benefitted from the talent and skill of Mrs. Virginia Thompson who also assumed responsibility for organizing and supervising the assistant editors, Miss Carol Conti, Mrs. Margaret Darland, and Miss Jean Kettler. We are extremely grateful to all the editors for the knowledge and patience they have devoted to these manuscripts.

In the actual publication of the material the University of California Press maintained their tradition of excellence. The typesetting was performed by the staff of Oliver Burridge Filmsetting Ltd., in Sussex, England.

The meetings of the Sixth Symposium were saddened by the absence of two of our long time friends and regular participants, William Feller and Alfréd Rényi. Professors J. L. Doob and Mark Kac were kind enough to write a short appreciation of Feller. For a similar appreciation of Rényi, we are indebted to Professor L. Schmetterer. The texts appear at the beginning of the second volume.

L.LC. J.N. E.L.S.

CONTENTS

Information and Control

PAUL LÉVY
1886–1971

At a time when the composition of the present volume was already far advanced, we heard, with sorrow, that Paul Lévy died on December 15, 1971. The printing schedule did not permit us to present here a memorial article appropriate to the unique stature of Lévy in the probabilistic field. The following text offers only a few indications on Lévy's career and his work. A more extensive article, with a bibliography, is expected to appear in the first issue of the *Annals of Probability*.

We are greatly indebted to Mrs. Lévy and Professor Laurent Schwartz for communicating to us documents on Paul Lévy's career.

Ed.

Paul Lévy was born in Paris on September 15, 1886, in a family with a strong traditional link with mathematics. Both his father, Lucien Lévy, and his grandfather were professors of mathematics.

In 1904, Paul Lévy secured first place at the competitive examination for entrance to the École Normale Superieure. He also placed second in the École Polytechnique examination. For various reasons, Lévy elected to join the École Polytechnique, which then enjoyed the majestic presence of Henri Poincaré. This was a surprising choice for someone interested in mathematical research, but it may by chance have saved Lévy's life by placing him in a less vulnerable position during the first world war. After leaving Polytechnique, Lévy became an engineering student at the École des Mines (Paris) and finally, in 1910, Professor at the École des Mines of Saint Étienne. Paul Lévy's doctoral thesis was accepted in 1911 by a jury consisting of Poincaré, Hadamard, and E. Picard. He joined the École Polytechnique in 1913 and replaced G. Humbert as Professeur d'Analyse in 1920. Lévy kept this position until 1959.

In spite of poor health, Lévy continued an uninterrupted stream of mathematical activity almost up to the time of his death on December 15, 1971. This activity resulted in the publication of about 260 scientific papers. In addition, Lévy published four scientific monographs as follows:

(i) *Leçons d'Analyse Fonctionnelle*, Gauthier Villars, Paris, 1922 (439 pages), 2nd edition, 1951, with the title, *Problèmes Concrets d'Analyse Fonctionnelle.*

(ii) *Calcul des Probabilités*, Gauthier Villars, Paris, 1925 (350 pages).

(iii) *Théorie de l'Addition des Variables Aléatoires*, Gauthier Villars, Paris, 1937 (17 + 328 pages), 2nd edition, 1954.

(iv) *Processus Stochastiques et Mouvement Brownien*, Gauthier Villars, Paris, 1948 (438 pages), 2nd edition, 1965.

Paul Lévy also published in 1935 and in 1964 Notices on his scientific work and in 1970 an autobiographical volume, *Quelques Aspects de la Pensée d'un Mathematicien* (Paris, A. Blanchard, 1970, 222 pages). This last book gives a charming and very candid description of Lévy's mathematical thoughts from his independent discovery in 1902 of von Koch's tangentless continuous curve up to his hesitant philosophical speculations on Paul Cohen's proof of the indecidability of the continuum hypothesis.

Paul Lévy's own work has had immense influence especially on the developments in the field of probability and stochastic processes. His magnificent *Théorie de l'Addition des Variables Aléatoires* was published 35 years ago. Since then the field of probability has undergone a period of unbridled expansion. In spite of this the reader who will take the trouble to ponder Lévy's writings of the thirties will still find them permeated with unbelievable and yet unexhausted treasures. These writings, in beautiful fluent classical French, do not have the dry formal structure younger mathematicians have been trained to expect. Lévy, with his powerful intuition, seemed to be able to "see" the sample functions of stochastic processes or the fluctuations in a sequence of random variables. They were intimate friends which he described, pausing only from time to time to state a more formal proposition. This does not make his proofs any less rigorous than the more usual ones, but it may prevent the hurried reader from assimilating the richness of the thought and even prompted Fréchet to comment that "Your results are more or less complex according to one's own perspective."

A description of the specifics of Lévy's work would be worth a volume itself. We shall give here only a few comments on some of the highlights.

Paul Lévy's early work, represented for instance by his doctoral thesis in 1911 and culminating in the 1922 volume on *Analyse Fonctionnelle*, revolves around the extension to an infinity of dimensions of the classical theorems relative to first and second order partial differential equations. This was influenced by Volterra's style of study of "functions of lines" and Hadamard's questions concerning the manner in which Green's functions depend on the contours of the domain. Paul Lévy was able to extend many of the results relative to first order partial differential equations, but the second order equations led to problems of a very different nature. The study of the Laplacian in Hilbert space led both Gâteaux and Lévy to introduce independently the idea of mean values taken on balls or other convex subsets of Hilbert space. Gâteaux, who was killed at the beginning of World War I, left unpublished manuscripts. Hadamard gave Lévy the task of rewriting them for publication. Some of the unsolved questions raised by these papers were combined with Lévy's own ideas and at least partially solved in the 1922 volume. This work of Lévy is not well known today, perhaps because functional analysis took around that time a rather more abstract direction with the introduction of Banach spaces and the ensuing emphasis on the general theory of linear operations, and perhaps because the extension of the Laplacian to Hilbert space by Lévy and Gâteaux turned out to be different from that needed in quantum field theory. The rather deep aspects of the geometry of Hilbert space studied by Lévy still remain to be fully explored. One can mention, as part of the incidental contributions of Lévy, his 1919 description of Lebesgue measure on the infinite dimensional cube carried out without knowledge of

Fréchet's 1915 paper on general σ-additive set functions and independently of the concurrent work of P. J. Daniell.

In 1919 Lévy was requested to give three lectures on the Calculus of Probabilities. This incident was to change radically the direction of Lévy's work and the field of probability itself. The general shape of the field at that time was not particularly resplendent, and the texts available to Lévy were even worse, ignoring in particular the works of Chebyshev and Liapounov. In this state of affairs, Lévy set out to prove the proposition suggested by Laplace and Poincaré that an error which is a sum of many independent terms will have a distribution close to the Gaussian, unless the maximum term yields a substantial contribution. With the exception of a conjecture which was to be proved by Cramér in 1936, Lévy achieved his stated goal in 1934.

As a first step, Lévy had to rethink what was meant by random variables and their distributions. For the description of these latter he introduces, independently of von Mises, the idea of cumulative distribution function and also an essentially measure-theoretic description relying on ordered countable families of finite partitions. At about the same time Lévy starts using, under the name of characteristic function, the Fourier transform of a probability measure, gives an inversion formula, and proves that the relation between Fourier transform and probability measure is bicontinuous. This theorem now very familiar, is better appreciated if one remembers that it does not extend to signed measures, that the correspondence is not at all uniformly continuous, and that the only inversion formulas available at that time were encumbered by various restrictions.

With this tool, Lévy proceeds to prove various versions of the central limit theorem and a number of propositions relative to symmetric or asymmetric stable laws. Some of the results including refinements of the previous work of G. Pólya and a version of Lindeberg's proof of the central limit theorem, can be found in Lévy's *Calcul des Probabilités* of 1925. The book also contains a chapter on applications to the kinetic theory of gases and even some discussion of the use of trimmed means for estimation when the errors are not in the domain of attraction of the Gaussian distribution.

It is curious to note that in 1925 Lévy discusses attraction to stable distributions, but considers convergence to the Gaussian distribution exclusively for sums of variables which possess second moments This restriction is notably absent ten years later in Lévy's work and in Feller's work.

After 1925 Lévy continues to elaborate on the behavior of sums of independent variables, rediscovering and improving in particular Khinchin's law of the iterated logarithm. However, the most spectacular contributions to the subject can be found in three papers: the 1931 paper in *Studia Mathematica*, the voluminous paper in *Journal de Mathématiques*, Vol. 14 (1935), and the two part paper in the *Bulletin des Sciences Mathématiques*, Vol. 59 (1935).

The method, used by Lévy in 1931, of bounding the oscillations of successive sums of independent random variables by the dispersion of the last term and through symmetrization remains one of the most powerful available in this domain. In the independent case, the 1935 papers give necessary and sufficient conditions for a sum of variables to have a distribution close to the Gaussian under the assumption that individually the terms are small. An equivalent result was obtained independently and published at approximately the same date by W. Feller. However, Lévy's description of the problem goes much deeper. He attempts to see what happens if the summands are not individually small and on this occasion formulates and elaborates the consequences of the conjecture that if $X + Y$ is Gaussian so are X and Y. Cramér proved the validity of the conjecture in 1936, thereby clearing up the situation and prompting Lévy to write his memorable monograph of 1937.

This seemed to complete the program Lévy had set for himself in 1919. However, in the meantime, Lévy had enlarged the scope of the investigation to the case of dependent variables. The main results available in this domain were those of S. Bernstein (1927). Lévy attacked the problem with such a flood of new ideas that even today the dust has not settled and it is not unusual to find papers which reprove particular cases of Lévy's results of 1935. It is in these 1935 papers that the convergence theorems for series of independent variables are extended to the martingale case and that the central limit theorem is extended to martingales.

For the latter, Lévy considers martingales differences and then conditional variances. He measures "time" according to the sums of these conditional variances and first proves a theorem concerning martingales stopped at fixed "time." He then proceeds to show that the result remains valid for other stopping rules, provided that the variables so obtained do not differ too much from the fixed "time" sums. For a very recent result in this direction, the reader may consult a paper by A. Dvoretzky in the present *Proceedings* (Volume II). (The names "martingale" and "stopping times" do not occur in these papers. Doob, to whom many of the fundamental results on martingales are due, borrowed that name from J. Ville's analysis of gambling systems.)

A glance at any modern text on probability or any of the standard journals will convince the reader that martingales have now penetrated the bulk of new developments.

Since Paul Lévy had first relied on characteristic functions for the proof of limit theorems, it is worth mentioning that curiously enough the 1937 volume makes very little use of this tool in proving the usual limit theorems. The only place where the Fourier transform appears essentially is in the Cramér-Lévy theorem. Paul Lévy mentioned several times that this was one of the very few instances where he could not obtain a proof by following his intuition. The fact that a nonvanishing entire characteristic function of order two must be Gaussian carries little probabilistic flavor and at the present time no intuitive proof exists.

On the contrary, Lévy's proof of the general central limit theorem retains a lot of intuitive appeal. To make it rigorous, Lévy had first shown that the space of probability measures on the line may be metrized in such a way that if two variables X and Y are such that $\Pr\{|X - Y| > \varepsilon\} \leqq \varepsilon$, then they have distributions differing by at most ε. Prohorov extended this result to separable metric spaces in 1953 and Strassen showed in 1965 that the property Lévy requested of his distance can in fact be taken as a definition of Prohorov's distance.

When looking at sums of independent variables Lévy separates the values they take into a set of small values and a set of large values. Through the use of concentration inequalities, he shows that it is legitimate to treat the two sets as if they were independent. Kolmogorov revived and improved this method around 1955. He and his school have shown that it leads to many deep approximation results.

In 1938, prompted by a question of J. Marcinkiewicz, Lévy returns to the study of the Brownian motion process and publishes two fundamental memoirs on the subject in *Compositio Mathematica* and in the *American Journal of Mathematics*. In these memoirs, one finds the description of the sets of zeros of ordinary Brownian motion, the distributions of first passage times, and their relation to the increasing stable process of exponent 1/2, the arcsine law for percentage of time spent above a level, and a number of other related results including the first definitions and theorems on what is now called the "local time" of the process. The *Compositio* memoir deals only with one dimensional processes.

The *American Journal of Mathematics* memoir considers a particle undergoing Brownian motion in the plane. There, Lévy introduces stochastic integrals to give relations satisfied by the area between the Brownian curve and its chord, and gives results on the measure of the Brownian curve. Many of these results appear in his book of 1948.

Lévy's ideas on "local times" and on stochastic integrals were soon noticed by K. Itô whose book with H. P. McKean (*Diffusion Processes and Their Sample Paths*, Springer-Verlag, Berlin, 1965) can be consulted for more recent development.

Lévy obtained in the early fifties several other groups of results on the measure of n dimensional Brownian motion and on the canonical representations of Laplacian random functions. Part of this, together with a mention of the beautiful results of Dvoretzky, Erdös, and Kakutani, was added to the 1965 edition of the book.

The preceding memoirs refer to stochastic processes which may be multidimensional, or even Hilbert valued, but where the indexing set T is the real line. Around 1945, Lévy starts the study of a very different object. Now the variables are real valued, but the index set T becomes an n dimensional space, or a sphere, or a Hilbert space. The underlying space T has then a metric structure and Lévy requires that the expectation $E|X(s) - X(t)|^2$ be the distance between s and t. The work described in the 1948 volume concerns mostly continuity properties of such a Gaussian process, including iterated logarithm laws.

By 1955 Lévy undertakes a deeper study of the processes and of their averages on spheres of varying radius and shows in particular that if T is infinite dimensional Hilbert space, the process is already determined by its values in arbitrarily small balls. The 1965 edition of *Processus Stochastiques* gives a summary of the main results obtained by Lévy between 1955 and 1963, including a modification of an integral representation formula of Chentsov and mention of improvements on Lévy's results due to T. Hida and H. P. McKean.

Another important part of Lévy's work started in 1950 as a result of a conversation with K. L. Chung. The 1951 memoir in the *Annales de l'École Normale Superieure* gives a classification of the states of Markov processes with a countable state space, together with constructions of various Markov processes exhibiting all sorts of curious and seemingly pathological behavior. It is a peculiar trait of Lévy's psychology that since his mind would follow the development in time of the trajectories, he could not conceive that one would dare call "Markov process" a process which was not strongly Markovian. This should be kept in mind while reading this memoir, which in spite of difficulties pointed out by K. L. Chung still gives a wonderful intuitive view of the situation.

The above does not exhaust Lévy's contributions to Probability, nor does it give justice to Lévy's mathematical work in other directions. Paul Lévy seems to have been the first to represent a stochastic process by a curve in Hilbert space. He also initiated the study of composition of random variables under laws other than addition and was led in this way to look at diffusion processes on a circle or a sphere. In pure analysis his early work includes several notes on the theory of functions of real or complex variables. Perhaps one of his most striking contributions to analysis is the famous theorem, often called theorem of Wiener-Lévy, according to which if a Fourier series converges absolutely to, say f, and if ϕ is analytic on the range of f, then the series of $\phi(f)$ also converges.

It is obviously not possible to give a complete description of the work here. The interested reader may find it pleasant and profitable to read Paul Lévy's own candid description in his 1970 essay on "Quelques aspects de la pensée d'un mathematicien." The probabilistically inclined reader will certainly find much food for thought in Paul Lévy's original papers. Their author is no longer with us, but his works will remain one of the monuments of the Calculus of Probability.

L. Le Cam

POINT PROCESSES AND FIRST PASSAGE PROBLEMS

YU. K. BELYAYEV
MOSCOW STATE UNIVERSITY, U.S.S.R.

1. Introduction and basic definitions

In many domains of application of probability theory, it becomes necessary to study various properties of random formations of points. Random sequences of arrival times in queuing systems were studied by C. Palm [17], A. Hinčin [9], F. Zitek [23], D. König, K. Matthes and K. Nawrotzki [11] and others. Statistical radiotechnica is also a source of similar problems. Here it became necessary to study the set of times corresponding to the crossing of a fixed level by a random signal (S. Rice [19], V. Tihonov [22]). It is of interest to study random point formations on the plane, on surfaces, and so forth. In this paper, we describe a general approach which makes it possible to investigate a wide class of random point sets, generated by random processes and fields, from a common point of view.

The theory of random point sets and the random streams which correspond to them can be regarded as a special branch of the theory of random processes. However, this branch can lay claim to a certain degree of independence. Specific concepts and methods of investigation arise in it. If we are considering a random sequence of points on the line, then its definition is easily reduced to the problem of specifying a suitably defined random process. In a general setting, such an approach is frequently insufficient.

We shall consider an adequately general scheme for defining random point sets and random streams (Belyayev [2], [4]). Let $[T, \mathscr{M}_T]$ be a measurable space of values of a parameter $t \in T$, where $\mathscr{M}_T$ is the σ-algebra of measurable sets, and $[\Omega, \mathscr{F}_\Omega, P]$ is the initial probability space of elementary events $\omega \in \Omega$.

DEFINITION 1.1. *By a random stream $\eta(\Delta)$ on $[T, \mathscr{M}_T]$ is meant a random function with domain $\mathscr{M}_T$ (we denote an element of $\mathscr{M}_T$ by Δ), $\eta(\Delta) = 0, 1, \cdots, \infty$, satisfying the relation $\eta(\cup_i \Delta_i) = \Sigma_i\, \eta(\Delta_i)$ for every countable sequence $\Delta_i \in \mathscr{M}_T$ for which $\Delta_i \cap \Delta_j = \varnothing$ whenever $i \neq j$.*

DEFINITION 1.2. *A random point set defined on $[T, \mathscr{M}_T]$ is a function $S = S(\omega)$ defined on Ω, whose values are subsets of T, and such that for every $\Delta \in \mathscr{M}_T$ the number of points in $S \cap \Delta$, denoted by $\eta(\Delta)$, is a random variable. The system of all such random variables $\eta(\Delta)$ is called the random stream generated by the random point set S.*

Interesting examples of random point sets arise in problems involving level crossings. If ξ_t is a real valued random process with continuous sample functions, $T = R^1$, $\mathscr{M}_T$ is the family of Borel sets of R^1, then $S_u = \{t: \xi_t = u\}$, that is, the trace of the process on the level u, is a random point set. The set $S_0 = \{t: \nabla\zeta_t = 0\}$—the set of stationary points of a random field ζ_t on $T = R^m$ whose sample functions are continuously differentiable—is also a random point set. Here

$$\nabla\zeta_t = \left(\frac{\partial\zeta_t}{\partial t_1}, \cdots, \frac{\partial\zeta_t}{\partial t_m}\right)', \qquad t = (t_1, \cdots, t_m)' \in T, \tag{1.1}$$

where ′ denotes transposition of vectors or matrices. One can give many examples of a variety of random point sets defined on the trajectories of random processes and fields.

Let $\mathscr{D}(\Delta)$ be the family of all possible partitions $d(\Delta)$ of a set $\Delta \in \mathscr{M}_T$, that is, all countable collections $d(\Delta) = \{\Delta_\alpha\}$, $\Delta_\alpha \in \mathscr{M}_T$, $\Delta_{\alpha_1} \cap \Delta_{\alpha_2} = \varnothing$, $\alpha_1 \neq \alpha_2$. For a random stream $\eta(\Delta)$, we introduce the set function

$$\lambda(\Delta) = \sup_{d(\Delta)\in\mathscr{D}(\Delta)} \sum_{\Delta_\alpha \in d(\Delta)} P\{\eta(\Delta_\alpha) > 0\}. \tag{1.2}$$

THEOREM 1.1. *The set function defined by* (1.2) *is a measure on* $\mathscr{M}_T$.

PROOF. The property $\lambda(\Delta) \geqq 0$ is obvious. If $\Delta = \cup_i \Delta_i$, $\Delta_{i_1} \cap \Delta_{i_2} = \varnothing$, $i_1 \neq i_2$, $\Delta_i \in \mathscr{M}_T$, then for $d(\Delta) = \cup_i d(\Delta_i)$, $d(\Delta_i) = \{\Delta_{i,\alpha}\}$, we have

$$\lambda(\Delta) \geqq \sum_i \sum_{\Delta_{i,\alpha}\in d(\Delta_i)} P\{\eta(\Delta_{i,\alpha}) > 0\}. \tag{1.3}$$

Now the $d(\Delta_i)$ can be chosen arbitrarily; therefore it follows from (1.3) that $\lambda(\Delta) \geqq \Sigma_i \lambda(\Delta_i)$. It is hardly more complicated to prove the opposite inequality.

Theorem 1.1 makes it possible to introduce a useful numerical characteristic, which generalizes the concept of the parameter of a random stream given by Hinčin [9].

DEFINITION 1.3. *The measure* $\lambda(\Delta)$, *defined by the relation* (1.2), *is called the parametric measure of the random stream* $\eta(\Delta)$. *The principal measure of* $\eta(\Delta)$ *is defined by the relation* $\mu(\Delta) = \mathbf{E}\eta(\Delta)$.

We shall call a system of subsets, $\mathscr{C} = \{\Delta_{n,k}\}$, $n, k = 1, 2, \cdots$, of the space T a fundamental system, if (1) $\Delta_{n,k} \in \mathscr{M}_T$, (2) $\Delta_{n,k} \cap \Delta_{n,\ell} = \varnothing$ for $k \neq \ell$, (3) $\Delta_{n,k} = \cup_{i\in I_{n,k}} \Delta_{n+1,i}$, $I_{n,k} \subset \{1, 2, \cdots\}$, (4) for any $t_1, t_2 \in T$, $t_1 \neq t_2$, there exists $n = n(t_1, t_2)$ and positive integers $i_1 \neq i_2$ such that $t_1 \in \Delta_{n,i_1}$, $t_2 \in \Delta_{n,i_2}$, (5) the σ-algebra generated by the family $\mathscr{C}$ coincides with $\mathscr{M}_T$. The following assertion is a natural generalization of the well-known result of V. S. Koroljuk (see [9]) that the parameter and the intensity of a stationary ordinary random stream coincide.

THEOREM 1.2. *If the random stream* $\eta(\Delta)$ *on* $[T, \mathscr{M}_T]$ *is generated by a random point set and a fundamental system* $\mathscr{C}$ *exists, then* $\lambda(\Delta) \equiv \mu(\Delta)$, $\Delta \in \mathscr{M}_T$.

PROOF. The proof is based upon the construction of a sequence of random functions

$$(1.4) \qquad \tilde{\eta}_n(\Delta_{n,k}) = \begin{cases} 1, & \eta(\Delta_{n,k}) > 0, \\ 0, & \eta(\Delta_{n,k}) = 0, \end{cases}$$

$\tilde{\eta}_n(\bigcup_{i\in I} \Delta_{n,k}) = \Sigma_{i\in I}\, \tilde{\eta}_n(\Delta_{n,i})$, $I \subset \{1, 2, \cdots\}$, which is nondecreasing and converges to $\eta(\Delta)$ from below.

Let us consider $[T^k, \mathscr{M}_T^k]$, the direct product of k copies of the measure space $[T, \mathscr{M}_T]$, whose points are of the form $(t_1, \cdots, t_k)$, $t_1, \cdots, t_k \in T$. Let S be a random point set on $[T, \mathscr{M}_T]$, and $\eta(\Delta)$ the random stream generated by S. We construct the random set S^{*k} on $[T^k, \mathscr{M}_T^k]$, taking for its points all those of the form $(s_1, \cdots, s_k) \in T^k$, where $s_i \in S$ and $s_i \neq s_j$ for $i \neq j$. We denote by $\eta^{*k}(\Delta^k)$ the random stream generated by S^{*k}.

Definition 1.4. *The k-parametric (k-principal) measure $\lambda_k(\Delta^k)$ ($\mu_k(\Delta^k)$) of the random stream $\eta(\Delta)$ generated by a random point set S is defined as the parametric (principal) measure of the stream $\eta^{*k}(\Delta^k)$ on $[T^k, \mathscr{M}_T^k]$.*

If T has a fundamental system, then we obtain as a corollary of Theorem 1.2 that $\lambda_k(\Delta^k) \equiv \mu_k(\Delta^k)$, that is,

$$(1.5) \qquad \lambda_k(\Delta^k) = \mathbf{E}\eta^{*k}(\Delta^k), \qquad \Delta^k \in \mathscr{M}_T^k.$$

From (1.5) and the definition of S^{*k}, we obtain the possibility of calculating various moments of properties of random point sets. The appropriate result can be stated in the form of a lemma.

Lemma 1.1. *If $\eta(\Delta)$ is a random stream generated by a random point set on a space $[T, \mathscr{M}_T]$ with a fundamental system $\mathscr{C}$, then the value of the k-parametric measure on a rectangular set $\tilde{\Delta}^k = \Delta_1 \times \cdots \times \Delta_1 \times \Delta_2 \times \cdots \times \Delta_\ell \times \cdots \times \Delta_\ell$, where Δ_1 is repeated k_1 times, $\cdots$, Δ_ℓ is repeated k_ℓ times, $k = k_1 + \cdots + k_\ell$, $\Delta_i \in \mathscr{M}_T$, $i \in [1, \ell]$, and $\Delta_i \cap \Delta_j = \varnothing$ for $i \neq j$, is given by*

$$(1.6) \qquad \lambda_k(\tilde{\Delta}^k) = \mathbf{E}\left\{\prod_{i=1}^{\ell} \prod_{j=1}^{k_i} [\eta(\Delta_i) - j + 1]^+\right\},$$

where $[x]^+ = \max(x, 0)$.

It follows from Lemma 1.1 that for a cubical set $\tilde{\Delta}^k = \Delta \times \cdots \times \Delta$, $\Delta \in \mathscr{M}_T$, the value $\lambda_k(\tilde{\Delta}^k)$ equals the kth factorial moment of the number of points $\eta(\Delta)$. If $k_i = 1$ and $\Delta_i \cap \Delta_j = \varnothing$, $i, j \in [1, \ell]$, $i \neq j$, then it follows from (1.6) that $\lambda_k(\Delta_1 \times \Delta_2 \times \cdots \times \Delta_\ell) = \mathbf{E}\{\Pi_{i=1}^{\ell}\, \eta(\Delta_i)\}$, which for $k = 2$ enables us to compute the covariance of the number of points falling into nonintersecting sets in terms of the 2-parametric measure. We especially single out the possibility of computing the higher moments of characteristics in terms of the parametric measure, due to the fact that in many problems it is possible to compute the value of the parametric measure in explicit analytic form.

If a measure $\nu(\Delta)$ defined on $[T, \mathscr{M}_T]$ is such that $\lambda_k(\Delta^k)$ is absolutely continuous with respect to the product measure $\nu^k = \nu \times \cdots \times \nu$, then

$$(1.7) \qquad \lambda_k(\Delta^k) = \int_{(t_1, \cdots, t_k) \in \Delta^k} \lambda_k(t_1, \cdots, t_k)\nu(dt_1) \cdots \nu(dt_k),$$

where $\lambda_k(t_1, \cdots, t_k)$ is called the k-parameter function of the random stream or random point set with respect to the measure ν.

Thus, with every random point set S on $[T, \mathscr{M}_T]$ we can, generally speaking, associate the infinite sequence of k-parametric measures, which in the presence of a fundamental system $\mathscr{C}$ enables us to find the moments of higher order. One of the important unsolved problems is whether or not the sequence of parametric measures $\lambda_k(\Delta^k)$ determines all probabilistic characteristics of S.

Under very general conditions on random streams which are homogeneous with respect to a group of motions in the space T, the parametric measure is proportional to the corresponding Haar measure, which generalizes the well-known result of Hinčin [9] on the existence of the parameter for a stationary random stream on $T = R^1$ (see Belyayev [4]).

2. Special systems of conditional probabilities

In a large number of applied problems, it becomes necessary to calculate the conditional probabilities of various events, when the conditioning event is the appearance in a random point set of a point with given coordinates. That this problem is not entirely simple is attested to by the fact that a number of papers have been devoted to the question of methods for determining the distribution function of the length of the interval between successive events from a stationary stream on the line (McFadden [15], Ryll-Nardzewski [20], Matthes [14], Cramér and Leadbetter [7], and others).

We shall assume that the random point set S considered below is defined on $[T, \mathscr{M}_T]$, which we shall assume has a fundamental system $\mathscr{C}$, and that the parametric measure is given by $\lambda(\Delta) = \int_\Delta \lambda(t)\nu(dt)$, where $\nu(\Delta)$ is some measure on $\mathscr{M}_T$. We denote the corresponding random stream by $\eta(\Delta, S)$. A random point set $S' \subset S$ generates a substream $\eta(\Delta, S') \leqq \eta(\Delta, S)$. The parametric measure $\lambda'(\Delta)$ of S' is also absolutely continuous with respect to $\nu(\Delta)$, with density $\lambda'(t) \leqq \lambda(t)$ almost everywhere relative to the measure $\nu(\Delta)$. The passage to $S' \subset S$ can be carried out by means of a screening (thinning out) operation, which is defined in the following way. With every point $t \in T$, we associate an event A_t or its complement $\bar{A}_t$, that is, we assume that it is known for each $\omega \in \Omega$ whether the event A_t or $\bar{A}_t$ occurred. We consider (for each ω) the subset $S(\mathscr{A}) \subset S$ consisting of those points $s \in S$ for which the event A_s occurs, assuming that this screening process yields a random point set $S(\mathscr{A})$. Since the parametric measures of the random streams $\eta(\Delta, S(\mathscr{A}))$ and $\eta(\Delta, S)$ satisfy the inequality $\eta(\Delta, S(\mathscr{A})) \leqq \eta(\Delta, S)$, their parameter functions $\lambda(t)$ and $\lambda(t, A_t)$ relative to the measure $\nu(\Delta)$ satisfy almost everywhere the inequality

$$\lambda(t, A_t) \leqq \lambda(t). \tag{2.1}$$

In most of the problems which the author has investigated, the inequality (2.1) can be extended to all values of $t \in T$. Assuming that (2.1) holds for all $t \in T$, we arrive at the following definition.

DEFINITION 2.1. *The conditional probability of the event A_t, the condition being that $t \in S$, is defined by the relation*

$$P^*(A_t | t \in S) = \frac{\lambda(t, A_t)}{\lambda(t)}. \tag{2.2}$$

If the random streams $\eta(\Delta, S(\mathscr{A}))$ and $\eta(\Delta, S)$ are homogeneous with respect to a group of motions $G = \{g\}$ in the space T with invariant measure $\nu(\Delta)$, and $\lambda(t) = \lambda$, $\lambda(t, A_t) = \lambda(gt, A_{gt}) = \lambda(A)$, then

$$P^*(A_t | t \in S) = \frac{\lambda(A)}{\lambda}. \tag{2.3}$$

The probabilities introduced by relations (2.1) and (2.2) turn out to be the natural ones in many concrete problems. In particular for ergodic random processes, the probabilities defined in this way, using the random set of crossings of a fixed level, have a simple statistical interpretation.

As an example, one can consider the case $T = R^1$, $\mathscr{M}_T$ the Borel sets, $\nu(\Delta)$ Lebesgue measure, $G = \{g\}$ the group of translations $t \to t + g$, $A_{k,t}(v) = \{\eta[(t, t + v)]\} = k$; then in accordance with (2.2),

$$\varphi_k(v) = \frac{\lambda(A_{k,t}(v))}{\lambda} \tag{2.4}$$

is the conditional probability that in a time interval of length v exactly k points from S will appear, the condition being that the initial moment of the interval also belongs to S. The $\varphi_k(v)$ are the well-known Palm–Hinčin functions.

Another particular example of this kind would be deducing that the derivative of a stationary Gaussian process at the moments when the trajectory leaves the level u has a Rayleigh distribution. However, it is important for us now to clarify that the method of screening given random streams in a suitable way enables us to define new probability distributions on the space of sample functions of a random process or field. We will therefore consider a more general scheme. This scheme can frequently prove useful in defining conditional probabilities. Suppose that a random point set S is defined on $[T, \mathscr{M}_T]$, where $T = T_1 \times T_2$, $\mathscr{M}_T = \mathscr{M}_{T_1} \times \mathscr{M}_{T_2}$, the $\mathscr{M}_{T_i}$ have fundamental systems $\mathscr{C}_i$, the $\nu_i(\Delta_i)$ are measures on the $\mathscr{M}_{T_i}$, and the parametric measure of S is given by

$$\lambda(\Delta) = \int_\Delta \lambda_2(t_1, t_2)\nu_1(dt_1)\nu_2(dt_2). \tag{2.5}$$

We assume further that with probability 1 all the points $s_\alpha = (s_{1,\alpha}, s_{2,\alpha}) \in S$ have different first coordinate, that is, $\alpha_1 \neq \alpha_2$ implies $s_{1,\alpha_1} \neq s_{1,\alpha_2}$. Considering the random point set $S_1 = \{s_{1,\alpha}\}$, consisting of the coordinates $s_{1,\alpha}$ of all points $s_\alpha = (s_{1,\alpha}, s_{2,\alpha}) \in S$, on the space $[T_1, \mathscr{M}_{T_1}]$, we find that its parametric measure is given by

$$\lambda_1(\Delta_1) = \int_{\Delta_1} \lambda_1(t_1)\nu_1(dt_1), \qquad \Delta_1 \in \mathscr{M}_{T_1}, \tag{2.6}$$

where $\lambda_1(t_1) = \int_{T_2} \lambda_2(t_1, t_2)\nu_2(dt_2)$. Thus, in accordance with (2.2), the conditional probability that for a point $(s_{1,\alpha}, s_{2,\alpha}) \in S$ we will have $s_{2,\alpha} \in \Delta_2 \in \mathscr{M}_{T_2}$, the condition being that $s_{1,\alpha} = t_1$, is given by

$$P^*(s_{2,\alpha} \in \Delta_2 | s_{1,\alpha} = t_1) = \frac{\int_{\Delta_2} \lambda_2(t_1, t_2)\nu_2(dt_2)}{\lambda_1(t_1)}. \tag{2.7}$$

Let us return to the problem of determining the distribution of the derivative, at the moment of leaving level u, of the trajectory of a stationary differentiable Gaussian process ξ_t, $\mathbf{E}\xi_t = 0$, $\mathbf{E}\xi_t^2 = 1$, $\mathbf{E}\dot{\xi}_t^2 = \lambda_2$. Here $T_1 = R^1$, $T_2 = R^+ = [0, \infty)$, and the $\mathscr{M}_{T_i}$ are the algebras of Borel sets. The point $s_{1,\alpha} = \tau$ is the instant of leaving the level u, and $s_{2,\alpha} = \dot{x}$ is the value $\dot{\xi} = \dot{x}$. Here

$$\lambda_2(\tau, \dot{x}) = \frac{\dot{x}}{2\pi\lambda_2^{1/2}} \exp\left\{-\frac{\dot{x}^2}{2\lambda_2} - \frac{u^2}{2}\right\} \tag{2.8}$$

with respect to Lebesgue measure in T. Consequently, in accordance with (2.7),

$$P^*(\dot{\xi}_\tau > v) = \exp\left\{-\frac{v^2}{2\lambda_2}\right\}, \tag{2.9}$$

that is, the conditional distribution of the values of the derivative $\dot{\xi}_\tau$ of a stationary Gaussian process is a Rayleigh distribution ([22]).

Let us consider a pair of correlated random objects (S, ζ_t), where S is a random point set on a space $[T, \mathscr{M}_T]$ having a fundamental system $\mathscr{C}$, and ζ_t is a random process (field) with values from a phase space $[\mathscr{E}, \mathscr{F}_{\mathscr{E}}]$. Examples of such pairs would be trajectories of random processes ζ_t, with the random point set S consisting of those values of the coordinate t for which the trajectory of the process has a local maximum, a saddle point, and so forth. The operation of screening the points of S described above can be carried out for every elementary event $\omega \in \Omega$ on the basis of the observed trajectory $\zeta_t = \zeta_t(\omega)$. For example, with every point $t \in T$ we can associate a collection of points $t_i = t_i(t) \in T$, $i \in [1, m]$. We form a new set S' by including a point $s \in S$ in S' whenever $\zeta_{t_i} \in A_i \in \mathscr{F}_{\mathscr{E}}$, and excluding s from S' when the contrary takes place. If $\lambda(t)$ is the parameter function of the random point set S relative to some measure $\nu(\Delta)$, $\Delta \in \mathscr{M}_T$, and S' is a random point set with parameter function $\lambda(t, t_1, \cdots, t_m, A_1, \cdots, A_m) \leqq \lambda(t)$, then one can define the conditional probability

$$P^*(\zeta_{t_i} \in A_i | t \in S) = \frac{\lambda(t, t_1, \cdots, t_m, A_1, \cdots, A_m)}{\lambda(t)}, \tag{2.10}$$

which corresponds to the probability that $\zeta_{t_i} \in A_i$, given that $t \in S$. Regarding (2.10) as a family of finite dimensional probabilities, one can pose the problem of constructing new probability measures in the space of sample functions of ζ_t. Of course there are difficulties here, analogous to those which arise in the construction of conditional probability measures in function spaces. However, the approach itself has definite interest and makes it possible to obtain interesting results.

As an example, which in fact makes use of this approach, we can mention a paper of the author and V. Nosko [5] on the determination of the asymptotic distributions of the duration of excursions of a stationary Gaussian process and of its envelope. In the paper [5], to study the envelope we take $T = R^1 \times \Gamma_r$, $\mathscr{E} = R^2 = \{x = (x_1, x_2)\}$. Let $\zeta_t = (\xi_{1,t}, \xi_{2,t})$ be a two dimensional stationary Gaussian process, $S = \{(\tau, x)\}$, where τ is an exit time and $x \in \Gamma_r$ is the exit point of the trajectory of ζ_t from the circle $\Gamma_r = \{x : x_1^2 + x_2^2 = r^2\}$. The components of the process ζ_t are

$$\xi_{1,t} = \int_0^\infty \cos \lambda t \, du(\lambda) + \int_0^\infty \sin \lambda t \, dv(\lambda), \tag{2.11}$$
$$\xi_{2,t} = \int_0^\infty \sin \lambda t \, du(\lambda) - \int_0^\infty \cos \lambda t \, dv(\lambda),$$

where $u(\lambda)$ and $v(\lambda)$ are mutually independent stationary Gaussian processes with orthogonal increments;

$$\mathbf{E}[du(\lambda)]^2 = \mathbf{E}[dv(\lambda)]^2 = dF(\lambda), \tag{2.12}$$

$$\int_0^\infty \lambda^2[\log(1 + \lambda)]^{1+\varepsilon} dF(\lambda) < \infty, \qquad \varepsilon > 0.$$

To determine a probability measure on the space of trajectories which exit at time t from the point $x \in \Gamma_r$, we put $t_i(t) = t + t_i$, $A_i \subset R^2$, $i \in [1, m]$; then

$$\lambda(t, x, t_1, \cdots, t_m, A_1, \cdots, A_m) \tag{2.13}$$
$$= \int_{x_i \in A_i, i \in [1,m]} \mathbf{E}\{(n(x)\dot{\zeta}_t)^+ \mid \zeta_t = x \in \Gamma_r, \zeta_{t_i} = x_i, i \in [1, m]\}$$
$$\cdot\, p_{t,t+t_1,\cdots,t+t_m}(x, x_1, \cdots, x_m)\, dx_1 \cdots dx_m, \qquad x_i \in R^2.$$

The parameter function of the random point set $S = \{(t, x)\} \subset R^1 \times \Gamma_r$ with respect to the measure $\nu = \nu_1 \times \nu_2$, where ν_1 is Lebesgue measure on R^1 and ν_2 is Lebesgue measure on Γ_r, is given by

$$\lambda(t, x) = \mathbf{E}\{(n(x)\dot{\zeta}_t)^+ \mid \zeta_t = x \in \Gamma_r\}\, p_t(x). \tag{2.14}$$

Here $n(x)$ is the normal vector of the circle Γ_r at the point x, while $p_t(x)$ and $p_{t,t+t_1,\cdots,t+t_m}(x, x_1, \cdots, x_m)$ are the probability densities for the values $\zeta_t = x$ and $\zeta_{t+t_i} = x_i$, $i \in [1, m]$. The conditional probabilities are constructed from (2.13) and (2.14) in accordance with (2.10). In [7] the probabilities obtained in this way are called ergodic probabilities. A study of them for $r \uparrow \infty$ leads to the following result ([5]):

THEOREM 2.1. *If ζ_t is a two dimensional stationary Gaussian process which satisfies the conditions enumerated above, then the limit of the distribution of the duration Δ of an excursion of the envelope, that is, $(\xi_{1,t}^2 + \xi_{2,t}^2)^{1/2}$, above the level r, has the following form in terms of ergodic probabilities:*

$$\lim_{r\uparrow\infty} P^*\{r\Delta > v\} = \exp\left\{-\frac{\lambda_2}{8\lambda_0^2}(xv)^2\right\}, \tag{2.15}$$

where

$$\lambda_k = \int_0^\infty \lambda^k \, dF(\lambda), \qquad x^2 = 1 - \frac{\lambda_1^2}{\lambda_0\lambda_2}. \tag{2.16}$$

3. Random point sets generated by random fields

With every trajectory of a random field ζ_t, $t \in R^m$, continuously differentiable with probability one, one can associate various random point sets. The problems which arise here are more diverse than problems concerning the crossing of levels by trajectories of random processes (see Longuet–Higgins [12], [13]). If we proceed along the path of direct generalization of level crossing problems, involving the trajectories of random processes, to the case of random fields, then we arrive at the necessity of studying random point sets $S_u = \{t: \zeta_t = u\}$, which for $m = 2$ form a family of contours (level lines) in the plane. Problems related to the study of the distribution of the number of such contours, their lengths, and so forth, present difficulties. The approach discussed in Sections 1 and 2 can be used to study local objects, while the level lines formed from the curves $\zeta_t = u$ are not local objects. However, here one can study random point sets which are introduced on the level lines in a special way.

A typical example is the study of the random point set S_0 of stationary points of a random field ζ_t, $t \in R^m$, that is, the set of points at which the gradient of the field equals zero. The study of subsets of S_0, such as the sets S_+ of local maxima (bursts), S_- of local minima and S_c of saddle points, deserves attention. If we are studying the reflection from a random surface of a trajectory of ζ_t, then it is natural to study the random set of shines, whose definition is given in [2]. We shall restrict ourselves here to a study of the properties of the random point set S_0.

Let us assume that ζ_t is a random field, $t \in R^m$, which is twice continuously differentiable with probability 1, $S_0 = \{t: \nabla\zeta_t = 0\}$ is the set of stationary points, where

$$\nabla\zeta_t = \left(\frac{\partial\zeta_t}{\partial t_1}, \cdots, \frac{\partial\zeta_t}{\partial t_m}\right)', \qquad t = (t_1, \cdots, t_m)'. \tag{3.1}$$

We denote by $p_t(x)$ the probability density of the values of the vector $\nabla\zeta_t = x$. We will call the joint distribution of the values of $\|\partial^2\zeta_t/\partial t_i\partial t_j\|$, $i, j \in [1, m]$ *determinantly nondegenerate* (det-nondegenerate) if, for every $x \in R^m$,

$$P\left\{\det\left\|\frac{\partial^2\zeta_t}{\partial t_i\partial t_j}\right\| = 0 \middle| \nabla\zeta_t = x\right\} = 0. \tag{3.2}$$

THEOREM 3.1 ([3]). *Suppose that the homogeneous random field ζ_t is twice continuously differentiable with probability one, $p_t(x) \leqq C < \infty$, and the distribu-*

tion of $\|\partial^2\zeta_t/\partial t_i\partial t_j\|$ *is det-nondegenerate. Then with probability one we have:* (1) *for* $\tau \in S_0$, $\det\|\partial^2\zeta_\tau/\partial t_i\partial t_j\| \neq 0$; (2) *every bounded region contains a finite number of points* $\tau \in S_0$.

PROOF. It is sufficient to study the random point set $S_0 \cap K_m$, where $K_m = \{t: 0 \leqq t_i \leqq 1, i \in [1, m]\}$ is the unit cube. We introduce the following notation for the modulus of continuity of a random field and of its first and second derivatives:

$$\tag{3.3} \begin{aligned} \omega_\zeta(h) &= \sup_{t', t'' \in K_m, |t'-t''|<h} |\zeta_{t'} - \zeta_{t''}|, \\ \omega_i(h) &= \sup_{t', t'' \in K_m, |t'-t''|<h} \left|\frac{\partial\zeta_{t'}}{\partial t_i} - \frac{\partial\zeta_{t''}}{\partial t_i}\right|, \\ \omega_{i,j}(h) &= \sup_{t', t'' \in K_m, |t'-t''|<h} \left|\frac{\partial^2\zeta_{t'}}{\partial t_i\partial t_j} - \frac{\partial^2\zeta_{t''}}{\partial t_i\partial t_j}\right|, \\ |t'-t''| &= \left(\sum_{i=1}^m (t_i' - t_i'')^2\right)^{1/2} \end{aligned}$$

Since the field is twice continuously differentiable, for any $\varepsilon > 0$ we can choose a continuous function $\omega_\varepsilon(h)$, $\omega_\varepsilon(h)\downarrow 0$ for $h \downarrow 0$, and a constant C_ε, $0 < C_\varepsilon < \infty$, such that for the event

$$\tag{3.4} \begin{aligned} E_\varepsilon = \Bigg\{ &\max_{i,j} \sup_{t \in K_m} \left|\frac{\partial^2\zeta_t}{\partial t_i\partial t_j}\right| \leqq C_\varepsilon, \\ &\max_{i \in [1,m]} \omega_i(h) \leqq mC_\varepsilon h, \ \max_{i,j \in [1,m]} \omega_{i,j}(h) \leqq \omega_\varepsilon(h), \text{ for all } h \leqq \sqrt{m} \Bigg\} \end{aligned}$$

we have $P\{E_\varepsilon\} > 1 - \varepsilon$.

We consider the sequence of decompositions of the cube K_m into 2^{mn} cubes:

$$\tag{3.5} \begin{aligned} K_{n,\bar{k}} &= \left\{t: \frac{k_i}{2^n} \leqq t_i \leqq \frac{k_i+1}{2^n}, i \in [1, m]\right\}, \\ t_{\bar{k}} &= \left(\frac{k_1}{2^n}, \cdots, \frac{k_m}{2^n}\right)', \ \bar{k} = (k_1, \cdots, k_m). \end{aligned}$$

Let G be the event that there exists $\tau \in S_0 \cap K_m$ for which $\det\|\partial^2\zeta_\tau/\partial t_i\partial t_j\| = 0$. In the same way we denote by $G_{n,\bar{k}}$ the similar event when $\tau \in S_0 \cap K_{n,\bar{k}}$. Since $G = \bigcup_{\bar{k}} G_{n,\bar{k}}$, we have

$$\tag{3.6} P\{G\} \leqq \sum_{\bar{k}} P\{G_{n,\bar{k}} \cap E_\varepsilon\} + P\{\bar{E}_\varepsilon\},$$

where $\bar{E}_\varepsilon$ is the complement of E_ε. Using (3.4), we obtain

$$\tag{3.7} \{G_{n,\bar{k}} \cap E_\varepsilon\} \subset \left\{\max_{i \in [1,m]} \left|\frac{\partial\zeta_t}{\partial t_i}\right| \leqq C_\varepsilon \frac{1}{2^n}, \left|\det\left\|\frac{\partial^2\zeta_t}{\partial t_i\partial t_j}\right\|\right| \leqq (m!)^2 C_\varepsilon^{m-1}\omega_\varepsilon\left(\frac{1}{2^n}\right)\right\}.$$

In terms of ordinary conditional probabilities, we obtain from (3.7)

$$(3.8)\qquad P\{G_{n,\bar{k}} \cap E_\varepsilon\} \leqq \int_{|x_i| \leqq C_\varepsilon 2^{-n}, i \in [1,m]} P\left\{\left|\det\left\|\frac{\partial^2 \zeta_t}{\partial t_i \partial t_j}\right\|\right| \leqq (m!)^2 C_\varepsilon^{m-1} \omega_\varepsilon\left(\frac{1}{2^n}\right) \middle| \nabla\zeta(t_k) = x\right\} p_t(x)\, dx.$$

Since $p_{t_k}(x) \leqq C$, and the conditional probability in the integrand in (3.8) is bounded by unity and tends to zero as $n \to \infty$ for any x, then by Lebesgue's theorem on passing to the limit inside the integral we obtain, for any $\delta > 0$ and $n \geqq n(\delta)$,

$$(3.9)\qquad P\{G_{n,\bar{k}} \cap E_\varepsilon\} \leqq \delta\left(\frac{1}{2^n}\right)^m.$$

From (3.6) to (3.9), we obtain

$$(3.10)\qquad P\{G\} \leqq 2^{mn}\delta\left(\frac{1}{2^n}\right)^m + \varepsilon.$$

Since δ and ε can be chosen as small as desired, it follows that $P\{G\} = 0$, which corresponds to the first part of the theorem.

We prove the second part of the theorem by assuming that in some cube the number of points from S_0 is infinite. This assumption implies that the event G has occurred, whose probability, as was just proven, is zero.

COROLLARY 3.1. *If ζ_t is a twice continuously differentiable (with probability 1) homogeneous Gaussian field for which the joint distribution of $\partial^2\zeta_t/\partial t_i\partial t_j$, $i, j \in [1, m]$ is nondegenerate, then the assertion of Theorem 3.1 holds.*

The proof follows from the mutual independence of the random variables $\partial\zeta_t/\partial t_k$ and $\partial^2\zeta_t/\partial t_i\partial t_j$ for any values of $i, j, k \in [1, m]$.

It is probable that the assertion of Corollary 3.1 can be strengthened somewhat by replacing the assumption of the twice continuous differentiability with probability 1 of ζ_t with the assumption of twice continuous differentiability of ζ_t in quadratic mean. One can formulate yet another assertion, similar to an assertion of Bulinskaja [6] on the absence of tangency in the problem of level crossings.

THEOREM 3.2 ([3]). *Let ζ_t be a twice continuously differentiable random field which has bounded probability densities for the quantities $\zeta_t = u$, $\nabla\zeta_t = \dot{u}$, and $p_t(u, \dot{u}) < C$. Then with probability 1 there are no stationary points corresponding to a given level r, that is, for $S_{0,r} = \{\tau : \nabla\zeta_\tau = 0, \zeta_\tau = r\}$ we have $P\{S_{0,r} \neq \varnothing\} = 0$.*

The proof can be carried out analogously to the proof of Theorem 3.1.

We remark that for $m = 2$, to transverse crossings of the level line $\zeta_t = r$ there correspond saddle points with height r. Since these are absent, the following alternative must prevail: a connected solution of $\zeta_t = r$, $t \in R^2$, is either a bounded isolated contour, possibly containing within itself another contour, or else it is a curve passing out of any bounded region. Presumably, for homogeneous Gaussian fields the boundedness of the contours satisfying the equation $\zeta_t = r$ is typical.

The k-parameter function which is needed to calculate the moments of functions (the mean, variance and so on) of random point sets generated by random fields, can be obtained in explicit form under weak restrictions. These restrictions, imposed on the field ζ_t, are conveniently formulated in the form of the following complex of conditions.

The complex of conditions $C_{\zeta,k}$. We assume that the real field ζ_t, $t \in R^m$, is with probability 1 twice continuously differentiable, and that the joint distributions of the quantities $\zeta_{t^i} = u_i$; $\nabla\zeta_{t^i} = \dot{u}^i$, $\nabla\zeta_{t^i}\nabla' = \|\ddot{u}^i_{\ell,r}\|$, at the points $t^i \in R^m$ have probability distributions $p_{t^1,\cdots,t^k}(u', \dot{U}, \ddot{Z})$ which are jointly continuous in all the variables, where $u' = (u_1, \cdots, u_k)$, $\dot{U} = \|\dot{u}^i_j\|$, $\ddot{Z} = \|\ddot{u}^i_{\ell,r}\|$, $j, \ell, r \in [1, m]$, $i \in [1, k]$. We also assume that in any bounded region $C \subset R^m$ the moduli of continuity $\omega_{i,j}(h)$ of the fields $\partial^2\zeta_t/\partial t_i\partial t_j$ satisfy the condition

(3.11) $$P\{\omega_{i,j}(h) > \varepsilon\} = o(h^{km}).$$

The following theorem gives conditions which are sufficient for (3.11) to hold.

THEOREM 3.3. *If for a real separable random field* ζ_t, $t, \tilde{h} \in R^m$,

(3.12) $$\sup_{|\tilde{h}_i| \leqq h, i \in [1,m]} P\{|\zeta_{t+\tilde{h}} - \zeta_t| > \varepsilon(h)\} \leqq g(h),$$

where the functions $\varepsilon(h)$ *and* $g(h)$ *satisfy the conditions*

(3.13) $$\sum_{n=1}^{\infty} 2^{m2^{n+1}} g(2^{-2^n}) < \infty, \qquad \sum_{n=1}^{\infty} \varepsilon(2^{-2^n}) < \infty,$$

then for the function

(3.14) $$\psi(h) = \sum_{n=j(h)}^{\infty} 2^{m2^{n+1}} g(2^{-2^n}), \quad 2^{-2^{j(h)+1}} \leqq h < 2^{-2^{j(h)}},$$

we have for $h \downarrow 0$

(3.15) $$P\{\omega_\zeta(h) > \varepsilon\} \leqq \psi(h).$$

The proof basically follows the method of A. N. Kolmogorov, which is presented in a paper of Slutsky [21], and also the method of Dudley [8]. The partitioning sets have the form

$$T_n = \left\{\left(\frac{k_1}{2^{2^n}}, \cdots, \frac{k_m}{2^{2^n}}\right), 0 \leqq k_i \leqq 2^{2^r}, i \in [1, m]\right\},$$

(3.16) $$T_{n,\bar{k}} = T_n \cap K_{n,\bar{k}},$$

$$K_{n,\bar{k}} = \left\{t: \frac{k_i}{2^{2^n}} \leqq t_i < \frac{k_i + 1}{2^{2^n}}, i \in [1, m]\right\},$$

and the basis events are

(3.17) $$A_{n,\bar{k},t} = \left\{\left|\zeta\left(\frac{k_1}{2^{2^n}}, \cdots, \frac{k_m}{2^{2^n}}\right) - \zeta(t)\right| \leqq \varepsilon(2^{-2^n})\right\},$$

where $\bar{k} = (k_1, \cdots, k_m)$, $\zeta(t) = \zeta_t$, $t \in T_{n+1,\bar{k}}$.

Verifying the hypotheses of Theorem 3.3 for suitably chosen functions $\varepsilon(h)$ and $g(h)$, we easily obtain the following corollary from (3.15).

COROLLARY 3.2. *If ζ_t, $t \in R^m$, is a separable Gaussian field, twice differentiable in quadratic mean, for which*

$$\max_{i,j \in [1,m]} \mathbf{E} \left| \frac{\partial^2 \zeta_{t+\tilde{h}}}{\partial t_i \partial t_j} - \frac{\partial^2 \zeta_t}{\partial t_i \partial t_j} \right|^2 \leqq \frac{C}{|\log |\tilde{h}||^{1+\varepsilon}}, \qquad C > 0, \varepsilon > 0, \tag{3.18}$$

then with probability one the fields $\partial^2 \zeta_t / \partial t_i \partial t_j$ have continuous trajectories for which (3.11) *is satisfied for any $\varepsilon > 0$ and any integer $\ell = km$. The complex of conditions $C_{\zeta,k}$, $k = 1, 2, \cdots$, holds for ζ_t.*

We will denote the positive (negative) definiteness of a matrix A by $A \succ 0$ ($A \prec 0$), and also put $\nabla \zeta_t \nabla' = \|\partial^2 \zeta_t / \partial t_i \partial t_j\|$. We introduce the following functions:

$$I_+(\nabla \zeta_t \nabla') = \begin{cases} 1, \nabla \zeta_t \nabla' \prec 0, \\ 0, \nabla \zeta_t \nabla' \nprec 0, \end{cases} \qquad I_-(\nabla \zeta_t \nabla') = \begin{cases} 1, \nabla \zeta_t \nabla' \succ 0, \\ 0, \nabla \zeta_t \nabla' \nsucc 0. \end{cases} \tag{3.19}$$

THEOREM 3.4. *If the complex of conditions $C_{\zeta,k}$ is satisfied, then the k-parameter functions with respect to Lebesgue measure in R^m of the random point sets of stationary points $S_0 = \{\tau\}$, of bursts $S_+ = \{\tau\}$, and of local minima $S_- = \{\tau\}$ generated by the random field ζ_t, $t \in R^m$, at which $\zeta_\tau > u$, are given respectively by*

$$\begin{aligned} \lambda_u^0(t^1, \cdots, t^k) &= \int_{u_i \geqq u, i \in [1,k]} \mathbf{E} \left\{ \prod_{i=1}^k |\det \nabla \zeta_{t^i} \nabla'| \,\middle|\, \zeta_{t^i} = u_i, \nabla \zeta_{t^i} = 0, i \in [1, k] \right\} \\ &\qquad p_{t^1 \cdots t^k}(u_1, \cdots, u_k, \dot{0})\, du_1 \cdots du_k, \end{aligned} \tag{3.20}$$

$$\begin{aligned} \lambda_u^+(t^1, \cdots, t^k) &= \int_{u_i \geqq u, i \in [1,k]} \mathbf{E} \left\{ \prod_{i=1}^k |\det \nabla \zeta_{t^i} \nabla'| I_+(\nabla \zeta_{t^i} \nabla') \,\middle|\, \zeta_{t^i} = u, \nabla \zeta_{t^i} = 0, i \in [1, k] \right\} \\ &\qquad p_{t^1 \cdots t^k}(u_1, \cdots, u_k, \dot{0})\, du_1 \cdots du_k, \end{aligned} \tag{3.21}$$

$$\begin{aligned} \lambda_u^-(t^1, \cdots, t^k) &= \int_{u_i \geqq u, i \in [1,k]} \mathbf{E} \left\{ \prod_{i=1}^k |\det \nabla \zeta_{t^i} \nabla'| I_-(\nabla \zeta_{t^i} \nabla') \,\middle|\, \zeta_{t^i} = u, \nabla \zeta_{t^i} = 0, i \in [1, k] \right\} \\ &\qquad p_{t^1 \cdots t^k}(u_1, \cdots, u_k, \dot{0})\, du_1 \cdots du_k. \end{aligned} \tag{3.22}$$

The proof is carried out by constructing upper and lower bounds for the probabilities of the events that there lies in the cube

$$\Delta_h(t) = \left\{ s : |t_i - s_i| \leqq \frac{h}{2}, i \in [1, m], t, s \in R^m \right\}, \qquad h \downarrow 0, \tag{3.23}$$

at least one point τ from S_0, S_+, S_-, at which $\zeta_\tau \geqq u$. In this connection, one first considers those subsets of S_0, S_+, S_- for which

$$\max_{i,j} \left|\frac{\partial^2 \zeta_\tau}{\partial t_i \partial t_j}\right| \leqq A, \ |\det \nabla\zeta_\tau \nabla'| \geqq \varepsilon. \tag{3.24}$$

Using the properties of the moduli of continuity $\omega_{i,j}(h)$, expressed by (3.11), we obtain the necessary bounds, following which we let $\varepsilon \downarrow 0$, $A \uparrow \infty$. The proof is not difficult, but because the formulas are cumbersome it would take up a lot of space, and will therefore not be given.

The relations (3.20) through (3.22) could be used as the basis for obtaining numerical results for concrete random fields. In this connection, it seems natural to use the method of statistical modeling in order to compute the integrals in (3.20) through (3.22). The corresponding programs have been tested for $m = 2, 3$, $k = 1, 2$. For $k = 1$ for a homogeneous Gaussian field satisfying $C_{\zeta,1}$, the principal terms in an asymptotic expansion of $\lambda_u^+(t)$ and $\lambda_u^0(t)$ for $u \uparrow \infty$ are found in a paper by Nosko [16]. It turns out that

$$\begin{aligned} \lambda_u^0(t) &= \lambda_u^+(t)\left[1 + O\left(\frac{1}{u}\right)\right] \\ &= \frac{[\det \Lambda^{(1)}]^{1/2}}{(2\pi)^{(m+1)/2}\lambda_0^{(2m-1)/2}} u^{m-1} \exp\left\{\frac{-u^2}{2\lambda_0}\right\}\left[1 + O\left(\frac{1}{u}\right)\right], \end{aligned} \tag{3.25}$$

$$\mathbf{E}\zeta_t = 0, \qquad \mathbf{E}\zeta_t^2 = \lambda_0, \qquad \Lambda^{(1)} = \mathbf{E}\nabla\zeta_t(\nabla\zeta_t)'.$$

It follows from (3.25) that for $u \uparrow \infty$ the parameter functions of the random point set S_0 and its subset S_+ are equivalent. This corresponds to the intuitive assumption that flattenings of the field, that is, points τ where $\nabla\zeta_\tau = 0$, at high levels correspond, as a rule, to local maxima. This is related to the fact that high saddle points and high local minima can appear only close to excursions having a complicated structure, for example near two-vertex bursts or vertices having hollows resembling craters, and so forth. Thus, for normal fields high vertices of a complicated structure are encountered infrequently.

Numerical calculations carried out for a homogeneous isotropic Gaussian field ζ_t, $t \in R^2$, with covariance function $R(x) = \exp\{-x_1^2 - x_2^2\}$ have shown that already at the level $u = \sqrt{3}$ the remainder term $O(1/u)$ does not have a practical influence. Thus, it may be hoped that calculations with the formula (3.25) for moderate values of u can be carried out on the basis of the principal term of the asymptotic expansion.

Using the method presented in Section 2 for determining the special conditional probabilities, it is easy to obtain the distribution of high bursts $\zeta_\tau = u + m_u$, given that a burst with height greater than u occurred at $\tau = t$. In particular, for $u \uparrow \infty$ we obtain from (3.25):

COROLLARY 3.3. *The limit of the special conditional distribution of the height of bursts of a homogeneous Gaussian field ζ_t, $t \in R^m$, satisfying $C_{\zeta,1}$, given that the height of the burst is greater than u, is given by*

$$\lim_{u\uparrow\infty} P\{um_u > v\} = \exp\left\{\frac{-v}{\lambda_0}\right\}, \tag{3.26}$$

for $\mathbf{E}\zeta_t = 0$, $\mathbf{E}\zeta_t^2 = \lambda_0$.

Study of the structure of a field ζ_t in the "neighborhood" of a burst also can be carried out by means of the special conditional probabilities. Here with every family of points $t^i \in R^m$ and families of sets $A_i \subset R^1$, $B_i \subset R^m$, $C_i \subset R^{m(m+1)/2}$, and so forth, one can associate the random point subset S'_+ of those bursts of the homogeneous field ζ_t such that if $\tau \in S_+$, then $\tau \in S'_+$ if $\zeta_\tau \in A_0$, $\zeta_{\tau+t^i} \in A_i$, $\nabla\zeta_{\tau+t^i} \in B_i$, $\nabla\zeta_{\tau+t^i}\nabla' \in C_i$, $i \in [1, n]$. If the conditions $C_{\zeta,k}$, $k = 1, 2, \cdots$ are satisfied, then the parameter function of S'_+ is given by

$$\lambda^+_{t^1\cdots t^n}(A_0, A_i, B_i, C_i, i\in[1,n]) = \int_{G_0} |\det \|\ddot{u}^0_{j,\ell}\|| p_{0,t^1,\cdots,t^n}(u, \dot{U}, \ddot{Z})\, du\, d\ddot{Z}, \tag{3.27}$$

where

$$\begin{aligned} G_0 = \{u_0 \in A_0,\ u_i \in A_i,\ (\dot{u}^i_1, \cdots, \dot{u}^i_m) \in B_i,\ \|\ddot{u}^i_{j,\ell}\| \in C_i,\ i \in [1, n], \\ \ell, j \in [1, m],\ \|\ddot{u}^0_{j,\ell}\| \prec 0\}, \end{aligned} \tag{3.28}$$

$$du = du_0\, du_1 \cdots du_n, \qquad \dot{U} = \|\dot{u}^i_j\|, \qquad \ddot{Z} = \|\ddot{u}^i_{j,\ell}\|, \qquad \dot{u}^0_j = 0.$$

The special conditional probabilities in the space of trajectories of the field having bursts at the point $\tau = 0$, of height $h \in A_0$, such that $\zeta_{t^i} \in A_i$, and so forth, are introduced as the ratios

$$P^*_{t^1\cdots t^n}(A_i, B_i, C_i, i \in [1, n] | A_0) = \frac{\lambda^+_{t^1\cdots t^n}(A_0, A_i, B_i, C_i, i \in [1, n])}{\lambda^+_0(A_0)}. \tag{3.29}$$

Interesting results were obtained in a paper by Nosko [16] on the structure of excursions of homogeneous Gaussian random fields above unboundedly increasing levels $u \uparrow \infty$. It was shown that when the complex of conditions $C_{\zeta,k}$, $k = 1, 2, \cdots$, is satisfied, then with probability arbitrarily close to unity an excursion of a trajectory of the field ζ_t, $t \in R^m$, containing a burst at the point τ, can be approximated within $o(1/u)$ by that part of the second order surface

$$z = u + \tfrac{1}{2}(t - \tau)'\Lambda_u(t - \tau) \tag{3.30}$$

lying on the plane $z = u$ in the space R^{m+1} of points $(t_1, \cdots, t_m, z)$, where

$$\Lambda_u = u \cdot \| -\lambda_{i,j}/\lambda_0 \|, \qquad \lambda_{i,j} = \mathbf{E}\frac{\partial\zeta_t}{\partial t_i}\frac{\partial\zeta_t}{\partial t_j}. \tag{3.31}$$

This result was obtained by introducing in the space of trajectories special conditional probabilities analogous to (3.29). For the particular case $m = 2$, the excursions are approximated by segments of an elliptical paraboloid. We mention here the following result.

THEOREM 3.5 (Nosko [16]). *Let ζ_t, $t \in R^2$, be a homogeneous Gaussian field satisfying the complex of conditions $C_{\zeta,k}$, $k = 1, 2, \cdots$. Then for $u \uparrow \infty$ the special conditional probabilities of the following functionals of excursions above the level u:*

the maximum m_u of the excursion, its cross sectional area S_u and its volume V_u, satisfy the relations

$$\lim_{u\uparrow\infty} P^*\{um_u > v\} = \lim_{u\uparrow\infty} P^*\{\gamma u S_u^2 > v\} = \lim_{u\uparrow\infty} P^*\{(2\gamma u^3 V_u)^{1/2} > v\} = \exp\left\{-\frac{v}{\lambda_0}\right\}, \tag{3.32}$$

where

$$\gamma = (2\pi\lambda_0)^{-1}\{\lambda_{1,1}\lambda_{2,2} - \lambda_{1,2}^2\}^{1/2}. \tag{3.33}$$

4. Estimation of the distribution of the maximum of a random field

Problems concerning the distribution of the maximum of a random field ζ_t when $t \in G$, where G is a region or surface in R^m, are interesting. An attempt to obtain the exact solution for such problems meets with considerable difficulties. The situation, however, is made easier by the fact that frequently one only needs to find an estimate for the probability of a trajectory of the field going above a high level u. In such cases, one can use an asymptotic approach [1]. For example, if ζ_t, $t \in R^m$, is a real Gaussian random field, and S_+ is the random point set of bursts of ζ_t above the level u, then for a region $G \subset R^m$ with a smooth boundary ∂G we have

$$P\{\sup_{t\in G} \zeta_t > u\} \leqq \mathbf{E}\eta_u^+(G) + \mathbf{E}\eta_u^+(\partial G), \tag{4.1}$$

where $\mathbf{E}\eta_u^+(G)$ is the average number of bursts in $S_+ \cap G$, and $\mathbf{E}\eta_u^+(\partial G)$ is the average number of bursts above the level u of the field ζ_t, $t \in \partial G$. One can show that if the complex of conditions $C_{\zeta,1}$ is satisfied, then the principal term in an asymptotic expression for the right side of (4.1) for $u \uparrow \infty$ has the form $\lambda_u^0 V$, where V is the volume of the region G and λ_u^0 can be found from (3.25). Thus, the level u_α for which $P\{\sup_{t\in G} \zeta_t > u_\alpha\} \approx \alpha$ can be found from the simple transcendental equation

$$VC_m\left(\frac{K_\alpha}{\sqrt{2\lambda_0}}\right)^{m-1} \exp\left\{-\left(\frac{K_\alpha}{\sqrt{2\lambda_0}}\right)^2\right\} = \alpha, \tag{4.2}$$

where

$$C_m = \frac{[\det \Lambda^{(1)}]^{1/2}}{2\pi^{(m+1)/2}\lambda_0^{m/2}}. \tag{4.3}$$

The results of the numerical computations which were mentioned in Section 3 show that, presumably, the u_α obtained from (4.2) gives good results if $u_\alpha/\sqrt{\lambda_0} > 3$. However, the problem of estimating the error which occurs here has not yet been solved.

If one assumes that the random stream of bursts above an unboundedly increasing level u is Poisson, then following the method of Cramér [7] one can obtain that for $G \uparrow R^m$

$$(4.4)\qquad \lim_{V=V(G)\uparrow\infty} P\left\{\sup_{x\in G}\frac{\zeta_x}{\sqrt{\lambda_0}} < \sqrt{2\log V} + \frac{(m-1)\log\log V}{2\sqrt{2\log V}} + \frac{(m-1)\log(\sqrt{2\lambda_0}) + \log C_m + z}{\sqrt{2\log V}}\right\} = \exp\{-\exp\{-z\}\}.$$

As G increases, one has to assume that ∂G is regular, that is, the volume V_d of the set of points whose distance from G does not exceed d is such that $V_d/V \to 0$. The relation (4.4) also can be used to calculate the critical level u_α. Denoting by z_β the quantile of the level β, $\exp\{-\exp\{-z_\beta\}\} = \beta$, $0 < \beta < 1$, we obtain from (4.4)

$$(4.5)\qquad \sqrt{\lambda_0}u_\alpha \approx \sqrt{2\log V} + \frac{(m-1)\log\log V}{2\sqrt{2\log V}} + \frac{(m-1)\log(\sqrt{2\lambda_0}) + \log C_m + z_{1-\alpha}}{\sqrt{2\log V}}.$$

It should be kept in mind that for small α ($\alpha \leqq 0.05$) and small values of V, calculation of the critical level by means of (4.5) can yield a nonmonotone dependence upon V. The method of calculation by means of (4.2) appears to be preferable.

By methods similar to those mentioned in Section 3, we have derived expressions for the parameter functions of the random set of critical points on the level lines of fields defined on manifolds [18] in R^m [1]. These expressions can be used to estimate the distribution of the maximum of a field defined over a surface.

REFERENCES

[1] YU. K. BELYAYEV, "Distribution of the maximum of a random field and its application to problems of reliability," *Izv. Akad. Nauk SSSR, Tehn. Kibernet.*, No. 2 (1970), pp. 77–84.

[2] ———, "New results and generalizations of problems of crossing type," in the Russian edition only of H. Cramér and M. Leadbetter, *Stationary Stochastic Processes*, Moscow, Izdat. "Mir," 1969, pp. 341–378.

[3] ———, "The random sets of bursts and shines of random fields," *Proceedings USSR–Japan Symposium on Probability*, Novosibirsk, 1969, pp. 11–18.

[4] ———, "On the characteristics of a random flow," *Theor. Probability Appl.*, Vol. 13 (1968), pp. 543–544.

[5] YU. K. BELYAYEV and V. P. NOSKO, "Characteristics of excursions above a high level for a Gaussian process and its envelope," *Theor. Probability Appl.*, Vol. 14 (1969), pp. 296–309.

[6] E. V. BULINSKAYA, "On the mean number of crossings of a level by a stationary Gaussian process," *Theor. Probability Appl.*, Vol. 6 (1961), pp. 435–438.

[7] H. CRAMÉR and M. LEADBETTER, *Stationary and Related Stochastic Processes; Sample Function Properties and Their Applications*, New York, Wiley, 1967.

[8] R. M. DUDLEY, "Gaussian processes on several parameters," *Ann. Math. Statist.*, Vol. 36 (1965), pp. 771–788.

[9] A. YA. HINČIN, *Papers on Queuing Theory*, Moscow, Fizmatgiz, 1963.

[10] M. KAC and D. SLEPIAN, "Large excursions of a Gaussian process," *Ann. Math. Statist.*, Vol. 30 (1959), pp. 1215–1228.

[11] D. KÖNIG, K. MATTHES, and K. NAWROTZKI, "Verallgemeinerungen der Erlangschen und Engsetschen formeln, eine Methode in der Bedienungstheorie," *Angewandte Mathematik und Mechanik*, Vol. 5 (1967), pp. 1–123.

[12] M. S. LONGUET-HIGGINS, "The statistical analysis of a random, moving surface," *Philos. Trans. Roy. Soc. London Ser. A*, Vol. 249 (1957), pp. 321–387.

[13] ———, "The statistical geometry of random surfaces," *Hydraulic Instability*, Symposium in Applied Mathematics, Vol. 13 (1962), pp. 105–143.

[14] K. MATTHES, "Stationäre zufällige Punktfolgen," *Jahresbericht d. Deutsch. Math. Verein*, Vol. 66 (1963/4), pp. 66–79.

[15] J. A. MCFADDEN, "On the lengths of intervals in a stationary point process," *J. Roy. Statist. Soc. Ser. B*, Vol. 24 (1962), pp. 364–382.

[16] V. P. NOSKO, "The characteristics of excursions of Gaussian homogeneous fields above a high level," *Proceedings USSR–Japan Symposium on Probability*, Novosibirsk, 1969.

[17] C. PALM, "Intensitätsschwankungen in Fernsprechverkehr," *Ericsson Technics*, Vol. 44 (1943), pp. 1–189.

[18] P. K. RAŠEVSKIĬ, *Riemannian Geometry and Tensor Analysis*, Moscow, Gostehizdat, 1953.

[19] S. O. RICE, "The mathematical analysis of random noise," *Bell System Tech. J.*, Vol. 23 (1944), pp. 282–332; Vol. 24 (1945), pp. 46–156.

[20] C. RYLL-NARDZEWSKI, "Remarks on processes of calls," *Proceedings of the Fourth Berkeley Symposium on Mathematical Statistics and Probability*, Berkeley and Los Angeles, University of California Press, 1961, Vol. 2, pp. 66–79.

[21] E. SLUTSKY, "Qualche proposizioni relativa alla teoria delle funzioni aleatorie," *Giorn. Ist. Ital. Attuari*, Vol. 8 (1937), pp. 183–199.

[22] V. I. TIHONOV, *Excursions of Random Processes*, Moscow, Izdat. "Nauka," 1970.

[23] F. ZÍTEK, "On the theory of ordinary streams," *Czechoslovak Math. J.*, Vol. 8 (1958), pp. 448–459. (In Russian, summary in German.)

LIMIT THEOREMS FOR RANDOM WALKS WITH BOUNDARIES

A. A. BOROVKOV
INSTITUTE OF MATHEMATICS
NOVOSIBIRSK, U.S.S.R.

1. Introduction

In this review, we consider boundary problems for random walks generated by sums of independent items and some of their generalizations.

Let $\xi_1, \xi_2, \cdots$ be identically distributed independent random variables with distribution frunction $F(x)$. Let $S_0 = 0$, $S_n = \Sigma_{k=1}^n \xi_k$ with $n = 1, 2, \cdots$. We shall study the properties of the random trajectory formed by the sums $S_0, S_1, S_2, \cdots$. Let n be an integer parameter and let $g_n^{\pm}(t)$ be two functions on the real axis with the following properties:

$$(1.1) \qquad g_n^+(0) > 0 > g_n^-(0), \qquad g_n^+(t) > g_n^-(t), \qquad t \geqq 0.$$

We shall denote by G_n the part of the halfplane $(t \geqq 0, x)$ which lies between these two curves. In the same halfplane (t, x), let us consider the trajectory formed by the points

$$(1.2) \qquad \left(\frac{k}{n}, S_k\right), \qquad k = 0, 1, 2, \cdots$$

One of the main boundary functionals of trajectory (1.2) is the time η_G at which it leaves the region G_n:

$$(1.3) \qquad \eta_G = \min\left\{\frac{k}{n} : \left(\frac{k}{n}, S_k\right) \notin G_n\right\}.$$

We shall define the value of the first jump χ_G across the boundary of the region G_n by the equalities

$$(1.4) \qquad \chi_G = S_{\eta_G} - g_n^+(\eta_G) \quad \text{or} \quad \chi_G = S_{\eta_G} - g_n^-(\eta_G),$$

depending on whether trajectory (1.2) crosses the upper or lower boundary of the region G_n. Note that in general the random variables η_G and χ_G are not defined on the whole space of elementary events. We put $\eta_G = \infty$, where η_G remains undefined. We shall not define the random value χ_G on the set $\{\eta_G = \infty\}$.

Problems variously connected with distributions of the functionals η_G and χ_G will be called boundary problems for random walks. It is well known that these problems play an important part in mathematical statistics (in sequential analysis, nonparametric methods, and so forth) in queueing theory, and in other similar fields of mathematics. For instance, the classical problem of sequential analysis leads to the elucidation of probabilities of the type

(1.5) $$P(\chi_G > 0; \eta_G < \infty),$$

for boundaries $g_n^{\pm}$ which are straight line boundaries. This is the probability of the event that the trajectory (1.2) crosses the upper boundary earlier than the lower one. The distribution of the maximum, $\bar{S}_n = \max_{0 \leqq k \leqq n} S_k$, is of great interest in queueing theory. It is evident that this distribution is also connected with boundary problems because

(1.6) $$P(\bar{S}_n > x) = P(\eta_G < 1),$$

where the region G_n is formed by the straight lines $g_n^+(t) = x$ and $g_n^-(t) = -\infty$. One could give examples of still more complicated applied and theoretical problems leading to boundary problems.

Considering limit theorems for boundary problems (that is, methods of the approximate calculation of distribution η_G, χ_G, $\bar{S}_n$ when $n \to \infty$), we shall distinguish rather conditionally between the regions of *normal* and *large* deviations. If the limit values of the probabilities in question are nondegenerate (different from 0 and 1), then we shall refer to all these cases as problems about normal deviations. The remaining cases will be referred to as problems on large deviations.

2. Normal deviations

The most important problem occurs when

(2.1) $$E\xi_k = 0, \qquad D\xi_k < \infty, \qquad g_n^{\pm}(t) = \sqrt{n}\, g^{\pm}(t).$$

Here the $g^{\pm}(t)$ possess the properties (1.1) and are continuous. Without restricting generality, we may assume $\mathscr{D}\xi_k$ to be equal to 1. It is well known that in this case

(2.2) $$\begin{aligned} \lim_{n\to\infty} P(\eta_G < v) &= p(v, g^+, g^-) \\ &= 1 - P\big(g^-(t) < w(t) < g^+(t); 0 \leqq t \leqq v\big), \end{aligned}$$

where $w(t)$ is a standard Brownian motion process.

Similar limits also exist for probabilities of the more general type,

(2.3) $$P\left(\eta_G < v, \chi_G > 0, \frac{S_{[nv]}}{\sqrt{n}} \in \Delta\right).$$

These limits are equal to the probabilities $p_1(v, g^+, g^-, \Delta)$ that the trajectory $w(t)$ during time v left the band

(2.4) $$G = \{(t, x): g^-(t) < x < g^+(t)\}$$

across the upper boundary and that the value $w(v)$ belongs to the interval Δ.

It is also possible to show that when $y > 0$,

$$\lim_{n\to\infty} P(\eta_G < v, \chi_G > y) = p_1(v, g^+, g^-, R)P(y), \tag{2.5}$$

where $R = (0, \infty)$ and $P(y) = \lim_{n\to\infty} P(\chi_{\bar{G}_n} > y)$ for the regions $\bar{G}_n$ generated by the straight lines $g_n^+(t) = n$ and $g_n^-(t) = -\infty$. The latter is the distribution of the first jump of sequence $\{S_k\}$ across the infinitely remote positive barrier.

Assertion (2.2) was obtained by Kolmogorov in 1931. Since that time, it has become clear that the problem of estimating the rate of convergence of distributions (2.2) and (2.3) to their limits is very complicated. The following result by S. V. Nagaev [1] is the most general in this direction.

THEOREM 2.1. *Let conditions (2.1) hold, let $c_3 = E|\xi_k|^3 < \infty$, and assume that the functions $g^\pm(t)$ satisfy the Lipschitz condition with constant L. Then an absolute constant A exists such that*

$$|P(\eta_G < v) - p(v, g^+, g^-)| < \frac{Ac_3^2(L+1)}{\sqrt{n}}. \tag{2.6}$$

The method of proving this theorem allows one to obtain the same estimates for the rate of convergence to their limits of the probabilities (2.3).

Obtaining the asymptotic expansions in powers of $1/\sqrt{n}$, requires more special assumptions concerning the distribution $F(x)$ and the form of the boundaries $g^\pm(t)$. One may find a rather comprehensive review of the achievements obtained in this field before 1964 in the paper by A. A. Borovkov and V. S. Koroljuk [3]. For instance, as asymptotic expansion of $P(\eta_G < v)$ of the form

$$\sum_{k=0}^{s} p_k n^{-k/2} + o(n^{-s/2}) \tag{2.7}$$

is possible if $E|\xi|^{2s+6} < \infty$, and the density of distribution $F(x)$ and the boundaries $g^\pm(t)$ have a sufficiently high degree of smoothness (Koroljuk).

The assumptions for one straight boundary, $g_n^+(t) = x = x(n)$, are more economical. If $F(t)$ has an absolutely continuous component and $E \exp\{\lambda\xi_k\} < \infty$ for $|\lambda| < \varepsilon$ for some $\varepsilon > 0$, then probabilities (2.3) admit the full asymptotic expansion in powers of x/n and $1/\sqrt{n}$ (Borovkov). If, in the last assertion, instead of finiteness of $E \exp\{\lambda\xi_k\}$, we demand only the existence of a finite number of moments, then the following result obtained recently by Nagaev will hold [4].

THEOREM 2.2. *Let $F(t)$ have an absolutely continuous component and suppose that $c_s = E|\xi_k|^s < \infty$, for $s > 3$. Then $(g^+(t) = x, g^-(t) = -\infty)$*

$$\begin{aligned} P(\eta_G > 1) &= P(\bar{S}_n < x\sqrt{n}) \\ &= \sqrt{\frac{2}{\pi}} \int_0^x \exp\{-\tfrac{1}{2}t^2\}\, dt + \exp\{-\tfrac{1}{2}x^2\} \sum_{j=1}^{s-3} \Pi_j(x) n^{-j/2} \\ &\quad + O(\min[n^{-1/2}, (1 + x^{1-s})n^{-(s-2)/2} \log^2 n]), \end{aligned} \tag{2.8}$$

where $\Pi_j(x)$ are some polynomials in x.

A similar but somewhat weaker result would hold if we required that

$$\limsup_{|\lambda|\to\infty} |E \exp\{i\lambda\xi_k\}| < 1, \tag{2.9}$$

instead of the existence of an absolutely continuous component.

The nature of the coefficients of polynomials Π_j is rather complicated ([4], [5]).

In the assertion mentioned above, we considered the event $\{\eta_G > v\}$ for $v = 1$. In this regard, we remark that the generality achieved by considering the events $\{\eta_G > v\}$ for arbitrary finite v is illusory. One can restrict oneself to the value $v = 1$. Indeed, $n\eta_G$ is an integer, but for integer nv, the event $\{\eta_G > v\}$ can be written as $\{\eta_G > 1\}$ for a new value of parameter $\tilde{n} = nv$ and new functions $\tilde{g}^{\pm}(t) = g^{\pm}(tv)v^{-1/2}$.

3. Large deviations

We arrive at the problem of large deviations if, for example, $E\xi_k = 0$, $D\xi_k = 1$, $g_n^{\pm}(t) = x(n)g^{\pm}(t)$, $x(n)/\sqrt{n} \to \infty$ with $n \to \infty$. In real problems, one can usually reduce determination of the asymptotic behavior of probabilities of the form

$$P\left(\eta_G < 1, \chi_G < 0, \frac{S_n}{x} \in \Delta\right), \qquad x = x(n), \tag{3.1}$$

or

$$P\left(\eta_G < 1, \chi_G > 0, \frac{S_n}{x} \in \Delta\right), \qquad x = x(n), \tag{3.2}$$

to the same boundary problems but with one boundary. For example, if $g^+(t) > 0$ and $g^-(t) < 0$, it is easy to see that

$$\begin{aligned} P(\eta_G < 1) \sim P\left\{\max_{k \leqq n}\left\{S_k - xg^+\left(\frac{k}{n}\right)\right\} > 0\right\} \\ + P\left\{\min_{k \leqq n}\left\{S_k - xg^-\left(\frac{k}{n}\right)\right\} < 0\right\}. \end{aligned} \tag{3.3}$$

Here, the relationship $a_n \sim b_n$ means that $a_n/b_n \to 1$ for $n \to \infty$. Accordingly, we shall restrict ourselves in this section to the case of one boundary when $g^-(t) = -\infty$.

Depending on the type of function $g^+(t) = g(t)$, when evaluating $P(\eta_G < 1)$ asymptotically, we arrive at two qualitatively different types of problems.

The first type of problem arises when $g(t) > 0$. In this case $P(\eta_G \leqq 1) \to 0$ with $n \to \infty$. Asymptotics of this probability are investigated in detail in [6]. The so called *level curves* play the most important part in their description. In order to describe these curves, we introduce the *deviation function*

(3.4) $$\Lambda(\alpha) = -\inf(-\lambda\alpha + \log\varphi(\lambda)), \qquad \varphi(\lambda) = E\exp\{\lambda\xi_k\}.$$

The function $\Lambda(\alpha) \geqq 0$ is defined for all real α. We denote

(3.5) $$\lambda_+ = \sup\{\lambda: \varphi(\lambda) < \infty\}, \qquad \lambda_- = \inf\{\lambda: \varphi(\lambda) < \infty\},$$
$$\alpha_+ = \lim_{\lambda\uparrow\lambda_+}[\log\varphi(\lambda)]', \qquad \alpha_- = \lim_{\lambda\downarrow\lambda_-}(\log\varphi(\lambda))'.$$

If $\lambda(\alpha)$ is a point where $\inf(-\lambda\alpha + \log\varphi(\lambda))$ is achieved, then $\Lambda(\alpha)$ may be expressed as

(3.6) $$\Lambda(\alpha) = \int_a^\alpha \lambda(u)\,du, \qquad a = E\xi_k.$$

From this it easily follows that $\Lambda(\alpha)$ is a convex function, achieving its minimum equal to 0 at $\alpha = a$. In the regions $(-\infty, \alpha_-)$, (α_+, ∞), the function $\lambda(\alpha)$ is constant and equals $\lambda_\mp$ respectively. Thus, the function $\Lambda(\alpha)$ is analytic in each of three regions, $(-\infty, \alpha_-)$, (α_-, α_+), (α_+, ∞). Discontinuity, for example, at the point α_+ is possible only if $\lambda_+ = \infty$, $\alpha_+ < \infty$ (the variable ξ_k is bounded from above by the value α_+) and $P(\xi_k = \alpha_k) > 0$. If the $\lambda_\pm$ are finite, then $\Lambda(\alpha)$ together with its first derivatives will be continuous at the points $\alpha_\pm$.

It is not difficult to find the expansion of the function $\Lambda(\alpha)$ valid in the neighborhood of the point $\alpha = a$, with coefficients of $(\alpha - a)^k$ defined by k semi-invariants of the distribution ξ.

For example, for the normal distribution $\varphi(\lambda) = \exp\{\frac{1}{2}\lambda^2\}$, $\alpha_\pm = \pm\infty$ and $\lambda(\alpha) = \alpha$.

For the Bernoulli scheme, $\varphi(\lambda) = pe^\lambda + qe^{-\lambda}$,

(3.7) $$\alpha_\pm = \pm 1, \qquad \lambda(\alpha) = \tfrac{1}{2}\log\frac{q(1+\alpha)}{p(1-\alpha)}.$$

For the centered Poisson distribution with parameter μ,

(3.8) $$\alpha_+ = \infty, \qquad \alpha_- = -\mu, \qquad \lambda(\alpha) = \log\frac{\mu+\alpha}{\mu}.$$

The probabilistic meaning of the deviation function is given by the equality

(3.9) $$\Lambda(\alpha) = -\lim_{\Delta_\alpha\to 0}\lim_{n\to\infty}\frac{1}{n}\log P\left(\frac{S_r}{n} \in \Delta_\alpha\right),$$

where Δ_α is a shrinking neighborhood of the point α.

The "inversion formula,"

(3.10) $$\varphi(\lambda) = \exp\left\{-\int_0^\lambda \theta(t)\,dt\right\},$$

is also valid, where $\theta(t)$ is the inverse function of

(3.11) $$t = \lambda(\theta) = \frac{\partial\Lambda(\theta)}{\partial\theta}.$$

In what follows, we assume that $a = E\xi_k = 0$, $D\xi_k = 1$, $\lambda_+ > 0$, $\lambda_- < 0$. We call the positive solution (if it exists) for the functional equation

$$(3.12) \qquad t\Lambda\left(\frac{a}{t}\right) = \Lambda(\tau), \qquad 0 < t \leqq 1,$$

the *level curve* $a_\tau(t)$, depending on parameter τ.

If ξ is such that $\Lambda(\alpha_+) = \infty$, then this equation has for each τ a unique solution $\alpha_\tau(t)$, which is a convex increasing function.

If $\Lambda(\alpha_+) < \infty$, then $a_\tau(t)$, possesses the indicated properties only if

$$(3.13) \qquad t \in (t_\tau, 1), \qquad t_\tau = \frac{\Lambda(\tau)}{\Lambda(\alpha_+)}.$$

When $t \in (0, t_\tau)$, the function $a_\tau(t)$ is to be defined as a segment of the straight line connecting the point (0, 0) with the end of the curve $a_\tau(t)$ at the point t_τ.

The functions $\alpha_\tau(t)$ are also increasing functions of τ. For τ small, $a_\tau(t) \sim \tau\sqrt{t}$. (The exact equality, $a_\tau(t) = \tau\sqrt{t}$, is true only for the normal distribution law.)

We may return now to the asymptotic description of $P(\eta_G < 1)$. To simplify the formulations of the results, we put $x(n) = n$ (if $x = \varepsilon n$, one may consider the function $g^*(t) = \varepsilon g(t)$ and $x^*(n) = n$). The main role is played by the maximal value of the parameter $\tau = \tau_g$ at which the curve $a_\tau = a_{\tau_g}$ first intersects the curve $g(t)$ as τ increases. The important fact is how this contact happened.

The following theorem is true.

THEOREM 3.1. *Let the set B of the points t, where the values $a_{\tau_g}(t)$ and $g(t)$ coincide, be contained in the interval $(t_{\tau_g}, 1)$ and be such that* mes $B > 0$. *Then*

$$(3.14) \qquad P(\eta_G < 1, \chi_G < y) \sim \phi_1(F, y, B)\sqrt{n}\exp\{-n\Lambda(\tau_g)\},$$

where $\phi_1(F, y, B)$ is a functional of known form.

Now let the point of contact v be a single one and in the neighborhood of this contact let the function $g(t)$ be p times differentiable. Let q be the number of initial derivatives of the functions $g(t)$ and $a_{\tau_g}(t)$ which are identical. Assume $p > q + 1$, then

$$(3.15) \qquad P(\eta_G < 1, \chi_G < y) \sim \phi_2(F, y, v, g)n^{1/2 - 1/(q+1)}\exp\{-n\Lambda(\tau_g)\},$$

where $\phi_2 > 0$ is also a known functional. However, unlike ϕ_1, the functional ϕ_2 has a local character with respect to g, since it depends only on $q + 1$ derivatives of the function $g(t)$ at the point v. When $x = o(n)$, the points of contact v of the functions $(x/n)g(t)$ and $a_{\tau_{xg/n}}(t)$ are, in general, mobile (they change with n) and the formulations of the results become more delicate, although the character of the dependence of the asymptotics on the value τ_g and on the type of contact remains as before.

The case of several points of contact is easily reduced to the case of a single point, considered in Theorem 3.1.

It is possible to make the analogous asymptotic analysis for the probabilities

$$P\left(\eta_G < 1, \chi_G < y, \frac{S_n}{x} \in \Delta\right). \tag{3.16}$$

The main factor here is the mutual disposition of curve $g(t)$ and the level curves; however, the level curves themselves are defined differently and already depend on two parameters rather than on one (see [6]).

4. Large deviations, special case

Consider the special case with a straight line boundary $g(t) = d + bt$. With the help of transformation of reversion it is always possible here to reduce the problem on distribution η_G to the investigation of the properties of $\bar{S}_n = \max_{0 \leqq k \leqq n} S_k$. The interesting case is that in which it happens that $E\xi_k = a < 0$ after the reversion, and therefore $\bar{S} = \bar{S}_\infty < \infty$ with probability 1. As mentioned above, a complete analysis of the asymptotic properties of the $\bar{S}_n$ distribution is in [5]. To those properties we add the following remarks (see [5], [7]). (The symbols c denotes different constants.)

THEOREM 4.1. *Suppose that $F(x)$ has an absolutely continuous component, and that there exists a root $q > 0$ of the equation $\varphi(\lambda) = 1$. Further, let $x/n \sim \alpha$, $\alpha_0 = \varphi'(q)$. Then if $\alpha < \alpha_0$, for a certain $\varepsilon > 0$,*

$$P(\bar{S}_n > x) \sim P(\bar{S} > x) = c \exp\{-qx\}(1 + o(\exp\{-\varepsilon x\})). \tag{4.1}$$

If $\alpha_0 < \alpha < \alpha_+$, then

$$P(\bar{S}_n > x) \sim c_1(\alpha) P(S_n > x) \sim \frac{c_2(\alpha)}{\sqrt{n}} \exp\left\{-n\Lambda\left(\frac{x}{n}\right)\right\}. \tag{4.2}$$

If $\alpha = \alpha_0(\Lambda(\alpha_0) = q)$, then the transition from one sort of asymptotics to another occurs with the help of the normal distribution function ϕ: for $u = (\sqrt{n}/\sigma)(x/n - a_0) = o(n^{1/6})$,

$$P(\bar{S}_n > x) = P(\bar{S} > x)\left[\phi(u) + \exp\{-\tfrac{1}{2}u^2\} \sum_{j=1}^{\infty} \frac{\Pi_{3j-1}(x)}{n^{j/2}}\right], \tag{4.3}$$

where σ is the variance of the distribution $F_q(A) = \int_A \exp\{qx\}\, dF(x)$, and $\Pi_k(u)$ are polynomials of degree k.

Since the asymptotic behavior of $P(S_n > x)$ has been studied well, to describe the asymptotic behavior of $P(\bar{S}_n > x) > P(S_n > x)$ in the general case (when the conditions of Theorem 4.1 are not satisfied), it is sufficient to know (1) the rate of the convergence of distributions $\bar{S}_n$ and $\bar{S}$, and (2) the asymptotic behavior of $\bar{S}$. The answer to the first question is in Theorem 4.2.

THEOREM 4.2. *For any λ such that $\varphi(\lambda) \leqq 1$,*

$$P(\bar{S} > x) - P(\bar{S}_n > x) \leqq \exp\{-\lambda x\}\varphi^n(\lambda). \tag{4.4}$$

(*One can write* $\exp\{-n\Lambda(x/n)\}$ *on the right side of* (4.4) *if* $\varphi[\lambda(x/n)] \leqq 1$.)

If $E\xi_k = a < 0$, $D\xi_k = 1$, *and* $c_m^+ = E[\max(0, \xi)]^m < \infty$ *for* $m \geqq 2$, *then* $\bar{c}_{m-1} \equiv E\bar{S}^{m-1} < \infty$ *and for all* $n \geqq n_0$,

$$P(\bar{S} > x) - P(\bar{S}_n > x) = \frac{2^{m+1} c_m^+ e}{|a|^m n^{m-1}} + \frac{\bar{c}_{m-1}}{(x + \frac{1}{2}|a|n)^{m-1}}, \tag{4.5}$$

where $n_0 = n_0(m, c_m)$ *is known explicitly.*

Concerning the asymptotic properties of the distribution of $\bar{S}$, it is well known that if $\lambda_+ > 0$, $\varphi(\lambda_+) \geqq 1$, then $P(\bar{S} > x) \sim c_1 e^{-qx}$ with $x \to \infty$ (at $\lambda_+ = q$ we also suppose that $\varphi'(\lambda_+) < \infty$.) Supplement this assertion by the following one.

THEOREM 4.3. *If* $\varphi(\lambda_+) < 1$ *or* $\lambda_+ = 0$, $\varphi'(0) = a > -\infty$, *then* $P(\bar{S} > x) \sim c_2 H(x)$, *where* $c_2 = |a|^{-1}$ *for* $\lambda_+ = 0$, *and where* $H(x) = \int_x^\infty (1 - F(t))\, dt$.

It is assumed here that for any $0 < h \leqq 1$ *and* $t \to \infty$,

$$\exp\{h\lambda_+\}\frac{H(t+h)}{H(t)} \to 1, \qquad 0 < \frac{\exp\{tb\lambda_+\}H(tb)}{\exp\{t\lambda_+\}H(t)} < c(b) < \infty. \tag{4.6}$$

Concerning the distribution of $\bar{S}$, we shall also note that there exists a class of distributions $\mathscr{R}$ everywhere dense in the sense of weak convergence in the set of all distributions, and such that for $F \in \mathscr{R}$ the distribution of $\bar{S}$ can be found in explicit form. This fact is used in conjunction with the "continuity theorem" showing when the nearness of F_1 and F_2 (in the sense of weak convergence) implies the nearness of corresponding suprema. The latter will hold if the value

$$\int_{-\infty}^{\infty} a(t)|F_1(t) - F_2(t)|\, dt \tag{4.7}$$

is small for a function $a(t) > 0$ such that $a(t) = 1$ when $t > 0$, $\int_{-\infty}^0 a(t)\, dt < \infty$. By itself, weak convergence, $F_1 \Rightarrow F_2$, is not sufficient for the convergence of suprema distributions.

The class $\mathscr{R}$, mentioned above, contains all the distributions for which either $\varphi_+(\lambda) = E(e^{\lambda\xi}; \xi \geqq 0)$ or $\varphi_-(\lambda) = E(e^{\lambda\xi}; \xi \leqq 0)$ are rational functions.

Results close to the latter were also obtained by H. J. Rossberg [8], [9] under some special conditions.

5. The second type of problem: problems on large deviations

Now let us return to the general boundary problems for large deviations. In Section 3, we considered the case where $g(t) > 0$, $t \in [0, 1]$. If $\inf_{(0,1)} g(t) < 0$, then $P(\eta_G \leqq 1) \to 1$, when $n \to \infty$ (here again $E\xi_k = 0$) and we shall investigate the rate of convergence to zero of the complementary probability $P(\eta_G > 1)$. The asymptotic nature of this probability appears to be quite different and significantly more complicated. Here one can find only the asymptotic behavior for the *logarithm* of this probability $P(\eta_G > 1)$.

Suppose $g(t)$ has no discontinuities of the second kind and for every t, $g(t) = \min\big(g(t-0), g(t+0)\big)$. Let t_0 be the point where $\inf_{(0,1)} g(t) = g(t_0)$ and let B_0 be the set of points $(0, 0)$, $\big(t, y > g(t)\big)$ for $0 \leqq t \leqq t_0$. We denote by $h(t)$ the lower boundary of the convex closure of the set B_0. The curve $h(t)$ evidently realizes the shortest path from point $(0, 0)$ to point $\big(t_0, g(t_0)\big)$ which does not intersect B_0. This function will be convex and absolutely continuous. If we put $h(t) = g(t_0)$ on $[t_0, 1]$, then $h(t)$ will keep this property. For such functions $h(t)$, the functional

$$W(h) = \int_0^1 \Lambda\left(\frac{dh}{dt}\right) dt = \int_0^{t_0} \Lambda\left(\frac{dh}{dt}\right) dt \tag{5.1}$$

is defined. First let $x = x(n) \sim n$ when $n \to \infty$. (As already noted, one can reduce the case $x \sim \varepsilon n$ to $x \sim n$ by changing the boundary.)

THEOREM 5.1. *If $h'(0) > -\infty$ and the interval $\big(h'(0), 0\big)$ contains no point of discontinuity of the function $\Lambda(\alpha)$, then* $\log P(\eta_G > 1) \sim -nW(h)$.

To describe the asymptotic form of the distribution η_G in the problem with a fixed terminal value, that is, the probability of the event $\{\eta_G > 1, S_n/x \in \Delta_b\}$ (where Δ_b is the neighborhood of the point b), it is necessary to construct the shortest path $h_b(t)$ connecting the points $(0, 0)$ with $(1, b)$ and not intersecting the set B_1 of the points $\big(t, y > g(t)\big), 0 \leqq t \leqq 1$. (This path will evidently coincide with $h(t)$ in the segment $[0, t_0]$ if $bt_0 \geqq g(t_0)$.)

THEOREM 5.2. *If Δ_b is a shrinking neighborhood of the point b,*

$$\lim_{\Delta_b \to 0} \lim_{n \to \infty} \frac{1}{n} \log P\left(\eta_G > 1, \frac{S_n}{x} \in \Delta_b\right) = -W(h_b). \tag{5.2}$$

If ξ_k is lattice-like and $P(S_n = bx) > 0$, then from the very beginning we may take a single point b as a Δ_b.

Now let us consider the case $x = o(n)$. Introduce the functional

$$V(h) = \int_0^1 \left(\frac{dh}{dt}\right)^2 dt = \int_0^1 \frac{dh}{dt}\, dh(t). \tag{5.3}$$

THEOREM 5.3. *If $x/n \to 0$, $x(n \log n)^{-1/2} \to \infty$, with $|\lambda_\pm| > 0$, then*

$$\log P(\eta_G > 1) \sim -\frac{x^2}{2n} V(h). \tag{5.4}$$

The analogous result is true for $P(\eta_G > 1, S_n/x \in \Delta_b)$.

These theorems, obtained in [10], show what the *exponential* part in the probability $P(\eta_G > 1)$ is. As for the *power* factor, we may conclude on the basis of different examples that it may be of any form depending on the corresponding function $g(t)$. It is possible to calculate this factor only for very special kinds of boundaries $g(t)$, for example, for the broken lines consisting of a finite number of straight line segments.

6. Generalization

It is natural to consider the next generalization of these problems. For example, some asymptotic problems of testing theory for statistical hypotheses lead to this. Denote by $S_n(t)$ a random broken line connecting the points $(k/n, S_k/x)$ for $k = 0, 1, \cdots, n$. In the previous parts, the asymptotics of

$$\log P\big(S_n(t) \in G\big) \tag{6.1}$$

were investigated when G is the set of functions $f(t) < g(t)$, $t \in [0, 1]$.

The question concerns the behavior of (6.1) when $n \to \infty$ if G is the arbitrary open set in the space $C(0, 1)$.

The following theorems hold here [10]. Denote by $\bar{G}$ the closure of G in the metric of space $C(0, 1)$ and by $E \subset C(0, 1)$ the set of all absolute continuous functions $f(t)$, $f(0) = 0$, for which there exists a finite number of intervals $\Delta_1, \Delta_2, \cdots, \Delta_r; \cup \bar{\Delta}_j = [0, 1]$, where $f'(t)$ is monotone and bounded. Obviously, $\Lambda(f'(t))$ is Riemann integrable for $f \in E$. Further, let

$$\begin{aligned} W(G) &= \inf_{f \in E \cap G} W(f) = \inf_{f \in E \cap G} \int_0^1 \Lambda(f'(t))\, dt, \\ V(G) &= \inf_{f \in E \cap G} V(f) = \inf_{f \in E \cap G} \int_0^1 (f'(t))^2\, dt. \end{aligned} \tag{6.2}$$

THEOREM 6.1. *Let $x \sim n$ and $W(G) = W(\bar{G})$. If $|\lambda_\pm| = \infty$, then*

$$\log P\big(S_n(t) \in G\big) \sim -nW(G). \tag{6.3}$$

THEOREM 6.2. *Let $x/n \to 0$, $x(n \log n)^{-1/2} \to \infty$, $V(G) = V(\bar{G})$. If $|\lambda_\pm| > 0$, then*

$$\log P\big(S_n(t) \in G\big) \sim -\frac{x^2}{2n} V(G). \tag{6.4}$$

The condition $|\lambda_\pm| = \infty$ in Theorem 6.1 implies that $\varphi(\lambda)$ is an entire function. We are quite sure that this requirement is unnecessary and that the theorem will hold with finite $\lambda_\pm$, but this is only a conjecture. We can give here only one sufficient condition. Namely, if $W(G) = W(\bar{G})$ and there exists a compact K in $C(0, 1)$ such that when $n \to \infty$

$$\log P(S_n \in G \cap K) > \log P\big(S_n(t) \in G\big) + o\left(\frac{x^2}{n}\right), \tag{6.5}$$

then (6.3) is true.

Relations (6.3) and (6.4) remain valid in problems with a fixed terminal value (see Theorem 5.2), but it is necessary to take the infimum of the functionals W, V for $f \in E_b \cap G$, where $E_b \in E$ contains only the functions for which $f(1) = b$.

Many of the results given remain valid also for the processes with independent increments. B. A. Rogozin [11] extends to these cases the theorems on asymptotic expansions and large deviations for the maximum $\bar{S}_n(t)$, where

$$\bar{S}_n(t) = \sup_{0 \leq u \leq t} S_n(u), \qquad S_n(t) = \frac{X(nt)}{x}, \tag{6.6}$$

and $X(u)$ is a process with independent increments, satisfying rather weak conditions.

If $X(u)$ is the sum of a Wiener process and of a generalized Poisson process, that is, if

$$\psi(\lambda) = \log E \exp \{\lambda X(1)\} = \lambda\theta - \tfrac{1}{2}\sigma^2\lambda^2 + \beta \int (\exp\{\lambda u\} - 1)\, dN(u), \tag{6.7}$$

where $N(u)$ is a distribution function, $\beta \geqq 0$, then the assertions of Theorem 6.1 and 6.2 remain completely valid, where G denotes now the set from the space $D(0, 1)$ containing with f its ρ_c neighborhood $\rho_c(f, g) = \sup_{[0,1]}|f(t) - g(t)|$. The closure $\bar{G}$ is used in the sense of ρ_c convergence. Then if $W(G) = W(\bar{G})$, $\psi(\lambda)$ is an integral function, and for $x \sim n$, we have

$$\log P(S_n(t) \in G) \sim -nW(G). \tag{6.8}$$

The case where $x = o(n)$ is similar. From this theorem one can derive, in particular, the theorem of I. N. Sanov [12] on large deviations of empirical distribution functions. Since this theorem was obtained by us under somewhat different conditions, we shall give it here. Let $F_n(t)$ be an empirical function constructed by n independent observations of a random variable with the continuous distribution function $F(t)$. And let G be a measurable ρ_c open set in $D(0, 1)$. Put

$$W_F(G) = \inf_{f \subset G \cap E_F} \int \log \frac{df}{dF}\, df, \qquad V_F(G) = \inf_{g \in E_F \cap G} \int \frac{dg}{dF}\, dg, \tag{6.9}$$

where E_F is the set of distribution functions g absolutely continuous with respect to F, and such that there exists a finite number of intervals $\Delta_1, \cdots, \Delta_r$, $\cup \bar{\Delta}_j = [-\infty, \infty]$, where dg/dF is monotone and bounded; $\bar{G}$ denotes the ρ_c closure of G.

THEOREM 6.3. *If* $W_F(G) = W_F(\bar{G})$, *then* $\log P(F_n(t) \in G) \sim -nW_F(G)$.

The following affirmation is also valid.

THEOREM 6.4. *If* $x/n \to 0$, $x(n \log n)^{-1/2} \to \infty$, $V_F(G) = V_F(\bar{G})$, *then*

$$\log P\left(F_n(t) - F(t) \in \frac{x}{n} G \right) \sim \frac{x^2}{2n}(1 - V_F(G)). \tag{6.10}$$

Theorems of this kind are essential in mathematical statistics.

REFERENCES

[1] A. N. KOLMOGOROV, "Eine Verallgemeinerung des Laplace-Ljapounoffschen Satzes," *Izv. Akad. Nauk. SSSR Otdel.*, (1931), p. 959.

[2] S. V. NAGAEV, "On the speed of convergence in a boundary problem, I, II," *Teor. Verojatnost. i Primenen.*, Vol. 15 (1970). pp. 179–199; pp. 419–441.

[3] A. A. Borovkov and V. S. Koroliuk, "On the results of an asymptotic analysis for problems with boundaries," *Teor. Verojatnost. i Primenen.*, Vol. 10 (1965), pp. 255–266.

[4] S. V. Nagaev, "Asymptotic expansions of the distribution functions of the maximum of sums of independent, identically distributed random variables," *Sibirsk. Mat. Ž.*, Vol. 11 (1970), pp. 381–406.

[5] A. A. Borovkov, "New limit theorems for boundary problems for sums of independent summands," *Sibirsk. Mat. Ž.*, Vol. 3 (1962), pp. 645–694.

[6] ———, "Analysis of large deviations for boundary problems with arbitrary boundaries, I, II," *Sibirsk. Mat. Ž.*, Vol. 5 (1964), pp. 253–284; pp. 750–767.

[7] ———, "On factorization identities and properties of the distribution of the supremum of successive sums," *Teor. Verjatnost. i Primenen.*, Vol. 15 (1970), pp. 377–418.

[8] H. J. Rossberg, "Eine neue Methode zur Behandlung der Integralgleichung von Lindley and ihrer Verallgemeinenung durch Finch," *Elektron. Informations verarb. und Kybernet.*, Vol. 3 (1967), pp. 215–238.

[9] ———, "Über die Verteilung von Wartezeiten," *Math. Nachr.*, Vol. 30 (1965), pp. 1–16.

[10] A. A. Borovkov, "Boundary problems for random walks and large deviations in function spaces," *Teor. Verojatnost. i Primenen.*, Vol. 12 (1967), pp. 635–654.

[11] B. A. Rogozin, "Distribution of the maximum of processes with independent increments," *Sibirsk. Mat. Ž.*, Vol. 10 (1969), pp. 1334–1363.

[12] I. N. Sanov, "On the probability of large deviations of random variables," *Mat. Sb.*, Vol. 42 (1957), pp. 11–44. (English translation, *Select. Trans. Math. Statist. Prob.*, Vol. 1 (1961), pp. 213–244.)

THE RANGE OF RANDOM WALK

NARESH C. JAIN and WILLIAM E. PRUITT
UNIVERSITY OF MINNESOTA

1. Introduction

Let $\{X_n, n \geqq 1\}$ be a sequence of independent identically distributed random variables, defined on a probability space $(\Omega, \mathscr{F}, P)$, which take values in the d dimensional integer lattice E_d. The sequence $\{S_n, n \geqq 0\}$ defined by $S_0 = 0$ and $S_n = \Sigma_{k=1}^n X_k$ is called a random walk. The range of the random walk, denoted by R_n, is the cardinality of the set $\{S_0, S_1, \cdots, S_n\}$; it is the number of distinct points visited by the random walk up to time n. Our object here is to study the asymptotic behavior of R_n. Two specific problems are considered:

(i) Does $R_n/ER_n \to 1$ a.s.? If so, this will be called the strong law for R_n.

(ii) Does $(R_n - ER_n)(\text{Var } R_n)^{-1/2}$ converge in distribution? If so, this will be called the central limit theorem for R_n.

The random walk may take place on a proper subgroup of E_d. In this case, the subgroup is isomorphic to some E_k for $k \leqq d$; if $k < d$, then the transformation should be made and the problem considered in k dimensions. We will assume throughout the paper that this reduction has been made, if necessary, and that d is the genuine dimension of the random walk.

Dvoretzky and Erdös [2] proved the strong law for the range of simple random walk for $d \geqq 2$. (Simple random walk is one for which the distribution of X_1 assigns probability $(2d)^{-1}$ to each of the $2d$ neighbors of the origin.) Their method was to obtain a somewhat crude estimate of Var R_n and then use the Chebyshev inequality. While this worked fairly easily for $d \geqq 3$, they had to work much harder for $d = 2$. By a rather sophisticated technique, they managed to improve the required probability estimate enough to obtain the proof.

Let $p = P[S_1 \neq 0, S_2 \neq 0, \cdots]$. The random walk is called transient if $p > 0$ and recurrent otherwise. Using a very elegant technique Kesten, Spitzer, and Whitman ([12], p. 38) proved that for *all* random walks $R_n/n \to p$ a.s. For transient random walks $ER_n \sim pn$, so that their result includes the strong law for all transient random walks.

There are recurrent random walks only if the dimension is one or two. In [7], we attempted to prove the strong law for R_n for the general recurrent random walk in two dimensions, but we succeeded only partially. Our method there was to imitate the proof of Dvoretzky and Erdös, that is, to obtain an estimate for Var R_n and then to improve the probability estimate by their methods. In Section 3, we will prove the strong law for R_n for *all* two dimensional recurrent random walks by an essentially different technique. We use a very delicate

Research was supported in part by the National Science Foundation Grant GP-21188.

method for estimating Var R_n, and the estimate is good enough so that once it is available the strong law follows in a fairly straightforward manner. In Section 4, we will show that if $EX_1 = 0$, $E|X_1|^2 < \infty$, and $d = 2$, then Var $R_n \sim cn^2/\log^4 n$ for some positive constant c. This should be compared with the Dvoretzky-Erdös bound which was $O(n^2 \log \log n/\log^3 n)$.

The second problem was first considered by Jain and Orey [6] who showed that if the random walk is strongly transient with $p < 1$, then Var $R_n \sim cn$ and the central limit theorem applies with the limit law normal. (The random walk is strongly transient if $\Sigma_{n=1}^{\infty} \Sigma_{j=n}^{\infty} P[S_j = 0] < \infty$.) Note that the case $p = 1$ is not interesting, since then $R_n = n + 1$ a.s. Thus, criteria for strong transience are of interest, since they will also imply the central limit theorem. All random walks with $d \geqq 5$ are strongly transient and if $EX_1 = 0$ and $E|X_1|^2 < \infty$, the random walk is strongly transient if and only if $d \geqq 5$. In Section 5, we give two additional sufficient conditions for strong transience. The first is that the random walk is aperiodic and not irreducible and the second is that $EX_1 \neq 0$ and $E|X_1|^2 < \infty$. These are both valid regardless of the dimension.

In [8], we considered the central limit theorem for random walks which were transient but not strongly transient. We proved that if $d = 4$, then Var $R_n \sim cn$ and the central limit theorem applies with a normal limit law. The growth of the variance should be compared with the Dvoretzky-Erdös bound which was $O(n \log n)$. For $d = 3$, we proved that Var $R_n = O(n \log n)$ in general. Under the additional assumption that $EX_1 = 0$ and $E|X_1|^2 < \infty$, we proved that Var $R_n \sim cn \log n$ and that the central limit theorem still is valid with a normal limit law. The Dvoretzky-Erdös bound in three dimensions was $O(n^{3/2})$.

The case of one dimension is rather unique. If Var $X_1 < \infty$, the random walk is either strongly transient if $EX_1 \neq 0$, or recurrent if $EX_1 = 0$. In the first case, the strong law and central limit theorem are both known for R_n as we have mentioned above. However, when $EX_1 = 0$ the situation is quite different and we shall prove in Section 6 that $n^{-1/2}R_n$ converges in distribution to a proper law. This implies that it is impossible to have a strong law, for $(R_n - \alpha_n)/\beta_n$ cannot converge even in probability to a nonzero constant for any sequences $\{\alpha_n\}$, $\{\beta_n\}$ except in the trivial case that R_n/β_n already converges to zero in probability.

It would be interesting to know whether the central limit theorem is valid for R_n when $d = 2$, at least when $EX_1 = 0$ and $E|X_1|^2 < \infty$, in which case the behavior of Var R_n is known. Another question is whether there is a limit law in general for R_n when $d = 3$. It is known that there is a limit law when $EX_1 = 0$ and $E|X_1|^2 < \infty$, or when the random walk is strongly transient. We have also said nothing about random walk on the line with Var $X_1 = \infty$. Of course, it is known that the strong law holds if the random walk is transient and the central limit theorem holds if it is strongly transient but it would be nice to have more information than this. One can also ask more delicate questions about the growth of R_n in general, that is, to obtain better upper and lower envelopes for R_n than are given by the strong law. Some results of this type are mentioned in [2].

2. Preliminaries

It will be convenient to think of the random walk as a Markov chain and we will use some of the terminology of general Markov chains. For $x \in E_d$, the random walk starting at x will refer to the random walk with $S_0 = x$ and $S_n = x + \Sigma_{k=1}^{n} X_k$. The notation $P_x[\cdot]$ will be used to denote probabilities of events related to this random walk; when $x = 0$, we will simply use $P[\cdot]$. Thus, for $n \geqq 0$ and $x, y \in E_d$, we let

$$P^n(x, y) = P_x[S_n = y] = P[S_n = y - x], \tag{2.1}$$

and note that $P^n(x, y) = P^n(0, y - x)$. For transient random walk the Green function is defined by $G(x, y) = \Sigma_{k=0}^{\infty} P^k(x, y)$. For an arbitrary set H of lattice points, T_H will denote the first hitting time of H, that is,

$$T_H = \min \{k \geqq 1 : S_k \in H\}; \tag{2.2}$$

if there are no positive integers k with $S_k \in H$, then $T_H = \infty$. If H consists of a single point x, we will write T_x instead of $T_{\{x\}}$. The taboo probabilities are defined by

$$P_H^n(x, y) = P_x[S_n = y, T_H \geqq n] \tag{2.3}$$

for $n \geqq 1$. We will use u_n for $P^n(0, 0)$, f_n for $P_0^n(0, 0)$, and

$$r_n = P[T_0 > n] = p + \sum_{k=n+1}^{\infty} f_k. \tag{2.4}$$

Another equation which is satisfied is $\Sigma_{k=0}^{n} u_k r_{n-k} = 1$; since r_n is monotone, it follows that

$$r_n \leqq \left(\sum_{k=0}^{n} u_k \right)^{-1}. \tag{2.5}$$

Kesten and Spitzer [10] proved that for any two dimensional random walk, r_n is slowly varying and this will be quite useful. We will need the following simple observation about slowly varying functions that decrease.

LEMMA 2.1. *Let $\{\ell_n\}$ be slowly varying and decreasing. Then there is a positive constant c such that if $j \leqq n$, then $j\ell_j \leqq cn\ell_n$. In particular, this implies for any two dimensional random walk there is a c such that $jr_j^4 \leqq cnr_n^4$ for $j \leqq n$.*

PROOF. Since ℓ_n is slowly varying, there exists an integer N such that for $n \geqq N$, $\ell_{2^n}/\ell_{2^{n+1}} < 2$. Let $2^N \leqq j \leqq n$; then there are integers β, γ with $\beta \leqq \gamma$ such that $2^\beta \leqq j < 2^{\beta+1}$ and $2^\gamma \leqq n < 2^{\gamma+1}$. Since ℓ_n is decreasing,

$$j\ell_j \leqq 2^{\beta+1}\ell_{2^\beta} \leqq 2 \cdot 2^{\gamma+1}\ell_{2^{\gamma+1}} \leqq 4n\ell_n. \tag{2.6}$$

To cover the cases with $j < 2^N$, we replace the 4 by a possibly larger constant c.

For any random walk in two dimensions with $p < 1$, there is a positive constant A such that

$$P^n(0, x) \leqq An^{-1} \tag{2.7}$$

for all $x \in E_2$ and $n \geqq 1$. This is a standard estimate; it is proved under these conditions in [8]. Another standard result which we shall use is that for any $n \geqq 1$ and $x \in E_d$,

$$P_0^n(0, x) = P_x^n(0, x). \tag{2.8}$$

This is proved by considering the dual or reversed random walk.

Lemma 2.2 is proved in [8] for $\gamma = 0, 1$. Although the general proof is the same, we shall give it since it is short and we will need some of the intermediate steps in Section 4.

LEMMA 2.2. *For* $\gamma \geqq 0$,

$$\sum_{k=1}^{m} P_0^k(0, x) r_{m-k}^{\gamma} \leqq \sum_{k=1}^{m} P^k(0, x) r_{m-k}^{\gamma+1}. \tag{2.9}$$

Equality holds for $\gamma = 0$.

PROOF. By considering the first return to zero,

$$P^k(0, x) = P_0^k(0, x) + \sum_{j=1}^{k-1} f_j P^{k-j}(0, x), \tag{2.10}$$

so that

$$\sum_{k=1}^{m} P_0^k(0, x) r_{m-k}^{\gamma} = \sum_{k=1}^{m} P^k(0, x) \left[r_{m-k}^{\gamma} - \sum_{j=1}^{m-k} f_j r_{m-k-j}^{\gamma} \right]. \tag{2.11}$$

The proof is completed by observing that

$$r_{m-k}^{\gamma} - \sum_{j=1}^{m-k} f_j r_{m-k-j}^{\gamma} = r_{m-k}^{\gamma+1} + \sum_{j=1}^{m-k} f_j [r_{m-k}^{\gamma} - r_{m-k-j}^{\gamma}] \tag{2.12}$$

and then using the monotonicity of r_n.

3. The strong law for R_n in the plane

The result that we will prove in this section is

THEOREM 3.1. *For any recurrent random walk in two dimensions,* $R_n/ER_n \to 1$ a.s. *as* $n \to \infty$.

The main part of the proof is to find an estimate of the form $\operatorname{Var} R_n = O(\varphi(n))$, where φ is a function with the property that for every given $\alpha > 1$ there is a sequence of positive integers $\{n_k\}$ such that $n_{k+1}/n_k \to \alpha$ and $\sum_{k=1}^{\infty} \varphi(n_k) n_k^{-2} r_{n_k}^{-2} < \infty$. Once we have such an estimate for the variance the proof of Theorem 3.1 can be finished as in [2] by the following argument. Since $ER_n \sim n r_n$, we have by Chebyshev's inequality, for every $\varepsilon > 0$,

$$P[|R_n - ER_n| \geqq \varepsilon ER_n] = O(\varphi(n) n^{-2} r_n^{-2}). \tag{3.1}$$

Let $\alpha > 1$ and $\{n_k\}$ be as above. Then by the Borel-Cantelli lemma, $R_n/ER_n \to 1$ a.s. along this subsequence. To fill in between, by the monotonicity of R_n, for $n_k \leqq n < n_{k+1}$,

$$\frac{R_{n_k}}{ER_{n_k}}\frac{ER_{n_k}}{ER_{n_{k+1}}} \leq \frac{R_n}{ER_n} \leq \frac{R_{n_{k+1}}}{ER_{n_{k+1}}}\frac{ER_{n_{k+1}}}{ER_{n_k}}, \tag{3.2}$$

and we know that the lower and upper bounds converge to α^{-1} and α, respectively, since $ER_n \sim nr_n$ and r_n is slowly varying. But α can be chosen as close to 1 as we please, so this is sufficient.

The remainder of this section will be devoted to obtaining the desired estimate for Var R_n.

THEOREM 3.2. *For any recurrent random walk in two dimensions*, Var $R_n = O(\varphi(8n))$ where

$$\varphi(n) = n + nr_n^4 \sum_{k=1}^{n} ku_k \log\left(\frac{n}{k}\right). \tag{3.3}$$

REMARK 3.1. Since ku_k is bounded, it follows from the theorem that Var $R_n = O(n^2 r_n^4)$. This is a very good bound for the case of simple random walk, but it is very poor in general.

PROOF. Let $Z_0 = 1$ and for $k \geqq 1$ let

$$Z_k = I[S_k \neq S_{k-1}, \cdots, S_k \neq S_0]. \tag{3.4}$$

Then $R_n = \Sigma_{k=0}^n Z_k$ and so

$$\text{Var } R_n = \sum_{k=1}^{n} \text{Var } Z_k + 2\sum_{i=1}^{n}\sum_{j=i+1}^{n} \text{Cov }(Z_i, Z_j). \tag{3.5}$$

The first sum is $O(n)$, so we will concentrate on the second one. By reversing the time parameter, it is easy to see that $EZ_i = r_i$ and that for $i < j$

$$EZ_iZ_j = \sum_{x\neq 0} P_0^{j-i}(0, x)P_x[T_x > i, T_0 > i]. \tag{3.6}$$

Hence, we can write

$$\text{Cov }(Z_i, Z_j) = EZ_iZ_j - r_ir_j = \sum_{x\neq 0} P_0^{j-i}(0, x)b_i(x), \tag{3.7}$$

where

$$\begin{aligned} b_i(x) &= P_x[T_x > i, T_0 > i] - P_x[T_x > i]P_x[T_0 > i] \\ &= P_x[T_x \leqq i, T_0 \leqq i] - P_x[T_x \leqq i]P_x[T_0 \leqq i]. \end{aligned} \tag{3.8}$$

Using the last expression for $b_i(x)$, we will find a more useful form for it. We have

$$\begin{aligned} P_x[T_x < T_0 \leqq i] &= \sum_{\nu=1}^{i} P_x[\nu = T_x < T_0 \leqq i] \\ &= \sum_{\nu=1}^{i}\sum_{\xi=1}^{i-\nu} P_{x0}^{\nu}(x, x)P_0^{\xi}(x, 0) \\ &= \sum_{\xi=1}^{i}\sum_{\nu=1}^{i-\xi} P_{x0}^{\nu}(x, x)P_0^{\xi}(x, 0). \end{aligned} \tag{3.9}$$

Similarly,

$$P_x[T_0 < T_x \leqq i] = \sum_{\nu=1}^{i} \sum_{\xi=1}^{i-\nu} P_{x0}^{\xi}(x, 0) P_x^{\nu}(0, x). \tag{3.10}$$

Considering the first visit to x,

$$\begin{aligned} \sum_{\xi=1}^{i-\nu} P_0^{\xi}(x, 0) &= \sum_{\xi=1}^{i-\nu} P_{x0}^{\xi}(x, 0) + \sum_{\xi=1}^{i-\nu} \sum_{\eta=1}^{\xi-1} P_{x0}^{\eta}(x, x) P_0^{\xi-\eta}(x, 0) \\ &= \sum_{\xi=1}^{i-\nu} P_{x0}^{\xi}(x, 0) + \sum_{\beta=1}^{i-\nu} \sum_{\eta=1}^{i-\nu-\beta} P_{x0}^{\eta}(x, x) P_0^{\beta}(x, 0). \end{aligned} \tag{3.11}$$

Substituting in (3.10) for $\Sigma_{\xi=1}^{i-\nu} P_{x0}^{\xi}(x, 0)$ from this expression, we get

$$P_x[T_0 < T_x \leqq i] = \sum_{\nu=1}^{i} \sum_{\xi=1}^{i-\nu} P_x^{\nu}(0, x) P_0^{\xi}(x, 0) \left[1 - \sum_{\eta=1}^{i-\nu-\xi} P_{x0}^{\eta}(x, x) \right]. \tag{3.12}$$

Changing the order of summation and combining with (3.9), we have

$$\begin{aligned} &P_x[T_x \leqq i, T_0 \leqq i] \\ &= \sum_{\xi=1}^{i} \sum_{\nu=1}^{i-\xi} P_0^{\xi}(x, 0) \left[P_{x0}^{\nu}(x, x) + P_x^{\nu}(0, x) \left\{ 1 - \sum_{\eta=1}^{i-\nu-\xi} P_{x0}^{\eta}(x, x) \right\} \right]. \end{aligned} \tag{3.13}$$

By considering the first visit to zero,

$$\sum_{\nu=1}^{i-\xi} P_x^{\nu}(x, x) = \sum_{\nu=1}^{i-\xi} P_{x0}^{\nu}(x, x) + \sum_{\beta=1}^{i-\xi} \sum_{\eta=1}^{i-\xi-\beta} P_{x0}^{\eta}(x, 0) P_x^{\beta}(0, x). \tag{3.14}$$

We substitute the expression this gives for $\Sigma_{\nu=1}^{i-\xi} P_{x0}^{\nu}(x, x)$ in the last expression. Since for $x \neq 0$,

$$\begin{aligned} &1 - \sum_{\eta=1}^{i-\nu-\xi} \{ P_{x0}^{\eta}(x, x) + P_{x0}^{\eta}(x, 0) \} \\ &\qquad = P_x[T_x > i - \nu - \xi, T_0 > i - \nu - \xi], \end{aligned} \tag{3.15}$$

we have

$$\begin{aligned} &P_x[T_x \leqq i, T_0 \leqq i] \\ &= \sum_{\xi=1}^{i} \sum_{\nu=1}^{i-\xi} P_0^{\xi}(x, 0) \{ f_\nu + P_x^{\nu}(0, x) P_x[T_x > i - \nu - \xi, T_0 > i - \nu - \xi] \}. \end{aligned} \tag{3.16}$$

But since

$$P_x[T_0 \leqq i] P_x[T_x \leqq i] = \sum_{\xi=1}^{i} P_0^{\xi}(x, 0) \sum_{\nu=1}^{i} f_\nu, \tag{3.17}$$

we obtain the cancellation we need in the expression for $b_i(x)$:

$$(3.18)\qquad b_i(x) = \sum_{\xi=1}^{i} P_0^{\xi}(x, 0) \Bigg\{ -\sum_{\nu=i-\xi+1}^{i} f_\nu + \sum_{\nu=1}^{i-\xi} P_x^{\nu}(0, x) P_x[T_x > i - \nu - \xi, T_0 > i - \nu - \xi] \Bigg\}.$$

At this point, we can afford to be somewhat crude and use the bound

$$(3.19)\qquad b_i(x) \leqq \sum_{\xi=1}^{i} \sum_{\nu=1}^{i-\xi} P_0^{\xi}(x, 0) P_x^{\nu}(0, x) r_{i-\nu-\xi}.$$

Using (2.8) and Lemma 2.2, we obtain

$$(3.20)\qquad b_i(x) \leqq \sum_{\xi=1}^{i} \sum_{\nu=1}^{i-\xi} P^{\xi}(x, 0) P^{\nu}(0, x) r_{i-\nu-\xi}^3.$$

Letting $\lambda = j - i$ in (3.7), we see that

$$(3.21)\qquad \begin{aligned} C_n &= \sum_{i=1}^{n} \sum_{j=i+1}^{n} \operatorname{Cov}(Z_i, Z_j) \\ &\leqq \sum_{i=1}^{n} \sum_{\lambda=1}^{n-i} \sum_{\xi=1}^{i} \sum_{\nu=1}^{i-\xi} \sum_{x \neq 0} P_0^{\lambda}(0, x) P^{\xi}(x, 0) P^{\nu}(0, x) r_{i-\nu-\xi}^3 \\ &= \sum_{i=1}^{n} \sum_{\lambda=1}^{n-i} \sum_{\xi=1}^{i} \sum_{\nu=1}^{i-\xi} \sum_{x \neq 0} P^{\lambda}(0, x) P^{\xi}(x, 0) P^{\nu}(0, x) r_{i-\nu-\xi}^3 r_{n-i-\lambda}, \end{aligned}$$

where Lemma 2.2 has again been used at the last step. We want to sum first on i; the relevant part is

$$(3.22)\qquad \sum_{i=\xi+\nu}^{n-\lambda} r_{i-\nu-\xi}^3 r_{n-i-\lambda}.$$

This is a convolution and since r_n is slowly varying and decreasing this is dominated by a constant times $(n - \lambda - \xi - \nu) r_{n-\lambda-\xi-\nu}^4$, which in turn is dominated by cnr_n^4 by Lemma 2.1. Thus, we have $C_n \leqq cnr_n^4 D_n$, say, where

$$(3.23)\qquad D_n = \sum_{\lambda=1}^{n} \sum_{\xi=1}^{n-\lambda} \sum_{\nu=1}^{n-\lambda-\xi} \sum_{x \neq 0} P^{\lambda}(0, x) P^{\xi}(x, 0) P^{\nu}(0, x).$$

Since this expression is symmetric in λ and ν, we need only consider $\nu \geqq \lambda$. Using the uniform estimate (2.7) on $P^{\nu}(0, x)$, we see that

$$D_n \leqq 2A \sum_{\xi=1}^{n} \sum_{\lambda=1}^{[(n-\xi)/2]} \sum_{\nu=\lambda}^{n-\xi-\lambda} \sum_{x\neq 0} P^{\lambda}(0, x)P^{\xi}(x, 0)\nu^{-1} \tag{3.24}$$

$$\leqq 2A \sum_{\xi=1}^{n} \sum_{\lambda=1}^{[(n-\xi)/2]} u_{\lambda+\xi}[\log e(n - \xi - \lambda) - \log \lambda]$$

$$\leqq 2A \sum_{\xi=1}^{n} \sum_{\lambda=1}^{n-\xi} u_{\lambda+\xi}[\log en - \log \lambda]$$

$$\leqq 2A \sum_{\beta=1}^{n} \sum_{\lambda=1}^{\beta} u_{\beta}[\log en - \log \lambda].$$

Next we use the inequality

$$\sum_{\lambda=1}^{\beta} \log \lambda \geqq \int_0^{\beta} \log y\, dy = \beta \log\left(\frac{\beta}{e}\right) \tag{3.25}$$

to obtain

$$D_n \leqq 2A \sum_{\beta=1}^{n} \beta u_{\beta} \log\left(\frac{e^2 n}{\beta}\right) \leqq 2A \sum_{\beta=1}^{8n} \beta u_{\beta} \log\left(\frac{8n}{\beta}\right). \tag{3.26}$$

Recalling that $C_n \leqq cnr_n^4 D_n$, this completes the proof of the theorem.

In order to complete the proof of Theorem 3.1, we must show that φ has the desired summability property. To do this, let $\psi(n) = \varphi(n)n^{-2}r_n^{-2}$, $n'_k = [\alpha^k]$ for $k \geqq 0$, and $n_k = 8n'_k$. We need only show that $\Sigma_k\, \psi(n_k) < \infty$. We note first that since $n^{-1}r_n^{-2} \leqq n^{-1/2}$ for large n, this is trivial for the contribution of the term n to $\varphi(n)$, so we need only consider the second term in $\varphi(n)$. From this point on the term n in $\varphi(n)$ will be ignored. Define an increasing sequence of integers by

$$m_j = \min\{k : u_0 + \cdots + u_k \geqq j\} \tag{3.27}$$

for $j \geqq 1$. Since $\Sigma_{k=0}^{\infty} u_k$ diverges and $u_k \to 0$, we have that the m_j are defined for all j and

$$\sum_{k=m_j+1}^{m_{j+1}} u_k \to 1 \quad \text{as} \quad j \to \infty. \tag{3.28}$$

Define

$$v_j = \sum_{k=n_j+1}^{n_{j+1}} u_k; \tag{3.29}$$

by (2.7),

$$v_j \leqq A n_j^{-1}(n_{j+1} - n_j) \sim A(\alpha - 1), \tag{3.30}$$

so that the v_j are bounded. For $m_j < n_i \leqq m_{j+1}$, we have by (2.5),

$$(3.31)\qquad \psi(n_i) \leqq n_i^{-1}j^{-2}\sum_{k=1}^{n_i} ku_k \log\left(\frac{n_i}{k}\right)$$

$$= n_i^{-1}j^{-2}\sum_{\beta=0}^{i-1}\sum_{k=n_\beta+1}^{n_{\beta+1}} ku_k \log\left(\frac{n_i}{k}\right) + O(n_i^{-1}\log n_i)$$

$$\leqq j^{-2}\sum_{\beta=0}^{i-1}\sum_{k=n_\beta+1}^{n_{\beta+1}}\left(\frac{n_{\beta+1}}{n_i}\right) u_k \log\left(\frac{n_i}{n_\beta}\right) + O(in_i^{-1})$$

$$\leqq cj^{-2}\sum_{\beta=0}^{i-1}\left(\frac{n_{\beta+1}}{n_i}\right) v_\beta(i-\beta) + O(in_i^{-1}),$$

where the error term is to take care of the sum for $k = 1, \cdots, 8$. Since this error term is summable, we will drop it. Now let

$$(3.32)\qquad I_j = \{i: m_j < n_i \leqq m_{j+1}\}.$$

Then

$$(3.33)\qquad \sum_{i\in I_j}\psi(n_i) \leqq cj^{-2}\sum_{i\in I_j}\sum_{\beta=0}^{i-1}\left(\frac{n_{\beta+1}}{n_i}\right) v_\beta(i-\beta)$$

and so it will suffice to show that the double sum is bounded uniformly in j. We change the order of summation and write ($\beta < I_j$ means $\beta \in \{i: n_i \leqq m_j\}$)

$$(3.34)\qquad \sum_{\beta<I_j}\sum_{i\in I_j}\left(\frac{n_{\beta+1}}{n_i}\right) v_\beta(i-\beta) + \sum_{\beta\in I_j}\sum_{i\in I_j, i>\beta}\left(\frac{n_{\beta+1}}{n_i}\right) v_\beta(i-\beta).$$

Since v_β is bounded and $n_i = 8[\alpha^i]$, this is bounded by

$$(3.35)\qquad c\sum_{i\in I_j}\sum_{\beta<I_j}\alpha^{\beta+1-i}(i-\beta) + c\sum_{\beta\in I_j} v_\beta.$$

By (3.28) and (3.30), the second sum is dominated for large j by

$$(3.36)\qquad \sum_{k=m_j+1}^{m_{j+1}} u_k + A(\alpha-1) \leqq 2 + A(\alpha-1).$$

Hence, it remains to look at the double sum. This is of the form

$$(3.37)\qquad \sum_{i=\lambda}^{\nu}\sum_{\beta=0}^{\lambda-1}(i-\beta)\alpha^{\beta+1-i} \leqq \sum_{k=1}^{\infty} k^2\alpha^{1-k} < \infty,$$

independent of λ and ν and hence of j. This proves that φ has the desired summability property and therefore completes the proof of Theorem 3.1.

4. The variance of R_n in the plane

We start this section with a theorem which gives the asymptotic behavior of f_n for all strongly aperiodic random walks in two dimensions with $EX_1 = 0$

and $E|X_1|^2 < \infty$. The result will be used to establish the asymptotic behavior of Var R_n in this case, but it is also of some independent interest since it establishes for this special case Kesten's conjecture [11] that

$$\lim_{n\to\infty} f_n^{-1} \sum_{k=0}^{n} f_k f_{n-k} = 2 \tag{4.1}$$

for all strongly aperiodic recurrent random walks. In his paper, Kesten found the asymptotic behavior of f_n and thereby verified (4.1) for one dimensional recurrent random walks with X_1 in the domain of attraction of a symmetric stable law. Some of the other results of [11] are also valid for the two dimensional random walks we are considering since they follow from (4.1).

THEOREM 4.1. *For a strongly aperiodic, two dimensional random walk with* $EX_1 = 0$ *and* $E|X_1|^2 < \infty$,

$$f_n \sim \frac{c_1}{n \log^2 n}, \tag{4.2}$$

where $c_1 = 2\pi|Q|^{1/2}$ *and* Q *is the covariance matrix of* X_1.

REMARK 4.1. The asymptotic behavior of r_n and also of $\Sigma_{j=1}^{n} jf_j$ can be obtained directly from Tauberian theorems, but this method does not give information about f_n since this sequence is not known to be monotone.

PROOF. Let $\gamma = [n/\log^3 n]$ and $\beta = n - 2\gamma$, and write

$$f_n = \sum_{x\neq 0} \sum_{y\neq 0} P_0^\gamma(0, x)P_0^\beta(x, y)P_0^\gamma(y, 0). \tag{4.3}$$

First we shall show that the error made by neglecting the taboo on the middle factor is small. Consider

$$\begin{aligned} 0 &\leqq P^\beta(x, y) - P_0^\beta(x, y) = P_x[T_0 \leqq \beta, S_\beta = y] \\ &\leqq \sum_{i=1}^{[\beta/2]} P_0^i(x, 0)P^{\beta-i}(0, y) + \sum_{i=[\beta/2]}^{\beta-1} P^i(x, 0)P_0^{\beta-i}(0, y), \end{aligned} \tag{4.4}$$

where the terms are classified according to the first time zero is hit if that is prior to $\beta/2$, but the last time if it occurs after $\beta/2$. This distinction is important. The estimate $f_n = O(1/n \log^2 n)$ which follows from (4.3) with $\gamma = [n/3]$ will be needed below; this is in Kesten and Spitzer [10]. Then

$$\begin{aligned} &\Big|f_n - \sum_{x\neq 0} \sum_{y\neq 0} P_0^\gamma(0, x)P^\beta(x, y)P_0^\gamma(y, 0)\Big| \\ &\leqq \sum_{i=1}^{[\beta/2]} \sum_{x\neq 0} \sum_{y\neq 0} P_0^\gamma(0, x)P_0^i(x, 0)P^{\beta-i}(0, y)P_0^\gamma(y, 0) \\ &\qquad + \sum_{i=[\beta/2]}^{\beta-1} \sum_{x\neq 0} \sum_{y\neq 0} P_0^\gamma(0, x)P^i(x, 0)P_0^{\beta-i}(0, y)P_0^\gamma(y, 0) \end{aligned} \tag{4.5}$$

$$= \sum_{i=1}^{[\beta/2]} \sum_{y \neq 0} f_{\gamma+i} P^{\beta-i}(0, y) P_0^{\gamma}(y, 0) + \sum_{i=[\beta/2]}^{\beta-1} \sum_{x \neq 0} P_0^{\gamma}(0, x) P^{i}(x, 0) f_{\beta-i+\gamma}$$

$$\leqq A \sum_{i=1}^{[\beta/2]} f_{\gamma+i}(\beta - i)^{-1} r_{\gamma} + A \sum_{i=[\beta/2]}^{\beta-1} r_{\gamma} i^{-1} f_{\beta-i+\gamma}$$

$$= O\left(\beta^{-1} r_{\gamma} \sum_{i=1}^{[\beta/2]} f_{\gamma+i}\right) = O\left(\beta^{-1} r_{\gamma} (\log \gamma)^{-2} \sum_{i=1}^{\beta} (\gamma + i)^{-1}\right)$$

$$= O\left(\frac{\log \log n}{n \log^3 n}\right).$$

With c_1 as in the statement of the theorem, the local limit theorem ([12], p. 77) gives

$$\left| P^{\beta}(x, y) - \frac{1}{c_1 \beta} \right| \leqq \frac{\varepsilon}{\beta} + O(\beta^{-2} |y - x|^2) \tag{4.6}$$

uniformly in x, y for β sufficiently large. Now

$$\beta^{-2} \sum_{x \neq 0} \sum_{y \neq 0} P_0^{\gamma}(0, x) |y - x|^2 P_0^{\gamma}(y, 0) \tag{4.7}$$

$$\leqq \beta^{-2} \sum_{x \neq 0} P^{\gamma}(0, x) \{\gamma E|X_1|^2 + |x|^2\}$$

$$\leqq 2\gamma \beta^{-2} E|X_1|^2 = O\left(\frac{1}{n \log^3 n}\right)$$

The proof is now complete for

$$\sum_{x \neq 0} \sum_{y \neq 0} P_0^{\gamma}(0, x) \frac{1}{c_1 \beta} P_0^{\gamma}(y, 0) = \frac{1}{c_1 \beta} r_{\gamma}^2 \sim \frac{c_1}{n \log^2 n} \tag{4.8}$$

the last step being a consequence of the known limit $r_n \sim c_1/\log n$. (See Lemma 2.3 of [7].)

THEOREM 4.2. *For two dimensional random walk with* $EX_1 = 0$ *and* $E|X_1|^2 < \infty$,

$$\operatorname{Var} R_n \sim \frac{c_2 n^2}{\log^4 n}, \tag{4.9}$$

where $c_2 = 8\pi^2 K|Q|$, *the covariance matrix of* X_1 *is* Q, *and*

$$K = -\int_0^1 \frac{\log w}{1 - w + w^2} dw + \frac{1}{2} - \frac{\pi^2}{12} = 0\ 84948659 \cdots \tag{4.10}$$

Note, however, that if the random walk takes place on a proper subgroup of E_2, *then a transformation should be made so that it will take place on all of* E_2 *and* Q *should be the covariance matrix for this transformed random walk.*

PROOF. The first step is to make a transformation, if necessary, so that the random walk does not take place on a proper subgroup. This does not change R_n and the fact that Q may change has been allowed for in the statement of the theorem. We shall assume for now that the random walk is strongly aperiodic and show at the end how to remove this assumption. From (3.5) and (3.7), we have that

$$\text{Var } R_n = 2 \sum_{i=1}^{n} \sum_{\lambda=1}^{n-i} \sum_{x \neq 0} P_0^{\lambda}(0, x) b_i(x) + O(n). \tag{4.11}$$

The expression we will use for $b_i(x)$ is given in (3.18). The contribution to the variance from the negative part of $b_i(x)$ will be considered first. This part will be denoted $-V_n$ and the positive part U_n so that $\text{Var } R_n = U_n - V_n + O(n)$. Now

$$\begin{aligned} V_n &= 2 \sum_{i=1}^{n} \sum_{\lambda=1}^{n-i} \sum_{x \neq 0} \sum_{\xi=1}^{i} \sum_{\nu=i-\xi+1}^{i} P_0^{\lambda}(0, x) P_0^{\xi}(x, 0) f_\nu \\ &= 2 \sum_{i=1}^{n} \sum_{\lambda=1}^{n-i} \sum_{\xi=1}^{i} \sum_{\nu=i-\xi+1}^{i} f_{\lambda+\xi} f_\nu \\ &\sim 2c_1^2 \sum_{i=2}^{n} \sum_{\xi=2}^{i-2} \frac{\log (n - i + \xi/\xi)}{\log (n - i + \xi) \log \xi} \frac{\log (i/i - \xi)}{\log i \log (i - \xi)}, \end{aligned} \tag{4.12}$$

where Theorem 4.1 has been used at the last step. Since the summands are bounded by 1, the contribution for $\xi \leqq n/\log^5 n$ is $O(n^2/\log^5 n)$, while for $\xi > n/\log^5 n$, one can replace $\log \xi$ with $\log n$. The same method applies to the other log terms in the denominator by considering where i, $i - \xi$, and $n - i + \xi$ are respectively $\leqq n/\log^5 n$. Thus,

$$\begin{aligned} V_n &\sim 2c_1^2 (\log n)^{-4} \sum_{i=2}^{n} \sum_{\xi=2}^{i-2} \log \frac{n - i + \xi}{\xi} \log \frac{i}{i - \xi} \\ &\sim 2c_1^2 n^2 (\log n)^{-4} \int_0^1 \int_0^z \log \frac{1 - z + y}{y} \log \frac{z}{z - y} \, dy \, dz \\ &= 2c_1^2 n^2 (\log n)^{-4} \frac{1}{2}\left(\frac{\pi^2}{6} - 1\right), \end{aligned} \tag{4.13}$$

since the double integral can be evaluated. Now we must consider the positive contribution. In (3.18), write

$$\begin{aligned} P_x[T_x > i - \nu - \xi, T_0 > i - \nu - \xi] \\ = r_{i-\nu-\xi} - P_x[T_x > i - \nu - \xi, T_0 \leqq i - \nu - \xi]; \end{aligned} \tag{4.14}$$

the next step will be to show that using the last of these probabilities in (3.18) will lead to a term of smaller order. To obtain a bound, note that by thinking of η as the last time zero is hit prior to $i - \nu - \xi$,

$$P_x[T_x > i - \nu - \xi, T_0 \leqq i - \nu - \xi] \leqq \sum_{\eta=1}^{i-\nu-\xi} P_x^{\eta}(x, 0) r_{i-\nu-\xi-\eta}. \tag{4.15}$$

Substituting this bound in (3.18) and (4.11), we see that this contribution to the variance is of order

$$(4.16) \quad \sum_{i=1}^{n} \sum_{\lambda=1}^{n-i} \sum_{x \neq 0} \sum_{\xi+\nu+\eta \leq i} P_0^\lambda(0, x) P_0^\xi(x, 0) P_x^\nu(0, x) P_x^\eta(x, 0) r_{i-\nu-\xi-\eta}$$

$$\leq \sum_{i=1}^{n} \sum_{\lambda=1}^{n-i} \sum_{x \neq 0} \sum_{\xi+\nu+\eta \leq i} P^\lambda(0, x) P^\xi(x, 0) P^\nu(0, x) P^\eta(x, 0) r_{n-i-\lambda} r_{i-\nu-\xi-\eta}^4$$

$$= O\left(n r_n^5 \sum_{\lambda=1}^{n} \sum_{x \neq 0} \sum_{\xi+\nu+\eta \leq n-\lambda} P^\lambda(0, x) P^\xi(x, 0) P^\nu(0, x) P^\eta(x, 0) \right),$$

where Lemma 2.2 has been applied four times at the second step and the i sum has been moved inside and Lemma 2.1 used at the last step. We will now show that the multiple sum is of order n. Since it is symmetric in λ and ν, in ξ and η, a bound is

$$(4.17) \quad 4A^2 \sum_{\lambda=1}^{n} \sum_{\xi=1}^{n} \sum_{x \neq 0} \sum_{\nu=\lambda}^{n} \sum_{\eta=\xi}^{n} P^\lambda(0, x) P^\xi(x, 0) \nu^{-1} \eta^{-1}$$

$$\leq 4A^2 \sum_{\lambda=1}^{n} \sum_{\xi=1}^{n} u_{\lambda+\xi} \log \frac{en}{\lambda} \log \frac{en}{\xi}$$

$$\leq 4A^3 \sum_{k=1}^{2n} \sum_{\xi=1}^{k-1} \frac{1}{k} \log \frac{en}{k-\xi} \log \frac{en}{\xi}$$

$$= O\left(\sum_{k=1}^{2n} \left(\log \frac{e^2 n}{k} \right)^2 \right) = O(n).$$

Thus, we have for the positive contribution to the variance

$$(4.18) \quad U_n = 2 \sum_{i=1}^{n} \sum_{\lambda=1}^{n-i} \sum_{x \neq 0} \sum_{\xi=1}^{i} \sum_{\nu=1}^{i-\xi} P_0^\lambda(0, x) P_0^\xi(x, 0) P_x^\nu(0, x) r_{i-\nu-\xi} + O(n^2 r_n^5).$$

We will now apply Lemma 2.2 three times to the main term here to obtain

$$(4.19) \quad 2 \sum_{i=1}^{n} \sum_{\lambda=1}^{n-i} \sum_{x \neq 0} \sum_{\xi=1}^{i} \sum_{\nu=1}^{i-\xi} P^\lambda(0, x) P^\xi(x, 0) P^\nu(0, x) r_{n-i-\lambda} r_{i-\nu-\xi}^3.$$

But since this is now the leading term, we need to examine the error made in the application of the lemma. To this end, note that for $\gamma \geq 1$,

$$(4.20) \quad 0 \leq \sum_{j=1}^{m-k} f_j(r_{m-k-j}^\gamma - r_{m-k}^\gamma) \leq \gamma \sum_{j=1}^{m-k} f_j r_{m-k-j}^{\gamma-1}(r_{m-k-j} - r_{m-k}).$$

Now, for $1 \leq j \leq (m-k)/2$, by Theorem 4.1,

$$(4.21) \quad r_{m-k-j} - r_{m-k} = \sum_{\beta=m-k-j+1}^{m-k} f_\beta = O\left(\frac{j}{m-k} r_{m-k}^2 \right),$$

and so

$$\sum_{j=1}^{m-k} f_j(r_{m-k-j}^{\gamma} - r_{m-k}^{\gamma}) \tag{4.22}$$

$$= O\left(\sum_{j=1}^{(m-k)/2} jf_j(m-k)^{-1}r_{m-k}^{\gamma+1} + \sum_{j=(m-k)/2}^{m-k} f_j r_{m-k-j}^{\gamma}\right)$$

$$= O(r_{m-k}^{\gamma+3} + r_{m-k}^{\gamma+2}) = O(r_{m-k}^{\gamma+2}).$$

Utilizing this in (2.12) and (2.11) yields

$$\sum_{k=1}^{m} P_0^k(0, x)r_{m-k}^{\gamma} = \sum_{k=1}^{m} P^k(0, x)[r_{m-k}^{\gamma+1} + O(r_{m-k}^{\gamma+2})]. \tag{4.23}$$

Thus, the error introduced by Lemma 2.2 will always lead to an extra factor of an r and this will ultimately give a term of order $n^2/\log^5 n$. We now return to the estimation of the main term (4.19). By summing first on i, we obtain

$$U_n = 2\sum_{k=3}^{n} a_k b_{n-k} + O\left(\frac{n^2}{\log^5 n}\right), \tag{4.24}$$

where

$$a_k = \sum\sum_{\xi+\nu+\lambda=k} \sum_{x\neq 0} P^{\lambda}(0, x)P^{\xi}(x, 0)P^{\nu}(0, x), \tag{4.25}$$

$$b_k = \sum_{j=0}^{k} r_j r_{k-j}^3 \sim kr_k^4. \tag{4.26}$$

We will prove next that $a_k \to a$; this will imply that

$$U_n \sim 2a\sum_{k=3}^{n} b_{n-k} \sim an^2 r_n^4 \sim \frac{ac_1^4 n^2}{\log^4 n}. \tag{4.27}$$

To show that $a_k \to a$, the local limit theorem ([12], pp. 77 and 79) will again be used. Write

$$P^{\beta}(0, x) = Q_{\beta}(x) + o(E_{\beta}(x)), \tag{4.28}$$

uniformly in x as $\beta \to \infty$, where

$$Q_{\beta}(x) = (2\pi\beta)^{-1}|Q|^{-1/2} \exp\left\{\frac{-x\cdot Q^{-1}x}{2\beta}\right\}, \tag{4.29}$$

$$E_{\beta}(x) = \min\{\beta^{-1}, |x|^{-2}\}. \tag{4.30}$$

Since it is clear that $Q_{\beta}(x) = O(E_{\beta}(x))$, the sum of all the error terms will go to zero if

$$\sum\sum_{\xi+\nu+\lambda=k} \sum_{x\neq 0} E_{\lambda}(x)E_{\xi}(x)E_{\nu}(x) \tag{4.31}$$

is bounded. Since this is symmetric in ξ, ν, λ, we may as well assume that $\xi \leqq \nu \leqq \lambda$. This means that $\lambda \geqq k/3$, and so (4.31) is bounded by a constant times

$$k^{-1}\sum_{\xi=1}^{k}\sum_{\nu=1}^{k}\sum_{|x|^2\leq k}E_\xi(x)E_\nu(x)+k^2\sum_{|x|^2\geq k}|x|^{-6} \tag{4.32}$$

$$\leq k^{-1}\sum_{|x|^2\leq k}\left(\log\frac{e^2k}{|x|^2}\right)^2+O(1)$$

$$=O\left[k^{-1}\sum_{j=1}^{k^{1/2}}j\left(2\log\frac{ek^{1/2}}{j}\right)^2\right]=O(1).$$

Now we must examine the leading terms. Letting

$$\mu=\xi^{-1}+\nu^{-1}+(k-\xi-\nu)^{-1}, \qquad y=x\mu^{1/2}, \tag{4.33}$$

we can write

$$a_k=\sum\sum_{\xi+\nu+\lambda=k}\sum_{x\neq 0}Q_\lambda(x)Q_\xi(x)Q_\nu(x)+o(1) \tag{4.34}$$

$$=(2\pi)^{-2}|Q|^{-1}\sum_{\xi=1}^{k-2}\sum_{\nu=1}^{k-\xi-1}\sum_{y}\{\xi\nu(k-\xi-\nu)\}^{-1}Q_1(y)+o(1)$$

$$\sim c_1^{-2}\sum_{\xi=1}^{k-2}\sum_{\nu=1}^{k-\xi-1}\{\xi\nu(k-\xi-\nu)\mu\}^{-1}\sim 2c_1^{-2}K_1,$$

where (see [5], p. 533)

$$K_1=\frac{1}{2}\int_0^1\int_0^{1-z}\frac{1}{z-z^2+y-y^2-yz}\,dydz=-\int_0^1\frac{\log w}{1-w+w^2}\,dw \tag{4.35}$$

$$=1.17195361935\cdots.$$

Recalling (4.13) and (4.27),

$$\operatorname{Var}R_n\sim 2c_1^2\left(K_1+\frac{1}{2}-\frac{\pi^2}{12}\right)n^2(\log n)^{-4}. \tag{4.36}$$

Thus, the proof of the theorem is complete for the case when the random walk is strongly aperiodic. To make the transition to the general case, let S_n denote the original random walk which has been made aperiodic by a transformation and let $P(x, y)$ denote its transition function. Then the random walk S_n' with transition function

$$P'(x,y)=\tfrac{1}{2}\delta(x,y)+\tfrac{1}{2}P(x,y) \tag{4.37}$$

will be strongly aperiodic. One can describe the paths of the new random walk by flipping a coin and each time remaining stationary if the coin falls heads and moving according to the original transition function if it falls tails. The range of the S_n' random walk will be denoted by R_n'. By conditioning on the sequence of heads and tails, it is clear that

$$ER_n'=\sum_{k=0}^{n}\binom{n}{k}2^{-n}ER_k, \qquad E(R_n')^2=\sum_{k=0}^{n}\binom{n}{k}2^{-n}ER_k^2. \tag{4.38}$$

For even n,

$$(4.39)\qquad \begin{aligned}|ER_{n/2}^2 - ER_k^2| &= \Big|\sum_{0 \leqq i,j \leqq n/2}\sum EZ_iZ_j - \sum_{0 \leqq i,j \leqq k}\sum EZ_iZ_j\Big| \\ &\leqq n|n - 2k|.\end{aligned}$$

Thus, by (4.38),

$$(4.40)\qquad \begin{aligned}|E(R_n')^2 - ER_{n/2}^2| &\leqq \sum_{|n-2k| \leqq n^{2/3}} \binom{n}{k} 2^{-n} n|n - 2k| \\ &\quad + \sum_{|n-2k| > n^{2/3}} \binom{n}{k} 2^{-n}|ER_k^2 - ER_{n/2}^2| \\ &\leqq n^{5/3} + n^2 P\left[\left|Y - \frac{n}{2}\right| \geqq \frac{1}{2} n^{2/3}\right],\end{aligned}$$

where Y is a binomial random variable. By Chebyshev's inequality this probability is $O(n^{-1/3})$. Thus, we have

$$(4.41)\qquad E(R_n')^2 = ER_{n/2}^2 + O(n^{5/3}).$$

The same estimate can be obtained for $(ER_n')^2$ and $(ER_{n/2})^2$ in essentially the same way. Therefore,

$$(4.42)\qquad \operatorname{Var} R_n \sim \operatorname{Var} R_{2n}' \sim \frac{8\pi^2 K|Q'|4n^2}{\log^4 n}$$

by the strongly aperiodic case. But since $Q' = Q/2$, it follows that $|Q'| = |Q|/4$ and the constant has the right form.

5. Sufficient conditions for strong transience

The following theorems give simple sufficient conditions for a random walk to be strongly transient. As we already mentioned in the introduction, the results of [6] apply to the range of such random walks.

First we need to introduce some terminology. A nonempty set $F \subset E_d$ is said to be closed if $\Sigma_{y \in F} P(x, y) = 1$ for every $x \in F$. The random walk is called irreducible if the only closed set is E_d. This is equivalent to saying that every lattice point can be reached from every other lattice point. The assumption that the random walk is aperiodic (that is, does not take place on a proper subgroup of E_d) which we have made without any real loss of generality is equivalent in the present terminology to the assumption that no two closed sets are disjoint.

THEOREM 5.1. *Every aperiodic random walk that is not irreducible is strongly transient.*

PROOF. Let $F_1 = \{x\colon G(0, x) > 0\}$ and $F_0 = \{x\colon G(x, 0) = 0\}$. Since the random walk is not irreducible, F_1 is a proper subset of E_d and hence F_0 is nonempty. The sets F_1 and F_0 are both closed and by the aperiodicity must have a common element x. Then there is a k such that

$$(5.1) \qquad P^k(0, x) = \varepsilon > 0, \qquad G(x, 0) = 0.$$

We may also assume that $\varepsilon < 1$, for if $\varepsilon = 1$, then $u_n = 0$ for $n \geqq k$ by (5.1) and the random walk is clearly strongly transient. Since $P^k(0, y) \leqq 1 - \varepsilon$ for $y \neq x$,

$$(5.2) \qquad P^k(0, y) \leqq \alpha = \max(\varepsilon, 1 - \varepsilon) \qquad \text{for all} \quad y \in E_d.$$

Now suppose that $z \in F_1$. Since $G(x + z, z) = G(x, 0) = 0$ and $G(0, z) > 0$, it follows that $G(x + z, 0) = 0$ or $x + z \in F_0$. Thus, for $z \in F_1$,

$$(5.3) \qquad \sum_{y \notin F_0} P^k(z, y) \leqq 1 - P^k(z, x + z) = 1 - \varepsilon \leqq \alpha.$$

Now suppose for all $z \in F_1 \cap F_0^c$ that $P^{nk}(z, 0) \leqq \alpha^n$; this is true for $n = 1$ by (5.2). For $z \in F_1 \cap F_0^c$, then

$$(5.4) \qquad P^{(n+1)k}(z, 0) = \sum_{y \in F_1 \cap F_0^c} P^k(z, y) P^{nk}(y, 0) \leqq \alpha^n \sum_{y \notin F_0} P^k(z, y) \leqq \alpha^{n+1}$$

by the induction hypothesis and (5.3). If $0 \in F_0$, the random walk is trivially strongly transient so we may assume that $0 \in F_1 \cap F_0^c$. Thus, we have $u_{nk} \leqq \alpha^n$ and for $0 < r < k$

$$(5.5) \qquad u_{nk+r} = \sum_{y \in F_1 \cap F_0^c} P^r(0, y) P^{nk}(y, 0) \leqq \alpha^n;$$

this is clearly sufficient for strong transience.

THEOREM 5.2. *A random walk with $EX_1 \neq 0$ and $E|X_1|^2 < \infty$ is strongly transient regardless of the dimension.*

PROOF. Since $EX_1 \neq 0$, there must be at least one component of X_1 with nonzero expectation. Furthermore, whenever the original random walk visits 0, so will the component ones so that $u_n \leqq u_n'$ where u_n' refers to one of the component random walks. Thus, if the component random walk is strongly transient, the original random walk will be as well. This reduces the problem to the one dimensional case. Let S_n be a random walk with $EX_1 \neq 0, EX_1^2 < \infty$, and $d = 1$. If $\operatorname{Var} X_1 = 0$, the random walk is degenerate and $EX_1 \neq 0$ implies $u_n = 0$ for all $n > 0$. We may then assume that $0 < \operatorname{Var} X_1 < \infty$ and by considering $-S_n$, if necessary, that $\mu = EX_1 > 0$. Now

$$(5.6) \qquad \sum_{n=0}^{\infty} \sum_{k=0}^{\infty} u_{n+k} = \sum_{n=0}^{\infty} \sum_{k=0}^{\infty} \sum_{x} P^n(0, x) P^k(x, 0) = \sum_{x} G(0, x) G(x, 0)$$

$$= G^2(0, 0) + 2 \sum_{x > 0} G(0, x) G(x, 0).$$

Thus, the random walk is strongly transient if and only if the last sum converges. If the random walk is aperiodic, the renewal theorem asserts that $\lim_{x \to \infty} G(0, x) = \mu^{-1} > 0$. If it is not aperiodic, the limit will still be positive if we restrict x to the subgroup. In either case, the random walk is strongly transient if and only if $\Sigma_{x > 0} G(x, 0)$ converges. But

$$\sum_{x>0} G(x, 0) = \sum_{x<0} G(0, x) = \sum_{x<0} \sum_{n=0}^{\infty} P^n(0, x) \tag{5.7}$$

$$= \sum_{n=0}^{\infty} P[S_n < 0] \leqq \sum_{n=0}^{\infty} P[|n^{-1}(S_n - n\mu)| \geqq \mu].$$

This series converges provided $\operatorname{Var} X_1 < \infty$ by a result of Erdös [3]. (Also, see [9].)

REMARK 5.1. The condition $E|X_1|^2 < \infty$ is essential in Theorem 5.2 even though the random walk is transient once EX_1 exists and is nonzero. To see that it is essential, note that the proof shows that a necessary and sufficient condition for strong transience in one dimension when $EX_1 = \mu > 0$ is the convergence of the series $\Sigma_n P[S_n < 0]$. It is easy to construct examples for which this series diverges once the requirement $EX_1^2 < \infty$ is dropped.

6. The range of one dimensional random walk

Let $\{S_n\}$ be a one dimensional random walk with $EX_1^2 < \infty$. If $EX_1 \neq 0$, then by Theorem 5.2 the random walk is strongly transient. The results of [6] then apply and the central limit theorem is valid provided only that the range R_n does not grow deterministically. The case of $EX_1 = 0$ is somewhat different and we shall deal with it now.

THEOREM 6.1. *Let $\{S_n\}$ be a one dimensional random walk with $EX_1 = 0$ and $0 < \operatorname{Var} X_1 = \sigma^2 < \infty$. Then R_n/ER_n converges in distribution to a proper law. The limit law is that of*

$$\left(\frac{\pi}{8}\right)^{1/2} \left\{\max_{0 \leqq t \leqq 1} Y(t) - \min_{0 \leqq t \leqq 1} Y(t)\right\}, \tag{6.1}$$

where $Y(t)$ is standard one dimensional Brownian motion.

PROOF. By making a transformation, if necessary, we may assume that the random walk is aperiodic since the transformation does not change R_n. Now $r_n \sim \sigma(2/\pi n)^{1/2}$ (see [12], p. 381), and

$$ER_n = \sum_{k=0}^{n} r_k \sim \sigma\left(\frac{8}{\pi}\right)^{1/2} n^{1/2}. \tag{6.2}$$

Let

$$M_n = \max_{0 \leqq j \leqq n} S_j, \qquad m_n = \min_{0 \leqq j \leqq n} S_j. \tag{6.3}$$

It follows readily from the results in [4] (see also [12], p. 232) that $EM_n \sim \sigma(2/\pi)^{1/2} n^{1/2}$ and so

$$EM_n - Em_n \sim \sigma\left(\frac{8}{\pi}\right)^{1/2} n^{1/2} \sim ER_n. \tag{6.4}$$

Since $M_n - m_n + 1 - R_n \geqq 0$ and

$$P[M_n - m_n + 1 - R_n \geqq \varepsilon\, ER_n] \leqq \frac{1 + EM_n - Em_n - ER_n}{\varepsilon\, ER_n} \to 0 \tag{6.5}$$

as $n \to \infty$, it follows that

$$\frac{M_n - m_n - R_n}{ER_n} \xrightarrow{P} 0. \tag{6.6}$$

Thus, R_n/ER_n and $(M_n - m_n)/ER_n$ have the same asymptotic distribution, and the limit behavior of the latter quantity follows from Donsker's invariance principle [1].

Theorem 6.1 is different from the other limit theorems for the range in that the limit law is not normal. We are also using a different scheme of normalization, but this difference is only apparent as we can now show that the standard deviation of R_n grows at the same rate as ER_n. Thus, using the usual normalization will only effect a scale change and translation on the limit law.

THEOREM 6.2. *Let $\{S_n\}$ be a one dimensional random walk with $EX_1 = 0$ and $0 < \operatorname{Var} X_1 < \infty$. Then there is a positive constant c such that* $\operatorname{Var} R_n \sim cn$.

PROOF. We know from the last theorem that $n^{-1/2}R_n$ converges in distribution and we will show that $\{n^{-1}R_n^2\}$ is uniformly integrable. This will prove the theorem with c being the variance of the limit law for $n^{-1/2}R_n$. Note that if $i \leqq j \leqq k$,

$$\begin{aligned} EZ_iZ_jZ_k &\leqq P[S_k \neq S_{k-1}, \cdots, S_k \neq S_j; S_j \neq S_{j-1}, \cdots, \\ &\qquad S_j \neq S_i; S_i \neq S_{i-1}, \cdots, S_i \neq 0] \\ &= r_{k-j}r_{j-i}r_i, \end{aligned} \tag{6.7}$$

and so

$$ER_n^3 \leqq 6 \sum_{i=0}^{n} \sum_{j=i}^{n} \sum_{k=j}^{n} r_{k-j}r_{j-i}r_i = O(n^{3/2}). \tag{6.8}$$

Thus,

$$\int_{[n^{-1}R_n^2 \geqq M]} n^{-1}R_n^2\, dP \leqq M^{-1/2} \int n^{-3/2}R_n^3\, dP = O(M^{-1/2}) \tag{6.9}$$

uniformly in n.

REMARK 6.1. In the same way, one can show that $n^{-k/2}ER_n^k$ converges to the kth moment of the limit law, which exists for each $k \geqq 1$.

Added in proof. W. Feller (*Ann. Math. Statist.*, Vol. 22 (1951), pp. 427–432) considered the asymptotic distribution of $(M_n - m_n)n^{-1/2}$ for $d = 1$. He also obtained a series expansion for the density of the random variable in (6.1).

REFERENCES

[1] M. D. DONSKER, "An invariance principle for certain probability limit theorems," *Mem. Amer. Math. Soc.*, No. 6 (1951), pp. 1–12.

[2] A. DVORETZKY and P. ERDÖS, "Some problems on random walk in space," *Proceedings of the Second Berkeley Symposium on Mathematical Statistics and Probability*, Berkeley and Los Angeles, University of California Press, 1951, pp. 353–367.

[3] P. ERDÖS, "On a theorem of Hsu and Robbins," *Ann. Math. Statist.*, Vol. 20 (1949), pp. 286–291.

[4] P. ERDÖS and M. KAC, "On certain limit theorems of the theory of probability," *Bull. Amer. Math. Soc.*, Vol. 52 (1946), pp. 292–302.

[5] I. S. GRADSHTEYN and I. M. RYZHIK, *Table of Integrals, Series, and Products*, New York and London, Academic Press, 1965.

[6] N. C. JAIN and S. OREY, "On the range of random walk," *Israel J. Math.*, Vol. 6 (1968), pp. 373–380.

[7] N. C. JAIN and W. E. PRUITT, "The range of recurrent random walk in the plane," *Z. Wahrscheinlichkeitstheorie und Verw. Gebiete*, Vol. 16 (1970), pp. 279–292.

[8] ———, "The range of transient random walk," *J. Analyse Math.*, Vol. 24 (1971), pp. 369–393.

[9] M. L. KATZ, "The probability in the tail of a distribution," *Ann. Math. Statist.*, Vol. 34 (1963), pp. 312–318.

[10] H. KESTEN and F. SPITZER, "Ratio theorems for random walks I," *J. Analyse Math.*, Vol. 11 (1963), pp. 285–322.

[11] H. KESTEN, "Ratio theorems for random walks II," *J. Analyse Math.*, Vol. 11 (1963), pp. 323–379.

[12] F. SPITZER, *Principles of Random Walk*, Princeton, Van Nostrand, 1964.

ON THE LAW OF THE ITERATED LOGARITHM FOR MAXIMA AND MINIMA

H. ROBBINS and D. SIEGMUND
COLUMBIA UNIVERSITY and BROOKHAVEN NATIONAL LABORATORY

1. Introduction and summary

Let $w(t)$, $0 \leqq t \leqq \infty$, denote a standard Wiener process. The general law of the iterated logarithm (see [6], p. 21) says that if g is a positive function such that $g(t)/\sqrt{t}$ is ultimately nondecreasing, then

$$P\{w(t) \geqq g(t) \text{ i.o. } t \uparrow \infty\} \tag{1.1}$$

equals 0 or 1, according as

$$\int_1^\infty \frac{g(t)}{t^{3/2}} \exp\left\{-\frac{1}{2}\frac{g^2(t)}{t}\right\} dt < \infty \text{ or } = \infty. \tag{1.2}$$

(The notation i.o. $t\uparrow\infty$ $(t\downarrow 0)$ means for arbitrarily large (small) t.) In particular, for $k \geqq 3$ and

$$g(t) = \left[2t\left(\log_2 t + \tfrac{3}{2}\log_3 t + \sum_{i=4}^{k} \log_i t + (1+\delta)\log_{k+1} t\right)\right]^{1/2}, \tag{1.3}$$

the probability (1.1) is 0 or 1 according as $\delta > 0$ or $\delta \leqq 0$. (We write $\log_2 = \log\log$, $e_2 = e^e$, and so on.)

For applications in statistics it is of interest to compute as accurately as possible

$$P\{w(t) \geqq g(t) \text{ for some } t \geqq \tau\} \tag{1.4}$$

for functions g for which this probability is < 1; that is, functions for which (1.2) converges (see [3], [10], [12]). In [11], we gave a method for computing (1.4) exactly for a certain class of functions g. A sketch of this method follows. Since $\exp\{\theta w(t) - \frac{1}{2}\theta^2 t\}$, $0 \leqq t < \infty$, is a martingale for each θ, Fubini's theorem shows that $\int_0^\infty \exp\{\theta w(t) - \frac{1}{2}\theta^2 t\}\, dF(\theta)$, $0 \leqq t < \infty$, is also a martingale for any σ-finite measure F on $(0, \infty)$. Let

Research supported by NIH Grant 1-R01-GM-16895-02, ONR Grant N00014-67-A-0108-0018, and in part by the United States Atomic Energy Commission.

$$f(x, t) = \int_0^\infty \exp\left\{\theta x - \frac{\theta^2 t}{2}\right\} dF(\theta), \tag{1.5}$$

and for each $t \geqq 0$ and $\varepsilon > 0$ let $A(t, \varepsilon)$ be the solution of

$$f(x, t) = \varepsilon. \tag{1.6}$$

Then

$$\begin{aligned} P\{w(t) \geqq A(t, \varepsilon) \text{ for some } t \geqq \tau\} \\ = P\{f(w(t), t) \geqq \varepsilon \text{ for some } t \geqq \tau\}, \end{aligned} \tag{1.7}$$

and we use an elementary martingale equality to evaluate the right side of (1.7). The relation of $w(t)$ to the sequence of sums of i.i.d. random variables with mean 0 and variance 1 then permits the asymptotic evaluation of boundary crossing probabilities for partial sums.

In view of (1.3) a choice of F of particular interest is, for $\delta > 0$,

$$dF(\theta) = \begin{cases} \left[\theta\left(\log\frac{1}{\theta}\right)\cdots\left(\log_{k-1}\frac{1}{\theta}\right)\left(\log_k\frac{1}{\theta}\right)^{1+\delta}\right]^{-1} d\theta & \text{for } \theta \leqq \dfrac{1}{e_k}, \\ 0 & \text{otherwise,} \end{cases} \tag{1.8}$$

for which it is shown in [11] that for any $\varepsilon > 1/\delta$,

$$\begin{aligned} A(t, \varepsilon) = \Bigg[2t\Bigg(\log_2 t + \tfrac{3}{2}\log_3 t + \sum_{i=4}^{k}\log_i t \\ + (1 + \delta)\log_{k+1} t + \log\frac{\varepsilon}{2\sqrt{\pi}} + o(1)\Bigg)\Bigg]^{1/2} \end{aligned} \tag{1.9}$$

as t $\to \infty$ and

$$P\{w(t) \geqq A(t, \varepsilon) \text{ for some } t \geqq 0\} = \frac{1}{\delta\varepsilon}. \tag{1.10}$$

The purpose of this paper is to obtain analogous results for maxima and minima of sequences $x_1, x_2, \cdots$ of i.i.d. random variables. We begin in Section 2 by establishing an analogue of the criterion (1.2) for a law of the iterated logarithm for sample minima. In Section 3, we give an application of this result to a conjecture of Darling and Erdös [2]. In Sections 4 and 5, we introduce a continuous time process v_t, $0 < t < \infty$, related to min $(x_1, \cdots, x_n)$ in much the same way that $w(t)$ is related to $x_1 + \cdots + x_n$, and apply the methods of [11] to the study of this process. In spite of the dissimilarity in the behavior of v_t and $w(t)$, the measure F defined by (1.8) plays the same role for v_t as for $w(t)$.

2. The law of the iterated logarithm for minima of uniform variables

THEOREM 1. *Let* $u_1, u_2, \cdots$ *be independent and uniform on* $(0, 1)$, *and let* $V_n = \min(u_1, \cdots, u_n)$. *Let* (c_n) *be any sequence of positive numbers. Then:*

(i) *if* $c_n/n \downarrow$ *for all sufficiently large* n, *then* $P\{nV_n \leqq c_n \text{ i.o.}\} = 0$ *or* 1 *according as*

$$\sum_1^\infty \frac{c_n}{n} \tag{2.1}$$

converges or diverges;

(ii) *if* $c_n/n \downarrow$ *and* $c_n \uparrow$ *for all sufficiently large* n, *then* $P\{nV_n \geqq c_n \text{ i.o.}\} = 0$ *or* 1 *according as*

$$\sum_1^\infty \frac{c_n}{n} e^{-c_n} \tag{2.2}$$

converges or diverges.

COROLLARY 1. *For* $k \geqq 3$,

$$P\left\{nV_n \geqq \log_2 n + 2\log_3 n + \sum_{i=4}^{k} \log_i n + (1 + \delta)\log_{k+1} n \text{ i.o.}\right\} \tag{2.3}$$

is equal to 0 *or* 1 *according as* $\delta > 0$ *or* $\delta \leqq 0$.

REMARK 2.1. The proof of (i) is an immediate consequence of the Borel–Cantelli lemma and the fact that if c_n/n is ultimately decreasing, then $V_n \leqq c_n/n$ i.o. if and only if $u_n \leqq c_n/n$ i.o. The proof of (ii) is much harder and will be given below.

REMARK 2.2. If $M_n = \max(u_1, \cdots, u_n)$, then Theorem 1 holds with V_n replaced by $1 - M_n$.

REMARK 2.3. Under different regularity conditions on the sequence (c_n), Ville [13] has shown that if (2.2) converges, then $P\{nV_n \geqq c_n \text{ i.o.}\} = 0$. His approach is similar to the one we take in Section 4. Pickands [8] has also obtained some results in the direction of Theorem 1.

REMARK 2.4. The condition in (ii) that c_n be ultimately increasing is bothersome in some applications (see Remark 2.5 below), but it cannot be dropped completely. For example, if $c_n = 1/n$, then both (2.1) and (2.2) converge. Hence by (i), $P\{nV_n \geqq c_n$ for all sufficiently large $n\} = 1$, which is incompatible with the conclusion of (ii) applied to the same sequence c_n.

REMARK 2.5. Let $x_1, x_2, \cdots$ be independent random variables with a common continuous distribution function F. Since $u_n = F(x_n)$ is uniform on $(0, 1)$ and

$$\begin{aligned} F[\min(x_1, \cdots, x_n)] &= \min[F(x_1), \cdots, F(x_n)] \\ &= \min(u_1, \cdots, u_n) = V_n, \end{aligned} \tag{2.4}$$

Theorem 1 implies a law of the iterated logarithm for $\min(x_1, \cdots, x_n)$. In particular, (ii) implies that if (a_n) is any sequence of numbers such that a_n is ultimately decreasing and $nF(a_n)$ is ultimately increasing, then $P\{\min(x_1, \cdots,$

$x_m) \geqq a_n$ i.o.$\} = 0$ or 1 according as

$$\sum_1^\infty F(a_n) \exp\{-nF(a_n)\} < \infty \text{ or } = \infty. \tag{2.5}$$

The condition that $nF(a_n)$ be ultimately increasing may be difficult to verify for a given F, and hence, it is worth observing (as will become apparent in the proof below) that the condition in (ii) that $c_n(= nF(a_n))$ be ultimately increasing may be replaced by the growth condition

$$\liminf_{n\to\infty} \frac{c_n}{\log_2 n} \geqq 1. \tag{2.6}$$

Moreover, it follows *a fortiori* that if $c_n \leqq (\geqq)\, c_n'$ and $P\{nV_n \geqq c_n' \text{ i.o.}\} = 1(0)$, then $P\{nV_n \geqq c_n \text{ i.o.}\} = 1(0)$. Hence, (ii) may be applied indirectly to some sequences (c_n) which satisfy neither the monotonicity conditions of (ii) nor the growth condition (2.6).

REMARK 2.6. Let $x_1, x_2, \cdots$ be independent $N(0, 1)$ random variables with distribution function $\Phi(x) = \int_{-\infty}^x \varphi(y)\, dy$, where $\varphi(y) = (2\pi)^{-1/2} \exp\{-\frac{1}{2}y^2\}$. For $k \geqq 3$ and δ arbitrary, let

$$a_n = -\left[2 \log \frac{n}{2\sqrt{\pi}} - \log_2 n - 2 \log\left(\log_2 n + 2 \log_3 n + \sum_{i=4}^k \log_i n + (1 + \delta) \log_{k+1} n\right)\right]^{1/2}. \tag{2.7}$$

From the fact that

$$\Phi(x) = \frac{1}{|x|} \varphi(x)\left[1 + O\left(\frac{1}{x^2}\right)\right], \qquad \text{as } x \to -\infty, \tag{2.8}$$

it can be shown that $c_n = n\Phi(a_n)$ satisfies (2.6), and hence, by (ii) and the preceding remark, that $P\{\min(x_1, \cdots, x_n) \geqq a_n \text{ i.o.}\} = 0$ or 1 according as $\delta > 0$ or $\delta \leqq 0$. Alternatively, it is possible using (2.8) to replace the criterion (2.5) by one involving the normal density φ; the argument of Lemma 8 below (together with (2.5), (2.8) and Remark 2.2) shows that if (a_n) is any ultimately increasing sequence of positive numbers such that $n a_n^{-1} \varphi(a_n)$ is ultimately increasing, then $P\{\max(x_1, \cdots, x_n) \leqq a_n \text{ i.o.}\} = 0$ or 1 according as

$$\sum_1^\infty \frac{\varphi(a_n)}{a_n} \exp\left\{-n \frac{\varphi(a_n)}{a_n}\right\} \tag{2.9}$$

converges or diverges.

The truth of (ii) follows from Theorem 2 and from Lemma 8 below which shows that the conditions of (ii) imply those of Theorem 2.

THEOREM 2. *Let* $\alpha > 0$ *and* $n_k = \exp\{\alpha k/\log k\}$, $k = 2, 3, \cdots$, *and assume that* c_n/n *is ultimately decreasing.*

(i) *If*

$$\sum_k \exp\{-c_{[n_k]}\} \tag{2.10}$$

converges for some α, *then* $P\{nV_n \geqq c_n \text{ i.o.}\} = 0$.

(ii) *If* (2.6) *holds and* (2.10) *diverges for some* α, *then* $P\{nV_n \geqq c_n \text{ i.o.}\} = 1$.

As usual, $[x]$ denotes the largest integer $\leqq x$. To avoid burdensome detail in the proof, we have ignored the difference between n_k and $[n_k]$.

PROOF. For (i), suppose that (2.10) converges for some α. By replacing c_n by $\min(c_n, 2\log_2 n)$, we may assume, without loss of generality, that

$$c_n \leqq 2\log_2 n. \tag{2.11}$$

By the Borel–Cantelli lemma, it suffices to show that

$$\sum_k P\{nV_n \geqq c_n \text{ for some } n_k < n \leqq n_{k+1}\} < \infty, \tag{2.12}$$

and hence, by the monotonicity of V_n and the ultimate monotonicity of c_n/n, to show that

$$\sum_k P\left\{V_{n_k} \geqq \frac{c_{n_{k+1}}}{n_{k+1}}\right\} < \infty. \tag{2.13}$$

But

$$\begin{aligned} \log P\left\{V_{n_k} \geqq \frac{c_{n_{k+1}}}{n_{k+1}}\right\} &= \log\left(1 - \frac{c_{n_{k+1}}}{n_{k+1}}\right)^{n_k} \leqq -\frac{n_k}{n_{k+1}}c_{n_{k+1}} \\ &\leqq -c_{n_{k+1}}\exp\left\{\frac{\alpha k}{\log(k+1)} - \frac{\alpha(k+1)}{\log(k+1)}\right\} \\ &= -c_{n_{k+1}}\exp\left\{\frac{-\alpha}{\log(k+1)}\right\} \\ &\leqq -c_{n_{k+1}}\left(1 - \frac{\alpha}{\log(k+1)}\right) \\ &\leqq -c_{n_{k+1}} + 2\alpha + o(1), \end{aligned} \tag{2.14}$$

where the last inequality follows from (2.11). Inequality 2.13 now follows immediately from the convergence of (2.10).

For (ii), assume that (2.6) holds and that the series (2.10) diverges for some α. Let $c'_n = \min(c_n, 2\log_2 n)$. Then $\Sigma_k \exp\{-c'_{n_k}\} \geqq \Sigma_k \exp\{-c_{n_k}\} = \infty$, and since by the first part of the theorem

$$P\{nV_n \geqq 2\log_2 n \text{ i.o.}\} = 0, \tag{2.15}$$

it follows that with no loss of generality we may again assume that (2.11) holds.

Let $A_k = \{n_k V_{n_k} \geqq c_{n_k}\}$. By Kolmogorov's 0–1 law, $P(\cap_{m=1}^{\infty} \cup_{k=m}^{\infty} A_k) = 0$ or 1, and hence, to show that infinitely many of the events A_k occur with probability 1, it suffices to show that for all k_0

$$P\left(\bigcup_{k=k_0}^{\infty} A_k\right) \geqq \frac{1}{8}. \tag{2.16}$$

Let $k_1 > k_0$ and for $k_0 \leqq k \leqq k_1$ let $B_k = A_k \cap A_{k+1}^c \cap \cdots \cap A_{k_1}^c$. Then

$$\bigcup_{k=k_0}^{\infty} A_k \supset \bigcup_{k=k_0}^{k_1} A_k = \bigcup_{k=k_0}^{k_1} B_k, \tag{2.17}$$

and the events $B_{k_0}, B_{k_0+1}, \cdots, B_{k_1}$ are disjoint. Hence, to prove (2.16), it suffices to show that there exists a $k_1 > k_0$, k_1 depending on k_0, such that

$$\sum_{k=k_0}^{k_1} P(B_k) \geqq \frac{1}{8}. \tag{2.18}$$

But for each $k_0 \leqq k \leqq k_1$,

$$B_k = A_k \cap \left\{V_{n_r}^{n_k} < \frac{c_{n_r}}{n_r} \text{ for all } k < r \leqq k_1\right\}, \tag{2.19}$$

where we have set $V_j^i = \min_{i<n\leqq j} u_n$. Hence, by the independence of the u_n,

$$\begin{aligned} P(B_k) &= P(A_k) P\left\{V_{n_r}^{n_k} < \frac{c_{n_r}}{n_r} \text{ for all } k < r \leqq k_1\right\} \\ &\geqq P(A_k)\left(1 - \sum_{r=k+1}^{k_1} P\left\{V_{n_r}^{n_k} \geqq \frac{c_{n_r}}{n_r}\right\}\right). \end{aligned} \tag{2.20}$$

It is easy to see from (2.6) that as $k \to \infty$, $P(A_k) \to 0$, and from Lemma 1 below, $\Sigma_k P(A_k) = \infty$. Hence, there exists a number K_0 (to be further specified below) such that for any $k_0 \geqq K_0$ and for some $k_1 > k_0$,

$$\frac{1}{4} \leqq \sum_{k=k_0}^{k_1} P(A_k) \leqq \frac{5}{16}. \tag{2.21}$$

It follows from (2.20) and (2.21) that to prove (2.18) it suffices to show that

$$\sup_{k_0 \leqq k \leqq k_1} \sum_{r=k+1}^{k_1} P\left\{V_{n_r}^{n_k} \geqq \frac{c_{n_r}}{n_r}\right\} \leqq \frac{1}{2}. \tag{2.22}$$

It will be shown in Lemma 8 below that if (2.6) holds, then (2.10) converges or diverges simultaneously for all values of α, and hence, it suffices to prove (2.22) for one value of α. This will be done in Lemmas 2 through 7 below, completing the proof of the theorem.

LEMMA 1. $\Sigma_k P(A_k) = \infty$.

Proof. Since $\log(1 - x) \geqq -x - x^2$ for all sufficiently small positive x, we have from (2.11) as $k \to \infty$,

$$\log P(A_k) = n_k \log\left(1 - \frac{c_{n_k}}{n_k}\right) \geqq n_k\left(-\frac{c_{n_k}}{n_k} - \frac{c_{n_k}^2}{n_k^2}\right) \geqq -c_{n_k} + o(1). \tag{2.23}$$

The lemma now follows from the divergence of (2.10).

In Lemmas 2 through 7 below, $\alpha > 1$ and $0 < \lambda < 1$ will be fixed numbers satisfying

$$\lambda > \left(\frac{5}{6}\right)^{1/2}, \tag{2.24}$$

and

$$\exp\{\alpha\lambda^3\} > 17. \tag{2.25}$$

Lemma 2. *There exists a number K_1 such that for all $k \geqq K_1$ and $r > k$,*

$$\frac{n_k}{n_r} \leqq \exp\left\{\frac{-\lambda\alpha(r - k)}{\log r}\right\}. \tag{2.26}$$

Proof. Let $v = r - k$. Then since $\log(1 + x) \leqq x$,

$$\log\frac{n_k}{n_r} = \frac{\alpha k}{\log k} - \frac{\alpha(k + v)}{\log(k + v)} = \frac{\alpha k \log(1 + v/k)}{\log k \log(k + v)} - \frac{\alpha v}{\log(k + v)} \tag{2.27}$$

$$\leqq -\frac{\alpha v}{\log(k + v)}\left(1 - \frac{1}{\log k}\right) \leqq -\lambda\alpha\frac{(r - k)}{\log r}$$

for $k \geqq K_1$, provided $\log K_1 \geqq (1 - \lambda)^{-1}$.

Lemma 3. *For each k let $r_1 = r_1(k)$ be the largest integer r such that $r - k < (\log r)^{1/2}$. There exists a number K_2 such that for all $k \geqq K_2$ and $k < r \leqq r_1$,*

$$P\left\{V_{n_r}^{n_k} \geqq \frac{c_{n_r}}{n_r}\right\} \leqq \exp\{-\alpha\lambda^3(r - k)\}. \tag{2.28}$$

Proof. From the inequality $1 - x \leqq e^{-x}$ and Lemma 1, for $r > k \geqq K_1$, we obtain

$$P\left\{V_{n_r}^{n_k} \geqq \frac{c_{n_r}}{n_r}\right\} = \left(1 - \frac{c_{n_r}}{n_r}\right)^{n_r - n_k} \leqq \exp\left\{-\left(1 - \frac{n_k}{n_r}\right)c_{n_r}\right\} \tag{2.29}$$

$$\leqq \exp\left\{-\left(1 - \exp\left\{-\lambda\alpha\left(\frac{r - k}{\log r}\right)\right\}\right)c_{n_r}\right\}.$$

For all sufficiently small positive x, $1 - e^{-x} \geqq \lambda x$. Hence, there exists K_3 so large that for all $k \geqq K_3$ and $k < r \leqq r_1$,

$$1 - \exp\left\{-\lambda\alpha\left(\frac{r - k}{\log r}\right)\right\} \geqq \alpha\lambda^2\left(\frac{r - k}{\log r}\right). \tag{2.30}$$

Finally, by (2.6), there exists K_4 so large that for $r > K_4$

$$c_{n_r} \geqq \lambda \log r. \tag{2.31}$$

With $K_2 = \max(K_1, K_3, K_4)$, the lemma follows from (2.29), (2.30), and (2.31).

LEMMA 4. *For each* $k \geqq K_2$ *and* $r > r_1$,

$$P\left\{V_{n_r}^{n_k} \geqq \frac{c_{n_r}}{n_r}\right\} \leqq \exp\{-\lambda^3 (\log r)^{1/2}\}. \tag{2.32}$$

PROOF. From (2.29) and (2.31), we have

$$\begin{aligned} P\left\{V_{n_r}^{n_k} \geqq \frac{c_{n_r}}{n_r}\right\} &\leqq \exp\left\{-\left(1 - \exp\left\{-\lambda \frac{r-k}{\log r}\right\}\right) c_{n_r}\right\} \\ &\leqq \exp\{-(1 - \exp\{-\lambda (\log r)^{-1/2}\})\, \lambda \log r\} \\ &\leqq \exp\{-\lambda^3 (\log r)^{1/2}\}. \end{aligned} \tag{2.33}$$

LEMMA 5. *For each* k, *let* $r_2 = r_2(k)$ *be the least integer* $r > k$ *such that* $r \geqq k + (\log r)^2$. *Then for all* $k > K_1$ *and* $r \geqq r_2$, $n_k/n_r \leqq 1/r$.

PROOF. By (2.27), for $k \geqq K_1$ and $r \geqq r_2$,

$$\log \frac{n_k}{n_r} \leqq -\lambda\alpha\left(\frac{r-k}{\log r}\right) \leqq -\lambda\alpha \log r < -\log r \tag{2.34}$$

LEMMA 6. *There exists a number* K_5 *such that for all* $k \geqq K_5$ *and* $r > r_2$,

$$P\left\{V_{n_r}^{n_k} \geqq \frac{c_{n_r}}{n_r}\right\} \leqq \frac{1}{\lambda^2} P(A_r). \tag{2.35}$$

PROOF. From (2.23), we have

$$P(A_r) \geqq \lambda \exp\{-c_{n_r}\} \tag{2.36}$$

for all $r \geqq$ some K_6. Hence, by Lemma 5, (2.11), (2.29), and (2.36), we have, for all $k \geqq K_5 \geqq \max(K_1, K_6)$ and $r \geqq r_2$,

$$\begin{aligned} P\left\{V_{n_r}^{n_k} \geqq \frac{c_{n_r}}{n_r}\right\} &\leqq \exp\left\{-\left(1 - \frac{n_k}{n_r}\right) c_{n_r}\right\} \\ &\leqq \frac{1}{\lambda} P(A_r) \exp\left\{\frac{2 \log_2 n_r}{r}\right\} \leqq \frac{1}{\lambda^2} P(A_r). \end{aligned} \tag{2.37}$$

Note that $r_2(k) \sim k$ as $k \to \infty$. Let K_7 be so large that for all $k \geqq K_7$,

$$r_2(k) \leqq 2k \tag{2.38}$$

and

$$8(\log k)^2 \exp\{-\lambda^3 (\log k)^{1/2}\} < \tfrac{1}{16}. \tag{2.39}$$

LEMMA 7. *For* $K_0 = \max(K_1, \cdots, K_7)$ *and all* $k_0 \geqq K_0$, *(2.22) holds.*

PROOF. For all $k_0 \leqq k \leqq k_1$,

$$\sum_{r=k+1}^{k_1} P\left\{V_{n_r}^{n_k} \geq \frac{c_{n_r}}{n_r}\right\} \leqq \sum_{r=k+1}^{r_1} + \sum_{r=r_1+1}^{r_2} + \sum_{r=r_2+1}^{k_1}, \tag{2.40}$$

which by Lemmas 3, 4, and 6, equations (2.21), (2.24), (2.25), (2.38), and (2.39), does not exceed

$$\sum_{r=k+1}^{\infty} \exp\{-\alpha\lambda^3(r-k)\} + 2(\log r_2)^2 \exp\{-\lambda^3(\log k)^{1/2}\} + \frac{1}{\lambda^2}\sum_{r=r_2}^{k_1} P(A_r) \tag{2.41}$$

$$\leqq \frac{\exp\{-\alpha\lambda^3\}}{1-\exp\{-\alpha\lambda^3\}} + 2(2\log k)^2 \exp\{-\lambda^3(\log k)^{1/2}\} + \frac{1}{\lambda^2}\left(\frac{5}{16}\right)$$

$$\leqq \frac{1}{16} + \frac{1}{16} + \frac{6}{16} = \frac{1}{2}.$$

The following lemma shows that the conditions of Theorem 1(ii) imply those of Theorem 2. Note that the condition that (c_n) be ultimately increasing is used only to show that (c_n) may without loss of generality be assumed to satisfy (2.6). This substantiates Remark (2.5) above.

LEMMA 8. *Let (c_n) be any sequence of positive numbers such that c_n/n is ultimately decreasing and either (c_n) is ultimately increasing or* (2.6) *holds. Then* (2.2) *converges if and only if* (2.10) *converges for all $\alpha > 0$.*

PROOF. First observe that without loss of generality we may assume that (2.11) holds. In fact, if c_n is ultimately increasing, so is $c_n' = \min(c_n, 2\log_2 n)$, while if (2.6) holds then it also holds for c_n', and it is easy to see that replacing c_n by c_n' does not alter the convergence or divergence of either (2.2) or (2.10).

We next show that with no loss of generality it may be assumed that (2.6) holds. Suppose that $c_n \leqq c_{n+1}$ for all $n \geqq n_0$. If $\lim_{n\to\infty} c_n < \infty$, then (2.2) and (2.10) both diverge and continue to do so if c_n is replaced by $c_n' = \max(c_n, \log_2 n)$. Suppose on the other hand that $c_n \to \infty$. Since xe^{-x} is decreasing for large x, we have

$$\sum_{n_0}^{n} \frac{c_k}{k} e^{-c_k} \geqq c_n e^{-c_n} \sum_{n_0}^{n} k^{-1} \geqq c_n e^{-c_n} \log n - O(1), \tag{2.42}$$

which $\to \infty$ along any subsequence n' for which $c_{n'} \leqq \log_2 n'$. Hence, if (2.2) converges, (2.6) holds. If (2.2) diverges, we see from (2.42) that we may replace c_n by $c_n' = \max(c_n, \log_2 n)$ and maintain divergence, so in this case as well we may assume that (2.6) holds.

It remains to prove the lemma under the assumption

$$\tfrac{1}{2}\log_2 n \leqq c_n \leqq 2\log_2 n. \tag{2.43}$$

Now

$$\left(1 - \frac{n_k}{n_{k+1}}\right) \sim 1 - \exp\left\{-\frac{\alpha}{\log k}\right\} \sim \frac{\alpha}{\log k} \sim \left(\frac{n_{k+1}}{n_k} - 1\right) \quad \text{as } k \to \infty, \tag{2.44}$$

and hence, from (2.43),

$$(2.45)\qquad c_{n_{k+1}}\left(1-\frac{n_k}{n_{k+1}}\right),\qquad c_{n_k}\left(\frac{n_{k+1}}{n_k}-1\right)$$

are bounded away from 0 and ∞. Since c_n/n is decreasing for large n, if (2.2) diverges, we have

$$(2.46)\quad \infty=\sum_k\sum_{n_k<n\le n_{k+1}}\frac{c_n}{n}\exp\left\{\frac{-c_n}{n}n\right\}\le\sum_k\left(\frac{c_{n_k}}{n_k}\exp\left\{-\frac{c_{n_{k+1}}}{n_{k+1}}n_k\right\}\right)(n_{k+1}-n_k)$$
$$\le\sum_k c_{n_k}\left(\frac{n_{k+1}}{n_k}-1\right)\exp\left\{-c_{n_{k+1}}+c_{n_{k+1}}\left(1-\frac{n_k}{n_{k+1}}\right)\right\}$$
$$\le \text{const.}\sum_k\exp\{-c_{n_{k+1}}\}.$$

The case in which (2.2) converges is treated similarly.

3. A conjecture of Darling and Erdös

In [2] Darling and Erdös obtained the limiting distribution of

$$(3.1)\qquad y_t=\max_{0\le\tau\le t}\frac{w(\tau)}{(\tau+1)^{1/2}}\qquad\text{as } t\to\infty.$$

(This question was suggested by an inequality in [9] concerning the statistical consequences of "optional stopping.") They also conjectured an iterated logarithm law for the process y_t, namely:

(a) there exists a constant $c_1>0$ such that

$$(3.2)\qquad P\left\{y_t\ge(2\log_2 t)^{1/2}+\frac{\log_3 t}{2(2\log_2 t)^{1/2}}+\frac{(c_1+\delta)\log_4 t}{(2\log_2 t)^{1/2}}\text{ i.o. } t\uparrow\infty\right\}$$
$$=0\text{ or }1\qquad\text{according as }\delta>0\text{ or }\delta<0;$$

and

(b) there exists a constant $c_2>0$ such that

$$(3.3)\qquad P\left\{y_t\le(2\log_2 t)^{1/2}+\frac{\log_3 t}{2(2\log_2 t)^{1/2}}-\frac{(c_2+\delta)\log_4 t}{(2\log_2 t)^{1/2}}\text{ i.o. } t\uparrow\infty\right\}$$
$$=0\text{ or }1\qquad\text{according as }\delta>0\text{ or }\delta<0.$$

Since y_t is increasing in t, it follows that for any ultimately increasing function $\psi(t)$, $y_t\ge\psi(t)$ for arbitrarily large t if and only if $w(t)\ge(t+1)^{1/2}\psi(t)$ for arbitrarily large t (see Remark 2.1). Hence, from (1.3) we see that the probability (3.2) is 1 for all c_1 and δ; that is, conjecture (a) is false. A correct version of (a) is

$$(3.4)\qquad P\left\{y_t\ge(2\log_2 t)^{1/2}+\frac{3\log_3 t}{2(2\log_2 t)^{1/2}}+\frac{(1+\delta)\log_4 t}{(2\log_2 t)^{1/2}}\text{ i.o. } t\uparrow\infty\right\}$$
$$=0\text{ or }1\qquad\text{according as }\delta>0\text{ or }\delta<0.$$

Conjecture (b) is correct, but more difficult to prove. In this section, we shall use Theorem 1(ii) to verify (b) and identify the constant c_2 as 1. We shall only sketch the proof, which relies greatly on the method of Motoo [6]. (Motoo's method illuminates the entire paper [2] as well as the relation of Theorem 1(i) to (a) and Theorem 1(ii) to (b).)

Let

$$U(t) = e^{-t}w(e^{2t} - 1). \tag{3.5}$$

It is easy to verify that

$$P\{U(t + s)\varepsilon\, dx \mid U(s) = u\} = \varphi\left(\frac{x - ue^{-t}}{(1 - e^{-2t})^{1/2}}\right)\frac{dx}{(1 - e^{-2t})^{1/2}}, \tag{3.6}$$

and hence, that $U(t)$, $0 \leqq t < \infty$, is a Markov process with stationary transition probabilities and infinitesimal generator

$$Df(x) = f''(x) - xf'(x). \tag{3.7}$$

(This $U(t)$ is the Ornstein–Uhlenbeck process with $U(0) = 0$.) To prove (b) with $c_2 = 1$, it suffices to show that

$$P\{\max_{0 \leqq \tau \leqq t} U(\tau) \leqq (2 \log t + \log_2 t - (2 + \delta) \log_3 t)^{1/2} \text{ i.o. } t \uparrow \infty\} = 0 \text{ or } 1 \quad \text{according as } \delta > 0 \text{ or } \delta < 0. \tag{3.8}$$

Define $T_0 = 0$ and for each $n = 1, 2, \cdots,$

$$\begin{aligned} T_{2n-1} &= \inf\{t: t > T_{2n-2}, U(t) = 1\}, \\ T_{2n} &= \inf\{t: t > T_{2n-1}, U(t) = 0\}. \end{aligned} \tag{3.9}$$

It may be shown that $\gamma = ET_2 < \infty$, and since $T_2 - T_0$, $T_4 - T_2, \cdots$ are independent and identically distributed, it follows from the strong law of large numbers that

$$P\left\{\frac{T_{2n}}{n} \to \gamma\right\} = 1. \tag{3.10}$$

Let $x_n = \max_{T_{2n-2} \leqq t < T_{2n}} U(t)$, $n = 1, 2, \cdots$. Then $x_1, x_2, \cdots$ are i.i.d., and for $a > 1$, $P\{x_n > a\}$ is the probability that the process $U(t)$ starting from 1 reaches the level a before it reaches 0. From (3.7) and standard diffusion theory it follows that $P\{x_n > a\} = g(1)$, where $g(x)$ satisfies $g''(x) - xg(x) = 0$, $0 < x < a$, subject to the boundary conditions $g(0) = 0$, $g(a) = 1$. Hence,

$$P\{x_n > a\} = \frac{\int_0^1 \exp\{\frac{1}{2}y^2\}\, dy}{\int_0^a \exp\{\frac{1}{2}y^2\}\, dy} = \eta a \exp\{-\tfrac{1}{2}a^2\}\left(1 + O\left(\frac{1}{a^2}\right)\right) \tag{3.11}$$

as $a \to \infty$, where we have put $\eta = \int_0^1 \exp\{\frac{1}{2}y^2\}\, dy$. Let $\delta > 0$ and

$$\psi(t) = [2 \log t + \log_2 t - (2 + \delta) \log_3 t]^{1/2}. \tag{3.12}$$

Since ψ is ultimately increasing, we have by (3.10) for any $\varepsilon > 0$,

$$\begin{aligned} P\{\max_{0 \leqq \tau \leqq t} U(t) \leqq \psi(t) \text{ i.o. } t \uparrow \infty\} &\leqq P\{\max_{0 \leqq t \leqq T_{2n}} U(t) \leqq \psi(T_{2n+2}) \text{ i.o.}\} \\ &\leqq P\{\max_{1 \leqq k \leqq n} x_k \leqq \psi(n(\gamma + \varepsilon)) \text{ i.o.}\}. \end{aligned} \tag{3.13}$$

It follows from Remarks 2.2 and 2.5 and some straightforward calculation using (3.11) that

$$P\{\max_{0 \leqq \tau \leqq t} U(t) \leqq \psi(t) \text{ i.o. } t \uparrow \infty\}, \tag{3.14}$$

that is, the probability (3.8), is 0 for $\delta > 0$. A similar argument shows that (3.8) is 1 for $\delta > 0$.

Motoo's method of proof of the criterion (1.2) for the Wiener process [6] is essentially a combination of the preceding argument with Theorem 1 (i) instead of Theorem 1 (ii). It is interesting to note that neither Motoo's nor our argument requires knowledge of the exact value of $\gamma = ET_2$ or of the constant η appearing in (3.11). A more careful analysis shows that under certain regularity conditions on the function ψ,

$$\begin{aligned} &P\{\max_{0 \leqq \tau \leqq t} U(\tau) \leqq \psi(t) \text{ i.o. } t \uparrow \infty\}, \\ &\qquad = 0 \text{ or } 1, \text{ according as } \int_1^\infty f(\psi(t)) \exp\left\{-\frac{t}{\sqrt{2\pi}} f(\psi(t))\right\} dt \end{aligned} \tag{3.15}$$

is convergent or divergent, where we have set $f(x) = x \exp\{-\frac{1}{2}x^2\}$. However, establishing this deeper criterion requires knowledge of the constants γ and η and in particular that $\gamma/\eta = (2\pi)^{1/2}$. (It is interesting to observe that if we had defined the stopping times T_n in our proof in terms of 0 and an arbitrary number $b > 0$, then γ and η would depend on b, but the ratio γ/η would not.) It is also necessary to sharpen (3.10) to, say,

$$P\left\{\frac{T_{2n} - n\gamma}{n^{1/2} \log n} \to 0\right\} = 1, \tag{3.16}$$

which is a consequence of the fact that $ET_2^2 < \infty$ and the usual proof of the strong law of large numbers using Kolmogorov's inequality and Kronecker's lemma (see [7]). We omit the details.

4. A continuous time extremal process

In this section, we introduce a continuous time process v_t which bears more or less the same relation to the process V_n as the Wiener process does to the sequence of partial sums of i.i.d. mean 0, variance 1, random variables, and give boundary crossing probabilities for this process analogous to those of [11] for the Wiener process.

Consider the sequence of processes $nV_{[nt]}$, $0 \leqq t < \infty$. For any $0 = t_0 < t_1 < t_2 < \cdots < t_r$ and $a_1 \geqq a_2 \geqq \cdots \geqq a_r > 0$, we have as $n \to \infty$

$$(4.1) \qquad P\{nV_{[nt_1]} \geqq a_1, \cdots, nV_{[nt_r]} \geqq a_r\} = \prod_{i=1}^{r} \left(1 - \frac{a_i}{n}\right)^{[nt_i]-[nt_{i-1}]} \to \exp\left\{-\sum_{i=1}^{n} a_i(t_i - t_{i-1})\right\}.$$

This suggests defining v_t, $0 < t < \infty$, by the following consistent family of joint distributions: for $0 = t_0 < t_1 < \cdots < t_r$ and $-\infty < a_i < \infty$, $i = 1, 2, \cdots, r$,

$$(4.2) \quad P\{v_{t_1} \geqq a_1, \cdots, v_{t_r} \geqq a_r\} = \exp\left\{-\sum_{i=1}^{r} \max(a_i, \cdots, a_r)^+ (t_i - t_{i-1})\right\}.$$

By Kolmogorov's consistency theorem, there exists a process, say $\tilde{v}_t$, having the finite dimensional joint distributions given by (4.2). Defining $v_t = \lim_{s \downarrow t} \tilde{v}_s$, where s runs through rationals greater than t, we obtain a process having the same finite dimensional joint distributions as $\tilde{v}_t$ and in addition right continuous, decreasing sample paths. We shall call any such process a *standard extremal process*.

It is easy to see from (4.2) that the process v_t, $0 < t < \infty$, is Markovian with stationary transition probability

$$(4.3) \quad P\{v_t \geqq a_2 | v_\tau = a_1\} = \begin{cases} \exp\{-a_2(t - \tau)\} & \text{for } a_1 \geqq a_2 \geqq 0, \\ 0 & \text{for } 0 \leqq a_1 < a_2. \end{cases}$$

For each $\tau > 0$, let $h = h(\tau) = \inf\{t: t > \tau, v_t < v_\tau\}$. Then $\{h > t\} = \{v_t = v_\tau\}$, and hence, by (4.3),

$$(4.4) \qquad P\{h > t | v_\tau = a\} = \exp\{-a(t - \tau)\}.$$

Also for $a_2 < a_1$,

$$\begin{aligned} (4.5) \qquad & P\{v_h \leqq a_2, t < h < t + \delta | v_\tau = a_1\} \\ &= P\{v_{t+\delta} \leqq a_2, t < h < t + \delta | v_\tau = a_1\} + o(\delta) \\ &= P\{v_{t+\delta} \leqq a_2, v_t \geqq a_1 | v_\tau = a_1\} + o(\delta) \\ &= (1 - \exp\{-\delta a_2\}) \exp\{-(t - \tau)a_1\} + o(\delta), \end{aligned}$$

so

$$(4.6) \qquad P\{v_h \leqq a_2 | v_\tau = a_1\} = \int_\tau^\infty P\{v_h \leqq a_2, h \in dt | v_\tau = a_1\} = \frac{a_2}{a_1}.$$

It follows that the sample paths of the process v_t, $0 < t < \infty$, may be described as follows: for any $\tau > 0$, if the process is in the state a at time τ, it remains there for a random length of time having a negative exponential distribution with parameter a and then jumps to a point uniformly distributed on $(0, a)$. By (4.2), $P\{v_{0+} = +\infty\} = 1$, and with probability 1 there are infinitely many jumps in each neighborhood of $t = 0$. Except for the behavior of v_t near $t = 0$, this description is analogous to that of the discrete time process V_n, which holds in each state a a random length of time which is geometrically distributed with parameter a and then moves to a point uniformly distributed on $(0, a)$. If $x_1, x_2, \cdots$ are i.i.d. with $P\{x_i \geqq x\} = e^{-x}$, and $v_n^* = \min(x_1, \cdots, x_n)$, then the process v_t interpolates the process v_n^* in the sense that the two sequences v_n and v_n^*, $n = 1, 2, \cdots$, have the same joint distributions.

Trivial modifications of the proof of Theorem 1 (ii) prove:

THEOREM 3. *If $c(t) \geqq 0$ is ultimately increasing and $c(t)/t$ is ultimately decreasing, then $P\{v_t \geqq c(t)/t, \text{ i.o. } t \uparrow \infty\} = 0$ or 1, according as $\int_1^\infty (c(t)/t)e^{-c(t)}\,dt$ converges or diverges.*

Since $P\{v_{0_+} = +\infty\} = 1$, it is of interest to obtain a description of the rate of growth of v_t as $t \downarrow 0$. A law of the iterated logarithm for v_t as $t \downarrow 0$ follows from Theorem 3 and the following inversion theorem.

THEOREM 4. *For each $v > 0$, let $T(v) = \sup\{t : v_t \geqq v\}$. The process $T(v)$, $0 < v < \infty$, is a standard extremal process.*

PROOF. The fact that the sample paths of $T(v)$, $0 < v < \infty$ are decreasing, right continuous step functions follows at once from the corresponding properties of v_t, $0 < t < \infty$. Hence, it suffices to show that $T(v)$ and v_t have the same finite dimensional joint distributions. For $0 < u < v$ and $\tau > t > 0$, except for a set of probability 0,

$$\{T(u) \geqq \tau, T(v) \geqq t\} = \{v_t \geqq v, v_\tau \geqq u\}, \tag{4.7}$$

and hence, by (4.2),

$$\begin{aligned} P\{T(u) \geqq \tau, T(v) \geqq t\} \\ = \exp\{-tv - (\tau - t)u\} = \exp\{-u\tau - (v - u)t\}. \end{aligned} \tag{4.8}$$

The general case of an arbitrary finite number of time points $u, v, \cdots, z$ follows by the same argument.

For any strictly decreasing function ψ defined on $(0, \infty)$, $v_t \geqq \psi(t)$ i.o. $t \downarrow 0$ if and only if $T(v) > \psi^{-1}(v)$ i.o. $v \uparrow \infty$. Hence, by Theorem 4,

$$P\{v_t \geqq \psi(t) \text{ i.o. } t \downarrow 0\} = P\{v_t > \psi^{-1}(t) \text{ i.o. } t \uparrow \infty\}. \tag{4.9}$$

For example, by Theorem 3 and (4.9), we have

$$P\left\{\limsup_{t \to 0} \frac{tv_t}{\log_2 \frac{1}{t}} = 1\right\} = 1. \tag{4.10}$$

We now show that the method of [11] yields boundary crossing probabilities for the process v_t, $0 < t < \infty$, analogous to those obtained there for the Wiener process.

Let F denote a measure on $(0, \infty)$ assigning finite measure to bounded intervals, and define for $x > 0$, $t \geqq 0$, $\varepsilon > 0$,

$$f(x, t) = \int_{\{0<y\leqq x\}} e^{yt}\, dF(y),$$
(4.11)
$$A(t, \varepsilon) = \inf\{x : f(x, t) \geqq \varepsilon\} (= \infty \text{ if } f(x, t) < \varepsilon \text{ for all } x).$$

It is easily seen that for all $t \geqq 0$, $x \geqq A(t, \varepsilon)$ if and only if $f(x, t) \geqq \varepsilon$, and that if $\tau_0 = \inf\{\tau : A(\tau, \varepsilon) < \infty\}$, then $A(\cdot, \varepsilon)$ is continuous and decreasing on (τ_0, ∞) and $A(\tau_0, \varepsilon) = \lim_{t\downarrow\tau_0} A(t, \varepsilon)$. Moreover, if $F\{A(t, \varepsilon)\} = 0$, then $f(A(t, \varepsilon), t) = \varepsilon$.

Let $\mathscr{F}_t = \mathscr{B}(v_\tau, \tau \leqq t)$. Since $\{I[v_t \geqq y]e^{yt}, \mathscr{F}_t, 0 \leqq t < \infty\}$ is a martingale for each $y > 0$, as may be verified by direct computation using (4.3), it follows from Fubini's theorem that $\{f(v_t, t), \mathscr{F}_t, 0 < t < \infty\}$ is also a martingale.

THEOREM 5. *For any* $\varepsilon > F\{(0, \infty)\}$

$$P\{v_t \geqq A(t, \varepsilon) \text{ for some } t > 0\} = \frac{F\{(0, \infty)\}}{\varepsilon}. \tag{4.12}$$

For each $\tau > 0$,

$$P\{v_t \geqq A(t, \varepsilon) \text{ for some } t \geqq \tau\} = \exp\{-\tau A(\tau, \varepsilon)\} \tag{4.13}$$
$$+ \varepsilon^{-1}\left[F\{(0, A(\tau, \varepsilon))\} - \exp\{-\tau A(\tau, \varepsilon)\}\int_{\{0<y<A(\tau,\varepsilon)\}} e^{\tau y}\, dF(y)\right]$$
$$\left(= \varepsilon^{-1}F\{(0, A(\tau, \varepsilon))\} \text{ if } F\{A(\tau, \varepsilon)\} = 0\right).$$

PROOF. The parenthetical part of (4.13) follows at once from the preceding line and the observation that if $F\{A(\tau, \varepsilon)\} = 0$, then

$$\int_{\{0<y<A(\tau,\varepsilon)\}} e^{\tau y}\, dF(y) = f(A(\tau, \varepsilon), \tau) = \varepsilon; \tag{4.14}$$

equation (4.12) follows from the parenthetical part of (4.13), by letting $\tau\downarrow\tau_0 = \inf\{t : A(t, \varepsilon) < \infty\}$ through any sequence of values such that $F\{A(\tau, \varepsilon)\} = 0$. The proof of (4.13) follows from Lemma 1 of [11], Remark (d) at the end of Section 3 of [11], and Lemma 9 below (see the proof of Theorem 1 of [11]).

LEMMA 9. *The function* $f(v_t, t)$ *tends to* 0 *in probability.*

PROOF. Let $c > 0$. By the weak convergence of the family $F_t\{\cdot\} = F\{(\cdot)\cap(0, c]/t\}$ to the 0 measure,

$$\int_{\{0<y\leqq c/t\}} e^{yt}\, dF(y) = \int_{\{0<y\leqq c\}} e^{y}\, dF\left(\frac{y}{t}\right) \to 0 \tag{4.15}$$

as $t \to \infty$. Hence, for any $\varepsilon > 0$, for all t sufficiently large,

$$P\{f(v_t, t) \geqq \varepsilon\} \leqq P\left\{v_t \geqq \frac{c}{t}\right\} \doteq e^{-c}, \tag{4.16}$$

which can be made arbitrarily small by taking c sufficiently large.

5. Asymptotic expansions for $A(t, \varepsilon)$

If the measure F of the preceding section is taken to be Lebesgue measure on $(0, \infty)$, it is easily seen from (4.11) that

$$A(t, \varepsilon) = \frac{1}{t} \log (1 + \varepsilon t) = \frac{1}{t} \left[\log t + \log \varepsilon + O\left(\frac{1}{t}\right) \right] \tag{5.1}$$

as $t \to \infty$. By Theorem 3 there exist functions $g(t) \sim \log_2 t/t$ as $t \to \infty$ such that $P\{v_t \geqq g(t) \text{ for some } t > 0\} < 1$, and it is natural to ask whether we can find boundaries with this rate of growth to which Theorem 5 applies.

THEOREM 6. *If F is defined by* (1.8), *then for* $k = 2$

$$A(t, \varepsilon) = \frac{1}{t} [\log_2 t + (2 + \delta) \log_3 t + \log \varepsilon + o(1)] \tag{5.2}$$

as $t \to \infty$, *while for* $k \geqq 3$,

$$A(t, \varepsilon) = \frac{1}{t} \left[\log_2 t + 2 \log_3 t + \sum_{i=4}^{k} \log_i t + (1 + \delta) \log_{k+1} t + \log \varepsilon + o(1) \right] \tag{5.3}$$

as $t \to \infty$.

(See equation (10) of [11] which describes the corresponding result for the Wiener process.)

To prove (5.2), let F be given by (1.8) with $k = 2$ and let $F'(y) = dF/dy$. (The proof of (5.3) when F is given by (1.8) with $k \geqq 3$ is similar and will be omitted.) It follows easily from (4.11) that

$$A = A(t, \varepsilon) \to 0 \tag{5.4}$$

as $t \to \infty$. Similarly,

$$tA \to \infty. \tag{5.5}$$

In fact, if there exists a number C such that $tA < C$ along a sequence of t values, then

$$\varepsilon \leqq \int_0^C e^z \, dF\left(\frac{z}{t}\right). \tag{5.6}$$

But $F(z/t) \to 0$ as $t \to \infty$ for all $z > 0$, and hence, $\int_0^C e^z \, dF(z/t) \to 0$ as $t \to \infty$, contradicting our supposition. Since F' is decreasing in a neighborhood of the origin, we have by (5.4) for all t sufficiently large,

$$\varepsilon = f(A, t) = \int_0^A e^{yt} F'(y) \, dy \geqq \frac{F'(A)(e^{tA} - 1)}{t}, \tag{5.7}$$

and hence, by (1.8) and (5.5),

$$\limsup_{t\to\infty} \frac{e^{tA}}{tA \log \dfrac{1}{A}\left(\log_2 \dfrac{1}{A}\right)^{1+\delta}} \leqq \varepsilon. \tag{5.8}$$

Rewriting (5.8) as

$$A \leqq \frac{1}{t}\left[\log \varepsilon t + \log A + \log_2 \frac{1}{A} + (1+\delta)\log_3 \frac{1}{A} + o(1)\right], \tag{5.9}$$

we see by (5.4) and (5.5) that $A \leqq (\log t/t)[1 + o(1)]$, and hence $\log A \leqq \log_2 t - \log t + o(1)$. Since (5.5) implies *a fortiori* that for all sufficiently large t we have $1/A \leqq t$, and hence

$$\log_k 1/A \leqq \log_k t, \tag{5.10}$$

we have from (5.9),

$$A \leqq \frac{1}{t}[\log \varepsilon + 2\log_2 t + (1+\delta)\log_3 t + o(1)]. \tag{5.11}$$

Thus $\log A \leqq -\log t + \log_3 t + O(1)$, which by substituting once more in (5.9) yields $A \leqq \log_2 t/t\,(1 + o(1))$ and hence

$$\log A \leqq -\log t + \log_3 t + o(1). \tag{5.12}$$

Finally, substituting (5.12) and (5.10) in (5.9), yields one half of (5.2), to wit

$$A \leqq \frac{1}{t}[\log_2 t + (2+\delta)\log_3 t + \log \varepsilon + o(1)]. \tag{5.13}$$

In particular,

$$\limsup_{t\to\infty} \frac{tA}{\log_2 t} \leqq 1. \tag{5.14}$$

To prove (5.13) with the inequality reversed let $0 < b < c < 1$. From (4.11), we have for all t sufficiently large

$$\begin{aligned} \varepsilon &= \left(\int_0^{bA} + \int_{bA}^{cA} + \int_{cA}^{A}\right) e^{yt}F'(y)\,dy \\ &\leqq e^{btA}F(bA) + \frac{e^{ctA}}{t}F'(bA) + \frac{e^{tA}}{t}F'(cA) \\ &\leqq \frac{e^{btA}}{\delta\left(\log \dfrac{1}{2A}\right)^{\delta}} + \frac{e^{ctA}}{btA \log \dfrac{1}{A}\left(\log \dfrac{1}{2A}\right)^{1+\delta}} + \frac{e^{tA}}{ctA \log \dfrac{1}{A}\left(\log \dfrac{1}{2A}\right)^{1+\delta}}. \end{aligned} \tag{5.15}$$

Let $b = 1/\log_2 t$. Then from (5.4), (5.5), (5.14), and (5.15), we obtain for large t

$$\varepsilon \leqq o(1) + \frac{e^{tA}}{\log 1/A} \leqq o(1) + \frac{e^{tA}}{\frac{1}{2}\log t}, \tag{5.16}$$

and hence,

$$\frac{\log_2 t}{t} = O(A). \tag{5.17}$$

From (5.14) and (5.17), it follows that for all η sufficiently small,

$$\eta \leqq \frac{tA}{\log_2 t} \leqq 1 + \eta \tag{5.18}$$

for all sufficiently large t. Using (5.18), we see that the second term on the right side of (5.15) is majorized by $\exp\{c(1 + \eta)\log_2 t\}/(\eta/2 \log t)$ for large t, which converges to 0 as $t \to \infty$ for η so small that $c(1 + \eta) < 1$. Hence, letting $t \to \infty$, then $c \to 1$ in (5.15), we have

$$\liminf_{t\to\infty} \frac{e^{tA}}{tA \log \frac{1}{A}\left(\log_2 \frac{1}{A}\right)^{1+\delta}} \geqq \varepsilon. \tag{5.19}$$

The reverse of inequality (5.13) now follows from (5.19) by an argument similar to that which led from (5.8) to (5.13). This completes the proof of (5.2).

6. Remarks

REMARK 6.1. Extremal processes in continuous time have been studied by Dwass [4] and Lamperti [5]. Lamperti proved an invariance theorem which is helpful in the proof of (6.2) below.

REMARK 6.2. Theorem 2(i) of [11] states that if g is a positive continuous function such that $g(t)/t^{1/2}$ is ultimately increasing and (1.2) converges, then for each $\tau > 0$,

$$P\{w(t) \geqq g(t) \text{ for some } t \geqq \tau\} = \lim_{m\to\infty} P\{S_n \geqq m^{1/2} g(n/m) \text{ for some } n \geqq \tau m\}, \tag{6.1}$$

where $S_n = x_1 + \cdots + x_n$, and $x_1, x_2, \cdots$ is any sequence of i.i.d. random variables having $Ex_1 = 0$, $Ex_1^2 = 1$. An analogous limit theorem for minima of uniform random variables is that if g is continuous and decreasing on some interval $(\tau_0, \infty)([\tau_0, \infty))$ and $\equiv \infty$ on $[0, \tau_0]([0, \tau_0))$, and if (2.10) converges with $c_n = ng(n)$, then for each $\tau > 0$,

$$P\{v_t \geqq g(t) \text{ for some } t \geqq \tau\} = \lim_{m\to\infty} P\left\{V_n \geqq \frac{1}{m} g\left(\frac{n}{m}\right) \text{ for some } n \geqq \tau m\right\}. \tag{6.2}$$

(As in Theorem 2(ii) of [11] a similar result holds if in (6.2) we replace $t \geqq \tau$ by $t > 0$ and $n \geqq \tau m$ by $n \geqq 1$.) The proof is similar in spirit to the proof of Theorem 2 of [11], but the details are much simpler.

REMARK 6.3. Using the probability integral transform, one can immediately obtain analogous results for random variables having arbitrary continuous distributions. (See Remark 2.5.) For example, if $x_1, x_2, \cdots$ are i.i.d. with $P\{x_i \geqq x\} = e^{-x}(x > 0)$ and if $g = e^{-h}$ satisfies the conditions of Remark 6.2 above, then the left side of (6.2) equals

$$\lim_{m \to \infty} P\left\{ \max_{1 \leqq k \leqq n} x_k \leqq \log m + h\left(\frac{n}{m}\right) \quad \text{for some } n \geqq \tau m \right\}. \tag{6.3}$$

If the x are standard normal random variables, the left side of (6.2) equals

$$\lim_{m \to \infty} P\left\{ \max_{1 \leqq k \leqq n} x_k \leqq 2^{1/2}\left[\log m + h\left(\frac{n}{m}\right) - \tfrac{1}{2}\log\left(\log m + h\left(\frac{n}{m}\right)\right) - \log 2\sqrt{\pi}\right]^{1/2} \quad \text{for some } n \geqq \tau m \right\}. \tag{6.4}$$

REMARK 6.4. It is possible to give a proof of Theorem 3 which is in the spirit of Motoo's proof [6] of the law of the iterated logarithm for the Wiener process. In fact, it may be shown that $\{x_t \equiv e^t v_{e^t}, 0 \leqq t < \infty\}$ is a positive recurrent Markov process, and since the sample paths of this process are continuous and increasing except for jumps in the negative direction, Motoo's method (as sketched in Section 3) applies with minor changes. To complete the argument it is necessary to compute (at least asymptotically as $a \to \infty$)

$$P\{x(T) \geqq a \mid x(0) = 1\}, \tag{6.5}$$

where $T = \inf\{t : x(t) \notin [1, a)\}$. Since the generator A of the process $x(t)$ is given by

$$Af(x) = xf'(x) + \int_0^x (f(u) - f(x))\, du \tag{6.6}$$

and since $p_a(x) \equiv P\{x(T) \geqq a \mid x(0) = x\}$ satisfies $Ap_a(x) = 0, 1 < x < a$, subject to $p_a(a) = 1$ and $p_a(x) = 0, 0 < x < 1$, it may be shown that

$$p_a(x) = \frac{e + \displaystyle\int_1^x \frac{e^u}{u}\, du}{e + \displaystyle\int_1^a \frac{e^u}{u}\, du}, \qquad 1 \leqq x \leqq a, \tag{6.7}$$

and hence $p_a(1) \sim ae^{-a}$ as $a \to \infty$.

Extensions of this approach are being investigated by Mr. J. Frankel.

REMARK 6.5. Minor changes in the method of proof of Theorem 1 applied to $\max_{0 \leqq s \leqq t}|w(s)|$ yield Chung's law of the iterated logarithm [1]; to wit: if ultimately $c(t) \uparrow$ and $c(t)/t \downarrow$, then

$$P\left\{ \max_{0 \leqq s \leqq t} |w(s)| \leqq \left(\frac{t}{c(t)}\right)^{1/2} \text{ for arbitrarily large } t \right\} = 0 \text{ or } 1, \tag{6.8}$$

according as

$$\int^{\infty} \frac{c(t)}{t} \exp\left\{-\frac{\pi^2}{8} c(t)\right\} dt < \infty \text{ or } = \infty. \tag{6.9}$$

The required computations are virtually identical with those of Lemmas 2 through 7 in light of the observation that for $0 < \tau < t$, $0 < y < x$,

$$P\{\max_{0 \leqq s \leqq t} |w(s)| \leqq x | \max_{0 \leqq s \leqq \tau} |w(s)| \leqq y\} \leqq P\{\max_{0 \leqq s \leqq t - \tau} |w(s)| \leqq x\}. \tag{6.10}$$

REFERENCES

[1] K. L. Chung, "On the maximum partial sums of sequences of independent random variables," *Trans. Amer. Math. Soc.*, Vol. 64 (1948), pp. 205–233.

[2] D. Darling and P. Erdös, "A limit theorem for the maximum of normalized sums of independent random variables," *Duke Math. J.*, Vol. 23 (1956), pp. 143–155.

[3] D. Darling and H. Robbins, "Iterated logarithm inequalities," *Proc. Nat. Acad. Sci. U.S.A.*, Vol. 57 (1967), pp. 1188–1192.

[4] M. Dwass, "Extremal processes," *Ann. Math. Statist.*, Vol. 35 (1964), pp. 1718–1725.

[5] J. Lamperti, "On extreme order statistics," *Ann. Math. Statist.*, Vol. 35 (1964), pp. 1726–1737.

[6] M. Motoo, "Proof of the law of the iterated logarithm through diffusion equation," *Ann. Inst. Statist. Math.*, Vol. 10 (1959), pp. 21–28.

[7] J. Neveu, *Mathematical Foundations of the Calculus of Probability*, San Francisco, Holden-Day, 1965.

[8] J. Pickands III, "Sample sequences of maxima," *Ann. Math. Statist.*, Vol. 38 (1967), pp. 1570–1575.

[9] H. Robbins, "Some aspects of the sequential design of experiments," *Bull. Amer. Math. Soc.*, Vol. 58 (1952), pp. 527–535.

[10] ———, "Statistical methods related to the law of the iterated logarithm," *Ann. Math. Statist.*, Vol. 41 (1970), pp. 1397–1409.

[11] H. Robbins and D. Siegmund, "Boundary crossing probabilities for the Wiener process and sample sums," *Ann. Math. Statist.*, Vol. 41 (1970), pp. 1410–1429.

[12] ———, "Confidence sequences and interminable tests," *Bull. Inst. Internat. Statist.*, Vol. 37 (1970), pp. 379–387.

[13] J. Ville, *Étude Critique de la Notion de Collectif*, Paris, Gauthier-Villars, 1939.

ASYMPTOTIC DISTRIBUTION OF THE MOMENT OF FIRST CROSSING OF A HIGH LEVEL BY A BIRTH AND DEATH PROCESS

A. D. SOLOVIEV
MOSCOW STATE UNIVERSITY, U.S.S.R.

1. Statement of the problem

In many applications of probability theory an essential role is played by birth and death processes, which is the name given to homogeneous Markov processes with a finite or countable number of states, which we denote by $0, 1, \cdots, n, \cdots$, in which an instantaneous transition is only possible between adjacent states. The probabilities $P_n(t) = P\{\xi(t) = n\}$ of these states satisfy the system of differential equations (see [2])

$$P_n'(t) = \lambda_{n-1}P_{n-1}(t) - (\lambda_n + \mu_n)P_n(t) + \mu_{n+1}P_{n+1}(t) \tag{1.1}$$

$n = 0, 1, \cdots$, where $\lambda_{-1} = \mu_0 = 0$.

If the number of states is finite and equals N, then $\lambda_N = \mu_{N+1} = 0$. It is also assumed that all the other parameters λ_n and μ_n are positive. Let us consider the random variable $\tau_{k,n}$, $k < n$, the passage time from state k to state n:

$$\tau_{k,n} = \inf\{t: \xi(t) = n, t > 0 | \xi(0) = k\}. \tag{1.2}$$

The random variables $\tau_{k,n}$ are of considerable interest in reliability theory, where birth and death processes describe the behavior of storage systems with replace ments. If the states $0, 1, \cdots, n - 1$, correspond to functioning states of a system, and other states correspond to nonfunctioning states of a system, then the random variable $\tau_{k,n}$ may be regarded as the length of time that the system works without a failure, if it starts in state k. Most often the state $\xi(t)$ is taken to be the number of nonfunctioning elements, at time t, in some system, and it is assumed that at the starting time the system was completely functioning, that is, $\xi(0) = 0$. Therefore, the study of the random variables $\tau_{0,n}$ is of greatest interest. Let us assume that our process has a stationary distribution. As is known [2], for this it is necessary and sufficient that the following conditions be satisfied:

$$\sum_{n=0}^{\infty} \theta_n < \infty, \qquad \sum_{n=0}^{\infty} \frac{1}{\lambda_n\theta_n} = \infty, \tag{1.3}$$

where

$$\theta_0 = 1, \qquad \theta_n = \frac{\lambda_0 \lambda_1 \cdots \lambda_{n-1}}{\mu_1 \mu_2 \cdots \mu_n} \tag{1.4}$$

(for the case of a finite number of states, a stationary distribution always exists).

For such a process the time intervals $\tau_1^+, \tau_2^+, \cdots, \tau_m^+, \cdots$, for which $\xi(t) < n$ will alternate with the intervals $\tau_1^-, \tau_2^-, \cdots, \tau_m^-, \cdots$ for which $\xi(t) \geqq n$. Since the process is Markovian, the lengths of all these intervals are independent, and from the existence of a stationary distribution it follows that τ_m^+ and τ_m^- are proper variables. Further it is not hard to see that at the beginning of every interval τ_m^+, $m \geqq 2$, the process will be in state $n - 1$, and therefore the distribution of any one of the variables τ_m^+, $m = 2$, coincides with the distribution of the variable $\tau_{n-1,n}$. Thus, in reliability problems $\tau_{n-1,n}$ can be regarded as the length of time during which the system works without failure or, as engineers say, the work per failure.

Let us assume now that at the initial moment the process is in a stationary regime, that is, $P\{\xi(0) = k\} = p_k$, where the $p_k = \theta_k p_0$ are the stationary probabilities. For reliability theory another random variable, which we denote by τ_n, is of interest:

$$\tau_n = \inf \{t \colon \xi(t) \geqq n, t \geqq 0\}. \tag{1.5}$$

This variable can be interpreted as the amount of time that the system works without failure in a stationary regime. Since the intervals τ_m^+ and τ_m^- form an extended alternating renewal process [1], the distribution of the variable τ_n is related to the distribution of the variable $\tau_{n-1,n}$ in the following way:

$$P\{\tau_n > t | \tau_n > 0\} = \frac{\int_t^\infty P\{\tau_{n-1,n} > x\}\, dx}{\int_0^\infty P\{\tau_{n-1,n} > x\}\, dx}. \tag{1.6}$$

Using ordinary methods, it is not hard to find the precise distributions of the variables $\tau_{k,n}$ and τ_n, which will be done below. However, computation with the exact formulas is very cumbersome. On the other hand, in applications the attainment of the level n denotes, as a rule, an undesirable event (an absence of demand in queuing theory, and a failure of the system in reliability theory), and therefore the parameters λ_i and μ_i and the level n are usually such that a crossing of the level n is "infrequent," that is, the level n is high. In such a situation, it is natural to investigate the asymptotic behavior of the variables $\tau_{k,n}$ and, as a consequence, to obtain approximate formulas for their distributions. In [5] it is shown that for fixed λ_i and μ_i, as $n \to \infty$ the distribution function of the appropriately normalized variable $\tau_{k,n}$ converges to the function

$$1 - ae^{-x}, \qquad x > 0, \quad 0 < a \leqq 1. \tag{1.7}$$

However, this result fails to satisfy us for two reasons.

First, in problems of reliability theory the level n (which is usually the number of elements in reserve) is almost never large. On the other hand, the parameters λ_i and μ_i most often depend upon the number n. For example, for the case of an immediate reserve with one maintenance unit [4] $\lambda_i = (n - i)\lambda$, $\mu_i = \mu$. Therefore, the following formulation of the problem is more natural and more general. Suppose that the parameters λ_i and μ_i and the level n vary in an arbitrary way; find conditions under which the distribution of the normalized variables $\tau_{k,n}$ converges to the distribution (1.7).

Second, any limit theorem which does not contain an estimate of the rate of convergence cannot, strictly speaking, be used for approximate calculations. Therefore, for applications it is highly essential to obtain constructive bounds on the rate of convergence, which do not contain the symbols $o(\cdot)$ and $O(\cdot)$. If F is the exact distribution, and Φ is the limit distribution, then a bound of the following type would seem to be ideal:

$$\|F - \Phi\| \leqq \varepsilon(\lambda_i, \mu_i, n) \text{ and } \|F - \Phi\| \sim \varepsilon(\lambda_i, \mu_i, n) \quad \text{as } \varepsilon(\lambda_i, \mu_i, n) \to 0. \tag{1.8}$$

An unimprovable bound of this kind gives simultaneously necessary and sufficient conditions for the convergence of F to Φ. It is estimates of just this kind which will be obtained in this paper. We will restrict ourselves to the study of the asymptotic behavior of the variables $\tau_{0,n}$, $\tau_{n-1,n}$ and τ_n since, as was shown, they are the variables of greatest interest in reliability theory.

2. The exact distributions of the variables $\tau_{0,n}$, $\tau_{n-1,n}$ and τ_n

We introduce the notation

$$\begin{aligned} P\{\tau_{0,n} < t\} &= F_n(t), \\ P\{\tau_{n-1,n} < t\} &= \Phi_n(t), \\ P\{\tau_n < t\} &= G_n(t). \end{aligned} \tag{2.1}$$

The exact distribution of the variable $\tau_{0,n}$ is found in [4]:

$$F_n(t) = \frac{1}{2\pi i}\int_{a-i\infty}^{a+i\infty} \frac{e^{zt}}{z\Delta_n(z)}\,dz, \qquad a > 0, \tag{2.2}$$

where the polynomials $\Delta_n(z) = 1 + \Delta_{n,1}z + \Delta_{n,2}z^2 + \cdots + \Delta_{n,n}z^n$ are determined from the recursion relation

$$\Delta_{k+1}(z) = \left(1 + \frac{\mu_k}{\lambda_k} + \frac{z}{\lambda_k}\right)\Delta_k(z) - \frac{\mu_k}{\lambda_k}\Delta_{k-1}(z). \tag{2.3}$$

Since $\tau_{0,n} = \tau_{0,1} + \cdots + \tau_{n-1,n}$ and since the terms are independent as a result of the process being Markovian, using elementary properties of the Laplace transform, we obtain

$$\Phi_n(t) = \frac{1}{2\pi i} \int_{a-i\infty}^{a+i\infty} \frac{\Delta_{n-1}(z)e^{zt}}{z\Delta_n(z)}\, dz, \qquad a > 0. \tag{2.4}$$

From formulas (2.2), (2.3), and (2.4), it is not hard to determine the expectation of $\tau_{0,n}$ and $\tau_{n-1,n}$:

$$\begin{aligned} M\tau_{0,n} &= T_{0,n} = \Delta_{n,1} = \sum_{k=0}^{n-1} \frac{1}{\lambda_k \theta_k} \sum_{s=0}^{k} \theta_s \\ M\tau_{n-1,n} &= T_{n-1,n} = T_{0,n} - T_{0,n-1} = \frac{1}{\lambda_{n-1}\theta_{n-1}} \sum_{k=0}^{n-1} \theta_k. \end{aligned} \tag{2.5}$$

To find the distribution of τ_n, we use formula (1.6):

$$1 - G_n(t) = P\{\tau_n > t\} = \frac{\sum_{k=0}^{n-1} p_k}{T_{n-1,n}} \int_t^{\infty} [1 - \Phi_n(x)]\, dx, \tag{2.6}$$

where the p_k are the stationary probabilities of the states.

Hence, again using the properties of the Laplace transform, we obtain

$$\begin{aligned} G_n(t) &= 1 - \mathscr{P}_n + \frac{\mathscr{P}_n}{2\pi i} \int_{a-i\infty}^{a+i\infty} \frac{\Delta_n(z) - \Delta_{n-1}(z)}{z^2 \Delta_n(z) T_{n-1,n}} e^{zt}\, dz \\ &= 1 - \mathscr{P}_n + \frac{\mathscr{P}_n}{2\pi i} \int_{a-i\infty}^{a+i\infty} \frac{\delta_n(z)\, e^{zt}}{z\Delta_n(z)}\, dz, \qquad a > 0, \end{aligned} \tag{2.7}$$

where

$$\begin{aligned} \mathscr{P}_n &= \sum_{k=0}^{n-1} p_k, \\ \delta_n(z) &= \frac{\Delta_n(z) - \Delta_{n-1}(z)}{zT_{n-1,n}} = 1 + \delta_{n,1} z + \cdots + \delta_{n,n-1} z^{n-1}. \end{aligned} \tag{2.8}$$

3. Properties of the polynomials $\Delta_n(z)$ and $\delta_n(z)$

LEMMA 3.1. *The roots of the polynomials $\Delta_n(z)$ and $\delta_n(z)$ have the following properties:*

(i) *they are simple and negative;*

(ii) *between any two adjacent roots of $\Delta_n(z)$ there lies a root of $\Delta_{n-1}(z)$;*

(iii) *between any two adjacent roots of $\Delta_n(z)$ there lies a root of $\delta_n(z)$.*

PROOF. The first and second assertions concerning the polynomials $\Delta_n(z)$ follow from the theory of orthogonal polynomials [6]. Let $-z'_1, -z'_2, \cdots, -z'_{n-1}$ be the roots of $\Delta_{n-1}(z)$, and $-z''_1, -z''_2, \cdots, -z''_n$ be the roots of $\Delta_n(z)$, numbered in the order of increasing modulus. Since

$$0 < z_1'' < z_1' < z_2'' < z_2' < \cdots < z_{n-1}' < z_n'', \tag{3.1}$$

the polynomial

$$\delta_n(z) = \frac{\Delta_n(z) - \Delta_{n-1}(z)}{zT_{n-1,n}} \tag{3.2}$$

is positive at the points $-z_1''$, $-z_3''$, $-z_5''$, $\cdots$, and negative at the points $-z_2''$, $-z_4''$, $-z_6''$, $\cdots$, from which it follows that all the roots of $\delta_n(z)$ are negative, and that between any two roots of $\Delta_n(z)$ there lies a root of $\delta_n(z)$. This proves the lemma.

Let us now introduce the normalized polynomials

$$\begin{aligned} \Delta_n\left(\frac{z}{\Delta_{n,1}}\right) &= 1 + z + a_{n,2}z^2 + \cdots + a_{n,n}z^n, \\ \delta_n\left(\frac{z}{\Delta_{n,1}}\right) &= 1 + b_{n,1}z + b_{n,2}z^2 + \cdots + b_{n,n-1}z^{n-1}. \end{aligned} \tag{3.3}$$

It follows from (2.3) that

$$\Delta_{n+1}(z) - \Delta_n(z) = z\frac{\sum_{k=0}^{n} \theta_k \Delta_k(z)}{\lambda_n \theta_n}, \tag{3.4}$$

from which we obtain the following recursion relation for the coefficients $\Delta_{n,k}$:

$$\Delta_{n,k} = \sum_{k=0}^{n-1} \frac{1}{\lambda_k \theta_k} \sum_{s=0}^{k} \theta_s \Delta_{s,k-1}. \tag{3.5}$$

Hence, the coefficients $a_{n,k}$ and $b_{n,k}$ of the normalized polynomials are given by

$$a_{n,k} = \frac{\Delta_{n,k}}{\Delta_{n,1}^k}, \qquad b_{n,k} = \frac{\Delta_{n,k+1} - \Delta_{n-1,k+1}}{\Delta_{n,1} - \Delta_{n-1,1}}. \tag{3.6}$$

Let us write these polynomials in the form

$$\begin{aligned} \Delta_n\left(\frac{z}{\Delta_{n,1}}\right) &= (1 + \alpha_0 z)(1 + \alpha_1 z)\cdots(1 + \alpha_{n-1}z), \\ \delta_n\left(\frac{z}{\Delta_{n,1}}\right) &= (1 + \beta_1 z)(1 + \beta_2 z)\cdots(1 + \beta_{n-1}z). \end{aligned} \tag{3.7}$$

If we assume that the numbers α_i and β_i are numbered in decreasing order, then as follows from Lemma 3.1

$$\alpha_0 > \beta_1 > \alpha_1 > \beta_2 > \alpha_2 > \cdots > \beta_{n-1} > \alpha_{n-1}. \tag{3.8}$$

We now make some estimates of the α_i and β_i.

LEMMA 3.2. *If we put* $\alpha = 1 - \alpha_0 = \alpha_1 + \cdots + \alpha_{n-1}$ *and* $\beta = \beta_1 + \beta_2 + \cdots + \beta_{n-1}$, *then*

$$\alpha < \beta, \tag{3.9}$$

and for $a_{n,2} < \frac{1}{4}$, *we have*

$$1 - (1 - 2a_{n,2})^{1/2} < \alpha < \frac{1 - (1 - 4a_{n,2})^{1/2}}{2}. \tag{3.10}$$

PROOF. Property (3.9) follows from (3.8). From (3.7), it follows that

$$\begin{aligned} \alpha_0 + \alpha_1 + \cdots + \alpha_{n-1} &= 1, \\ \alpha_0^2 + \alpha_1^2 + \cdots + \alpha_{n-1}^2 &= 1 - 2a_{n,2}; \end{aligned} \tag{3.11}$$

whence

$$1 - 2a_{n,2} < \alpha_0(\alpha_0 + \cdots + \alpha_{n-1}) = 1 - \alpha, \tag{3.12}$$

that is, $\alpha < \frac{1}{2}$. Further,

$$1 - 2a_{n,2} < \alpha_0^2 + (\alpha_1 + \cdots + \alpha_{n-1})^2 = \alpha^2 + (1 - \alpha)^2 \tag{3.13}$$

or

$$\alpha^2 - \alpha + a_{n,2} > 0. \tag{3.14}$$

Solving this inequality and keeping in mind that $\alpha < \frac{1}{2}$, we obtain

$$\alpha < \tfrac{1}{2}[1 - (1 - 4a_{n,2})^{1/2}]. \tag{3.15}$$

On the other hand,

$$\alpha_0^2 = (1 - \alpha)^2 < 1 - 2a_{n,2}, \tag{3.16}$$

that is, $\alpha > 1 - (1 - 2a_{n,2})^{1/2}$. This proves the lemma.

COROLLARY 3.1. *It follows from Lemma* 3.2 *that for* $a_{n,2} \to 0$

$$\alpha = a_{n,2} + O(a_{n,2}^2). \tag{3.17}$$

LEMMA 3.3. *For any* $k \geqq 1$ *the following inequalities hold:*

$$a_{n,k} \leqq \frac{\alpha^{k-1}}{(k - 1)!}, \tag{3.18}$$

$$b_{n,k} \leqq \frac{\beta^k}{k!}. \tag{3.19}$$

PROOF. We shall say that the series $A(z) = \Sigma_{n=0}^{\infty} a_n z^n$ is a majorant of the series $B(z) = \Sigma_{n=0}^{\infty} b_n z^n$, if $|b_n| \leqq a_n$ for every n. We will write this relation as $B(z) \ll A(z)$. It is easily seen that the following holds. If $B_1(z) \ll A_1(z)$ and

$B_2(z) \ll A_2(z)$, then

$$B_1(z) \cdot B_2(z) \ll A_1(z) \cdot A_2(z). \tag{3.20}$$

This property will be applied in order to estimate the coefficients $a_{n,k}$ and $b_{n,k}$:

$$\Delta_n\left(\frac{z}{\Delta_{n,1}}\right) = (1 + \alpha_0 z)(1 + \alpha_1 z) \cdots (1 + \alpha_{n-1} z) \ll (1 + \alpha_0 z)e^{\alpha z}. \tag{3.21}$$

Hence,

$$a_{n,k} \leqq \frac{\alpha^k}{k!} + \frac{(1 - \alpha)\alpha^{k-1}}{(k - 1)!} \leqq \frac{\alpha^{k-1}}{(k - 1)!}. \tag{3.22}$$

It is simpler yet to bound the $b_{n,k}$:

$$\delta_n\left(\frac{z}{\Delta_{n,1}}\right) = (1 + \beta_1 z) \cdots (1 + \beta_{n-1} z) \ll e^{\beta z}, \tag{3.23}$$

that is, $b_{n,k} \leqq \beta^k/k!$. This proves the lemma.

COROLLARY 3.2. *From inequality* (3.18) *and relation* (3.17) *it follows that for* $a_{n,2} \to 0$, *we have*

$$a_{n,k} = O\left(a_{n,2}^{k-1}\right). \tag{3.24}$$

4. Asymptotic behavior of the variable $\tau_{0,n}$

We will consider the normalized variable $\xi_n = \tau_{0,n}/T_{0,n}$. It follows from (2.2) that

$$P\{\xi_n < x\} = \frac{1}{2\pi i}\int_{a-i\infty}^{a+i\infty} \frac{e^{zx}\, dz}{z(1 + \alpha_0 z)(1 + \alpha_1 z) \cdots (1 + \alpha_{n-1} z)}. \tag{4.1}$$

THEOREM 4.1. *For* $a_{n,2} < \frac{1}{4}$, *we have the inequality*

$$\max_{0 \leqq t < \infty}\left|F_n(t) - 1 + \exp\left\{-\frac{t}{T_{0,n}}\right\}\right| \leqq \frac{\alpha}{1 - \alpha} \leqq \frac{1 - (1 - 4a_{n,2})^{1/2}}{1 + (1 - 4a_{n,2})^{1/2}}. \tag{4.2}$$

PROOF. It is evident from (4.1) that $\xi_n = \eta_0 + \eta_1 + \cdots + \eta_{n-1}$, where the η_i are independent and distributed according to an exponential law

$$P\{\eta_i > x\} = \exp\left\{-\frac{x}{\alpha_i}\right\}, \qquad i = 0, 1, \cdots, n - 1. \tag{4.3}$$

Let $P\{\eta_1 + \cdots + \eta_{n-1} > x\} = f_n(x)$. Then

$$(4.4)\qquad \begin{aligned} \varepsilon_n(x) &= F_n(T_n x) - 1 + e^{-x} \\ &= \frac{1}{\alpha_0}\int_0^x \exp\left\{-\frac{x-u}{\alpha_0}\right\}[1 - f_n(u)]\,du - 1 + e^{-x} \\ &= e^{-x} - \exp\left\{-\frac{x}{\alpha_0}\right\} - \frac{1}{\alpha_0}\int_0^x \exp\left\{-\frac{x-u}{\alpha_0}\right\} f_n(u)\,du. \end{aligned}$$

We bound this quantity from above and below:

$$(4.5)\qquad \begin{aligned} \varepsilon_n(x) &\leqq e^{-x} - \exp\left\{-\frac{x}{\alpha_0}\right\} \leqq \frac{\alpha}{e(1-\alpha)}, \\ \varepsilon_n(x) &\geqq -\frac{1}{\alpha_0}\int_0^x \exp\left\{-\frac{x-u}{\alpha_0}\right\} f_n(u)\,du \geqq -\frac{1}{\alpha_0}\int_0^\infty f_n(u)\,du \\ &= \frac{1}{\alpha_0} M(\eta_1 + \cdots + \eta_{n-1}) = -\frac{\alpha}{1-\alpha}. \end{aligned}$$

The assertion of the theorem follows from these bounds and inequality (3.10).

COROLLARY 4.1. *In order that*

$$(4.6)\qquad \lim P\left\{\frac{\tau_{0,n}}{T_{0,n}} > x\right\} = e^{-x},$$

it is necessary and sufficient that $a_{n,2} \to 0$.

The necessity is evident from formula (4.1), and the sufficiency follows from inequality (4.2).

REMARK 4.1. Estimating the difference $\varepsilon_n(x)$, for example, at the point $x = \sqrt{\alpha}$, it is not hard to show that for $a_{n,2} \to 0$

$$(4.7)\qquad \max_{0 \leqq t < \infty} \left| F_n(t) - 1 + \exp\left\{-\frac{t}{T_{0,n}}\right\}\right| \sim a_{n,2},$$

and thus inequality (4.2) is asymptotically exact. We also note that from (3.6) it is easy to find an explicit expression for $a_{n,2}$:

$$(4.8)\qquad a_{n,2} = \frac{1}{T_{0,n}^2} \sum_{k=0}^{n-1} \frac{1}{\lambda_k \theta_k} \sum_{s=0}^{k} \theta_s T_{0,s},$$

and therefore inequality (4.2) enables us to easily and precisely estimate the rate of convergence in (4.6).

An even more precise approximation for the distribution $F_n(t)$ yields:

THEOREM 4.2. *For* $a_{n,2} < \frac{1}{4}$, *we have the equality*

$$(4.9)\qquad F_n(t) = 1 - \frac{\lambda'}{\lambda}\exp\left\{-\frac{\lambda t}{T_{0,n}}\right\} + \theta\,\frac{2a_{n,2}}{1 - 4a_{n,2}}\exp\left\{-\frac{t}{2a_{n,2}T_{0,n}}\right\},$$

where $0 < \theta < 1$ *and*

$$(4.10)\qquad \lambda = \frac{1}{\alpha_0} = 1 + \sum_{m=1}^{\infty} \frac{1}{m!} \frac{d^{m-1}}{dw^{m-1}} [\varphi_n^m(-w)]_{w=1},$$

$$(4.11)\qquad \lambda' = \frac{T_{0,n}}{\Delta_n'\left(-\dfrac{\lambda}{T_{0,n}}\right)} = \sum_{m=0}^{\infty} \frac{1}{m!} \frac{d^m}{dw^m} [\varphi_n^m(-w)]_{w=1}$$

and

$$(4.12)\qquad \varphi_n(z) = \Delta_n\left(\frac{z}{T_{0,n}}\right) - 1 - z.$$

PROOF. We move the path of integration in the integral in (4.1) to the left so as to separate out the residues at the points $z = 0$ and $z = -1/\alpha_0 = -\lambda$:

$$(4.13)\qquad F_n(T_{0,n}x) = 1 - \frac{e^{-\lambda x}}{\dfrac{\lambda}{T_{0,n}} \Delta_n'\left(-\dfrac{\lambda}{T_{0,n}}\right)} + R_n,$$

where

$$(4.14)\qquad R_n = \frac{1}{2\pi i} \int_{-b-i\infty}^{-b+i\infty} \frac{e^{zx}\, dz}{z(1 + \alpha_0 z) \cdots (1 + \alpha_{n-1} z)}, \qquad \frac{1}{\alpha_0} < b < \frac{1}{\alpha_1}.$$

Bounding R_n in terms of the maximum modulus of the integrand, we obtain

$$(4.15)\qquad |R_n| \leqq \frac{1}{2\pi} \int_{-\infty}^{\infty} \frac{e^{-bx}\, ds}{(b^2 + s^2)^{1/2} [(b\alpha_0 - 1)^2 + \alpha_0^2 s^2]^{1/2} (1 - \alpha b)}$$

$$= \frac{e^{-bx}}{2\pi(b\alpha_0 - 1)(1 - b\alpha)} \int_{-\infty}^{\infty} \frac{d(s/b)}{[1 + (s/b)^2]^{1/2} [1 + \alpha_0^2 s^2/(b\alpha_0 - 1)^2]^{1/2}}$$

$$\leqq \frac{e^{-bx}}{2(b\alpha_0 - 1)(1 - b\alpha)}.$$

Putting $b = [2\alpha(1 - \alpha)]^{-1}$, we find that

$$(4.16)\qquad |R_n| \leqq \frac{2\alpha\alpha_0 \exp\left\{-\dfrac{x}{2\alpha\alpha_0}\right\}}{1 - 4\alpha\alpha_0} < \frac{2a_{n,2}}{1 - 4a_{n,2}} \exp\left\{-\frac{x}{2a_{n,2}}\right\}$$

since $\alpha\alpha_0 < a_{n,2}$. We now write the residue at the point $z = -1/\alpha_0 = -\lambda$ in the form

$$(4.17)\qquad \frac{1}{2\pi i} \int_{|z+\lambda| = \rho} \frac{e^{zx}\, dx}{z[1 + z + \varphi_n(z)]},$$

where ρ is sufficiently small, and

$$\varphi_n(z) = a_{n,2}z^2 + a_{n,3}z^3 + \cdots + a_{n,n}z^n. \tag{4.18}$$

One could expand the integrand in powers of $\varphi_n(z)$ and obtain the corresponding expansion for the residue. It has the form $\Sigma_{k=0}^{\infty} \pi_k(x)e^{-x}$, where $\pi_k(x)$ is a polynomial of degree k. By virtue of the inequalities (3.18), this series will be an asymptotic one. However, for the residue that we are considering, one can obtain a more convenient representation. We have

$$\frac{1}{2\pi i}\int_{|z+\lambda|=\rho} \frac{e^{zx}\,dz}{z[1 + z + \varphi_n(z)]} = \frac{\lambda'}{\lambda} e^{-\lambda x}, \tag{4.19}$$

where λ satisfies the equation $1 = \lambda - \varphi_n(-\lambda)$, and $\lambda' = [1 + \varphi_n'(-\lambda)]^{-1}$. Now consider the function $w = z - \varphi_n(-z)$, and denote its inverse by $z = \psi(w)$ It is easily seen that $\lambda = \psi(1)$ and $\lambda' = \psi'(1)$. We expand the function $\psi'(w)$ in a series:

$$\begin{aligned} \psi'(w) &= \frac{1}{2\pi i}\int_C \frac{d\psi(\zeta)}{\zeta - w} = \frac{1}{2\pi i}\int_{C_1} \frac{dz}{z - w - \varphi_n(-z)} \\ &= \sum_{m=0}^{\infty} \frac{1}{2\pi i}\int_{C_1} \frac{\varphi_n^m(-z)}{(z - w)^{m+1}}\,dz = \sum_{m=0}^{\infty} \frac{1}{m!}\frac{d^m}{dw^m}[\varphi_n^m(-w)]. \end{aligned} \tag{4.20}$$

Hence,

$$\psi(w) = w + \sum_{m=1}^{\infty} \frac{1}{m!}\frac{d^{m-1}}{dw^{m-1}}[\varphi_n^m(-w)]. \tag{4.21}$$

Letting $w = 1$ in these series, we obtain the desired expansions (4.10), (4.11). This proves the theorem.

REMARK 4.2. Using inequality (3.18), it is not hard to show by a simple estimate using the maximum modulus that the mth terms of these series do not exceed the quantity $2[2(e^{2\alpha} - 1)]^m$, from which, taking (3.10) into account, it follows that the series (4.10), (4.11) converge for $a_{n,2} < 0.2$.

5. The asymptotic behavior of the variables τ_n

THEOREM 5.1. *For $\beta < (e - 1)/e$ one has the inequality*

$$\max_{0 \leqq t < \infty} \left| G_n(t) - 1 + \mathscr{P}_n \exp\left\{-\frac{t}{T_{0,n}}\right\}\right| \leqq \beta\mathscr{P}_n. \tag{5.1}$$

PROOF. As follows from (2.7),

$$\begin{aligned} G_n(T_{0,n}x) &= P\left\{\frac{\tau_n}{T_{0,n}} < x\right\} \\ &= 1 - \mathscr{P}_n + \frac{\mathscr{P}_n}{2\pi i}\int_{a-i\infty}^{a+i\infty} \frac{(1 + \beta_1 z)\cdots(1 + \beta_{n-1}z)e^{zx}}{z(1 + \alpha_0 z)\cdots(1 + \alpha_{n-1}z)}\,dz \end{aligned} \tag{5.2}$$

$$= 1 - \mathscr{P}_n \exp\left\{-\frac{x}{\alpha_0}\right\} + \mathscr{P}_n \frac{1}{\alpha_0}\int_0^x \exp\left\{-\frac{x-u}{\alpha_0}\right\}[g_n(u) - 1]\,du,$$

where

$$g_n(u) = \frac{1}{2\pi i}\int_{a-i\infty}^{a+i\infty} \frac{(1+\beta_1 z)\cdots(1+\beta_{n-1}z)e^{zu}}{z(1+\alpha_1 z)\cdots(1+\alpha_{n-1}z)}\,dz = 1 + \sum_{k=1}^{n-1}\frac{A_k}{\alpha_k}\exp\left\{-\frac{u}{\alpha_k}\right\}. \tag{5.3}$$

From inequalities (3.8), we easily obtain

$$A_k = \frac{\left(1-\frac{\beta_1}{\alpha_k}\right)\cdots\left(1-\frac{\beta_{n-1}}{\alpha_k}\right)}{-\frac{1}{\alpha_k}\left(1-\frac{\alpha_1}{\alpha_k}\right)\cdots\left(1-\frac{\alpha_{k-1}}{\alpha_k}\right)\left(1-\frac{\alpha_{k+1}}{\alpha_k}\right)\cdots\left(1-\frac{\alpha_{n-1}}{\alpha_k}\right)} > 0. \tag{5.4}$$

Consequently,

$$\left|\frac{1}{\alpha_0}\int_0^x \exp\left\{-\frac{x-u}{\alpha_0}\right\}[g_n(u)-1]\,du\right| < \frac{1}{\alpha_0}\int_0^\infty [g_n(u)-1]\,du = \frac{1}{\alpha_0}\sum_{k=1}^{n-1} A_k = \frac{\beta-\alpha}{1-\alpha}. \tag{5.5}$$

We can now bound the difference:

$$0 < G_n(T_0 x) - 1 + \mathscr{P}_n e^{-x} = \mathscr{P}_n\left\{e^{-x} - e^{-x/\alpha} + \frac{1}{\alpha_0}\int_0^x \exp\left\{-\frac{x-u}{\alpha_0}\right\}[g_n(u)-1]\,du\right\} < \left[\frac{\alpha}{e(1-\alpha)} + \frac{\beta-\alpha}{1-\alpha}\right]\mathscr{P}_n. \tag{5.6}$$

If $\beta < (e-1)/e = 0.63\cdots$, then the last expression does not exceed $\beta\mathscr{P}_n$, from which we obtain the assertion of the theorem.

Corollary 5.1. *From the theorem just proved, it follows that*

$$\lim_{\beta\to 0} P\left\{\frac{\tau_n}{T_{0,n}} > x \,\middle|\, \tau_n > 0\right\} = e^{-x}. \tag{5.7}$$

REMARK 5.1. It is not hard to show that for $x = 1$

$$G_n(T_{0,n}) - 1 + \mathscr{P}_n e^{-1} \sim \frac{\beta}{e} \mathscr{P}_n, \tag{5.8}$$

from which it follows that for $\beta \to 0$ the bound (5.1) is exact, as concerns order.

REMARK 5.2. Proceeding as in Theorem 4.2, it is not difficult to obtain a more precise bound. For $\beta < \frac{1}{2}$

$$G_n(t) = 1 - \mathscr{P}_n \frac{\lambda''}{\lambda} \exp\left\{-\frac{\lambda t}{T_{0,n}}\right\} + \theta \frac{\beta}{1 - 2\beta} \exp\left\{-\frac{t}{\beta T_{0,n}}\right\}, \quad |\theta| < 1, \tag{5.9}$$

where λ is defined by the series (4.10) and

$$\lambda'' = \sum_{m=0}^{\infty} \frac{1}{m!} \frac{d^m}{dw^m} \left[\delta_n\left(-\frac{w}{T_{0,n}}\right) \varphi_n^m\left(-\frac{w}{T_{0,n}}\right)\right]_{w=1}. \tag{5.10}$$

It is easy to show that the mth term of this series is of order β^m.

6. The asymptotic behavior of the variables $\tau_{n-1,n}$

THEOREM 6.1. *For any $t \geqq 0$, we have the inequality*

$$\begin{aligned} &\left|\Phi_n(t) - 1 + \frac{T_{n-1,n}}{T_{0,n}} \exp\left\{-\frac{t}{T_{0,n}}\right\}\right| \\ &\qquad \leqq \left[1 - \frac{T_{n-1,n}(1 - \alpha - \beta)}{T_{0,n}(1 - \alpha)}\right] \exp\left\{-\frac{t}{\alpha T_{0,n}}\right\} + \left|\frac{\beta}{1 - \alpha}\right|. \end{aligned} \tag{6.1}$$

PROOF. It follows from (2.4) that

$$\begin{aligned} P\left\{\frac{\tau_{n-1,n}}{T_{0,n}} > x\right\} &= 1 - \Phi_n(T_{0,n}x) \\ &= \frac{T_{n-1,n}}{T_{0,n}} \frac{1}{2\pi i} \int_{a-i\infty}^{a+i\infty} \frac{\delta_n\left(\frac{z}{T_{0,n}}\right) e^{zx}}{\Delta_n\left(\frac{z}{T_{0,n}}\right)} dz, \qquad a > 0. \end{aligned} \tag{6.2}$$

We consider the integral

$$\begin{aligned} h_n(x) &= \frac{1}{2\pi i} \int_{a-i\infty}^{a+i\infty} \frac{\delta_n\left(\frac{z}{T_{0,n}}\right) e^{zx}}{\Delta_n\left(\frac{z}{T_{0,n}}\right)} dz \\ &= B_0 e^{-x/\alpha_0} + B_1 e^{-x/\alpha_1} + \cdots + B_{n-1} e^{-x/\alpha_{n-1}}. \end{aligned} \tag{6.3}$$

Making use of the inequalities (3.8), it is easily shown that $B_k > 0$. Moreover,

$$\text{(6.4)} \qquad \frac{1}{1-\alpha} > B_0 = \frac{\left(1 - \frac{\beta_1}{\alpha_0}\right) \cdots \left(1 - \frac{\beta_{n-1}}{\alpha_0}\right)}{\alpha_0\left(1 - \frac{\alpha_1}{\alpha_0}\right) \cdots \left(1 - \frac{\alpha_{n-1}}{\alpha_0}\right)} > \frac{1 - \alpha - \beta}{1 - \alpha}.$$

It is not hard to see that

$$\text{(6.5)} \qquad B_0 + B_1 + \cdots + B_{n-1} = h_n(0) = \frac{T_{0,n}}{T_{n-1,n}}.$$

Further,

$$\text{(6.6)} \qquad \begin{aligned} |h_n(x) - e^{-x}| &\leqq |B_0 e^{-x/\alpha_0} - e^{-x}| + |h_n(x) - B_0 e^{-x/\alpha_0}| \\ &\leqq |B_0 - 1| + (B_1 + \cdots + B_{n-1})\, e^{-x/\alpha_1} \\ &\leqq \frac{\beta}{1-\alpha} + \left[\frac{T_{0,n}}{T_{n-1,n}} - \frac{1-\alpha-\beta}{1-\alpha}\right] e^{-x/\alpha}, \end{aligned}$$

from which we obtain the assertion of the theorem.

COROLLARY 6.1. *From the theorem just proved, it follows that for any* $x > 0$

$$\text{(6.7)} \qquad \lim_{\beta \to 0} \left[\Phi_n(T_{0,n}x) - 1 + \frac{T_{n-1,n}}{T_{0,n}} e^{-x}\right] = 0;$$

however, this convergence, generally speaking, is not uniform in the neighborhood of zero. In order that the limit in (6.7) be uniform, it is necessary and sufficient that

$$\text{(6.8)} \qquad \lim_{\beta \to 0} \frac{T_{n-1,n}}{T_{0,n}} = 1.$$

7. A class of infinitely divisible laws

We have earlier obtained conditions under which the distributions of the suitably normalized variables $\tau_{0,n}$, τ_n and $\tau_{n-1,n}$ converge to an exponential distribution (possibly with a jump at zero). It would be natural to go further and attempt to find other limiting distributions for these variables. If, passing to the limit, one varies the level n as well as the parameters λ_k and μ_k, then the problem is empty, since by choosing these parameters appropriately we can always make the roots of the polynomials $\Delta_n(z)$ equal to any preassigned negative numbers. Therefore, we assume that the parameters λ_k and μ_k are fixed, and that the level $n \to \infty$.

For this statement of the problem, the following holds.

THEOREM 7.1. *Suppose that the parameters* λ_k *and* μ_k *satisfy the conditions*

$$\text{(7.1)} \qquad \lim_{n \to \infty} T_{0,n} = +\infty,$$

$$\lim_{n\to\infty} \frac{T_{0,n+1}}{T_{0,n}} = 1, \tag{7.2}$$

$$\lim_{n\to\infty} \frac{T_{0,n}}{\lambda_n T_{n,n+1}^2} = \gamma, \qquad 0 \leqq \gamma \leqq \infty. \tag{7.3}$$

Then

$$\lim_{n\to\infty} P\left\{\frac{\tau_{0,n}}{T_{0,n}} < x\right\} = \frac{1}{2\pi i}\int_{a-i\infty}^{a+i\infty} \frac{e^{zx}}{z\Phi(z, \gamma)}\, dz, \tag{7.4}$$

where

$$\Phi(z, \gamma) = \sum_{k=0}^{\infty} \frac{\Gamma(\gamma)(\gamma z)^k}{k!\Gamma(k + \gamma)}. \tag{7.5}$$

PROOF. We shall prove that for any k, $\lim_{n\to\infty} a_{n,k} = a_k$ exists. From formula (3.6) it is not difficult to find an explicit expression for $a_{n,k}$:

$$a_{n,k} = \frac{\Delta_{n,k}}{\Delta_{n,1}^k} = \frac{1}{T_{0,n}^k} \sum_{\ell=0}^{n-1} \frac{1}{\lambda_\ell \theta_\ell} \sum_{s=0}^{\ell} \theta_s a_{s,k-1} T_{0,s}^{k-1}. \tag{7.6}$$

We will prove the existence of the limit by induction on k. For $k = 1$, we have $a_{n,1} = 1$ and the assertion is trivial. Let us assume that $\lim_{n\to\infty} a_{n,k-1} = a_{k-1}$ exists. Since $T_{0,n} \uparrow \infty$, to find the limit of the $a_{n,k}$, we can apply a theorem of Stolz [3] (the finite-difference analog of L'Hospital's rule):

$$\begin{aligned} \lim_{n\to\infty} a_{n,k} &= \lim_{n\to\infty} \frac{\sum_{\ell=0}^{n-1} \frac{1}{\lambda_\ell \theta_\ell} \sum_{s=0}^{\ell} \theta_s a_{s,k-1} T_{0,s}^{k-1}}{T_{0,n}^k} \\ &= \lim_{n\to\infty} \frac{\sum_{s=0}^{n} \theta_s a_{s,k-1} T_{0,s}^{k-1}}{k T_{0,n+1}^{k-1} \sum_{s=0}^{n} \theta_s} \\ &= \lim_{n\to\infty} \frac{\theta_n a_{n,k-1} T_{0,n}}{k[(k-1)T_{0,n}^{k-2} T_{n,n+1}^2 \lambda_n \theta_n + T_{0,n}^{k-1}\theta_n]} \\ &= \lim_{n\to\infty} \frac{a_{n,k-1}}{k\left[(k-1)\frac{\lambda_n T_{n,n+1}^2}{T_{0,n}} + 1\right]} = \frac{\gamma a_{k-1}}{k(k - 1 + \gamma)} = a_k. \end{aligned} \tag{7.7}$$

We have twice applied Stolz's theorem, using here the condition (7.1). Thus

$$a_k = \frac{\gamma a_{k-1}}{k(k - 1 + \gamma)} = \frac{\gamma^k \Gamma(\gamma)}{k!\Gamma(k + \gamma)}, \tag{7.8}$$

from which it follows, taking account of the inequality (3.18), that

$$\lim_{n\to\infty} \Delta_n\left(\frac{z}{T_{0,n}}\right) = \sum_{k=0}^{\infty} \frac{\Gamma(\gamma)(\gamma z)^k}{k!\,\Gamma(k+\gamma)} = \Phi(z, \gamma) \tag{7.9}$$

uniformly on any finite interval. Therefore,

$$\lim_{n\to\infty} P\left\{\frac{\tau_{0,n}}{T_{0,n}} < x\right\} = \frac{1}{2\pi i}\int_{a-i\infty}^{a+i\infty} \frac{e^{zx}}{z\Phi(z, \gamma)}\, dz, \tag{7.10}$$

which proves the theorem.

REMARK 7.1. It follows from the hypothesis of the theorem that the terms in the sum $\tau_{0,n} = \tau_{0,1} + \tau_{1,2} + \cdots + \tau_{n-1,n}$ are uniformly small. Therefore, the distribution (7.4) is infinitely divisible.

REMARK 7.2. For the extreme values, we have

$$\Phi(z, 0) = 1 + z, \qquad \Phi(z, \infty) = e^z, \tag{7.11}$$

which corresponds to the exponential and the identity distribution.

REMARK 7.3. In a similar way, one can show that if the conditions (7.1), (7.2), (7.3) are satisfied, then

$$\lim_{n\to\infty} \delta_n\left(\frac{z}{T_{0,n}}\right) = \Phi'_z(z, \gamma), \tag{7.12}$$

from which it follows that

$$\lim_{n\to\infty} P\left\{\frac{\tau_n}{T_{0,n}} < x\right\} = \frac{1}{2\pi i}\int_{a-i\infty}^{a+i\infty} \frac{\Phi'_z(z, \gamma)}{z\Phi(z, \gamma)}\, e^{zx}\, dz, \tag{7.13}$$

and

$$\lim_{n\to\infty} P\left\{\frac{\tau_{n-1,n}}{T_{0,n}} > x\right\}\frac{T_{0,n}}{T_{n-1,n}} = \frac{1}{2\pi i}\int_{a-i\infty}^{a+i\infty} \frac{\Phi'_z(z, \gamma)}{\Phi(z, \gamma)}\, e^{zx}\, dz. \tag{7.14}$$

In concluding this paper, we note that the results of Section 4 (bounds on the distribution of $\tau_{0,n}$) are obviously valid also for an arbitrary Markov process with a finite number of states for which

$$M \exp\{-z\tau_{0,n}\} = \frac{1}{P(z)}, \tag{7.15}$$

where $P(z)$ is a polynomial, all of whose roots are negative. The equality (7.15) will, for example, be satisfied for a process in which instantaneous transitions upward can only occur by one state at a time. In the general case, where

$$M \exp\{-z\tau_{0,n}\} = \frac{Q(z)}{P(z)} \tag{7.16}$$

and the roots of the polynomial $P(z)$ are negative, the asymptotic behavior of the variables $\tau_{0,n}$ can be investigated by our methods, if certain restrictions are imposed upon the growth of the polynomial $Q(z)$.

REFERENCES

[1] D. R. Cox, *Renewal Theory*, London, Methuen; New York, Wiley, 1962.
[2] W. Feller, *An Introduction to Probability Theory and Its Applications*, Vol. 1, New York, Wiley, 1968 (3rd ed.).
[3] G. M. Fichtenholz, *Course in Differential and Integral Calculus*, Moscow and Leningrad, Fizmatgiz, 1963. (In Russian.)
[4] B. V. Gnedenko, Yu. K. Belyayev, and A. D. Soloviev, *Mathematical Methods of Reliability Theory* (edited by Richard Barlow), New York, Academic Press, 1969.
[5] S. Karlin and J. L. McGregor, "The differential equations of birth-and-death processes, and the Stieltjes moment problem," *Trans. Amer. Math. Soc.*, Vol. 85 (1957), pp. 489–546.
[6] I. P. Natanson, *Constructive Function Theory* (translated by J. R. Schulenberger), New York, Ungar, 1965.

MARTIN BOUNDARIES OF RANDOM WALKS ON LOCALLY COMPACT GROUPS

R. AZENCOTT
UNIVERSITY OF CALIFORNIA, BERKELEY
and
P. CARTIER
I.H.E.S., BURES-SUR-YVETTE

Introduction

Let G be a separable locally compact space and let (X_t), t in T, be a transient Markov process with values in G, where T is either the set of positive integers (discrete time) or the set of positive real numbers (continuous time). Let (Q^t) be the semigroup of transition kernels of (X_t). Let f and λ be, respectively, a positive Borel function on G and a positive measure on the Borel σ-field of G. Call f (respectively, λ) excessive if $Q^t f \leqq f$ and $\lim_{t\to 0} Q^t f = f$ (respectively, $\lambda Q^t \leqq \lambda$), and invariant if $Q^t f = f$ (respectively, $\lambda Q^t = \lambda$).

Around 1955, the early studies of excessive functions of a Markov process centered around two problems: the relations between Brownian motion and Newtonian potential theory, and the behavior of the trajectories of the process (X_t) as $t \to +\infty$. The latter approach can be traced back to D. Blackwell ([4], 1955) who noticed the link between bounded invariant functions and the subsets of G in which (X_t) stays, from some finite time on, with positive probability. The importance of these 'sojourn' sets became clear after W. Feller's magistral article ([24], 1956), where they are used to construct (discrete T and G) a compactification $G \cup F$ of G such that each bounded invariant function f extends continuously to $G \cup F$ and is uniquely determined by its values on the Feller boundary F.

The other approach was initiated by two papers of J. L. Doob: a study of the behavior of subharmonic functions along Brownian paths ([16], 1954), and a probabilistic approach to the potential theory of the heat equation ([17], 1955). The relation between potential theory and general (transient) Markov processes was completely clarified by G. Hunt soon after ([32], 1957–1958).

These two trends of thought each found their expression in Doob's work ([18], 1959) which revived the methods used by R. Martin ([39], 1941) in his classical study of harmonic functions. In this article, Doob constructed (for

The work of R. Azencott was partially supported by NSF Grant GP–24490.

discrete T and G) a compactification of G by essentially adjoining to G a set of extreme invariant functions, obtained the integral representation of excessive functions in terms of extreme excessive functions, and proved the basic results about the almost sure convergence of (X_t) to the Martin boundary, as $t \to +\infty$. Hunt ([33], 1960) introduced new methods of achieving Doob's results, in particular the reversal of the sense of time for Markov processes.

It soon became clear that Feller's boundary is almost always too large (D. Kendall [34], 1960, J. Feldman [26], 1962), and in later studies the approach of Doob and Hunt has prevailed. The extension of potential theoretic results to the recurrent case began around 1960–1961, first for the case of random walks (K. Itô and H. J. McKean, and F. Spitzer) and then for the case of Markov chains (J. G. Kemeny and J. L. Snell). A very lucid exposition of the main ideas of boundary and potential theory for Markov chains (for discrete T and G) was given by J. Neveu ([46], 1964), using G. Choquet's results on convex cones ([11], 1956) to obtain the integral representation of excessive functions. In [36] (1965), H. Kunita and T. Watanabe treated the case of continuous time and general state space. They considered two processes in duality with respect to a measure on the state space, a situation whose importance had been recognized earlier by G. A. Hunt [32] and P. A. Meyer (thesis). The construction of the (exit) boundary uses, then, the compact caps of the cone of coexcessive measures (that is, measures which are excessive with respect to the dual process).

Meyer ([42], 1968) gave a new presentation of their results using Choquet's integral representation theorem.

When the state space G is a topological group, a class of Markov processes is naturally linked to the group structure, the random walks on G. We now restrict our attention to this situation. In the abelian case, the first significant result was reached by G. Choquet and J. Deny ([12], 1960), who showed that the extreme invariant functions are essentially characters of the group. For the particular case when $G = Z$, group of integers, this was noticed simultaneously by J. L. Doob, J. L. Snell, and R. Williamson ([22], 1960).

For the case $G = Z^n$, the boundaries of random walks were then carefully described by P. Hennequin ([31], thesis 1962) and the asymptotic behavior of the Green function at the boundary was obtained by P. Ney and F. Spitzer ([47], 1966). The basic results of the potential theory of random walks on Z^n, exposed by Spitzer ([49], 1964) were soon extended to countable abelian groups in a joint paper with H. Kesten ([35], 1966).

For the nonabelian case, the extreme invariant functions were obtained first for finitely generated groups by E. Dynkin and M. Maljutov ([23], 1961). A major advance was made by H. Furstenberg ([27], 1963) giving an integral representation of bounded invariant functions for random walks on semisimple connected Lie groups, containing as a particular case the classical Poisson formula relative to harmonic functions on a disc. He obtained partial results ([29], 1965) about the cone of all nonnegative invariant functions for the same class of groups. The main results of [27] were extended by R. Azencott ([3], 1970) to a larger class of groups.

We have tried to outline briefly the evolution of the main currents of ideas relevant to our work. Lack of space has forced us to make a number of sizeable omissions: the essential influences of "abstract" potential theory, the important applications of the theory of excessive functions, and the rich material concerning the recurrent case have been barely alluded to.

The "Poisson formula" obtained by Furstenberg [27] involves a family of explicitly described, compact homogeneous spaces of G, the "Poisson spaces" of G. One question arises naturally: what are the relations between the Poisson spaces of G and the Martin boundaries of random walks of G? This problem is solved in the last part of the present work. We study the case of a transient random walk on a locally compact separable group; there is then a natural random walk in duality with the first one with respect to any right invariant Haar measure on G. Let $\bar{U}$ be the potential kernel of the dual random walk. To any function $r \geqq 0$ on G such that $0 < \bar{U}r < \infty$ ("reference" function), we associate as in [36] a continuous one to one map from G into a compact cap of the cone of coexcessive measures. As in Neveu [46] and Meyer [42], Choquet's theorem gives, then, the integral representation of coexcessive measures. The classical, Martin type compactifications of G have been abandoned here, mainly because G does not in general act continuously on these spaces. Even a strong restriction on the type of reference function used ("adapted" reference function, see Sections 16 and 17) only insures an action of G on part of the boundary. The only favorable case seems to be the one when the closed semigroup generated in G by the support of the law of the random walk is large enough (see Section 8).

We have preferred to imbed G in the sets of *rays* of the cone of coexcessive measures obtaining thus a Hausdorff space $G \cup B$ with countable base, and an "intrinsic boundary" B. This space is neither metrizable nor compact, in general, but the slight measure theoretic technicalities required by this situation are balanced by the fact that G acts continuously on $G \cup B$. We also obtain intrinsic formulations (independent of the reference function r) for the main classical results: integral representation of coexcessive measures and convergence to the boundary.

We then prove a "Poisson formula" for bounded invariant functions, also in intrinsic form, and use it to essentially identify the Poisson space and the "active part" of the intrinsic boundary.

Although reference functions have been eliminated in the formulation of the main results, they have been used in many proofs, and it is, in fact, possible to present the whole question in the more classical setting of Martin compactifications, provided only "adapted" reference functions are used (see Section 17). We also point out that our proofs of the basic (nonintrinsic) results on integral representation and convergence to the boundary follow classical patterns, essentially those outlined by Neveu [46].

Part A. Preliminaries

In this part, the main conventions are stated and a description of the random walks on a group is given. A derivation of the recurrence criterion for such random walks is proposed, after K. L. Chung and W. H. J. Fuchs [15]. The boundary theory described is nontrivial only in the case of transient random walks (see Section 8) which will be considered exclusively in later parts.

1. Notations and conventions

1.1. *Measure theory.* Let E be a Hausdorff space. The smallest σ-algebra containing the class of open subsets of E is denoted by $\mathscr{B}(E)$ and its elements are called the *Borel subsets* of E. We shall consider exclusively real valued functions on E, with nonnegative infinite values included. For instance, $b^+(E)$ denotes the set of all such functions which are Borel measurable (that is, $\mathscr{B}(E)$ measurable) and $C_c(E)$ denotes the functions which are finite at every point, continuous, and have compact support.

A *measure* μ on E is a σ-additive mapping from $\mathscr{B}(E)$ into $[0, +\infty]$ and μ is a *probability* measure if, moreover, $\mu(E) = 1$. The unit point mass at a point x of E is a measure denoted ε_x. We use the notations $\langle \mu, f \rangle$ and $\int_E f(x)\mu(dx)$ for the integral of a function f in $b^+(E)$ with respect to the measure μ; such an integral may be infinite. By definition of ε_x, one gets $f(x) = \langle \varepsilon_x, f \rangle$ for any f in $b^+(E)$.

The measure μ on E is called a *Radon measure* if it enjoys the following properties:

(a) (*local finiteness*) every point of E has an open neighborhood V such that $\mu(V)$ is finite;

(b) (*inner regularity*) for every Borel subset A of E, the number $\mu(A)$ is the L.U.B. of the numbers $\mu(K)$ where K runs over the class of compact subsets of A.

When μ is a Radon measure, $\mu(K)$ is finite whenever K is compact and among the closed subsets of E whose complement is μ null there is a smallest one called the *support* of μ. When E is a separable locally compact space, local finiteness means that $\mu(K)$ is finite for K compact, and it implies inner regularity [8].

Let E and E' be Hausdorff spaces. A *kernel* Q from E into E' is a mapping from $b^+(E')$ into $b^+(E)$ such that $Q(\Sigma_{n=1}^{\infty} f'_n) = \Sigma_{n=1}^{\infty} Qf'_n$ for f'_n in $b^+(E')$, $n \geqq 1$. The kernel Q is called *markovian* if and only if $Q1 = 1$. Let μ be a measure on E; then there exists a unique measure μQ on E' such that $\langle \mu Q, f' \rangle = \langle \mu, Qf' \rangle$ for each f' in $b^+(E')$. In particular, to Q there corresponds a map q from E into the set of measures on E' given by $q(x) = \varepsilon_x Q$. Then $Qf'(x) = \langle q(x), f' \rangle$ for f' in $b^+(E')$. If μ is a measure on E, one gets

$$\mu Q(A') = \int_E q(x)(A')\mu(dx) \tag{1.1}$$

for each Borel subset A' of E'; we shall abbreviate this relation as $\mu Q = \int_E q(x)\mu(dx)$. Finally, let Q' be a kernel from E' into another Hausdorff space E''.

The *composite kernel* QQ' from E into E'' is defined by $(QQ')f'' = Q(Q'f'')$ for f'' in $b^+(E'')$. By duality, one gets $\mu(QQ') = (\mu Q)Q'$ for each measure μ on E.

1.2. *Topological groups.* Let G be a separable locally compact group. The *convolution* $\mu * \mu'$ of two measures μ and μ' on G is the image of the product measure $\mu \otimes \mu'$ by the multiplication map $(g, g') \mapsto gg'$ from $G \times G$ into G. Note the following integral formula

$$\langle \mu * \mu', f \rangle = \int_G \int_G f(xy)\mu(dx)\mu'(dy), \qquad f \text{ in } b^+(G). \tag{1.2}$$

The n-fold convolution $\mu * \cdots * \mu$ shall be abbreviated as μ^n. For any measure μ on G, the *opposite measure* $\bar{\mu}$ is defined by $\bar{\mu}(A) = \mu(A^{-1})$ for each Borel subset A of G.

A *right invariant Haar measure* m on G is a nonzero Radon measure such that $m(Ag) = m(A)$ for A in $\mathscr{B}(G)$ and g in G. It is unique up to multiplication by a positive real number and there exists a continuous function Δ on G, the *module function of* G, such that $m(g^{-1}A) = \Delta(g)m(A)$ for g in G and A in $\mathscr{B}(G)$.

By a G *space*, we mean a Hausdorff space E upon which G acts continuously from the left. The group G acts on measures on E by $(g\mu)(A) = \mu(g^{-1}A)$ for g in G and A in $\mathscr{B}(E)$. In particular, G acts on itself by left translations, and hence on the measures on G. One gets from the definitions the relations $g\mu = \varepsilon_g * \mu$ and $gm = \Delta(g)m$ for g in G, μ a measure on G, and m a right invariant Haar measure on G.

A probability measure μ on G is *spread out* if it satisfies the following equivalent conditions:

(a) there is an integer $n \geqq 0$ such that the n-fold convolution μ^n is nonsingular with respect to a right invariant Haar measure;

(b) there is an integer $n \geqq 0$, a right invariant Haar measure m, and a nonempty open set V in G such that $\mu^n(A) \geqq m(A)$ for any Borel subset A of V.

This is the case for instance if μ is absolutely continuous with respect to a Haar measure (see [3] for a study of this notion).

2. Sums of independent random variables

In this and the next section, G denotes a separable locally compact group and μ a probability measure on G.

Let us denote by $(Z_n)_{n \geqq 1}$ an independent sequence of G valued random variables with the common probability law μ, and define $S_0 = e$ (the unit element of G) and $S_n = Z_1 \cdots Z_n$ for $n \geqq 1$. The canonical sample space $(\Omega, \mathscr{B}(\Omega), \mathbf{P})$ for the process $(Z_n)_{n \geqq 1}$ is described as follows: Ω is the topological product space $\Pi_{n=1}^{\infty} G_n$, where $G_n = G$ for each n, $\mathscr{B}(\Omega)$ is the class of Borel subsets of Ω, and $\mathbf{P} = \otimes_{n=1}^{\infty} \mu_n$ where $\mu_n = \mu$ for each n. Moreover, Z_n is the projection of Ω onto its nth factor. Since the topology of G is countably generated, $\mathscr{B}(\Omega)$ is the smallest σ-algebra for which the projections Z_n are measurable (as functions with values in $(G, \mathscr{B}(G))$).

Let G_μ be the smallest closed subgroup of G containing the support of μ. For every $n \geqq 0$, the support of μ^n is contained in G_μ; if μ is spread out on G, the support of μ^n has an inner point for some $n \geqq 0$, and hence G_μ has some inner point, that is, G_μ is open. Consequently, one gets $G_\mu = G$ if G is connected and μ spread out on G (for instance μ absolutely continuous with respect to a Haar measure). In any case, the random variables Z_n and S_n take almost surely their values in G_μ, and we can consider $(Z_n)_{n \geqq 1}$ and $(S_n)_{n \geqq 0}$ as random processes carried by the separable locally compact group G_μ.

We shall say that an element g of G is *μ recurrent* if and only if each neighborhood of g is hit infinitely often by almost every path of the process $(S_n)_{n \geqq 0}$. The following theorem is an easy generalization of the results of Chung and Fuchs [15]. According to the previous remarks, there is no real loss of generality in assuming $G_\mu = G$ and this hypothesis simplifies the enunciation.

THEOREM 2.1. *Assume that there is no proper closed subgroup of G containing the support of μ. Let us define the measure $\pi = \Sigma_{n=0}^{\infty} \mu^n$ on G. There is the following dichotomy:*

(i) (*transient case*) *no element of G is μ recurrent and π is a Radon measure;*

(ii) (*recurrent case*) *every element of G is μ recurrent and $\pi(V)$ is infinite for every nonempty open set V in G.*

PROOF. Let R be the set of μ recurrent elements. We shall prove that R is equal to $\varnothing$ or to G. For that purpose, we introduce the set S of all elements g of G enjoying the following property:

for every open neighborhood V of g there exists an integer $n \geqq 0$ such that $\mathbf{P}[S_n \in V] > 0$.

The complement of S in G is the largest open set U such that $\mathbf{P}[S_n \in U] = 0$ for all $n \geqq 0$, hence contains no μ recurrent point. It follows that S is closed and contains R.

We prove next the inclusion $S^{-1}R \subset R$. Indeed, let g be in S and h be in R, and let U be an open neighborhood of $g^{-1}h$. By continuity of the operation in G, we can find open neighborhoods V and W, of g and h, respectively, such that $V^{-1}W \subset U$. By definition of S, there exists an integer $k \geqq 0$ such that the event $A = [S_k \in V]$ has positive probability. Let A' be the set of all ω in A such that there exist infinitely many integers $n \geqq 0$ such that $S_{n+k}(\omega) \in W$. One gets $A' \in \mathscr{B}(\Omega)$, and since W is an open neighborhood of the μ recurrent point h, one has $\mathbf{P}[A'] = \mathbf{P}[A] > 0$. Put $S'_n = S_k^{-1}S_{n+k}$ for $n \geqq 0$; the process $(S'_n)_{n \geqq 0}$ is then independent from $Z_1, \cdots, Z_k$, hence from A', and for every ω in A' the relation $S_{n+k}(\omega) \in W$ entails $S'_n(\omega) = S_k(\omega)^{-1}S_{n+k}(\omega) \in V^{-1}W \subset U$. It follows that almost every path of the process $(S'_n)_{n \geqq 0}$ hits U infinitely often, and since the process $(S'_n)_{n \geqq 0}$ has the same law as $(S_n)_{n \geqq 0}$ and U is an arbitrary open neighborhood of $g^{-1}h$, it follows that $g^{-1}h$ is μ recurrent and we are through.

Assume now R nonempty. From $R \subset S$ and $S^{-1}R \subset R$, one gets $R^{-1}R \subset R$, that is, R is a subgroup of G. Consequently, e belongs to R; hence, $S^{-1} = S^{-1}e \subset S^{-1}R \subset R = R^{-1}$, that is, $S \subset R$. Finally, $S = R$ is a closed subgroup of G and since μ is the probability law of $S_1 = Z_1$, its support is contained in

$S = R$. Hence, $R = G$.

Assume that $\pi(V)$ is finite for some nonempty open set V in G. Since $\pi(V) = \Sigma_{n=0}^{\infty} \mathbf{P}[S_n \in V]$, it follows from the Borel-Cantelli lemma that almost every path of the process $(S_n)_{n \geqq 0}$ hits V only finitely many times. Consequently, no point of V is μ recurrent and from above there is no μ recurrent point at all.

Therefore, in the case $R = G$, one gets $\pi(V) = +\infty$ for every nonempty open set V in G. When $R = \varnothing$, one can use the reasoning of Chung and Fuchs ([15], p. 4) to show that π is a Radon measure; one needs only to note that there exists a left invariant metric defining the topology of G. *Q.E.D.*

The previous proof gives a useful criterion for transient processes. Indeed, call a subset Γ of G a semigroup, if it contains the unit element e of G and is closed under multiplication. Denote by Γ_μ the smallest closed semigroup containing the support of μ. It is easy to see that the support of μ^n is the closure of the set of products $g_1 \cdots g_n$ for $g_1, \cdots, g_n$ running over the support of μ. Since μ^n is the probability law of S_n, it is easy to see that Γ_μ is the set denoted S in the proof of Theorem 2.1. We have seen that R nonempty entails $S = R = G$. We see therefore that *the inequality $\Gamma_\mu \neq G_\mu$ can occur in the transient case only.* When G is the additive real group, the inequality $\Gamma_\mu \neq G_\mu$ means that the probability law μ of the elementary step is supported by either one of the two half lines bounded by 0, and such a one sided process is necessarily transient. There are obvious geometric generalizations of this case.

3. Description of the random walk of law μ

We shall keep the previous notation. For every g in G, the *random walk of law μ starting at g* is the process $(gS_n)_{n \geqq 0}$. More generally, let α be a probability measure on G. The *random walk of law μ and initial distribution* α is the process $(X_n)_{n \geqq 0}$, of the form $X_n = X_0 S_n$, where X_0 is any G valued random variable with probability law α independent of the process $(Z_n)_{n \geqq 0}$.

The canonical sample space for these processes is $(W, \mathscr{B}(W))$, where W is the topological product space $\Pi_{n=0}^{\infty} G_n$ with $G_n = G$ for every $n \geqq 0$, and X_n is the projection $W \mapsto G$ on the nth factor. We denote by $\mathbf{P}^g$ the probability law of the random walk of law μ starting at g, that is, the image of $\mathbf{P}$ by the continuous mapping $(g_1, g_2, \cdots, g_n, \cdots) \mapsto (g, gg_1, gg_1g_2, \cdots, gg_1 \cdots g_n, \cdots)$ from Ω to W. Similarly, one denotes by $\mathbf{P}^\alpha$ the probability law of the random walk of law μ and initial distribution α. It is easily shown that for any A in $\mathscr{B}(W)$, the function $g \mapsto \mathbf{P}^g[A]$ is Borel measurable on G and that

$$\mathbf{P}^\alpha[A] = \int_G \mathbf{P}^g[A]\alpha(dg). \tag{3.1}$$

We shall use this formula as a definition of the measure $\mathbf{P}^\alpha$ on W whenever α is a Radon measure on G. We denote by $\mathbf{E}^g$ and $\mathbf{E}^\alpha$ the expectation functionals corresponding respectively to $\mathbf{P}^g$ and $\mathbf{P}^\alpha$. If the function f on W is nonnegative and Borel measurable, the function $g \mapsto \mathbf{E}^g[f]$ is Borel measurable on G and one gets the formula

$$\mathbf{E}^\alpha[f] = \int_G \mathbf{E}^g[f]\alpha(dg). \tag{3.2}$$

This formula reduces to (3.1) when f is the indicator function I_A of the Borel set A.

The *transition kernel* of the random walk is the kernel Q on G defined by

$$Qf(g) = \int_G f(gh)\mu(dh), \qquad f \text{ in } b^+(G), g \text{ in } G, \tag{3.3}$$

and the *shift* is the kernel θ on W defined by

$$\theta F(g_0, g_1, \cdots, g_n, \cdots) = F(g_1, g_2, \cdots, g_{n+1}, \cdots), \qquad F \text{ in } b^+(W). \tag{3.4}$$

The *Markov property* of the random walk is expressed by the relation $\mathbf{E}^\alpha[F \cdot f(X_{n+1})] = \mathbf{E}^\alpha[F \cdot Qf(X_n)]$, where F depends only on $X_0, \cdots, X_n$. By induction on n, one gets

$$\mathbf{E}^\alpha[f_0(X_0) \cdots f_n(X_n)] = \langle \alpha, f_0 Q f_1 Q \cdots f_{n-1} Q f_n \rangle \tag{3.5}$$

for $f_0, \cdots, f_n$ in $b^+(G)$. Specializing $f_0, \cdots, f_n$ to indicator functions in (3.5) gives

$$\mathbf{P}^\alpha[X_0 \in A_0, \cdots, X_n \in A_n] = \langle \alpha, I_{A_0} Q I_{A_1} Q \cdots I_{A_{n-1}} Q I_{A_n} \rangle \tag{3.6}$$

for any finite sequence of Borel subsets $A_0, \cdots, A_n$ of G.

Part B. Construction of the intrinsic boundary

Here are our main assumptions: *G is a separable locally compact group and μ a probability measure on G*; we assume that $\pi = \Sigma_{n=0}^\infty \mu^n$ is a *Radon measure* (transient case). This part is devoted to an elementary study of the potential theory associated with π and to the construction of the intrinsic boundary corresponding to μ. Finally, we shall prove a certain number of theorems asserting the existence of integral representations.

4. Excessive measures and excessive functions

Let $\bar{\mu}$ and $\bar{\pi}$ be the opposite measures of μ and π. Define the kernels $\bar{Q}$ and $\bar{U}$ on G by the formulas

$$\bar{Q}f(g) = \int_G f(gh)\bar{\mu}(dh), \qquad \bar{U}f(g) = \int_G f(gh)\bar{\pi}(dh) \tag{4.1}$$

for f in $b^+(G)$ and g in G. For a measure λ, one gets dually

$$\lambda\bar{Q} = \lambda * \bar{\mu}, \qquad \lambda\bar{U} = \lambda * \bar{\pi}. \tag{4.2}$$

The *potential kernel* $\bar{U}$ is defined in terms of $\bar{Q}$ by $\bar{U} = \Sigma_{n=0}^{\infty} \bar{Q}^n$, or more precisely

(4.3) $$\bar{U}f = \sum_{n=0}^{\infty} \bar{Q}^n f, \qquad \lambda\bar{U} = \sum_{n=0}^{\infty} \lambda\bar{Q}^n$$

for f in $b^+(G)$ and any measure λ on G.

The transition kernels Q and $\bar{Q}$ of the random walks of law μ and $\bar{\mu}$ are *in duality* with respect to any right invariant Haar measure m, that is, they satisfy the (easily checked) identity

(4.4) $$\langle m, f \cdot Qf' \rangle = \langle m, f' \cdot \bar{Q}f \rangle$$

for any f, f' in $b^+(G)$.

DEFINITION 4.1. *A function f on G is called excessive (respectively, invariant), if $f \in b^+(G)$ and $Qf \leqq f$ (respectively, $Qf = f$). Any bounded Borel function such that $Qf = f$ will also be called invariant. Let λ be a measure on G; one calls λ excessive (invariant), if it is a Radon measure and $\lambda\bar{Q} \leqq \lambda (\lambda\bar{Q} = \lambda)$. One calls λ a potential, if there exists a measure α such that $\lambda = \alpha\bar{U}$.*

According to the classical terminology, our excessive measures (respectively, potentials) should be called coexcessive (respectively, copotentials), since the kernels Q and $\bar{Q}$ are in duality (see [36]). Since the construction of the boundary involves only the excessive measures in our sense, there is little inconvenience if we delete the prefix co. We shall come back to the study of excessive functions in Part D. For the moment, we simply note that if an excessive function f is m locally integrable, the measure $f \cdot m$ is excessive (a direct consequence of (4.4)). We also remark that m is an invariant measure.

We shall denote by $\mathscr{E}$ the class of excessive measures and by $\mathscr{I}$ the class of invariant measures. Both are convex cones, that is, closed under addition and multiplication by a nonnegative real number. Any invariant measure is excessive; if a Radon measure is a potential, it is excessive according to the following consequence of (4.3),

(4.5) $$\alpha\bar{U} = \alpha + (\alpha\bar{U})\bar{Q}.$$

In the following, we shall denote by $\mathscr{M}$ the space of Radon measures on G endowed with the vague topology, that is, the coarsest topology making continuous the real valued functionals $\lambda \mapsto \langle \lambda, f \rangle$ for f in $C_c^+(G)$. We now gather the main algebraic and topological properties of the cone $\mathscr{E}$ of excessive measures. The proofs follow well-known patterns (see, for instance, [46]) and have been included here for the sake of completeness only.

THEOREM 4.1. (i) *Let λ be an excessive measure. There exist two Radon measures α and β on G such that $\lambda = \alpha\bar{U} + \beta$ and $\beta\bar{Q} = \beta$ (Riesz decomposition). The measures α and β are uniquely determined by λ; indeed, $\alpha = \lambda - \lambda\bar{Q}$ and the decreasing sequence $(\lambda\bar{Q}^n)_{n \geqq 0}$ tends to β. Moreover, β is the largest among the invariant measures majorized by λ.*

(ii) *The convex cone $\mathscr{E}$ is a lattice for its intrinsic order.*

PROOF. (i) Since λ is a Radon measure and $\lambda\bar{Q} \leqq \lambda$, there exists a Radon measure α such that $\lambda = \alpha + \lambda\bar{Q}$. By induction, one gets $\lambda\bar{Q}^{n+1} \leqq \lambda\bar{Q}^n$ for $n \geqq 0$, and thus there exists the limit $\beta = \lim_{n\to\infty} \lambda\bar{Q}^n$. From the definition of α, one gets

$$\lambda = \alpha + \alpha\bar{Q} + \cdots + \alpha\bar{Q}^{n-1} + \lambda\bar{Q}^n \tag{4.6}$$

by induction on $n \geqq 1$. By going to the limit in (4.6), one gets $\lambda = \alpha\bar{U} + \beta$ as required.

Let us show that β is invariant. Substituting $\lambda = \alpha\bar{U} + \beta$ into the relation $\lambda = \alpha + \lambda\bar{Q}$ gives

$$\alpha\bar{U} + \beta = \alpha + (\alpha\bar{U} + \beta)\bar{Q} = \alpha\bar{U} + \beta\bar{Q} \tag{4.7}$$

by (4.5). Cancelling out $\alpha\bar{U}$ gives $\beta = \beta\bar{Q}$. It is clear that the invariant measure β is majorized by λ. Furthermore, if β' is invariant and $\beta' \leqq \lambda$, one gets $\beta' = \beta'\bar{Q}^n \leqq \lambda\bar{Q}^n$ for any integer $n \geqq 0$, hence, $\beta' \leqq \beta$ by going to the limit.

Let α' and β' be Radon measures such that $\lambda = \alpha'\bar{U} + \beta'$ and $\beta'\bar{Q} = \beta'$. From (4.5), one gets $\lambda = \alpha' + (\alpha'\bar{U})\bar{Q} + \beta'\bar{Q} = \alpha' + \lambda\bar{Q}$; hence $\alpha' = \alpha$, and therefore $\beta' = \beta$.

(ii) We denote by $\lambda_1 \succ \lambda_2$ the intrinsic order in the convex cone $\mathscr{E}$. By definition, this relation means the existence of a measure λ_3 in $\mathscr{E}$ such that $\lambda_1 = \lambda_2 + \lambda_3$. According to (i), write $\lambda_i = \alpha_i\bar{U} + \beta_i$ with β_i invariant for $i = 1, 2$. It is immediate that $\lambda_1 \succ \lambda_2$ is equivalent to $\alpha_1 \geqq \alpha_2$ and $\beta_1 \geqq \beta_2$ (note that $\beta_1 \geqq \beta_2$ implies that $\beta_1 - \beta_2$ is an invariant measure).

With the previous notations, denote by α (respectively, β) the largest among the Radon measures that are majorized in the usual sense by α_1 and α_2 (respectively, β_1 and β_2). The existence of α and β is well known ([7], p. 53). For $i = 1, 2$ one gets $\beta \leqq \beta_i$; hence $\beta\bar{Q} \leqq \beta_i\bar{Q} = \beta_i$. By definition of β, we have $\beta\bar{Q} \leqq \beta$. By (i), there is a largest invariant measure γ majorized in the usual sense by the excessive measure β (that is, by β_1 and β_2) namely, $\gamma = \lim_{n\to\infty} \beta\bar{Q}^n$. It is then immediate that $\lambda_1 \wedge \lambda_2 = \alpha\bar{U} + \gamma$ is the G.L.B. of λ_1 and λ_2 in $(\mathscr{E}, \succ)$.

Finally, from $\lambda_1 \wedge \lambda_2 \prec \lambda_1 \prec \lambda_1 + \lambda_2$, one deduces the existence of an excessive measure $\lambda_1 \vee \lambda_2$ such that $\lambda_1 + \lambda_2 = (\lambda_1 \wedge \lambda_2) + (\lambda_1 \vee \lambda_2)$. The proof that $\lambda_1 \vee \lambda_2$ is the L.U.B. of λ_1 and λ_2 in $(\mathscr{E}, \succ)$ is then straightforward. *Q.E.D.*

THEOREM 4.2. *The convex cone $\mathscr{E}$ of excessive measures is closed in the space $\mathscr{M}$ of all Radon measures on G. Moreover, any excessive measure is the limit of an increasing sequence of potentials.*

PROOF. Let f in $C_c^+(G)$. It is well known that $\bar{Q}f$ is a continuous function on G; hence, $\langle\alpha\bar{Q}, f\rangle = \langle\alpha, \bar{Q}f\rangle$ is the L.U.B. of the numbers $\langle\alpha, g\rangle$ for g in $C_c^+(G)$ and $g \leqq \bar{Q}f$, whatever be the Radon measure α. Hence, $\mathscr{E}$ is singled out from $\mathscr{M}$ by the set of inequalities $\langle\alpha, f\rangle \geqq \langle\alpha, g\rangle$ for f and g in $C_c^+(G)$ such that $g \leqq \bar{Q}f$. Each of these inequalities defines a vaguely closed set in $\mathscr{M}$; thus, $\mathscr{E}$ is vaguely closed in $\mathscr{M}$.

Let λ be any excessive measure with Riesz decomposition $\lambda = \alpha\bar{U} + \beta$. For any compact subset K of G, the reduite β_K of β on K is a potential such that $I_K \cdot \beta \leqq \beta_K \leqq \beta$ and $\beta_K \leqq \beta_L$ for K contained in L (the properties of the reduites needed here are derived again in Part C).

Since G is a separable locally compact space, we can find an increasing sequence $(K_n)_{n\geqq 1}$ of compact subsets of G such that $G = \bigcup_{n=1}^{\infty} K_n$ and the sequence of potentials $\alpha\bar{U} + \beta_{K_n}$ is increasing and clearly tends to λ. *Q.E.D.*

5. Intrinsic boundary of G

As before, let $\mathscr{E}$ stand for the space of excessive measures on G with the vague topology. The *ray* generated by a measure $\lambda \neq 0$ in $\mathscr{E}$ is as usual the set of all measures $t \cdot \lambda$, where t runs over the positive real numbers. The rays form a partition of the open subspace $\mathscr{E} - \{0\}$ of $\mathscr{E}$. We shall denote by $\mathscr{R}$ the set of all rays endowed with the topology obtained by considering it as a quotient space of $\mathscr{E} - \{0\}$. The ray D is called *extreme* if and only if the relations $\lambda \prec \lambda'$ and $\lambda' \in D$ imply $\lambda \in D$ for any measure $\lambda \neq 0$ in $\mathscr{E}'$. Let $\mathscr{S}$ be the set of all extreme rays and B the subset of $\mathscr{S}$ consisting of the extreme rays, all of whose elements are invariant measures. Finally, for every g in G, let $i(g)$ be the ray generated by the potential $\varepsilon_g \bar{U} = g\bar{\pi}$.

LEMMA 5.1. *The mapping i is injective and $\mathscr{S}$ is the disjoint union of $i(G)$ and B.*

PROOF. Let g and g' in G be such that $i(g) = i(g')$. There exists therefore a real number $t > 0$ such that $\varepsilon_{g'}\bar{U} = t \cdot \varepsilon_g \bar{U}$; hence, $\varepsilon_{g'} = t \cdot \varepsilon_g$ by Theorem 4.1, (i). This last relation is possible only if $t = 1$ and $g = g'$; hence, i is injective.

Let λ be a nonzero excessive measure with Riesz decomposition $\lambda = \alpha\bar{U} + \beta$. Since $\alpha\bar{U} \prec \lambda$ and $\beta \prec \lambda$, the ray generated by λ can be extreme only if $\alpha\bar{U}$ or β vanishes; that is, if λ is a potential or an invariant measure. Finally, the potential $\alpha\bar{U}$ generates an extreme ray if and only if every measure α' with $\alpha' \leqq \alpha$ is proportional to α; it is well known that this means that α is a point measure. *Q.E.D.*

DEFINITION 5.1. *The intrinsic boundary of G (with respect to μ) is the set B of extreme rays in $\mathscr{E}$ consisting of invariant measures. The intrinsic completion of G (with respect to μ) is the disjoint union $\hat{G}$ of G and B.*

We extend the map $i\colon G \mapsto \mathscr{R}$ to a map $j\colon \hat{G} \mapsto \mathscr{R}$ by $j(x) = x$ for x in B. By definition a set U in $\hat{G}$ is called open if there exist open sets V in G and V' in $\mathscr{R}$ such that $U = V \cup j^{-1}(V')$. The axioms for a topology are easily checked (use the continuity of $i\colon G \mapsto \mathscr{R}$); hence, $\hat{G}$ becomes a topological space. Moreover, by Lemma 5.1, j is a continuous bijection from $\hat{G}$ onto $\mathscr{S}$ (but not necessarily a homeomorphism); furthermore, G with its given topology and B with the topology induced from $\mathscr{R}$ are subspaces of $\hat{G}$ with G open and B closed.

Now we let G operate on $\hat{G}$. For g in G one gets $g(\lambda\bar{Q}) = (g\lambda)\bar{Q}$ (λ in $\mathscr{M}$); hence, the map $\lambda \mapsto g\lambda$ leaves both $\mathscr{E}$ and $\mathscr{I}$ invariant. The group G operates therefore by automorphisms of the convex cone $\mathscr{E}$; hence, it operates on the set $\mathscr{R}$ of rays in $\mathscr{E}$. It is clear that $\mathscr{S}$ and B are invariant under G and that $g \cdot i(g') =$

$i(gg')$ for g, g' in G. The action of G on $\hat{G}$ is given by the left translations on G and the previous action on B, in such a way that the bijection $j: \hat{G} \mapsto \mathcal{S}$ is compatible with the operations of G.

LEMMA 5.2. *The intrinsic completion $\hat{G}$ is a Hausdorff space having a countable base of open sets, and G acts continuously on $\hat{G}$.*

PROOF. By construction, one has a continuous injective map $j: G \mapsto \mathcal{R}$; hence, to show that $\hat{G}$ is Hausdorff, it suffices to show that $\mathcal{R}$ is Hausdorff. The equivalence relation defined in $\mathcal{E} - \{0\}$ by the partition in rays is clearly open, and its graph is closed. Hence, the quotient space $\mathcal{R}$ is Hausdorff. The natural projection $q: \mathcal{E} - \{0\} \mapsto \mathcal{R}$ is open, and $\mathcal{E}$ being a subspace of the separable metrizable space $\mathcal{M}$, has a countable base of open sets: hence, the topology of $\mathcal{R}$ has a countable base. The definition of the topology of $\hat{G}$ implies then immediately that $\hat{G}$ has a countable base of open sets.

Let us prove now that G acts continuously upon the convex cone $\mathcal{E}$ (upon $\mathcal{M}$, indeed!). We have to show that for any f in $C_c^+(G)$, the numerical function F defined on $G \times \mathcal{E}$ by

$$F(g, \lambda) = \langle g \cdot \lambda, f \rangle = \int_G f(gx)\lambda(dx) \tag{5.1}$$

is continuous. Let $\varepsilon > 0$ and (g_0, λ_0) in $G \times \mathcal{E}$ be fixed. Let U be a compact neighborhood of g_0 and S be the (compact) support of f; the set $L = U^{-1}S$ is then compact in G and one can choose a function f' in $C_c^+(G)$ taking the constant value 1 on L. Also, let c be a real number such that $c > \langle \lambda_0, f' \rangle$. Since f is left uniformly continuous, there exists a compact neighborhood V of g_0 contained in U such that

$$|f(gx) - f(g_0x)| \leqq \frac{\varepsilon}{2c} \tag{5.2}$$

for g in V and x in G. The left side of this inequality vanishes for x off L (for g fixed in V). Hence, we can strengthen (5.2) as follows:

$$|f(gx) - f(g_0x)| \leqq \frac{\varepsilon \cdot f'(x)}{2c}, \qquad g \text{ in } V,\ x \text{ in } G. \tag{5.3}$$

The function f'' defined by $f''(x) = f(g_0x)$, x in G, is in $C_c^+(G)$. Hence, the set of measures λ in $\mathcal{E}$ satisfying the inequalities

$$\langle \lambda, f' \rangle < c, \qquad |\langle \lambda, f'' \rangle - \langle \lambda_0, f'' \rangle| < \frac{\varepsilon}{2} \tag{5.4}$$

is an open neighborhood W of λ_0 in $\mathcal{E}$. For g in V and λ in W, one gets

$$\begin{aligned} |F(g, \lambda) - F(g_0, \lambda)| &\leqq |F(g, \lambda) - F(g_0, \lambda)| + |F(g_0, \lambda) - F(g_0, \lambda_0)| \\ &\leqq \int_G |f(gx) - f(g_0x)|\lambda(dx) + |\langle \lambda, f'' \rangle - \langle \lambda_0, f'' \rangle| \\ &\leqq \frac{\varepsilon \langle \lambda, f' \rangle}{2c} + \frac{\varepsilon}{2} < \varepsilon \end{aligned} \tag{5.5}$$

by using (5.1), (5.3), and (5.4). The continuity of F is therefore established.

If q is the canonical mapping from $\mathscr{E} - \{0\}$ onto $\mathscr{R}$, one has a commutative diagram

$$(5.6) \qquad \begin{array}{ccc} G \times \mathscr{E} & \xrightarrow{\;m\;} & \mathscr{E} \\ {\scriptstyle Id_G \times q}\downarrow & & \downarrow{\scriptstyle q} \\ G \times \mathscr{R} & \xrightarrow{\;m'\;} & \mathscr{R} \end{array}$$

with $m(g, \lambda) = g \cdot \lambda$ and $m'(g, x) = g \cdot x$ for g in G, λ in $\mathscr{E}$ and x in $\mathscr{R}$. We have shown that m is continuous. Clearly, q is surjective and open; hence, $Id_G \times q$ is surjective and open and it follows from (5.6) that m' is continuous. Thus, G acts continuously upon $\mathscr{R}$ and, *a fortiori*, upon the stable subspace $\mathscr{S}$ of extreme rays in $\mathscr{E}$. This fact implies readily that G operates continuously upon $\hat{G}$. *Q.E.D.*

6. Integral representation of excessive measures

We shall need an ancillary notion, that of a reference function.

DEFINITION 6.1. *A reference function on G is a continuous function r on G such that $r(g) > 0$ for each g in G and the potential $\bar{U}r$ is a finite continuous function on G.*

LEMMA 6.1. *For any Radon measure λ on G there exists a bounded reference function r such that $\langle \lambda, r \rangle$ is finite.*

PROOF. Since G is a separable locally compact space, there exists an increasing sequence $(f_n)_{n \geqq 1}$ in $C_c^+(G)$ with limit 1 at any point of G.

The potential $\bar{U}f$ of any function f in $C_c^+(G)$ is a continuous function on G, and hence bounded on the support of f; by the maximum principle ([40], p. 228) the function $\bar{U}f$ is therefore bounded on G.

Then let c_n be the maximum among the numbers 1, $\langle \lambda, f_n \rangle$ and $\sup_{g \in G} \bar{U}f_n(g)$. It is an easy matter to check that $r = \sum_{n=1}^{\infty} c_n^{-1} 2^{-n} f_n$ is the required reference function. *Q.E.D.*

We shall now apply Choquet's theory of integral representations to the convex cone of excessive measures. Let r be any continuous function on G with positive values and let $\mathscr{E}_r$ be the set of λ in $\mathscr{E}$ such that $\langle \lambda, r \rangle \leqq 1$. It is immediately verified that $\mathscr{E}_r$ is a *cap* in $\mathscr{E}$, that is, a convex subset containing 0 with convex complement in $\mathscr{E}$. Furthermore, let Σ_r denote the set of nonzero extreme points in $\mathscr{E}_r$, that is, the set of excessive measures λ such that $\langle \lambda, r \rangle = 1$ which generate extreme rays. We claim that the cap $\mathscr{E}_r$ *is vaguely compact*: indeed, the inequality $\langle \lambda, r \rangle \leqq 1$ is equivalent to the set of inequalities $\langle \lambda, f \rangle \leqq 1$ for f in $C_c^+(G)$ majorized by r; hence $\mathscr{E}_r$ is vaguely closed. Furthermore, for f in $C_c^+(G)$, the positive continuous function r has a positive minimum on the compact support of f; hence, there exists a constant $c > 0$ such that $f \leqq c \cdot r$. This last inequality implies $\langle \lambda, f \rangle \leqq c$ for any λ in $\mathscr{E}_r$ and the compactness of $\mathscr{E}_r$ follows from Tychonov's theorem. Finally, from Lemma 6.1, it follows that *$\mathscr{E}$ is the union of its compact caps $\mathscr{E}_r$, where r runs over the reference functions.*

Let r be a reference function and λ be an excessive measure such that $\langle \lambda, r\rangle$ be finite. By Choquet's theorem ([7], [40]) *there exists a bounded measure* δ_r *on* Σ_r *such that* $\lambda = \int_{\Sigma_r} \sigma \cdot \delta_r(d\sigma)$ *and such a measure is unique* since the convex cone $\mathscr{E}$ is a lattice for its intrinsic order. Let g be in G. By assumption, $\bar{U}r(g) = \langle g\bar{\pi}, r\rangle$ is finite; hence, there exists in Σ_r a unique measure generating the extreme ray $i(g)$, namely, $k_r(g) = \bar{U}r(g)^{-1} \cdot g\bar{\pi}$. For f in $C_c^+(G)$, one gets $\langle k_r(g), f\rangle = \bar{U}f(g)/\bar{U}r(g)$ for any g in G; since the functions $\bar{U}f$ and $\bar{U}r$ are continuous on G, it follows that k_r is a vaguely continuous map from G into Σ_r. Furthermore, k_r is injective since i is injective (Lemma 5.1) and Lusin's theorem ([5], p. 135) applies, since G is a separable locally compact space: k_r is a Borel isomorphism of G onto a Borel subset $k_r(G)$ of Σ_r. Therefore, any measure on $k_r(G)$ lifts uniquely to a measure on G and, for instance, the restriction of δ_r to $k_r(G)$ lifts to a bounded measure γ on G. Let $\alpha = (\bar{U}r)^{-1} \cdot \gamma$; by an easy calculation, one gets

$$\lambda = \alpha\bar{U} + \int_{B_r} \sigma \cdot \delta_r(d\sigma) \tag{6.1}$$

where $B_r = \Sigma_r - k_r(G)$. By Lemma 5.1, B_r consists of invariant measures, and therefore the integral in (6.1) represents an invariant measure. Thus, in (6.1) we have the Riesz decomposition of λ. This decomposition corresponds to the decomposition of Σ_r into $k_r(G)$ and B_r and this supports the heuristic view that an invariant measure is the potential of a charge located at the boundary (here B_r is the boundary).

We summarize our discussion in the following theorem.

THEOREM 6.1. *Let r be a reference function and λ be an invariant measure such that $\langle \lambda, r\rangle$ is finite. Let B_r be the set of the invariant measures σ, such that $\langle \sigma, r\rangle = 1$, which generates an extreme ray in $\mathscr{E}$. Then there exists a unique bounded measure δ_r on B_r such that $\lambda - \int_{B_r} \sigma \cdot \delta_r(d\sigma)$.*

7. Intrinsic integral representations

The results derived in the previous section depend on the choice of a reference function r. We now show how to switch to the intrinsic boundary and get intrinsic formulations.

LEMMA 7.1. *Let λ be an invariant measure on G and r a reference function such that $\langle \lambda, r\rangle$ is finite. Define B_r and δ_r as in Theorem 6.1. There exists on $G \times B_r$ a unique Radon measure $\Theta_{\lambda,r}$ taking the following values on the rectangle sets:*

$$\Theta_{\lambda,r}(A \times A') = \int_{A'} \sigma(A)\delta_r(d\sigma), \qquad A \text{ in } \mathscr{B}(G),\ A' \text{ in } \mathscr{B}(B_r). \tag{7.1}$$

Moreover, $\Theta_{\lambda,r}$ projects onto the measure λ on G.

PROOF. Since G is a separable locally compact space, each vaguely compact set of Radon measures on G is metrizable. In particular, $\mathscr{E}_r$ is metrizable. Since k_r is a continuous map from G into $\mathscr{E}_r$ and G is a countable union of compact subsets, $k_r(G)$ is the union of a sequence of compact subsets of $\mathscr{E}_r$. It is known ([40], p. 282) that the set Σ_r of extreme points of $\mathscr{E}_r$ is a countable intersection of open subsets of $\mathscr{E}_r$; hence, $B_r = \Sigma_r - k_r(G)$ has the same property. It follows ([5], p. 123) that G and B_r are Polish spaces, that is, homeomorphic to complete separable metric spaces. Hence, the topologies of G and B_r are countably generated and the σ-algebra $\mathscr{B}(G \times B_r)$ is generated by the rectangle sets $A \times A'$, where A is in $\mathscr{B}(G)$ and A' in $\mathscr{B}(B_r)$.

Let f be in $C_c^+(G)$; by definition of the vague topology, the function $\sigma \mapsto \langle\sigma, f\rangle$ on B_r is continuous. By a familiar argument of monotone classes, it follows that the mapping $\sigma \mapsto \langle\sigma, f\rangle$ is Borel measurable on B_r for any f in $b^+(G)$. One defines therefore a Markovian kernel K_r from B_r into G by

$$K_r f(\sigma) = \langle\sigma, rf\rangle, \qquad \sigma \text{ in } B_r, f \text{ in } b^+(G). \tag{7.2}$$

We can now use a construction familiar from the theory of Markov processes. From the bounded measure δ_r on B_r and the Markovian kernel K_r from B_r into G, one derives a bounded measure Θ on $G \times B_r$ characterized by the following relation:

$$\Theta(A \times A') = \langle\delta_r, I_{A'} \cdot K_r I_A\rangle, \qquad A \text{ in } \mathscr{B}(G), A' \text{ in } \mathscr{B}(B_r). \tag{7.3}$$

Since $G \times B_r$ is a Polish space, the bounded measure Θ on it is a Radon measure by Prokhorov's theorem ([8], p. 49). The measure $\Theta_{\lambda,r}$ on $G \times B_r$, product of Θ by the locally bounded continuous function $(g, \sigma) \mapsto r(g)^{-1}$ is therefore a Radon measure.

Equation (7.1) is readily checked. Moreover, one deduces the relation

$$\Theta_{\lambda,r}(A \times B_r) = \int_{B_r} \sigma(A)\delta_r(d\sigma) = \lambda(A) \tag{7.4}$$

as a particular case of (7.1); hence, $\Theta_{\lambda,r}$ projects onto the measure λ on G. Finally, since the σ-algebra $\mathscr{B}(G \times B_r)$ is generated by the rectangle sets, there is at most one measure taking given values on the rectangle sets, hence the uniqueness of $\Theta_{\lambda,r}$. *Q.E.D.*

LEMMA 7.2. *Let λ be an invariant measure on G. For each reference function r, let q_r be the continuous mapping of $G \times B_r$ into $G \times B$ which sends (g, σ) into (g, x), where x is the ray generated by σ. There exists a Radon measure Θ_λ on $G \times B$ with the following property: for each reference function r such that $\langle\lambda, r\rangle$ is finite, the image by q_r of the measure $\Theta_{\lambda,r}$ on $G \times B_r$ defined in Lemma 7.1 is equal to Θ_λ. Moreover, Θ_λ projects onto the measure λ on G.*

PROOF. Let r be a reference function such that $\langle\lambda, r\rangle$ is finite. We denote by Λ_r the image of $\Theta_{\lambda,r}$ by q_r. For any compact subset K of G one gets

$$\Lambda_r(K \times B) = \Theta_{\lambda,r}(K \times B_r) = \lambda(K) < \infty. \tag{7.5}$$

Hence, Λ_r is locally finite and projects onto λ. The inner regularity of $\Theta_{\lambda,r}$ and the continuity of q_r imply inner regularity for Λ_r. Thus, Λ_r is a Radon measure on $G \times B$. If s is any reference function such that $\langle\lambda, s\rangle$ is finite, then $r + s$ is a reference function and $\langle \lambda, r + s \rangle$ is finite. Therefore the proof of the lemma will be achieved if one establishes the equality $\Lambda_r = \Lambda_s$ in the case $r \leqq s$.

From now on, fix two reference functions r and s such that $r \leqq s$ and $\langle \lambda, s \rangle$ is finite. Again, using Lusin's theorem, one sees that the set B'_r of extreme rays generated by the measures belonging to B_r is a Borel subset of B and that B'_r is Borel isomorphic to B_r under the natural map. The construction of Λ_r can be rephrased as follows: for each x in B'_r let $k_r(x)$ be the unique measure σ in the ray x such that $\langle \sigma, r \rangle = 1$; there exists a unique measure δ'_r on B'_r such that $\lambda = \int_{B'_r} k_r(x)\delta'_r(dx)$ and then Λ_r is given by

(7.6) $$\Lambda_r = \int_{B'_r} (k_r(x) \otimes \varepsilon_x)\delta'_r(dx).$$

Since $r \leqq s$, one gets $B'_r \supset B'_s$ and there exists a function f in $b^+(B'_s)$ such that

(7.7) $$k_s(x) = f(x) \cdot k_r(x), \qquad x \text{ in } B'_s,$$

namely, $f(x) = \langle k_s(x), r \rangle$ for x in B'_s. We have then

(7.8) $$\lambda = \int_{B'_s} k_s(x)\delta'_s(dx) = \int_{B'_s} k_r(x) \cdot f(x)\delta'_s(dx),$$

and by the uniqueness of δ'_r one concludes that δ'_r is carried by B'_s and that $\delta'_r(dx) = f(x) \cdot \delta'_s(dx)$ on B'_s. The proof of $\Lambda_r = \Lambda_s$ follows then by a trivial calculation from the definition (7.6) of Λ_r and the corresponding relation for Λ_s. *Q.E.D.*

To summarize, we have attached to any invariant measure λ on G a Radon measure Θ_λ on $G \times B$ with projection λ onto the first factor space. The projection of Θ_λ onto the second factor space is not σ-finite in general, and before disintegrating Θ_λ with respect to the second projection, we have to replace it by an equivalent bounded measure. This is achieved with the help of a reference function r such that $\langle \lambda, r \rangle$ is finite, the result being given by (7.6), namely, $\Theta_\lambda = \int_{B'_r} (k_r(x) \otimes \varepsilon_x)\delta'_r(dx)$. On the other hand, the first projection of Θ_λ being the Radon measure λ, we could appeal to general results ([8], p. 39) to get a disintegration of Θ_λ with respect to the first projection. Such a disintegration is unique up to null sets only, but fortunately we can achieve a very smooth result in an important particular case. The probabilistic significance of the measures Θ_λ and γ will appear in the next part (Theorem 12.2 and 12.3).

LEMMA 7.3. *Let m be a right invariant Haar measure on G. There exists a unique Radon probability measure γ on B such that*

(7.9) $$\Theta_m = \int_G (\varepsilon_g \otimes g \cdot \gamma)m(dg).$$

PROOF. The group G acts upon $G \times B$ by $g \cdot (g', x) = (gg', g \cdot x)$. We shall first establish the relation

(7.10) $$g \cdot \Theta_m = \Delta(g) \cdot \Theta_m, \qquad g \text{ in } G,$$

where Δ is the module function of G. Indeed, let r be a reference function such that $\langle m, r\rangle = 1$ and let B'_r and $k_r(x)$ be as in the proof of Lemma 7.2. There exists a unique probability measure μ_r on B'_r such that

$$(7.11)\qquad m = \int_{B'_r} k_r(x)\mu_r(dx).$$

Then one gets

$$(7.12)\qquad \Theta_m = \int_{B'_r} (k_r(x) \otimes \varepsilon_x)\mu_r(dx).$$

Let g be in G. It is immediate that the relation $s(x) = \Delta(g)\cdot r(g^{-1}x)$ (for x in G) defines a reference function s such that $\langle m, s\rangle = 1$. One gets easily

$$(7.13)\qquad gk_r(x) = \Delta(g)\cdot k_s(gx), \qquad x \text{ in } B.$$

Transforming (7.11) by g one finds

$$(7.14)\qquad \Delta(g)\cdot m = \int_{B'_r} \Delta(g)\cdot k_s(gx)\mu_r(dx),$$

since $gm = \Delta(g)\cdot m$. From the uniqueness of the integral representation of an invariant measure and from $gB'_r = B'_s$, one concludes that g transforms the probability measure μ_r on B'_r into the probability measure μ_s on B'_s. We act now upon (7.12) with g and get

$$(7.15)\qquad g\Theta_m = \int_{B'_r} (gk_r(x) \otimes g\varepsilon_x)\mu_r(dx) = \int_{B'_r} \Delta(g)\cdot(k_s(gx) \otimes \varepsilon_{gx})\mu_r(dx)$$

$$= \Delta(g)\int_{B'_s} (k_s(y) \otimes \varepsilon_y)\mu_s(dy) = \Delta(g)\cdot\Theta_m.$$

Hence, the sought after relation (7.10) follows.

The function Δ_1 on $G \times B$ defined by $\Delta_1(h, x) = \Delta(h)$ is continuous and locally bounded. Therefore, $\Delta_1\cdot\Theta_m$ is a Radon measure on $G \times B$. We denote by α the image of $\Delta_1\cdot\Theta_m$ by the homeomorphism $(h, x) \mapsto (h, h^{-1}x)$ of $G \times B$ with itself. For any function F in $b^+(G \times B)$, one gets

$$(7.16)\qquad \int_{G\times B} F(h, x)\alpha(dh, dx) = \int_{G\times B} \Delta(h)\cdot F(h, h^{-1}x)\Theta_m(dh, dx).$$

In the same manner, (7.10) is made explicit by the following transformation formula

$$(7.17)\qquad \int_{G\times B} F(gh, gx)\Theta_m(dh, dx) = \Delta(g)\int_{G\times B} F(h, x)\Theta_m(dh, dx)$$

for any g in G. By an easy calculation, one deduces from (7.16) and (7.17) the following transformation formula for α

$$\int_{G\times B} F(gh, x)\alpha(dh, dx) = \int_{G\times B} F(h, x)\alpha(dh, dx), \tag{7.18}$$

where F is in $b^+(G \times B)$ and g in G.

From (7.16), one deduces that the projection of α onto the first factor of $G \times B$ is equal to $\Delta\cdot m$, and from (7.18), one recovers the well known fact that $\Delta\cdot m$ is a *left invariant* Haar measure. By specializing (7.18), one gets

$$\alpha(gA \times A') = \alpha(A \times A'), \qquad A \text{ in } \mathscr{B}(G), A' \text{ in } \mathscr{B}(B), g \text{ in } G. \tag{7.19}$$

For fixed A' in $\mathscr{B}(B)$, the mapping $A \mapsto \alpha(A \times A')$ of $\mathscr{B}(G)$ into $[0, +\infty]$ is therefore a *left invariant* Radon measure on G. From the uniqueness of Haar measure, one gets the existence of a functional γ on $\mathscr{B}(B)$ such that

$$\alpha(A \times A') = (\Delta\cdot m)(A)\cdot\gamma(A'), \qquad A \text{ in } \mathscr{B}(G), A' \text{ in } \mathscr{B}(B). \tag{7.20}$$

It then follows easily that γ is a Radon probability measure on B and that (7.20) is equivalent to the relation $\alpha = (\Delta\cdot m) \otimes \gamma$. Using (7.16) and Fubini's theorem, one gets finally the following integration formula

$$\langle\Theta_m, F\rangle = \int_G m(dg) \int_B F(g, gx)\gamma(dx), \qquad F \text{ in } b^+(G \times B), \tag{7.21}$$

which is nothing other than the sought after formula (7.9).

It remains to prove that (7.9) characterizes γ uniquely. Let γ' be any Radon probability measure on B such that $\Theta_m = \int_G (\varepsilon_g \otimes g\cdot\gamma')m(dg)$. Making this relation more explicit, one gets

$$\langle\Theta_m, F\rangle = \int_G m(dg) \int_B F(g, gx)\gamma'(dx), \qquad F \text{ in } b^+(G \times B) \tag{7.22}$$

by analogy with (7.21). Using (7.16), one gets $\alpha = (\Delta\cdot m) \otimes \gamma'$; hence, finally $\gamma' = \gamma$. *Q.E.D.*

8. Additional remarks

8.1. *Smoothness of the intrinsic boundary.* The intrinsic boundary B of G (with respect to μ) may seem very large. Since the space of rays in the cone $\mathscr{M} - \{0\}$ of *all positive Radon measures* on G is regular if and only if G is compact, it is highly plausible that B is not always a metrizable space, although we have no nontrivial counter examples (that is, such that $G_\mu = G$). Nevertheless, since the topology of B has a countable base, each compact subset of B is metrizable. It follows that any bounded Radon measure on B is carried by a countable union T of metrizable compact subsets of B. Since G is also a countable union of metrizable compact subsets and G acts continuously upon B, one can even assume that T is stable under G. This applies, for instance, to the measures δ'_r and γ defined above; it follows that the measure Θ_λ is carried by $G \times T$, where T is a subset of B with the previous properties. In summary, the measures we have to work with have all the desirable smoothness.

8.2. *Martin compactification.* Call any continuous nonnegative function r on G (not necessarily positive) such that $\bar{U}r$ is a positive continuous function on G a *generalized reference function.* Classically (*see* [46], [36], for instance), to each generalized reference function r is associated the Martin compactification G_r of G, which is characterized up to homeomorphism by the following properties:

(a) the space G_r is compact and metrizable;

(b) G, with its topology, is an open dense subset of G_r;

(c) for f in $C_c^+(G)$ the function $\bar{U}f/\bar{U}r$ on G extends uniquely to a continuous function L_f on G_r and these functions L_f separate the points of $G_r - G$.

The theorems of existence of an integral representation can be described in terms of G_r. But the main disadvantage of the space G_r is that the action of G on G does not in general extend to a continuous action of G on G_r. The best that can be achieved in general is to obtain a continuous action of G on a Borel subset of G_r, large enough to permit the integral representation of excessive measures; this necessitates the use of reference functions of a special type (see Section 16). We are nevertheless going to describe one case where the Martin compactification seems preferable to the intrinsic boundary of G; let Γ be the support of π; this is also the smallest closed semigroup in G containing the support of μ. One shows easily the equivalence of the two following properties:

(a′) there exists a compact subset K of G such that $G = K \cdot \Gamma$;

(b′) there exists a function r in $C_c^+(G)$ such that $\bar{U}r(g) > 0$ for any g in G, that is, there exists a generalized reference function r having compact support.

Let us assume that (a′) and (b′) hold. Then, the Martin compactification G_r associated with the generalized reference functions r with compact support are all homeomorphic to a metrizable compact space G^* on which G acts continuously. We sketch the construction of G^*. For any r in $C_c^+(G)$, let N_r be the set of all excessive measures λ such that $\langle \lambda, r \rangle = 1$. One first shows that $\bar{U}r > 0$ implies $\langle \lambda, r \rangle > 0$ for each excessive measure $\lambda \neq 0$ and that N_r is vaguely compact. Hence, if $\bar{U}r > 0$, any ray contains one and only one point in the vaguely compact set N_r; this implies that the space $\mathscr{R}$ of rays is compact and metrizable. Call $G_\infty = G \cup \{\infty\}$ the Alexandrov compactification of G and define a map q from G into $G_\infty \times \mathscr{R}$ by $q(g) = (g, i(g))$. There is then a closed subset B^* of $\mathscr{R}$ such that $\overline{q(G)} - q(G) = \{\infty\} \times B^*$. One defines G^* as the disjoint union of G and B^*, one extends q to a bijection q' of G^* onto $\overline{q(G)}$ by mapping any x in B^* into (∞, x), and one gives G^* the topology that makes q' a homeomorphism. It is straightforward to check (a) and (b). If f and r' are in $C_c^+(G)$ and if $\bar{U}r' > 0$, the boundary value of $\bar{U}f/\bar{U}r'$ at x, for x in B^*, is defined as the number $\langle \lambda, f \rangle / \langle \lambda, r' \rangle$, where λ is any representative of the ray x, and the extended function $\bar{U}f/\bar{U}r'$ is continuous on G^*, which proves (c). Hence, G^* is homeomorphic to $G_{r'}$ for any $r' \in C_c^+(G)$ such that $\bar{U}r'$ be continuous and > 0.

The previous construction of G^* shows that the action of G upon itself by left translations extends to a continuous action of G upon G^*. Moreover, using the fact that any excessive measure is the limit of an increasing sequence of potentials, one shows that B is contained in B^* and this fact allows one to consider the intrinsic completion $\hat{G}$ of G as a dense subspace of G^*, namely, a countable intersection of open subsets.

8.3. *Recurrent case.* Let us assume that there exists no proper closed subgroup of G containing the support of μ and that $\Sigma_{n=0}^{\infty} \mu^n(V)$ is infinite for every nonempty open subset V of G (see Theorem 2.1). We shall show that any Radon measure λ such that $\lambda \bar{Q} \leqq \lambda$ is right invariant, hence that the convex cone $\mathscr{E}$ has just one ray. Indeed, let f be in $C_c^+(G)$ and F the nonnegative continuous function on G defined by $F(g) = \int_G f(yg^{-1})\lambda(dy)$. The following calculation shows that $QF \leqq F$:

$$
\begin{aligned}
(8.1) \qquad QF(g) &= \int_G F(gx)\mu(dx) = \int_G F(gx^{-1})\bar{\mu}(dx) \\
&= \int_G \bar{\mu}(dx) \int_G f(yxg^{-1})\lambda(dy) \\
&= \int_G f(zg^{-1})(\lambda * \bar{\mu})(dz) \leqq \int_G f(zg^{-1})\lambda(dz) = F(g).
\end{aligned}
$$

From $QF \leqq F$, it follows that F is constant ([1]; [46], p. 64) hence that λ is right invariant.

It is now clear why the methods used in this part cannot provide nontrivial boundaries in the recurrent case.

Part C. Convergence to the Boundary

Our assumptions are the same as for Part B. The scope of this part is primarily probabilistic. We shall devote ourselves to the proof of several limit theorems giving the asymptotic behavior of the random walk of law μ on G.

9. Relativization

Let λ be an invariant measure. Probabilistically, the relativized process associated with λ is defined as follows. From $\lambda * \bar{\mu} = \lambda$, one gets the existence of a bilateral random walk (Y_n) with n running over the integers of both signs, where each random variable Y_n has λ as distribution and the elementary steps $Y_{n-1}^{-1} Y_n$ are independent with the same probability law $\bar{\mu}$. Then the relativized process is $(Y_{-n})_{n \geqq 0}$ by definition. In the sequel, we shall need only the distribution Π^{λ} of this process in the path space W and we proceed to give a direct construction of Π^{λ}.

PROPOSITION 9.1. *Let λ be an invariant measure. There exists on the path space W with projections X_n, $n \geqq 0$, a unique Radon measure Π^λ such that*

$$\Pi^\lambda[X_0 \in A_0, \cdots, X_n \in A_n] = \langle \lambda, I_{A_n}\bar{Q}I_{A_{n-1}} \cdots I_{A_1}\bar{Q}I_{A_0}\rangle \tag{9.1}$$

holds whatever be the integer $n \geqq 0$ and the Borel subsets $A_0, \cdots, A_n$ of G. Moreover, if m is a right invariant Haar measure, Π^m is equal to $\mathbf{P}^m$.

PROOF. It is well known that two measures on W which agree on the cylinder sets $A_0 \times A_1 \times \cdots \times A_n \times G \times G \times \cdots$ are equal; hence, there can be at most one measure Π^λ for which (9.1) obtains.

For each integer $n \geqq 0$, let $W_n = G \times \cdots \times G$ ($n + 1$ factors) and let Π_n be the image of the Radon measure $\lambda \otimes \bar{\mu} \otimes \cdots \otimes \bar{\mu}$ (n factors $\bar{\mu}$) by the homeomorphism of W_n with itself which maps a point $(g_0, g_1, \cdots, g_n)$ onto the point with ith coordinate equal to $g_0 g_1 \cdots g_{n-i}$ for $0 \leqq i \leqq n$. Now let $f_0, f_1, \cdots, f_n$ in $b^+(G)$; the integral of $f_0 \otimes f_1 \otimes \cdots \otimes f_n$ with respect to Π_n is then equal to

$$J = \int_G \cdots \int_G f_0(g_0 g_1 \cdots g_{n-1} g_n) f_1(g_0 g_1 \cdots g_{n-1}), \tag{9.2}$$
$$\cdots f_{n-1}(g_0 g_1) f_n(g_0) \lambda(dg_0) \bar{\mu}(dg_1) \cdots \bar{\mu}(dg_n).$$

Assume $n \geqq 1$. In the previous integral only the first factor contains g_n. Hence, integrating first with respect to g_n and using formula (4.1) defining $\bar{Q}f$, we get a similar integral with the sequence of $n + 1$ functions $f_0, f_1, \cdots, f_n$ replaced by the sequence of n functions $f_1(\bar{Q}f_0), f_2, \cdots, f_n$. By induction on n, one gets

$$\langle \Pi_n, f_0 \otimes \cdots \otimes f_n\rangle = \langle \lambda, f_n \bar{Q} f_{n-1} \bar{Q} \cdots f_1 \bar{Q} f_0\rangle. \tag{9.3}$$

Let $r > 0$ be a continuous function on G such that $\langle \lambda, r\rangle = 1$ and let Π_n^r be the product of the measure Π_n on W_n by the continuous function $(g_0, \cdots, g_n) \mapsto r(g_0)$. Then Π_0^r is the probability measure $r \cdot \lambda$ on $W_0 = G$. For $f_0, \cdots, f_n$ in $b^+(G)$, one gets

$$\langle \Pi_n^r, f_0 \otimes \cdots \otimes f_n\rangle = \langle \lambda, f_n \bar{Q} f_{n-1} \bar{Q} \cdots f_1 \bar{Q}(f_0 r)\rangle \tag{9.4}$$

from (9.3) and $\lambda\bar{Q} = \lambda$ implies

$$\langle \Pi_n^r, f_0 \otimes \cdots \otimes f_{n-1} \otimes 1\rangle = \langle \Pi_{n-1}^r, f_0 \otimes \cdots \otimes f_{n-1}\rangle \tag{9.5}$$

whenever $n \geqq 1$. Otherwise stated, the projection of Π_n^r onto the first n factors of W_n is equal to Π_{n-1}^r, and since Π_0^r is a probability measure, it follows that Π_n^r is a probability Radon measure for each $n \geqq 0$. By Kolmogorov's theorem ([8], p. 54), there exists a unique Radon probability measure $\Pi^{\lambda,r}$ on W whose projection onto the product W_n of the first $n + 1$ factors is equal to Π_n^r for each $n \geqq 0$. As a final step, define Π^λ as the Radon measure on W product of the Radon measure $\Pi^{\lambda,r}$ with the continuous locally bounded function $r(X_0)^{-1}$. From (9.4), one gets

$$\langle \Pi^{\lambda,r}, f_0(X_0) \cdots f_n(X_n)\rangle = \langle \lambda, f_n \bar{Q} f_{n-1} \bar{Q} \cdots f_1 \bar{Q}(f_0 r)\rangle. \tag{9.6}$$

Hence,

$$(9.7)\qquad \langle \Pi^{\lambda}, f_0(X_0)\cdots f_n(X_n)\rangle = \langle \lambda, f_n\bar{Q}f_{n-1}\bar{Q}\cdots f_1\bar{Q}f_0\rangle$$

whatever the integer $n \geqq 0$ and the functions $f_0, \cdots, f_n$ in $b^+(G)$. The sought after relation (9.1) is the particular case of (9.7), where $f_0, \cdots, f_n$ are indicator functions.

A glance at (3.5) and (9.7) shows that, using the duality between Q and $\bar{Q}$ (relation (4.4)), the proof of $\Pi^m = \mathbf{P}^m$ is reduced to a straightforward induction on n. *Q.E.D.*

In the following, we shall use without further comment the notation Π^{λ} and the symbol $\mathbf{H}^{\lambda}$ for the integral defined by Π^{λ}. For r in $b^+(G)$, we shall denote by $\Pi^{\lambda,r}$ the product of the measure Π^{λ} on W by the function $r(X_0)$; the integral corresponding to $\Pi^{\lambda,r}$ will be denoted by $\mathbf{H}^{\lambda,r}$.

REMARK. The customary definition of relativized processes works for invariant measures of the form $f\cdot m$ only, where $Qf = f$. Such a process is defined as the Markov process with initial distribution $f\cdot m$ and transition kernel Q^f given by

$$(9.8)\qquad Q^f u = \begin{cases} f^{-1}Q(fu) & \text{on the set } [f > 0], \\ 0 & \text{elsewhere,} \end{cases}$$

for u in $b^+(G)$. The following calculation using (4.4) and the readily verified relation $f\cdot Q^f u = Q(fu)$, shows that our definition agrees with the previous description:

$$\begin{aligned}(9.9)\qquad \mathbf{H}^{f\cdot m}&[f_0(X_0)\cdots f_n(X_n)] \\ &= \langle m, ff_n\bar{Q}f_{n-1}\bar{Q}\cdots f_1\bar{Q}f_0\rangle = \langle m, f_0Qf_1Q\cdots f_{n-1}Q(ff_n)\rangle \\ &= \langle m, ff_0Q^ff_1Q^f\cdots f_{n-1}Q^ff_n\rangle = \langle f\cdot m, f_0Q^ff_1Q^f\cdots f_{n-1}Q^ff_n\rangle.\end{aligned}$$

Similarly, for any invariant measure λ, it is easy to show the existence of a transition kernel Q_{λ} in duality with $\bar{Q}$ with respect to λ (that is, $\langle \lambda, f\cdot\bar{Q}f'\rangle = \langle \lambda, f'\cdot Q_{\lambda}f\rangle$) such that the relativized process associated with λ is the Markov process with initial distribution λ and transition kernel Q_{λ}. We point out that Q_{λ} is not necessarily unique.

10. Reduites of measures

Let λ be an invariant measure and K a compact subset of G. First, we shall prove the transient character of the relativized process. Indeed, let r be a reference function such that $\langle \lambda, r\rangle = 1$. From (9.6) one gets

$$(10.1)\qquad \sum_{n=0}^{\infty} \Pi^{\lambda,r}[X_n \in K] = \sum_{n=0}^{\infty} \langle \lambda, I_K\cdot\bar{Q}^n r\rangle = \int_K \bar{U}r(g)\lambda(dg).$$

The last integral is finite because the continuous function $\bar{U}r$ is bounded on the compact set K and $\lambda(K)$ is finite. Since $r > 0$, the measures Π^{λ} and $\Pi^{\lambda,r}$ have the same null sets. From the Borel-Cantelli lemma, one concludes that Π^λ *almost no path in W hits K infinitely often.*

Define W_K as the set of all w in W such that the set of integers $n \geqq 0$ for which $X_n(w)$ belongs to K is finite and nonempty. For w in W_K, one denotes $t_K(w)$ the largest among the integers n such that $X_n(w) \in K$. That is, t_K is the last time that the process is in K.

The *reduite* of λ on K is the measure on G defined by $\lambda_K(A) = \Pi^\lambda[X_0 \in A, W_K]$ for A in $\mathscr{B}(G)$. The following lemma states some elementary properties of the reduites.

LEMMA 10.1. *The reduite λ_K is a potential and $I_K \cdot \lambda \leqq \lambda_K \leqq \lambda$.*

PROOF. By definition, one has

$$\lambda_K(A) = \Pi^\lambda[X_0 \in A, W_K] \leqq \Pi^\lambda[X_0 \in A] = \lambda(A), \qquad A \text{ in } \mathscr{B}(G). \tag{10.2}$$

Hence, $\lambda_K \leqq \lambda$. Moreover, whenever A is contained in K the event $[X_0 \in A]$ is contained up to a Π^λ null set in W_K because of the transient character proved above. We therefore have equality everywhere in the previous calculation, and hence $I_K\lambda \leqq \lambda_K$.

To prove that λ_K is a potential, we need the following formula

$$\mathbf{H}^\lambda[f(X_0)\cdot\theta^n F] = \mathbf{H}^\lambda[\bar{Q}^n f(X_0)\cdot F] \tag{10.3}$$

for f in $b^+(G)$ and F in $b^+(W)$. An easy induction reduces the proof of (10.3) to the proof of the particular case $n = 1$; in this case, we can content ourselves with taking F of the form $f_0(X_0)\cdots f_n(X_n)$, where $f_0, \cdots, f_n$ are in $b^+(G)$ and the sought after relation follows immediately from (9.7).

Define the measure α on G by $\alpha(A) = \Pi^\lambda[X_0 \in A, t_K = 0]$ for A in $\mathscr{B}(G)$ and let J be the indicator of the event $[t_K = 0]$. It is clear that $\theta^n J$ is the indicator of the event $[t_K = n]$ for each integer $n \geqq 0$. Hence, the indicator Φ of W_K is $\Sigma_{n=0}^\infty \theta^n J$. For f in $b^+(G)$, one therefore gets

$$\begin{aligned}\langle\lambda_K, f\rangle &= \mathbf{H}^\lambda[f(X_0)\cdot\Phi] = \sum_{n=0}^\infty \mathbf{H}^\lambda[f(X_0)\cdot\theta^n J] = \sum_{n=0}^\infty \mathbf{H}^\lambda[\bar{Q}^n f(X_0)\cdot J] \\ &= \mathbf{H}^\lambda[\bar{U}f(X_0)\cdot J] = \langle\alpha, \bar{U}f\rangle\end{aligned} \tag{10.4}$$

by using (10.3). Hence, $\lambda_K = \alpha\bar{U}$ is a potential as promised. *Q.E.D.*

Finally, let us consider a compact subset L of G containing K. It is immediate that W_K is contained in W_L up to a Π^λ null set and that $t_K(w) \leqq t_L(w)$ for w in $W_K \cap W_L$. Moreover, $\lambda_K \leqq \lambda_L$.

11. The basic convergence lemma

The following result is the main ingredient to prove convergence of the random walk to the boundary. It is an extension of a theorem of Doob [18] who treated the case of Markov chains with discrete state space. The arrangement of our proof follows rather closely Hunt [33] and Neveu [46].

THEOREM 11.1. *Let λ be an invariant measure, r a reference function such that $\langle \lambda, r \rangle = 1$ and f in $C_c^+(G)$. For every $n \geqq 0$ define the real valued random variable F_n by $F_n = \bar{U}f(X_n)/\bar{U}r(X_n)$. Then the sequence $(F_n)_{n \geqq 0}$ ends Π^λ almost surely to a random variable F_∞ such that $\mathbf{H}^{\lambda, r}[F_\infty] = \langle \lambda, f \rangle$.*

We shall subdivide the proof into several parts.

(A) Let K be a compact subset of G and $t = t_K$; for each integer $n \geqq 0$, we denote by T_n the set of paths w in W_K such that $t(w) \geqq n$, and define the real valued random variable F_n^* by

$$F_n^*(w) = \begin{cases} F_{t(w)-n}(w) & \text{if } w \in T_n, \\ 0 & \text{otherwise.} \end{cases} \tag{11.1}$$

Also, the G valued random variables X_{t-i} are defined on $T_i \supset T_n$ for $0 \leqq i \leqq n$; let $\mathscr{A}_n$ be the smallest σ-algebra of subsets of T_n containing the sets $[X_{t-i} \in A] \cap T_n$ for $0 \leqq i \leqq n$ and A in $\mathscr{B}(G)$. Furthermore, we let $\mathscr{A}_n^*$ be the σ-algebra consisting of the Borel subsets A of W such that $A \cap T_n$ belongs to $\mathscr{A}_n$.

LEMMA 11.1. *The sequence $(F_n^*)_{n \geqq 0}$ is a supermartingale with respect to the increasing family $(\mathscr{A}_n^*)_{n \geqq 0}$ of σ-algebras and the probability measure $\Pi^{\lambda, r}$.*

We fix an integer $n \geqq 0$. Let L be any $\mathscr{A}_n^*$ measurable function on W with values in $[0, +\infty]$. By definition of $\mathscr{A}_n^*$, there exists a function L' in $b^+(G \times \cdots \times G)$ ($n + 1$ factors G) such that $L = L'(X_{t-n}, \cdots, X_{t-1}, X_t)$ on T_n. Let J be the indicator of the event $[t = 0]$ and h be the continuous nonnegative function $\bar{U}f/\bar{U}r$ on G and let $L'' = L'(X_0, \cdots, X_n) \cdot \theta^n J$. On the set T_n, the function F_n^* coincides with $h(X_{t-n})$; hence, the function $F_n^* \cdot L$ coincides with $h(X_{p-n}) \cdot L'(X_{p-n}, \cdots, X_{p-1}, X_p)$ on the set $[t = p]$ for any integer $p \geqq n$ and vanishes outside $T_n = \bigcup_{p \geqq n} [t = p]$. Otherwise stated, one has

$$\begin{aligned} F_n^* \cdot L &= \sum_{p=n}^{\infty} h(X_{p-n}) L'(X_{p-n}, \cdots, X_{p-1}, X_p) \cdot \theta^p J \\ &= \sum_{q=0}^{\infty} \theta^q [h(X_0) \cdot L'']. \end{aligned} \tag{11.2}$$

Since F_{n+1}^* is zero outside T_{n+1}, one gets by a similar reasoning the relation

$$\begin{aligned} F_{n+1}^* \cdot L &= \sum_{p=n+1}^{\infty} h(X_{p-n-1}) \cdot L'(X_{p-n}, \cdots, X_{p-1}, X_p) \cdot \theta^p J \\ &= \sum_{q=0}^{\infty} \theta^q [h(X_0) \cdot \theta L'']. \end{aligned} \tag{11.3}$$

Using (10.3), one derives the formula

$$\mathbf{H}^{\lambda, r}\left[\sum_{q=0}^{\infty} \theta^q R\right] = \mathbf{H}^\lambda[\bar{U}r(X_0) \cdot R], \qquad R \text{ in } b^+(W), \tag{11.4}$$

and from the above representation of $F_n^* \cdot L$, one gets

$$\mathbf{H}^{\lambda, r}[F_n^* \cdot L] = \mathbf{H}^\lambda[\bar{U}r(X_0) \cdot h(X_0) \cdot L''] = \mathbf{H}^\lambda[\bar{U}f(X_0) \cdot L'']. \tag{11.5}$$

In order to get a similar expression for $F^*_{n+1} \cdot L$, we just need to replace L'' by $\theta L''$. Once again using (10.3), we get

$$\text{(11.6)} \qquad \mathbf{H}^{\lambda,r}[F^*_{n+1} \cdot L] = \mathbf{H}^{\lambda}[\bar{U}f(X_0) \cdot \theta L''] = \mathbf{H}^{\lambda}[\bar{Q}\bar{U}f(X_0) \cdot L''] \leqq \mathbf{H}^{\lambda}[\bar{U}f(X_0) \cdot L''] = \mathbf{H}^{\lambda,r}[F^*_n \cdot L].$$

Therefore, we have the inequality $\mathbf{H}^{\lambda,r}[F^*_{n+1} \cdot L] \leqq \mathbf{H}^{\lambda,r}[F^*_n \cdot L]$ for any $\mathscr{A}^*_n$ measurable function L on W with values in $[0, +\infty]$. Since F^*_n is obviously $\mathscr{A}^*_n$ measurable, the lemma is proved.

(B) Let a and b be two rational numbers with $0 < a < b$ and let N be the (random) number of upward crossings of $[a, b]$ by the random sequence $(F_n)_{n \geqq 0}$. Let K and t be as in (A) and define N^*_K as the (random) number of downward crossings of $[a, b]$ by the random sequence $(F^*_n)_{n \geqq 0}$, that is, the number of upward crossings by $(F_n)_{n \geqq 0}$ of $[a, b]$ in the random interval $[0, t]$. According to the classical Doob inequality for a nonnegative supermartingale, one has

$$\text{(11.7)} \qquad (b - a) \cdot \mathbf{H}^{\lambda,r}[N^*_K] \leqq \mathbf{H}^{\lambda,r}[F^*_0].$$

To compute $\mathbf{H}^{\lambda,r}[F^*_0]$, it suffices to let $n = 0$ and $L = 1$ in (11.5), which yields

$$\text{(11.8)} \qquad \mathbf{H}^{\lambda,r}[F^*_0] = \mathbf{H}^{\lambda}[\bar{U}f(X_0) \cdot J].$$

Define the measure α on G by $\alpha(A) = \Pi^{\lambda}[X_0 \in A, t = 0]$; then we have

$$\text{(11.9)} \qquad \mathbf{H}^{\lambda}[\bar{U}f(X_0) \cdot J] = \langle \alpha, \bar{U}f \rangle = \langle \alpha\bar{U}, f \rangle = \langle \lambda_K, f \rangle$$

because $\alpha\bar{U}$ is equal to λ_K by the proof of Lemma 10.1. Since $\lambda_K \leqq \lambda$, we conclude from the relations (11.7) to (11.9) the following inequality

$$\text{(11.10)} \qquad (b - a) \cdot \mathbf{H}^{\lambda,r}[N^*_K] \leqq \langle \lambda, f \rangle.$$

Since G is a separable locally compact space, we can find an increasing sequence $(K_p)_{p \geqq 0}$ of compact subsets of G such that $G = \bigcup_{p=0}^{\infty} K_p$. Because a path is doomed to meet at least one of the compact sets K_p, the transient character shows that up to Π^{λ} null sets $(W_{K_p})_{p \geqq 0}$ is an increasing sequence of Borel subsets of W, whose union exhausts W.

Moreover, for w in W_{K_p}, the sequence $\big(t_{K_q}(w)\big)_{q \geqq p}$ increases Π^{λ} almost surely without bound. Hence, the sequence of random variables $(N^*_{K_p})_{p \geqq 0}$ increases to N. Going to the limit in (11.10), we get

$$\text{(11.11)} \qquad (b - a) \cdot \mathbf{H}^{\lambda,r}[N] \leqq \langle \lambda, f \rangle.$$

Hence, whatever be a and b, the number N of upward crossings of $[a, b]$ by $(F_n)_{n \geqq 0}$ is Π^{λ} almost surely finite and the random sequence $(F_n)_{n \geqq 0}$ converges Π^{λ} almost surely (note that Π^{λ} and $\Pi^{\lambda,r}$ have the same null sets).

(C) Define $F_{\infty} = \lim_{n \to \infty} F_n$. Take the sequence $(K_p)_{p \geqq 0}$ as previously and define the random variables R_p as follows

$$\text{(11.12)} \qquad R_p(w) = \begin{cases} F_{t_p(w)}(w) & \text{for } w \text{ in } W_{K_p}, \\ 0 & \text{otherwise,} \end{cases}$$

where t_p is the exit time associated with the compact set K_p. By the previous remarks about W_{K_p} and the relation $\lim_{p\to\infty} t_p = \infty$ Π^λ almost surely, one gets $F_\infty = \lim_{p\to\infty} R_p$ (Π^λ almost surely). For $K = K_p$, the random variable denoted F_0^* in (B) is nothing else than R_p, and by (11.8) and (11.9), one gets

$$\mathbf{H}^{\lambda,r}[R_p] = \langle \lambda_{K_p}, f\rangle. \tag{11.13}$$

The positive continuous function r on G has a positive minimum on the compact support of f. Hence, there exists a constant $c > 0$ such that $f \leqq c \cdot r$. It follows that $\bar{U}f/\bar{U}r$, and hence F_n and R_p are bounded by c. By the bounded convergence theorem, one gets

$$\mathbf{H}^{\lambda,r}[F_\infty] = \lim_{p\to\infty} \mathbf{H}^{\lambda,r}[R_p] = \lim_{p\to\infty} \langle \lambda_{K_p}, f\rangle \tag{11.14}$$

from (11.13). Since the sequence of measures $(\lambda_{K_p})_{p\geqq 0}$ tends increasingly to λ (see Lemma 10.1 and the proof of Theorem 4.2), the number $\langle \lambda_{K_p}, f\rangle$ tends to $\langle \lambda, f\rangle$ as p tends to infinity. Finally, one gets the desired relation $\mathbf{H}^{\lambda,r}[F_\infty] = \langle \lambda, f\rangle$. *Q.E.D.*

12. Convergence to the boundary

We come to the core of this part and establish three convergence theorems. The first two deal with the relativized process and have an ancillary character.

Recall notation from Section 6. If r is a reference function and $\mathscr{E}_r$ is the set of all excessive measures λ such that $\langle \lambda, r\rangle \leqq 1$, the vaguely continuous map k_r from G into $\mathscr{E}_r$ is defined by $k_r(g) = \bar{U}r(g)^{-1} \cdot g\bar{\pi}$ for g in G. Moreover, Σ_r is the set of nonzero extreme points of the convex set $\mathscr{E}_r$ and $B_r = \Sigma_r - k_r(G)$.

THEOREM 12.1. *Let λ be an invariant measure and r a reference function such that $\langle \lambda, r\rangle = 1$. Then there exists a random element X in B_r such that $k_r(X_n)$ tends Π^λ almost surely to X. Moreover, for each Borel subset A of B_r, one has $\Pi^{\lambda,r}[X \in A] = \delta_r(A)$ where δ_r is the unique probability measure on B_r such that $\lambda = \int_{B_r} \sigma \cdot \delta_r(d\sigma)$. Finally, for λ in B_r, the relation $X = \lambda$ holds Π^λ almost surely.*

PROOF. By definition, one has $\langle k_r(X_n), f\rangle = \bar{U}f(X_n)/\bar{U}r(X_n)$ for f in $C_c^+(G)$ and $n \geqq 0$. Moreover, let D be a countable dense subset of $C_c^+(G)$ (uniform convergence on G). Then a sequence of elements λ_n of $\mathscr{E}_r$ has a limit in $\mathscr{E}_r$ if and only if $\langle \lambda_n, f\rangle$ has a limit for each f in D, and a mapping T from W into $\mathscr{E}_r$ is Borel measurable if and only if the numerical function $\langle T, f\rangle$ is Borel measurable for each f in D.

From these remarks and Theorem 11.1, one gets the existence of a random element X in $\mathscr{E}_r$ defined on the path space W such that $\lim_{n\to\infty} k_r(X_n) = X$ holds Π^λ almost surely and that $\mathbf{H}^{\lambda,r}[\langle X, f\rangle] = \langle \lambda, f\rangle$ holds for each f in D. This last relation can also be written

$$\lambda = \int_{\mathscr{E}_r} \sigma \cdot \nu(d\sigma), \tag{12.1}$$

where ν is the probability measure on $\mathscr{E}_r$ defined by $\nu(A) = \Pi^{\lambda,r}[X \in A]$ for A in $\mathscr{B}(\mathscr{E}_r)$. If λ is in B_r, there can be no nontrivial representation of the form (12.1). Hence, $\nu = \varepsilon_\lambda$, that is, $X = \lambda$ holds Π^λ almost surely.

From (9.6) and the definition of δ_r, one gets immediately

$$\Pi^{\lambda,r}[E] = \int_{B_r} \Pi^{\sigma,r}[E]\delta_r(d\sigma), \qquad E \text{ in } \mathscr{B}(W). \tag{12.2}$$

We have already shown $\Pi^{\sigma,r}[X = \sigma] = 1$ for σ in B_r. Hence, $\Pi^{\sigma,r}[X \in B_r] = 1$ for each σ in B_r. By (12.2), one therefore gets $\nu(B_r) = \Pi^{\lambda,r}[X \in B_r] = 1$. From (12.1), one gets $\lambda = \int_{B_r} \sigma \cdot \nu(d\sigma)$ and finally $\nu = \delta_r$. *Q.E.D.*

REMARK. Using (9.1) instead of (9.6), one gets the integral formula

$$\Pi^\lambda[E] = \int_{B_r} \Pi^\sigma[E]\delta_r(d\sigma) \qquad E \text{ in } \mathscr{B}(W) \tag{12.3}$$

instead of (12.2). For σ in B_r, we know that $X = \sigma$ holds Π^σ almost surely. Therefore,

$$\Pi^\sigma[X_0 \in A, X \in A'] = \sigma(A)\varepsilon_\sigma(A') \tag{12.4}$$

for A in $\mathscr{B}(G)$ and A' in $\mathscr{B}(B_r)$. The last two formulas give

$$\Pi^\lambda[X_0 \in A, X \in A'] = \int_{A'} \sigma(A)\delta_r(d\sigma) = \Theta_{\lambda,r}(A \times A'). \tag{12.5}$$

This gives us the probabilistic meaning of the measure $\Theta_{\lambda,r}$ on $G \times B_r$ defined by Lemma 7.1. Indeed, one gets

$$\Pi^\lambda[(X_0, X) \in C] = \Theta_{\lambda,r}(C), \qquad C \text{ in } \mathscr{B}(G \times B_r). \tag{12.6}$$

With the previous notations, let p_r be the canonical continuous map from B_r into the intrinsic boundary B of G, namely, $p_r(\lambda)$ is the ray generated by λ. If $(g_n)_{n \geqq 0}$ is a sequence of points of G and σ a point in B_r, the relation $\lim_{n\to\infty} k_r(g_n) = \sigma$ in $\mathscr{E}_r$ implies $\lim_{n\to\infty} g_n = p_r(\sigma)$ in $\hat{G}$. Define the random element X_∞ in $\hat{G}$ by $X_\infty = p_r(X)$. Using Theorem 12.1, the previous remark, and Lemma 7.2, one gets the following result immediately.

THEOREM 12.2. *Let λ be an invariant measure. There exists a random element X_∞ in the intrinsic boundary B defined over the sample space W such that the relation $\lim_{n\to\infty} X_n = X_\infty$ holds Π^λ almost surely in the space $\hat{G}$. If the measure λ generates the extreme ray x, then $X_\infty = x$ holds Π^λ almost surely. Furthermore, for any Borel subset C of $G \times B$, one gets*

$$\Pi^\lambda[(X_0, X_\infty) \in C] = \Theta_\lambda(C), \tag{12.7}$$

where the measure Θ_λ on $G \times B$ has been defined in Lemma 7.2.

We are now ready to prove our main theorem about the convergence to the boundary of the random walk of law μ on G.

THEOREM 12.3. *Let* $(Z_n)_{n \geqq 1}$ *be an independent sequence of* G *valued random elements with the common probability law* μ *and let* $S_0 = e$, $S_n = Z_1 \cdots Z_n$ *for* $n \geqq 1$. *There exists a random element* S_∞ *in the intrinsic boundary* B *of* G (with respect to μ) *such that* S_n *tends to* S_∞ *almost surely in the extended space* $\hat{G} = G \cup B$. *Moreover the probability law of* S_∞ *is the probability measure* γ *on* B *defined by Lemma* 7.3.

PROOF. Roughly speaking, the almost sure convergence of S_n to a point in B is obtained as follows. Take any random element Z in G with distribution a right invariant Haar measure m, independent from the process $(Z_n)_{n \geqq 1}$. Then the process $(ZS_n)_{n \geqq 0}$ has the distribution $\mathbf{P}^m = \Pi^m$ in the path space W. Hence by Theorem 12.2, it converges almost surely to the boundary B. By Fubini's theorem, for m almost any sample value g of Z, the process $(gS_n)_{n \geqq 0}$ converges almost surely to the boundary; since G acts by homeomorphisms upon $\hat{G}$, the process $(S_n)_{n \geqq 0}$ converges almost surely to the boundary B.

The previous argument is marred by some measurability difficulties; indeed, we don't know that $\hat{G}$ is a metrizable space. Hence, the measurability of the limit S_∞ is not ensured *a priori*. One could be tempted to work in $\mathscr{E}_r$ for some reference function r, but there the invariance under G is lost. We shall now repeat the previous reasoning taking more care of the measurability questions.

In the sample space Ω of the process $(Z_n)_{n \geqq 1}$, let us distinguish the part Ω_1 consisting of the sample points ω such that $S_n(\omega)$ converges in $\hat{G}$ to a point in B, to be denoted by $S_\infty(\omega)$. In the same way, W_1 is the set of paths w in W converging in $\hat{G}$ to a point in B, to be denoted by $X_\infty(w)$. There is a homeomorphism Φ of $G \times \Omega$ with W characterized by the following relation

$$X_n(\Phi(g, \omega)) = g \cdot S_n(\omega), \qquad n \geqq 0, g \text{ in } G, \omega \text{ in } \Omega. \tag{12.8}$$

Since G acts by homeomorphisms upon $\hat{G}$, one gets $W_1 = \Phi(G \times \Omega_1)$ and

$$X_\infty(\Phi(g, \omega)) = g \cdot S_\infty(\omega), \qquad g \text{ in } G, \omega \text{ in } \Omega_1. \tag{12.9}$$

By Theorem 12.2 with $\lambda = m$, there exists a Borel subset W_2 of W_1 such that $\Pi^m[W - W_2] = 0$ and that X_∞ induces a Borel measurable map from W_2 into B. Since Φ is a homeomorphism of $G \times \Omega$ with W transforming the measure $m \otimes \mathbf{P}$ into $\mathbf{P}^m = \Pi^m$, by Fubini's theorem, one gets

$$0 = \Pi^m[W - W_2] = \int_G \mathbf{P}[\Omega - \Omega_g] m(dg), \tag{12.10}$$

where Ω_g is the set of ω in Ω such that $\Phi(g, \omega) \in W_2$. Hence, there is at least a point g_0 such that $\mathbf{P}[\Omega - \Omega_{g_0}] = 0$. Thus, $\Omega^* = \Omega_{g_0}$ is a Borel subset of Ω such that $\mathbf{P}[\Omega^*] = 1$ and S_∞ is a Borel measurable map from Ω^* into B such that $\lim_{n \to \infty} S_n(\omega) = S_\infty(\omega)$ for any ω in Ω^*.

It remains to identify the probability law γ of S_∞ in B. Let us modify S_∞ by giving it some fixed value $b \in B$ in $\Omega - \Omega^*$. We modify X_∞ so that (12.9) remains valid. Let F in $b^+(G \times B)$ and $L = F(X_0, X_\infty)$. According to (12.7), one gets

$$\mathbf{E}^m[L] = \mathbf{H}^m[L] = \langle \Theta_m, F \rangle. \tag{12.11}$$

Moreover, from (12.9) and (12.8) one infers $L\Phi(g, \omega) = F(g, g \cdot S_\infty(\omega))$, and since Φ transforms $m \otimes \mathbf{P}$ into $\mathbf{P}^m$, one gets

$$\text{(12.12)} \quad \mathbf{E}^m[L] = \int_G \int_\Omega F(g, g \cdot S_\infty(\omega)) m(dg) \mathbf{P}(d\omega) = \int_G \int_B F(g, g \cdot x) m(dg) \gamma(dx).$$

Comparing (12.11) and (12.12) gives

$$\text{(12.13)} \qquad \langle \Theta_m, F \rangle = \int_G \int_B F(g, g \cdot x) m(dg) \gamma(dx)$$

for an arbitrary F in $b^+(G \times B)$, that is, $\Theta_m = \int_G (\varepsilon_g \otimes g \cdot \gamma) m(dg)$. Hence, γ has the characteristic property stated in Lemma 7.3. *Q.E.D.*

The study of *bounded* invariant functions (Section 15) involves only a part of the boundary B which we now describe.

DEFINITION 12.1. *Let r be a reference function such that $\langle m, r \rangle = 1$. Let μ_r be the image of $\Pi^{m,r} = \mathbf{P}^{r \cdot m}$ by X_∞. We call the active part N of the boundary B the (closed) support in B of the probability measure μ_r. The space N does not depend on r and is invariant by G.*

If r and r' are reference functions such that $\langle m, r \rangle = \langle m, r' \rangle = 1$, we see by (3.1) that the probability measures $\mathbf{P}^{r \cdot m}$ and $\mathbf{P}^{r' \cdot m}$ are equivalent. Hence, μ_r and $\mu_{r'}$ are equivalent, and consequently have the same support. It is obvious, by Theorem 12.1, that μ_r is the measure occuring in (7.11); the proof of Lemma 7.3 shows then that $g\mu_r = \mu_s$, where s is another reference function such that $\langle m, s \rangle = 1$. We then have $\mu_r \sim \mu_s \sim g\mu_r$. The measure μ_r is hence *quasi-invariant* (equivalent to its translates by elements of G) and *a fortiori*, its support N is invariant by G.

To justify the terminology "active part", we note that the limit $X_\infty = \lim_{n \to \infty} X_n$ exists and belongs to the active part N of the boundary, $\mathbf{P}^g$ a.s., for *each* g in G. The proof is completely similar to the proof of Theorem 12.3; we simply have to modify the definitions of Ω_1 and W_1: namely, Ω_1 is the set of ω in Ω such that $S_n(\omega)$ converges in $\hat{G}$ to a point in the active part N of B. A similar definition is used for W_1. Since N is invariant by G, we still have $\Phi(G \times \Omega_1) = W_1$. We thus obtain $\mathbf{P}[S_\infty = \lim_{n \to \infty} S_n$ and S_∞ in $N] = 1$, and using (12.8) and $g \cdot N = N$, we get $\mathbf{P}^g[X_\infty = \lim_{n \to \infty} X_n$ and X_∞ in $N] = 1$, for each g in G.

The interest of the notion of the active part of the boundary lies essentially in the fact that in many cases (see Section 17), N can be determined completely, while the boundary B remains unknown.

13. Additional remarks

13.1. From Theorem 12.2, one deduces that any point in B is the limit in $\hat{G}$ of some sequence of points in G. This could be proved directly by a purely analytical argument. Indeed, from the fact that any excessive measure is the limit of an increasing sequence of potentials (Theorem 4.2), one infers easily that $\mathscr{E}_r$ is the closed convex hull of $k_r(G)$ whatever the reference function r is.

By a classical result, any extreme point of $\mathscr{E}_r$ is of the form $\lambda = \lim_{n\to\infty} k_r(g_n)$. Hence, $x = \lim_{n\to\infty} g_n$ in $\hat{G}$ for the ray x generated by λ. Since any point in B is the ray generated by an extreme point of $\mathscr{E}_r$ for some suitable reference function r, we are through.

13.2. Let r be a *generalized* reference function (see Section 8). One shows easily that $\langle\lambda, r\rangle > 0$ for any excessive measure $\lambda \neq 0$, that the set $\mathscr{E}_r$ of excessive measures λ with $\langle\lambda, r\rangle \leqq 1$ is vaguely compact and that the measures Π^λ and $\Pi^{\lambda,r}$ have the same null sets when $\langle\lambda, r\rangle$ is finite. Using these remarks, one checks that the proofs of Theorem 6.1, Lemmas 7.1 to 7.3, Theorems 11.1 and 12.1 remain valid when r is a generalized reference function.

We have refrained from making this generalization because we feel that the reference functions and the compact spaces $\mathscr{E}_r$ are only auxiliary tools and that the ultimate concern is with the intrinsic boundary B. The most interesting generalized reference functions are those with compact support; but, if they exist at all, the convex cone $\mathscr{E}$ has a compact basis and it is better to work directly with the Martin compactification G^* of Section 8.2 without having recourse to the reference functions.

Part D. Bounded Invariant Functions

In this part, we give an integral representation of the bounded invariant functions analogous to the representation obtained by Furstenberg [27] and we use it to compare the Poisson space of μ to the intrinsic boundary of G.

14. Invariant functions

We have seen in Section 4 that if an excessive function f is locally m integrable, the measure $f\cdot m$ is excessive. This is particularly interesting in the case when μ is spread out since we have the following lemma.

Lemma 14.1. *Assume that μ is spread out on G. An excessive function f is locally m integrable if and only if f is m almost everywhere finite.*

Proof. Let f be an excessive function finite on the complement of an m null set A. Since μ is spread out there exists a nonempty open subset V of G, an integer n, and a real number $c > 0$ such that μ^n majorizes $c\cdot m$ on V. Let h be in G; since $m(A) = 0$, there is a g such that $g \in hV^{-1}$ and g is not in A. We have

$$\infty > f(g) \geqq \langle g\mu^n, f\rangle \geqq c\langle gm, I_{gV}\cdot f\rangle \geqq c\Delta(g)\langle m, I_{gV}\cdot f\rangle. \tag{14.1}$$

Since gV is a neighborhood of h and h is arbitrary, f is locally m integrable. The converse is obvious. *Q.E.D.*

When the support of μ is contained in no proper closed subgroup of G, an invariant function is in $L_2(G)$ only if it is m a.e. constant, and hence m a.e. zero when G is not compact (see U. Grenander, *Probabilities of Algebraic Structures*, p. 58).

Now let f be an invariant function in $L_1(G)$. The measure $f\cdot m = \lambda$ is an invariant bounded measure; for any $f' \in C_c^+(G)$, the function f_1 defined by

$f_1(g) = \langle g\lambda, f' \rangle$ is a continuous bounded invariant function. This remark is used (in [3] and [28]) to deduce a representation of the bounded invariant measures from the integral representation of the bounded invariant functions. We recall ([3], Proposition 1.6) that *when μ is spread out, the bounded invariant functions are continuous*. We call H the Banach space of all bounded invariant functions (with the norm of uniform convergence).

15. Integral representation of bounded invariant functions

We assume in this section that μ is spread out. Let r be a reference function such that $\langle m, r \rangle = 1$, let N be the active part of the intrinsic boundary B, and let μ_r be the quasi-invariant measure on N, image of $\mathbf{P}^{r\cdot m}$ by X_∞ (Definition 12.1). Note that the null sets of μ_r are independent of r, so that the Banach space $L_\infty(N, \mu_r)$ does not depend on r.

THEOREM 15.1. *Assume that μ is spread out. There exists an isometry $f \mapsto \hat{f}$ from the Banach space H of bounded invariant functions onto $L_\infty(N, \mu_r)$ such that*

$$\hat{f}(X_\infty) = \lim_{n\to\infty} f(X_n), \qquad \mathbf{P}^{r\cdot m} \text{ a.s.,} \tag{15.1}$$

and

$$f(g) = \langle g\gamma, \hat{f} \rangle, \qquad g \text{ in } G, \tag{15.2}$$

where γ is the probability measure on N occuring in Theorem 12.3 *and Lemma* 7.3.

PROOF. We recall the notation of Sections 6 and 7; B_r is the set of the extreme invariant measures σ such that $\langle \sigma, r \rangle = 1$; B_r' is the corresponding Borel subset of rays in B, and the natural map $p_r: B_r \mapsto B_r'$ is a Borel isomorphism. By Theorem 6.1, there is a unique probability measure δ_r such that

$$m = \int_{B_r} \sigma \cdot \delta_r(d\sigma), \tag{15.3}$$

and, taking account of (7.11), we have $p_r(\delta_r) = \mu_r$.

Let f be a bounded invariant function; let λ be the invariant measure $f\cdot m$; assume first $f \geqq 0$. Since $\langle \lambda, r \rangle$ is finite, by Theorem 6.1 there is a unique bounded measure β_r on B_r such that $\lambda = \int_{B_r} \sigma \cdot \beta_r(d\sigma)$. This result is readily extended to the case when f is not positive by writing $f = (f + \|f\|) - \|f\|$. The measure $\|f\|m - \lambda$ is a positive invariant measure. Hence, by Theorem 6.1, $\|f\|\delta_r - \beta_r$ is a positive measure. There is then a function f_r in $L_\infty(B_r, \delta_r)$ such that $\beta_r = f_r \cdot \delta_r$ and $\|f_r\|_\infty \leqq \|f\|$. On the other hand,

$$\begin{aligned} \lambda = f\cdot m &= \int_{B_r} \sigma f_r(\sigma)\, \delta_r(d\sigma) \\ &\leqq \|f_r\|_\infty \int_{B_r} \sigma \cdot \delta_r(d\sigma) = \|f_r\|_\infty \cdot m. \end{aligned} \tag{15.4}$$

Since f is continuous, (15.4) implies $\|f\| \leqq \|f_r\|_\infty$, and finally $\|f\| = \|f_r\|_\infty$. The map $f \mapsto f_r$ clearly defines an isometry from H onto $L_\infty(B_r, \delta_r)$. Since $L_\infty(B_r, \delta_r)$ is isometric to $L_\infty(N, \mu_r)$ by the map $f_r \mapsto f'_r = f_r \circ p_r^{-1}$, we have an isometry $f \mapsto f'_r$ from H onto $L_\infty(N, \mu_r)$. We shall see, in fact, that the equivalence class of f'_r in $L_\infty(N, \mu_r)$ does not depend on r.

As in Section 12, define $X\colon W \mapsto B_r$ by $X = \lim_{n\to\infty} k_r(X_n)$ if the limit exists in $\mathscr{E}_r$, and X arbitrary elsewhere. For any function F in $b^+(W)$, we have by (12.2) and (15.3),

$$\mathbf{H}^{m,r}[F] = \int_{B_r} \delta_r(d\sigma)\mathbf{H}^{\sigma,r}[F]. \tag{15.5}$$

Let h be a function in $b^+(G)$; applying (15.5), we get

$$\mathbf{H}^{m,r}[h(X_n)f_r(X)] = \int_{B_r} \delta_r(d\sigma)\mathbf{H}^{\sigma,r}[h(X_n)f_r(X)]. \tag{15.6}$$

Since, by Theorem 12.1, $\Pi^{\sigma,r}[X = \sigma] = 1$ for σ in B_r, (15.6) becomes

$$\mathbf{H}^{m,r}[h(X_n)f_r(X)] = \int_{B_r} \delta_r(d\sigma)f_r(\sigma)\mathbf{H}^{\sigma,r}[h(X_n)]. \tag{15.7}$$

From (9.6), we obtain

$$\mathbf{H}^{\sigma,r}[h(X_n)] = \langle\sigma, h\cdot\bar{Q}^n r\rangle. \tag{15.8}$$

Using (15.8) and (15.4), we transform (15.7) into

$$\mathbf{H}^{m,r}[h(X_n)f_r(X)] = \int_G m(dg)h(g)\bar{Q}^n r(g)f(g). \tag{15.9}$$

In the particular case $f = 1$ (and hence $f_r = 1$), (15.9) yields

$$\mathbf{H}^{m,r}[h(X_n)] = \int_G m(dg)h(g)\bar{Q}^n r(g), \tag{15.10}$$

which shows that the distribution of X_n for the law $\Pi^{m,r}$ is $\bar{Q}^n r\cdot m$. We can then rewrite (15.9) as

$$\mathbf{H}^{m,r}[f_r(X)\,|\,X_n] = f(X_n), \qquad \Pi^{m,r} \text{ a.s.} \tag{15.11}$$

The left side is a bounded martingale and $f_r(X)$ is measurable with respect to the σ-algebra generated by the X_n. Hence, we have

$$\lim_{n\to\infty} \mathbf{H}^{m,r}[f_r(X)\,|\,X_n] = f_r(X), \qquad \Pi^{m,r} \text{ a.s.,} \tag{15.12}$$

which combined with (15.11), implies

$$f_r(X) = \lim_{n\to\infty} f(X_n), \qquad \Pi^{m,r} \text{ a.s.} \tag{15.13}$$

The continuity of p_r and Theorems 12.1 and 12.2 show that

$$p_r(X) = X_\infty, \qquad \Pi^m \text{ a.s.,} \tag{15.14}$$

for any reference function r.

Let s be another reference function such that $\langle m, s\rangle = 1$. Since Π^m, $\Pi^{m,r}$, and $\Pi^{m,s}$ have the same null sets, equations (15.14) and (15.13) imply

$$f'_r(X_\infty) = f'_s(X_\infty), \qquad \Pi^{m,r} \text{ a.s.,} \tag{15.15}$$

(taking account of the definitions $f'_r = f_r \circ p_r^{-1}$ and $f'_s = f_s \circ p_s^{-1}$). The image of $\Pi^{m,r} = \mathbf{P}^{r\cdot m}$ by X_∞ is μ_r. Hence, f'_r and f'_s define the same element of $L_\infty(N, \mu_r)$. We now call $\hat{f}$ the equivalence class (independent of r) of f'_r in $L_\infty(N, \mu_r)$; the equality (15.13) readily implies (15.1), since $f_r(X) = \hat{f}(X_\infty)$, $\Pi^{m,r}$ a.s. Taking account of (3.1), (15.1) implies

$$\hat{f}(X_\infty) = \lim_{n\to\infty} f(X_n), \qquad \mathbf{P}^g \text{ a.s.,} \tag{15.16}$$

for m almost every g in G. Since f is bounded, (15.16) gives

$$\lim_{n\to\infty} \mathbf{E}^g[f(X_n)] = \mathbf{E}^g[\hat{f}(X_\infty)], \qquad m \text{ a.e. } g \text{ in } G. \tag{15.17}$$

We have, since f is invariant,

$$f(g) = Q^n f(g) = \mathbf{E}^g[f(X_n)], \qquad g \text{ in } G. \tag{15.18}$$

From Theorem 12.3, we see that the image of $\mathbf{P}^e$ by X_∞ is the probability measure γ; this shows, by (12.9) and (12.8) that the image of $\mathbf{P}^g$ by X_∞ is $g\gamma$. We can now deduce, from (15.17) and (15.18),

$$f(g) = \langle g\gamma, \hat{f}\rangle, \qquad m \text{ a.e. } g \text{ in } G. \tag{15.19}$$

The definition of X_∞ shows that $F = \hat{f}(X_\infty)$ is shift invariant (see (3.4)), that is, $\theta F = F$. If we define

$$h(g) = \langle g\gamma, \hat{f}\rangle = \mathbf{E}^g[F], \tag{15.20}$$

using the Markov property, we get

$$Qh(g) = \mathbf{E}^g[h(X_1)] = \mathbf{E}^g[\mathbf{E}^{X_1}[F]] = \mathbf{E}^g[\mathbf{E}^g[\theta F \mid X_1]] = \mathbf{E}^g[\theta F] = h(g). \tag{15.21}$$

Since h is bounded and invariant, it is continuous (see Section 14); since f has the same properties, we see that (15.19) implies (15.2). *Q.E.D.*

We now study formula (15.2) in more detail.

PROPOSITION 15.1. *Assume that μ is spread out. There is a Borel positive function u_r on $G \times N$ such that*

$$\frac{d(g\gamma)}{d\mu_r}(x) = u_r(g, x) \qquad g \text{ in } G, x \text{ in } N. \tag{15.22}$$

For μ_r almost every x, the measure $u_r(\cdot, x)\cdot m$ is extreme and invariant. The measure δ_r on B_r such that $m = \int_{B_r} \sigma\cdot\delta_r(d\sigma)$ is carried by the set of extreme invariant measures σ such that $\sigma \ll m$. Any bounded invariant function f has the representation

$$f(g) = \int_N u_r(g, x)\hat{f}(x)\, \mu_r(dx), \qquad g \text{ in } G; \tag{15.23}$$

moreover, if $\mu \ll m$, u_r *can be chosen such that* $u_r(\cdot\,, x)$ *is an invariant function for* μ_r *almost every* x.

PROOF. Let $\hat{f}$ be any bounded Borel function on N and f the corresponding invariant function. From (15.2), we see that $f = 0$ if and only if for each g in G, $\hat{f} = 0$, $g\gamma$ a.e. By (15.4), $f = 0$ if and only if $f_r = 0$, δ_r a.e., that is to say, if and only if $\hat{f} = 0$, μ_r a.e. Hence, $\hat{f} = 0$, μ_r a.e. if and only if $\hat{f} = 0$, $g\gamma$ a.e. In particular, $g\gamma \ll \mu_r$ for each g in G. Since μ_r is carried by a countable union of metrizable compact sets (see Section 8), there is a positive Borel function u_r on $G \times N$ such that (15.22) is satisfied. Equation (15.23) is then an immediate consequence of (15.2). We have $Qf = f$, which implies, by (15.23) and by the fact that $\hat{f}$ is arbitrary in $L_\infty(N, \mu_r)$,

$$\int \mu(dh) u_r(gh, x) = u_r(g, x), \qquad \mu_r \text{ a.e. } x, \tag{15.24}$$

for each g in G. Applying Fubini's theorem to the set of pairs (g, x) in $G \times N$ for which (15.24) holds, we see that for μ_r a.e. x the function $u_r(\cdot\,, x)$ satisfies

$$Qu_r(\cdot\,, x) = u_r(\cdot\,, x), \qquad m \text{ a.e.} \tag{15.25}$$

Equation (4.4) shows then by duality that the measure $u_r(\cdot, x)\cdot m$ on G is invariant for μ_r a.e. x. Note that if $\mu \ll m$, (15.25) implies $Q^2 u_r(\cdot\,, x) = Qu_r(\cdot\,, x)$ everywhere, so that we can replace $u_r(\cdot\,, x)$ by the invariant function $Qu_r(\cdot\,, x)$. Coming back to the general case, call $\varphi(x)$ the invariant measure $u_r(\cdot\,, x)\cdot m$. We get from (15.23)

$$f\cdot m = \int_N \varphi(x)\hat{f}(x)\mu_r(dx). \tag{15.26}$$

Since $\mu_r = p_r(\delta_r)$ and $\hat{f} = f_r \circ p_r^{-1}$, we have

$$f\cdot m = \int_{B_r} \varphi \circ p_r(\sigma) f_r(\sigma)\delta_r(d\sigma). \tag{15.27}$$

Comparing with (15.4), we see that since f_r is arbitrary, we must have $\varphi \circ p_r(\sigma) = \sigma$, for δ_r a.e. σ. Hence, we have $\sigma \ll m$ for δ_r a.e. σ and $\varphi(x)$ is an extreme invariant measure for μ_r a.e. x. *Q.E.D.*

16. A special type of reference function

To prove the main result of this section, we shall have to use compact caps $\mathscr{E}_r$ of the cone of excessive measures, such that the subcone of $\mathscr{E}$ generated by $\mathscr{E}_r$ is stable by G, and on which G acts with some sort of uniformity. The appropriate reference functions are constructed below.

PROPOSITION 16.1. *For any probability measure μ on G (transient case) there is a reference function r such that*

$$\lim_{g \to e} \frac{r(gh) - r(h)}{r(h)} = 0 \tag{16.1}$$

uniformly in h and such that $\langle m, r \rangle = 1$.

We shall need a lemma.

LEMMA 16.1. *Let Δ be the modular function of G. There is a bounded Radon measure η on G, an open subgroup G_0 of G, and a finite positive function ε on G_0 such that*

(i) *$\varepsilon(g)$ tends to 0 when g tends to e,*
(ii) *$\langle \eta, \Delta \rangle = 1$,*
(iii) *$|\langle \eta g, f \rangle - \langle \eta, f \rangle| \leqq \varepsilon(g) \langle \eta, f \rangle$,*

for any g in G_0, f in $b^+(G)$.

PROOF. Assume first that G is a Lie group having a finite number of connected components. Let d be a left invariant distance on G. It is known ([30], p. 75) that there is a positive number k such that for $b > k$ the function $\Delta(g) \exp\{-bd(e, g)\}$ is in $L_1(G, m)$, and such that $\Delta(g) \leqq k \exp\{kd(e, g)\}$ for g in G. Define the measure η on G by $\eta(dg) = C \exp\{-bd(e, g)\} \Delta(g) m(dg)$. For b large enough, η is bounded, and $\langle \eta, \Delta \rangle$ is finite; we then choose C such that $\langle \eta, \Delta \rangle = 1$. The proof of (iii) is readily deduced from the following elementary inequality: for g in G and h in G,

$$|\exp\{-bd(e, hg)\} - \exp\{-bd(e, h)\}| \leqq \varepsilon(g) \exp\{-bd(e, h)\}, \tag{16.2}$$

where $\varepsilon(g) = \exp\{bd(e, g)\} - 1$.

Assume now that the quotient of G by its connected component is compact. There exists ([43], pp. 153 and 175) a normal compact subgroup K of G such that $G_1 = G/K$ is a Lie group with a finite number of connected components. Let m_K be the normed Haar measure on K; for each f in $C_c^+(G)$, the function $\bar{f}(g) = \langle m_K g, f \rangle$ can be considered as a function on G_1. Let η_1 be a measure on G_1 satisfying (i), (ii), (iii). Define η on G by $\langle \eta, f \rangle = \langle \eta_1, \bar{f} \rangle$. It is readily checked that η satisfies (i), (ii), (iii) with $G_0 = G$.

Finally, in the general case, G contains an open subgroup G_1 such that the quotient of G_1 by its connected component is compact ([43], pp. 153 and 175). Write G as a disjoint union $G = \bigcup_{n \geqq 0} g_n G_1$. Let η_1 be a measure on G_1 satisfying (i), (ii), (iii). Define η on G by $\eta = \Sigma_{n \geqq 0} c_n g_n \eta_1$, where $\Sigma_{n \geqq 0} c_n < \infty$ and $\Sigma_{n \geqq 0} c_n \Delta(g_n) = 1$, which implies (ii). Then, one checks (iii) with $G_0 = G_1$. *Q.E.D.*

PROOF OF PROPOSITION 16.1. The proof of Lemma 6.1 shows the existence of a bounded reference function s such that $\langle m, s \rangle = 1$ and such that $\bar{U}s$ is bounded. Let η be the measure on G obtained in Lemma 16.1. Define r by

$$r(g) = \langle \eta g, s \rangle, \qquad g \text{ in } G. \tag{16.3}$$

We have $\bar{U}r(g) = \langle \eta g, \bar{U}s\rangle$, since $\bar{U}$ commutes with left translations on G. Since s and $\bar{U}s$ are both continuous, bounded and positive, it is clear that r and $\bar{U}r$ have the same properties. Hence, r is a reference function. We have

$$\langle m, r\rangle = \int \eta(dg)\langle gm, s\rangle = \langle \eta, \Delta\rangle = 1. \tag{16.4}$$

For g in G_0, h in G, we have, by (16.3),

$$|r(gh) - r(h)| = |\langle \eta g, R_h s\rangle - \langle \eta, R_h s\rangle|, \tag{16.5}$$

where $R_h s(g) = s(gh)$. Applying inequality (iii) from Lemma 16.1, we get $|r(gh) - r(h)| \leqq \varepsilon(g) r(h)$, for g in G_0 and h in G and the proof is completed.

The following result describes the action of G on B_r when r satisfies (16.1).

PROPOSITION 16.2. *Let r be a reference function satisfying* (16.1) *and such that* $\langle m, r\rangle = 1$. *Some open subgroup G_0 of G acts then on $\mathscr{E}_r - \{0\}$ by*

$$T_g(\sigma) = \frac{\langle \sigma, r\rangle}{\langle g\sigma, r\rangle} g\sigma, \tag{16.6}$$

and we have

$$\lim_{g\to e, g\in G_0} T_g(\sigma) = \sigma, \tag{16.7}$$

uniformly for σ in B_r. There is a Borel subset Σ of B_r such that the whole group G acts continuously on Σ by (16.6) *and such that* $\delta_r(\Sigma) = 1$, *where δ_r is the measure on B_r such that* $m = \int_{B_r} \sigma \cdot \delta_r(d\sigma)$.

PROOF. For σ in $\mathscr{E}_r$, define the function F_σ on G by

$$F_\sigma(g) = \langle g\sigma, r\rangle, \qquad g \text{ in } G. \tag{16.8}$$

By the proofs of Proposition 16.1, there exists an open subgroup G_0 of G such that if $F_\sigma(g)$ is finite for some g in G, F_σ is finite on $G_0 g$; moreover, (16.1) shows then that F_σ is continuous at g. We thus see that on each coset $G_0 g$, the function F_σ is either finite and continuous or identically infinite. This has two pleasant consequences used below:

(a) if F_σ is m a.e. finite, F_σ is everywhere finite and continuous;

(b) if we choose a sequence (g_n) such that G is a disjoint union of the sets $G_0 g_n$, we see that F_σ is everywhere finite and continuous if and only if $F_\sigma(g_n)$ is finite for all n.

Also, since σ is in $\mathscr{E}_r$, $F_\sigma(e) = 1$, and all the F_σ are finite and continuous on G_0. It is then easy to check that (16.6) defines a continuous action of G_0 on $\mathscr{E}_r - \{0\}$, which obviously leaves globally invariant the set B_r of extreme invariant points of $\mathscr{E}_r - \{0\}$.

Formula (16.1) implies readily that the (F_σ), σ in $\mathscr{E}_r - \{0\}$, are equicontinuous at e, and it is then trivial to check (16.7)—where the restriction σ in B_r is essential.

Let $\dot{\Sigma}$ be the set of all σ in B_r such that F_σ is everywhere finite and continuous. Conclusion (b) above shows clearly that Σ is a Borel set. From $\langle gm, r\rangle = \Delta(g)$

and $gm = \int_{B_r} g\sigma \cdot \delta_r(d\sigma)$ (a consequence of (15.3)), we obtain that, for each g in G, $\langle g\sigma, r\rangle$ is finite δ_r a.e. By Fubini's theorem, this implies that, for δ_r a.e. σ in B_r, the function F_σ is m a.e. finite on G. Using conclusion (a) above, we then see that $\delta_r(\Sigma) = 1$. It is then immediate that (16.6) defines a continuous action of G on Σ. *Q.E.D.*

17. Intrinsic boundary and Poisson space

In this section, *we assume that μ is spread out*. As before, we call H the Banach space of bounded invariant functions. Let H_u be the closed subspace of H consisting of those f in H which are *left uniformly continuous* on G. A construction due to Furstenberg ([27], [3]) shows the existence of a compact G space Π (depending on μ), a probability measure ν on Π, and an isometry $f \mapsto \bar{f}$ from H_u onto $C(\Pi)$ (space of continuous functions on Π) such that

$$f(g) = \langle g\nu, \bar{f}\rangle, \qquad g \text{ in } G; \tag{17.1}$$

Π and ν are called Poisson space and Poisson kernel of μ, respectively. They have been studied extensively in [27] and [3] and in a large number of cases Π is known explicitely. We are going to compare Π and the intrinsic boundary.

We recall briefly the construction of Π (see [3]). For every Borel function f on G, we define a measurable function $F = t(f)$ on the sample space W by

$$F(W) = \begin{cases} \lim\limits_{n\to\infty} f(X_n(w)) & \text{if the limit exists,} \\ 0 & \text{elsewhere.} \end{cases} \tag{17.2}$$

Two functions on W are considered as equivalent if and only if they are $\mathbf{P}^g$ a.s. equal, for each g in G; we call $\bar{t}(f)$ the equivalence class of $t(f)$. The set $\{\bar{t}(f) \mid f \text{ in } H_u\}$ is a C^* algebra A for the norm

$$\|\bar{t}(f)\| = \sup_{g\in G} \|t(f)\|_{L_\infty(W,\mathbf{P}^g)} \tag{17.3}$$

The map $\bar{t}$ is an isometry from H_u onto A; the Poisson space Π is the spectrum of A.

THEOREM 17.1. *Let μ be a probability measure on the locally compact separable group G, and let Π and ν be the Poisson space and Poisson kernel of μ. Assume that μ is spread out, and that the random walk of law μ is transient. Let N be the active part of the intrinsic boundary B of G, and γ the measure on N occurring in Theorem* 12.3. *Then there is a Borel subset Π_1 of Π, invariant by G, such that $\nu(\Pi_1) = 1$, and a continuous map ψ from Π_1 to N, commuting with the action of G, such that $\psi(\nu) = \gamma$. If, moreover, Π is a homogeneous space of G, the map ψ is in fact a homeomorphism from Π onto N, commuting with the action of G.*

PROOF. Let r be a reference function *as in Proposition* 16.1. (The notation is that of Section 15.) Let q be the isometry from $L_\infty(N, \mu_r)$ onto H such that

$$f(g) = q(\hat{f})(g) = \langle g\gamma, \hat{f}\rangle, \qquad g \text{ in } G, \hat{f} \text{ in } L_\infty(N, \mu_r). \tag{17.4}$$

Let M be the compact support of δ_r in $\mathscr{E}_r$. Since $\delta_r(B_r) = 1$, the space $C(M)$ of continuous functions on M is naturally identified to a Banach subalgebra of $L_\infty(B_r, \delta_r)$, which we denote by $C_1(M)$. Any function f_r in $C_1(M)$ is clearly uniformly continuous on B_r. Taking account of (16.7), we see that

$$\lim_{g \to e, g \in G_0} \|f_r \circ T_g - f_r\| = 0, \qquad f_r \text{ in } C_1(M). \tag{17.5}$$

The isometry s from $L_\infty(B_r, \delta_r)$ onto $L_\infty(N, \mu_r)$ defined by $\hat{f} = s(f_r) = f_r \circ p_r^{-1}$ maps $C_1(M)$ onto a Banach subalgebra $C_2(M)$ of $L_\infty(N, \mu_r)$, and we rewrite (17.5) as

$$\lim_{g \to e, g \in G_0} \|\hat{f}(g\cdot) - \hat{f}(\cdot)\| = 0, \qquad \hat{f} \text{ in } C_2(M). \tag{17.6}$$

By (17.4), the invariant function corresponding to $\hat{f}(g\cdot)$ is $f(g\cdot)$. Since q is an isometry, $\|f(g\cdot) - f(\cdot)\| = \|\hat{f}(g\cdot) - \hat{f}(\cdot)\|$ and (17.6) shows that if $\hat{f}$ is in $C_2(M)$, f is left uniformly continuous. Hence, q maps $C_2(M)$ into H_u.

From (15.1), we get

$$\hat{f}(X_\infty) = \lim_{n \to \infty} f(X_n) = t \circ q(\hat{f}), \qquad \mathbf{P}^{r \cdot m} \text{ a.s.} \tag{17.7}$$

Let $F_i = \bar{t} \circ q(\hat{f}_i)$ for $i = 1, 2$, with $\hat{f}_i$ in $C_2(M)$. Since $\bar{t}$ is an isometry from H_u onto A and since A is an algebra, there is an f in H_u such that $F_1 F_2 = \bar{t}(f)$. We then have in A

$$\bar{t} \circ q(\hat{f}) = \bar{t}(f) = F_1 F_2 = \bar{t} \circ q(\hat{f}_1) \cdot \bar{t} \circ q(\hat{f}_2), \tag{17.8}$$

which by definition, is equivalent to

$$t \circ q(\hat{f}) = t \circ q(\hat{f}_1) \cdot t \circ q(\hat{f}_2), \tag{17.9}$$

$\mathbf{P}^g$ a.s., for each g in G.

Combining equations (17.7) and (17.9), we get

$$\hat{f}(X_\infty) = \hat{f}_1(X_\infty) \hat{f}_2(X_\infty), \qquad \mathbf{P}^{r \cdot m} \text{ a.s.} \tag{17.10}$$

The image $\mathbf{P}^{r \cdot m}$ by X_∞ being μ_r, we see that $\hat{f} = \hat{f}_1 \hat{f}_2$ in $L_\infty(N, \mu_r)$. But (17.8) now becomes

$$\bar{t} \circ q(\hat{f}_1 \hat{f}_2) = \bar{t} \circ q(\hat{f}_1) \cdot \bar{t} \circ q(\hat{f}_2), \qquad \hat{f}_1, \hat{f}_2 \text{ in } C_2(M). \tag{17.11}$$

Hence, the isometry $\bar{t} \circ q$ is a homomorphism of algebras from $C_2(M)$ into A, such that $\bar{t} \circ q(1) = 1$. Since $C(M) \mapsto C_1(M)$ and $C_1(M) \mapsto C_2(M)$ are isomorphisms of Banach algebras (preserving 1), we have a homomorphism of $C(M)$ into A (preserving 1). By duality, we obtain a continuous mapping Φ from the spectrum Π of A onto M.

Let $\hat{f}$ be a function in $C_1(M)$ and f the corresponding bounded invariant function (left uniformly continuous); the function f_r on B_r is the restriction to B_r of a continuous function f_r' on M; define $f_r' \circ \Phi = \bar{f}$. For each g in G, we have

$$f(g) = \langle gv, \bar{f} \rangle = \langle \Phi(gv), f_r' \rangle. \tag{17.12}$$

But we also have, by (15.2) and $\hat{f} = f_r \circ p_r^{-1}$,

$$f(g) = \langle g\gamma, \hat{f} \rangle = \langle p_r^{-1}(g\gamma), f_r \rangle. \tag{17.13}$$

Finally, for each g in G,

$$\langle \Phi(g\nu), f_r' \rangle = \langle I_{B_r} \cdot p_r^{-1}(g\gamma), f_r' \rangle. \tag{17.14}$$

Since the function f_r' is arbitrary in $C(M)$, we conclude that

$$\Phi(g\nu) = I_{B_r} \cdot p_r^{-1}(g\gamma) \tag{17.15}$$

for each g in G.

Let Σ be the subset of B_r defined in Proposition 16.2. We have seen in the proof of Proposition 15.1 that $g\gamma \ll \mu_r$ for each g in G. Since $\delta_r(\Sigma) = 1$ and $\delta_r = p_r^{-1}(\mu_r)$, we see that Σ has measure one for $p_r^{-1}(g\gamma)$, for each g in G. Hence, by (17.15), we have

$$\Phi(g\nu) = I_\Sigma \cdot p_r^{-1}(g\gamma) \tag{17.16}$$

for each g in G.

Define $\Pi' = \Phi^{-1}(\Sigma)$. From (17.16), we deduce $g\nu(\Pi') = 1$ for each g in G. The map $\psi = p_r \circ \Phi$ is obviously defined and continuous on Π' and maps Π' into the active part N of the boundary. We then rewrite (17.16) as

$$\psi(g\nu) = g\gamma \tag{17.17}$$

for each g in G.

We now show that ψ commutes with the action of G (which is not obvious since $C(M)$ is not invariant by G *a priori*). Let x in Π' and g in G be such that gx is in Π'. It is known ([3], p. 13) that the measure ν on Π is "contractile," that is, any point mass on Π belongs to the vague closure of the set $(h\nu)_{h \in G}$. Let $(h_i)_{i \in I}$ be a net in G such that $\lim_I (h_i\nu) = \varepsilon_x$ (point mass at x), which implies since Π is a G space, $\lim_I (gh_i\nu) = \varepsilon_{gx}$. Assume that $g\psi(x)$ and $\psi(gx)$ are distinct points of N. Since N is Hausdorff and since G acts continuously on N, we can find in N neighborhoods A of $\psi(x)$ and B of $\psi(gx)$ such that gA and B are disjoint. Then $\psi^{-1}(A)$ is a neighborhood of x in Π'. Hence, $\psi^{-1}(A) = C \cap \Pi'$, where C is a neighborhood of x in Π. The vague convergence of $(h_i\nu)$ to ε_x in the compact space Π implies the existence of i_0 in I such that for $i > i_0$, $h_i\nu(C) > \frac{2}{3}$. Since $h_i\nu(\Pi') = 1$, we then have $h_i\nu[\psi^{-1}(A)] > \frac{2}{3}$ for $i > i_0$, that is, by (17.17), $h_i\gamma(A) > \frac{2}{3}$. Similarly, we find i_1 in I such that for $i > i_1$, $gh_i\gamma(B) > \frac{2}{3}$. Choosing i larger than i_0 and i_1, we have

$$1 \geqq gh_i\gamma(gA) + gh_i\gamma(B) = h_i\gamma(A) + gh_i\gamma(B), \tag{17.18}$$

an obvious contradiction since the last term is larger than $\frac{4}{3}$. We have proved

$$\psi(gx) = g\psi(x) \qquad x, gx \text{ in } \Pi'. \tag{17.19}$$

Let (g_n) be a dense sequence in G, containing the identity of G, and let $\Pi_1 = \bigcap_n g_n^{-1}\Pi'$. Since $g\nu(\Pi') = 1$ for any g in G, we have $g\nu(\Pi_1) = 1$ for each g in G, and by (17.19) we have

$$\psi(g_n x) = g_n \psi(x) \tag{17.20}$$

for any n, for x in Π_1. By (16.6), we see that $p_r(T_g \sigma) = g p_r(\sigma)$ for any σ in B_r. Hence, (17.20) implies

$$\Phi(g_n x) = T_{g_n}\Phi(x) \tag{17.21}$$

for any n, for x in Π_1. But x in Π_1 implies $\Phi(x)$ in Σ, which by Proposition 16.2 implies that $T_g\Phi(x)$ is a continuous function of g, for any fixed x in Π_1. Obviously, $\Phi(gx)$ has the same property, and from (17.21) we get

$$\Phi(gx) = T_g\Phi(x) \tag{17.22}$$

for any g, any x in Π_1. Since $T_g(B_r) = B_r$, we see that $G\Pi_1$ is included in Π'. *A fortiori* $G\Pi_1$ is included in $g_n^{-1}\Pi'$ for any g_n, which implies $G\Pi_1 = \Pi_1$. By composition with p_r, from (17.22) we can now deduce

$$\psi(gx) = g\psi(x) \tag{17.23}$$

for any g in G, any x in Π_1.

We now assume that Π is a homogeneous space of G. Then we obviously have $\Pi_1 = \Pi$. Hence, $\psi(\Pi)$ is a compact subset of N; on the other hand, $\psi(\Pi) = p_r(\Phi(\Pi_1)) = p_r(\Sigma)$ so that $\mu_r[\psi(\Pi)] = \delta_r(\Sigma) = 1$. But N is the closed support of μ_r, so that $N = \psi(\Pi)$, and ψ is a continuous map from Π onto N commuting with the action of G. We now show that ψ is an isomorphism. Let $\bar{f}$ be a continuous function on Π and let f be the corresponding left uniformly continuous, bounded invariant function defined by (17.1). By Theorem 15.1, there is an $\hat{f}$ in $L_\infty(N, \mu_r)$ such that $f(g) = \langle g\gamma, \hat{f}\rangle$, for g in G. Using (17.17), we then have

$$\langle g\nu, \bar{f}\rangle = f(g) = \langle g\gamma, \hat{f}\rangle = \langle \psi(g\nu), \hat{f}\rangle = \langle g\nu, \hat{f}\circ\psi\rangle, \qquad g \text{ in } G. \tag{17.24}$$

By Theorem 1.3 in [3], this implies $\bar{f} = \hat{f}\circ\psi$, ε a.e. on Π, where ε is any quasi-invariant measure on Π. Define $V(g) = \bar{f}(gx)$ and $F(g) = \hat{f}\circ\psi(gx)$ for g in G, where x is an arbitrary fixed point in Π. Then V is continuous on G and $V = F$ m a.e. on G. Let G' be the stability group of $\psi(x)$ in G and let h be in G'. We have $F(gh) = F(g)$ for each g in G. Hence, $V(gh) = V(g)$ for m a.e. g in G; the continuity of V implies then $V(gh) = V(g)$ for all g in G. In particular, we see that $\bar{f}(hx) = \bar{f}(x)$ for each h in G'. Since $\bar{f}$ is arbitrary in $C(\Pi)$, we obtain $G'x = x$, and G' is included in the stability group of x; the converse inclusion is obvious since ψ commutes with the action of G. Hence, x and $\psi(x)$ have the same stability groups. Then ψ must be a bijection (Π and N are homogeneous spaces), and Π being compact, ψ is a homeomorphism. *Q.E.D.*

REMARKS. The case where Π is a homogeneous space of G has been studied in [27] and [3]. It occurs in particular if G is a semisimple connected Lie group or if G is a compact extension of a solvable group of a certain type (including the nilpotent groups, (see [3] for details)). In both cases, the space Π is known explicitly.

Let us call reference function *adapted* to G any generalized reference function r (see Section 8) such that the functions (F_σ), σ in $\mathscr{E}_r$, defined by (16.8) are finite and equicontinuous at e (as functions on some open subgroup G_0, of G). The reference functions constructed in Proposition 16.1, as well as the generalized reference functions with compact support (see Section 8) are adapted to G. When r is adapted to G, G_0, acts uniformly on B_r (see (16.7) and (16.6)) and it is possible to obtain a result analogous to Theorem 17.1 in terms of the Martin compactification G_r (see Section 8), that is, to identify the Poisson space and the active part of G_r (modulo null sets which are empty in the homogeneous case).

REFERENCES

[1] J. Azema, M. Kaplan-Duflo, and D. Revuz, "Chaînes de Markov récurrentes," unpublished expository paper.

[2] R. Azencott, "Espaces de Poisson et frontière de Martin," *C. R. Acad. Sci. Paris, Ser. A*, Vol. 268 (1969), pp. 1406–1409.

[3] ———, Espaces de Poisson des Groupes Localement Compacts, Lecture Notes in Mathematics, Vol. 148, Berlin-Heidelberg-New York, Springer-Verlag, 1970.

[4] D. Blackwell, "On transient Markov processes with a countable number of states and stationary transition probabilities," *Ann. Math. Statist.*, Vol. 26 (1955), pp. 654–658.

[5] N. Bourbaki, *Topologie Générale*, Chapter 9, Paris, Hermann, 1958 (2nd ed.).

[6] ———, *Espaces Vectoriels Topologiques*, Chapters 1 and 2, Paris, Hermann, 1966 (2nd ed.)

[7] ———, *Intégration*, Chapters 1 through 4, Paris, Hermann, 1965 (2nd ed.).

[8] ———, *Intégration*, Chapter 9, Paris, Hermann, 1969.

[9] L. Breiman, "On transient Markov chains with application to the uniqueness problem for Markov processes," *Ann. Math. Statist.*, Vol. 28 (1957), pp. 499–503.

[10] ———, "Transient atomic chains with a denumerable number of states," *Ann. Math. Statist.*, Vol. 29 (1958), pp. 212–218.

[11] G. Choquet, "Existence et unicité des représentations intégrales au moyen des points extrémaux dans les cônes convexes," *Séminaire Bourbaki*, Exp. 139, New York, Benjamin, 1956.

[12] G. Choquet and J. Deny, "Sur l'équation de convolution $\mu = \mu * \sigma$," *C. R. Acad. Sci. Paris, Ser. A*, Vol. 250 (1960), pp. 799–801.

[13] K. L. Chung, "On the boundary theory for Markov Chains, I, II," *Acta Math.*, Vol. 110 (1963), pp. 19–77, Vol. 115 (1966), pp. 111–163.

[14] ———, *Markov Chains with Stationary Transition Probability*, Berlin-New York-Heidelberg, Springer-Verlag, 1960.

[15] K. L. Chung and W. H. J. Fuchs, "On the distribution of values of sums of random variables," *Mem. Amer. Math. Soc.*, Vol. 6 (1951), pp. 1–12.

[16] J. L. Doob, "Semi-martingales and subharmonic functions," *Trans. Amer. Math. Soc.*, Vol. 77 (1954), pp. 86–121.

[17] ———, "A probabilistic approach to the heat equation," *Trans. Amer. Math. Soc.*, Vol. 80 (1955), pp. 216–280.

[18] ———, "Discrete potential theory and boundaries," *J. Math. Mech.*, Vol. 8 (1959), pp. 433–458; correction, p. 993.

[19] ———, "Compactification of the discrete state space of a Markov process," *Z. Wahrscheinlichkeitstheorie verw. Gebiete*, Vol. 10 (1968), pp. 236–251.

[20] ———, "Probability theory and the first boundary value problem," *Illinois J. Math.*, Vol. 2 (1958), pp. 19–36.

[21] ———, "Conditional brownian motion and the boundary limits of harmonic functions," *Bull Soc. Math. France*, Vol. 85 (1957), pp. 431–458.

[22] J. L. Doob, J. L. Snell, and R. Williamson, "Applications of boundary theory to sums of independent random variables," *Contributions to Probability and Statistics*, Stanford, Stanford University Press, 1960, pp. 182–197.

[23] E. Dynkin and M. Maljutov, "Random walks on a group with a finite number of generators," *Soviet Math. Dokl.*, Vol. 2 (1961), pp. 399–402.

[24] W. Feller, "Boundaries induced by non-negative matrices," *Trans. Amer. Math. Soc.*, Vol. 83 (1956), pp. 19–54.

[25] ———, "On boundaries and lateral conditions for the Kolmogoroff differential equations," *Ann. of Math.*, Vol. 65 (1957), pp. 527–570.

[26] J. Feldman, "Feller and Martin boundaries for countable sets," *Illinois J. Math.*, Vol. 6 (1962), pp. 357–366.

[27] H. Furstenberg, "A Poisson formula for semisimple Lie groups," *Ann. of Math.*, Vol. 77 (1963), pp. 335–386.

[28] ———, "Noncommuting random products," *Trans. Amer. Math. Soc.*, Vol. 108 (1963), pp. 377–428.

[29] ———, "Translation invariant cones of functions on semisimple Lie groups," *Bull. Amer. Math. Soc.*, Vol. 71 (1965), pp. 271–326.

[30] L. Gårding, "Vecteurs analytiques dans les représentations des groupes de Lie," *Bull. Soc. Math. France*, Vol. 88 (1960), pp. 73–93.

[31] P. Hennequin, "Processus de Markov en cascade," *Ann. Inst. H. Poincaré, Sect. B.*, Vol. 19 (1963), pp. 109–196.

[32] G. A. Hunt, "Markoff processes and potentials, I, II, III," *Illinois J. Math.*, Vol. 1 (1957), pp. 44–93; pp. 316–369; Vol. 2 (1958), pp. 151–213.

[33] ———, "Markoff chains and Martin boundaries," *Illinois J. Math.*, Vol. 4 (1960), pp. 313–340.

[34] D. Kendall, "Hyperstonian spaces associated with Markov chains," *Proc. London Math. Soc.*, Vol. 10 (1960), pp. 67–87.

[35] H. Kesten and F. Spitzer, "Random walks on countable infinite abelian groups," *Acta Math.*, Vol. 113–114 (1965), pp. 237–265.

[36] H. Kunita and T. Watanabe, "Markov processes and Martin boundaries, I," *Illinois J. Math.*, Vol. 9 (1965), pp. 485–526.

[37] ———, "On certain reversed processes and their applications to potential theory and boundary theory," *J. Math. Mech.*, Vol. 15 (1966), pp. 393–434.

[38] ———, "Some theorems concerning resolvents over locally compact spaces," *Proceedings of the Fifth Berkeley Symposium on Mathematical Statistics and Probability*, Berkeley and Los Angeles, University of California Press, 1967, Vol. 2, Part 2, pp. 131–164.

[39] R. S. Martin, "Minimal positive harmonic functions," *Trans. Amer. Math. Soc.*, Vol. 49 (1941), pp. 137–172.

[40] P. A. Meyer, *Probabilités et Potentiel*, Paris, Hermann, 1966.

[41] ———, *Processus de Markov*, Lecture Notes in Mathematics, Vol. 26, Berlin-Heidelberg-New York, Springer-Verlag, 1967.

[42] ———, *Processus de Markov: la Frontière de Martin*, Lecture Notes in Mathematics, Vol. 77, Berlin-Heidelberg-New York, Springer-Verlag, 1968.

[43] D. Montgomery and L. Zippin, *Topological Transformation Groups*," New York, Interscience, 1955.

[44] J. Neveu, "Lattice methods and submarkovian processes," *Proceedings of the Fourth Berkeley Symposium on Mathematical Statistics and Probability*, Berkeley and Los Angeles, University of California Press, 1961, Vol. 1, pp. 347–391.

[45] ———, "Sur les états d'entrée et les états fictifs d'un processus de Markov," *Ann. Inst. H. Poincaré*, Vol. 17 (1962), pp. 324–337.
[46] ———, "Chaînes de Markov et théorie du potentiel," *Ann. Univ. Clermont*, Vol. 24 (1964), pp. 37–89.
[47] P. Ney and F. Spitzer, "The Martin boundary for random walks," *Trans. Amer. Math. Soc.*, Vol. 121 (1966), pp. 116–132.
[48] D. Ray, "Resolvents, transition functions and strongly Markovian processes," *Ann. of Math.*, Vol. 70 (1959), pp. 43–78.
[49] F. Spitzer, *Principles of Random Walk*, New York, Van Nostrand, 1964.
[50] J. B. Walsh, "The Martin boundary and completion of Markov chains," *Z. Wahrscheinlichkeitstheorie verw. Gebiete*, Vol. 14 (1970), pp. 169–188.

THE STRUCTURE OF A MARKOV CHAIN

J. L. DOOB
UNIVERSITY OF ILLINOIS

1. Introduction

Let p be a standard transition function on the set I of integers, that is, a function from $(0, \infty) \times I \times I$ into $[0, 1]$ satisfying

$$\sum_j p(t, i, j) = 1, \qquad p(s + t, i, k) = \sum_j p(s, i, j)p(t, j, k), \tag{1.1}$$

together with the continuity condition $\lim_{t\to 0} p(t, i, i) = 1$. Let f be an absolute probability function, that is, a function from $(0, \infty) \times I$ into $[0, 1]$, satisfying

$$\sum_j f(t, j) = 1, \qquad f(s + t, j) = \sum_i f(s, i)p(t, i, j). \tag{1.2}$$

Let L be an arbitrary set containing I as a subset. There is then a Markov process $\{x(t), t > 0\}$ with state space L having the specified transition and absolute probability functions. The notation $x(t)$ will always refer to the tth random variable of such a process, and the process will be called "smooth" if L is topological and if almost every sample function is right continuous with left limits on $(0, \infty)$. Note that this condition does not require the existence of a right limit at 0. For each $t > 0$ the random variable $x(t)$ almost surely has its values in I, but it has been known since Ray's work [11] in 1959 that L and the process can be chosen to make the process and properly chosen extensions of the transition function have desirable smoothness properties. One can always choose L to be an entrance space in the sense of [4]; that is, one can choose L to satisfy the following conditions:

(a) L is a Borel subset of a compact metric space in which I is dense;

(b) for every absolute probability function f there is a smooth corresponding process with state space L;

(c) for every integer j, $p(\cdot, \cdot, j)$ has a continuous extension to $(0, \infty) \times L$ and (1.1) is satisfied for i allowed to be any point of L;

(d) if ξ is in L and if $\{x(t), t > 0\}$ is a smooth process with absolute probability function given by $f(t, i) = p(t, \xi, i)$, then $x(0+)$ exists (and is in L) almost surely.

In the following, i, j, k are integers and ξ, η are points of a specified entrance space. The probability measure determined by a smooth process with $f(t, i) =$

$p(t, \xi, i)$ will be denoted by P_ξ. If ξ is not a branch point, $P_\xi\{x(0+) = \xi\} = 1$ and we define $x(0) = \xi$. The process will then be called a smooth process with initial state ξ. There are many entrance spaces, and none can be described as best for all purposes. Any full analysis of Markov chains must include an analysis of the possible entrance spaces for a given transition function.

The present paper is devoted to showing how, after ramifying any given entrance space in order to get certain functions continuous, the resolvent of a process can be expressed in terms of the resolvent of the process killed at a certain type of terminal time together with certain extra operators. The case of interest is when all states are stable. The technique has been used for Hunt processes, but unfortunately under the hypothesis (too restrictive in the present context) that there are no branch points. The necessary theorems on additive functionals when the related Markov processes may have branch points, apparently, have not been stated in the literature, but the extension to the branch point case seems straightforward.

2. Stable case

We shall be most interested in the "stable case" by which we mean that for every i

$$-p'_{i,i}(0) = \sum_{j \neq i} p'_{i,j}(0) < \infty. \tag{2.1}$$

As usual, we write q_i for $-p'_{i,i}(0)$ and $q_{i,j}$ for $p'_{i,j}(0)$ when $j \neq i$.

In the stable case, any smooth process with initial state an integer proceeds at first by jumps. More precisely, if the initial state is i, there is an exponential holding time with expected value $1/q_i$; then there is a jump to j with probability $q_{i,j}/q_i$; then there is an exponential holding time with expected value $1/q_j$; and so on. If T_1 ($\leqq \infty$) is the first explosion time, that is, the supremum of these jump times, the process is integer valued to time T_1. Many papers have been devoted to the character of the paths after time T_1 when $T_1 < \infty$.

In terms of transition functions, the problem is the following. Define $\bar{p}$ by

$$\bar{p}(t, i, j) = P_i\{x(t) = j, T_1 > t\}. \tag{2.2}$$

Then $p \geqq \bar{p}$, with equality only under special restrictions on the matrix $(q_{i,j})$, and the problem is to find the class of possible transition functions p for a given $\bar{p}$, that is, for a given matrix $(q_{i,j})$. In earlier papers, it has always been assumed that the number of ways sample functions can "go to infinity" at T_1 is discrete (usually finite) in some suitable sense. See, for example, the recent literature [2], [5], [12]. If an entrance space has been chosen as state space, one can describe the process at time T_1: the distribution of $x(T_1)$ if $x(T_1 -) = \xi$ is the branching distribution $p(0+, \xi, \cdot)$, that is, the limiting distribution of $p(t, \xi, \cdot)$ when $t \to 0$. (See [4].) If ξ is not a branch point, and only then, this distribution is supported by the singleton $\{\xi\}$. If $p(0+, \xi, \cdot)$ is supported by I for every ξ, one can continue from T_1 using the matrix $(q_{i,j})$; and in fact, in this case $(q_{i,j})$ together with the branching distributions determine the process completely. If the branching distributions are not restricted as stated, we shall see below how the process can be continued—using only $(q_{i,j})$, the state space topology, and the branching distributions—to a time we designate as T_∞.

For an important class of stopping times T, including T_∞, we shall show that if a transition probability p^T is defined like $\bar{p}$ in equation (2.2) but using T instead of T_1, then any given entrance space can be ramified into one in which $p^T(t, \cdot, j)$ has a continuous extension to the space, and the corresponding resolvent operator will then take bounded functions into continuous functions. This property will be used in expressing the resolvent operator of the given process in terms of the resolvent operator of the process killed at T, together with certain other operators.

3. Terminal times and corresponding excessive functions

We assume the usual background of σ-algebras and so on for Hunt processes (using in the discussion of smooth processes, for example, the space of right continuous functions with left limits from $[0, \infty)$ into the entrance state space less the branch points (where branch points are allowed as left limits), identifying the value of the function at 0 with $x(0+)$. The obvious changes are to be made on the rare occasions when the absolute probability function is not chosen to ensure the existence of $x(0+)$. In particular, we shall use the translation operator θ_t. Note that our smooth processes are not necessarily quasi left continuous except in a slightly extended sense, but have the strong Markov property (see [3], [4]).

Let L be an entrance space. A terminal time will be called "admissible" if the following three conditions are satisfied:

(a) T is perfect (see [1]);

(b) for every nonbranch point ξ, $\lim_{n\to\infty} \theta_{s_n} T = 0$, P_ξ almost everywhere where $T = 0$, whenever $\{s_n, n \geqq 1\}$ is a sequence of positive numbers with limit 0;

(c) $P_\xi\{T > 0\} = 1$ if ξ is in I.

According to the Blumenthal zero-one law, the probability in (c) must be either 0 or 1 if ξ is not a branch point.

If ξ is not a branch point and $t > 0$, define

$$p^T(t, \xi, k) = P_\xi\{x(t) = k, T > t\}. \tag{3.1}$$

Then the Chapman–Kolmogorov equation system

$$\begin{aligned} p^T(s + t, \xi, k) &= \sum_j P_\xi\{x(s + t) = k, x(s) = j, T > s, \theta_s T > t\} \\ &= \sum_j E_\xi\{P\{x(s + t) = k, \theta_s T > t \mid \mathscr{F}(s)\}\, 1_{T > s, x(s) = j}\} \\ &= \sum_j p^T(s, \xi, j) p^T(t, j, k) \end{aligned} \tag{3.2}$$

is satisfied.

If ξ is a branch point, we make the obvious definition corresponding to starting the process at ξ and having it jump at once:

$$p^T(t, \xi, k) = \int_L p^T(t, \eta, k)p(0+, \xi, d\eta), \tag{3.3}$$

where we use the fact that $p(0+, \xi, \cdot)$ is supported by the set of nonbranch points. The Chapman–Kolmogorov equation system is satisfied by p^T for every initial state in L. If $s < t$,

$$\sum_j p(s, \xi, j)p^T(t - s, j, k) = P_\xi\{x(t) = k, \theta_s T > t - s\} \geqq p^T(t, \xi, k) \tag{3.4}$$

and the left side has limit $p^T(t, \xi, k)$ when $s \to 0$. Then the function $p(\cdot, \cdot, k) - p^T(\cdot, \cdot, k)$ is space time excessive (see [4]). Instead of $p(\cdot, \cdot, k)$, any other function on $(0, \infty) \times L$ which dominates $p^T(\cdot, \cdot, k)$ and is an exit law can be used, for example, the constant function 1.

Define R, R^T by

$$R_{\xi,j}(\lambda) = \int_0^\infty e^{-\lambda s}p(s, \xi, j)\, ds, \qquad R^T_{\xi,j}(\lambda) = \int_0^\infty e^{-\lambda s}p^T(s, \xi, j)\, ds \tag{3.5}$$

for $\lambda > 0$. Then $R_{\cdot,j}(\lambda)$ is λ excessive and what we have proved about $p - p^T$ implies that $R_{\cdot,j}(\lambda) - R^T_{\cdot,j}(\lambda)$ is also λ excessive.

A terminal state Δ can be introduced in the usual way to make p^T stochastic, and this function, with second and third arguments restricted to $I^\Delta = I \cup \{\Delta\}$ is then a standard transition function relative to I^Δ.

4. Terminal time examples

Throughout the rest of this paper we consider only the stable case. For a function which is right continuous with left limits from $[0, \infty)$ into an entrance space, we call a point of $[0, \infty)$ an "explosion point" of that function if it is a limit point of discontinuities. The set Γ of explosion points is closed. Because of the nature of the sample functions of smooth processes in the stable case, the set Γ is, stated roughly, determined by a sample function in such a way that Γ is independent of the state space. More precisely, consider a process $\{x(t), t \text{ rational} > 0\}$ with the given transition function and some absolute probability function. We take the state space as I (untopologized). Almost every sample function is identically constant in a neighborhood of each rational parameter value. Define an explosion time for a sample function as any real number with the property that in every neighborhood of the number there are infinitely many maximal constancy intervals. If L is any entrance space, the process can be extended into a smooth process with state space L by defining $x(t)$ for irrational $t > 0$ as $x(t+)$ when this limit exists in the L topology. For almost every sample function, the set Γ is then the same whether defined in terms of the original or in terms of the extended process.

Define a terminal time T_1 extending the definition in Section 2, as the infimum of points in $\Gamma - \{0\}$, (or ∞, if there is no such point). The qualification in parentheses will be omitted in further definitions. Then T_1 is admissible. Note that $x(T_1)$ may be integer valued and that $\theta_{T_1} T_1$ need not be 0.

Define $T_{1,1} = T_1$. If α is a countable ordinal for which $T_{1,\beta}$ is defined when $\beta < \alpha$, define $T_{1,\alpha} = \sup_{\beta<\alpha} T_{1,\beta}$ if α is a limit ordinal and $T_{1,\alpha} = T_{1,\alpha-1} + \theta_{T_{1,\alpha-1}} T_1$ otherwise. There is a first countable α, depending on the sample function, with $T_{1,\alpha} = T_{1,\alpha+1} = \cdots$; using this α, we define $T_{1,\infty} = T_{1,\alpha}$. Then $T_{1,\infty}$ is a terminal time, because this method applied to any terminal time yields a terminal time. The admissible terminal time $T_2 = \lim_{s\to 0} \theta_s T_{1,\infty}$ is the first limit point from the right of explosion times. The method of obtaining T_2 from T_1 is now applied to T_2 to obtain an admissible terminal time T_3, and so on. More precisely, if T_α is already defined for α a countable ordinal $T_{\alpha+1}$ is obtained from T_α as T_2 was from T_1. If T_β is defined for $\beta < \alpha$, where α is a (countable) limit ordinal, T_α is defined as $\lim_{s\to 0} \theta_s (\sup_{\beta<\alpha} T_\beta)$. A standard argument shows that for some countable ordinal α, depending on the sample function, $T_\alpha = T_{\alpha+1}$, so that $\theta_{T_\alpha} T_\alpha = 0$, and that, for any assignment of probability measures to the Markov process in question, there is an index α independent of the sample function, such that $T_\alpha = T_{\alpha+1}$ almost surely. Thus, if T_∞ is defined as T_α for α large enough to make $T_\alpha = T_{\alpha+1}$, we obtain a stopping time with $\theta_{T_\infty} T_\infty = 0$. Every terminal time we have obtained here is defined in terms of explosion times and is therefore meaningful for every entrance space, and the meanings are the same for all entrance spaces just as those of T_1 are.

Obviously, p^{T_1} is determined completely by the matrix $(q_{i,j})$ if the initial state is an integer. If the initial state is neither an integer nor a branch point, the continuity properties of space time excessive functions [4] imply that

$$p^{T_1}(t, \xi, j) = \limsup_{i\to\xi} p^{T_1}(t, i, j). \tag{4.1}$$

If the initial state is a branch point, $p^{T_1}(t, \xi, \cdot)$ is determined by (3.3). Thus, p^{T_1} is determined by $(q_{i,j})$, the branching distributions, and the entrance space topology. Evidently these same elements also determine $p^{T_2}, \cdots, p^{T_\infty}$.

5. Ramification of an entrance space

Let K_0 be an entrance space, a Borel subset of the compact metric space K, with metric ρ. Let T be an admissible terminal time defined in terms of the explosion points of sample functions. Then T has a well-defined meaning for every entrance space, in fact, a meaning independent of the entrance space (see Section 4). Enlarge K to K^Δ by the adjunction of an isolated state Δ, setting $\rho(\xi, \Delta) = 1$ and $p(t, \Delta, \Delta) = 1$. Let ρ^* be a metric on $I^\Delta = I \cup \{\Delta\}$, chosen in such a way that if K^* is the completion of I^Δ under ρ^* there is a Borel subset K_0^* of K^* which is an entrance space for the restriction of p^T to $(0, \infty) \times I^\Delta \times I^\Delta$. Define R, R^T as in Section 3. It is known (see [3], for example) that ρ^* can be chosen so that $\{i_n, n \geqq 1\}$ is a Cauchy sequence if and only if $\{R^T_{i_n,j}(\lambda), n \geqq 1\}$ is a Cauchy sequence for all j and all $\lambda > 0$, equivalently for λ in a countable dense subset of $(0, \infty)$. We assume that such a choice has been made.

Now let $\hat{K}$ be the completion of I^Δ in the metric $\rho + \rho^*$. Then $\hat{K}$ is a compact metric space in which I^Δ is dense and there is a unique continuous map α $[\alpha^*]$ from $\hat{K}$ onto K^Δ $[K^*]$ leaving I^Δ invariant. Since $\hat{K}$ can also be considered the completion of K^Δ $[K^*]$ in the metric $\rho + \rho^*$, K^Δ and K^* can be thought of as subsets of $\hat{K}$.

Define

$$\begin{aligned} &\hat{K}_0 = \alpha^{-1}(K_0^\Delta) \cap \alpha^{*-1}(K_0^*) \\ &\hat{p}(t, \xi, j) = p(t, \alpha(\xi), j), \qquad \hat{p}_T(t, \xi, j) = p^T(t, \alpha^*(\xi), j), \qquad \xi \in \hat{K}_0, \end{aligned} \tag{5.1}$$

so that $\hat{p} = p$ and $\hat{p}_T = p^T$ on $(0, \infty) \times I^\Delta \times I^\Delta$. For each j, $\hat{p}(\cdot, \cdot, j)$ and $\hat{p}_T(\cdot, \cdot, j)$ are continuous on $(0, \infty) \times \hat{K}_0$. (In discussing these functions, Δ is considered an honorary integer.) Moreover, $\hat{p}$ and $\hat{p}_T$ are stochastic transition functions satisfying the Chapman–Kolmogorov equation system.

Let $\{x(t), t \text{ rational} > 0\}$ be a process with state space I and an integer initial state. Almost every sample function of this process has right and left limits at all real positive ($\geqq 0$) times in the K topology. Furthermore, the fact that $R_{\cdot,j}(\lambda)$ and $R_{\cdot,j}(\lambda) - R^T_{\cdot,j}(\lambda)$ are both λ excessive on K (see Section 3) implies that almost every sample function of the $R^T_{x(t)j}(\lambda)$ process has right and left limits at all real positive times in the K^* topology. We conclude that almost every $x(\cdot)$ process sample function has right and left limits at all real positive times in the K^* topology and therefore also in the $\hat{K}$ topology.

Define $\hat{x}(t) = x(t+)$ (limit in the $\hat{K}$ topology). The process $\{\hat{x}(t), t \geqq 0\}$ is a smooth process with state space $\hat{K}_0$ and the image process $\{\alpha[\hat{x}(t)], t \geqq 0\}$ is a smooth process with state space K_0 and the same initial state. The three processes $\{\alpha[\hat{x}(t)], t \geqq 0\}$, $\{\alpha^*[\hat{x}(t)], t \geqq 0\}$, $\{\hat{x}(t), t \geqq 0\}$, when made identically Δ at times $\geqq T$ have identical finite dimensional distributions. More generally, the corresponding discussion goes through for any absolute probability function, for example by starting processes at time $1/n$ and then making $n \to \infty$. Thus, we have the situation where K_0 satisfies conditions (a), (b), (c) of an entrance space relative to the restrictions to $(0, \infty) \times I^\Delta \times I^\Delta$ of both p and p^T. That is, in the terminology of [4], $\hat{K}_0$ is an entrance adapted space for these two restrictions. According to [4], there is then a Borel subset L of $\hat{K}_0$ which is an entrance space for the first restriction and therefore for the second restriction (because if $\hat{x}(0+)$ exists this right limit also exists when the process is made identically Δ at the times $\geqq T$). We observe that we have now identified $\hat{p}_T$ with $\hat{p}^T$.

We have thus found an entrance space L on which the function $\hat{p}^T(\cdot, \cdot, j)$ is continuous as desired. A trivial modification of this discussion would yield the corresponding result simultaneously for countably many admissible terminal times T. Thus, for example, there is a space which is simultaneously an entrance space for every T_j in Section 4. Moreover, the corresponding reversed processes can be handled at the same time, for example, to make the final state space also an entrance-exit space.

6. Continuation of Markov chains

Analyses of the continuation of a Hunt type Markov process from the first time a set is hit have been made by many authors (for example recently by Dynkin [7], Motoo [9], and Okabe [10]). The corresponding analysis of Markov chains in the stable case has been limited to the analysis of what happens after the first explosion time, with strong restrictions on the number of ways sample functions can "go to ∞" at this time (see, for example, Chung [2] and the more recent Shih [12]). The techniques used in the analysis of Hunt process continuation have not been applied to the stable chain case. We shall now make such an application, following Dynkin [7] (who treated Hunt processes) with appropriate modifications. A new difficulty is the fact that no attack on chains can avoid the possibility of branch points, whereas the no-branch-point hypothesis has not been thought improperly restrictive in the Hunt theory and the associated theory of additive functionals.

Let L be an entrance space for our transition function in the stable case and let T be an admissible terminal time depending on the explosion points with the additional property that $\theta_T T = 0$. For example, T_∞, as defined in Section 4, is such a terminal time. Then p^T is well defined and we suppose that L has been ramified if necessary to make $p^T(\cdot, \cdot, j)$ continuous for every choice of j including the terminal state. The function $R^T_{\cdot,j}(\lambda)$ is then continuous on L. More generally, if f is a bounded function on L (or even merely on I) and if the obvious operational notation is used, the function $R^T(\lambda)f$ is continuous on L. This follows, for example, from the particular case just mentioned and the fact that the function

$$\sum_j R^T_{\cdot,j}(\lambda) = \frac{1}{\lambda} - R_{\cdot,\Delta}(\lambda) \tag{6.1}$$

is continuous, so that the sum on the left converges uniformly on compacta.

Let F be the closed set of points ξ at which $P_\xi\{T = 0\} = 1$, that is, at which $\Sigma_j p^T(t, \xi, j) = 0$ for all $t > 0$, or equivalently at which $\lambda R^T_{\xi,\Delta}(\lambda) = 1$ for every $\lambda > 0$. Let F_b be the set of branch points in F. Since $\theta_T T = 0$, for any smooth process $\{x(t), t > 0\}$ the random variable $x(T)$ has its values almost surely in F where $0 < T < \infty$, including 0 if $x(0+)$ exists and $x(0)$ is defined as this right limit. In fact, the distribution of $x(T)$ is supported by $F - F_b$, since almost no sample path hits a branch point.

If S is the hitting time of F, then $T = S + \theta_S T$ where $T > S$, whereas $\theta_S T = 0$ by definition of F. Thus, $T = S$ is the hitting time of F, and the condition $\theta_T T = 0$, combined with condition (b) for an admissible stopping time, implies that every point ξ of F is regular for F. Then $p(0+, \xi, F) = 1$ for

ξ in F. For a smooth process, if T is the limit of an increasing sequence $\{S_n, n \geqq 1\}$ of stopping times then $p[0+, x(T-), F] = 1$ almost everywhere where $S_n < T < \infty$ for all n. In particular if $T = T_\infty$ as defined in Section 4, then $x(T-) \in F$ almost everywhere where $0 < T < \infty$.

Since $\Sigma_j R^T_{\xi,j}(\cdot)$ is the Laplace transform of a monotone decreasing function, the function $\lambda \rightsquigarrow \lambda \Sigma_j R^T_{\xi,j}(\lambda)$ is an increasing function. Then if α is a strictly positive number, to be retained thoughout the following, and if $K_\lambda f = R^T(\lambda)f/R^T(\alpha)1$ off F, then off F

$$|K_\lambda f| \leqq \begin{cases} \sup |f| & \text{if } \lambda \geqq \alpha, \\ \dfrac{\alpha}{\lambda} \sup |f| & \text{if } \lambda \leqq \alpha. \end{cases} \tag{6.2}$$

Let $\{x(t), t \geqq 0\}$ be a smooth process with $x(0) = \xi$, where we allow ξ to be arbitrary; smoothness at 0 in this context means that $x(0+)$ exists, but is almost certainly ξ if and only if ξ is not a branch point. Then, since almost every sample function is right continuous on $(0, \infty)$ (and we ignore the exceptional sample functions below), the parameter set for which a sample path lies in the open set $L - F$ is the union of disjoint (maximal) intervals open on the right. We denote the endpoints of the intervals generically by γ, δ. Here $\gamma + \theta_\gamma T = \delta$. Since almost every sample function is integer valued for Lebesgue almost every parameter value, these intervals cover Lebesgue almost every point of $(0, \infty)$. In this property, the context is simpler than that in [7], [9], [10]. As Dynkin pointed out in [6], for a smooth process and bounded f, under certain hypotheses satisfied in the present study, $K_\lambda f$ has a right limit at every left endpoint γ, for almost every sample function. We shall use this fact.

Let f be a bounded function on L (only its values on I are relevant). Let h be a bounded continuous function on $L - F$ with the property that for almost every sample function of a smooth process $\lim_{s\to 0} h[x(\gamma + s)]$ exists for all γ. We denote this limit by $h[x(\gamma)]^+$. In a similar context [7] Dynkin proved that

$$\begin{aligned} E_\xi\left\{\sum \exp\{-\lambda\gamma\}\, h[x(\gamma)]^+ \int_\gamma^\delta f[x(t)] \exp\{-\mu(t-\gamma)\}\, dt\right\} \\ = E_\xi\left\{\sum \exp\{-\lambda\gamma\}\, (hK_\mu f)[x(\gamma)]^+ \int_\gamma^\delta \exp\{-\alpha(t-\gamma)\}\, dt\right\}. \end{aligned} \tag{6.3}$$

Here λ and μ are strictly positive and the sum here and below is over all strictly positive γ. (Dynkin's proof yields (6.3), although his stated hypotheses are more special than ours.) Define ϕ_h^λ by

$$\phi_h^\lambda(\xi) = E_\xi\left\{\sum \exp\{-\lambda\gamma\}\, h[x(\gamma)]^+ \int_\gamma^\delta \exp\{-\alpha(t-\gamma)\}\, dt\right\}. \tag{6.4}$$

It is easy to see that this function is λ excessive and Dynkin's proof [7] that ϕ_h^λ is a regular λ potential is valid here. Thus, applying the Šur-Meyer theorem, we find as Dynkin did, that there is a continuous additive functional A_h^λ such that

$$\phi_h^\lambda(\xi) = E_\xi\left\{\int_0^\infty \exp\{-\lambda t\}\, dA_h^\lambda(t)\right\}. \tag{6.5}$$

Dynkin's proof in [7] that A_h^λ does not depend on λ is valid in our context and we omit the superscript from now on. We shall write A instead of A_1. When $\lambda = \alpha$ and $h = 1$, we find

$$\phi_1^\alpha(\xi) = E_\xi\{e^{-\alpha T}\}\alpha^{-1} = E_\xi\left\{\int_0^\infty e^{-\alpha t}\, dA(t)\right\}. \tag{6.6}$$

Hence,

$$E_\xi\{e^{-\alpha T}\phi_1^\alpha[x(T)]\} = E_\xi\{e^{-\alpha T}\}\alpha^{-1} = E_\xi\left\{\int_{(T,\infty)} e^{-\alpha t}\, dA(t)\right\}, \tag{6.7}$$

where we have used the fact that $\theta_T T = 0$ and must make the obvious conventions if T is not finite. Comparing (6.7) with (6.6), we see that $A(T) = 0$, P_ξ almost surely for all ξ. Thus (see [1]), A is supported by F; more precisely, the measure $dA(t)$ is P_ξ almost surely supported by the set of t with $x(t)$ in F.

Define for g a bounded Borel measurable function on L,

$$\nu^\lambda(\xi, g) = E_\xi\left\{\int_0^\infty e^{-\lambda t} g[x(t)]\, dA(t)\right\}. \tag{6.8}$$

Then $\nu^\lambda(\xi, \cdot)$ defines a measure, and $\nu^\lambda(\xi, g)$ is the integral of g with respect to this measure. Below, "null set" is a set of $\nu^\lambda(\xi, \cdot)$ measure 0 for all ξ. This condition is independent of $\lambda > 0$. For example, $L - F$ is a null set.

By hypothesis, the space L is a Borel subset of a compact metric space L'; it is convenient to introduce L' at this point in order to follow Dynkin in [7]. By a linearity argument, he shows (translating his result into our context) that if ξ is not in some null set then there is a measure $b(\xi, \cdot)$ of Borel subsets of L' with $b(\xi, L') \leqq 1$ such that if h is continuous on L', $b(\cdot, h)$ is Borel measurable and $dA_h(t) = b[x(t)]\, dA(t)$. Thus, for such a choice of h, and hence for every bounded Borel measurable h on L',

$$\begin{aligned} E_\xi&\left\{\sum \exp\{-\lambda\gamma\}\, h[x(\gamma)] \int_\gamma^\delta \exp\{-\alpha(t-\gamma)\}\, dt\right\} \\ &= E_\xi\left\{\int_0^\infty \exp\{-\lambda t\}\, b[x(t), h]\, dA(t)\right\}. \end{aligned} \tag{6.9}$$

If h is the indicator function of L and $\lambda = \alpha$, (6.7) together with (6.9) imply that $b(\xi, L) = 1$ for ξ not in a null set. In other words for ξ not in a null set, $b(\xi, \cdot)$ is a probability measure supported by L, and we can drop L' again. If f is bounded on L, (6.3) yields

$$(6.10)\qquad E_\xi\left\{\int_T^\infty \exp\{-\lambda t\} f[x(t)]\,dt\right\}$$
$$= E_\xi\left\{\sum \exp\{-\lambda\gamma\}\int_\gamma^\delta \exp\{-\lambda(t-\gamma)\} f[x(t)]\,dt\right\}$$
$$= E_\xi\left\{\sum \exp\{-\lambda\gamma\}\,(K_\lambda f)[x(\gamma)]^+ \int_\gamma^\delta \exp\{-\alpha(t-\gamma)\}\,dt\right\}.$$

If $K_\lambda f$ were defined bounded and Borel measurable on L, with $(K_\lambda f)[x(\gamma)] = (K_\lambda f)[x(\gamma)]^+$, or if $h[x(\gamma)]$ in (6.9) could be replaced by $h[x(\gamma)]^+$, then we could identify $K_\lambda f$ with h in (6.9) to get

$$(6.11)\quad E_\xi\left\{\int_T^\infty \exp\{-\lambda t\} f[x(t)]\,dt\right\} = E_\xi\left\{\int_0^\infty \exp\{-\lambda t\}\, b[x(t), K_\lambda f]\,dA(t)\right\}$$
$$= \nu^\lambda[\xi, b(\cdot, K_\lambda f)].$$

In order to get some version of (6.11), one can ramify $L - F$ to a space on which $K_\lambda f$ has a continuous extension [7]. The measure $\nu^\lambda(\xi, \cdot)$ is then a measure on this new space. With this interpretation of (6.11), the representation of the resolvent R in terms of R^T, ν, K_λ is now trivial:

$$(6.12)\qquad [R(\lambda)f](\xi) = E_\xi\left\{\int_0^\infty \exp\{-\lambda t\} f[x(t)]\,dt\right\}$$
$$= [R^T(\lambda)f](\xi) + \nu^\lambda[\xi, b(\cdot, K_\lambda f)].$$

The continuous additive functional A determines the "boundary" process on F as usual.

REFERENCES

[1] R. M. BLUMENTHAL and R. K. GETOOR, *Markov Processes and Potential Theory*, New York, Academic Press, 1968.

[2] KAI LAI CHUNG, *Lectures on Boundary Theory for Markov Chains*, Annals of Mathematics Studies, Princeton, Princeton University, 1970.

[3] J. L. DOOB, "Compactification of the discrete state space of a Markov process," *Z. Wahrscheinlichkeitstheorie und Verw. Gebiete*, Vol. 10 (1968), pp. 236–251.

[4] ———, "State spaces for Markov chains," *Trans. Amer. Math. Soc.*, Vol. 149 (1970), pp. 279–305.

[5] E. B. DYNKIN, "General boundary conditions for Markov processes with countably many states," *Teor. Verojatnost. i Primenen.*, Vol. 12 (1967), pp. 222–257. (In Russian.)

[6] ———, "On the starting points of wanderings of Markov processes," *Teor. Verojatnost. i Primenen.*, Vol. 13 (1968), pp. 490–493. (In Russian.)

[7] ———, "On the continuation of a Markov process," *Teor. Verojatnost. i Primenen.*, Vol. 13 (1968), pp. 708–713. (In Russian.)
[8] Paul A. Meyer, *Probability and Potentials*, Waltham, Blaisdell, 1966.
[9] Minoru Motoo, "Application of additive functionals to the boundary problem of Markov processes (Lévy's system of U-processes)," *Proceedings of the Fifth Berkeley Symposium on Mathematical Statistics and Probability*, Berkeley and Los Angeles, University of California Press, Vol. 2, Part 2, 1967, pp. 75–110.
[10] Yasunori Okabe, "The resolution of an irregularity of boundary points in the boundary problem for Markov processes," *J. Math. Soc. Japan*, Vol. 22 (1970), pp. 47–104.
[11] Daniel Ray, "Resolvents, transition functions, and strongly Markovian processes," *Ann. Math.*, Vol. 70 (1959), pp. 43–72.
[12] C. T. Shih, "Construction of all Markov chains having a fixed minimal chain," to appear.

CLASSICAL POTENTIAL THEORY AND BROWNIAN MOTION

SIDNEY C. PORT and CHARLES J. STONE
UNIVERSITY OF CALIFORNIA, LOS ANGELES

1. Introduction

Pioneering work of Doob, Kac, and Kakutani showed that there was a beautiful and deep connection between certain problems in the study of Brownian motion and those of classical potential theory. This work stimulated much research on the theory of Markov processes. In spite of all this work, however, there doesn't appear anywhere in the literature any reasonably complete treatment of the connection between potential theory and Brownian motion. In this paper and its companion "Logarithmic Potentials and Planar Brownian Motion" which follows in this volume, we present this connection in a way that is both elementary and essentially selfcontained. Our treatment here is not complete and will be expanded upon in a forthcoming monograph.

This paper, being basically expository in nature, contains essentially nothing new. Its novelty (if any) consists in the treatment given to the topics discussed. In one place, however, we do seem to have a result that is new. This is in finding all bounded solutions of the modified Dirichlet problem for any arbitrary open set G, and in giving a necessary and sufficient condition for there to be a unique such solution.

In this paper, we consider a Brownian motion process X_t in $n \geqq 2$ dimensional Euclidean space R^n. Let B be a Borel set and set

$$(1.1) \qquad V_B = \inf\{t > 0 : X_t \in B\}, \qquad V_B = \infty \text{ if } X_t \notin B \text{ for all } t > 0.$$

In Section 2, we present preliminary facts about Brownian motion that are needed to develop classical potential theory from a probabilistic point of view. A set B is called polar if $P_x(V_B < \infty) \equiv 0$. A point x is called regular for B if $P_x(V_B = 0) = 1$. In Section 3, we prove that the set of points in B that are not regular for B is a polar set. In this section, we also show that points are polar and gather together a few more facts of a technical nature that are needed for work in the later sections. The Dirichlet problem for an arbitrary open set G is discussed in Section 4.

Starting with Section 5 and throughout the remainder of this paper, we assume that we are dealing with Brownian motion in $n \geqq 3$ dimensions. (The planar

The preparation of this paper was supported in part by NSF Grant GP-17868.

case is discussed in the companion paper.) In Section 5, we introduce Newtonian potentials and show how they are related to the Brownian motion process. We also prove that $|X_t| \to \infty$ with probability one as $t \to \infty$ and prove the extended maximum principle for potentials.

The remaining two sections are devoted to developing the notion of capacitary measure and potentials from a probabilistic view point.

2. Preliminaries

A Brownian motion process in $n \geqq 2$ dimensional Euclidean space R^n is a stochastic process X_t, $0 \leqq t < \infty$, having the following properties: (i) for each $t > 0$ and $h \geqq 0$, $X_{t+h} - X_h$ has the normal density $\mathrm{p}(t, x) = (2\pi t)^{-n/2} \exp\{-|x|^2/2t\}$; and (ii) for $0 < t_1 < t_2 < \cdots < t_n < \infty$, $X_{t_1} - X_0, X_{t_2} - X_{t_1}, \cdots, X_{t_n} - X_{t_{n-1}}$ are independent random variables. It is well known that a version of this process can be selected so that the sample functions $t \to X_t$ are continuous with probability one. Henceforth, we will always assume that our process has this continuity property.

The distribution of X_t depends on the distribution of X_0. We let $P_x(\cdot)$ denote the probability of $\cdot$ given $X_0 = x$ and we let E_x denote expectation relative to P_x.

Let B be any Borel set. The *first hitting* time V_B of B is defined by $V_B = \inf\{t > 0 : X_t \in B\}$ if $X_t \in B$ for some $t > 0$. If $X_t \notin B$ for all $t > 0$, we define $V_B = \infty$. When $V_B < \infty$, the *first hitting place* in B is the random variable X_{V_B}.

In the sequel, we will need to know some continuity properties of $P_x(V_B \leqq t)$. These are given by:

PROPOSITION 2.1. *The function $P_x(V_B \leqq t)$ is a continuous function in t for $t > 0$ for fixed x and a lower semicontinuous function in x for fixed $t > 0$.*

PROOF. Suppose for some $t > 0$, $P_x(V_B = t) > 0$. Then for any h, $0 < h < t$,

$$0 < P_x(V_B = t) \leqq \int_{R^n} p(h, y - x) P_y(V_B = t - h)\, dy, \tag{2.1}$$

so

$$\int_{R^n} P_y(V_B = t - h)\, dy > 0 \tag{2.2}$$

for all h, $0 < h < t$. But this is impossible because for any $r > 0$,

$$\int_{|y| \leqq r} P_y(V_B \leqq t)\, dy \leqq \int_{|y| \leqq r} dy < \infty. \tag{2.3}$$

Thus the measure

$$\int_{|y| \leqq r} P_y(V_B \in dt)\, dy \tag{2.4}$$

can have only countably many atoms.

To establish the lower semicontinuity in x for fixed $t > 0$ note that

$$(2.5) \qquad \int_{R^n} p(h, y - x) P_y(V_B < t - h)\, dy = P_x\big(X_s \in B \text{ for some } s \in (h, t)\big)$$

is a continuous function in x that increases to $P_x(V_B < t)$ as $h \downarrow 0$. Thus, $P_x(V_B < t) = P_x(V_B \leqq t)$ is lower semicontinuous in x. This establishes the proposition.

The next fact we prove tells us that the mean time to leave any *bounded* set B is finite.

PROPOSITION 2.2. *Let B be relatively compact. Then* $\sup_x E_x V_{B^c} < \infty$.

PROOF. If $x \in (\bar{B})^c$ then $P_x(V_{B^c} = 0) = 1$, so only points $x \in \bar{B}$ need be considered. Let $t > 0$. Then

$$(2.6) \qquad \inf_{x \in \bar{B}} P_x(X_t \in B^c) = \delta > 0,$$

and thus for any $x \in \bar{B}$,

$$(2.7) \qquad P_x(V_{B^c} \leqq t) \geqq P_x(X_t \in B^c) \geqq \delta.$$

Hence, for $x \in \bar{B}$,

$$(2.8) \qquad P_x(V_{B^c} > t) \leqq 1 - \delta.$$

But then

$$(2.9) \qquad P_x(V_{B^c} > 2t) = \int_{\bar{B}} P_x(V_{B^c} > t, X_t \in dz) P_z(V_{B^c} > t) \leqq (1 - \delta)^2,$$

and by induction, for $x \in \bar{B}$,

$$(2.10) \qquad P_x(V_{B^c} > nt) \leqq (1 - \delta)^n.$$

Hence,

$$(2.11) \qquad \sup_{x \in \bar{B}} E_x V_{B^c} \leqq \frac{t}{\delta} < \infty,$$

as desired.

Let A and B be any Borel sets. Then clearly

$$(2.12) \qquad P_x(X_t \in A) = P_x(V_B \leqq t, X_t \in A) + P_x(V_B > t, X_t \in A).$$

Now

$$(2.13) \qquad P_x(V_B \leqq t, X_t \in A)$$
$$= \int_{0-}^{t} \int_{\bar{B}} P_x(X_t \in A \mid V_B = s, X_{V_B} = z) P_x(V_B \in ds, X_{V_B} \in dz).$$

Brownian motion, along with a large variety of Markov processes, possesses the important property that for $s \leqq t$,

$$P_x(X_t \in A \mid V_B = s, X_{V_B} = z) = P_z(X_{t-s} \in A). \tag{2.14}$$

This property is a consequence of the *strong Markov property*. It is certainly intuitively plausible that (2.14) should hold and a rigorous proof is not difficult to supply. Since the proof would involve us in a more thorough discussion of the measure theoretic structure of a Brownian motion process than we care to go into in this paper, we will omit the proof. (The interested reader should consult Chapter I of [1] for a complete discussion of the strong Markov property.)

Using (2.14), we obtain from (2.12) that

$$P_x(X_t \in A) - \int_{0-}^{t} \int_{\bar{B}} P_x(V_B \in ds, X_{V_B} \in dz) P_z(X_{t-s} \in A) = P_x(V_B > t, X_t \in A). \tag{2.15}$$

The left side of (2.15) has

$$p(t, y - x) - \int_{0-}^{t} \int_{\bar{B}} P_x(V_B \in ds, X_{V_B} \in dz) p(t - s, y - z) \tag{2.16}$$

as a density. Fatou's lemma shows that

$$\int_{0-}^{t} \int_{\bar{B}} P_x(V_B \in ds, X_{V_B} \in dz) p(t - s, y - z) \tag{2.17}$$

is a lower semicontinuous function in y. As the left side of (2.15) is absolutely continuous, the measure $P_x(V_B > t, X_t \in dy)$ is also absolutely continuous. Let $q(t, x, y)$ be any density of $P_x(V_B > t, X_t \in dy)$. Then

$$p(t, y - x) - \int_{0-}^{t} \int_{\bar{B}} P_x(V_B \in ds, X_{V_B} \in dz) p(t - s, y - z) = q(t, x, y) \qquad \text{a.e. } y. \tag{2.18}$$

Consequently, the left side of (2.18) is $\geqq 0$ a.e. y and being upper semicontinuous, it is $\geqq 0$ for all y. We may therefore use the left side of (2.18) to define a density for $P_x(V_B > t, X_t \in dy)$ for all y. Denote this density by $q_B(t, x, y)$. Then for *all* y,

$$p(t, y - x) - \int_{0-}^{t} \int_{\bar{B}} P_x(V_B \in ds, X_{V_B} \in dz) p(t - s, y - z) = q_B(t, x, y). \tag{2.19}$$

Henceforth in this paper, q_B will always denote this density.

The densities $p(t, x)$ have the semigroup property:

$$p(t + s, y - x) = \int_{R^n} p(t, z - x) p(s, y - z)\, dz. \tag{2.20}$$

Define P^t on the bounded or nonnegative measurable functions by

$$P^t f(x) = \int_{R^n} p(t, y - x) f(y)\, dy = E_x f(X_t). \tag{2.21}$$

Then

$$P^{t+s}f = P^t(P^s f). \tag{2.22}$$

Since Brownian motion is a Markov process,

$$\begin{aligned} P_x(V_B > t + s, X_{t+s} \in A) \\ = \int_{B^c} P_x(V_B > t, X_t \in dz) P_z(V_B > s, X_s \in A), \end{aligned} \tag{2.23}$$

and thus for almost all y

$$q_B(t + s, x, y) = \int_{B^c} q_B(t, x, z) q_B(s, z, y)\, dz. \tag{2.24}$$

We will now show that (2.24) holds for *all* $y \in R^n$. To do this note first that by using (2.20) and (2.19), it is easily verified that

$$\lim_{\varepsilon \downarrow 0} \int_{R^n} q_B(t - \varepsilon, x, z) p(\varepsilon, y - z)\, dz = q_B(t, x, y). \tag{2.25}$$

Using this and the fact that (2.24) holds for a.e. y, we see that (2.24) holds for all y.

Let $\lambda > 0$. (Henceforth, λ will always denote a positive real number.) Define

$$\begin{aligned} g^\lambda(x) &= \int_0^\infty p(t, x)\, e^{-\lambda t}\, dt, \\ g_B^\lambda(x, y) &= \int_0^\infty q_B(t, x, y)\, e^{-\lambda t}\, dt, \end{aligned} \tag{2.26}$$

and

$$\Pi_B^\lambda(x, dz) = \int_0^\infty e^{-\lambda t} P_x(V_B \in dt, X_{V_B} \in dz). \tag{2.27}$$

The quantities corresponding to these for $\lambda = 0$ are denoted by the same symbol without the λ (for example, Π_B^λ for $\lambda = 0$ is Π_B). For future reference, we note that $\Pi_B^\lambda(x, dz)$ is supported on $\bar{B}$ and

$$\int_{\bar{B}} \Pi_B^\lambda(x, dz) f(x) = E_x\big(\exp\{-\lambda V_B\} f(X_{V_B})\big). \tag{2.28}$$

Also,

$$\int_{R^n} g_B^\lambda(x, y) f(y)\, dy = E_x \int_0^{V_B} f(X_t)\, e^{-\lambda t}\, dt. \tag{2.29}$$

Observe that $g^\lambda(0) = \infty$, but otherwise $g^\lambda(x)$ is a continuous function that vanishes as $x \to \infty$. Since $p(t, y - x) \geqq q_B(t, x, y)$, we see that $g^\lambda(y - x) \geqq g_B^\lambda(x, y)$. It follows from (2.19) that

$$g^\lambda(y - x) = \int_{\bar{B}} \Pi_B^\lambda(x, dz) g^\lambda(y - z) + g_B^\lambda(x, y). \tag{2.30}$$

This equation will play a major role in our development.

Now $g^\lambda(x) = g^\lambda(-x)$, so $g^\lambda(y - x) = g^\lambda(x - y)$. It can be shown that *$g_B^\lambda(x, y)$ is a symmetric function in x and y*. A proof of this fact can be obtained by a fairly simple probabilistic argument. We will not prove this fact here but refer the reader to Chapter VI of [1] for a complete discussion.

Using (2.30), we see that the symmetry of g_B^λ implies that

$$\int_{\bar{B}} \Pi_B^\lambda(x, dz) g^\lambda(y - z) = \int_{\bar{B}} \Pi_B^\lambda(y, dz) g^\lambda(x - z). \tag{2.31}$$

This relation, as well as the symmetry of g_B^λ, will be of constant use in our development of potential theory.

Let τ be an orthogonal transformation. Then $p(t, \tau x) = p(t, x)$. It follows from this fact that if τ is a rotation about a point a and S is a sphere of center a, then $P_a(X_{V_S} \in A) = P_a(X_{V_S} \in \tau A)$ for any Borel subset $A \subset S$. Hence, $P_a(X_{V_S} \in dy)$ must be the uniform distribution on S. Let $\sigma_r(a, dy)$ denote the uniform distribution on the sphere of center a and radius r. Then $P_a(X_{V_S} \in dy) = \sigma_r(a, dy)$ when S is the sphere of center a and radius r. Using this fact, we will now prove a fact that will be used several times in the sequel.

PROPOSITION 2.3. *Let f be a bounded function such that $f = P^t f$ for all $t > 0$. Then f is a constant.*

PROOF. Since $f = P^t f$ for all $t > 0$, we see that f is continuous and that $f(x) = \lambda \int_0^\infty g^\lambda(y - x) f(y)\, dy$. Let S be the complement of the ball of center a and radius r. Using (2.30), we see that

$$f(x) = \lambda \int_0^\infty g_S^\lambda(x, y) f(y)\, dy + \Pi_S^\lambda f(x). \tag{2.32}$$

Now

$$\int_0^\infty g_S^\lambda(x, y)|f(y)|\, dy \leqq \|f\|_\infty \int_0^\infty e^{-\lambda t} P_x(V_S > t)\, dt. \tag{2.33}$$

Since S^c is bounded, $E_x V_S < \infty$, so

$$\lim_{\lambda \downarrow 0} \int_0^\infty e^{-\lambda t} P_x(V_S > t)\, dt = \int_0^\infty P_x(V_S > t)\, dt = E_x V_S < \infty. \tag{2.34}$$

Letting $\lambda \downarrow 0$ in (2.32), we see that $f(x) = \Pi_S f(x)$. In particular, for $x = a$, we see that

$$f(a) = E_a f(X_{V_S}) = \int_{\partial S} f(y) \sigma_r(a, dy). \tag{2.35}$$

Thus, f is its average over every sphere. Hence, f is harmonic, and as it is bounded it must be a constant. This establishes the proposition.

3. Regular points

Let B be a Borel set. It can be shown by a simple measure theoretic argument that $P_x(V_B = 0)$ is either 0 or 1. (See Chapter I of [1].) A point x is called *regular for B if* this probability is 1. Otherwise x is said to be *irregular for B*. Let B^r denote the set of regular points for B. Clearly, $\mathring{B} \subset B^r \subset \bar{B}$. A simple sufficient condition for regularity is the following.

PROPOSITION 3.1 (Poincaré's test). *Let k be a truncated cone of vertex 0, radius r_0 and angle opening α. Then $x \in B^r$ if $x + k \subset B$.*

PROOF. Clearly,

$$(3.1)\quad \begin{aligned} P_x(V_B \leqq t) &\geqq P_x(X_t \in B) \geqq P_x(X_t \in k + x) = P_0(X_t \in k) \\ &= (2\pi t)^{-n/2} \int_k \exp\left\{-\frac{|x|^2}{2t}\right\} dx = \alpha(2\pi t)^{-n/2} \int_0^{r_0} \exp\left\{-\frac{r^2}{2t}\right\} r^{n-1}\, dr \\ &\geqq \delta > 0 \end{aligned}$$

for all $1 \geqq t > 0$. Thus, $P_x(V_B = 0) > 0$ and hence $P_x(V_B = 0) = 1$, as desired.

Using (2.19), we see that if $x \in B^r$, then $g_B^\lambda(x, y) = 0$ for all y. By symmetry, $g_B^\lambda(x, y) = 0$ for $y \in B^r$.

A simple but important device will be employed in many proofs. Let B be a closed set and let B_n, $n \geqq 1$, be a family of closed sets such that $B_1 \supset \mathring{B}_1 \supset B_2 \supset \cdots$, $\bigcap_n B_n = \bigcap_n \mathring{B}_n = B$.

PROPOSITION 3.2. *Let B and B_n be as described above. Then $P_x(V_{B_n} \uparrow V_B) = 1$ for $x \in B^c \cup B^r$.*

PROOF. Clearly, the V_{B_n} are nondecreasing, and thus $V_{B_n} \uparrow V \leqq V_B$. If $V = \infty$, then $V_B = \infty$. On the other hand if $V < \infty$, then $X_{V_{B_n}} \to X_V \in \bigcap_n B_n = B$. Thus, $V \geqq V_B$ whenever $V > 0$. Clearly, $P_x(V > 0) = 1$ when $x \in B^c$. If $x \in B^r$ then $P_x(V = V_B = 0) = 1$. This establishes the proposition.

For our later work, we will need the following simple corollary of Propositions 3.1 and 3.2.

COROLLARY 3.1. *Let G be a nonempty open set. Then there is an increasing sequence G_n of open sets with compact closures contained in G such that $G_1 \subset \bar{G}_1 \subset G_2 \subset \cdots$, $\bigcup_n G_n = G$ and such that each point of ∂G_n is regular for G_n^c. The times $V_{\partial G_n}$, $n \geqq 1$, are such that $P_x(V_{\partial G_n} \uparrow V_{\partial G} = 1$ for all $x \in G$.*

PROOF. Let k_n be compact sets such that $k_1 \subset k_2 \subset$ and $\bigcup_n k_n = G$. Cover k_1 by a finite number of open balls whose union D is such that $\bar{D}$ is contained in G. By increasing the radii of some balls if necessary, we can conclude from Poincaré's test that each point of ∂D is regular for $\bar{D}^c$. Let $G_1 = D$. Thus, each point of ∂G_1 is regular for G_1^c and $\bar{G}_1 \subset G$. Apply the same procedure to $\bar{G}_1 \cup k_2$, and so forth. Clearly, $G_1 \subset G_2 \subset \cdots$, $\bigcup_n G_n = G$, so $G_1^c \supset (G_1^c)^\circ \supset \cdots$, and $\bigcap_n G_n^c = G^c$. Using Proposition 3.2, $P_x(V_{G_n^c} \uparrow V_{G^c}) = 1$ for all $x \in G$. But $P_x(V_{\partial G} = V_{G^c}) = 1$, $x \in G$ and $P_x(V_{\partial G_n} = V_{G_n^c}) = 1$ for $x \in G_n$, so $P_x(V_{\partial G_n} \uparrow V_{\partial G}) = 1$, $x \in G$, as desired.

A set B is called *polar* if $P_x(V_B < \infty) \equiv 0$. Clearly, such sets are negligible since no Brownian motion process can ever hit such a set in positive time. Later when we introduce the notion of capacity as given in classical potential theory we will see that a polar set and a set of capacity 0 are equivalent.

Let B be a Borel set. If $x \in \mathring{B}$ then x is a regular point of B, while $x \in (\bar{B})^c$ is irregular for B. Thus, only points on ∂B are in question. An important fact is that the points in B that are irregular for B constitute a polar set, that is, $(B^r)^c \cap B$ is a polar set. We will prove this fact here for B a closed set. To carry out the proof of this fact we will need some preliminary facts, some of which are of interest in their own right.

We first show that $(B^r)^c \cap B$ is a Borel set having measure 0.

LEMMA 3.1. *Let B be a Borel set. Then B^r is a G_δ set and $D = (B^r)^c \cap B$ has measure* 0.

PROOF. A point x is regular if and only if $P_x(V_B = 0) = 1$. Thus,

$$B^r = \{x: P_x(V_B = 0) = 1\} = \bigcap_{n=1}^{\infty} \left\{x: P_x\left(V_B \leqq \frac{1}{n}\right) > 1 - \frac{1}{n}\right\}. \tag{3.2}$$

By Proposition 2.1, $P_x(V_B \leqq 1/n)$ is a lower semicontinuous function so $\{x: P_x(V_B \leqq 1/n) > 1 - 1/n\}$ is open. To see that D has measure 0, we can proceed as follows. Let $A \subset D$ be relatively compact and note that

$$P_x(X_t \in A) \leqq P_x(V_A \leqq t) \leqq P_x(V_B \leqq t), \tag{3.3}$$

so for $x \in A$,

$$\limsup_{t \downarrow 0} P_x(X_t \in A) \leqq \lim_{t \downarrow 0} P_x(V_B \leqq t) = 0. \tag{3.4}$$

Thus,

$$\lim_{t \downarrow 0} \int_A P_x(X_t \in A)\, dx = 0. \tag{3.5}$$

The function

$$f(x) = \int_{R^n} 1_A(z + x) 1_A(z)\, dz \tag{3.6}$$

is a continuous function and

$$\int_A P_x(X_t \in A)\, dx = E_0 f(X_t). \tag{3.7}$$

Hence,

$$f(0) = \lim_{t \downarrow 0} E_x f(X_t) = \lim_{t \downarrow 0} \int_A P_x(X_t \in A) = 0. \tag{3.8}$$

But $f(0) = |A|$, so $|A| = 0$.

Our next result shows that $E_x(\exp\{-\lambda V_B\})$ is the "λ potential" of the measure μ_B^λ defined by

$$\mu_B^\lambda(A) = \lambda \int_{R^n} \Pi_B^\lambda(x, A)\, dx. \tag{3.9}$$

LEMMA 3.2. *Let B be any Borel set. Then*

$$E_x(\exp\{-\lambda V_B\}) = \int_{R^n} g^\lambda(y - x)\mu_B^\lambda(dy). \tag{3.10}$$

PROOF. Integrating both sides of (2.30) on x over R^n, we see that

$$1 = \int_{\bar{B}} \mu_B^\lambda(dz) g^\lambda(y - z) + \lambda \int_{R^n} g_B^\lambda(x, y)\, dx. \tag{3.11}$$

But by the symmetry of g_B^λ, we see that

$$\int_{R^n} g_B^\lambda(x, y)\, dx = \int_{R^n} g_B^\lambda(y, x)\, dx = [1 - E_y(\exp\{-\lambda V_B\})]\lambda^{-1}. \tag{3.12}$$

This establishes the lemma.

REMARK. Since $g^\lambda(x)$ is bounded away from 0 on compacts, it follows from (3.10) that $\mu_B^\lambda(\bar{B}) < \infty$ whenever $\bar{B}$ is compact.

LEMMA 3.3. *Let B be a compact set and let B_n, $n \geqq 1$, be compacts such that $B_1 \supset \mathring{B}_1 \supset B_2 \supset \cdots$, $\bigcap_n \mathring{B}_n = \bigcap_n B_n = B$. Then the total mass $C^\lambda(B_n)$ of the measure $\mu_{B_n}^\lambda$ converges to the total mass $C^\lambda(B)$ of μ_B^λ as $n \to \infty$.*

PROOF. By Proposition 3.2, for $x \in B^r \cup B^c$,

$$E_x(\exp\{-\lambda V_{B_n}\}) \downarrow E_x(\exp\{-\lambda V_B\}). \tag{3.13}$$

By Lemma 3.1, $(B^r)^c \cap B = (B^r \cup B^c)^c$ has measure 0, so (3.13) holds for a.e. x. Thus, by monotone convergence and (3.10),

$$C^\lambda(B_n) = \lambda \int_{R^n} E_x(\exp\{-\lambda V_{B_n}\})\, dx \downarrow \lambda \int_{R^n} E_x(\exp\{-\lambda V_B\})\, dx = C^\lambda(B). \tag{3.14}$$

This establishes the lemma.

LEMMA 3.4. *Suppose k is a compact set such that* $\sup_{x\in R^n} E_x(\exp\{-\lambda V_k\}) = \beta < 1$. *Then k is polar.*

PROOF. Let k_n be compacts containing k such that $k_1 \supset \mathring{k}_1 \supset k_2 \supset \cdots$, $\bigcap_n k_n = \bigcap_n \mathring{k}_n = k$. Then each point in k is a regular point of k_n for all n. On the one hand (3.10) shows that

$$\int_{R^n}\int_{R^n} g^\lambda(y - x)\mu_k^\lambda(dy)\mu_{k_n}^\lambda(dx) \tag{3.15}$$

$$= \int_{R^n} E_x(\exp\{-\lambda V_k\})\mu_{k_n}^\lambda(dx) \leqq \beta C^\lambda(k_n).$$

On the other hand,

$$(3.16)\qquad \int_{R^n}\int_{R^n} g^\lambda(y - x)\mu_k^\lambda(dy)\mu_{k_n}^\lambda(dx) = \int_{R^n} E_y(\exp\{-\lambda V_{k_n}\})\mu_k^\lambda(dy) = C^\lambda(k).$$

Thus, $C^\lambda(k) \leqq \beta C^\lambda(k_n)$. By Lemma 3.3, $C^\lambda(k_n)\downarrow C^\lambda(k)$, and thus $C^\lambda(k) \leqq \beta C^\lambda(k)$. Hence, $C^\lambda(k) = 0$. But using (3.10), we then see that $E_x(\exp\{-\lambda V_k\}) \equiv 0$, and thus k is polar, as desired.

Let μ be a finite measure having compact support k. Let $G^\lambda\mu(x) = \int_k g^\lambda(y - x)\mu(dy)$. Our next lemma shows that $G^\lambda\mu$ satisfies a maximum principle.

LEMMA 3.5. *Let μ be a finite measure having compact support k and let $M = \sup_{x\in k} G^\lambda\mu(x)$. Then $G^\lambda\mu(x) \leqq M$ for all $x \in R^n$.*

PROOF. If $M = \infty$ there is nothing to prove so assume $M < \infty$. Let $\varepsilon > 0$ be given and let $A = \{x: G^\lambda\mu(x) < M + \varepsilon\}$. Each point of A must be a regular point of A. To see this, suppose $x_0 \in A$ and x_0 is irregular for A. Then $P_{x_0}(V_A = 0) = 0$, and thus

$$(3.17)\qquad \lim_{t\downarrow 0} P_{x_0}(X_t \in A) \leqq \lim_{t\downarrow 0} P_{x_0}(V_A \leqq t) = 0.$$

However,

$$(3.18)\qquad G^\lambda\mu(x_0) \geqq e^{-\lambda t}P^t G^\lambda\mu(x_0) \geqq e^{-\lambda t}\int_{A^c} p(t, y - x_0)G^\lambda\mu(y)\,dy \geqq (M + \varepsilon)\,e^{-\lambda t}P_{x_0}(X_t \in A^c).$$

By (3.17), $\lim_{t\downarrow 0} P_{x_0}(X_t \in A^c)\,e^{-\lambda t} = 1$, and thus $G^\lambda\mu(x_0) \geqq M + \varepsilon$, a contradiction. Hence, $A \subset A^r$. Since $k \subset A \subset A^r$ and $g_A^\lambda(x, y) = 0$ for all $y \in A^r$, we see that $\int_k g_A^\lambda(x, y)\mu(dy) = 0$. By (2.30), we then see that

$$(3.19)\qquad G^\lambda\mu(x) = \int_{\bar{A}} \Pi_A^\lambda(x, dz)G^\lambda\mu(z), \qquad x \in R^n.$$

The function $G^\lambda\mu(z)$ is lower semicontinuous in z (since it is the limit of the increasing sequence $e^{-\lambda/n}P^{1/n}G^\lambda\mu(z)$, $n \geqq 1$, of continuous functions). Consequently, $\{z: G^\lambda\mu(z) \leqq M + \varepsilon\}$ is a closed set. Clearly $\bar{A} \subset \{z: G^\lambda\mu(z) \leqq M + \varepsilon\}$, and thus by (3.19), $G^\lambda\mu(x) \leqq M + \varepsilon$ for all $x \in R^n$. As ε is arbitrary, $G^\lambda\mu(x) \leqq M$ for all x. This establishes the lemma.

COROLLARY 3.2. *Let A be a closed set such that $A^r = \varnothing$. Then A is polar.*

PROOF. It suffices to consider the case when A is compact. The function $E_x(\exp\{-\lambda V_A\})$ is lower semicontinuous. Indeed, $P_x(V_A \leqq t)$ is lower semicontinuous in x and thus by Fatou's lemma,

$$\text{(3.20)}\quad \liminf_{x\to x_0} E_x(\exp\{-\lambda V_A\} = \liminf_{x\to x_0} \lambda \int_0^\infty P_x(V_A \leqq t)\, e^{-\lambda t}\, dt$$
$$\geqq \lambda \int_0^\infty \liminf_{x\to x_0} P_x(V_A \leqq t)\, e^{-\lambda t}\, dt$$
$$\geqq \lambda \int_0^\infty P_{x_0}(V_A \leqq t)\, e^{-\lambda t}\, dt = E_{x_0}(\exp\{-\lambda V_A\}).$$

Let

$$\text{(3.21)}\qquad A_n = \left\{x: E_x(\exp\{-\lambda V_A\}) \leqq 1 - \frac{1}{n}\right\} \cap A.$$

Then $A_n \subset A$ is compact and by Lemma 3.2, for $x \in A_n$,

$$\text{(3.22)}\qquad G^\lambda \mu^\lambda_{A_n}(x) = E_x(\exp\{-\lambda V_{A_n}\}) \leqq E_x(\exp\{-\lambda V_A\}) \leqq 1 - \frac{1}{n}.$$

By Lemmas 3.2 and 3.5, we then see that $E_x(\exp\{-\lambda V_{A_n}\}) \leqq 1 - 1/n$ for all $x \in R^n$, and thus by Lemma 3.4, A_n is polar. Since $A = \bigcup_n A_n$, A is polar.

We may now prove the following basic theorem.

THEOREM 3.1. *Let B be any closed set. Then $(B^r)^c \cap B$ is polar.*

PROOF. It suffices to consider B compact. Let

$$\text{(3.23)}\qquad D_n = \left\{x: E_x(\exp\{-\lambda V_B\}) \leqq 1 - \frac{1}{n}\right\},$$

and let $B_n = B \cap D_n$. Then

$$\text{(3.24)}\qquad (B^r)^c \cap B = \bigcup_{n=1}^\infty B_n.$$

Since $B_n \subset B$, $E_x(\exp\{-\lambda V_{B_n}\}) \leqq E_x(\exp\{-\lambda V_B\})$. If x is irregular for B then $E_x(\exp\{-\lambda V_{B_n}\}) < 1$ so x is irregular for B_n. Suppose x is regular for B. Then $x \in D_n^c$, and as D_n is closed, x is irregular for D_n and hence also for B_n. Thus, $B_n^r = \varnothing$. Corollary 3.2 then implies that each B_n is polar, and thus $(B^r)^c \cap B$ is polar. This establishes the theorem.

We conclude this section by pointing out a simple corollary of Lemma 3.2. Though its proof is trivial it is important enough to be stated as

THEOREM 3.2. *A one point set is a polar set.*

PROOF. Using (3.10) for the set $B = \{a\}$, we see that

$$\text{(3.25)}\qquad E_x(\exp\{-\lambda V_{\{a\}}\}) = g^\lambda(a - x)\, \Pi^\lambda_{\{a\}}(a).$$

Since $E_a(\exp\{-\lambda V_{\{a\}}\}) \leqq 1$ and $g^\lambda(0) = \infty$, it must be that $\Pi^\lambda_{\{a\}}(a) = 0$. But then $E_x(\exp\{-\lambda V_{\{a\}}\}) \equiv 0$ so $P_x(V_{\{a\}} < \infty) \equiv 0$.

4. Dirichlet problem

Let G be an open set. The *classical Dirichlet problem* for G with boundary function φ is as follows. Given φ on ∂G find f harmonic in G and continuous on $\overline{G}$ such that $f = \varphi$ on ∂G. In general, even when φ is restricted to be a bounded continuous function at each point of ∂G, this problem may have no solution. If solutions do exist, then unless G is bounded, they may not be unique. The *modified Dirichlet problem* eases the continuity requirements on $\overline{G}$ by allowing the function f (which must still be harmonic in G) to be discontinuous at an exceptional set of points on ∂G. One of the nicest connections between Brownian motion and potential theory is the elegant and simple treatment it allows for the modified Dirichlet problem.

DEFINITION 4.1. *Let G be an open set. A point $x_0 \in \partial G$ is said to be nonsingular if $\lim_{x \in G, x \to x_0} \Pi_{\partial G}\, \varphi(x) = \varphi(x_0)$ for all bounded functions that are continuous on ∂G. Otherwise a point $x_0 \in \partial G$ is called singular.*

THEOREM 4.1. *Let φ be a bounded measurable function defined on ∂G. Then $\Pi_{\partial G}\varphi(x)$ is harmonic in G. Moreover, if $x_0 \in \partial G$ is a point of continuity of φ and is also a regular point of G^c, then $\lim_{x \in G, x \to x_0} \Pi_{\partial G}\varphi(x) = \varphi(x_0)$.*

PROOF. It is easily seen that $P^t \Pi_{\partial G}\, \varphi(x)$ is continuous in x for all $t > 0$ and that $\lim_{t \to 0} P^t \Pi_{\partial G}\, \varphi(x) = \Pi_{\partial G}\, \varphi(x)$ uniformly on compact subsets of G. We conclude that $\Pi_{\partial G}\, \varphi(x)$, $x \in G$, is continuous. Let $x \in G$ and let S_r be a ball of center x and radius r such that $\overline{S}_r \subset G$. Then clearly the process starting from x must first hit ∂S_r in positive time before it can hit ∂G. Thus,

$$\Pi_{\partial G}\, \varphi(x) = \int_{\partial S_r} \Pi_{\partial S_r}(x, dz)\, \Pi_{\partial G}\, \varphi(z). \tag{4.1}$$

But as argued in Section 2, $\Pi_{\partial S_r}(x, dz) = \sigma_r(x, dz)$. Thus,

$$\Pi_{\partial G}\, \varphi(x) = \int_{\partial S_r} \Pi_{\partial G}\, \varphi(z) \sigma_r(x, dz), \tag{4.2}$$

so $\Pi_{\partial G}\, \varphi$ is indeed harmonic at x.

Suppose now that $x_0 \in \partial G$ is both a point of continuity of φ and a regular point of G^c. Given $\delta > 0$ and $\varepsilon > 0$, we can find a t_0 such that $P\big(\sup_{t \leqq t_0} |X_t - X_0| \leqq \frac{1}{2}\delta\big) > 1 - \varepsilon$. Let N be any neighborhood of x_0, and let $D(\delta)$ be a closed ball of center x_0 and radius $\delta \subset N$. Let $B = G^c$. Then for $x \in D(\frac{1}{2}\delta)$,

$$\begin{aligned} P_x(V_B \leqq t_0, X_{V_B} \notin N) & \\ &\leqq P\big(|X_t - X_0| > \tfrac{1}{2}\delta \text{ for some } t \leqq t_0\big) \leqq \varepsilon. \end{aligned} \tag{4.3}$$

Thus, for $x \in D(\frac{1}{2}\delta)$,

$$\begin{aligned} P_x(V_B \leqq t_0, X_{V_B} \in N) &= P_x(V_B \leqq t_0) - P_x(V_B \leqq t_0, X_{V_B} \notin N) \\ &\geqq P_x(V_B \leqq t_0) - \varepsilon. \end{aligned} \tag{4.4}$$

Since x_0 is regular for B and $P_x(V_B \leqq t)$ is a lower semicontinuous function in x, we see that

(4.5) $$1 = P_{x_0}(V_B \leqq t_0) \leqq \liminf_{x \to x_0} P_x(V_B \leqq t_0) \leqq 1.$$

Thus, by (4.4), $1 \geqq \liminf_{x \to x_0} P_x(V_B \leqq t_0, X_{V_B} \in N) \geqq 1 - \varepsilon$ and as ε is arbitrary we see that

(4.6) $$\lim_{x \to x_0} P_x(V_B \leqq t_0, X_{V_B} \in N) = 1.$$

But

(4.7) $$1 \geqq \Pi_B(x, N) \geqq P_x(V_B \leqq t_0, X_{V_B} \in N),$$

so for any neighborhood N of x_0

(4.8) $$\lim_{x \to x_0} \Pi_B(x, N^c) = 0.$$

Moreover, as $P_x(V_B \leqq t_0) \leqq P_x(V_B < \infty)$, equation (4.4) shows that

(4.9) $$\lim_{x \to x_0} P_x(V_B < \infty) = 1.$$

Now as φ is continuous at x_0, we can choose a neighborhood N of x_0 such that $|\varphi(x) - \varphi(x_0)| \leqq \varepsilon$, $x \in N$. But

(4.10) $$\begin{aligned} |\Pi_B \varphi(x) - \varphi(x_0)| &\leqq \int_N \Pi_B(x, dz)|\varphi(z) - \varphi(x_0)| \\ &\quad + \int_{N^c} \Pi_B(x, dz)|\varphi(z) - \varphi(x_0)| + \varphi(x_0)P_x(V_B = \infty) \\ &\leqq \varepsilon + 2\|\varphi\| \Pi_B(x, N^c) + \varphi(x_0)P_x(V_B = \infty). \end{aligned}$$

Using (4.8) and (4.9), we see that

(4.11) $$\lim_{x \to x_0} \Pi_B \varphi(x) = \varphi(x_0).$$

Finally, if $x \in G$, then $P_x(V_{\partial G} = V_B) = 1$. Thus (4.11) shows that

(4.12) $$\lim_{x \to x_0, x \in G} \Pi_{\partial G} \varphi(x) = \varphi(x_0),$$

as desired. This establishes the theorem.

COROLLARY 4.1. *A point $x_0 \in \partial G$ is nonsingular if and only if it is a regular point for G^c. The set of singular points of ∂G is thus a polar F^σ set.*

PROOF. If $x_0 \in (G^c)^r \cap \partial G$, then Theorem 4.1 shows that x_0 is a nonsingular point. Suppose now that x_0 is a nonsingular point. Then $\Pi_{\partial G}(x, dy)$ converges weakly to the unit mass $\varepsilon_{x_0}(dy)$ at x_0. Thus, given any $\varepsilon > 0$ and any neighborhood N of x_0, we can find a closed ball S of center $x_0 \subset N$ such that $P_x(X_{V_{\partial G}} \in N) \geqq 1 - \varepsilon$, $x \in S \cap G$. But

(4.13) $$P_{x_0}(X_{V_{\partial G}} \in N) = P_{x_0}(X_{V_{\partial G}} \in N; V_{\partial G} \leqq V_{\partial S}) + P_{x_0}(X_{V_{\partial G}} \in N; V_{\partial G} > V_{\partial S}).$$

Since $S \subset N$, $P_{x_0}(X_{V_{\partial G}} \in N \mid V_{\partial G} \leqq V_{\partial S}) = 1$, so

$$P_{x_0}(X_{V_{\partial G}} \in N;\ V_{\partial G} \leqq V_{\partial S}) = P_{x_0}(V_{\partial G} \leqq V_{\partial S}). \tag{4.14}$$

Also,

$$\begin{aligned} P_{x_0}(X_{V_{\partial G}} \in N \mid V_{\partial G} > V_{\partial S}) &= \int_{\partial S} P_{x_0}(V_{\partial G} > V_{\partial S};\ X_{V_{\partial S}} \in dz) P_z(X_{V_{\partial G}} \in N) \\ &\geqq (1 - \varepsilon) P_{x_0}(V_{\partial G} > V_{\partial S}). \end{aligned} \tag{4.15}$$

Thus, $P_{x_0}(X_{V_{\partial G}} \in N) \geqq 1 - \varepsilon$.

Hence, $\Pi_{\partial G}(x_0, dy)$ is the unit mass at x_0 so $P_{x_0}(X_{V_{\partial G}} = x_0) = 1$. But Theorem 3.2 shows that $\{x_0\}$ is a polar set. Thus, $P_{x_0}\{X_{V_{\partial G}} = x_0\} = 1$ can only be true if $P_{x_0}(V_{\partial G} = 0) = 1$, so x_0 is a regular point of ∂G. Since $\partial G \subset G^c$, x_0 is then also a regular point for G^c. This establishes the theorem.

Theorem 4.1 and its corollary show at once that for φ a bounded continuous function on ∂G, the function $\Pi_{\partial G}\,\varphi$ is a solution to the Dirichlet problem for G with boundary function φ provided each point of ∂G is nonsingular.

From Theorem 3.1 and Corollary 4.1, we know that the set of singular points of ∂G is a polar set. Let N be any polar set that contains all singular points of ∂G. Let φ be a bounded function on ∂G that is continuous at each point of $N^c \cap \partial G$. The *modified Dirichlet problem* consists in finding a function f harmonic on G and continuous on $G \cup (\partial G \cap N^c)$ such that $f = \varphi$ on ∂G. We know that $\Pi_{\partial G}\,\varphi$ is a solution to the modified problem. If we choose $\varphi \equiv 1$, we see that $P_x(V_{\partial G} < \infty)$ is a solution. Thus, $P_x(V_{\partial G} = \infty) = 1 - P_x(V_{\partial G} < \infty)$ is a solution to the modified problem with boundary function 0. Our principal goal in the remainder of this section is to show that $\Pi_{\partial G}\,\varphi(x) + \alpha P_x(V_{\partial G} = \infty)$ are the only bounded solutions to the modified problem.

We will start our investigation with a bounded G.

THEOREM 4.2. *Let G be a bounded open set and let N be a polar set that contains the singular points of ∂G. Suppose φ is a bounded function on ∂G that is continuous on $\partial G \cap N^c$. Then $\Pi_{\partial G}\,\varphi$ is the unique bounded solution to the modified Dirichlet problem for G with boundary function φ.*

PROOF. Suppose first that all points on ∂G are nonsingular and that the exceptional set N is empty. Then the modified problem becomes just the classical Dirichlet problem. Suppose f is any solution. Then $f - \Pi_{\partial G}\,\varphi = h$ vanishes on ∂G, is harmonic on G, and continuous on $\bar{G}$. The maximum principle then tells us that h vanishes on G so $f = \Pi_{\partial G}\,\varphi$ on G.

Consider now an arbitrary bounded open set G and allow an exceptional polar set N containing the singular points of ∂G. By Corollaries 3.1 and 4.1, we can exhaust G by an increasing sequence of open sets $G_1 \subset \bar{G}_1 \subset G_2 \subset \cdots$, $\bigcup_n G_n = G$, such that all points of ∂G_n are nonsingular, and such that $P_x(V_{\partial G_n} \uparrow V_{\partial G}) = 1$ for all $x \in G$. Let f be a bounded solution to the modified problem on G. Then f is continuous on $\bar{G}_n$ and harmonic on G_n so it is a solution to the classical Dirichlet problem on G_n with boundary function f on ∂G_n. By what was proved above, $\Pi_{\partial G_n} f$ is the unique solution to this problem, so

$$f(x) = \Pi_{\partial G_n} f(x), \qquad x \in G_n. \tag{4.16}$$

Now as $\bar{G}$ is compact, Proposition 2.2 shows that $E_x V_{\bar{G}^c} < \infty$, so certainly $P_x(V_{G^c} < \infty) = 1$. But for any $x \in G$, $P_x(V_{G^c} = V_{\partial G}) = 1$, so $P_x(V_{\partial G} < \infty) = 1$ for all $x \in G$. Since $\bar{G}_n \subset G$, it must also be that $P_x(V_{\partial G_n} < \infty) = 1$ for all $x \in G_n$. Fix $x \in G$. Then

$$\Pi_{\partial G_n} f(x) = E_x[f(X_{V_{\partial G_n}})]. \tag{4.17}$$

Now $P_x(V_{\partial G_n} \uparrow V_{\partial G}) = 1$ and as $P_x(V_{\partial G} < \infty) = 1$, we see that $P_x(\lim_{n\to\infty} X_{V_{\partial G_n}} = X_{V_{\partial G}}) = 1$. But $P_x(X_{V_{\partial G}} \in N) = 0$ and f is continuous on $(\partial G) \cap N^c$ so

$$P_x\big(\lim_{n\to\infty} f(X_{V_{\partial G_n}}) = \varphi(X_{V_{\partial G}})\big) = 1. \tag{4.18}$$

Since f is bounded,

$$\lim_{n\to\infty} E_x f(X_{V_{\partial G_n}}) = E_x[\lim_{n\to\infty} f(X_{V_{\partial G_n}})] = E_x \varphi(X_{V_{\partial G}}). \tag{4.19}$$

Using (4.16) and (4.19), we see that for any $x \in G$, $f(x) = \Pi_{\partial G}\, \varphi(x)$ as desired.

COROLLARY 4.2. *Let G be a bounded open set. The classical Dirichlet problem has a solution for all continuous boundary functions φ if and only if ∂G has no singular points. In that case $\Pi_{\partial G}\, \varphi$ is the unique solution with boundary function φ.*

PROOF. Since $\bar{G}$ is compact, a solution f of the classical Dirichlet problem for G for φ continuous on ∂G is automatically a bounded solution for the modified problem with N being the set of all singular points of ∂G. But then $f(x) = \Pi_{\partial G}\, \varphi(x)$, $x \in G$. As f is a classical solution, $\lim_{x\to x_0} \Pi_{\partial G}\, \varphi(x_0) = \varphi(x_0)$. Since this is true for all φ continuous on ∂G, we see that all points in ∂G are nonsingular. This establishes the corollary.

To handle the case when G is unbounded, we will require some preliminary information on the process X_t stopped when it hits G^c (for typographical simplicity we shall put $G^c = F$). Let $Y_t = X_{t \wedge V_F}$ and set ${}_F P^t f(x) = E_x f(Y_t)$, and

$$Q_F^t f(x) = E_x[f(X_t), V_F > t] = \int_G q_F(t, x, y) f(y)\, dy. \tag{4.20}$$

Note that for $x \in G$, $P_x(V_{\partial G} = V_F) = 1$, so that for $x \in G$,

$${}_F P^t f(x) = Q_F^t f(x) + E_x[f(X_{V_{\partial G}}); V_{\partial G} \leqq t]. \tag{4.21}$$

We say that a function defined on $\bar{G}$ is invariant for ${}_F P^t$ on G if

$${}_F P^t f(x) = f(x), \qquad x \in G. \tag{4.22}$$

Similarly, a function f defined on G is Q_F^t invariant if $f(x) = Q_F^t f(x)$, $x \in G$.

LEMMA 4.1. *For any bounded function* φ *on* ∂G, $\Pi_{\partial G}\,\varphi(x) + \alpha P_x(V_{\partial G} = \infty)$ *is a bounded* ${}_FP^t$ *invariant function on* G. *Conversely, every bounded* ${}_FP^t$ *invariant function on* G *is of this form.*

PROOF. Let φ be bounded on ∂G and let $x \in G$. Then by (4.21),

$$(4.23)\qquad {}_FP^t\,\Pi_{\partial G}\,\varphi(x) = Q_F^t\,\Pi_{\partial G}\,\varphi(x) + E_x[\Pi_{\partial G}\,\varphi(X_{V_{\partial G}});\,V_{\partial G} \leqq t].$$

By Theorem 3.1, $P_x(X_{V_{\partial G}} \in (\partial G)^r) = 1$. Since $\Pi_{\partial G}\,(x, dy)$ is the unit mass at x if $x \in (\partial G)^r$, we see that $\Pi_{\partial G}\,\varphi(X_{V_{\partial G}}) = \varphi(X_{V_{\partial G}})$ with probability one, and thus the second term on the right in (4.23) is just $E_x[\varphi(X_{V_{\partial G}})\colon V_{\partial G} \leqq t]$. The first term is just

$$(4.24)\qquad Q_F^t\,\Pi_{\partial G}\,\varphi(x) = E_x[\varphi(X_{V_{\partial G}});\,t < V_{\partial G} < \infty].$$

Hence, the right side of (4.23) is just $\Pi_{\partial G}\,\varphi(x)$. Thus, for any φ, $\Pi_{\partial G}\,\varphi$ is ${}_FP^t$ invariant. In particular, for $\varphi \equiv 1$, we see that $P_x(V_{\partial G} < \infty)$ is ${}_FP^t$ invariant, and thus as 1 is clearly ${}_FP^t$ invariant so is $1 - P_x(V_{\partial G} < \infty) = P_x(V_{\partial G} = \infty)$. This shows that $\Pi_{\partial G}\,\varphi + \alpha P_x(V_{\partial G} = \infty)$ is ${}_FP^t$ invariant. It is clearly bounded if φ is bounded.

Suppose now that f is any bounded ${}_FP^t$ invariant function on G. Since constants are ${}_FP^t$ invariant, we can assume that $f \geqq 0$. Then

$$(4.25)\qquad E_x[f(X_{V_{\partial G}});\,V_{\partial G} \leqq t] \uparrow \Pi_{\partial G}\,f(x), \qquad t \to \infty.$$

By (4.21), we then see that $Q_F^t\,f$ is decreasing as $t \to \infty$. Let h denote its limit. Then for $x \in G$, $f(x) = h(x) + \Pi_{\partial G}\,f(x)$.

By dominated convergence and the semigroup property of Q_F^t, we see that for $x \in G$

$$(4.26)\qquad \begin{aligned} Q_F^t h(x) &= Q_F^t[\lim_{s\to\infty} Q_F^s f](x) = \lim_{s\to\infty} Q_F^t[Q_F^s f](x) \\ &= \lim_{s\to\infty} Q_F^{t+s} f(x) = h(x), \end{aligned}$$

so h is Q_F^t invariant. Note that $Q_F^t h(x) = 0$ for a.e. $x \in F$. Thus, if we define $h(x)$ to be 0 in $G^c(F)$, we see that

$$(4.27)\qquad P^{t+s}h(x) = P^t(P^s h)(x) \geqq P^t(Q_F^s h)(x) = P^t h(x),$$

so $P^t h(x)$ is increasing in t. As $P^t h(x) \leqq \sup_x h(x)$, $\lim_{t\to\infty} P^t h(x) = h_1(x)$ exists. The monotone convergence theorem then shows that $P^t h_1(x) = h_1(x)$ for all $x \in R^n$ and all $t > 0$. Thus, h_1 must be a constant. (See Proposition 2.3.) Denote this constant by α.

Now

$$(4.28)\qquad \begin{aligned} P^t h(x) &= Q_F^t h(x) + E_x[h(X_t);\,V_F \leqq t] \\ &= h(x) + E_x[h(X_t);\,V_F \leqq t], \end{aligned}$$

so taking the limit as $t \to \infty$ we see that

$$\alpha - h(x) = \lim_{t\to\infty} E_x[h(X_t);\ V_F \leqq t] = h_2(x). \tag{4.29}$$

Now $h_2(x) \leqq [\sup_x h(x)]P_x(V_F < \infty)$, so

$$Q_F^s h_2(x) \leqq [\sup_x h(x)]P_x(s < V_F < \infty). \tag{4.30}$$

Thus, $\lim_{s\to\infty} Q_F^s h_2(x) = 0$. But then, as $Q_F^s \alpha = \alpha P_x(V_F > s)$, we see from (4.29) that

$$\alpha P_x(V_F > s) = h(x) + Q_F^s h_2(x). \tag{4.31}$$

Letting $s \to \infty$, we see that $\alpha P_x(V_F = \infty) = h(x)$. Since for $x \in G$, $P_x(V_F = \infty) = P_x(V_{\partial G} = \infty)$ the lemma is proved.

We can now establish the following theorem.

THEOREM 4.3. *Let G be any open set and let N be a polar set that contains the singular points of ∂G. Suppose φ is bounded on ∂G and continuous on $(\partial G) \cap N^c$. Then the only bounded solutions f to the modified Dirichlet problem for G with boundary function φ are*

$$f(x) = \Pi_{\partial G}\, \varphi(x) + \alpha P_x(V_{\partial G} = \infty), \tag{4.32}$$

for α an arbitrary constant. Conversely, every such function is a solution of the modified problem.

PROOF. By Theorem 4.1 and Corollary 4.1, we already know that $\Pi_{\partial G}\, \varphi + \alpha P_x(V_{\partial G} = \infty)$ is a bounded solution. Suppose f is any other bounded solution. Let S_r be the open ball of center 0 and radius r and let $G_r = G \cap S_r$. Consider the modified Dirichlet problem on G_r with boundary function f. Then clearly f as a function on G_r is a solution. Since G_r is bounded, Theorem 4.2 tells us that

$$f(x) = \Pi_{\partial G_r} f(x), \qquad x \in G_r. \tag{4.33}$$

By Lemma 4.1, $\Pi_{\partial G_r} f$ is a bounded ${}_{F_r}P^t$ invariant function on G_r, where we use $(G_r)^c = F_r$ for typographical reasons. Then for any $t > 0$ and $x \in G_r$,

$$\begin{aligned} f(x) = \Pi_{\partial G_r} f(x) &= Q_{F_r}^t\, \Pi_{\partial G_r} f(x) + E_x[\Pi_{\partial G_r} f(x_{V_{\partial G_r}});\ V_{\partial G_r} \leqq t] \\ &= \int_{G_r} P_x(V_{F_r} > t,\ X_t \in dy) f(y) + E_x[f(X_{V_{\partial G_r}});\ V_{\partial G_r} \leqq t]. \end{aligned} \tag{4.34}$$

Now $V_{\partial S_r} \uparrow \infty$ with probability one, and thus for $x \in G$

$$P_x(V_{\partial G_r} = V_{\partial G} \text{ for all sufficiently large } r \text{ whenever } V_{\partial G} < \infty) = 1. \tag{4.35}$$

If $V_{\partial G} = \infty$ then $V_{\partial G_r} \uparrow \infty$. Hence, in every case when $x \in G$, $P_x(\lim_r V_{\partial G_r} = V_{\partial G}) = 1$. Then (4.21) and (4.34) show that as $r \to \infty$, for $x \in G$,

$$(4.36) \qquad \begin{aligned} f(x) &= \int_G P_x(V_F > t, X_t \in dy) f(y) + E_x[f(X_{V_{\partial G}}); V_{\partial G} \leqq t] \\ &= {}_F P^t f(x). \end{aligned}$$

Thus, f is ${}_F P^t$ invariant on G. Lemma 4.1 then shows that $f(x) = \Pi_{\partial G} f + \alpha P_x(V_{\partial G} = \infty)$, $x \in G$. Since f is defined to be φ on ∂G, we see that $f(x) = \Pi_{\partial G}\, \varphi(x) + \alpha P_x(V_B = \infty)$ as desired. This establishes the theorem.

As a simple application of this theorem we prove the following extension property of harmonic functions relative to polar sets.

COROLLARY 4.3. *Let G be an open set. Suppose f is locally bounded and harmonic on G except perhaps on a relatively closed polar subset N. Then f extends uniquely to a harmonic function on G.*

PROOF. Let S be an open ball such that $\bar{S} \subset G$ and let $G_1 = S \cap N^c$. Then each point on ∂G_1 not in N is regular for G_1^c and f is continuous at each point of ∂G_1 not in N. Thus, as f is bounded on G_1 it is the unique bounded solution to the modified Dirichlet problem for ∂G_1 with boundary function f. But $\Pi_{\partial S} f$ is harmonic on S and assumes boundary value $f(x_0)$ at each point of ∂S not in N. Thus, it too is a bounded solution to the modified Dirichlet problem for G_1. Therefore, $f(x) = \Pi_{\partial S} f(x)$, $x \in G_1$. This shows that f can be extended to be a harmonic function on S and thus $\lim_{x \to x_0} f(x) = f(x_0)$ must exist for each $x_0 \in S$. Since S can be any open ball $\subset G$, f extends everywhere in G as a harmonic function. Define $f^*(x)$ to be $f(x)$ for $x \in G \cap N^c$ and define $f^*(x_0) = \lim_{x \to x_0} f(x)$ for $x_0 \in N$. Then $f^*(x)$ is harmonic on G and agrees with f on $G \cap N^c$. Since N has measure 0, f^* is the unique such function.

5. Newtonian potentials

Throughout the remainder of this paper we will consider Brownian motion in R^n, $n \geqq 3$. The planar case will be treated in our companion paper in this volume.

An easy computation shows that when $n \geqq 3$,

$$(5.1) \qquad \begin{aligned} \lim_{\lambda \downarrow 0} g^\lambda(x) &= \int_0^\infty p(t, x)\, dt \\ &= \int_0^\infty (2\pi t)^{-n/2} \exp\left\{\frac{-|x|^2}{2t}\right\} dt = \frac{\Gamma\left(\frac{n}{2} - 1\right)}{2\pi^{n/2}} |x|^{2-n}. \end{aligned}$$

where the convergence is uniform in compacts not containing 0. For a function f or measure μ define Gf and $G\mu$ by

$$(5.2) \qquad Gf(x) = \int_{R^n} g(y - x) f(y)\, dy,$$

and

$$G\mu(x) = \int_{R^n} g(y - x)\mu(dy), \tag{5.3}$$

respectively. Gf is called *the potential of f* and $G\mu$ the potential of μ. One easily checks that Gf is a continuous function vanishing at ∞ whenever f is a bounded measurable function with compact support and that $G\mu(x)$ is lower semicontinuous and superharmonic whenever μ is a finite measure. It is useful to know that the potential of μ determines μ whenever $G\mu$ is sufficiently finite.

THEOREM 5.1. *If μ is a measure such that $G\mu < \infty$ a.e. then $G\mu$ determines μ.*

PROOF. Suppose μ and ν are two measures such that $G\mu$ and $G\nu < \infty$ a.e. and $G\mu = G\nu$ a.e. Then $P^tG\mu = P^tG\nu$, and for any point x for which $G\mu < \infty$ we have

$$G\mu(x) - P^tG\mu(x) = \int_0^t P^s\mu(x)\,ds, \tag{5.4}$$

where $P^s\mu(x) = \int_{R^n} p(s, y - x)\mu(dy)$. Thus, if x is such that $G\mu(x) < \infty$ and $G\nu(x) < \infty$, then

$$\int_0^t P^s\mu(x)\,ds = \int_0^t P^s\nu(x)\,ds, \tag{5.5}$$

so (5.5) holds for a.e. x. Let h be a bounded, nonnegative function having compact support such that $0 < \int_{R^n} G\mu(x)h(x)\,dx < \infty$ and set $g = Gh$. Then g is a bounded strictly positive continuous function and

$$\int_{R^n} g(x)\mu(dx) = \int_{R^n} G\mu(x)h(x)\,dx = \int_{R^n} G\nu(x)h(x)\,dx \tag{5.6}$$
$$= \int_{R^n} g(x)\nu(dx) < \infty.$$

Let f be any continuous function, $0 \leqq f \leqq 1$. Observe that

$$\int_{R^n} f(x)g(x)P^s\mu(x)\,dx = \int_{R^n} \mu(dx)P^s(fg)(x) \tag{5.7}$$

and

$$\int_{R^n} f(x)g(x)P^s\nu(x)\,dx = \int_{R^n} \nu(dx)P^s(fg)(x). \tag{5.8}$$

Using (5.5), we then see that

$$\frac{1}{t}\int_{R^n} \mu(dx)\int_0^t P^s(fg)(x)\,ds = \frac{1}{t}\int_{R^n} \nu(dx)\int_0^t P^s(fg)(x)\,ds. \tag{5.9}$$

Since fg is a bounded continuous function, $P^sfg \to fg$ as $s \downarrow 0$, and $P^sfg \leqq P^sg = P^sGh \leqq Gh = g$, and by (5.6) g is both μ and ν integrable. Hence, using (5.9) and dominated convergence, we see upon letting $t \downarrow 0$ in (5.9) that

$$\int_{R^n} \mu(dx) g(x) f(x) = \int_{R^n} \nu(dx) g(x) f(x). \tag{5.10}$$

Since f can be any bounded continuous function, (5.10) shows that $\mu(dx)g(x) = \nu(dx)g(x)$ and as $g > 0$ for all x it must be that $\mu(dx) = \nu(dx)$. This establishes the theorem.

A useful fact about Brownian motion in R^n, $n \geqq 3$, is the following proposition.

PROPOSITION 5.1. *Let B be any bounded set. Then*

$$\lim_{t \to \infty} P_x(X_s \in B \text{ for some } s > t) = 0, \tag{5.11}$$

or equivalently, $P_x(\lim_{t \to \infty} |X_t| = \infty) = 1$.

PROOF. Let k be a compact set of positive measure such that

$$\inf_{0 \leqq s \leqq 1} \inf_{z \in \bar{B}} P_z(X_s \in k) = \delta > 0. \tag{5.12}$$

By integrating from 0 to $t + 1$ on both sides of (2.15) and then integrating by parts, it follows that

$$\begin{aligned} \int_0^{t+1} P_x(X_s \in k)\, ds &\geqq \int_{\bar{B}} \int_t^{t+1} P_x(V_B \leqq s, X_{V_B} \in dz) P_z(X_{t+1-s} \in k)\, ds \\ &\geqq \delta P_x(V_B \leqq t). \end{aligned} \tag{5.13}$$

Letting $t \uparrow \infty$, we see that $\int_k g(y - x)\, dy \geqq \delta P_x(V_B < \infty)$. But

$$\begin{aligned} P_x(X_s \in B \text{ for some } s > t) &= \int_{R^n} p(t, z - x) P_z(V_B < \infty)\, dz \\ &\leqq \delta^{-1} \int_{R^n} p(t, z - x) \left[\int_k g(y - z)\, dy \right] dz \\ &= \delta^{-1} \int_t^{\infty} P^s 1_k(x)\, ds, \end{aligned} \tag{5.14}$$

where 1_k is the indicator function of k. Since

$$\lim_{t \uparrow \infty} \int_t^{\infty} P^s 1_k(x) = 0, \tag{5.15}$$

Proposition 5.1 holds.

Using (2.30) and monotone convergence, we see that for any Borel set B,

$$g(y - x) = \int_{\bar{B}} \Pi_B(x, dz) g(y - z) + g_B(x, y). \tag{5.16}$$

It is quite easy to prove the following theorem.

THEOREM 5.2. *The function g_B has the following properties:*

(i) $g_B \geqq 0$;

(ii) $g_B(x, y) = g_B(y, x)$;

(iii) $g_B(x, y) < \infty$ *for* $x \neq y$ *and* $g_B(x, x) = \infty$ *for* $x \in (\bar{B})^c$;

(iv) *for fixed* x, $g_B(x, \cdot)$ *is upper semicontinuous and subharmonic on* $R^n - \{x\}$;

(v) *for fixed* x, $g_B(x, y)$ *is harmonic in* $y \in (\bar{B})^c - \{x\}$;

(vi) *for fixed* x, $g_B(x, y) - g(y - x)$ *is harmonic in* $y \in (\bar{B})^c$;

(vii) $\lim_{y \to y_0} g_B(x, y) = g_B(x, y_0) = 0$ *if* $y_0 \in B^r$.

PROOF. Properties (i) and (ii) follow from the fact that they are true for $g_B^\lambda \uparrow g_B$, $\lambda \downarrow 0$. Properties (iii) to (vi) follow at once from (5.16) and the fact that

$$\int_{\bar{B}} \Pi_B x, dz) g(y - z), \tag{5.17}$$

as a function of y is lower semicontinuous, superharmonic on R^n, and harmonic on $(\bar{B})^c$. Finally, to see that (vii) is true note that if $y_0 \in B^r$, then $g_B^\lambda(x, y_0) = g_B^\lambda(y_0, x) \equiv 0$, so $g_B(x, y_0) = 0$. But by (iv)

$$0 \leqq \limsup_{y \to y_0} g_B(x, y) \leqq g_B(x, y_0) = 0. \tag{5.18}$$

Let G be an open set. *The Green function* g_G^* of G is the smallest nonnegative function h defined on $G \times G$ such that $h(x, y) - g(y - x)$ is harmonic in y.

An important connection between potential theory and Brownian motion is that g_{G^c} as a function on $G \times G$ is the Green function.

THEOREM 5.3. *The Green function of the open set G is the function g_{G^c} restricted to $G \times G$.*

PROOF. The proof of Theorem 5.2 tells us that g_{G^c} has the required properties so it is only necessary to show that g_{G^c} restricted to $G \times G$ is the smallest such function. Suppose g^* is another function having the required properties. Consider first the case when every point of ∂G is regular for G^c and G is bounded. Then for any $y_0 \in \partial G$ we see by Theorem 5.2 (vii) that

$$\liminf_{y \to y_0} [g^*(x, y) - g_{G^c}(x, y)] = \liminf_{y \to y_0} g^*(x, y) \geqq 0. \tag{5.19}$$

Since $g^*(x, y) - g_{G^c}(x, y)$ is harmonic in G, the minimum principle tells us that

$$g^*(x, y) - g_{G^c}(x, y) \geqq 0 \qquad \text{for all } y \in G. \tag{5.20}$$

Suppose now that G is any open set. By Corollary 3.1, we can find open sets $G_n \subset G$, $G_1 \subset \bar{G}_1 \subset G_2 \subset \cdots$, $\cup_n G_n = G$, such that each point of ∂G_n is regular for G_n^c and such that $P_x(V_{\partial G_n} \uparrow V_{\partial G}) = 1$ for all $x \in G$. The function g^* viewed as a function on $G_n \times G_n$ has the required properties, and thus by what has just been proven

$$g^*(x, y) \geqq g_{G_n^c}(x, y), \qquad x, y \in G_n. \tag{5.21}$$

Now the functions $g_{G_n^c}$, $n \geqq 1$, are increasing for $x, y \in G$. Indeed, since each point of G_n^c is regular for G_n^c, $0 = g_{G_n^c}(x, y) \leqq g_{G_{n+1}^c}(x, y)$, if either $x \in G_n^c$ or $y \in G_n^c$. On the other hand if x and y are both in G_n, then as $g_{G_{n+1}^c}(x, y)$ has the required properties on $G_{n+1} \supset G_n$ it does so on G_n. Thus, by what was proved above, $g_{G_n^c}(x, y) \leqq g_{G_{n+1}^c}(x, y)$, $x, y \in G_n$. We will finish the proof by showing that the limit, $\lim_n g_{G_n^c}$, agrees with g_{G^c} on $G \times G$. If $x = y$, then $g_{G_n^c}(x, x) = \infty$ for all sufficiently large n and so does $g_{G^c}(x, x)$ so the desired result holds in this case. Suppose $x \neq y$. Then (5.16) shows that

$$g_{G_n^c}(x, y) = g(y - x) - \int_{G_n^c} \Pi_{G_n^c}(x, dz) g(y - z). \tag{5.22}$$

If x and $y \in G$, then x and $y \in G_n$ for all $n \geqq n_0$ for some n_0. Assume $x, y \in G_n$. Then

$$\begin{aligned} \int_{G_n^c} \Pi_{G_n^c}(x, dz) g(y - z) &= E_x g(y - X_{V_{G_n^c}}) = E_x g(y - X_{V_{\partial G_n}}) \\ &= E_x[g(y - X_{V_{\partial G_n}}); V_{\partial G} < \infty] \\ &\quad + E_x[g(y - X_{V_{\partial G_n}}); V_{\partial G} = \infty]. \end{aligned} \tag{5.23}$$

Since $g(y - x)$ is a bounded continuous function in x on G_n^c for $y \in G_n$ and $P_x(X_{V_{\partial G_n}} \to X_{V_{\partial G}} \mid V_{\partial G} < \infty) = 1$, we see that

$$\begin{aligned} \lim_{n \to \infty} E_x[g(y - X_{V_{\partial G_n}}); V_{\partial G} < \infty] &= E_x[g(y - X_{V_{\partial G}}); V_{\partial G} < \infty] \\ &= E_x[g(y - X_{V_{G^c}}); V_{G^c} < \infty] \\ &= \int_{G^c} \Pi_{G^c}(x, dz) g(y - z). \end{aligned} \tag{5.24}$$

Moreover, by Proposition 5.1, $P_x(\lim_{t \to \infty} |X_t| = \infty) = 1$. Since $P_x(V_{\partial G_n} \uparrow \infty \mid V_{\partial G} = \infty) = 1$ and $\lim_{|x| \to \infty} g(y - x) = 0$, it follows by dominated convergence that

$$\lim_{n \to \infty} E_x[g(y - X_{V_{\partial G_n}}); V_{\partial G} = \infty] = 0. \tag{5.25}$$

Thus, from (5.22) and (5.23), we see that for any $x, y \in G$,

$$\lim_{n \to \infty} g_{G_n^c}(x, y) = g(y - x) - \int_{G^c} \Pi_{G^c}(x, dz) g(y - z). \tag{5.26}$$

But by (5.16) the right side of (5.26) is just $g_{G^c}(x, y)$ as desired. This completes the proof.

THEOREM 5.4. *Let μ be a finite measure having support B. Let N be a subset of B such that $\mu(N) = 0$. If $G\mu(x) \leqq M < \infty$ for all $x \in B \cap N^c$, then $\sup_x G\mu(x) \leqq M$.*

PROOF. Choose $\varepsilon > 0$ and let

$$A = \{x: G\mu(x) < M + \varepsilon\}. \tag{5.27}$$

Suppose for some $x_0 \in A$, x_0 is irregular for A. Then $P_{x_0}(V_A = 0) = 0$, and thus

$$\lim_{t\to 0} P_{x_0}(X_t \in A) \leqq \lim_{t\to 0} P_{x_0}(V_A \leqq t) = 0. \tag{5.28}$$

Observe that

$$G\mu(x_0) \geqq P^t G\mu(x_0) \geqq \int_{A^c} p_t(y - x) G\mu(y)\, dy \geqq (M + \varepsilon) P_{x_0}(X_t \in A^c). \tag{5.29}$$

Thus, $G\mu(x_0) \geqq M + \varepsilon$, a contradiction. Therefore, each point of A is regular for A. Using (5.16), we see that

$$G\mu(x) = \int_{\bar{A}} \Pi_A\,(x, dz) G\mu(z) + \int_B g_A(x, y)\mu(dy). \tag{5.30}$$

Now as $\mu(N) = 0$,

$$\int_B g_A(x, y)\mu(dy) = \int_{B\cap N^c} g_A(x, y)\mu(dy). \tag{5.31}$$

But $B \cap N^c \subset A$, and each point of A is regular for A, so $g_A(x, y) = 0$ for all $y \in B \cap N^c$. Thus, we see that

$$G\mu(x) = \int_{\bar{A}} \Pi_A\,(x, dz) G\mu(z). \tag{5.32}$$

Now $G\mu(x)$ is a lower semicontinuous function, and thus $\{x\colon G\mu(x) \leqq M + \varepsilon\}$ is closed. Hence, $\bar{A} \subset \{x\colon G\mu(x) \leqq M + \varepsilon\}$, and thus (5.32) shows that $G\mu(x) \leqq M + \varepsilon$. As ε is arbitrary, $G\mu(x) \leqq M$ as desired.

6. Equilibrium measure

Let S_r be the closed ball of center 0 and radius r and let $G = S_r^c$. The hitting distribution $\Pi_{S_r}\,(x, dy)$ of S_r is easily found. Indeed, for $x \in S_r$, $\Pi_{S_r}\,(x, dy)$ is just the unit mass at x while for $x \in G$, $\Pi_{S_r}\,(x, dy) = \Pi_{\partial G}(x, dy)$. Since S_r is compact

$$\lim_{|x|\to\infty} P_x(V_{S_r} = \infty) = \lim_{|x|\to\infty} P_x(V_{\partial G} = \infty) = 1. \tag{6.1}$$

Thus, for any continuous function φ on ∂G, $\Pi_{\partial G}\,\varphi$ is the unique bounded solution to the Dirichlet problem for G with boundary function φ that vanishes at ∞. It is easily checked that

$$h(x) = \int_{\partial G} r^{n-2}\big||x|^2 - r^2\big||\xi - x|^{-n}\varphi(\xi)\sigma_r(d\xi), \tag{6.2}$$

is a bounded harmonic function on G taking values φ on ∂G and vanishes at ∞. Thus, for $x \in G$,

$$\Pi_{S_r}\,(x, d\xi) = r^{n-2}\big||x|^2 - r^2\big||\xi - x|^{-n}\, d\sigma_r(\xi). \tag{6.3}$$

It follows at once from (6.3) that $\Pi_{S_r}(x, d\xi)g(x)^{-1}$ converges strongly as $|x| \to \infty$ to $k^{-1}r^{n-2}\sigma_r$, where $k = \Gamma((n/2) - 1)/2\pi^{n/2}$.

Let B be any relatively compact set and let S_r be a closed ball of center 0 and radius r that contains $\bar{B}$ in its interior. Then for $|x| > r$

$$\Pi_B(x, A) = \int_{\partial S_r} \Pi_{\partial S_r}(x, d\xi)\, \Pi_B(\xi, A), \tag{6.4}$$

and thus for any Borel set A,

$$\lim_{|x| \to \infty} \frac{\Pi_B(x, A)}{g(x)} = \int_{\partial S_r} k^{-1}\sigma_r(d\xi)\, \Pi_B(\xi, A). \tag{6.5}$$

We have thus established the following important theorem.

THEOREM 6.1. *Let B be any relatively compact set. Then the measure*

$$\mu_B(dy) = \lim_{|x| \to \infty} \frac{\Pi_B(x, dy)}{g(x)} \tag{6.6}$$

exists in the sense of strong convergence of measures. For any ball S_r of center 0 and radius r containing $\bar{B}$ in its interior

$$\mu_B(dy) = \int_{\partial S_r} \Pi_B(\xi, dy) k^{-1}\sigma_r(d\xi). \tag{6.7}$$

DEFINITION 6.1. *The measure μ_B is called the equilibrium measure of B and its total mass $C(B)$ is called the capacity of B.*

Since $\Pi_B(x, N) \equiv 0$ whenever N is a polar set, we see that $\mu_B(N) = 0$ for any polar set. It is also clear that μ_B is concentrated in the outer boundary of B.

By use of probability theory, we have directly defined an equilibrium measure and capacity for any relatively compact Borel set. We must now show that this is consistent with the definitions usually given in potential theory. The equilibrium measure (also called the capacitory measure) is usually defined only for compact sets and in the following manner.

Let $\mathcal{M}(B)$ denote all *nonzero* bounded measures having compact support contained in B whose potentials are bounded above by 1. When B is compact it is then shown that there is a unique measure γ_B supported on B such that $G\gamma_B = \sup_{\mu \in \mathcal{M}(B)} G\mu$. The measure γ_B is what is usually called the capacitory measure of B and its total mass the capacity. We will now show that $\gamma_B = \mu_B$. (In the classical theory of potentials capacitory measures are only defined for compact sets B.)

As a first step towards this goal we will show the following important theorem.

THEOREM 6.2. *Let B be a relatively compact set. Then $P_x(V_B < \infty) = G\mu_B(x)$.*

PROOF. By (5.16),

$$g(y - x) = \int_{\bar{B}} \Pi_B(x, dz) g(y - z) + g_B(x, y). \tag{6.8}$$

Since $g(y - x)/g(y) \to 1$ as $|y| \to \infty$ uniformly on compacts, it follows from (6.8) that $\lim_{|y|\to\infty} g_B(x, y)g(y)^{-1} = P_x(V_B = \infty)$, the convergence being uniform on compacts. By symmetry,

$$\lim_{|x|\to\infty} \frac{g_B(x, y)}{g(x)} = P_y(V_B = \infty), \tag{6.9}$$

uniformly in y on compacts. Let f be any nonnegative bounded measurable function having compact support. Then $Gf(z)$ is a bounded function. From (6.9), we see that

$$\lim_{|x|\to\infty} \int_{R^n} \frac{g_B(x, y)}{g(x)} f(y)\, dy = \int_{R^n} P_y(V_B = \infty) f(y)\, dy, \tag{6.10}$$

while by Theorem 6.1,

$$\lim_{|x|\to\infty} \int_{\bar{B}} \frac{\Pi_B\,(x, dz)Gf(z)}{g(x)} = \int_{\bar{B}} \mu_B(dz)Gf(z). \tag{6.11}$$

Thus, using (6.8), we see that

$$\int_{\bar{B}} \mu_B(dz)Gf(z) = \int_{R^n} P_y(V_B < \infty) f(y)\, dy. \tag{6.12}$$

As f is an arbitrary bounded function having compact support, it follows from (6.12) that for a.e. y,

$$G\mu_B(y) = P_y(V_B < \infty). \tag{6.13}$$

Let $\varphi_B(y) = P_y(V_B < \infty)$. Then $P^t\varphi_B(y) = P_y(X_s \in B \text{ for some } s > t)$ increases to $\varphi_B(y)$ as $t \downarrow 0$. Also

$$P^t G\mu_B = G\mu_B - \int_0^t P^s \mu_B\, ds \tag{6.14}$$

increases to $G\mu_B$ as $t\downarrow 0$. From (6.13), we see that $P^t G\mu_B = P^t\varphi_B$, and thus letting $t\downarrow 0$, we see that (6.13) holds for all y. This establishes the theorem.

It follows from Theorem 6.2 that $C(B) = 0$ if and only if B is a polar set. We can now easily show that $\gamma_B = \mu_B$.

THEOREM 6.3. *Let B be a compact set. Then $P_x(V_B < \infty) = \sup_{\mu\in\mathcal{M}(B)} G\mu(x)$.*

PROOF. Since B is compact, we can find compact sets B_n such that $B \subset \mathring{B}_n$ for all n and $B_1 \supset \mathring{B}_1 \supset B_2 \supset \cdots$, $\bigcap_n B_n = \bigcap_n \mathring{B}_n = B$. By Proposition 3.2, $P_x(V_{B_n} \uparrow V_B) = 1$ for $x \in B^c \cup B^r$. Thus, for $x \in B^c \cup B^r$ and f a continuous function

$$\begin{aligned} \lim_n \Pi_{B_n} f(x) &= \lim_n E_x[f(X_{V_{B_n}}); V_B < \infty] + \lim_n E_x[f(X_{V_{B_n}}); V_{B_n} < \infty, V_B = \infty] \\ &= \Pi_B f(x). \end{aligned} \tag{6.15}$$

In particular, by taking f continuous with compact support and equal to 1 on B_1, we see that $P_x(V_{B_n} < \infty) \downarrow P_x(V_B < \infty)$ for $x \in B^r \cup B^c$. Since $(B^c \cup B^r)^c$ has measure 0, $P_x(V_{B_n} < \infty) \downarrow P_x(V_B < \infty)$ a.e. Let $\mu \in \mathcal{M}(B)$. Then as each point of B is a regular point of B_n (since $B \subset B_n^0$), $\int_B g_{B_n}(x, y)\mu(dy) = 0$, and thus by (6.8),

$$(6.16) \qquad G\mu(x) = \int_{B_n} \Pi_{B_n}(x, dz) G\mu(z) \leqq \Pi_{B_n}(x, B_n) = P_x(V_{B_n} < \infty).$$

Thus, for each $x \in B^c \cup B^r$ we see upon letting $n \to \infty$ that

$$(6.17) \qquad G\mu(x) \leqq P_x(V_B < \infty) = \varphi_B(x),$$

so that (6.17) is valid for a.e. x. Hence, for all x, $P^t G\mu(x) \leqq P^t \varphi_B(x)$, and thus letting $t \downarrow 0$ we see that (6.17) holds for *all* x. Using this and the fact that $\mu_B \in \mathcal{M}(B)$, we see that Theorem 6.3 holds.

Theorems 6.2 and 6.3 and our uniqueness theorem (Theorem 5.1) show that $\gamma_B = \mu_B$ when B is compact.

An immediate consequence of Theorems 6.2 and 6.3 is the following corollary.

COROLLARY 6.1. *Let B be relatively compact. Then for any $\mu \in \mathcal{M}(B)$, $\mu(R^n) \leqq C(B)$.*

PROOF. By Theorems 6.2 and 6.3, we know that

$$(6.18) \qquad \int_{\bar{B}} \frac{g(y - x)}{g(x)} \mu(dy) \leqq \int_{\bar{B}} \frac{g(y - x)}{g(x)} \mu_B(dy).$$

Since $g(y - x) g(x)^{-1} \to 1$ as $|x| \to \infty$ uniformly on compacts, we see by letting $|x| \to \infty$ that $\mu(R^n) = \mu(\bar{B}) \leqq \mu_B(\bar{B}) = C(B)$, as desired.

Let U be any open set. We can then find compact sets $k_n \subset U$ such that $k_1 \subset k_2 \subset \cdots$, $\cup_n k_n = U$. Since $X_t \in U$ if and only if $X_t \in k_n$ for all sufficiently large n, $P_x(V_{k_n} \downarrow V_U) = 1$ for all x, so $P_x(V_{k_n} < \infty) \uparrow P_x(V_U < \infty)$. By Theorem 6.2, $P_x(V_{k_n} < \infty) = G\mu_{k_n}(x)$. Thus, $G\mu_{k_n}(x) \uparrow P_x(V_U < \infty)$. But as $\mu_{k_n} \in \mathcal{M}(U)$,

$$(6.19) \qquad P_x(V_U < \infty) \leqq \sup_{\mu \in \mathcal{M}(U)} G\mu(x).$$

On the other hand if $\mu \in \mathcal{M}(U)$ has compact support k, then Theorems 6.2 and 6.3 show that

$$(6.20) \qquad G\mu(x) \leqq G\mu_k(x) = P_x(V_k < \infty) \leqq P_x(V_U < \infty).$$

Thus, $P_x(V_U < \infty)$ is the *smallest majorant* of potentials of measures in $\mathcal{M}(U)$. This characterization of $P_x(V_U < \infty)$ together with the one given for compact sets by Theorem 6.3 shows that $P_x(V_B < \infty)$ for B an open set or a compact set is the electrostatic potential of B for such sets.

The capacity function $C(\cdot)$ defined for all relatively compact sets has the following properties.

THEOREM 6.4. *Let A and B be relatively compact. Then:*

(i) $C(A) \leqq C(B)$, if $A \subset B$;

(ii) $C(A \cup B) \leqq C(A) + C(B) - C(A \cap B)$;

(iii) $C(A + a) = C(A)$;

(iv) $C(-A) = C(A)$;

(v) $C(rA) = r^{n-2}C(A)$;

(vi) *if B is open and $\bar{B}$ compact*, $C(B) = \sup\{C(k): k \subset B, k \text{ compact}\}$;

(vii) *if B is compact*, $C(B) = \inf\{C(U): U \supset B, U \text{ open}, \bar{U} \text{ compact}\}$.

PROOF. Using Theorem 6.2 and the fact that $g(y - x)/g(x) \to 1$ uniformly on compacts as $|x| \to \infty$, we see that for any relatively compact set B

$$C(B) = \lim_{|x|\to\infty} \frac{P_x(V_B < \infty)}{g(x)} = \int_{\partial S} P_\xi(V_B < \infty)k^{-1}\sigma_r(d\xi), \tag{6.21}$$

for any ball of center 0 and radius r containing $\bar{B}$ in its interior. Thus, to establish (i) to (v), it is only necessary to establish the appropriate inequalities for $P_x(V_B < \infty)$. Hence, (i) to (v) follow from:

(a) $P_x(V_A < \infty) \leqq P_x(V_B < \infty)$, $A \subset B$;

(b) $P_x(V_{A\cap B} < \infty) \leqq P_x(V_A < \infty, V_B < \infty) = P_x(V_A < \infty) + P_x(V_B < \infty) - P_x(V_{A\cup B} < \infty)$;

(c) $P_x(V_A < \infty) = P_{x+a}(V_{A+a} < \infty)$;

(d) $P_x(V_A < \infty) = P_x(V_{-A} < \infty)$;

(e) $P_x(V_A < \infty) = P_{rx}(V_{rA} < \infty)$.

To prove (vi), let k_n be compact sets $\subset B$ such that $k_1 \subset k_2 \subset \cdots$ and $\bigcup_n k_n = B$. Then $P_x(V_{k_n} < \infty)\uparrow P_x(V_B < \infty)$. Let D be an open relatively compact set containing $\bar{B}$. Then

$$\begin{aligned} C(k_n) &\\ &= \int_{R^n} P_x(V_D < \infty)\mu_{k_n}(dx) \\ &= \int_{\bar{D}} G\mu_{k_n}(x)\mu_D(dx)\uparrow\int_{\bar{D}} P_x(V_B < \infty)\mu_D(dx) \\ &= \int_{\bar{B}} G\mu_D(x)\mu_B(dx) = \int_{\bar{B}} P_x(V_D < \infty)\mu_B(dx) = C(B). \end{aligned} \tag{6.22}$$

To establish (vii), let U_n be open, relatively compact, and such that $U_1 \supset \bar{U}_2 \supset U_2 \supset \cdots$, $\bigcap_n U_n = \bigcap_n \bar{U}_n = B$. Choose S_r to be an open ball of center 0 and radius r that contains $\bar{U}_1$ in its interior. Then for $\xi \in \partial S_r$, $P_\xi(V_{U_n} < \infty)\downarrow P_\xi(V_B < \infty)$, and thus

$$C(U_n) = \int_{\partial S_r} P_\xi(V_{U_n} < \infty)k^{-1}\sigma_r(d\xi)\downarrow\int_{\partial S_r} P_\xi(V_B < \infty)k^{-1}\sigma_r(d\xi) = C(B). \tag{6.23}$$

This establishes the theorem.

Let $C^*(B) = C(B)$ if B is a compact set. For any open set U define $C^*(U)$ as $\sup \{C(k): k \subset U, k \text{ compact}\}$. We say that a set B is capacitable if $\sup \{C^*(k): k \subset B, k \text{ compact}\} = \inf \{C^*(U): U \supset B, U \text{ open}\}$. For a capacitable set define $C^*(B)$ as $\sup \{C(k): k \subset B, k \text{ compact}\}$. Property (vi) shows that if U is relatively compact then $C^*(U) = C(U)$. This fact, together with (i) and (ii) shows that $C(\cdot)$ is a Choquet capacity (see, for example, [1]) on the compact sets. By Choquet's capacity theorem, C^* is its unique extension to the Borel sets and every Borel set is capacitable.

Now for a relatively compact set B, we have already defined its capacity by $C(B)$. To see that $C^*(B) = C(B)$ note that if $k \subset B$, k compact then $C(k) \leqq C(B)$, and thus

$$C^*(B) = \sup \{C(k): k \subset B. k \text{ compact}\} \leqq C(B). \tag{6.24}$$

Also if U is open and relatively compact then $C(B) \leqq C(U)$, and thus

$$C^*(B) = \inf \{C(U); U \supset B, U \text{ open}\} \geqq C(B). \tag{6.25}$$

Thus, $C^*(B) = C(B)$.

Now that we have the capacity defined for all Borel sets, let us point out that B has capacity 0 if and only if every compact subset of B has capacity 0. (We could have used this property to define sets of capacity 0 directly.) Our next two results charactize polar sets.

THEOREM 6.5. *Let B be any Borel set. Then*

(i) *B is polar if and only if every compact subset of B is polar;*

(ii) *B is polar if and only if $C^*(B) = 0$.*

PROOF. Clearly, if B is polar, then so is every compact subset of B. On the other hand, if every compact subset of B is polar, then for any relatively compact $A \subset B$, $C(A) = \sup \{C(k): k \subset A, k \text{ compact}\} = 0$, so A is polar. Thus, B must be polar. Therefore (i) holds. Similarly, if B is polar, then every compact subset is polar so $C^*(B) = \sup \{C(k): k \subset B, k \text{ compact}\} = 0$. Conversely, if $C^*(B) = 0$, then $C(k) = 0$ for all compact sets $k \subset B$, and thus by (i) B is polar. This establishes the theorem.

COROLLARY 6.2. *Let B be any Borel set. Then B is polar if and only if $\mathscr{M}(B) = \varnothing$. Equivalently, B is polar if and only if $\sup_x G\mu(x) = \infty$ for any bounded nonzero measure having compact support $\subset B$.*

PROOF. Suppose there is a nonzero measure μ having compact support $k \subset B$ such that $\sup_x G\mu(x) \leqq M < \infty$. The measure μ/M then belongs to $\mathscr{M}(k) \subset \mathscr{M}(B)$ and clearly

$$P_x(V_B < \infty) \geqq P_x(V_k < \infty) \geqq G\mu(x). \tag{6.26}$$

Since μ is nonzero, $G\mu(x) > 0$ for some x, and thus $P_x(V_B < \infty) > 0$ for some x. Hence, B is not polar. On the other hand if B is not polar then some compact subset k of B must also be nonpolar. Hence, $P_x(V_k < \infty) > 0$ for some x. Since $\mu_k \in \mathscr{M}(k)$, $\mathscr{M}(k) \neq \varnothing$, and $G\mu_k(x) \leqq 1$. This establishes the corollary.

7. Equilibrium sets

So far we have discussed equilibrium measures only for relatively compact sets. In this section, we will examine to what extent these notions go over to unbounded sets. We start our discussion with the following theorem.

THEOREM 7.1. *Let B be a Borel set. Then either $P_x(V_B < \infty) \equiv 1$ or $P_x(X_s \in B$ for some $s > t) \to 0$ as $t \to \infty$.*

PROOF. Let $\varphi(x) = P_x(V_B < \infty)$. Then $P^t\varphi$ is a decreasing function in t as $t \to \infty$. Let $r(x) = \lim_{t\uparrow\infty} P^t\varphi(x)$ and set $h = \varphi - r$. Clearly, $P^th(x)\downarrow 0$ as $t \to \infty$. Using dominated convergence and the semigroup property of P^t, it easily follows that $r = P^tr$ for all $t > 0$. Thus, $r(x) \equiv \alpha$ for some constant α (see Proposition 2.3). Hence, $\varphi = \alpha + h$. Now

$$P_x(t < V_B < \infty) = \int_{R^n} q_B(t, x, y)\varphi(y)\,dy \geqq \alpha P_x(V_B > t). \tag{7.1}$$

Letting $t\uparrow\infty$, we see that $\alpha P_x(V_B = \infty) \leqq 0$. Thus, either $\alpha = 0$ or $P_x(V_B = \infty) \equiv 0$. In the first case $\varphi = h$ and in the second case $P_x(V_B < \infty) \equiv 1$ as desired.

DEFINITION 7.1. *A Borel set B is called recurrent if $P_x(V_B < \infty) \equiv 1$; it is called transient if $P_x(X_s \subset B$ for some $s > t)\downarrow 0$.*

By Theorem 7.1, we know that every Borel set is either transient or recurrent.

Our next result extends Theorem 6.2 from relatively compact sets to transient sets.

THEOREM 7.2. *Let B be a transient set. Then there is a unique Radon measure μ_B such that $P_x(V_B < \infty) = G\mu_B(x)$. The measure μ_B is concentrated on ∂B. Moreover, if B_m, $m \geqq 1$, is any sequence of relatively compact sets such that $B_1 \subset B_2 \subset \cdots$, $\bigcup_m B_m = B$, then $G\mu_{B_m}\uparrow G\mu_B$ as $m \to \infty$ and the measures μ_{B_m} converge vaguely to μ_B.*

PROOF. Let B_m, $m \geqq 1$, be as in the statement of the theorem. Then $\bigcup_m [V_{B_m} < \infty] = [V_B < \infty]$, and thus $P_x(V_{B_m} < \infty)\uparrow P_x(V_B < \infty)$. By Theorem 6.2,

$$G\mu_{B_m}(x) = P_x(V_{B_m} < \infty) \leqq P_x(V_B < \infty) = \varphi_B(x). \tag{7.2}$$

Let k be any compact set. Since $\inf_{y\in k} g(y - x) = \delta(x) > 0$, it follows from (7.2) that

$$\delta(x)\mu_{B_m}(k) \leqq \int_k g(y - x)\mu_{B_m}(dy) \leqq G\mu_{B_m}(x) \leqq \varphi_B(x). \tag{7.3}$$

Thus, $\sup_m \mu_{B_m}(k) < \infty$. Consequently, there is a subsequence, $\mu_{B'_m}$, of the measures μ_{B_m} that converge vaguely to some measure μ_B.

Let $f \geqq 0$ be continuous with compact support, and let S_r be the closed ball of center 0 and radius r. Then, as $g_{S_r^c}(x, y) \equiv 0$ for each $x \in S_r^c$, we see that

$$\text{(7.4)} \qquad \int_{S_r^c} \mu_{B'_m}(dx) Gf(x) = \int_{S_r^c} \mu_{B'_m}(dx)\, \Pi_{S_r^c}\, Gf(x) \leqq \int_{R^n} \mu_{B'_m}(dx)\, \Pi_{S_r^c}\, Gf(x).$$

By letting $\lambda \downarrow 0$ in (2.31), we see that

$$\text{(7.5)} \qquad \int_{R^n} \Pi_{S_r^c}\,(x, dz) g(y - z) = \int_{R^n} \Pi_{S_r^c}\,(y, dz) g(x - z).$$

Using this fact, we compute

$$\begin{aligned}\text{(7.6)} \qquad \int_{R^n} \mu_{B'_m}(dx)\, \Pi_{S_r^c}\, Gf(x) &= \int_{R^n}\int_{R^n}\int_{R^n} \mu_{B'_m}(dx)\, \Pi_{S_r^c}\,(x, dz) g(y - z) f(y)\, dy \\ &= \int_{R^n}\int_{R^n}\int_{R^n} \mu_{B'_m}(dx) f(y)\, dy\, \Pi_{S_r^c}(y, dz) g(x - z) \\ &= \int_{R^n} \Pi_{S_r^c}\, G\mu_{B'_m}(y) f(y)\, dy.\end{aligned}$$

Using (7.2), (7.4), and (7.6), we see that

$$\text{(7.7)} \qquad \int_{S_r^c} Gf(x) \mu_{B'_m}(dx) \leqq \int_{R^n} \Pi_{S_r^c}\, \varphi_B(y) f(y)\, dy.$$

Now for any $t > 0$.

$$\text{(7.8)} \qquad \Pi_{S_r^c}\, \varphi_B(y) = E_y[\varphi_B(X_{V_{S_r^c}})] \leqq P_y(V_{S_r^c} \leqq t) + P^t \varphi_B(y).$$

Since B is a transient set, $P^t \varphi_B \downarrow 0$ as $t \uparrow \infty$. In addition, $P_y(V_{S_r^c} \leqq t) \downarrow 0$ as $r \uparrow \infty$ because $P_y(\lim_{r \to \infty} V_{S_r^c} = \infty) = 1$. It follows from these two facts that

$$\text{(7.9)} \qquad \lim_{r \to \infty} \Pi_{S_r^c}\, \varphi_B(y) = 0.$$

Hence, by (7.7) and (7.9), we see that given any $\varepsilon > 0$ there is an $r_0 < \infty$ such that for $r \geqq r_0$,

$$\text{(7.10)} \qquad \sup_m \int_{S_r^c} \mu_{B'_m}(dx) Gf(x) \leqq \varepsilon.$$

Using the fact that Gf is a bounded continuous function, we see that

$$\text{(7.11)} \qquad \lim_{r\to\infty}\left[\lim_{m\to\infty}\int_{S_r} Gf(x)\mu_{B'_m}(dx)\right] = \lim_{r\to\infty}\int_{S_r} Gf(x)\mu_B(dx)$$

$$= \int_{R^n} Gf(x)\mu_B(dx)$$

$$= \int_{R^n} G\mu_B(x)f(x)\,dx.$$

Moreover, monotone convergence shows that

$$\text{(7.12)} \quad \lim_{m\to\infty}\int_{R^n} Gf(x)\mu_{B'_m}(dx) = \lim_{m\to\infty}\int_{R^n} G\mu_{B'_m}(x)f(x)\,dx = \int_{R^n}\varphi_B(x)f(x)\,dx.$$

It follows easily from (7.10) through (7.12) that

$$\text{(7.13)} \qquad \int_{R^n} G\mu_B(x)f(x)\,dx = \int_{R^n}\varphi_B(x)f(x)\,dx.$$

Since f can be any nonnegative continuous function with compact support, (7.13) implies that for a.e. x,

$$\text{(7.14)} \qquad G\mu_B(x) = \varphi_B(x).$$

From (7.14), we see that $P^t G\mu_B(x) = P^t\varphi_B(x)$ for all x. Since $P^t G\mu_B \uparrow G\mu_B$ and $P^t\varphi_B \uparrow \varphi_B$ as $t \downarrow 0$, it follows that (7.14) holds for *all* x.

Suppose $\mu_{B''_m}$ is another subsequence of the measures μ_{B_m} that converge vaguely to a measure μ'_B. The same argument as used above will again show that $G\mu'_B = \varphi_B$. By Theorem 5.1, it must then be that $\mu'_B = \mu_B$. Thus, the measures μ_{B_m} converge vaguely to μ_B. Theorem 5.1 tells us that μ_B is the unique measure whose potential is φ_B. To see that μ_B is concentrated on ∂B, we can proceed as follows. Let S_m be the closed ball of center 0 and radius m, and let $B_m = B \cap S_m$. Then $B_1 \subset B_2 \subset \cdots$, and $\bigcup_m B_m = B$. Thus, the measures μ_{B_m} converge vaguely to μ_B. Since μ_{B_m} is concentrated on ∂B_m and each interior point of B is an interior point of B_m for m sufficiently large, μ_B must be concentrated on ∂B. This completes the proof.

We will now show that the total mass of the measure μ_B in Theorem 7.2 is $C^*(B)$. To do this, we will need the following

PROPOSITION 7.1. *Let B be a transient set. Suppose $A \subset B$. Then $\mu_A(R^n) \leqq \mu_B(R^n)$.*

PROOF. Let D_m be a family of relatively compact sets that increase to R^n. Using Theorem 7.2, we then see that

$$(7.15)\qquad \begin{aligned}\int_{R^n} \mu_A(dx) G\mu_{D_m}(x) &= \int_{R^n} \mu_{D_m}(dx) G\mu_A(x) = \int_{R^n} \mu_{D_m}(dx) P_x(V_A < \infty)\\ &\leqq \int_{R^n} \mu_{D_m}(dx) P_x(V_B < \infty) = \int_{R^n} \mu_{D_m}(dx) G\mu_B(x)\\ &= \int_{R^n} \mu_B(dx) G\mu_{D_m}(x).\end{aligned}$$

Since

$$(7.16)\qquad G\mu_{D_m}(x) = P_x(V_{D_m} < \infty) \uparrow 1$$

as $m \uparrow \infty$, it follows from (7.15) (by monotone convergence) that $\mu_A(R^n) \leqq \mu_B(R^n)$, as desired.

THEOREM 7.3. *Let B be a transient set. Then $\mu_B(R^n) = C^*(B)$. Moreover, if B_m, $m \geqq 1$, is a sequence of relatively compact subsets of B such that $B_1 \subset B_2 \subset \cdots$, $\bigcup_m B_m = B$, then $C(B_m) \uparrow C^*(B)$.*

PROOF. Let B_m, $m \geqq 1$, be as in the statement of the theorem. Then by Proposition 7.1,

$$(7.17)\qquad C(B_1) \leqq C(B_2) \leqq \cdots \leqq \mu_B(R^n).$$

On the other hand, let f_r, $r \geqq 1$, be continuous functions with compact support such that $0 \leqq f_r \leqq 1$ and such that $f_r \uparrow 1$ as $r \uparrow \infty$. Then

$$(7.18)\qquad C(B_m) \geqq \int_{R^n} f_r(x) \mu_{B_m}(dx),$$

and thus by Theorem 7.2,

$$(7.19)\qquad \liminf_{m \to \infty} C(B_m) \geqq \int_{R^n} \mu_B(dx) f_r(x).$$

Letting $r \uparrow \infty$, we see that

$$(7.20)\qquad \liminf_{m \to \infty} C(B_m) \geqq \mu_B(R^n).$$

Hence, $C(B_m) \uparrow \mu_B(R^n)$.

From our results in Section 6, we know that

$$(7.21)\qquad C(B_m) = \sup \{C(k): k \subset B_m, k \text{ compact}\}.$$

Suppose $\mu_B(R^n) = \infty$. Given any $N > 0$, we can then find an m such that $C(B_m) \geqq 2N$. From (7.21), we see that we can find a compact set $k \subset B_m$ such that $C(k) \geqq C(B_m) - N$. Thus, $C(k) \geqq N$. Hence, $\sup \{C(k): k \subset B, k \text{ compact}\} = C^*(B) = \infty$. Suppose now that $\mu_B(R^n) < \infty$, and let $\varepsilon > 0$ be given. We can then choose m such that $C(B_m) \geqq \mu(R^n) - \varepsilon$. From (7.21), we see that there is a compact set $k \subset B_m$ such that $C(k) \geqq C(B_m) - \varepsilon$, and thus $C(k) \geqq \mu(R^n) - 2\varepsilon$. Hence,

$$C^*(B) = \sup \{C(k): k \subset B, k \text{ compact}\} = \mu_B(R^n). \tag{7.22}$$

This establishes the theorem.

Let B be a closed set. As usual, let $\mathcal{M}(B)$ denote all nonzero bounded measures having compact support $\subset B$ whose potentials are bounded above by 1.

An important link between probability theory and potential theory is the following

THEOREM 7.4. *Let B be a closed set. Then*

$$\sup \{G\mu(x): \mu \in \mathcal{M}(B)\} = P_x(V_B < \infty). \tag{7.23}$$

PROOF. Since B is closed, we can find compact sets $B_n \subset B$, $B_1 \subset B_2 \subset \cdots$, $\bigcup_n B_n = B$. Then clearly, $P_x(V_{B_n} < \infty) \uparrow P_x(V_B < \infty)$, $n \to \infty$. If $\mu \in \mathcal{M}(B)$ has compact support $k \subset B_n$, then by Theorems 6.2 and 6.3,

$$G\mu(x) \leqq G\mu_{B_n}(x) \leqq P_x(V_{B_n} < \infty) \leqq P_x(V_B < \infty), \tag{7.24}$$

and thus

$$\sup \{G\mu(x): \mu \in \mathcal{M}(B)\} \leqq P_x(V_B < \infty). \tag{7.25}$$

But $G\mu_{B_n}(x) = P_x(V_{B_n} < \infty) \uparrow P_x(V_B < \infty)$, so

$$\sup \{G\mu(x): \mu \in \mathcal{M}(B)\} = P_x(V_B < \infty). \tag{7.26}$$

This establishes the theorem.

COROLLARY 7.1. *A closed set B is transient if and only if there is a Radon measure μ_B supported on ∂B such that*

$$G\mu_B(x) = \sup \{G\mu(x): \mu \in \mathcal{M}(B)\}. \tag{7.27}$$

PROOF. This follows at once from Theorems 7.2 and 7.4.

If B is a transient set the measure μ_B is called the *equilibrium measure* of B and its potential is called the *equilibrium potential*, just as in the case of a compact set. The total mass of μ_B is the capacity of B. Theorem 7.3 shows this is consistent with the extension of $C(\cdot)$ from the relatively compact sets.

We are now in a position to state our results on the Dirichlet problem for G in analytical terms. Note that for $x \in G$, $P_x(V_{\partial G} = \infty) = P_x(V_{G^c} = \infty)$. We want to know when $P_x(V_{\partial G} = \infty) = 0$ for all $x \in G$. Suppose this is the case. Then G^c must be a recurrent set. For suppose $P_x(V_{G^c} < \infty) = 1$ for all $x \in G$. It is always true that $P_x(V_{G^c} < \infty) = 1$ for all $x \in G^c$ except perhaps at the points in G^c that are irregular. But these exceptional points form a set of measure 0, and thus $P_x(V_{G^c} < \infty) = 1$ a.e. Since $P^t\varphi_{G^c} \uparrow \varphi_{G^c}$ as $t \downarrow 0$, we then see that $P_x(V_{G^c} < \infty) = 1$ for all x, so G^c is recurrent. Conversely, if G^c is recurrent then $P_x(V_{G^c} < \infty) \equiv 1$, so $P_x(V_{\partial G} = \infty) = 0$ for all $x \in G$. Thus we have the following.

THEOREM 7.5. *Let G be an open set. The modified Dirichlet problem for ∂G with boundary function φ has $\Pi_{\partial G}\,\varphi$ as its unique bounded solution if and only if G^c is a recurrent set. If G^c is a transient set then a constant multiple of $P_x(V_{G^c} = \infty)$ can be added to the solution $\Pi_{\partial G}\,\varphi$. These constitute the only bounded solutions to the problem.*

REFERENCE

[1] R. BLUMENTHAL and R. GETOOR, *Markov Processes and Potential Theory*, New York, Academic Press, 1968.

LOGARITHMIC POTENTIALS AND PLANAR BROWNIAN MOTION

SIDNEY C. PORT and CHARLES J. STONE
UNIVERSITY OF CALIFORNIA, LOS ANGELES

In this paper we continue our discussion of the connection between potential theory and Brownian motion begun in "Classical Potential Theory and Brownian Motion" that also appears in this Symposium volume. Throughout this paper, we will be dealing with a two dimensional Brownian motion process. We will continue numbering the sections from where we left off in the previous paper.

8. Planar Brownian motion

In Section 5, we saw that for a Brownian motion process in $n \geqq 3$ dimensions, $P_x(\lim_{t\to\infty} |X_t| = \infty) = 1$ for all x. In sharp contrast to this situation, a planar Brownian motion is certain to hit any nonpolar set.

THEOREM 8.1. *Let B be a Borel set. Then $P_x(V_B < \infty)$ is either identically* 1 *or identically* 0.

PROOF. A simple computation shows that for any $x \in R^2$, $\int_0^t p(s, x)\, ds \uparrow \infty$ as $t \uparrow \infty$. Thus, for any nonnegative function f having nonzero integral,

$$\lim_{t\to\infty} \int_0^t P^s f(x)\, ds = \infty. \tag{8.1}$$

Let $\varphi(x) = P_x(V_B < \infty)$. Then for any $h > 0$,

$$0 \leqq \int_0^t P^s(\varphi - P^h\varphi)\, ds = \int_0^h P^s\varphi\, ds - \int_t^{t+h} P^s\varphi\, ds \leqq 2h. \tag{8.2}$$

Letting $t \uparrow \infty$, we see that

$$0 \leqq \int_0^\infty P^s(\varphi - P^h\varphi)\, ds \leqq 2h. \tag{8.3}$$

But then it must be that $\varphi = P^h\varphi$ a.e. Since $P^t\varphi \uparrow \varphi$ as $t \downarrow 0$ and $P^t(P^h\varphi) \uparrow P^h\varphi$ as $t \downarrow 0$, it follows that $\varphi(x) = P^h\varphi(x)$ for all x. Using Proposition 2.3, we see that $\varphi(x) \equiv \alpha$ for some constant α. Now

$$P_x(t < V_B < \infty) = \int_{R^2} q_B(t, x, y)\varphi(y)\, dy = \alpha P_x(V_B > t). \tag{8.4}$$

Research supported in part by NSF Grant GP-17868.

Letting $t \uparrow \infty$, we see that $\alpha P_x(V_B = \infty) = 0$. Thus, either $P_x(V_B = \infty) \equiv 0$ or $\alpha = 0$. In the first case $\varphi(x) \equiv 1$, while in the second case $\varphi(x) \equiv 0$. This establishes the proposition.

The difference between planar Brownian motion and Brownian motion in $n \geqq 3$ dimensions has its analytical counterpart in potential theory. We will now show that the potentials associated with planar Brownian motion are logarithmic potentials.

Let 1 denote the point (1, 0) and let $a^\lambda(x) = g^\lambda(1) - g^\lambda(x)$. Using (2.30), we see that for $x \neq y$,

$$a^\lambda(y - x) = \int_{\bar{B}} \Pi_B^\lambda(x, dz)a^\lambda(y - z) - g_B^\lambda(x, y) + L_B^\lambda(x), \tag{8.5}$$

where

$$L_B^\lambda(x) = g^\lambda(1)[1 - E_x(\exp\{-\lambda V_B\})]. \tag{8.6}$$

Now

$$a^\lambda(x) = \int_0^\infty e^{-\lambda t}[p(t, 1) - p(t, x)]\, dt. \tag{8.7}$$

If $|x| \geqq 1$, then $p(t, 1) - p(t, x) \geqq 0$; so for $|x| \geqq 1$, $a^\lambda(x)$ is increasing. On the other hand, for $|x| < 1$, $p(t, 1) - p(t, x) < 0$ so $-a^\lambda(x)$ is increasing. In either case

$$\lim_{\lambda \downarrow 0} a^\lambda(x) = \int_0^\infty [p(t, 1) - p(t, x)]\, dt = \frac{1}{\pi} \log |x|, \tag{8.8}$$

and the convergence is uniform on any compact set not containing 0. For simplicity, we set $a(x) = (1/\pi) \log |x|$.

Our principle result in this section will be to establish the following theorem.

THEOREM 8.2. *Let B be a nonpolar set. Then $g_B(x, y) < \infty$ for $x \neq y$ and $\lim_{\lambda \downarrow 0} L_B^\lambda(x) = L_B(x)$ exists and is finite for all x. Moreover, for $x \neq y$,*

$$a(y - x) - \int_{\bar{B}} \Pi_B(x, dz)a(y - z) = -g_B(x, y) + L_B(x). \tag{8.9}$$

Before getting on with the proof, we observe first if $E_x V_B < \infty$ for all x then $L_B(x) \equiv 0$. Indeed,

$$L_B^\lambda(x) = \lambda g^\lambda(1) \left[\frac{1 - E_x(\exp\{-\lambda V_B\})}{\lambda}\right]; \tag{8.10}$$

so if $E_x V_B < \infty$, then the expression in the brackets converges to $E_x V_B < \infty$, while $\lambda g^\lambda(1) \to 0$ as $\lambda \to 0$. In particular by Proposition 2.2, $L_B(x) \equiv 0$ whenever B^c is relatively compact.

The proof of Theorem 8.2 is long and will be divided into several lemmas.

LEMMA 8.1. *Suppose B is relatively compact. Then*

$$\lim_{\lambda \downarrow 0} \int_{\bar{B}} \Pi_B^\lambda(x, dz)a^\lambda(y - z) = \int_{\bar{B}} \Pi_B(x, dz)a(y - z). \tag{8.11}$$

PROOF. If B is polar, there is nothing to prove since both Π_B^λ and Π_B are the zero measure, so suppose B is nonpolar. If $y \notin \bar{B}$, then as $\lambda \downarrow 0$, $a^\lambda(y - z)$ converges to $a(y - z)$ uniformly in $z \in \bar{B}$, and thus (8.11) holds in this case. Suppose $y \in \bar{B}$. Let D_ε be the open disk of center y and radius $\varepsilon < 1$. We can write

$$(8.12) \qquad \int_{\bar{B}} \Pi_B^\lambda(x, dz)a^\lambda(y - z)$$
$$= \int_{\bar{B} \cap D_\varepsilon^c} \Pi_B^\lambda(x, dz)a^\lambda(y - z) + \int_{\bar{B} \cap D_\varepsilon} \Pi_B^\lambda(x, dz)a^\lambda(y - z).$$

Since

$$(8.13) \qquad \int_{\bar{B} \cap D_\varepsilon^c} \Pi_B^\lambda(x, dz)a^\lambda(y - z) = E_x[\exp\{-\lambda V_B\}\, a^\lambda(y - X_{V_B})1_{D_\varepsilon^c}(X_{V_B})]$$

and $a^\lambda(y - z)$ converges to $a(y - z)$ uniformly for $z \in \bar{B} \cap D_\varepsilon^c$, we see that

$$(8.14) \qquad \lim_{\lambda \downarrow 0} \int_{\bar{B} \cap D_\varepsilon^c} \Pi_B^\lambda(x, dz)a^\lambda(y - z) = E_x[a(y - X_{V_B})1_{D_\varepsilon^c}(X_{V_B})].$$

On the other hand,

$$(8.15) \qquad -\int_{\bar{B} \cap D_\varepsilon} \Pi_B^\lambda(x, dz)a^\lambda(y - z)$$
$$= E_x[\exp\{-\lambda V_B\}[-a^\lambda(y - X_{V_B})]1_{D_\varepsilon}(X_{V_B}).$$

Since

$$(8.16) \qquad -\exp\{-\lambda V_B\}a^\lambda(y - X_{V_B})1_{D_\varepsilon}(X_{V_B}) \uparrow -a(y - X_{V_B})1_{D\varepsilon}(X_{V_B}),$$

monotone convergence shows that

$$(8.17) \qquad \lim_{\lambda \downarrow 0} \int_{\bar{B} \cap D_\varepsilon} \Pi_B^\lambda(x, dz)a^\lambda(y - z) = E_x[a(y - X_{V_B})1_{D\varepsilon}(X_{V_B})] \geqq -\infty.$$

As

$$(8.18) \qquad -\infty < E_x[a(y - X_{V_B})1_{D_\varepsilon^c}(X_{V_B}) < +\infty,$$

we see from (8.14) and (8.17) that (8.11) holds. This establishes the lemma.

LEMMA 8.2. *Suppose B is a nonpolar relatively compact set. Let A be a compact set of positive measure such that $A \cap \bar{B} = \varnothing$. Then*

$$(8.19) \qquad \lim_{\lambda \downarrow 0} \int_A g_B^\lambda(x, y)\, dy = E_x \int_0^{V_B} 1_A(X_t)\, dt < \infty.$$

PROOF. For a given x_0 there is a $t_0 > 1$ such that

$$(8.20) \qquad P_{x_0}(X_s \in B \text{ for some } s \in (1, t_0))$$
$$= \int_{R^2} p(1, y - x_0)P_y(V_B \leqq t_0 - 1)\, dy > 0,$$

since otherwise

$$\int_{R^n} p(1, y - x_0)P_y(V_B < \infty)\, dy = P_{x_0}(V_B < \infty) = 0. \tag{8.21}$$

Thus, there must be a compact set F of positive measure such that $P_y(V_B \leqq t_0 - 1) > 0$ for all $y \in F$. As $p(1, x)$ is a strictly positive continuous function, it follows that

$$\inf_{x\in A} P_x(V_B \leqq t_0) \geqq \inf_{x\in A} \int_F p(1, y - x)P_y(V_B \leqq t_0 - 1) = \delta > 0. \tag{8.22}$$

Let $I_j = [jt_0, (j + 1)t_0)$ and let $C = \{t\colon X_t \in A,\ V_B > t\}$. Define the index set Γ by $j \in \Gamma$ if and only if $I_j \cap C \neq \varnothing$, and enumerate Γ increasing order by $j_1 < j_2 < \cdots$. Define the times $T_1 < T_2 < \cdots$ as follows:

$$T_1 = \inf\{t\colon t \in C\}\ (= \infty \text{ if there is no such } t) \tag{8.23}$$

and

$$T_{n+1} = \inf\{t\colon t \in C \cap [j_n t_0, \infty\}\ (= \infty \text{ if there is no such } t). \tag{8.24}$$

Let $N \leqq \infty$ denote the number of indices in Γ. Then

$$\begin{aligned} P_x(N > n, N \leqq n + 2) &= P_x(T_n < \infty, T_{n+2} = \infty) \\ &\geqq P_x(T_n < \infty, V_B \leqq T_n + t_0) \\ &= \int_A P_x(T_n < \infty, X_{T_n} \in dz)P_z(V_B \leqq t_0) \\ &\geqq \delta P_x(T_n < \infty) = \delta P_x(N > n), \end{aligned} \tag{8.25}$$

so $P_x(N > n + 2) \leqq (1 - \delta)P_x(N > n)$. Thus, $E_xN < \infty$, and hence

$$E_x \int_0^{V_B} 1_A(X_t)\, dt = E_x|C| \leqq E_x\Big|\bigcup_{j\in\Gamma} I_j\Big| = t_0E_xN < \infty, \tag{8.26}$$

as desired. This establishes Lemma 8.2.

We can now prove Theorem 8.2 when B is a relatively compact set.

LEMMA 8.3. *Suppose B is a nonplanar relatively compact set. Then Theorem 8.2 holds.*

PROOF. Let A be as in Lemma 8.2. Since

$$\int_A a^\lambda(y - x)\, dy \to \int_A a(y - x)\, dy \tag{8.27}$$

uniformly in x on compacts, we see that

$$\lim_{\lambda\downarrow 0} \int_{\bar B} \Pi_B^\lambda(x, dz) \int_A a^\lambda(y - z)\, dy = \int_{\bar B} \Pi_B(x, dz) \int_A a(y - z)\, dy. \tag{8.28}$$

By Lemma 8.2,

$$\int_A g_B^\lambda(x, y)\,dy \uparrow \int_A g_B(x, y)\,dy < \infty. \tag{8.29}$$

By (8.5)

$$|A|L_B^\lambda(x) = \int_A a^\lambda(y - x)\,dy - \int_{\bar{B}} \Pi_B^\lambda(x, dz)\int_A a^\lambda(y - z)\,dy + \int_A g_B^\lambda(x, y)\,dy. \tag{8.30}$$

Since the right side has a finite limit as $\lambda \downarrow 0$, we see that $L_B^\lambda(x)$ must have a finite limit as $\lambda \downarrow 0$. Call this limit function $L_B(x)$. By (8.5),

$$-g_B^\lambda(x, y) = a^\lambda(y - x) - \int_{\bar{B}} \Pi_B^\lambda(x, dz)a^\lambda(y - z) - L_B^\lambda(x), \tag{8.31}$$

and thus for $x \neq y$,

$$\lim_{\lambda\downarrow 0}\left[-g_B^\lambda(x, y) + \int_{\bar{B}} \Pi_B^\lambda(x, dz)a^\lambda(y - z)\right] = a(y - x) - L_B(x) \tag{8.32}$$

is finite. Now $0 \leq g_B^\lambda(x, y) \uparrow g_B(x, y) \leqq +\infty$ and by Lemma 8.1

$$\lim_{\lambda\downarrow 0}\int_{\bar{B}} \Pi_B^\lambda(x, dz)a^\lambda(y - z) = \int_{\bar{B}} \Pi_B(x, dz)a(y - z), \tag{8.33}$$

and

$$-\infty \leqq \int_{\bar{B}} \Pi_B(x, dz)a(y - z) < +\infty. \tag{8.34}$$

Thus for $x \neq y$, we see that $g_B(x, y) < +\infty$, $\int_{\bar{B}} \Pi_B(x, dz)a(y - z) > -\infty$, and that (8.9) holds.

To handle the unbounded case we need two additional lemmas.

LEMMA 8.4. *Suppose $A \subset B$. Then $g_A(x, y) \geqq g_B(x, y)$ for all x, y.*

PROOF. Since $A \subset B$, $V_A \geqq V_B$ and thus for any Borel set F,

$$P_x(V_A > t, X_t \in F) \geqq P_x(V_B > t, X_t \in F). \tag{8.35}$$

Hence (in the notation of Section 2) for a.e. y,

$$q_A(t, x, y) \geqq q_B(t, x, y). \tag{8.36}$$

But then

$$\int_{R^2} q_A(t - \varepsilon, x, z)p(\varepsilon, y - z)\,dz \geqq \int_{R^2} q_B(t - \varepsilon, x, z)p(\varepsilon, y - z)\,dz, \tag{8.37}$$

and thus letting $\varepsilon \downarrow 0$ (see Section 2), it follows that (8.36) holds for all y.

Integrating on t, we see that

$$(8.38)\qquad g_A(x, y) = \int_0^\infty q_A(t, x, y)\,dt \geqq \int_0^\infty q_B(t, x, y)\,dt = g_B(x, y)$$

as desired.

LEMMA 8.5. *Let B be any Borel set. Then for $x \neq y$,*

$$(8.39)\qquad \lim_{\lambda \downarrow 0} \int_{\bar{B}} \Pi_B^\lambda(x, dz) a^\lambda(y - z) = \int_{\bar{B}} \Pi_B(x, dz) a(y - z) > -\infty.$$

PROOF. If B is polar, there is nothing to prove so suppose B is nonpolar. Suppose $y \in (\bar{B})^c$. Let D_r be a disk of center 0 and radius $r > 1 + |y|$. Then $|y - z| > 1$ for $z \in D_r^c$. Hence, by monotone convergence

$$(8.40)\qquad \int_{\bar{B} \cap D_r^c} \Pi_B^\lambda(x, dz) a^\lambda(y - z) = E_x[\exp\{-\lambda V_B\} a^\lambda(y - X_{V_B}) 1_{D_r^c}(X_{V_B})]$$

$$\uparrow E_x[a(y - X_{V_B}) 1_{D_r^c}(X_{V_B})] = \int_{\bar{B} \cap D_r^c} \Pi_B(x, dz) a(y - z) \geqq 0.$$

On the other hand, $a^\lambda(y - z) \to a(y - z)$ uniformly in $z \in \bar{B} \cap D_r$ so

$$(8.41)\qquad \lim_{\lambda \downarrow 0} \int_{\bar{B} \cap D_r} \Pi_B^\lambda(x, dz) a^\lambda(y - z) = \int_{\bar{B} \cap D_r} \Pi_B(x, dz) a(y - z),$$

and

$$(8.42)\qquad -\infty < \int_{\bar{B} \cap D_r} \Pi_B(x, dz) a(y - z) < \infty.$$

Thus, (8.39) holds for $y \notin \bar{B}$. Suppose $y \in \bar{B}$ and let A_ε be the disk of center y and radius $\varepsilon < 1$. We can write

$$(8.43)\qquad \int_{\bar{B}} \Pi_B^\lambda(x, dz) a^\lambda(y - z) = \int_{D_r \cap A_\varepsilon} \Pi_B^\lambda(x, dz) a^\lambda(y - z)$$
$$+ \int_{D_r \cap A_\varepsilon^c} \Pi_B^\lambda(x, dz) a^\lambda(y - z)$$
$$+ \int_{D_r^c} \Pi_B^\lambda(x, dz) a^\lambda(y - z).$$

Choose r so large that $B \cap D_r$ is nonpolar and such that $|y - z| > 1$ for $z \in D_r^c$. The second term on the right of (8.43) converges to

$$(8.44)\qquad -\infty < \int_{D_r \cap A_\varepsilon^c} \Pi_B(x, dz) a(y - z) < \infty,$$

as $\lambda \downarrow 0$, since $a^\lambda(y - z) \to a(y - z)$ uniformly for $z \in \bar{B} \cap D_r \cap A_\varepsilon^c$. Let $B_r = B \cap D_r$, and note that $\Pi_{B_r}(x, dz) \geqq \Pi_B(x, dz)$ for $z \in D_r$. Since $-a^\lambda(y - z) \uparrow -a(y - z)$, $z \in A_\varepsilon$, it follows from the monotone convergence theorem that

$$(8.45) \quad -\int_{D_r \cap A_\varepsilon} \Pi_B^\lambda(x, dz) a^\lambda(y - z)$$
$$= E_x(\exp\{-\lambda V_B\}[-a^\lambda(y - X_{V_B}) 1_{A_\varepsilon}(X_{V_B}) 1_{D_r}(X_{V_B}))$$
$$\uparrow E_x[-a(y - X_{V_B}) 1_{A_\varepsilon}(X_{V_B}) 1_{D_r}(X_{V_B})]$$
$$= -\int_{D_r \cap A_\varepsilon} \Pi_B(x, dz) a(y - z).$$

Since $a(y - z) \leqq 0$ for $z \in A_\varepsilon$, we see that

$$(8.46) \quad \int_{D_r \cap A_\varepsilon} \Pi_B(x, dz) a(y - z) \geqq \int_{D_r \cap A_\varepsilon} \Pi_{B_r}(x, dz) a(y - z).$$

By Lemma 8.3,

$$(8.47) \quad \int_{\bar{B}_r} \Pi_{B_r}(x, dz) a(y - z) > -\infty,$$

and it is clear that

$$(8.48) \quad \infty > \int_{\bar{B}_r \cap (D_r \cap A_\varepsilon)^c} \Pi_{B_r}(x, dz) a(y - z) > -\infty.$$

Hence,

$$(8.49) \quad \lim_{\lambda \downarrow 0} \int_{D_r \cap A_\varepsilon} \Pi_B^\lambda(x, dz) a^\lambda(y - z) = \int_{D_r \cap A_\varepsilon} \Pi_B(x, dz) a(y - z) > -\infty.$$

Finally, as in the case when $y \in (\bar{B})^c$, monotone convergence shows that

$$(8.50) \quad \lim_{\lambda \downarrow 0} \int_{D_r^c} \Pi_B^\lambda(x, dz) a^\lambda(y - z) = \int_{D_r^c} \Pi_B(x, dz) a(y - z) \geqq 0.$$

Thus, using (8.43), we see that (8.39) holds for $y \in \bar{B}$. This completes the proof.

We may now easily establish Theorem 8.2.

PROOF OF THEOREM 8.2. Since B is nonpolar some relatively compact subset $A \subset B$ must be nonpolar. But then, by Lemma 8.3, $g_B(x, y) \leqq g_A(x, y) < \infty$ for $x \neq y$. Also

$$(8.51) \quad L_B^\lambda(x) = g^\lambda(1)[1 - E_x(\exp\{-\lambda V_B\})]$$
$$\leq g^\lambda(1)[1 - E_x(\exp\{-\lambda V_A\})] = L_A^\lambda(x),$$

so

$$(8.52) \quad \limsup_{\lambda \downarrow 0} L_B^\lambda(x) \leqq L_A(x) < \infty.$$

Using (8.5), we see that for $x \neq y$,

$$\lim_{\lambda \downarrow 0} \left[\int_{\bar{B}} \Pi_B^\lambda(x, dz) a^\lambda(y - z) + L_B^\lambda(x) \right] = a(y - x) + g_B(x, y) \tag{8.53}$$

has a finite limit.

Using (8.52) and Lemma 8.5, we see that

$$\lim_{\lambda \downarrow 0} \int_{\bar{B}} \Pi_B^\lambda(x, dz) a^\lambda(y - z) = \int_{\bar{B}} \Pi_B(x, dz) a(y - z), \tag{8.54}$$

must be finite for $x \neq y$. Thus, it must be that $\lim_{\lambda \downarrow 0} L_B^\lambda(x) = L_B(x)$ exists and that (8.9) is satisfied. This establishes the theorem.

One of the main applications of Theorem 8.2 is to show that $g_B(x, y) < \infty$, $x \neq y$, $x, y \in B^c$, when B is a compact nonpolar set. As we will see in the next section, g_B restricted to $B^c \times B^c$ (for B a closed set) is just the Green function of B^c. Now if one is interested only in showing that $g_B(x, y) < \infty$ for $x \neq y$, $x, y \in B^c$ and in showing that (8.9) is valid for such x, y when B is a nonpolar compact set then the proof of the theorem can be considerably shortened. In fact, all one needs is Lemma 8.2 and the simple parts of Lemmas 8.1 and 8.3.

A simple consequence of Theorem 8.2 is the following result.

THEOREM 8.3. *Let B be a nonpolar Borel set. Then*

$$\lim_{t \to \infty} \log(t) P_x(V_B > t) = 2\pi L_B(x). \tag{8.55}$$

PROOF. One easily checks that

$$\lambda g^\lambda(1) \sim \frac{1}{2\pi} \lambda \log\left(\frac{1}{\lambda}\right), \qquad \lambda \downarrow 0. \tag{8.56}$$

The theorem follows from this and the fact that

$$\lim_{\lambda \downarrow 0} \lambda g^\lambda(1) \left[\frac{1 - E_x(\exp\{-\lambda V_B\})}{\lambda} \right] \equiv L_B(x) \tag{8.57}$$

exists by well-known Tauberian theorems.

Of course Theorem 8.3 is uninteresting when $E_x V_B < \infty$ since then $L_B(x) \equiv 0$. When B is relatively compact, however, $L_B(x) > 0$ for all $|x|$ sufficiently large. In Section 10, we will see that $L_B(x) = \lim_{|y| \to \infty} g_B(x, y)$ whenever B is relatively compact.

9. Green function

Now that we have Theorem 8.2 at our disposal, we can easily derive the properties of $g_B(x, y)$ for a nonpolar set B.

THEOREM 9.1. *Let B be nonpolar. Then*

(i) $0 \leqq g_B(x, y) < \infty$ *for* $x \neq y$,
(ii) $g_B(x, y) = g_B(y, x)$,

(iii) $g_B(x, y) + a(y - x)$ *is harmonic in* y *on* $(\bar{B})^c$,
(iv) $g_B(x, y)$ *is harmonic in* y *on* $(\bar{B})^c - \{x\}$,
(v) $g_B(x, y)$ *is subharmonic in* y *and upper semicontinuous in* y *on* $R^2 - \{x\}$,
(vi) $\lim_{y \to y_0} g_B(x, y) = g_B(x, y_0) = 0, y_0 \in B^r$.

PROOF. Part (i) is part of Theorem 8.2. Part (ii) follows from the fact that it is true for g_B^λ and letting $\lambda \downarrow 0$. Parts (iii) to (v) follow from (8.9) and the fact that $a(y - x)$ is harmonic in $y \neq x$ and that for any finite measure μ, $\int_{\bar{B}} \log |y - x| \mu(dx)$ is an upper semicontinuous and subharmonic function that is harmonic in $(\bar{B})^c$. Finally, (vi) follows from the upper semicontinuity and the fact that $g_B^\lambda(x, y_0) = 0$ for all λ and x.

An open set G *is called Greenian if* there exists a function $g(x, y)$ on $G \times G$ such that $g(x, y) + a(y - x)$ is harmonic in y on G. If G is Greenian the smallest such function is called the Green function of G.

From our work in Section 5, we know that in dimension $n \geqq 3$ every open set is Greenian and that g_{G^c} restricted to $G \times G$ is its Green function. Based on this, and Theorem 9.1, we would fully expect that the following holds in the planar case.

THEOREM 9.2. *An open set* $G \subset R^2$ *is Greenian if and only if* G^c *is nonpolar. In that case* g_{G^c} *restricted to* $G \times G$ *is the Green function of* G.

PROOF. Suppose first that G is a bounded open set such that each point of ∂G is regular for G^c. Then clearly G^c is nonpolar so Theorem 9.1 shows that g_{G^c} restricted to $G \times G$ is Greenian. To see that it is the smallest of the Greenian functions, suppose g is another such function. Then by property (vi) of Theorem 9.1,

$$\liminf_{y \to y_0} [g(x, y) - g_{G^c}(x, y)] = \liminf_{y \to y_0} g(x, y) \geqq 0. \tag{9.1}$$

Thus, by the minimum principle, $g(x, y) - g_{G^c}(x, y) \geqq 0$ on G.

Suppose now that G is any open subset of R^2. By Corollary 3.1, we can find bounded open sets $G_1 \subset \bar{G}_1 \subset G_2 \subset \cdots$, $\cup_n G_n = G$, such that each point of ∂G_n is regular for G_n^c and such that $P_x(V_{\partial G_n} \uparrow V_{\partial G}) = 1$ for all $x \in G$.

By Lemma 8.4, the sequence $g_{G_n^c}$ is increasing and bounded above by $g_{G^c}(x, y)$. Since $g_{G_n^c}(x, y) + a(y - x)$ is harmonic in $y \in G$ for fixed $x \subset G_n$, Harnack's theorem tells us that the limit function $g^*(x, y) \leqq g_{G^c}(x, y)$ is such that in a given component of G it is either identically infinite or a harmonic function on $G - \{x\}$. Let $f \geqq 0$ be arbitrary. Then for any $x \in G$,

$$E_x \int_0^{V_{\partial G_n}} f(X_t)\, dt = E_x \int_0^{V_{G_n^c}} f(X_t)\, dt \uparrow E_x \int_0^{V_{G^c}} f(X_t)\, dt, \qquad n \uparrow \infty. \tag{9.2}$$

But

$$E_x \int_0^{V_{G_n^c}} f(X_t)\, dt = \int_{R^2} g_{G_n^c}(x, y) f(y)\, dy, \tag{9.3}$$

so monotone convergence gives that

$$\int_{R^2} g_{G_n^c}(x, y)f(y)\,dy \uparrow \int_{R^2} g^*(x, y)f(y)\,dy, \tag{9.4}$$

as $n \uparrow \infty$. Thus,

$$\int_{R^2} g^*(x, y)f(y)\,dy = \int_{R^2} g_{G^c}(x, y)f(y)\,dy. \tag{9.5}$$

Suppose G^c is nonpolar. Then as $g^*(x, y) \leqq g_{G^c}(x, y)$, $g^*(x, y) < \infty$ for $x \neq y$, and thus $g^*(x, y) + a(y - x)$ is harmonic in y on G. Since the function f in (9.5) is any nonnegative function, (9.5) implies that $g^*(x, y) = g_{G^c}(x, y)$ a.e. y. Since $g_{G^c}(x, y) + a(y - x)$ is harmonic in y, we see that $g^* = g_{G^c}$ on $G \times G$. Suppose g is any function having the required properties. Then this function restricted to G_n also has the required properties so by what has already been proved, $g(x, y) \geqq g_{G_n^c}(x, y)$, $x, y \in G_n$. Thus,

$$g_{G^c}(x, y) = \lim_n g_{G_n^c}(x, y) \leqq g(x, y). \tag{9.6}$$

Hence, g_{G^c} restricted to $G \times G$ is the Green function of G. Finally, suppose G^c is polar. Then G cannot be Greenian. Indeed, if it were Greenian, then let g be a function with the required properties. But then, as argued above

$$g^*(x, y) \leqq g(x, y) < \infty, \qquad x \neq y. \tag{9.7}$$

But (9.4) shows that $g^*(x, y) = \infty$ for a.e. y, a contradiction. This completes the proof.

10. Logarithmic potentials

Let μ be a bounded measure having compact support k. The function $\varphi_\mu(x) = -\int_k a(y - x)\mu(dy)$ is called *the potential* of μ. One easily verifies that $\varphi_\mu(x)$ is a lower semicontinuous function that is superharmonic on R^2 and harmonic on k^c.

Let B be a nonpolar relatively compact set. Since $a(y - x) - a(y) \to 0$ as $|y| \to \infty$, uniformly in x on compacts, it follows at once from (8.9) that, uniformly in x on compacts,

$$\lim_{|y| \to \infty} g_B(x, y) = L_B(x). \tag{10.1}$$

As $g_B(x, y) = g_B(y, x)$, we see that

$$\lim_{|x| \to \infty} g_B(x, y) = L_B(y). \tag{10.2}$$

Let D_r be the closed disk of center 0 and radius r. The hitting measure, $\Pi_{D_r}(x, dy)$, is just the unit mass at x for $x \in D_r$. For $x \in D_r^c$ and φ a continuous function on ∂D_r, $\Pi_{D_r}\varphi(x) = \Pi_{\partial D_r}\varphi(x)$ is the unique bounded solution to the

Dirichlet problem for D_r^c with boundary function φ. As is well known from texts on complex variables or partial differential equations, the solution to this Dirichlet problem is provided by the Poisson integral. Thus,

$$\Pi_{D_r}(x, dy) = \frac{[|x|^2 - r^2]}{|y - x|^2} \sigma_r(0, dy), \qquad x \in D_r^c. \tag{10.3}$$

It follows from (10.3) that

$$\lim_{|x| \to \infty} \Pi_{D_r}(x, dy) = \sigma_r(0, dy), \tag{10.4}$$

in the sense of strong convergence of measures.

Let B be any nonpolar relatively compact set and let D_r be a disk of center 0 and radius r that contains $\bar{B}$ in its interior. Then for any bounded function φ

$$\Pi_B\varphi(x) = \int_{\partial D_r} \Pi_{D_r}(x, dz)\Pi_B\varphi(z), \qquad x \in D_r^c. \tag{10.5}$$

Hence,

$$\lim_{|x| \to \infty} \Pi_B\varphi(x) = \int_{\partial D_r} \Pi_B\varphi(z)\sigma_r(0, dz). \tag{10.6}$$

Let

$$\mu_B(dy) = \int_{\partial D_r} \Pi_B(z, dy)\sigma_r(0, dz). \tag{10.7}$$

Equation (10.6) shows that $\mu_B(dy) = \lim_{|x| \to \infty} \Pi_B(x, dy)$ in the sense of strong convergence of measures. We have thus proved the following important result.

THEOREM 10.1. *Let B be a nonpolar relatively compact set. Then $g_B(x, y) \to L_B(y)$ as $|x| \to \infty$, and $\Pi_B(x, dy) \to \mu_B(dy)$ as $|x| \to \infty$ in the sense of strong convergence.*

The measure μ_B has the obvious probabilistic significance as the hitting probability of B starting from infinity. We will show that μ_B should be considered in potential theoretic terms as the equilibrium measure of B.

THEOREM 10.2. *Let B be a nonpolar relatively compact set. Then*

$$\lim_{|x| \to \infty} [L_B(x) - a(x)] = k(B) \tag{10.8}$$

exists and is finite. Moreover,

$$\varphi_{\mu_B}(x) = k(B) - L_B(x). \tag{10.9}$$

PROOF. Suppose $y \notin \bar{B}$. Then $a(y - z)$ is bounded for $z \in \bar{B}$, and thus by Theorem 10.2,

$$\lim_{|x| \to \infty} \int_{\bar{B}} \Pi_B(x, dz)a(y - z) = \int_{\bar{B}} \mu_B(dz)a(y - z) \tag{10.10}$$

exists and is finite. Using (8.9), we see that

$$a(y - x) - a(x) - \int_{\bar{B}} \Pi_B(x, dz)a(y - z) + g_B(x, y) = L_B(x) - a(x). \tag{10.11}$$

Using (10.10) and (10.2), we see that the left side converges to $\varphi_{\mu_B}(y) + L_B(y)$ as $|x| \to \infty$. Hence, the right side must have a finite limit. This establishes (10.8) and (10.9) for $y \in (\bar{B})^c$. Suppose $y \in \bar{B}$. Then (8.9) shows

$$\lim_{|x| \to \infty} \int_{\bar{B}} \Pi_B(x, dz)a(y - z) = L_B(y) - k(B). \tag{10.12}$$

The function $g_B(\xi, y) + L_B(\xi) - a(y - \xi)$ is clearly bounded in ξ if $|\xi| > r$ for some sufficiently large r. But then by (8.9),

$$\int_{\bar{B}} \Pi_B(\xi, dz)a(y - z) = g_B(\xi, y) + L_B(\xi) - a(y - \xi) \tag{10.13}$$

is bounded in ξ for $|\xi| > r$. Now for any closed disk D_r of center 0 and radius r containing $\bar{B}$ in its interior,

$$\int_{\bar{B}} \Pi_B(x, dz)a(y - z) = \int_{\partial D_r} \Pi_{D_r}(x, d\xi) \int_{\bar{B}} \Pi_B(\xi, dz)a(y - z), \qquad x \notin \bar{D}. \tag{10.14}$$

Thus, by Theorem 10.1 and equation (10.7),

$$\lim_{|x| \to \infty} \int_{\bar{B}} \Pi_B(x, dz)a(y - z) = \int_{\partial D_r} \sigma_r(0, d\xi) \int_{\bar{B}} \Pi_B(\xi, dz)a(y - z) = \int_{\bar{B}} \mu_B(dz)a(y - z). \tag{10.15}$$

This establishes the theorem for $y \in \bar{B}$ and thereby completes the proof.

DEFINITION 10.1. *Let B be a nonpolar relatively compact set. The measure μ_B in Theorem* 10.1 *is called the* equilibrium measure *of B. The constant $k(B)$ in Theorem* 10.2 *is called the* Robin's constant *of B and the potential of μ_B is called the* equilibrium potential *of B.*

For a relatively compact polar set, we define $k(B) = +\infty$. Theorem 10.3 given below will show that this is the natural definition of $k(B)$ for a polar set.

PROPOSITION 10.1. *Let A and B be two relatively compact sets such that $A \subset B$. Then $k(A) \geqq k(B)$.*

PROOF. If B is polar then A must also be polar. In this case $k(A) = k(B) = +\infty$. Suppose B is nonpolar. If A is polar, then $k(A) = +\infty$ and $k(B) < \infty$ so the proposition is valid. Suppose A is also nonpolar. Then $L_A \geqq L_B$ (since $L_A^\lambda \geqq L_B^\lambda$) and (10.8) shows that $k(A) \geqq k(B)$.

THEOREM 10.3. *Let B be a compact set. Then*

$$k(B) = \sup \{k(U): U \text{ open}, U \supset B \text{ and } \bar{U} \text{ compact}\}. \tag{10.16}$$

On the other hand, if U is an open relatively compact set, then

$$k(U) = \inf \{k(A) : A \text{ compact}, A \subset U\}. \tag{10.17}$$

PROOF. Suppose B is compact. Let B_n, $n \geqq 1$ be relatively compact open sets such that $B_1 \supset \bar{B}_2 \supset B_2 \supset \cdots, \bigcap_n B_n = \bigcap_n \bar{B}_n = B$. Then as was shown in Proposition 3.2, $P_x(V_{B_n} \uparrow V_B) = 1$ for all $x \in B^c \cup B^r$. Suppose that B is polar. Let $f \geqq 0$ be continuous with compact support and have integral 1 and set $Af(z) = \int_{R^2} a(y - z)f(y)\, dy$. Then for any $x \in B^c$,

$$\int_{R_n} g_{B_n}(x, y)f(y)\, dy = E_x \int_0^{V_{B_n}} f(X_t)\, dt \uparrow \infty, \qquad n \to \infty. \tag{10.18}$$

Now $Af(x)$ is a continuous function, and thus for all n

$$\left| \int_{\bar{B}_n} \Pi_{B_n}(x, dz) Af(z) \right| \leqq \sup_{z \in \bar{B}_1} |Af(z)| = M < \infty. \tag{10.19}$$

Using (8.9), we see that

$$\begin{aligned} \left| - \int_{R^n} g_{B_n}(x, y)f(y)\, dy + L_{B_n}(x) \right| &= \left| Af(x) - \int_{\bar{B}_n} \Pi_{B_n}(x, dz) Af(z) \right| \\ &\leqq |Af(x)| + M < \infty. \end{aligned} \tag{10.20}$$

Thus, using (10.18), we see that $L_{B_n}(x) \uparrow \infty$ for each $x \in B^r$. By (10.9),

$$\int_{R^2} \varphi_{\mu_{B_n}}(x) f(x)\, dx + \int_{R^2} L_{B_n}(x) f(x)\, dx = k(B_n). \tag{10.21}$$

But

$$\left| \int_{R^2} \varphi_{\mu_{B_n}}(x) f(x)\, dx \right| = \left| \int_{R^2} Af(x) \mu_{B_n}(dx) \right| \leqq M < \infty, \tag{10.22}$$

and $\int_{R^2} L_{B_n}(x) f(x)\, dx \uparrow \infty$. Thus, $k(B_n) \uparrow \infty$. This establishes (10.16) when B is polar.

Suppose now that B is nonpolar. Let f and Af be as before. Then for $x \in B^c \cup B^r$,

$$\begin{aligned} \int_{R^n} g_{B_n}(x, y) f(y)\, dy &= E_x \int_0^{V_{B_n}} f(X_t)\, dt \uparrow E_x \int_0^{V_B} f(X_t)\, dt \\ &= \int_{R2} g_B(x, y) f(y)\, dy. \end{aligned} \tag{10.23}$$

Also as $P_x(X_{V_{B_n}} \in \bar{B}_1) = 1$ for all n.

$$\begin{aligned} \lim_{n \to \infty} \int_{\bar{B}_n} \Pi_{B_n}(x, dz) Af(z) &= \lim_{n \to \infty} E_x Af(X_{V_{B_n}}) \\ &= E_x Af(X_{V_B}) = \int_{\bar{B}} \Pi_B(x, dz) Af(z). \end{aligned} \tag{10.24}$$

Hence by (8.9), for $x \in B^c \cup B^r$,

$$(10.25)\quad \lim_{n\to\infty} L_{B_n}(x)$$

$$= \lim_{n\to\infty}\left[Af(x) - \int_{\bar{B}_n} \Pi_{B_n}(x, dz)Af(z) + \int_{R^2} g_{B_n}(x, y)f(y)\,dy\right]$$

$$= Af(x) - \int_{\bar{B}} \Pi_B(x, dz)Af(z) + \int_{R^2} g_B(x, y)f(y)\,dy = L_B(x).$$

Since $(B^r)^c \cap B$ has measure 0, we see that

$$(10.26)\qquad \int_{R^2} L_{B_n}(x)f(x)\,dx \uparrow \int_{R^2} L_B(x)f(x)\,dx.$$

Now if D is a disk of center 0 containing $\bar{B}$ in its interior,

$$(10.27)\qquad \lim_{n\to\infty}\int_{R^2}\mu_{B_n}(dx)Af(x) = \lim_{n\to\infty}\int_{\partial D}\sigma(0, d\xi)E_\xi Af(X_{V_{B_n}})$$

$$= \int_{\partial D}\sigma(0, d\xi)E_\xi Af(X_{V_B})$$

$$= \int_{R^2}\mu_B(dx)Af(x).$$

Hence, using this fact, (10.26), and (10.21), we see that $k(B_n) \uparrow k(B)$. This establishes (10.16). To prove (10.17), note that we can find compacts $A_n \subset U$ such that $A_1 \subset A_2 \subset \cdots$, $\cup_n A_n = U$. But then $P_x(V_{A_n} \downarrow V_U) = 1$ for all x. The remainder of the proof of (10.17) is similar to the proof of (10.16) for B a non polar set. We omit these details.

Let A and B be two Borel sets. Then,

$$(10.28)\qquad P_x(V_{A\cap B} \leqq t) \leqq P_x(V_A \leqq t, V_B \leqq t)$$
$$\leqq P_x(V_A \leqq t) + P_x(V_B \leqq t) - P_x(V_{A\cup B} \leqq t).$$

Thus, $P_x(V_{A\cap B} > t) \geqq P_x(V_A > t) + P_x(V_B > t) - P_x(V_{A\cup B} > t)$. It follows from this and (8.6) that

$$(10.29)\qquad L^\lambda_{A\cap B}(x) \geqq L^\lambda_A(x) + L^\lambda_B(x) - L^\lambda_{A\cup B}(x).$$

Letting $\lambda \downarrow 0$, we see that whenever A and B are nonpolar

$$(10.30)\qquad L_{A\cap B}(x) + L_{A\cup B}(x) \geqq L_A(x) + L_B(x).$$

If we take $L_k(x) \equiv \infty$ whenever k is polar, then (10.30) is valid for all sets. Using (10.8), we see that for relatively compact sets

$$(10.31)\qquad -k(A\cup B) + [-k(A\cap B)] \leqq -k(A) + (-k(B)).$$

Also as $L_A^\lambda(x) \geqq L_B^\lambda(x)$ for $A \subset B$, we see that $L_A(x) \geqq L_B(x)$, so for $A \subset B$,

$$-k(A) \leqq -k(B). \tag{10.32}$$

Define $k^*(B) = -k(B)$ if B is compact and define $k^*(U) = \sup\{k^*(B): B \subset U, B \text{ compact}\}$ if U is open. Then (10.16), (10.17), (10.31), and (10.32) show that $k^*(\cdot)$ is a Choquet capacity on the compact sets. The extension theorem of Choquet then implies that for any Borel set B,

$$\sup\{k^*(A): A \subset B, A \text{ compact}\} = \inf\{k^*(U): U \supset B, U \text{ open}\}, \tag{10.33}$$

and that the common value $k^*(B)$ is such that $k^*(A) \leqq k^*(B)$, $A \subset B$ and

$$k^*(A \cup B) + k^*(A \cap B) \leqq k^*(A) + k^*(B). \tag{10.34}$$

Let B be any relatively compact set. Then for any compact set $A \subset B$ and any open set $U \supset B$, $-k(A) \leqq -k(B) \leqq -k(U)$. Thus,

$$k^*(B) \leqq -k(B) \leqq k^*(B). \tag{10.35}$$

Equation (10.35) tells us that whenever B is relatively compact, $k(B) = -k^*(B)$. For a general Borel set, define $k(B)$ to be $-k^*(B)$ and define the capacity $C(B)$ of B to be $e^{-k(B)}$. Our discussion above has shown the following result.

THEOREM 10.4. *Let B be any relatively compact set. Then*

$$\inf\{k(A): A \subset B, A \text{ compact}\} = k(B) = \sup\{k(U): U \text{ open}, U \supset B\}. \tag{10.36}$$

COROLLARY 10.1. *A Borel set B has capacity 0 if and only if it is polar.*

PROOF. The set B has capacity 0 if and only if $k(B) = \infty$. We have already shown that for a relatively compact set B this is the case if and only if B is polar. If B is not relatively compact and $k(B) = \infty$, then (10.36) shows that $k(A) = \infty$ for every compact subset of B, and thus (again by (10.36)) $k(D) = \infty$ for every relatively compact subset D of B. But then every relatively compact subset of B is polar, and as B is a countable union of relatively compact sets, B must be polar. Conversely, if B is polar, every compact subset of B is polar, so for any compact subset $A \subset B$, $k(A) = \infty$. Thus, (10.36) shows that $k(B) = \infty$. This establishes the corollary.

If B is a relatively compact set in dimension $n \geqq 3$, then as was shown in Section 6, B has capacity 0 (that is, B is polar) if and only if the only finite measure μ having support on B with a bounded potential is the 0 measure. For potentials in the plane we have the following analog.

THEOREM 10.5. *A relatively compact set B is polar if and only if* $\sup_x \varphi_\mu(x) = +\infty$ *for every nonzero finite measure μ having support on B.*

PROOF. Let A be a relatively compact open set containing $\bar{B}$. Let $a_N(x) = a(x)$ if $a(x) \geqq -N$ and let $a_N(x) = -N$ if $a(x) < -N$, $N > 0$. Then clearly

$$-\int_{R^2} \mu(dx) \int_{R^2} a_N(y - x)\mu_A(dy) = -\int_{R^2} \mu_A(dy) \int_{R^2} a_N(y - x)\mu(dx). \tag{10.37}$$

By monotone convergence as $N \uparrow \infty$, we see that

$$\int_{\bar{B}} \mu(dx)\varphi_{\mu_A}(x) = \int_{\bar{A}} \mu_A(dx)\varphi_\mu(x). \tag{10.38}$$

Since $\bar{B} \subset A$ equation (10.9) shows that the left side is $\mu(\bar{B})k(A)$. Thus, $\mu(\bar{B})k(A) \leqq \sup_x \varphi_\mu(x)$, and hence by Theorem 10.4, $\mu(\bar{B})k(B) \leqq \sup_x \varphi_\mu(x)$. If B is polar then $k(B) = +\infty$. On the other hand, if $\sup_x \varphi_\mu(x) = +\infty$ for all nonzero μ supported on $\bar{B}$, then B must be polar. For, if B were nonpolar then the equilibrium measure μ_B of B would be a probability measure on B whose potential would be $\leqq k(B) < +\infty$ everywhere, a contradiction. This establishes the theorem.

REMARK. By using the maximum principle for potentials φ_μ (see, for example, Hille, *Analytic Function Theory*, Vol. II), we replace $\sup_x$ by $\sup_{x \in \bar{B}}$ in Theorem 10.5.

POTENTIAL OPERATORS FOR MARKOV PROCESSES

KEN-ITI SATO
TOKYO UNIVERSITY OF EDUCATION

1. Introduction

Yosida's definition of potential operators for semigroups [17] makes it possible to deal with transient Markov processes and a class of recurrent Markov processes in a unified operator theoretical way. In this paper, we prove some general properties of his potential operators, show which Markov processes admit the potential operators, and investigate the cases of processes with stationary independent increments as typical examples.

Let T_t be a strongly continuous semigroup of linear operators on a Banach space $\mathscr{B}$ satisfying

$$\sup_{t \geqq 0} \|T_t\| < \infty, \tag{1.1}$$

with infinitesimal generator A and resolvent

$$J_\lambda = (\lambda - A)^{-1}, \qquad \lambda > 0. \tag{1.2}$$

Following Yosida, we define potential operator V for the semigroup by

$$Vf = s \lim_{\lambda \to 0} J_\lambda f, \tag{1.3}$$

when and only when the limit exists for f in a dense subset of $\mathscr{B}$. The domain $\mathscr{D}(V)$ is the collection of f such that the limit exists. We will give conditions for the existence of the potential operator (Theorem 2.2) and prove some general properties (Theorem 2.3), summarizing Yosida's results [17], [19] with a few results added. The relation with other definitions of potential operators is shown in Theorem 2.4. In Section 3, we consider the case where $\mathscr{B}$ is the Banach space $C_0(S)$ of real valued continuous functions on S vanishing at infinity, S being a locally compact Hausdorff space with a countable base, and T_t is a semigroup induced by a Markov process transition probability. We will prove that the semigroup admits a potential operator if the Markov process is either transient or null recurrent, and that it does not admit a potential operator if the process is positive recurrent. Processes with stationary independent increments on Euclidean spaces are examined in Section 4. The fact that they admit potential operators (Theorem 4.1) is a generalization of Yosida's result [18] on Brownian motions. The domain and the representation of potential operators are investigated for Brownian motions, stable processes, and some other processes in Section 5. We return in Section 6 to a general situation and consider generalization of maximum principles for classical potential operators to operators in Banach lattices. New types of maximum principles are introduced for the adjoints of potential operators.

Several works have been done recently on potential operators of recurrent Markov processes ([7], [10], and others). Authors use different definitions of potential operators. It seems that an advantage for Yosida's potential operators lies in their direct connection with infinitesimal generators.

2. Potential operators for semigroups on Banach spaces

Let $\mathscr{B}$ be a Banach space, and $\mathscr{B}^*$ be its adjoint space. We use the notation $(\varphi, f) = \varphi(f)$ for $\varphi \in \mathscr{B}^*$ and $f \in \mathscr{B}$. The limit in the strong, weak, or weak* convergence is denoted by s lim, w-lim, or w^* lim, respectively. By *dense*, we mean strongly dense. We say that a subset $\mathscr{M}$ of $\mathscr{B}^*$ is *w^* dense* if for each $\varphi \in \mathscr{B}^*$ there is a sequence $\{\varphi_n\}$ in $\mathscr{M}$ such that φ_n weakly* converges to φ. Thus w^* denseness implies denseness in the sense of weak* topology. The symbols $\mathscr{D}$, $\mathscr{R}$, and $\mathscr{N}$ mean domain, range, and null space of an operator. In this section, $\{T_t; t \geqq 0\}$ is always a strongly continuous semigroup of linear operators on $\mathscr{B}$ satisfying (1.1), A is its infinitesimal generator, and J_λ is the resolvent operator (1.2). It is known that A has dense domain and determines the semigroup uniquely and that

$$J_\lambda f = \int_0^\infty e^{-\lambda t}\, T_t f\, dt, \tag{2.1}$$

(see [3] or [16]). Let T_t^*, J_λ^*, and A^* be the adjoint operators of T_t, J_λ, and A, respectively. The following theorem has a preliminary character, but is interesting in itself.

THEOREM 2.1.

(i) *The semigroup $\{T_t^*; t \geqq 0\}$ is a weakly* continuous semigroup on $\mathscr{B}^*$.*

(ii) *The operator A^* has w^* dense domain and*

$$A^*\psi = w^* \lim_{t\to 0} t^{-1}(T_t^*\psi - \psi), \qquad \psi \in \mathscr{D}(A^*). \tag{2.2}$$

Conversely, if the right side of (2.2) *exists, then $\psi \in \mathscr{D}(A^*)$.*

(iii) *The following relations hold*

$$J_\lambda^* = (\lambda - A^*)^{-1}, \tag{2.3}$$

$$J_\lambda^*\varphi = \int_0^\infty e^{-\lambda t}\, T_t^*\varphi\, dt, \qquad \varphi \in \mathscr{B}^*. \tag{2.4}$$

(iv) *The operator A^* determines T_t uniquely.*

The integral in (2.4) is defined to be the element $\psi \in \mathscr{B}^*$ which satisfies

$$(\psi, f) = \int_0^\infty e^{-\lambda t} (T_t^* \varphi, f)\, dt, \qquad f \in \mathscr{B}. \tag{2.5}$$

PROOF. Part (i) is obvious. Equation (2.3) is a consequence of (1.2), (see [16], p. 224). The domain of A^* is w^* dense since $\lambda J_\lambda^* \varphi$ converges to φ weakly* as $\lambda \to \infty$ and $\lambda J_\lambda^* \varphi \in \mathscr{D}(A^*)$. If φ is the right side of (2.2), then $J_\lambda^*(\lambda\psi - \varphi) = \psi$, and hence $\psi \in \mathscr{D}(A^*)$ and $A^*\psi = \varphi$, because we have

$$\begin{aligned}(J_\lambda^* \varphi, f) &= \lim_{t \to 0} t^{-1}(T_t^* \psi - \psi, J_\lambda f) = (\psi, AJ_\lambda f) \\ &= (\psi, \lambda J_\lambda f - f) = (\lambda J_\lambda^* \psi - \psi, f)\end{aligned} \tag{2.6}$$

for any $f \in \mathscr{B}$. If ψ is the right side of (2.4), then

$$\begin{aligned}(T_t^* \psi, f) &= \int_0^\infty e^{-\lambda s} (T_s^* \varphi, T_t f)\, ds = \int_0^\infty e^{-\lambda s} (T_{t+s}^* \varphi, f)\, ds \\ &= e^{\lambda t} \int_t^\infty e^{-\lambda s} (T_s^* \varphi, f)\, ds\end{aligned} \tag{2.7}$$

for any f, which implies $t^{-1}(T_t^*\psi - \psi, f) \to (\lambda\psi - \varphi, f)$, and hence $\psi \in \mathscr{D}(A^*)$ and $A^*\psi = \lambda\psi - \varphi$. This shows (2.4) by (2.3). In order to finish the proof of (2.2), note that any $\psi \in \mathscr{D}(A^*)$ is represented as $\psi = J_\lambda^* \varphi$ by (2.3), and hence $w^* \lim t^{-1}(T_t^*\psi - \psi)$ exists by (2.4) and the above argument. The operator A^* determines T_t uniquely since A^* determines J_λ^* by (2.3) and J_λ^* determines J_λ. The proof is complete.

The potential operator defined in Section 1 does not always exist. But we have simple criteria for its existence.

THEOREM 2.2. *The following conditions are equivalent:*

(a) *T_t admits a potential operator;*

(b) *$\mathscr{R}(A)$ is dense;*

(c) *$\lambda J_\lambda f \to 0$ $(\lambda \to 0)$ strongly for all f;*

(d) *$\lambda J_\lambda f \to 0$ $(\lambda \to 0)$ weakly for all f;*

(e) *$t^{-1} \int_0^t T_s f\, ds \to 0$ $(t \to \infty)$ strongly for all f;*

(f) *$t^{-1} \int_0^t T_s f\, ds \to 0$ $(t \to \infty)$ weakly for all f;*

(g) *A^* is one to one;*

(h) *$\lambda J_\lambda^* \varphi \to 0$ $(\lambda \to 0)$ weakly* for all φ;*

(i) *$t^{-1} \int_0^t T_s^* \varphi\, ds \to 0$ $(t \to \infty)$ weakly* for all φ.*

The equivalence of the first four conditions is proved by Yosida [17]. He introduces also the condition (h) in [19]. Conditions (e), (f), (i) are new. The following condition is also equivalent, though apparently weaker: for each f in a dense set in $\mathscr{B}$ there is a sequence $\{\lambda_n\}$ decreasing to 0 such that $\lambda_n J_{\lambda_n} f$ converges to 0 weakly (see [17]).

PROOF. That (d) is equivalent to (h) is evident. The equivalence of (f) and (i) is also evident, since we have

$$\left(t^{-1}\int_0^t T_s^*\varphi\, ds, f\right) = \left(\varphi, t^{-1}\int_0^t T_s f\, ds\right). \tag{2.8}$$

Equivalence of (b) and (g) is of general character ([16], p. 224). Using

$$\lambda\|J_\lambda\| \leqq M, \tag{2.9}$$

where M is the bound of $\|T_t\|$, we get the implication (a) $\Rightarrow$ (c); (c) implies (d); (d) implies (b) because the closure of $\mathscr{R}(A)$ is closed in weak topology by the Hahn-Banach theorem ([16], p. 125) and $AJ_\lambda f = \lambda J_\lambda f - f \to f$ weakly as $\lambda \to 0$; (b) implies (c), since if $f = Au$ then

$$\|\lambda J_\lambda f\| = \|\lambda(\lambda J_\lambda - 1)u\| \leqq \lambda(1 + M)\|u\| \to 0, \tag{2.10}$$

and since we can use (2.9) and (b) for general f. On the other hand, (b) and (c) together imply (a) because $J_\lambda Au = \lambda J_\lambda u - u \to -u$ strongly. Thus, (a), (b), (c), (d) are equivalent. If $f = Au$, then

$$t^{-1}\int_0^t T_s f\, ds = t^{-1}\int_0^t AT_s u\, ds = t^{-1}(T_t u - u) \to 0, \qquad t \to \infty. \tag{2.11}$$

Since we have

$$\left\|t^{-1}\int_0^t T_s f\, ds\right\| \leqq M\|f\|, \qquad f \in \mathscr{B}, \tag{2.12}$$

(b) implies (e); (e) $\Rightarrow$ (f) is evident. If (f) holds, then

$$t^{-1}\int_0^t (\varphi, T_s f)\, ds = \left(\varphi, t^{-1}\int_0^t T_s f\, ds\right) \to 0, \qquad t \to \infty, \tag{2.13}$$

for each φ and f, which implies

$$\lambda\int_0^\infty e^{-\lambda s}(\varphi, T_s f)\, ds \to 0, \qquad \lambda \to 0, \tag{2.14}$$

by the Abelian theorem for Laplace transforms ([15], p. 181), that is, the condition (d). The proof is complete.

THEOREM 2.3. *Suppose that* $\{T_t;\, t \geqq 0\}$ *admits a potential operator* V *and let* V^* *be the adjoint operator of* V. *Then,* A, A^*, V, V^* *are all one to one,* $V = -A^{-1}$, *and* $V^* = -(A^*)^{-1}$. *Subspaces* $\mathscr{D}(V) = \mathscr{R}(A)$ *and* $\mathscr{R}(V) = \mathscr{D}(A)$ *are both dense in* $\mathscr{B}$; *similarly,* $\mathscr{D}(V^*) = \mathscr{R}(A^*)$ *and* $\mathscr{R}(V^*) = \mathscr{D}(A^*)$ *are both* w^* *dense in* $\mathscr{B}^*$. *Furthermore,*

$$V^*\varphi = w^* \lim_{\lambda \to 0} J_\lambda^*\varphi, \qquad \varphi \in \mathscr{D}(V^*) \tag{2.15}$$

holds. The collection of φ *such that the limit in the right side of* (2.15) *exists coincides with* $\mathscr{D}(V^*)$.

The theorem is proved by Yosida [17], [19]. We give the proof for completeness.

PROOF. If $u \in \mathscr{D}(A)$, then $J_\lambda Au = \lambda J_\lambda u - u \to -u$ strongly by Theorem 2.2 (c), and hence $Au \in \mathscr{D}(V)$ and $-VAu = u$. If $f \in \mathscr{D}(V)$, then $AJ_\lambda f \to -f$ strongly likewise, and hence $Vf \in \mathscr{D}(A)$ and $-AVf = f$ by the closedness of A. Thus, $\mathscr{N}(A) = \mathscr{N}(V) = \{0\}$ and $V = -A^{-1}$. This implies $\mathscr{N}(A^*) = \mathscr{N}(V^*) = \{0\}$ and $V^* = -(A^*)^{-1}$ by [16], p. 224. If $J_\lambda^*\varphi$ has weak* limit ψ as $\lambda \to 0$, then it follows from

$$(2.16) \qquad (\varphi - \lambda J_\lambda^*\varphi, Vf) = (-A^*J_\lambda^*\varphi, Vf) = (J_\lambda^*\varphi, f),$$

that $(\varphi, Vf) = (\psi, f)$ for all $f \in \mathscr{D}(V)$, which means $\varphi \in \mathscr{D}(V^*)$ and $V^*\varphi = \psi$. Conversely, if $\varphi \in \mathscr{D}(V^*)$ and $V^*\varphi = \psi$, then $J_\lambda^*\varphi = -J_\lambda^*A^*\psi = \psi - \lambda J_\lambda^*\psi$ converges to ψ weakly*. The remaining assertions are trivial consequences of the previous theorems.

COROLLARY 2.1. *The operator V determines $\{T_t\}$ uniquely, and so does V^*.*

THEOREM 2.4. *Suppose that $\{T_t\}$ admits a potential operator V.*

(i) *The following five conditions are equivalent:*

(a) *$f \in \mathscr{D}(V)$ and $Vf = u$;*

(b) *$J_\lambda f \to u (\lambda \to 0)$ weakly;*

(c) *$(1 - \lambda J_\lambda)u = J_\lambda f$ for some $\lambda > 0$;*

(d) *$(1 - \lambda J_\lambda)u = J_\lambda f$ for all $\lambda > 0$;*

(e) *$(1 - T_t)u = \int_0^t T_s f\, ds$ for all $t \geqq 0$.*

(ii) *In order that $f \in \mathscr{D}(V)$, $Vf = u$, and $w \lim T_t u\ (t \to \infty)$ exists, it is necessary and sufficient that*

$$(2.17) \qquad u = w \lim_{t \to \infty} \int_0^t T_s f\, ds.$$

(iii) *Assertion* (ii) *remains valid with w lim replaced by s lim.*

Equivalence of (d) and (e) is observed by Kondō, and he studies solution of (d) and (e) in an extended sense for recurrent Markov processes [7].

PROOF. The implication (a) ⇒ (b) is trivial. If $u = w \lim J_\mu f$, then by the resolvent equation

$$(2.18) \qquad J_\lambda f - J_\mu f + (\lambda - \mu)J_\lambda J_\mu f = 0,$$

we have (d). Thus, (b) implies (d). The implication (d) ⇒ (c) is trivial. If (c) holds, then it follows from (2.18) that

$$(2.19) \qquad J_\mu f = J_\lambda f + (\lambda - \mu)J_\mu(1 - \lambda J_\lambda)u = J_\lambda f - (\lambda - \mu)J_\mu AJ_\lambda u,$$

which strongly converges to $J_\lambda f + \lambda J_\lambda u = u$, and we have (*a*). Also, we see equivalence of (a) and (e), noting that $f = -Au$. For (ii) let u be defined by (2.17). Then u satisfies (e), and hence (a). According to (e) and (2.17), $T_t u$ weakly converges to 0 as $t \to \infty$. Conversely, let $Vf = u$ and $T_t u \to v$ weakly as $t \to \infty$. Then we have $T_s v = v$ for every s, and hence $v = 0$ by $\mathscr{N}(A) = \{0\}$. Thus, (2.17) follows from (e). Replacing w lim by s lim, we get the proof of (iii).

A set $\mathscr{M}$ is called a *core* of a closed operator T, if $\mathscr{M} \subset \mathscr{D}(T)$ and if the smallest closed extension of $T | \mathscr{M}$ coincides with T, where $T | \mathscr{M}$ is the restriction of T to $\mathscr{M}$ (see [6], p. 166). The notion of core is important, because if $\mathscr{M}$ is a core of the potential operator V, then $V | \mathscr{M}$ determines the semigroup. Note that V is a closed operator since A is closed. Although it is usually difficult to find explicit expression of V, it is sometimes possible to find the expression on some core. See Section 5 for examples. If $\mathscr{M}$ is a core of T, then $\mathscr{M}$ is dense in $\mathscr{D}(T)$ and $T(\mathscr{M})$ is dense in $\mathscr{R}(T)$. But the converse is not true in general. We can prove the following assertion: let T be a closed linear operator and let $\mathscr{M}$ be a linear subspace of $\mathscr{D}(T)$. Suppose that for each $f \in \mathscr{D}(T)$ there are a sequence $\{f_n\}$ in $\mathscr{M}$ and an element $g \in \mathscr{B}$ such that $w \lim f_n = f$ and $w \lim Tf_n = g$. Then, $\mathscr{M}$ is a core of T.

3. Potential operators for Markov process semigroups

Let S be a locally compact Hausdorff space with a countable base. Let $C_0(S)$ be the Banach space of real valued continuous functions on S vanishing at infinity if S is not compact, or the Banach space of real valued continuous functions on S if S is compact. We denote the collection of continuous functions with compact supports by $C_K(S)$ or C_K, and the collection of nonnegative functions in C_K by C_K^+. Let $\{T_t;\, t \geqq 0\}$ be a strongly continuous semigroup of positive linear operators on $C_0(S)$ with norm $\|T_t\| \leqq 1$. There corresponds to $\{T_t\}$ a right continuous, time homogeneous Markov process on S with transition probability $P(t, x, dy)$ such that

$$T_t f(x) = \int_S f(y) P(t, x, dy). \tag{3.1}$$

We call the Markov process (or the semigroup) *recurrent* if

$$\int_0^\infty P(t, x, U)\, dt = \infty \tag{3.2}$$

for all x and all open neighborhoods U of x, and *transient* if

$$\int_0^\infty P(t, x, K)\, dt < \infty \tag{3.3}$$

for all x and all compact K. It is *null recurrent* if it is recurrent and

$$\lim_{t \to \infty} P(t, x, K) = 0 \tag{3.4}$$

for all x and all compact K, and *positive recurrent* if

$$\liminf_{t \to \infty} P(t, x, U) > 0 \tag{3.5}$$

for all x and all open neighborhood U of x. We will show the relations of these notions with the existence of a potential operator. In applying the theorems in

Section 2, note that $C_0^*(S)$ is the space of signed measures with bounded variation normed by the total variation, and that $\{f_n\}$ converges weakly to f in $C_0(S)$ if and only if $f_n(x)$ converges to $f(x)$ pointwise and $\sup_n \|f_n\| < \infty$.

THEOREM 3.1. *If $\{T_t\}$ is transient, then it admits a potential operator.*

PROOF. Since C_K is dense, it suffices to show that $t^{-1} \int_0^t T_s f(x)\, ds$ tends to 0 pointwise as $t \to \infty$ for $f \in C_K$ (Theorem 2.2). But this is easily seen because $\int_0^\infty T_s|f|(x)\, ds$ is finite.

THEOREM 3.2. *If $\{T_t\}$ is null recurrent or, more generally, if* (3.4) *holds, then it admits a potential operator V and*

$$Vf = w \lim_{t\to\infty} \int_0^t T_s f\, ds. \tag{3.6}$$

PROOF. It follows from (3.4) that

$$w \lim_{t\to\infty} T_t f = 0 \tag{3.7}$$

for all $f \in C_K$, hence for all $f \in C_0(S)$. The theorem is then obtained from Theorems 2.2 and 2.4.

THEOREM 3.3. *If $\{T_t\}$ is positive recurrent, or if the process has a finite invariant measure, then the potential operator does not exist.*

PROOF. Property (3.5) implies that condition (f) of Theorem 2.2 does not hold. Existence of a finite invariant measure φ contradicts condition (i) of the same theorem since $T^*\varphi = \varphi \neq 0$.

REMARK. If S is a countable set with discrete topology and all points communicate with each other, then the following three conditions for the process are equivalent: to admit a potential operator; to be transient or null recurrent; to have no finite invariant measure. In fact, transience, null recurrence, and positive recurrence cover all possibilities in this case, and the process is positive recurrent if and only if it has a finite invariant measure [2].

If the process is transient, C_K is not necessarily contained in $\mathscr{D}(V)$. But we have:

THEOREM 3.4. *Suppose that $\{T_t\}$ admits a potential operator V and that $C_K \subset \mathscr{D}(V)$. Then, it is transient, C_K is a core of V, and*

$$Vf = s \lim_{t\to\infty} \int_0^t T_s f\, ds. \tag{3.8}$$

PROOF. If $f \in C_K^+$, then

$$\int_0^\infty T_t f(x)\, dt = \lim_{\lambda\to 0} \int_0^\infty e^{-\lambda t} T_t f(x)\, dt = Vf(x) < \infty. \tag{3.9}$$

Hence, the process is transient and we have (3.8) for C_K^+ by applying Dini's theorem to the one point compactification of S. Let V_0 be the smallest closed extension of $V | C_K$. We have to prove $V_0 = V$. First, let us show that if $f \in C_K^+$, then $J_\lambda f \in \mathscr{D}(V_0)$. In fact, let $u = J_\lambda Vf$ and let $\{g_n\}$ be an increasing sequence in

C_K^+ which converges strongly to $J_\lambda f$. Since we have

$$\int_0^\infty T_t J_\lambda f(x)\,dt = \int_0^\infty \int_0^\infty e^{-\lambda s} T_{t+s} f(x)\,ds\,dt = \int_0^\infty e^{-\lambda s} T_s V f(x)\,ds = u, \tag{3.10}$$

Vg_n tends strongly to u by Dini's theorem, and hence $J_\lambda f \in \mathscr{D}(V_0)$. Also, we have

$$\lambda V_0 J_\lambda f + J_\lambda f = V_0 f \tag{3.11}$$

for $f \in C_K$. It follows from (3.11) that $\mathscr{R}(V_0)$ is dense. We see that if $f \in \mathscr{D}(V_0)$, then $J_\lambda f \in \mathscr{D}(V_0)$ and (3.11) holds. Thus, $\mathscr{R}(\lambda V_0 + 1)$ contains $\mathscr{R}(V_0)$, and hence is dense. If $g_n = (\lambda V_0 + 1)f_n \to g$ strongly, then $f_n = g_n - \lambda J_\lambda g_n \to g - \lambda J_\lambda g$ and hence $g \in \mathscr{R}(\lambda V_0 + 1)$ by closedness of V_0. The whole space is thus $\mathscr{R}(\lambda V_0 + 1)$. On the other hand, $\lambda V + 1$ is a one to one mapping and an extension of $\lambda V_0 + 1$, whence $V = V_0$. If $f \in C_K$, then it follows from (3.8) and Theorem 2.4 that $T_t V f \to 0$ strongly as $t \to \infty$. Since $V(C_K)$ is dense, $T_t g \to 0$ strongly for all $g \in C_0$, and hence we have (3.8), completing the proof.

If the process is conservative (that is, $P(t, x, S) = 1$ for all t and x), then V is unbounded. In fact, if V is bounded, then $0 \leqq J_\lambda f_n \leqq V f_n \leqq \|V\|$ for $0 \leqq f_n \leqq 1$ and, letting $f_n(x)$ increase to 1, we should get $\lambda^{-1} \leqq \|V\|$ for any $\lambda > 0$, which is absurd. This is in contrast to the fact that there are many bounded infinitesimal generators.

4. Potential operators for processes with stationary independent increments

Let $X_t(\omega)$, $t \geqq 0$, be a right continuous stochastic process on R^N starting at the origin with stationary independent increments defined on a probability space $(\Omega, \mathscr{F}, P)$. The process $x + X_t(\omega)$ is a Markov process starting at x. Its transition operator carries $C_0(R^N)$ into itself and forms a strongly continuous semigroup T_t, which commutes with any translation L_y defined by $L_y f(x) = f(x + y)$. Conversely, every strongly continuous positive semigroup T_t on $C_0(R^N)$ with norm $\|T_t\| = 1$ which commutes with translations is induced in this way. We denote the totality of infinitesimal generators of such semigroups on $C_0(R^N)$ by $\mathbf{G}_N$.

THEOREM 4.1. *The infinitesimal generator $A \in \mathbf{G}_N$ admits a potential operator, except if A is the zero operator.*

This fact, which generalizes [18], is a consequence of Section 3, since the process is transient or null recurrent. But we will prove this theorem from an estimation of $\|T_t\|$ (Theorem 4.3). Theorem 4.2 gives the representation of infinitesimal generators $\mathbf{G}_N$, and our proof of Theorem 4.3 which makes use of Theorem 4.2 and decomposition of semigroups may be of some interest. In the following, $D_i = \partial/\partial x_i$, $D_{i,j} = \partial^2/\partial x_i \partial x_j$, and C_K^∞ is the set of C^∞ functions with compact supports; $S(f)$ is the support of function f.

THEOREM 4.2. *Let $A \in \mathbf{G}_N$. Then,*
(i) $C_K^\infty \subset \mathcal{D}(A)$;
(ii) C_K^∞ *is a core of* A;
(iii) *for any* $u \in C_K^\infty$, Au *is of the form*

$$Au(x) = \sum_{i,j=1}^{N} a_{i,j}D_{i,j}u(x) + \sum_{i=1}^{N} b_i D_i u(x) + \int_{R^N\backslash\{0\}} \left[u(x+y) - u(x) - \chi_U(y)\sum_{i=1}^{N} y_i D_i u(x)\right] n(dy), \tag{4.1}$$

where $a_{i,j}$ *and* b_i *are constants,* $(a_{i,j})$ *is a symmetric positive semidefinite matrix,* χ_U *is the indicator function of the open unit ball* U, *and* n *is a measure on* $R^N\backslash\{0\}$ *satisfying*

$$n(R^N\backslash U) < \infty, \qquad \int_{U\backslash\{0\}} |y|^2 n(dy) < \infty. \tag{4.2}$$

Conversely, if A *satisfies* (iii) *and* $\mathcal{D}(A) = C_K^\infty$, *then* A *is closable in* $C_0(R^N)$ *and the smallest closed extension of* A *is a member of* $\mathbf{G}_N$.

The measure n is called *Lévy measure*. A proof is found in [11]. Hunt [4] has a similar theorem.

THEOREM 4.3. *Let $A \in \mathbf{G}_N$.*
(i) *If* $a_{i,j} \neq 0$ *for some* i, j, *then*

$$\|T_t f\| \leqq (4\pi\alpha t)^{-1/2}\|f\| \operatorname{diam} S(f) \tag{4.3}$$

for any $f \in C_K$ *and* $t > 0$, *where* α *is the maximum eigenvalue of* $(a_{i,j})$.

(ii) *If the Lévy measure* n *does not identically vanish, then*

$$\|T_t f\| \leqq e(2\pi\beta t)^{-1/2}\|f\|[1 + \varepsilon^{-1} \operatorname{diam} S(f)] \tag{4.4}$$

for any $f \in C_K$, $t > 0$, *and* $\varepsilon > 0$, *where*

$$\beta = \beta(\varepsilon) = \max_{1 \leqq i \leqq N} \max\{n(\{y; y_i \geqq \varepsilon\}), n(\{y; y_i \leqq -\varepsilon\})\}. \tag{4.5}$$

In the proof, we use the following lemma, which is easily proved. This is closely connected with Theorem 1 of Trotter [14].

LEMMA 4.1. *Let* $A^{(1)}, A^{(2)} \in \mathbf{G}_N$ *and let* $T_t^{(1)}$ *and* $T_t^{(2)}$ *be respective generated semigroups. Then,* $T_t^{(1)}$ *and* $T_s^{(2)}$ *commute for any* t *and* s. *Let* $T_t^{(3)} = T_t^{(2)}T_t^{(1)}$. *Then* $T_t^{(3)}$ *is a strongly continuous semigroup with infinitesimal generator* $A^{(3)} \in \mathbf{G}_N$ *and* $A^{(3)}u = A^{(1)}u + A^{(2)}u$ *for* $u \in C_K^\infty$.

PROOF OF THEOREM 4.3.

(i) By an orthogonal transformation of R^N, we can and shall assume that the matrix $(a_{i,j})$ is diagonal and $a_{1,1} = \alpha$. Let $A^{(1)}$ and $A^{(2)}$ be members of $\mathbf{G}_N$ such that $A^{(1)}u = \alpha D_{1,1}u$ and $A^{(2)}u = Au - A^{(1)}u$ for $u \in C_K^\infty$ (Theorem 4.2). By using

Lemma 4.1, we have $\|T_t f\| = \|T_t^{(2)} T_t^{(1)} f\| \leqq \|T_t^{(1)} f\|$, which implies (4.3) since

$$(4.6) \qquad T_t^{(1)} f(x) = \int_{R^1} \frac{\exp\{-y_1^2/4\alpha t\}}{(4\pi\alpha t)^{1/2}} f(x_1 + y_1, x_2, \cdots, x_N)\, dy_1.$$

(ii) Let us consider the case where $\beta = n(\{y; y_1 \geqq \varepsilon\})$. The other cases are treated in the same manner. Let $n^{(1)}$ be a measure defined by $n^{(1)}(E) = n(E \cap \{y; y_1 \geqq \varepsilon\})$, let $A^{(1)}$ be a bounded operator defined by

$$(4.7) \qquad A^{(1)} u(x) = \int [u(x + y) - u(x)] n^{(1)}(dy),$$

and let $A^{(2)} = A - A^{(1)}$. By virtue of Lemma 4.1, it is enough to prove the estimate (4.4) with T_t replaced by $T_t^{(1)}$ generated by $A^{(1)}$. Let $Y_t(\omega)$, $t \geqq 0$, be a Poisson process with paths being right continuous step functions and $EY_t = \beta t$. Let $\{Z_k(\omega); k = 1, 2, \cdots\}$ be independent identically distributed R^N valued random variables independent of the process $\{Y_t; t \geqq 0\}$, each Z_k having distribution $\beta^{-1} n^{(1)}$. Let $S_0 = 0$ and $S_k(\omega) = \Sigma_{j=1}^k Z_j(\omega)$. Then, $S_{Y_t(\omega)}(\omega)$ is a process with stationary independent increments which induces the semigroup $T_t^{(1)}$. We have

$$(4.8) \qquad T_t^{(1)} f(x) = \sum_{k=0}^{\infty} P(Y_t = k) Ef(x + S_k),$$

and hence

$$(4.9) \qquad |T_t^{(1)} f(x)| \leqq e^{-\beta t} \left(\max_{k \geqq 0} \frac{(\beta t)^k}{k!} \right) \|f\| \sum_{k=0}^{\infty} P(S_k \in S(f) - x).$$

We have

$$(4.10) \qquad \sum_{k=0}^{\infty} P(S_k \in S(f) - x) \leqq E \sum_{k=0}^{\infty} \chi_{K_1 - x_1}(S_{k,1}) \leqq 1 + \varepsilon^{-1} \operatorname{diam} S(f),$$

where K_1 is the projection of $S(f)$ into the first coordinate, and $S_{k,1} = \Sigma_{j=1}^k Z_{j,1}$, $Z_{j,1}$ being the first coordinate of Z_j. Note that $Z_{j,1} \geqq \varepsilon$ with probability one. Let $\rho(k) = (k!)^{-1} e^{-k} (k + 1)^{k + 1/2}$. Since we have

$$(4.11) \qquad \sup_{k \geqq 0} \rho(k) = \lim_{k \to \infty} \rho(k) = e(2\pi)^{-1/2},$$

by elementary calculus (it will suffice to recall the proof of Stirling's formula), we have

$$(4.12) \qquad e^{-\beta t} \left(\max_{k \geqq 0} \frac{(\beta t)^k}{k!} \right) = e^{-\beta t} \frac{(\beta t)^{[\beta t]}}{[\beta t]!} \leqq (\beta t)^{-1/2} \rho([\beta t]) \leqq e(2\pi\beta t)^{-1/2},$$

which, combined with (4.9) and (4.10), proves (4.4). The proof is complete.

PROOF OF THEOREM 4.1. In the cases described in Theorem 4.3, $T_t f$ strongly converges to 0 as $t \to \infty$ for every $f \in C_0(R^N)$, which implies the existence of a potential operator by Theorem 2.2. There remains the case where $Au = \sum_{i=1}^{N} b_i D_i u$ for $u \in C_K^\infty$ and $b_i \neq 0$ for some i. In this case, the semigroup is $T_t f(x) = f(x_1 + b_1 t, \cdots, x_N + b_N t)$, and hence $T_t f$ weakly converges to 0 as $t \to \infty$ for every f. This also suffices for our conclusion by Theorem 2.2.

We give some results on cores of the potential operator V of $A \in \mathbf{G}_N$.

THEOREM 4.4. *The collection $\mathscr{M}$ of $f \in \mathscr{D}(V)$ such that f and Vf are integrable is a core of V. If $f \in \mathscr{M}$, then*

$$\int_{R^N} f(x)\, dx = 0. \tag{4.13}$$

PROOF. It is easy to see that if $g \in C_0$ is integrable, then $J_\lambda g$ is integrable and

$$\lambda \int J_\lambda g(x)\, dx = \int g(x)\, dx. \tag{4.14}$$

If $f \in \mathscr{M}$, then we get (4.13) from (4.14) by letting $g = f + \lambda V f$, $f = g - \lambda J_\lambda g$. Since C_K^∞ is a core of A (Theorem 4.2), the set $\mathscr{M}_1$ of $f \in \mathscr{D}(V)$ such that Vf belongs to C_K^∞ is a core of V. We claim that Au is integrable if $u \in C_K^\infty$, which will imply $\mathscr{M}_1 \subset \mathscr{M}$ and completes the proof. Let $K = S(u)$. In the expression (4.1) of Au, the first two terms have compact supports, and the integral

$$\iint_{R^N \times R^N} \left| u(x + y) - u(x) - \chi_U(y) \sum_{i=1}^{N} y_i D_i u(x) \right| dx n(dy) \tag{4.15}$$

is finite because the integral over each of the following four sets is finite: $x \in K$ and $y \in U$; $x \in K$ and $y \in R^N \backslash U$; $x + y \in K$ and $y \in U$; $x + y \in K$ and $y \in R^N \backslash U$.

THEOREM 4.5. *If $n(R^N \backslash B_a) = 0$ for some $B_a = \{x; |x| \leqq a\}$, then $C_K \cap \mathscr{D}(V)$ is a core of V.*

PROOF. The condition implies $Au \in C_K$ if $u \in C_K^\infty$. Hence, the set $\mathscr{M}_1$ defined in the preceding proof is a subset of $C_K \cap \mathscr{D}(V)$.

An open question is whether $C_K \cap \mathscr{D}(V)$ is a core of V without any assumption on Lévy measure.

It should be noted that a paper by Port and Stone [10] includes the following result: let Σ be the collection of points x such that for each open neighborhood U of x there is a $t > 0$ satisfying $P(X_t \in U) > 0$, assume that the closed group generated by Σ is R^N, and consider transient cases. In case $N = 1$ and X_t has finite nonzero mean, a function f in C_K belongs to $\mathscr{D}(V)$ if and only if it satisfies (4.13). Otherwise all functions in C_K belong to $\mathscr{D}(V)$.

5. Examples

EXAMPLE 5.1. *Brownian motion.* Let $A \in \mathbf{G}_N$ be such that

$$Au(x) = \sum_{i=1}^{N} D_{i,i} u(x), \qquad u \in C_K^\infty, \tag{5.1}$$

and V be the corresponding potential operator.

Let $N = 1$. Then, a function f in $C_0(R^1)$ belongs to $\mathscr{D}(V)$ if and only if

$$\int_{-\infty}^{\infty} f(x)\,dx = 0, \tag{5.2}$$

and

$$\lim_{a,b\to\infty}\left[\int_{-a}^{b} xf(x)\,dx - a\int_{-\infty}^{-a} f(x)\,dx + b\int_{b}^{\infty} f(x)\,dx\right] = 0. \tag{5.3}$$

If $f \in \mathscr{D}(V)$, then

$$Vf(x) = \lim_{a\to\infty} 2^{-1}\left[-\int_{-a}^{a} |y - x| f(y)\,dy + a\int_{-a}^{a} f(y)\,dy\right]. \tag{5.4}$$

Here integrals with infinite endpoints are understood as Riemann improper integrals.

COROLLARY 5.1. *Consider the case where $xf(x) \in L^1$. A necessary and sufficient condition for $f \in \mathscr{D}(V)$ is*

$$\int_R xf(x)\,dx = \int_R f(x)\,dx = 0. \tag{5.5}$$

If $f \in \mathscr{D}(V)$, then

$$Vf(x) = -\frac{1}{2}\int_R |y - x| f(y)\,dy. \tag{5.6}$$

Note that there are functions f in $\mathscr{D}(V)$ for which $xf(x)$ is not an L^1 function. Consider, for example, $f = u''$, letting $u(x)$ equal $x^\alpha \sin x$, $-1 \leqq \alpha < 0$, for large x.

COROLLARY 5.2. *Let I be a closed interval and let $f \in \mathscr{D}(V)$. Then, $S(f) \subset I$ if and only if $S(Vf) \subset I$.*

PROOF. It is known that $\mathscr{D}(A)$ is the collection of $u \in C_0(R^1)$ such that u is of class C^2 and $u'' \in C_0(R^1)$. We have $Au = u''$ for all $u \in \mathscr{D}(A)$. Let $f \in \mathscr{D}(V)$. It follows that $u = Vf \in \mathscr{D}(A)$, $f = -u''$, and $u' \in C_0(R^1)$, whence (5.2). Equation (5.3) follows from

$$\int_{-a}^{b} xf(x)\,dx = a\int_{-\infty}^{-a} xf(x)\,dx - b\int_{b}^{\infty} xf(x)\,dx + u(-a) - u(b). \tag{5.7}$$

In order to prove the converse, let f be a function in $C_0(R^1)$ satisfying (5.2) and (5.3). Let

$$g_1(x) = \int_x^{\infty} f(y)\,dy, \qquad g_2(x) = \int_{-\infty}^{x} f(y)\,dy, \tag{5.8}$$

$$h_1(x) = \lim_{b\to\infty}\left[\int_x^b yf(y)\,dy + b\int_b^{\infty} f(y)\,dy\right], \tag{5.9}$$

$$h_2(x) = \lim_{a \to \infty} \left[\int_{-a}^{x} yf(y)\,dy - a \int_{-\infty}^{-a} f(y)\,dy \right]. \tag{5.10}$$

All of these exist and $g_1 + g_2 = 0$, $h_1 + h_2 = 0$. Define u by the right side of (5.4). Then we have

$$u(x) = xg_1(x) - h_1(x) = -xg_2(x) + h_2(x). \tag{5.11}$$

Since $xg_1(x) - h_1(x)$ is

$$\int_{x_0}^{x} yf(y)\,dy + x \int_{x}^{\infty} f(y)\,dy - \lim_{b \to \infty} \left[\int_{x_0}^{b} yf(y)\,dy + b \int_{b}^{\infty} f(y)\,dy \right] \tag{5.12}$$

for fixed x_0, we have $xg_1(x) - h_1(x) \to 0$ as $x \to \infty$. Similarly, $xg_2(x) - h_2(x) \to 0$ as $x \to -\infty$, and hence $u \in C_0$. Since $u'' = -f$, it follows that $u \in \mathscr{D}(A)$ and $Au = -f$.

For higher dimensions, the following are classical results. For $N = 2$, a function f in C_K belongs to $\mathscr{D}(V)$ if and only if f has integral null. If $f \in C_K \cap \mathscr{D}(V)$, then

$$Vf(x) = -\frac{1}{2\pi} \int_{R^2} f(y) \log |y - x|\,dy. \tag{5.13}$$

For $N \geqq 3$, C_K is contained in $\mathscr{D}(V)$, and

$$Vf(x) = \frac{\Gamma(N/2)}{2(N-2)\pi^{N/2}} \int_{R^N} \frac{f(y)}{|y - x|^{N-2}}\,dy \tag{5.14}$$

for $f \in C_K$. We do not know the complete description of $\mathscr{D}(V)$ for $N \geqq 2$, but we have the following partial result for $N = 2$. Let $\mathscr{M}$ be the collection of $f(x)$ in $C_0(R^2)$ which depends only on $|x|$. Let B_a be the closed disc with radius a centered at the origin. Then, $f \in \mathscr{D}(V)$ if and only if

$$\lim_{a \to \infty} \left[-\int_{B_a} f(y) \log |y|\,dy + (\log a) \int_{B_a} f(y)\,dy \right] \tag{5.15}$$

exists and is finite, provided that $f \in \mathscr{M}$. If $f \in \mathscr{D}(V) \cap \mathscr{M}$, then

$$Vf(x) = \lim_{a \to \infty} \frac{1}{2\pi} \left[-\int_{B_a} \log |y - x| f(y)\,dy + \log a \int_{B_a} f(y)\,dy \right]. \tag{5.16}$$

The relations $S(f) \subset B_a$ and $S(Vf) \subset B_a$ are equivalent, provided that $f \in \mathscr{D}(V) \cap \mathscr{M}$. The proof, which makes use of Green functions for discs, is omitted. In any dimension, $C_K \cap \mathscr{D}(V)$ is a core of V by Theorem 4.5.

In the sequel, we denote by $\mathscr{M}_0$ the collection of functions in $C_K(R^1)$ with integral null.

EXAMPLE 5.2. *Brownian motion with drift.* Suppose $A \in \mathbf{G}_1$ and $Au = u'' + bu'$, $b \neq 0$, for $u \in C_K^\infty$. The process is transient. Suppose $b > 0$ for simplicity. Then, it is easy to see that $C_K \cap \mathscr{D}(V) = \mathscr{M}_0$, that $\mathscr{M}_0$ is a core of V, and that

$$Vf(x) = \frac{1}{b}\left[\int_{-\infty}^{x} f(y) \exp\{b(y - x)\}\, dy + \int_{x}^{\infty} f(y)\, dy\right] \tag{5.17}$$

for $f \in \mathscr{M}_0$.

EXAMPLE 5.3. *Deterministic motion.* Suppose $A \in \mathbf{G}_1$ and $Au = bu'$, $b \neq 0$, for $u \in C_K^\infty$. It is trivial that $C_K \cap \mathscr{D}(V) = \mathscr{M}_0$, which is a core of V, and that

$$Vf(x) = \frac{1}{b}\int_{x}^{\infty} f(y)\, dy \tag{5.18}$$

for $f \in \mathscr{M}_0$, provided that $b > 0$.

EXAMPLE 5.4. *Stable processes.* Let X_t be a one dimensional stable process with index α. Excluding deterministic motions and normalizing time scale, we have the characteristic function $E \exp\{i\zeta X_t\}$ of the form

$$\exp\{-t\zeta^2\}, \tag{5.19}$$

if $\alpha = 2$,

$$\exp\left\{-t|\zeta|^\alpha\left(1 + i\beta \tan\frac{\pi\alpha}{2} \operatorname{sgn} \zeta\right)\right\}, \qquad -1 \leqq \beta \leqq 1, \tag{5.20}$$

if $0 < \alpha < 1$ or $1 < \alpha < 2$, and

$$\exp\{-t|\zeta|(1 - i\gamma \operatorname{sgn} \zeta)\}, \qquad -\infty < \gamma < \infty, \tag{5.21}$$

if $\alpha = 1$. Equation (5.19) is the Brownian motion examined above. The process is recurrent if $1 \leqq \alpha \leqq 2$, and transient if $0 < \alpha < 1$.

If $1 < \alpha < 2$, then $C_K \cap \mathscr{D}(V) = \mathscr{M}_0$, $\mathscr{M}_0$ is a core of V, and

$$Vf(x) = \int_R k(y - x) f(y)\, dy, \qquad f \in C_K \cap \mathscr{D}(V), \tag{5.22}$$

$$k(x) = \frac{|x|^{\alpha - 1}(1 - \beta \operatorname{sgn} x)}{2(1 + h^2)\Gamma(\alpha) \cos\frac{\pi\alpha}{2}}, \qquad h = \beta \tan\frac{\pi\alpha}{2}. \tag{5.23}$$

Indeed, the fact $C_K \cap \mathscr{D}(V) = \mathscr{M}_0$ and the expressions (5.22), (5.23) are proved by Port [9]. His proof can be simplified by systematic use of Theorem 2.4 (ii). The proof that $\mathscr{M}_0$ is a core of V is as follows. First, note that if $u \in C_K^\infty$, then $Au(x) = O(|x|^{-\alpha-1})$ as $|x| \to \infty$, and hence Au has integral null by Theorem 4.4, because the Lévy measure has density $c_1 x^{-\alpha-1}$ for $x > 0$ and $c_2|x|^{-\alpha-1}$ for $x < 0$ with some nonnegative constants c_1, c_2. Therefore, since $\{Au;\ u \in C_K^\infty\}$ is a core of V, it suffices to show that for each function $f \in C_0$ with integral null satisfying

$f(x) = O(|x|^{-\alpha-1})$ as $|x| \to \infty$ there exists a sequence $\{f_n\}$ in $\mathcal{M}_0$ such that $f_n \to f$ weakly and $Vf_n \to u$ weakly for some $u \in C_0$. Let

$$u(x) = \int k(y - x)f(y)\,dy = \int [k(y - x) - k(-x)]f(y)\,dy. \tag{5.24}$$

Since there is an estimate

$$|k(y - x) - k(-x)| \leqq \text{const}\,|y|^{\alpha-1} \tag{5.25}$$

([9], p. 146) and since $k(y - x) - k(x)$ tends to 0 as $|x| \to \infty$, u belongs to C_0. Choose a function $g \in C_K^+$ such that $g(x) = 1$ on $[-1, 1]$, and let $f_n(x) = (a_n + f(x))g(x/n)$, choosing a_n in such a way that f_n has integral null. It follows that $a_n = O(n^{-\alpha-1})$. Hence, $\|f_n - f\| \to 0$ as $n \to \infty$ and $f_n(x) = O(|x|^{-\alpha-1})$ uniformly in n as $|x| \to \infty$. Using (5.22) for Vf_n, we see that $\|Vf_n - u\| \to 0$, completing the proof.

If $\alpha = 1$, then we have similarly $C_K \cap \mathcal{D}(V) = \mathcal{M}_0$ and (5.22) *with*

$$k(x) = \frac{-\log|x| + 2^{-1}\pi\gamma\,\text{sgn}\,x}{\pi(1 + \gamma^2)} \tag{5.26}$$

and $\mathcal{M}_0$ is a core of V.

To prove this, let $p(t, x)$ be the distribution density (Cauchy) of X_t and let

$$q(t, x) = \int_0^t (p(s, x) - p(s, 1))\,ds. \tag{5.27}$$

We have

$$q(t, x) = \frac{1}{2\pi(1 + \gamma^2)}\left(\log\frac{t^2 + (x - \gamma t)^2}{t^2 + (1 - \gamma t)^2} - 2\log|x|\right) + g(t, x), \tag{5.28}$$

$$\lim_{t\to\infty} q(t, x) = k(x) - \gamma[2(1 + \gamma^2)]^{-1} \tag{5.29}$$

with $k(x)$ defined by (5.26) and $g(t, x)$ a bounded function. Hence, we have estimate

$$|q(t, x)| \leq c_1\big|\log|x|\big| + c_2, \tag{5.30}$$

$$|q(t, x + y) - q(t, x)| \leqq c_3\big|\log|1 + y/x|\big| + c_4 \tag{5.31}$$

for $x \neq 0, -y$ with c_i independent of x, y, t. Let $f \in \mathcal{M}_0$ and let u be defined by (5.24). Then we see that $u \in C_0$. Using (5.30) and (5.31), we can prove that

$$\begin{aligned}\int_0^t T_s f(x)\,ds &= \int_R q(t, y - x)f(y)\,dy \\ &= \int_R (q(t, y - x) - q(t, -x))f(y)\,dt,\end{aligned} \tag{5.32}$$

and that (5.32) is bounded in t and x and tends pointwise to $u(x)$ as $t \to \infty$. Hence, $f \in \mathscr{D}(V)$ and $Vf = u$ by Theorem 2.4 (ii). Conversely, if $f \in C_K \cap \mathscr{D}(V)$, then $\int_0^t T_s f(x)\, ds$ is bounded by the same theorem, and hence f must have integral null because

$$(5.33) \qquad \int_0^t T_s f(0)\, ds = \int q(t, y) f(y)\, dy + \int f(y)\, dy \int_0^t p(s, 1)\, ds.$$

The fact that $\mathscr{M}_0$ is a core of V is proved by the same idea as in case $1 < \alpha < 2$. Details are omitted.

If $0 < \alpha < 1$, *then* $C_K \subset \mathscr{D}(V)$, C_K *is a core of* V, *and we have* (5.22) *with* $k(x)$ *defined by* (5.23).

In fact, we have

$$(5.34) \qquad \int_0^\infty p(t, x)\, dt = k(x), \qquad x \neq 0$$

for distribution density $p(t, x)$ of X_t and so $\int_0^t T_s f\, ds$ converges weakly to the right side of (5.22) as $t \to \infty$ if $f \in C_K$. Thus, we can apply Theorem 2.4 (ii). Finally, C_K is a core by Theorem 3.4.

EXAMPLE 5.5. *Poisson process.* If $Au(x) = u(x + \ell) - u(x)$ with $\ell \neq 0$, then

$$(5.35) \qquad Vf(x) = \sum_{k=0}^{\infty} f(x + k\ell), \qquad f \in C_K \cap \mathscr{D}(V),$$

and $C_K \cap \mathscr{D}(V)$ is the collection of $f \in C_K$ such that $\int f(x)\varphi(x)\, dx = 0$ for all $\varphi \in \Phi$, where Φ is the set of continuous periodic functions with period ℓ.

EXAMPLE 5.6. *Symmetrized Poisson process.* If $Au(x) = u(x + \ell) + u(x - \ell) - 2u(x)$ with $\ell \neq 0$, then

$$(5.36) \qquad Vf(x) = -\frac{1}{2} \sum_{k=-\infty}^{\infty} |k| f(x + k\ell), \qquad f \in C_K \cap \mathscr{D}(V),$$

and $C_K \cap \mathscr{D}(V)$ is the set of $f \in C_K$ such that

$$(5.37) \qquad \int f(x)\varphi(x)\, dx = \int f(x) x \varphi(x)\, dx = 0$$

for all $\varphi \in \Phi$, where Φ is the same as in 5.5.

EXAMPLE 5.7. *Brownian motion on* $[0, \infty)$ *with reflecting boundary condition.* This is null recurrent and the potential operator V in $C_0([0, \infty))$ is

$$(5.38) \qquad Vf(x) = -\frac{1}{2} \int_0^\infty (|y - x| + |y + x|) f(y)\, dy$$

for $f \in C_K([0, \infty)) \cap \mathscr{D}(V)$. The function $f \in C_K([0, \infty))$ belongs to $\mathscr{D}(V)$ if and only if f has null integral. No other condition appears.

EXAMPLE 5.8. *Brownian motion with reflecting boundary condition on a strip* $S = R^1 \times [0, 1]$. This is also null recurrent. Kimio Kazi [6a] found the potential operator V for this process. If $f \in C_K \cap \mathscr{D}(V)$, then

$$Vf(x_1, x_2) = \int_S k(x_1, x_2; y_1, y_2) f(y_1, y_2)\, dy_1 dy_2, \tag{5.39}$$

$$\begin{aligned} k(x_1, x_2; y_1, y_2) \\ = -2^{-1}|y_1 - x_1| + k_1(x_1, x_2; y_1, y_2) + k_1(x_1, -x_2; y_1, y_2), \end{aligned} \tag{5.40}$$

$$\begin{aligned} k_1(x_1, x_2; y_1, y_2) \\ = -(4\pi)^{-1} \log \big(1 - 2 \exp\{-\pi|y_1 - x_1|\}\cdot\cos \pi|y_2 - x_2| \\ + \exp\{-2\pi|y_1 - x_1|\}\big). \end{aligned} \tag{5.41}$$

A function $f \in C_K(S)$ belongs to $\mathscr{D}(V)$ if and only if

$$\int_S f(x_1, x_2)\, dx_1 dx_2 = \int_S x_1 f(x_1, x_2)\, dx_1 dx_2 = 0. \tag{5.42}$$

The kernel k is approximately equal to that of the one dimensional Brownian motion if $|y_1 - x_1|$ is large, but has logarithmic singularity when (x_1, x_2) and (y_1, y_2) are close. We do not know such a nice expression of potential kernel for the similar process on $R^2 \times [0, 1]$.

6. Some properties of V and V^*

Let $\mathscr{B}$ be a Banach lattice (see [1] or [16] for definition). The symbols f^+, f^-, and $|f|$ mean $f \vee 0$, $-(f \wedge 0)$, and $f \vee (-f)$, respectively. The adjoint space $\mathscr{B}^*$ becomes a Banach lattice in the naturally induced order where $\varphi \leqq \psi$ means that $(\varphi, f) \leq (\psi, f)$ for all $f \in \mathscr{B}$ such that $f \geqq 0$ ([1], p. 368). We call $\{T_t\}$ an *M semigroup* on $\mathscr{B}$ if it is a strongly continuous semigroup of positive linear operators with norm $\leqq 1$. We will show characteristic properties of potential operators of M semigroups and their adjoint operators.

We use a functional on $\mathscr{B} \times \mathscr{B}$ defined by

$$\rho(f, g) = \lim_{\varepsilon \to 0+} \varepsilon^{-1}(\|(f + \varepsilon g)^+\| - \|f^+\|). \tag{6.1}$$

This functional is introduced in [12] with notation $\varphi_0(f, g)$, and shown to have several nice properties. We define $\rho(\varphi, \psi)$ on $\mathscr{B}^* \times \mathscr{B}^*$ in the same way.

THEOREM 6.1. *Let* $\{T_t\}$ *be a strongly continuous semigroup on* $\mathscr{B}$ *satisfying condition* (1.1) *and admitting a potential operator* V. *Then, the following conditions are equivalent:*

(a) T_t *is an* M *semigroup;*
(b) $\rho(Vf, f) \geqq 0$ *for all* $f \in \mathscr{D}(V)$;
(c) $\|(Vf)^+\| \leqq \|(Vf + \varepsilon f)^+\|$ *for all* $f \in \mathscr{D}(V)$ *and* $\varepsilon > 0$;

(d) $\rho(V^*\varphi, \varphi) \geqq 0$ *for all* $\varphi \in \mathscr{D}(V^*)$;

(e) $\|(V^*\varphi)^+\| \leqq \|(V^*\varphi + \varepsilon\varphi)^+\|$ *for all* $\varphi \in \mathscr{D}(V^*)$ *and* $\varepsilon > 0$.

PROOF. The family $\{T_t\}$ is an M semigroup if and only if J_λ is positive and has norm $\leqq \lambda^{-1}$ for all $\lambda > 0$. Hence, the equivalence of (a), (b), (c) is proved in [13]. Likewise, both (d) and (e) are equivalent to the fact that J_λ^* is positive with norm $\leqq \lambda^{-1}$ for all $\lambda > 0$. The last property is equivalent to the same property for J_λ.

If we rewrite the above conditions by using the infinitesimal generator A and its adjoint A^* instead of V and V^*, then the theorem is true even in case $\{T_t\}$ does not admit a potential operator.

In the case $\mathscr{B} = C_0(S)$, we have

$$\rho(f, g) = \max_{x \in K(f^+)} g(x), \qquad f^+ \neq 0, \tag{6.2}$$

where $K(f^+) = \{x; f^+(x) = \|f^+\|\}$. (If $f^+ = 0$, then $\rho(f, g) \geqq 0$ for all g by definition (6.1). Hence, the inequality in (b) is trivial if $Vf \leqq 0$. Similar remarks apply to the other conditions.) Therefore, condition (b) in $C_0(S)$ is the weak principle of positive maximum studied by Hunt [4]. We can prove

$$\rho(\varphi, \psi) = \psi_\varphi^c(S_\varphi^+) + \|(\psi_\varphi^s)^+\| \tag{6.3}$$

for all signed measures $\varphi, \psi \in C_0^*(S)$, where ψ_φ^c and ψ_φ^s are, respectively, the absolutely continuous part and the singular part of ψ with respect to $|\varphi|$, and S_φ^+ is the positive set in the Hahn decomposition of S relative to φ. Thus, condition (d) seems to be a new kind of maximum principle for adjoint potential operators.

In order that a given operator V be the potential operator of an M semigroup, some properties of V^* are decisive, as Yosida [20] suggests. Let us define

$$\tau(\varphi, \psi) = \lim_{\varepsilon \to 0+} \varepsilon^{-1}(\|\varphi + \varepsilon\psi\| - \|\varphi\|). \tag{6.4}$$

The following theorem is a consequence of [20] and Theorem 6.1.

THEOREM 6.2. *Let* V *be a closed linear operator satisfying condition* (b) *of Theorem* 6.1 *with domain and range both dense in* $\mathscr{B}$. *Then the following are equivalent:*

(a) *there exists an* M *semigroup with potential* V;

(b) V^* *satisfies condition* (d) *of Theorem* 6.1;

(c) V^* *satisfies condition* (e) *of Theorem* 6.1;

(d) $\tau(V^*\varphi, \varphi) \geqq 0$ *for all* $\varphi \in \mathscr{D}(V^*)$;

(e) $\|V^*\varphi\| \leqq \|V^*\varphi + \varepsilon\varphi\|$ *for all* $\varphi \in \mathscr{D}(V^*)$ *and* $\varepsilon > 0$;

(*f*) $V^* + \varepsilon$ *is one to one for all* $\varepsilon > 0$.

Actually, (d) and (e) are equivalent in general [12]. If we replace *closed* in this theorem by *one to one*, V is closable ([12]) and the theorem remains true with V in (a) replaced by the smallest closed extension of V. In the case $\mathscr{B}^* =$

$C_0^*(S)$, we can prove

$$\tau(\varphi, \psi) = \psi_\varphi^c(S_\varphi^+) - \psi_\varphi^c(S_\varphi^-) + \|\psi_\varphi^s\|. \tag{6.5}$$

So, (d) is another kind of maximum principle for V^*.

If $\mathscr{B}$ is a sublattice of a vector lattice $\tilde{\mathscr{B}}$ and we require that T_t preserves e majorization (that is, $f \leqq e$ implies $T_t f \leqq e$) for a fixed $e \in \tilde{\mathscr{B}}$, then a property which replaces condition (b) of Theorem 6.1 is investigated in [8] and [12].

REFERENCES

[1] G. Birkhoff, *Lattice Theory*, American Mathematical Society Colloquium Publication, Vol. 25, Providence, 1967 (3rd ed.).

[2] K. L. Chung, *Markov Chains with Stationary Transition Probabilities*, New York, Springer, 1967 (2nd ed.).

[3] N. Dunford and J. T. Schwartz, *Linear Operators*, Part 1, New York, Interscience, 1958.

[4] G. A. Hunt, "Semigroups of measures on Lie groups," *Trans. Amer. Math. Soc.*, Vol. 81 (1956), pp. 264–293.

[5] ———, "Markov processes and potentials II," *Illinois J. Math.*, Vol. 1 (1957), pp. 316–369.

[6] T. Kato, *Perturbation Theory for Linear Operators*, New York, Springer, 1966.

[6a] K. Kazi, "Potential operator for Brownian motion process on a strip with reflecting boundary condition," *Sci. Rep. Tokyo Kyoiku Daigaku, Sect. A*, Vol. 1 (1971), pp. 53–56.

[7] R. Kondō, "On weak potential operators for recurrent Markov processes," *J. Math. Kyoto Univ.*, Vol. 11 (1971), pp. 11–44.

[8] H. Kunita, "Sub-Markov semi-groups in Banach lattices," *Proceedings of the International Conference on Functional Analysis and Related Topics*, Tokyo, University of Tokyo Press, 1970, pp. 332–343.

[9] S. C. Port, "Potentials associated with recurrent stable processes," *Markov Processes and Potential Theory*, New York–London–Sydney, Wiley, 1967, pp. 135–163.

[10] S. C. Port and C. J. Stone, "Infinitely divisible processes and their potential theory (First part)," *Ann. Inst. Fourier (Grenoble)*, Vol. 21, No. 2 (1971), pp. 157–275. Second part to appear.

[11] K. Sato, "Semigroups and Markov processes," Lecture notes, University of Minnesota, 1968.

[12] ———, "On dispersive operators in Banach lattices," *Pacific J. Math.*, Vol. 33 (1970), pp. 429–443.

[13] ———, "Positive pseudo-resolvents in Banach lattices," *J. Fac. Sci. Univ. Tokyo Sect. I.*, Vol. 17 (1970), pp. 305–313.

[14] H. F. Trotter, "On the product of semi-groups of operators," *Proc. Amer. Math. Soc.*, Vol. 10 (1959), pp. 545–551.

[15] D. V. Widder, *The Laplace Transform*, Princeton, Princeton University Press, 1941.

[16] K. Yosida, *Functional Analysis*, Berlin–Heidelberg–New York, Springer, 1965.

[17] ———, "The existence of the potential operator associated with an equicontinuous semi-group of class (C_0)," *Studia Math.*, Vol. 31 (1968), pp. 531–533.

[18] ———, "On the potential operators associated with Brownian motions," *J. Analyse Math.*, Vol. 23, (1970), pp. 461–465.

[19] ———, "On the existence of abstract potential operators and the principle of majoration associated with them," Preprint Series, Research Institute for Mathematical Sciences, Kyoto University, 1970.

[20] ———, "A characterization of abstract potential operators," in same Preprint Series as [19], 1970.

APPROXIMATION OF CONTINUOUS ADDITIVE FUNCTIONALS

R. K. GETOOR
UNIVERSITY OF CALIFORNIA, SAN DIEGO

1. Introduction

The purpose of this exposition is to give correct proofs of two well known and reasonably important propositions concerning continuous additive functionals. We adopt the terminology and notation of [1] throughout. We fix once and for all a standard process $X = (\Omega, \mathscr{F}, \mathscr{F}_t, X_t, \theta_t, P^x)$ with state space E. (See (I–9.2); all such references are to [1].)

The following two theorems are important facts about continuous additive functionals (CAF's) of such a process. (See (IV–2.21) or [2].)

THEOREM 1. *Let A be a* CAF *of X. Then $A = \Sigma_{n=1}^{\infty} A^n$ where each A^n is a* CAF *of X having a bounded one potential.*

Making use of Theorem 1, one can establish the following result. (See (V–2.1) or [2].)

THEOREM 2. *Suppose that X has a reference measure (that is, satisfies the hypothesis of absolute continuity). Then every* CAF *of X is equivalent to a perfect* CAF.

Unfortunately, the proofs known to me of Theorem 1 are not convincing. For example, the "proof" in [1] goes as follows. Let A be a CAF of X. Define

$$\varphi(x) = E^x \int_0^{\infty} e^{-t} e^{-A_t}\, dt. \tag{1.1}$$

Clearly, $0 < \varphi \leqq 1$ and φ is universally measurable; actually it is not difficult to see that φ is nearly Borel, but this is not required. Let $R = \inf\{t: A_t = \infty\}$. Then it is easy to check that R is a terminal time and that $P^x(R > 0) = 1$ for all x. Obviously, $\varphi(x) = E^x \int_0^R e^{-t} e^{-A_t}\, dt$. Now if T is any stopping time,

$$\begin{aligned} E^x\{e^{-T}\varphi(X_T); T < R\} &= E^x\left\{e^{-T} \int_0^{R\circ\theta_T} e^{-t} e^{-A_t\circ\theta_T}\, dt; T < R\right\} \\ &= E^x\left\{e^{A_T} \int_T^R e^{-u} e^{-A_u}\, du; R < T\right\}, \end{aligned} \tag{1.2}$$

This research was partially supported by the Air Force Office of Scientific Research, Office of Aerospace Research, United States Air Force, under AFOSR Grant AF-AFOSR 1261-67.

and so using T15, Chapter VII, [3], one finds

$$(1.3)\qquad \begin{aligned} U_A^1\varphi(x) &= E^x \int_0^\infty e^{-t}\varphi(X_t)\,dA_t \\ &= E^x \int_0^\infty e^{-t}\varphi(X_t) I_{[0,R)}(t)\,dA_t \\ &= E^x \int_0^R e^{A_t} \int_t^R e^{-u}e^{-A_u}\,du\,dA_t \\ &= E^x \int_0^R e^{-u}e^{-A_u}\left(\int_0^u e^{A_t}\,dA_t\right) du \\ &= E^x \int_0^R e^{-u}(1 - e^{-A_u})\,du \leqq 1. \end{aligned}$$

Next let f_n be the indicator function of $\{1/(n + 1) < \varphi \leqq 1/n\}$ for $n \geqq 1$. Clearly, $\Sigma f_n = 1$ and so if we define $A_t^n = \int_0^t f_n(X_s)\,dA_s$, then $\Sigma A^n = A$. Also,

$$(1.4)\qquad \begin{aligned} E^x \int_0^\infty e^{-t}\,dA_t^n &= E^x \int_0 e^{-t} f_n(X_t)\,dA_t \\ &\leqq (n + 1)E^x \int_0^\infty e^{-t}\varphi(X_t)\,dA_t \leqq n + 1. \end{aligned}$$

Consequently, each A^n is a CAF of X with a bounded one potential.

The *joker*, of course, comes in this last sentence; namely, although $t \to A_t^n$ is continuous almost surely, A^n need not be an additive functional. To see the issue fix n and let $B = A^n$ and $f = f_n$. Then

$$(1.5)\qquad B_{t+s} = B_t + \int_0^s f(X_u)\circ\theta_t\,d_u A_{u+t}.$$

Now $A_{u+t} = A_t + A_u \circ \theta_t$ and so if $A_t < \infty$, $dA_{u+t} = d(A_u \circ \theta_t)$ which yields

$$(1.6)\qquad B_{t+s} = B_t + B_s \circ \theta_t$$

if $A_t < \infty$. Obviously, (1.6) holds if $B_t = \infty$, but there is no reason for (1.6) to hold on $\{A_t = \infty; B_t < \infty\}$. If $A_t = \infty$, then $A_{u+t} = \infty$ for all u and so $dA_{u+t} = 0$. Therefore, although (1.6) need not hold, at least

$$(1.7)\qquad B_{t+s} \leqq B_t + B_s \circ \theta_t.$$

However, something of value can be salvaged from this discussion. Let f_n and A^n be as above. Note that

$$(1.8)\qquad A_t^n = \int_0^t f_n(X_s)\,dA_s = \int_0^{t \wedge R} f_n(X_s)\,dA_s$$

since dA_s puts no mass on the interval $[R, \infty]$. In particular, each A^n is a CAF of (X, R) with a bounded one potential; recall that for B to be an additive functional of (X, R) we only require that $B_{t+s} = B_t + B_s \circ \theta_t$ almost surely on $\{R > t\}$ and that B is continuous at R and constant on $[R, \infty]$. Thus, we have proved the following lemma.

LEMMA 1.1. *Let A be a* CAF *of X. Then $A = \Sigma_{n=1}^{\infty} A^n$, where each A^n is a* CAF *of (X, R) having a bounded one potential.*

Most likely Lemma 1.1 would suffice in many situations. Still it is of interest to know that Theorem 1 is valid. The main purpose of this note is to present a proof of Theorem 1. It is not at all surprising that Lemma 1.1 will be used in our argument. Once Theorem 1 is established Theorem 2 follows as in [1]. However, because our proof of Theorem 1 is rather long, there is some interest in giving a direct proof of Theorem 2 which avoids an appeal to Theorem 1. We present such a proof in Section 2.

Although our proof of Theorem 1 is rather involved, all of the ideas and techniques that we will need are contained in Section V–5 of [1]. Since these techniques are of some interest in themselves and not particularly well known, it is perhaps worthwhile to present them here in a situation that is substantially simpler than that of Section V–5 of [1]. Consequently, we will give complete details even though this necessitates repeating certain arguments given in [1].

The key fact that we need is the following interesting result which is essentially (V–5.12).

THEOREM 3. *Let T be the hitting time of a finely open nearly Borel set and let A be a* CAF *of (X, T) with a bounded one potential. Let $\eta < 1$ and let $K = \{x: E^x(e^{-T}) < \eta\}$. Then there is a* CAF, *B of X with a bounded one potential such that for every x and $f \in \mathcal{E}_{\dagger}^*$ which vanishes off K, we have*

$$E^x \int_0^T e^{-t} f(X_t)\, dA_t = E^x \int_0^T e^{-t} f(X_t)\, dB_t. \tag{1.9}$$

Most likely this theorem is true for an arbitrary exact terminal time T, but our proof makes use of the fact that T is the hitting time of a finely open set. Of course, one could easily abstract the property of T needed for the proof to go through, but this would be of very little interest.

As mentioned before, Section 2 is devoted to a proof of Theorem 2. In Section 3 we prove Theorem 1 assuming Theorem 3, while in Section 4 we prove Theorem 3.

2. Proof of Theorem 2

We begin with some preliminary facts that will also be used in Section 3. We fix an additive functional A of X and for the moment we assume only that A has no infinite discontinuity. We assume without loss of generality that $t \to A_t(\omega)$ is right continuous and nondecreasing for all ω. Recall that $A_0 = 0$ and $t \to A_t$ is continuous at ζ. We will usually omit the phrase "almost surely" in our discussions. Let $R = \inf\{t: A_t = \infty\}$. By right continuity $A_R = \infty$ if $R < \infty$ and since A has no infinite discontinuity, A is continuous at R if $R < \infty$. Of course, A is continuous at R if $R = \infty$ because $A_\infty = \lim_{t\uparrow\infty} A_t$ by convention. It is easy to see that R is a terminal time and that $P^x(R > 0) = 1$ for all x. Therefore, R is an exact terminal time. Let $R_n = \inf\{t: A_t \geqq n\}$. Then each R_n is a stopping time and $R_n < R$ when $R < \infty$ because A has no infinite discontinuity. Clearly, $\{R_n\}$ is increasing. Let $T = \lim R_n \leqq R$. Since $A(R_n) \geqq n$ on $\{R_n < \infty\}$, it is clear that $A(T) = \infty$ on $\{T < \infty\}$. Consequently, $T = R$. Thus, $\{R_n\}$ is an increasing sequence of stopping times with limit R and $R_n < R$ for all n if $R < \infty$. Let $\psi(x) = E^x(e^{-R})$. Because R is an exact terminal time ψ is 1-excessive and $0 \leqq \psi < 1$. Let $E_n = \{\psi > 1 - 1/n\}$. Then each E_n is a finely open nearly Borel set, and the E_n decrease to the empty set. Finally, let T_n be the hitting time of E_n. The following lemma is well known. Since a more general and considerably more complicated version is given in [1], we will give the proof here even though only very standard techniques are involved.

LEMMA 2.1. *Using the above notation* $T_n \leqq R$, $\lim T_n = R$, *and* $T_n < R$ *if* $R < \infty$.

PROOF. By the usual supermartingale considerations $e^{-R_n}\psi(X_{R_n}) \to e^{-R} L$ where $0 \leqq L \leqq 1$ and since R is a strong terminal time, we have, for any $\Gamma \in \mathscr{F}_{R_k}$ and $n \geqq k$,

$$E^x\{e^{-R_n}\psi(X_{R_n}); \Gamma; R_n < R\} = E^x\{e^{-R}; \Gamma; R_n < R\}. \tag{2.1}$$

Letting $n \to \infty$, we obtain

$$E^x\{e^{-R}L; \Gamma; R_n < R, \forall\, n\} = E^x\{e^{-R}; \Gamma; R_n < R, \forall\, n\} \tag{2.2}$$

for all $\Gamma \in \vee \mathscr{F}_{R_k}$. Let $\Gamma = \{R < \infty\} \in \vee \mathscr{F}_{R_k}$. Since $R_n < R$ if $R < \infty$, we see that $L = \lim \psi(X_{R_n}) = 1$ if $R < \infty$ and since ψ is 1-excessive, this yields $\lim_{t\uparrow R} \psi(X_t) = 1$ if $R < \infty$.

Now if $0 \leqq t < T_n$, $\psi(X_t) \leqq 1 - 1/n$, and consequently $T_n < R$ if $R < \infty$ because $\lim_{t\uparrow R} \psi(X_t) = 1$. Hence, $T_n \leqq R$ and $T_n < R$ if $R < \infty$. Also, $\psi(X_{T_n}) \geqq 1 - 1/n$ if $T_n < \infty$ and so

$$E^x\{e^{-(R-T_n)}; T_n < R\} = E^x\{\psi(X_{T_n}); T_n < R\} \geqq \left(1 - \frac{1}{n}\right) P^x(T_n < R). \tag{2.3}$$

Letting $n \to \infty$, we see that $\lim T_n = R$ on $\{T_n < R; \forall\, n\}$. But $\lim T_n = R$ on $\{T_n = R \text{ for some } n\}$, and so Lemma 2.1 is established.

The importance of Lemma 2.1 is that the T_n are *hitting* times of finely open sets and hence are *perfect exact* terminal times.

We now are ready to prove Theorem 2. We assume that A is a CAF of X and we will use the notation developed above. Define $B^n_t = A(t \wedge T_n)$. Then each B^n is a CAF of (X, T_n) and B^n is finite on $[0, T_n)$; this is clear if $R < \infty$ because then $T_n < R$ and it is true *a priori* if $R = \infty$. But $I_{[0,T_n)}(t)$ is a perfect multiplicative functional of X and so it follows from (V–2.1) that each B^n is perfect. (The proof of (V–2.1) is valid for all CAF's of (X, M) which are finite on $[0, S)$ where

$S = \inf\{t\colon M_t = 0\}$.) As a result for each n there exists $\Lambda_n \in \mathscr{F}$ with $P^x(\Lambda_n) = 0$ for all x such that if $\omega \notin \Lambda_n$, $B^n_{t+s} = B^n_t + B^n_s \circ \theta_t I_{[0,T_n)}(t)$ identically in t and s. Let $\Lambda_0 = \{\lim T_n \neq R\}$ and $\Lambda = \bigcup_{n \geqq 0} \Lambda_n$. The proof of Theorem 2 is completed by observing that

$$(2.4) \qquad \{A_{u+t} \neq A_t + A_u \circ \theta_t \text{ for some } t \text{ and } u\} \subset \Lambda.$$

3. Proof of Theorem 1

Let A be a CAF of X. Then by Lemma 1.1 we can write $A = \Sigma A^n$ where each A^n is a CAF of (X, R) with a bounded one potential.

LEMMA 3.1. *Let B be a CAF of (X, R) with a bounded one potential. Then there exist CAF's B^n of X, each having a bounded one potential such that $B_t = \Sigma B^n_t$ if $t < R$.*

Before coming to the proof of this lemma, let us use it to prove Theorem 1. Applying Lemma 3.1 to each A^n, we have

$$(3.1) \qquad A_t = \sum_n A^n_t = \sum_n \sum_k A^{n,k}_t \qquad \text{if} \quad t < R,$$

where each $A^{n,k}$ is a CAF of X with a bounded one potential. But if $t \geqq R$, $A_t = \infty$, and since the double sum in (3.1) is monotone in t, it also must be infinite if $t \geqq R$. Thus, (3.1) holds for all t establishing Theorem 1.

It remains to prove Lemma 3.1. We do this assuming Theorem 3 which will be proved in Section 4. As in Section 2 let $\psi(x) = E^x(e^{-R})$ and let T_n be the hitting time of the finely open set $E_n = \{\psi > 1 - 1/n\}$. Then $T_n \uparrow R$ according to Lemma 2.1. Let $G_n = \{\psi \leqq 1 - 1/n\}$ and let $\varphi_n(x) = E^x(e^{-T_n})$. Next define $K^{n,k} = \{\varphi_n < 1 - 1/k\}$. It is immediate that $K^{n,k}$ increases with both n and k, and so if we let $K_n = K^{n,n}$ then $K_n \subset G_n$ for each n and $\bigcup K_n = E$. Now $t \to B(t \wedge T_n)$ is a CAF of (X, T_n) with a bounded one potential and so by Theorem 3 there exists a CAF, C^n, of X with a bounded one potential such that if $f \in \mathscr{E}^*_+$ and vanishes off K_n then

$$(3.2) \qquad E^x \int_0^{T_n} e^{-t} f(X_t)\, dB_t = E^x \int_0^{T_n} e^{-t} f(X_t)\, dC^n_t.$$

We need the following compatibility relationship: if $f \geqq 0$ vanishes off K_n, then for all m

$$(3.3) \qquad E^x \int_0^{T_m} e^{-t} f(X_t)\, dC^n_t = E^x \int_0^{T_m} e^{-t} f(X_t)\, dB_t.$$

Suppose firstly that $m < n$. It follows from (3.2) that

$$(3.4) \qquad \bar{B}_t = \int_0^{t \wedge T_n} I_{K_n}(X_u)\, dB_u, \qquad \bar{C}_t = \int_0^{t \wedge T_n} I_{K_n}(X_u)\, dC^n_u$$

define CAF's of (X, T_n) with the same bounded one potential. Consequently, by the uniqueness theorem for CAF'S, $\bar{B} = \bar{C}$ (that is, $\bar{B}$ and $\bar{C}$ are equivalent). But $T_m \leqq T_n$ and hence (3.3) holds if $m < n$.

Next suppose that $m > n$. Then $K_n \subset G_n \subset G_m$. Recall that $E_m = E - G_m$ and T_m is the hitting time of E_m. Let S be the hitting time $K_n \cup E_m$ and define stopping times as follows: $S_0 = 0$,

$$S_{2k+1} = S_{2k} + T_n \circ \theta_{S_{2k}}, S_{2k+2} = S_{2k+1} + S \circ \theta_{S_{2k+2}}, \tag{3.5}$$

for $k \geqq 0$. Then $\{S_k\}$ forms an increasing sequence of stopping times and since E_m is finely open, $S_k \leqq T_m$ for all k. Also, $X(S_{2k}) \in K_n$ if $S_{2k} < T_m$ and using the definition of K_n this yields

$$\begin{aligned} E^x\{e^{-S_{2k+1}}; S_{2k+1} < T_m\} &\leqq E^x\{\exp\{-(S_{2k} + T_n \circ \theta_{S_{2k}})\}; S_{2k} < T_m\} \\ &\leqq (1 - 1/n)E^x\{e^{-S_{2k}}; S_{2k} < T_m\} \\ &\leqq (1 - 1/n)E^x\{e^{-S_{2k-1}}; S_{2k-1} < T_m\}. \end{aligned} \tag{3.6}$$

Consequently, $\lim S_k = T_m$. But f vanishes off K_n and $X_t \notin K_n$ if $S_{2k+1} \leqq t < S_{2k+2}$. As a result using (3.2), we obtain

$$\begin{aligned} E^x \int_0^{T_m} e^{-t} f(X_t)\, dB_t &= \sum_{k=0}^{\infty} E^x \int_{S_{2k}}^{S_{2k+1}} e^{-t} f(X_t)\, dB_t \\ &= \sum_{k=0}^{\infty} E^x \left\{ e^{S_{2k}} E^{X(S_{2k})} \int_0^{T_n} e^{-t} f(X_t)\, dB_t \right\} \\ &= \sum_{k=0}^{\infty} E^x \left\{ e^{-S_{2k}} E^{X(S_{2k})} \int_0^{T_n} e^{-t} f(X_t)\, dC_t^n \right\} \\ &= E^x \int_0^{T_m} e^{-t} f(X_t)\, dC_t^n. \end{aligned} \tag{3.7}$$

Thus, (3.3) is established since it reduces to (3.2) when $m = n$.

Now disjoint the K_n: $J_1 = K_1, \cdots, J_n = K_n - \bigcup_{j<n} K_j$. Thus, $\{J_n\}$ is a disjoint sequence of nearly Borel sets such that $\bigcup J_n = E$ and $J_n \subset K_n$ for each n. Define

$$B_t^n = \int_0^t I_{J_n}(X_s)\, dC_s^n. \tag{3.8}$$

Each B^n is a CAF of X with a bounded one potential. Let $C_t = \Sigma B_t^n$ and let

$f \in \mathscr{E}_+^*$. Then for each n

$$(3.9) \qquad E^x \int_0^{T_n} e^{-t} f(X_t)\, dC_t = \sum_k E^x \int_0^{T_n} e^{-t}(f I_{J_k})(X_t)\, dC_t^k$$
$$= \sum_k E^x \int_0^{T_n} e^{-t}(f I_{J_k})(X_t)\, dB_t$$
$$= E^x \int_0^{T_n} e^{-t} f(X_t)\, dB_t,$$

and letting $n \to \infty$, we obtain

$$(3.10) \qquad E^x \int_0^R e^{-t} f(X_t)\, dC_t = E^x \int_0^R e^{-t} f(X_t)\, dB_t.$$

Since $R > 0$ almost surely, this implies that $t \to C_t$ is finite on $[0, R)$ and it is then easy to see that C is a CAF of X. Once again the uniqueness theorem for CAF's tells us that $B_t = C_t$ if $t < R$. But $C = \Sigma B^n$ where each B^n is a CAF of X with a bounded one potential, and so Lemma 3.1 is established.

4. Proof of Theorem 3

The proof of Theorem 3 is rather long and so we will break it up into several lemmas. We refer the reader to Section 1 for the statement of Theorem 3. We begin with some notation that will be used throughout the proof. Let G be the finely open set such that $T = T_G$. Let $\psi(x) = E^x(e^{-T})$. Then $K = \{\psi < \eta\}$ where $\eta < 1$ and $K \subset \{\psi \leqq \eta\} \subset E - G$. Define $T_0 = 0$ and for $n \geqq 0$

$$(4.1) \qquad T_{2n+1} = T_{2n} + T \circ \theta_{T_{2n}}, \qquad T_{2n+2} = T_{2n+1} + T_K \circ \theta_{T_{2+1}}.$$

Thus, $\{T_n\}$ is an increasing sequence of stopping times, and for any x and $n \geqq 1$

$$(4.2) \qquad E^x\{e^{-T_{2n+1}}; T_{2n} < \infty\} = E^x\{e^{-T_{2n}} \psi(X_{T_{2n}}); T_{2n} < \infty\}$$
$$\leqq \eta E^x\{e^{-T_{2n}}; T_{2n} < \infty\}$$
$$\leqq \eta E^x\{e^{-T_{2n-1}}; T_{2n-2} < \infty\}$$

because $\psi(X_{T_{2n}}) \leqq \eta$ if $T_{2n} < \infty$ and $n \geqq 1$. As a result $\lim T_n = \infty$.

Suppose for the moment that there is a CAF, B of X for which the conclusion of Theorem 3 holds. If we define

$$(4.3) \qquad u(x) = E^x \int_0^T e^{-t} I_K(X_t)\, dA_t = U_A^1 I_K(x),$$

then because $X_t \notin K$ if $T_{2n-1} \leqq t < T_{2n}$ we can compute $U_B^1 I_K(x)$ as follows

$$(4.4)\qquad \begin{aligned} U_B^1 I_K(x) &= E^x \int_0^\infty e^{-t} I_K(X_t)\, dB_t \\ &= \sum_{n=0}^\infty E^x \int_{T_{2n}}^{T_{2n+1}} e^{-t} I_K(X_t)\, dB_t \\ &= \sum_{n=0}^\infty E^x \left\{ e^{-T_{2n}} E^{X(T_{2n})} \int_0^T e^{-t} I_K(X_t)\, dB_t \right\} \\ &= \sum_{n=0}^\infty E^x \{ e^{-T_{2n}} u(X_{T_{2n}}) \}. \end{aligned}$$

The main part of the proof of Theorem 3 consists in showing that if we *define*

$$(4.5)\qquad w(x) = \sum_{n=0}^\infty E^x \{ e^{-T_{2n}} u(X_{T_{2n}}) \},$$

then w is a regular one potential of X, and hence the one potential of CAF of X. By hypothesis, u is bounded and since

$$(4.6)\qquad w(x) \leqq \|u\| \sum_{n=0}^\infty E^x(e^{-T_{2n}}) \leqq \|u\| \sum_{n=0}^\infty \eta^n < \infty,$$

w is also bounded.

LEMMA 4.1. *Let K be as above. Then $w = P_K^1 w$.*

PROOF. For typographical simplicity let $Q = T_K$. Then

$$(4.7)\qquad \begin{aligned} P_K^1 w(x) &= E^x \{ e^{-Q} w(X_Q) \} \\ &= \sum_{n=0}^\infty E^x \{ \exp\{ -(Q + T_{2n} \circ \theta_Q) \} u(X_{Q + T_{2n} \circ \theta_Q}) \}. \end{aligned}$$

Break each summand into an integral over $\{Q < T_1\}$ and over $\{Q \geqq T_1\}$. A straightforward induction argument shows that if $k \geqq 1$, $Q + T_k \circ \theta_Q = T_k$ on $\{Q < T_1\}$. On the other hand if $Q \geqq T_1$, then $Q = T_2$. But then $Q + T_1 \circ \theta_Q = T_2 + T \circ \theta_{T_2} = T_3$ and again one sees by induction that for $k \geqq 0$, $Q + T_k \circ \theta_Q = T_{k+2}$ if $Q \geqq T_1$. Consequently,

$$(4.8)\qquad P_K^1 w(x) = E^x \{ e^{-Q} u(X_Q);\, Q < T_1 \} + \sum_{n=1}^\infty E^x \{ e^{-T_{2n}} u(X_{T_{2n}}) \}.$$

Therefore,

$$(4.9)\qquad w(x) - P_K^1 w(x) = u(x) - E^x \{ e^{-Q} u(X_Q);\, Q < T_1 \}.$$

But $T_1 = T$, $Q = T_K$, and using the definition of u (see (4.3)), we obtain

$$(4.10)\qquad E^x \{ e^{-Q} u(X_Q);\, Q < T_1 \} = E^x \int_0^T e^{-t} I_K(X_t)\, dA_t = u(x).$$

Therefore, $w = P_K^1 w$, completing the proof of Lemma 4.1.

LEMMA 4.2. *If J is any compact set, then $P_J^1 w \leqq w$.*

PROOF. Let $S = T_J + Q \circ \theta_{T_J}$ where $Q = T_K$ as before. Now $X_S \in K \cup K^r$ if $S < \infty$. But $X_t \notin K \cup K^r$ if $T_{2n+1} \leqq t < T_{2n+2}$, and so $\{S < \infty\} = \bigcup_n \{T_{2n} \leqq S < T_{2n+1}\}$. Also, it is easy to check by induction that for $k \geqq 0$, $T_{k+2} = T_2 + T_k \circ \theta_{T_2}$. Hence,

$$w(x) = u(x) + \sum_{n=1}^{\infty} E^x\{e^{-T_{2n}} u(X_{T_{2n}})\} \tag{4.11}$$
$$= u(x) + E^x\{e^{-T_2} w(X_{T_2})\}.$$

Again one checks that for $k \geqq 1$, $S + T_k \circ \theta_S = T_k$ if $S < T_1$. Now $\{S < T_1\} \in \mathscr{F}_{T_1} \subset \mathscr{F}_{T_2}$ and so

$$E^x\{e^{-S} w(X_S); S < T_1\} \tag{4.12}$$
$$= E^x\{e^{-S} u(X_S); S < T_1\} + E^x\{e^{-T_2} w(X_{T_2}); S < T_1\}.$$

Using (4.11) and the fact that u is one (X, T) excessive, we obtain

$$E^x\{e^{-S} w(X_S); S < T_1\} + E^x\{e^{-T_2} w(X_{T_2}); S \geqq T_1\} \leqq w(x). \tag{4.13}$$

We next prove by induction that for all $n \geqq 1$.

$$w(x) \geqq E^x\{e^{-S} w(X_S); S < T_{2n}\} + E^x\{e^{-T_{2n}} w(X_{T_{2n}}); S \geq T_{2n}\}. \tag{4.14}$$

If $n = 1$, this reduces to (4.13) because S lies in some interval $[T_{2k}, T_{2k+1})$ when S is finite. Assume (4.14) for a fixed value of n. The second summand on the right side of (4.14) may be written

$$E^x\{e^{-S} w(X_S): S = T_{2n}\} + E^x\{e^{-T_{2n}} w(X_{T_{2n}}); S > T_{2n}\}. \tag{4.15}$$

It is immediate that if $T_{2n} < S$ then $T_{2n-1} < T_J$. Recall that $S = T_J + Q \circ \theta_{T_J}$ and $T_{2n} = T_{2n-1} + Q \circ \theta_{T_{2n-1}}$. But this together with the fact that K is finely open implies that $T_{2n} < T_J$ if $T_{2n} < S$. Consequently, $T_{2n} + S \circ \theta_{T_{2n}} = S$ if $T_{2n} < S$. Combining these observations with (4.13), we obtain

$$E^x\{e^{-T_{2n}} w(X_{T_{2n}}); S > T_{2n}\} \tag{4.16}$$
$$\geqq E^x\{e^{-T_{2n}} E^{X(T_{2n})}[e^{-S} w(X_S); S < T_1]; S > T_{2n}\}$$
$$+ E^x\{e^{-T_{2n}} E^{X(T_{2n})}[e^{-T_2} w(X_{T_2}); S \geqq T_1]; S > T_{2n}\}$$
$$= E^x\{e^{-S} w(X_S); T_{2n} < S < T_{2n+1}\}$$
$$+ E^x\{e^{-T_{2n+2}} w(X_{T_{2n+2}}); S \geqq T_{2n+1}\}.$$

But $\{T_{2n} < S < T_{2n+1}\} = \{T_{2n} < S < T_{2n+2}\}$ and $\{S \geqq T_{2n+1}\} = \{S \geqq T_{2n+2}\}$. As a result (4.14) holds with n replaced by $n + 1$, and hence it holds for all $n \geqq 1$. Now $\lim T_n = \infty$ and so letting $n \to \infty$ in (4.14), we obtain $w \geqq P_S^1 w$. But $P_S^1 w = P_J^1 P_K^1 w = P_J^1 w$ since $w = P_K^1 w$ by Lemma 4.1, completing the proof of Lemma 4.2.

LEMMA 4.3. *The function w is 1-excessive.*

PROOF. In light of Lemma 4.2 and Dynkin's theorem (II–5.3), it will suffice to show that $\liminf_{t \downarrow 0} P_t^1 w(x) \geqq w(x)$ for all x. Suppose first of all that x is not regular for K. Then almost surely P^x, $t + Q \circ \theta_t = Q$ for t sufficiently small, and since $w = P_K^1 w$ this yields

$$\begin{aligned}(4.17)\qquad \lim_{t \to 0} P_t^1 w(x) &= \lim_{t \to 0} P_t^1 P_K^1 w(x)\\ &= \lim_{t \to 0} E^x\{\exp\{-(t + Q \circ \theta_t)\} w(X_{t + Q \circ \theta_t})\}\\ &= P_K^1 w(x) = w(x).\end{aligned}$$

Suppose on the other hand that x is regular for K. Then $P^x(t < T) \to 1$ as $t \to 0$ and so using (4.11) with $T = T_1$,

$$\begin{aligned}(4.18)\qquad P_t^1 w(x) &\geqq E^x\{e^{-t} w(X_t); t < T\}\\ &= E^x\{e^{-t} u(X_t); t < T\} + E^x\{e^{-T_2} w(X_{T_2}); t < T\}.\end{aligned}$$

Because u is $1 - (X, T)$ excessive this approaches $u(x) + E^x\{e^{-T_2} w(X_{T_2})\} = w(x)$ as $t \to 0$, completing the proof of Lemma 4.3.

LEMMA 4.4. *The function w is a regular one potential.*

PROOF. We must show that if $\{S_n\}$ is an increasing sequence of stopping times with limit S, then $P_{S_n}^1 w \to P_S^1 w$. It follows from (IV–3.6) and (IV–3.8) that we need consider only the case $S_n = T_{B_n}$ where $\{B_n\}$ is a decreasing sequence of nearly Borel sets. In particular each S_n is a strong terminal time and consequently so is their limit S. In checking that $P_{S_n}^1 w(x) \to P_S^1 w(x)$, we may assume that $P^x(S_n > 0) = 1$ since if $S_n = 0$ for all n the conclusion is obvious. Now fix x and let

$$(4.19)\qquad a_{n,k} = E^x\{e^{-S_n} w(X_{S_n}); T_k < S_n \leqq T_{k+1}\}$$

and

$$(4.20)\qquad a_k = E^x\{e^{-S} w(X_S); T_k < S \leqq T_{k+1}\}.$$

Then $P_{S_n}^1 w(x) = \Sigma_k a_{n,k}$ and $P_S^1 w(x) = \Sigma_k a_k$. It will suffice to show that for each k, $a_{n,k} \to a_k$ as $n \to \infty$ because $\Sigma_{k \geqq N} a_{n,k} \leqq \|w\| E^x(e^{-T_N}) \to 0$ as $N \to \infty$. Suppose first of all that k is even, say $k = 2j$. If R is any strong terminal time then on $\{T_{2j} < R \leqq T_{2j+1}\}$ we have $R = T_{2j} + R \circ \theta_{T_{2j}}$, and also because T is the hitting time of a finely open set $R + T_2 \circ \theta_R = T_{2j+2}$. Now using (4.11), we obtain for any strong terminal time R

$$\begin{aligned}(4.21)\qquad & E^x\{e^{-R} w(X_R); T_{2j} < R \leqq T_{2j+1}\}\\ &= E^x\{e^{-R} u(X_R); T_{2j} < R; R \circ \theta_{T_{2j}} \leqq T \circ \theta_{T_{2j}}\}\\ &\quad + E^x\{e^{-R} E^{X(R)}[e^{-T_2} w(X_{T_2})]; T_{2j} < R \leqq T_{2j+1}\}\\ &= E^x\{e^{-T_{2j}} E^{X(T_{2j})}[e^{-R} u(X_R); R \leqq T]; T_{2j} < R\}\\ &\quad + E^x\{e^{-T_{2j+2}} w(X_{T_{2j+2}}); T_{2j} < R \leqq T_{2j+1}\}.\end{aligned}$$

In (4.21), we may replace R by either S_n or S. Observe that the set $\{T_{2j} < S_n\}$ approaches the set $\{T_{2j} < S\}$ as $n \to \infty$ and that $\{T_{2j} < S_n \leqq T_{2j+1}\}$ approaches $\{T_{2j} < S \leqq T_{2j+1}\}$ as $n \to \infty$. Now u is a regular one potential of (X, T) since it is the one potential of a CAF of (X, T), and $u(X_T) = 0$ because X_T is regular for G; recall $T = T_G$ with G finely open. As a result for any y

$$(4.22) \qquad \begin{aligned} E^y\{e^{-S_n}u(X_{S_n}); S_n \leqq T\} &= E^y\{e^{-S_n}u(X_{S_n}); S_n < T\} \\ &\to E^y\{e^{-S}u(X_S); S < T\} \\ &= E^y\{e^{-S}u(X_S); S \leqq T\}, \end{aligned}$$

as $n \to \infty$. Consequently, $a_{n,2j} \to a_{2j}$ as $n \to \infty$. Next consider the case in which k is odd, say $k = 2j + 1$. Using the fact that $w = P_K^1 w$, we obtain

$$(4.23) \qquad a_{n,2j+1} = E^x\{\exp\{-S_n + T_K \circ \theta_{S_n}\}\, w(X_{S_n + T_K \circ \theta_{S_n}});\, T_{2j+1} < S_n \leqq T_{2j+2}\}$$

and a similar expression for a_{2j+1} with S_n replaced by S. But on $\{T_{2j+1} < S_n \leqq T_{2j+2}\}$ we have $S_n + T_K \circ \theta_{S_n} = T_{2j+2}$ while on $\{T_{2j+1} < S \leqq T_{2j+2}\}$, $S + T_K \circ \theta_S = T_{2j+2}$ because K is finely open. From this and the fact that $S_n \uparrow S$, it is immediate that $a_{n,2j+1} \to a_{2j+1}$ as $n \to \infty$. This completes the proof of Lemma 4.4.

We are now prepared to complete the proof of Theorem 3. Since w is a regular one potential there is a CAF, B of X such that $w = U_B^1 1$, that is, w is the one potential of B. Now $D_t = B_{t \wedge T}$ is a CAF of (X, T) and

$$(4.24) \qquad U_D^1 1(x) = E^x \int_0^T e^{-t}\, dB_t = w(x) - E^x\{e^{-T_1} w(X_{T_1})\}.$$

From Lemma 4.1

$$(4.25) \qquad E^x\{e^{-T_1} w(X_{T_1})\} = E^x\{e^{-T_2} w(X_{T_2})\},$$

and so by (4.11), $U_D^1 1 = u$. Hence, D and $t \to \int_0^t I_K(X_u)\, dA_u$ are equivalent CAF's of (X, T). Therefore, $E^x \int_0^T e^{-t} f(X_t)\, dA_t = E^x \int_0^T e^{-t} f(X_t)\, dB_t$ if f vanishes off K, completing the proof of Theorem 3.

REFERENCES

[1] R. M. Blumenthal and R. K. Getoor, *Markov Processes and Potential Theory*, New York, Academic Press, 1968.

[2] C. Doléans-Dade, "Fonctionnelles additives parfaites," *Séminaire de Probabilités II Université de Strasbourg, Lecture Notes in Mathematics*, Vol. 51, Berlin-Heidelberg-New York, Springer-Verlag, 1968.

[3] P. A. Meyer, *Probability and Potentials*, Waltham, Blaisdell, 1966.

POISSON POINT PROCESSES ATTACHED TO MARKOV PROCESSES

KIYOSI ITÔ
CORNELL UNIVERSITY

1. Introduction

The notion of point processes with values in a general space was formulated by K. Matthes [4]. A point process is called *Poisson*, if it is σ-discrete and is a renewal process. We will prove in this paper that such a process can be characterized by a measure on the space of values, called the *characteristic measure*.

Let X be a standard Markov process with the state space S. Fix a state $a \in S$ and suppose that a is recurrent state for X. Let $S(t)$ be the inverse local time of X at a. By defining $Y(t)$ to be the excursion of X in $(S(t-), S(t+))$ for the t value such that $S(t+) > S(t-)$, we shall obtain a point process called the *excursion point process* with values in the space of paths. Using the strong Markov property of X, we can prove that Y is a Poisson point process. Its characteristic measure, called the *excursion law*, is a σ-finite measure on the space of paths. Although it may be an infinite measure, the conditional measure, when the values of the path up to time t is assigned, is equal to the probability law of the path of the process X starting at the value of the path at t and stopped at the hitting time for a. Using this idea, we can determine the class of all possible standard Markov processes whose stopped process at the hitting time for a is a given one.

We presented this idea in our lecture at Kyoto in 1969 [3] and gave the *integral representation* of the excursion law to discuss the jumping-in case in which the excursion starts outside a. P. A. Meyer [5] discussed the general case in which continuous entering may be possible by introducing the *entrance law*. In our present paper, we will prove the *integral representation* of the excursion law in terms of the *extremal excursion laws* for the general case. It is not difficult to parametrize the extremal excursion laws by the entrance Martin boundary points for the stopped process and to determine the generator of X, though we shall not discuss it here.

E. B. Dynkin and A. A. Yushkevich [1], [2] discussed a very general extension problem which includes our problem as a special case. We shall deal with their case from our viewpoint. The excursion point process defined similarly is no longer Poisson but will be called Markov. It seems useful to study point processes of Markov type in general.

This work was supported by a National Science Foundation Grant GP-19658.

2. Point functions

Let U be an abstract Borel space associated with a σ-algebra $B(U)$ on U whose member is called a Borel subset of U. Let T_+ be the open half line $(0, \infty)$ which is called the *time interval*. The product space $T_+ \times U$ is considered as a Borel space associated with the product σ-algebra $B(T_+ \times U)$ of the topological σ-algebra $B(T_+)$ on T_+ and the σ-algebra $B(U)$ on U.

A *point function* $p: T_+ \to U$ is defined to be a map from a countable set $D_p \subset T$ into U. The space of all point functions: $T_+ \to U$ is denoted by $\Pi = \Pi(T_+^-, U)$. For $p \in \Pi$ and $E \in B(T_+ \times U)$, we denote by $N(E, p)$ the number of the time points $t \in D_p$ for which $(t, p(t)) \in E$. The space Π is regarded as a Borel space associated with the σ-algebra $B(\Pi)$ generated by the sets: $\{p \in \Pi: N(E, p) = k\}$, $E \in B(T_+ \times U)$, $k = 0, 1, 2, \cdots$.

We will introduce several operations in Π. Let $E \in B(T_+ \times U)$. The *restriction* $p|E$ of p to E is defined to be the point function g such that

$$D_g = \{t \in D_p : (t, p(t)) \in E\} \tag{2.1}$$

and $g(t) = p(t)$ for $t \in D_g$.

Let $V \in B(U)$. The *range restriction* $p|_r V$ of p to V is defined to be the restriction $p|T_+ \times V$. Let $S \in B(T_+)$. The *domain restriction* $p|_d S$ is defined to be the restriction $p|S \times U$. Let $s \in T_+$. The *stopped point function* $\alpha_s p$ of p at s is defined to be the domain restriction of p to $(0, s]$ and the *shifted point function* $g = \theta_s p$ is defined by $D_g = \{t: t + s \in D_p\}$ and $g(t) = p(t + s)$.

Let p_n for $n = 1, 2, \cdots$, be a sequence of point processes. If the D_{p_n}, $n = 1, 2, \cdots$, are disjoint, then the direct sum $p = \Sigma_n p_n$ is defined by $D_p = \cup_n D_{p_n}$ and $p(t) = p_n(t)$ for $t \in D_{p_n}$.

3. Point processes

Let Π be the space of all point functions: $T_+ \to U$ as in Section 2. Let (Ω, P) be a probability space, where P is a complete probability measure on Ω. A map $Y: \Omega \to \Pi$ is called a *point process* if it is measurable. The image of ω by Y is denoted Y_ω and is called the sample point function of Y corresponding to ω. The value of Y_ω at t if t belongs to the domain of Y_ω is denoted by $Y_\omega(t)$. It follows from the definition that $N(E, Y_\omega)$ is measurable in ω and is therefore a random variable on (Ω, P) if $E \in B(T \times U)$. The probability law of Y is clearly a probability measure on $(\Pi, B(\Pi))$.

The process Y is called *discrete* if $N((0, t) \times U, Y) < \infty$ a.s. for every $t < \infty$. It is called *σ-discrete* if for every t we can find $U_n = U_n(t) \in B(U)$, $n = 1, 2, \cdots$ such that $N((0, t) \times U_n, Y) < \infty$ a.s. for every n and that $U = \cup_n U_n$.

The process Y is called "*renewal*" if for every $t < \infty$ we have that $\alpha_t Y$ and $\theta_t Y$ are independent and that $\theta_t Y$ has the same probability law as Y for every t. By virtue of the following theorem, we call a σ-discrete and renewal point process a *Poisson point process*.

THEOREM 3.1. *Let Y be σ-discrete and renewal. Then $N(E_i, Y), i = 1, 2, \cdots, k$ are independent and Poisson distributed for every finite disjoint system $\{E_i\} \subset B(T_+ \times U)$.*

REMARK. A probability measure concentrated at ∞ is regarded as a special case of Poisson measure with mean ∞.

PROOF. Because of the σ-discreteness of Y and the second condition of the renewal property, we can take $\{U_n\}$ independent of t such that $N((0, t) \times U_n, Y) < \infty$ a.s. for every t and such that $U = \cup_n U_n$. It can be also assumed that U_n increases with n. Since

$$N(E, Y) = \lim_n N(E \cap (T_+ \times U_n), Y), \tag{3.1}$$

and since the Poisson property and the independence property are inherited by the limit of random variables, we can assume that all E_i are included in $T_+ \times U_n$. By the renewal property of Y, we can easily see that $Z(t) \equiv N((0, t) \times U_n, Y)$ is a stochastic process with stationary independent increments and that its sample function increases only with jumps $=1$, a.s. This $Z(t)$ is therefore a Poisson process. This implies that $Z(t)$ is continuous in probability.

Now set $Z_i(t) = N(E_i, (0, t) \times U_n)$ and $Z^*(t) = \Sigma_i iZ_i(t)$. Since $Z^*(t) - Z^*(s) \leqq k(Z(t) - Z(S))$ for $s < t$, Z^* is continuous in probability with Z. The process Z^* has independent increments and its sample function increases with jumps. Thus, Z^* is a Lévy process and $Z_i(t)$ is the number of jumps $= i$ of Z^* up to time t. The $\{Z_i(t)\}_i$ are therefore independent and Poisson distributed. Since

$$N(E_i, Y) = \lim_{t\to\infty} Z_i(t), \tag{3.2}$$

the proof is completed.

As an immediate result of Theorem 2.1 we have the following theorem.

THEOREM 3.2. *In order for Y to be a Poisson point process, it is necessary and sufficient that it be the sum of independent discrete Poisson point processes.*

4. Characteristic measure

Let Y be a Poisson point process defined in Section 3 and set $m(E) = E_P(N(E, Y))$, where $E_P =$ expectation. Then m is a measure on $T_+ \times U$ which is shift invariant in the time direction because of the second condition in the renewal property of Y. By the σ-discreteness of Y, we have $m((0, t) \times U_n) < \infty$ for the U_n introduced in the proof of Theorem 3.1. This implies that m is σ-finite. Therefore, m is the product measure of the Lebesgue measure on T_+ and a unique σ-finite measure n on U. The measure n is called the *characteristic measure* n of Y by virtue of the following theorem.

THEOREM 4.1. *The probability law of a Poisson point process Y is determined by its characteristic measure n.*

PROOF. The measure n determines the joint distribution of $N(E_i, Y)$, $i = 1, 2, \cdots, k$ for disjoint $E_i \in B(T_+ \times U)$ by Theorem 3.1. Since $N(E, Y)$ is additive in E, it is also true for nondisjoint E_i. This completes the proof.

The following theorem shows that any arbitrary σ-finite measure on U induces a Poisson point process.

THEOREM 4.2. *Let n be a σ-finite measure on U. Then there exists a Poisson point process whose characteristic measure is n.*

Before proving this, we will study the structure of a Poisson point process whose characteristic measure is finite.

THEOREM 4.3. *A Poisson point process is discrete if and only if its characteristic measure is finite.*

PROOF. Observe that the number of $s \in D_Y \cap (0, t)$ is $N((0, t) \times U, Y)$.

THEOREM 4.4A. *Let Y be a discrete Poisson point process with characteristic measure n. (Then n is a finite measure by the previous theorem.) Let D_Y be $\tau_1(\omega)$, $\tau_2(\omega), \cdots$, and let $\xi_i(\omega) = Y_\omega(\tau_i(\omega))$. Then we have the following:*

(i) $\tau_i - \tau_{i-1}$, $i = 1, 2, \cdots$, $(\tau_0 = 0)$, $\xi_1, \xi_2, \cdots$ *are independent;*

(ii) $\tau_i - \tau_{i-1}$ *is exponentially distributed with mean* $1/n(U)$, *that is,* $P(\tau_i - \tau_{i-1} > t) = e^{-tn(U)}$;

(iii) $P(\xi_i \in A) = n(A)/n(U)$, $A \in B(U)$.

PROOF. Let $\alpha_i > 0$ and $V_i \in B(U)$, $i = 1, 2, \cdots, k$. Write $\phi_p(t)$ for $([pt] + 1)/p$, $[t]$ being the greatest integer $\leqq t$. Then we have

$$\tag{4.1} \begin{aligned} E_P\left[\exp\left\{-\sum_{i=1}^{k}\alpha_i\tau_i\right\}, \xi_i \in V_i, i = 1, 2, \cdots, k\right] &= \lim_{p\to\infty} E_P\left[\exp\left\{-\sum_{i=1}^{k}\alpha_i\phi_p(\tau_i)\right\}, \xi_i \in V_i, \tau_i - \tau_{i-1} > \frac{1}{p}, \right. \\ &\qquad\qquad \left. i = 1, 2, \cdots, k\right] \\ &= \lim_{p\to\infty} \sum_{0<\nu_1<\cdots<\nu_k} \exp\left\{-\sum_{i=1}^{k}\frac{\alpha_i\nu_i}{p}\right\} P\left(\xi_i \in V_i, \frac{\nu_i - 1}{p} < \tau_i \leqq \frac{\nu_i}{p}, \right. \\ &\qquad\qquad \left. i = 1, 2, \cdots, k\right). \end{aligned}$$

The event in the second factor can be expressed as

$$\tag{4.2} \left.\begin{aligned} &N\left(Y, \left(\frac{\nu_i - 1}{p}, \frac{\nu_i}{p}\right] \times V_i\right) = 1 \\ &N\left(Y, \left(\frac{\nu_i - 1}{p}, \frac{\nu_i}{p}\right] \times (U - V_i)\right) = 0 \end{aligned}\right\} \quad \text{for } i = 1, 2, \cdots, k,$$

$$\tag{4.3} N\left(Y, \left(\frac{\nu - 1}{p}, \frac{\nu}{p}\right] \times U\right) = 0 \qquad \text{for } \nu \neq \nu_1, \cdots, \nu_k, \nu \leqq \nu_k.$$

Using Theorem 3.1, we can easily see that its probability is

$$\exp\left\{\frac{-v_k n(U)}{p}\right\}\left(\frac{1}{p}\right)^k \prod_{i=1}^{k} n(V_i). \tag{4.4}$$

Therefore, the above limit is expressed as an integral form

$$\begin{aligned} E_P\left[\exp\left\{-\sum_{i=1}^{k}\alpha_i\tau_i\right\}, \xi_i \in V_i, i = 1, 2, \cdots, k\right] \\ = \prod_{i=1}^{k} n(V_i) \int\cdots\int_{0<t_1<\cdots<t_k} \exp\left\{-\sum_{i=1}^{k}\alpha_i t_i - t_k n(U)\right\} dt_1 \cdots dt_k. \end{aligned} \tag{4.5}$$

Changing variables in the integral, we obtain

$$\begin{aligned} E_P\left(\exp\left\{-\sum_{i=1}^{k}\beta_i(\tau_i - \tau_{i-1})\right\}, \xi_i \in V_i, i = 1, 2, \cdots, k\right) \\ = \prod_{i=1}^{k} n(V_i) \int\cdots\int_{s_1\cdots s_k>0} \exp\left\{-\sum_{i=1}^{k}\beta_i s_i - \left(\sum_{i=1}^{k} s_i\right) n(U)\right\} ds_1 \cdots ds_k \end{aligned} \tag{4.6}$$

for $\beta_1 > \beta_2 > \cdots > \beta_k > 0$. This is true for $\beta_1, \beta_2, \cdots, \beta_k > 0$ by analytic continuation in β_i. It is now easy to complete the proof.

This theorem suggests a method to construct a Poisson point process whose characteristic measure is a given finite measure.

THEOREM 4.4B. *Let n be a finite measure on U. Suppose that $\sigma_1, \sigma_2, \cdots, \xi_1, \xi_2, \cdots$ are independent and that*

$$P(\sigma_i > t) = \exp\{-tn(U)\}, \qquad P(\xi_i \in A) = \frac{n(A)}{n(U)}. \tag{4.7}$$

Define a point process Y by

$$D_Y = \{\sigma_1, \sigma_1 + \sigma_2, \cdots\}, \qquad Y(\sigma_1 + \cdots + \sigma_k) = \xi_k. \tag{4.8}$$

Then Y is a Poisson point process with characteristic measure $= n$.

Now we will prove Theorem 4.2. The case $n(U) < \infty$ has been discussed above. If $n(U) = \infty$, then we have a disjoint countable decomposition of U: $U = \cup_i U_i$, $n(U_i) < \infty$. Let $n_i(A) = n(A \cap U_i)$. Then n_i is a finite measure on U and we have a Poisson point process Y_i with characteristic measure n_i. We can assume that $Y_1, Y_2, \cdots,$ are independent. First we will remark that

D_{Y_i}, $i = 1, 2, \cdots$, are disjoint a.s. In fact for $i \neq j$,

(4.9)

$$P(D_{Y_i} \cap D_{Y_j} \cap (0, t) \neq \varnothing)$$
$$\leqq \sum_{k=1}^{p} P\left(N\left(U \times \left(\frac{(k-1)t}{p}, \frac{kt}{p}\right], Y_i\right) \neq 0, N\left(U \times \left(\frac{(k-1)t}{p}, \frac{kt}{p}\right], Y_j\right) \neq 0\right)$$
$$\leqq t^2 n_i(U) n_j(U) \left(\frac{1}{k}\right)^2 k \to 0,$$

as $k \to \infty$. By letting $t \to \infty$, we have $P(D_{Y_i} \cap D_{Y_j} \neq \varnothing) = 0$ for i, j fixed. Therefore, $D_{Y_1}, D_{Y_2}, \cdots$, are disjoint a.s.

Let Y be the sum of $Y_1, Y_2, \cdots$, (see the end of Section 1). It is easy to show that Y is a Poisson point process whose characteristic measure is n.

Let $\varphi: T \times U \rightsquigarrow [0, \infty)$ be measurable $B(T \times U)/B[0, \infty)$. Then we have the following result.

THEOREM 4.5. *With the convention that* $\exp\{-\infty\} = 0$, *we have*

$$E_P[\exp\{-\alpha \sum_{t \in D_Y} \varphi(t, Y_t)\}] = \exp\left\{\iint_{T_+ \times U} (\exp\{-\alpha\varphi(t, u)\} - 1)\, dt\, n(du)\right\}. \tag{4.10}$$

PROOF. If φ is a simple function, this follows from Theorem 3.1 and the definition of n. We can derive the general case by taking limits.

Let Φ be a random variable with values in $[0, \infty]$. The condition $\Phi < \infty$ a.s. is equivalent to

$$\lim_{\alpha \to 0+} E(\exp\{-\alpha\Phi\}) = 1. \tag{4.11}$$

Using this fact, we get the following theorem from Theorem 4.5.

THEOREM 4.6. *The condition*

$$\sum_{t \in D_Y} \varphi(t, Y_t) < \infty \text{ a.s.} \tag{4.12}$$

is equivalent to

$$\iint_{T \times U} \varphi(t, u) \wedge 1\, dt\, n(du) < \infty. \tag{4.13}$$

REMARK. This condition is also equivalent to

$$\iint_{T \times U} (1 - \exp\{-\varphi(t, u)\})\, dt\, n(du) < \infty. \tag{4.14}$$

THEOREM 4.7. *Let Y be a Poisson point process with values in U. The range restriction Y^* of Y to a set $U^* \in B(U)$ is also a Poisson point process whose characteristic measure is the restriction of that of Y to U^*.*

5. The strong renewal property of Poisson point processes

Let Y be a Poisson point process and $B_t(Y)$ be the σ-algebra generated by the stopped process $\alpha_t Y$. It is easy to see that $B_t(Y)$ is right continuous, that is, $B_t(Y) = \bigcap_{s>t} B_s(Y)$ a.s., where two σ-algebras are said to be equal a.s. if every member of one σ-algebra differs from a member of the other by a null set.

Let σ be a stopping time with respect to the increasing family $B_t(Y)$, $t \in T_+$. Suppose that $\sigma < \infty$ a.s. Then we have the following.

THEOREM 5.1 (Strong Renewal Property). *The process Y has the strong renewal property:*

(i) *$\alpha_\sigma Y$ and $\theta_\sigma Y$ are independent;*

(ii) *$\theta_\sigma Y$ has the same probability law as Y.*

The idea of the proof is the same as that of the proof of the strong Markov property in the theory of Markov processes and is omitted.

6. The recurrent extension of a Markov process at a fixed state

Let S be a locally compact metric space. Let δ stand for the *cemetery*, an extra point to be added to S as an isolated point. Denote the topological σ-algebra on S by $B(S)$ and let T stand for the closed half line $[0, \infty)$ with the topological σ-algebra $B(T)$.

Let U stand for the space of all right continuous functions: $T \to S \cup \{\delta\}$ with left limits. Let $B(U)$ denote the σ-algebra on U generated by the cylinder Borel subsets of U. It is the same as the topological σ-algebra with respect to the Skorohod topology in U. A member of U is often called a *path*. The hitting time for a, the stopped path at t, and the shifted path at t are denoted by $\sigma_a(u)$, $\alpha_t(u)$, and $\theta_t(u)$ as usual. To avoid typographical difficulty, we write $\alpha_a(u)$ for the stopped path at $\sigma_a(u)$.

Let $X = (X_t, P_b)$ be a standard Markov process with the state space S, where P_b denotes the probability law of the path starting at b which is clearly a probability measure on $(U, B(U))$, completed if necessary.

The process X stopped at the hitting time σ_a for a is also a standard Markov process which is denoted by $\alpha_a X$. The state a is a trap for $\alpha_a X$.

Let $X^0 = (X_t^0, P_b^0)$ be a standard Markov process with a trap at a. Any standard Markov process $X = (X_t, P_b)$ with the state space S is called a *recurrent extension* of X^0 at a if $\alpha_a X$ is equivalent to X^0 and if a is a *recurrent state* for X, that is, $P_a(\sigma_a < \infty) = 1$. The process X^0 itself is a recurrent extension of X^0, but there are many other extensions. Our problem is to determine all possible recurrent extensions of X^0. We will exclude the trivial extension X^0 from our discussion.

Let X be a recurrent extension of X^0. We will exclude the trivial extension. Since $P_a(\sigma_a < \infty) = 1$, we have two cases.

Case 1 (Discrete Visiting Case). $P_a(0 < \sigma_a < \infty) = 1$. In this case the visiting times of the path of X at a form a discrete set.

Case 2. $P_a(\sigma_a = 0) = 1$. In this case we have exactly one additive functional

$A(t)$ such that

$$E_b(e^{-\sigma_a}) = E_b\left(\int_0^\infty e^{-t}\, dA(t)\right). \tag{6.1}$$

This function $A(t)$ is called the Blumenthal–Getoor *local time* of X at a. The path of $A(t)$ increases continuously from 0 to ∞ with t. This case is divided into two subcases. Let τ_a denote the exit time from a.

Case 2(a) (Exponential Holding Case). τ_a is exponentially distributed with finite and positive mean.

Case 2(b). (Instantaneous Case). $P_a(\tau_a = 0) = 1$.

We will define the excursion process Y of X with respect to P_a. In case 1, Y is a sequence of random variables with values in U, $Y_k = \alpha_{\sigma(k)}(\theta_{\sigma(k-1)}X)$, where $\sigma(0) = 0$ and $\sigma(k)(k > 0)$ is the kth hitting time for a. Since Y_k is a random variable with values in U for each k, it is a stochastic process and

$$Y_k(t) = \begin{cases} X(\sigma(k-1) + t), & 0 \leqq t < \sigma(k) - \sigma(k-1), \\ a, & t \geqq \sigma(k) - \sigma(k-1). \end{cases} \tag{6.2}$$

By the strong Markov property of X at $\sigma(k)$, we can easily prove that all Y_k have the same probability law. In Case 2, Y is a point process with values in U:

$$\begin{aligned} D_Y &= \{s : S(s+) - S(s-) > 0\}, & S(t) &= A^{-1}(t), \\ Y_s &= \alpha_a(\theta_{S(s-)}X), & s &\in D_Y; \end{aligned} \tag{6.3}$$

the second equation can be written as

$$Y_s(t) = \begin{cases} X(S(s-) + t), & 0 \leqq t < S(s+) - S(s-), \\ a, & t \geqq S(s+) - S(s-). \end{cases} \tag{6.4}$$

The process Y is discrete in Case 2(a), but not in Case 2(b). Even in Case 2(b) Y is σ-discrete, because $S(s+) - S(s-) > 1/k$ is possible only for a finite number of s values in every finite time interval. Y is also renewal, as we can prove by the strong Markov property of X. Therefore, the excursion process is a Poisson point process in Case 2.

The *excursion law* of X at a is defined to be the common probability law of Y_k in Case 1 and the characteristic measure of Y in Case 2.

THEOREM 6.1. *The excursion law n of X and the probability laws $\{P_b^0\}_{b \neq a}$ determine the probability laws $\{P_b\}$ of X.*

PROOF. Because of the strong Markov property, it is enough to prove that P_a is determined by n and $\{P_b^0\}_{b \neq a}$. Since n determines Y, this is obvious in Case 1. We will discuss Case 2. Since $S(s)$ is a subordinator, that is, an increasing homogeneous Lévy process, we have

$$S(s-) = ms + \sum_{t \leqq s} \big(S(t-) - S(t-)\big) = ms + \sum_{t \leqq s, t \in D_y} \sigma_a(Y_t), \tag{6.5}$$

m being a nonnegative constant, called the *delay coefficient* of X at a. Therefore, S is determined by m and Y. Since $S(s)$ is increasing and since we have

$$X_t = \begin{cases} Y_s(t - S(s-)), & S(s-) \leqq t < S(s+), \\ a, & S(s-) = t = S(s-), \end{cases} \tag{6.6}$$

a.s. (P_a), the probability law P_a is determined by m and the probability law of Y. But the latter is determined by n. Since m is also determined by n by the theorem below, P_a is determined by n.

THEOREM 6.2. *In Case 2, we have*

$$m = 1 - \int_U (1 - \exp\{-\sigma_a(u)\}) n(du). \tag{6.7}$$

PROOF. By the definition of the local time, we have

$$\begin{aligned} 1 = E_a(\exp\{-\sigma_a\}) &= E_a\left(\int_0^\infty \exp\{-t\}\, dA(t)\right) \\ &= E_a\left(\int_0^\infty \exp\{-S(s)\}\, ds\right) = \int_0^\infty E_a(\exp\{-S(s)\})\, ds \\ &= \int_0^\infty \exp\{-ms\} E_a\left[\exp\left\{-\sum_{t \leqq s, t \in D_y} \sigma_a(Y_t)\right\}\right] ds \\ &= \int_0^\infty \exp\{-ms\} \exp\left\{-s \int_U (1 - \exp\{-\sigma_a(u)\}) n(du)\right\} ds \end{aligned} \tag{6.8}$$

by Theorem 4.5, which equals $[m - \int_U (1 - \exp\{-\sigma_a(u)\}) n(du)]^{-1}$. This completes the proof.

THEOREM 6.3. *The excursion law satisfies the following conditions:*

(i) *n is concentrated on the set $U' \equiv \{u : 0 < \sigma_a(u) < \infty,\ u(t) = a$ for $t \geqq \sigma_a(u)\}$;*

(ii) *$n\{u \in U : u(0) \notin V(a)\} < \infty$ for every neighborhood $V(a)$ of a;*

(ii′) *$\int_U (1 - \exp\{-\sigma_a(u)\}) n(du) \leqq 1$;*

(iii) *$n\{u : \sigma_a(u) > t,\ u \in \Lambda_t,\ \theta_t u \in M\} = \int_{\Lambda_t \cap (\sigma_a > t)} P^0_{u(t)}(M) n(du)$ for $t > 0$, $\Lambda_t \in B_t(U)$ and $M \in B(U)$;*

(iii′) *$n(u\colon u(0) \in B,\ u \in M) = \int_{u(0) \in B} P^0_{u(0)}(M) n(du)$ for $B \in B(S - \{a\})$ and $M \in B(U)$.*

PROOF. Condition (i) is obvious. Condition (ii′) is obvious in Case 1 and it follows at once from Theorem 6.2 in Case 2. Condition (ii) is obvious in Case 1. To prove it in Case 2, consider the restriction Y^* of Y to $U^* = \{u : u(0) \notin V(a)\}$. Then $t \in D_{Y^*}$ implies $X(S(t-)-) = a$ and $X(S(t-)+) \notin V(a)$. Since the set of such t values is discrete, Y^* is discrete. But Y^* is a Poisson point process with the characteristic measure $= n|U^*$ (Theorem 4.7). Therefore, the total measure of $n|U^*$ is finite and so we have $n(U^*) < \infty$. Condition (iii) is obvious and condition (iii′) is trivial in Case 1.

To prove (iii) in Case 2, consider the measures (Meyer's entrance law):

$$r_t(B) = n\{u : \sigma_a(u) > t, u(t) \in B\} \qquad B \in B(S), t > 0. \tag{6.9}$$

First we will prove that for $B \in B(S)$ and $M \in B(U)$,

$$n\{n : \sigma_a(u) > t, u(t) \in B, \theta_t u \in M\} = \int_B r_t(db) P_b(M). \tag{6.10}$$

Let $V = \{u \in U : \sigma_a(u) > t\}$ and $Z = Y|_r V$. Since $s \in D_Z$ implies $S(s+) - S(s-) > t$, the set of such s values is discrete. Therefore, Z is a discrete Poisson point process with the characteristic measure $= n \mid V$. This implies that $n(V) < \infty$. Let τ be the first time in D_Y. By Theorem 4.4A we have

$$P_a(\alpha_a(\theta_{S(\tau-)}X) \in M') = \frac{n(M' \cap V)}{n(V)}, \qquad M' \in B(U). \tag{6.11}$$

Setting $M' = \{u : \sigma_a(u) > t, u(t) \in B, \theta_t u \in M\}$, $M \in B(U)$, we have

$$\begin{aligned} n\{u &: \sigma_a(u) > t, u(t) \in B, \theta_t u \in M\} \\ &= n(V) P_a(\sigma_a(\theta_{S(\tau-)}X) > t, X_{S(\tau-)+t}B, \alpha_a(\theta_{S(\tau-)+t}(X)) \in M). \end{aligned} \tag{6.12}$$

Since $S(\tau-) + t = \inf\{\alpha > 0 : X_s \neq a \text{ for } \alpha - t < s \leqq \alpha\}$, $S(\tau-) + t$ is a stopping time for X. Since $\sigma_a(\theta_{S(\tau-)}X) > t$ is the same as $X_{S(\tau-)+t} \neq a$, this event is measurable $(B_{S(\tau-)+t})$. Therefore, we have

$$\begin{aligned} n\{u &: \sigma_a(u) > t, u(t) \in B, \theta_t u \in M\} \\ &= n(V) \int_S P_a(\sigma_a(\theta_{S(\tau-)}X) > t, X_{S(\tau-)+t} \in db) P_b(X \in M). \end{aligned} \tag{6.13}$$

Setting $M = \{u : u(0) \in B\}$, we have

$$\begin{aligned} r_t(B) &\equiv n\{u : \sigma_a(u) > t, u(t) \in B\} \\ &= n(V) P_a(\sigma_a(\theta_{S(\tau-)}X) > t, X_{S(\tau-)+t} \in B). \end{aligned} \tag{6.14}$$

Putting this in the above formula, we obtain (6.9). Equation (6.9) can be written as

$$n\{u : \sigma_a(u) > t, u(t) \in B, \theta_t u \in M\} = \int_{(u(t) \in B) \cap (\sigma_a(u) > t)} P_{u(t)}(M) n(du). \tag{6.15}$$

Using this and the Markov property of X and noticing that $\sigma_a(u) > t$ implies $\sigma_a(u) > s$ and $\sigma_a(u) = s + \sigma_a(\theta_s u)$ for $s \leqq t$, we get

$$\begin{aligned} n\{u &: \sigma_a(u) > t, u(t_1) \in B_1, \cdots, u(t_k) \in B_k, \theta_t u \in M\} \\ &= \int_{u(t_1) \in B_1, \cdots, u(t_k) \in B_k, \sigma_a(u) > t} P_{u(t)}(M) n(du), \end{aligned} \tag{6.16}$$

which implies (iii). A similar and even simpler argument shows (iii′).

THEOREM 6.4. *In order for a σ-finite measure n on U to be the characteristic measure of a recurrent extension of X^0 it is necessary that n satisfies the following conditions in addition to* (i), (ii), (ii′), (iii) *and* (iii′) *in Theorem* 6.3*:*

Case 1 (*Discrete Visiting Case*). *n is a probability measure concentrated on $U^a \equiv \{u \in U' : u(0) = a\}$;*

Case 2(*a*) (*Exponential Holding Case*). *n is a finite but not identically zero measure concentrated on $U^+ \equiv \{u \in U' : u(0) \neq a\}$ such that $\int_U (1 - \exp\{-\sigma_a(u)\})n(du) < 1$;*

Case 2(*b*) (*Instantaneous Case*). *n is an infinite measure such that $n(U^a) = 0$ or ∞.*

PROOF. First we will prove the necessity of the conditions. In Case 1, n is the probability law of the path of X^0 starting at a and is therefore concentrated in U^a. In Case 2, n is proportional to the probability law (with respect to P_a) of $\alpha_a(\theta_\tau X)$, when τ is the exit time from a. Since $\infty > \tau > 0$ a.s., $X(\tau) \equiv \alpha_a(\theta_\tau X)(0) \neq a$ by the strong Markov property of X. This implies that n is concentrated in U^+. Since the local time of X at a is proportional to the actual visiting time in this case, the delay coefficient m must be positive. Thus, we have the inequality. In Case 3, n must be an infinite measure, because the excursion process Y is not discrete. If $0 < n(U^a) < \infty$, then $n(U^+) = \infty$ and $Y^a \equiv Y|_r U^a$ would be a discrete Poisson point process. Let τ be the first time in D_{Y^a}. Then $S(\tau-)$ would be a stopping time for X, that is,

$$S(\tau-) = \inf\{t : X_t = a \text{ and there exists } t' > t \text{ for all } s \in (t, t')X_s \neq a\} \tag{6.17}$$

and the strong Markov property of X would be violated at $S(\tau-)$.

For the proof of the sufficiency, it is enough to construct the path of X starting at a with the excursion measure $= n$. First we construct the Poisson point process Y with the characteristic measure $= n$ by Theorem 4.2. It is easy to construct the path of X starting at a whose excursion process has the same probability law as Y by reversing the procedure of deriving the excursion process from a Markov process.

7. The integral representation of the excursion law

Let X^0 be a standard process on S with a trap at $a \in S$ and $\mathscr{E}(X^0)$ be the set of all σ-finite measures on U satisfying the five conditions (i), (ii), (ii′), (iii) and (iii′) in Theorem 6.3. Define the norm $n \in \mathscr{E}(X^0)$ by

$$\|n\| = \int_U (1 - \exp\{-\sigma_a(u)\})n(du). \tag{7.1}$$

Let $\mathscr{E}_1(X^0)$ denote the set of all $n \in \mathscr{E}(X^0)$ with $\|n\| = 1$. Clearly, $\mathscr{E}_1(X^0)$ is a convex set. This suggests that any $n \in \mathscr{E}(X^0)$ has an integral representation in terms of extremal ones. The measure $n \in \mathscr{E}_1(X^0)$ is called *extremal* if

$$n = c_1 u_1 + c_2 u_2 (c_1, c_2 > 0, c_1 + c_2 = 1, u_1, u_2 \in \mathscr{E}_1(X)) \tag{7.2}$$

implies $u_1 = u_2 = u$.

Suppose that n is concentrated on $U^+ = \{u \in U' : u(0) \neq a\}$. Then condition (iii′) implies

$$n(\cdot) = \int_{b \neq a} k(db) P_b(\cdot), \; k(B) = n\{u : u(0) \in B\}. \tag{7.3}$$

Since $P_b(\cdot)$, $b \neq a$, satisfies all conditions (i) to (iii′), it belongs to $\mathscr{E}(X^0)$, and therefore we have

$$n_b \equiv \frac{P_b(\cdot)}{\|P_b\|} = \frac{P_b(\cdot)}{E_b(1 - \exp\{-\sigma_a\})} \in \mathscr{E}_1(X). \tag{7.4}$$

Therefore (7.3) can be written as

$$n(\cdot) = \int_{b \in S - \{a\}} \lambda(db) n_b(\cdot), \qquad \lambda(db) = E_b(1 - \exp\{-\sigma_a\}) k(db). \tag{7.5}$$

Since $k(S - V(a)) < \infty$ for every neighborhood $V(a)$ of a and

$$\int_{S - \{a\}} E_b(1 - \exp\{\sigma_a\}) k(db) = 1 \tag{7.6}$$

by (ii) and $\|u\| = 1$, λ satisfies

$$\int_{S - V(a)} \frac{\lambda(db)}{E_b(1 - \exp\{-\sigma_a\})} < \infty \qquad \text{for every } V(a) \tag{7.7}$$

and $\lambda(S - \{a\}) = 1$.

Using (7.5) and noticing that $n_b = c_1 n_1 + c_2 n_2$, $c_1, c_2 > 0$, implies that both n_1 and n_2 are concentrated on $U^b \equiv \{u \in U' : u(0) = b\}$, we can easily prove that n_b, $b \neq a$, is extremal.

Suppose that $u \in \mathscr{E}_1(X^0)$ and $B \in B(S)$ and set

$$U^B = \{u \in U: u(0) \in B\}, \qquad n^B = n \,|\, U^B. \tag{7.8}$$

If $n^B \neq 0$, then $n^B/\|n^B\| \in \mathscr{E}_1(X^0)$. Using this fact, we can easily see that if $n \in \mathscr{E}_1(X^0)$ is extremal, then n is concentrated in U^b for some $b \in S$. If $b \neq a$, then $n = n_b$ by (7.5). However, there are many extremal ones concentrated in U^a. Let N^a be the set of all such extremal ones and write $B(N^a)$ for the σ-algebra on N^a generated by the sets $\{v \in N^a : v(\Lambda) < c\}$, $\Lambda \in B(U)$, $c > 0$.

If n is concentrated on U^a, then $n(\cdot) = \int_{N^a} v(\cdot) \lambda(dv)$, where λ is a probability measure on N^a. Once this is proved, we can easily obtain the following theorem.

THEOREM 7.1. *The measure $n \in \mathscr{E}_1(X^0)$ can be expressed uniquely as*

$$n(\cdot) = \int_{b \neq a} n_b(\cdot) \lambda(db) + \int_{N^a} v(\cdot) \lambda(dv) \tag{7.9}$$

with a probability measure λ on $(S - \{a\}) \cup N^a$.

PROOF. Since n is concentrated on U^a, we will regard n as a measure on U^a from now on. Let Ω be the product space $T \times U^a$ associated with the product σ-algebra. We introduce a probability measure Q on Ω by

$$(7.10) \qquad Q(d\tau du) = 1_{\tau < \sigma_a(u)} \exp\{-\tau\}\, d\tau\, n(du),$$

where $1_{\tau<\sigma_a(u)}$ denotes the indicator of the set $\{(\tau, u) \in \Omega : \tau < \sigma_a(u)\}$. It is easy to see that $Q(\Omega) = \|n\| = 1$. We will use the same notation for a measure and the integral based on it, for example $Q(f) = \int_\Omega f(\omega) Q(d\omega)$. We also use the same notation for a σ-algebra and for the class of all bounded real functions measurable with respect to it.

Consider a stochastic process $Z_t(\omega)$ on (Ω, Q) defined by

$$(7.11) \qquad Z_t(\omega) = Z_t(\tau, u) = \begin{cases} u(t) & \text{for } t < \tau, \\ a & \text{for } t \geqq \tau, \end{cases}$$

and let $B_t(Z)$ denote the σ-algebra generated by Z_s, $s \leqq t$. We will first prove that

$$(7.12) \qquad Q[1_{t<\tau} g(\theta_t u) | B_t(Z)] = 1_{t<\iota} n_{u(t)}((1 - \exp\{-\sigma_a\})g) \qquad \text{a.s. } (Q)$$

for every $g \in B(U^a)$. For this purpose it is enough to prove that

$$(7.13) \qquad Q[f_1(Z(t_1)) \cdots f_n(Z(t_n)) 1_{t<\tau} g(\theta_t u)]$$
$$= Q[f_1(Z(t_1)) \cdots f_n(Z(t_n)) 1_{t<\tau} n_{u(t)}((1 - \exp\{-\sigma_a\})g)],$$

for $t_1 < t_2 < \cdots < t_n \leqq t$ and $f_i \in B(S_i)$. The $Z(t_i)$ can be replaced by $u(t_i)$ in the above equation because of the factor $1_{t<\tau}$. It is therefore enough to prove that

$$(7.14) \qquad Q[f(u) 1_{t<\tau} g(\theta_t u)] = Q[f(u) 1_{t<\tau} n_{u(t)}(1 - \exp\{-\sigma_a) g]$$

for every $f \in B_t(U)$. Integrating by $d\tau$ and using property (iii), we can prove that both sides are equal to

$$(7.15) \qquad \int f(u) 1_{t<\sigma_a} \exp\{-t\} E^0_{u(t)}((1 - \exp\{-\sigma_a\})g) n(du).$$

The set U^a is a Borel subset of U with respect to the Skorohod topology whose topological σ-algebra is the same as $B(U)$. Therefore letting $B_{0+}(Z) = \bigcap_{t>0} B_t(Z)$, we can define on $(U, B(U))$ the *conditional probability measure* $\tilde{Q}(\cdot | B_{0+}(Z))$ of the random variable $\omega = (\tau, u) \rightsquigarrow u$. Define a measure ν_ω on $(U, B(U))$ depending on ω by

$$(7.16) \qquad \nu_\omega(du) = \frac{1}{1 - \exp\{-\sigma_a(u)\}} \tilde{Q}(du | B_{0+}).$$

Then we have

$$(7.17) \qquad \nu_\omega(g) = Q\left(\frac{g(u)}{1 - \exp\{-\sigma_a(u)\}} \middle| B_{0+} \right) \qquad \text{a.s. } (Q).$$

The measure ν_ω is clearly an N^a valued function on Ω, measurable with respect to $B_{0+}(Z)$. Let λ be the probability law of this random variable.

Let $g \in B(U^a)$. Then we have

$$\begin{aligned}(7.18)\qquad \int_{N^a} \nu(g)\lambda(d\nu) &= Q(\nu_\omega(g)) = Q\left(\frac{g(u)}{1 - \exp\{-\sigma_a(u)\}}\right) \\ &= \int_{U^a}\int_0^{\sigma_a(u)} e^{-\tau}\,d\tau \frac{g(u)}{1-\exp\{-\sigma_a(u)\}} n(du) \\ &= n(g).\end{aligned}$$

To complete the proof, we need only prove $\nu_\omega \in \mathscr{E}_1(X^0)$. The only difficult condition we have to verify is (iii), that is,

$$(7.19)\qquad \nu_\omega(f(u)g(\theta_t u)) = \nu_\omega(f(u)E^0_{u(t)}(g))$$

for $t > 0$, $f \in B_t(U^a)$ and $g \in B(U^a)$. We have to prove this except on a null set which is independent of t, f and g. First fix $t > 0$. Since ν_ω and P^0_b are measures, it is enough to prove (7.19) in case f and g are of the following form:

$$(7.20)\qquad \begin{aligned} f(u) &= f_1(u(t_1))f_2(u(t_2))\cdots f_k(u(t_k)), &\quad 0 < t_1 < \cdots < t_k, \\ g(u) &= g_1(u(s_1))g_2(u(s_2))\cdots g_m(u(s_m)), &\quad 0 < s_1 < \cdots < s_m,\end{aligned}$$

where t_i and s_j are taken from a fixed countable dense subset of $(0, \infty)$ and f_i and g_j are taken from a fixed countable number of bounded continuous functions on S which form a ring generating the σ-algebra on S. Take $\delta < t_1$ and write f_δ for $f \circ \theta_{-\delta}$. Then we have

$$\begin{aligned}(7.21)\qquad \nu_\omega(f(u)g(\theta_t u)) &= Q\left(\frac{f(u)g(\theta_t u)}{1 - \exp\{-\sigma_a(u)\}}\middle| B_{0+}(Z)\right) \\ &= \lim_{\delta \to 0+} Q\left(1_{\delta<\tau}\frac{f_\delta(\theta_\delta u)g(\theta_{t-\delta}\theta_\delta u)}{1 - \exp\{-\sigma_a(\theta_\delta(u))\}}\middle| B_{0+}(Z)\right).\end{aligned}$$

Noticing that $Q(\cdot|B_{0+}(Z)) = Q(Q(\cdot|B_t(Z))|B_{0+}(Z))$ and using (7.14), we can see that the expression above is

$$\begin{aligned}(7.22)\qquad &\lim_{\delta\to 0+} Q(1_{\delta<\tau}n_{u(\delta)}(f_\delta(u)g(\theta_{t-\delta}u))|B_{0+}(Z)) \\ &\qquad = \lim_{\delta\to 0+} Q(1_{\delta<\tau}n_{u(\delta)}(f_\delta(u)\cdot E_{u(t-\delta)}(g))|B_{0+}(Z)).\end{aligned}$$

By a similar argument, we have

$$\begin{aligned}(7.23)\qquad \nu_\omega(f(u)E_{u(t)}(g)) &= Q\left(\frac{f(u)E_{u(t)}(g)}{1 - \exp\{-\sigma_a(u)\}}\middle| B_{0+}\right) \\ &= \lim_{\delta\to 0+} Q(1_{\delta<\tau}n_{u(\delta)}(f_\delta(u)\cdot E_{u(t-\delta)}(g))|B_{0+}(Z)).\end{aligned}$$

Thus, (7.19) is proved for such f and g except on a null ω set. Since there are a countable number of possible pairs of (f, g), equation (7.19) holds for every

(f, g) except on a null ω set which may depend on t. If (7.19) is true for t, then it is true for every $t' > t$ by the Markov property of X. It is therefore enough to verify (7.19) only for a sequence $t_k \downarrow 0$. Thus, the exceptional ω set can be taken independently of t.

◇ ◇ ◇ ◇ ◇

We wish to express our gratitude to K. L. Chung and R. K. Getoor for their valuable suggestions.

REFERENCES

[1] E. B. DYNKIN, "On excursions of Markov processes," *Theor. Probability Appl.*, Vol. 13 (1968), pp. 672–676.

[2] E. B. DYNKIN and A. A. YUSHKEVICH, "On the starting points of incursions of Markov processes," *Theor. Probability Appl.*, Vol. 13 (1968), pp. 469–470.

[3] K. ITÔ, "Poisson point processes and their application to Markov processes," Department of Mathematics, Kyoto University, mimeographed notes, 1969.

[4] K. MATTHES, "Stationäre zufällige Punktmengen, I," *Jber. Deutsch. Math.-Verein.*, Vol. 66 (1963), pp. 69–79.

[5] P. A. MEYER, "Processus de Poisson ponctuels, d'après K. Itô," *Séminaire de Probabilités*, No. 5, 1969–1970, Lecture Notes in Mathematics, Berlin–Heidelberg–New York, Springer-Verlag, 1971.

REGENERATIVE PHENOMENA AND THE CHARACTERIZATION OF MARKOV TRANSITION PROBABILITIES

J. F. C. KINGMAN
OXFORD UNIVERSITY

1. Markov chains in continuous time

It is the object of this paper to draw together certain lines of research which during the last decade have grown out of the problem of characterizing the functions which can arise as transition probabilities of continuous time Markov chains. This problem is now solved (see Sections 8 and 9), although as usual its solution has thrown up further problems which demand attention.

The evolution of a Markov chain X_t in continuous time, with stationary transition probabilities, on a countable state space S, is as usual [2] described by the functions

$$p_{i,j}(t) = \mathbf{P}(X_{s+t} = j \mid X_s = i) \tag{1.1}$$

for $i, j \in S$, $t > 0$. These necessarily satisfy the conditions

$$p_{i,j}(t) \geqq 0, \qquad \sum_{j \in S} p_{i,j}(t) = 1, \tag{1.2}$$

and

$$p_{i,j}(t + u) = \sum_{k \in S} p_{i,k}(t) p_{k,j}(u), \tag{1.3}$$

to which it is usual to add the continuity condition

$$\lim_{t \to 0} p_{i,j}(t) = p_{i,j}(0) = \delta_{i,j}. \tag{1.4}$$

Conversely, given any array $(p_{i,j};\ i, j \in S)$ of functions satisfying (1.2) and (1.3), a Markov chain X_t can be constructed so as to satisfy (1.1).

It is therefore not surprising that a substantial part of the theory of Markov chains should be concerned with the consequences of (1.2), (1.3), and (1.4) for the functions $p_{i,j}$. It is possible to regard this as a problem in pure analysis, but those methods that have proved most powerful have had strong probabilistic motivation. The following list of typical results, taken from [2], will illustrate the achievements of this part of the theory (they are arranged in roughly increasing order of difficulty):

(I) $p_{i,i}(t) > 0$ *for all* $t > 0$;

(II) $p_{i,j}$ *is uniformly continuous on* $[0, \infty)$;

(III) (Doob) *the limit* $q_i = \lim_{t \to 0} t^{-1}\{1 - p_{i,i}(t)\}$ *exists in* $0 \leqq q_i \leqq \infty$, *and* $p_{i,i}(t) \geqq e^{-q_i t}$.

(IV) (Kolmogorov) *the limit* $\pi_{i,j} = \lim_{t \to \infty} p_{i,j}(t)$ *exists* (and a good deal is known about its properties);

(V) (Kolmogorov) *the finite limit* $q_{i,j} = \lim_{t \to 0} t^{-1} p_{i,j}(t)$ *exists when* $i \neq j$;

(VI) (Austin-Ornstein) *the function* $p_{i,j}$ *is either identically zero or always positive on* $(0, \infty)$;

(VII) (Austin-Ornstein-Chung) *the function* $p_{i,j}$ *is continuously differentiable in* $(0, \infty)$ *for* $i = j$ *and in* $[0, \infty)$ *for* $i \neq j$.

This last result cannot be substantially improved; Yuskevitch showed that $p_{i,j}$ need not have a second derivative, and Smith showed that $p'_{i,i}$ need not be continuous at the origin.

In the light of results such as these, the question at once arises [13] of characterizing the functions $p_{i,j}$. It is clear from (1.4) that there are two distinct cases to be considered, according as $i = j$ (the *diagonal* case) or $i \neq j$ (the *nondiagonal* case).

The natural first step in approaching such a problem is to ask whether it can be solved in the usually simpler situation of discrete time, when the variables t, u in (1.2) and (1.3) take only integer values. (Condition (1.4) then has no force.) The answer was given long ago by Feller and Chung in terms of the notion of a *renewal sequence*. A sequence $(u_n; n \geqq 0)$ is called a renewal sequence if there exists a sequence $(f_n; n \geqq 1)$ satisfying

$$f_n \geqq 0, \qquad \sum_{n=1}^{\infty} f_n \leqq 1, \tag{1.5}$$

and such that (u_n) is determined recursively by the equations

$$u_0 = 1, \qquad u_n = \sum_{r=1}^{n} f_r u_{n-r}, \qquad n \geqq 1. \tag{1.6}$$

Then the results of Feller and Chung may be summarized as follows.

THEOREM 1.1 [6]. *A sequence* (a_n) *can be expressed in the form*

$$a_n = p_{i,i}(n), \tag{1.7}$$

for some discrete Markov chain if and only if (a_n) *is a renewal sequence. A sequence* (b_n) *can be expressed in the form*

$$b_n = p_{i,j}(n), \qquad i \neq j, \tag{1.8}$$

if and only if there exists a renewal sequence (u_n) *and a sequence* (f'_n) *satisfying* (1.5) *such that*

$$b_n = \sum_{r=1}^{n} f'_r u_{n-r}. \tag{1.9}$$

It should be noted that the solution for the diagonal case is effective in the sense that, given a sequence (a_n), we can test whether it is a renewal sequence by using (1.6) to compute the corresponding f_n. This is not however true of the nondiagonal case; the representation (1.9) is far from unique, and I know of no sure way of deciding of a sequence (b_n) whether it can be expressed in this form. This feature will persist in the much more difficult solution of the continuous time problem.

2. Regenerative phenomena

It is a notable feature of some (but not all) of the arguments used in Markov chain theory that they really only concern one or two states of the set S. Thus if interest centers on one particular state i, it is often possible to lump all the remaining states together in a single state "not i." More precisely, it may suffice to consider, not X_t, but the process

$$Z_t = \varphi(X_t), \tag{2.1}$$

where $\varphi(i) = 1$, $\varphi(j) = 0, j \neq i$.

The process Z_t is in general non-Markovian, but it has a simple structure governed by the function $p_{i,i}$. If $\mathbf{P}_i$ denotes probability conditional upon $\{X_0 = i\}$, then for $0 = t_0 < t_1 < t_2 < \cdots < t_n$,

$$\begin{aligned} \mathbf{P}_i(Z_{t_r} &= 1; r = 1, 2, \cdots, n) \\ &= \mathbf{P}_i(X_{t_r} = i; r = 1, 2, \cdots, n) \\ &= \prod_{r=1}^{n} p_{i,i}(t_r - t_{r-1}). \end{aligned} \tag{2.2}$$

This suggests the following definition.

A *regenerative phenomenon* with *p-function* p is a stochastic process $(Z_t; t > 0)$ taking values 0 and 1, such that for $0 = t_0 < t_1 < t_2 < \cdots < t_n$,

$$\mathbf{P}(Z_{t_r} = 1; r = 1, 2, \cdots, n) = \prod_{r=1}^{n} p(t_r - t_{r-1}). \tag{2.3}$$

A function $p\colon (0, \infty) \rightsquigarrow [0, 1]$ is called a *p-function* if there is a regenerative phenomenon having p as p-function.

The left side of (2.3) is of course equal to

$$\mathbf{E}(Z_{t_1} Z_{t_2} \cdots Z_{t_n}), \tag{2.4}$$

and thus the p-function p determines the expectation of any linear combination of products of values of the process Z. In particular, p determines

$$\mathbf{P}(Z_{t_r} = \alpha_r; r = 1, 2, \cdots, n) = (-1)^{\Sigma \alpha_r} \mathbf{E}\left\{ \sum_{r=1}^{n} (1 - \alpha_r - Z_{t_r}) \right\}, \tag{2.5}$$

whenever $\alpha_r = 0$ or 1, so that the finite dimensional distributions of Z are known once p has been specified.

It is of course necessary that when (2.5) is calculated the result should be nonnegative. This requirement imposes, for each n, 2^n inequalities on the function p, though it turns out that all but 2 of these are consequences of those for smaller values of n. Thus p satisfies, for each $n \geqq 1$, a pair of inequalities, which may be written, for $n = 1$,

$$0 \leqq p(t) \leqq 1, \tag{2.6}$$

for $n = 2$,

$$p(t)p(u) \leqq p(t + u) \leqq 1 - p(t) + p(t)p(u), \tag{2.7}$$

for $n = 3$,

$$\begin{aligned} p(t)p(u + v) + p(t + u)p(v) - p(t)p(u)p(v) \\ \leqq p(t + u + v) \\ \leqq 1 - p(t) - p(t + u) + p(t)p(u) + p(t)p(u + v) \\ + p(t + u)p(v) - p(t)p(u)p(v), \end{aligned} \tag{2.8}$$

and so on. Conversely, the Daniell-Kolmogorov theorem establishes the existence of the process Z whenever this infinite family of functional inequalities is satisfied, showing that these inequalities are both necessary and sufficient for p to be a p-function.

For any $h > 0$, the events $E_n = \{Z_{nh} = 1\}$ form a recurrent event in the sense of Feller [6], since for $0 = r_0 < r_1 < \cdots < r_k$,

$$\mathbf{P}\left(\bigcap_{k=1}^{n} E_{r_k}\right) = \prod_{k=1}^{n} p\{(r_k - r_{k-1})h\}. \tag{2.9}$$

Thus, the sequence $\big(p(nh)\big)$ is a renewal sequence. This simple remark is one of the most powerful tools in the theory of p-functions.

In view of (2.2), we can now say that (2.1) defines a regenerative phenomenon with p-function $p_{i,i}$. Thus, *any diagonal Markov transition function* $p_{i,i}$ *is a* p*-function*. Theorem 1.1 might suggest that the converse ought to be true; regrettably it is not.

3. Standard p-functions

A p-function is called *standard* (by analogy with Chung's terminology for chains satisfying (1.4)) if

$$\lim_{t \to 0} p(t) = 1, \tag{3.1}$$

and the class of standard p-functions is denoted by $\mathscr{P}$. Then (1.4) shows that the p-function $p_{i,i}$ is standard, and if $\mathscr{PM}$ denotes the class of all diagonal Markov transition functions then

$$\mathscr{PM} \subseteqq \mathscr{P}. \tag{3.2}$$

If we combine (3.1) and (2.7) with the fact that $(p(nh))$ is a renewal sequence, a number of the simpler results known for $\mathscr{P}\mathscr{M}$ can be proved in the wider class $\mathscr{P}$. Thus, the following theorem is proved by quite elementary arguments.

THEOREM 3.1 [13]. *If p is any standard p-function, then*

(i) *p is uniformly continuous and strictly positive on* $(0, \infty)$,

(ii) *the limit*

$$p(\infty) = \lim_{t\to\infty} p(t) \tag{3.3}$$

exists, and

(iii) *the limit*

$$q = \lim_{t\to 0} t^{-1}\{1 - p(t)\} \tag{3.4}$$

exists in $0 \leqq q \leqq \infty$ *and*

$$p(t) \geqq e^{-qt}. \tag{3.5}$$

Because $(p(nh))$ is a renewal sequence, there must exist a sequence $(f_n(h))$ with

$$f_n(h) \geqq 0, \qquad \sum_{n=1}^{\infty} f_n(h) \leqq 1, \tag{3.6}$$

such that (1.6) holds with $f_n = f_n(h)$, $u_n = p(nh)$. It follows easily that, for $|z| < 1$,

$$\sum_{n=0}^{\infty} p(nh)z^n = \left\{1 - \sum_{n=1}^{\infty} f_n(h)z^n\right\}^{-1}. \tag{3.7}$$

Now fix $\theta > 0$, and set $z = e^{-\theta h}$ in (3.6). If the left side is multiplied by h it converges, as $h \to 0$; to the Laplace transform

$$r(\theta) = \int_0^{\infty} p(t)e^{-\theta t}\,dt \tag{3.8}$$

of p. The limiting behavior of the right side may be examined using the Helly compactness theorem, and a rather technical argument then leads to the following basic characterization of the class $\mathscr{P}$.

THEOREM 3.2 [13]. *If p is any standard p-function, there is a unique measure* μ *on the Borel subsets of* $(0, \infty]$ *with*

$$\int_{(0,\infty]} (1 - e^{-x})\mu(dx) < \infty, \tag{3.9}$$

such that, for all $\theta > 0$,

$$\int_0^{\infty} p(t)e^{-\theta t}\,dt = \left\{\theta + \int_{(0,\infty]} (1 - e^{-\theta x})\mu(dx)\right\}^{-1}. \tag{3.10}$$

Conversely, if μ is any Borel measure on $(0, \infty]$ *satisfying* (3.9), *there exists a unique continuous function p satisfying* (3.10), *and p is a standard p-function.*

Thus, (3.10) sets up a one to one correspondence between $\mathscr{P}$ and the class of measures μ satisfying (3.9). The limits whose existence is asserted in Theorem 3.1 are simply expressed in terms of μ;

$$p(\infty) = \left\{1 + \int x\mu(dx)\right\}^{-1}, \tag{3.11}$$

$$q = \mu(0, \infty]. \tag{3.12}$$

A regenerative phenomenon with $q < \infty$ is called *stable*, and one with $q = \infty$ *instantaneous*.

As an example, suppose that μ concentrates all its mass q at a single point a, $0 < a < \infty$. Then

$$r(\theta) = (\theta + q - qe^{-\theta a})^{-1}, \tag{3.13}$$

which inverts to give

$$p(t) = \sum_{n=0}^{[t/a]} \pi_n\{q(t - na)\}, \tag{3.14}$$

where π_n denotes the Poisson probability

$$\pi_n(\lambda) = \frac{\lambda^n e^{-\lambda}}{n!}. \tag{3.15}$$

This is an oscillating p-function, converging to the limit $p(\infty) = (1 + qa)^{-1}$. It is differentiable everywhere except at the point $t = a$, where it has left and right derivatives

$$D_-p(a) = -qe^{-qa}, \qquad D_+p(a) = q - qe^{-qa}. \tag{3.16}$$

In view of the theorem cited as (VII) in Section 1, the p-function (3.14) cannot therefore come from a Markov chain. Thus, *the inclusion* (3.2) *is strict.*

The differentiability behavior of (3.14) is entirely typical of that of p-functions for which the corresponding measures μ have atoms in $(0, \infty)$. In fact, let $m(t) = \mu(t, \infty]$, so that m is finite, nonincreasing and right continuous in $(0, \infty)$, and integrable on $(0, 1)$ (because of (3.9)). If m_n denotes the n-fold convolution of m with itself, we have the following result.

THEOREM 3.3 [15]. *The series*

$$b(t) = \sum_{n=1}^{\infty} (-1)^{n-1} m_n(t) \tag{3.17}$$

is uniformly absolutely convergent on compact intervals in $(0, \infty)$, *and all terms except possibly the first are continuous. The equation*

$$p(t) = 1 - \int_0^t b(u)\, du \tag{3.18}$$

holds, and shows that p has finite right and left derivatives at all points t in $0 < t < \infty$, and that

$$D_+p(t) - D_-p(t) = \mu\{t\}. \tag{3.19}$$

In particular, p is continuously differentiable in $(0, \infty)$ if and only if μ has no atoms in $(0, \infty)$.

Thus, the (diagonal case of the) Austin-Ornstein differentiability theorem is equivalent to the statement that, for p in $\mathscr{P}\mathscr{M}$, the measure μ has no atoms, except perhaps at ∞. This result will be considerably strengthened in Section 7.

The fundamental formula (3.10) has a number of other important uses. It can for example be used to examine the rate of convergence of $p(t)$ to $p(\infty)$, to establish the Volterra equation

$$p(t) = 1 - \int_0^t p(t - u)\mu(u, \infty]\, du, \tag{3.20}$$

(of which (3.17) and (3.18) describe an iterative solution), and to generalize a theorem of Kendall [8] by showing that every function in $\mathscr{P}$ admits a Fourier representation of the form

$$p(t) = p(\infty) + \int_0^\infty \varphi(\omega) \cos \omega t\, d\omega, \tag{3.21}$$

where $\varphi \geqq 0$. For these results and others, the reader is referred to [13] and [15].

4. Additive processes

The right side of (3.10) strongly suggests a connection with the theory of additive processes (processes with stationary independent increments), a connection which in the Markov case was exploited by Lévy [25]. If Z is a regenerative phenomenon with standard p-function p, it is easy to check that

$$\lim_{h \to 0} \mathbf{P}(Z_{t+h} \neq Z_t) = 0, \tag{4.1}$$

so that Z has a measurable version, and it makes sense to consider the process

$$\tau_t = \int_0^t Z_u\, du. \tag{4.2}$$

Since the sample functions of τ are continuous and nondecreasing, there exists an inverse process T defined by

$$T_t = \inf\{s > 0;\ \tau_s \geqq t\}. \tag{4.3}$$

Then [13] T is an additive process, with Lévy-Khinchin representation

$$\log \mathbf{E}(e^{-\theta T_t}) = -t\left\{\theta + \int_{(0,\infty]} (1 - e^{-\theta x})\mu(dx)\right\}^{-1}. \tag{4.4}$$

The term $-t\theta$ signifies a constant drift, so that

$$T_t = t + \xi_t, \tag{4.5}$$

where the additive process ξ increases in jumps.

If $q = \mu(0, \infty] < \infty$, the jumps of ξ occur at the points of a Poisson process of rate q, the height of each jump having distribution function

$$F(x) = q^{-1}\mu(0, x]. \tag{4.6}$$

Translating this back into a description of Z, it shows that the sample functions of (a separable version of) Z are step functions. The lengths of the intervals of constancy are independent random variables, those with $Z_t = 1$ having distribution function $1 - e^{-qx}$, and those with $Z_t = 0$ having distribution function F.

When $q = \infty$, the jumps of ξ are dense, and the sample function behavior of Z becomes much more complex. A version can be chosen in which the set $\{t; Z_t = 0\}$ is a countable union of intervals, but the complement $\{t; Z_t = 1\}$ can never be so, and instead resembles a Cantor set (though of positive measure). The measure μ determines the lengths of the component intervals of the former set, in the sense that, for $c > 0$, the intervals of length greater than c are well ordered, with distribution function

$$F_c(x) = \frac{\mu(c, x]}{\mu(c, \infty]}. \tag{4.7}$$

The problem of choosing a suitable version of Z in the instantaneous case has been considered (in unpublished lectures) by Kendall, who remarks that if T is a right continuous, strong Markov, additive process satisfying (4.4), then

$$\{t; Z_t = 1\} = \{T_u; u \geqq 0\} \tag{4.8}$$

defines a convenient version of the regenerative phenomenon whose p-function is given by (3.10). This construction permits the calculation of some useful distributions. For instance, the backward recurrence time

$$\beta_t = \inf\{u > 0; Z_{t-u} = 1\} \tag{4.9}$$

has distribution

$$\begin{aligned} \mathbf{P}\{\beta_t = 0\} &= p(t), \\ \mathbf{P}\{\beta_t \in (u, u + du)\} &= p(t - u)\mu(u, \infty]\cdot du. \end{aligned} \tag{4.10}$$

In fact, β is a Markov process, and

$$\{t; \tilde{Z}_t = 1\} = \{t; \beta_t = 0\} \tag{4.11}$$

defines a version $\tilde{Z}$ of Z. For related work, raising the possibility of a "strong regenerative property," see [7].

5. Markov measures

Returning now to the problem of describing the class $\mathscr{PM}$ of diagonal Markov transition functions $p_{i,i}$, we first exhibit a large subclass of $\mathscr{PM}$. Let (u_n) be any renewal sequence; then by Theorem 1.1 there exists a discrete time Markov chain with $u_n = p_{a,a}(n)$ for one state a. If c is any positive constant, consider the functions

$$\tilde{p}_{i,j}(t) = \sum_{n=0}^{\infty} p_{i,j}(n)\pi_n(ct), \tag{5.1}$$

where π_n is the Poisson probability (3.15). An elementary computation shows that these satisfy (1.2), (1.3), and (1.4), so that the function $\tilde{p}_{a,a}$ belongs to $\mathscr{PM}$. Thus, for any renewal sequence (u_n) and any $c > 0$, $\mathscr{PM}$ contains the function

$$p(t) = \sum_{n=0}^{\infty} u_n \pi_n(ct). \tag{5.2}$$

The class of functions of the form (5.2) is denoted by $\mathscr{Q}$. Since $\mathscr{Q}$ contains only stable p-functions, it cannot exhaust $\mathscr{PM}$, so that

$$\mathscr{Q} \subset \mathscr{PM} \subset \mathscr{P}. \tag{5.3}$$

For any p in $\mathscr{P}$, and any positive integer k, the sequence $(p(nk^{-1}))$ is a renewal sequence, so that

$$p_k(t) = \sum_{n=0}^{\infty} p(nk^{-1})\pi_n(kt) \tag{5.4}$$

belongs to $\mathscr{Q}$. It is a simple consequence of the weak law of large numbers that, for all t,

$$p(t) = \lim_{k\to\infty} p_k(t), \tag{5.5}$$

so that every function in $\mathscr{P}$ is the pointwise limit of a sequence of functions in $\mathscr{Q}$.

It is useful to express this fact in more formal topological language. Every p-function is a function from $(0, \infty)$ into $[0, 1]$, and may therefore be regarded as an element of the product space

$$\Pi = [0, 1]^{(0,\infty)}, \tag{5.6}$$

whose product topology is compact by Tychonov's theorem. The set of p-functions (standard or not), being defined by the inequalities (2.6), (2.7), (2.8), $\cdots$, is clearly closed in Π, and thus inherits a compact Hausdorff topology. In this topology $\mathscr{P}$ is not closed (consider the sequence $p_n(t) = e^{-nt}$), and the subspace topology on $\mathscr{P}$, though Hausdorff and indeed metrizable [5], is not compact. Equation (5.5) shows that $\mathscr{Q}$, and *a fortiori* $\mathscr{PM}$, is dense in $\mathscr{P}$. It is this fact that makes the identification of $\mathscr{PM}$ as a subset of $\mathscr{P}$ a somewhat delicate matter.

In the one to one correspondence set up by Theorem 3.2 between $\mathscr{P}$ and the class of measures satisfying (3.9), the latter class inherits a topology from that of $\mathscr{P}$. This topology has been identified by Kendall [10] and Davidson [5]; a sequence (p_n) in $\mathscr{P}$ converges to p in $\mathscr{P}$ if and only if the corresponding measures μ_n, μ satisfy

$$\lim_{n\to\infty} \int_{(0,\infty]} (1 - e^{-x})\varphi(x)\mu_n(dx) = \int_{(0,\infty]} (1 - e^{-x})\varphi(x)\mu(dx) \tag{5.7}$$

for every bounded continuous function φ on $(0, \infty]$.

In the correspondence (3.10), the subset $\mathscr{PM}$ of $\mathscr{P}$ corresponds to a proper (albeit dense in the sense of (5.7)) subset of the class of measures satisfying (3.9). It will be convenient to describe the members of this subset simply as *Markov measures*, so that $\mathscr{PM}$ is determined once the Markov measures have been characterized. In view of the discussion in Section 4, this amounts to the characterization of the possible distributions of lengths of excursions from a particular state in a Markov chain.

Examples of Markov measures can be obtained by computing the measure μ corresponding to p-functions of the form (5.2). It is not difficult to calculate that, if (f_n) is the sequence related to (u_n) by (1.6), then

$$\begin{aligned} \mu(dt) &= c^2\, dt \sum_{n=2}^{\infty} f_n \pi_{n-2}(ct), \\ \mu\{\infty\} &= c\left\{1 - \sum_{n=1}^{\infty} f_r\right\}. \end{aligned} \tag{5.8}$$

The point to note about this measure is that, apart from a possible atom at ∞, it has a density with respect to Lebesgue measure of the form $e^{-ct}P(t)$, where P is a power series with nonnegative coefficients.

More general examples can be constructed by a technique due to Yuskevitch ([26], see also [9]). Consider first a Markov chain of the special type constructed by Lévy [25], on the nonnegative integers, in which any excursion from 0 only visits one other state. The states $i \geqq 1$ must be stable, but 0 may be instantaneous. Fix positive integers b_i, m_i, $i = 1, 2, \cdots$, and construct a new chain on the state space,

$$\{0\} \cup \{(i, \alpha); i = 1, 2, \cdots, \alpha = 0, 1, 2, \cdots, m_i\}, \tag{5.9}$$

by replacing a sojourn in i by a progress through the states $(i, 0)$, $(i, 1)$, $\cdots$, (i, m_i) and return to 0, the time spent in (i, α) having probability density $b_i e^{-b_i t}$. It is not difficult to show that, in this new chain, the measure μ corresponding to the p-function $p_{0,0}$ has density in $(0, \infty)$ of the form

$$h(t) = \sum_{i=1}^{\infty} c_i t^{m_i} e^{-b_i t}, \tag{5.10}$$

where c_i is a nonnegative constant depending on the parameters of the original chain.

By a suitable choice of these parameters [17], one can obtain any measure satisfying (3.9) whose density is of the form

$$h(t) = \sum_{m=0}^{\infty} \sum_{n=1}^{\infty} a_{m,n} t^m e^{-nt}, \qquad a_{m,n} \geqq 0. \tag{5.11}$$

Any such measure is therefore a Markov measure, regardless of the value of $\mu\{\infty\}$. It thus becomes urgent to know which functions of a positive variable t can be expressed in the form (5.11); the answer to a slightly more general question is given by the following theorem.

THEOREM 5.1 [23]. *A function* h: $(0, \infty) \rightarrow [0, \infty]$, *not identically zero, is expressible in the form*

$$h(t) = \sum_{m=0}^{\infty} \sum_{n=0}^{\infty} a_{m,n} t^m e^{-nt} \tag{5.12}$$

with $a_{m,n} \geqq 0$ *if and only if it is lower semicontinuous and satisfies*

$$h(t) \geqq a t^m e^{-nt} \tag{5.13}$$

for some $a > 0$ *and some nonnegative integers* m, n.

Thus, the Yuskevitch construction shows that μ is a Markov measure whenever it has a density h in $(0, \infty)$ which is lower semicontinuous and satisfies (5.13). We shall see that these sufficient conditions come very close to being necessary.

6. Quasi-Markov chains

The notion of regenerative phenomenon is an abstraction of the process obtained by lumping together all states of a Markov chain except one. For some purposes, however, this is too drastic; if for instance one is interested in the transition probability $p_{i,j}$, $i \neq j$, then both states i and j should retain their identities. This suggests the following more general definition ([14], [16]).

A *quasi-Markov chain* of order N is a stochastic process Z_t, $t \geqq 0$, taking values $0, 1, 2, \cdots, N$, in such a way that, for $0 = t_0 < t_1 < t_2 < \cdots < t_n$ and $\alpha_0, \alpha_1, \alpha_2, \cdots, \alpha_n \in \{1, 2, \cdots, N\}$,

$$\mathbf{P}\{Z_{t_k} = \alpha_k (k = 1, 2, \cdots, n) \mid Z_0 = \alpha_0\} = \prod_{k=1}^{n} p_{\alpha_{k-1}, \alpha_k}(t_k - t_{k-1}). \tag{6.1}$$

Here the functions $p_{\alpha,\beta}$, which determine the finite dimensional distributions of Z so long as $Z_0 \neq 0$, are the elements of a matrix

$$\mathbf{p}(t) = (p_{\alpha,\beta}(t);\ \alpha, \beta = 1, 2, \cdots, N), \tag{6.2}$$

called the p-matrix of the chain. It is important to note that (6.1) is *not* required to hold if $\alpha_k = 0$ for any k; the state 0 is anomalous. A quasi-Markov chain of order 1 is essentially a regenerative phenomenon. A quasi-Markov chain is said to be standard if

$$\lim_{t \to 0} \mathbf{p}(t) = \mathbf{I}. \tag{6.3}$$

the identity matrix of order N.

If X_t is a Markov chain, and $i_1, i_2, \cdots, i_N$ are any N distinct states, then it is immediate that the process

$$Z_t = \psi(X_t), \tag{6.4}$$

where

$$\psi(i_\alpha) = \alpha, \qquad \psi(j) = 0, \qquad j \notin \{i_1, i_2, \cdots, i_N\}, \tag{6.5}$$

is a standard quasi-Markov chain with p-matrix

$$\big(p_{i_\alpha, i_\beta}(t);\ \alpha, \beta = 1, 2, \cdots, N\big). \tag{6.6}$$

The analysis described in Sections 2 and 3 can be extended to the more general situation of a quasi-Markov chain; for the details see [14]. For present purposes it suffices to quote the main characterization theorem which generalizes Theorem 3.2.

THEOREM 6.1 [14]. *In order that a continuous $(N \times N)$ matrix valued function $\mathbf{p}(t)$, $t > 0$, be the p-matrix of a standard quasi-Markov chain, it is necessary and sufficient that its Laplace transform*

$$\mathbf{r}(\theta) = \int_0^\infty \mathbf{p}(t) e^{-\theta t}\, dt \tag{6.7}$$

should have, for all $\theta > 0$, an inverse

$$\mathbf{r}(\theta)^{-1} = \big(r^{\alpha, \beta}(\theta)\big), \tag{6.8}$$

with

$$r^{\alpha, \alpha}(\theta) = \theta + a_\alpha + \int_{(0, \infty)} (1 - e^{-\theta x}) \mu_\alpha(dx), \tag{6.9}$$

and (for $\alpha \neq \beta$)

$$r^{\alpha, \beta}(\theta) = - \int_{[0, \infty)} e^{-\theta x} \mu_{\alpha, \beta}(dx), \tag{6.10}$$

where μ_α is a Borel measure on $(0, \infty)$ satisfying (3.9), $\mu_{\alpha, \beta}$ a totally finite Borel measure on $[0, \infty)$, and, for all α,

$$\sum_{\beta \neq \alpha} \mu_{\alpha, \beta}[0, \infty) \leqq a_\alpha. \tag{6.11}$$

It is possible to give probabilistic interpretations of the quantities occurring in these formulae. Comparing (6.9) with (3.10) it appears that $1/r^{\alpha,\alpha}(\theta)$ is the Laplace transform of a p-function p_α; this turns out to be a taboo function in the sense of Chung [2],

$$p_\alpha(t) = {}_H p_{\alpha,\alpha}(t), \quad H = \{\beta; 1 \leqq \beta \leqq N, \beta \neq \alpha\}. \tag{6.12}$$

The measure $\mu_{\alpha,\beta}$ has the following interpretation, for a suitably regular version of Z. Scan a sample function of Z for the first times $\sigma \leqq \tau$ with

$$Z_{\sigma-} = \alpha, \qquad Z_{\tau+} = \beta, \qquad Z_t = 0, \qquad \sigma < t < \tau. \tag{6.13}$$

Then $\mu_{\alpha,\beta}$ is a multiple of the distribution of $(\tau - \sigma)$.

The formulae of Theorem 6.1 take on particularly useful forms when $N = 2$. Inverting the (2×2) matrix $\mathbf{r}(\theta)^{-1}$ directly, we have

$$\mathbf{r}(\theta) = \det[\mathbf{r}(\theta)]\begin{pmatrix} r^{2,2}(\theta) & -r^{1,2}(\theta) \\ -r^{2,1}(\theta) & r^{1,1}(\theta) \end{pmatrix}, \tag{6.14}$$

so that

$$\frac{r_{1,2}(\theta)}{r_{2,2}(\theta)} = -\frac{r^{1,2}(\theta)}{r^{1,1}(\theta)} = r_1(\theta)\int e^{-\theta x}\mu_{1,2}(dx). \tag{6.15}$$

Using (6.12), this implies that

$$p_{1,2}(t) = \int_0^t f_{1,2}(s)p_{2,2}(t-s)\,ds, \tag{6.16}$$

where

$$f_{1,2}(t) = \int_0^t {}_2p_{1,1}(t-u)\mu_{1,2}(du). \tag{6.17}$$

Applying this in particular to the quasi-Markov chain obtained from the Markov chain by lumping together all states except i and j, we obtain the equation

$$p_{i,j}(t) = \int_0^t f_{i,j}(s)p_{j,j}(t-s)\,ds, \tag{6.18}$$

where

$$f_{i,j}(t) = \int_0^t {}_jp_{i,i}(t-u)\mu_{i,j}(du). \tag{6.19}$$

Equation (6.18) is the celebrated *first passage decomposition* (usually proved by quite different methods [2]); the fact that $f_{i,j}$ itself has the decomposition (6.19) is crucial. Condition (6.11) is in this context equivalent to the inequality

$$\int_0^\infty f_{i,j}(t)\,dt = \int_0^\infty {}_jp_{i,i}(t)\,dt\cdot\mu_{i,j}[0,\infty) \leqq 1. \tag{6.20}$$

If in the above argument we had started with $r_{1,2}(\theta)/r_{1,1}(\theta)$ instead of

$r_{1,2}(\theta)/r_{2,2}(\theta)$, we should have obtained the *last exit decomposition*,

$$p_{i,j}(t) = \int_0^t p_{i,i}(t - s)g_{i,j}(s)\,ds, \tag{6.21}$$

where

$$g_{i,j}(t) = \int_0^t {}_ip_{j,j}(t - u)\mu_{i,j}(du). \tag{6.22}$$

Such identities are naturally written in convolution notation:

$$\begin{aligned} f_{i,j} &= {}_jp_{i,i} * d\mu_{i,j}, \qquad & p_{i,j} &= f_{i,j} * p_{j,j} = {}_jp_{i,i} * d\mu_{i,j} * p_{j,j}, \\ g_{i,j} &= {}_ip_{j,j} * d\mu_{i,j}, \qquad & p_{i,j} &= g_{i,j} * p_{i,i} = {}_ip_{j,j} * d\mu_{i,j} * p_{i,i}. \end{aligned} \tag{6.23}$$

7. Properties of Markov measures

The first passage and last exit decompositions can be combined, by an argument shown to me by Professor Reuter, to give important positive information about the measure μ associated by Theorem 3.2 with the p-function $p_{i,i}$ in a Markov chain. By (1.3),

$$p_{i,i}(t + u) = p_{i,i}(t)p_{i,i}(u) + \sum_{j \neq i} p_{i,j}(t)p_{j,i}(u), \tag{7.1}$$

so that, using (6.18) and (6.21),

$$\begin{aligned} p_{i,i}(t + u) - p_{i,i}(t)p_{i,i}(u) \\ = \sum_{j \neq i} \int_0^t p_{i,i}(t - s)g_{i,j}(s)\,ds \int_0^u f_{j,i}(v)p_{i,i}(u - v)\,dv. \end{aligned} \tag{7.2}$$

Multiplying by $e^{-\alpha t - \beta u}$ and integrating over $t > 0$, $u > 0$, we have

$$\frac{r_{i,i}(\alpha) - r_{i,i}(\beta)}{\beta - \alpha} - r_{i,i}(\alpha)r_{i,i}(\beta) = \sum_{j \neq i} r_{i,i}(\alpha)\hat{g}_{i,j}(\alpha)\hat{f}_{j,i}(\beta)r_{i,i}(\beta), \tag{7.3}$$

where $\alpha, \beta > 0$, $\alpha \neq \beta$, and $\hat{\varphi}$ denotes the Laplace transform of the function φ. Using (3.10) to express $r_{i,i}$ in terms of the measure μ, and simplifying, we have

$$\int_{(0,\infty)} \frac{e^{-\alpha x} - e^{-\beta x}}{\beta - \alpha}\,\mu(dx) = \sum_{j \neq i} \hat{g}_{i,j}(\alpha)\hat{f}_{j,i}(\beta). \tag{7.4}$$

It is not difficult to see that this implies that μ has a density h in $(0, \infty)$, and that

$$h(t + u) = \sum_{j \neq i} g_{i,j}(t)f_{j,i}(u), \tag{7.5}$$

for almost all (t, u). That μ is absolutely continuous was suggested without proof by Lévy [25]; in view of the remarks following Theorem 3.3, it implies at once the continuous differentiability of $p_{i,i}$.

If in (7.5) we replace t by $(t - u)$ and integrate with respect to $u \in (0, t)$, we obtain the formula

$$th(t) = \sum_{j \neq i} \int_0^t g_{i,j}(t - u) f_{j,i}(u)\, du, \tag{7.6}$$

for almost all t. Thus, one version of the density of μ is given by

$$h(t) = t^{-1} \sum_{j \neq i} k_j(t), \tag{7.7}$$

where

$$k_j = g_{i,j} * f_{j,i} = {}_i p_{j,j} * d\mu_{i,j} * d\mu_{j,i} * {}_i p_{j,j}. \tag{7.8}$$

Now for any p-function p, $p^{(2)} = p * p$ is a continuous function on $(0, \infty)$ with $\lim_{t \to 0} p^{(2)}(t) = 0$. It follows easily that $k_j = {}_i p_{j,j}{}^{(2)} * d\mu_{i,j} * d\mu_{j,i}$ is continuous, and hence that the density h in (7.7) is lower semicontinuous in $(0, \infty)$.

Now suppose that h is not identically zero. Then there must exist $j \neq i$ for which k_j is not identically zero, and for this j (7.8) shows that neither $\mu_{i,j}$ nor $\mu_{j,i}$ can be identically zero. The p-function ${}_i p_{j,j}$ satisfies, as a simple consequence of the left inequality of (2.7), the inequality

$${}_i p_{j,j}(t) \geqq e^{-\alpha t}, \tag{7.9}$$

for some α and all sufficiently large t. If (7.9) is substituted into (7.8) and (7.7), it follows that, for some β, and all sufficiently large t, $h(t) \geqq e^{-\beta t}$.

A further result follows from the Austin-Ornstein positivity theorem cited as (VI) in Section 1. With the particular value of j chosen above, (6.18) and (6.19) show that neither $p_{i,j}$ nor $p_{j,i}$ vanishes identically. The theorem then asserts that, for all $t > 0$, $p_{i,j}(t) > 0$, $p_{j,i}(t) > 0$. Using (6.18) and (6.19) again, it follows that, for all $\varepsilon > 0$,

$$\mu_{i,j}[0, \varepsilon) > 0, \qquad \mu_{j,i}[0, \varepsilon) > 0. \tag{7.10}$$

Substituting (7.10) into (7.8) and (7.7), we have $h(t) > 0$, $t > 0$.

Combining all these results, we have a set of necessary conditions for a measure μ to correspond (in (3.10)) to a function in $\mathcal{PM}$.

THEOREM 7.1. *Every Markov measure μ has (as well as a possible atom at infinity) a lower semicontinuous density h on $(0, \infty)$ which is either identically zero or satisfies (for some β) the inequalities*

$$\begin{aligned} h(t) &> 0 && \text{for all } t > 0, \\ h(t) &\geqq e^{-\beta t} && \text{for sufficiently large } t. \end{aligned} \tag{7.11}$$

8. The solution of the Markov characterization problem

If Theorems (5.1) and (7.1) are compared, it will be seen that the gap between the sufficient and the necessary conditions, for μ to be a Markov measure, is just the gap between the inequalities (5.13) and (7.11). In fact, the gap is narrower than might at first appear, since a positive lower semicontinuous function is bounded away from zero on every compact interval. It follows easily that *a lower semicontinuous function* h *satisfying* (7.11) *satisfies* (5.12) *if and only if, for some integer* m,

$$h(t) \geqq t^m \tag{8.1}$$

for all sufficiently small t.

All that remains therefore to complete the characterization of $\mathscr{PM}$ is to determine to what extent (8.1) is necessary. In fact, it is not necessary at all, as was shown in [23] by calculating the measure μ for an *escalator* [2], an infinite string of states through which the process moves in a finite time (that is, a divergent pure birth process with instantaneous return). For such a chain, it is easy [21] to see that $h(t) \to 0$ as $t \to 0$ faster than any monomial. Moreover, it is possible (and essential) to go much further than this, and to prove the following result.

THEOREM 8.1 [23]. *Let* ω *be any positive continuous function on* $(0, 1]$. *Then there exists an escalator with the property that the probability density* h *of the time spent in traversing it satisfies*

$$h(t) \leqq \omega(t), \qquad 0 < t \leqq 1. \tag{8.2}$$

With this as a tool, the solution now proceeds with only technical difficulties. In the Yuskevitch construction, each finite string of states is replaced by a suitably chosen escalator, and it is then shown to be possible to realize any positive lower semicontinuous function h satisfying (7.11). In other words, the converse of Theorem 7.1 is true. Collecting the various results together, we therefore have the following fundamental theorem.

THEOREM 8.2 [23]. *A continuous function* $p(t)\,(t > 0)$ *can be expressed in the form* $p(t) = p_{i,i}(t)$ *in some Markov chain if and only if its Laplace transform can be expressed in the*

$$\int_0^\infty p(t)e^{-\theta t}\,dt = \left\{\theta + a + \int_0^\infty (1 - e^{-\theta t})h(t)\,dt\right\}^{-1}, \tag{8.3}$$

where $a \geqq 0$ *and* h *is a lower semicontinuous function which is either identically zero or satisfies the inequalities* (7.11).

Because of the way the theorem has been proved, one can in fact say rather more, for it shows that any function in $\mathscr{PM}$ can be realized in a special sort of chain, a *bouquet of escalators*. If a person moves through the state space according to such a chain, his wanderings can be described as follows. Starting at the distinguished state i, he is presented with a choice of escalators. He chooses one, ascends it, and on reaching the top after an infinite number of jumps returns at once to i, where he again has the opportunity to choose. Notice that in this chain all the states, except perhaps i, are stable.

Now suppose the chain is modified by providing each state except i with a cinema showing a film of a totally instantaneous chain (independent for each state, and for each visit to a state) which the person is required to watch while waiting in the state. If the specification is enlarged to include the state of the film, the result is another chain in which every state, except perhaps i, is instantaneous.

It follows that the function $p_{i,i}$ contains no information about the stability or otherwise of the other states of the chain. This is in line with the experience of authors who have constructed pathological Markov chains (for example, [11]), who have found that extreme sample function irregularity is consistent with simple functional forms for the individual transition probabilities.

9. The nondiagonal problem

Theorem 8.2 is the complete continuous time analogue of the first half of Theorem 1.1, and it therefore remains to find the analogue of the second, non-diagonal, part. The key to this problem lies in equations (6.18) and ((6.19), which combine to give

$$p_{i,j} = {}_j p_{i,i} * d\mu_{i,j} * p_{j,j}. \tag{9.1}$$

In this decomposition, the first and third members clearly belong to $\mathscr{PM}$. The totally finite measure $\mu_{i,j}$ has, as noted in Section 6, a probabilistic interpretation as a multiple of the distribution of $(\tau - \sigma)$, where (if $X_0 = i$),

$$\begin{aligned} \tau &= \inf \{t; X_t = j\}, \\ \sigma &= \sup \{t < \tau; X_t = i\}. \end{aligned} \tag{9.2}$$

From this, it is not difficult to show [17] that, apart from a possible atom $b = \mu_{i,j}\{0\}$, μ is a Markov measure.

It is shown in [17] that these conditions are sufficient as well as necessary, in the sense of the following theorem.

THEOREM 9.1 [17]. *A function $q(t)$, $t > 0$, can be expressed in the form $q(t) = p_{i,j}(t)$, $i \neq j$, in some Markov chain if and only if it can be written*

$$q = bp_1 * p_2 + p_1 * h * p_2, \tag{9.3}$$

where b is a nonnegative constant, p_1 and p_2 belong to $\mathscr{PM}$, h is a lower semi-continuous function, either identically zero or satisfying (7.11), *and*

$$\left\{ b + \int_0^\infty h(t)\, dt \right\} \int_0^\infty p_1(t)\, dt \leqq 1. \tag{9.4}$$

As in the discrete time case, there appears to be no canonical form of the decomposition (9.3), and no effective way of deciding, of a given function q, whether it is expressible in the form (9.3). The effective description of the nondiagonal Markov transition probabilities therefore remains an open problem, presumably more difficult than the corresponding discrete time problem.

If this problem can be solved, it will probably be easy to characterize functions of the form

$$p_{i,A}(t) = \sum_{j \in A} p_{i,j}(t), \qquad A \subseteq S, \tag{9.5}$$

and to solve the corresponding problem [19] for purely discontinuous Markov processes on uncountable state spaces.

10. Multiplicative properties of p-functions

If p_1 and p_2 are p-functions, we can construct corresponding regenerative phenomena Z^1 and Z^2 on distinct probability spaces Ω_1, Ω_2. Then the process Z defined on the product space $\Omega_1 \times \Omega_2$, with the product probability measure, by

$$Z_t(\omega_1, \omega_2) = Z_t^1(\omega_1) Z_t^2(\omega_2), \tag{10.1}$$

is a regenerative phenomenon with p-function

$$p(t) = p_1(t) p_2(t). \tag{10.2}$$

Thus, the product of two p-functions is itself a p-function.

If p_1 and p_2 are standard, then so is $p = p_1 p_2$, so that $\mathscr{P}$ is a commutative Hausdorff topological semigroup, with identity e given by $e(t) \equiv 1$. The arithmetical properties of this semigroup have been extensively studied by Kendall [10] and Davidson [5]. In particular, Kendall observed that there is a strong resemblance between $\mathscr{P}$ and the convolution semigroup $\mathscr{W}$ of probability measures on the line, an observation which is the starting point of the theory of *delphic semigroups*. This theory, like the classical theory of $\mathscr{W}$, leans heavily on the concept of an *infinitely divisible* element of the semigroup, one which can be expressed in the form

$$p = (p_n)^n, \tag{10.3}$$

for every integer $n \geqq 2$, where p_n belongs to the semigroup.

It was observed in [16] that $\mathscr{P}$ contains every continuous function p with $0 < p(t) \leqq p(0) = 1$ such that

$$\varphi(t) = -\log p(t) \tag{10.4}$$

is concave. Such p-functions are clearly infinitely divisible, for we may take

$$p_n(t) = \exp\{-n^{-1}\varphi(t)\}. \tag{10.5}$$

Kendall showed that they are the only infinitely divisible elements of $\mathscr{P}$.

It is very natural to ask which of the infinitely divisible p-functions belong to $\mathscr{P}\mathscr{M}$. In principle this question is answered by Theorem 8.2, but this is difficult to apply since the measure μ depends in a very complicated way on φ (see [22]).

A more useful answer is given by the following theorem (of which the regrettably complex proof will be published elsewhere).

THEOREM 10.1. *Let p be an infinitely divisible element of $\mathscr{P}$, and write the concave function φ in the form*

$$\varphi(t) = \int_{(0,\infty]} \min(t, x)\lambda(dx), \tag{10.6}$$

λ being a measure on $(0, \infty]$ with

$$\int_{(0,\infty]} \min(1, x)\lambda(dx) < \infty. \tag{10.7}$$

Then p belongs to $\mathscr{PM}$ if λ has a lower semicontinuous density in $(0, \infty)$, and $\lambda(0, \varepsilon) > 0$ for all $\varepsilon > 0$ (unless $\lambda(0, \infty) = 0$).

A special class of infinitely divisible p-functions is the class of completely monotonic functions, that is, those expressible in the form

$$p(t) = \int_{[0,\infty)} e^{-tx}\nu(dx) \tag{10.8}$$

for probability measures ν on $[0, \infty)$. These all belong to $\mathscr{PM}$; indeed, it was proved in [18] that they are exactly the diagonal transition probabilities of *reversible* chains. The problem of characterizing the nondiagonal transition functions of reversible chains remains open.

There are a number of other open problems in the multiplicative theory of $\mathscr{P}$, often motivated by the corresponding problems for the classical semigroup $\mathscr{W}$. To take just one example, Davidson has conjectured that, if p is any infinitely divisible element of $\mathscr{P}$ which is not of the form $p(t) = e^{-qt}$, then p has a factor which is not infinitely divisible.

11. Inequalities for p-functions

Freedman has asked the following question. If in a Markov chain, for some $i \in S$, $t > 0$, $c > 0$, it is known that

$$p_{i,i}(t) \geqq c, \tag{11.1}$$

can one give a lower bound for $p_{i,i}(s)$ $(s < t)$? He has given a partial answer to this question (in joint work with Blackwell [1]; the result was independently discovered by Davidson [5]) in the form of the inequality

$$p_{i,i}(s) \geqq \tfrac{1}{2} + (c - \tfrac{3}{4})^{1/2}, \tag{11.2}$$

as long as $c > \frac{3}{4}$.

The elegant proof of this inequality uses only the right inequality of (2.7); it is therefore true of all p in $\mathscr{P}$ that, if $c > \frac{3}{4}$, then

$$p(t) \geqq c \text{ implies } p(s) \geqq \tfrac{1}{2} + (c - \tfrac{3}{4})^{1/2} \tag{11.3}$$

for $s < t$. The result can indeed be extended, since $\mathscr{P}$ is not closed in the product space Π, to all functions in its closure $\bar{\mathscr{P}}$. Now the elements of $\bar{\mathscr{P}}$ are p-functions (since the set of p-functions is closed in Π) and those in $\bar{\mathscr{P}} - \mathscr{P}$ might be called *semistandard*. The p-functions which are not standard have been studied in [20], where the following rather deep theorem is proved.

THEOREM 11.1 [20]. *Every p-function satisfies one and only one of the following four conditions*:

(i) *p is standard*;

(ii) *there exists a constant $a \in (0, 1)$ and a standard p-function $\bar{p}$ such that* $p(t) = a\bar{p}(t)$;

(iii) $p(t) = 0$ *for almost all t*;

(iv) *p is not Lebesgue measurable.*

The functions of type (ii) are easy to handle, and are all semistandard since

$$a\bar{p}(t) = \lim_{n\to\infty} \exp\{-\min[nt, -\log a]\}\bar{p}(t). \tag{11.4}$$

Those of type (iv) are sufficiently pathological to ignore, though it is conceivable that some may perhaps be semistandard. The functions of type (iii) require however more attention; they include for instance the functions of the form

$$p(t) = \begin{cases} u_t & t \text{ integral}, \\ 0 & \text{otherwise}, \end{cases} \tag{11.5}$$

where (u_n) is any renewal sequence.

If p is any semistandard p-function of type (iii), then $p(t) \leqq \frac{3}{4}$ for all $t > 0$, since otherwise (11.3) would imply that $p(s) > 0$ for all $s \in (0, t)$. Hence,

$$\gamma = \sup\{p(t); t > 0, p \text{ semistandard of type (iii)}\} \tag{11.6}$$

satisfies

$$\gamma \leqq \tfrac{3}{4}. \tag{11.7}$$

To obtain an inequality in the other direction, consider the standard p-function (3.14), set $a = 1 - b/q$, and let $q \to \infty$ for fixed b. The result is the semistandard p-function of the form (11.5), with

$$u_n = \pi_n(nb). \tag{11.8}$$

This is greatest when $n = b = 1$, when it has the value e^{-1}, so that

$$\gamma \geqq e^{-1}. \tag{11.9}$$

All the evidence suggests that in fact $\gamma = e^{-1}$, but no proof is known. Any proof would almost certainly imply a substantial improvement on the Davidson-

Blackwell-Freedman inequality (11.2), and might even describe the subsets of the plane defined by

$$\Gamma_1 = \{(p(s), p(t)); p \in \mathscr{P}\mathscr{M}\},$$
(11.10) $$\Gamma_2 = \{(p(s), p(t)); p \in \mathscr{P}\},$$
$$\Gamma_3 = \{(p(s), p(t)); p \in \overline{\mathscr{P}}\} = \overline{\Gamma}_1.$$

12. Approximate regenerative phenomena

It is appropriate to end on a tentative note. The idea of a regenerative phenomenon is intended to describe the situation in which a Markov process returns to its starting point for a set of time points of positive measure. This is natural when handling countable state spaces, and is sometimes relevant [19] in more general situations. But many Markov processes on continuous state spaces do not return to their starting point, or do so only on a set of time instants of zero measure. For these it is more natural to think in terms of return to small neighborhoods of the starting point. Thus, instead of a single process Z_t one has a family of processes Z_t^N corresponding to neighborhoods N, and a corresponding family of "approximate p-functions" p^N. Whether there is too little structure here for a valuable theory only time will tell, but any such development would, in effect, be an abstract version of the theory of local time for diffusion processes.

A very similar situation occurs in the boundary theory of Markov chains ([3], [4]). Here one is led to an equation very like (3.10), but missing the crucial "drift" term:

(12.1) $$\int e^{-\theta t}p(dt) - \left\{\int (1 - e^{-\theta x})\mu(dx)\right\}^{-1}.$$

It is no longer possible to assert that P is absolutely continuous, and the complexity of the resulting theory shows very clearly how much reliance is placed on the drift term θ in (3.10). For example, even the proper formulation of the Volterra equation which generalizes (3.20) requires analysis of formidable depth [12]. The problem is nevertheless mentioned here, in the hope that an approximate regenerative theory, of comparable scope to the exact one related in this paper, might one day be found (see the remarks of Professor Chung and Dr. Williams in the discussion of [16]).

REFERENCES

[1] D. Blackwell and D. Freedman, "On the local behavior of Markov transition probabilities," *Ann. Math. Statist.*, Vol. 39 (1968), pp. 2123–2127.

[2] K. L. Chung, *Markov Chains with Stationary Transition Probabilities*, Berlin and New York, Springer, 1967.

[3] ———, "On the boundary theory for Markov chains," *Acta Math.*, Vol. 110 (1963), pp. 19–77.

[4] ———, "On the boundary theory for Markov chains II," *Acta Math.*, Vol. 115 (1966), pp. 111–163.

[5] R. DAVIDSON, "Arithmetic and other properties of certain delphic semigroups," *Z. Wahrscheinlichkeitstheorie und Verw. Gebiete*, Vol. 10 (1968), pp. 120–172.

[6] W. FELLER, *An Introduction to Probability Theory and Its Applications*, Vol. 1, New York, Wiley, 1957.

[7] J. HOFFMAN-JORGENSEN, "Markov sets," *Preprint Series*, No. 10, Matematisk Institut, Aarhus, 1967–8.

[8] D. G. KENDALL, "Unitary dilations of one-parameter semigroups of Markov transition operators, and the corresponding integral representations for Markov processes with a countable infinity of states," *Proc. London Math. Soc.*, Vol. 9 (1959), pp. 417–431.

[9] ———, "Some recent developments in the theory of denumerable Markov processes," *Transactions of the Fourth Prague Conference on Information Theory, Statistical Decision Functions and Random Processes*, Prague, 1967, pp. 11–17.

[10] ———, "Delphic semi-groups, infinitely divisible regenerative phenomena, and the arithmetic of p-functions," *Z. Wahrscheinlichkeitstheorie und Verw. Gebiete*, Vol. 9 (1968), pp. 163–195.

[11] D. G. KENDALL and G. E. H. REUTER, "Some pathological Markov processes with a denumerable infinity of states and the associated semi-groups of operators on 1," *Proceedings of the International Congress of Mathematicians* (1954, Amsterdam), Vol. 3, 1956, pp. 377–415.

[12] H. KESTEN, "Hitting probabilities of single points for processes with stationary independent increments," *Mem. Amer. Math. Soc.*, No. 93 (1969), pp. 1–129.

[13] J. F. C. KINGMAN, "The stochastic theory of regenerative events," *Z. Wahrscheinlichkeitstheorie und Verw. Gebiete*, Vol. 2 (1964), pp. 180–224.

[14] ———, "Linked systems of regenerative events," *Proc. London Math. Soc.*, Vol. 15 (1965), pp. 125–150.

[15] ———, "Some further analytical results in the theory of regenerative events," *J. Math. Anal. Appl.*, Vol. 11 (1965), pp. 422–433.

[16] ———, "An approach to the study of Markov processes," *J. Roy. Statist. Soc., Ser. B*, Vol. 28 (1966), pp. 417–447.

[17] ———, "Markov transition probabilities I," *Z. Wahrscheinlichkeitstheorie und Verw. Gebiete*, Vol 6 (1967), pp. 248–270.

[18] ———, "Markov transition probabilities II; completely monotonic functions," *Z. Wahrscheinlichkeitstheorie und Verw. Gebiete*, Vol. 9 (1967), pp. 1 9.

[19] ———, "Markov transition probabilities III; general state spaces," *Z. Wahrscheinlichkeitstheorie und Verw. Gebiete*, Vol. 10 (1968), pp. 87–101.

[20] ———, "On measurable p-functions," *Z. Wahrscheinlichkeitstheorie und Verw. Gebiete*, Vol. 11 (1968), pp. 1–8.

[21] ———, "Markov transition probabilities IV; recurrence time distributions," *Z. Wahrscheinlichkeitstheorie und Verw. Gebiete*, Vol. 11 (1968), pp. 9–17.

[22] ———, "An application of the theory of regenerative phenomena," *Proc. Camb. Phil. Soc.*, Vol. 68 (1970), pp. 697–701.

[23] ———, "Markov transition probabilities V," *Z. Wahrscheinlichkeitstheorie und Verw. Gebiete*, Vol. 17 (1971), pp. 89–103.

[24] ———, "A class of positive-definite functions," *Problems in Analysis (A Symposium in Honor of Solomon Bochner)* (edited by R. C. Gunning), Princeton, 1970, pp. 93–110.

[25] P. LÉVY, "Systèmes markoviens et stationnaires. Cas dénombrable," *Ann. Sci. Ecole Norm. Sup.*, Vol. 68 (1951), pp. 327–381.

[26] A. A. YUSKEVITCH, "On differentiability of transition functions of homogeneous Markov processes with a countable number of states," *Učen. Zap. Moskov. Gos. Univ.*, Vol. 186, Mat. 9 (1959), pp. 141–160.

STOCHASTIC DIFFERENTIAL EQUATIONS AND MODELS OF RANDOM PROCESSES

E. J. McSHANE
UNIVERSITY OF VIRGINIA

1. Description of a desirable model

Let us suppose that we are investigating a system whose state can be adequately specified by n real numbers $x^1, \cdots, x^n$. We shall suppose that by some acceptable scientific theory it is predicted that, in the absence of disturbances from outside the system, the x^i develop in time in accordance with certain differential equations,

$$\dot{x}^i = g_0^i(t, x), \qquad i = 1, \cdots, n. \tag{1.1}$$

If there are disturbances or noises, $n^1(t), \cdots, n^r(t)$, the underlying theory of such systems will often permit us to conclude that

$$\dot{x}^i = g_0^i(t, x) + \sum_{\rho=1}^{r} g_\rho^i(t, x) n^\rho(t), \qquad i = 1, \cdots, n, \tag{1.2}$$

where g_ρ^i is the sensitivity of the ith coordinate to the ρth noise. However in the underlying theory, equation (1.2) will usually have a limited domain of applicability; in particular, we could not usually retain confidence in the trustworthiness of (1.2) if the noise were unbounded. But for sufficiently well-behaved bounded noises we can rewrite (1.2) in the form

$$dx^i = g_0^i(t, x)\, dt + \sum_\rho g_\rho^i(t, x)\, dz^\rho, \tag{1.3}$$

or

$$x^i(t) = x_0^i + \int_a^t g_0^i[s, x(s)]\, ds + \sum_\rho \int_a^t g_\rho^i[s, x(s)]\, dz^\rho(s), \tag{1.4}$$

where

$$z^\rho(t) = z^\rho(a) + \int_a^t n^\rho(s)\, ds; \tag{1.5}$$

Research supported in part by the U.S. Army Research Office under Grant ARO-DG-1005.

with bounded n^ρ or Lipschitzian z^ρ, these are solvable by traditional methods, and (perhaps with still stronger requirements on the z^ρ) will describe the evolution of the system with as much certainty as the underlying scientific theory of such systems permits.

Usually however, we are interested, not in the response of the system to specified noises z^ρ, but in statistical properties of the responses of the system to random noises. As is well known, this causes a dilemma. The processes z^ρ most amenable to probabilistic study are martingales, especially the Wiener process and closely related processes. But these have almost surely non-Lipschitzian sample functions and lie outside the domain of applicability of the scientific theory that led to (1.4). The integrals with respect to z^ρ in (1.4) cannot even be interpreted as Riemann-Stieltjes or Lebesgue-Stieltjes integrals. Interpreting them as Itô integrals restores meaning to all terms in (1.4), but gives no ground for confidence that the solution (1.4) will continue to represent the time development of the system. It is a familiar fact that the uncritical use of (1.4) can lead to mismatches between system and model that are often considered paradoxical.

E. Wong and M. Zakai have made a major contribution [8], [9] to the removal of these "paradoxes." Suppose that we are studying a system which, for Lipschitzian disturbances z^ρ, is governed by (1.4) with $n = r = 1$. For notational simplicity we omit the superscripts on x^0, z^ρ, g^i_ρ, and so forth. Let z be a Brownian motion process on an interval $[a, b]$. Let π be a finite set of numbers $t_1, \cdots, t_{k+1}$ with

$$(1.6)\qquad a = t_1 < t_2 < \cdots < t_{k+1} = b,$$

and let Z be the process whose sample paths coincide with those of z at the t_j and are linear between them. Then the solutions X of (1.4) with Z in place of z, that is, the solutions of the ordinary equations

$$(1.7)\qquad X(t) = x_0 + \int_a^t g_0[s, X(s)] + \int_a^t g_1[s, X(s)]\, dZ(s),$$

are random variables; and as the mesh of π (that is, $\max\,[t_{j+1} - t_j]$) tends to 0, the X converge in quadratic mean to a limit x. But this limit is not the solution of (1.4), but of

$$(1.8)\qquad x(t) = x_0 + \int_a^t g_0[s, x(s)]\, ds + \int_a^t g_1[s, x(s)]\, dz(s) + \frac{1}{2}\int_a^t g_1[s, x(s)] g_{1,x}[s, x(s)]\, ds.$$

(Wong and Zakai have also established this for a more general class of disturbances than Brownian motion processes, see [9].)

These results of Wong and Zakai show us, at least in some important cases, how to model systems affected by noise. If for Lipschitzian disturbances the system evolves according to (1.4) (with subscripts and superscripts suppressed), then if the physically admissible Lipschitzian distances are idealized to Brownian motion processes equation (1.4) should be replaced by (1.8). If this is done, the solution of (1.8) will be close in quadratic mean to the solutions of (1.4) for at least some Lipschitzian disturbances with finite dimensional distributions close to those of the Brownian motion idealization.

Nevertheless, it is at least inconvenient, as well as aesthetically unsatisfying, to have different equations for different types of disturbances. It would be preferable to have a theory of integration that would apply both to processes with Lipschitzian sample functions, to Brownian motion and to other martingales that have so often proved useful; and correspondingly, it would be preferable to have a method of modeling systems that is consistent with the basic model (1.4) when the disturbances are Lipschitzian and gives "nearly" the same result when a Lipschitzian disturbance is replaced by a martingale type idealization that is in some reasonable sense "close" to it. More specifically, we shall seek to replace (1.4) by another set of so called differential equations (really integral equations) with the following desirable properties.

(a) *Inclusiveness.* The integrals in the equations should be defined for some recognizable class of processes z^ρ, large enough to include all processes with Lipschitzian sample paths and also to include all Brownian motion and such modifications of Brownian motion as have been useful in applications.

(b) *Consistency.* For Lipschitzian disturbances, the solutions of the equations should coincide with the solutions of the equations (1.4) that are given to us (for smooth disturbances) by the scientific theory of the system.

(c) *Stability.* This property is not easy to describe precisely. Suppose that we have introduced some sort of topology in the space of random processes, so that the convergence of a sequence of processes $z_1, z_2 \cdots$ to a limit process z is meaningful and is in principle experimentally verifiable, with the customary allowance for experimental error. Then, under unexcessive restrictions, if processes $(z_j^1, \cdots, z_j^r)$ converge to $(z^1, \cdots, z^r)$, the solutions $(x_j^1, \cdots, x_j^n)$ corresponding to the z_j^ρ should also converge to the solutions $(x^1, \cdots, x^n)$ corresponding to the limit $(z^1, \cdots, z^r)$. As a special case, if $n = r = 1$, the solution of the equation when z is Brownian motion should coincide with the solution of Wong-Zakai equation (1.8).

In order to develop such a theory, we must define, for a class of processes with the inclusiveness property (a) the types of integrals needed in the equations; we must develop a calculus for these integrals that will permit us to study differential equations; we must specify the differential equations of our model; we must show that these differential equations are solvable; and we must show that their solutions possess the consistency property (b) and the stability property (c). The remainder of this paper is an outline of the steps in this program.

During the Sixth Berkeley Symposium, I had the pleasure and profit of several conversations with Professor Eugene Wong. In particular, the present version of Theorem 9.1 owes its existence to his tactfully expressed dissatisfaction with an earlier version in which weaker conclusions were drawn from stronger hypotheses.

2. Definition of the integral

To avoid repetition, we henceforth suppose that $(\Omega, \mathscr{A}, P)$ is a probability triple, that T is a set of real numbers, that $[a, b]$ is a closed interval contained in T, and also that

$$(2.1) \qquad f = (f(\tau, \omega): \tau \in T, \omega \in \Omega), \qquad Z^k = (z^k(t, \omega): t \in [a, b], \omega \in \Omega),$$

$k = 1, \cdots, q$, are real stochastic processes on T and on $[a, b]$, respectively. By a *partition* of $[a, b]$ (with evaluation points in T) we shall mean a finite set

$$(2.2) \qquad \Pi = (t_1, \cdots, t_{\ell+1}; \tau_1, \cdots, \tau_\ell)$$

of real numbers such that

$$(2.3) \qquad a = t_1 \leqq t_2 \leqq \cdots \leqq t_{\ell+1} = b$$

and $\tau_i \in T$, $i = 1, \cdots, \ell$. The t_i are called the division points of Π, and the τ_i the partition points of Π. (We usually omit the words "with evaluation points in T.")

Apart from notation, the partitions Π with $\tau_i = t_i$ were used one hundred and fifty years ago by Cauchy to define the integral of a continuous function; so we shall call them *Cauchy partitions*. Partitions with τ_i in $[t_i, t_{i+1}]$ for each i were used by Riemann, and we shall call them *Riemann partitions*. But for use with stochastic processes it proves highly advantageous to use partitions such that $t_i \geqq \tau_i$, $i = 1, \cdots, \ell$, and these we shall call *belated partitions*.

If Π is a Cauchy, or Riemann, or belated partition, with notation (2.2), we define

$$(2.4) \qquad \text{mesh } \Pi = \max \{t_{j+1} - \min \{t_j, \tau_j\} : j = 1, \cdots, \ell\}.$$

Corresponding to the processes (2.1) and the partition (2.2), we define the *Riemann sum* $S(\Pi; f, z^1, \cdots, z^q)$ to be the random variable (r.v.) whose value at ω (in Ω) is given by

$$(2.5) \qquad S(\Pi; f, z^1, \cdots, z^q)(\omega) = \sum_{i=1}^{\ell} \left\{ f(\tau_i; \omega) \prod_{k=1}^{g} [z^k(t_{i+1}\omega) - z^k(t_i, \omega)] \right\}.$$

We can now define the family of integrals that we shall use in our models.

DEFINITION 2.1. *The process f has a* belated integral *with respect to $(z^1, \cdots, z^q)$ over $[a, b]$ if, Π being restricted to the class of belated partitions of $[a, b]$, there is an r. v. J such that $S(\Pi; f, z^1, \cdots, z^q)$ converges in probability to J as mesh Π tends to 0. Every such J is called a* weak version *of the integral, and is denoted (possibly ambiguously) by*

$$(2.6) \qquad (w) \int_a^b f(t, \omega)\, dz^1(t, \omega), \cdots, dz^q(t, \omega).$$

Such a J is a strong version *of the integral, and is denoted by*

$$(2.7) \qquad \int_a^b f(t, \omega)\, dz^1(t, \omega), \cdots, dz^q(t, \omega),$$

if for each ω_0 *in* Ω *such that the limit (with notation* (2.2))

$$(2.8) \qquad \ell(\omega_0) = \lim_{\text{mesh }\Pi\to 0} f(t_i, \omega_0) \prod_{k=1}^{g} [z^k(t_{i+1}, \omega_0) - z^k(t_i, \omega_0)]$$

exists it is true that $J(\omega_0) = \ell(\omega_0)$.

As usual, we omit the ω when convenient. It is quite easy to show that if f is integrable with respect to $(z^1, \cdots, z^q)$, a strong version of the integral exists.

3. The stochastic model

If the sample functions of f are bounded and those of the z^k are Lipschitzian, there is no difficulty in proving that if $q > 1$, then

$$(3.1) \qquad \int_a^b f(t)\, dz^1(t) \cdots dz^q(t) = 0.$$

Suppose then that the functions f^i and g^i_ρ and the derivatives of the latter with respect to the x^i are continuous. By (3.1), if the sample functions x^i all satisfy (1.4) and the functions

$$(3.2) \qquad g^i_{\rho,\sigma}(x, t), \qquad i = 1, \cdots, n;\ \rho, \sigma = 1, \cdots, r;\ t \in T,\ x \in R^n$$

are continuous, then the integrals

$$(3.3) \qquad \sum_{\rho,\sigma} \int_a^t g^i_{\rho,\sigma}[x(s), s]\, dz^\rho(s)\, dz^\sigma(s)$$

exist and are zero for all i in $\{1, \cdots, n\}$ and t in T. Hence, the x^i also satisfy

$$(3.4) \qquad x^i(t) = x^i(a) + \int_a^t g^i_0[x(s), s]\, ds + \sum_\rho \int_a^t g^i_\rho[x(s), s]\, dz^\rho(s) + \tfrac{1}{2} \sum_{\rho,\sigma} \int_a^t g^i_{\rho,\sigma}[x(s), s]\, dz^\rho(s)\, dz^\sigma(s),$$

$i = 1, \cdots, n;\ a \leqq t \leqq b$, the integrals either being computed for each sample curve or understood as strict versions of belated integrals. No matter how we choose the (continuous) functions (3.2), we obtain the consistency property (b).

But soon we shall show that the belated integrals can be defined for a class of processes large enough to possess the inclusiveness property (a). When this larger class of z^ρ is permitted, the integrals in (3.3) no longer all vanish, and the stability property (c) does not hold for all choices of functions (3.2). In fact, it is far from clear that it will hold for *any* such functions. We make the choice

$$(3.5) \qquad g^i_{\rho,\sigma}(x, t) = \tfrac{1}{2} \sum_{j=1}^{n} g^i_{\rho,x^j}(x, t) g^j_\sigma(x, t), \qquad i = 1, \cdots, n.$$

This is our selection principle. We do *not* consider that we have added a correction term (3.5) to the "standard" equation (1.4). Rather, from the aggregate of all equations (3.4), we have selected the one specified by (3.5) instead of the simplest looking one with all functions (3.2) equal to 0. All the equations (3.4) are equally in accord with the underlying theory that gave us equations (1.4), assuming as before that this theory has been established only for Lipschitzian z^ρ. But setting the functions (3.2) equal to 0 gives us merely typographical simplicity, while (as we shall ultimately show) the choice (3.5) gives us, at least under some restrictions, the much more important virtue of stability.

4. Principal existence theorem for the belated integral

Throughout this paper, note T will denote a set of real numbers and $[a, b]$ a closed interval contained in T. Moreover, the symbol $F_\tau (\tau \in T)$ will always denote a σ-subalgebra of $\mathscr{A}$, and we shall always assume *if τ and σ are in T and $\sigma \leqq \tau$, then $F_\sigma \subseteqq F_\tau$.*

For the sake of brevity, if x is a process defined on some subset D_x of T, we shall use the expression "x is F. measurable" to mean "for each t in D_x, $(x(t, \cdot)$ is F_t measurable." Furthermore, to avoid complicated typography we use $F(\tau)$ to denote F_τ whenever convenient; in particular, we write $F(t_j)$ instead of writing the t_j as a subscript to the F.

The processes $(z^1, \cdots, z^r)$ that play the principal role in our theory are those processes on $[a, b]$ that satisfy the following conditions.

CONDITION 4.1. *Each $z^\rho (\rho = 1, \cdots, r)$ is F. measurable, and there exist positive numbers K and δ and a positive integer q such that if $\rho \in \{1, \cdots, r\}$ and $a \leqq s \leqq t \leqq b$ and $t - s < \delta$, then a.s.*

$$\begin{aligned} &|E([z^\rho(t) - z^\rho(s)] \mid F_s)| \leq K(t - s), \\ &E([z^\rho(t) - z^\rho(s)]^{2k} \mid F_s) \leqq K(t - s), \qquad k = 1, \cdots, q. \end{aligned} \tag{4.1}$$

If x is a vector in R^n, we define $|x| = [\Sigma_1^n (x^i)^2]^{1/2}$; if $(x(\omega); \omega \in \Omega)$ is an n vector valued r.v., we define $\|x\| = E(|x|^2)^{1/2}$ whenever this expectation exists.

At this stage we observe that the existence of the integral in Definition 2.1 can be proved, for $q = 1$, under much weaker hypotheses than Condition 4.1; for $q \geqq 2$ considerable weakening is also possible, though not as much as for $q = 1$. But the gain in generality is bought at a high price in simplicity. With Condition 4.1, we already have as much inclusiveness as was asked for in (a) and Definition 2.1 is only slightly more complicated than the standard definition of the Riemann integral. Greater inclusiveness would require introducing concepts and methods too sophisticated for some potential users, and would not justify its cost.

The next lemma is an essential element of several later proofs. Its proof differs only trivially from that of Lemma 1 in [5].

LEMMA 4.1. *Let $F_1, \cdots, F_m$ be σ-subalgebras of $\mathscr{A}$ with $F_1 \subseteqq F_2 \subseteqq \cdots \subseteqq F_m$. Let $u_1, \cdots, u_m$ and $\Delta_1, \cdots, \Delta_m$ be r.v. with finite second moments such that for* *each k in $\{1, \cdots, m\}$, all u_j with $j \leqq k$ and all Δ_j with $j < k$ are F_k measurable. Let C_j, D_j for $j = 1, \cdots, m$, be numbers such that a.s.*

$$|E(\Delta_j | F_j)| \leqq C_j, E(\Delta_j^2 | F_j) \leqq D_j. \tag{4.2}$$

Then

$$\left\| \sum_{j=1}^{m} u_j \Delta_j \right\| \leqq 2 \sum_{j=1}^{m} C_j \|u_j\| + \left\{ \sum_{j=1}^{m} D_j \|u_j\|^2 \right\}^{1/2}. \tag{4.3}$$

It is convenient to state a frequently used corollary.

COROLLARY 4.1. *Let Condition 4.1 be satisfied, and let Π (with notation (2.2)) be a partition of $[a, b]$. For each j in $\{1, \cdots, \ell\}$, let u_j be an $F[t_j]$ measurable r.v. with finite second moment. Then*

$$\left\| \sum_{j=1}^{\ell} u_j \prod_{k=1}^{q} [z^k(t_{j+1}) - z^k(t_j)] \right\| \leqq B \left\{ \sum_{j=1}^{\ell} \|u_j\|^2 (t_{j+1} - t_j) \right\}^{1/2}, \tag{4.4}$$

where $B = 2K(b - a)^{1/2} + K^{1/2}$.

PROOF. We define

$$\begin{aligned} \Delta_j &= \prod_{k=1}^{q} [2^k(t_{j+1}) - z^k(t_j)], \\ C_j &= D_j = K(t_{j+1} - t_j). \end{aligned} \tag{4.5}$$

The hypotheses of Lemma 4.1 are satisfied, so

$$\begin{aligned} \left\| \sum_{j=1}^{\ell} u_j \Delta_j \right\| \leqq 2K \sum_{j=1}^{\ell} \{\|u_j\| (t_{j+1} - t_j)\}^{1/2} \{(t_{j+1} - t_j)\}^{1/2} \\ + \left\{ \sum_{j=1}^{\ell} \|u_j\|^2 K(t_{j+1} - t_j) \right\}^{1/2}. \end{aligned} \tag{4.6}$$

Applying the Cauchy-Buniakowsky-Schwarz inequality to the first sum in the right member yields the desired conclusion.

We can now state and prove an existence theorem of particular importance in the rest of this paper. In this theorem the integrand is assumed to have the following rather strong continuity property:

(d) *f is bounded in L_2 norm on T, and is continuous in L_2 norm at almost all points of $[a, b]$.*

For such integrands we have the following theorem.

THEOREM 4.1. *Let $z^1, \cdots, z^q$ satisfy Condition 4.1. Let $(f(\tau) : \tau \in T)$ be F. measurable and satisfy* (d). *Then for every subinterval $[c, e]$ of $[a, b]$, f has a (belated stochastic) integral over $[c, e]$, and this integral has an F_e measurable version. Moreover, for every $\varepsilon > 0$ there is a δ' with $0 < \delta' < \delta$ such that, for all subintervals $[c, e]$ of $[a, b]$ and all belated partitions of $[c, e]$ with mesh $\Pi < \delta'$*

it is true that

$$\|S(\Pi; f, z^1, \cdots, z^q) - \int_c^e f(t)\, dz^1(t) \cdots dz^q(t)\| < \varepsilon. \tag{4.7}$$

PROOF. Consider the case in which f is continuous in L_2 norm on T. If Π' and Π'' are belated partitions

$$\begin{aligned} \Pi' &= (t_1, \cdots, t_{\ell+1}; \tau'_1, \cdots, \tau'_\ell), \\ \Pi'' &= (t_1, \cdots, t_{\ell+1}; \tau''_1, \cdots, \tau''_\ell), \end{aligned} \tag{4.8}$$

of $[c, e]$ with the same division points,

$$\begin{aligned} &S(\Pi'; f, z^1, \cdots, z^q) - S(\Pi''; f, z^1, \cdots, z^q) \\ &\qquad = \sum_{j=1}^{\ell} [f(\tau'_j) - f(\tau''_j)] \prod_{k=1}^{q} [z^k(t_{j+1}) - z^k(t_j)]. \end{aligned} \tag{4.9}$$

By Corollary 4.1,

$$\begin{aligned} &\|S(\Pi'; f, z^1, \cdots, z^q) - S(\Pi''; f, z^1, \cdots, z^q)\| \\ &\qquad \leqq B(e - c)^{1/2} \max_j \|f(\tau'_j) - f(\tau''_j)\|. \end{aligned} \tag{4.10}$$

For $q = 1$, the restriction to Π' and Π'' with the same division points is easily removed. Given two partitions

$$\begin{aligned} \Pi' &= (t'_1, \cdots, t'_{\gamma+1}; \tau'_1, \cdots, \tau'_\gamma), \\ \Pi'' &= (t''_1, \cdots, t''_{n+1}; \tau''_1, \cdots, \tau''_n) \end{aligned} \tag{4.11}$$

of $[c, e]$, we say that the latter is obtained from the former by *adjunction of division points* if each t'_i is one of the t''_j, and if $[t''_j, t''_{j+1}] \subseteqq [t'_i, t'_{i+1}]$ then $\tau''_j = \tau'_i$. In this case it is easily seen that

$$S(\Pi'; f, z^1) = S(\Pi''; f, z^1). \tag{4.12}$$

If Π' and Π'' are any two belated partitions of $[c, e]$, in computing their Riemann sums (for $q = 1$) there is thus no loss of generality in supposing that Π' and Π'' have the same division points, so

$$\|S(\Pi''; f, z^1) - S(\Pi''; f, z^1)\| \leqq B(e - c)^{1/2} \max_i \|f(\tau'_i) - f(\tau''_i)\|, \tag{4.13}$$

and this can be made arbitrarily small by restricting Π' and Π'' to have small mesh. So the Riemann sums converge in L_2 norm and mesh Π tends to 0, and Theorem 4.1 holds if $q = 1$ and f is continuous in L_2 norm. The latter restriction can be removed by much the same devices as in the case of the ordinary Riemann integral.

If $q > 1$ and Π'' is obtained from Π' by adjunction of division points, the analogue of (4.12) fails. For example, if $q = 2$ and each $[t_j, t_{j+1}]$ of Π' contains in its interior either a single division point of Π'' (which we then call s_i) or no such point (in which case we define s_i to be t_i), we readily calculate

$$(4.14)\qquad S(\Pi', f, z^1, z^2) - S(\Pi''; f, z^1, z^2)$$
$$= \sum_{j=1}^{\ell} f(\tau_j)\{[z^1(s_j) - z^1(t_j)][z^2(t_{j+1}) - z^2(s_j)]$$
$$+ [z^2(s_j) - z^2(t_j)][z^1(t_{j+1}) - z^1(s_j)]\},$$

which is not in general 0. However, we can find a useful estimate of its norm. Define $\mu(\Pi') = \max_j[t_{j+1} - t_j]$. Since we have assumed Condition 4.1 holds, we find

$$(4.15)\qquad |E([z^1(s_j) - z^1(t_j)][z^2(t_{j+1}) - z^2(s_j)]\,|\,F(t_j))|$$
$$= |E(z^1(s_j) - z^1(t_j)E([z^2(t_{j+1}) - z^2(s_j)]\,|\,F(s_j))\,|\,F(t_j))|$$
$$\leqq K(t_{j+1} - s_j)E(|z^1(s_j) - z^1(t_j)|\,|\,F(t_j))$$
$$\leqq K(t_{j+1} - s_j)[E([z^1(s_j) - z^1(t_j)]^2\,|\,F(t_j))]^{1/2}$$
$$\leqq K^{3/2}\mu(\Pi')^{1/2}(t_{j+1} - t_j).$$

The same estimate holds with z^1 and z^2 interchanged. Similarly,

$$(4.16)\qquad E([z^1(s_j) - z^1(t_j)]^2[z^2(t_{j+1}) - z^2(s_j)]^2\,|\,F(t_j))$$
$$\leqq K^2\mu(\Pi')(t_{j+1} - t_j),$$

and likewise with z^1 and z^2 interchanged. We now apply Lemma 4.1 to each of the two sums in (4.14); by (4.15) and (4.16), we obtain

$$(4.17)\qquad \|S(\Pi'; f, z^1, z^2) - S(\Pi''; f, z^1, z^2)\| \leqq C[\mu(\Pi')]^{1/2},$$

where

$$(4.18)\qquad C = 2K \sup\{\|f(\tau)\| : \tau \in T\}[2K^{1/2}(b - a) + (b - a)^{1/2}].$$

If $q > 2$, the estimate (4.17) (with a different c) is still valid, the proof is not essentially different but the details are more tedious.

We shall repeatedly use the following procedure.

PROCEDURE 4.1. *Given a partition* $\Pi = (t_1, \cdots, t_{\ell+1}; \tau_1, \cdots, \tau_\ell)$, *we adjoin to* Π *as new division points the midpoints of all those intervals* $[t_j, t_{j+1}]$ *such that* $t_{j+1} - t_j \geqq \frac{1}{2}\mu(\Pi)$.

We form a sequence of partitions

$$(4.19)\qquad \Pi'_0 = \Pi', \Pi'_1, \Pi'_2, \cdots, \Pi'_\alpha,$$

each formed from the preceding by applying Procedure 4.1. We carry it to a large enough α so that no interval of the original Π' remains unsubdivided; then each interval of Π'_α will have length at least $\frac{1}{2}\mu(\Pi'_\alpha)$. Since

$$(4.20)\qquad \mu(\Pi'_k) = 2^{-k}\mu(\Pi'),$$

we may also suppose that $\mu(\Pi'_\alpha)$ is less than half the length of the smallest interval in Π''. Next, starting with Π'', we form the sequence

$$(4.21) \qquad \Pi''_0 = \Pi'', \Pi''_1, \Pi''_2, \cdots, \Pi''_\beta$$

by repeated application of Procedure 4.1. We can and do choose β so that

$$(4.22) \qquad 2^{-1/2}\mu(\Pi'_\alpha) \leqq \mu(\Pi''_\beta) \leqq 2^{1/2}\mu(\Pi'_\alpha);$$

this is possible by the analogue of (4.20). By (4.17) and (4.20),

$$(4.23) \qquad \begin{aligned} &\|S(\Pi'_\alpha; f, z^1, z^2) - S(\Pi'; f, z^1, z^2)\| \\ &\quad \leqq \sum_{n=0}^{\alpha-1} C[\mu(\Pi'_h)]^{1/2} \\ &\quad \leqq (2 + 2^{1/2})C[\mu(\Pi')]^{1/2}. \end{aligned}$$

Similarly,

$$(4.24) \qquad \begin{aligned} &\|S(\Pi''_\beta; f, z^1, z^2) - S(\Pi''; f, z^1, z^2)\| \\ &\quad \leqq (2 + 2^{1/2})C[\mu(\Pi'')]^{1/2}. \end{aligned}$$

Every interval in Π'_α has length at least $\frac{1}{2}\mu(\Pi'_\alpha)$, and likewise for Π''_β. So every interval in Π'_α has length at least $\mu(\Pi''_\beta)/2^{3/2}$, and vice versa. Thus, each interval of Π'_α contains at most three division points of Π''_β. We can adjoin these to Π'_α in two stages, obtaining a partition Π'_* such that (by (4.17))

$$(4.25) \qquad \begin{aligned} &\|S(\Pi'_*, f, z^1, z^2) - S(\Pi'_\alpha; f, z^1, z^2)\| \\ &\quad \leqq 2C[\mu(\Pi')]^{1/2}. \end{aligned}$$

Similarly, we can adjoin the division points of Π'_ρ to Π''_β in at most two stages, obtaining a partition Π''_* such that

$$(4.26) \qquad \begin{aligned} &\|S(\Pi''_*; f, z^1, z^2) - S(\Pi''_\beta; f, z^1, z^2)\| \\ &\quad \leqq 2C[\mu(\Pi'')]^{1/2}. \end{aligned}$$

Now Π'_* and Π''_* have the same division points. By (4.10), (4.19), (4.20), (4.21), and (4.22), for L_2 continuous f, the Riemann sums for Π', for Π'_*, for Π''_* and for Π'' have differences (in the order named) that have L_2 norms which are arbitrarily small if mesh Π' and mesh Π'' are small. This implies that $S(\Pi; f, z^1, \cdots, z^q)$ converges in L_2 norm as mesh Π tends to 0, and the integral exists. The uniform closeness of Riemann sum to integral follows from the fact that all estimates of L_2 norms were uniformly valid; and we could have used F_e everywhere in place of $\mathscr{A}$ without changing anything, which would give us an F_e measurable integral.

5. An estimate, and a second existence theorem

Suppose that the hypotheses of Theorem 4.1 are satisfied, and that Π (with notation (2.2)) is a belated partition of a subinterval $[c, e]$ of $[a, b]$. By Corollary 4.1,

$$\|S(\Pi; f, z^1, \cdots, z^q)\| \leqq B\left\{\sum_{j=1}^{\ell} \|f(\tau_j)\|^2(t_{j+1} - t_j)\right\}^{1/2}. \tag{5.1}$$

But by the special case of Theorem 4.1 in which Ω contains a single point and $q = 1$ and $z(t) = t$, the belated integral of $\|f\|$ with respect to t exists. Since Cauchy partitions are both Riemann partitions and belated partitions, the belated integral of $\|f\|$ is its Riemann integral, and by letting mesh Π tend to 0 we obtain from (5.1)

$$\left\|\int_c^e f(t)\, dz^1(t) \cdots dz^q(t)\right\| \leqq B\left\{\int_c^e \|f(t)\|^2\, dt\right\}^{1/2}. \tag{5.2}$$

If T is an interval, and f is $F.$ measurable and $(f(\tau, \omega)\colon \omega \in t, \omega \in \Omega)$ is $dt\, dP$ measurable on $T \times \Omega$, and

$$\int_T E[f(\tau)^2]\, d\tau < \infty, \tag{5.3}$$

it is possible to find (as in [1], p. 440) a sequence of bounded processes $f_1, f_2, \cdots$, satisfying the hypotheses of Theorem 4.1 such that

$$\lim_{n \to \infty} \int_a^b E(|f_n - f|^2)\, d\tau = 0; \tag{5.4}$$

in fact, by a slight modification of the construction in [1] we may choose f_n that are continuous in L_2 norm. Then by (5.2) the integrals

$$\int_c^e f_n(t)\, dz^1(\tau) \cdots dz^q(\tau), \qquad n = 1, 2, 3, \cdots, \tag{5.5}$$

form a Cauchy sequence in $L_2(\Omega, P)$, and hence have a limit in that space. We can accept this limit as the definition of the integral of f with respect to $(z^1, \cdots, z^q)$, thus extending the class of integrable functions so as to have the same sort of closure properties as the Itô integral. Such properties are valuable in many investigations. But in this paper we have no need of them, so we pursue this no farther.

In Definition 2.1, we used the concept of convergence in probability. In Theorem 4.1, we obtained more: the Riemann sums converged in L_2 norm. There is an intermediate kind of convergence that is sometimes encountered, that we shall call *uniform convergence in near L_2 norm*. We define it in the setting of functions of partitions, although it evidently can be applied to more general limit processes. (In the definition, 1_A denotes the indicator function of the set A.)

DEFINITION 5.1. *Assume that to each subinterval* $[c, e]$ *of* $[a, b]$ *and to each belated partition* Π *of* $[c, e]$ *there corresponds an r.v.* $x(\Pi, [c, e])$ *and an r.v.* $x_0([c, e])$. *Then* $x(\Pi, [c, e])$ converges to $x_0([c, e])$ in near L_2 norm, uniformly on subintervals $[c, e]$ of $[a, b]$, *if to each positive* ε *there corresponds a positive* δ *and a subset* A *of* Ω *with* $P(A) > 1 - \varepsilon$ *such that for all* $[c, e] \subseteqq [a, b]$ *and all* Π *with mesh* $\Pi < \delta$,

$$\|1_A[x(\Pi, [c, e]) - x_0([c, e])]\| < \varepsilon. \tag{5.6}$$

Clearly this implies uniform convergence in the metric of convergence in probability, and is implied by uniform convergence in L_2 norm.

Many processes f possess the following important property:

(e) f *is separable, and with probability* 1 *the sample function* $[f(\tau, \omega): \tau \in T]$ *is bounded*. (Note that the bound is not assumed to be independent of ω.)

For integrands f with this property, we can prove the following theorem.

THEOREM 5.1. *Let Condition* 4.1 *be satisfied. Let* $[f(\tau): \tau \in T]$ *satisfy* (e), *and be* $F_{\cdot}$ *measurable, and be continuous in probability at almost all points of* $[a, b]$. *Then for every subinterval* $[c, e]$ *of* $[a, b]$, f *has a belated integral with respect to* $(z^1, \cdots, z^q)$ *over* $[c, e]$, *and this integral has an* F_e *measurable version. Moreover, the Riemann sums* $S(\Pi; f, z^1, \cdots, z^q)$ *corresponding to belated partitions* Π *of* $[c, e]$ *converge to the integral over* $[c, e]$ *uniformly in near* L_2 *norm as mesh* $\Pi \to 0$.

PROOF. Let $S \subseteqq T$ be a separate set for f. There is a subset Λ of Ω with $P\Lambda = 0$ such that for every open interval I and every ω in $\Omega - \Lambda$, the functions $[f(\tau, \omega): \tau \in I \cap T]$ and $[f(\tau, \omega): \tau \in I \cap S]$ have equal suprema and equal infima. Let ε be positive. For each positive N we define $A_N(\tau)$ $(\tau \in T)$ to be the set of all ω in Ω such that $|f(s, \omega)| \leqq N$ *for* $s = \tau$ *and for all* $s < \tau$ *in* S. This is F_τ measurable, and $P[A_N(\tau)]$ is nonincreasing. By (e), we can choose N large enough so that

$$P[A_N(b)] > 1 - \varepsilon. \tag{5.7}$$

Let $\phi_N(\tau, \cdot)$ be the indicator function of $A_N(\tau)$. Then by definition of A_N, $f\phi_N$ is bounded. It is also fairly obviously F_τ measurable at each τ in T, and it is continuous in probability (hence, being bounded, it is continuous in L_2 norm) except on the union of the null set of discontinuities of f and the countable set of discontinuities of $P(A_N(\tau))$. So, by Theorem 4.1 for every $[c, e] \leqq [a, b]$ the Riemann sums

$$S(\Pi; f\phi_N, z^1, \cdots, z^q) \tag{5.8}$$

converge as mesh $\Pi \to 0$ to the integral of $f\phi_N$ over $[c, e]$, uniformly with respect to $[c, e]$. But the sums (5.8) coincide on $A_N(b) - \Lambda$ with

$$S(\Pi; f, z^1, \cdots, z^2). \tag{5.9}$$

From this and (5.7), it follows readily that the sums (5.9) converge in near L_2 norm to a limit, which is by definition the integral of f over $[c, e]$; and the convergence is uniform with respect to $[c, e]$.

6. Examples

Suppose first that z^1 and z^q are both the same Wiener process w. Then if $a \leqq s \leqq t \leqq b$, $w(t) - w(s)$ is independent of F_s. Let Π (with the usual notation (2.2)) be a belated partition, and define

$$(6.1) \qquad \Delta_j = [w(t_{j+1}) - w(t_j)]^2 - (t_{j+1} - t_j), \qquad j = 1, \cdots, \ell.$$

Then $E(\Delta_j | F(t_j)) = 0$, $E(\Delta_j^2 | F(t_j)) = 2(t_{j-1} - t_j)^2$.

If f satisfies the hypotheses of Theorem 4.1 by Lemma 4.1,

$$(6.2) \qquad \left\| \sum_{j=1}^{\ell} f(t_j)\Delta_j \right\| \leqq \left\{ \sum_{j=1}^{\ell} 2\|f(t_j)\|^2(t_{j+1} - t_j)^2 \right\}^{1/2},$$

which tends to 0 with mesh Π. By Theorem 4.1, $S(\Pi; f, z^1, z^2)$ has a limit as mesh $\Pi \to 0$; by (6.1) and (6.2), $S(\Pi; f, t)$ has the same limit. So if f satisfies the hypotheses of Theorem 4.1, we have

$$(6.3) \qquad \int_a^b f(t)\,(dw)^2 = \int_a^b f(t)\,dt.$$

This also holds if f satisfies the hypotheses of Theorem 5.1.

The next lemma is useful because it often permits us to discard integrals with several dz^ρ. It applies to disturbances that satisfy the following condition.

Condition 6.1. *To each positive ε there corresponds a positive δ and a set $A \subseteqq \Omega$ with $P(A) > 1 - \varepsilon$ such that if $a \leqq s \leqq t \leqq b$ and $t - s < \delta$,*

$$(6.4) \qquad |z^\rho(t, \omega) - z^\rho(s, \omega)| < \varepsilon(t - s)^{1/3}$$

for all ω in A.

For example, by a well-known theorem of Kolmogorov (see Neveu [7], p. 97) z^ρ satisfies Condition 6.1 if there is a constant K such that

$$(6.5) \qquad E([z^\rho(t) - z^\rho(s)]^8) \leqq K(t - s)^4, \qquad a \leqq s \leqq t \leqq b.$$

Theorem 6.1. *Let the hypotheses of Theorem 4.1 or Theorem 5.1 be satisfied. If $q \geqq 3$, and $z^1, \cdots, z^q$, satisfy Lemma 6.1, then*

$$(6.6) \qquad \int_c^e f(t)\,dz^1(t) \cdots dz^q(t) = 0.$$

Proof. Suppose first that $|f(\tau, \omega)|$ has an upper bound N on $T \times \Omega$. Let ε be positive, and let δ and A serve for all z^k in Condition 6.1. If Π (with notation (2.2)) is a belated partition with mesh $\Pi < \min(1, \delta)$, for all ω in A we have

$$(6.7) \qquad \left| \sum_{j=1}^{\rho} f(\tau_j, \omega) \prod_{k=1}^{q} (z^k(t_{j+1}, \omega) - z^k(t_j, \omega)) \right| \leqq N\varepsilon^q(b - a).$$

So the Riemann sums converge in near L_2 norm to 0, and the integral is 0.

If f satisfies the hypotheses of Theorem 4.1, for each positive N, we define

$$f_N(\tau, \omega) = \begin{cases} f(\tau, \omega) & \text{if} \quad -N \leqq f(\tau, \omega) \leqq N, \\ N & \text{if} \quad f(\tau, \omega) > N, \\ -N & \text{if} \quad f(\tau, \omega) < -N. \end{cases} \tag{6.8}$$

By the proof just completed, the integral of f_N is 0 for all N. So by (5.2),

$$\left\| \int_c^e f(t)\, dz^1(t) \cdots dz^q(t) \right\| \leqq B \left\{ \int_c^e \|f(t) - f_N(t)\|^2\, dt \right\}^{1/2}. \tag{6.9}$$

The right member tends to 0 as $N \to \infty$, so the left member is 0.

If the hypotheses of Theorem 5.1 are satisfied, with the notation of that theorem, $f\phi_N$ has integral 0 for all N, so the integral of f is 0.

There are other useful sets of conditions that eliminate integrals, but we will confine ourselves to two simple cases.

THEOREM 6.2. *If the hypotheses of Theorem* 4.1 *or of Theorem* 5.1 *hold with* $q \geqq 2$, *and* $z^q(t) = t$, *then* $\int_a^b f(t)\, dz^1 \cdots dz^q = 0$.

PROOF. With Π as in (2.2), define

$$\Delta_j^k = z^k(t_{j+1}) - z^k(t_j). \tag{6.10}$$

Then

$$\left| E\left(\prod_{k=1}^{q} \Delta_j^k \,|\, F_{t_j} \right) \right| = (t_{j+1} - t_j) \left| E\left(\prod_{k=1}^{q-1} \Delta_j^k \,|\, F_{t_j} \right) \right|, \tag{6.11}$$

$$E\left(\left[\prod_{k=1}^{q} \Delta_j^k \right]^2 | F_{t_j} \right) = (t_{j+1} - t_j)^2 E\left(\left[\prod_{k=1}^{q-1} \Delta_j^k \right]^2 | F_{t_j} \right). \tag{6.12}$$

If $q = 2$, the right members of (6.11) and (6.12) do not exceed $K(t_{j+1} - t_j)^2$ and $K(t_{j+1} - t_j)^3$, respectively, by Condition 4.1; so, under the hypotheses of Theorem 4.1, Lemma 4.1 assures us that $\|S(\Pi; f, z^1, \cdots, z^q)\|$ tends to 0 with mesh Π. If the hypotheses of Theorem 5.1 hold, the conclusion is established by the use of the functions ϕ_N of Theorem 5.1.

If $q > 2$, let r and s be integers at most $\frac{1}{2}q$ with $r + s = q - 1$. Then

$$E\left(\prod_{k=1}^{q-1} \Delta_j^k | F_{t_j} \right) \leqq \left\{ E\left(\left[\prod_{k=1}^{r} \Delta_j^k \right]^2 | F_{t_j} \right) \right\}^{1/2} \left\{ E\left(\left[\prod_{k=r+1}^{q-1} \Delta_j^k \right]^2 | F_{t_j} \right) \right\}^{1/2}. \tag{6.13}$$

Since $[\Pi_1^r \Delta_j^k]^2 \leqq \Sigma_1^r [\Delta_j^k]^{2r}$ and $2^r \leqq q$, by Condition 4.1, the first factor in the right member of (6.13) does not exceed a constant multiple of $(t_{j+1} - t_j)^{1/2}$. The same is true of the second factor, so the left member of (6.11) does not exceed a multiple of $(t_{j+1} - t_j)^2$. Likewise the left member of (6.12) does not exceed a multiple of $(t_{j+1} - t_j)^3$.

The rest of the proof is as for $q = 2$.

THEOREM 6.3. *Let z^1 and z^2 be processes such that if $a \leqq s \leqq t \leqq b$, then $z^1(t) - z^1(s)$ and $z^2(t) - z^2(s)$ are conditionally independent as conditioned by F_s. Let the hypotheses of Theorem* 4.1 *hold. Then*

$$\int_a^b f(t)\, dz^1(t)\, dz^2(t) = 0. \tag{6.14}$$

PROOF. We use the same notation as in the preceding proof. Then

$$\begin{aligned} |E(\Delta_j, z^1\Delta_j z^2 \,|\, F_s(t_j))| &= |E(\Delta, z^1 \,|\, F(t_j))E(\Delta_j z^2 \,|\, F(t_j))| \\ &\leqq K^2(t_{j+1} - t_j)^2, \end{aligned} \tag{6.15}$$

$$\begin{aligned} E([\Delta_j z^1 \Delta_j z^2]^2 \,|\, F(t_j)) &= E([\Delta_j z^1]^2 \,|\, F(t_j))E([\Delta_j z^2]^2 \,|\, F(t_j)) \\ &\leqq K^2(t_{j+1} - t_j)^2. \end{aligned} \tag{6.16}$$

By Lemma 4.1, $\|S(\Pi; f, z^1, z^2)\|$ tends to 0 with mesh Π.

7. Existence theorem for a functional equation

If the hypotheses of Theorem 4.1 are satisfied and we define a process F on $[a, b]$ by setting

$$F(t) = \int_a^t f(s)\, dz^1(s) \cdots dz^q(s), \qquad t \in [a, b], \tag{7.1}$$

we know by Theorem 4.1 that $F(t)$ has finite second moment and can be chosen $F_.$ measurable. By (5.2), we know that it satisfies a Hölder condition of exponent 1/2 in L_2 norm. Processes with these properties occur often enough in succeeding pages to justify giving them a name.

DEFINITION 7.1. *Let $H_{1/2}(T, F_.)$ be the class of all (real or vector valued) processes x on T such that for all t in T, $x(t)$ is F_t measurable and $E(|x(t)|^2) < \infty$, and there is a number H^* such that if s and t are in T,*

$$\|x(t) - x(s)\| \leqq H^*(t - s)^{1/2}. \tag{7.2}$$

COROLLARY 7.1. *If the hypotheses of Theorem* 4.1 *are satisfied and F is defined by* (7.1), *F belongs to $H_{1/2}([a, b], F_.)$.*

Instead of restricting ourselves to stochastic "differential equations" such as (3.3), we shall discuss a class of functional equations

$$x^i(\tau) = y^i(\tau), \qquad \tau \in T, \tau \leqq a, \tag{7.3}$$

$$\begin{aligned} x^i(t) = y^i(t) &+ \int_a^t g_0^i(s, x(s))\, ds \\ &+ \sum_{\mathscr{C}} \int_a^t g^i_{\rho,\sigma,\cdots,\phi}(s, x(s))\, dz^\rho(s)\, dz^\sigma(s) \cdots dz^\phi(s), \end{aligned} \tag{7.4}$$

$$i = 1, \cdots, n;\ a \leqq t \leqq b,$$

where the letters denote members of $\{1, \cdots, r\}$, and $\mathscr{C}$ is a finite set of finite ordered sequences $(\rho, \sigma, \cdots, \phi)$ of members of $\{1, \cdots, r\}$. The functions $g_0^i, g_{\rho,\sigma,\cdots,\phi}^i$ will be called coefficients. We shall make the following assumptions.

ASSUMPTION 7.1. *The class $\mathscr{P}$ is a linear class of n-vector valued processes on T that contains $H_{1/2}(T, F_\cdot)$, and is closed under uniform convergence in L_2 norm.*

ASSUMPTION 7.2. *Each coefficient g is defined on $T \times \mathscr{P}$, and for fixed x in $\mathscr{P}$, $g(\cdot, x)$ is bounded in L_2 norm on T and is continuous in L_2 norm at almost all points of $[a, b]$.*

ASSUMPTION 7.3. *If F is a σ-subalgebra of $\mathscr{A}$, and $t \in T$, and x is a process in $\mathscr{P}$ such that $x(\tau)$ is F measurable for all $\tau \leqq t$ in T, then $g(t, x)$ is also F measurable.*

For Theorem 7.1 it would be adequate to choose $H_{1/2}(t, F_\cdot)$ for $\mathscr{P}$. However, in the case of differential equations a little more latitude is convenient. Suppose that G_0^i and $G_{\rho,\sigma,\cdots,\phi}^i$ are functions on $T \times \mathbf{R}^n$ such that, for a certain subset N_0 of T with Lebesgue measure 0 and a certain positive L, it is true that G_0^i and $G_{\rho,\sigma,\cdots,\phi}^i$ are continuous in all variables at all points (t, x) with $t \in T - N_0$ and $x \in \mathbf{R}^n$, and for all t in T and x_1, x_2 in $\mathbf{R}^n$

$$|G_0^i(t, x_1) - G_0^i(t, x_2)| \leqq L|x_1 - x_2|, \tag{7.5}$$

and likewise for the $G_{\rho,\sigma,\cdots,\phi}^i$. Then for all processes x with finite second moments we can define

$$g_0^i(t, x) = G_0^i(t, x(t)), \tag{7.6}$$

$$g_{\rho,\sigma,\cdots,\phi}^i(t, x) = G_{\rho,\sigma,\cdots,\phi}^i(t, x(t)), \qquad t \in T, \tag{7.7}$$

and Assumption 7.3 is satisfied. To attain Assumptions 7.1 and 7.2 also, we can make the following assumption.

ASSUMPTION 7.4. *$\mathscr{P}$ is the class of all processes bounded in L_2 norm on T and continuous in L_2 norm at almost all points of $[a, b]$.*

We can simplify notation a little by defining $z^0(t) = t$ for all real t, and adjoining the one element sequence (0) to $\mathscr{C}$. With this understanding, equations (7.3) and (7.4) take the notationally simpler form

$$x^i(\tau) = y^i(\tau), \qquad \tau \in T, \tau \leqq a, \tag{7.8}$$

$$x^i(t) = y^i(t) + \sum_{\mathscr{C}} \int_a^t g_{\rho,\sigma,\cdots,\phi}^i(s, x(s))\, dz^\rho(s) \cdots dz^\phi(s), \tag{7.9}$$

$$t = 1, \cdots, n; a \leqq t \leqq b.$$

THEOREM 7.1. *Let the coefficients in* (7.8) *and* (7.9) *satisfy Assumptions* 7.2 *and* 7.3, *and let the z^ρ satisfy Condition* 4.1. *Assume also that there exists a positive L such that if x_1 and x_2 are in $\mathscr{P}$ and $t \in [a, b]$, for each coefficient g in* (7.8) *and* (7.9) *it is true that*

$$\|g(t, x_1) - g(t, x_2)\| \leqq L \sup\{\|x_1(s) - x_2(s)\|: s \in T, s \leqq t\}. \tag{7.10}$$

Let y belong to $\mathscr{P}$ and be $F_.$ measurable. Then there is an $F_.$ measurable process $x(\cdot)$ in $\mathscr{P}$ such that $x^i(\tau) = y^i(\tau)$, $\tau \in T$, $\tau \leqq a$, and (7.8) holds for $a \leqq t \leqq b$. If $y(\cdot) \in H_{1/2}(T, F.)$, so does $x(\cdot)$. Moreover, if x_1 is any $F.$ measurable process satisfying (7.8) and (7.9) then $P[x_1(t) = x(t)] = 1$ for all t in T.

PROOF. Hypothesis (7.10) guarantees that the coefficients are nonanticipative; if x_1 and x_2 belong to $\mathscr{P}$ and $x_1(\tau) = x_2(\tau)$ if $\tau \in T$ and $\tau \leqq t$, then $g(t, x_1) = g(t, x_2)$. If x is defined only on the part of T in $(-\infty, t]$ and has an extension $\tilde{x}$ to T that belongs to $\mathscr{P}$, by (7.10) all such extensions $\tilde{x}$ give the same value to $g(t, x)$. To simplify notation, we shall define $g(t, x)$ to mean that common value.

We use Picard's method. We define $x_0 = y$, and then successively

$$x^i_{k+1}(\tau) = y^i(\tau), \qquad \tau \in T, \tau \leqq a, \tag{7.11}$$

$$x^i_{k+1}(\tau) = y^i(\tau) + \sum_{\mathscr{C}} \int_a^\tau g^i_{\rho,\sigma,\cdots,\phi}(s, x_k)\, dz^\rho\, dz^\sigma \cdots dz^\phi. \tag{7.12}$$

By hypothesis x_0 is in $\mathscr{P}$. If we assume x_k in that class, the integrands in (7.11) and (7.12) satisfy the hypotheses of Theorem 4.1 by Corollary 7.1, x^i_{k+1} belongs to $H_{1/2}(T, F.)$. Thus, (7.11) and (7.12) define x_k for $k = 0, 1, 2, 3, \cdots$. Define, for every process x on T and every t in T,

$$N(t, x) = \sup\{\|x(\tau)\| \mid \tau \in T, \tau \leqq t\}. \tag{7.13}$$

If $g^i_{\rho,\sigma,\cdots,\phi}$ is one of the coefficients in (7.8) and (7.9) and $k \geqq 1$, by (5.2) and hypothesis (7.10),

$$\begin{aligned} &\left\| \int_a^t \{g^i_{\rho,\sigma,\cdots,\phi}(s, x_k) - g^i_{\rho,\sigma,\cdots,\phi}(s, x_{k+1})\}\, dz^\rho\, dz^\sigma \cdots dz^\phi \right\| \\ &\quad \leqq B\left\{\int_a^t \|g^i_{\rho,\sigma,\cdots,\phi}(s, x_k) - g^i_{\rho,\sigma,\cdots,\phi}(s, x_{k-1})\|^2\, ds\right\}^{1/2} \\ &\quad \leqq B\left\{\int_a^t L^2 N(s, x_k - x_{k-1})^2\, ds\right\}^{1/2}. \end{aligned} \tag{7.14}$$

Let B_0 be the product of n, B, L^2 and the number of sequences in the set $\mathscr{C}$. By (7.11), (7.12), and (7.14),

$$\|x_{k+1}(t) - x_k(t)\| \leqq B_0\left\{\int_a^t N(s, x_k - x_{k-1})\, ds\right\}^{1/2}. \tag{7.15}$$

Since this estimate is still valid if we replace t in the left member by any smaller member of T.

$$N(t, x_{k+1} - x_k)^2 \leqq B_0^2\left\{\int_a^t N(s, x_k - x_{k-1})^2\, ds\right\}. \tag{7.16}$$

We can now prove by induction (with $x_{-1} \equiv 0$)

$$N(t, x_k - x_{k-1})^2 \leqq \{\sup \|y(t)\|^2\} B_0^{2k} \frac{(t - a)^k}{k!}, \qquad k = 0, 1, 2, \cdots, \tag{7.17}$$

For $k = 0$, this is simply the statement $N(t, y)^2 \leqq \sup \|y(t)\|^2$. If (7.17) holds for k, by (7.16),

$$\begin{aligned} N(t, x_{k+1} - x_k)^2 &\leqq B_0^2 \left\{ \int_a^t (\sup \|y\|^2) \left[\frac{B_0^{2k}}{k!} \right] (s - a)^k \, ds \right\} \\ &= \sup \|y\|^2 \left[\frac{B_0^{2k+2}}{(k+1)!} \right] (t - a)^{k+1}, \end{aligned} \tag{7.18}$$

so (7.17) holds for all nonnegative integers k. It follows at once that the sums

$$x_k = \sum_{k=0}^{h} (x_k - x_{k-1}) \tag{7.19}$$

converge uniformly in L_2 norm to a limit, which we call x. This limit belongs to $H_{1/2}(T, F.)$, and by (7.11) and (7.12) it satisfies (7.8) and (7.9).

If x' and x'' both satisfy (7.8) and (7.9) and are $F.$ measurable, just as we proved (7.16) we can prove

$$N(t, x' - x'')^2 \leqq B_0^2 \left\{ \int_a^t N(s, x' - x'')^2 \, ds \right\}^{1/2}. \tag{7.20}$$

The only solution of this is $N(s, x' - x'') = 0$, which completes the proof.

8. Cauchy-Maruyama approximations

G. Maruyama [4] has extended the well-known Cauchy (or Euler) method of constructing polygonal approximate solutions, proceeding successively from each vertex to the next, to the stochastic differential equations (1.3). It is easy to extend this procedure still further to equations of the form of (7.8) and (7.9). Given any Cauchy partition

$$\Pi = (t_1, \cdots, t_{\ell+1}; t_1, \cdots, t_\ell) \tag{8.1}$$

of $[a, b]$, we first define $\bar{x}(\tau) = y(\tau)$ for all τ in T with $\tau \leqq a$. Then $\bar{x}$ having been defined for all τ in T with $\tau \leqq t_j$, we define it on $(t_j, t_{j+1}]$ by setting

$$\begin{aligned} \bar{x}^i(\tau) = \bar{x}^i(t_j) &+ y^i(t) - y^i(t_j) \\ &+ \sum_{\mathscr{C}} g^i_{\rho,\sigma,\cdots,\phi}(t_j, \bar{x})[z^\rho(t) - z^\rho(t_j)] \cdots [z^\phi(t) - z^\phi(t_j)]. \end{aligned} \tag{8.2}$$

(Notice that the coefficients are defined, even though $\bar{x}$ has been defined only up to t_j, by the first paragraph of the proof of Theorem 7.1.)

We can prove that under the hypotheses of Theorem 7.1 these Cauchy-Maruyama functions x converge to the solution x of (7.8) and (7.9) uniformly in L_2 norm as mesh $\Pi \to 0$. But in Sections 9 and 11, we shall need a different approximation, in which we shall permit a small departure from equality in (8.2).

We shall suppose that $\mathscr{P}$ is defined by Assumption 7.4, and we shall adopt the abbreviations

$$(8.3)\qquad \Delta_j t = t_{j+1} - t_j,\ \Delta_j y = y(t_{j+1}) - y(t_j),\ \Delta_j z^\rho = z^\rho(t_{j+1}) - z^\rho(t_j).$$

Suppose now that to each Cauchy partition Π (with notation (8.1)) there corresponds a process $\bar{x}$ with the following properties:

(f) $\bar{x}(\tau) = y(\tau),\ \tau \in T,\ \tau \leqq a$;

(g) *to each positive* ε *there corresponds a positive* δ *such that, if mesh* $\Pi < \delta$

$$(8.4)\qquad \begin{aligned} &\|\bar{x}(t) - \bar{x}(t_j) - y(t) + y(t_j) \\ &\quad - \Sigma\, g^i_{\rho,\sigma,\cdots,\phi}(t_j, \bar{x})[z^\rho(t) - z^\rho(t_j)] \cdots [z^\phi(t) - z^\phi(t_j)]\| \\ &\quad \leqq \varepsilon(1 + \sup\{\|\bar{x}(\tau)\| : \tau \in T,\ \tau \leqq t\})(t_j \leqq t < t_{j+1}), \end{aligned}$$

$$(8.5)\qquad \begin{aligned} &\|\bar{x}(t_{j+1}) - \bar{x}(t_j) - \Delta_j y - \Sigma\, g_{\rho,\sigma,\cdots,\phi}(t_j, \bar{x})\Delta_j z^\rho \cdots \Delta_j z^\rho\| \\ &\quad \leqq \varepsilon(1 + \sup\{\|\bar{x}(\tau)\| : \tau \in T,\ \tau \leqq t_{j+1}\})\Delta_j t; \end{aligned}$$

(h) *if* $\tau \in T$ *and* $\tau \leqq t_j$, $\bar{x}(\tau)$ *is* $F[t_j]$ *measurable.*

(The Cauchy-Maruyama functions (8.3) clearly satisfy these requirements.) We can then prove the following theorem.

THEOREM 8.1. *Let the hypotheses of Theorem* 7.1 *hold. Assume that to each Cauchy partition* Π *of* $[a, b]$ *there corresponds a process* $\bar{x}$ *in* $\mathscr{P}$ *such that* (f), (g), *and* (h) *hold. Then as mesh* Π *tends to zero,* $\bar{x}$ *converges in* L_2 *norm, uniformly on* T, *to the solution* x *of* (7.8) *and* (7.9).

PROOF. By Theorem 4.1, the solution x of (7.8) and (7.9) exists, and we can and do choose it to be $F_{\cdot}$ measurable. Let Π be a Cauchy partition, with notation (2.1); let t be a point of $[a, b]$; and define

$$(8.6)\qquad t_k = \text{largest number in set } \{t_1, \cdots, t_\ell\} \cap (-\infty, t],$$

$$(8.7)\qquad N(t) = \sup\{\|\bar{x}(\tau) - x(\tau)\| : \tau \in T,\ \tau \leqq t_k\},$$

$$(8.8)\qquad \begin{aligned} x^i(t) = y^i(t) &+ \sum_{j=1}^{k-1} g^i_{\rho,\sigma,\cdots,\phi}(t_j, x)\Delta_j z^\rho \cdots \Delta_j z \\ &+ \Sigma\, g^i_{\rho,\cdots,\phi}(t_k, x)[z^\rho(t) - z^\rho(t_k)] \cdots [z^\phi(t) - z^\phi(t_k)]. \end{aligned}$$

(For $\tau \in t,\ \tau \leqq a$ we take $X(\tau) = y(\tau)$.)

Let $M - 1$ be an upper bound for $\|x(\tau)\|$ on T, and let ε be any number such that $0 < 2(1 + b - a)\varepsilon < 1$. By Theorem 4.1, there is a positive δ_1 such that if mesh $\Pi < \delta_1$,

$$(8.9)\qquad \|X(\tau) - x(\tau)\| < \varepsilon(\tau \in T).$$

With δ as in (g), we let Π be any belated partition such that

$$\text{mesh } \Pi < \min \{\delta, \delta_1\}, \tag{8.10}$$

Then, with t_k defined by (8.6),

$$\begin{aligned}
\bar{x}^i(t) - x^i(t) \\
&= X^i(t) - x^i(t) \\
&+ \sum_{j=1}^{k-1} \{\bar{x}^i(t_{j+1}) - \bar{x}^i(t_j) - \Delta_j y^i - \sum_{\mathscr{C}} g^i_{\rho,\cdots,\phi}(t_j, \bar{x})\Delta_j z^\rho \cdots \Delta_j z^\phi\} \\
&+ \bar{x}^i(t) - \bar{x}^i(t_k) - [y^i(t) - y^i(t_k)] \\
&- \sum_{\mathscr{C}} g^i_{\rho,\cdots,\phi}(t_k, \bar{x})[z^\rho(t) - z^\rho(t_k)] \cdots [z^\phi(t) - z^\phi(t_k)] \\
&+ \sum_{j=1}^{k-1} \sum_{\mathscr{C}} [g^i_{\rho,\cdots,\phi}(t_j, \bar{x}) - g^i_{\rho,\cdots,\phi}(t_j, x)]\Delta_j z^\rho \cdots \Delta_j z^\phi \\
&+ \sum_{\mathscr{C}} [g^i_{\rho,\cdots,\phi}(t_k, \bar{x}) - g^i_{\rho,\cdots,\phi}(t_k, x)][z^\rho(t) - z^\rho(t_k)] \cdots \\
&\qquad [z^\phi(t) - z^\rho(t_k)].
\end{aligned} \tag{8.11}$$

By hypothesis (7.10) of Theorem 7.1 with (5.1),

$$\begin{aligned}
&\Bigg\| \sum_{j=1}^{k-1} [g^i_{\rho,\cdots,\phi}(t_j, \bar{x}) - g^i_{\rho,\cdots,\phi}(t_j, x)]\Delta_j z^\rho \cdots \Delta_j z^\phi \\
&\quad + [g^i_{\rho,\cdots,\phi}(t_k, \bar{x}) - g^i_{\rho,\cdots,\phi}(t_k, x)][z^\rho(t) - z^\rho(t_k)] \cdots [z^\phi(t) - z^\phi(t_k)] \Bigg\| \\
&\leqq B\left\{\sum_{j=1}^{k-1} L^2(N(t_j))^2 \Delta_j t + L^2(N(t_k))^2[t - t_k]\right\}^{1/2} \\
&= BL\left\{\int_a^t N(s)^2\, ds\right\}^{1/2}.
\end{aligned} \tag{8.12}$$

Since $M - 1$ is an upper bound for $\|x\|$, by (8.7), $\|\bar{x}(t)\| \leqq M - 1 + N(t)$, and the right members of (8.4) and (8.5) are at most $\varepsilon(M + N(t))$, $\varepsilon(M + N(t_{j+1}))$, respectively. So if C is the number of members of the set $\mathscr{C}$, from (g), (8.9), (8.10), (8.11), and (8.12), we deduce

$$\begin{aligned}
&\|\bar{x}(t) - x(t)\| \\
&\leqq \varepsilon + \varepsilon\{M + N(t_k)\}(t_k - a) + \varepsilon\{M + N(t)\} + CBL\left\{\int_a^t N(s)^2\, ds\right\}^{1/2}.
\end{aligned} \tag{8.13}$$

The right member is a nondecreasing function of t, so (8.13) remains valid if we replace t_k in the right member by t and then replace t by any larger number, or equivalently replace t in the left member by any smaller number. So

$$N(t) \leqq \varepsilon[1 + M(1 + b - a)] + \varepsilon(1 + b - a)N(t) + CBL\left\{\int_a^t N(s)^2\,ds\right\}^{1/2}. \tag{8.14}$$

By the fact we have chosen ε such that $0 < 2(1 + b - a)\varepsilon < 1$, this implies

$$N(t) \leqq 2\varepsilon[1 + M(1 + b - a)] + 2CBL\left\{\int_a^t N(s)^2\,ds\right\}^{1/2} \tag{8.15}$$

To condense notation, we write

$$\begin{aligned} P &= 2[1 + M(1 + b - a)], \\ Q &= 2CBL. \end{aligned} \tag{8.16}$$

Then from (8.15) we can deduce that

$$N(t) \leqq 2\varepsilon P \exp\{(2P^2Q^2[t - a])(a \leqq t \leqq b)\}, \tag{8.17}$$

for (8.17) holds at $t = a$. If it does not hold everywhere in $[a, b]$, there is a first point t_0 at which it fails. Then it holds on $[a, t_0]$, so by (8.15),

$$\begin{aligned} N(t_0) &\leqq \varepsilon P + Q\left[\int_a^t 4\varepsilon^2P^2 \exp\{4P^2Q^2[s - a]\}\,ds\}\right]^{1/2} \\ &= \varepsilon P + \varepsilon[\exp\{4P^2Q^2[t_0 - a]\} - 1]^{1/2} \\ &< 2\varepsilon P \exp\{2P^2Q^2[t_0 - a]\}, \end{aligned} \tag{8.18}$$

contradicting the assumption that (8.17) fails at t_0. Since for every positive ε, (8.17) holds whenever (8.10) does, $\|\bar{x}(t) - x(t)\|$ converges uniformly to 0 as mesh $\Pi \to 0$, which completes the proof.

The only use made of the enlarged class defined by Assumption 7.4 was to guarantee that the coefficients $g^i_{\rho,\cdots,\phi}(t_j, \bar{x})$ are defined. If the $\bar{x}$ are the Cauchy-Maruyama functions defined by (8.2), they are in $H_{1/2}[T, F_.]$, and we can use this for our class $\mathscr{P}$, abandoning Assumption 7.4.

9. Stochastic differential equations and related ordinary equations

We shall now revert back to stochastic differential equations like those in Sections 1 to 3, in which the coefficients g^i_ρ, and so forth, are functions of t and $x(t)$ and independent of earlier values of $x(\tau)$. We suppose that these have the properties ascribed to the coefficients G^i_0, and so forth, as stated after Assumption 7.3; but we use g instead of G. Moreover, we take T to be the same as $[a, b]$, and $y(t)$ is simply an initial value x_0, which is an F_a measurable r.v. For such equations we shall show that, with the definition in (3.5), equations of the form (3.4) have the stability property that for a rather large class of processes Z^ρ

interpolated in the z^ρ and having piecewise smooth sample paths, the solutions of (3.3) with Z^ρ tend to those with z^ρ, uniformly in near L_2 norm.

To avoid inordinately long formulae, we change the notation somewhat. We define

$$\begin{aligned} &x^0(t) = t, \qquad z^0(t) = t, && -\infty < t < \infty \\ &g_0^0(x) = 1, \qquad g_1^0(x) = \cdots = g_r^0(x) = 0, && x \in \mathbf{R}^{n+1}. \end{aligned} \tag{9.1}$$

The variables α, β will always have range $\{0, \cdots, n\}$, and ρ, σ, τ will have range $\{0, \cdots, r\}$. A summation sign such as Σ_σ or Σ_ρ will denote the sum over the whole range of that variable. Also, Π will always denote a Cauchy partition with notation (8.1), and t_j will denote a division point of Π. An equation such as $u_\rho^i = v_\rho^i$ will always be understood to hold for all i, ρ in the range of those variables, unless some other range is expressly specified.

For all x^0 in $[a, b]$ and $(x^1, \cdots, x^n)$ in $\mathbf{R}^n$ we define

$$g_{\rho,\sigma}^i(x) = \sum_\alpha \frac{\partial g_\rho^i(x)}{\partial x^\alpha} g_\sigma^\alpha(x), \tag{9.2}$$

provided that the indicated derivatives exist. Equations (1.3) now take the form

$$x^i(t) = x_0^i + \sum_\rho \int_a^t g_\rho^i(x(s))\, dz^\sigma, \tag{9.3}$$

where the initial value x_0 is always assumed to be an F_a measurable r.v. The analogue of (3.4), with (3.5), is

$$x^i(t) = x_0^i + \sum_\rho \int_a^t g_\rho^i(x(s))\, dz^\rho(s) + \tfrac{1}{2} \sum_\rho \int_{\sigma a}^{t} g_{\rho,\sigma}^i(x(s))\, dz^\rho(s)\, dz^\sigma(s). \tag{9.4}$$

This is not identical with (3.4), for the last sum contains terms with $\rho = 0$ or $\sigma = 0$, and (3.4) does not. However, by Theorem 6.2, all such integrals vanish for all processes z^ρ that we shall consider. Furthermore, even if ρ and σ are positive, the definition (9.2) contains a term (with $\alpha = 0$) which is lacking in (3.5). But by (9.1) this term is 0. So the solutions of (9.4) are the same as those of (3.4) with (3.5) for all processes z^ρ that we shall permit.

In Theorem 7.1, we assumed that the coefficients were Lipschitzian in $x(\cdot)$, and merely almost everywhere continuous in t. To simplify proofs, we now replace this by a somewhat unnecessarily strong substitute:

ASSUMPTION 9.1. *The functions g^i are continuously differentiable on the set of x with $a \leqq x^0 \leqq b$; and there is a positive L such that, if x and x'' are both in that set,*

$$\begin{aligned} |g_\rho^i(x') - g_\rho^i(x'')| &\leqq L|x' - x''|, \\ |g_{\rho,\sigma}^i(x') - g_{\rho,\sigma}^i(x'')| &\leqq L|x' - x''|. \end{aligned} \tag{9.5}$$

Instead of restricting ourselves to linear interpolation as mentioned in Section 1, we shall permit certain other kinds. Let

(9.6) $$\phi_\rho(t): 0 \leqq t \leqq 1, \qquad \rho = 0, 1, \cdots, r,$$

be Lipschitzian functions such that

(9.7) $$\phi_\rho(0) = 0, \qquad \phi_\rho(1) = 1$$

and

(9.8) $$\phi_0(t) = t, \qquad 0 \leqq t \leqq 1.$$

Then for each z^ρ and each Cauchy partition Π, we define functions Z^ρ by setting

(9.9) $$Z^\rho(t, \omega) = z^\rho(t_j, \omega) + \phi_\rho\left(\frac{t - t_j}{t_{j+1} - t_j}\right)[z^\rho(t_{j+1}) - z^\rho(t_j)], \qquad t_j \leqq t \leqq t_{j+1}.$$

In particular,

(9.10) $$Z^0(t) = t, \qquad a \leqq t \leqq b.$$

We define

(9.11) $$J_{\rho,\sigma} = \int_0^1 [1 - \phi_\rho(s)]\dot{\phi}_\sigma(s)\, ds.$$

Our principal stability theorem, which we now state, overlaps considerably with the results of Wong and Zakai ([8], [9]). Although the present methods are different, Theorem 9.1 obviously owes its existence to those previous results. Besides this, the present version of Theorem 9.1 replaces an earlier version with stronger hypotheses because Professor Wong pointed out the desirability of improvement.

THEOREM 9.1. *Let Assumption* 9.1 *hold, and let the* z^ρ *satisfy Condition* 6.1. *Let* $\phi_0, \cdots, \phi_r$ *have the properties described above. Assume that for each* ρ *and* σ *in* $\{0, 1, \cdots, r\}$, *either:*

(i) *to each* $\varepsilon > 0$ *corresponds a* $\delta > 0$ *such that if* $a \leqq s \leqq t \leqq b$ *and* $t - s < \delta$ *then a.s.,*

(9.12) $$\begin{aligned} |E([z^\rho(t) - z^\rho(s)][z^\sigma(t) - z^\sigma(s)] \mid F_s)| &\leqq \varepsilon(t - s), \\ E([z^\rho(t) - z^\rho(s)]^2[z^\sigma(t) - z^\sigma(s)]^2 \mid F_s) &\leqq \varepsilon(t - s), \end{aligned}$$

or else

(ii) $J_{\rho,\sigma} = 1/2$.

Then, as mesh $\Pi \to 0$, *the solution* X *of*

(9.13) $$X^i(t) = x_0^i + \sum_\rho \int_a^t g_\rho^i(X(s))\dot{Z}^\rho(s)\, ds, \qquad a \leqq t \leqq b$$

converges uniformly in near L_2 *norm on* $[a, b]$ *to the solution* x *of*

$$(9.14)\qquad x(t) = x_0 + \sum_{\rho} \int_a^t g_{\rho}^i(x(s))\, dz^{\rho}(s) + \tfrac{1}{2} \sum_{\rho,\sigma} \int_a^t g_{\rho,\sigma}^i(x(s))\, dz^{\rho}(s)\, dz^{\sigma}(s),$$

$$a \leqq t \leqq b.$$

PROOF. Observe that if $\rho = \sigma$, condition (ii) is satisfied, while if $\rho = 0$ or $\sigma = 0$ condition (i) holds.

The solution x of (9.14) also satisfies

$$(9.15)\qquad x^i(t) = x_0^i + \sum_{\rho} \int_a^t g_{\rho}^i(x(s))\, dz^{\rho}(s) + \sum_{\rho,\sigma} J_{\rho,\sigma} \int_a^t g_{\rho,\sigma}^i(x(s))\, dz^{\rho}(s)\, dz^{\sigma}(s),$$

$$a \leqq t \leqq b,$$

since those integrals in (9.15) with coefficients $J_{\rho,\sigma} \neq 1/2$ all vanish by (5.2).

We again define $\Delta_j t$ and $\Delta_j z^{\rho}$ by (8.3). Let ε be positive, and let δ and A correspond to ε for all the z^{ρ} as in Condition 6.1. Let Π be a Cauchy partition with mesh $\Pi < \delta$. For each k in $\{1, \cdots, \ell\}$, we define A_k to be the set of ω in Ω such that the inequalities

$$(9.16)\qquad |\Delta_j z^{\rho}(\omega)| \leqq \varepsilon(\Delta_j t)^{1/3}, \qquad \rho = 1, \cdots, r$$

all hold for $j = 1, \cdots, k$; then $A_{\ell} \supseteqq A$. Corresponding to Π, we now define a process $\bar{x}$ as follows. First, $\bar{x}(a) = x_0$. Then, $\bar{x}(t_j, \omega)$ having been defined, we define

$$(9.17)\qquad \bar{x}^i(t_{j+1}, \omega) = X^i(t_{j+1}, \omega) \qquad \text{if} \quad \omega \in A_j,$$

$$(9.18)\qquad \bar{x}(t_{j+1}, \omega) = \bar{x}^i(t_j, \omega) + \sum_{\rho} g_{\rho}^i(\bar{x}(t_j, \omega))\Delta_j z^{\rho} + \sum_{\rho,\sigma} J_{\rho,\sigma} g_{\rho,\sigma}^i(\bar{x}(t_j, \omega))\Delta_j z^{\rho}\Delta_j z^{\sigma} \qquad \text{if} \quad \omega \in \Omega - A_j.$$

In either case we define

$$(9.19)\qquad \bar{x}^i(t, \omega) = \bar{x}^i(t_j, \omega) \qquad t_j \leqq t < t_{j+1}.$$

The set A_j defined by (9.16) is $F[t_{j+1}]$ measurable, and $Z^{\rho}(t)$ is a linear function of $z^{\rho}(t_j)$ and $z^{\rho}(t_{j+1})$ for $t_j \leqq t \leqq t_{j+1}$; so by (9.13) $X(t_{j+1})$ is a continuous function of the $z^{\rho}(t_h)$ for $h = 1, \cdots, j + 1$, and is $F[t_{j+1}]$ measurable. So by (9.17) and (9.18), $\bar{x}(t_{j+1})$ is $F[t_{j+1}]$ measurable, and hypothesis (h) is satisfied. So is (f); and (8.4) follows readily from (9.19) and Condition 4.1.

If $\omega \in A_j$, from (9.17) we obtain by integration by parts in (9.13),

$$(9.20)\qquad \bar{x}^i(t_{j+1}) - \bar{x}^i(t_j) - \sum_{\rho} g_{\rho}^i(\bar{x}(t_j))\Delta_j z^{\rho} - \sum_{\rho,\sigma} g_{\rho,\sigma}^i(\bar{x}(t_j)) \int_{t_j}^{t_{j+1}} [Z^{\rho}(t_{j+1}) - Z^{\rho}(s)]\dot{Z}^{\sigma}(s)\, ds$$

$$= \sum_{\rho,\sigma} \int_{t_j}^{t_{j+1}} [g_{\rho,\sigma}^i[X(s)] - g_{\rho,\sigma}^i[\bar{x}(t_j)]][Z^{\rho}(t_{j+1}) - Z^{\rho}(s)]\dot{Z}^{\sigma}(s)\, ds.$$

For the rest of this proof C_1, C_2, and so forth, will denote positive numbers whose values are determined by the numbers $n, r, K, L, g_\rho^i(0)$ and $\sup|\dot{\phi}_\rho(t)|$; we omit the easy but uninspiring computation of the expressions for the C_i.

For t in $[t_j, t_{j+1}]$ and ω in A_j, we define

$$N(t, \omega) = \sup\{|X(\tau, \omega) - \bar{x}(t_j, \omega)| : t_j \leqq \tau \leqq t\}. \tag{9.21}$$

Then by Assumption 9.1,

$$\begin{aligned} |g_\rho^i(X(t, \omega))| &\leqq |g_\rho^i(0)| + |g_\rho^i(\bar{x}(t_j, \omega)) - g_\rho^i(0)| \\ &\quad + |g_\rho^i(X(t, \omega)) - g_\rho^i(\bar{x}(t_j, \omega))| \\ &\leqq C_1 + L|\bar{x}(t_j, \omega)| + LN(t, \omega). \end{aligned} \tag{9.22}$$

Hence by (9.17) and (9.16),

$$\begin{aligned} |X^i(t, \omega) - \bar{x}^i(t_j, \omega)| &\leqq \sum_\rho \int_{t_j}^t |g_\rho^i(X(s))||\dot{Z}^\rho(s)|\, ds \\ &\leqq \varepsilon[C_2 + C_3|\bar{x}(t_j, \omega)| + C_4 N(t, \omega)](\Delta_j t)^{1/3}. \end{aligned} \tag{9.23}$$

This remains valid if in the left member we replace t by any number τ in $[t_j, t]$, so

$$N(t, \omega) \leqq \varepsilon[C_5 + C_6|\bar{x}(t_j, \omega)| + C_7 N(t, \omega)](\Delta_j t)^{1/3}. \tag{9.24}$$

Since C_7 does not depend on ε, we may and shall restrict our attention to ε such that

$$0 < \varepsilon < \frac{1}{2C_7}(b - a)^{-1/3}. \tag{9.25}$$

Then from (9.24), we obtain

$$N(t, \omega) \leqq 2\varepsilon[C_5 + C_6|\bar{x}(t_j, \omega)|](\Delta_j t)^{1/3}. \tag{9.26}$$

From this, with (9.20) and (9.21),

$$\begin{aligned} \Big|\bar{x}^i(t_{j+1}) - \bar{x}^i(t_j) - \sum_\rho g_\rho^i(\bar{x}(t_j))\Delta_j z^\rho - \sum_{\rho,\sigma} J_{\rho,\sigma} g_{\rho,\sigma}^i(\bar{x}(t_j))\Delta_j z^\rho \Delta_j z^\sigma\Big| \\ \leqq (C_8 + C_9|\bar{x}^i(t_j)|)\varepsilon\Delta_j t. \end{aligned} \tag{9.27}$$

If $\omega \in \Omega - A_j$, this is trivial; the left member of (9.27) is 0 by definition.

By (9.27),

$$\begin{aligned} \Big\|\bar{x}(t_{j+1}) - \bar{x}(t_j) - \sum_\rho g_\rho^i(\bar{x}(t_j))\Delta_j z^\rho - \sum_{\rho,\sigma} J_{\rho,\sigma} g_{\rho,\sigma}^i(\bar{x}(t_j))\Delta_j z^\rho \Delta_j z^\sigma\Big\| \\ \leqq \|C_8\varepsilon\Delta t\| + \|C_9\varepsilon\Delta t\bar{x}(t_j)\|; \end{aligned} \tag{9.28}$$

so (8.5) is satisfied, and by Theorem 8.1, $\bar{x}$ converges to x uniformly in L_2 norm as mesh $\Pi \to 0$.

If $\omega \in A_j$, $\bar{x}(t_j)$ was defined to be $X(t_j)$, and by (9.21) and (9.26),

$$|X^i(t) - \bar{x}^i(t)| \leqq 2\varepsilon[C_5 + C_6|\bar{x}^i(t)|](\Delta_j t)^{1/3}. \tag{9.29}$$

Since $\bar{x}$ converges uniformly in L_2 norm to x, its L_2 norm is bounded, and (9.29) implies that if mesh Π is small,

$$(9.30)\qquad \left\{\int_A |X(t, \omega) - \bar{x}(t, \omega)|^2 P(d\omega)\right\}^{1/2} \leqq C_{10}\varepsilon(\text{mesh }\Pi)^{1/3}.$$

This, with the uniform convergence of $\bar{x}$ to x in L_2 norm, shows that X tends to x uniformly in near L_2 norm as mesh Π tends to 0.

10. Stability, and its limitations

Theorem 9.1 can be regarded as a statement about stability of the solutions of (9.14). Since the last set of stochastic integrals in (9.14) vanish for Lipschitzian z^ρ and z^σ, the theorem informs us that, on the family of disturbances z^ρ consisting of one process satisfying Condition 4.1 and Lemma 6.1 together with all processes interpolated in the z^ρ in accordance with Theorem 9.1, the solutions of (9.14) depend, in a stable or continuous manner, on the disturbances. By estimating the closeness of all approximations we could extend this to a larger collection of z^ρ, all satisfying Condition 4.1 and Condition 6.1 with the same constants, together with all disturbances interpolated in them as in Theorem 9.1.

It would be desirable to permit another kind of interpolation often encountered in applications, in which the z^ρ are approximated by functions Z^ρ that coincide with z^ρ at evenly spaced t_j and have derivatives whose Fourier transforms vanish outside some finite interval. Theorem 9.1 gives us a feeble substitute for this. Let Ψ be infinitely differentiable and nondecreasing on $(-\infty, \infty)$, with $\Psi(t) = 0$ if $t \leqq 0$ and $\Psi(t) = 1$ if $t \geqq 1$. We choose

$$(10.1)\qquad \phi_0(t) = t, \qquad \phi_\rho(t) = \Psi(t) \qquad \rho = 1, \cdots, r,$$

and for each Π, we interpolate Z^ρ in z^ρ using the ϕ_ρ and extend Z^ρ by constancy on $(-\infty, a]$ and $[b, \infty)$. Then the Z^ρ have Fourier transforms that tend to 0 at $\pm\infty$ faster than any negative power of the independent variable, and the Z^ρ satisfy the requirements of Theorem 9.1.

For the case $n = r = 1$, with z^1 a Wiener process, Wong and Zakai [9] have proved a theorem that shows the possibility of using Z^ρ whose derivatives have bounded spectra. Omitting superscripts i and ρ, let $Z_1, Z_2, \cdots$, be a sequence of processes on $[a, b]$ such that with probability 1, $Z_n(t, \omega)$ tends pointwise to $z(t, \omega)$ and has a bounded piecewise continuous derivative, and such that there are finite valued processes n_0, k such that a.s.

$$(10.2)\qquad |Z_n(t, \omega)| \leqq k(\omega), \qquad a \leqq t \leqq b, \textit{ if } n \geqq n_0(\omega).$$

Wong and Zakai then showed that, under essentially the same hypotheses on g_1 and g_{11} as in Theorem 9.1 the solutions X of (9.13) converge almost surely to the solution x of (9.14) (in which we can replace $dzdz$ by dt, by (6.3)).

This theorem does not extend to the case $n = r = 2$, even when z^1 and z^2 are independent Brownian motions, as the following example shows. Consider the stochastic differential equations

$$(10.3)\qquad x^1(t) = \int_0^t dz^1(s), \qquad x^2(t) = \int_0^t x^1(s)\, dz^2(s), \qquad 0 \leqq t \leqq 1,$$

in which z^1 and z^2 are independent standard Wiener processes. Let ψ_1 and ψ_2 be infinitely differentiable nondecreasing functions on $(-\infty, \infty)$ such that

$$(10.4)\qquad \psi(t) = \begin{cases} 0 & \text{if } t \leqq 0, \\ 1 & \text{if } t \geqq 1/2, \end{cases}$$

$$(10.5)\qquad \psi(t) = \begin{cases} 0 & \text{if } t \leqq 1/2, \\ 1 & \text{if } t \geqq 1. \end{cases}$$

Then

$$(10.6)\qquad \begin{aligned} &\int_0^1 [1 - \psi_1(s)]\dot{\psi}_2(s)\, ds = 0, \\ &\int_0^1 [1 - \psi_2(s)]\dot{\psi}_1(s)\, ds = 1. \end{aligned}$$

Given a Cauchy partition Π, with the usual notation (8.1), for $\rho = 1, 2$ and for $t_j \leqq t \leqq t_{j+1}$, we define

$$(10.7)\qquad \begin{aligned} Z^\rho(t, \omega) &= z^\rho(t, \omega) + \psi_\rho\left(\frac{t - t_j}{\Delta_j t}\right) \Delta_j z^\rho(\omega), \\ \tilde{Z}^\rho(t, \omega) &= z^\rho(t, \omega) + \psi_{3-\rho}\left(\frac{t - t_j}{\Delta_j t}\right) \Delta_j z^\rho(\omega), \end{aligned}$$

if

$$(10.8)\qquad \Delta_j z^1(\omega)\Delta_j z^2(\omega) \geqq 0,$$

and we define $Z^\rho(t, \omega)$ and $\tilde{Z}^\rho(t, \omega)$ by (10.7) with the right members interchanged if (10.8) is false. We extend Z^ρ and $\tilde{Z}^\rho$ by constancy on $(-\infty, a]$ and on $[b, \infty)$. Then Z^ρ and $\tilde{Z}^\rho$, $\rho = 1, 2$, are infinitely differentiable, have the same bounds as z^ρ, and with probability 1 converge to z^ρ uniformly on $[a, b]$. By (9.20) and (10.6), for the corresponding solutions of (10.3) we have

$$(10.9)\qquad \begin{aligned} X^1(t_{j+1}) - X(t_j) &= \Delta_j z^1, \\ X^2(t_{j+1}) - X^2(t_j) &= X^1(t_j)\Delta_j z^2 + (\Delta_j z^1 \Delta_j z^2)^+, \end{aligned}$$

and

$$(10.10)\qquad \begin{aligned} \tilde{X}^1(t_{j+1}) - \tilde{X}^1(t_j) &= \Delta_j z^1, \\ \tilde{X}^2(t_{j+1}) - \tilde{X}^2(t_j) &= \tilde{X}^1(t_j)\Delta_j z^2 - (\Delta_j z^1 \Delta_j z^2)^-. \end{aligned}$$

Define

$$\xi^\rho = X^\rho - \tilde{X}^\rho, \qquad \rho = 1, 2; \tag{10.11}$$

then from (10.9) and (10.10)

$$\xi^1(t_{j+1}) - \xi^1(t_j) = 0, \tag{10.12}$$

$$\xi^2(t_{j+1}) - \xi^2(t_j) = \xi^1(t_j) + |\Delta_j z^1 \Delta_j z^2|. \tag{10.13}$$

From (10.12), $\xi^1(t_j) = 0$ for all j, so by (10.13)

$$\xi^2(1) = \sum_j |\Delta_j z^1 \Delta_j z^2|. \tag{10.14}$$

Since the $\Delta_j z^1$ and $\Delta_j z^2$ are independent normal r.v., $\Delta_j z^\rho$ having mean 0 and variance $\Delta_j t$, we readily compute

$$E(\xi^2(1)) = \sum_j \frac{2\Delta_j t}{\pi} = \frac{2}{\pi}, \tag{10.15}$$

and

$$\begin{aligned} \operatorname{Var} \xi^2(1) &\leqq E\Big(\sum_j [\Delta_j z^1]^2 [\Delta_j z^2]^2\Big) \\ &= \sum_j (\Delta_j t)^2, \end{aligned} \tag{10.16}$$

which tends to 0 with mesh Π. So $X^2(1) - \tilde{X}^2(1)$ tends in L_2 norm to $2/\pi$ as mesh Π tends to 0, and it is impossible that $X^2(1)$ and $\tilde{X}^2(1)$ both tend to the same limit $x^2(1)$, a.s., or even in probability.

The example shows the inherent limitations on stability of models. With such a simple system as (10.3), when mesh Π is small, the results of linear interpolation in z^ρ and of the interpolation (10.9) in z^ρ will have differences that are uniformly arbitrarily small for almost all ω. Yet the solutions X of the ordinary equations (10.4) and (10.5) corresponding to those two practically indistinguishable disturbances will not be arbitrarily close to each other in L_2 norm. Hence, no "selection principle" can possibly provide a model that is consistent, inclusive enough to include Lipschitzian processes and Brownian motions, and so thoroughly stable as to yield practically indistinguishable solutions corresponding to practically indistinguishable disturbances. The limited stability described in Theorem 9.1 may be about as much as we can attain.

Perhaps we are studying the problem from the wrong end. As mentioned in Section 1, if we wish to stay in the domain of trustworthiness of classical scientific theories, we should hold to Lipschitzian disturbances. Idealizations to martingales are made for mathematical convenience, and they depart from the Lipschitzian case so far that no martingale can have a.s. Lipschitzian sample paths unless the sample paths are a.s. constant (see Fisk, [3]).

In Theorem 9.1, and in the theorems of Wong and Zakai, the idealization is the starting point, and it is approximated by the Lipschitzian Z^ρ. Since it is the

Lipschitzian case that is presented to us by the outside world, it would seem more significant to find how well we can approximate the processes of classical theory by our idealizations, rather than the reverse. But this would appear to be a difficult undertaking.

11. A Runge-Kutta type of approximation

The Cauchy (or Euler) polygons are useful in the theory of ordinary differential equations, but for computation they are much inferior to the Runge-Kutta approximations. As adapted to equations (9.3), this method can be described thus. Given a partition Π (with the usual notation (8.1)), we define $y(a) = x_0$, and then define $y(t_2), \cdots$ successively as follows. From $y(t_j)$ we first compute, as in (8.2), the value of

$$y^i(t_j) + \sum_\rho g^i_\rho[y(t_j)]\Delta_j z^\rho. \tag{11.1}$$

But instead of using the g^i_ρ corresponding to these sums as coefficients for the next step, as in the Cauchy-Maruyama process, we average them with the $g^i_\rho(y(t_j))$ to furnish a second approximation to the values of the g^i_ρ for use in estimating $y(t_{j+1})$. Thus, we have

$$\begin{aligned} y(t_{j+1}) = y(t_j) &+ \tfrac{1}{2}\sum_\rho [g^i_\rho(y(t_j))\Delta_j z^\rho] \\ &+ \tfrac{1}{2}\sum_\rho g^i_\rho[y(t_j) + \sum_\sigma g_\sigma(y(t_j))\Delta_j z^\sigma]\Delta_j z^\rho. \end{aligned} \tag{11.2}$$

The values of y at points interior to intervals $[t_k, t_{k+1}]$ are of secondary interest; we could, for example, define them by linear interpolation.

The preservation of a formula or an algorithm is a much less basic stability property than that discussed in the preceding section. Nevertheless, it is to some extent significant, as well as computationally convenient, that if we try to approximate solutions of (9.3) by the Runge-Kutta method for processes satisfying the hypotheses of Theorem 9.1 and one more continuity condition Assumption 11.1, the approximations will converge, not to the solution of (9.3), but to the solution of (9.14). This in a sense gives added recommendation to our "selection principle." But besides this, it permits us to use a well-known computation procedure to approximate the solution of (9.14), whether the z^ρ are Lipschitzian or are martingales or any other processes satisfying the hypotheses of Theorem 9.1, without having to interpolate to find the Z^ρ and without having to solve equations (9.13).

Equation (11.2) may be regarded as the first step in the iterative solution of

$$y(t_{j+1}) = y(t_j) + \tfrac{1}{2}\sum_\rho \{g^i_\rho[y(t_j)] + g^i_\rho[y(t_{j+1})]\}\Delta_j z^\rho. \tag{11.3}$$

To guarantee the convergence of such an iterative process, it is desirable and usual to make assumptions that guarantee that the successive corrections form a diminishing sequence. One such assumption, for the present problem, is the following.

ASSUMPTION 11.1. *There are positive numbers* δ_1, L_1 *such that if* x_1 *and* x_2 *are points of* $\mathbf{R}^{n+1}$ *with* x_1^0 *in* $[a, b]$ *and* $|x_2 - x_1| \leqq \delta_1(1 + |x_1|)$, *then*

$$|g^i_{\rho,x^\alpha}(x_2) - g^i_{\rho,x^\alpha}(x_1)| \leqq \frac{L_1|x_1 - x_2|}{1 + |x_1 - x_2|}, \tag{11.4}$$

This rather strong uniform continuity requirement will be further discussed after proving the next theorem.

THEOREM 11.1. *Let the* z^ρ *and* g^i_ρ *satisfy the hypotheses of Theorem* 9.1, *and also satisfy Assumption* 11.1. *For each Cauchy partition* Π *of* $[a, b]$, *let* y *be the process determined by the Runge-Kutta process* (11.2), *with linear interpolation between the division points of* Π. *Then as mesh* $\Pi \to 0$, y *converges on* $[a, b]$ *uniformly in near* L_2 *norm to the solution* x *of* (9.14).

PROOF. Let ε be positive, and let δ and A correspond to ε for all the z^ρ as in Condition 6.1. We define the sets $A_1, \cdots, A_\ell$ as in the sentence containing (9.16) and we define a process $\bar{x}$ corresponding to Π as follows. First, $\bar{x}(a) = x_0$. Next, $\bar{x}(t_j)$ having been defined, we define $\bar{x}(t_{j+1})$ by

$$\bar{x}(t_{j+1}, \omega) = y(t_{j+1}, \omega) \qquad \text{if} \quad \omega \,\varepsilon\, A_j, \tag{11.5}$$

$$\begin{aligned}\bar{x}(t_{j+1}, \omega) = \bar{x}(t_j, \omega) &+ \sum_\rho g^i_\rho(\bar{x}(t_j, \omega))\Delta_j z^\rho \\ &+ \tfrac{1}{2}\sum_{\rho,\sigma} g^i_{\rho,\sigma}(\bar{x}(t_j, \omega))\Delta_j z^\rho \Delta_j z^\sigma \\ &\qquad\qquad \text{if} \quad \omega \in \Omega - A_j.\end{aligned} \tag{11.6}$$

Finally, we define

$$\bar{x}^i(t, \omega) = \bar{x}^i(t_j, \omega), \qquad t_j \leqq t < t_{j+1}. \tag{11.7}$$

It is easy to verify that these $\bar{x}$ satisfy conditions (f), (h), and (8.4). For ω in A_j, by applying the theorem of the mean to (11.2), we find

$$\begin{aligned}y^i(t_{j+1}) = y^i(t_j) &+ \tfrac{1}{2}\sum_\rho g^i_\rho(y(t_j))\Delta_j z^\rho \\ &+ \tfrac{1}{2}\sum_\rho \{g^i_\rho(y(t_j)) + \sum_\sigma g^i_{\rho,\sigma}(\eta)\Delta_j z^\sigma\}\Delta_j z^\rho,\end{aligned} \tag{11.8}$$

where

$$\eta^i = y^i(t_j) + \theta^i_j(\omega)\sum_\tau g^i_\tau(y(t_j))\Delta_j z^\tau \tag{11.9}$$

for some $\theta_j^i(\omega)$ in $(0, 1)$. As in Section 9, we use C_1, C_2, and so forth, to denote numbers determined by the data of the problem. Then, by Assumption 11.1, (11.9) implies

$$(11.10) \qquad |g^i_{\rho,\sigma}(\eta) - g^i_{\rho,\sigma}(y(t_j))| \leqq (C_1 + C_2|y(t_j)|)\varepsilon(\Delta_1 t)^{1/3}.$$

From this and (11.8), since $\omega \in A_j$,

$$(11.11) \quad |\bar{x}^i(t_{j+1}) - \bar{x}^i(t_j) - \sum_\rho g^i_\rho(\bar{x}(t_j))\Delta_j z^\rho - \tfrac{1}{2}\sum_{\rho,\sigma} g^i_{\rho,\sigma}(\bar{x}(t_j))\Delta_j z^\rho \Delta_j z^\sigma| \leqq (C_3 + C_4|\bar{x}(t_j)|)\varepsilon^3\Delta_j t.$$

This also holds if $\omega \in \Omega - A_j$, since then the left member is 0 by definition. By taking the expectation of the square of the left member of (11.11), we find, as in (9.28), that (8.5) also is satisfied.

Now by Theorem 8.1, $\bar{x}$ converges to the solution x of (9.14) uniformly in L_2 norm, as mesh $\Pi \to 0$. Since x is continuous in L_2 norm and the $x(t_j)$ are uniformly close in L_2 norm to $x(t_j)$ at all division points of Π if mesh Π is small, it is easy to see that if we modify $\bar{x}$ by retaining its values at the t_j but interpolating linearly between them, the modified process also converges to x uniformly in L_2 norm. But this modified process coincides with the Runge-Kutta process y for all ω in A, and so the proof is complete.

Assumption 11.1 is strong, but from an experimental or computational point of view it can be tolerated. Ordinarily there will be some bound B on the norms of the x that interest us. In an experiment, points with $|x| > B$ will make the points too far away to be involved in the process under investigation; in computation, B could be a bound on the numbers within the machine's capacity. If we replace the $g^i_\rho(x)$ by other functions $G^i_\rho(x)$ that satisfy Assumption 11.1 and coincide with $g^i_\rho(x)$ whenever $|x| \leqq B$, the solution of (9.14) with coefficients G^i_ρ will coincide with the solution of (9.13) as written unless the solution somewhere has norm greater than B: and unless the probability of this is negligibly small, we face worse troubles than the mere nonconvergence of the Runge-Kutta procedure.

Professor H. Rubin informs me that Dr. Donald Fisk, in his doctoral dissertation at Michigan State University, has defined and studied a stochastic integral which is the limit of "trapezoidal rule" approximations, the values of the integrand at the beginning and end of each interval $[t_j, t_{j+1}]$ being averaged. (Professor Rubin also furnished reference [2].) Existence theorems are established for integrals $\int f dz$ in which z is a quasi-martingale (see Fisk, [3]). I have not had the opportunity of seeing this dissertation, but it is evident that the application of Fisk's integral to differential equations (1.3) must be closely related to the procedure described in Theorem 11.1, and even more closely related to the process mentioned in (11.3).

REFERENCES

[1] J. L. DOOB, *Stochastic Processes*, New York, Wiley, 1953.
[2] D. L. FISK, "Quasi-Martingales and stochastic integrals," Technical Report No. 1, Dept. Statistics, Michigan State University, 1963.
[3] ———, "Quasi-Martingales," *Trans. Amer. Math. Soc.*, Vol. 120 (1965), pp. 369–389.
[4] G. MARUYAMA, "Continuous Markov processes and stochastic equations," *Rend. Circ. Mat. Palermo*, Vol. 4 (1955), pp. 48–90.
[5] E. J. MCSHANE, "Stochastic integrals and stochastic functional equations," *SIAM J. Appl. Math.*, Vol. 17 (1969), pp. 287–306.
[6] ———, "Toward a stochastic calculus," *Proc. Nat. Acad. Sci., U.S.A.*, Vol. 63 (1969), Part I, pp. 275–280; Part II, pp. 1084–1087.
[7] J. NEVEU, *Mathematical Foundations of the Calculus of Probability* (translated by A. Feinstein), San Francisco, Holden-Day, 1965.
[8] E. WONG and M. ZAKAI, "On the relation between ordinary and stochastic differential equations," *Internat. J. Engrg. Sci.*, Vol. 3 (1965), pp. 213–229.
[9] ———, "On the convergence of ordinary integrals to stochastic integrals," *Ann. Math. Statist.*, Vol. 36 (1965), pp. 1560–1564.

BIRTH AND DEATH OF MARKOV PROCESSES

P. A. MEYER
UNIVERSITY OF STRASBOURG
R. T. SMYTHE
UNIVERSITY OF WASHINGTON
J. B. WALSH
UNIVERSITY OF STRASBOURG

1. Introduction

We must start with some basic notation and definitions. To avoid stopping the process at time 0, we shall deal with one particular situation, leaving it to the specialist to check whether our conclusions remain true when all hypotheses are deleted. Let E be a locally compact space with countable base. Some $\Delta \in E$ has been singled out for infamous purposes. Let Ω be the set of all mappings $\omega : \mathbf{R}_+ \rightsquigarrow E$ which are right continuous and possess a "lifetime" ζ (possibly 0 or $+\infty$), namely,

(1.1) $$\omega(t) \neq \Delta \text{ for } t < \zeta(\omega), \qquad \omega(t) = \Delta \text{ for } t > \zeta(\omega).$$

We set as usual $X_t(\omega) = \omega(t)$, $X_\infty(\omega) = \Delta$, and provide Ω with the natural family of σ-fields $(\mathscr{F}_t^0)_{t \leq \infty}$ of the process (X_t). Given now a Hunt transition semigroup $(P_t)_{t \geqq 0}$ on E, with Δ as an absorbing point, we can define as usual measures P^μ, P^x on Ω, for which the process (X_t) is Markovian, with the transition semigroup (P_t) and initial measures μ, ε_x. The assumptions concerning left limits in the definition of Hunt processes will be superfluous most of the time. We postpone all other definitions to the main text.

We are interested in operations on the sample paths which preserve the homogeneous Markov character of the process, with possible alteration of the semigroup. Known examples of these operations are: turning a set into an absorbing barrier; restarting the process at a stopping time (we shall use the terminology "optional r.v." rather than "stopping time"); killing the process at a terminal time; reversing time at an L time (L times are called cooptional random variables below); clock changing relative to a continuous additive functional. Our purpose here consists in giving two more examples of such transformations.

Let us say informally that a positive random variable R is a *birth time* for the process if the process $(X_{R+t})_{t \geqq 0}$ starting at time R is, for every law P^μ, a homogeneous Markov process (its transition semigroup may depend on R, but not on μ). Similarly, replacing the process starting at R by the process killed at R, we get the notion of a *death time*. Then two of the preceding examples can be stated as follows:

(a) optional times are birth times,

(b) terminal times are death times.

We are going to prove here that:

(c) coterminal times are birth times,

(d) cooptional times are death times.

Thus, for instance, the last exit time from a Borel set, which was historically the first example of an L time (or cooptional r.v.), turns out to be the model for coterminal times also. According to (c) and (d), it is both a birth time and a death time.

Properties (c) and (d) are dual to (a) and (b), since time reversal from ζ exchanges optional for cooptional, terminal for coterminal and birth for death. Unfortunately, full proofs using time reversal would be very cumbersome, owing to the impossibility of reversing on $\{\zeta = \infty\}$. Therefore, we have used time reversal only as a guide to intuition.

2. Cooptional random variables and killing

Let us complete our notations. As usual, $\mathscr{F}^{\mu}$ denotes the σ-field $\mathscr{F}^0$ completed with respect to P^{μ}, and $\mathscr{F}_t^{\mu}$ is $\mathscr{F}_t^0$ augmented with all subsets of $\mathscr{F}^{\mu}$ of measure 0; we set $\mathscr{F} = \bigcap_{\mu} \mathscr{F}^{\mu}$, $\mathscr{F}_t = \bigcap_{\mu} \mathscr{F}_t^{\mu}$. It is well known that the family $(\mathscr{F}_t)$ is right continuous.

The *shift operator* θ_t is defined as usual by $X_s(\theta_t\omega) = X_{s+t}(\omega)$ for all $s \geqq 0$, and the *killing operator* k_t is defined by

$$X_s(k_t\omega) = \begin{cases} X_s(\omega) & \text{if } s < t, \\ \Delta & \text{if } s \geqq t. \end{cases} \tag{2.1}$$

We have $k_s \circ k_t = k_{s \wedge t}$.

DEFINITION 2.1. *A positive random variable L on $(\Omega, \mathscr{F})$ is cooptional if and only if we have identically in $t \geqq 0$*

$$L(\theta_t\omega) = \big(L(\omega) - t\big)^+. \tag{2.2}$$

One generally allows a negligible set on which (2.2) doesn't hold. For the sake of simplicity, we demand a true identity. If (2.2) isn't an identity, one can generally restrict Ω to some shift invariant and killing invariant subset Ω_0 of full measure on which (2.2) holds identically, and the extension of our results becomes trivial.

The most classical examples of cooptional times are *last exit times*: if A is Borel in E, one sets $L_A(\omega) = \sup\{t \geqq 0 : X_t(\omega) \in A\}$ with the usual convention that $\sup(\varnothing) = 0$. The terminology "cooptional" requires some justification, provided by the following remark. Let T be a random variable on $(\Omega, \mathscr{F}^0)$, such that $0 \leqq T \leqq \zeta$. Then T is optional relative to the family $(\mathscr{F}_{t+}^0)$ if and only if the identity $T \circ k_t = T \wedge t$ holds, and this is dual to (2.2) by time reversal at ζ.

We collect in the following proposition some useful and simple results on cooptional random variables.

PROPOSITION 2.1. (i) *If L is cooptional, so is $(L - t)^+$ for $t \geqq 0$.*

(ii) *If L and L' are cooptional, so are $L \wedge L'$ and $L \vee L'$.*

(iii) *Let L be cooptional, and $\mathcal{G}_L$ be the set of all $A \in \mathcal{F}$ such that for every $u \geqq 0$, $A \cap \{L > u\} = \theta_u^{-1}(A) \cap \{L > u\}$. Then $\mathcal{G}_L$ is a σ-field, and X_L is $\mathcal{G}_L$ measurable (and also $\{L = \infty\} \in \mathcal{G}_L$). If $A \in \mathcal{G}_L$, the random variable L^A defined by $L^A = L$ on A, $L^A = 0$ on A^c is cooptional.*

The proofs are very easy, and we give no details.

DEFINITION 2.2. *The excessive function associated to the cooptional random variable L is $c_L(\cdot) = P^{\cdot}\{L > 0\}$.*

We have $P_t c_L(x) = P^x\{L \circ \theta_t > 0\} = P^x\{L > t\}$. This implies immediately that c_L is excessive. We shall write c instead of c_L, to simplify notation, and define as usual the c path semigroup Q_t as

$$Q_t(x, dy) = \begin{cases} \dfrac{1}{c(x)} P_t(x, dy) c(y) & \text{if } c(x) \neq 0, \\ \varepsilon_\Delta(dy) & \text{if } c(x) = 0. \end{cases} \tag{2.3}$$

We give now the main theorem of this section. The result is quite easy, but it has one surprising feature: we are *killing* the process (X_t), and get a new process (Y_t) whose transition semigroup *is not dominated* by (P_t). This is shocking at first, but becomes quite easy to understand if one notes that if (X_t) starts at x, then the initial measure of (Y_t) is not ε_x, but $c(x)\varepsilon_x + (1 - c(x))\varepsilon_\Delta$.

THEOREM 2.1. *Let L be cooptional, and (Y_t) be the process (X_t) killed at time L*

$$Y_t(\omega) = \begin{cases} X_t(\omega) & \text{if } t < L(\omega), \\ \Delta & \text{if } t \geqq L(\omega). \end{cases} \tag{2.4}$$

If Ω is given the measure P^μ, the process (Y_t) is Markovian with (Q_t) as transition semigroup, and $c \cdot \mu + \langle \mu, 1 - c \rangle \varepsilon_\Delta$ as initial measure.

PROOF. Let us define the σ-field $\mathcal{H}_t$ as the set of all $A \in \mathcal{F}$ such that there exists $A_t \in \mathcal{F}_t$ satisfying $A \cap \{t < L\} = A_t \cap \{t < L\}$. The family $(\mathcal{H}_s)$ is increasing, and Y_t is $\mathcal{H}_t$ measurable.

Let s and t be two epochs such that $s < t$, ϕ be a bounded $\mathcal{H}_s$ measurable random variable, and f be Borel and bounded on E, equal to 0 at Δ. The theorem amounts to saying that

$$E^\mu[\phi \cdot f \circ Y_t] = E^\mu[\phi \cdot Q_{t-s}(Y_s, f)]. \tag{2.5}$$

Let us compute the left side. It is equal to

$$(2.6)\qquad E^{\mu}[\phi\cdot f\circ X_t\cdot I_{\{t<L\}}] = E^{\mu}[\phi_s\cdot f\circ X_t\cdot I_{\{t<L\}}] = E^{\mu}[\phi_s\cdot f\circ X_t\cdot I_{\{L\circ\theta_t>0\}}],$$

where ϕ_s is (according to the definition of $\mathscr{H}_s$) some $\mathscr{F}_s$ measurable and bounded r.v. equal to ϕ on $\{s < L\}$. Taking conditional expectations with respect to $\mathscr{F}_t$, we get

$$(2.7)\qquad E^{\mu}[\phi_s\cdot f\circ X_t\cdot P^{X_t}\{L > 0\}] = E^{\mu}[\phi_s\cdot((fc)\circ X_t)].$$

Taking conditional expectations with respect to $\mathscr{F}_s$, we write this last integral as $E^{\mu}[\phi_s\cdot P_{t-s}(X_s, fc)]$. On the other hand, $P_{t-s}(fc) = 0$ on $\{c = 0\}$ since the function c is excessive, and the expectation can be written

$$(2.8)\qquad E^{\mu}[\phi_s\cdot I_{\{c>0\}}\cdot P_{t-s}(X_s, fc)] = E^{\mu}[\phi_s\cdot c\circ X_s\cdot Q_{t-s}(X_s, f)]$$

On the other hand, the right side of (2.5) is equal to

$$(2.9)\qquad \begin{aligned} E^{\mu}[\phi\cdot Q_{t-s}(X_s, f)\cdot I_{\{s<L\}}] &= E^{\mu}[\phi_s\cdot Q_{t-s}(X_s, f)\cdot I_{\{s<L\}}] \\ &= E^{\mu}[\phi_s\cdot Q_{t-s}(X_s, f)\cdot c\circ X_s] \end{aligned}$$

(by the same reasoning as above). The theorem is proved.

We shall leave the main subject for a moment, to discuss the relation between the existence of cooptional times and the transience of the process. Let us say that a cooptional time L is *trivial* if for all x either $P^x\{L = 0\} = 1$, or $P^x\{L = \infty\} = 1$, and denote by U the potential operator $\int_0^\infty P_t\,dt$. We say that an excessive function h is *nearly constant* if for all x $P^x\{h\circ X_t = h(x)$ for all $t\} = 1$.

THEOREM 2.2. *The following statements are equivalent:*

(i) *$U(x, B)$ takes on only the values* 0 *and* ∞*;*

(ii) *all cooptional times are trivial.*

PROOF. Assume (i), and let h be excessive, bounded by 1. Then h cannot have a potential part, and therefore is invariant. From martingale theory, $h(x) = E^x[h_\infty]$, $h\circ X_t = E[h_\infty|\mathscr{F}_t]$, where $h_\infty = \lim_{s\to\infty} h\circ X_s$. Now $1 - h^2$ is excessive, hence invariant; therefore, $h^2(x) = E^x[h_\infty^2]$ and h_∞ must be P^x a.s. equal to $h(x)$, and h is nearly constant. This part of the reasoning is well known.

Let L be cooptional, and assume that $P^x\{L = \infty\} < 1$. Set $h(x) = P^x\{L = \infty\}$. This is an invariant function, and $h_\infty = I_{\{L=\infty\}}$. Since h_∞ is a constant P^x a.s., and is not a.s. 1, $h_\infty = 0$ P^x a.s. Then the excessive function $c_L = P^{\cdot}\{L > 0\}$ is such that $P_t c_L \to 0$ at x as $t\to\infty$. Therefore at x, $c_L = \lim_{t\to 0} U((c_L - P_t c_L)/t)$; since U takes on the values 0 and ∞ only, $c_L(x) = 0$. Finally, $P^x\{L = 0\} = 1$.

Conversely, assume that for some $x \in E$ and $B \subset E$, we have $0 < U(x, B) < \infty$. Then $P_t(x, U(I_B)) \to 0$ as $t \to \infty$. Hence, $U(X_t, B) \to 0$ P^x a.s. as $t \to 0$.

Define L as the last exit time from $\{U(\cdot, B) > \varepsilon\}$; then for $\varepsilon > 0$, L is nontrivial in the stronger sense that $P^x\{0 < L < \infty\} \neq 0$. Theorem 2.2 is proved.

3. Some trivialities about fields

In Section 5, it will be useful to define $\mathscr{F}_R$ for random variables R which are not optional. This has already been done by K. L. Chung and J. L. Doob in a well-known paper, but their definition is not convenient for our purposes. Therefore, we are going to shift for a while to "general abstract nonsense." This section may be safely skipped, except for a glance at the results.

DEFINITION 3.1. *Let R be any positive r.v. on $(\Omega, \mathscr{F})$. We define $\mathscr{F}_R$ as the set of all $A \in \mathscr{F}$ such that for every $t \geqq 0$ there exists $A_t \in \mathscr{F}_t$ such that $A \cap \{R < t\} = A_t \cap \{R < t\}$. We say that R is honest if R is $\mathscr{F}_R$ measurable.*

In more general situations $A \in \mathscr{F}$ should be replaced by $A \in \mathscr{F}_\infty$. If R is optional, Definition 3.1 is equivalent to the standard definition of $\mathscr{F}_R$.

Here are some obvious remarks. A real valued r.v. Y is $\mathscr{F}_R$ measurable if and only if for every t there exists some $\mathscr{F}_t$ measurable r.v. Y_t such that $Y = Y_t$ on $\{R < t\}$. Then for any optional T there exists some $\mathscr{F}_T$ measurable Y_T such that $Y = Y_T$ on $\{R < T\}$ (start with a countably valued T, and pass to the limit, using the right continuity of the family $(\mathscr{F}_t)$). Also, $\{R < T\}$ can be replaced by $\{R \leqq T\}$. Finally, note that the family $(\mathscr{F}_{R+t})_{t \geqq 0}$ is increasing and right continuous.

The random variable R is honest if and only if, for every t, R is equal on $\{R < t\}$ to some $\mathscr{F}_t$ measurable r.v. R_t. This allows a simple characterization of $\mathscr{F}_R$ when R is honest.

PROPOSITION 3.1. *Let R be honest. Then a random variable Y is $\mathscr{F}_R$ measurable if and only if there exists a progressively measurable process (Y_t) such that $Y = Y_R$ on $\{R < \infty\}$.*

PROOF. The condition is necessary: indeed, for every s rational choose some r.v. Y_s^1, which is $\mathscr{F}_s$ measurable, such that $Y = Y_s^1$ on $\{R < s\}$. Then for every t set $Y_t = \liminf_{s \downarrow t, s > t} Y_s^1$ where s is rational; this is a progressive process, and $Y = Y_R$ on $\{R < \infty\}$.

Conversely, if (Y_t) is progressive, and $Y = Y_R$, then Y is equal on $\{R < t\}$ to Y_{R_t}, which is $\mathscr{F}_t$ measurable. Therefore, Y is $\mathscr{F}_R$ measurable.

We do not need anything more about honest random variables. Let us just quote an amusing result. Consider the abstract situation of a right continuous family $(\mathscr{F}_t)$, such that $\mathscr{F}_0$ contains all sets of measure 0 for the basic measure P. Then R is honest if and only if there exists a well-measurable subset H of $\mathbf{R}_+ \times \Omega$ such that R is the *end of* H:

$$R(\omega) = \sup \{t : (t, \omega) \in H\}. \tag{3.1}$$

The class of all random variables which are ends of *predictable* sets is strictly smaller than that of honest random variables and particularly interesting, as shown by a recent paper of Azema.

4. Terminal and coterminal times, duality

DEFINITION 4.1. *A random variable* $L \geqq 0$ *on* $(\Omega, \mathscr{F})$ *is a coterminal time if*

(i) $L \circ \theta_s = (L - s)^+$ *for every* s,

(ii) $L \circ k_s = L$ *on* $\{L < s\}$,

(iii) $L \circ k_s \leqq s$ *for every* s.

The first property means that L is cooptional. The second property implies that L is honest, and would mean precisely honestness if the σ-fields had not been completed. The third property is equivalent to the inequality $L \leqq \zeta$: indeed, applying (iii) with $s = \zeta$, we get that $L \leqq \zeta$, and conversely $L \leqq \zeta$ implies $L \circ k_s \leqq \zeta \circ k_s \leqq s$.

Examples of coterminal times are: the lifetime ζ and last exit times from sets; last exit times of the left limit process (X_{t-}) from sets; last jumps of one given kind, and so forth.

We state some obvious properties of coterminal times.

PROPOSITION 4.1. *Let L be a coterminal time. Then:*

(i) $L \circ k_s \leqq L$ *for every* s;

(ii) $s < t$ *implies* $L \circ k_s \leqq L \circ k_t$;

(iii) $s < t$, $L \circ k_s = 0$, *and* $L \circ k_t > 0$ *imply* $L \circ k_t \geqq s$;

(iv) *set* $L' = \sup_{s<\infty} L \circ k_s$. *Then* L' *is a coterminal time,* $L' \leqq L$, *with equality on* $\{L < \infty\}$, *and* $L \circ k_t = L' \circ k_t$ *for every finite* t.

PROOF. (i) If $s \leqq L$, then $L \circ k_s \leqq s \leqq L$ from Definition 4.1 (iii). If $L < s$, $L \circ k_s = L$ from Definition 4.1 (ii).

(ii) Composition of (i) with k_t gives $L \circ k_s = L \circ k_s \circ k_t \leqq L \circ k_t$.

(iii) Assume the contrary: $L \circ k_t < s$. Then from Definition 4.1 (ii) $L \circ k_s \circ k_t = L \circ k_t$, a contradiction since the right side is greater than 0, while the left side is equal to $L \circ k_s = 0$.

(iv) According to (ii), this "sup" is really a lim. We first have $L' \leqq L$ from (i). If $L < \infty$, take $s > L$ and apply Definition 4.1 (ii), we find $L' \geqq L \circ k_s = L$, so L and L' are equal on $\{L < \infty\}$. We have for finite t $L' \circ k_t = \lim_{s\to\infty} L \circ k_s \circ k_t = L \circ k_t$ (take $s > t$).

We must prove that L' is a coterminal time. We have seen that Definition 4.1 (iii) is equivalent to $L' \leqq \zeta$; since $L' \leqq L$ and L satisfies Definition 4.1 (iii), the property holds. Next $(L \circ k_u) \circ \theta_s = L \circ \theta_s \circ k_{u-s}$ if $u > s$, and this is equal to $(L \circ k_{u-s} - s)^+$ from Definition 4.1 (i). Letting $u \to \infty$, we get property Definition 4.1 (i) for L'. Let us finally prove Definition 4.1 (ii): assume $L' < s$, then we have $L \circ k_u < s$ if u is finite; hence, from Definition 4.1 (ii) $L \circ k_u = L \circ k_s \circ k_u = L \circ k_s$ if $u > s$. Letting $u \to \infty$, we get $L' = L' \circ k_s$.

DEFINITION 4.2. *The coterminal time L is exact if* $L = \sup_{s<\infty} L \circ k_s$.

If L is not exact, then the coterminal time L' we have just considered is exact, and differs from L only on $\{L = \infty\}$. Thus, exactness is not much of a restriction.

We are going now to construct the terminal time associated with the coterminal time L.

DEFINITION 4.3. *The time* T_L *(or simply* T*) is the function on* Ω *defined by*

$$T_L(\omega) = \inf\{t > 0: L(k_t\omega) > 0\}. \tag{4.1}$$

Note that $T_L = T_{L'}$.

All the results on T_L which may be needed below are collected in the following proposition.

PROPOSITION 4.2. *The function $T = T_L$ satisfies the following relations:*

(i) *T is an optional r.v.;*

(ii) *$T \leqq L$ on $\{L > 0\}$, $T = \infty$ on $\{L = 0\}$;*

(iii) *$T \circ \theta_t = \infty$ is equivalent to $L' \leqq t$; hence, $L' = \sup \{t : T \circ \theta_t < \infty\}$;*

(iv) *$T \circ k_s = \infty$ on $\{T > s\}$;*

(v) *T is a perfect, exact terminal time;*

(vi) *$a < L' \leqq b$ is equivalent to $T \circ \theta_a \leqq b - a$, $T \circ \theta_b = \infty$.*

PROOF. (i) The inequality $T < t$ holds if and only if there exists some rational $r < t$ such that $L \circ k_r > 0$. This implies that T is an optional r.v.

(ii) If $L = 0$, then $L \circ k_t \leqq L$ is 0 for all t, and hence $T = \infty$. The inequality $T \leqq L$ is obvious on $L = \infty$. If $0 < L < \infty$, take $t > L$; therefore $L = L \circ k_t$ from Definition 4.1 (ii), and since $L > 0$ we have $T \leqq t$, and finally $T \leqq L$.

(iii) Using Definition 4.1 (i),

$$(4.2) \qquad T \circ \theta_t = \inf \{u > 0 : L \circ k_u \circ \theta_t > 0\} = \inf \{u > 0 : L \circ \theta_t \circ k_{u+t} > 0\}$$
$$= \inf \{u > 0 : L \circ k_{u+t} > t\},$$

so $(T \circ \theta_t = \infty) \Leftrightarrow (L \circ k_{u+t} \leqq t \text{ for all } t) \Leftrightarrow (L' \leqq t)$.

(iv) is obvious, from the definition of T.

(v) A *terminal time* is an optional r.v. S such that for every t, $S = t + S \circ \theta_t$ a.s. on $\{S > t\}$. In the definition which is most commonly used, the exceptional set may depend on t; if it can be chosen independent of t, S is a *perfect* terminal time (in our case the exceptional set will turn out to be empty). Also S is *exact* if $t + S \circ \theta_t \to S$ a.s. as $t \to 0$.

From the proof of (iii), we have

$$(4.3) \qquad t + T \circ \theta_t = \inf \{w > t : L \circ k_w > t\} = \inf \{w > 0 : L \circ k_w > t\}$$

(since $w \geqq L \circ k_w$, according to Definition 4.1 (iii)). Assume that $t < T$, and choose u such that $t < u < T$. Then $L \circ k_u = 0$, and from Proposition 4.1 (iii), $L \circ k_w > 0$ implies $L \circ k_w \geqq u > t$. Therefore, on $t < T$ we have

$$(4.4) \qquad t + T \circ \theta_t = \inf \{w > 0 : L \circ k_w > 0\} = T,$$

and this means that T is a perfect terminal time without exceptional set.

Let us prove that T is exact. We always have

$$(4.5) \qquad t + T \circ \theta_t = \inf \{w > 0 : L \circ k_w > t\}$$
$$\geqq \inf \{w > 0 : L \circ k_w > 0\} = T.$$

There is equality on $\{T = \infty\}$. On $\{T < \infty\}$ we have $T < T + u$ for every $u > 0$. Therefore, $L \circ k_{T+u} > 0$, and for t small enough $t + T \circ \theta_t < T + u$; hence, $T = \lim_{t \to \infty} t + T \circ \theta_t$.

(vi) The relation $a < L' \leqq b$ is equivalent to $T \circ \theta_a < \infty$, $T \circ \theta_b = \infty$ according to (iii), but $(T \circ \theta_a \leqq b - a, T \circ \theta_b = \infty) \Rightarrow (T \circ \theta_a < \infty, T \circ \theta_b = \infty) \Rightarrow (T \circ \theta_a \leqq b - a, T \circ \theta_b = \infty)$ since $\infty > T \circ \theta_a > b - a$ implies $(b - a) + T \circ \theta_{b-a} \circ \theta_a = T \circ \theta_a < \infty$, according to (v), and therefore $T \circ \theta_b < \infty$.

We have now everything we need to prove that coterminal times are birth times. However, only one side of the duality between terminal and coterminal times appears in Propositions 4.1 and 4.2, and we must give the other side.

PROPOSITION 4.3. *Let T be a random variable on $(\Omega, \mathscr{F})$ which satisfies the following properties:*

(i) $T < s \Rightarrow T \circ k_s = T$;

(ii) $s < T \Rightarrow T \circ k_s = \infty$;

(iii) $s < T \Rightarrow T - s = T \circ \theta_s$.

Set L_T (or simply L) $= \sup \{t : T \circ \theta_t < \infty\}$. Then L is an exact coterminal time, and $T_L = T$ if and only if T is exact, that is $t + T \circ \theta_t \downarrow T$ as $t \downarrow 0$.

PROOF. Let us show first that $T < s \Leftrightarrow T \circ k_s < s$. The implication $\Rightarrow$ is obvious from (i). Conversely, if $T \circ k_s < r < s$, (i) implies that $T \circ k_r \circ k_s = T \circ k_r < r$, which excludes the possibility that $T > r$ according to (ii). Therefore, $T \leqq r < s$. As a consequence, we have that $\{T < s\} \in \mathscr{F}_s$; hence, T is optional and a terminal time according to (iii). Note that T is not an arbitrary terminal time, because of property (ii): ζ, for instance, is a terminal time and $s < \zeta$ implies $\zeta \circ k_s = s$, not ∞.

It follows at once from (iii) that $s + T \circ \theta_s \geqq T$, and then that $t + T \circ \theta_t$ increases in t. Let us set $T' = \lim_{t \to 0} t + T \circ \theta_t$. We have $T' \geqq T$, with $T' = T$ on $\{T > 0\}$. If $s < T'$, then $s < T \circ \theta_t$ for t small enough, and (iii) gives $s + T \circ \theta_{s+t} = T \circ \theta_t$; letting $t \to 0$, we get $s + T' \circ \theta_s = T'$. Hence, T' satisfies (iii). If $s < T'$, then $s < t + T \circ \theta_t$ for all t; hence $T \circ k_{s-t} \circ \theta_t = \infty$ from (ii), or $T \circ \theta_t \circ k_s = \infty$. Finally, $T' \circ k_s = \infty$. If $T' < s$, then $t + T \circ \theta_t < s$ for t small enough. Hence, $T \circ \theta_t = T \circ k_{s-t} \circ \theta_t$ from (i); and letting $t \to 0$, we find again that $T' \circ k_s = T'$, that is, T' satisfies (i). Thus, T' is an exact terminal time. Note that $L_T = L_{T'}$.

Let us prove that L is an exact coterminal time. We have

$$L \circ \theta_s = \sup \{t : T \circ \theta_{t+s} < \infty\} = (L - s)^+. \tag{4.6}$$

Hence, L satisfies Definition 4.1 (i). Next, assume that T is not identically 0 (in which case everything would be trivial); denote by $[\Delta]$ the constant sample function equal to Δ, choose some ω such that $T(\omega) > 0$ and some $s < T(\omega)$. Then $T(k_s\omega) = \infty > s$ from (ii), and from (iii) $T(\theta_s(k_s\omega)) = \infty$; otherwise stated, $T([\Delta]) = \infty$.

Then we have

$$L \circ k_s = \sup \{t : T \circ \theta_t \circ k_s < \infty\} = \sup \{t : T \circ k_{(s-t)^+} \circ \theta_t < \infty\}. \tag{4.7}$$

If $t = s$, we have $T([\Delta]) = \infty$. Hence, $L \circ k_s \leqq s$, that is, Definition 4.1 (iii) is true. Let us remark that $T \leqq T \circ k_s$ for all s: if we had $T > r > T \circ k_s$ for some r, we would have from (i) $T \circ k_r = \infty$, from (ii) $T \circ k_r \circ k_s = T \circ k_s < \infty$. Hence,

$r > s$ and $T > s$; thus, $T \circ k_s = \infty$ from (ii), which is a contradiction. This implies that $L \circ k_s \leqq L$. On the other hand, if $s > L$, choose some s' such that $s > s' > L$. Then $T \circ \theta_{s'} = \infty$. Hence, $r < L$ implies $T \circ \theta_r < \infty$, which implies $T \circ \theta_r \leqq s' - r < s - r$ from (iii). Therefore from (i), we get $T \circ k_{s-r} \circ \theta_r = T \circ \theta_r$, or $T \circ \theta_r \circ k_s = T \circ \theta_r < \infty$; and finally, $r \leqq L \circ k_s$. Otherwise stated, $L = L \circ k_s$ and property (ii) of Definition 4.1 is proved.

If $L = \infty$, we have $T \circ \theta_t < \infty$ for all t. Given t, choose $w > t + T \circ \theta_t$. Then $w - t > T \circ \theta_t$ and from (i) $T \circ k_{w-t} \circ \theta_t = T \circ \theta_t < \infty$. Hence, $T \circ \theta_t \circ k_w < \infty$, and $L \circ k_w \geqq t$. Since t is arbitrary, we have also $L' = \infty$ and L is exact.

We finish with the proof that $T_L = T$. We have

(4.8) $$T_L = \inf\{t > 0 : L \circ k_t > 0\} = \inf\{t > 0 : \text{for some } s \quad T \circ \theta_s \circ k_t < \infty\}$$
$$= \inf\{t > 0 : T' \circ k_t < \infty\} = \inf\{t > 0 : t > T'\} = T'.$$

5. Birth of the process at a coterminal time

We consider a coterminal time L, exact, and its associated terminal time T. Let us define a new semigroup (K_t) by killing at time T,

(5.1) $$K_t(x, f) = E^x[f \circ X_t \cdot I_{\{t < T\}}].$$

It has Δ as an absorbing point (since $L([\Delta]) = 0$, $T([\Delta]) = +\infty$), but it does not satisfy $K_t 1 = 1$. That does not matter.

Let us define also $g(x) = P^x\{L = 0\} = P^x\{T = \infty\}$ for all $x \in E$, including Δ (see Proposition 4.2 (iii)). The following computation shows that g is *invariant* for (K_t):

(5.2) $$K_t(x, g) = P^x\{T \circ \theta_t = \infty, t < T\} = P^x\{T - t = \infty, t < T\}$$
$$= P^x\{T = \infty\} = g(x).$$

We can therefore define the conditioned semigroup,

(5.3) $$K_t^g(x, dy) = \begin{cases} \dfrac{1}{g(x)} K_t(x, dy) g(y) & \text{if } g(x) \neq 0, \\ \varepsilon_\Delta(dy) & \text{if } g(x) = 0. \end{cases}$$

If $g(x) \neq 0$, this is a probability measure carried by $\{g > 0\}$.

We shall need the following elementary property of the semigroup (K_t^g). Let $(Y_s)_{s \geqq 0}$ be any right continuous Markov process with (K_s^g) as its transition semigroup, and let f be a bounded continuous function on E. Set $j = K_t^g f$. Then the process (Y_s) is strong Markov, and the process $(j \circ Y_s)_{s \geqq 0}$ is a.s. right continuous. These properties belong to the folklore of Markov processes, the second statement being a particular case of the so-called "Feller property of the semigroup in the fine topology." They are true for all right continuous strong Markov semigroups (see the bibliographical comments at the end of the paper).

We state and prove now our main result. The restriction to the open interval $(0, \infty)$ is essential.

THEOREM 5.1. *The process* $(X_{L+t})_{t>0}$ *is a strong Markov process with respect to the family* $(\mathscr{F}_{L+t})$, *with* (K_t^g) *as transition semigroup.*

PROOF. Implicit in the statement is the hypothesis that Ω has been provided with some law P^μ; we drop μ from our notation, however.

Let L_n be the dyadic approximation of L from above, that is, $L_n = k2^{-n}$ if and only if $(k-1)2^{-n} < L \leqq k2^{-n}$, $k \geqq 1$, and $L_n = 0$ on $\{L = 0\}$. We are going to prove that, for f continuous and bounded on E and $0 < s < t$,

$$E[f \circ X_{L_n+t} | \mathscr{F}_{L_n+s}] = K_{t-s}^g(X_{L_n+s}, f) \tag{5.4}$$

(the fields are defined as in Definition 3.1, L, L_n, $L_n + s$, $L_n + t$ being honest; X_{L_n+s} is $\mathscr{F}_{L_n+s}$ measurable according to Proposition 4.2). This relation will imply Theorem 5.1. Indeed, each process $(X_{L_n+u})_{u>0}$ is a right continuous Markov process with (K_u^g) as transition semigroup according to (5.4). Using the Feller property above, $K_t^g f$ is a.s. right continuous on the sample paths of every process (X_{L_n+u}), and therefore as $n \to \infty$, on the sample paths of $(X_{L+u})_{u>0}$. If $H \in \mathscr{F}_{L+s} \subset \mathscr{F}_{L_n+s}$, the relation

$$\int_H f \circ X_{L_n+t}\, dP = \int_H K_{t-s}^g f \circ X_{L_n+s}\, dP \tag{5.5}$$

passes nicely to the limit as $n \to \infty$, giving the result we seek.

Let us therefore prove (5.4). Take $H \in \mathscr{F}_{L_n+s}$. Then letting $J_k = \{k2^{-n} < L \leqq (k+1)2^{-n}\}$,

$$\int_H f \circ X_{L_n+t}\, dP = \int_{H, L=0} f \circ X_t\, dP + \sum_{k=0}^{\infty} \int_{HnJ_k} f \circ X_{(k+1)2^{-n}}\, dP. \tag{5.6}$$

Since $H \in \mathscr{F}_{L_n} + s$, and we are on $\{L \leqq (k+1)2^{-n}\}$, therefore on $\{L_n + s \leqq (k+1)2^{-n} + s\}$, we may replace H by some $H' \in \mathscr{F}_{(k+1)2^{-n}+s}$. The kth term of the sum can be written

$$E[f \circ X_{(k+1)2^{-n}+t}, H', T \circ \theta_{k2^{-n}} \leqq 2^{-n}, T \circ \theta_{(k+1)2^{-n}} = \infty\}. \tag{5.7}$$

We write $\{T \circ \theta_{(k+1)2^{-n}} = \infty\}$ as $\{T \circ \theta_{(k+1)2^{-n}} > t\} \cap \{T \circ \theta_{(k+1)2^{-n}+t} = \infty\}$ and take conditional expectations with respect to $\mathscr{F}_{(k+1)2^{-n}+t}$. We get

$$\begin{aligned} E[f \circ X_{(k+1)2^{-n}+t}, H', T \circ \theta_{k2^{-n}} \leqq 2^{-n}, T \circ \theta_{(k+1)2^{-n}} > t, g \circ X_{(k+1)2^{-n}+t}] \\ = E[H', T \circ \theta_{k2^{-n}} \leqq 2^{-n}, T \circ \theta_{(k+1)2^{-n}} > s, \\ T \circ \theta_{(k+1)2^{-n}+s} > t - s, (fg) \circ X_{(k+1)2^{-n}+t}]. \end{aligned} \tag{5.8}$$

Taking now conditional expectations with respect to $\mathscr{F}_{(k+1)2^{-n}+s}$, equation (5.8) is equal to

$$E[H', T \circ \theta_{k2^{-n}} \leqq 2^{-n}, T \circ \theta_{(k+1)2^{-n}} > s, K_{t-s}(X_{(k+1)2^{-n}+s}, fg)]. \tag{5.9}$$

We replace $K_{t-s}(fg)$ by $gK^g_{t-s}(f)$, a legitimate step since $K_{t-s}(fg) = 0$ on $\{g = 0\}$, and we get

$$E[H', T\circ\theta_{k2^{-n}} \leqq 2^{-n}, T\circ\theta_{(k+1)2^{-n}} > s, g\circ X_{(k+1)2^{-n}+s}, K^g_{t-s}f\circ X_{(k+1)2^{-n}+s}] \tag{5.10}$$

which is equal to

$$E[H', k2^{-n} < L \leqq (k+1)2^{-n}, K^g_{t-s}f\circ X_{(k+1)2^{-n}+s}], \tag{5.11}$$

according to a reasoning similar to the first transformation of this proof. We now replace H' by H, sum over k (the first term on the right side of (5.6) needs a similar, but slightly different treatment), and we get

$$\int_H f\circ X_{L_n+t}\, dP = \int_H K^g_{t-s}f\circ X_{L_n+s}\, dP, \tag{5.12}$$

which is equivalent to (5.4).

REMARK. An easy computation shows that

$$K^g_t(x, f) = \frac{1}{g(x)} Q_t(x, fg) = \frac{E^x[f\circ X_t, T < \infty]}{P^x\{T < \infty\}}. \tag{5.13}$$

Thus, (K^g_t) appears as the transition semigroup conditioned by the fact that $T = \infty$. For instance, if L is the last hitting time of A, (K^g_t) is the semigroup of the original process conditioned not to hit A. This is reasonably intuitive.

REFERENCES

[1] J. AZEMA, M. DUFLO, and D. REVUZ, "Propriétés relatives des processus de Markov récurrents," *Z. Wahrscheinlichkeitstheorie und Verw. Gebiete*, Vol. 13 (1969), pp. 286–314.

[2] R. M. BLUMENTHAL and R. K. GETOOR, *Markov Processes and Potential Theory*, New York and London, Academic Press, 1968.

[3] K. L. CHUNG and J. L. DOOB, "Fields, optionality and measurability," *Amer. J. Math.*, Vol. 87 (1965) pp. 397–424.

[4] K. L. CHUNG and J. B. WALSH, "To reverse a Markov process," *Acta Math.*, Vol. 123 (1970), pp. 225–251.

[5] E. B. DYNKIN, *Markov Processes I, II* (translated by J. Fabius, V. Greenberg, A. Maitra, and G. Majone), New York, Academic Press; Berlin–Göttingen–Heidelberg, Springer-Verlag, 1965.

[6] P. A. MEYER, "Le retournement du temps d'après Chung et Walsh," *Séminaire de probabilités V*, Université de Strasbourg Lecture Notes in Mathematics, Heidelberg, Springer-Verlag, 1971.

[7] M. NAGASAWA, "Time reversions of Markov processes," *Nagoya Math. J.*, Vol. 24 (1964), pp. 177–204.

[8] D. REVUZ, oral communication.

[9] J. B. WALSH, "Some remarks on the Feller Property," *Ann. Math. Statist.*, Vol. 41 (1970), pp. 1672–1683.

STOCHASTIC INTEGRALS AND PROCESSES WITH STATIONARY INDEPENDENT INCREMENTS

P. WARWICK MILLAR
UNIVERSITY OF CALIFORNIA, BERKELEY

1. Introduction

Let $X = \{X(t); 0 \leqq t \leqq 1\}$ be a real valued stochastic process with stationary independent increments and right continuous paths $X(0) = 0$. The characteristic function of $X(t)$ then has the form $\exp\{t\psi(u)\}$, where

$$(1.1) \qquad \psi(u) = iug - \tfrac{1}{2}\sigma^2 u^2 + \int \left[e^{iux} - 1 - \frac{iux}{1 + x^2} \right] \nu(dx).$$

The measure ν is called the Lévy measure of X, and ψ is called the exponent. It will be assumed throughout that $\sigma^2 = 0$. The index $\beta(X)$ of the process X is

$$(1.2) \qquad \beta(X) = \inf \left\{ p > 0; \int_{|x| < 1} |x|^p \nu(dx) < \infty \right\}.$$

If $\int_{|x|<1} |x| \nu(dx) < \infty$, then by subtracting a linear term from X one may write the exponent ψ as

$$(1.3) \qquad \psi(u) = \int [1 - e^{iux}] \nu(dx);$$

it will be assumed from now on that the exponent is in this form whenever $\int_{|x|<1} |x| \nu(dx) < \infty$.

This paper studies the sample function behavior of processes $Y = \{Y(t); 0 \leqq t \leqq 1\}$, where $Y(t)$ has the form $Y(t) = \int_0^t v(s)\, dX(s)$ and where $v = \{v(s); 0 \leqq s \leqq 1\}$ is a stochastic process of a special type described below. Section 2 contains a development of the theory of such stochastic integrals, together with conventions and notations prerequisite for the rest of the paper. The construction of the stochastic integral is made to depend on an inequality of L. E. Dubins and J. L. Savage, and has applications to more general theories of stochastic integration. In Section 3 a local limit theorem is proved. If $|v(s)| \leqq 1$ and if $p > \beta(X)$, then $|Y(t)| t^{-1/p}$ converges to zero a.s. as $t \downarrow 0$. This generalizes (with different proof) a result known for the case $v(s) \equiv 1$ (see [2]). To state the results of Section 4, let $\pi_n: 0 = t_{n,1} < \cdots < t_{n,k_n} = 1$ be a sequence of partitions of $[0, 1]$ satisfying

$$(1.4) \qquad \lim_{n \to \infty} \max_k [t_{n,k+1} - t_{n,k}] = 0;$$

Work supported by NSF Grant GP-15283.

and let

$$V(\pi_n, Y, p) = \sum_k |Y(t_{n,k+1}) - Y(t_{n,k})|^p. \tag{1.5}$$

If $|v(s)| \leqq 1$, and if $\int_{|x|<1} |x|^p \nu(dx) < \infty$, then $V(\pi_n, Y, p)$ converges in probability to $\Sigma_{0 \leqq s \leqq 1} |v(s)|^p |j(X, s)|^p$, where $j(X, s) = X(s) - X(s-)$. This generalizes and improves (with different proof) the result obtained in [18] under the additional assumptions $p > \beta(X)$ and $v(s) \equiv 1$. Moreover, if the Lévy measure is concentrated on a finite interval, convergence is L_r for every $r < \infty$. Section 5 is devoted to the concept of zero jumps. A zero jump is experienced by X at time t if either $X(t, \omega) < 0 < X(t-, \omega)$ or $X(t-, \omega) < 0 < X(t, \omega)$. This concept appears to be of some interest in describing the sample function behavior of X. For example, as shown in Section 5, X may jump over zero infinitely often as $t \downarrow 0$, yet without ever hitting zero itself. Section 5 begins by establishing a stochastic integral formula analogous to the famous Itô formula for the case of Brownian motion. This is eventually made to yield an inequality relating the zero jumps up to time t to other more tractable quantities. This development draws on the results established in preceding sections.

2. Stochastic integral

Throughout this paper, the notation $X = \{X(t); 0 \leqq t \leqq 1\}$ will always denote a process with stationary independent increments, as in Section 1. Let $X^a(t)$ be the sum up to time t of all the jumps of X having absolute magnitude greater than a. Define $X_a(t) = X(t) - X^a(t)$. Then X^a and X_a are independent processes with independent increments. The process X_a has moments of all orders, its Lévy measure is concentrated on $(-a, a)$, and $\{X_a(t) - ct; t \geqq 0\}$ is a martingale if $c = EX_a(1)$. The exponent of X^a is

$$\psi^a(u) = \int_{|x|>a} [e^{iux} - 1]\nu(dx), \tag{2.1}$$

and that of X_a is $\psi_a(u) = \psi(u) - \psi^a(u)$. In the interval $[0, 1]$, the path $X(\cdot, \omega)$, for each ω, will experience only a finite number of jumps exceeding a. Hence, if a is large, $X(t) = X_a(t)$, $0 \leqq t \leqq 1$, for all ω in a set Ω_a, where $\Omega_a \uparrow 1$ as $a \to \infty$. For this reason we can (and will) often replace the process X in arguments below with the truncated process X_a. (See also [18], where this point of view is further explained.) Let us proceed now with the development of the stochastic integral.

Let $\{F(t); t \geqq 0\}$ denote the family of sigma fields given by $F(t) = F\{X(s): s \leqq t\}$. All processes are henceforth automatically assumed to be adapted to the family $F(t)$. Let G be the sigma field on $[0, \infty) \times \Omega$ generated by all the processes with left continuous paths. A process $v = \{v(t); t \geqq 0\}$ is "previsible" if it is measurable when regarded as a map of $([0, \infty) \times \Omega, G) \to$

$((-\infty, \infty), B)$. Here B denotes the Borel sigma field. A previsible process v is a step function if

$$v(t) = \sum_{1 \leqq k \leqq n} I_{(t_k, t_{k+1}]}(t) v_k, \qquad 0 \leqq t_1 \leqq \cdots \leqq t_n \leqq 1, \tag{2.2}$$

and v_k is an $F(t_k)$ measurable random variable. (For any set A, I_A will always denote the indicator function of A.) If v is a step function, then the stochastic integral of v with respect to X is the right continuous process $Y = \{Y(t); 0 \leqq t \leqq 1\}$ defined by

$$Y(t) = \int_0^t v(s)\, dX(s) = v_1[X(t_2) - X(t_1)] + \cdots + v_{k(t)}[X(t) - X(t_{k(t)})], \tag{2.3}$$

where $k(t) = \max\{k : t_k < t\}$.

The problem now is to extend this definition to more general processes v. There are several ways of doing this. The following approach, which is applicable in much wider contexts (see the remark following the proof of Theorem 2.2), extends the version by H. McKean, Jr. [13] of Itô's original approach to Brownian motion through the use of different martingale inequalities. For the present paper, it was desirable to have the integral defined in a canonical manner as an a.s. limit uniformly in t, $0 \leqq t \leqq 1$. For other methods, consult P. A. Meyer [15]. The present method yields in a natural way a larger class of v that may serve as integrands than do the methods of [15].

Let us begin with an inequality of Dubins and Savage [10]. Let $\{F_n ; n \geqq 0\}$ be an increasing family of sigma fields; let $\{d_k ; k \geqq 1\}$ be a sequence of martingale differences, and let $\{w_n ; n \geqq 1\}$ be a sequence of random variables such that w_k is F_{k-1} measurable. Finally, let $h_n = E(d_n^2 | F_{n-1})$. Then, assuming d_k is F_k measurable for $k \geqq 1$,

$$P\left\{\sum_{k=1}^n d_k \geqq a \sum_{k=1}^n h_k + b \text{ for some } n\right\} \leqq (1 + ab)^{-1}. \tag{2.4}$$

This bound is known to be sharp, and implies that

$$P\left\{\left|\sum_{k=1}^n d_k\right| \geqq a \sum_{k=1}^n h_k + b; \text{ for some } n\right\} \leqq 2(1 + ab)^{-1}. \tag{2.5}$$

It then follows that

$$P\left\{\left|\sum_{k=1}^n w_k d_k\right| \geqq a \sum_{k=1}^n w_k^2 h_k + b \;\text{ for some } n\right\} \leqq 2(1 + ab)^{-1}. \tag{2.6}$$

Hence, if v is any step function, and X is any integrable process with stationary independent increments, then

$$(2.7)\qquad P\left\{\left|\int_0^t v(s)\,d[X(s) - cs]\right| \geqq a\int_0^t v(s)^2 r\,ds + b;\ \text{for some } t,\ 0 \leqq t \leqq 1\right\} \leqq 2(1 + ab)^{-1},$$

where $c = EX(1)$ and $r = \text{Var}\,X(1)$; or

$$(2.8)\qquad P\left\{\left|\int_0^t v(s)\,dX(s)\right| \geqq ar\int_0^t v(s)^2\,ds + |c|\int_0^t |v(s)|\,ds + b;\ \text{for some } t\right\} \leqq 2(1 + ab)^{-1}.$$

The following theorem gives the existence of the stochastic integral.

THEOREM 2.1. *Let v_n be a sequence of step functions, and let v be a measurable process, not necessarily previsible, such that*

$$(2.9)\qquad P\left\{\int_0^1 [v(s) - v_n(s)]^2\,ds \leqq 2^{-n};\ n\uparrow\infty\right\} = 1.$$

Then the sequence of step function integrals $\int_0^t v_n(s)\,dX(s)$ converges for every ω, uniformly in t, $0 \leqq t \leqq 1$. Any process v satisfying

$$(2.10)\qquad P\left\{\int_0^1 [v(s)]^2\,ds < \infty\right\} = 1$$

may be represented as a limit of step functions as in (2.9).

Of course, one defines

$$(2.11)\qquad Y(t) = \int_0^t v(s)\,dX(s) = \lim_{n\to\infty}\int_0^t v_n(s)\,dX(s),$$

and so the existence of $Y(t)$ is established as an a.s. limit for all $t \in [0, 1]$, simultaneously. By virtue of the uniform convergence, one automatically has the following corollary.

COROLLARY 2.1. *The process $\{Y(t) : t \geqq 0\}$ is right continuous.*

PROOF OF THEOREM 2.1. Because of the facts stated at the beginning of this section, there will be no loss in generality in assuming the Lévy measure of X concentrated on a finite interval $[-a, a]$ (If

$$(2.12)\qquad \Omega_a = \{\omega : X(t) = X_a(t);\ 0 \leqq t \leqq 1\},$$

then $\Omega_a \uparrow \Omega$ as $a \uparrow \infty$; the step function integrals relative to X and to X_a are then exactly the same for each $\omega \in \Omega_a$.) Let $c = EX(1)$ and $r = \text{Var}\,X(1)$. Suppose first that v_n is a sequence of step functions such that

$$(2.13)\qquad P\left\{\int_0^1 [v_n(s)]^2\,ds \leqq 2^{-n};\ n\uparrow\infty\right\} = 1.$$

Then from (2.8);

$$\left\{\sup_{0 \leqq t \leqq 1} \left| \int_0^t v_n(s)\, dX(s) \right| \leqq 2^{-n/3};\ n \uparrow \infty \right\} = 1. \tag{2.14}$$

For, let $a_n = [2^n n^4]^{1/2}$ and $b_n = 2^{-n/2}$, so that $(1 + a_n b_n)^{-1}(1 + n^2)^{-1}$ is the general term of a convergent series. Then from the Borel–Cantelli lemma,

$$P\left\{\left| \int_0^t v_n(s)\, dX(s) \right| \geqq a_n r \int_0^t [v_n(s)]^2\, ds + |c| \int_0^t |v_n(s)|\, ds + b_n;\right.$$
$$\left.\text{for some } t,\ n \uparrow \infty \right\} = 0. \tag{2.15}$$

But $\int_0^t v_n^2(s)\, ds \leqq \int_0^1 v_n^2(s)\, ds \leqq 2^{-n}$ for all but a finite number of n, and so also

$$\int_0^t |v_n(s)|\, ds \leqq \left[\int_0^1 v_n^2(s)\, ds\right]^{1/2} \leqq 2^{-n/2} \tag{2.16}$$

for all but a finite number of n, implying that

$$P\left\{\sup_{0 \leqq t \leqq 1} \left| \int_0^t v_n(s)\, dX(s) \right| \geqq 2^{-n/2}(n^2 r + |c| + 1);\ n \uparrow \infty \right\} = 0. \tag{2.17}$$

Therefore, (2.14) holds.

Next, let $\{v_n\}$ be a sequence of step functions satisfying (2.9). By (2.14), $\sup_{0 \leqq t \leqq 1} \left|\int_0^t (v_{n+1} - v_n)\, dX\right|$ goes to zero geometrically fast, so that $\int_0^t v_n(s)\, dX(s)$ converges uniformly in t for almost every ω. This completes the proof of the existence of $\int_0^t v(s)\, dX(s)$. The last statement of the theorem is proved as in McKean's book [13].

Remark 2.1. Let $\{X(t), t \geqq 0\}$ be an L_2 martingale, and $A(t)$ the natural increasing process associated with it. These concepts are defined in [14]. The analogue of (2.7) then becomes, if v is a step function,

$$P\left\{\left| \int_0^t v(s)\, dX(s) \right| \geqq a \int_0^t v^2(s)\, dA(s) + b;\ \text{for some } t \right\} \leqq 2(1 + ab)^{-1}. \tag{2.18}$$

Using the argument of Theorem 2.1, one can obtain the existence of $\int_0^t v(s)\, dX(s)$ as an a.s. limit, uniformly in t, whenever there exist step functions v_n such that

$$P\left\{\int_0^1 [v(s) - v_n(s)]^2\, dA(s) < 2^{-n};\ n \uparrow \infty \right\} = 1. \tag{2.19}$$

In particular, one can see directly that this is possible whenever

$$P\left\{\int_0^1 v^2(s)\, dA(s) < \infty \right\} = 1 \tag{2.20}$$

and A has continuous paths (that is, X is quasileft continuous), and so establishes the existence of $\int_0^t v(s)\, dX(s)$ for all v satisfying (2.20). This seems to be an improvement on P. Courrège's approach to this problem [7].

The process $Y(t) = \int_0^t v(s)\,dX(s)$ is now defined for any process v satisfying (2.9). One can then enquire of the existence of the stochastic integral $\int_0^t w(s)\,dY(s)$. If w is a step function, then

$$\begin{aligned}(2.21)\qquad \int w(s)\,dY(s) &= \sum w_k I_{(t_k, t_{k+1}]}[Y(_{k+1}) - Y(t_k)] \\ &= \sum w_k I_{(t_k, t_{k+1}]} \int_{t_k}^{t_{k+1}} v(s)\,dX(s) = \int w(s)v(s)\,dX(s).\end{aligned}$$

Using the development of Theorem 2.1, one checks that if $P\{\int_0^1 w^2(s)\,ds < \infty\} = 1$, then there are step functions w_n such that

$$(2.22)\qquad P\left\{\int_0^1 [w(s) - w_n(s)]^2\,ds < 2^{-n};\, n \uparrow \infty\right\} = 1,$$

and $\int_0^t w_n(s)\,dY(s)$ converges to $\int_0^t w(s)v(s)\,dX(s)$ for each ω, uniformly in $t, 0 \leqq t \leqq 1$.

Several properties of the process $Y(t) = \int_0^t v(s)\,dX(s)$ will now be recorded for future reference.

PROPERTY 2.1. *Suppose $X = X_1 + X_2$, where the X_i are processes with independent increments. Then*

$$(2.23)\qquad \int_0^t v(s)\,dX(s) = \int_0^t v(s)\,dX_1(s) + \int_0^t v(s)\,dX_2(s), \qquad 0 \leqq t \leqq 1.$$

PROPERTY 2.2. *Let the Lévy measure of X be concentrated on a finite interval. There is a constant K depending only on $EX(1)$ and* Var $X(1)$ *such that*

$$(2.24)\qquad E\left[\int_0^t v(s)\,dX(s)\right]^2 \leqq KE\int_0^t v^2(s)\,ds,$$

$$(2.25)\qquad E\left[\sup_{0 \leqq t \leqq 1} \int_0^t v(s)\,dX(s)\right]^2 \leqq KE\int_0^1 v^2(s)\,ds.$$

This follows upon observing that the following hold for step functions v:

$$(2.26)\qquad E\left|\int_0^t v(s)\,d[X(s) - cs]\right|^2 = rE\int_0^t v^2(s)\,ds,$$

and

$$(2.27)\qquad E\left|\sup_{0 \leqq t \leqq 1} \int_0^t v(s)\,d[X(s) - cs]\right|^2 \leqq 2rE\int_0^t v^2(s)\,ds.$$

These, of course are well-known martingale results. Here, as usual, $c = EX(1)$ and $r = \operatorname{Var} X(1)$.

PROPERTY 2.3. *If the Lévy measure of X is concentrated on a finite interval and* (2.9) *holds, then*

$$P\left\{\left|\int_0^t v(s)\,dX(s)\right| \geqq ar\int_0^t v^2(s)\,ds + |c|\int_0^t |v(s)|\,ds + b;\ \text{for some } t\right\} \leqq 2(1+ab)^{-1}. \tag{2.28}$$

PROPERTY 2.4. *If the Lévy measure of X is concentrated on a finite interval, if* $c = EX(1)$, *and if* $|v(t)| \leqq 1$, $0 \leqq t \leqq 1$, *then for* $p \geqq 1$,

$$E\sup_{0\leqq s\leqq t}\left|\int_0^t v(s)\,dX(s)\right|^p \leqq c_p E|X(t)|^p + c_p ct^p. \tag{2.29}$$

Here c_p is a constant depending on p only.

If $p > 1$, Theorem 9 of D. Burkholder [5] (see also [16]) implies that

$$E\sup_{0\leqq s\leqq t}\left|\int_0^t v(s)\,d[X(s) - cs]\right|^p \leqq c_p E|X(t) - ct|^p \tag{2.30}$$

whenever v is a step function. That (2.30) holds for $p = 1$ follows from Theorem 1.1 of [17]. Property 2.4 is a consequence of (2.30).

PROPERTY 2.5. *Suppose that the Lévy measure of X is concentrated on* $[-a, a]$ *and that* (2.9) *holds. If*

$$\psi(u) = iug + \int_{|x|<a}\left[e^{iux} - 1 - \frac{iux}{1+x^2}\right]\nu(dx), \tag{2.31}$$

let X^n *be the process with exponent*

$$iug + \int_{s_n<|x|<a}\left[e^{iux} - i - \frac{iux}{1+x^2}\right]\nu(dx). \tag{2.32}$$

Let Y^n *be the process*

$$Y^n(t) = \int_0^t v(s)\,dX^n(s). \tag{2.33}$$

Choose s_n *so that*

$$\int_{|x|\leqq s_n} |x|^2\nu(dx) < 2^{-n}. \tag{2.34}$$

Then for each ω, $Y^n(t) \to Y(t)$, *uniformly for* $0 \leqq t \leqq 1$.

PROOF. The process $X(t) - X^n(t)$ has stationary independent increments. According to Property 2.3, if $c_n = E[X(1) - X^n(1)]$ and $r_n = \text{Var}[X(1) - X^n(1)]$, then for constants a_n, b_n,

$$(2.35)\quad P\left\{\sup_{0 \leqq t \leqq 1} \left|\int_0^t v(s)\, dX(s) - \int_0^t v(s)\, dX^n(s)\right| > a_n r_n \int_0^t v^2(s)\, ds + |c_n| \int_0^t |v(s)|\, ds + b_n\right\} \leqq 2(1 + a_n b_n)^{-1}.$$

Since the exponent of the process $X(t) - X^n(t)$ is

$$(2.36)\quad \int_{0 < |x| \leqq s_n} \left[e^{iux} - 1 - \frac{iux}{1 + x^2}\right] \nu(dx),$$

$r_n \leqq 2^{-n}$ and $c_n \leqq 2^{-n/2}$. Set $a_n = n^3$ and $b_n = n^{-1}$, and conclude Property 2.5 from the Borel–Cantelli theorem.

PROPERTY 2.6. *Suppose almost all paths of X are of bounded variation. (A necessary and sufficient condition for this is $\int_{|x|<1} |x| \nu(dx) < \infty$.) Let v be previsible and satisfy* (2.10). *Let $L - \int_0^t v(s)\, dX(s)$ be the ordinary Lebesgue–Stieltjes integral (calculated for each ω), and $\int_0^t v(s)\, dX(s)$ the stochastic integral. Then*

$$(2.37)\quad \int_0^t v(s)\, dX(s) = L - \int_0^t v(s)\, dX(s) \qquad \text{a.s., } 0 \leqq t \leqq 1,$$

(the exceptional set does not depend on t).

PROOF. Suppose without any loss that the Lévy measure of X is on a finite interval. Let C be the class of bounded step functions and H the class of bounded previsible processes for which (2.37) holds. Then C is a vector space closed under $\wedge$, and H is a vector space of real functions on $\Omega \times [0, 1]$ containing C. Let $v_n \in H$, $0 \leqq v_n \leqq M$, and $v_n \uparrow v$. From Property 2.2,

$$(2.38)\quad \lim_{n\to\infty} \int_0^t v_n(s)\, dX(s) = \int_0^t v(s)\, dX(s)$$

in L_2 and from monotone convergence,

$$(2.39)\quad \lim_{n\to\infty} L - \int_0^t v_n(s)\, dX(s) = L - \int_0^t v(s)\, dX(s)$$

a.e. and in L_2. Hence, there is a null set A such that if $\omega \notin A$, then

$$(2.40)\quad \int_0^t v(s)\, dX(s) = L - \int_0^t v(s)\, dX(s)$$

for every rational t. Since both sides are right continuous, equality holds for all t, and $\omega \notin A$; it follows that H is closed under monotone limits. Therefore, H contains all bounded previsible v by T20 of Meyer [14], or rather the remark following it. Next, let v be nonnegative, previsible. There then exist bounded previsible v_n such that $v_n \uparrow v$, and we can choose v_n so that

$$P\left\{\int_0^1 (v - v_n)^2\, ds < 2^{-n},\ n \uparrow \infty\right\} = 1. \tag{2.41}$$

Arguing with Property 2.3 in a manner now familiar, one sees that for almost all ω, $\int_0^t v_n(s)\, dX(s)$ converges to $\int_0^t v(s)\, dX(s)$, uniformly in t. Since

$$\int_0^t v_n(s)\, dX(s) = L - \int_0^t v_n(s)\, dX(s) \tag{2.42}$$

by what has already been proved, it need be proved only that

$$\lim_{n\to\infty} L - \int_0^t v_n(s)\, dX(s) = L - \int_0^t v(s)\, dX(s) \qquad \text{a.e.} \tag{2.43}$$

Write $X = X_1 - X_2$, where X_1, X_2 are independent processes with stationary independent increments, each with increasing paths. Since $0 \leqq v_n \uparrow v$, clearly,

$$\lim_{n\to\infty} L - \int_0^t v_n(s)\, dX_i(s) = L - \int_0^t v(s)\, dX_i(s) \leqq \infty, \qquad i = 1, 2, \tag{2.44}$$

for every ω, so it is necessary only to show that $L - \int_0^t v(s)\, dX_i(s) < \infty$. But by what has already been proved,

$$\infty > \lim_{n\to\infty} \int_0^t v_n(s)\, dX_i(s) = \lim L - \int_0^t v_n(s)\, dX_i(s), \tag{2.45}$$

since X_i is a process with stationary independent increments. For general previsible v satisfying (2.9), write $v = v^+ - v^-$.

REMARK 2.2. A theorem of this nature has already been proved for L_2 martingales by Meyer and C. Doléans-Dade; the first half of the present proof follows theirs [8].

The following property is a convenient technicality.

PROPERTY 2.7. *Let v be a (measurable) process satisfying* (2.10), *and let v_n be a sequence of step functions satisfying* (2.9). *Then there is a previsible process $\hat{v}$ such that*

$$P\left\{\int_0^1 [\hat{v}(s) - v_n(s)]^2\, ds < 2^{-n},\ n \uparrow \infty\right\} = 1. \tag{2.46}$$

In particular, the stochastic integrals $\int_0^t v(s)\, dX(s)$ and $\int_0^t \hat{v}(s)\, dX(s)$ will be the same.

It follows from Property 2.7 that in discussion of stochastic integrals, there is no loss in generality in assuming v previsible. Accordingly, this assumption will be in effect throughout the remainder of the paper.

3. Local limit theorems

The goal of this section is to prove the following theorem.

THEOREM 3.1. *Let* $X = \{X(t); 0 \leqq t\}$ *be a process with stationary, independent increments and index* $\beta(X)$. *Let* $v = \{v(t); 0 \leqq t\}$ *be a stochastic process with* $\sup_{0 \leqq t \leqq 1} |v(t)| \leqq 1$. *Let* $Y = \{Y(t); 0 \leqq t\}$ *be the process* $Y(t) = \int_0^t v(s)\, dX(s)$. *If* $p > \beta(X)$, *then as* $t \downarrow 0$,

$$\frac{Y(t)}{t^{1/p}} \to 0 \qquad \text{a.s.} \tag{3.1}$$

This theorem is known for the case $v(s) \equiv 1$. The first proof was given by R. M. Blumenthal and R. K. Getoor [2]; in [18], a second proof is obtained as a consequence of a more general theorem. The method of proof below is new, even for the case $v \equiv 1$; since stochastic integration destroys stationarity and independence of increments, the methods used up to now had to be replaced in order to obtain the present theorem. A standard truncation argument shows that the hypothesis $|v(t)| \leqq 1$, all t, can be replaced by the hypothesis that almost all paths of v be bounded on finite intervals.

Before proceeding, it will be convenient to record as lemmas two results proved in [18] (Theorems 3.1 and 2.1, respectively).

LEMMA 3.1. *If* $\int_{|x|<1} |x|^p \nu(dx) < \infty$, *then*

$$\sum_{s \leqq 1} |j(X, s)|^p < \infty \qquad \text{a.s.} \tag{3.2}$$

Here, $j(X, s) = X(s) - X(s-)$, as usual.

LEMMA 3.2. *Let* ν *be concentrated on a finite interval and* $\int_{|x|<1} |x|^p \nu(dx) < \infty$. *Then if* $0 < p \leqq 2$,

$$E|X(t)|^p \leqq Ct, \qquad 0 \leqq t \leqq 1. \tag{3.3}$$

The constant

$$C \leqq 2^p \left[\int |x|^p \nu(dx) + e t^{p-1} \right] \tag{3.4}$$

if $p > 1$ *and* $e = EX(1)$; *if* $p \leqq 1$, *then* $C \leqq 2 \int |x|^p \nu(dx)$.

PROOF OF THEOREM 3.1. The proof falls into several parts.

(a) Suppose that $1 < p < 2$. Without loss of generality suppose the Lévy measure is concentrated on $[-a, a]$. Moreover, there is also no loss in assuming that X is a martingale. For, if X is not a martingale and if $c = EX(1)$, then $X(t) - ct$ and $Y(t) - c \int_0^t v(s)\, ds$ are martingales; and, since $p > 1$,

$$\frac{\left|\int_0^t v(s)\, ds\right|}{t^{1/p}} \leqq t^{(p-1)/p} \to 0 \qquad \text{a.s.} \tag{3.5}$$

as $t \to 0$.

Next, pick p' subject to $\beta(X) < p' < p, p' > 1$. Then if $\varepsilon > 0$,

$$P\{|Y(2^{-n})| > \varepsilon 2^{-n/p}\} \leqq E|Y(2^{-n})|^{p'}\varepsilon^{-p'}2^{np'/p} \leqq CE|X(2^{-n})|^{p'}\varepsilon^{-p'}2^{np'/p}, \leqq C\varepsilon^{-p'}2^{-n[1-(p'/p)]}, \tag{3.6}$$

using (2.29) and (3.3). Hence,

$$\sum_n P\{|Y(2^{-n})| > \varepsilon 2^{-n/p}\} < \infty, \tag{3.7}$$

so that

$$\lim_{n\to\infty} |Y(2^{-n})|2^{n/p} = 0 \qquad \text{a.s.} \tag{3.8}$$

by the Borel–Cantelli theorem. Moreover,

$$\sup_{t:2^{-n-1}\leqq t\leqq 2^{-n}} |Y(t) - Y(2^{-n-1})|2^{n/p} \to 0 \qquad \text{a.s.} \tag{3.9}$$

as $n \to \infty$. This is verified as follows. Since $\{Y(t) - Y(2^{-n-1}), t \geqq 2^{-n-1}\}$ is a martingale, Doob's submartingale maximal inequality applied to the submartingale $\{|Y(t) - Y(2^{-n-1})|^{p'}; t \geqq 2^{-n-1}\}$ yields

$$\begin{aligned} P\{&\sup_{t:2^{-n-1}\leqq t\leqq 2^{-n}} |Y(t) - Y(2^{-n-1})| > \varepsilon 2^{-n/p}\} \\ &= P\{\sup_{t:2^{-n-1}\leqq t\leqq 2^{-n}} |Y(t) - Y(2^{-n-1})|^{p'} > \varepsilon^{p'}2^{-np'/p}\} \\ &\leqq E|Y(2^{-n}) - Y(2^{-n-1})|^{p'}\varepsilon^{-p'}2^{np'/p}. \end{aligned} \tag{3.10}$$

By inequality (2.29) (Property 2.4) and Lemma 3.2, the last expression does not exceed

$$C[2^{-n} - 2^{-n-1}]2^{np'/p} = C2^{-n[1-(p'/p)]}. \tag{3.11}$$

The Borel–Cantelli theorem then yields (3.9).

Finally, one deduces that $\lim_{t\to 0} |Y(t)|t^{-1/p} = 0$ a.s. for the case $1 < p < 2$ by means of the following calculation. If t is given, then there exists an integer $n(n = n(t))$ such that $2^{-n-1} \leqq t \leqq 2^{-n}$. Then

$$\begin{aligned} &|Y(t)|t^{-1/p} \\ &\leqq |Y(2^{-n-1})|2^{(n+1)/p} + \sup_{t:2^{-n-1}\leqq t\leqq 2^{-n}} |Y(t) - Y(2^{-n})|2^{(n+1)/p}. \end{aligned} \tag{3.12}$$

By (3.8) and (3.9) the right side goes to zero as $t \to 0$ ($n \to \infty$).

(b) Next suppose $\beta(X) = 2$. Using the same inequalities of (a), one sees that

$$P\{|Y(2^{-n})| > \varepsilon n 2^{-n/2}\} \leqq Cn^{-2}, \tag{3.13}$$

implying by the Borel–Cantelli lemma that $|Y(2^{-n})|2^{n/2}n^{-1} \to 0$ a.s. Hence, if $p > 2$, $|Y(2^{-n})|2^{n/p} \to 0$ a.s. Similarly, it is easy to see that

$$\sup_{t: 2^{-n-1} \leqq t \leqq 2^{-n}} |Y(t) - Y(2^{-n-1})|2^{n/p} \to 0 \qquad \text{a.s.} \tag{3.14}$$

as $n \to \infty$, and the argument is completed as in (a).

(c) Suppose $p \leqq 1$. According to the conventions of Section 1, the exponent of X is of the form $\int [e^{iux} - 1]\nu(dx)$. This implies that $X(t) = T(t) - S(t)$, where T and S are independent processes, each with increasing paths. It is then enough to prove the result for $X(t) = T(t)$. But in this case, $|\int_0^t v(s)\, dX(s)| \leqq |X(t)|$, so the result follows from the known fact for X.

REMARK 3.1. A modification of the argument of Millar [18] shows that if we assume only

$$\int_{|x|<1} |x|^p \nu(dx) < \infty, \qquad |v(t)| \leqq 1 \qquad \text{for all } t, \tag{3.15}$$

then $|\int v(s)\, dX(s)|t^{-1/p} \to 0$ in probability. Condition (3.15) is slightly weaker than $p > \beta(X)$, and it is known that under condition (3.15) one cannot assert a.s. convergence in general.

REMARK 3.2. If $|v(t, \omega)| > \delta$ all (t, ω), and if $p < \beta(X)$, then

$$\limsup_{t \to 0} \left| \int_0^t v(s)\, dX(s) \right| t^{-1/p} = +\infty \text{ a.s.} \tag{3.16}$$

The proof is similar to the proof of the case $v \equiv 1$ in [18], and is omitted.

4. Variation of sample functions

Throughout this section, $\{\pi_n; n \geqq 1\}$ will be a sequence of partitions of $[0, 1]$ with $\pi_n: 0 = t_{n,1} < \cdots < t_{n,k_n} = 1$, and will satisfy

$$\lim_{n \to \infty} \max_k [t_{n,k+1} - t_{n,k}] = 0. \tag{4.1}$$

The partition π_{n+1} is not necessarily a refinement of π_n. If $Y = \{Y(t); 0 \leqq t \leqq 1\}$ is a stochastic process, define

$$V(\pi_n, Y, p) = V_n(Y, p) = \sum_k |Y(t_{n,k+1}) - Y(t_{n,k})|^p. \tag{4.2}$$

THEOREM 4.1. *Let* $v = \{v(t); t \geqq 0\}$ *satisfy* $\sup_t |v(t)| \leqq 1$. *Let* $X = \{X(t); t \geqq 0\}$ *be a process with stationary independent increments such that* $\int_{|x|<1} |x|^p \nu(dx) < \infty$. *Let* $Y(t) = \int_0^t v(s)\, dX(s)$. *Then* $V(\pi_n, Y, p)$ *converges in probability to* $\Sigma_{0 \leqq s \leqq 1} |v(s)|^p |j(X, s)|^p < \infty$, *where* $j(X, s) = X(s) - X(s-)$.

This theorem constitutes an improvement of Theorem 3.2 of [18], which asserts the convergence in probability under the hypothesis $v \equiv 1$ and the somewhat stronger assumption that $p > \beta(X)$. The present proof is different and rather simpler than the proof of [18]. It will reveal that in the case $p \leqq 1$, almost everywhere convergence of $V_n(Y, p)$ holds, even if π_{n+1} is not a refinement of π_n; this is a further improvement of the results of [18].

PROOF. (a) The case $p \leqq 1$. Since v is previsible and $p \leqq 1$, $\int_0^t v(s)\,dX(s)$ is an ordinary Lebesgue–Stieltjes integral for each ω (Property 6, Section 2), so that in fact

$$(4.3)\qquad \int_0^t v(s)\,dX(s) = \sum_{0 \leqq s \leqq t} v(s)[X(s) - X(s-)].$$

(According to the conventions of Section 1, X will have no "linear part" if $p \leqq 1$.) It is clear that

$$(4.4)\qquad \liminf_n V_n \geqq \sum_{0 \leqq s \leqq 1} |v(s)|^p |j(X, s)|^p.$$

However, since

$$(4.5)\qquad V_n(Y, p) = \sum_k \left| \int_{t_{n,k}}^{t_{n,k+1}} v(s)\,dX(s) \right|^p,$$

and since

$$(4.6)\qquad \begin{aligned} \left| \int_{t_{n,k}}^{t_{n,k+1}} v(s)\,dX(s) \right|^p &= \Big| \sum_{t_{n,k} \leqq s \leqq t_{n,k+1}} v(s) j(X, s) \Big|^p \\ &\leqq \Big[\sum_{t_{n,k} \leqq s \leqq t_{n,k+1}} |v(s)|\,|j(X, s)| \Big]^p \\ &\leqq \sum_{t_{n,k} \leqq s \leqq t_{n,k+1}} |v(s)|^p |j(X, s)|^p \end{aligned}$$

(because $p \leqq 1$), it follows that

$$(4.7)\qquad V_n(Y, p) \leqq \sum_{0 \leqq s \leqq 1} |v(s)|^p |j(X, s)|^p$$

for all n. This proves (a), since $|v(s)| \leqq 1$ and $\Sigma_{0 \leqq s \leqq 1} |j(X, s)|^p$ is finite a.s. by Lemma 3.1.

(b) Consider the case $p > 1$. Suppose without loss of generality that the Lévy measure of X concentrates on $[-1, 1]$. If the exponent of X is of the form

$$(4.8)\qquad \psi(u) = iug + \int_{|x| \leqq 1} [e^{iux} - 1 - iux/(1 + x^2)]\,\nu(dx),$$

then write $X = X_n + X^n$, where $X^n = \{X^n(t); t \geqq 0\}$ is the process with stationary independent increments having exponent

$$(4.9)\qquad iug + \int_{1 \geqq |x| \geqq (1/n)} \left[e^{iux} - 1 - \frac{iux}{1 + x^2} \right] \nu(dx).$$

Then

$$(4.10)\qquad Y(t) = \int_0^t v(s)\,dX(s) = \int_0^t v(s)\,dX_n(s) + \int_0^t v(s)\,dX^n(s) = Y_n(t) + Y^n(t),$$

and also

$$\sum_{0 \leqq s \leqq 1} |v(s)|^p |j(X, s)|^p = \sum_{0 \leqq s \leqq 1} |v(s)|^p |j(X_n, s)|^p + \sum_{0 \leqq s \leqq 1} |v(s)|^p |j(X^n, s)|^p = J_n + J^n. \tag{4.11}$$

Then

$$\begin{aligned} E|V_j(Y, p) - V_j(Y^n, p)| &= E\Big|\sum_k |Y_n(t_{j,k+1}) - Y_n(t_{j,k}) + \{Y^n(t_{j,k+1}) - Y^n(t_{j,k})\}|^p \\ &\qquad - \sum_k |Y^n(t_{j,k+1}) - Y^n(t_{j,k})|^p\Big| \\ &\leqq pE \sum_k |Y^n(t_{j,k+1}) - Y^n(t_{j,k})|^{p-1} |Y_n(t_{j,k+1}) - Y_n(t_{j,k})| \\ &\qquad + pEV_j(Y_n, p) \\ &\leqq p \sum_k E^{(p-1)/p} |Y^n(t_{j,k+1}) - Y^n(t_{j,k})|^p E^{1/p} |Y_n(t_{j,k+1}) - Y_n(t_{j,k})|^p \\ &\qquad + pEV_j(Y_n, p). \end{aligned} \tag{4.12}$$

In the calculation (4.12), we have used the following elementary inequality where $1 < p \leqq 2$ and $0 < s < 1$,

$$\begin{aligned} \big| |x + y|^p - |x|^p \big| &= p|x + sy|^{p-1} |y| \\ &\leqq p(|x|^{p-1} + |y|^{p-1})|y| \\ &= p|x|^{p-1}|y| + p|y|^p. \end{aligned} \tag{4.13}$$

Next, observe that if $e_n = EX_n(1)$ then by (2.29)

$$\begin{aligned} E|Y_n(t) - Y_n(s)|^p &\leqq c_p E|X_n(t) - X_n(s)|^p + c_p |e_n|^p |t - s|^p \\ &\leqq K\left[\int_{|x| < (1/n)} |x|^p \nu(dx) + |e_n|^p |t - s|^{p-1}\right](t - s), \end{aligned} \tag{4.14}$$

using Lemma 3.2, and where K is an absolute constant. Similarly, if $e^n = EX^n(1)$, then

$$\begin{aligned} E|Y^n(t) - Y^n(s)|^p &\leqq K\left[\int_{(1/n) \leqq |x| < 1} |x|^p \nu(dx) + |e^n|^p |t - s|^{p-1}\right](t - s). \\ &\leqq M|t - s|, \end{aligned} \tag{4.15}$$

where M is a finite constant independent of n. Let

$$\|\pi_j\| = \max_k |t_{j,k+1} - t_{j,k}|, \qquad b = \sup_n \{|e_n|^p\} < \infty; \tag{4.16}$$

and let

$$c_n(j) = K\left[\int_{|x|<(1/n)} |x|^p \nu(dx) + b\|\pi_j\|^{p-1}\right]. \tag{4.17}$$

Then, using (4.14) and (4.15) in (4.12), we have

$$\begin{aligned} E|V_j(Y, p) - V_j(Y^n, p)| &\leqq p \sum_k M^{(p-1)/p}(t_{j,k+1} - t_{j,k})^{(p-1)/p}[c_n(j)]^{1/p}(t_{j,k+1} - t_{j,k})^{1/p} \\ &\leqq \text{const } c_n(j)^{1/p} \equiv Cc_n(j)^{1/p}. \end{aligned} \tag{4.18}$$

Since p > 1, it is obvious that

$$V_j(Y^n, p) \to J^n \qquad \text{a.s.} \tag{4.19}$$

as $j \to \infty$, because X^n is piecewise linear with a finite number of jumps in any interval. Let $\varepsilon > 0$. Then

$$\begin{aligned} &P\{|V_j(Y) - J| > \varepsilon\} \\ &\leqq P\{|V_j(Y^n) - J^n| > \tfrac{1}{3}\varepsilon\} + P\{|V_j(Y) - V_j(Y^n)| > \tfrac{1}{3}\varepsilon\} + P\{J_n > \tfrac{1}{3}\varepsilon\}. \end{aligned} \tag{4.20}$$

As $n \to \infty$, the third term above goes to zero. The middle term is dominated by $C \int_{|x|<1/n} |x|^p \nu(dx) + C\|\pi_j\|^{p-1}$, because of (4.18) and Chebyshev's inequality. Therefore, if δ is given, one may choose n so large that

$$P\{|V_j(Y) - J| > \varepsilon\} \leqq P\{|V_j(Y^n) - J^n| > \tfrac{1}{3}\varepsilon\} + C\|\pi_j\|\varepsilon^{-1} + \delta. \tag{4.21}$$

Letting $j \to \infty$ and applying (4.19), now completes the proof that $V_j(Y)$ converges to J in probability.

THEOREM 4.2. *If the Lévy measure of X concentrates on a finite interval, and if* $\int_{|x|<1} |x|^p \nu(dx) < \infty$, *then* $V_n(Y, p)$ *converges in* L_r *norm for every* r, $1 \leqq r < \infty$. *(The process Y satisfies the hypotheses of Theorem 4.1.)*

In order to prove this, it will be convenient to establish the following lemma. Let $\{d_k; k \geqq 1\}$ be a sequence of martingale differences, with $E|d_k|^p < \infty$, and let $f_n = \sum_{k=1}^n d_k$.

LEMMA 4.1. *There are positive constants C, c depending on p only such that for every* $n \geqq 1$:

(i) *if* $1 \leqq p \leqq 2$, *then* $E|f_n|^p \leqq C \sum_{k=1}^n E|d_k|^p$;

(ii) *if* $p \geqq 2$, *then* $E|f_n|^p \geqq c \sum_{k=1}^n E|d_k|^p$.

PROOF. Part (i) is known (see C. G. Esseen and B. von Bahr [11]). Part (ii) is established by a simple modification of S. D. Chatterji's proof [6] of the Esseen–von Bahr result. Since

$$\inf_x \left\{\frac{(|1 + x|^p - 1 - px)}{|x|^p}\right\} = c_p > 0, \tag{4.22}$$

it follows that

$$|A + B|^p \geqq |A|^p + c_p|B|^p - \operatorname{sgn}[A]|A|^{p-1}B, \tag{4.23}$$

so that

$$E|f_{n+1}|^p = E|f_n + d_{n+1}|^p \geqq E|f_n|^p + c_pE|d_{n+1}|^p. \tag{4.24}$$

The result now follows by induction.

LEMMA 4.2. *If the Lévy measure of X is concentrated on a finite interval, if $\int_{|x|<1} |x|^p \nu(dx) < \infty$ for some p, $0 < p \leqq 2$, and if $\sup_t |v(t)| \leqq 1$, then*

$$E\left\{\left|\int_s^t v(u)\, dX(u)\right|^p \middle| F(s)\right\} \leqq c_p(t - s), \qquad 0 \leqq t - s \leqq 1. \tag{4.25}$$

PROOF. Let $\{d_k; k \geqq 1\}$ be as in Lemma 4.1; let $\{v_k; k \geqq 1\}$ be a sequence of bounded random variables, $|v_k| \leqq 1$ and v_k measurable with respect to $F_{k-1} = F(d_1, \cdots, d_{k-1})$. Results of Burkholder [5] then imply if $p > 1$ that $E|\Sigma_{k=1}^n v_k d_k|^p \leqq c_pE|\Sigma_{k=1}^n d_k|^p$. Let m be given and $A \in F_m$. Then

$$\int_A \left|\sum_{m+1}^n v_k d_k\right|^p = \int \left|\sum_{m+1}^n (v_k I_A) d_k\right|^p \leqq c_p \int_A \left|\sum_{m+1}^n d_k\right|^p, \tag{4.26}$$

implying that

$$E\left\{\left|\sum_{m+1}^n v_k d_k\right|^p \middle| F_m\right\} \leqq c_pE\left\{\left|\sum_{m+1}^n d_k\right|^p \middle| F_m\right\}. \tag{4.27}$$

If X is a martingale, and if v is a step function, it now follows that

$$E\left\{\left|\int_s^t v(u)\, dX(u)\right|^p \middle| F(s)\right\} \leqq c_pE\{|X(t) - X(s)|^p | F(s)\} \leqq c_p(t - s), \tag{4.28}$$

using Lemma 3.2. Conclusion (4.25) now follows for all bounded v by passage to the limit. To handle the case when X is not a martingale, consider $X(t) - tEX(1)$. The case $p \leqq 1$ is easy.

PROOF OF THEOREM 4.2. To show convergence in L_r, $1 \leqq r < 2$, it suffices to show $\sup_j E[V_j(Y, p)]^2 < \infty$. The case $1 < p \leqq 2$ will be discussed first. Here there is no loss in generality in assuming that X is a martingale (subtract off a multiple of t if necessary). Then, in obvious notation,

$$\begin{aligned} E[V_j(Y, p)]^2 &= E\left[\sum_k |Y(t_{j,k+1}) - Y(t_{j,k})|^p\right]^2 \\ &= E\left[\sum_k |\Delta_k|^p\right]^2 \\ &= E\sum_k |\Delta_k|^{2p} + E\sum_{i \neq k} |\Delta_k|^p|\Delta_i|^p. \end{aligned} \tag{4.29}$$

But $E\,\Sigma\,|\Delta_k|^{2p} \leqq cE|Y(1)|^{2p} < \infty$, from Lemma 4.1(ii). Also, suppressing the

j in $t_{j,k}$, if $i < k$,

$$(4.30)\qquad E|\Delta_k|^p|\Delta_i|^p = E|Y(t_{i+1}) - Y(t_i)|^p E\{|Y(t_{k+1}) - Y(t_k)|^p\,|F(t_{i+1})\}$$

and

$$(4.31)\qquad E\{|Y(t_{k+1}) - Y(t_k)|^p|F(t_{i+1})\} \leqq c(t_{k+1} - t_k),$$

using Lemma 4.2. Therefore,

$$(4.32)\qquad E\sum_{i\neq k}|\Delta_k|^p|\Delta_i|^p \leqq c\sum_{i\neq k}(t_{i+1} - t_i)(t_{k+1} - t_k) \leqq c.$$

It follows that $\sup_j E[V_j(Y,p)]^2 < \infty$, proving convergence in L_r, $1 \leqq r < 2$. One proves convergence in L_r, $1 \leqq p < 2n$ by showing in a similar manner that $\sup_j E[V_j(Y,p)]^{2n} < \infty$.

If $p \leqq 1$, then from the proof of part (a) of Theorem 4.1, $V_j(Y,p) \leqq \Sigma_{0\leqq s\leqq 1}|j(X,s)|^p$. The Lévy measure of the latter random variable is concentrated on a finite interval (see [18]), so it has moments of all orders.

Let us conclude this section with a few remarks about the a.s. convergence of $V_j(Y,p)$. The case where $v \equiv 1$ was discussed in [18], where some open problems were listed. For general v, we give only the following rather limited results (compare with [4] and [21]).

THEOREM 4.3. *Assume* $\sup_t |v(t)| \leqq 1$ *and* $Y(t) = \int_0^t v(s)\,dX(s)$. *If* $\Sigma_n[\Sigma_k(t_{n,k+1} - t_{n,k})^2]^{1/2} < \infty$, *then* $V_j(Y,2)$ *converges a.s.*

PROOF. Assume without loss of generality that the Lévy measure concentrates on a finite interval. Then if $c = EX(1)$, $X(t) - ct$ and $Y'(t) = Y(t) - c\int_0^t v(s)\,ds$ are martingales. Moreover,

$$(4.33)\qquad \begin{aligned}\sum_k [Y(t_{n,k+1}) - Y(t_{n,k})]^2 &= \sum_k [Y'(t_{n,k+1}) - Y'(t_{n,k})]^2 \\ &\quad + \sum_k c^2\left[\int_{t_{n,k}}^{t_{n,k+1}} v(s)\,ds\right]^2 \\ &\quad + 2c\sum_k [Y'(t_{n,k+1}) - Y'(t_{n,k})]\int_{t_{n,k}}^{t_{n,k+1}} v(s)\,ds.\end{aligned}$$

Since the second term on the right obviously converges to zero a.s. as $n \to \infty$, and since the expectation of the third term is less than

$$(4.34)\qquad \begin{aligned}E\sum_k{}^{1/2}|Y'(t_{n,k+1}) - Y'(t_{n,k})|^2\sum_k{}^{1/2}(t_{n,k+1} - t_{n,k})^2 \\ \leqq \text{const}\sum_k{}^{1/2}(t_{n,k+1} - t_{n,k})^2\end{aligned}$$

(by Lemma 3.2), it follows from the Borel–Cantelli theorem that there is no loss in generality in assuming that X and Y are martingales.

In Section 5, the following formula will be established:

$$\sum_{0 \leqq s \leqq t} [v(s)]^2 [j(X, s)]^2 = \left[\int_0^t v(s)\, dX(s) \right]^2 - 2 \int_0^t v(s) Y(s)\, dX(s). \tag{4.35}$$

Assuming (4.35) and using in the third equality that X is a martingale,

$$\begin{aligned} (4.36) \quad & E \left| V_n(Y, 2) - \sum_{0 \leqq s \leqq 1} [v(s)]^2 [j(X, s)]^2 \right|^2 \\ & = E \left| \sum_k \left\{ \left[\int_{t_{n,k}}^{t_{n,k+1}} v(s)\, dX(s) \right]^2 - \sum_{t_{n,k} \leqq s \leqq t_{n,k+1}} [v(s)]^2 [j(X, s)]^2 \right\} \right|^2 \\ & = 4E \left| \sum_k \int_{t_{n,k}}^{t_{n,k+1}} v(s) [Y(s) - Y(t_{n,k})]\, dX(s) \right|^2 \\ & = 4 \sum_k E \left| \int_{t_{n,k}}^{t_{n,k+1}} v(s) [Y(s) - Y(t_{n,k})]\, dX(s) \right|^2 \\ & = 4c \sum_k \int_{t_{n,k}}^{t_{n,k+1}} E v^2(s) [Y(s) - Y(t_{n,k})]^2\, ds \\ & \leqq \text{const} \sum_k (t_{n,k+1} - t_{n,k})^2. \end{aligned}$$

The result now follows from the Borel–Cantelli theorem. It is clear from the proof that if X is known to be a martingale, or if every truncation of X is a martingale (which happens if X is symmetric), then one needs only $\Sigma_n \Sigma_k (t_{n,k+1} - t_{n,k})^2 < \infty$.

5. Zero jumps

In this section, some of the preceding ideas will be used to study certain sample function properties of X itself, in particular the zero jumps (defined below) of the process X. Before doing this, however, it is necessary to have an analogue of Itô's formula for processes $Y(t) = \int_0^t v(s)\, dX(s)$.

THEOREM 5.1. *Suppose* $\int_{|x| < 1} |x|^p \nu(dx) < \infty$, *and* $\sup_t |v(s)| \leqq 1$. *Let F be a real function such that (i) F' exists and is continuous; (ii) for every neighborhood N of 0 there is a constant M (depending on N perhaps) such that*

$$|F(x + h) - F(x) - F'(x)h| \leqq M h^p, \qquad x \in N, x + h \in N. \tag{5.1}$$

Then

$$\begin{aligned} (5.2) \quad F(Y(t)) = & \int_0^t F'[Y(s-)]\, dY(s) \\ & + \sum_{0 \leqq s \leqq t} \{F[Y(s)] - F[Y(s-)] - F'[Y(s-)] j(Y, s)\}. \end{aligned}$$

Theorems of this type have been established in great generality for quasi-martingales (of which Y is an example) by Meyer [15], but under more restrictive hypotheses on F (F is assumed twice continuously differentiable). The special nature of the process Y and the theory of sample variation in Section 4 permit the stronger conclusion here. Theorem 5.1 yields, for example, the following conclusion.

COROLLARY 5.1. *Let Y satisfy the hypotheses of Theorem* 5.1, *and assume in addition that $p > 1$. Then $|Y(t)|^p$ is a quasimartingale.*

Also, choosing $F(x) = x^2$ yields formula (4.35), *required in the preceding section.*

PROOF. As usual, one may suppose that the Lévy measure of X is on $[-1, 1]$. Let X^n, Y^n be defined as in the proof of Theorem 4.1, part (b). Then $Y^n(t) = \int_0^t v(s)\, dX^n(s)$ is an ordinary Lebesgue–Stieltjes integral for each ω. Let $S_0 \equiv 0$, and $S_1 < S_2 < \cdots$ be an enumeration of the jumps of Y^n (since X^n has only a finite number of jumps in any finite interval, the same is true of Y^n). If t is fixed, let $T_k = \min\{S_k, t\}$. Then

$$
\begin{aligned}
(5.3)\quad F[Y^n(t)] &= \sum_k \{F[Y^n(T_{k+1}-)] - F[Y^n(T_k)]\} + \sum_k F[Y^n(T_k)] - F[Y^n(T_k-)] \\
&= \sum_k \{F[Y^n(T_{k+1}-)] - F[Y^n(T_k)]\} + \sum_k F'[Y^n(T_k-)]j(Y^n, T_k) \\
&\qquad + \sum_k \{F[Y^n(T_k)] - F[Y^n(T_k-)] - F'[Y^n(T_k-)]j(Y^n, T_k)\} \\
&= \sum_k \int_{(T_k, T_{k+1})} F'[Y^n(s)]\, dY^n(s) + \sum_k F'[Y^n(T_k-)]j(Y^n, T_k) \\
&\qquad + \sum_{0 \leqq s \leqq t} \{F[Y^n(s)] - F[Y^n(s-)] - F'[Y^n(s-)]j(Y^n, s)\} \\
&= \int_0^t F'[Y^n(s-)]\, dY^n(s) \\
&\qquad + \sum_{0 \leqq s \leqq t} \{F[Y^n(s)] - F[Y^n(s-)] - F'[Y^n(s-)]j(Y^n, s)\},
\end{aligned}
$$

using the fact that Y^n is continuous on (T_k, T_{k+1}). By choosing a subsequence if necessary, we may suppose that for each ω the paths of Y^n converge uniformly for $0 \leqq t \leqq 1$ to the paths of Y (see Property 2.5). For each ω, the paths of Y are bounded on $[0, 1]$. Therefore, there exists $M(\omega)$ such that

$$
\begin{aligned}
(5.4)\quad &|F[Y^n(s)] - F[Y^n(s-)] - F'[Y^n(s-)]j(Y^n, s)| \\
&\qquad \leqq M(\omega)|j(Y^n, s)|^p \leqq M(\omega)|j(X, s)|^p.
\end{aligned}
$$

Since $\Sigma_{0 \leqq s \leqq 1} |j(X, s)|^p < \infty$ (Lemma 3.1), it follows that

$$
(5.5)\quad \sum_{0 \leqq s \leqq t} F[Y^n(s)] - F[Y^n(s-)] - F'[Y^n(s-)]j(Y^n, s)
$$

converges for the given ω to

$$
(5.6)\quad \sum_{0 \leqq s \leqq t} F[Y(s)] - F[Y(s-)] - F'[Y(s-)]j(Y, s),
$$

a process having paths of bounded variation. Finally, one checks that for each ω, $\int_0^t F'[Y^n(s-)]\,dY^n(s)$ converges to $\int_0^t F'[Y(s-)]\,dY(s)$; by taking a further subsequence if necessary, one can ensure that for each ω the convergence is uniform in t, $0 \leqq t \leqq 1$.

REMARK 5.1. The proof shows that if the Lévy measure of X is concentrated on a set bounded away from zero, then the conclusion of Theorem 5.1 holds with no further hypotheses other than the existence and continuity of F'.

DEFINITION 5.1. *A process X is said to have a zero jump at s if either $X(s) \leqq 0 < X(s-)$ or $X(s) > 0 \geqq X(s-)$. The notation $Z(t) = \Sigma'_{0 \leqq s \leqq t} |X(s)|$ will denote the sum of all $|X(s)|$, $s \leqq t$, over only those s at which a zero jump occurs.*

The rest of this section will be devoted to a study of the process $Z(t)$. The asymmetry in the definition of zero jump was introduced in order to keep the statement of Theorem 5.2 simple. In all cases of real interest (specifically when $\nu(R) = \infty$), the definition can be replaced by the condition that either $X(s) < 0 < X(s-)$ or $X(s-) < 0 < X(s)$, as shown in the proof of Theorem 5.2.

Of course if may happen that the process Z is rather trivial, for example when X is a subordinator. On the other hand, many examples exist of processes X which experience infinitely many zero jumps as t varies in any interval of the form $[0, \varepsilon]$, $\varepsilon > 0$. For example, if the Lévy measure of X is concentrated on $(0, \infty)$, then X has only upward jumps; however, if $\int_{-\infty}^{\infty} [\lambda + Re\,\psi(x)]^{-1}\,dx < \infty$ for all positive λ, then it is known that $\inf\{t > 0;\, X(t) = 0\} = 0$ a.e. (Blumenthal and Getoor, [3], p. 64). It is easy to see that this implies that X has infinitely many zero jumps. In such cases, it is not even clear *a priori* that the process $Z(t)$ is finite—this will be the conclusion of Theorem 5.2, where a stochastic upper bound for Z is derived.

It is interesting to notice also that as $t \downarrow 0$, a process may jump over zero infinitely often, but without hitting 0 itself infinitely often. Here is an example. Let X be the symmetric Cauchy process. It is then known that 0 is not regular for $\{0\}$; that is, the process does not pass through 0 infinitely often as $t \downarrow 0$ (see Port [19]). However, it is also known (see Blumenthal and Getoor [2], or Millar [18]) that

$$\limsup_{t \to 0} |X(t)| t^{-1/2} = +\infty, \tag{5.7}$$

implying by symmetry and the zero-one law that

$$\limsup_{t \to 0} X(t) t^{-1/2} = +\infty \tag{5.8}$$

and

$$\liminf_{t \to 0} X(t) t^{-1/2} = -\infty. \tag{5.9}$$

In particular, X must pass from above to below zero infinitely often as $t \downarrow 0$. Since the process does not hit zero while doing this, it therefore must jump over 0 infinitely often.

THEOREM 5.2. *The process sum of jumps* $Z(t)$ *satisfies the relation*

$$Z(t) \leqq X(t)^+ - \int_0^t I_{(0,\infty)}[X(s-)]\, dX(s). \tag{5.10}$$

PROOF. Suppose without loss of generality that the Lévy measure concentrates on $[-1, 1]$.

(a) Suppose that the exponent of X is of the form

$$\psi(u) = iug + \int [e^{iux} - 1]\nu(dx), \tag{5.11}$$

where $\nu(R) < \infty$. Then the paths of X are piecewise linear (see, for example, [12], p. 274). Let $\varepsilon > 0$ and $F(x) = \int_0^x (\varepsilon \wedge y)^+ \, dy$. By Theorem 5.1 (or rather, Remark 5.1),

$$\begin{aligned} F[X(t)] &= \int_0^t F'[X(s-)]\, dX(s) \\ &\quad + \sum_{0 \leqq s \leqq t} F[X(s)] - F[X(s-)] - F'[X(s-)]j(X, s), \end{aligned} \tag{5.12}$$

where the sum on the right is only a finite sum for each ω, and the integral on the right is an ordinary Lebesgue–Stieltjes integral (Property 2.6). Divide (5.2) by ε, and let $\varepsilon \downarrow 0$. Then

$$\varepsilon^{-1}F[X(t)] = \varepsilon^{-1} \int_0^{X(t)} (\varepsilon \wedge y)^+ \, dy, \tag{5.13}$$

so that

$$\begin{aligned} \lim_{\varepsilon \to 0} \varepsilon^{-1}F[X(t)] &= 0, \text{ if } X(t) \leqq 0 \\ &= \lim_{\varepsilon \to 0} \left[\varepsilon^{-1} \int_0^{\varepsilon} y \, dy + \varepsilon^{-1} \int_{\varepsilon}^{X(t)} \varepsilon \, dy \right] = X(t), \end{aligned} \tag{5.14}$$

if $X(t) > 0$. That is, $\lim_{\varepsilon \to 0} \varepsilon^{-1}F[X(t)] = X(t)^+$. Since $\int_0^t F'[X(s-)]\, dX(s) = \int_0^t [\varepsilon \wedge X(s-)]^+ \, dX(s)$ is a Lebesgue–Stieltjes integral, and since $0 \leq (\varepsilon \wedge X(s-))^+ \varepsilon^{-1} \leqq 1$ with

$$\lim_{\varepsilon \downarrow 0} \varepsilon^{-1}[\varepsilon \wedge X(s-)]^+ = I_{(0,\infty)}[X(s-)], \tag{5.15}$$

we may apply the dominated convergence theorem to obtain

$$\lim_{\varepsilon \downarrow 0} \int_0^t F'[X(s-)]\, dX(s) = \int_0^t I_{(0,\infty)}[X(s-)]\, dX(s). \tag{5.16}$$

Finally, to evaluate the remaining term, it suffices to evaluate only

$$\lim_{\varepsilon \downarrow 0} \left[\int_{X(s-)}^{X(s)} (\varepsilon \wedge y)^+ \, dy - [\varepsilon \wedge X(s-)]^+ j(X, s) \right] \tag{5.17}$$

for each s, since there is involved only a finite sum. There are several cases. If

$X(s) > 0 \geqq X(s-)$, and if $\varepsilon < X(s)$, then

$$\varepsilon^{-1}\left[\int_{X(s-)}^{X(s)} (\varepsilon \wedge y)^+ \, dy - [\varepsilon \wedge X(s-)]^+ j(X, s)\right] \tag{5.18}$$

$$= \varepsilon^{-1}\left[\int_0^{\varepsilon} y \, dy + \int_{\varepsilon}^{X(s)} \varepsilon \, dy\right] \to X(s).$$

Similarly, if $X(s) \leqq 0 < X(s-)$, then the resulting limit is $-X(s)$. If either $X(s), X(s-) > 0$, or $X(s), X(s-) < 0$, then the resulting limit is 0. Hence, considering all cases, one finds that

$$\lim_{\varepsilon \downarrow 0}\left[\int_{X(s-)}^{X(s)} (\varepsilon \wedge y)^+ \, dy - [\varepsilon \wedge X(s-)]^+ j(X, s)\right] = |X(s)| \tag{5.19}$$

if there is a zero jump at s, and the limit is zero if not. This establishes the theorem for the special case. Note that in this case, the theorem is true with equality.

(b) The process X is said to have a strict zero jump at s if either $X(s-) < 0 < X(s)$ or $X(s) < 0 < X(s-)$. Suppose that $\nu(R) = \infty$. Then for almost every ω, all the zero jumps of the path $X(t)$ are strict. Here is the proof of this fact. Let $\{T_k, k \geqq 1\}$ be a sequence of stopping times that enumerates all the jumps of X. For example, let

$$T_{n,1} = \inf\left\{t > 0; |X(t) - X(t-)| \in \left[\frac{1}{n}, \frac{1}{n-1}\right)\right\}, \tag{5.20}$$

and

$$T_{n,k+1} = \inf\left\{t > T_{n,k}; |X(t) - X(t-)| \in \left[\frac{1}{n}, \frac{1}{n-1}\right)\right\}. \tag{5.21}$$

Then $\{T_{n,k}; n \geqq 1, k \geqq 1\}$ enumerates the jumps.

The desired assertion will follow if we show that $X(T_k)$ and $X(T_k-)$ are nonatomic. Let $\varepsilon > 0$. The process $Y(t) = X(t + \varepsilon) - X(\varepsilon)$ is independent of $X(\varepsilon)$, and $Y \sim X$. Let $\{S_n; n \geqq 1\}$ be an enumeration of the jumps of $\{Y(t)\}$. Then

$$\begin{aligned} &P\{\varepsilon < T_k, X(T_k) = 0\} \\ &\quad = \sum_j P\{\varepsilon < T_k, T_k = S_j + \varepsilon, X(\varepsilon) + [X(S_j + \varepsilon) - X(\varepsilon)] = 0\} \\ &\quad = \sum_j P\{\varepsilon < T_k, T_k = S_j + \varepsilon, X(\varepsilon) + Y(S_j) = 0\}. \end{aligned} \tag{5.22}$$

But $P\{\varepsilon < T_k, T_k = S_j + \varepsilon, X(\varepsilon) + Y(S_j) = 0\} \leqq P\{X(\varepsilon) + Y(S_j) = 0\}$, and $X(\varepsilon)$ is independent of $Y(S_j)$. Also, since $\nu(R) = \infty$, $X(\varepsilon)$ is nonatomic (A. Wintner and P. Hartman [20]; see also J. R. Blum and M. Rosenblatt [1]).

Since it is well known that the convolution of two measures is nonatomic if at least one of them is, $P\{X(\varepsilon) + Y(S_j) = 0\} = 0$, and so $P\{\varepsilon < T_k, X(T_k) = 0\} = 0$. Let $\varepsilon \downarrow 0$ to get the result, since $P\{T_k > 0\} = 1$ (see the construction of T_k above). The case of $X(T_k-)$ is treated similarly.

(c) Theorem 5.2 will now be verified under the assumption $\nu(R) = \infty$, and this will complete the proof. By part (b), one need consider only strict zero jumps. Let X^n be the process defined in the proof of Theorem 4.1, part (b). By choosing a subsequence if necessary, one may assume that for each ω the paths of X^n on $[0, 1]$ converge uniformly to those of X (see Property 2.5). By part (a) of the present proof,

$$\sum_{0 \leq s \leq t}{}' |X^n(s)| = X^n(t)^+ - \int_0^t I_{(0,\infty)}[X^n(s-)]\, dX^n(s). \tag{5.23}$$

Now let $n \to \infty$ (through the subsequence, if necessary).

Then for every ω, $X^n(t)^+ \to X(t)^+$, $0 \leq t \leq 1$. Also, for each t,

$$\int_0^t I_{(0,\infty)}[X^n(s-)]\, dX^n(s) \to \int_0^t I_{(0,\infty)}[X(s-)]\, dX(s) \tag{5.24}$$

in L_2. For,

$$\begin{aligned} &\left\| \int_0^t I_{(0,\infty)}[X^n(s-)]\, dX^n(s) - \int_0^t I_{(0,\infty)}[X(s-)]\, dX(s) \right\|_2 \\ &\leq \left\| \int_0^t I_{(0,\infty)}[X^n(s-)]\, dX^n(s) - \int_0^t I_{(0,\infty)}[X^n(s-)]\, dX(s) \right\|_2 \\ &\quad + \left\| \int_0^t \{I_{(0,\infty)}[X^n(s-)] - I_{(0,\infty)}[X(s-)]\}\, dX(s) \right\| \\ &= S_1 + S_2. \end{aligned} \tag{5.25}$$

By Property 2.2, $S_2^2 \leq K \int_0^1 E\{I_{(0,\infty)}[X^n(s-)] - I_{(0,\infty)}[X(s-)]\}^2\, dt$. Moreover, for each ω, $I_{(0,\infty)}[X^n(s-)] \to I_{(0,\infty)}[X(s-)]$ for every fixed s (recall, for every ω, $X^n(s) \to X(s)$ uniformly for $0 \leq s \leq 1$), except possibly those s at which $X(s-) = 0$. But if $\nu(R) = \infty$, these s have Lebesgue measure 0, so $I_{(0,\infty)}[X^n(s-)] \to I_{(0,\infty)}[X(s-)]$ for almost all $(\omega, s)\,(dP \times ds)$. Therefore, by dominated convergence, $S_2 \to 0$. Also

$$S_1 = \left\| \int_0^t I_{(0,\infty)}[X^n(s-)]\, d[X^n(s) - X(s)] \right\|_2 \leq \|X^n(t) - X(t)\|_2 \to 0 \tag{5.26}$$

as $n \to \infty$. In fact, by using the stronger inequality given in Property 2.2, it is easy to see that we have the stronger result:

$$\sup_{0 \leq t \leq 1} \left| \int_0^t I_{(0,\infty)}[X^n(s-)]\, dX^n(s) - \int_0^t I_{(0,\infty)}[X(s-)]\, dX(s) \right| \to 0 \tag{5.27}$$

in L_2, so by choosing a further subsequence if necessary one obtains for each ω

$$\int_0^t I_{(0,\infty)}[X^n(s-)]\,dX^n(s) \to \int_0^t I_{(0,\infty)}[X(s-)]\,dX(s), \tag{5.28}$$

uniformly in t, $0 \leqq t \leqq 1$. Assume from now on that $n \to \infty$ through this subsequence.

Finally, consider $\lim_{n\to\infty} \Sigma'_{0\leqq s\leqq t} |X^n(s)|$. Let

$$T_{n,1} = \inf\left\{t > 0;\ X \text{ has a zero jump at } t, \text{ and } |j(X, t)| \in \left[\frac{1}{n}, \frac{1}{n-1}\right)\right\}, \tag{5.29}$$

$$T_{n,k+1} = \inf\left\{t > T_{n,k};\ X \text{ has a zero jump at } t \text{ and } |j(X, t)| \in \left[\frac{1}{n}, \frac{1}{n-1}\right)\right\}. \tag{5.30}$$

Then $\{T_{n,k};\ n \geqq 1,\ k \geqq 1\}$ enumerates all the zero jumps of X; let $\{T_k;\ k \geqq 1\}$ be a list of the $\{T_{n,k}\}$. By part (b), either $X(T_k) < 0 < X(T_k-)$ or $X(T_k-) < 0 < X(T_k)$ (for a specified ω). It therefore follows from the uniform convergence that if T_k is the time of a zero jump for X, then T_k is also the time of a (strict) zero jump for X^n, for all sufficiently large n (how large will depend on ω). Therefore, for each ω, if $T_k(\omega) \leqq t$, then

$$|X(T_k)| \leqq \liminf_{n\to\infty} \sum_{0\leqq s\leqq t}' |X^n(s)| \leqq X(t)^+ - \int_0^t I_{(0,\infty)}[X(s-)]\,dX(s). \tag{5.31}$$

This clearly continues to hold if we replace the left side by any finite sum $\Sigma_{k=1}^n |X(T_k)|$. The limit on the right is taken through a subsequence for which the limit will exist for every ω and uniformly in t, $0 \leqq t \leqq 1$. The formula of Theorem 5.2 therefore holds for almost all ω, and all t, $0 \leqq t \leqq 1$, the exceptional ω set not depending on t. This completes the proof.

REMARK 5.2. Presumably the inequality of Theorem 5.2 may be replaced by equality, but I have not been able to show this. Notice that the case $\nu(R) = \infty$ could not be treated by taking a limit in the formula (5.2) directly, since the function $F(x) = \int_0^x (\varepsilon \wedge y)^+\,dy$ does not satisfy the hypothesis of Theorem 5.1. The formula of Theorem 5.2 should be compared to the formula attributed to Tanaka (see McKean [13]) for the local time of Brownian motion.

The process $Z(t)$ is not local time (Z does not have continuous paths). However, it is natural to wonder whether, if $Z(t)$ were smoothed out appropriately, the result would be similar to local time (whenever local time exists).

Note added in proof. Equality in Theorem 5.2 can be established by using the theory of Lévy systems developed by S. Watanabe ("On discontinuous additive functionals and Lévy measures of a Markov process," *Japan J. Math.*, Vol. 34 (1964), pp. 53–70). Connections between the process Z and local time are contained in recent work of Getoor and Millar ("Some limit theorems for local time," to appear in *Compositio Math.*).

REFERENCES

[1] J. R. Blum and M. Rosenblatt, "On the structure of infinitely divisible distributions," *Pacific J. Math.*, Vol. 9 (1959), pp. 1–7.

[2] R. M. Blumenthal and R. K. Getoor, "Sample functions of stochastic processes with stationary independent increments," *J. Math. Mech.*, Vol. 10 (1961), pp. 493–516.

[3] ———, "Local times for Markov processes," *Z. Wahrscheinlichkeitstheorie und Verw. Gebiete*, Vol. 3 (1964), pp. 50–74.

[4] G. A. Brosamler, "Quadratic variation of potentials and harmonic functions," *Trans. Amer. Math. Soc.*, Vol. 149 (1970), pp. 243–257.

[5] D. L. Burkholder, "Martingale transforms," *Ann. Math. Statist.*, Vol. 37 (1966), pp. 1494–1504.

[6] S. D. Chatterji, "An L^p convergence theorem," *Ann. Math. Statist.*, Vol. 40 (1969), pp. 1068–1070.

[7] P. Courrège, "Intégrales stochastiques et martingales de carré intégrable," *Seminaire de Théorie du Potentiel, 7 ième année*, Institut Henri Poincaré, Secrétariat mathématique, Paris, 1963.

[8] C. Doléans-Dade and P. A. Meyer, "Intégrales stochastiques par rapport aux martingales locales," *Séminaire de Probabilités* IV, *Université de Strasbourg*, Lecture Notes in Mathematics, Vol. 124, Heidelberg, Springer, 1970.

[9] J. L. Doob, *Stochastic Processes*, New York, Wiley, 1953.

[10] L. E. Dubins and L. J. Savage, "A Tchebycheff-like inequality for stochastic processes," *Proc. Nat. Acad. Sci., U.S.A.*, Vol. 53 (1965), pp. 274–275.

[11] C. G. Esseen and B. von Bahr, "Inequalities for the rth absolute moment of a sum of random variables, $1 \leq r \leq 2$," *Ann. Math. Statist.*, Vol. 36 (1965), pp. 299–303.

[12] I. I. Gikhman and A. V. Skorokhod, *Introduction to the Theory of Random Processes*, Philadelphia, W. B. Saunders, 1969.

[13] H. P. McKean, Jr., *Stochastic Integrals*, New York, Academic Press, 1969.

[14] P. A. Meyer, *Probability and Potentials*, Waltham, Blaisdell Pub., 1966.

[15] ———, "Intégrales stochastiques, I, II, III, IV," *Séminaire de Probabilités* I, *Université de Strasbourg*, Lecture Notes in Mathematics, Vol. 39, Heidelberg, Springer, 1967.

[16] P. W. Millar, "Martingale integrals," *Trans. Amer. Math. Soc.*, Vol. 133 (1968), pp. 145–168.

[17] ———, "Martingales with independent increments," *Ann. Math. Statist.*, Vol. 40 (1969), pp. 1033–1041.

[18] ———, "Path behavior of processes with stationary independent increments," *Z. Wahrscheinlichkeitstheorie und Verw. Gebiete*, Vol. 17 (1971), pp. 53–73.

[19] S. C. Port, "Hitting times and potentials for recurrent stable processes," *J. Analyse Math.*, Vol. 20 (1968), pp. 371–395.

[20] A. Wintner and P. Hartman, "On the infinitesimal generator of integral convolutions," *Amer. J. Math.*, Vol. 64 (1942), pp. 273–298.

[21] E. Wong and M. Zakai, "The oscillation of stochastic integrals," *Z. Wahrscheinlichkeitstheorie und Verw. Gebiete*, Vol. 4 (1965), pp. 103–112.

ON THE SUPPORT OF DIFFUSION PROCESSES WITH APPLICATIONS TO THE STRONG MAXIMUM PRINCIPLE

DANIEL W. STROOCK and S. R. S. VARADHAN
COURANT INSTITUTE, NEW YORK UNIVERSITY

1. Introduction

Let $a\colon [0, \infty) \times R^d \rightsquigarrow S_d$ and $b\colon [0, \infty) \times R^d \rightsquigarrow R^d$ be bounded continuous functions, where S_d denotes the class of symmetric, nonnegative definite $d \times d$ matrices. From a and b form the operator

$$L_t = \frac{1}{2} \sum_{i,j=1}^{d} a^{ij}(t, x) \frac{\partial^2}{\partial x_i \partial x_j} + \sum_{i=1}^{d} b_i(t, x) \frac{\partial}{\partial x_i}. \tag{1.1}$$

A strong maximal principle for the operator $(\partial/\partial t) + L_t$ is a statement of the form: "for each open $\mathscr{G} \subseteqq [0, \infty) \times R^d$ and each $(t_0, x_0) \in \mathscr{G}$ there is a set $\mathscr{G}(t_0, x_0) \subseteqq \mathscr{G}$ with the property that $(\partial f/\partial t) + L_t f \geqq 0$ on $\mathscr{G}(t_0, x_0)$ and $f(t_0, x_0) = \sup_{\mathscr{G}(t_0, x_0)} f(t, x)$ imply $f \equiv f(t_0, x_0)$ on $\mathscr{G}(t_0, x_0)$." Of course, in order for a strong maximum principle to be very interesting it must describe the set $\mathscr{G}(t_0, x_0)$. Further, it should be possible to show that $\mathscr{G}(t_0, x_0)$ is maximal. That is, one wants to know that if $(t_1, x_1) \in \mathscr{G} - \mathscr{G}(t_0, x_0)$, then there is an f satisfying $(\partial f/\partial t) + L_t f \geqq 0$ on $\mathscr{G}$ (perhaps in a generalized sense) such that $f(t_0, x_0) = \sup f(t, x)$, and $f(t_1, x_1) < f(t_0, x_0)$.

In the case when $a(t, x)$ is positive definite for all (t, x), L. Nirenberg [6] has shown that $\mathscr{G}(t_0, x_0)$ can be taken as the closure in $\mathscr{G}$ of the set of $(t_1, x_1) \in \mathscr{G} \cap ([t_0, \infty) \times R^d)$ such that there exists a continuous map $\phi\colon [t_0, t_1] \rightsquigarrow R^d$ with the properties that $\phi(t_0) = x_0$, $\phi(t_1) = x_1$, and $(t, \phi(t)) \in \mathscr{G}$ for all $t \in (t_0, t_1)$. We will give a probabilistic proof of the Nirenberg maximum principle in Section 3. Moreover, we will also prove there that Nirenberg's $\mathscr{G}(t_0, x_0)$ is maximal in the desired sense.

If a is only nonnegative definite, the problem of finding a suitable maximum principle is more difficult. Results in this direction have been proved by J.-M. Bony [1] and C. D. Hill [3]. Both of these authors employ a modification of the technique originally introduced by E. Hopf for elliptic operators and later adapted by Nirenberg for parabolic ones. The major drawback to Bony's

Results obtained at the Courant Institute of Mathematical Sciences, New York University, this research was sponsored by the U.S. Air Force Office of Scientific Research. Contract AF-49(638)-1719.

work is his restrictive assumptions on the coefficients of L_t. As we will point out later, our own approach has not removed all his restrictions.

Before going into the details, we will conclude this section with an outline of our method. Suppose P_{t_0,x_0} is a probability measure on $C([t_0, \infty), R^d)$ with the properties that $P_{t_0,x_0}(x(t_0) = x_0) = 1$ and

$$\int_{t_0}^{t} \left(\frac{\partial}{\partial u} + L_u\right) f(u, x(u))\, du \tag{1.2}$$

is a martingale for all $f \in C_0^\infty([t_0, \infty) \times R^d)$ (here and in what follows $C_0^\infty(S)$ denotes the space of infinitely differentiable functions having compact support in S). Given an open $\mathscr{G} \subseteqq [0, \infty) \times R^d$, let $\tau = \inf\{t \geqq t_0 : (t_0, x(t)) \notin \mathscr{G}\}$. Then it is easy to see that $f(t \wedge \tau, x(t \wedge \tau))$ is a submartingale if $f \in C_b^{1,2}(\mathscr{G})$ and $(\partial f/\partial t) + L_t \geqq 0$. (We use $C_b^{1,2}(S)$ to denote the class of bounded f having one bounded continuous t derivative and two bounded continuous x derivatives on S.) Hence,

$$f(t_0, x_0) \leqq E^{P_{t_0,x_0}}[f(t \wedge \tau, x(t \wedge \tau))]. \tag{1.3}$$

In particular, if $f(t_0, x_0) = \sup_{\mathscr{G}} f(t, x)$, then $f(t_1, x_1) = f(t_0, x_0)$ at all (t_1, x_1) for which there exists a path $\phi \in \text{supp}\,(P_{t_0,x_0})$ such that $\phi(t_1) = x_1$ and $(t, \phi(t)) \in \mathscr{G}$ for $t \in [t_0, t_1]$. Thus, for example, if

$$\text{supp}\,(P_{t_0,x_0}) = \{\phi \in C([0, \infty), R^d) : \phi(t_0) = x_0\}, \tag{1.4}$$

then $(\partial/\partial t) + L_t$ satisfies the Nirenberg maximum principle. What we are going to do is study the measure P_{t_0,x_0} and try to describe its support.

2. Background material

In this section, we discuss diffusions from the point of view introduced in [9]. Our notation throughout is the same as it was in that paper. Namely,

$$\begin{aligned} &\Omega = C([0, \infty), R^d), \\ &x(t, \omega) = x_t(\omega) \text{ is the position of } \omega \text{ at time } t, \\ &\mathscr{M}_t^s = \mathscr{B}[x_u : s \leqq u \leqq t], \qquad \mathscr{M}^s = \mathscr{B}[x_u : u \geqq s]. \end{aligned} \tag{2.1}$$

In order to discuss the weak convergence of measures on $\langle \Omega, \mathscr{M}^s \rangle$, we will sometimes think of a measure on $\langle \Omega, \mathscr{M}^s \rangle$ as defined on $C([s, \infty), R^d)$. A useful criterion for the relative compactness of a set Γ of probability measures P on $\langle \Omega, \mathscr{M}^s \rangle$ is the following:

$$\lim_{R \to \infty} \sup_{P \in \Gamma} P(|x(s)| \geqq R) = 0, \tag{2.2}$$

$$\sup_{P \in \Gamma} E^P[|x(t_2) - x(t_1)|^4] \leqq C_T(t_2 - t_1)^2, \quad s \leqq t_1 \leqq t_2 \leqq T, \quad T > 0. \tag{2.3}$$

A proof of this fact may be found in [7].

A function η on $[s, \infty) \times \Omega$ into a measurable space is said to be s *nonanticipating* if η is $\mathscr{B}_{[s,\infty]} \times \mathscr{M}^s$ measurable and $\eta(t)$ is $\mathscr{M}_t^s$ measurable for each $t \geqq s$. If P is a probability measure on $\langle\Omega, \mathscr{M}^s\rangle$, then η is a P *martingale* if η is complex valued, s nonanticipating, and

$$E^P[\eta(t_2) \mid \mathscr{M}_{t_1}^s] = \eta(t_1) \qquad \text{a.s. } P \tag{2.4}$$

for $s \leqq t_1 < t_2$.

Let $a: [0, \infty) \times R^d \to S_d$ and $b: [0, \infty) \times R^d \to R^d$ be bounded and measurable. Define

$$L_t = \frac{1}{2} \sum_{i,j=1}^{d} a^{ij}(t, x) + \sum_{i=1}^{d} b_i(t, x). \tag{2.5}$$

A probability measure P on $\langle\Omega, \mathscr{M}^{t_0}\rangle$ is said to solve the *martingale problem for L_t starting at (t_0, x_0) if $P(x(t_0) = x_0) = 1$ and $f(x(t)) - \int_{t_0}^t L_s f(x(u))\,du$ is a P* martingale for all $f \in C_0^\infty(R^d)$. In [9] it was shown that if a is continuous and $a(t, x)$ is positive definite for all (t, x), then there is exactly one solution $P_{t,x}$ to the martingale problem for L_t starting at (t, x). Moreover, we proved there that the family $\{P_{t,x}: (t, x) \in [0, \infty) \times R^d\}$ forms a strong Feller, strong Markov process. The purpose of the present section is to extend this result to the case when a and b are smooth in x and $a(t, x)$ is only nonnegative definite. The idea is to reduce the martingale problem to a stochastic differential equation which can be solved by the techniques introduced by K. Itô in [4]. For purposes of easy reference, we state here the following theorem, whose proof may be found in [10].

THEOREM 2.1. *Let $a: [t_0, \infty) \times \Omega \to S_d$ and $b: [t_0, \infty) \times \Omega \to R^d$ be bounded t_0 nonanticipating functions and define*

$$L_t = \frac{1}{2} \sum_{i,j=1}^{d} a^{ij}(t) \frac{\partial^2}{\partial x_i \partial x_j} + \sum_{i=1}^{d} b_i(t) \frac{\partial}{\partial x_i}. \tag{2.6}$$

Suppose $\alpha: [t_0, \infty) \times \Omega \to R^d$ is a continuous (in time) t_0 nonanticipating function and that P is a probability measure on $\langle\Omega, \mathscr{M}^{t_0}\rangle$. Then the following are equivalent:

(i) *$f(\alpha(t)) - \int_{t_0}^t L_u f(\alpha(u))\,du$ is a P martingale for all $f \in C_0^\infty(R^d)$;*

(ii) *$f(t, \alpha(t)) - \int_{t_0}^t [(\partial/\partial u) + L_u] f(u, \alpha(u))\,du$ is a P martingale for all $f \in C_b^{1,2}([t_0, \infty) \times R^d)$;*

(iii) *$\exp\{\langle\theta, \alpha(t) - \alpha(t_0) - \int_{t_0}^t b(u)\,du\rangle - \frac{1}{2}\int_{t_0}^t \langle\theta, a(u)\,\theta\rangle\,du\}$ is a P martingale for all $\theta \in R^d$.*

Moreover, if P satisfies one of these and if $\bar{\alpha}(t) - \int_0^t b(u)\,du$, then $d\bar{\alpha}(t)$ stochastic integrals $\int_{t_0}^t \langle\theta(u), d\bar{\alpha}(u)\rangle$ can be defined when the integrand $\theta: [t_0, \infty) \times \Omega \to R^d$ is t_0 nonanticipating and satisfies

$$E^P\left[\int_{t_0}^t \langle\theta(u), a(u)\theta(u)\rangle\,du\right] < \infty, \qquad t \geqq t_0. \tag{2.7}$$

The process $\int_{t_0}^t \langle\theta(u), d\bar{\alpha}(u)\rangle$ *is a continuous P martingale; and if* $\langle\theta(u), a(u)\theta(u)\rangle$ *is bounded, then*

$$\exp\left\{\int_{t_0}^t \langle\theta(u), d\bar{\alpha}(u)\rangle - \frac{1}{2}\int_{t_0}^t \langle\theta(u), a(u)\theta(u)\rangle\, du\right\} \tag{2.8}$$

is a P martingale. Finally, for $f \in C_b^{1,2}([t_0, \infty) \times R^d)$, *one has*

$$f(t, \alpha(t)) - f(t_0, \alpha(t_0)) = \int_{t_0}^t \langle\nabla_x f(u, \alpha(u)), d\bar{\alpha}(u)\rangle + \int_{t_0}^t \left(\frac{\partial}{\partial u} + L_u\right) f(u, \alpha(u))\, du, \tag{2.9}$$

where $\nabla_x f(t, x) = [(\partial f/\partial x_1)(t, x), \cdots, (\partial f/\partial x_d)(t, x)]$.

In [9], we showed that if the a in Theorem 2.1 is uniformly positive definite, then $\beta(t) = \int_{t_0}^t a^{-1/2}(u)\, d\bar{\alpha}(u)$ is a P Brownian motion (that is, $P(\beta(t_0) = 0) = 1$ and $\exp\{\langle\theta, \beta(t)\rangle - \frac{1}{2}|\theta|^2(t - t_0)\}$ is a P martingale for all $\theta \in R^d$), where $a^{1/2}$ is the positive definite, symmetric square root of a. Hence,

$$\alpha(t) - \alpha(t_0) = \int_{t_0}^t a^{1/2}(u)\, d\beta(u) + \int_{t_0}^t b(u)\, du \quad \text{a.s. } P \tag{2.10}$$

for some P Brownian motion β. We will now extend this result to nonnegative definite a. As we will see, this entails enlarging the sample space.

THEOREM 2.2. *Let a, b, and* α *be as in Theorem 2.1* and *suppose P is a probability measure of* $\langle\Omega, \mathscr{M}^{t_0}\rangle$ *satisfying one of the conditions* (i), (ii), *or* (iii) *of that theorem. Assume* $\sigma: [t_0, \infty) \times \Omega \rightsquigarrow R^d \times R^d$ *is* t_0 *nonanticipating and satisfies* $a(u) = \sigma(u)\sigma^*(u)$. *Then there is an extension* $\hat{P}$ *of P to* $\langle\Omega \times \Omega, \mathscr{M}^{t_0} \times \mathscr{M}^{t_0}\rangle$ *and a* $\hat{P}$ *Brownian motion* $\hat{\beta}$ *such that*

$$\alpha(t) - \alpha(t_0) = \int_{t_0}^t \sigma(u)\, d\hat{\beta}(u) + \int_{t_0}^t b(u)\, du \quad \text{a.s. } \hat{P}. \tag{2.11}$$

PROOF. It suffices to treat the case when $\sigma = a^{1/2}$. Indeed, if $a = \sigma\sigma^*$ and $U = (a + \varepsilon I)^{-1/2}\sigma$, then $U_\varepsilon \to U_0$ as $\varepsilon \downarrow 0$, where U_0 is an orthogonal transformation such that $a^{1/2} = U_0\sigma$. Hence, if

$$\alpha(t) - \alpha(t_0) = \int_{t_0}^t a^{1/2}(u)\, d\hat{\beta}(u) + \int_{t_0}^t b(u)\, du \quad \text{a.s. } \hat{P}, \tag{2.12}$$

then

$$\alpha(t) - \alpha(t_0) = \int_{t_0}^t \sigma(u)\, d\hat{\hat{\beta}}(u) + \int_{t_0}^t b(u)\, du \quad \text{a.s. } \hat{P}, \tag{2.13}$$

where $\hat{\hat{\beta}}(t) = \int_{t_0}^t U_0^*(u)\, d\hat{\beta}(u)$ is again a $\hat{P}$ Brownian motion.

To prove the assertion when $\sigma = a^{1/2}$, define $\tilde{a}(u) = \lim_{\varepsilon\downarrow 0} a^{1/2}(u)(a(u) + \varepsilon I)^{-1}$. Then $\tilde{a}(u)a^{1/2}(u) = a^{1/2}(u)\tilde{a}(u) = E_R(u)$, where $E_R(u)$ is the orthogonal projection onto the range of $a(u)$. Let $\beta(t)$ be a W Brownian motion on $\langle \Omega, \mathscr{M}^{t_0} \rangle$ and define $\hat{P} = P \times W$. Then, by Theorem 2.1,

$$\hat{\beta}(t) = \int_{t_0}^{t} \tilde{a}(u)\, d\bar{\alpha}(u) + \int_{t_0}^{t} E_N(u)\, d\beta(u) \tag{2.14}$$

is a $\hat{P}$ Brownian motion, where $E_N(u) = I - E_R(u)$. Moreover,

$$\begin{aligned} \int_{t_0}^{t} a^{1/2}(u)\, d\hat{\beta}(u) &= \int_{t_0}^{t} E_R(u)\, d\bar{\alpha}(u) + \int_{t_0}^{t} a^{1/2}(u) E_N(u)\, d\beta(u) \\ &= \int_{t_0}^{t} E_R(u)\, d\bar{\alpha}(u), \end{aligned} \tag{2.15}$$

and

$$\begin{aligned} E^{\hat{P}}\left[\left|\bar{\alpha}(t) - \bar{\alpha}(t_0) - \int_{t_0}^{t} E_R(u)\, d\bar{\alpha}(u)\right|^2\right] &= E^{\hat{P}}\left[\left|\int_{t_0}^{t} E_N(u)\, d\bar{\alpha}(u)\right|^2\right] \\ &= E^{\hat{P}}\left[\int_{t_0}^{t} \operatorname{tr}\big(E_N(u) a(u) E_N(u)\big)\, du\right] = 0, \end{aligned} \tag{2.16}$$

where tr means trace. *Q.E.D.*

Theorem 2.2 is the multidimensional analogue of Theorem 5.3 in J. L. Doob's book [2].

THEOREM 2.3. *Let $a: [0, \infty) \times R^d \rightsquigarrow S_d$ and $b: [0, \infty) \times R^d \rightsquigarrow R^d$ be bounded measurable functions. Assume that there is a bounded measurable $\sigma: [0, \infty) \times R^d \rightsquigarrow S_d$ such that $a = \sigma\sigma^*$ and*

$$\sup_{0 \leqq t \leqq T, |x| + |y| \leqq R} \big(\|\sigma(t, x) - \sigma(t, y)\| + |b(t, x) - b(t, y)|\big) \leqq C(T, R)|x - y| \tag{2.17}$$

for all $T, R > 0$. Then for each (t_0, x_0) there is exactly one solution P_{t_0, x_0} to the martingale problem for

$$L_t = \frac{1}{2} \sum_{i,j=1}^{d} a^{ij}(t, x) \frac{\partial^2}{\partial x_i \partial x_j} + \sum_{i=1}^{d} b_i(t, x) \frac{\partial}{\partial x_i} \tag{2.18}$$

starting at (t_0, x_0). Moreover, the family $\{P_{t_0, x_0}: (t_0, x_0) \in [0, \infty) \times R^d\}$ forms a Feller continuous, strong Markov process.

PROOF. We will only prove the first assertion, since the second one follows by standard methods used in [9]. Moreover, we will restrict ourselves to the case when $C(T, R)$ is independent of T and R, because the general case can then be handled by the techniques employed in Theorem 5.6 of [9].

To prove existence, let β be a W Brownian motion on $\langle \Omega, \mathscr{M}^{t_0} \rangle$. Define $\xi_0(t) \equiv x_0$ and

$$\xi_{n+1}(t) = x_0 + \int_{t_0}^{t} \sigma\big(u, \xi_n(u)\big)\, d\beta(u) + \int_{t_0}^{t} b\big(u, \xi_n(u)\big)\, du. \tag{2.19}$$

Following H. P. McKean [5], it is easy to show that

$$W\Big(\sup_{t_0 \leqq t \leqq T} |\xi_n(t) - \xi(t)| > \varepsilon \Big) \to 0 \tag{2.20}$$

for all $T > t_0$, where

$$\xi(t) = x_0 + \int_{t_0}^{t} \sigma\big(u, \xi(u)\big)\, d\beta(u) + \int_{t_0}^{t} b\big(u, \xi(u)\big)\, du. \tag{2.21}$$

Letting P be the distribution of $\xi(t)$ (that is, P is defined on $\langle \Omega, \mathscr{M}^{t_0} \rangle$ by

$$P\big(x(t_1) \in \Gamma_1, \cdots, x(t_n) \in \Gamma_n\big) = W\big(\xi(t_1) \in \Gamma_1, \cdots, \xi(t_n) \in \Gamma_n\big), \tag{2.22}$$

and using Theorem 2.1, one sees that P solves the martingale problem for L_t starting at (t_0, x_0).

Turning to uniqueness, suppose that P is a solution. Choose $\hat{P}$ and $\hat{\beta}(t)$ as in Theorem 2.2. Using the technique of the preceding paragraph, we can find $\xi_n(t)$ such that

$$\xi_{n+1}(t) = x_0 + \int_{t_0}^{t} a^{1/2}\big(u, \xi_n(u)\big)\, d\hat{\beta}(u) + \int_{t_0}^{t} b\big(u, \xi_n(u)\big)\, du, \tag{2.23}$$

and $\xi_n(t) \to x(t)$ in probability uniformly on finite intervals. Because the distribution of each $\xi_n(t)$ is uniquely determined, P is unique. *Q.E.D.*

We next state another theorem for reference purposes. Its proof can be found in [9].

THEOREM 2.4 (Cameron-Martin). *Let $\beta(t)$ be a W Brownian motion on $\langle \Omega, \mathscr{M}^{t_0} \rangle$ and suppose $c: [t_0, \infty) \times \Omega \twoheadrightarrow R^d$ is bounded and t_0 nonanticipating. Then*

$$R(t) = \exp\left\{ \int_{t_0}^{t} \langle c(u), d\beta(u) \rangle - \frac{1}{2} \int_{t_0}^{t} |c(u)|^2\, du \right\} \tag{2.24}$$

is a W martingale. In particular, there is a unique probability measure Q on $\langle \Omega, \mathscr{M}^{t_0} \rangle$ such that

$$\frac{dQ}{dW} = R(t) \text{ on } \mathscr{M}_t^{t_0}, \qquad t \geqq t_0. \tag{2.25}$$

Moreover, $\bar{\beta}(t) = \beta(t) - \int_{t_0}^{t} c(u)\, du$ is a Brownian motion.

COROLLARY 2.1. *Let a, b, L_t, and σ be as in Theorem 2.3, and let $c: [0, \infty) \times R^d \twoheadrightarrow R^d$ be bounded and measurable. Suppose P_{t_0, x_0} is the solution to the martingale problem for L_t starting at (t_0, x_0), and choose $\hat{P}_{t_0, x_0}$ and $\hat{\beta}(t)$ as in Theorem 2.2 so that*

$$(2.26) \qquad x(t) = x_0 + \int_{t_0}^t \sigma(u, x(u))\, d\hat{\beta}(u) + \int_{t_0}^t b(u, x(u))\, du \qquad a.s.\ \hat{\Gamma}_{t_0, x_0}.$$

Define

$$(2.27) \qquad R(t) = E^{P_{t_0,x_0}}\left[\exp\left\{\int_{t_0}^t \langle c(u, x(u)), d\hat{\beta}(u)\rangle - \frac{1}{2}\int_{t_0}^t |c(u)|^2\, du\right\}\middle| \mathscr{M}_t^{t_0}\right],$$

and determine Q_{t_0,x_0} *by the relations*

$$(2.28) \qquad \frac{dQ_{t_0,x_0}}{dP_{t_0,x_0}} = R(t) \text{ on } \mathscr{M}_t^{t_0}, \qquad t \geqq t_0.$$

Then Q_{t_0,x_0} *is the only solution to the martingale problem for*

$$(2.29) \qquad L_t^c = \frac{1}{2}\sum_{i,j=1}^d a^{ij}(t, x)\frac{\partial^2}{\partial x_i \partial x_j} + \sum_{i=1}^d (b + \sigma c)_i(t, x)\frac{\partial}{\partial x_i}$$

starting at (t_0, x_0). *Moreover, the family* $\{Q_{t,x}: (t, x) \in [0, \infty) \times R^d\}$ *forms a strong Markov process which is Feller continuous when c is continuous in* x.

REMARK 2.1. If $a: [0, \infty) \times R^d \rightsquigarrow S_d$ is bounded and measurable, then

$$(2.30) \qquad \sup_{t \leqq T, |x| + |y| \leqq R} \|a^{1/2}(t, x) - a^{1/2}(t, y)\| \leqq C(T, R)|x - y|,$$

if $a(t, x)$ is twice continuous differentiable in x and

$$(2.31) \qquad \max_{1 \leqq i,j \leqq d} \sup_{t \leqq T, |x| + |y| \leqq R} \left\| \frac{\partial^2 a}{\partial x_i \partial x_j}(t, x)\right\| \leqq C(T, R).$$

This fact is proved by R. S. Philips and L. Sarason [8].

REMARK 2.2. Unfortunately there is no nice criterion on $a: [0, \infty) \times R^d \rightsquigarrow S_d$ which guarantees the existence of a smooth $\sigma: [0, \infty) \times R^d \rightsquigarrow R^d \otimes R^d$ such that $a = \sigma\sigma^*$. Nonetheless, we will often assume in what follows that L_t can be written in the form

$$(2.32) \qquad L_t = \tfrac{1}{2}\sigma^* \nabla_x \cdot \sigma^* \nabla_x + b \cdot \nabla_x,$$

where

$$(2.33) \qquad \sigma^* \nabla_x \cdot \sigma^* \nabla_x = \sum_{i,j=1}^d \sigma^{i\ell}\frac{\partial}{\partial x_i}\left(\sigma^{j\ell}\frac{\partial}{\partial x_j}\right).$$

(Here, and in what follows, repeated indices are summed.) Notice that (2.32) can be written as

$$(2.34) \qquad L_t = \frac{1}{2}\sum_{i,j=1}^d a^{ij}\frac{\partial^2}{\partial x_i \partial x_j} + \sum_{i=1}^d \left(b + \frac{1}{2}\sigma'\sigma\right)_i \frac{\partial}{\partial x_i},$$

where $a = \sigma\sigma^*$ and the vector $\sigma'\sigma$ is defined by

$$(2.35) \qquad (\sigma'\sigma)_i = \sigma^{ij}_{,\ell}\sigma^{\ell j}.$$

(We have used here, and will continue to use, the notation $f_{,\ell}$ to stand for $\partial f/\partial x_\ell$.) What specific assumptions are made about the smoothness σ will depend on our immediate needs. But in any case, it will be necessary to assume that σ is once differentiable in x in order to even define $\sigma^*\nabla_x \cdot \sigma^*\nabla_x$.

3. The nondegenerate case

Throughout this section $a: [0, \infty) \times R^d \rightsquigarrow S_d$ will be bounded, continuous, positive definite valued function and $b: [0, \infty) \times R^d \rightsquigarrow R^d$ will be bounded and measurable. We will use P_{t_0, x_0} to denote the unique solution to the martingale problem for

$$L_t = \frac{1}{2} \sum_{i,j=1}^{d} a^{ij}(t, x) \frac{\partial^2}{\partial x_i \partial x_j} + \sum_{i=1}^{d} b_i(t, x) \frac{\partial}{\partial x_i} \tag{3.1}$$

starting at (t_0, x_0), and $P^0_{t_0, x_0}$ will denote the unique solution for

$$L^0_t = \frac{1}{2} \sum_{i,j=1}^{d} a^{ij}(t, x) \frac{\partial^2}{\partial x_i \partial x_j} \tag{3.2}$$

starting at (t_0, x_0). Our aim is to prove that

$$\mathrm{supp}\,(P_{t_0, x_0}) = \Omega(t_0, x_0), \tag{3.3}$$

where $\Omega(t_0, x_0)$ is the set of $\omega \in \Omega$ such that $x(t_0, \omega) = x_0$.

Before proceeding, we make two simplifying observations. First, since P_{t_0, x_0} and $P^0_{t_0, x_0}$ are equivalent (that is, mutually absolutely continuous) on $\mathscr{M}_t^{t_0}$ for all $t \geqq t_0$, it suffices to work with $P^0_{t_0, x_0}$. Second, by an obvious transformation, we can always assume that $t_0 = 0$ and $x_0 = 0$. Thus, what we need to show is that

$$\mathrm{supp}\,(P_0) = \Omega_0, \tag{3.4}$$

where $P_0 = P^0_{0,0}$ and $\Omega_0 = \Omega(0, 0)$.

LEMMA 3.1. *Let $\phi: [0, \infty) \rightsquigarrow R^d$ be once continuously differentiable such that $\phi(0) = 0$. Then for all $T > 0$ and $\varepsilon > 0$, $P_0(\|x(t) - \phi(t)\|_T^0 < \varepsilon) > 0$, where $\|\cdot\|_T^s$ denotes the sup norm on the interval $[s, T]$.*

PROOF. Let $\psi(t) = \chi_{[0,T]}(t)\dot{\phi}(t)$ and define Q_0 by

$$\frac{dQ_0}{dP_0} = R(t) = \exp\left\{\int_0^t \langle a^{-1}(u, x(u)), \psi(u)\, dx(u)\rangle - \frac{1}{2}\int_0^t \langle \psi(u), a^{-1}(u, x(u)), \psi(u)\rangle\, du\right\} \tag{3.5}$$

on $\mathscr{M}_t^0$, $t \geqq 0$. Then Q_0 is the unique solution to the martingale problem for

$$L_t^\psi = \frac{1}{2} \sum_{i,j=1}^{d} a^{ij}(t, x) \frac{\partial^2}{\partial x_i \partial x_j} + \sum_{i=1}^{d} \psi_i(t) \frac{\partial}{\partial x_i} \tag{3.6}$$

starting at (0, 0). In particular, there is a Q_0 Brownian motion $\beta(t)$ such that

$$(3.7) \qquad \bar{x}(t) = \int_0^t a^{1/2}(u, x(u))\, d\beta(u),$$

where $\bar{x}(t) = x(t) - \int_0^t \psi(u)\,du = x(t) - \phi(t \wedge T)$. Hence, by Theorem 1 in [11], $Q_0(\|\bar{x}(t)\|_T^0 < \varepsilon) > 0$ for all $\varepsilon > 0$. Since Q_0 and P_0 are equivalent on $\mathcal{M}_T^0$, this implies $P_0(\|x(t) - \phi(t)\|_T^0 < \varepsilon) > 0$ for all $\varepsilon > 0$. *Q.E.D.*

Using Lemma 3.1, we see that supp(P_0) contains all differentiable paths which start at 0. Because these are dense in Ω_0, equation (3.4) is now proved. We have therefore proved the following theorem.

THEOREM 3.1. *If P_{t_0,x_0} is the solution to the martingale problem for L_t starting at (t_0, x_0), then equation* (3.3) *holds.*

As we saw in Section 1, equation (3.3) implies Nirenberg's strong maximum principle. In fact, it implies more. Let $\mathcal{G}$ be an open set in $[0, \infty) \times R^d$ and let $(t_0, x_0) \in \mathcal{G}$. Define $\mathcal{H}_{L_t}^-(t_0, x_0, \mathcal{G})$ to be the set of $f: \mathcal{G} \rightsquigarrow R \cup \{-\infty\}$ which are upper semicontinuous, bounded above, and have the property that $f(t \wedge \tau, x(t \wedge \tau))$ is a P_{t_0,x_0} submartingale, that is,

$$(3.8) \qquad f(t_1 \wedge \tau, x(t_1 \wedge \tau)) \leqq E^{P_{t_0,x_0}}[f(t_2 \wedge \tau, x(t_2 \wedge \tau)) | \mathcal{M}_{t_1 \wedge \tau}^{t_0}], \qquad t_0 \leqq t_1 \leqq t_2,$$

where $\tau = \inf\{t \geqq t_0: (t, x(t)) \notin \mathcal{G}\}$. Note that $\mathcal{H}_{L_t}^-(t_0, x_0, \mathcal{G})$ is closed under nonincreasing limits, multiplication by nonnegative constants, and maximums (that is, if $f_1, f_2 \in \mathcal{H}_{L_t}^-(t_0, x_0, \mathcal{G})$, then so is $f_1 \vee f_2$). Next define (t_0, x_0) to be the closure in $\mathcal{G}$ of the set of $(t_1, x_1) \in \mathcal{G} \cap ([t_0, \infty) \times R^d)$ such that $x_0 = \phi(t_0)$ and $x_1 = \phi(t_1)$ for some $\phi \in C([t_0, t_1], R^d)$ satisfying $(t, \phi(t)) \in \mathcal{G}$, $t \in [t_0, t_1]$. Observe that if $f \in C_b^{1,2}(\mathcal{G})$ and $((\partial/\partial t) + L_t)f \geqq 0$ on $\mathcal{G}(t_0, x_0)$, then $f \in \mathcal{H}_{L_t}^-(t_0, x_0)$. Finally, define

$$(3.9) \qquad \mathcal{H}_{L_t}^-(\mathcal{G}) = \bigcap_{\mathcal{G}} \mathcal{H}_{L_t}^-(t_0, x_0, \mathcal{G}).$$

THEOREM 3.2. *If $f \in \mathcal{H}_{L_t}^-(t_0, x_0, \mathcal{G})$ and $f(t_0, x_0) = \sup f(t, x)$, then $f \equiv f(t_0, x_0)$ on $\mathcal{G}(t_0, x_0)$. Moreover, if $(t_1, x_1) \in \mathcal{G} - \mathcal{G}(t_0, x_0)$, then there is an $f \in \mathcal{H}_{L_t}^-(\mathcal{G})$ such that $f(t_0, x_0) = \sup f(t, x)$ and $f(t_1, x_1) < f(t_0, x_0)$.*

PROOF. The first assertion follows from the argument given in Section 1. To prove the second assertion, let $(t_1, x_1) \in \mathcal{G} - \mathcal{G}(t_0, x_0)$. If $t_1 < t_0$, take $f(t, x) = t \wedge t_0$. If $t_1 \geqq t_0$, choose an open neighborhood N of (t_1, x_1) such that $N \subseteqq \mathcal{G}$ and $N \cap \mathcal{G}(t_0, x_0) = \phi$. Let $h \in C_0^\infty(N)$ such that $-1 \leqq h \leqq 0$ and $h(t_1, x_1) = -1$. Define

$$(3.10) \qquad f(t, x) = E^{P_{t,x}}\left[\int_t^{\tau_t} e^{-u} h(u, x(u))\, du\right],$$

where $\tau_t = \inf\{u \geqq t: (u, x(u)) \notin \mathscr{G}\}$. Since $\{P_{t,x}: (t, x) \in [0, \infty) \times R^d\}$ is strongly Feller continuous, f is continuous. Moreover,

$$(3.11) \qquad \begin{aligned} &E^{P_{s,x}}[f(t \wedge \tau_s, x(t \wedge \tau_s))] \\ &\quad = E^{P_{s,x}}\left[\chi_{\tau_s > t} \int_t^{\tau_t} e^{-u} h(u, x(u))\, du\right] \geqq f(s, x), \end{aligned}$$

for $(s, x) \in \mathscr{G}$ and $t \geqq s$; and therefore

$$(3.12) \qquad \begin{aligned} &E^{P_{s,x}}[f(t_2 \wedge \tau_s, x(t_2 \wedge \tau_s)) | \mathscr{M}^s_{t_1 \wedge \tau_s}] \\ &\quad = E^{P_{t_1 \wedge \tau_s, x(t_1 \wedge \tau_s)}}[f(t_2 \wedge \tau_{t_1}, x(t_2 \wedge \tau_{t_1}))] \\ &\quad \geqq f(t_1 \wedge \tau_3, x(t_1 \wedge \tau_s)) \end{aligned}$$

for $s \leqq t_1 \leqq t_2$. This proves that $f \in \mathscr{H}^-_{L_t}(\mathscr{G})$. Clearly, $f \leqq 0$, $f(t_0, x_0) = 0$, and $f(t_1, x_1) < 0$. *Q.E.D.*

REMARK 3.1. It is important to know in what sense $\mathscr{H}^-_{L_t}(\mathscr{G})$ is an extension of the class of $f \in C^{1,2}_b(\mathscr{G})$ satisfying $(\partial f/\partial t) + L_t f \geqq 0$ on $\mathscr{G}$. Using the estimates obtained in [9], one can show that $\mathscr{H}^-_{L_t}(\mathscr{G})$ contains the class of $f \in W^{1,2}_p(\mathscr{G})$ (see [9] for the definition of $W^{1,2}$) satisfying $(\partial f/\partial t) + L_t f \geqq 0$ when $p > (d + 2)/2$. To give a complete analytic description of $\mathscr{H}^-_{L_t}(\mathscr{G})$, consider the transition function $\hat{P}(t, (s, x), \cdot)$ defined by

$$(3.13) \qquad \hat{P}(t, (s, x), \Delta \times \Gamma) = \chi_\Delta(s + t) P_{s,t}(x(s + t) \wedge \tau_s) \in \Gamma)$$

for $\Delta \in \mathscr{B}_{[0,\infty)}$ and $\Gamma \in \mathscr{B}[\overline{\mathscr{G}}]$, where $\tau_s = \inf\{t \geqq s: (t, x(t)) \notin \mathscr{G}\}$. It is easy to see that

$$(3.14) \qquad \hat{P}(t_1 + t_2, (s, x), \cdot) = \int_0^\infty \int_{\overline{\mathscr{G}}} \hat{P}(t_1, (s, x), du \times dy) \hat{P}(t_2, (u, y), \cdot).$$

Thus, we can define a Markov semigroup $\{\hat{T}_t\}_{t \geqq 0}$ on $\mathscr{B}(\mathscr{G})$ by setting

$$(3.15) \qquad \hat{T}_t f(s, x) = \int_0^\infty \int_{\overline{\mathscr{G}}} \hat{P}(t, (s, x), du \times dy) f(u, y).$$

Furthermore, $\{\hat{T}_t\}_{t \geqq 0}$ is the only Markov semigroup having the property that

$$(3.16) \qquad \hat{T}_t f(s, x) - f(s, x) = \int_0^t \left(\hat{T}_u\left[\chi_{\mathscr{G}} \cdot \left(\frac{\partial f}{\partial u} + L_u f\right)\right]\right)(u, x)\, du$$

for all $f \in C^{1,2}_b(\overline{\mathscr{G}})$. Finally, $\mathscr{H}^-_{L_t}(\mathscr{G})$ coincides with the class of f on $\overline{\mathscr{G}}$ such that f is bounded above and $\hat{T}_t f(s, x) \downarrow f(s, x)$ for $(s, x) \in \mathscr{G}$. When $L_t = \frac{1}{2}\Delta$, it is easy to see that $\mathscr{H}^-_{L_t}(\mathscr{G})$ is just the class of subparabolic functions on $\mathscr{G}$.

4. The degenerate case, part I

Let $\sigma: [0, \infty) \times R^d \rightsquigarrow R^d \otimes R^d$ and $b: [0, \infty) \times R^d \rightsquigarrow R^d$ be bounded measurable functions. In this section, we will assume that b and the first spatial derivatives of σ are uniformly Lipschitz continuous in x. From σ and b, we form the operator

$$L_t = \tfrac{1}{2}\sigma^*\nabla_x \cdot \sigma^*\nabla_x + b \cdot \nabla_x, \tag{4.1}$$

by the prescription given in Remark 2.2. Under the above assumptions, we know from Theorem 2.3 that there is exactly one solution P_{t_0,x_0} to the martingale problem for L_t starting at (t_0, x_0). The purpose of this section is to prove that

$$\operatorname{supp}(P_{t_0,x_0}) \subseteq \overline{\mathscr{S}_{\sigma,b}(t_0, x_0)}, \tag{4.2}$$

where $\overline{\mathscr{S}_{\sigma,b}(t_0, x_0)}$ is the class of $\phi \in C([0, \infty), R^d)$ for which there exists a piecewise constant $\psi\colon [t_0, \infty) \to R^d$ such that

$$\phi(t) = x_0 + \int_{t_0}^{t} \sigma(u, \phi(u))\psi(u)\, du + \int_{t_0}^{t} b(u, \phi(u))\, du, \qquad t \geqq t_0. \tag{4.3}$$

Clearly, it suffices to treat the case when $t_0 = 0$ and $x_0 = 0$.

Let $\beta(t)$ be a W Brownian motion on $\langle \Omega, \mathscr{M}^0 \rangle$. Given $n \geqq 0$, define $t_n = [2^n t]/(2^n)$, $t_n^+ = ([2^n t] + 1)/2^n$, and

$$\dot{\beta}^{(n)}(t) = 2^n(\beta(t_n^+) - \beta(t_n)). \tag{4.4}$$

Next, let $\xi^{(n)}(t)$ be the stochastic process determined by the ordinary integral equation

$$\xi^{(n)}(t) = \int_0^t \sigma(u, \xi^{(n)}(u))\dot{\beta}^{(n)}(u)\, du + \int_0^t b(u, \xi^{(n)}(u))\, du; \tag{4.5}$$

and denote by P_n the distribution of $\xi^{(n)}(t)$. Clearly, $\operatorname{supp}(P_n) \subseteq \overline{\mathscr{S}_{\sigma,b}(0, 0)}$ for all $n \geqq 0$. Hence, if we show that P_n tends weakly to $P_{0,0}$ as $n \to \infty$, then it will follow that $\operatorname{supp}(P_{0,0}) \subseteq \overline{\mathscr{S}_{\sigma,b}(0, 0)}$. Thus, we must prove that $P_n \Rightarrow P_{0,0}$. Results of this sort are familiar in various branches of applied mathematics when $d = 1$ (see E. Wong and M. Zakai [12]). However, to the best of our knowledge, the proof which follows is the first complete one for $d > 1$.

The procedure which we will use consists of two steps. The first of these is to prove that $\{P_n\}_{n \geqq 1}$ is relatively weakly compact. Once this is done, we will then show that every convergent subsequence of $\{P_n\}_{n \geqq 1}$ converges to a solution of the martingale problem for L_t starting at $(0, 0)$. For convenience in writing, we will use the following notation:

$$\eta^{(n)}(t) = \int_0^t \sigma(u, \xi^{(n)}(u))\dot{\beta}^{(n)}(u)\, du, \tag{4.6}$$

$$\alpha^{(n)}(t) = \sigma\left(t, \xi^{(n)}\left(\frac{[2^n t]}{2^n}\right)\right)\dot{\beta}^{(n)}(t), \tag{4.7}$$

(4.8)
$$(\sigma'\sigma)_i^{\ell,\ell'}(t, x) = \left(\frac{\partial}{\partial x_j}\sigma^{i\ell}(t, x)\right)\sigma^{j\ell'}(t, x),$$

(4.9)
$$\Delta_k^{(n)} = \beta\left(\frac{k+1}{2^n}\right) - \beta\left(\frac{k}{2^n}\right).$$

LEMMA 4.1. *The set* $\{P_n\}_{n \geqq 1}$ *is relatively weakly compact.*

PROOF. It suffices to prove that

(4.10)
$$\sup_n E^{P_n}[|x(t) - x(s)|^4] \leqq C_T|t_2 - s|^2, \quad 0 \leqq s \leqq t \leqq T, \quad T > 0.$$

To do this, first observe that

(4.11)
$$E^{P_n}|x(t) - x(s)|^4 \leqq 8\left(E^W[|\eta^{(n)}(t) - \eta^{(n)}(s)|^4] + E^W\left[\left|\int_s^t b(u, \xi^{(n)}(u))\,du\right|^4\right]\right).$$

Hence, it suffices to examine

(4.12)
$$E^W[|\eta^{(n)}(t) - \eta^{(n)}(s)|^4].$$

But

(4.13)
$$\eta^{(n)}(t) - \eta^{(n)}(s) = \int_s^t \alpha^{(n)}(u)\,du + \int_s^t du \int_{u_n}^u dv(\sigma'\sigma)^{\ell,\ell'}(u, \xi^{(n)}(v))\dot{\beta}_\ell^{(n)}(v)\dot{\beta}_{\ell'}^{(n)}(v),$$

and

(4.14)
$$E^W\left[\left|\int_s^t \alpha^{(n)}(u)\,du\right|^4\right] = E^W\left[\left|\int_s^t \sigma(u, \xi^{(n)}(u_n))\dot{\beta}^{(n)}(u)\,du\right|^4\right] = E^W\left[\left|\int_{s_n}^{t_n} \sigma^{(n)}(u)\,d\beta(u)\right|^4\right] \leqq C_1(t - s)^2,$$

where

(4.15)
$$\sigma^{(n)}(u) = 2^n \int_{u_n \vee s}^{u_n^+ \wedge t} \sigma(v, \xi^{(n)}(u_n))\,dv.$$

Finally,

(4.16)
$$\begin{aligned} E^W&\left[\left|\int_s^t du \int_{u_n}^u dv(\sigma'\sigma)^{\ell,\ell'}(u, \xi^{(n)}(v))\dot{\beta}_\ell^{(n)}(v)\dot{\beta}_{\ell'}^{(n)}(v)\right|^4\right] \\ &\leqq (t - s)^3 E^W\left[\int_s^t du\left|\int_{u_n}^u dv(\sigma'\sigma)^{\ell,\ell'}(u, \xi^{(n)}(v))\dot{\beta}_\ell^{(n)}(v)\dot{\beta}_{\ell'}^{(n)}(v)\right|^4\right] \\ &\leqq C_2(t - s)^3 \sum_{k=[2^n s]}^{[2^n t]} 2^{8n} \int_{k/2^n}^{k+1/2^n} du\left(u - \frac{k}{2^n}\right)^4 E^W[|\Delta_k^{(n)}|^8] \\ &\leqq C_3(t - s)^3. \end{aligned}$$

Q.E.D.

We now have to show that if P is the limit of a convergent subsequence of $\{P_n\}$, then

$$E^P[F\cdot(f(x(t))-f(x(s)))] = E^P\left[F\cdot\int_s^t L_u f(x(u))\,du\right] \tag{4.17}$$

for all $f\in C_0^\infty(R^d)$, $0\leq s<t$, and bounded $\mathscr{M}_s^0$ measurable $F\colon\Omega\to R$. Clearly, it will suffice to do this when s and t have the form $k/2^N$ and F is continuous as well as bounded and $\mathscr{M}_s^0$ measurable. For the sake of convenience, we will use $\{P_n\}_{n\geq 1}$ to denote the subsequence which converges to P. Observe that

$$\begin{aligned} E^{P_n}&[F\cdot(f(x(t))-f(x(s)))] \\ &= E^{P_n}\left[F\cdot\int_s^t\langle\nabla_x f(x(u)), b(u,x(u))\rangle\,du\right] \\ &\quad + E^W\left[F\cdot\int_s^t\langle\nabla_x f(\xi^{(n)}(u)), \alpha^{(n)}(u)\rangle\,du\right] \\ &\quad + E^W\left[F\cdot\int_s^t\langle\nabla_x f(\xi^{(n)}(u)), \dot\eta^{(n)}(u)-\alpha^{(n)}(u)\rangle\,du\right] \\ &= I_1^{(n)}+I_2^{(n)}+I_3^{(n)}. \end{aligned} \tag{4.18}$$

Obviously, $I_1^{(n)}\to E^P[F\cdot\int_s^t\langle\nabla_x f(x(u)), b(u,x(u))\rangle\,du]$. Before examining the limiting behavior of the other two terms, we need the following simple lemma.

LEMMA 4.2. *If ϕ and ψ are bounded measurable functions on $[s,t]$, then*

$$2^n\int_s^t du\,\phi(u)\int_{u_n}^u dv\,\psi(v)\to\frac{1}{2}\int_s^t\phi(u)\psi(u)\,du. \tag{4.19}$$

PROOF. Let $\psi_n(u)=2^n\int_{u_n}^u\psi(v)\,dv$. Then $\{\psi_n\}_{n\geq 1}$ is obviously relatively compact in the weak topology on $L^1([s,t])$ induced by $L^\infty([s,t])$. Hence, it suffices to prove the result when ϕ is continuous. Next observe that

$$2^n\int_s^t du\,\phi(u)\int_{u_n}^u dv\,\psi(v) = 2^n\int_s^t du\,\psi(u)\int_u^{u_n^+}dv\,\phi(v). \tag{4.20}$$

Thus, by the argument just used, we may assume that both ϕ and ψ are continuous. But the result is obvious for continuous ϕ and ψ, and so we are done. Q.E.D.

LEMMA 4.3. *Let $L_u^0=\frac{1}{2}(\sigma\sigma^*)^{ij}(t,x)(\partial^2/\partial x_i\partial x_j)$. Then*

$$I_2^{(n)}\to E^P\left[F\cdot\int_s^t L_u^0 f(x(u))\,du\right]. \tag{4.21}$$

PROOF. We will use $H(x)$ to denote the Hessian matrix of f. Note that

$$E^W[\alpha^{(n)}(u)\,|\,\mathscr{M}_{u_n}]=0, \tag{4.22}$$

and therefore

$$(4.23)\qquad I_2^{(n)} = E^W\left[F\cdot\int_s^t \langle \nabla_x f(\xi^{(n)}(u)) - \nabla_x f(\xi^{(n)}(u_n)), \alpha^{(n)}(u)\rangle\, du\right]$$

$$= E^W\left[F\cdot\int_u^t du\int_{u_n}^u dv\langle \dot\eta^{(n)}(v), H(\xi^{(n)}(v))\alpha^{(n)}(u)\rangle\right]$$

$$+ E^W\left[F\cdot\int_s^t du\int_{u_n}^u dv\langle b(v, \xi^{(n)}(v)), H(\xi^{(n)}(v))\alpha^{(n)}(u)\rangle\right]$$

$$= J_1^{(n)} + J_2^{(n)}.$$

Clearly, $|J_2^{(n)}| \to 0$ and

$$(4.24)\qquad J_1^{(n)} = E^W\left[F\cdot\int_s^t du\int_{u_n}^u dv\langle \alpha^{(n)}(v), H(\xi^{(n)}(v)\alpha^{(n)}(u)\rangle\right]$$

$$+ E^W\Bigg[F\cdot\int_s^t du\int_{u_n}^u dv\int_{u_n}^v dw\langle(\sigma'\sigma)^{\ell,\ell'}(v, \xi^{(n)}(w))\dot\beta_\ell^{(n)}\dot\beta_{\ell'}^{(n)},$$

$$H[\xi^{(n)}(v)]\alpha^{(n)}(u)\rangle\Bigg]$$

$$= J_3^{(n)} + J_4^{(n)}.$$

Again, it is obvious that $|J_4^{(n)} \to 0$ and that

$$(4.25)\qquad J_3^{(n)} = E^W\left[F\cdot\int_s^t du\int_{u_n}^u dv\langle \alpha^{(n)}(v), H(\xi^{(n)}(u_n))\alpha^{(n)}(u)\rangle\right]$$

$$+ E^W\left[F\cdot\int_s^t du\int_{u_n}^u dv\langle \alpha^{(n)}(v), \big(H(\xi^{(n)}(v)) - H(\xi^{(n)}(u_n))\big)\alpha^{(n)}(u)\rangle\right]$$

$$= J_5^{(n)} + J_6^{(n)}.$$

Since $|J_6^{(n)}| \to 0$, it remains to examine $J_5^{(n)}$. Clearly,

$$(4.26)\qquad J_5^{(n)} = E^W\left[2^n F\cdot\int_s^t du\int_{u_n}^u dv \operatorname{tr}\left[\sigma^*(v, \xi^{(n)}(v_n))H(\xi^{(n)}(u_n))\sigma(u, \xi^{(n)}(u_n))\right]\right]$$

$$= E^{P_n}\left[2^n F\cdot\int_s^t du\int_{u_n}^u dv \operatorname{tr}\left[\sigma^*(v, x(v_n))H(x(u_n))\right]\sigma(u, x(u_n))\right].$$

Hence, by Lemma 4.2 and the fact that $P_n \Rightarrow P$, we have that

$$(4.27)\qquad J_5^{(n)} \to \frac{1}{2}E^P\left[F\cdot\int_s^t \operatorname{tr}\left[\sigma^*(u, v(u))H(x(u))\sigma(u, x(u))\right] du\right].$$

Q.E.D.

LEMMA 4.4. *Let* $(\sigma'\sigma)_i(t, x) = (\sigma'\sigma)_i^{\ell\ell}(t, x)$. *Then*

$$(4.28)\qquad I_3^{(n)} \to \frac{1}{2}E^P\left[F\cdot\int_s^t \nabla_x f(x(u)), (\sigma'\sigma)(u, x(u))\, du\right].$$

PROOF. Note that

$$(4.29)\quad I_3^{(n)} = E^W\left[F \cdot \int_s^t du \int_{u_n}^u dv\, \nabla_x f(\xi^{(n)}(u)), (\sigma\sigma')^{\ell\ell'}(u, \xi^{(n)}(v))\dot\beta_\ell(v)\dot\beta_{\ell'}(v)\right]$$

$$= E^W\left[F \cdot \int_s^t \nabla_x f(\xi^{(n)}(u_n)), (\sigma\sigma')(u, \xi^{(n)}(u_n))(u - u_n)\, du\right]$$

$$+ E^W\Big[F \cdot \int_s^t du \int_{u_n}^u dv\, \nabla_x f(\xi^{(n)}(u_n)),$$

$$[(\sigma\sigma')^{\ell\ell'}(u, \xi^{(n)}(v)) - (\sigma\sigma')^{\ell\ell'}(u, \xi^{(n)}(u_n))]\dot\beta_\ell(v)\dot\beta_{\ell'}(v)\Big]$$

$$+ E^W\Big[F \cdot \int_s^t du \int_{u_n}^u dv \langle \nabla_x f(\xi^{(n)}(u)) - \nabla_x f(\xi^{(n)}(u_n)),$$

$$(\sigma'\sigma)^{\ell\ell'}(u, \xi^{(n)}(v))\dot\beta_\ell(v)\dot\beta_{\ell'}(v)\rangle\Big]$$

$$= J_1^{(n)} + J_2^{(n)} + J_3^{(n)}.$$

Clearly, $|J_2^{(n)}|$ and $|J_3^{(n)}|$ tend to zero. Moreover,

$$(4.30)\quad J_1^{(n)} = E^{P_n}\left[F \cdot \int_s^t \langle \nabla_x f(x(u_n)), (\sigma\sigma')(u, x(u_n))\rangle (u - u_n)\, du\right]$$

$$\to \frac{1}{2} E^P\left[F \cdot \int_s^t \nabla_x f(x(u)), (\sigma\sigma')(u, x(u))\, du\right].$$

Q.E.D.

THEOREM 4.1. *Let P_n be the distribution of the process $\xi^{(n)}(t)$ defined in (4.5). Then P_n converges weakly to $P_{0,0}$ as $n \to \infty$. In particular, equation (4.2) is valid for all (t_0, x_0).*

COROLLARY 4.1. *If $c\colon [0, \infty) \times R^d \to R^d$ is bounded and measurable, then for each (t_0, x_0) the unique solution $P^c_{t_0,x_0}$ to the martingale problem for*

$$(4.31)\quad L_t^c = \tfrac{1}{2}\sigma^*\nabla_x \cdot \sigma^*\nabla_x + (b + \sigma c)\cdot \nabla_x$$

starting at (t_0, x_0) satisfies

$$(4.32)\quad \operatorname{supp}(P^c_{t_0,x_0}) \subseteqq \overline{\mathscr{S}_{\sigma,b}(t_0, x_0)}.$$

PROOF. According to Corollary 2.1, $P^c_{t_0,x_0}$ and P_{t_0,x_0} are equivalent on $\mathscr{M}_t^{t_0}$ for all $t \geqq t_0$. Hence, $\operatorname{supp}(P^c_{t_0,x_0}) = \operatorname{supp}(P_{t_0,x_0})$. *Q.E.D.*

REMARK 4.1. Let $a_i\colon [0, \infty) \times R^d \to S_d$, $i = 1, 2$, and $b\colon [0, \infty) \times R^d \to R^d$ be bounded measurable functions which are uniformly Lipschitz continuous in x. Define $\mathscr{S}_{a_i,b}(t_0, x_0)$ to be the class of $\phi \in C([0, \infty), R^d)$ for which there exists a piecewise constant $\psi\colon [0, \infty) \times R^d \to R^d$ such that

$$(4.33)\quad \phi(t) = x_0 + \int_{t_0}^t a_i(u, \phi(u))\psi(u)\, du + \int_{t_0}^t b(u, \phi(u))\, du, \qquad t \geqq t_0.$$

Then Range $(a_1(t, x)) \subseteqq$ Range $(a_2(t, x))$, $(t, x) \in [t_0, \infty) \times R^d$, implies

$$\overline{\mathscr{S}_{a_1,b}(t_0, x_0)} \subseteqq \overline{\mathscr{S}_{a_2,b}(t_0, x_0)}. \tag{4.34}$$

REMARK 4.2. Suppose $a: [0, \infty) \times R^d \rightsquigarrow S_d$ satisfies the conditions stated in Remark 2.1. Then for each (t_0, x_0) there is exactly one solution P_{t_0,x_0} to the martingale problem for $L_t = \frac{1}{2}\nabla_x \cdot (a\nabla_x) + b \cdot \nabla_x$, where

$$\nabla_x \cdot (a\nabla_x) = \sum_{i,j=1}^{d} \frac{\partial}{\partial x_i}\left(a^{ij}(t, x)\frac{\partial}{\partial x_j}\right) \tag{4.35}$$

and $b: [0, \infty) \times R^d \rightsquigarrow R^d$ is bounded, measurable, and uniformly Lipschitz continuous in s. Moreover, if $a^{1/2}(t, x)$ possesses first spatial derivatives which are uniformly Lipschitz continuous in x, then L_t can be written in the form

$$L_t = \tfrac{1}{2}a^{1/2}\nabla_x \cdot a^{1/2}\nabla_x + (b + a^{1/2}c) \cdot \nabla_x; \tag{4.36}$$

and therefore

$$\operatorname{supp}(P_{t_0,x_0}) \subseteqq \overline{\mathscr{S}_{a^{1/2},b}(t_0, x_0)}. \tag{4.37}$$

By Remark 4.1, $\overline{\mathscr{S}_{a^{1/2},b}(t_0, x_0)} = \overline{\mathscr{S}_{a,b}(t_0, x_0)}$, and so we have

$$\operatorname{supp}(P_{t_0,x_0}) \subseteqq \overline{\mathscr{S}_{a,b}(t_0, x_0)}. \tag{4.38}$$

Using a localization procedure, it is possible to prove (4.38) under the assumptions on a stated in Remark 2.1, without any further conditions on $a^{1/2}$.

REMARK 4.3. Suppose $\sigma: [0, \infty) \times R^d \rightsquigarrow R^d \times R^d$ and $b: [0, \infty) \times R^d \rightsquigarrow R^d$ are bounded infinitely differentiable functions. Define vector fields

$$X_\ell = \sum_{i=1}^{d} \sigma^{i\ell}(t, x)\frac{\partial}{\partial x_i}, \qquad 1 \leqq \ell \leqq d, \tag{4.39}$$

and

$$Y = \sum_{i=1}^{d} b_i(t, x)\frac{\partial}{\partial x_i}. \tag{4.40}$$

Then

$$\tfrac{1}{2}\sigma^*\nabla_x \cdot \sigma^*\nabla_x + b \cdot \nabla_x = \tfrac{1}{2}\sum_{\ell=1}^{d} X_\ell^2 + Y. \tag{4.41}$$

Using the techniques of Bony [1], one can show that $\overline{\mathscr{S}_{\sigma,b}(t_0, x_0)}$ contains all $\phi \in C([0, \infty), R^d)$ such that

$$\phi(t) = x_0 + \int_{t_0}^{t} Z(u, \phi(u))\,du + \int_{t_0}^{t} Y(u, \phi(u))\,du, \qquad t \geqq t_0, \tag{4.42}$$

where Z is an element of the Lie algebra $\mathscr{L}(X_1, \cdots, X_d)$ generated by $X_1, \cdots, X_d$. In particular, if $\mathscr{L}(X_1, \cdots, X_d)$ has rank d at every point, then $\overline{\mathscr{S}_{\sigma,b}(t_0, x_0)}$ coincides with the set of $\phi \in C([0, \infty), R^d)$ such that $\phi(t_0) = x_0$.

5. The degenerate case, part II

Throughout this section $\sigma: [0, \infty) \times R^d \rightsquigarrow R^d \otimes R^d$ will satisfy $\sigma^{ij} \in C_b^{1,2}([0, \infty) \times R^d)$, $1 \leqq i, j \leqq d$, and $b: [0, \infty) \times R^d \rightsquigarrow R^d$ will be a bounded measurable function which is uniformly Lipschitz continuous in x. For each (t_0, x_0), P_{t_0,x_0} will denote the unique solution to the martingale problem for $L_t = \frac{1}{2}\sigma^*\nabla_x \cdot \sigma^*\nabla_x + b \cdot \nabla_x$, starting at (t_0, x_0). Our aim is to prove that

$$\text{supp}\,(P_{t_0,x_0}) = \overline{\mathscr{S}_{\sigma,b}(t_0, x_0)}, \tag{5.1}$$

where $\mathscr{S}_{\sigma,b}(t_0, x_0)$ is defined as in Section 4. In view of Theorem 4.1, equation (5.1) will be proved once we have shown that for all $T > t_0$, $\varepsilon > 0$, and ϕ in a dense subset of $\mathscr{S}_{\sigma,b}(t_0, x_0)$,

$$P_{t_0,x_0}\left(\|x(t) - \phi(t)\|_T^{t_0} < \varepsilon\right) > 0 \tag{5.2}$$

and $x_0 = 0$.

Using Theorem 2.2, one sees that the desired result is equivalent to proving that, for a dense set of $\phi \in \mathscr{S}_{\sigma,b}(0, 0)$,

$$W\left(\|\eta(t) - \phi(t)\|_T^0 < \varepsilon\right) > 0, \qquad T > 0,\ \varepsilon > 0, \tag{5.3}$$

where

$$\eta(t) = \int_0^t \sigma(u, \eta(u))\, d\beta(u) + \int_0^t \tilde{b}(u, \eta(u))\, du \qquad \text{a.s. } W, \tag{5.4}$$

$\beta(t)$ being a W Brownian motion and $\tilde{b}$ standing for $b + \frac{1}{2}\sigma'\sigma$ (see equation (2.35) for the definition of $\sigma'\sigma$). Actually, what we will show is more; namely, we will prove that if $\psi \in C^2([0, \infty) \times R^d)$, $\psi(0) = 0$, and

$$\phi(t) = x_0 + \int_0^t \sigma(u, \phi(u))\dot{\psi}(u)\, du + \int_0^t b(u, \phi(u))\, du, \tag{5.5}$$

then, for all $\varepsilon > 0$ and $T > 0$,

$$W\left(\|\eta(t) - \phi(t)\|_T^0 < \varepsilon \,\middle|\, |||\beta(t) - \psi(t)|||_T^0 < \delta\right) \to 1 \tag{5.6}$$

as $\delta \downarrow 0$, where $|||\alpha(t)|||_T^0 = \max_{1 \leqq i \leqq d} \|\alpha_i(t)\|_T^0$. We will first prove (5.6) when $\psi \equiv 0$.

After some easy manipulations involving Itô's formula (see McKean [5]), one can show that (5.4) is equivalent to

$$\eta(t) = \int_0^t b(u, \eta(u))\, du + \sigma(t, \eta(t))\beta(t) - \int_0^t \left[\left(\frac{\partial}{\partial u} + L_u\right)\sigma\right](u, \eta(u))\beta(u)\, du - \Delta(t), \tag{5.7}$$

where

$$\Delta_i(t) = \sum_{j \neq k} \int_t^t (\sigma^{ij}_\ell \sigma^{\ell k})(u, \eta(u))\beta_j(u)\, d\beta_k(u) + \sum_j \int_0^t (\sigma^{ij}_\ell \sigma^{\ell j})(u, \eta(u))\, d\beta_j^2(u). \tag{5.8}$$

Thus, in order to prove that

$$W(\|\eta(t) - \phi(t)\|_T^0 < \varepsilon \,|\, \|\|\beta(t)\|\|_T^0 < \delta) \to 1 \tag{5.9}$$

as $\delta \downarrow 0$, where $\phi(t) = \int_0^t b(u, \phi(u))\, du$, it suffices to show that

$$W(\|\Delta(t)\|_T^0 < \varepsilon \,|\, \|\|\beta(t)\|\|_T^0 < \delta) \to 1 \tag{5.10}$$

as $\delta \downarrow 0$.

LEMMA 5.1. *Let $\eta(t)$ be given by (5.4) and suppose $f\colon [0, \infty) \times R^d \rightsquigarrow R$ is bounded and uniformly continuous. Then for all $\varepsilon > 0$*

$$W\left(\left\| \int_0^t f(u, \eta(u))\, d\beta_i^2(u) \right\|_T^0 < \varepsilon \,\middle|\, \|\|\beta(t)\|\|_T^0 < \delta\right) \to 1 \tag{5.11}$$

as $\delta \downarrow 0$, and

$$W\left(\left\| \int_0^t f(u, \eta(u))\beta_i(u)\, d\beta_j(u) \right\|_T^0 < \varepsilon \,\middle|\, \|\|\beta(t)\|\|_T^0 < \delta\right) \to 1 \tag{5.12}$$

as $\delta \downarrow 0$, where $1 \leqq i \leqq d$ and $j \neq i$.

PROOF. We will first prove (5.11) under the assumption that $f \in C_b^\infty([0, \infty) \times R^d)$. Applying Itô's formula, we have

$$\begin{aligned} \int_0^t f(u, \eta(u))\, d\beta_i^2(u) &= f(t, \eta(t))\beta_i^2(t) - \int_0^t \beta_i^2(u)(f_u + L_u f)(u, \eta(u))\, du \\ &\quad - 2\int_0^t \beta_i(u)(\sigma^* \nabla_x f)_i(u, \eta(u))\, du \\ &\quad - \int_0^t \beta_i^2(u)\langle(\sigma^* \nabla_x f)(u, \eta(u)), d\beta(u)\rangle. \end{aligned} \tag{5.13}$$

Clearly, the first three terms on the right tend to 0 as $\|\|\beta(t)\|\|_T^0 \to 0$. Moreover, by standard estimates,

$$\begin{aligned} W\left(\left\| \int_0^t \beta_i^2(u)\langle(\sigma^* \nabla_x f)(u, \eta(u)), d\beta(u)\rangle \right\|_T^0 > \varepsilon, \|\|\beta(t)\|\|_T^0 < \delta\right) \\ \leqq A_1 \exp\left\{\frac{-B_1\varepsilon^2}{\delta^4 T}\right\} \end{aligned} \tag{5.14}$$

and

$$W(\|\beta(t)\|_T^0 < \delta) \geqq A_2 \exp\left\{\frac{-B_2 T}{\delta^2}\right\}. \tag{5.15}$$

Hence, (5.11) is proved in the case when $f \in C_b^\infty([0, \infty) \times R^d)$. Now suppose f is uniformly continuous and choose $\{f_n\}_0^\infty \subseteq C_b^\infty([0, \infty) \times R^d)$ so that $f_n \to f$ uniformly. Then

$$\begin{aligned}(5.16)\qquad & \int_0^t f(u, \eta(u))\, d\beta_i^2(u) - \int_0^t f_n(u, \eta(u))\, d\beta_i^2(u) \\ &= 2\int_0^t (f - f_n)(y, \eta(u))\beta_i(u)\, d\beta_i(u) + \int_0^t (f - f_n)(u, \eta(u))\, du.\end{aligned}$$

Given $\varepsilon > 0$, and $T > 0$, it is clear that for large n,

$$\begin{aligned}(5.17)\qquad & W\left(\left\| \int_0^t f(u, \eta(u))\, d\beta_i^2(u) - \int_0^t f_0(u, \eta(u))\, d\beta_i^2(u) \right\|_T^0 > \varepsilon, \|\|\beta(t)\|\|_T^0 < \delta\right) \\ &\leqq W\left(\left\| \int_0^t (f - f_n)(u, \eta(u))\beta_i(u)\, d\beta_i(u) \right\|_T^0 > \tfrac{1}{2}\varepsilon, \|\|\beta(t)\|\|_T^0 < \delta\right) \\ &\leqq A \exp\left\{\frac{-B_1 \varepsilon^2}{\|f - f_n\|\delta^2 T}\right\}.\end{aligned}$$

Hence, for all $\varepsilon > 0$ and $\alpha > 0$, there is an $n(\varepsilon, \alpha)$ such that

$$\begin{aligned}(5.18)\qquad & W\left(\left\| \int_0^t f(u, \eta(u))\, d\beta_i^2(u) \right.\right. \\ &\qquad \left.\left. - \int_0^t f_n(u, \eta(u))\, d\beta_i^2(u) \right\|_T^0 > \varepsilon \,\middle|\, \|\|\beta(t)\|\|_T^0 < \delta\right) < \alpha\end{aligned}$$

independent of $0 < \delta \leqq 1$. Combining this with our original result, (5.11) follows.

Next we will prove (5.12) under the assumption that $f \in C_b^\infty([0, \infty) \times R^d)$. Again the general result follows from this by the technique used above. For convenience, we will let $\xi(t) = \int_0^t \beta_i(u)\, d\beta_j(u)$. Using Itô's formula, we have

$$\begin{aligned}(5.19)\qquad & \int_0^t f(u, \eta(u))\beta_i(u)\, d\beta_j(u) \\ &= f(t, \eta(t))\zeta(t) - \int_0^t \xi(u)(f_n + L_n f)(u, \eta(u))\, du \\ &\quad - \int_0^t (\sigma^* \nabla_x f)_j(u, \eta(u))\beta_i(u)\, du \\ &\quad - \int_0^t \xi(u)\langle(\sigma^* \nabla_x f)(u, \eta(u)), d\beta(u)\rangle.\end{aligned}$$

Applying Theorem A.1 of the Appendix, we see that only the last term on the right need be examined. Let $\phi = (\sigma^* \nabla_x f)_\ell$. Then

$$\begin{aligned}(5.20)\qquad \int_0^t \xi(u)(\sigma^*\nabla_x f)_\ell(u, \eta(u))\, d\beta_\ell(u) &= \int_0^t \phi(u, \eta(u))\, d(\xi(u)\beta_\ell(u)) \\ &\quad - \int_0^t \phi(u, \eta(u))\beta_i(u)\beta_\ell(u)\, d\beta_j(u) - \delta_{j,\ell}\int_0^t \phi(u, \eta(u))\beta_i(u)\, du.\end{aligned}$$

The third term on the right gives no trouble. Moreover, the second term can be handled by the estimates used to prove (5.11). Thus, we need only worry about the first term. But

$$\begin{aligned}(5.21)\qquad \int_0^t \phi(u, \eta(u))\, d(\xi(u)\beta_\ell(u)) &= \phi(t, \eta(t))\xi(t)\beta_\ell(t) - \int_0^t \xi(u)\beta_\ell(u)(\phi_u + L_u\phi)(u, \eta(u))\, du \\ &\quad - \int_0^t \xi(u)(\sigma^*\nabla_x)_\ell(u, \eta(u))\, du \\ &\quad - \int_0^t \beta_\ell(u)\beta_i(u)(\sigma^*\nabla_x\phi)_j(u, \eta(u))\, du \\ &\quad - \int_0^t \langle \xi(u)\beta_\ell(u)(\sigma^*\nabla_x\phi)(u, \eta(u)), d\beta(u)\rangle.\end{aligned}$$

Again, only the final term need be examined. Using $\gamma(t)$ to denote this term, we have

$$\begin{aligned}(5.22)\qquad W(\|\gamma(t)\|_T^0 \geqq \varepsilon, \|\!|\beta(t)\|\!|_T^0 < \delta) &= W(\|\gamma(t)\|_T^0 \geqq \varepsilon, \|\!|\beta(t)\|\!|_T^0 < \delta, \|\xi(t)\|_T^0 < M\delta) \\ &\quad + W(\|\gamma(t)\|_T^0 \geqq \varepsilon, \|\!|\beta(t)\|\!|_T^0 < \delta, \|\xi(t)\|_T^0 \geqq M\delta).\end{aligned}$$

By the standard estimates,

$$(5.23)\qquad W(\|\gamma(t)\|_T^0 \geqq \varepsilon, \|\beta(t)\|_T^0 \leqq \delta, \|\xi(t)\|_T^0 < M\delta) \leqq A \exp\left\{-\frac{B\varepsilon^2}{M^2\delta^4}\right\}$$

By Theorem A.1,

$$(5.24)\qquad \lim_{M\to\infty}\ \sup_{0<\delta\leqq 1} \frac{W(\|\gamma(t)\|_T^0 \geqq \varepsilon, \|\!|\beta(t)\|\!|_T^0 < \delta, \|\xi(t)\|_T^0 \geqq M\delta)}{W(\|\!|\beta(t)\|\!|_T^0 < \delta)} = 0.$$

Combining these, we see that

$$(5.25)\qquad W(\|\gamma(t)\|_T^0 \geqq \varepsilon \mid \|\!|\beta(t)\|\!|_T^0 < \delta) \to 0$$

as $\delta \downarrow 0$. *Q.E.D.*

Clearly, (5.10) is an immediate consequence of Lemma 5.1. Hence, we have proved (5.6) when $\psi \equiv 0$.

THEOREM 5.1. *Let $\eta(t)$ be given by*

$$\eta(t) = x_0 + \int_{t_0}^{t} \sigma(u, \eta(u))\, d\beta(u) + \int_{t_0}^{t} \tilde{b}(u, \eta(u))\, du \quad a.s.\ W, \qquad t \geqq t_0, \tag{5.26}$$

where $\beta(t)$ is a W Brownian motion. Given $\psi \in C^2([0, \infty), R^d)$ satisfying $\psi(t_0) = 0$, define $\phi(t)$ by

$$\phi(t) = x_0 + \int_{t_0}^{t} \sigma(u, \phi(u))\dot{\psi}(u)\, du + \int_{t_0}^{t} b(u, \phi(u))\, du, \qquad t \geqq t_0. \tag{5.27}$$

Then for all $\varepsilon > 0$

$$W(\|\eta(t) - \phi(t)\|_T^{t_0} < \varepsilon \mid \|\beta(t) - \psi(t)\|_T^{t_0} < \delta) \to 1 \tag{5.28}$$

as $\delta \downarrow 0$.

PROOF. We may assume that $t_0 = 0$ and $x_0 = 0$. The case when $\psi \equiv 0$ has just been proved. To handle the general case, define $\overline{W}$ so that

$$\frac{d\overline{W}}{dW} = R(t) = \exp\left\{\int_0^t \langle \dot{\psi}(u), d\beta(u)\rangle - \frac{1}{2}\int_0^t |\dot{\psi}(u)|^2\, du\right\}, \tag{5.29}$$

on $\mathscr{M}_t^0$, $t \geq 0$. Then, by Theorem 2.4, $\overline{\beta}(t) = \beta(t) - \psi(t)$ is a $\overline{W}$ Brownian motion and

$$\eta(t) = \int_0^t \sigma(u, \eta(u))\, d\overline{\beta}(u) + \int_0^t \tilde{c}(u, \eta(u))\, du \qquad \text{a.s. } \overline{W}, \qquad t \geqq 0, \tag{5.30}$$

where $c = b + \sigma\dot{\psi}$ and $\tilde{c} = c + \frac{1}{2}\sigma'\sigma$. Hence,

$$\overline{W}(\|\eta(t) - \phi(t)\|_T^0 < \varepsilon \mid \|\overline{\beta}(t)\|_T^0 < \delta) \to 1 \tag{5.31}$$

as $\delta \downarrow 0$, where $\phi(t) = \int_0^t c(u, \phi(u))\, du$. But this means that

$$\begin{aligned}
&\lim_{\delta\downarrow 0} W(\|\eta(t) - \phi(t)\|_T^0 < \varepsilon \mid \|\beta(t) - \psi(t)\|_T^0 < \delta) \\
&= \lim_{\delta\downarrow 0} \frac{W(\|\eta(t) - \phi(t)\|_T^0 < \varepsilon,\ \|\beta(t) - \psi(t)\|_T^0 < \delta)}{E^W[R(T)\chi_{[0,\varepsilon)}(\|\eta(t) - \phi(t)\|_T^0)\chi_{[0,\delta)}(\|\beta(t) - \psi(t)\|_T^0)]} \\
&\qquad \cdot \frac{E^W[R(T)\chi_{[0,\delta)}(\|\beta(t) - \psi(t)\|_T^0)]}{W(\|\beta(t) - \psi(t)\|_T^0 < \delta)} \\
&= 1,
\end{aligned} \tag{5.32}$$

since

$$R(t) = \exp\left\{\dot{\psi}(t)\beta(t) - \int_0^t \langle \beta(u), \ddot{\psi}(u)\rangle\, du - \frac{1}{2}\int_0^t |\dot{\psi}(u)|^2\, du\right\}, \tag{5.33}$$

is a continuous functional of β. *Q.E.D.*

THEOREM 5.2. *Let* $c: [0, \infty) \times R^d \rightsquigarrow R^d$ *be bounded and measurable, and set*

$$L_t = \tfrac{1}{2}\sigma^* \nabla_x \cdot \sigma^* \nabla_x + (b + \sigma c) \cdot \nabla_x \tag{5.34}$$

(as before). If P_{t_0,x_0} *is the solution to the martingale problem for* (5.34) *starting at* (t_0, x_0), *then equation* (5.1) *obtains.*

PROOF. By Corollary 2.1, we may take $c \equiv 0$. Choose $\hat{P}_{t_0,x_0}$ and $\hat{\beta}(t)$ as in Theorem 2.2. Then

$$x(t) + \int_{t_0}^{t} \sigma(u, x(u))\, d\hat{\beta}(u) + \int_{t_0}^{t} \tilde{b}(u, x(u))\, du \quad \text{a.s. } \hat{P}_{t_0,x_0}, \qquad t \geqq t_0. \tag{5.35}$$

Hence, by Theorem 5.1, for all $T \geqq t_0$ and $\varepsilon > 0$,

$$\hat{P}_{t_0,x_0}(\|x(t) - (t)\|_T^{t_0} < | \|\hat{\beta}(t) - \psi(t)\|_T^{t_0} < \delta) \to 1 \tag{5.36}$$

as $\delta \downarrow 0$, where $\psi \in C^2([t_0, \infty), R^d)$ satisfies $\psi(t_0) = 0$ and (5.27). By Theorem 3.1, $\hat{P}_{t_0,x_0}(\|\hat{\beta}(t) - \psi(t)\|_T^{t_0} < \delta) > 0$ for all $T > t_0$ and $\delta > 0$. In particular, supp (P_{t_0,x_0}) contains a dense subset of $\mathscr{S}_{\sigma,b}(t_0, x_0)$.

REMARK 5.1. Let $\langle X, \rho \rangle$ be a metric space and suppose that μ is a probability measure on X. Given a μ measurable transformation $T: X \rightsquigarrow X$, we say that $x \in X$ is a *continuity point* of the transformation T if for all $\varepsilon > 0$,

$$\lim_{\delta \downarrow 0} \frac{\mu(T^{-1}(B(T(x), \varepsilon)) \cap B(x, \delta))}{\mu(B(x, \delta))} = 1, \tag{5.37}$$

where $B(y, \alpha) = \{z \in X: \rho(y, z) < \alpha\}$. In the terminology of continuity points, Theorem 5.1 says the following. Define $T: C^2([0, \infty), R^d) \rightsquigarrow \Omega$ so that $T(\psi) = \phi$, where

$$\phi(t) = \int_0^t \sigma(u, \phi(u))\dot{\psi}(u)\, du + \int_0^t b(u, \phi(u))\, du, \qquad t \geqq 0, \tag{5.38}$$

and let W be Wiener measure on Ω (that is, $x(t)$ is a W Brownian motion). Then the transformation $\hat{T}$ such that $\hat{T}(\omega)(t) = \eta(t, \omega)$, where

$$\eta(t) = \int_0^t \sigma(u, \eta(u))\, dx(u) + \int_0^t \tilde{b}(u, \eta(u))\, du \quad \text{a.s. } W, \tag{5.39}$$

is a W measurable extension of T of Ω with the property that all elements in $C^2([0, \infty), R^d)$ are continuity points of $\hat{T}$.

REMARK 5.2. It seems likely that the hypotheses under which Theorem 5.1 was proved are close to the best possible. However, the authors believe that Theorem 5.2 is valid under weaker assumptions. In particular, it seems likely that if $a: [0, \infty) \times R^d \rightsquigarrow S_d$ satisfies $a^{ij} \in C_b^{1,2}([0, \infty) \times R^d)$, $1 \leqq i, j \leqq d$ and $b: [0, \infty) \times R^d \rightsquigarrow R^d$ is a bounded, measurable function which is uniformly Lipschitz continuous in x, then

$$\text{supp } (P_{t_0,x_0}) = \overline{\mathscr{S}_{a,b}(t_0, x_0)}, \tag{5.40}$$

where P_{t_0,x_0} is the solution to the martingale problem for $L_t = \frac{1}{2}\nabla_x \cdot (a\nabla_x) + b\cdot\nabla_x$ starting at (t_0, x_0).

6. Applications

Let σ and b be as in Section 5 and define $\mathscr{S}_{\sigma,b}(t_0, x_0)$, $(t_0, x_0) \in [0, \infty) \times R^d$, accordingly. Given an open $\mathscr{G} \subseteq [0, \infty) \times R^d$, take $\mathscr{G}_{\sigma,b}(t_0, x_0)$ to be the closure in $\mathscr{G}$ of $\{(t, \phi(t)) : t \geqq t_0, \phi \in \mathscr{S}_{\sigma,b}(t_0, x_0)$, and $(s, \phi(s)) \in \mathscr{G}$ for $t_0 \leqq s \leqq t\}$. Let $c: [0, \infty) \times R^d \to R^d$ be bounded, measurable, and continuous in x. Denote by P_{t_0,x_0} the solution to the martingale problem for $L_t = \frac{1}{2}\sigma^*\nabla_x \cdot \sigma^*\nabla_x + (b + \sigma c)\cdot\nabla_x$, starting at (t_0, x_0). We will use $\mathscr{H}_{L_t}^-(t_0, x_0, \mathscr{G})$ to denote the class of upper semicontinuous functions f on $\mathscr{G}$ into $R \cup \{-\infty\}$ which are bounded above and have the property that $f(t \wedge \tau, x(t \wedge \tau))$, $t \geqq t_0$, is a P_{t_0,x_0} submartingale, where $\tau = \inf\{t \geqq t_0 : (t, x(t)) \notin \mathscr{G}\}$. We use $\mathscr{H}_{L_t}^-(\mathscr{G})$ to denote the class $\bigcap_{\mathscr{G}} \mathscr{H}_{L_t}^-(t_0, x_0, \mathscr{G})$. Observe that $\mathscr{H}_{L_t}^-(t_0, x_0, \mathscr{G})$ and $\mathscr{H}_{L_t}^-(\mathscr{G})$ have the same closure properties as the analogously defined classes in Section 3. Further, note that if $f \in C_b^{1,2}(\mathscr{G})$, then $f \in \mathscr{H}_{L_t}^-(t_0, x_0, \mathscr{G})$ if and only if $(\partial f/\partial t) + L_t f \geqq 0$ on $\mathscr{G}(t_0, x_0)$.

THEOREM 6.1. *If $f \in \mathscr{H}_{L_t}^-(t_0, x_0, \mathscr{G})$ and $f(t_0, x_0) = \sup_{\mathscr{G}} f(t, x)$, then $f \equiv f(t_0, x_0)$ on $\mathscr{G}(t_0, x_0)$. Moreover, if $(t_1, x_1) \in \mathscr{G} - \mathscr{G}(t_0, x_0)$, then there is an $f \in \mathscr{H}_{L_t}^-(\mathscr{G})$ such that $f(t_0, x_0) = \sup_{\mathscr{G}} f(t, x)$ and $f(t_1, x_1) < f(t_0, x_0)$.*

PROOF. The proof is exactly the same as that of Theorem 3.2. The only difference lies in the fact that the $P_{t,x}$ are no longer strongly Feller continuous and therefore $E^{P_{t,x}}[\int_t^{\tau_t} e^{-u}h(u, x(u))\, du]$ will not, in general, be continuous. Nonetheless, if $h \leqq 0$, then it will still be upper semicontinuous, by virtue of the Feller continuity of the $P_{t,x}$. *Q.E.D.*

REMARK 6.1. The class $\mathscr{H}_{L_t}^-(\mathscr{G})$ admits the same semigroup interpretation as we gave in Remark 3.1, only it is no longer true that every measurable function f which is bounded above and satisfies $\hat{T}_t f \downarrow f$ is upper semicontinuous in the ordinary sense. However, it will be upper semicontinuous in the "intrinsic topology" of the time-space process; and one can still show that such an f will be constant on the intrinsic closure of $\{(t, \phi(t)) : t \geqq t_0, \phi \in \mathscr{S}_{\sigma,b}(t_0, x_0)$, and $(s, \phi(s)) \in \mathscr{G}$ for $t_0 \leqq s \leqq t\}$ if $f(t_0, x_0) = \sup_{\mathscr{G}} f(t, x)$.

◇ ◇ ◇ ◇ ◇

APPENDIX

Let $\beta(t)$ be a W Brownian motion and set $\xi(t) = \int_0^t \beta_i(u)\, d\beta_j(u)$, where $i \neq j$. The purpose of this section is to prove that for all $T > 0$,

$$\text{(A.1)} \qquad \lim_{M\to\infty} \inf_{0<\delta\leqq 1} W(\|\xi(t)\|_T^0 < M\delta \mid \|\|\beta(t)\|\|_T^0 < \delta) = 1.$$

It is clear that we may assume that $d = 2$, $i = 1$, and $j = 2$.

For each $(s, x) \in [0, \infty) \times R^2$, let $W_{s,x}$ be Wiener measure starting at (s, x) (that is, $W_{s,x}$ is the measure on $\langle \Omega, \mathscr{M}^s \rangle$ such that $x(t) - x$ is a $W_{s,x}$ Brownian motion). Define

$$\tau_s = \inf \{t \geqq s : |x_1(t)| \vee |x_2(t)| \geqq \delta\}, \tag{A.2}$$

and let

$$Q_{s,x}(A) = W_{s,x}(A \mid \{\tau_s > T\}), \qquad A \in \mathscr{M}^s. \tag{A.3}$$

LEMMA A.1. *Let* $g(s, x) = W_{s,x}(\tau_s > T)$ *and* $h(s, a) = W_{s,(a,0)}(\|x_1(t)\|_T^s < \delta)$. *Then* $g(s, x) = h(s, x_1) \cdot h(s, x_2)$, *and* $h \in C^\infty[(0, T) \times (-\delta, \delta)]$. *Next, let* $b(s, a) = h_a(s, a)/h(s, a)$ *in* $(0, T) \times (-\delta, \delta)$. *Then* $b(s, \cdot)$ *is nonincreasing and* $b(s, -a) = -b(s, a)$. *Finally, define* $B(s, x) = (b(s, x_1), b(s, x_2))$. *Then for all* $s \in [0, \tau)$ *and* x *such that* $|x_1| \vee |x_2| < \delta$,

$$\beta(t) = x(t) - x - \int_s^{t \wedge \tau} B(u, x(u))\, du \tag{A.4}$$

is a $Q_{s,x}$ *Brownian motion.*

PROOF. The first assertion is trivial. To prove the second assertion, note that $h_s + \frac{1}{2}h_{aa} = 0$ and that $h_s \geqq 0$. Hence,

$$b_a(s, a) = \frac{hh_{aa} - h_a^2}{h^2} \leqq 0, \tag{A.5}$$

and so $b(s, \cdot)$ is nonincreasing. Also, note that $h(s, -a) = h(s, a)$, and therefore $h_a(s, -a) = -h_a(s, a)$. Hence, $b(s, -a) = -b(s, a)$.

Finally, to prove the last assertion, let $\{\delta_n\}_1^\infty \subseteqq (0, \delta)$ such that $\delta_n \uparrow \delta$ and define

$$\tau_s^{(n)} = (\inf \{t \geqq s : |x_1(t)| \vee |x_2(t)| \geqq \delta_n\}) \wedge T. \tag{A.6}$$

Then, by Itô's formula,

$$\begin{aligned} &g(t \wedge \tau_s^{(n)}, x(t \wedge \tau_s^{(n)})) - g(s, x) \\ &\qquad = \int_s^{t \wedge \tau_s^{(n)}} g(u, x(u)) \langle B(u, x(u)), dx(u) \rangle \quad \text{a.s. } W_{s,x}. \end{aligned} \tag{A.7}$$

Hence,

$$\begin{aligned} &\frac{g(t \wedge \tau_s^{(n)}, x(t \wedge \tau_s^{(n)}))}{g(s, x)} \\ &\quad = R^{(n)}(t) \\ &\quad = \exp \left\{ \int_s^{t \wedge \tau_s^{(n)}} \langle B(u, x(u)), dx(u) \rangle - \frac{1}{2} \int_s^{t \wedge \tau_s^{(n)}} |B(u, x(u))|^2\, du \right\} \\ &\qquad\qquad \text{a.s. } W_{s,x}. \end{aligned} \tag{A.8}$$

Given $A \in \mathscr{M}^s_{t \wedge \tau_s^{(n)}}$, we now have

$$\text{(A.9)} \qquad Q_{s,x}(A) = \frac{W_{s,x}(A \cap \{\tau_s > T\})}{g(s, x)} = \frac{E^{W_{s,x}}[\chi_A g(t \wedge \tau_s^{(n)}, x(t \wedge \tau_s^{(n)})]}{g(s, x)}$$

$$= E^{W_{s,x}}[R^{(n)}(t)\chi_A].$$

Thus, by Theorem 2.1,

$$\text{(A.10)} \qquad X_\theta^{(n)}(t)$$

$$= \exp\left\{\langle\theta, x(t \wedge \tau_s^{(n)}) - x - \int_s^{t \wedge \tau_s^{(n)}} B(u, x(u))\, du\rangle - \tfrac{1}{2}|\theta|^2(t \wedge \tau_s^{(n)} - s)\right\},$$

is a $Q_{s,x}$ martingale for all $\theta \in R^2$. Since $\tau_s^{(n)} \uparrow T$ as $n \to \infty$ and

$$\text{(A.11)} \qquad E^{Q_{s,x}}[|X_\theta^{(n)}(t)|^2] = E^{Q_{s,x}}[X_{2\theta}^{(n)}(t) \exp\{|\theta|^2(t \wedge \tau_s^{(n)} - s)\}]$$
$$\leqq \exp\{|\theta^{2(T-s)}\},$$

it follows that

$$\text{(A.12)} \qquad \exp\left\{\langle\theta, x(t \wedge T) - x - \int_s^{t \wedge T} B(u, x(u))\, du\rangle - \tfrac{1}{2}|\theta|^2(t \wedge T - s)\right\},$$

is a $Q_{s,x}$ martingale for all $\theta \in R^2$. Combining these, we have that

$$\text{(A.13)} \qquad x(t) - x - \int_s^{t \wedge T} B(u, x(u))\, du$$

is a $Q_{s,x}$ Brownian motion. *Q.E.D.*

LEMMA A.2. *Let* $\xi(t) = \int_0^t x_1(u)\, dx_2(u)$. *Then*

$$\text{(A.14)} \qquad \sup_{0 \leqq s \leqq T, |x_1| \vee |x_2| < \delta} E^{Q_{s,x}}[|\xi(T) - \xi(s)|^2] \leqq C\delta^2.$$

PROOF. Note that by Lemma A.1

$$\text{(A.15)} \qquad \xi(t) = \int_s^t x_1(u)\, d\beta_1(u) + \int_s^{t \wedge T} x_1(u) b(u, x_2(u))\, du,$$

where $\beta_1(t) = x_1(t) - x_1 - \int_s^{t \wedge T} b(u, x_1(u))\, du$ is a one dimensional $Q_{s,x}$ Brownian motion. Hence,

$$\text{(A.16)} \qquad E^{Q_{s,x}}[|\xi(T) - \xi(s)|^2]$$

$$\leqq 2E^{Q_{s,x}}\left[\left|\int_s^T x_1(u)\, d\beta_1(u)\right|^2\right] + 2E^{Q_{s,x}}\left[\left|\int_s^T x_1(u) b(u, x_2(u))\, du\right|^2\right].$$

Since $x_1(u) < \delta$ a.s. $Q_{s,x}$ for $s \leqq u \leqq T$,

$$\text{(A.17)} \qquad E^{Q_{s,x}}\left[\left|\int_s^T x_1(u)\, d\beta_1(u)\right|^2\right] \leqq \delta^2(T - s).$$

Using the independence of $x_1(\cdot)$ and $x_2(\cdot)$ under $Q_{s,x}$, we have

$$\text{(A.18)} \quad E^{Q_{s,x}}\left[\left|\int_s^T x_1(u)b(u, x_2(u))\,du\right|^2\right] = 2 \iint\limits_{s \leqq u_1 \leqq u_2 \leqq T} E^{Q_{s,x}}[x_1(u_1)x_1(u_2)] \cdot E^{Q_{s,x}} \cdot [b(u_1, x_2(u_1))b(u_2, x_2(u_2))]\,du_1 du_2.$$

Observe that

$$\text{(A.19)} \quad E^{Q_{s,x}}[b(u_1, x_2(u_1))b(u_1, x_2(u_2))] = E^{Q_{s,x}}[b(u_1, x_2(u_1))E^{Q_{u_1,(0,x_2(u_1))}}[b(u_2, x_2(u_2))]].$$

Since $(b(u, \cdot)$ is nonincreasing and antisymmetric, it is easy to see that $v(u_1, \cdot) = E^{Q_{u_1,(0,\cdot)}}[b(u_2, x_2(u_2))]$ has the same properties. In particular,

$$\text{(A.20)} \quad E^{Q_{s,x}}[b(u_1, x_2(u_1))b(u, x_2(u_2))] \geqq 0.$$

Hence,

$$\text{(A.21)} \quad E^{Q_{s,x}}\left[\left|\int_s^T x_1(u)b(u, x_2(u))\,du\right|^2\right] \leqq \delta^2 E^{Q_{s,x}}\left[\left|\int_s^T b(u, x_2(u))\,du\right|^2\right] = \delta^2 E^{Q_{s,x}}[|\beta_2(T) - x_2(T) - x_2|^2] \leqq C\delta^2.$$

Q.E.D.

THEOREM A.1. *Define* $\xi(t)$ *as in Lemma* A.2 *and let* $W = W_{0,0}$. *Then*

$$\text{(A.22)} \quad \lim_{M \to \infty} \inf_{0 < \delta \leqq 1} W(\|\xi(t)\|_T^0 < M\delta \,|\, \|x(t)\|_T^0 < \delta) = 1.$$

PROOF. Let σ be a stopping time which is dominated by T. Then for $A \in \mathscr{M}_\sigma^0$,

$$\begin{aligned} \text{(A.23)} \quad & E^W[\chi_{A\cap\{\tau_s > T\}}|\xi(T) - \xi(\sigma)|^2] \\ &= E^W[\chi_{A\cap\{\tau_s > \sigma\}}E^{W_{\sigma,x(\sigma)}}(\chi_{\{\tau_0 > T\}}|\xi(T) - \xi(\sigma)|^2)] \\ &= E^W[\chi_{A\cap\{\tau_s > \sigma\}}W_{\sigma,x(\sigma)}(\tau_\sigma > T)E^{Q_{\sigma,x(\sigma)}}[|\xi(T) - \xi(\sigma)|^2]] \\ &\leqq C\delta^2 W(A \cap \{\tau_s > T\}). \end{aligned}$$

Thus if $Q = Q_{0,0}$, then

$$\text{(A.24)} \quad E^Q[|\xi(T) - \xi(\sigma)|^2 \,|\, \mathscr{M}_\sigma^0] \leqq C\delta^2.$$

Now let $\sigma_\ell = (\inf\{t \geqq 0 : |\xi(t)| \geqq \ell\delta) \wedge T$. Then

$$\begin{aligned} \text{(A.25)} \quad & Q(\{\sigma_{2\ell} < T\} \cap \{|\xi(T)| \leqq \ell\delta\}) \\ &\leqq Q(\{\sigma_{2\ell} < T\} \cap \{|\xi(T) - \xi(\sigma_{2\ell})| \geqq \ell\delta\}) \\ &\leqq \frac{C}{\ell^2} Q(\sigma_{2\ell} < T). \end{aligned}$$

Thus,

$$Q(\sigma_{2\ell} < T) \leqq \frac{Q(|\xi(T)| > \ell\delta)}{1 - (C/\ell^2)} \leqq \frac{C}{\ell^2}\left(1 - \frac{C}{\ell^2}\right)^{-1}. \tag{A.26}$$

Q.E.D.

REFERENCES

[1] J. M. Bony, "Principe du maximum, unicité du problème de Cauchy et inégalité de Harnack pour les operateurs elliptiques degenérés," *Ann. Inst. Fourier*, Vol. 19 (1969), pp. 277–304.

[2] J. L. Doob, *Stochastic Processes*, New York, Wiley, 1953.

[3] C. D. Hill, "A sharp maximum principle for degenerate elliptic-parabolic equations," *J. Math. and Mech.*, Vol. 20 (1970), pp. 213–228.

[4] K. Itô, "On stochastic differential equations," *Mem. Amer. Math. Soc.*, No. 4 (1951).

[5] H. P. McKean, *Stochastic Integrals*, New York, Academic Press, 1969.

[6] L. Nirenberg, "A strong maximum principle for parabolic equations," *Comm. Pure Appl. Math.*, Vol. 6 (1953), pp. 167–177.

[7] Yu. V. Prokhorov, "Convergence of random processes and limit theorems," *Theory Probability Appl.*, Vol. 1 (1956), pp. 118–122.

[8] R. S. Philips and L. Sarason, "Elliptic-parabolic equations of the second order," *J. Math. Mech.*, Vol. 17 (1968), pp. 891–917.

[9] D. W. Stroock and S. R. S. Varadhan, "Diffusion processes with continuous coefficients," *Comm. Pure Appl. Math.* II, Vol. 22 (1969), pp. 345–400; 479–530.

[10] ———, "Diffusion processes with boundary conditions," *Comm. Pure Appl. Math.*, Vol. 24 (1971), pp. 147–225.

[11] D. W. Stroock, "On the growth of stochastic integrals," *Z. Wahrscheinlichkeitstheorie und Verw. Gebiete*, Vol. 18 (1971), pp. 340–344.

[12] E. Wong and M. Zakai, "Riemann-Stieltjes approximation of stochastic integrals," *Z. Wahrscheinlichkeitstheorie und Verw. Gebiete*, Vol. 12 (1969), pp. 87–97.

DIFFUSION PROCESSES

DANIEL W. STROOCK and S. R. S. VARADHAN
COURANT INSTITUTE, NEW YORK UNIVERSITY

1. Introduction

One of the major problems in the theory of diffusion processes is to construct the process for a given set of diffusion coefficients. A diffusion process in R^d is hopefully determined by the two sets of coefficients

$$(1.1)\qquad \begin{aligned} a &= a(t, x) = \{a_{ij}(t, x)\}, && 1 \leqq i, j \leqq d,\ t \in [0, \infty),\ x \in R^d, \\ b &= b(t, x) = \{b_j(t, x)\}, && 1 \leqq j \leqq d,\ t \in [0, \infty),\ x \in R^d. \end{aligned}$$

Here a is a positive semidefinite symmetric matrix for each t and x, and b is a d vector for each t and x. There are various ways of describing exactly what we mean by a diffusion process corresponding to the specified set of coefficients. We shall adopt the following approach.

Let Ω be the space of R^d valued continuous functions on $[0, \infty)$. The value of a function $\omega = x(\cdot)$ in Ω at time t will be denoted by $x(t)$. The σ-field generated by $x(s)$ for $t_1 \leqq s \leqq t_2$ will be denoted by $M_{t_2}^{t_1}$. If $t_1 = 0$, we will denote this by M_{t_2} and by M^{t_1} in case $t_2 = \infty$, where M is the σ-field generated by $x(s)$ for $0 \leqq s < \infty$. The space Ω can be viewed as a complete separable metric space, with uniform convergence on bounded intervals defining the topology. Then M is the Borel σ-field in Ω. A stochastic process with values in R^d, defined for $t \geqq t_0$, is a probability measure on (Ω, M^{t_0}).

Given the coefficients $\{a_{ij}(t, x)\}$ and $\{b_j(t, x)\}$, we define an operator L_t acting on functions $f(x) \in C_0^\infty(R^d)$ by

$$(1.2)\qquad (L_t f)(x) = \frac{1}{2} \sum a_{ij}(t, x) \frac{\partial^2 f}{\partial x_i \partial x_j} + \sum b_j(t, x) \frac{\partial f}{\partial x_j}.$$

We say that a measure P is a solution to the Martingale problem corresponding to the given coefficients, starting at time t_0 from the point x_0 if

(a) P is a probability measure on (Ω, M^{t_0}) such that $P[x(t_0) = x_0] = 1$, and

(b) for each $f \in C_0^\infty(R^d)$, $f(x(t)) - \int_{t_0}^t (L_s f)(x(s))\, ds$ is a martingale relative to $(\Omega, M_t^{t_0}, P)$.

Under suitable conditions on the coefficients a and b, one should attempt to answer the following questions:

(1) For each t_0 and x_0, does a solution P_{t_0, x_0} exist?

Results obtained at the Courant Institute of Mathematical Sciences, New York University; this research was sponsored by the US Air Force Office of Scientific Research, Contract AF-49(638)-1719.

(2) Is it unique?

(3) Is the solution a strong Markov process?

(4) Does the solution depend continuously, in some suitable sense, on t_0, x_0, a and b?

(5) If a diffusion process can be constructed in some other natural manner, does it coincide with the above construction?

Let us take up question (5) first. There are at least two other possible ways of constructing diffusion processes for given coefficients. If we use the stochastic differential equations of Itô, we take a matrix $\sigma(t, x)$ such that

$$\sigma(t, x)\sigma^*(t, x) = a(t, x). \tag{1.3}$$

If $\beta(t)$ is Brownian motion in d dimensions, one can set up the stochastic differential equation

$$dx(t) = \sigma\big(t, x(t)\big)\, d\beta(t) + b\big(t, x(t)\big)\, dt. \tag{1.4}$$

If σ and b are assumed to be bounded and uniformly (with respect to t) Lipschitz continuous in x, one can show that the above equation has a unique solution $x(t)$ for $t \geqq t_0$, for each initial condition $x(t_0) = x_0$. The process so obtained is the diffusion process corresponding to a and b starting from the point x_0 at time t_0.

One can prove that if the solution to the stochastic differential equation exists, then the solution to the martingale problem also exists. Moreover, if the former is unique, then so is the latter. Furthermore, one shows that the two solutions are the same in the sense that the solution to the martingale problem is the distribution of the solution to the stochastic differential equation.

Another possibility is to use the theory of partial differential equations. We consider the fundamental solution $p(s, x, t, y)$ of the equation

$$\frac{\partial p}{\partial s} + \frac{1}{2} \sum a_{ij}(s, x) \frac{\partial^2 p}{\partial x_i \partial x_j} + \sum b_j(s, x) \frac{\partial p}{\partial x_j} = 0, \tag{1.5}$$

where $p(s, x, t, y)$ serves as the transition probability density of a Markov process. This is taken as the diffusion process corresponding to a and b. One can verify that, in this case also, the solution to the martingale problem exists, is unique, and coincides with the process constructed through the fundamental solution. The existence of a fundamental solution is proved under the assumption that a and b are bounded, satisfy a Hölder condition and a is uniformly positive definite.

2. Existence and uniqueness: the general case

Let us now assume that $a(t, x)$ is bounded continuous and positive definite for each t and x. Assume $b(t, x)$ bounded and measurable. Under these assumptions we can answer the questions (1) through (4) raised in the introduction, affirmatively.

THEOREM 2.1. *The solution $P_{s,x}$ to the martingale problem, starting from the point x at some s, exists and is unique for every s and x. Moreover, $P_{s,x}$ is a strong Markov process with transition probabilities $P(s, x, t, A) = P_{s,x}[x(t) \in A]$. Further, the solution depends continuously on s, x, $a(\cdot, \cdot)$, and $b(\cdot, \cdot)$ in the following sense: if $a_n(\cdot, \cdot) \to a(\cdot, \cdot)$ uniformly on compact sets, $b_n(\cdot, \cdot) \to b(\cdot, \cdot)$ in measure, $s_n \to s$, $x_n \to x$ and if $a_n(\cdot, \cdot)$, $b_n(\cdot, \cdot)$ are uniformly bounded, then the solution P_n corresponding to a_n, b_n starting from x_n at time s_n converges weakly to the solution P for the limiting coefficients starting from x at time s.*

This theorem can be found in [2] and [3].

3. Boundary conditions

Let us suppose that $G \subset R^d$ is a smooth region. More precisely, there is a function $\phi(x) \in C_b^2(R^d)$ such that

$$G = [x: \phi(x) > 0],$$
$$\delta G = [x: \phi(x) = 0], \tag{3.1}$$
$$\|\nabla G\| \geqq 1 \text{ on } \delta G.$$

As before we shall assume that a is bounded, positive definite for each t and x and is continuous on $[0, \infty) \times \bar{G}$. Assume b is bounded and measurable. One can check that some sort of boundary condition on δG is needed to describe what happens to the process when it reaches the boundary δG. A class of these boundary conditions are of the form

$$\rho(s, x)\frac{\partial}{\partial s} + \sum_{j=1}^{d} \gamma_j(s, x)\frac{\partial}{\partial x_j} = 0, \tag{3.2}$$

where ρ and γ are suitable function on the boundary $[0, \infty) \times \delta G$. We shall assume the following regarding ρ and γ:

(i) γ is Lipschitz continuous in t and x, is bounded, and there is a constant $\beta > 0$, such that $\langle \gamma, \nabla\phi \rangle \geqq \beta > 0$; for all $t, x \in [0, \infty) \times \delta G$;

(ii) either ρ is identically zero or it is everywhere positive, is bounded, and satisfies a Lipschitz condition in t and x.

We formulate the problem in the following manner. A solution corresponding to a, b, ρ, and γ, starting a time t_0 from the point x_0, is a measure P on (Ω, M^{t_0}), where now $\Omega = C[[0, \infty), \bar{G}]$ and M^{t_0} is the natural σ field as before. The measure P is such that

(a) $P[x(t_0) = x_0] = 1$, and

(b) for every $u \in C_0^\infty[[0, \infty) \times \bar{G}]$ with $\rho(\partial u/\partial s) + \gamma, \nabla u \geqq 0$ on $[0, \infty) \times \delta G$,

$$u(t, x(t)) - \int_{t_0}^{t} (U_s + L_s u)(s, x(s))\chi_G(x(s))\, ds, \tag{3.3}$$

is a submartingale relative to $(\Omega, M_t^{t_0}, P)$. We will call P, if it exists, a solution to the submartingale problem.

THEOREM 3.1. *Under the assumptions* (a) *and* (b) *on* ρ *and* γ, *along with the assumptions on* a, b *mentioned in the earlier section, for each* $x \in \overline{G}$ *and* $s \geqq 0$, *there is a unique solution* $P_{s,x}$ *to the submartingale problem. The solution is a strong Markov process. Moreover, the solution depends continuously on* s, x, a, b, ρ, *and* γ.

REMARK 3.1. In the homogeneous case, that is, when a, b, ρ, and γ are independent of t, assumption (b) on ρ can be replaced by the assumption that ρ is bounded, continuous and nonnegative.

These results can be found in [4].

4. Invariance principle

Using the results of Sections 2 and 3, one can prove general theorems concerning the convergence of Markov chains to diffusion processes. The conditions for convergence are very natural and involve essentially the first two moments. For simplicity, we shall treat the homogeneous case and assume further that $d = 1$. The general case is quite similar. Let us suppose that there is no boundary.

For each $\delta > 0$, $\pi_\delta(x, dy)$ is the transition probability of a Markov chain with R as its state space. The transitions of the chain occur in multiples of time δ. Let us define

$$(4.1)\qquad \begin{aligned} b_\delta(x) &= \frac{1}{\delta}\int (y - x)\pi_\delta(x, dy),\\ a_\delta(x) &= \frac{1}{\delta}\int (y - x)^2\pi_\delta(x, dy),\\ \theta_\delta(x) &= \frac{1}{\delta}\int |y - x|^\mu\pi_\delta(x, dy), \qquad \mu > 2. \end{aligned}$$

We assume:

(a) $|b_\delta(x)| \leqq C$ and $b_\delta(x)$ converges as δ tends to zero to a continuous limit $b(x)$ uniformly on bounded intervals;

(b) $|a_\delta(x)| \leqq C$ and $a_\delta(x)$ converges as δ tends to zero to a continuous limit $a(x)$ uniformly on bounded intervals;

(c) $\theta_\delta(x) \to 0$, as δ tends to zero, uniformly on bounded intervals;

(d) the solution to the martingale problem for any starting point, corresponding to the coefficients a and b is unique.

THEOREM 4.1. *Under the above assumptions the Markov chain converges as* δ *tends to zero, to the diffusion corresponding to* a *and* b.

The general case for the d dimensional case, where a and b could depend on t is treated in [3]. There are similar results when a boundary is involved and these can be found in [4].

5. Special case: $d = 2$

The method of the main theorem mentioned in Section 2, involves several steps. The hypothesis of continuity of $a(t, x)$ is superfluous. What one needs is that the discontinuities of $a(t, x)$ should be small compared to the eigenvalues of $a(t, x)$. The actual constants are hard to follow through in general. But when $d = 2$, in the homogeneous case, this leads to the following theorem.

THEOREM 5.1. *If $a(x)$ is uniformly positive definite on compact sets and if trace $a(x) \equiv 1$, then for any bounded measurable $b(x)$, existence and uniqueness hold for the martingale problem.*

By a random time change, the condition that trace $a(x) \equiv 1$ can be removed and replaced by the boundedness of a. Once uniqueness is established, one shows by a standard reasoning that the process is strongly Markovian. With a little more work, one can establish the continuous dependence of the solution on the coefficients and the starting point.

These results can be found essentially in [1].

6. Special case: $d = 1$

When $d = 1$, the results are even stronger. If $a(t, x)$ is bounded measurable and uniformly positive on compact sets and if $b(t, x)$ is bounded and measurable, then the solution to the martingale problem exists and is unique for every starting point. Since a proof has not appeared anywhere, we will give a quick sketch. From the way uniqueness is proved in [2] for the general case the basic step is a perturbation in L_p. When $d = 1$, p can be taken to be 2. We write the operator

$$\frac{1}{2}\,a(t, x)\,\frac{\partial}{\partial x^2} + \frac{\partial}{\partial t}, \tag{6.1}$$

as

$$\frac{1}{2}\,\ell\,\frac{\partial}{\partial x^2} + \frac{1}{2}\big(a(t, x) - \ell\big)\frac{\partial}{\partial x^2} + \frac{\partial}{\partial t}, \tag{6.2}$$

where ℓ is chosen to be a suitable large number. For this to work in L_2, where the norms of the various operators are explicitly computable, it suffices that

$$\big|a(t, x) - \ell\big| \leqq \ell' < \ell. \tag{6.3}$$

If a is such that $0 < \alpha_1 \leqq a \leqq \alpha_2 < \infty$, then ℓ can be taken as α_2 and ℓ' can be $\alpha_2 - \alpha_1$. Since uniqueness is purely a local property, our assertion follows.

We fix $0 < \alpha_1 < \alpha_2 < \infty$ and consider the totality of all bounded measurable $a(t, x)$ satisfying $\alpha_1 \leqq a(t, x) \leqq \alpha_2$. For simplicity, we shall assume that b is identically zero. By existence and uniqueness mentioned above, we know that there are transition probabilities $\{p_a(s, x, t, dy)\}$ depending on a. As a varies over the above class, one can check that $\{p_a\}$ varies over a compact family. The convergence notion is weak convergence in dy which is uniform over compact sets of s, x and t. The limits are all again transition probabilities corresponding to some a in the same class.

This means that there is a notion of "weak" convergence such that $a_n \to a$ "weakly" if and only if $p_{a_n} \to p_a$ as described above. It will of course be very interesting to know precisely what this convergence is. There are two special cases worth noticing. If $a_n(t, x)$ are purely functions of x, then $p_{a_n} \to p_a$ if and only if

$$\int_{\ell_1}^{\ell_2} \frac{dx}{a_n(x)} \to \int_{\ell_1}^{\ell_2} \frac{dx}{a(x)} \tag{6.4}$$

for $-\infty < \ell_1 < \ell_2 < \infty$. If $a_n(t, x)$ are purely functions of t, then $p_{a_n} \to p_a$ if and only if

$$\int_{\ell_1}^{\ell_2} a_n(s)\, ds \to \int_{\ell_1}^{\ell_2} a(s)\, ds \tag{6.5}$$

for $0 \leqq \ell_1 \leqq \ell_2 < \infty$.

For the general case when a_n depends on both t and x, these special cases provide conflicting clues. The problem is more involved than one expects to begin with.

7. An example

Let us define the coefficients

$$a_+^h(x) = \begin{cases} \alpha & \text{if } \left[\dfrac{x}{h}\right] \text{ is even,} \\ \beta & \text{if } \left[\dfrac{x}{h}\right] \text{ is odd,} \end{cases} \tag{7.1}$$

and

$$a_-^h(x) = \begin{cases} \alpha & \text{if } \left[\dfrac{x}{h}\right] \text{ is odd,} \\ \beta & \text{if } \left[\dfrac{x}{h}\right] \text{ is even,} \end{cases} \tag{7.2}$$

$\beta > \alpha > 0$. (For $x < 0$, $[x] = -[-x] - 1$.)
Let $\pi_\pm^h(t, x, dy)$ be the transition probabilities corresponding to the homogeneous coefficients $a_\pm^h(x)$. Let us define

$$a^{k,h}(t, x) = \begin{cases} a_+^h(x) & \text{if } \left[\dfrac{t}{k}\right] \text{ is even,} \\ a_-^h(x) & \text{if } \left[\dfrac{t}{k}\right] \text{ is odd.} \end{cases} \tag{7.3}$$

Let $\pi^{k,h}(s, x, t, dy)$ be the transition probability corresponding to $a^{k,h}$. We denote by $P^{k,h}_{s,x}$, the process corresponding to $a^{k,h}$ starting off at time s from x.

THEOREM 7.1. *Let h_n and k_n tend to zero such that $\rho_n = k_n h_n^{-2}$ converges to a limit ρ, $0 \leqq \rho \leqq \infty$. Then π^{k_n,h_n} converges to a limit π which is Brownian motion*

$$\pi(s, x, t, dy) = [2\pi(t - s)\sigma^2]^{-1/2} \exp\left\{-\frac{|y - x|^2}{2\sigma^2(t - s)}\right\}, \tag{7.4}$$

where $\sigma^2 = \sigma^2(\rho)$ is a continuous function of ρ with $\sigma^2(0) = \frac{1}{2}(\alpha + \beta)$ and $\sigma^2(\infty) = 2\alpha\beta/(\alpha + \beta)$.

PROOF. There are always convergent subsequences because of compactness. The limits are uniform in s, x, and t, so that the periodicity of $\pi^{k,h}$ leads to the invariance of the limit π, with respect to space and time translations. Therefore, any possible limit π is Brownian motion. We only have to compute σ^2.

Let us denote by π_n the transition probability π^{k_n,h_n}, by a_n the corresponding coefficient and by P_n the process starting from 0 at time 0. Let us define

$$\sigma_n^2 = E^{P_n}|x(1)|^2 = E^{P_n} \int_0^1 a_n(s, x(s))\, ds. \tag{7.5}$$

The computation of σ_n^2 really involves an idea of how much time, on the average, the process spends in the regions where $a = \alpha$ and $a = \beta$. We shall suppose that N is such that $2Nk_n = 1$. This involves at most an error of magnitude k_n in the computation of σ_n^2. Instead of considering the process $x(s)$, let us consider the process

$$y(s) = h_n^{-1} x(k_n s). \tag{7.6}$$

We denote by Q_n the measure corresponding to it. The generator of the $y(s)$ process can be written as

$$\frac{1}{2}\,\rho_n a(t, x) \frac{\partial^2}{\partial x^2}, \tag{7.7}$$

where $p_n = k_n h_n^{-1}$ and $a(t, x) = a^{1,1}(t, x)$. Now for the y process it is a question of how much time is spent in regions $\{a = \alpha\}$ and $\{a = \beta\}$ relatively up to time $2N$. Since the problem is periodic, we consider the reduced problem on the circle. A^+ and A^- are the upper and lower semicircles. π^+ and π^- are the transition probabilities for the homogeneous processes corresponding to $a^+(x)$ and $a^-(x)$, where

$$a^+(x) = \begin{cases} \alpha & \text{on } A^+, \\ \beta & \text{on } A^-, \end{cases} \tag{7.8}$$

and

$$a^-(x) = \begin{cases} \beta & \text{on } A^+, \\ \alpha & \text{on } A^-. \end{cases} \tag{7.9}$$

If $\pi^{\pm}(t)$ denotes the semigroup corresponding to $a^{\pm}$, then the ρ_n in front of the generator changes $\pi^{\pm}(t)$ to $\pi^{\pm}(t\rho_n)$. We can now write

$$\begin{aligned}
(7.10)\qquad \sigma_n^2 &= E^{P_n}\left[\int_0^1 a_n(s, x(s))\,ds\right] \\
&= k_n E^{Q_n}\left[\int_0^{2N} a(s, y(s))\,ds\right] \\
&= k_n \sum_{r=1}^{N} E^{Q_n} \int_{(2r-1)}^{2r} a(s, y(s))\,ds + k_n \sum_{r=0}^{N-1} E^{Q_n} \int_{2r}^{(2r+1)} a(s, y(s))\,ds \\
&= k_n \sum_{r=1}^{N} E^{Q_n} \int_{(2r-1)}^{2r} a^-(y(s))\,ds + k_n \sum_{r=0}^{N-1} E^{Q_n} \int_{2r}^{(2r+1)} a^+(y(s))\,ds.
\end{aligned}$$

Since k_n is nearly $(2N)^{-1}$, it suffices to look at $E^{Q_n} \int_{2r}^{(2r+1)} a^+(y(s))\,ds$ and $E^{Q_n} \int_{(2r-1)}^{2r} a^-(y(s))\,ds$ for large r and n.

We have

$$(7.11)\qquad E^{Q_n} \int_{2r}^{(2r+1)} a^+(y(s))\,ds = [\pi^+(\rho_n)\pi^-(\rho_n)]^r \left(\int_0^1 \pi^+(s\rho_n)a^+\,ds\right)$$

and

$$(7.12)\qquad E^{Q_n} \int_{(2r-1)}^{2r} a^-(y(s))\,ds = [\pi^-(\rho_n)\pi^+(\rho_n)]^{r-1} \left(\int_0^1 \pi^-(s\rho_n)a^-\,ds\right).$$

As ρ_n tends to ρ and $n \to \infty$, it is clear that we have to look at the invariant measures μ_ρ^+ and μ_ρ^- solving $\mu_\rho^+\pi^+(\rho)\pi^-(\rho) = \mu_\rho^+$ and $\mu_\rho^-\pi^-(\rho)\pi^+(\rho) = \mu_\rho^-$. We can then write

$$(7.13)\qquad \lim_{n\to\infty} \sigma_n^2 = \sigma^2 = \frac{1}{2}\left[\int_0^1 \mu_\rho^+\pi^+(s\rho)a^+\,ds + \int_0^1 \mu_\rho^-\pi^-(s\rho)a^-\,ds\right],$$

where σ^2 is of course only a function of ρ. By standard techniques, one can justify all the steps and even compute $\sigma^2(\rho)$ when $\rho = 0$ or ∞.

REFERENCES

[1] N. V. Krylov, "On the first boundary value problem for elliptic equations of second order," *Differentsial'nye Uravneniya*, Vol. 3 (1967), pp. 315–326. (In Russian.)

[2] D. W. Stroock and S. R. S. Varadhan, "Diffusion processes with continuous coefficients I," *Comm. Pure Appl. Math.*, Vol. 22 (1969), pp. 345–400.

[3] ———, "Diffusion processes with continuous coefficients II," *Comm. Pure Appl. Math.*, Vol. 22 (1969), pp. 479–530.

[4] ———, "Diffusion processes with boundary conditions," *Comm. Pure Appl. Math.*, Vol. 24 (1971), pp. 147–225.

RANDOM FIELDS OF SEGMENTS AND RANDOM MOSAICS ON A PLANE

R. V. AMBARTSUMIAN

Mathematics and Mechanics Institute

Armenian Academy of Sciences

1. Introduction

The position of an undirected segment of straight line of length τ in a Euclidean plane is determined by the triple coordinate $X = (x, y, \varphi)$, where x and y are the cartesian coordinates of the center of the segment and φ is the angle made by the segment with the zero direction. Let $\mathscr{X}$ denote the phase space of segment coordinates, that is, the layer in the three dimensional Euclidean space defined by the inequalities $-\infty < x < \infty$, $-\infty < y < \infty$, $0 < \varphi < \pi$.

Let $\mathscr{X}$ be a subset of the phase space $\mathscr{X}$ and let τ be a positive real function defined for $X \in \mathscr{X}$. Assign a length $\tau(X)$ to the segment which occupies the position X. This defines a certain set J of segments in the plane. We shall write $J = [\mathscr{X}; \tau(X)]$.

Call **I** the set of all those J for which the number of segments which intersects every bounded subset of the plane is finite. Moreover, if $J \in \mathbf{I}$, then, by definition, any two segments of J either do not intersect or they intersect at a single point.

Take a Borel set B in the phase space and a Borel set $T \subset (0, \infty)$. Each such pair (B, T) defines a subset of **I**, namely, the set of those $J = [\mathscr{X}; \tau(X)]$ such that $\mathscr{X} \cap B$ contains exactly one point and such that $X_0 \in \mathscr{X} \cap B$ implies $\tau(X_0) \in T$. Also, for each B, consider the subset of **I** formed by those J such that $\mathscr{X} \cap B = \varnothing$. The sets just introduced will be called cylindrical subsets of **I**.

Let $\mathscr{B}$ denote the minimal σ-algebra generated by the cylindrical subsets of **I** and let $(\Omega, \mathscr{A}, \mu)$ be a probability space.

Definition 1. *A $(\mathscr{B}, \mathscr{A})$ measurable map $\omega \rightsquigarrow J(\omega)\Omega$ of Ω into* **I** *is called a random field of segments (r.f.s.) in the plane.*

If an r.f.s. $J(\omega)$ is given, then a probability measure P will be induced in $\mathscr{B}$, which we shall call the distribution of the r.f.s. $J(\omega)$.

The group of all Euclidean motions of a plane induces a group of transformations of $\mathscr{B}$ into itself (the group of motions of $\mathscr{B}$). An r.f.s. is called homogeneous and isotropic (h.i.r.f.s.), if its distribution is invariant with respect to the group of motions of $\mathscr{B}$. Only homogeneous and isotropic random fields of segments are examined herein.

An r.f.s. $J(\omega)$ is called a random mosaic if, with probability 1, $J(\omega)$ generates a mosaic, that is, a partition of the plane into convex, bounded polygons.

Let L be a fixed line on the plane and let $J(\omega)$ be an h.i.r.f.s. Let $\{\mathscr{P}_i\}$ be the set of points $J(\omega) \cap L$ and let ψ_i be the angle of intersection at the point $\mathscr{P}_i$. The set of pairs $\{\mathscr{P}_i, \psi_i\}$ defines on L a random labeled sequence of points (r.l.s.p.) in the sense of K. Matthes [1].

We call a star the common point of two or more closed segments which intersect at nonzero angles, taken together with the directions of the segments issuing therefrom.

The definition given above for the r.l.s.p. on the line L is acceptable since it is easy to show that if $J(\omega)$ is an h.i.r.f.s., then with probability 1 no stars enter into $J(\omega) \cap L$. With probability 1, none of the free ends of the segments comprising $J(\omega)$ lies on L.

It is also easy to see that in case $J(\omega)$ is an h.i.r.f.s., the distribution of the r.l.s.p. $\{\mathscr{P}_i, \psi_i\}$ is independent of the selection of line L, and possesses the property of stationarity.

Let us introduce the probability

$$P(\overline{d\tau}, \overline{d\psi}), \qquad \overline{d\tau} = d\tau_1, \cdots, d\tau_m, \quad \overline{d\psi} = d\psi_1, \cdots, d\psi_m, \tag{1.1}$$

that there will be just one intersection in the r.l.s.p. $\{\mathscr{P}_i, \psi_i\}$ in nonintersecting intervals of length $d\tau_i$, placed arbitrarily on the line L, and that the angle of the intersection occurring in $d\tau_i$ will lie in some interval of the opening $d\psi_i$, $i = 1, \cdots, m$.

Let us present several examples of a homogeneous and isotropic random field of segments (h.i.r.f.s.).

EXAMPLE 1. As $\mathscr{X}(\omega)$, let us select a random point field in $\mathscr{X}$ which is a contraction of a homogeneous random field of points having a Poisson distribution with parameter λ in the whole space (x, y, φ). Let us put $\tau_\omega(X) = a$, where a is a constant. Such an h.i.r.f.s. is called a field of segments of length a, scattered independently in the plane.

EXAMPLE 2. Let $\mathscr{F}$ be some figure formed by k segments. Let us select a segment from components of $\mathscr{F}$ and let us call it the leader. For simplicity, we assume that $\mathscr{F}$ is symmetric relative to the leading segment. To a fixed countable set of leading segments on the plane corresponds a well-defined arrangement of figures congruent to $\mathscr{F}$. If the set of leading segments arises from segments scattered independently over the plane, then the set of segments entering in the corresponding set of congruent figures will be an h.i.r.f.s.

EXAMPLE 3. Let us give a line on the (x, y) plane by the equation

$$x \cos \varphi + y \sin \varphi = p. \tag{1.2}$$

The coordinates (φ, p) of all possible lines fill a strip $0 < \varphi < \pi$, $p > 0$.

In this strip let $\mathscr{X}'(\omega)$ be a random field of points which is a contraction of a homogeneous Poisson random field of points in the whole (φ, p) plane. A set of lines $\mathscr{L}$ in the (x, y) plane corresponds to each set of points $\mathscr{X}'$ in the strip.

Therefore, a random set (field) of lines $\mathscr{L}(\omega)$ on the plane corresponds to $\mathscr{X}'(\omega)$. Each set of lines which does not contain parallels can be considered as a set J of segments; namely, we can consider each line $L \in \mathscr{L}$ as the union of segments into which L is divided by other lines belonging to $\mathscr{L}$. We thus obtain an h.i.r.f.s. $J(\omega)$ which is a random mosaic. We call this random mosaic the simplest.

EXAMPLE 4. Let us select a number p, $0 < p < 1$. Remove or retain each of the segments of the simplest random mosaic according to independent trials, with probabilities p and $1 - p$ of the outcomes. The set of segments which are retained forms an h.i.r.f.s.

EXAMPLE 5. Let us select a number $\varepsilon > 0$. Let us delete all segments of length less than ε in the simplest random mosaic, and let us shorten the remaining segments by $\frac{1}{2}\varepsilon$ at both ends. The set of segments retained and shortened evidently forms an h.i.r.f.s.

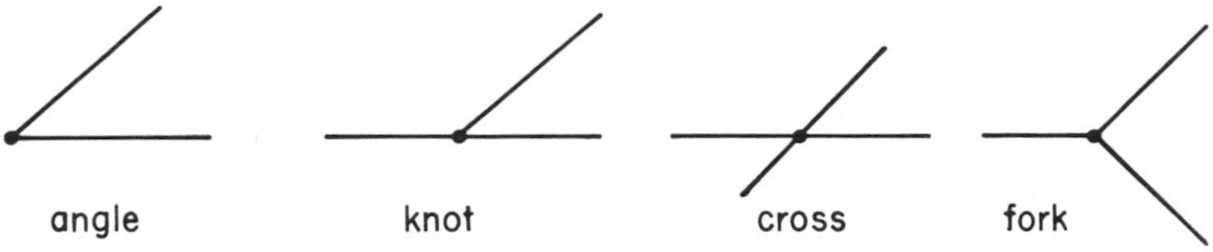

FIGURE 1

Considering the examples presented, we see that the r.f.s. of Example 4 has, with probability 1, stars of only the first three out of the four so-called simple kinds of stars (see Figure 1). With probability 1 the r.f.s. of Examples 1 and 3 only have stars of the cross kind, and Example 5 has no stars with probability 1. As concerns Example 2, by selecting the figure $\mathscr{F}$ in a suitable manner, an r.f.s. can be obtained with stars of any "complex" kind.

It is also easy to establish that for each r.f.s. of Examples 1, 3, 4, and 5 the r.l.s.p. $\{\mathscr{P}_i, \psi_i\}$ has a distribution of the form

$$P(\overline{d\tau}, \overline{d\psi}) = \prod_{i=1}^{m} \lambda d\tau_i \cdot \prod_{i=1}^{m} \tfrac{1}{2} \sin \psi_i \, d\psi_i. \tag{1.3}$$

In other words, for the r.f.s. $J(\omega)$ from Examples 1, 3, 4, and 5, the sequence of points of intersection $\{\mathscr{P}_i\}$ with any line L has a Poisson distribution, the sequence of angles of intersection $\{\psi_i\}$ is independent of the sequence of points of intersection $\{\mathscr{P}_i\}$, and the angles in the sequence $\{\psi_i\}$ are independent. The particular form of the density $\frac{1}{2} \sin \psi$ results from the homogeneity and isotropy of $J(\omega)$.

The present research was undertaken in order to clarify which kinds of stars are generally possible for h.i.r.f.s. for which the distribution $\{\mathscr{P}_i, \psi_i\}$ of the random labeled sequence of points (r.l.s.p.) of intersection with a line has the form (1.3), which is simplest in some sense. It is shown in Section 2 that compliance with (1.3) actually imposes strong constraints on the possible kinds of stars in the class of so-called regular h.i.r.f.s. Namely, it has been established that it follows from (1.3) that stars of a regular h.i.r.f.s. can only be of the four simple kinds (Figure 1) or missing entirely. Other kinds of stars are excluded by condition (1.3). However, let us note that this result cannot be considered final, since we have no examples of h.i.r.f.s. for which (1.3) is satisfied and which possess stars of the fork kind with positive probability.

Indeed, the result mentioned follows from weaker assumptions than (1.3) relative to the r.l.s.p. $\{\mathscr{P}_i, \psi_i\}$. Thus, according to Lemma 1, for this it is sufficient to demand that the distribution of points of intersection on L be a Poisson distribution (independent of the distribution of the sequence of angles $\{\psi_i\}$). Theorem 2 asserts that the same follows just from the fact of factorization

$$P(\overline{d\tau}, \overline{d\psi}) = p(\overline{d\tau}) \cdot \Lambda(\overline{d\psi}) \tag{1.4}$$

without any particular assumptions relative to the form of p and Λ.

Restriction to the class of regular h.i.r.f.s. is the basic hypothesis for the validity of the whole theory.

DEFINITION 2. *An h.i.r.f.s. $J(\omega)$ is called regular if the following two conditions are satisfied.*

(i) *There exists a number $\varepsilon > 0$ such that with probability 1, $J(\omega)$ will not contain parallel segments not lying on one line, and situated at a distance less than ε.*

(ii) *Let $\{\beta_i\}$ be the set of angles between all the pairs of segments of $J(\omega)$ (for fixed ω), which have a nonempty intersection with the unit circle centered at the origin. If two segments lie on the same line, then the angle between them is assumed to be π. There exists a number $\varepsilon > 0$ such that the mathematical expectation of the random quantities $\Sigma_\varepsilon [1 + (\pi - \beta_i) \cot \beta_i]$ and N_ε is finite. Here Σ_ε denotes the sum extended over those pairs of segments whose distance does not exceed ε, and N_ε is the number of nonzero components in this sum.*

The distance between segments in (i) and (ii) is understood to be the distance between sets.

Regular homogeneous and isotropic random mosaics (h.i.r.m.) are examined in Section 3. In this case the stars are the vertices of the mosaics, whereupon the possibilities of vertices of angle type and the lack of vertices are at once excluded. The solution of the problem of the possible kinds of vertices for regular h.i.r.m. satisfying (1.3) is given somewhat later. It excludes vertices of fork type. Let us note that this result remains true for some conditions much weaker than (1.3) on the distribution of the r.l.s.p. $\{\mathscr{P}_i, \psi_i\}$. It is sufficient to require compliance with

$$P(\overline{d\tau}, \overline{d\psi}) = p(\overline{d\tau}) \cdot \Pi \tfrac{1}{2} \sin \psi_i \, d\psi_i \tag{1.5}$$

without further specification of the form of p.

Combining this with known results from the theory of random fields of lines on a plane, one obtains the following uniqueness theorem.

THEOREM 1. *Let $J(\omega)$ be a regular, homogeneous, isotropic random mosaic which, with probability unity, has no nodes of knot type. Suppose that intersecting*

$J(\omega)$ with a straight line yields a random labeled sequence of intersections, where the sequence of angles $\{\psi_i\}$ is independent of the sequence of intersection points $\{\mathscr{P}_i\}$. Assume also that the angles $\{\psi_i\}$ are mutually independent. Then $J(\omega)$ is a mixture of the simplest mosaics.

By mixture of simplest random mosaics is understood the random mosaic of Example 3, where the parameter λ of the Poisson field of points in the strip $0 < \varphi < \pi$, $p > 0$, is itself a random quantity.

2. Random fields of segments

LEMMA 1. *Let $J(\omega)$ be a regular, homogeneous and isotropic field of segments. Let $p_k(\tau)$ be the probability that exactly k intersections with segments belonging to $J(\omega)$ will occur in a segment of length τ fixed in the plane. Then for each $k \geqq 2$ the limit $\lim_{\tau\to 0} p_k(\tau)/\tau^2 < \infty$ exists. The equality $\lim_{\tau \to 0} p_3(\tau)/\tau^2 = 0$ is the necessary and sufficient condition that, with probability 1, there will be no stars or only stars of the simplest kinds in $J(\omega)$.*

PROOF. Let dX denote an element of kinematic measure, $dX = dx\, dy\, d\varphi$. Let us introduce the function $\delta_k(X, \tau, \omega)$ which takes value unity if the segment of length τ and position X has exactly k intersections with $J(\omega)$ and takes value zero otherwise.

Let g be a star on the plane. Consider temporarily that the rays issuing from g are infinite. Let $\alpha_1, \cdots, \alpha_n$ be the successive angles between rays numbered serially in the counterclockwise direction around g (Figure 2). Let Z_i be the set

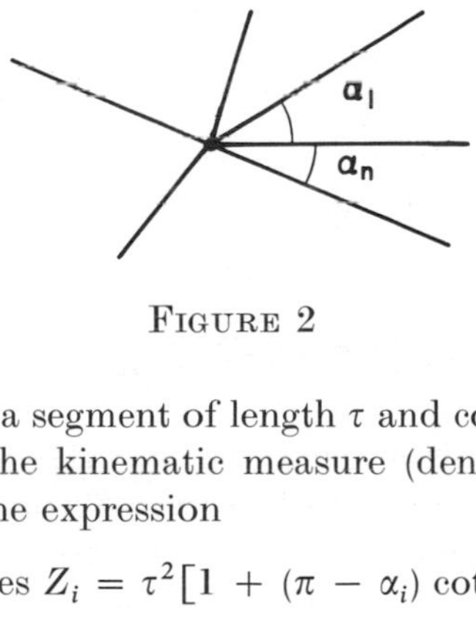

FIGURE 2

of positions X such that a segment of length τ and coordinate X intersects both sides of the angle α_i. The kinematic measure (denoted mes) of Z_i is known (see [2]) and given by the expression

$$\text{mes } Z_i = \tau^2[1 + (\pi - \alpha_i) \cot \alpha_i]. \tag{2.1}$$

For $k \geqq 2$, the set of positions of a moving segment of length τ for which exactly k intersections with rays issuing from g occur, is the set $Y_k = \{X;$ exactly $k - 1$ events among the Z_i, $i = 1, \cdots, n$, occur$\}$.

According to a known combinatorial formula,

$$\text{mes } Y_k = \sum_i (-1)^i \binom{k+i}{i} S_i, \tag{2.2}$$

where

$$S_i = \Sigma \text{ mes } Z_{k_1} Z_{k_2} \cdots Z_{k_i}, \tag{2.3}$$

the summation extending to all subsets of size i of the set $\{1, 2, \cdots, n\}$. Since each of the sets of the form $Z_{k_1} Z_{k_2} \cdots Z_{k_i}$ is evidently always a set of X for which a segment of length τ intersects two sides of some angle, the kinematic measure of these sets has the form (2.1), that is,

$$\text{mes } Z_{k_1} Z_{k_2} \cdots Z_{k_i} = c \cdot \tau^2, \qquad c \geqq 0. \tag{2.4}$$

We thus arrive at the deduction that

$$\text{mes } Y_k = C_k(g) \cdot \tau^2, \tag{2.5}$$

where $C_k(g) \geqq 0$ is independent of τ.

Let K be a circle of radius $\frac{1}{2}$ centered at the origin and let us fix $\omega \in \Omega$ such that there are no stars from $J(\omega)$ on the boundary of K. From the homogeneity and isotropy of the r.f.s. $J(\omega)$, it follows that the measure of the set of such ω equals 1. For all $\tau < \tau(\omega)$, where $\tau(\omega)$ depends only on the arrangement of the segments of $J(\omega)$ in the unit circle, we will have ($k \geqq 2$)

$$\iiint_{\Delta} \delta_k(X; \tau; \omega)\, dX = \tau^2 \sum_{g \in K} C_k(g), \tag{2.6}$$

where $\Delta = \{X$; the center of a segment with coordinate X belongs to the circle $K\}$, and the summation on the right is over all nodes of the lattice belonging to K. Therefore, we have for almost every $\omega \in \Omega$ and $k \geqq 2$

$$\lim_{\tau \to 0} \frac{1}{\tau^2} \iiint_{\Delta} \delta_k(X; \tau; \omega)\, dX = \sum_{g \in K} C_k(g). \tag{2.7}$$

At the same time it is easy to see that for all $k \geqq 2$ and $\tau < \varepsilon$ (see the definition of a regular h.i.r.f.s.),

$$\iiint_{\Delta} \delta_k(X; \tau; \omega)\, dX \leqq \sum_{\varepsilon} M_{ij}(\tau), \tag{2.8}$$

where $M_{ij}(\tau)$ is the kinematic measure of a set of segments of length τ, which yield intersections with the ith and jth segments from $J(\omega)$. The summation is over all pairs of segments from $J(\omega)$, which intersect the unit circle and are separated by a distance smaller than ε.

Let β denote the angle between these two segments, then we obtain

$$M_{ij}(\tau) \leqq \tau^2[1 + (\pi - \beta) \cot \beta]. \tag{2.9}$$

From (2.8) and (2.9), we obtain

$$\frac{1}{\tau^2} \iiint_{\Delta} \delta_k(X; \tau; \omega)\, dX \leqq \Sigma\, [1 + (\pi - \beta_i) \cot \beta_i]. \tag{2.10}$$

Since we have assumed that a summable function of ω is on the right side in (2.10), then by integrating (2.7) with respect to the measure μ, we obtain by means of the Lebesgue theorem on the passage to the limit under the integral sign

$$\lim_{\tau \to 0} \frac{1}{\tau^2} \mathscr{E} \iiint \delta_k(X; \tau; \omega)\, dX = \mathscr{E} \sum_{g \in K} C_k(g). \tag{2.11}$$

We denote integration with respect to the measure μ in Ω by the expectation symbol $\mathscr{E}$.

Because of the homogeneity and isotropy of the r.f.s. $J(\omega)$, one has $\mathscr{E}\delta_k(X; \tau; \omega) = p_k(\tau)$ identically in X, so that by Fubini's theorem it follows from (2.11) that

$$\text{mes}\, \Delta \cdot \lim_{\tau \to 0} \frac{1}{\tau^2} p_k(\tau) = \mathscr{E} \sum_{g \in K} C_k(g). \tag{2.12}$$

It is clear from (2.12) that $\lim_{\tau \to 0} p_3(\tau)/\tau^2 = 0$ if and only if the sum $\Sigma_{g \in K} C_3(g)$ vanishes with probability 1. This can occur when there are no stars in the domain K with probability 1; then it follows from the homogeneity and isotropy of $J(\omega)$ that with probability 1 there are no stars in the whole plane. If it is assumed that with positive probability there are stars in K, then there remains to assume that for each star $g \in K$ we have $C_3(g) = 0$.

However, it is easy to verify that $C_3(g) = 0$ if and only if g is a simple star. If with probability 1 there are only simple stars in K, then because of the homogeneity and isotropy of $J(\omega)$ the same is true for the whole plane. This completes the proof of Lemma 1.

Let W_τ denote the event that on a segment of length τ fixed on a plane there are exactly three intersections with segments of the r.f.s. $J(\omega)$, and that the sides intersect the segment τ at angles such that their continuations converge at a point.

LEMMA 2. *Let $J(\omega)$ be a regular, homogeneous, and isotropic random field of segments. If nonsimple stars are encountered with positive probability in $J(\omega)$, then*

$$\lim_{\tau \to 0} \frac{P(W_\tau)}{p_3(\tau)} = 1. \tag{2.13}$$

PROOF. Let us introduce the function that $\delta(X; \tau; \omega)$ equals one if the event W_τ holds for a segment of length τ with coordinate X, and $\delta(X; \tau; \omega)$ equals zero otherwise. Let us fix $\omega \in \Omega$. There is a number $\tau(\omega)$ such that for all $\tau < \tau(\omega)$ we have

$$\delta[X; \tau; \omega] = \delta_3(X; \tau; \omega) \quad \text{when } X \in \Delta, \tag{2.14}$$

and therefore, for all $\tau < \tau(\omega)$

$$\iiint_\Delta \delta(X; \tau; \omega)\, dX = \iiint_\Delta \delta_3(X; \tau; \omega)\, dX \tag{2.15}$$

from which we conclude that for each $\omega \in \Omega$ there exists a limit equal to

$$\lim_{\tau \to 0} \frac{1}{\tau^2} \iiint_{\Delta} \delta(X; \tau; \omega)\, dX = \lim_{\tau \to 0} \frac{1}{\tau^2} \iiint_{\Delta} \delta_3(X; \tau; \omega)\, dX. \tag{2.16}$$

Let us integrate (2.16) with respect to the measure μ in Ω. Since $J(\omega)$ is a regular h.i.r.f.s., the order of integration with respect to μ and the passage to the limit can be interchanged in both integrals. After reduction by mes Δ, and using Fubini's theorem, this yields

$$\lim_{\tau \to 0} \frac{1}{\tau^2} P(W_\tau) = \lim_{\tau \to 0} \frac{1}{\tau^2} p_3(\tau). \tag{2.17}$$

We have utilized the facts that $\mathscr{E}\delta(X; \tau; \omega)$ is independent of X, and

$$\mathscr{E}\delta(X; \tau; \omega) = P(W_\tau). \tag{2.18}$$

We have therefore found that there exists a finite limit

$$\lim_{\tau \to 0} \frac{P(W_\tau)}{p_3(\tau)} \cdot \frac{p_3(\tau)}{\tau^2} = \lim_{\tau \to 0} \frac{p_3(\tau)}{\tau^2} < \infty. \tag{2.19}$$

It follows that if $\lim_{\tau \to 0} p_3(\tau)/\tau^2 > 0$, then $\lim_{\tau \to 0} P(W_\tau)/p_3(\tau) = 1$, which together with Lemma 1 yields the proof of Lemma 2.

Let $x_1, x_2, x_3, \psi_1, \psi_2, \psi_3$, $(x_1 < x_2 < x_3)$ be the abscissas and the corresponding angles of three points of intersection of a segment of length τ fixed on a plane with segments from $J(\omega)$, under the condition that there are just three intersections on the fixed segment. Let us select the direction on the segment so that the inequality $x_2 - x_1 < x_3 - x_2$ will always be satisfied.

Let us put $u = (x_2 - x_1)/(x_3 - x_1)$. Let $F_\tau(u; \psi_1, \psi_2, \psi_3)$ be the joint distribution of the random variables $u, \psi_1, \psi_2, \psi_3$. Then evidently

$$\frac{P(W_\tau)}{p_3(\tau)} = \iiint_{V_\tau} dF_\tau(u; \psi_1, \psi_2, \psi_3), \tag{2.20}$$

where the integration is over that portion of the four dimensional space $u, \psi_1, \psi_2, \psi_3$ corresponding to the event W_τ.

The factorization

$$dF_\tau(u; \psi_1, \psi_2, \psi_3) = dF_\tau(u)\, d\Phi(\psi_1, \psi_2, \psi_3) \tag{2.21}$$

corresponds to the hypothesis of an independent sequence of angles. In this case we have,

$$\frac{P(W_\tau)}{p_3(\tau)} = \int_0^{1/2} dF_\tau(u) \int_{V(u)} d\Phi(\psi_1, \psi_2, \psi_3), \tag{2.22}$$

where $V(u)$ is some surface in the space (ψ_1, ψ_2, ψ_3). It is now natural to seek the limit $(\tau \to 0)$ of (2.22).

LEMMA 3. *Let $J(\omega)$ be a regular, homogeneous, and isotropic random field of segments possessing, with positive probability, nonsimple stars, and let $F_\tau(u)$ be a marginal distribution function of the random variable u. Then, F_τ converges weakly to some absolutely continuous distribution function $F(u)$ as $\tau \to 0$.*

PROOF. Let us consider a star g, which is not simple, and let us temporarily assume that rays issuing from g are continued infinitely. For those positions X of segments of length τ for which just three intersections with the rays issuing from g occur (we denote the set of such positions by $S_3(\tau; g)$), the ratio $u = (x_2 - x_1)/(x_3 - x_1)$ is defined. If it is considered that the coordinate X is distributed uniformly within $S_3(\tau; g)$, then u has some distribution function $F_g(u)$, where as follows from similarity considerations, $F_g(u)$ is independent of τ, and, as is easy to verify, is absolutely continuous.

Let us fix $\omega \in \Omega$, thereby, fixing the field of segments $J(\omega)$. Let us put

$$S_3(\tau; \omega) = \{X; \delta_3(X; \tau; \omega) \neq 0, X \in \Delta\}. \tag{2.23}$$

For $X \in S_3(\tau; \omega)$ the ratio $u = (x_2 - x_1)/(x_3 - x_1)$ is also defined, and if it is considered that X is distributed uniformly within $S_3(\tau; \omega)$, then u acquires a distribution function $F_\omega(u; \tau)$, which for sufficiently small τ satisfies the relation

$$F_\omega(u; \tau) = \frac{1}{\sum_{g \in K} C_3(g)} \sum_{g \in K} C_3(g) F_g(u). \tag{2.24}$$

Obviously, the right side is an absolutely continuous distribution function which for all $\omega \in \Omega$ equals $\lim_{\tau \to 0} F_\omega(u; \tau) = F_\omega(u)$.

In other words, we have shown that for each $\omega \in \Omega$, there exists limits

$$\begin{aligned} \lim_{\tau \to 0} \frac{1}{\tau^2} \iiint_\Delta \delta_3(u; \tau; \omega; X)\, dX \\ = F_\omega(u) \lim_{\tau \to 0} \frac{1}{\tau^2} \iiint_\Delta \delta_3(X; \tau; \omega)\, dX, \end{aligned} \tag{2.25}$$

where

$$\delta[u; \tau; \omega; X] = \begin{cases} 1 & \text{if } X \in S_3(\tau; \omega) \text{ and } (x_2 - x_1)/(x_3 - x_1) < u, \\ 0 & \text{otherwise.} \end{cases} \tag{2.26}$$

Integrating (2.25) with respect to the measure μ in the space Ω and taking into account that

$$\mathscr{E} \delta_3(u; \tau; \omega; X) - F_\tau(u) p_3(\tau), \tag{2.27}$$

we obtain, analogously to the proof of Lemma 1, that

$$\lim_{\tau \to 0} F_\tau(u) = \mathscr{E} \frac{1}{\sum_{g \in K} C_3(g)} \sum_{g \in K} C_3(g) F_g(u). \tag{2.28}$$

Since the right side is obviously absolutely continuous, Lemma 3 is proved.

THEOREM 2. *Let $J(\omega)$ be a regular, homogeneous, and isotropic random field* *of segments on a plane. If the sequences* $\{\mathscr{P}_i\}$ *and* $\{\psi_i\}$ *are independent in a random labeled point sequence* $\{\mathscr{P}_i, \psi_i\}$, *then the stars belonging to* $J(\omega)$ *can only be simple.*

PROOF. Let us assume on the contrary that with positive probability $J(\omega)$ contains stars different from the simple ones. Then by Lemma 2

$$\lim_{\tau \to 0} \frac{P(W_\tau)}{p_3(\tau)} = 1. \tag{2.29}$$

At the same time, according to Lemma 3, the same limit can be represented as

$$1 = \int_0^{1/2} f(u)\, du \int_{V(u)} d\Phi(\psi_1, \psi_2, \psi_3). \tag{2.30}$$

Hence, it follows that the functions $\int_{V(u)} d\Phi(\psi_1, \psi_2, \psi_3)$ equal one for almost all u for which $f(u) > 0$. But this is impossible since there are several such points u and since the surfaces $V(u)$ do not intersect for different values of u. The contradiction obtained proves Theorem 2.

3. Random mosaics

Theorem 2 can also be considered as a result concerning homogeneous and isotropic random mosaics (h.i.r.m.). Replacing the words "fields of segments" by "mosaics" and "stars" by "vertices" in the formulation of Theorem 2, it should be taken into account that of the simple stars only knots, forks, and crosses can be vertices.

THEOREM 3. *Let* $J(\omega)$ *be a regular, homogeneous, and isotropic random mosaic. If the sequences* $\{\mathscr{P}_i\}$ *and* $\{\psi_i\}$ *in the random labeled point sequence* $\{\mathscr{P}_i, \psi_i\}$ *of intersections of* $J(\omega)$ *with a line fixed on a plane are independent, the angles* ψ_i *are mutually independent, then the vertices of* $J(\omega)$ *can only be of knot and cross type.*

Let us precede the proof of Theorem 3 with still another lemma.

LEMMA 4. *Let* $J(\omega)$ *be a regular, homogeneous, and isotropic random mosaic which, with probability* 1, *possesses only simple vertices, and* $\nu_1 (\nu_2, \nu_3)$ *the mean number of vertices of knot (fork, cross, respectively) type per unit area. Let a segment of length* τ *be fixed on a plane. Let us introduce the random variables:* $\delta_2(\tau)$ *equals one, if two intersections of the segment with* $J(\omega)$ *occur; and* $\delta_2(\tau)$ *equals zero otherwise. Let* ψ_i, $i = 1, 2$, *be the angles of intersection on the segment* τ *defined when* $\delta_2(\tau) = 1$. *Then*

$$\begin{aligned} &\pi \lim_{\tau \to 0} \frac{1}{\tau^2} \mathscr{E}\delta_2(\tau)[1 + (\pi - |\psi_1 - \psi_2|) \cot |\psi_1 - \psi_2|]^{-1} \\ &\qquad = 2\nu_1 + 3\nu_2 + 4\nu_3, \\ &\pi \lim_{\tau \to 0} \frac{1}{\tau^2} \mathscr{E}\delta_2(\tau)|\psi_1 - \psi_2|[1 + (\pi - |\psi_1 - \psi_2|) \cot |\psi_1 - \psi_2|]^{-1} \\ &\qquad = \pi(\nu_1 + 2\nu_2 + 2\nu_3) \end{aligned} \tag{3.1}$$

where $\mathscr{E}$ *denotes the mathematical expectation.*

PROOF. Let us fix $\omega \in \Omega$. Let $M_1(\omega)(M_2, M_3)$ denote the number of vertices of the type knot (fork, cross, respectively) in the circle K.

Let a_i and a_j be two sides of the mosaic $J(\omega)$ issuing from the same vertex. Let us put $\delta(X; \tau; a_i, a_j)$ equal to one if a segment of length τ with coordinate X intersects a_i and a_j, and put $\delta(X; \tau; a_i, a_j)$ equal to zero otherwise.

We have

$$\delta_2(X; \tau; \omega) = \sum_{i<j} \delta(X; \tau; a_i; a_j). \tag{3.2}$$

It follows from (2.1) that for all sufficiently small values of τ

$$\frac{1}{\tau^2} \iiint \frac{\delta(X; \tau; a_i; a_j)\, dX}{1 + (\pi - |\psi_1 - \psi_2|) \cot |\psi_1 - \psi_2|} = 1 \tag{3.3}$$

and

$$\frac{1}{\tau^2} \iiint \frac{|\psi_1 - \psi_2| \delta(X; \tau; a_i; a_j)\, dX}{1 + (\pi - |\psi_1 - \psi_2|) \cot |\psi_1 - \psi_2|} = \alpha_{ij}, \tag{3.4}$$

where ψ_1 and ψ_2 are angles of intersection of a segment of length τ with coordinate X and sides in $J(\omega)$. Indeed, for all X for which $\delta(X; \tau; a_i, a_j) = 1$, the quantity $|\psi_1 - \psi_2|$ is constant and equals the angle α_{ij} between the sides $a_1, a_j \in J(\omega)$. By summation, we obtain from (3.3) and (3.4) that for all sufficiently small values of τ (ω is fixed)

$$\frac{1}{\tau^2} \iiint_{\Delta} \frac{\delta_2(X; \tau; \omega)\, dX}{1 + (\pi - |\psi_1 - \psi_2|) \cot |\psi_1 - \psi_2|} = 2M_1(\omega) + 3M_2(\omega) + 4M_3(\omega) \tag{3.5}$$

and

$$\frac{1}{\tau^2} \iiint_{\Delta} \frac{|\psi_1 - \psi_2| \delta_2(X; \tau; \omega)\, dX}{1 + (\pi - |\psi_1 - \psi_2|) \cot |\psi_1 - \psi_2|} = \pi M_1(\omega) + 2\pi M_2(\omega) + 2\pi M_3(\omega). \tag{3.6}$$

To clarify matters, let us note that the right side of the second equality represents the sum of the angles for those vertices of the mosaic $J(\omega)$ which lie in K. Indeed, the sum of these angles is equal to π for a vertex of knot type while it is equal to 2π for forks or cross type vertices.

For almost all $\omega \in \Omega$ the limit forms of (3.5) and (3.6) are

$$\lim_{\tau \to 0} \frac{1}{\tau^2} \iiint_{\Delta} \frac{\delta_2(X; \tau; \omega)\, dX}{1 + (\pi - |\psi_1 - \psi_2|) \cot |\psi_1 - \psi_2|} = 2M_1(\omega) + 3M_2(\omega) + 4M_3(\omega) \tag{3.7}$$

and

$$\text{(3.8)} \qquad \lim_{\tau \to 0} \frac{1}{\tau^2} \iiint\limits_{\Delta} \frac{|\psi_1 - \psi_2| \delta_2(X; \tau; \omega)\, dX}{1 + (\pi - |\psi_1 - \psi_2|) \cot |\psi_1 - \psi_2|}$$
$$= \pi\big(M_1(\omega) + 2M_2(\omega) + 2M_3(\omega)\big)$$

and can be integrated with respect to the measure μ in Ω. Since we assume that the random mosaic $J(\omega)$ is regular, and the estimates

$$\text{(3.9)} \qquad \frac{1}{\tau^2} \iiint\limits_{\Delta} \frac{\delta_2(X; \tau; \omega)\, dX}{1 + (\pi - |\psi_1 - \psi_2|) \cot |\psi_1 - \psi_2|} < N_\varepsilon(\omega)$$

and

$$\text{(3.10)} \qquad \frac{1}{\tau^2} \iiint\limits_{\Delta} \frac{|\psi_1 - \psi_2| \delta_2(X; \tau; \omega)\, dX}{1 + (\pi - |\psi_1 - \psi_2|) \cot |\psi_1 - \psi_2|} < \pi N_\varepsilon(\omega)$$

are valid for all $\tau < \varepsilon$, then by first applying the Lebesgue theorem on the passage to the limit under the integral sign to the integrals with respect to the measure μ from the left sides of (3.7) and (3.8), and then Fubini's theorem, we obtain

$$\text{(3.11)} \qquad \text{mes}\, \Delta \lim_{\tau \to 0} \frac{1}{\tau^2} \mathscr{E} \delta_2(\tau)[1 + (\pi - |\psi_1 - \psi_2|) \cot |\psi_1 - \psi_2|]^{-1}$$
$$= 2\overline{M_1(\omega)} + 3\overline{M_2} + 4\overline{M_3}$$

and

$$\text{(3.12)} \qquad \text{mes}\, \Delta \lim_{\varepsilon \to 0} \frac{1}{\tau^2} \mathscr{E} \delta_2(\tau)|\psi_1 - \psi_2|[1 + (\pi - |\psi_1 - \psi_2|) \cot |\psi_1 - \psi_2]^{-1}$$
$$= \pi(\overline{M_1} + 2\overline{M_2} + 2\overline{M_3}).$$

The assertion of Lemma 4 is now obtained from the relationship $\nu_i = \mathscr{E} M_i(\omega)$ [area of $K]^{-1}$ since mes $\Delta = \pi \cdot$ area of K. This proves Lemma 4.

Now, if $J(\omega)$ is an r.h.i.r.m. for which the sequences $\{\mathscr{P}_i\}$ and $\{\psi_i\}$ are independent in the r.l.s.p. $\{\mathscr{P}_i, \psi_i\}$, then the result of Lemma 4 becomes

$$\text{(3.13)} \qquad \pi \lim_{\tau \to 0} \frac{p_2(\tau)}{\tau^2} \mathscr{E}[1 + (\pi - |\psi_1 - \psi_2|) \cot |\psi_1 - \psi_2|]^{-1}$$
$$= 2\nu_1 + 3\nu_2 + 4\nu_3,$$

and

$$\text{(3.14)} \qquad \pi \lim_{\tau \to 0} \frac{p_2(\tau)}{\tau^2} \mathscr{E}|\psi_1 - \psi_2|[1 + (\pi - |\psi_1 - \psi_2|) \cot |\psi_1 - \psi_2|]^{-1}$$
$$= \pi(\nu_1 + 2\nu_2 + 2\nu_3).$$

Let us note that the ratio of the mathematical expectations

$$(3.15)\qquad a = \frac{\mathscr{E}|\psi_1 - \psi_2|[1 + (\pi - |\psi_1 - \psi_2|)\cot|\psi_1 - \psi_2|]^{-1}}{\mathscr{E}[1 + (\pi - |\psi_1 - \psi_2|)\cot|\psi_1 - \psi_2|]},$$

to which the sense of mean value of the angle for an "arbitrary" vertex of the polygons comprising the mosaic can be ascribed in this case, is easily evaluated if it is assumed that the angles ψ_1 and ψ_2 are independent, that is, that (1.5) holds. For this it is sufficient to consider the simplest random mosaic for which the angles ψ_1 and ψ_2 are independent, as we have already remarked. Indeed, in this case we should put $v_1 = 0$, $v_2 = 0$, after which we find by dividing (3.13) by (3.14), $a = \frac{1}{2}\pi$. Therefore, if it is assumed that the conditions of Theorem 3 are satisfied, then v_1, v_2, and v_3 satisfy the relationships

$$(3.16)\qquad 2v_1 + 3v_2 + 4v_3 = A, \qquad \pi v_1 + 2\pi v_2 + 2\pi v_3 = \tfrac{1}{2}\pi A,$$

for some A. Multiplying the first equality by $\frac{1}{2}\pi$ and subtracting this result from the second, we find that $v_2 = 0$. Hence, the assertion of Theorem 3 evidently follows.

Finally, let us consider the conditions of Theorem 1. Since as follows from Theorem 3, the possibility (with probability 1) of the existence of vertices of cross type in $J(\omega)$ is all that remains, then we arrive at the conclusion that $J(\omega)$, with probability 1, is formed by (infinite) straight lines. Therefore, there is a stationary point process on the line L, and through each of its points $\{\mathscr{P}_i\}$ we draw a random straight line with orientation ψ_i such that the ψ_i are mutually independent for different values of i, are independent of the point process $\{\mathscr{P}_i\}$, and have the common density $\frac{1}{2}\sin\psi$. These random lines indeed form the random mosaic $J(\omega)$. As has been shown in [3] from the condition that on every other line L' parallel to L the distribution of the point process of intersection with the random lines agrees with that for $\{\mathscr{P}_i\}$ on L (homogeneity and isotropy conditions of $J(\omega)$), it results that $\{\mathscr{P}_i\}$ is a mixture of stationary Poisson point sequences, in other words, a Poisson point sequence of random intensity.

From this the assertion of Theorem 3 follows in an evident manner.

REFERENCES

[1] K. Matthes, "Stationäre zufällige punktfolgen. I," *Jber. Deutsch. Math.-Verein.*, Vol. 66 (1963), pp. 66–79.

[2] L. A. Santaló, *Introduction to Integral Geometry*, Paris, Herman, 1953.

[3] R. Davidson, "Construction of line processes. Second order properties," *Izv. Akad. Nauk. Armjan. SSR Ser. Mat.*, Vol. 5 (1970).

NONHOMOGENEOUS POISSON FIELDS OF RANDOM LINES WITH APPLICATIONS TO TRAFFIC FLOW

HERBERT SOLOMON[1]
STANFORD UNIVERSITY
and
PETER C. C. WANG[2]
NAVAL POSTGRADUATE SCHOOL, MONTEREY

1. Introduction

This study was prompted by investigations of models of traffic flow on a highway through analyses of the structure and properties of Poisson fields of random lines in a plane. It is possible to view the trajectory of a car produced by its time and space coordinates on the highway as a straight line in that plane if the car travels at a constant speed once it enters the highway and then never leaves the highway. These traffic considerations plus the property of time invariance for traffic flow distributions lead to one model for traffic flow on a divided highway developed by Rényi [10]. This idealized model is simpler to study than the more realistic situation that provided Rényi's motivation and which he also subjects to analysis, namely, cars do lose time because of an overtaking of one car by another even on a divided highway with two lanes for traffic moving in one direction.

In his paper, Rényi found it convenient to start from the stochastic process of entrance times of the cars at a fixed point on the highway. Other authors start from the spatial process of cars distributed in locations along the highway at some fixed time according to some random law. The traffic flow results of Weiss and Herman [13] who study the spatial process for the idealized model are analogous to Rényi's results which stem from the temporal process. To demonstrate the equivalence of the two results, care must be taken to employ the appropriate measure in deriving distributions related to traffic flow. Both Rényi, and Weiss and Herman, achieved asymptotic results for traffic flow distributions. We will reproduce both results in Section 4 as special cases of our development of traffic flow models through the structure of random lines in the plane. It should be mentioned here that Brown [3] reconsiders Rényi's idealized model and derives exact distributions rather than asymptotic distributions for spatial and speed distributions of cars.

[1]This research partially supported by Contract No. FH11-7698 U.S. Department of Transportation.
[2]This research partially supported by Contract No. NRO42-268 U.S. Office of Naval Research.

To this point we have not been specific about the random processes governing entrance times of cars on a highway, positions of cars on a highway, and speed distributions of cars on the highway. The Poisson process is the assumed machinery governing car entrance times or equivalently car positions and the speed distributions for each car are assumed to be identically and independently distributed (i.i.d.). Starting from the spatial process, Breiman [2] considered the idealized model and proved that the Poisson process is the only process obeying the time invariance property—namely if at a time t_0, the spatial process is Poisson with specific parameter, and the speeds of the cars are i.i.d. with respect to each other and the positions of the cars at time t_0, then the process will have the same properties at any other time t.

Obviously, other results for Poisson processes can be germane to traffic flow situations and similarly this can be so for results in queueing processes. There is a vast literature in both subjects. However, it is pertinent to this exposition to mention some results for the $M/G/\infty$ queue. The highway can be regarded as the infinite server for each car suffers no delay when it enters the highway in our model; in addition the input is Poisson and the service time distribution is the distribution of the distances traveled by each car before it overtakes or is overtaken by another car on the highway or equivalently the distribution of the time expended until an overtaking occurs. In a paper on Markov processes, Kingman [6] arrives at a general formulation that can be reduced to our idealized traffic flow model or equivalently the $M/G/\infty$ queue. This produces the result that the distribution of cars on the highway is Poisson with parameter

$$\omega \int_0^\infty \frac{1}{v}\, dG(v). \tag{1.1}$$

where ω is the parameter of the Poisson process for cars entering the highway and $G(v)$ is the cumulative distribution function for the speed of each car. This result is employed by Rényi in his paper where he cites other authors who have produced it. It also falls out of the development in this paper and appears in Theorem 4.2.

Along these lines there is a recent paper by Brown [4] in which he discusses an estimation procedure for G in the $M/G/\infty$ queue for which data are kept only on the times cars enter or leave the highway without identification of cars (that is, no pairing of entering and departing times for any one car). This is a different model from the one to which we give central attention in this paper. In its representation in the time-space plane we would have one straight line going through an origin on the time axis or equivalently on the spatial axis to indicate an arbitrary car (or observer car) always traveling on the highway at some constant speed, v_0, but all the other cars would be indicated by line segments from whose lengths we could get distance traveled on highway (still assumed to be i.i.d.) and from whose orientation angle with the t axis we could get the speed of the car. This produces the problem of the distribution of the number of intersections made by the line segments with the fixed line, an interesting but unsolved problem in geometrical probability.

However, this serves to return us to the central issue of intersections of random lines in the plane and its relationship to traffic flow models. We now turn to the formulation where the arbitrary car and all other cars are indicated by straight lines in the plane. The number of intersections of the arbitrary line (observer car) by the other lines determines the number of overtakings of slower cars made by the observer car plus the number of times it was overtaken by faster cars. We are interested in this distribution and also in the distributions of faster car overtakings of the observer car and the overtaking of slower cars by the observer car.

The structure of random lines in the plane and the properties resulting from a specific structure are therefore pertinent to analyses of traffic flow for our idealized model. The notion of a homogeneous Poisson field of random lines in the plane and its consequences have been developed by Miles in several papers [7], [8]. Additional development for nonhomogeneous Poisson fields of random lines is required for study of traffic flow models. In subsequent sections, we provide a formulation for a nonhomogeneous Poisson field of random lines and develop its structure and characteristics. This makes it possible to provide a different proof of Rênyi's theorem and the Weiss and Herman result on traffic flow and allows for further understanding of traffic flow models. It also provides a format for viewing their results as special cases of a more general model. In fact, this model provides a unified treatment for viewing any aspect of the idealized traffic flow model.

2. Development

First we formalize the notion of straight lines distributed "at random" throughout the plane. We will describe the plane in terms of (t, x) coordinates for subsequently the t axis will be employed to register time of arrival of cars at a fixed point on a highway and the x axis will in similar fashion report on spatial positions of cars on a highway at a fixed point in time. Naturally the time invariance property will insure that the conditions will prevail at any point in time. Any line in the (t, x) plane can be represented as

$$p = t \cos \alpha + x \sin \alpha, \qquad -\infty < p < \infty, 0 \leq \alpha < \pi, \tag{2.1}$$

where p is the signed length of the perpendicular to the line from an arbitrary origin 0, and α is the angle this perpendicular makes with the t axis (Figure 1). Note that if the intersection of the perpendicular with the line is in the third or fourth quadrant, p is taken to be negative. A set of lines $\{(p_i, \alpha_i): i = 0, \pm 1, \pm 2, \cdots\}$ constitutes a Poisson field under the following conditions.

(1) The distances $\cdots \leqq p_{-2} \leqq p_{-1} \leqq p_0 \leqq p_1 \leqq p_2 \leqq \cdots$ of the lines from an arbitrary origin 0, arranged according to magnitude represent the coordinates of the events of a Poisson process with constant parameter, say λ. Thus, the

number of p_i in an interval of length L has a Poisson distribution with mean λL.

(2) The orientations α_i of each line with a fixed but arbitrary axis (say the t axis) in the plane are independent and obey a uniform distribution in the interval $[0, \pi)$.

Thus, a reasonable representation of random lines in the plane is that of the Poisson field. This definition of randomness for lines in the plane also has the property that the randomness is unaffected by the choice of origin or line to serve as t axis, since it can be demonstrated that except for a constant factor $\int dp d\alpha$ is the only invariant measure under the group of rotations and translations that transform the line (p, α) to the line (p', α'). We will return to this structure and its characteristics, but now we employ it as a point of departure to initiate discussion of a nonhomogeneous Poisson field of random lines. To achieve this we will relax condition (2) above and ask only that the α_i be identically and independently distributed (i.i.d.).

For ease in the algebra of our traffic flow models, we will employ instead of α_i an angle formed by the intersection of the t axis with a line in the plane and we label this θ (Figure 1); note that $v = \tan \theta$. Also we will only be concerned with

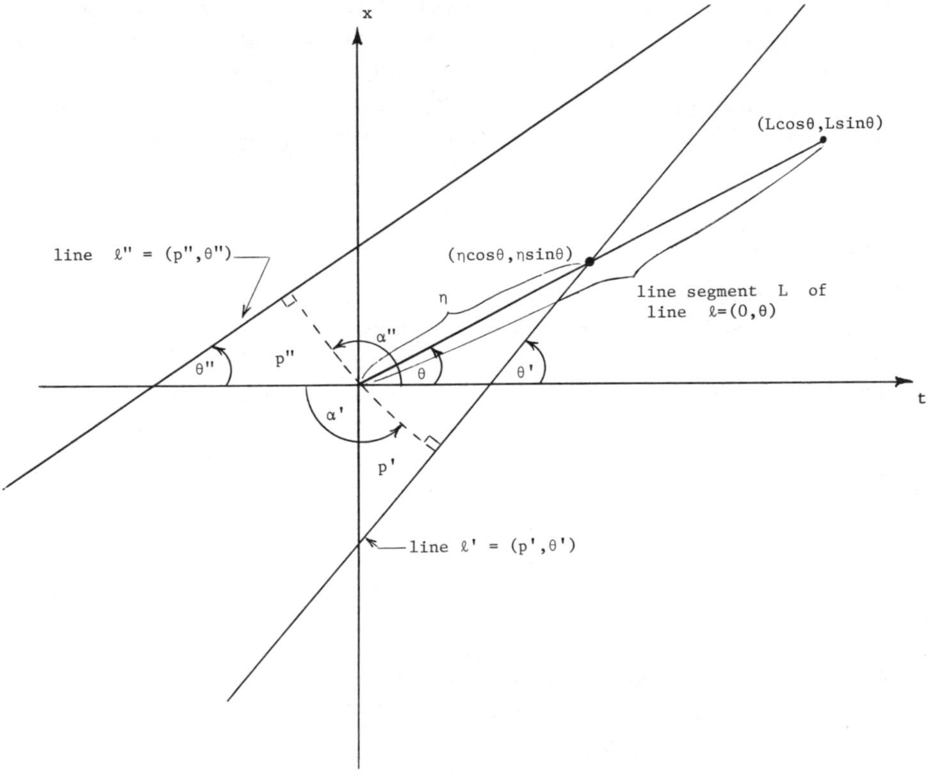

FIGURE 1

those lines where p_i falls in the second or fourth quadrant since this will yield all positive car velocities. The inclusion of the p_i in the first and third quadrant does not complicate the mathematical development, but they are not relevant. Thus, $\alpha = \frac{1}{2}\pi + \theta$ and the lines of interest will now be parametrized by (p, θ) where

$$p = -t \sin \theta + x \cos \theta, \qquad 0 \leqq \theta < \tfrac{1}{2}\pi. \tag{2.2}$$

Equation (2.2) takes care of the sign of p for it insures that p will be positive if it is in the second quadrant and negative in the fourth quadrant.

The set $\mathscr{L}$ of lines $\{(p_i, \alpha_i): i = 0, \pm 1, \pm 2, \cdots\}$ becomes a nonhomogeneous Poisson field if we require invariant measure only under translation, and we look into this situation because it will be helpful in our traffic flow models. Under this constraint, we now have the same conditions except that the orientation angles α_i of each line are i.i.d. random variables with common distribution function in the interval $[0, \pi)$. Thus, $\int dpd\alpha$ is no longer the appropriate measure. The diagram in Figure 1 delineates the situation where the origin can be arbitrarily chosen at any point on a specific and fixed t axis because invariance is preserved now only under translation.

The orientations θ_i are independent and identically distributed with common distribution F in the interval $[0, \frac{1}{2}\pi)$ and further the sequence of values $\langle\theta_i\rangle$ are independent of $\langle p_i\rangle$. This is equivalent to the statement that the velocities of cars, namely, $v_i = \tan \theta_i$ are independent and identically distributed with common distribution G on $[0, \infty)$ and thus $\langle v_i = \tan \theta_i\rangle$ are independent of $\langle p_i\rangle$.

When $\theta_0 = 0, p_0 = 0$, the traffic flow is characterized by a distribution of time intercepts on the t axis; when $\theta_0 = \frac{1}{2}\pi, p_0 = 0$, the traffic flow is characterized by a distribution of cars spaced along the x axis. For any other value of θ, the traffic flow is measured along a trajectory line. In the traffic literature, trajectories for low density traffic flow (no delays in overtaking) may be assumed to be linear in the time-space plane. Thus in any development, we must employ the appropriate measure to characterize distributions of traffic flow in such matters, for example, as distribution of number of overtakings. For our purposes where Poisson processes will be the underpinning for traffic flow in both spatial and temporal processes, the evaluation of the appropriate Poisson intensity parameter will be paramount as will be the relationships between these parameters for different measures.

The following exposition and the diagram in Figure 2 are included to make clear how the departure to nonhomogeneous Poisson fields occurs. Consider the (p, θ) plane. The homogeneous Poisson field occurs when points on the p axis follow a Poisson process with parameter λ independent of p and θ is uniformly distributed from 0 to $\frac{1}{2}\pi$. Given a fixed interval on p containing exactly n points, each follows a uniform distribution whose range is the length of the interval. When the interval is of unit length the density is dp. Similarly the density for θ is $d\theta$ and the joint density is $dpd\theta$ leading to $\int dpd\theta$ as the measure. This is invariant under rotations and translations. If θ_i is i.i.d. but not uniformly

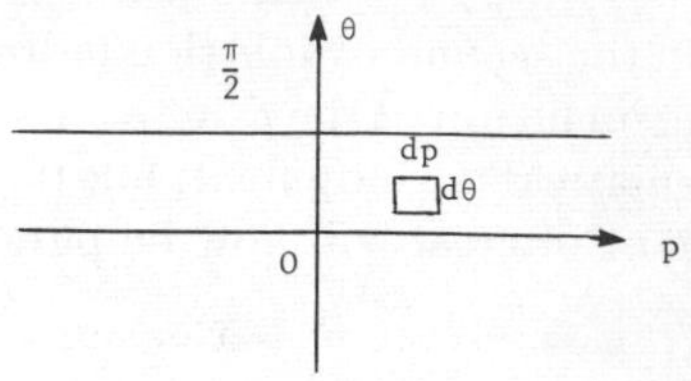

FIGURE 2

distributed, then the Poisson process for points of intersection along any trajectory line (p, θ) is maintained but the density for θ is no longer $d\theta$. Thus, any $dpd\theta$ rectangle as in Figure 2 will have the same measure only under translation on the p axis.

Also if θ is uniformly distributed but the points on the p axis follow a Poisson with parameter $\lambda(p)$, we obtain a nonhomogeneous Poisson field of random lines. If there are departures in the structure of both θ_i and p_i as listed above, then we obviously have a nonhomogeneous Poisson field of random lines where a Poisson process for points of intersections along any trajectory line (p, θ) will be maintained but the counting will be measured by the values of (p, θ) or equivalently the values of (t, x).

3. Basic results

The main results for nonhomogeneous Poisson fields of random lines $\mathscr{L}$ are discussed in this section. All sets under investigation are assumed to be measurable and events of probability zero are neglected. For instance, a possible realization of $\mathscr{L}$ is one in which there are no lines at all with probability zero and this is omitted. Many of the following results must be qualified by the phrase "with probability one"; however, this is often omitted for brevity. One basic feature of a special nonhomogeneous Poisson field of random lines, where p is Poisson with parameter λ and θ is i.i.d. is given in the following theorem.

THEOREM 3.1. *Points of intersections of such random lines $\mathscr{L}$ and any arbitrary and fixed line (p_0, θ_0) form a Poisson process with parameter $\lambda(\theta_0)$, where*

$$\lambda(\theta_0) = \lambda \cos \theta_0 \int_0^\infty |v - \tan \theta_0| (1 + v)^{-1/2}\, dG(v) \tag{3.1}$$

and λ is the parameter of the Poisson field of random lines and $v = \tan \theta$.

NOTE. The counting is done on the arbitrary and fixed line (p_0, θ_0) and of course $\langle v = \tan \theta \rangle$ for all the lines in $\mathscr{L}$. In our traffic model, a point of intersection on the line (p_0, θ_0) when represented in the coordinates of the time-space plane (t_0, x_0) may be viewed in the following traffic sense—t_0 is the actual time of car overtaking and x_0 is the actual spatial position where the car overtaking event occurs. This is developed more fully in Section 4.

PROOF. We note that the random mechanism in $\mathscr{L}$ is invariant under translation and is thus unaffected by the choice of origin; hence, we can have the arbitrary line go through the origin such that the segment of line $(0, \theta_0)$ with length L emanates from the origin and θ_0 is the fixed angle associated with this arbitrary line. Now denote η as the length of this segment measured from the origin to the point of intersection with another line $\ell' = (p', \theta') \in \mathscr{L}$ (see Figure 1). We can classify lines in $\mathscr{L}$ that intersect with line segment L into two groups, namely:

(1) for θ such that $0 < \theta < \theta_0$, $p > 0$ we have $\eta \sin(\theta_0 - \theta) = p$, if and only if

$$0 < \eta = p \csc(\theta_0 - \theta) < L. \tag{3.2}$$

(2) for θ such that $\theta_0 < \theta < \frac{1}{2}\pi$, $p < 0$ we have $\eta \sin(\theta - \theta_0) = -p$, if and only if

$$0 < \eta = -p \csc(\theta - \theta_0) < L. \tag{3.3}$$

Let N_L denote the number of lines in $\mathscr{L}$ intersecting the segment of length L. Then we will show that

$$Pr\{N_L = n\} = \exp\{-L\lambda\mu\} \frac{(L\lambda\mu)^n}{n!}, \tag{3.4}$$

and upon evaluation, that

$$\mu = \cos\theta_0 \int_0^\infty |v - \tan\theta_0| (1 + v^2)^{-1/2}\, dG(v). \tag{3.5}$$

Recall that $\cos\theta_0$ and $\tan\theta_0$ are constants depending on the θ_0 of the arbitrary line. Denote N_p the number of random lines whose signed distance p, to the origin is between $-L\sin\theta_0$ and $L\cos\theta_0$. Then

$$Pr\{N_L = n\} = \sum_{m=0}^{\infty} Pr\{N_L = n \mid N_p = m\} Pr\{N_p = m\}. \tag{3.6}$$

Clearly, no line can intersect the segment L unless its minimum distance to the origin is between $-L\sin\theta_0$ and $L\cos\theta_0$. Thus, N_p must be more than N_L; that is,

$$Pr\{N_L = n \mid N_p = m\} = 0 \quad \text{for } n > m, \tag{3.7}$$

and therefore

$$Pr\{N_L = n\} = \sum_{m=n}^{\infty} Pr\{N_L = n \mid N_p = m\}\, Pr\{N_p = m\}. \tag{3.8}$$

Let $\mu = Pr\{N_L = 1 \mid N_p = 1\}$. Then since the random lines are independent, that is, sequences $\langle v_i \rangle$ and $\langle p_i \rangle$ are independent, we have

$$Pr\{N_L = n \mid N_p = m\} = \binom{m}{n} \mu^n (1 - \mu)^{m-n} \quad \text{for } m \geqq n. \tag{3.9}$$

By the definition of random lines $\mathscr{L}$, we have

$$Pr\{N_p = m\} = \exp\{-\lambda L(\sin\theta_0 + \cos\theta_0)\}\frac{[\lambda L(\cos\theta_0 + \sin\theta_0)]^m}{m!}. \tag{3.10}$$

Thus,

$$\begin{aligned} Pr\{N_L = n\} &= \sum_{m=n}^{\infty}\binom{m}{n}\mu^n(1-\mu)^{m-n} \\ &\qquad \exp\{-\lambda L(\sin\theta_0 + \cos\theta_0)\}\frac{[\lambda L(\cos\theta_0 + \sin\theta_0)]^m}{m!} \\ &= \exp\{-\lambda L\mu(\sin\theta_0 + \cos\theta_0)\}\frac{[\lambda L\mu(\sin\theta_0 + \cos\theta_0)]^n}{n!}. \end{aligned} \tag{3.11}$$

Now we evaluate μ, the probability that a line whose minimum signed distance to the origin is between $-L\sin\theta_0$ and $L\cos\theta_0$, will intersect the segment L. Write

$$\begin{aligned} \mu &= Pr\{N_L = 1 \mid N_p = 1\} \\ &= Pr\{0 < \eta < L \mid -L\sin\theta_0 < p < L\cos\theta_0\} \\ &= Pr\{0 < \eta < L; 0 < \theta < \theta_0 \mid -L\sin\theta_0 < p < L\cos\theta_0\} \\ &\qquad + Pr\{0 < \eta < L; \theta_0 < \theta < \tfrac{1}{2}\pi \mid -L\sin\theta_0 < p < L\cos\theta_0\} \\ &= \frac{1}{L(\sin\theta_0 + \cos\theta_0)}\Bigg[\int_{-L\sin\theta_0}^{0} Pr\{0 < \eta < L, 0 < \theta < \theta_0 \mid p\}\,dp \\ &\qquad + \int_0^{L\cos\theta_0} Pr\{0 < \eta < L; \theta_0 < \theta < \tfrac{1}{2}\pi \mid p\}\,dp\Bigg]. \end{aligned} \tag{3.12}$$

Therefore, we have

$$\begin{aligned} L\mu(\cos\theta_0 + \sin\theta_0) &= \int_{-L\sin\theta_0}^{0} Pr\{0 < \eta < L; 0 < \theta < \theta_0 \mid p\}\,dp \\ &\qquad + \int_0^{L\cos\theta_0} Pr\{0 < \eta < L; \theta_0 < \theta < \tfrac{1}{2}\pi \mid p\}\,dp \\ &= L\cos\theta_0\Bigg[\int_0^{\tan\theta_0}(\tan\theta_0 - v)(1+v^2)^{-1/2}\,dG(v) \\ &\qquad + \int_{\tan\theta_0}^{\infty}(v - \tan\theta_0)(1+v^2)^{-1/2}\,dG(v)\Bigg]. \end{aligned} \tag{3.13}$$

Hence, we may conclude that

$$\mu = \frac{\cos\theta_0}{\cos\theta_0 + \sin\theta_0}\int_0^{\infty}|v - \tan\theta_0|(1+v^2)^{-1/2}\,dG(v), \tag{3.14}$$

and this in turn gives the result that

$$(3.15)\qquad Pr\,\{N_L = n\} = \exp\left\{-\lambda L \cos\theta_0 \int_0^\infty |v - \tan\theta_0|(1 + v^2)^{-1/2}\, dG(v)\right\}$$
$$\cdot\left[\lambda L \cos\theta_0 \int_0^\infty |v - \tan\theta_0|(1 + v^2)^{-1/2}\, dG(v)\right]^n (n!)^{-1}.$$

That is, N_L, the number of intersections with segment L from lines in $\mathscr{L}$, follows the Poisson distribution with parameter

$$(3.16)\qquad \lambda \cos\theta_0 \int_0^\infty |v - \tan\theta_0|(1 + v^2)^{-1/2}\, dG(v).$$

Let us denote N_L^+ the number of lines in $\mathscr{L}$ intersecting the segment L with $\theta_0 < \theta < \frac{1}{2}\pi$ and denote N_L^- the number of lines in $\mathscr{L}$ intersecting the segment L with $0 < \theta < \theta_0$. (Clearly, $N_L = N_L^+ + N_L^-$.) Theorem 3.1 permits us to state the following result immediately.

THEOREM 3.2. *Let $\langle\tau_i^-\rangle$ be points of intersections of random lines $\mathscr{L}$ with $\theta_0 < \theta < \frac{1}{2}\pi$ and any arbitrary line (p_0, θ_0) and let $\langle\tau_i^+\rangle$ be points of intersections of random lines $\mathscr{L}$ with $0 < \theta < \theta_0$ and line (p_0, θ_0). Then $\langle\tau_i^+\rangle$ and $\langle\tau_i^-\rangle$ form two independent Poisson processes, with parameters*

$$(3.17)\qquad \lambda^+(\theta_0) = \lambda \cos\theta_0 \int_0^{\tan\theta_0} (\tan\theta_0 - v)(1 + v^2)^{-1/2}\, dG(v)$$

and

$$(3.18)\qquad \lambda^-(\theta_0) = \lambda \cos\theta_0 \int_{\tan\theta_0}^\infty (v - \tan\theta_0)(1 + v^2)^{-1/2}\, dG(v).$$

In traffic terms, $\lambda^+(\theta_0)$ is the intensity of the Poisson process generated by the fixed car $K(p_0, \theta_0)$ overtaking slower cars and $\lambda^-(\theta_0)$ for faster cars overtaking $K(p_0, \theta_0)$. We remark here that if $\theta_0 = 0$, $p_0 = 0$, namely the t axis, then $\lambda^+(0) = 0$ and $\lambda^-(0) = \lambda \int_0^\infty v(1 + v^2)^{-1/2}\, dG(v)$, that is, if the random variable N_p is distributed according to the Poisson distribution with parameter λ, then the points of intersection of random lines $\mathscr{L}$ with the t axis form a point process distributed according to the Poisson distribution with parameter $\lambda \int_0^\infty v(1 + v^2)^{-1/2}\, dG(v)$. Similarly, if $\theta_0 = \frac{1}{2}\pi$, $p_0 = 0$, namely the x axis, then

$$(3.19)\qquad \lambda(\tfrac{1}{2}\pi) = \lambda \int_0^\infty (1 + v^2)^{-1/2}\, dG(v),$$

and the counting is done along the spatial axis. The corresponding $\lambda^+(\frac{1}{2}\pi) = \lambda(\frac{1}{2}\pi)$ and $\lambda^-(\frac{1}{2}\pi) = 0$. Hence, we have established the following result.

THEOREM 3.3. (i) *Points of intersection of the field $\mathscr{L}$ and the t axis (temporal counting) form a Poisson process with parameter λ_t where*

$$(3.20)\qquad \lambda_t = \lambda \int_0^\infty v(1 + v^2)^{-1/2}\, dG(v).$$

(ii) *Points of intersection of the field $\mathscr{L}$ and the x axis (spatial counting) form a Poisson process with parameter λ_x, where*

$$\lambda_x = \lambda \int_0^\infty (1 + v^2)^{-1/2}\, dG(v). \tag{3.21}$$

Let the sequence $\langle \tau_{1k} \rangle$ denote the instants when the arbitrary line (p_0, θ_0) segment of length L intersects random lines of $\mathscr{L}$ whose orientations θ_i belong to a given set Θ_1 and sequence $\langle \tau_{2k} \rangle$ denote the instants when the arbitrary line (p_0, θ_0) segment of length L intersects random lines of $\mathscr{L}$ whose orientations θ_i belong to a given set Θ_2. We now state a generalized version of the results in Theorem 3.4.

THEOREM 3.4. *If $\Theta_1 \cap \Theta_2 = \varnothing$, then the two sequences $\langle \tau_{1k} \rangle$ and $\langle \tau_{2k} \rangle$ form two independent Poisson processes with parameters $\lambda_1(\theta_0)$ and $\lambda_2(\theta_0)$, respectively, where*

$$\lambda_i(\theta_0) = \lambda \int_{\tan \Theta_i} |v - \tan \theta_0| (1 + v^2)^{-1/2}\, dG(v), \qquad i = 1, 2, \tag{3.22}$$

and

$$\tan \Theta_i = \{v \mid v = \tan \theta \text{ such that } \theta \in \Theta_i\}, \qquad i = 1, 2. \tag{3.23}$$

The details of the proof are omitted since it is essentially the proof used in Theorem 3.1.

In the next paragraphs, we establish some similar results employing $\langle \tau_i \rangle$ and $\langle \eta_i \rangle$, where the τ_i are the arrival times of cars on a highway measured from a fixed position, say $x = 0$, and η_i are corresponding positions of these cars on the highway at a fixed time, say t_i. Let us denote $\langle \tau_i \rangle$ the sequence of points of intersections of a given random family of lines $\mathscr{A}$ with the t axis. Let $\langle v_i = \tan \theta_i \rangle$ be the sequence of i.i.d. random variables. Those θ_i are the orientations of random lines in $\mathscr{A}$. Denote $\langle \eta_i \rangle$ the sequence of points of intersections of random lines in $\mathscr{A}$ with the x axis. We employ $\mathscr{A}$ instead of $\mathscr{L}$ for now we do not wish to assume as in $\mathscr{L}$ that $\langle p_i \rangle$ and $\langle v_i \rangle$ are independent sequences. In the following theorems, we will employ $\langle \eta_i \rangle$ and $\langle v_i \rangle$, or $\langle \tau_i \rangle$ and $\langle v_i \rangle$ as independent sequences and these assumptions will be made specific in each statement of the theorem. The results can be stated as follows.

THEOREM 3.5. *If $\langle \tau_i \rangle$ forms a Poisson process with parameter λ_τ and sequences $\langle \tau_i \rangle$ and $\langle v_i \rangle$ are independent, then $\langle p_i \rangle$ forms a Poisson process with parameter* $\lambda_\eta \int_0^\infty [(1 + v^2)^{1/2}/v]\, dG(v)$.

Similarly, we have:

THEOREM 3.6. *If $\langle \eta_i \rangle$ forms a Poisson process with parameter λ_η and sequence $\langle \eta_i \rangle$ and $\langle v_i \rangle$ are independent, then $\langle p_i \rangle$ forms a Poisson process with parameter* $\lambda_\eta \int_0^\infty (1 + v^2)^{1/2}\, dG(v)$.

We shall give a proof of Theorem 3.5. The proof of Theorem 3.6 is similar and hence is omitted.

Proof of Theorem 3.5. Based on the proof similar to that used in Theorem 3.1, we denote $N_t(c)$ the number of lines in $\mathscr{A}$ intersecting the t axis in an interval of length c and denote N_p for the number of lines in $\mathscr{A}$ whose p_i is bounded by $0 < p_i < p$. Then it is clear we want to show that

$$(3.24)\quad Pr\{N_p = n\} = \exp\left\{-\lambda p \int_0^\infty \frac{(1+v^2)^{1/2}}{v}\, dG(v)\right\}\left[\lambda p \int_0^\infty \frac{(1+v^2)^{1/2}}{v}\, dG(v)\right]^n (n!)^{-1}.$$

Following the previous development, we arrive at

$$(3.25)\qquad Pr\{N_p = n\} = \lim_{c\to\infty} \exp\{-\lambda_1 C\mu_c\} \frac{(\lambda_1 C\mu_c)^n}{n!}$$

and

$$(3.26)\qquad \mu_c = Pr\{N_p = 1 \mid N_t(c) = 1\}.$$

It remains to show that

$$(3.27)\qquad \lim_{c\to\infty} c\mu_c = p \int_0^\infty \frac{(1+v^2)^{1/2}}{v}\, dG(v),$$

$$(3.28)\qquad \mu_c - Pr\{N_p = 1 \mid N_t(c) - 1\} = \frac{1}{c}\int_0^c Pr\{N_p = 1 \mid \tau\}\, d\tau,$$

$$(3.29)\qquad c\mu_c = \int_0^c Pr\{N_p = 1 \mid \tau\}\, d\tau,$$

$$\begin{aligned}(3.30)\qquad \lim_{c\to\infty} c\mu_c &= \int_0^\infty Pr\{N_p = 1 \mid \tau\}\, d\tau \\ &= \int_0^\infty Pr\left\{0 < \theta < \sin^{-1}\frac{p}{\tau} \,\middle|\, \tau\right\} d\tau \\ &= \int_0^\infty \int_0^{p/\mathrm{sintan}^{-1} v} d\tau\, dG(v) \\ &= p \int_0^\infty \frac{(1+v^2)^{1/2}}{v}\, dG(v).\end{aligned}$$

This completes the proof of Theorem 3.5.

4. Traffic flow applications

The purpose of this section is to discuss a number of results that can be related to low density traffic flow models on an infinite highway in the light of the developments in Section 3. These models were initiated and developed principally in papers by Rényi [10], Weiss and Herman [13], and Breiman [2], sometimes without specific reference to low density traffic flow. Of the theorems presented in this section, some are known but all the proofs are new and developed in a unified manner.

Rényi [10] has developed and analyzed a model of traffic flow on a divided highway that extends in one direction out to infinity without traffic lights or other barriers. It is assumed that the speed of each car is constant but its value is governed by a random variable and passing is always achieved without delays. Assuming that the temporal distribution of cars is described by a Poisson process, Rényi obtained some asymptotic results for the spatial distribution of cars along the highway. In what follows, we shall reproduce Rényi's theorems and include other results dealing with low density traffic flow. It will also be demonstrated for Rényi's model, that if the spatial distribution of cars is assumed to obey a Poisson process then the temporal distribution of cars (that is, arrival times at some fixed position) is again a Poisson process. This new result establishes a crucial structural property of Rényi's model for low density traffic. In detail, the assumptions of Rényi's model are:

(a) instants $\langle t_i \rangle_{i=1}^{\infty}$ at which cars enter the highway at a fixed position form a homogeneous Poisson process with parameter ω;

(b) a car arriving at a certain point on the highway at instant t_i chooses a velocity V_i and then moves with this constant velocity; the random variables $\langle V_k \rangle$ are independently and identically distributed with distribution function $G(v) = Pr\{V \leqq v\}$ and sequences $\{V_k\}$ and $\langle t_k \rangle$ are independent;

(c) $\int_0^{\infty} (1/v)\, dG(v) < \infty$, that is, the mean value of $1/v$ is finite; without this condition a traffic jam would arise and make all traffic flow impossible;

(d) no delay in overtaking a car traveling at a slower speed when it is approached.

Suppose an arbitrary car $K(t_0, v_0)$ arrives at some fixed point of the highway at time t_0, where it assumes and maintains the fixed velocity v_0. Let $\langle t_k^+ \rangle$ denote the instants at which the car $K(t_0, v_0)$ overtakes slower cars and $\langle t_k^- \rangle$ denote instants at which car $K(t_0, v_0)$ is overtaken by faster cars. Rényi has obtained the following asymptotic result.

THEOREM 4.1. (Rényi). *The instants* $\{t_k^+\}$ *and* $\{t_k^-\}$ *form two independent homogeneous Poisson processes, with parameters:*

$$\omega^+(v_0) = \omega \int_0^{v_0} \frac{v_0 - v}{v}\, dG(v), \qquad \omega^-(v_0) = \omega \int_{v_0}^{\infty} \frac{v - v_0}{v}\, dG(v). \tag{4.1}$$

Rényi's proof of the above theorem is based on the following two properties of Poisson processes:

(A) if $\langle t_i \rangle$ are the instants of time when an event occurs in a homogeneous Poisson process with parameter ω, and $\zeta_1, \zeta_2, \cdots$, is a sequence of independent positive random variables, each having the same distribution $G(\zeta)$ and each is independent of the process $\{t_k\}$, then the time instants $t_k \zeta_k$, $k = 1, 2, \cdots$, also form a homogeneous Poisson process with density

$$\omega^* = \omega \int_0^\infty \frac{1}{\zeta}\, dG(\zeta); \tag{4.2}$$

(B) if a subsequence $\{t_{v_k}\}$ of the instants $\{t_k\}$, in which an event occurs in a Poisson process with density ω, is selected at random in such a way that for each j the probability of the event A_j that j should belong to the subsequence $\{v_k\}$ is equal to r, $0 < r < 1$, and the events A_j, $j = 1, 2, \cdots$, are independent, and if $\{t_{u_k}\}$ are the instants that are not selected, (that is, j belongs to the sequence $\{u_k\}$ if and only if it does not belong to the sequence $\{v_k\}$), then $\{t_{v_j}\}$ and $\{t_{u_k}\}$ are two independent Poisson processes with density ωr and $\omega(1 - r)$.

It is now known from a result of Wang [12] that property (B), in some sense, is a characteristic property for Poisson processes. We shall establish Theorem 4.1 without using property (A). Thus, it may be inferred that property (B) implies property (A).

Proof of Theorem 4.1. The trajectory of any car in the time-space diagram in the preceding section for Rényi's low density traffic model is realized by a straight line. Let us denote the trajectories of all cars on the highway as a set $\mathscr{A}$. Then it is clear that $\mathscr{A}$ possesses the properties of a nonhomogeneous Poisson field of random lines. Denote M_L^+ the number of lines in $\mathscr{A}$ that intersect segment L from below and M_p the number of lines in $\mathscr{A}$ whose arrival times are in $(0, t_0)$. Then

$$\begin{aligned} Pr\,\{M_L^+ = n\} &= \sum_{m=n}^{\infty} \binom{m}{n} \mu^n (1-\mu)^{m-n} \exp\{-\omega t_0\} \frac{(\omega t_0)^m}{m!} \\ &= \exp\{-\omega t_0 \mu\} \frac{(\omega t_0 \mu)^n}{n!}, \end{aligned} \tag{4.3}$$

where

$$\begin{aligned} \mu &= Pr\,\{M_L^+ = 1 \mid M_p = 1\} \\ &= Pr\left\{\theta \geqq \tan^{-1} \frac{x_0}{t_0 - p} \,\middle|\, 0 < p < t_0\right\} \\ &= \frac{1}{t_0} \int_0^{t_0} Pr\left\{V \geqq \frac{x_0}{t_0 - p}\right\} dp \\ &= \int_{v_0}^{\infty} \frac{v - v_0}{v}\, dG(v), \end{aligned} \tag{4.4}$$

and $v_0 = x_0/t_0$.

Similarly, we define M_L^- as the number of lines in $\mathscr{A}$ intersecting L from above and M_p^c as the number of lines in $\mathscr{A}$ whose arrival times fall in the interval $-c$ and 0, $c > 0$. We can compute

$$Pr\,\{M_L^- = n\} = \lim_{c \to \infty} \exp\{-\omega c \mu_c^*\} \frac{(\omega c \mu_c^*)^n}{n!} \tag{4.5}$$

where $\mu_c^* = Pr\{M_L^- = 1 | M_p^c = 1\}$ and

$$\lim_{c\to\infty} c\mu_c^* = \lim_{c\to\infty} cPr\{M_L^- = 1 | M_p^c = 1\} \tag{4.6}$$
$$= \lim \int_{-c}^{0} \int_{0}^{x_0/(t_0-p)} dG(v)\,dp$$
$$= \int_{-\infty}^{0} \int_{0}^{x_0/(t_0-p)} dG(v)\,dp$$
$$= \int_{0}^{v_0} \frac{x_0 - vt_0}{v}\,dG(v)$$
$$= t_0 \int_{0}^{v_0} \frac{v_0 - v}{v}\,dG(v).$$

Random variables M_L^- and M_L^+ are independent because the events involved come from disjoint intervals. This completes the proof of Theorem 4.1.

The counting interval employed in the above theorem is on the time axis. In what follows, a similar approach to the problem dealing with a spatial counting interval is employed and produces some interesting results.

Denote M^- the number of lines in $\mathscr{A}$ intersecting L from above and M^+ the number of lines from below. Denote M_{x_0} the number of lines in $\mathscr{A}$ whose spatial positions at $t = 0$ are between 0 and x_0 and similarly $M_{x_0}^c$ for 0 and $-c$. Let λ^* be the spatial density. Then we have

$$Pr\{M^- = n\} = \exp\{-\lambda^* x_0 \mu\} \frac{(\lambda^* x_0 \mu)^n}{n!}, \tag{4.7}$$

where

$$\mu = Pr\{M^- = 1 | M_{x_0} = 1\} \tag{4.8}$$
$$= \frac{t_0}{x_0} \int_{0}^{v_0} (v_0 - v)\,dG(v),$$

and

$$Pr\{M^+ = n\} = \lim_{c\to\infty} \sum_{m=n}^{\infty} Pr\{M^+ = n | M_{x_0}^c = m\} Pr\{M_{x_0}^c = m\} \tag{4.9}$$
$$= \lim_{c\to\infty} \exp\{-\lambda^* c_1^c\} \frac{(\lambda^* c\mu_1^c)^n}{n!}$$

where $\mu_1^c = Pr\{M^+ = 1 | M_{x_0}^c = 1\}$. It can be easily verified that

$$\lim_{c\to\infty} c\mu_1^c = t_0 \int_{v_0}^{\infty} (v - v_0)\,dG(v) \tag{4.10}$$

and random variables M^- and M^+ are independent. Now denote $M = M^+ + M^-$. We conclude that

$$(4.11)\qquad Pr\{M = n\} = \exp\left\{-\lambda^* t_0 \int_0^\infty |v_0 - v|\, dG(v)\right\} \cdot \left[\lambda^* t_0 \int_0^\infty |v_0 - v|\, dG(v)\right]^n (n!)^{-1}.$$

The above result appeared initially in the paper [13] by Weiss and Herman mentioned previously.

In the next paragraph, results are derived about the spatial distribution of vehicles if the temporal distribution (distribution of arrival times) is assumed to be Poisson.

Denote by S^+ the number of lines in $\mathscr{A}$ intersecting $(0, x_0)$ and $x_0 > 0$ at time zero and by S_c^+ the number of lines in $\mathscr{A}$ whose arrival times are in the interval $(-c, 0)$. We further denote by S^- the number of lines in $\mathscr{A}$ intersecting $(-x_0, 0)$, $x_0 > 0$ at time zero and by S_c^- the number of lines in $\mathscr{A}$ whose arrival times are in the interval $(0, c)$. Let us compute the quantities $Pr\{S^+ = n\}$, $Pr\{S^- = n\}$, and $Pr\{S = S^+ + S^- = n\}$. We have

$$(4.12)\qquad Pr\{S^+ = n\} = \lim_{c\to\infty} \exp\{-\omega c \mu_2\} \frac{(\omega c \mu_2)^n}{n!},$$

where ω is the temporal density and $\mu_2 = Pr\{S^+ = 1 | S_c^+ = 1\}$. It can be shown easily that

$$(4.13)\qquad \lim_{c\to\infty} c\mu_2 = x_0 \int_0^\infty \frac{1}{v}\, dG(v).$$

We conclude that

$$(4.14)\qquad Pr\{S^+ = n\} = \exp\left\{-\omega x_0 \int_0^\infty \frac{1}{v}\, dG(v)\right\}\left[\omega x_0 \int_0^\infty \frac{1}{v}\, dG(v)\right]^n (n!)^{-1}.$$

Similarly, we obtain $Pr\{S_n^- = n\} = Pr\{S^+ = n\}$ and

$$(4.15)\qquad Pr\{S = S^+ + S^- = n\} = \exp\left\{-2\omega x_0 \int_0^\infty \frac{1}{v}\, dG(v)\right\} \cdot \left[2\omega x_0 \int_0^\infty \frac{1}{v}\, dG(v)\right]^n (n!)^{-1}.$$

We can now summarize as follows.

THEOREM 4.2. *If $\langle t_i \rangle$ forms a Poisson process with parameter ω and sequences $\langle t_i \rangle$ and $\langle V_i \rangle$ are independent, then the locations of vehicles on the highway at time $t = 0$, namely $\langle x_i \rangle$, form a Poisson process with parameter $\omega \int_0^\infty (1/v)\, dG(v)$.*

THEOREM 4.3. *If $\langle x_i \rangle$ forms a Poisson process with parameter λ^* and sequences $\langle x_i \rangle$ and $\langle V_i \rangle$ are independent and $\langle x_i^+ \rangle$ denotes the positions at which the car $K(v_0)$ overtakes slower cars and $\langle x_i^- \rangle$ denotes the positions at which car $K(v_0)$ is overtaken by faster cars, then the two sequences $\langle x_i^+ \rangle$ and $\langle x_i^- \rangle$ form two independent (homogeneous) Poisson processes, with parameters*

$$(4.16)\qquad \lambda_{+}^{*}(v_0) = \lambda^* \int_0^{v_0} (v_0 - v)\, dG(v) \qquad \lambda_{-}^{*}(v_0) = \lambda^* \int_{v_0}^{\infty} (v - v_0)\, dG(v).$$

This result is analogous to the Rényi result which we developed as Theorem 4.1 except that the counting of overtakings is accomplished on the spatial axis rather than on the time axis. The next theorem provides results analogous to those in Theorem 4.2.

THEOREM 4.4. *If $\langle x_i \rangle$ forms a Poisson process with parameter λ^* and sequences $\langle x_i \rangle$ and $\langle V_i \rangle$ are independent and the $\langle V_i \rangle$ are i.i.d. random variables with common distribution $G(v) = Pr\{V \leqq v\}$, then the corresponding $\langle t_i \rangle$ arrival times at position $x = 0$ form a Poisson process with parameter $\lambda^* E(V)$, where $E(V) = \int_0^\infty v\, dG(v)$.*

PROOF. The proof is again based on the binomial mixing as presented in property (B) and hence details are omitted.

5. Concluding remarks

REMARK 5.1. On the basis of the work in the previous sections, it appears that we can view the main structural property for a nonhomogeneous Poisson field of random lines $\mathscr{L}$ in the following way. The point process obtained by the intersections of lines in the field $\mathscr{L}$ with any fixed line (p_0, θ_0) forms a Poisson process subject to the existence of the integral $\int_0^\infty h(v)\, dG(v)$ for some suitable $h(v)$, say $1/v$, $v(1 + v^2)^{-1/2}$, $(1 + v^2)^{-1/2}$, and others.

REMARK 5.2. In light of the statement in Theorem 3.2, we can offer the more general result below for which the proof is immediate and hence is omitted.

THEOREM 5.1. *Let $\Theta_1, \cdots, \Theta_m$ be m disjoint intervals on θ and let $P_1, \cdots, P_m$ be m intervals on p. Recall that $\mathscr{L}$ is the nonhomogeneous Poisson field defined previously, then the random variables $N(P_i, \Theta_i)$, $i = 1, 2, \cdots, m$, where $N(P_i, \Theta_i) = \{$no. of $(p, \theta) \in \mathscr{L}$ such that $p \in P_i$, $\theta \in \Theta_i\}$, $i = 1, 2, \cdots, m$, are m independent Poisson random variables. Consequently, the lines (p, θ) are points of a two dimensional nonhomogeneous Poisson process with parameter $\lambda(\Theta)$ that depends on θ.*

Consider the strips in the (p, θ) plane where $0 \leqq \theta < \frac{1}{2}\pi$ and $-\infty < p < \infty$; see Figure 2. Thus, the number of p in an interval of length I on the p axis whose θ are in the set Θ has a nonhomogeneous Poisson distribution with mean

$$(5.1)\qquad \lambda(\Theta) = \lambda I \int_{\Theta} dF,$$

where F is the c.d.f. on the random variable θ.

It is clear, for the homogeneous Poisson field of random lines where F is the uniform distribution, that $\int_\Theta dF$ equals the length of the interval measure for θ divided by $\frac{1}{2}\pi$. Hence, the parameter

$$(5.2)\qquad \lambda I \int_\Theta dF = \frac{2}{\pi} \lambda I \text{ (length of } \Theta),$$

and thus it depends only on the length of the set Θ and the length of I on the p axis. Thus, Theorem 5.1 holds for homogeneous Poisson fields of random lines and in this way it adds to Miles' results.

All results obtained in this paper should be capable of extension to other nonhomogeneous Poisson fields, say, where λ is the function of p, $\lambda = \lambda(p)$, or where $\lambda = \lambda(t, x)$.

REMARK 5.3. The results announced by Miles in [7] and [8] also fall out immediately from our development because there the orientations α_i are independent and uniformly, distributed, $0 \leqq \alpha_i < \pi$. Then the following result is immediate: the points of intersection of the random lines $\mathscr{A}$ and an arbitrary line (p_0, θ_0) form a Poisson process with parameter $2\lambda/\pi$.

REMARK 5.4. Based on the results stated in Theorem 3.3 (i) and Theorem 3.5, one might expect to get the following identity

$$\lambda_\tau = \lambda_\tau \int_0^\infty \frac{(1 + v^2)^{1/2}}{v} \, dG(v) \int_0^\infty v(1 + v^2)^{-1/2} \, dG(v). \tag{5.3}$$

But the identity is true if and only if the field $\mathscr{L}$ consists of parallel lines alone. This reduces the field of random lines $\mathscr{L}$ to the case initially studied by Goudsmit [5], who employed it as a first attempt to study random lines in the plane in connection with examining the randomness of tracks left in a cloud chamber by a particle.

REMARK 5.5. The structure and properties of random lines in the plane that are developed in this paper make it possible to review and extend results in still other applications. In a paper reporting on the pattern in a planar region of one species of vegetation with respect to another, Pielou [9] defined a random pattern as one in which the alternation between species along any line transect is Markovian. In a subsequent paper, Bartlett [1] indicated that she did not establish the existence of a two state planar process that could produce this Markovian property. Switzer [11] then demonstrated the existence of a finite state random process in the plane, namely, the homogeneous Poisson field or random lines with the property that alternation among states along any straight line is Markovian. In this paper, a whole class of finite state random processes in the plane that accomplishes this is presented by our results for nonhomogeneous Poisson fields of random lines.

REFERENCES

[1] M. S. BARTLETT, "A note on a spatial pattern," *Biometrics*, Vol. 20 (1964), pp. 891–892.

[2] L. BREIMAN, "The Poisson tendency in traffic distributions," *Ann. Math. Statist.*, Vol. 34 (1963), pp. 308–311.

[3] M. BROWN, "Some results on a traffic model of Rényi," *J. Appl. Prob.*, Vol. 6 (1969), pp. 293–300.

[4] ———, "An $M/G/\infty$ estimation problem," *Ann. Math. Statist.*, Vol. 41 (1970), pp. 651–654.

[5] S. A. GOUDSMIT, "Random distribution of lines in a plane," *Rev. Modern Phys.* Vol. 17 (1945), pp. 321–322.

[6] J. F. C. Kingman, "Markov population processes," *J. Appl. Prob.*, Vol. 6 (1969), pp. 1–18.
[7] R. E. Miles, "Random polygons determined by random lines in a plane," *Proc. Nat. Acad. Sci. U.S.A.*, Vol. 52 1964), pp. 901–907.
[8] ———, "Random polygons determined by random lines in a plane, II," *Proc. Nat. Acad. Sci. U.S.A.*, Vol. 52 (1964), pp. 1157–1160.
[9] E. C. Pielou, "The spatial pattern of two-phase patchworks of vegetation," *Biometrics*, Vol. 20 (1964), pp. 156–167.
[10] A. Rényi, "On two mathematical models of the traffic on a divided highway," *J. Appl. Prob.*, Vol. 1 (1964), pp. 311–320.
[11] P. Switzer, "A random set process in the plane with a Markovian property," *Ann. Math. Statist.*, Vol. 36 (1965), pp. 1859–1863.
[12] P. C. C. Wang, "A characterization of the Poisson distribution based on random splitting and random expanding," Stanford University Technical Report No. 158, (NR 042–067), 1970.
[13] G. Weiss and R. Herman, "Statistical properties of low-density traffic," *Quart. Appl. Math.*, Vol. 20 (1962), pp. 121–130.

MULTIVARIATE POINT PROCESSES

D. R. COX
Imperial College, University of London
and
P. A. W. LEWIS*
Imperial College, University of London
and
IBM Research Center

1. Introduction

We consider in this paper events of two or more types occurring in a one dimensional continuum, usually time. The classification of the events may be by a qualitative variable attached to each event, or by their arising in a common time scale but in different physical locations. Such multivariate point processes, or multitype series of events, are to be distinguished from events of one type occurring in an n dimensional continuum and considered, for example, by Bartlett [2]. It is of course possible to have multivariate point processes in, say, two dimensions, for example, the locations of accidents labelled by day of occurrence, but we do not consider this extension here.

Multivariate series of events arise in many contexts; the following are a few examples.

Example 1.1. Queues are a well-known situation in which bivariate point processes arise as the input and output, although interest in the joint properties of the input and output processes is fairly recent (for example, Daley [16] and Brown [7]). The two processes occur simultaneously in time. Many of the variants on simple queueing situations which have been considered give rise to more than two point processes.

Example 1.2. An important and rich source of multivariate point processes is neurophysiology (Perkel, Gerstein, and Moore [41]). Information is carried along nerve bundles by spikes which occur randomly in time. (The spikes are extremely narrow and, at least in many situations, their shape and height do not appear to vary or carry information.) The neuronal spike trains of different types may be observations at different locations with no definite knowledge of physical connection, or may be the inputs and outputs to nerve connections (neurons).

Example 1.3. When the events are crossings of a given level by a real valued stochastic process in continuous time, the up crossings and down crossings of the

*Support under a National Institutes of Health Special Fellowship (2–FO3–GM 38922–02) is gratefully acknowledged. Present address: Naval Postgraduate School, Monterey, California.

level constitute a very special bivariate point process in which the two types of events alternate (Leadbetter [27]). However, up crossings of two different levels produce a general type of bivariate point process which is of interest, for example, in reliability investigations.

EXAMPLE 1.4. In reliability studies over time, of continuously operating machines such as computers, the failures are more often than not labelled according to the part of the system in which they occurred, or according to some other qualitative characterization of the failure, for example, mechanical or electrical. One might also be interested in studying interactions between preventive maintenance points and failures occurring during normal operation. Again a comparison between failure patterns in separately located computers (Lewis [29]) might be of interest in determining whether some unknown common variable, such as temperature and/or humidity, influences reliability.

EXAMPLE 1.5. Cox [11] has considered the problem of analyzing events of two types in textile production. The two types of event may be breakdowns in the loom and faults in the cloth, or different types of breakdown of the loom. The continuum is length of thread rather than time.

EXAMPLE 1.6. In the analysis of electrocardiograms the trace is continuous, but both regular heart beats and various types of ectopic heart beats occur. It is therefore of interest to analyze electrocardiograms as bivariate event processes, even though defining the precise time of occurrence of the event (heartbeat) may present some problems.

EXAMPLE 1.7. Traffic studies are a rich source of multivariate point processes. Just two possibilities are that the events may be the passage of cars by a point on a road when the type of event is differentiated by direction of travel, or we may consider passage of cars past two different positions.

EXAMPLE 1.8. Finally, physical phenomena such as volcanoes or earthquakes (Vere-Jones [45], [46], [47]) may have distinguishing features of many kinds—generally highly compacted attributes of the process, for example, the general location of the origin of the earthquake.

Multivariate point processes can be regarded as very special cases of univariate point processes in which a real valued quantity is associated with each point event, that is, special cases of what Bartlett [3] has called, rather generally, line processes. In particular, if the real valued quantity takes only two possible values, we have in effect a bivariate process of events of two types.

Three broad types of problems arise for multivariate point processes. The first are general theoretical and structural problems of which the most outstanding is the problem of characterizing the dependence and interaction between a number of processes. This is the only general theoretical question we will consider in any detail; it is intimately connected with the statistical analysis of bivariate point processes.

The second type of problem is the calculation of the properties of special stochastic models suggested, for example, by physical considerations. This in general is a formidable task even for quite simple models.

Thirdly, there are problems of statistical analysis. These include:

(a) comparing rates in independent processes (Cox and Lewis [14], Chapter 9) from finite samples;

(b) assessing possible dependence between two processes from finite samples;

(c) determining, again from finite samples, the probabilistic structure of a mechanism which transforms one process into a second quite clearly dependent process.

The range of the problems will become clear in the main body of the paper. The topics considered are briefly as follows.

In Section 2, we give some notation and define various types of interevent sequences and counting processes which occur in bivariate point processes. Concepts such as independence of the marginal processes and stationarity and regularity of the complete, bivariate process are defined. The ideas of this section are illustrated by considering two independent renewal processes and also the semi-Markov process (Markov renewal process).

In Section 3, we study dependence and correlation in bivariate point processes, defining complete intensity functions and second order cross intensity functions and cross spectra, giving their relationship to covariance time surfaces. Doubly stochastic bivariate point processes are defined and their cross intensity function is given. Other simple models of bivariate point processes are defined through the complete intensity and cross intensity functions. In this way, various degrees of interaction between events in the bivariate process can be specified. A class of bivariate Markov interval processes is defined.

In Section 4, a simple delay model with marginal Poisson processes is considered in some detail. Other special physical models are considered briefly in Section 5.

General comments on bivariate Poisson processes are given at the end of Section 5; a bivariate Poisson process is defined simply as a bivariate point process whose marginal processes are Poisson processes.

Statistical procedures are considered in Section 6, including the estimation of second order cross intensity functions and cross spectra, as well as covariance time surfaces. Tests for dependence in general and particular situations are considered, and statistical procedures for some special processes are given.

Throughout, emphasis is placed on concepts rather than on mathematical details and a number of open questions are indicated. For the most part we deal with bivariate processes, that is, with events of two types; the generalization of more than two types of events is on the whole straightforward.

2. General definitions and ideas

2.1. *Regularity.* Throughout Section 2, we deal with bivariate processes, that is, processes of events of two types called type a and type b. The process of, say, type a events alone is called a *marginal process*.

In a univariate point process such as a marginal process in a bivariate point process, regularity is defined by requiring that in any interval of length Δt

$$Pr\{(\text{number events in } \Delta t) > 1\} = o(\Delta t). \tag{2.1}$$

Regularity is intuitively the nonoccurrence of multiple events.

For bivariate processes, we say the process is *marginally regular* if its marginal processes, considered as univariate point processes, are both regular. The bivariate process is said to be *regular* if the process of superposed marginal events is regular, that is, if the process of events regardless of type is regular. This type of regularity, of course, implies *marginal regularity*.

A simple, rather degenerate, bivariate process is obtained by taking three Poisson processes, say I, II, and III, and superposing processes I and II to obtain the events of type a and superposing II and III to obtain the events of type b (Marshall and Olkin [33]). Clearly, the bivariate process is marginally regular but not regular. However, if the events of type b are made up of process III events superposed with process II events delayed by a fixed amount, the resulting bivariate process is regular. A commonly used alternative to the word *regular* is *orderly*.

2.2. *Independence and stationarity.* Independence of the marginal processes in a bivariate process is intuitively defined as independence of the number of events (counts) in any two sets of intervals in the marginal processes. The more difficult problem of specifying dependence (and correlation) in the bivariate process is central to this paper and will be taken up in the next section.

In the sequel, we will be primarily concerned with *transient* or *stationary* bivariate point processes, as opposed to *nonhomogeneous* processes. The latter type of process is defined roughly as one with either an evolutionary or cyclic trend, whereas a *transient* process is roughly one whose probabilistic structure eventually becomes stationary (time invariant). There are a number of types of stationarity which need to be defined more carefully.

DEFINITION 2.1 (Simple stationarity). *Let* $N^{(a)}(t_1^{(1)}, t_1^{(1)} + \tau_1^{(1)})$ *be the number of events of type a in the interval* $(t_1^{(1)}, t_1^{(1)} + \tau_1^{(1)}]$ *and* $N^{(b)}(t_2^{(1)}, t_2^{(1)} + \tau_2^{(1)})$ *be the number of events of type b in the interval* $(t_2^{(1)}, t_2^{(1)} + \tau_2^{(1)}]$. *The bivariate point process is said to have* simple stationarity *if*

$$\begin{aligned} Pr\{N^{(a)}(t_1^{(1)}, t_1^{(1)} + \tau_1^{(1)}) = n^{(a)};\ N^{(b)}(t_2^{(1)}, t_2^{(1)} + \tau_2^{(1)}) = n^{(b)}\} \\ = Pr\{N^{(a)}(t_1^{(1)} + y, t_1^{(1)} + \tau_1^{(1)} + y) = n^{(a)}; \\ N^{(b)}(t_2^{(1)} + y, t_2^{(1)} + \tau_2^{(1)} + y) = n^{(b)}\}, \end{aligned} \tag{2.2}$$

for all $t_1, t_2, \tau_1^{(1)}, \tau_2^{(1)}, y > 0$.

In other words, the joint distribution of the number of type a events in a fixed interval and the number of type b events in another fixed interval is invariant under translation.

Simple stationarity of the bivariate process implies an analogous property for the individual marginal processes and for the superposed process.

In the sequel, we assume that for the marginal processes considered individually the probabilities of more than one type a event in τ_1 and more than one type b event in τ_2 are, respectively, $\rho_a\tau_1 + o(\tau_1)$ as $\tau_1 \to 0$ and $\rho_b\tau_2 + o(\tau_2)$ as $\tau_2 \to 0$, where ρ_a and ρ_b are finite.

Simple stationarity and these finiteness conditions imply that the univariate forward recurrence time relationships in the marginal processes and the pooled processes hold (Lawrance [25]).

If in addition the process is regular, Korolyuk's theorem implies that ρ_a, ρ_b, and $\rho_a + \rho_b$ are, respectively, the rates of events of types a, events of types b, and events regardless of type.

DEFINITION 2.2 (Second order stationarity). *By extension, we say that the bivariate point process has* second order stationarity (*weak stationarity*) *if the joint distribution of the number of type a events in two fixed intervals and the number of type b events in another two fixed intervals is invariant under translation.*

This type of stationarity is necessary in the sequel for the definition of a time invariant cross intensity function. Clearly, it implies second order stationarity for the marginal processes considered individually and for the superposed marginal processes.

DEFINITION 2.3 (Complete stationarity). *By extension,* complete stationarity *for a bivariate point process is invariant under translation for the joint distribution of counts in arbitrary numbers of intervals in each process.*

2.3. *Asynchronous counts and intervals.* In specifying stationarity, we did not mention the time origin or the method of starting the process. There are three main possibilities.

(i) The process is started at time $t = 0$ with initial conditions which produce stationarity, referred to as *stationary initial conditions.*

(ii) The process is transient and is considered beyond $t = 0$ as its start moves off to the left. The process then becomes stationary as the start moves to minus infinity. There is generally a specification of the state of the process at $t = 0$ known as the *stationary equilibrium conditions.*

Note that in both (i) and (ii) stationarity is defined by invariance under shifts to the right.

(iii) In a stationary point process, a time is specified without knowledge of the events and is taken to be the origin, $t = 0$. The time $t = 0$ is said to be an *arbitrary time* in the (stationary) process, selected by an *asynchronous sampling* of the process.

Now there is associated with the stationary bivariate point process a counting process $\mathbf{N}(t_1, t_2) = \{N^{(a)}(t_1), N^{(b)}(t_2)\}$, where

$$N^{(a)}(t_1) \text{ is the number of type } a \text{ events in } (0, t_1], \tag{2.3}$$
$$N^{(b)}(t_2) \text{ is the number of type } b \text{ events in } (0, t_2],$$

and a bivariate sequence of intervals $\{X^{(a)}(i), X^{(b)}(j)\}$, where, assuming regularity of the process, $X^{(a)}(1)$ is the forward recurrence time in the process of type a events (that is, the time from $t = 0$ to the first type a event), $X^{(a)}(2)$ is the time between the first and second type a events, and so forth; and the $\{X^{(b)}(j)\}$ sequence is defined similarly.

Note that for asynchronous sampling of a stationary process the indices i and j can take negative values; in particular, $\{X^{(a)}(-1), X^{(b)}(-1)\}$ are the bivariate backward recurrence times.

There is a fundamental relationship connecting the bivariate counting processes with the bivariate interval processes; this is a direct generalization of the relationship for the univariate case:

$$N^{(a)}(t_1) < n^{(a)}, \qquad N^{(b)}(t_2) < n^{(b)}, \tag{2.4}$$

if and only if

$$\begin{aligned} S^{(a)}(n^{(a)}) &= X^{(a)}(1) + \cdots + X^{(a)}(n^{(a)}) > t_1, \\ S^{(b)}(n^{(b)}) &= X^{(b)}(1) + \cdots + X^{(b)}(n^{(b)}) > t_2. \end{aligned} \tag{2.5}$$

Probability relationships are written down directly from these identities connecting the bivariate distribution of counts with the bivariate distribution of the sums of intervals $S^{(a)}(n^{(a)})$ and $S^{(b)}(n^{(b)})$.

Equations (2.4) and (2.5) can be used, for example, to prove the asymptotic bivariate normality of $\{N^{(a)}(t_1), N^{(b)}(t_2)\}$ for a broad class of bivariate point processes.

2.4. *Semisynchronous sampling.* In a univariate point process, *synchronous* sampling of the stationary process refers to the placement of the time origin at an arbitrary event and the examination of the counts and intervals following this arbitrary event (Cox and Lewis [14], Chapter 4, and McFadden [35]). In more precise terms (Leadbetter [27], [28], and Lawrance [24]), the notion of an arbitrary event in a stationary point process is the event $\{N(0, \tau) \geqq 1\}$ as τ tends to zero, and the distribution function $F(t)$ of the interval between the arbitrary event and the following event is defined to be

$$1 - F(t) = \lim_{\tau \to 0+} Pr\{N(\tau, \tau + t) = 0 | N(0, \tau) \geqq 1\}. \tag{2.6}$$

In bivariate point processes, the situation is more complex. Synchronous sampling of the marginal process of type a events produces *semisynchronous sampling* of the process of b events from an arbitrary a event, and *vice versa*.

The bivariate counting processes and intervals following these two types of sampling are denoted as follows:

(a) for semisynchronous sampling of b by a,

$N_a^{(a)}(t_1)$ is the number of type a events following an origin at a type a event;

$N_a^{(b)}(t_1)$ is the number of type b events following an origin at a type a event;

$\{X_a^{(a)}(i)\}$ is, for $i = 1$, the time from the origin at a type a event to the next event of type a, and for $i = 2, 3, \cdots$, the intervals between subsequent type a events;

$\{X_a^{(b)}(j)\}$ is, for $j = 1$, the time from the origin at a type a event to the first subsequent type b event, and for $j = 2, 3, \cdots$, the intervals between subsequent type b events;

(b) for semisynchronous sampling of a by b, the subscript becomes b instead of a, in the above expressions, indicating the nature of the origin.

Note that in general (Slivnyak [44]) the sequences $\{X_a^{(a)}(i)\}$ and $\{X_b^{(b)}(j)\}$, being the synchronous interval processes in the marginal point processes, are stationary, whereas $\{X_a^{(b)}(i)\}$ and $\{X_b^{(a)}(j)\}$, the semisynchronous intervals, are in general not stationary. Also for independent processes, the semisynchronous sequences are identical with the asynchronous sequences $\{X^{(a)}(i)\}$ and $\{X^{(b)}(j)\}$.

2.5. *Pooling and superposition of processes.* In discussing regularity, we referred to the superposition of the two marginal processes in the bivariate process. This is the univariate process of events of both types considered without specification of the event type and is referred to simply as the *superposed process.* Study of the superposed process of rate $\rho_a + \rho_b$ is an intimate part of the analysis of the bivariate process. Asynchronous sampling of the superimposed process gives counts and intervals denoted by $N^{(\cdot)}(t_1)$ and $\{X^{(\cdot)}(i)\}$, whereas synchronous sampling, that is, the process considered conditionally (in the Khinchin sense) on the existence of an event of an unspecified type at the origin, gives $N_\cdot^{(\cdot)}(t_1)$ and $\{X_\cdot^{(\cdot)}(i)\}$.

Semisynchronous sampling of the superposed process by events of type a or type b is also possible and the notation should be clear.

We call the superposed process with specification of the event type the *pooled* process. The original bivariate process can then be respecified in terms of the process

$$\{X_\cdot^{(\cdot)}(i), \qquad T_\cdot(i)\}, \tag{2.7}$$

where $T_\cdot(i)$ is a binary valued process indicating the type of the ith event after the origin in the superposed process with synchronous sampling. Clearly, the marginal processes of event types, that is, $\{T(i)\}$, $\{T_\cdot(i)\}$, $\{T_a(i)\}$ and $\{T_b(i)\}$ are themselves of interest. Note that they are in general not stationary processes for all types of sampling and are related to the processes defined in Sections 2.3 and 2.4. Thus, for example,

$$\{X_a^{(\cdot)}(1) \leqq x;\, T_a(1) = a\} \Leftrightarrow \{X_a^{(a)} < X_a^{(b)};\, X_a^{(a)} \leqq x\}, \tag{2.8}$$

with much more complicated statements relating events of higher index i. The binary sequence of event types has no counterpart in univariate point process.

Thus, there are many possible representations of a bivariate point process. Which is the most fruitful is likely to depend on the particular application.

As a very simple practical example of these representations, consider a generalization of the alternating renewal process. We have a sequence of positive random variables $W(1)$, $Z(1)$, $W(2)$, $Z(2)$, $\cdots$, representing operating and repair intervals in a machine. It is natural to assume that $W(i)$ and $Z(i)$ are mutually correlated but independent of other pairs of operating and repair times. Type a events, occurring at the end of the $W(i)$ variables, are machine failures. Type b events occur at the end of $Z(i)$ variables and represent times at which the machine goes back into service.

Specification of the process is straightforward and simple in terms of the pooled process variables $\{X_{\cdot}^{(\cdot)}(i), T_{\cdot}(i)\}$, $\{X_a^{(\cdot)}(i), T_a(i)\}$, and so forth. However, marginally the type b events are a renewal process, whereas the type a events are a nonrenewal, non-Markovian point process and the dependency structure expressed through the intervals in the marginals is complex.

2.6. *Successive semisynchronous sampling.* Finally, we mention the possibility of successive semisynchronous samples of the marginal process of type b events by a events. The origin is at an a event, as in ordinary semisynchronous sampling and connected with this a event is the time forward (or backward) to the next b event. Subsequent a events are associated with the times forward (or backward) to the next b event. It is not clear how generally useful this procedure is in studying bivariate point processes. It has been used, however, by Brown [7] in studying identifiability problems in $M/G/\infty$ queues; see also Section 6.4.

2.7. *Palm-Khinchin formulae.* In the theory of univariate point processes, there are relations connecting the distributions of sums of synchronous intervals and sums of asynchronous intervals. Similar relationships connect the synchronous and asynchronous counting processes. These relationships are sometimes called the Palm-Khinchin formulae and are given, for example, by Cox and Lewis ([14], Chapter 4).

The best known of these relations connects the distributions of the synchronous and asynchronous forward recurrence times in a stationary point process with finite rate ρ (Lawrance [25]). In the context of the marginal process of type a events,

$$\rho_a\{1 - F_{X_a^{(a)}}(t)\} = D_t^+ F_{X^{(a)}}(t), \tag{2.9}$$

where D_t^+ denotes a right derivative. For moments when the relevant moments exist we have

$$E\{(X^{(a)})^r\} = \frac{\rho_a E\{(X_a^{(a)})^{r+1}\}}{r + 1}. \tag{2.10}$$

Palm-Khinchin type formulae for bivariate point processes have been developed by Wisniewski [50], [51]. They are far more complex than those for univariate processes, both in terms of the number of relationships involved and in the analytical problems encountered. Thus, on the first point there are not only interval relationships, but also relationships between the probabilistic structures of the binary sequences $\{T(i)\}$ and $\{T_{\cdot}(i)\}$ and between the probabilistic structures of the binary sequence $\{T(i)\}$ and the binary sequences $\{T_a(i)\}$ and $\{T_b(i)\}$.

On the second point, the analytical problems are illustrated by the following argument. It is easily shown that an arbitrarily selected point in the (univariate) superposed process is of type a with probability $\rho_a/(\rho_a + \rho_b)$. Thus, any probabilistic statement about the variables $\{X_\cdot^{(\cdot)}(i), T_\cdot(i)\}$, say $g(X_\cdot^{(\cdot)}(1), T_\cdot(1), \cdots)$, is expressible in terms of the same probabilistic statement for $\{X_a^{(\cdot)}(i), T_a(i)\}$ and $\{X_b^{(\cdot)}(i), T_b(i)\}$,

$$g(X_\cdot^{(\cdot)}(1), T_\cdot(1), \cdots) = \frac{\rho_a}{\rho_a + \rho_b} g(X_a^{(\cdot)}(1), \cdots) + \frac{\rho_b}{\rho_a + \rho_b} g(X_b^{(\cdot)}(1), \cdots). \tag{2.11}$$

Now if a relationship between $g(X^{(\cdot)}(1), T(1), \cdots)$, and $g(X_\cdot^{(\cdot)}(1), T_\cdot(1), \cdots)$ exists, we can relate the asynchronous sequence to the two semisynchronous sequences through (2.11). But the usual univariate Palm-Khinchin formulae relate univariate distributions of *sums* of asynchronous intervals to univariate distributions of sums of synchronous intervals. Clearly, formulae relating *joint* properties of asynchronous intervals and types to joint properties of synchronous intervals $X_\cdot(i)$ and types $T_\cdot(i)$ are needed if one is, for example, to relate, through (2.11) and generalizations of (2.8), bivariate distributions of asynchronous forward recurrence times $\{X^{(a)}(1), X^{(b)}(1)\}$ to the bivariate distributions of the semisynchronous forward recurrence times $\{X_a^{(a)}(1), X_a^{(b)}(1)\}$ and $\{X_b^{(a)}(1), X_b^{(b)}(1)\}$.

Lawrance [26] has noted this need for extended Palm-Khinchin formulae and conjectured results in the univariate case.

Of Wisniewski's results [50], [51], we cite here only two moment formulae. These relate the moments of the joint asynchronous forward recurrence times $\{X^{(a)}(1), X^{(b)}(1)\}$ with the moments of both of the semisynchronous forward recurrence times, $\{X_a^{(a)}(1), X_a^{(b)}(1)\}$ and $\{X_b^{(a)}(1), X_b^{(b)}(1)\}$. The feature that probabilistic properties of both semisynchronous sequences are needed to determine probabilistic properties of the asynchronous sequence is characteristic of all these relationships, and follows from (2.11).

We have for the bivariate analogues to (2.10), for $r = 1$,

$$\begin{aligned} &\tfrac{1}{2}\rho_a E[\{X_a^{(a)}(1)\}^2] + \tfrac{1}{2}\rho_b E[\{X_b^{(b)}(1)\}^2] \\ &\quad = E\{X^{(a)}\} + E\{X^{(b)}\} \\ &\quad = \rho_a E\{X_a^{(a)}(1)\, X_a^{(b)}(1)\} + \rho_b E\{X_b^{(a)}(1)\, X_b^{(b)}(1)\} \end{aligned} \tag{2.12}$$

and

$$\begin{aligned} &12E\{X^{(a)}(1)\, X^{(b)}(1)\} \\ &\quad = \rho_a E[3X_a^{(a)}(1)\,\{X_a^{(b)}(1)\}^2 + 3\{X_a^{(a)}(1)\}^2\, X_a^{(b)}(1) - \{X_a^{(a)}(1)\}^3] \\ &\qquad + \rho_b E[3X_b^{(b)}(1)\,\{X_b^{(a)}(1)\}^2 + 3\{X_b^{(b)}(1)\}^2\, X_b^{(a)}(1) - \{X_b^{(b)}(1)\}^3]. \end{aligned} \tag{2.13}$$

The interesting feature of (2.12) is that the correlation between semisynchronous forward recurrence times is a function only of the properties of the marginal processes and not of the dependency structure of the bivariate point process. Moreover, (2.13) shows that if we use correlation between the asynchronous forward recurrence times as a measure of dependence in the bivariate point process, this dependence only affects the third order joint moments of the semisynchronous forward recurrence times.

2.8. *Examples.* To illustrate the definitions and concepts introduced above, we consider two very simple bivariate point processes. The analytical details developed here will be used in Section 6 in considering the statistical analysis of bivariate point processes.

EXAMPLE 2.1. *Independent renewal processes.* Consider two *delayed* renewal processes $\{X^{(a)}(1); X_a^{(a)}(i), i = 2, 3, \cdots\}$ and $\{X^{(b)}(1); X_b^{(b)}(j), j = 2, 3, \cdots\}$, where using a shortened notation,

$$G^{(a)}(x) = Pr\{X^{(a)}(1) \leqq x\} = \frac{\int_0^x \{1 - F^{(a)}(u)\}\, du}{E_a(X)}, \tag{2.14}$$

$$Pr\{X_a^{(a)}(i) \leqq x\} = F^{(a)}(x), \quad E_a(X) = \int_0^\infty x\, dF^{(a)}(x), \tag{2.15}$$

with similar definitions for the process of type b events. The distribution of the variable $X^{(a)}(1)$ in (2.14) and the analogous distribution for $X^{(b)}(1)$ are the stationary initial conditions for the marginal renewal processes, and clearly the independence of the processes implies that these distributions (jointly) give stationarity to the bivariate process. Because of independence there is no difference between semisynchronous and asynchronous sampling; the process is defined completely in terms of the properties of the asynchronous and synchronous intervals.

Properties of intervals in the superposed process, and properties of successive intervals and event types in the pooled process, that is, $\{X^{(\cdot)}(i), T(i)\}$ are very difficult to obtain explicitly. The sequences $X^{(\cdot)}(i)$ and $T(i)$ are neither stationary nor independent, but contain transient effects. We have for example

$$\begin{aligned} Pr\{X^{(\cdot)}(1) > x; T(1) = a\} &= Pr\{X^{(b)}(1) > X^{(a)}(1) > x\} \\ &= \int_x^\infty \{1 - G^{(b)}(y)\}\, dG^{(a)}(y) \end{aligned} \tag{2.16}$$

and, marginally,

$$Pr\{T(1) = a\} = \int_0^\infty \{1 - G^{(b)}(y)\}\, dG^{(a)}(y). \tag{2.17}$$

The only simple case is where the two renewal processes are Poisson processes with parameters ρ_a and ρ_b. Then, of course, $\{X^{(\cdot)}(i)\}$ is a Poisson process of rate $\rho_a + \rho_b$ and $T(i)$ is an independent binomial sequence

$$Pr\{T(1) = a\} = Pr\{T(1) = a | X^{(\cdot)} > x\} = \frac{\rho_a}{\rho_a + \rho_b}. \tag{2.18}$$

EXAMPLE 2.2. *Semi-Markov processes (Markov renewal processes).* The two state semi-Markov process is the simplest bivariate process with dependent structure and plays, in bivariate process theory, a role similar to that played in univariate process theory by the renewal process. It is, in a sense, the closest one gets in bivariate processes to a regenerative process. The process is defined in terms of the sequences $\{X_a^{(\cdot)}(i), T_a(i)\}$, and $\{X_b^{(\cdot)}(i), T_b(i)\}$, the type processes $\{T_a(i)\}$ and $\{T_b(i)\}$ being Markov chains with transition matrix

$$\mathbf{P} = \begin{pmatrix} p_{aa} & p_{ab} \\ p_{ba} & p_{bb} \end{pmatrix} = \begin{pmatrix} \alpha_1 & 1 - \alpha_1 \\ 1 - \alpha_2 & \alpha_2 \end{pmatrix}, \tag{2.19}$$

while the distributions of the random variables $X_a^{(\cdot)}(i)$, $X^{(\cdot)}(i)$, and $X_b^{(\cdot)}(i)$ depend only on the type of events at i and $(i - 1)$. Thus, illustrating the regenerative nature of the process, we define $F_{aa}(x)$ to be, for $i \geqq 2$,

$$\begin{aligned} F_{aa}(x) &= Pr\{X_a^{(\cdot)}(i) \leqq x \,|\, T_a(i) = a, T_a(i-1) = a\} \\ &= Pr\{X_\cdot^{(\cdot)}(i) \leqq x \,|\, T_\cdot(i) = a, T_\cdot(i-1) = a\} \\ &= Pr\{X^{(\cdot)}(i) \leqq x \,|\, T(i) = a, T(i-1) = a\} \end{aligned} \tag{2.20}$$

with equivalent definitions for $F_{ab}(x)$, $F_{ba}(x)$, $F_{bb}(x)$. Thus, the effect of the initial sampling disappears when the type of the first subsequent events is known.

The joint distributions of the time from the origin to the first event and the type of the first event ($i = 1$) are either quite arbitrary initial conditions, or initial conditions established by the kind of sampling involved at the origin and denoted by the subscript on the interval random variable. Thus for asynchronous sampling, we get stationary initial conditions which are specified by the joint distribution of $X^{(\cdot)}(i)$ and $T(1)$.

These stationary equilibrium conditions (Pyke and Schaufele [43]) are that $T(1) = a$ and $T(1) = b$ have probabilities p_a and p_b, where

$$\{p_a, p_b\} = \{p_a, p_b\}\,\mathbf{P} = \left\{\frac{1 - \alpha_2}{2 - \alpha_1 - \alpha_2}, \frac{1 - \alpha_1}{2 - \alpha_1 - \alpha_2}\right\}, \tag{2.21}$$

the equilibrium probabilities of the Markov chain, and the time from the origin to an event of type a has distribution function

$$\frac{p_{ba} \int_0^x R_{ba}(u)\,du}{E(X_b^{(b)}(1))} + \frac{p_{aa} \int_0^x R_{aa}(u)\,du}{E(X_a^{(a)}(1))}, \tag{2.22}$$

with a similar definition for the time to an event of type b.

Cinlar [9] has reviewed the properties of semi-Markov processes. Our view of these processes, being related to statistical problems arising in the analysis of bivariate point processes, will be somewhat different from the usual one. Thus, note that in the marginal processes the regenerative property of the semi-Markov process implies that the times between events of type a, $X_a^{(a)}(i)$, $i = 1, 2, \cdots$, are independent and identically distributed, as are the $X_b^{(b)}(j)$, $j = 1, 2, \cdots$. Therefore, the marginal processes are renewal processes and we say that the semi-Markov process is a *bivariate renewal process*. Since the types of successive renewals (events) form a Markov chain, the process is also called a Markov renewal process ([9], p. 130). However, the two marginal renewal processes together with the Markov chain of event types do not determine the process.

The dependency structure of this bivariate renewal process can also be examined through joint properties of forward recurrence times in the process. The joint forward recurrence times $\{X_a^{(a)}(1), X_a^{(b)}(1)\}$ for semisynchronous sampling of b by a are, in the terminology of semi-Markov process theory, the first passage times from state a to state a and from state a to state b, with similar definitions for $\{X_b^{(a)}(1), X_b^{(b)}(1)\}$. Denoting the marginal distributions of these random variables by $F_a^{(a)}(x)$ and so forth, we have the equations

$$F_a^{(a)}(x) = p_{aa}F_{aa}(x) + p_{ab}F_{ab}(x) * F_b^{(a)}(x), \tag{2.23}$$

$$F_a^{(b)}(x) = p_{ab}F_{ab}(x) + p_{aa}F_{aa}(x) * F_a^{(b)}(x), \tag{2.24}$$

$$F_b^{(a)}(x) = p_{ba}F_{ba}(x) + p_{bb}F_{bb}(x) * F_b^{(a)}(x), \tag{2.25}$$

$$F_b^{(b)}(x) = p_{bb}F_{bb}(x) + p_{ba}F_{ba}(x) * F_a^{(b)}(x), \tag{2.26}$$

where $*$ denotes Stieltjes convolution and $F_{aa}(x)$ and so forth, are defined in (2.20).

These equations can be solved using Laplace–Stieltjes transforms. Thus, if $\mathscr{F}_a^{(a)}(s)$ is the Laplace–Stieltjes transform of $F_a^{(a)}(x)$, and so forth, we get

$$\mathscr{F}_a^{(a)}(s) = \alpha_1 \mathscr{F}_{aa}(s) + \frac{(1 - \alpha_1)(1 - \alpha_2)\mathscr{F}_{ab}(s)\mathscr{F}_{ba}(s)}{1 - \alpha_2 \mathscr{F}_{bb}(s)}, \tag{2.27}$$

$$\mathscr{F}_b^{(b)}(s) = \alpha_1 \mathscr{F}_{bb}(s) + \frac{(1 - \alpha_1)(1 - \alpha_2)\mathscr{F}_{ab}(s)\mathscr{F}_{ba}(s)}{1 - \alpha_1 \mathscr{F}_{aa}(s)}. \tag{2.28}$$

From these results, we can write down joint forward recurrence time distributions using the regenerative properties of the process. For example,

$$\begin{aligned} (2.29)\quad & R_{X_a^{(a)}(1), X_a^{(b)}}(x_1, x_2) \\ &= Pr\{X_a^{(a)}(1) > x_1, X_a^{(b)} > x_2\} \\ &= \begin{cases} \alpha_1 R_{aa}(x_1) + (1 - \alpha_1)\left\{R_{ab}(x_1) + \displaystyle\int_{x_2}^{x_1} [1 - F_b^{(a)}(x_1 - u)]\, dF_{ab}(u)\right\} \\ \hfill \text{if } x_1 \geqq x_2, \\ (1 - \alpha_1)R_{ab}(x_2) + \alpha_1 \left\{R_{aa}(x_2) + \displaystyle\int_{x_1}^{x_2} [1 - F_a^{(b)}(x_2 - u)]\, dF_{aa}(u)\right\} \\ \hfill \text{if } x_2 \geqq x_1. \end{cases} \end{aligned}$$

It is actually much simpler, because of the regenerative nature of the process, to express results in terms of the order statistics and order types associated with $R(x_1, x_2)$. These aspects of the process are worked out in greater detail by Wisniewski [50].

Note that the process derived previously as two independent Poisson processes is a very particular form of semi-Markov process. The question then arises whether there are any other semi-Markov processes with Poisson marginals and the answer is clearly yes. For example, when $\alpha_1 = \alpha_2 = 0$ we have the special case of an alternating renewal process and choosing $F_{ab}(x)$ and $F_{ba}(x)$ to be distributions of random variables proportional to chi square variables with one degree of freedom gives Poisson marginals. The example shows in fact that one can produce any desired marginal renewal processes in a semi-Markov process, as is also clear from (2.27) and (2.28).

From equations such as (2.14) and (2.15), it can be shown that no bivariate process of independent renewal marginals is a semi-Markov process unless the marginals are also Poisson processes. The dependency structure in a semi-Markov process is actually better characterized by the second order cross intensity function, which we introduce in the next section, rather than by joint moments of forward recurrence times. This cross intensity function together with the two distributions of intervals in the marginal renewal processes (or equivalently the intensity functions of the marginal renewal processes) completely specifies the semi-Markov process.

3. Dependence and correlation in bivariate point processes

3.1. *Specification.* We now consider in more detail the specification of the structure of bivariate point processes. It is common in the study of particular stochastic processes to find that physically the same process can be specified in several equivalent but superficially different ways. A simple and familiar example is the stationary univariate Poisson process which can be specified as:

(a) a process in which the numbers of events in disjoint sets have independent Poisson distributions with means proportional to the measures of the sets;

(b) a renewal process with exponentially distributed intervals;

(c) a process in which the probability of an event in $(t, t + \Delta t]$ has an especially simple form, as $\Delta t \to 0$.

We call these three specifications, respectively, the counting, the interval, and the intensity specifications. Univariate point processes can in general be specified in these three ways, if the initial conditions are properly chosen.

While the counting specification (a) for bivariate point processes is in principle fundamental, it is often too complicated to be very fruitful. If the joint characteristic functional of the process, defined by an obvious generalization of the univariate case, can be obtained in a useful form, this does give a concise representation of all the joint distributions of counts; even then, such a characteristic functional would usually give little insight into the physical mechanism generating the process.

Often, special processes are most conveniently handled through some kind of interval specification, especially when this corresponds rather closely to the physical origin of the process. In particular, the two state semi-Markov process is most simply specified in this way, as shown in Section 2.8. The two main types of interval specifications discussed in Section 2 were the specifications in terms of the intervals in the marginal processes, or the intervals and event types in the pooled process. The latter is the basic specification for the semi-Markov process. Other processes, such as various kinds of inhibited processes and the bivariate Poisson process of Section 4.1 are specified rather less directly in terms of relations between intervals and event types.

However, in some ways the most convenient general specification is through the intensity. Denote by $\mathscr{H}_t$ the history of the process at time t, that is, a complete specification of the occurrences in $(-\infty, t]$ measured backwards from an origin at t, then two time points t', t'' have the same history if and only if the observed sequences $\{x^{(a)}(-1), x^{(a)}(-2), \cdots\}$, $\{x^{(b)}(-1), x^{(b)}(-2), \cdots\}$ are identical if measured from origins at t' and at t''.

Then a marginally regular process is specified by

$$\lambda^{(a)}(t; \mathscr{H}_t) = \lim_{\Delta t \to 0+} \frac{Pr\{N^{(a)}(t, t + \Delta t) \geqq 1 | \mathscr{H}_t\}}{\Delta t}, \tag{3.1}$$

$$\lambda^{(b)}(t; \mathscr{H}_t) = \lim_{\Delta t \to 0+} \frac{Pr\{N^{(b)}(t, t + \Delta t) \geqq 1 | \mathscr{H}_t\}}{\Delta t}, \tag{3.2}$$

$$\lambda^{(ab)}(t; \mathscr{H}_t) = \lim_{\Delta t \to 0+} \frac{Pr\{N^{(a)}(t, t + \Delta t) N^{(b)}(t, t + \Delta t) \geqq 1 | \mathscr{H}_t\}}{\Delta t}. \tag{3.3}$$

We call these functions the *complete intensity functions* of the process. If the process is regular $\lambda^{(ab)}(t; \mathscr{H}_t) = 0$. Given the functions (3.1) through (3.3) and some initial conditions, we can construct a discretized realization of the process, although this is, of course, a clumsy method of simulation if the interval specification is at all simple.

One advantage of the complete intensity specification is that one can generate families of models of increasing complexity by allowing more and more complex dependency on $\mathscr{H}_t$. This may be useful, for instance, in testing consistency of data with a given type of model, for example, a semi-Markov process. Further, if the main features of $\mathscr{H}_t$ that determine the intensity functions can be found, an appropriate type of model may be indicated.

As an example of a complete intensity specification, consider the two state semi-Markov process. Here the only aspects of $\mathscr{H}_t$ that are relevant, if at least one event has occurred before t, are the backward recurrence time to the previous event and the type of that event $\{x^{(\cdot)}(-1), t(-1)\}$. Any initial conditions disappear once one event has occurred. For convenience, we write the partial history as (u, a), if the preceding event is of type a and (u, b) if it is of type b. Then assuming that the process is regular, that is, that none of the interval distributions has an atom at zero, we have

$$\lambda^{(a)}\{t; (u, a)\} \equiv \lambda_a^{(a)}(u) = \frac{p_{aa} f_{aa}(u)}{p_{aa} R_{aa}(u) + p_{ab} R_{ab}(u)}, \tag{3.4}$$

$$\lambda^{(b)}\{t;\,(u,a)\} \equiv \lambda_a^{(b)}(u) = \frac{p_{ab}f_{ab}(u)}{p_{aa}R_{aa}(u) + p_{ab}R_{ab}(u)}, \tag{3.5}$$

with similar expressions defining $\lambda_b^{(a)}(u)$ and $\lambda_b^{(b)}(u)$ when the partial history is (u, b). These complete intensities are analogues of the hazard function, or age specific failure rate, which can be used to specify a univariate renewal process.

The semi-Markov process is characterized by the dependence on $\mathscr{H}_t$ being only on u and the type of the preceding event. We can generalize the semi-Markov process in many ways, for instance by allowing a dependence on both of the backward recurrence times $x^{(a)}(-1)$ and $x^{(b)}(-1)$; see Section 3.8.

3.2. *Properties of complete intensity functions.* We now consider briefly some properties of complete intensity functions. It is supposed for simplicity that the process is regular and that it is observed for a time long enough to allow initial conditions to be disregarded. For the semi-Markov process "long enough" is the occurrence of at least one event.

(i) If the process is completely stationary, as defined in Section 2.3, the intensity functions depend only on $\mathscr{H}_t$ and not on t.

(ii) Nonstationary generalizations of a given stationary process can be produced by inserting into the intensity a function either of t, for example, $e^{\gamma t}$, $\exp\{\gamma\cos(\omega_0 t + \phi)\}$, or of the numbers of events that have occurred since the start of the process.

(iii) The intensity specification of a stationary process is unique in the sense that if we have two different intensity specifications and can find a set of histories of nonzero probability such that, say, the first intensity specification gives greater intensity of events of type a than the second, then the two processes are distinguishable from suitable data. Note that this is not the same question as whether two different specifications containing unknown parameters are distinguishable.

(iv) The events of different types are independent if and only if $\lambda^{(a)}$ and $\lambda^{(b)}$ involve $\mathscr{H}_t$ only through the histories of the separate processes of events of type a and type b, denoted, respectively, by $\mathscr{H}_t^{(a)}$ and $\mathscr{H}_t^{(b)}$.

(v) We can call the process purely a dependent if both $\lambda^{(a)}$ and $\lambda^{(b)}$ depend on $\mathscr{H}_t$ only through $\mathscr{H}_t^{(a)}$. In many ways the simplest example of such a process is obtained when both intensities depend only on the backward recurrence time in the process of events of type a, that is, on the time $u^{(a)}$ measured back to the previous event of type a. Denote the intensities by $\lambda^{(a)}(u^{(a)})$ and $\lambda^{(b)}(u^{(a)})$. Then the events of type a form a renewal process; if in particular $\lambda^{(a)}(\cdot)$ is constant, the events of type a form a Poisson process. If simple functional forms are assumed for the intensities, the likelihood of data can be obtained in a fairly simple form and hence an efficient statistical analysis derived.

(vi) A different kind of purely a dependent process is derived from a shot noise process based on the a events, that is, by considering a stochastic process $Z^{(a)}(t)$ defined by

$$Z^{(a)}(t) = \int_0^\infty g(u)\,dN^{(a)}(t-u). \tag{3.6}$$

We then take $\lambda^{(b)}(t)$ and possibly also $\lambda^{(a)}(t)$ to depend only on $Z^{(a)}(t)$. In particular if $g(u) = 1(u < \Delta)$ and $g(u) = 0(u \geqq \Delta)$, the intensities depend only on the number of events of type a in $(t - \Delta, t)$. Hawkes [22] has considered some processes of this type.

(vii) The intensity functions look in one direction in time. This approach is therefore rather less suitable for processes in a spatial continuum, where there may be no reason for picking out one spatial direction rather than another.

(viii) Some simple processes, for example, the bivariate Poisson process of Section 4.1, have intensity specifications that appear quite difficult to obtain.

3.3. *Second order cross intensity functions.* In Section 3.2, we considered the complete intensity functions which specify probabilities of occurrence given the entire history $\mathscr{H}_t$. For some purposes, it is useful with stationary (second order) processes to be less ambitious and to consider probabilities of occurrence conditionally on much less information than the entire history $\mathscr{H}_t$. We then call the functions corresponding to (3.1) through (3.3) incomplete intensity functions. For example, using the notation of (3.4) and (3.5), both

$$\lim_{\Delta t \to 0+} Pr \frac{\{N^{(a)}(t, t + \Delta t) \geqq 1 \,|\, (u, a)\}}{\Delta t} \tag{3.7}$$

and the corresponding function when the last event is of type b are defined for any regular stationary process, even though they specify the process completely only for semi-Markov processes.

A particularly important incomplete intensity function is obtained when one conditions on the information that an event of specified type occurs at the time origin. Again for simplicity, we consider stationary regular processes. Write

$$h_a^{(a)}(t) = \lim_{\Delta t \to 0+} Pr \frac{\{N^{(a)}(t, t + \Delta t) \geqq 1 \,|\, \text{type } a \text{ event at } 0\}}{\Delta t}, \tag{3.8}$$

$$h_a^{(b)}(t) = \lim_{\Delta t \to 0+} Pr \frac{\{N^{(b)}(t, t + \Delta t) \geqq 1 \,|\, \text{type } a \text{ event at } 0\}}{\Delta t}, \tag{3.9}$$

with similar definitions for $h_b^{(a)}(t)$, $h_b^{(b)}(t)$ if an event of type b occurs at 0. We call the function (3.9) a second order cross intensity function. For nonregular processes, it may be helpful to introduce intensities conditionally on events of both types occurring at 0.

Note that the cross intensity functions $h_a^{(b)}(t)$ and $h_b^{(a)}(t)$ will contain Dirac delta functions as components if, for example, there is a nonzero probability that an event of type b will occur exactly τ away from a type a event. For the process to be regular, rather than merely marginally regular, the cross intensity functions must not contain delta functions at the origin.

Note too that $h_a^{(b)}(t)$ is well defined near $t = 0$ and will typically be continuous there.

If following an event of type a at the origin, subsequent events of type a are at times $S_a^{(a)}(1), S_a^{(a)}(2), \cdots$ and those of type b at $S_a^{(b)}(1), S_a^{(b)}(2), \cdots$, we can write

$$h_a^{(a)}(t) = \sum_{r=1}^{\infty} f_{S_a^{(a)}(r)}(t), \tag{3.10}$$

$$h_a^{(b)}(t) = \sum_{r=1}^{\infty} f_{S_a^{(b)}(r)}(t), \tag{3.11}$$

where $f_U(\cdot)$ is the probability density function of the random variable U.

If the process of type a events is a renewal process, (3.10) is a function familiar as the renewal density. For small t the contribution from $r = 1$ is likely to be dominant in all these functions.

The intensities are defined for all t. However, $h_a^{(a)}(t)$ and $h_b^{(b)}(t)$ are even functions of t. Further, it follows from the definition of conditional probability that

$$h_b^{(a)}(t)\rho_b = h_a^{(b)}(-t)\rho_a, \tag{3.12}$$

where ρ_a and ρ_b are the rates of the two processes. For processes without long term effects, we have that as $t \to \infty$ or $t \to -\infty$

$$h_a^{(a)}(t) \to \rho_a, \qquad h_b^{(a)}(t) \to \rho_a, \qquad h_b^{(b)}(t) \to \rho_b. \tag{3.13}$$

Sometimes it may be required to calculate the intensity function of the superposed process, that is, the process in which the type of event is disregarded. Given an event at the time origin, it has probability $\rho_a/(\rho_a + \rho_b)$ of being a type a, and hence the intensity of the superposed process is

$$h_{\cdot}^{(\cdot)}(t) = \frac{\rho_a}{\rho_a + \rho_b}\{h_a^{(a)}(t) + h_a^{(b)}(t)\} + \frac{\rho_b}{\rho_a + \rho_b}\{h_b^{(a)}(t) + h_b^{(b)}(t)\}. \tag{3.14}$$

This is a general formula for the intensity function of the superposition of two, possibly dependent, processes.

3.4. *Covariance densities.* For some purposes, it is slightly more convenient to work with covariance densities rather than with the second order intensity functions; see, for example, Bartlett [4]. To define the *cross covariance density*, we consider the random variables $N^{(a)}(0, \Delta' t)$ and $N^{(b)}(t, t + \Delta'' t)$ and define

$$\begin{aligned}\gamma_a^{(b)}(t) &= \lim_{\Delta' t, \Delta'' t \to 0+} \frac{\text{Cov}\,\{N^{(a)}(0, \Delta' t), N^{(b)}(t, t + \Delta'' t)\}}{\Delta' t\, \Delta'' t} \\ &= \rho_a h_a^{(b)}(t) - \rho_a \rho_b.\end{aligned} \tag{3.15}$$

It follows directly from (3.14) or from (3.12) and (3.15), that $\gamma_a^{(b)}(t) = \lambda_b^{(a)}(-t)$. Note that an *autocovariance density* such as $\gamma_a^{(a)}(t)$ can be written

$$\gamma_a^{(a)}(t) = \rho_a \delta(t) + \gamma_{a,\text{cont}}^{(a)}(t) = \rho_a \delta(t) + \rho_a\{h_b^{(a)}(t) - \rho_a\}, \tag{3.16}$$

where the second terms are continuous at $t = 0$ and $\delta(t)$ denotes the Dirac delta function.

We denote by $V^{(ab)}(t_1, t_2)$ the covariance between $N^{(a)}(t_1)$ and $N^{(b)}(t_2)$ in the stationary bivariate process. This is called the *covariance time surface*. Then

$$\begin{aligned} V^{(ab)}(t_1, t_2) &= \operatorname{Cov}\{N^{(a)}(t_1), N^{(b)}(t_2)\} \\ &= \operatorname{Cov}\left\{\int_0^{t_1} dN^{(a)}(u), \int_0^{t_2} dN^{(b)}(v)\right\} \\ &= \int_0^{t_1}\int_0^{t_2} \gamma_b^{(a)}(u - v)\, du\, dv. \end{aligned} \tag{3.17}$$

In the special case $t_1 = t_2 = t$, we write $V^{(ab)}(t, t) = V^{(ab)}(t)$. It follows from (3.17) that

$$V^{(ab)}(t) = \int_0^t (t - v)\{\gamma_b^{(a)}(v) + \gamma_a^{(b)}(v)\}\, dv. \tag{3.18}$$

Note that in (3.18) a delta function component at the origin can enter one but not both of the cross covariance densities. If, in (3.18), we take the special highly degenerate case when the type b and type a processes coincide point for point, we obtain the well-known variance time formula

$$\begin{aligned} V^{(aa)}(t) &= \operatorname{Var}\{N^{(a)}(t)\} \\ &= \rho_a t + 2\int_{0+}^t (t - v)\gamma_a^{(a)}(v)\, dv. \end{aligned} \tag{3.19}$$

An interesting question concerns the conditions under which a set of functions $\{\gamma_a^{(a)}(t), \gamma_b^{(a)}(t), \gamma_a^{(b)}(t), \gamma_b^{(b)}(t)\}$ can be the covariance densities of a bivariate point process. Now for all α and β

$$\begin{aligned} &\operatorname{Var}\{\alpha N^{(a)}(t_1) + \beta N^{(b)}(t_2)\} \\ &\qquad = \alpha^2 V^{(aa)}(t_1) + 2\alpha\beta V^{(ab)}(t_1, t_2) + \beta^2 V^{(bb)}(t_2) \\ &\qquad \geqq 0. \end{aligned} \tag{3.20}$$

Thus for all t_1 and t_2,

$$V^{(aa)}(t_1) \geqq 0, \qquad \{V^{(ab)}(t_1, t_2)\}^2 \leqq V^{(aa)}(t_1)V^{(bb)}(t_2), \qquad V^{(bb)}(t_2) \geqq 0. \tag{3.21}$$

The conditions (3.21) can be used to show that certain proposed functions cannot be covariance densities. It would be interesting to know whether corresponding to any functions satisfying (3.21) there always exists a corresponding stationary bivariate point process.

Nothing special is learned by letting t_1 and $t_2 \to 0$ in (3.21). If, however, we let $t_1 = t_2 \to \infty$, we have under weak conditions that

$$V^{(aa)}(t) \sim t\left\{\rho_a + 2\int_{0+}^{\infty} \gamma_a^{(a)}(v)\, dv\right\} = \rho_a t I^{(aa)}, \tag{3.22}$$

$$V^{(ab)}(t) \sim t\int_0^{\infty} \{\gamma_b^{(a)}(v) + \gamma_a^{(b)}(v)\}\, dv = (\rho_a\rho_b)^{1/2} t I^{(ab)}, \tag{3.23}$$

$$V^{(bb)}(t) \sim t\left\{\rho_b + 2\int_0^\infty \gamma_b^{(b)}(v)\,dv\right\} = \rho_b t I^{(bb)}, \tag{3.24}$$

where the right sides of these equations define three asymptotic measures of dispersion I. The conditions

$$I^{(aa)} \geqq 0, \qquad \{I^{(ab)}\}^2 \leqq I^{(aa)}I^{(bb)}, \qquad I^{(bb)} \geqq 0 \tag{3.25}$$

must, of course, be satisfied, in virtue of (3.21).

3.5. *Some special processes.* The second order intensity functions, or equivalently the covariance densities, are not the most natural means of representing the dependencies in a point process, if these dependencies take special account of the nearest events of either or both types. Thus for the semi-Markov process, the second order intensity functions satisfy integral equations; see, for example, Cox and Miller [15], pp. 352–356. The relation with the defining functions of the process is therefore indirect. Thus, while in principle the distributions defining the process could be estimated from data via the second order intensity functions, this would be a roundabout approach, and probably very inefficient.

We now discuss briefly two processes for which the second order intensity functions are more directly related to the underlying mechanism of the process.

Consider an arbitrary regular stationary process of events of type a. Let each event of type a be displaced by a random amount to form a corresponding event of type b; the displacements of different points are independent and identically distributed random variables with probability density function $p(\cdot)$. Denote the probability density function of the difference between two such random variables by $q(\cdot)$. Then a direct probability calculation for the limiting, stationary, process shows that (Cox [12])

$$h_a^{(b)}(t) = p(t) + \int_{-\infty}^{\infty} h_a^{(b)}(v)p(t - v)\,dv, \tag{3.26}$$

$$h_b^{(b)}(t) = \int_{-\infty}^{\infty} h_a^{(a)}(v)q(t - v)\,dv. \tag{3.27}$$

In particular, if the type a events form a Poisson process, $h_a^{(a)}(v) = \rho_a$, so that

$$h_a^{(b)}(t) = p(t) + \rho_a, \qquad h_b^{(b)}(t) = \rho_a. \tag{3.28}$$

The constancy of $h_b^{(b)}(t)$ is an immediate consequence of the easily proved fact that the type b events on their own form a Poisson process. The results (3.28) lead to quite direct methods of estimating $p(\cdot)$ from data and to tests of the adequacy of the model. For positive displacements the type a events could be the inputs to an $M/G/\infty$ queue, the type b events being the outputs. Generalizations of this delay process are considered in Sections 4 and 5.

As a second example, consider a *bivariate doubly stochastic Poisson process.* That is, we have an unobservable real valued (nonnegative) stationary bivariate process $\{\mathbf{\Lambda}(t)\} = \{\Lambda_a(t), \Lambda_b(t)\}$. Conditionally on the realized value of this process, we observe two independent nonstationary Poisson processes with rates, respectively, $\Lambda_a(t)$ for the type a events and $\Lambda_b(t)$ for the type b events. Then, by first arguing conditionally on the realized value of $\{\mathbf{\Lambda}(t)\}$, we have a stationary bivariate point process with

$$\begin{aligned} \gamma_a^{(a)}(t) &= \rho_a\delta(t) + c_\Lambda^{(aa)}(t), \qquad \gamma_b^{(a)}(t) = c_\Lambda^{(ab)}(t), \\ \gamma_b^{(b)}(t) &= \rho_b\delta(t) + c_\Lambda^{(bb)}(t), \end{aligned} \tag{3.29}$$

where $E\{\mathbf{\Lambda}(t)\} = (\rho_a, \rho_b)$ and the c_Λ are the auto and cross covariance functions of $\{\mathbf{\Lambda}(t)\}$.

Again there is a quite direct connection between the covariance densities and the underlying mechanism of the process. Two special cases are of interest. One is when $\Lambda_a(t) = \Lambda_b(t)\,\rho_a/\rho_b$, leading to some simplification of (3.29). Another special case is

$$\begin{aligned} \Lambda_a(t) &= \rho_a + R_a \cos(\omega_0 t + \theta + \Phi), \\ \Lambda_b(t) &= \rho_b + R_b \cos(\omega_0 t + \Phi). \end{aligned} \tag{3.30}$$

In this

$$\begin{aligned} &E(R_a) = E(R_b) = 0, \\ E(R_a^2) = \sigma_{aa}, \qquad &E(R_{ab}) = \sigma_{ab}, \qquad E(R_{bb}) = \sigma_{bb}, \end{aligned} \tag{3.31}$$

and the random variable Φ is uniformly distributed over $(0, 2\pi)$ independently of R_a and R_b. Further, ρ_a, ρ_b, ω_0, and θ are constants and, to keep the Λ nonnegative, $|R_a| \leqq \rho_a, |R_b| \leqq \rho_b$. This defines a stationary although nonergodic process $\{\mathbf{\Lambda}(t)\}$.

Specifications (3.30) and (3.31) yield

$$\begin{aligned} \gamma_a^{(a)}(t) &= \rho_a\delta(t) + \sigma_{aa}\cos(\omega_0 t), \qquad \gamma_b^{(a)}(t) = \sigma_{ab}\cos\{\omega_0(t + \theta)\}, \\ \gamma_b^{(b)}(t) &= \rho_b\delta(t) + \sigma_{bb}\cos(\omega_0 t). \end{aligned} \tag{3.32}$$

Of course this process is extremely special. Note, however, that fairly general processes with a sinusoidal component in the intensity can be produced by starting from the complete intensity functions of a stationary process and either adding a sinusoidal component or multiplying by the exponential of a sinusoidal component; the latter has the advantage of ensuring automatically a nonnegative complete intensity function.

3.6. *Spectral analysis of the counting process.* For Gaussian stationary stochastic processes, study of spectral properties is useful for three rather different reasons:

(a) the spectral representation of the process itself may be helpful;
(b) the spectral representation of the covariance matrix may be helpful;
(c) the effect on the process of a stationary linear operator is neatly expressed.

For point processes a general representation analogous to (a) has been discussed by Brillinger [6]. Bartlett [1] has given some interesting second order theory and applications in the univariate case. For doubly stochastic Poisson processes, which are of course very special, we can often use a full spectral representation for the defining $\{\mathbf{\Lambda}(t)\}$ process; indeed the $\{\mathbf{\Lambda}(t)\}$ process may be nearly Gaussian.

If we are content with a spectral analysis of the covariance density, we can write, in particular for the complex valued *cross spectral density function*,

$$g_b^{(a)}(\omega) = \frac{1}{2\pi} \int_{-\infty}^{\infty} e^{-i\omega t} \gamma_b^{(a)}(t)\, dt = g_a^{(b)}(-\omega). \tag{3.33}$$

Because of the mathematical equivalence between the covariance density and the spectral density, the previous general and particular results for covariance densities can all be expressed in terms of the spectral properties. While these will not be given in full here, note first that the measures of dispersion in (3.22) through (3.24) are given by

$$\begin{aligned} \rho_a I^{(aa)} &= 2\pi g_a^{(a)}(0), \qquad (\rho_a \rho_b)^{1/2} I^{(ab)} = 2\pi g_b^{(a)}(0), \\ \rho_b I^{(bb)} &= 2\pi g_b^{(b)}(0). \end{aligned} \tag{3.34}$$

All the results (3.26) through (3.32) can be expressed simply in terms of the spectral properties. Thus, from (3.29), the spectral analysis of the bivariate doubly stochastic Poisson process leads directly to the spectral properties of the process $\{\mathbf{\Lambda}(t)\}$ on subtracting the "white" Poisson spectra. Thus, spectral analysis of a doubly stochastic Poisson process is likely to be useful whenever the process $\{\mathbf{\Lambda}(t)\}$ has an enlightening spectral form. In the special case (3.32) where $\{\mathbf{\Lambda}(t)\}$ is sinusoidal,

$$g_a^{(a)}(\omega) = \frac{\rho_a}{2\pi} + \frac{\sigma_{aa}}{2\pi}\, \delta(\omega - \omega_0), \tag{3.35}$$

$$g_b^{(a)}(\omega) = \frac{\sigma_{ab}}{2\pi}\, e^{i\omega\theta}\, \delta(\omega - \omega_0). \tag{3.36}$$

The complex valued cross spectral density can be split in the usual way into real and imaginary components, which indicate the relative phases of the fluctuations in the processes of events of type a and type b. We can also define the coherency as $|g_b^{(a)}(\omega)|^2/\{g_a^{(a)}(\omega) g_b^{(b)}(\omega)\}$; for the doubly stochastic process driven by proportional intensities, $\Lambda_b(t) \propto \Lambda_a(t)$, and the coherency is one for all ω, provided that the "white" Poisson component is removed from the denominator.

The natural analogue for point processes of the stationary linear operators on a real valued process is random translation, summarized in (3.26) and (3.27). It follows directly from these equations and from the relation between covariance densities and intensity functions that

$$g_a^{(b)}(\omega) = \rho_a p^{\dagger}(\omega) + g_a^{(a)}(\omega) p^{\dagger}(\omega), \tag{3.37}$$

$$\left\{g_b^{(b)}(\omega) - \frac{\rho_b}{2\pi}\right\} = q^\dagger(\omega)\left\{g_a^{(a)}(\omega) - \frac{\rho_a}{2\pi}\right\}, \tag{3.38}$$

where $p^\dagger(\omega)$, $q^\dagger(\omega)$ are the Fourier transforms of $p(t)$ and $q(t)$. A more general type of random translation for bivariate processes is discussed in Section 5.

3.7. *Variance and covariance time functions.* For univariate point processes the covariance density or spectral functions are mathematically equivalent to the variance time function $V^{(aa)}(t)$ of (3.19), which gives as a function of t the variance of the number of events in an interval of length t. This function is useful for some kinds of statistical analysis; examination of its behavior for large t is equivalent analytically to looking at the low frequency part of the spectrum.

For bivariate point processes, it might be thought that the variance time function $V^{(aa)}(t)$, $V^{(bb)}(t)$, and the covariance time function $V^{(ab)}(t)$ of (3.18) are equivalent to the other second order specifications. This is not the case, however, because it is clear from (3.18) that only the combinations $\gamma_b^{(a)}\omega) + \gamma_a^{(b)}(\omega)$ can be found from $V^{(ab)}(t)$ and this is not enough to fix the cross covariance function of the process.

The cross covariance density can, however, be found from the covariance time surface, $V^{(ab)}(t_1, t_2)$ of (3.17).

The covariance time function and surface are useful for some rather special statistical purposes.

The variance time function of the superposed process is

$$V^{(\cdot\cdot)}(t) = V^{(aa)}(t) + 2V^{(ab)}(t) + V^{(bb)}(t); \tag{3.39}$$

this is equivalent to the relation (3.14) for intensity functions.

3.8. *Bivariate interval specifications; bivariate Markov interval processes.* As has been mentioned several times in this section, the second order intensity functions and their equivalents are most likely to be useful when the dependencies in the underlying mechanism do not specifically involve nearest neighbors, or other features of the process that are most naturally expressed serially, that is, through event number either in the pooled process or in the marginal processes rather than through real time.

For processes in which an interval specification is more appropriate, there are many ways of introducing functions wholly or partially specifying the dependency structure of the process. For a stationary univariate process we can consider the sequence of intervals between successive events as a stochastic process, indexed by serial number, that is, as a real valued process in discrete time. The second order properties are described by an autocovariance sequence which, say for events of type a, is

$$\gamma_x^{(aa)}(j) = \text{Cov}\,\{X_a^{(a)}(k), X_a^{(a)}(k + j)\}, \qquad j = 0, \pm 1, \cdots. \tag{3.40}$$

McFadden [35] has shown that the autocovariance sequence is related to the distribution of counts by the simple, although indirect, formula

$$\gamma_x^{(aa)}(j) = \frac{1}{\rho_a}\left[\int_0^\infty Pr\{N^{(a)}(t) = j\}\, dt - \frac{1}{\rho_a}\right]. \tag{3.41}$$

Stationarity of the $X_a^{(a)}(i)$ sequence is discussed in Slivnyak [44].

If the distribution of $N^{(a)}(t)$ is given by the probability generating function

$$\phi^{(a)}(z, t) = \sum_{j=0}^{\infty} z^j Pr\{N^{(a)}(t) = j\} \tag{3.42}$$

with Laplace transform $\phi^{(a)*}(z, s)$, it will be convenient to substitute in (3.41) the result

$$\int_0^\infty Pr\{N^{(a)}(t) = j\}\, dt = \left[\frac{1}{j!}\frac{\partial^j \phi^{(a)*}(z, s)}{\partial z^j}\right]_{z=0, s=0+}. \tag{3.43}$$

The sequence (3.40) and the analogous one for events of type b summarize the second order marginal properties. To study the joint properties of various kinds of intervals between events, the following are some of the possibilities.

(i) The two sets of intervals $\{X_a^{(a)}(r), X_b^{(b)}(r)\}$ may be considered as a bivariate process in discrete time, that is, we may use serial number in each process as a common discrete time scale. Cross covariances and cross spectra can then be defined in the usual way. While this may occasionally be fruitful, it is not a useful general approach, because for almost all physical models events in the two processes with a common serial number will be far apart in real time. Another problem is that if the process is sampled semisynchronously, say on a type a event, the sequence $X_a^{(b)}(r)$ is not a stationary sequence, although it will generally "converge" to the sequence $X_b^{(b)}(r)$. Again, sufficiently far out from the sampling point, events in the two processes with common serial number will be far apart in real time.

(ii) We may consider the intervals between successive events in the process taken regardless of type, that is, the superposed process. This gives a third covariance sequence, namely, $\gamma_x^{(\cdot\cdot)}(j)$. For particular processes this can be calculated from (3.41) applied to the pooled process, particularly if the joint distribution of the count $N^{(a)}(t)$ and $N^{(b)}(t)$ are available.

In fact, if the joint distribution of $\{N^{(a)}(t), N^{(b)}(t)\}$ is specified by the joint probability generating function $\phi^{(ab)}(z_a, z_b, t)$ with Laplace transform $\phi^{(ab)*}(z_a, z_b, s)$, we have from (3.41) and (3.43) that

$$\gamma_x^{(\cdot\cdot)}(j) = \frac{1}{\rho_a + \rho_b}\left\{\frac{1}{j!}\frac{\partial^j \phi^{(ab)*}(z, z, s)}{\partial z^j} - \frac{1}{\rho_a + \rho_b}\right\}, \tag{3.44}$$

the derivative being evaluated at $z = 0$, $s = 0+$.

A limitation of this approach is, however, that independence of the type a and type b events is not reflected in any simple general relation between the three covariance sequences; this is clear from (3.44). Consider also the process of two independent renewal processes of Section 2.3; the covariance sequence for the superposed sequence is complex and not directly informative.

(iii) The discussion of (i) and (ii) suggests that we consider some properties of the intervals in the pooled sequence, that is, the superposed process with the type of each event being distinguished. Possibilities and questions that arise include the following.

(a) The sequence of event types can be considered as a binary time series. In particular, it might be useful to construct a simple test of dependence of the two series based on the nonrandomness of the sequence of event types. Such a test would, however, at least in its simplest form, require the assumption that the marginal processes are Poisson processes.

(b) We can examine the distributions and in particular the means of the backward recurrence times from events of one type to those of the opposite type, that is, $X_b^{(a)}(-1)$ and $X_a^{(b)}(-1)$. If the two types of events are independently distributed, the two "mixed" recurrence times should have marginal distributions corresponding to the equilibrium recurrence time distributions in the marginal process of events of types a and b.

(c) A more symmetrical possibility similar in spirit to (b) is to examine the joint distribution of the two backward recurrence times measured from an arbitrary time origin, that is, of $X^{(a)}(-1)$ and $X^{(b)}(-1)$; the marginal distributions are, of course, the usual ones from univariate theory. If the events of the two types are independent, the two recurrence times are independently distributed with the distribution of the equilibrium recurrence times. Note, however, the discussion following (2.13). It would be possible to adapt (b) and (c) to take account of forward as well as of backward recurrence times.

(d) Probably the most useful general procedure for examining dependence in a bivariate process through intervals is to consider intensities conditional on the two separate asynchronous backward recurrence times. This is not quite analogous to the use of second order intensities of Section 3.3. Denote the realized backward recurrence times from an arbitrary time origin in the stationary process by u_a, u_b, that is, $u_a = x^{(a)}(-1)$, $u_b = x^{(b)}(-1)$. We then define the *serial intensity functions* for a stationary regular process by

$$\lambda^{(a)}(u_a, u_b) = \lim_{\Delta t \to 0+} \frac{Pr\{N^{(a)}(\Delta t) \geqq 1 \mid U_a = u_a, U_b = u_b\}}{\Delta t}, \tag{3.45}$$

with an analogous definition for $\lambda^{(b)}(u_a, u_b)$. These are, in a sense, third order rather than second order functions, since they involve occurrences at three points.

Now these two serial intensities are defined for all regular stationary processes, but they are complete intensity functions in the sense of Section 3.1 only for a very special class of process that we shall call *bivariate Markov interval processes*. These processes include semi-Markov processes and independent renewal processes; note, however, that in general the marginal processes associated with a bivariate Markov interval process are not univariate renewal processes (as with semi-Markov processes) and this is why we have not called the processes bivariate renewal processes.

As an example of this type of process, consider the alternating process of Section 2.5 with disjoint pairwise dependence. Denote the marginal distribution of the $W(i)$ by $G(x)$, and the conditional distribution of the $Z(i)$, given $W(i) = w$, by $F(z|w)$. If $\bar{G}(x) = 1 - G(x)$, $\bar{F}(z|w) = 1 - F(z|w)$, and the probability densities $g(x)$ and $f(z|w)$ exist, then

$$(3.46) \qquad \gamma^{(a)}(u_a, u_b) = \frac{g(u_a)}{\bar{G}(u_a)}, \qquad \gamma^{(b)}(u_a, u_b) = \frac{f(u_a|u_b - u_a)}{\bar{F}(u_a|u_b - u_a)}.$$

These are essentially hazard (failure rate) functions.

Thorough study of bivariate Markov interval processes would be of interest. The main properties can be obtained in principle because of the fairly simple Markov structure of the process. In particular if $p(u, v)$ denotes the bivariate probability density function of the backward recurrence times from an arbitrary time, that is, of (U_a, U_b) or $(X^{(a)}(-1), X^{(b)}(-1))$, then

$$(3.47) \qquad \frac{\partial p(u, v)}{\partial u} + \frac{\partial p(u, v)}{\partial v} = -\{\lambda^{(a)}(u, v) + \lambda^{(b)}(u, v)\}p(u, v),$$

and

$$(3.48) \qquad p(0, v) = \int_0^\infty p(u, v)\lambda^{(a)}(u, v)\, du, \qquad p(u, 0) = \int_0^\infty p(u, v)\lambda^{(b)}(u, v)\, dv.$$

From the normalized solution of these equations, some of the simpler properties of the process can be deduced.

More generally, (3.45) may be a useful semiqualitative summary of the local serial properties of a bivariate point process. It does not seem possible to deduce the properties of the marginal processes given just $\lambda^{(a)}(u_a, u_b)$ and $\lambda^{(b)}(u_a, u_b)$, except for very particular processes such as the Markov interval process. For the alternating process with pairwise disjoint dependence, we indicated this difficulty in Section 2.5.

4. A bivariate delayed Poisson process model with Poisson noise

In previous sections, we defined bivariate Poisson processes to be those processes whose marginal processes (processes of type a events and type b events) are Poisson processes. Bivariate Poisson processes with a dependency structure which is completely specified by the second order intensity function arise from semi-Markov (Markov renewal) processes. The complete intensity function is also particularly simple.

Other bivariate Poisson processes can be constructed and in the present section we examine in some detail one such process. Its physical specification is very simple, although the specification of its dependency structure via a complete intensity function is difficult. The details of the model also illustrate the definitions introduced in Section 2.

General considerations on bivariate Poisson processes will be given in the next section.

4.1. *Construction of the model.* Suppose we have an unobservable main or generating Poisson process of rate μ. Events from the main process are delayed (independently) by random amounts Y_a with common distribution $F_a(t)$ and superposed on a "noise" process which is Poisson with rate λ_a. The resulting process is the observed marginal process of type a events. Similarly, the events in the main process are delayed (independently) by random amounts with common distribution $F_b(t)$ and superposed with another independent noise process which is Poisson with rate λ_b. The resulting process is then the marginal process of type b events. It is not observed which type a and which type b events originate from common main events.

In what follows, we assume for simplicity that the two delays associated with each main point are independent and positive random variables. The process has a number of possible interpretations. One is as an immigration death process with immigration consisting of couples "arriving" and type a events being deaths of men and type b events being deaths of women. Other queueing or service situations should be evident. The Poisson noise processes are added for generality and because they lead to interesting complications in inference procedures. In particular applications, it might be known that one or both noise processes are absent.

Various special cases are of interest. Thus, if delays of both types are equal with probability one, we have the Marshall–Olkin process [34] mentioned in Section 2. Without the added noise and if delays on one side (say, the a event side) are zero with probability one, we have the delay process of Section 3.5 or, equivalently, an $M/G/\infty$ queue, where type a events are arrivals and type b events are departures. The noise process on the a event side would correspond to independent balking in the arrival process.

4.2. *Some simple properties of the model.* If we consider the transient process from its initiation, it is well known (for example, Cox and Lewis [14], p. 209) that the processes are nonhomogeneous Poisson processes with rates that are, respectively,

$$\rho_a(t_1) = \lambda_a + \mu F_a(t_1), \tag{4.1}$$

$$\rho_b(t_2) = \lambda_b + \mu F_b(t_2). \tag{4.2}$$

Furthermore, the superposed process is a generalized branching Poisson process whose properties are given by Lewis [30] and Vere-Jones [47]. Thus, at each point in the main or generating process there are, with probability $(\lambda_a + \lambda_b)/$

$(\lambda_a + \lambda_b + \mu)$, no subsidiary events, and, with probability $\mu/(\lambda_a + \lambda_b + \mu)$, two subsidiary events. In the second case, the two subsidiary events are independently displaced from the main or parent event by amounts having distributions $F_a(\cdot)$ and $F_b(\cdot)$.

It is also known (Doob [17]) that as $t \to \infty$, or the origin moves off to the right, the marginal processes become simple stationary Poisson processes of rates

$$\rho_a = \lambda_a + \mu, \qquad \rho_b = \lambda_b + \mu, \tag{4.3}$$

respectively, for any distributions $F_a(u)$ and $F_b(u)$. The superposed process is then a stationary generalized branching Poisson process of rate $\lambda_a + \lambda_b + 2\mu$. The bivariate process is unusual in this respect, since there are very few dependent point processes whose superposition has a simple structure. The properties of the process of event types $\{T(i)\}$ or $\{T.(i)\}$ are, however, by no means simple to obtain, as will be evident when we consider bivariate properties below. Note too that stationarity of the marginal and superposed process does not imply stationarity of the bivariate process. A counterexample will be given later when initial conditions are discussed.

Asymptotic results for the bivariate counting process $\{N^{(a)}(t_1), N^{(b)}(t_2)\}$ can be obtained by a simple generalization of the methods of Lewis [30]. If, for simplicity, $t_1 = t_2 = t$, the intuitive basis of the method is that when t is very large, the proportion of events that are delayed from the generating process until after t goes (in some sense) to zero and the process behaves as though all events are concentrated at their generating event, that is, like the Marshall–Olkin process. Thus,

$$E\{N^{(a)}(t)\} = \text{Var}\,\{N^{(a)}(t)\} = V^{(aa)}(t) \sim \rho_a t, \tag{4.4}$$

$$E\{N^{(b)}(t)\} = \text{Var}\,\{N^{(b)}(t)\} = V^{(bb)}(t) \sim \rho_b t, \tag{4.5}$$

$$\text{Var}\,\{N^{(\cdot)}(t)\} \sim (\rho_a + \rho_b + 2\mu)t, \tag{4.6}$$

and therefore

$$\text{Cov}\,\{N^{(a)}(t), N^{(b)}(t)\} = V^{(ab)}(t) \sim \mu t. \tag{4.7}$$

The asymptotic measures of dispersion $I^{(aa)}$, $I^{(ab)}$, and $I^{(bb)}$ defined in equations (3.22) to (3.24) are therefore 1, $\mu/(\rho_a\rho_b)^{1/2}$, and 1. Result (4.7) will be useful in a statistical analysis of the process. By similar methods (Lewis, [30]), one can establish the joint asymptotic normality of the bivariate counting process.

Another property of the process which is simple to derive is the second order cross intensity function (3.8) or the covariance density function (3.14). In fact because of the Poisson nature of the main process and the independence of the noise processes from the main process, there is a contribution to the covariance density only if the type b event is a delayed event and the event at τ is the same event appearing in the type a event process with its delay of Y_a. Thus using (3.10), we get the cross covariance function

$$\gamma_b^{(a)}(\tau) = \mu f_{Y_a - Y_b}(\tau) = \gamma_a^{(b)}(-\tau) \tag{4.8}$$

if $F_{Y_a}(\cdot)$ and $F_{Y_b}(\cdot)$ are absolutely continuous.

If $F_{Y_a}(\cdot)$ and $F_{Y_b}(\cdot)$ have jumps, there will be delta function components in the cross intensity. In particular, when Y_a and Y_b are zero with probability one, there is a delta function component at zero and the process is marginally regular but not regular.

Result (4.8) will be verified from the more detailed results we derive next for the asynchronously sampled, stationary bivariate process. For this we must first consider detailed results for the transient process.

4.3. *The transient counting process.* The number of events of type a in an interval $(0, t_1]$ following the start of the process is denoted by $N_0^{(a)}(t_1)$ and the number of events of type b in $(0, t_2]$ by $N_0^{(b)}(t_2)$.

Assume first that $t_2 \geqq t_1 > 0$.

Now if a main event occurs at time v in the interval $(0, t_1]$, then it contributes either one or no events to the type a event process in $(0, t_1]$ and one or no events to the type b event process in $(0, t_2]$. This bivariate binomial random variable has generating function

$$\begin{aligned} 1 &+ (1 - z_1)(1 - z_2)F_a(t_1 - v)F_b(t_2 - v) \\ &+ (z_2 - 1)F_b(t_2 - v) + (z_1 - 1)F_a(t_1 - v). \end{aligned} \tag{4.9}$$

Since we will be using the conditional properties of Poisson processes in our derivation, we require the time v to be uniformly distributed over $(0, t_1]$ and the resulting generating function for the contribution of each main point is obtained by integrating (4.9) with respect to v from 0 to t_1 and dividing by t_1. After some manipulation, this gives

$$\begin{aligned} 1 &+ \frac{(1 - z_1 - z_2 + z_1 z_2)}{t_1} \int_0^{t_1} F_a(v)F_b(t_2 - t_1 + v)\, dv \\ &+ \frac{(z_2 - 1)}{t_1} \int_0^{t_1} F_b(t_2 - t_1 + v)\, dv + \frac{(z_1 - 1)}{t_1} \int_0^{t_1} F_a(v)\, dv \\ &= Q(z_1, z_2, t_2, t_1). \end{aligned} \tag{4.10}$$

Now assume that there are k_1 events from the main Poisson process of rate μ in $(0, t_1]$, and k_2 main events in $(t_1, t_2]$. Then using the conditional properties of the Poisson process and the independence of the number of main events in $(0, t_1]$ and $(t_1, t_2]$, we get for the conditional generating function of $N_0^{(a)}(t_1)$ and $N_0^{(b)}(t_2)$

$$\begin{aligned} &\exp\{\lambda_a t_1(z_1 - 1) + \lambda_b t_2(z_2 - 1)\}\{Q(z_1, z_2, t_2, t_1)\}^{k_1} \\ &\cdot \left\{1 + \frac{(z_2 - 1)}{(t_2 - t_1)} \int_0^{t_2 - t_1} F_b(u)\, du\right\}^{k_2}. \end{aligned} \tag{4.11}$$

Removing the conditioning on the independently Poisson distributed number of events k_1 and k_2, we have for the logarithm of the joint generating function of $N_0^{(a)}(t_1)$ and $N_0^{(b)}(t_2)$

$$
\begin{aligned}
(4.12)\quad \psi_0(z_1, z_2; t_1, t_2) &= \log \phi(z_1, z_2; t_1, t_2) \\
&= \rho_a t_1(z_1 - 1) + \rho_b t_2(z_2 - 1) \\
&\quad - \mu(z_2 - 1)\int_0^{t_2} R_b(u)\,du - \mu(z_1 - 1)\int_0^{t_1} R_a(u)\,du \\
&\quad + \mu(1 - z_1 - z_2 + z_1 z_2)\int_0^{t_1} F_a(u)F_b(t_2 - t_1 + u)\,du,
\end{aligned}
$$

where $\rho_b = \lambda_a + \mu$, $\rho_b = \lambda_b + \mu$, $R_b(u) = 1 - F_b(u)$, and we still have $t_2 \geqq t_1$.

A similar derivation gives the result for $t_1 \geqq t_2$ and we can write for the general case

$$
\begin{aligned}
(4.13)\quad \psi_0(z_1, z_2; t_1, t_2) &= \rho_a t_1(z_1 - 1) + \rho_b t_2(z_2 - 1) \\
&\quad - \mu(z_2 - 1)\int_0^{t_2} R_b(v)\,dv - \mu(z_1 - 1)\int_0^{t_1} R_a(v)\,dv \\
&\quad + \mu(1 - z_1)(1 - z_2)\int_0^{\min(t_1, t_2)} F_a(t_1 - v)F_b(t_2 - v)\,dv.
\end{aligned}
$$

The expected numbers of events (4.1) and (4.2) in the marginal processes also come out of (4.13), as do the properties of the transient, generalized branching Poisson process obtained by superposing events of type a and type b. Moreover, when the random variables Y_a and Y_b have fixed values, $\psi_0(0, 0; t_1, t_2)$ gives the logarithm of the survivor function of the bivariate exponential distribution of Marshall and Olkin [33], [34].

Note that $\psi_0(z_1, z_2; t_1, t_2)$ is the generating function of a bivariate Poisson variate, that is, a bivariate distribution with Poisson marginals. It is, in fact, the bivariate form of the multivariate distribution which Dwass and Teicher [18] showed to be the only infinitely divisible Poisson distribution:

$$
(4.14)\qquad \phi(\mathbf{z}) = \exp\left\{\sum_{i=1}^{n} a_i(z_i - 1) + \sum_{i<j} a_{ij}(z_i - 1)(z_j - 1) + \cdots + a_{1,2,\cdots,n}\prod_{i=1}^{n}(z_i - 1)\right\}.
$$

However, since the coefficients in (4.13) depend on t_1 and t_2, the joint distribution of events of type a in two disjoint intervals and events of type b in another two disjoint intervals will not have the form (4.14). This is clearly only true for the highly degenerate Marshall–Olkin process of Section 2.

4.4. *The stationary asynchronous counting process.* To derive the properties of the generating function of counts in the stationary limiting process, or equivalently the asynchronously sampled stationary process, we consider first the number of events of type a in $(t, t_1]$ and of type b in $(t, t_2]$. Because of the independent interval properties of the main and noise Poisson processes of rates λ_a, λ_b, and μ, respectively, this is made up independently from noise and main events occurring in $(t, \max(t_1, t_2)]$, whose generating function is given by (4.13), and by main events occurring in $(0, t]$ and delayed into $(t, t_1]$ or $(t, t_2]$.

Consider, therefore, the generating function of the latter type of events. A main event at v in $(0, t]$ generates either one or no type a events in $(t, t_1]$ and either one or no events of type b in $(t, t_2]$. The generating function of this bivariate binomial random variable is

$$1 + (z_1 - 1)p_a + (z_2 - 1)p_b + (z_1 - 1)(z_2 - 1)p_a p_b, \tag{4.15}$$

where

$$p_a = p_a(1; t_1; t; v) = R_a(t - v) - R_a(t + t_1 - v) \tag{4.16}$$

and

$$p_b = p_b(1; t_2; t; v) = R_b(t - v) - R_b(t + t_2 - v). \tag{4.17}$$

If the start time v is assumed to be uniformly distributed over $(0, t]$, then the generating function becomes

$$1 + (z_1 - 1)\frac{\bar{p}_a}{t} + (z_2 - 1)\frac{\bar{p}_b}{t} + (z_1 - 1)(z_2 - 1)\frac{\overline{p_a p_b}}{t}, \tag{4.18}$$

where

$$\bar{p}_a = \int_0^t \{R_a(v) - R_a(v + t_1)\}\, dv, \tag{4.19}$$

$$\bar{p}_b = \int_0^t \{R_b(v) - R_b(v + t_2)\}\, dv, \tag{4.20}$$

$$\overline{p_a p_b} = \int_0^t \{R_a(v) - R_a(v + t_1)\}\{R_b(v) - R_b(v + t_2)\}\, dv. \tag{4.21}$$

It follows from (4.19) and (4.20) that if t_1 and t_2 are finite, we have, even if $E(Y_a)$ and $E(Y_b)$ are infinite,

$$\lim_{t\to\infty} \bar{p}_a = \int_0^{t_1} R_a(v)\, dv, \qquad \lim_{t\to\infty} \bar{p}_b = \int_0^{t_2} R_b(v)\, dv; \tag{4.22}$$

and since $\overline{p_a p_b} \leqq \bar{p}_b$ for all t, t_1, t_2, we have that

$$\lim_{t\to\infty} \overline{p_a p_b} = \int_0^{\infty} \{R_a(v) - R_a(v + t_1)\}\{R_b(v) - R_b(v + t_2)\}\, dv \tag{4.23}$$

exists for finite t_1 and t_2.

The results (4.19) through (4.23) are used, as in the derivation of (4.12), to obtain the cumulant generating function of the contribution of delayed events of type a and type b to $(t, t_1]$ and $(t, t_2]$ when $t \to \infty$. This is

$$(4.24)\quad \psi^+(z_1, z_2; t_1, t_2) = \mu(z_1 - 1)\int_0^{t_1} R_a(v)\,dv + \mu(z_2 - 1)\int_0^{t_2} R_b(v)\,dv + \mu(z_1 - 1)(z_2 - 1)\int_0^\infty \{R_a(v) - R_a(v + t_1)\}\{R_b(v) - R_b(v + t_2)\}\,dv.$$

Combined with (4.13), we have for the stationary bivariate process the result

$$(4.25)\quad \psi(z_1, z_2; t_1, t_2) = \rho_a t_1(z_1 - 1) + \rho_b t_2(z_2 - 1) + \mu(z_1 - 1)(z_2 - 1) \cdot\left[\int_0^{\min(t_1,t_2)} F_a(t_1 - v)F_b(t_2 - v)\,dv + \int_0^\infty \{R_a(v) - R_a(v + t_1)\}\{R_b(v) - R_b(v + t_2)\}\,dv\right].$$

Note that this is the cumulant generating function of a bivariate Poisson distribution and that the covariance time function (3.17) is the term in (4.25) multiplying $(z_1 - 1)(z_2 - 1)$;

$$(4.26)\quad V^{(ab)}(t_1, t_2) = \mu\int_0^{\min(t_1,t_2)} R_a(t - v_1)R_b(t_2 - v)\,dv + \mu\int_0^\infty \{R_a(v) - R_a(v + t_1)\}\{R_b(v) - R_b(v + t_2)\}\,dv.$$

Differentiation of this expression with respect to t_1 and t_2 gives, after some manipulation, the covariance density (4.8), as predicted by the general formula (3.17).

Thus, if the densities associated with $R_a(\cdot)$ and $R_b(\cdot)$ exist, we can express (4.25) as

$$(4.27)\quad \psi(z_1, z_2; t_1, t_2) = \rho_a t_1(z_1 - 1) + \rho_b t_2(z_2 - 1) + (z_1 - 1)(z_2 - 1)\mu\int_0^{t_1}\int_0^{t_2} f_{Y_a - Y_b}(u - v)\,du\,dv.$$

There are a number of alternative forms for and derivations of this distribution.

The behavior of $V^{(ab)}(t_1, t_2)$, although it is clearly a monotone nondecreasing function of both t_1 and t_2, is complex and will not be studied further here. In (4.7), we saw that along the line $t_1 = t_2 = t$ it is asymptotically μt.

We have also not established the complete stationarity of the limiting bivariate process; this follows from the fact that the delay depends only on the distance from the Poisson generating event, and can be established rigorously using bivariate characteristic functionals.

The complete intensity functions for this process (3.1) cannot be written down and although the second order intensity function is simple it does not specify the dependency structure of the process completely, as it does for the bivariate semi-Markov process. Note too that the cross covariance function (4.8) is always positive, so that there is in effect no inhibition of type a events by b events. In fact from the construction of the process, it is clear that just the opposite effect takes place. We examine the dependency structure of the delay process in more detail here by looking at the joint asynchronous forward recurrence time distribution. This distribution is of some interest in itself.

4.5. *The joint asynchronous forward recurrence times.* In the asynchronous process of the previous section, the time to the kth event of type a, $S^{(a)}(k)$, has a gamma distribution with parameter k and $S^{(b)}(h)$ has a gamma distribution with parameter h. Thus, the joint distribution of these random variables is a bivariate gamma distribution of mixed marginal parameters k and h which is obtained from the generating function (4.25) via the fundamental relationship (2.5). We consider only the joint forward recurrence times $S^{(a)}(1) = X^{(a)}(1)$ and $S^{(b)}(1) = X^{(b)}(1)$ which have a bivariate exponential distribution:

$$(4.28) \qquad \begin{aligned} R_{ab}(t_1, t_2) &= Pr\{X^{(a)}(1) > t_1, X^{(b)}(1) > t_2\} \\ &= \exp \{\psi(0, 0; t_1, t_2)\} \\ &= \exp \{-\rho_a t_1 - \rho_b t_2 + V^{(ab)}(t_1, t_2)\}. \end{aligned}$$

Clearly, this bivariate exponential distribution reduces to the distribution discussed by Marshall and Olkin [33] in the degenerate case when there are no delays (or fixed delays [34]). For no delays

$$(4.29) \qquad R_{ab}(t_1, t_2) = -\rho_a t_1 - \rho_b t_2 + \mu \min (t_1, t_2).$$

The bivariate exponential distribution (4.27) is not the same as the infinitely divisible exponential distribution discussed by Gaver [20], Moran and Vere-Jones [38], and others. Whenever the delay distributions $R_a(\cdot)$ and $R_b(\cdot)$ have jumps, $R_{ab}(t_1, t_2)$ will have singularities.

For the correlation coefficient, we have

$$(4.30) \qquad \rho_a \rho_b \operatorname{Corr} \{X^{(a)}(1), X^{(b)}(1)\} = \int_0^\infty \int_0^\infty R_{ab}(t_1, t_2)\, dt_1\, dt_2 - \frac{1}{\rho_a \rho_b}.$$

It is not possible to integrate this expression explicitly except in special cases. However, since $V^{(ab)}(t_1, t_2) \geqq 0$, we clearly have that the correlation coefficient is greater than zero.

For the special case (4.29) the correlation (4.30) is $1/\{(1 + (\lambda_a + \lambda_b)/\mu)\}$.

We do not pursue further here the properties of the process obtainable from the joint distribution of counts (4.25) of the synchronous counting process $\{N^{(a)}(t_1), N^{(b)}(t_2)\}$. However, it is useful to summarize what useful properties can be derived for this or any other bivariate process such as those given in the next section, from this bivariate distribution.

(i) The marginal generating functions ($z_1 = 0$ or $z_2 = 0$) give the correlation structure of the marginal interval process through equation (3.43). This is trivial for the delay process.

(ii) The generating function with $z_1 = z_2$ gives the correlation structure of the intervals in the superposed process through (3.44). For the delay process this is the interval correlation structure of a clustering (branching) Poisson process.

(iii) The covariance time surface and cross intensity and marginal intensity functions can be obtained. Again for the delay process this is trivial.

(iv) The joint distribution of the asynchronous forward recurrence times $\{X^{(a)}(1), X^{(b)}(1)\}$ can be calculated. Other functions of interest are the smaller and larger of $X^{(a)}(1)$ and $X^{(b)}(1)$, and the conditional distributions and expectations, for example, $E\{X^{(a)}(1)|X^{(b)}(1) = x\}$. The latter is difficult to obtain for the delay process, the regression being highly nonlinear.

(v) In principle, one can obtain not only the distributions of the smaller and larger of $X^{(a)}(1)$ and $X^{(b)}(1)$, but also the order type (jointly or marginally) since

$$Pr\{T(1) = a\} = Pr\{X^{(a)}(1) < X^{(b)}(1)\}. \tag{4.31}$$

(vi) It is not possible to obtain the complete distributions of types, for example, $Pr\{T(1) = a; T(2) = a\}$ from the bivariate distribution of asynchronous counts, since these counts are related to the sums of intervals by (2.5). For this information, we need more complete probability relationships, that is, for the pooled process $\{X^{(\cdot)}(1), T(1); X^{(\cdot)}(2), \cdots\}$. Note too that $\{T(i)\}$ is not a stationary binary sequence.

It is possible to obtain distributions of semisynchronous counting processes for the delay processes although we do not do this here. One reason for doing this is to obtain information on the distribution of the stationary sequences $T_{\cdot}(i)$. Thus,

$$Pr\{T_{\cdot}(0) = a, T_{\cdot}(1) = b\} = \frac{\rho_a}{\rho_a + \rho_b} Pr\{X_a^{(b)}(1) < X_a^{(a)}(1)\}, \tag{4.32}$$

and so forth, from which the correlation coefficient of lag one is obtained. It is not possible to carry the argument to lags of greater than one solely with joint distributions of sums of semisynchronous intervals.

4.6. *Stationary initial conditions.* We discuss here briefly the problem of obtaining stationary initial conditions for the delay process, since this has some bearing on the problems considered in this paper.

Note that for the marginal processes in the delayed Poisson process the numbers of events generated before t which are delayed beyond t have, if $E(Y_a) < \infty$ and $E(Y_b) < \infty$, Poisson distributions with parameters $\mu E(Y_a)$ and $\mu E(Y_b)$, respectively, when $t \to \infty$. Denote these random variables by $Z^{(a)}$ and $Z^{(b)}$. If the transient process of Section 4.3 is started with an additional number

Z_a of type a events which occur independently at distances $\bar{Y}_a(1), \cdots, \bar{Y}_a(Z_a)$ from the origin, where

$$Pr\{\bar{Y}_a(i) \leqq t\} = \int_0^t \frac{R_a(u)\,du}{\{E(Y_a)\}}, \tag{4.33}$$

and with an additional number Z_b of type b events which occur independently at distances $\bar{Y}_b(1), \cdots, \bar{Y}_b(z_b)$ from the origin, where the common distribution of the $\bar{Y}_b(j)$ is directly analogous to that of the $\bar{Y}_a(i)$, then the marginal processes are stationary Poisson processes (Lewis [30]). However, the bivariate process is not stationary. This can be verified, for instance, by obtaining the covariance density from the resulting generating function and noting that it depends on t_1 and t_2 separately, and not just on their difference.

In obtaining stationary initial conditions, the joint distribution of Z_a and Z_b is needed. Without going into the details of the limiting process, the generating function for these random variables is clearly (4.24) when $t_1 \to \infty$ and $t_2 \to \infty$. Thus,

$$\begin{aligned}(4.34)\quad \psi_{Z_a, Z_b}(z_1, z_2) &= (z_1 - 1)\mu E(Y_a) + (z_2 - 1)\mu E(Y_b) + (z_1 - 1)(z_2 - 1)\mu \\ &\quad \cdot E\{\min (Y_a, Y_b)\},\end{aligned}$$

where $E\{\min (Y_a, Y_b)\} = \int_0^\infty R_a(v)R_b(v)\,dv$. This is the generating function (4.14) of a bivariate Poisson distribution.

Further details of this model, including the complete stationary initial conditions, will be given in another paper.

5. Some other special processes

We discuss here briefly several important models for bivariate point processes. The specification of the models is through the structure of intervals and is based on direct physical considerations, unlike, say, the bivariate Markov process with its specification of degree of dependence through the complete intensity functions. At the end of the section we consider the general problem of specifying the form of bivariate Poisson processes.

5.1. *Single process subject to bivariate delays.* The bivariate delayed Poisson process of the previous section can be generalized in several ways. First, the delays Y_a and Y_b might be correlated since, for instance, in the example of a man and wife in a bivariate immigration death process, their residual lifetimes would be correlated. Again Y_a and Y_b may take both positive and negative values. The stationary analysis of the previous section goes through essentially unchanged although specifying initial conditions is difficult. The covariance function (4.8) is the same except that $f_{Y_a - Y_b}(t)$ is, of course, no longer a simple convolution.

Another extension is to consider main processes which are, say, regular stationary point processes with rate μ and intensity function $h_\mu(t)$. Then the cross intensity function for the bivariate process, $h_a^{(b)}(t)$, becomes

$$\frac{\lambda_a}{\lambda_a + \mu} \rho_b + \frac{\mu}{\lambda_a + \mu} \left\{ \lambda_b + f_{Y_a - Y_b}(t) + \int_{-\infty}^{\infty} h_\mu(u) f_{Y_a - Y_b}(t - u)\, du \right\}, \tag{5.1}$$

with a similar expression for $h_b^{(a)}(t)$. These should be compared with (3.26). Except when the main process is a renewal process, explicit results beyond the intensity function are difficult to obtain. For the renewal case an integral equation can be written down, as also for branching renewal processes (Lewis, [32]); from the integral equation higher moments of the bivariate counting process can be derived.

5.2. *Bivariate point process subject to delays.* Instead of having a univariate point process in which each point (say the ith) is delayed by two different amounts $Y_a(i)$ and $Y_b(i)$ to form the bivariate process, one can have a main bivariate point process in which the ith type a event is delayed by $Y_a(i)$ and the jth type b event is delayed by $Y_b(j)$, thus forming a new bivariate point process. This does not reduce to the bivariate delay process of Section 5.1 although it is conceptually similar.

The simplest illustration is where there is error (jitter) in recording the positions of the points. Usually the errors are taken to be independently distributed, although Y_a and Y_b may have different distributions. Another situation is an immigration death process with two different types of immigrants.

If the main process has cross intensities $\bar{h}_a^{(b)}(t)$ and $\bar{h}_b^{(a)}(t)$, then the delayed bivariate process (with no added Poisson noise) has cross intensity

$$h_a^{(b)}(t) = \int_{-\infty}^{\infty} \bar{h}_a^{(b)}(t - v) f_{Y_a - Y_b}(v)\, dv. \tag{5.2}$$

It will not be possible from data to separate properties of the jitter process from those of the underlying main process, unless strong special assumptions are made.

An interesting situation occurs when the main process is a semi-Markov process with marginal processes which are Poisson processes as, for example, in Section 2. Then the delayed bivariate point process is, in equilibrium, a bivariate Poisson process.

5.3. *Clustering processes.* Univariate clustering processes (Neyman and Scott [39], Vere-Jones [47], and Lewis [30]) are important. Each main event generates one subsidiary sequence of events and the subsidiary sequences have a finite number of points with probability one. The subsidiary processes are independent of one another but can be of quite general structure. When the subsidiary processes are finite renewal processes, the clustering process is known as a Bartlett–Lewis process; when the events are generated by independent delays from the initiating main event, the process is known as a Neyman–Scott cluster process.

The bivariate delay process and delayed bivariate process described in the previous two subsections are special cases of bivariate cluster processes and clearly both types of main process are possible for these cluster processes. As an example of a bivariate main process generating two different types of subsidiary process, Lewis [29] considered computer failure patterns and discussed the possibility of two types of subsidiary sequences, one generated by permanent component failures and the other by intermittent component failures.

There are many possibilities that will not be discussed here. Some general points of interest are, however, the following.

(i) When the main process is a univariate Poisson process, producing a bivariate clustering Poisson process, bivariate superposition of such processes again produces a bivariate clustering Poisson process. The process is thus infinitely divisible.

(ii) Both the marginal processes and the superposed processes are (generalized) cluster processes. Thus, we can use known results for these processes and expressions such as (3.39) and (3.14) to find variance time curves and cross intensities for the bivariate process. When the main process is a semi-Markov process, the marginal processes are clustering (or branching) renewal processes (Lewis, [32]).

(iii) The analysis in Section 4 can be used for these processes when the main process is a univariate Poisson process. Bivariate characteristic functionals are probably also useful.

5.4. *Selective inhibition.* A simple, realistic and analytically interesting model arises in neurophysiological contexts. We have two series of events, the first called the inhibitory series of events and the second the excitatory series of events, occurring on a common time scale. Each event in the inhibitory series blocks only the next excitatory event (and blocks it only if no following inhibitory event occurs before the excitatory event). This is the simplest of many possibilities.

Although only the sequence of noninhibited excitatory events (the responses) is usually studied, Lawrance has pointed out that there are a number of bivariate processes generated by this mechanism, in particular the inhibitory events and the responses [24], [25]. These may constitute the input and output to a neuron, and are the only pair we consider here. In particular, we take the excitatory process to be a Poisson process with rate ρ_a and the inhibiting process to be a renewal process with interevent probability distribution function $F_b(x)$. The response process has dependent intervals unless the inhibitory process also is a Poisson process.

When the excitatory process is a renewal process with interevent probability distribution function $F_a(x)$ and the inhibitory process is Poisson with rate ρ_b, the process of responses is a renewal process. This follows because the original renewal process is in effect being thinned at a rate depending only on the time since the last recorded response and such an operation preserves the renewal property. This bivariate renewal process is not a semi-Markov process, as can be seen by attempting to write down the complete intensity functions (3.4) and (3.5). The complete intensity functions become simple only for the trivariate process of inhibitory events, responses, and nonresponses.

Coleman and Gastwirth [10] have shown that it is possible to pick $F_a(x)$ so that the responses also form a Poisson process. The covariance density of this bivariate Poisson process can be obtained; it is always negative (personal communication, T. K. M. Wisniewski).

Other forms of selective inhibition can be postulated; some have been discussed by Coleman and Gastwirth [10]. Another possibility is the simultaneous inhibition, as above, of two excitatory processes by a single, unobservable inhibitory process. When the inhibitory process is Poisson and the excitatory processes are renewal processes, the two response processes are a bivariate renewal process.

There are, of course, many other neurophysiological models, generally more complicated than the selective inhibition models and many times involving the doubly stochastic mechanism discussed in Section 3.5. An interesting example is given by Walloe, Jansen, and Nygaard [48].

5.5. *General remarks on bivariate Poisson processes.* In this and previous sections, we have encountered several examples of bivariate Poisson processes, defined as bivariate point processes in which the marginal processes are Poisson processes.

(i) The degenerate Poisson process of Marshall and Olkin was discussed in Section 2.

(ii) The process in (i) is a special case of a broad family of bivariate Poisson processes generated by bivariate delays on univariate Poisson processes. Several other examples arise in considering delays on bivariate Poisson processes.

(iii) Semi-Markov processes have renewal marginals and a broad class of bivariate Poisson processes is obtained by choosing the marginal processes to be Poisson processes. Delays added to these particular semi-Markov processes again produce bivariate Poisson processes.

(iv) A rather special case arises when a Poisson process inhibits a renewal process.

Another example is mentioned because it illustrates the problem considered in Section 3.8 of specifying dependency structure in terms of the bivariate, discrete time sequence of marginal intervals. Thus, we can start the process and require that the intervals in the marginals with the same serial index be bivariate exponentials. Any bivariate exponential distribution may be used, such as (4.28) or those of Gaver [20], Plackett [42], Freund [19], and Griffiths [21]. The interval structure is stationary, as is the counting process of the marginals, which are Poisson processes. The bivariate counting process is, however, not stationary. It is not clear whether one gets the counting process to be stationary, as defined in Section 2, when moving away from the origin, but since the time lag between the dependent intervals increases indefinitely as $n \to \infty$ the process is degenerate and tends to almost independent Poisson processes a long time from the origin.

No general structure is known for bivariate Poisson processes. There follow some general comments and some open questions.

(i) The bivariate Poisson process as defined is infinitely divisible in that bivariate superposition produces bivariate Poisson processes. However, of the above models of bivariate Poisson processes, only the bivariate delayed Poisson process keeps the same dependency structure under bivariate superposition.

(ii) Does unlimited bivariate superposition produce two independent Poisson processes? The answer is, generally, yes (Cinlar [8]).

(iii) It can be shown that successive independent delays on bivariate Poisson processes (and most bivariate processes) produces in the limit a process of independent Poisson processes. This can be seen from (5.2) and (5.3), but needs bivariate characteristic functionals for a complete proof.

(iv) The numbers of events of the two types in an interval $(0, t]$ in a bivariate Poisson process have a bivariate Poisson distribution. Some general properties of such distributions are known (Dwass and Teicher [18]); the bivariate Poisson distribution (4.33) is the only infinitely divisible bivariate Poisson distribution. An open question of interest in investigating bivariate Poisson processes is whether, when Z_1, Z_2, and $Z_1 + Z_2$ have marginally Poisson distributions of means μ_1, μ_2, and $\mu_1 + \mu_2$, Z_1 and Z_2 are independent. If this is so, a bivariate Poisson process in which the superposed marginal process is a Poisson process must have the events of two types independent.

(v) The broad class of stationary bivariate Poisson processes arising from delay mechanisms have positive cross covariance densities, that is, no "inhibitory effect." For the semi-Markov process with Poisson marginals, it is an open question as to whether cross covariance densities which take on negative values exist. In particular, for the alternating renewal process with identical gamma distributions of index one for up and down times, the cross covariance is strictly positive. The only model which is known to produce a bivariate Poisson process with strictly negative covariance density is the Poisson inhibited renewal process described earlier in this section.

6. Statistical analysis

6.1. *General discussion.* We now consider in outline some of the statistical problems that arise in analyzing data from a bivariate point process. If a particular type of model is suggested by physical considerations, it will be required to estimate the parameters and test goodness of fit. In some applications, a fairly simple test of dependence between events of different types will be the primary requirement. In yet other cases, the estimation of such functions as the covariance densities will be required to give a general indication of the nature of the process, possibly leading to the suggestion of a more specific model. In all cases, the detection and elimination of nonstationarity may be required.

There is one important general distinction to be drawn, parallel to that between correlation and regression in the analysis of quantitative data. It may be that both types of event are to be treated symmetrically, and that in particular the stochastic character of both types needs analysis. This is broadly the attitude implicit in the previous sections. Alternatively one type of event, say b, may be causally dependent on previous a events, or it may be required to predict events of type b given information about the previous occurrences of events of both types. Then it will be sensible to examine the occurrence of the b's conditionally on the observed sequence of a's and not to consider the stochastic mechanism generating the a's; this is analogous to the treatment of the independent variable in regression analysis as "fixed." Note in particular that the pattern of the a's might be very nonstationary and yet if the mechanism generating the b's is stable, simple "stationary" analyses may be available.

In the rest of this section, we sketch a few of the statistical ideas required in analyzing this sort of data.

6.2. *Likelihood analyses.* If a particular probability model is indicated as the basis of the analysis when the model is specified except for unknown parameters, in principle it will be a good thing to obtain the likelihood of the data from which exactly or asymptotically optimum procedures of analysis can be derived, for example, by the method of maximum likelihood; of course, this presupposes that the usual theorems of maximum likelihood theory can be extended to cover such applications. Unfortunately, even for univariate point processes, there are relatively few models for which the likelihood can be obtained in a useful form. Thus, one is often driven to rather *ad hoc* procedures.

Here we note a few very particular processes for which the likelihood can be calculated.

In a semi-Markov process, the likelihood can be obtained as a product of a factor associated with the two state Markov chain and factors associated with the four distributions of duration; if sampling is for a fixed time there will be one "censored" duration. Moore and Pyke [36] have examined this in detail with particular reference to the asymptotic distributions obtained when sampling is for a fixed time, so that the numbers of intervals of various types are random variables.

A rather similar analysis can be applied to the bivariate Markov process of intervals of Section 3.8, although a more complex notation is necessary. Let

$$L^{(a)}(x; v, w) = \exp\left\{-\int_0^x \lambda^{(a)}(z + v, z + w)\, dz\right\}, \tag{6.1}$$

$$L^{(b)}(x; v, w) = \exp\left\{-\int_0^x \lambda^{(b)}(z + v, z + w)\, dz\right\}. \tag{6.2}$$

We can summarize the observations as a sequence of intervals between successive events in the pooled process, where the intervals are of type aa, ab, ba, or bb. We characterize each interval by its length x and by the backward recurrence time at the start of the interval measured to the event of opposite type. Denote this by v if measured to a type a event and by w if measured to a type b event.

Then the contribution to the likelihood of the length of the interval and the type of the event at the end of the interval is

(6.3) $\lambda^{(a)}(x, w + x)\, L^{(a)}(x; 0, w)\, L^{(b)}(x; 0, w)$ for an *aa* interval,

(6.4) $\lambda^{(b)}(x, w + x)\, L^{(a)}(x; 0, w)\, L^{(b)}(x; 0, w)$ for an *ab* interval,

(6.5) $\lambda^{(a)}(v + x, x)\ L^{(a)}(x; v, 0)\ L^{(b)}(x; v, 0)$ for a *ba* interval,

(6.6) $\lambda^{(b)}(v + x, x)\ L^{(a)}(x; v, 0)\ L^{(b)}(x; v, 0)$ for a *bb* interval.

Thus, once the intensities are specified parametrically the likelihood can be written down and, for example, maximized numerically.

Now the above discussion is for the "correlational" approach in which the two types of event are treated symmetrically. If, however, we treat the events of type b as the dependent process and argue conditionally on the observed sequence of events of type a, the analysis is simplified, in effect by replacing $\lambda^{(a)}(\cdot, \cdot)$ and $L^{(a)}(\cdot, \cdot, \cdot)$ by unity.

A particular case of interest is when the intensities are linear functions of their arguments. This, of course, precludes having the semi-Markov process as a special case.

A further example of when a likelihood analysis is feasible is provided by the bivariate sinusoidal Poisson process of (3.30) with R_a, R_b, and Φ regarded as unknown parameters. An analysis in terms of exponential family likelihoods is obtained by taking (3.20) to refer to the log intensity; for the univariate analysis see Lewis [31].

In the bivariate case, we reparametrize and have

$$\lambda_a(t) = \frac{\rho_a \exp\{R_a \cos(\omega_0 t + \theta + \Phi)\}}{I_0(R_a)} \tag{6.7}$$

and

$$\lambda_b(t) = \frac{\rho_b \exp\{R_b \cos(\omega_0 t + \Phi)\}}{I_0(R_b)}, \tag{6.8}$$

where $I_0(R_b)$ is a zero order modified Bessel function of the first kind. It is convenient to assume that observation on both processes is for a common fixed period t_0, where $\omega_0 t_0$ is an integral multiple of 2π, say $2\pi p$. Then $\int_0^{t_0} \lambda_a(u)\, du = \rho_a t_0$ and $\int_0^{t_0} \lambda_b(u)\, du = \rho_b t_0$.

If $n^{(a)}$ type a events are observed in $(0, t_0]$ at times $t_1^{(a)}, \cdots, t_{n^{(a)}}^{(a)}$ and $n^{(b)}$ type b events at times $t_1^{(b)}, \cdots, t_{n^{(b)}}^{(b)}$, then, using the likelihood for the nonhomogeneous bivariate Poisson process

$$\prod_{i=1}^{n^{(a)}} \lambda_a(t_i^{(a)}) \prod_{j=1}^{n^{(b)}} \lambda_b(t_j^{(b)}) \exp\{-\rho_a t_0 - \rho_b t_0\}, \tag{6.9}$$

we find that the set of sufficient statistics for $\{\rho_a, \rho_b, R_a \cos(\theta + \Phi), R_a \sin(\theta + \Phi), R_b \cos \Phi, R_b \sin \Phi\}$ are $\{n^{(a)}, n^{(b)}, \mathscr{A}_a(\omega_0), \mathscr{B}_a(\omega_0), \mathscr{A}_b(\omega_0), \mathscr{B}_b(\omega_0)\}$, where

$$(6.10) \qquad \mathscr{A}_a(\omega_0) = \sum_{i-1}^{n^{(a)}} \cos(\omega_0 t_i^{(a)}), \qquad \mathscr{B}_a(\omega_0) = \sum_{j=1}^{n^{(b)}} \sin(\omega_0 t_j^{(a)}),$$

with similar definitions for $\mathscr{A}_b(\omega_0)$ and $\mathscr{B}_b(\omega_0)$.

Typically, if $R_a = R_b = R$, maximum likelihood estimates of R and tests of $R = 0$ are based on monotone functions of $\mathscr{A}_a(\omega_0)$, $\mathscr{B}_a(\omega_0)$, $\mathscr{A}_b(\omega_0)$, and $\mathscr{B}_b(\omega_0)$. The estimation and testing procedures are formally equivalent to tests for directionality on a circle from two independent samples when the direction vector has a von Mises distribution (Watson and Williams, [49]).

Other trend analyses can be carried out with a similar type of likelihood analysis if the model is a nonhomogeneous bivariate Poisson process.

For most other special models, including quite simple ones such as the delayed Poisson process of Section 4, it does not seem possible to obtain the likelihood in usable form; it would be helpful to have ways of obtaining useful pseudo-likelihoods for such processes.

For testing goodness of fit, it may sometimes be possible to imbed the model under test in some richer family; for instance, agreement with a parametric semi-Markov model could be tested by fitting some more general bivariate Markov interval process and comparing the maximum likelihoods achieved. More usually, however, it will be a case of finding relatively *ad hoc* test statistics to examine various aspects of the model.

In situations in which the model of independent renewal processes or the semi-Markov model may be relevant, the following procedures are likely to be useful. To test consistency with an independent renewal process model, we may:

(a) examine for possible nonstationarity,

(b) test the marginal processes for consistency with a univariate renewal model (Cox and Lewis [14], Chapter 6),

(c) test for dependence using the estimates of the cross intensity given in the next section, or test that the event types do not have the first order Markov property.

If dependence is present, it may be natural to see whether the data are consistent with a semi-Markov process. (Note, however, that the family of independent renewal models is not contained in the family of semi Markov models.) To test for the adequacy of an assumed parametric semi-Markov model, we may, for example, proceed as follows:

(a) examine for possible nonstationarity,

(b) test the sequence of event types for the first order Markov property (Billingsley [5]),

(c) examine the distributional form of the four separate types of interval,

(d) examine the dependence of intervals on the preceding interval and the preceding event type.

6.3. *Estimation of intensities and associated functions.* If a likelihood based analysis is not feasible, we must use the more empirical approach of choosing aspects of the process thought to be particularly indicative of its structure and estimating these aspects from the data. In this way we may be able to obtain estimates of unknown parameters and tests of the adequacy of a proposed model.

In the following discussion, we assume that the process is stationary. With extensive data, it will be wise first to analyze the data in separate sections, pooling the results only if the sections are reasonably consistent.

The main aspects of the process likely to be useful as a basis for such procedures are the frequency distributions of intervals of various kinds, the second order functions of Section 3.3 through 3.7 and, the bivariate interval properties, in particular the serial intensity functions (3.42). As stressed in Section 3, it will often happen that one or other of the above aspects is directly related to the underlying mechanism of the process and hence is suitable for statistical analysis.

Estimation of the univariate second order functions does not need special discussion here. We therefore merely comment briefly on the estimation of the serial intensity functions and the cross properties; for the latter the procedures closely parallel the corresponding univariate estimation procedures.

6.3.1. *Cross intensity function.* To obtain a smoothed estimate of the cross intensity function $h_a^{(b)}(t)$, choose a grouping interval Δ and count the total number of times a type b event occurs a distance between t and $t + \Delta$ to the right of a type a event; let the random variable corresponding to this number be $R_a^{(b)}(t, t + \Delta)$. In practice, we form a histogram from all possible intervals between events of type a and events of type b. We now follow closely the argument of Cox and Lewis ([14], p. 122) writing, for observations over $(0, t_0)$,

$$(6.11)\qquad R_a^{(b)}(t, t + \Delta) = \left\{\int_{u=0}^{t_0-t-\Delta}\int_{x=t}^{t+\Delta} + \int_{u=t_0-t-\Delta}^{t_0-t}\int_{x=t}^{t_0-u}\right\} dN^{(a)}(u)\, dN^{(b)}(u + x).$$

Now for a stationary process

$$(6.12)\qquad E\{dN^{(a)}(u)\, dN^{(b)}(u + x)\} = \rho_a h_a^{(b)}(x)\, du\, dx,$$

and a direct calculation, plus the assumption that $h_a^{(b)}(x)$ varies little over $(t, t + \Delta)$, gives

$$(6.13)\qquad E\{R_a^{(b)}(t, t + \Delta)\} = (t_0 - t - \tfrac{1}{2}\Delta)\rho_a \int_t^{t+\Delta} h_a^{(b)}(x)\, dx,$$

thus leading to a nearly unbiased estimate of the integral of the cross intensity over $(t, t + \Delta)$.

If the type b events are distributed in a Poisson process independently of the type a events, we can find the exact moments of $R_a^{(b)}(t, t + \Delta)$, by arguing conditionally both on the number of type b events and on the whole observed process of type a events (see Section 6.4). To a first approximation, $R_a^{(b)}(t, t + \Delta)$

has (conditionally) a Poisson distribution of mean $n^{(a)}n^{(b)}\Delta/t_0$ provided that Δ is small and, in particular, that few type a events occur within Δ of one another. This provides the basis for a test of the strong null hypothesis that the type b events follow an independent Poisson process; it would be interesting to study the extent to which the test is distorted if the type b events are distributed independently of the type a events, although not in a Poisson process.

6.3.2. *Cross spectrum.* Estimation of the cross spectrum is based on the cross periodogram, defined as follows. For each marginal process, we define the finite Fourier–Stieltjes transforms of $N^{(a)}(t)$ and $N^{(b)}(t)$ (Cox and Lewis [14], p. 124) to be

$$H_{t_0}^{(a)}(\omega) = (2\pi t_0)^{-1/2} \sum_{\ell=1}^{n^{(a)}} \exp\{i\omega t_\ell^{(a)}\} = (2\pi t_0)^{-1/2}\{\mathscr{A}_{t_0}^{(a)}(\omega) + i\mathscr{B}_{t_0}^{(a)}(\omega)\}, \tag{6.14}$$

$$H_{t_0}^{(b)}(\omega) = (2\pi t_0)^{-1/2} \sum_{j=1}^{n^{(b)}} \exp\{i\omega t_j^{(b)}\} = (2\pi t_0)^{-1/2}\{\mathscr{A}_{t_0}^{(b)}(\omega) + i\mathscr{B}_{t_0}^{(b)}(\omega)\}. \tag{6.15}$$

The cross periodogram is then

$$\mathscr{I}_{t_0}^{(ab)}(\omega) = H_{t_0}^{(a)}(\omega)\, \bar{H}_{t_0}^{(b)}(\omega) \tag{6.16}$$

(Jenkins [23]). Thus, the estimates of the amplitude and phase of harmonic components of fixed frequencies in a nonhomogeneous bivariate Poisson model considered in the previous section are functions of the empirical spectral components. It can also be shown, as for the univariate case (Lewis [31]), that $\mathscr{I}_{t_0}^{(ab)}(\omega)$ is the Fourier transform of the unsmoothed estimator of the cross intensity function obtained from all possible intervals between events of type a and events of type b.

The distribution theory of $\mathscr{I}_{t_0}^{(ab)}(\omega)$ for independent Poisson processes follows simply from the conditional properties of the Poisson processes. Thus, we find that $A_{t_0}^{(a)}(\omega)$ and $B_{t_0}^{(a)}(\omega)$ have the (conditional) joint generating function

$$[I_0\{(\xi_a^2 + \xi_b^2)^{1/2}\}]^{n^{(a)}} \tag{6.17}$$

if $\omega t_0 = 2\pi p$, from which it can be shown, for example, that $\mathscr{A}_{t_0}^{(a)}(\omega)$ and $\mathscr{B}_{t_0}^{(b)}(\omega)$ go rapidly to independent normal random variables with means 0 and standard deviations $\frac{1}{2}t_0\rho_a$ as $n^{(a)}$ becomes large. Consequently, the real and imaginary components of the cross periodogram have double exponential distributions centered at zero with a variance which does not decrease as t_0 increases.

At two frequencies ω_1 and ω_2 such that $\omega_1 t_0 = 2\pi p_1$ and $\omega_2 t_0 = 2\pi p_2$, the real components of $\mathscr{I}_{t_0}^{(ab)}(\omega_1)$ and $\mathscr{I}_{t_0}^{(ab)}(\omega_2)$ are asymptotically uncorrelated, as are the imaginary components. Consequently, smoothing of the periodogram is required to get consistent estimates of the in phase and out of phase components of the cross spectrum. The problems of bias, smoothing, and computation of the spectral estimates are similar to those for the univariate case discussed in detail by Lewis [31].

Note that the smoothed intensity function or the smoothed spectral estimates can be used to estimate the delay probability density function in the one sided Poisson delay model (see equations (3.26) and (3.27)) and the difference of the delays in the two sided (bivariate) Poisson delay model. In the first case, the estimation procedure is probably much more efficient, in some sense, than the procedure discussed by Brown [7] unless the mean delay is much shorter than the mean time between events in the main Poisson process.

6.3.3. *Covariance time function.* Another problem that arises with the bivariate Poisson delay process is to test for the presence of the Poisson noise and to estimate the rate μ of the unobservable main process. Since the covariance time curve $V^{(ab)}(t) \sim \mu t$, we can estimate μ by estimating $V^{(ab)}(t)$ and also test for Poisson noise by comparing the estimated measures of dispersion $I^{(aa)}$, $I^{(bb)}$, and $I^{(ab)}$, defined in (3.22), (3.23), and (3.24). Care will be needed over possible nonstationarity.

The simplest method for estimating $V^{(ab)}(t)$ is to estimate the variance time curves $V^{(aa)}(t)$, $V^{(bb)}(t)$, and $V^{(\cdot\cdot)}(t)$ with the procedures given by Cox and Lewis ([14], Chapter 5) and to use (3.39) to give an estimate of $V^{(ab)}(t)$.

There is no evident reason for estimating the covariance time surface $C(t_1 t_2)$ along any line except $t_1 = t_2$.

6.3.4. *Serial intensity function.* Estimation of the serial intensity functions raises new problems, somewhat analogous to the analysis of life tables. Consider the estimation of $\lambda^{(a)}(u_a, u_b)$ of (3.42). One approach is to pass to discrete time, dividing the time axis into small intervals of length Δ. Each such interval is characterized by the values of (u_a, u_b) measured from the center of the interval if no type a event occurs within the interval, and by the values of (u_a, u_b) at the type a event in question if one such event occurs; we assume for simplicity of exposition that the occurrence of multiple type a events can be ignored. Thus, each time interval contributes a binary response plus the values of two explanatory variables (u_a, u_b); the procedure extends to the case of more than two explanatory variables, and to the situation in which multiple type a events occur within the intervals Δ.

We can now do one or both of the following:

(a) assume a simple functional form for the dependence on (u_a, u_b) of the probability $\lambda^{(a)}(u_a, u_b)\Delta$ of a type a event and fit by weighted least squares or maximum likelihood (Cox [13]);

(b) group into fairly coarse "cells" in the (u_a, u_b) plane and find the proportion of "successes" in each cell.

It is likely that standard methods based on an assumption of independent binomial trials are approximately applicable to such data and, if so, specific assumptions about the form of the serial intensities can be tested. In particular, we can test the hypothesis that the process is, say, purely a dependent, making the further assumption to begin with that the dependence is only on u_a.

By extensions of this method, that is, by bringing in dependencies on more aspects of the history at time t than merely u_a and u_b, it may be possible to build up empirically a fairly simple model for the process.

6.4. *Simple tests for dependence.* As noted previously, it may sometimes be required to construct simple tests of the null hypothesis that the type a and type b events are independent, as defined in Section 2.2. This may be done in various ways. Much the simplest situation arises when we consider the dependence of, say, the type b events on the type a events, argue conditionally on the observed type a process, and consider the strong null hypothesis that the type b events form an independent Poisson process. Then, conditionally on the total number of events of type b, the positions of the type b events, $t_1^{(b)}, \cdots, t_{n(a)}^{(b)}$ are independently and uniformly distributed over the period of observation. Thus in principle, the exact distribution of any test statistic can be obtained free of nuisance parameters.

The two simplest of the many possible test statistics are probably:

(a) particular ordinates of the cross intensity function, usually that near the origin; equivalently we can use the statistic $R_a^{(b)}(0, \Delta)$ of Section 6.3, directly;

(b) the sample mean recurrence time backwards from a type b event to the nearest preceding type a event.

The null distribution of $R_a^{(b)}(0, \Delta)$ can be found as follows. Place an interval of length Δ to the right of each type a event. (It is assumed for convenience that either there is a type a event at the origin, or that the position of the last type a event before the origin is available.) Let $\pi_0, \pi_1, \pi_2, \cdots, \pi_{n(a)}$ be the proportion of the observed interval $(0, t_0)$ covered jointly by $0, 1, 2, \cdots, n^{(a)}$ of these intervals Δ. Then, if there are $n^{(b)}$ events of type b in all, the null distribution of $R_a^{(b)}(0, \Delta)$ is that of the sum of $n^{(b)}$ independent random variables each taking the value i with probability π_i, $i = 1, \cdots, n^{(a)}$.

Similarly, for the second test statistic, we can find the null distribution as follows. Regard the sequence of intervals between successive type a events as a finite population $x = \{x_1, \cdots, x_N\}$, say. This includes the intervals from 0 to the first type a event and from the last type a event to t_0. If t_0 is preassigned, $N = n^{(a)} + 1$. Note that $\Sigma x_i = t_0$. Then the null distribution of the test statistic is that of the mean of n_b independent and identically distributed random variables each with probability density function

$$\frac{1}{t_0} \sum_{i=1}^{N} U(x; x_i), \tag{6.18}$$

where

$$U(x; x_i) = \begin{cases} 1, & 0 \leqq x \leqq x_i, \\ 0, & \text{otherwise}. \end{cases} \tag{6.19}$$

Thus, in particular, the null mean and variance of the test statistic are

$$\frac{\sum x_i^2}{2t_0}, \qquad \frac{1}{n^{(b)}} \left\{ \frac{1}{3t_0} \sum x_i^3 - \frac{(\sum x_i^2)^2}{4t_0^2} \right\}. \tag{6.20}$$

A strong central limit effect may be expected.

The tests derived here may be compared with similar ones in which the null distribution is derived by computer simulation, permuting at random the observed sequences of intervals (Perkel [40]; Moore, Perkel, and Segundo [37]; Perkel, Gerstein, and Moore [41]). In both types of procedure, it is not clear how satisfactory the tests are in practice as general tests of independence, when the type b process is not marginally Poisson. Note, however, that in order to obtain a null distribution for (a) and (b) above it is necessary to assume only that one of the marginal processes is a Poisson process.

If it is required to treat the two processes symmetrically, taking the null hypothesis that there are two mutually independent Poisson processes, there are many possibilities, including the use of the estimated cross spectral or cross intensity functions or of a two sample test based on the idea that, conditionally on $n^{(a)}$ and $n^{(b)}$, the times to events in the two processes are the order statistics from two independent populations of uniformly distributed random variables. Again, in the symmetrical case when both marginal processes are clearly not Poisson processes, tests of independence based on the cross spectrum are probably the best broad tests. For this purpose, investigation of the robustness of the distribution theory given in Section 6.3 would be worthwhile.

◇ ◇ ◇ ◇ ◇

We are indebted to Mr. T. K. M. Wisniewski, Dr. A. J. Lawrance, and Professor D. P. Gaver for helpful discussions during the growth of this paper.

REFERENCES

[1] M. S. Bartlett, "The spectral analysis of point processes," *J. Roy. Statist. Soc. Ser. B*, Vol. 25 (1963), pp. 264–296.

[2] ———, "The spectral analysis of two-dimensional point processes," *Biometrika*, Vol. 51 (1964), pp. 299–311.

[3] ———, "Line processes and their spectral analysis," *Proceedings of the Fifth Berkeley Symposium on Mathematical Statistics and Probability*, Berkeley and Los Angeles, University of California Press, 1967, Vol. 3, pp. 135–154.

[4] ———, *An Introduction to Stochastic Processes*, Cambridge, Cambridge University Press, 1966 (2nd ed.).

[5] P. Billingsley, "Statistical methods in Markov chains," *Ann. Math. Statist.*, Vol. 32 (1961), pp. 12–40.

[6] D. R. Brillinger, "The spectral analysis of stationary interval functions," *Proccedings of the Sixth Berkeley Symposium on Mathematical Statistics and Probability*, Berkeley and Los Angeles, University of California Press, 1972, Vol. 1, pp. 483–513.

[7] M. Brown, "An $M/G/\infty$ estimation problem," *Ann. Math. Statist.*, Vol. 41 (1970), pp. 651–654.

[8] E. Cinlar, "On the superposition of m-dimensional point processes," *J. Appl. Prob.*, Vol. 5 (1968), pp. 169–176.

[9] ———, "Markov renewal theory," *Adv. Appl. Prob.*, Vol. 1 (1969), pp. 123–187.

[10] R. Coleman and J. L. Gastwirth, "Some models for interaction of renewal processes related to neuron firing," *J. Appl. Prob.*, Vol. 6 (1969), pp. 38–58.

[11] D. R. Cox, "Some statistical methods connected with series of events," *J. Roy. Statist. Soc. Ser. B*, Vol. 17 (1955), pp. 129–164.

[12] ———, "Some models for series of events," *Bull. Inst. Internat. Statist.*, Vol. 40 (1963), 737–746.

[13] ———, *Analysis of Binary Data*, London, Methuen; New York, Barnes and Noble, 1970.

[14] D. R. Cox and P. A. W. Lewis, *The Statistical Analysis of Series of Events*, London, Methuen; New York, Barnes and Noble, 1966.

[15] D. R. Cox and H. D. Miller, *The Theory of Stochastic Processes*, London, Methuen, 1966.

[16] D. J. Daley, "The correlation structure of the output process of some single server queueing systems," *Ann. Math. Statist.*, Vol. 39 (1968), pp. 1007–1019.

[17] J. L. Doob, *Stochastic Processes*, New York, Wiley, 1953.

[18] M. Dwass and H. Teicher, "On infinitely divisible random vectors," *Ann. Math. Statist.*, Vol. 28 (1957), pp. 461–470.

[19] J. E. Freund, "A bivariate extension of the exponential distribution," *J. Amer. Statist. Assoc.*, Vol. 56 (1961), pp. 971–977.

[20] D. P. Gaver, "Multivariate gamma distributions generated by mixture," *Sankhyā Ser. A*, Vol. 32 (1970), pp. 123–126.

[21] R. C. Griffiths, "The canonical correlation coefficients of bivariate gamma distributions," *Ann. Math. Statist.*, Vol. 40 (1969), pp. 1401–1408.

[22] A. G. Hawkes, "Spectra of some self-exciting and mutually exciting point processes," *Biometrika*, Vol. 58 (1971), pp. 83–90.

[23] G. M. Jenkins, "Contribution to a discussion of paper by M. S. Bartlett," *J. Roy. Statist. Soc. Ser. B*, Vol. 25 (1963), pp. 290–292.

[24] A. J. Lawrance, "Selective interaction of a Poisson and renewal process: First-order stationary point results," *J. Appl. Prob.*, Vol. 7 (1970), pp. 359–372.

[25] ———, "Selective interaction of a stationary point process and a renewal process," *J. Appl. Prob.*, Vol. 7 (1970), pp. 483–489.

[26] ———, "Selective interaction of a Poisson and renewal process: The dependency structure of the intervals between responses," *J. Appl. Prob.*, Vol. 8 (1971), pp. 170–184.

[27] M. R. Leadbetter, "On streams of events and mixtures of streams," *J. Roy. Statist. Soc. Ser. B*, Vol. 28 (1966), pp. 218–227.

[28] ———, "On the distribution of times between events in a stationary stream of events," *J. Roy. Statist. Soc. Ser. B*, Vol. 31 (1969), pp. 295–302.

[29] P. A. W. Lewis, "A branching Poisson process model for the analysis of computer failure patterns," *J. Roy. Statist. Soc. Ser. B*, Vol. 26 (1964), pp. 398–456.

[30] ———, "Asymptotic properties and equilibrium conditions for branching Poisson processes," *J. Appl. Prob.*, Vol. 6 (1969), pp. 355–371.

[31] ———, "Remarks on the theory, computation and application of the spectral analysis of series of events," *J. Sound Vib.*, Vol. 12 (1970), pp. 353–375.

[32] ———, "Asymptotic properties of branching renewal processes," *J. Appl. Prob.*, to appear.

[33] A. W. Marshall and I. Olkin, "A multivariate exponential distribution," *J. Amer. Statist. Assoc.*, Vol. 62 (1967), pp. 30–44.

[34] ———, "A generalized bivariate exponential distribution," *J. Appl. Prob.*, Vol. 4 (1967), pp. 291–302.

[35] J. A. McFadden, "On the lengths of intervals in a stationary point process," *J. Roy. Statist. Soc. Ser. B*, Vol. 24 (1962), pp. 364–382.

[36] E. H. Moore and R. Pyke, "Estimation of the transition distributions of a Markov renewal process," *Ann. Inst. Statist. Math.*, Vol. 20 (1968), pp. 411–424.

[37] G. P. Moore, D. H. Perkel, and J. P. Segundo, "Statistical analysis and functional interpretation of neuronal spike data," *Ann. Rev. Psychology*, Vol. 28 (1966), pp. 493–522

[38] P. A. P. Moran and D. Vere-Jones, "The infinite divisibility of multivariate gamma distributions," *Sankhyā Ser. A*, Vol. 31 (1969), pp. 191–194.

[39] J. NEYMAN and E. L. SCOTT, "Statistical approach to problems of cosmology," *J. Roy. Statist. Soc. Ser. B*, Vol. 20 (1958), pp. 1–29.

[40] D. H. PERKEL, "Statistical techniques for detecting and classifying neuronal interactions," *Symposium on Information Processing in Sight Sensory Systems*, Pasadena, California Institute of Technology, 1965, pp. 216–238.

[41] D. H. PERKEL, G. L. GERSTEIN, and G. P. MOORE, "Neuronal spike trains and stochastic point processes II. Simultaneous spike trains," *Biophys. J.*, Vol. 7 (1967), pp. 419–440.

[42] R. L. PLACKETT, "A class of bivariate distributions," *J. Amer. Statist. Assoc.*, Vol. 60 (1965), pp. 516–522.

[43] R. PYKE and R. A. SCHAUFELE, "The existence and uniqueness of stationary measures for Markov renewal processes," *Ann. Math. Statist.*, Vol. 37 (1966), pp. 1439–1462.

[44] I. M. SLIVNYAK, "Some properties of stationary flows of homogeneous random events," *Theor. Probability Appl.*, Vol. 7 (1962), pp. 336–341.

[45] D. VERE-JONES, S. TURNOVSKY, and G. A. EIBY, "A statistical survey of earthquakes in the main seismic region of New Zealand. Part I. Time trends in the pattern of recorded activity," *New Zealand J. Geol. Geophys.*, Vol. 7 (1964), pp. 722–744.

[46] D. VERE-JONES and R. D. DAVIES, "A statistical survey of earthquakes in the main seismic region of New Zealand, Part II. Time series analysis," *New Zealand J. Geol. Geophys.*, Vol. 9 (1966), pp. 251–284.

[47] D. VERE-JONES, "Stochastic models for earthquake occurrence," *J. Roy. Statist. Soc. Ser. B*, Vol. 32 (1970), pp. 1–62.

[48] L. WALLOE, J. K. S. JANSEN, and K. NYGAARD, "A computer simulated model of a second order sensory neuron," *Kybernetik*, Vol. 6 (1969), pp. 130–140.

[49] G. S. WATSON and E. J. WILLIAMS, "On the construction of significance tests on the circle and the sphere," *Biometrika*, Vol. 43 (1956), pp. 344–352.

[50] T. K. M. WISNIEWSKI, "Forward recurrence time relations in bivariate point processes," *J. Appl. Prob.*, Vol. 9 (1972).

[51] ———, "Extended recurrence time relations in bivariate point processes," to appear.

ON BASIC RESULTS OF POINT PROCESS THEORY

M. R. LEADBETTER
UNIVERSITY OF NORTH CAROLINA

1. Introduction

There are many existing approaches to the theory of point processes. Some of these—following the original work of Khinchin [9] are "analytical" and others (for example, [15], [8]) quite abstract in nature. Here we will take a position somewhat in the middle in describing the development of some of the basic theory of point processes in a relatively general setting, but by using largely the simple techniques of proof described for the real line in [11]. We shall survey a number of known results—giving simple derivations of certain existing theorems (or their adaptations in our setting) and obtain some results which we believe to be new. Our framework for describing a general point process will be essentially that of Belyayev [2], while that for Section 4 concerning Palm distributions is developed from the approach of Matthes [14].

First we give the necessary background and notation. There are various essentially equivalent ways of defining the basic structure of a point process. For example, for point processes on the line, one may consider the space of integer valued functions $x(t)$ with $x(0) = 0$, which increase by a finite number of jumps in any finite interval. The events of the process then correspond to jumps of $x(t)$. One advantage of such a specification is that multiple events fit naturally into the framework.

To define point processes on an arbitrary space T, it is often appropriate to consider the "sample points" ω to be subsets of T. This is the point of view taken in [18], where each ω is itself a countable subset of the real line, the set of points "where events occur." Sometimes, however, a point process arises from some existing probabilistic situation (such as the zeros of a continuous parameter stochastic process) and one may wish to preserve the existing framework in the discussion. A convenient structure for this is the following, used in [2]. Let $(\Omega, \mathscr{F}, P)$ be a probability space and $(T, \mathscr{T})$ a measurable space (T is the space "in which the events will occur"). For each $\omega \in \Omega$, let S_ω be a subset of T. If for each $E \in \mathscr{T}$

$$N(E) = N_\omega(E) = \text{card } (E \cap S_\omega) \tag{1.1}$$

is a (possibly infinite valued) random variable, then S_ω is called a *random set* and the family $\{N(E) : E \in \mathscr{T}\}$ a *point process*. The "events" of the process are, of course, the points of S_ω.

The model may be generalized slightly to take account of *multiple events*—that is, the possible occurrence of more than one event at some $t \in T$. The definition of $N(E)$ as card $(E \cap S_\omega)$ shows that $N(E)$ is an integer valued measure (on the subsets of T) for each ω. As a measure, $N_\omega(\cdot)$ has its mass confined to S_ω and $N_\omega(\{t\}) = 1$ for each $t \in S_\omega$.

To allow multiple events, we may simply redefine $N_\omega(E)$ to be an integer valued measure with all its mass confined to S_ω, with $N_\omega(\{t\}) \geqq 1$ for each $t \in S_\omega$ and such that $N_\omega(E)$ is a random variable for each $E \in \mathscr{T}$. If $N_\omega(\{t\}) > 1$, we say a *multiple event* occurs at t. If there is zero probability that any $t \in S_\omega$ is multiple, we say the process is *without multiple events*.

If we say a process may have multiple events, we shall be referring to this framework and shall write $M(E)$ for the number of multiple events in E. In such a case, we shall write $N^*(E)$ for card $\{S_\omega \cap E\}$ and refer to $N^*(E)$ as the number of events in E *without regard to their multiplicities*. (Of course, $N(E)$ is the total number of events in E.)

In the manner just described, a point process may be regarded as a special type of *random measure*. This concept has been developed in considerable generality for stationary cases (see, for example, [15]), but this generality will not be pursued here.

Another method of taking account of multiple events is to replace each $t \in S_\omega$ by a pair (t, k_t), where k_t is a "mark" associated with t denoting the multiplicity. This again is capable of considerable generalization by considering rather arbitrary kinds of "marks" and the appropriate additional measure theoretic structure. These ideas have been developed by Matthes (see, for example, [14]) for stationary point processes on the real line and provide an elegant framework for obtaining results, for example, in relation to Palm distributions. In such cases, the marks are chosen to be highly dependent on the set S_ω (for example, translates of S_ω). At the same time, most results of interest can be obtained by using essentially these techniques, but without explicit reference to marks. Hence, we here use the framework previously explained.

For stationary point processes on the real line, there are several important basic theorems. Included among these are (writing $N(s, t)$ for the number of events in $(s, t]$):

(i) the theorem of Khinchin regarding the existence of the *parameter* $\lambda = \lim_{t\downarrow 0} Pr\{N(0, t) \geqq 1\}/t$;

(ii) Korolyuk's theorem which, in its sharpest form, says that for a stationary point process without multiple events, λ is equal to the *intensity* $\mu = \mathscr{E}N(0, 1)$ that is, the mean number of events per unit time; λ and μ may be infinite; if multiple events may occur, we replace μ by $\mathscr{E}N^*(0,1)$;

(iii) for the regular (orderly, ordinary) case (that is, when $Pr\{N(0, t) > 1\} = o(t)$ as $t \downarrow 0$) multiple events have probability zero;

(iv) "Dobrushin's lemma"—a converse to (iii)—stating that if $\lambda < \infty$ and multiple events have probability zero, then the process is regular.

Various analogues of these results have been studied for nonstationary point processes on the real line in [20], [4], [5], largely by using properties of Burkhill integrals. A clarifying and general viewpoint has been more recently given by Belyayev [2]. Specifically in [2], generalizations of the two constants λ, μ are made in terms of *measures* $\lambda(\cdot)$, $\mu(\cdot)$ on the space T, instead of in terms of point functions. The *principal measure* $\mu(\cdot)$ is defined (as customarily) on $\mathscr{T}$ simply by $\mu(E) = \mathscr{E}N(E)$—countable additivity of N guaranteeing countable additivity of $\mu(\cdot)$. On the other hand, the *parametric measure* $\lambda(\cdot)$ is defined in [2] by

$$\lambda(E) = \sup\left\{\sum_1^\infty Pr\,\{N(E_i) > 0\} : E_i \in T,\, E_i \text{ disjoint},\, \bigcup_1^\infty E_i = E\right\}. \tag{1.2}$$

It is easily shown that $\lambda(\cdot)$ is a measure and it is clear that $\lambda(E) \leqq \mu(E)$ for all $E \in \mathscr{T}$. For a stationary point process on the real line, $\lambda(E) = \lambda m(E)$ and $\mu(E) = \mu m(E)$, where m denotes Lebesgue measure.

Using these definitions, it is possible to extend the basic results quoted above to apply to point processes which may be nonstationary, on spaces T more general than the line (including any Euclidean space). These generalizations are systematically described in Section 2. In Section 3, stationarity is discussed in general terms (with particular reference to Khinchin's theorem) when T is a topological group. In both these sections, the general lines of development are those of [2], with adaptation of the results in presenting a somewhat different viewpoint, and with emphasis on simplicity of proofs obtained by direct analogy with those of [11].

Finally, in Section 4, we discuss some basic results relative to Palm distributions (and their expressions as limits of conditional probabilities), for stationary point processes on the real line. The approach is essentially that of [14] (without explicit reference to marks), again with emphasis on the simplicity of proofs obtained from the techniques of [11].

2. The basic general theorems

The notation already developed will be used throughout this section. We shall systematically obtain the generalizations of the basic theorems referred to in Section 1. This development follows the same general lines as [2] but with differences of detail and perspective.

All that is to be said in *general* relative to Khinchin's theorem concerning the existence of the intensity, is contained in Belyayev's definition of the parametric measure (1.2) given in Section 1. (For special cases, when T has a group structure and the point process is stationary, it is possible to say more that is directly analogous to the real line case—mention of this will be made later.)

It is shown in [2] that the truth of the generalized version of Korolyuk's theorem, namely, $\lambda(E) = \mu(E)$ for all $E \in \mathscr{T}$ (for a point process without multiple events or $\lambda(E) = \mathscr{E}N^*(E)$ if multiple events may occur), depends on the structure of T rather than on any stationarity assumption. The proof given directly generalizes that of [11] for stationary processes on the real line. This is most clearly seen for a nonstationary process on the real line ($\mathscr{T}$ then being the Borel sets). For then if E is an interval $(a, b]$, we may divide E into n equal subintervals E_{ni}, $i = 1, \cdots, n$, and write $\chi_{ni} = 1$ if $N(E_{ni}) \geqq 1$, $\chi_{ni} = 0$ otherwise. Assuming there are no multiple events, it is easily seen that $N_n = \Sigma_{i=1}^n \chi_{ni} \to N(E)$ with probability one, as $n \to \infty$, and hence by Fatou's lemma,

$$(2.1) \qquad \mu(E) \leqq \liminf \mathscr{E} N_n = \liminf \sum_{i=1}^{n} Pr\,\{\chi_{ni} = 1\} \leqq \lambda(E).$$

But $\lambda(E) \leqq \mu(E)$, and hence, $\lambda(E) = \mu(E)$ for all E of the form $(a, b]$. Thus, $\lambda(E) = \mu(E)$ for all Borel sets E, provided μ is σ-finite.

For the above proof to be useful when T is a more general space, we require T to have sets playing the role of intervals. A suitable definition of such a class of sets is given by Belyayev [2] and called a "fundamental system of dissecting sets" for T. Here we shall use a somewhat different definition to achieve the desired results. Specifically, we here say that a class $\mathscr{C} = \{E_{nk}: n, k = 1, 2 \cdots\}$ of sets $E_{nk} \in \mathscr{T}$ is a *dissecting system* for T if

(i) $\mathscr{C}$ is a "determining class" (see [3]) for σ-finite measures on $\mathscr{T}$; that is, two σ-finite measures equal on $\mathscr{C}$ are equal on $\mathscr{T}$ (for example, $\mathscr{C}$ may be a semiring generating $\mathscr{T}$);

(ii) for any given set $E \in \mathscr{C}$, there is corresponding to each $n = 1, 2, 3, \cdots$ a set I_n of integers such that

(a) E_{nk} are disjoint subsets of E for $k \in I_n$ with $E - \bigcup_{k \in I_n} E_{nk} \subset F_n \in \mathscr{T}$, where $F_n \downarrow \varnothing$, the empty set, as $n \to \infty$, and

(b) given any two points $t_1 \neq t_2$ of E, for all sufficiently large values of n (that is, for all $n \geqq$ some $n_0(t_1, t_2)$), there are sets E_{nk_1}, E_{nk_2}, $k_1, k_2 \in I_n (k_1 \neq k_2$, such that $t_1 \in E_{nk_1}$, $t_2 \in E_{nk_2}$.

For example, for the real line, we may take E_{nk} to be any interval $(a, b]$ with rational endpoints and of length $1/n$. We note also that the requirement in (ii) (a) that $\lim F_n = \varnothing$ may be replaced by $Pr\,\{N(\lim F_n) = 0\} = 1$, but this, of course, depends on the process as well as the structure of T.

The proof of Korolyuk's theorem given for the real line now generalizes at once to apply to a point process on a space T possessing a dissecting system. This is easily seen from the following lemma.

LEMMA 2.1. *Consider a point process on a space T possessing a dissecting system $C = \{E_{nk}\}$. With the above notation for $E \in C$, $k \in I_n$, write $\chi_{nk} = 1$ if $N(E_{nk}) > 0$, $\chi_{nk} = 0$ otherwise. Let*

$$(2.2) \qquad N_n = \sum_{k \in I_n} \chi_{nk}.$$

Then

$$(2.3) \qquad N_n \to N^*(E) \leqq \infty \text{ with probability one,}$$

as $n \to \infty$.

Further, if $\chi_{nk}^* = 1$ *when* $N(E_{nk}) > 1$ *and* $\chi_{nk}^* = 0$ *otherwise, and if* $N(E) < \infty$ *with probability one, then*

$$\sum_{k \in I_n} \chi_{nk}^* \to M(E) \text{ with probability one,} \tag{2.4}$$

as $n \to \infty$.

PROOF. It is clear that $N_n \leqq N(E)$. On the other hand, if $\infty \geqq N(E) \geqq m$, there are points $t_1, \cdots, t_m$, where events occur. For large n, these are eventually contained in different sets E_{nk} and hence $N_n \geqq m$. Equation (2.3) follows by combining these results.

The second part follows by noting that since (with probability one) only a finite number of distinct events occur, they are eventually contained in different E_{nk} sets when n is large, and thus for such n, $N_n^* = M(E)$.

By using the first part of this lemma, Korolyuk's theorem follows as for the real line by a simple application of Fatou's lemma. Stated specifically we have (see [2]):

THEOREM 2.1 (Generalized Korolyuk's theorem). *For a point process, with* σ*-finite principal measure, on a space* T *possessing a dissecting system, we have* $\lambda(E) = \mathscr{E}N^*(E)$ *for all* $E \in \mathscr{T}$. *In particular if there are no multiple events, then the principal and parametric measures coincide on* $\mathscr{T}$.

If $\lambda(\cdot)$ or $\mu(\cdot)$ is absolutely continuous with respect to some σ-finite measure ν on T, then under the above conditions the densities $d\lambda/d\nu$, $d\mu/d\nu$ coincide a.e. This reduces again to the usual statement of Korolyuk's theorem for stationary point processes on the line.

A point process on T is called regular (orderly, ordinary—cf [20], [5] and especially [2]) with respect to a dissecting system $\mathscr{C} = \{E_{nk}\}$ if

$$\lim_{n \to \infty} \sup_k \frac{Pr\{N(E_{nk}) > 1\}}{Pr\{N(E_{nk}) > 0\}} = 0. \tag{2.5}$$

(For simplicity, we shall always assume $Pr\{N(E_{nk}) > 0\} \neq 0$ for any n, k.) This definition applies to a point process which may conceivably have multiple events. However, the next result shows that in fact regularity precludes the occurrence of multiple events under simple conditions on T.

THEOREM 2.2. *Consider a point process (allowing multiple events) on a space* T *possessing a dissecting system* $\mathscr{C}$. *Suppose the process is regular and that there exist* $E_n \in \mathscr{C}$, $E_n \uparrow T$ *such that* $\lambda(E_n) < \infty$. *Then, with probability one, the process has no multiple events.*

PROOF. Let $E \in \mathscr{C}$ be such that $\lambda(E) < \infty$. Write again $M(E)$ for the number of *multiple* events in E. Then by Lemma 2.1, $M(E) = \lim_{n \to \infty} \Sigma_{k \in I_n} \chi_{nk}^*$ (with the usual notation), where χ_{nk}^* is one or zero according as $N(E_{nk}) > 1$ or not. Hence,

$$\begin{aligned} \mathscr{E}M(E) &\leqq \liminf_n \Big[\sum_{k \in I_n} Pr\{N(E_{nk}) > 1\}\Big] \\ &\leqq \liminf_n \left[\sup_k \frac{Pr\{N(E_{nk}) > 1\}}{Pr\{N(E_{nk}) > 0\}} \sum_{j \in I_n} Pr\{N(E_{nj}) > 0\}\right], \end{aligned} \tag{2.6}$$

which is zero since by regularity the first term in the braces tends to zero, and the sum does not exceed $\lambda(E) < \infty$. Since $\mathscr{E}M(E)$ is a measure on $\mathscr{T}$, $\mathscr{E}M(T) = \lim_{n\to\infty} \mathscr{E}M(E_n) = 0$, and hence, $M(T) = 0$ with probability one.

A converse result of Theorem 2.2 is "Dobrushin's lemma." A general form of this given in [2] assumes a "homogeneous" point process—for which T possesses a fundamental system $\mathscr{C}$ of dissecting sets such that $p_{nk}(0) = Pr\{N(E_{nk}) > 0\}$, $p_{nk}(1) = Pr\{N(E_{nk}) > 1\}$ are each dependent on n but not on k for $E_{nk} \in \mathscr{C}$. This assumption does not imply stationarity of the point process (indeed there may be no "translations" defined on T), but it may well be that the only interesting homogeneous processes are stationary ones. We give a less restricted result below. It may still be of greatest interest in the stationary case, but it does allow considerable variation in the quantities $p_{nk}(0)$, $p_{nk}(1)$ for fixed n.

Specifically to obtain Dobrushin's lemma, we shall assume the existence of a dissecting system $\mathscr{C} = \{E_{nk}\}$ for which there is a sequence $\{\theta_n\}$ of nonnegative real numbers, and a function $\phi(\theta) \to 0$ as $\theta \to 0$, such that for each n

$$\theta_n \leqq \frac{Pr\{N(E_{nk}) > 1\}}{Pr\{N(E_{nk}) > 0\}} \leqq \phi(\theta_n) \tag{2.7}$$

for all k.

THEOREM 2.3 (Generalized version of Dobrushin's lemma). *Consider a point process without multiple events, on a space T possessing a dissecting system $\mathscr{C}$ satisfying* (2.7). *Suppose $\lambda(E) < \infty$ for some $E \in \mathscr{C}$. Then the point process is regular.*

PROOF. Using the notation of Lemma 2.1, we have

$$\sum_{k\in I_n} \chi_{nk} \to N(E), \qquad \sum_{k\in I_n} \chi^*_{nk} \to 0 \tag{2.8}$$

with probability one, as $n \to \infty$. Since both sums are dominated by $N(E)$ and $\mathscr{E}N(E) = \mu(E) = \lambda(E) < \infty$, it follows by dominated convergence that

$$\sum_{k\in I_n} Pr\{N(E_{nk}) > 0\} = \mathscr{E}\{\sum_{k\in I_n} \chi_{nk}\} \to \mu(E) = \lambda(E), \tag{2.9}$$

and similarly that

$$\sum_{k\in I_n} Pr\{N(E_{nk}) > 1\} \to 0. \tag{2.10}$$

Hence by (2.7),

$$\theta_n \sum_{k\in I_n} Pr\{N(E_{nk}) > 0\} \leqq \sum_{k\in I_n} Pr\{N(E_{nk}) > 1\} \to 0, \tag{2.11}$$

and thus by (2.9), $\theta_n \to 0$ $(\lambda(E) \geqq Pr\{N(E) > 0\} > 0$ since $E \in \mathscr{C})$.

Finally, from (2.7) again,

$$\sup_k \left[\frac{Pr\{N(E_{nk}) > 1\}}{Pr\{N(E_{nk}) > 0\}}\right] \leqq \phi(\theta_n) \to 0 \tag{2.12}$$

as $n \to \infty$.

3. Stationarity generalities

A very great deal of literature exists relative to stationary point processes on the real line (see Section 4). One expects to be able to say less about stationary point processes on the plane or in R^n (see, for example, [6]). However, there is quite a good deal that may be said even when T is just assumed to be a (locally compact) topological group. In this section, we comment briefly on a few aspects of such results.

If T is a locally compact (Hausdorff) group, the natural σ-field $\mathscr{T}$ is the class of Borel sets—generated by the open sets of T. It is usually convenient to assume (and we here do) that T is also σ-compact, and then $\mathscr{T}$ is also generated by the compact sets of T. (It is, in fact, sometimes assumed that T is second countable; for example, [15]. While this additional assumption may be necessary for some purposes it does, however, imply that the group is also metrizable.)

For a point process on such a group T, stationarity may be defined in terms of the invariance of the joint distributions of $N(tE_1), \cdots, N(tE_n)$ for $t \in T$, where n is any fixed positive integer and the E_i are any fixed sets of $\mathscr{T}$ ($tE = \{ts\colon s \in E\}$, ts denoting the group operation). If T is not abelian this gives a concept of "left stationarity," "right stationarity" being correspondingly defined.

Under (say, left) stationarity, the principal and parametric measures $\mu(\cdot)$, $\lambda(\cdot)$ are (left) invariant Borel measures which are regular provided their values on compact sets are finite ([7], Theorem 64I)—which we will assume. Thus, $\lambda(\cdot)$ and $\mu(\cdot)$ are just constant multiples of the Haar measure $m(\cdot)$ on T, $\lambda(E) = \lambda m(E)$, $\mu(E) = \mu m(E)$, say, for all $E \in \mathscr{T}$, where λ and μ are constants, the *parameter* and the *intensity* of the stationary point process, respectively. Questions concerning the parameter and intensity in such a setting have been discussed to some extent in [1]. The general line of argument above is that of [2].

If in addition T possesses a dissecting system $\mathscr{C} = \{E_{nk}\}$ of, say, bounded sets (that is, having compact closures) and if a stationary point process on T is without multiple events, then Theorem 2.1 shows that $\lambda = \mu < \infty$. This is Korolyuk's theorem in the stationary case. Further, in such a case it is not unreasonable to suppose that $P\{N(E_{nk}) > 0\}$ and $m(E_{nk})$ are independent of k (which will hold if, for example, for fixed n the E_{nk} are translates of each other). Then using the notation of Theorem 2.3 we have, from the proof of that theorem,

$$r_n Pr\,\{N(E_{n0}) > 0\} \to \lambda(E) = \lambda m(E) \tag{3.1}$$

for $E \in \mathscr{C}$, where r_n is the (necessarily finite) number of integers in the set I_n and E_{n0} is any given E_{nk} for $k \in I_n$.

But since by definition of $\mathscr{C}$.

$$E - \bigcup_{k \in I_n} E_{nk} \subset F_n \downarrow \varnothing, \tag{3.2}$$

it follows that

$$r_n m(E_{n0}) = \sum_{k \in I_n} m(E_{nk}) \to m(E), \tag{3.3}$$

and hence, that

$$\frac{Pr\,\{N(E_{n0}) > 0\}}{m(E_{n0})} \to \lambda \tag{3.4}$$

as $n \to \infty$. It is this latter property that the parameter satisfies in Khinchin's existence theorem. We summarize this as a theorem. For convenience of statement, we will *here* call a dissecting system $\mathscr{C}$ *homogeneous* if the distribution of $N(E_{nk})$ and $m(E_{nk})$ do not depend on k for each fixed n.

THEOREM 3.1. *Consider a stationary point process without multiple events on locally compact group T. Suppose T is also σ-compact. Then there exist constants λ, μ such that $\lambda(E) = \lambda m(E)$, $\mu(E) = \mu m(E)$ for all $E \in \mathscr{T}$, where $m(\cdot)$ is the Haar measure of T, $0 \leqq \lambda \leqq \mu < \infty$.*

Suppose, in addition, that T has a homogeneous dissecting system $\mathscr{C} = \{E_{nk}\}$ of bounded sets E_{nk}. Then the point process is regular, $\lambda = \mu$ and

$$\lim_{n\to\infty} \frac{Pr\,\{N(E_{n0}) > 0\}}{m(E_{n0})} = \lambda, \tag{3.5}$$

where E_{n0} is any E_{nk}.

COROLLARY 3.1. *The stated results hold if the condition that $\mathscr{C}$ be a homogeneous dissecting system is replaced by the requirement that for each fixed n, the sets E_{nk} are all translates of each other.*

The above remarks have been concerned with a stationary process without multiple events. When multiple events are allowed, the appropriate generalizations of the real line results occur. For example, if E is a set of $\mathscr{T}$ with $\mu(E) < \infty$ (for example, E compact), and if $N_s(E)$ denotes the number of those events in E which have "multiplicity" $s = 1, 2, \cdots$, then $p_s = \mathscr{E}N_s(E)/\lambda(E)$ is a probability distribution on the integers $1, 2, \cdots$. we may interpret $\{p_s\}$ as the "probability that an event has multiplicity s." If in addition T has a homogeneous dissecting system $\mathscr{C} = \{E_{nk}\}$ and we choose $E \in \mathscr{C}$ with $\mu(E) < \infty$ writing $\chi_{nk}^s = 1$ if $N(E_{nk}) = s$ and $\chi_{nk}^s = 0$ otherwise, then, similarly to Lemma 2.1,

$$\sum_{k \in I_n} \chi_{nk}^s \to N_s(E) \tag{3.6}$$

as $n \to \infty$, with probability one. The familiar argument of taking expectations and using dominated convergence shows that $r_n Pr\,\{N(E_{n0}) = s\} \to \mathscr{E}N_s(E)$, where E_{n0} is any E_{nk} and r_n is the number of points in I_n. Similarly, $r_n Pr\,\{N(E_{n0}) \geqq 1\} \to \lambda(E)$. Thus,

$$p_s = \lim_{n\to\infty} Pr\,\{N(E_{n0}) = s \,|\, N(E_{n0}) \geqq 1\}, \tag{3.7}$$

giving intuitive justification to the description of p_s as the probability that an event has multiplicity s (under these assumptions p_s does not depend on E). Further questions of this type are considered in [16] when $T = R^n$. We note that the above calculation may also be considered as a special case of that in the next section concerning Palm distributions.

4. Concerning Palm distributions

For a stationary point process, the Palm distribution P_0 gives a precise meaning to the intuitive notion of conditional probability "given an event of the process occurred at some point (for example, $t = 0$)." When T is the real line, we may write (for certain sets $F \in \mathscr{F}$)

$$P_0(F) = \lim_{\delta \downarrow 0} Pr\{F \mid N((-\delta, 0)) \geqq 1\}. \tag{4.1}$$

That is, $P_0(F)$ is then the limit of the conditional probability of F given an event occurred in an interval near $t = 0$ as that interval shrinks. For example, if F denotes the occurrence of at least one event in the interval $(0, t]$ (that is, $\{N(0, t) \geqq 1\}$), then $P_0(F) = F_1(t)$, the distribution function for the time to the first event after time zero given an event occurred "at" time zero.

This kind of procedure for particular sets F was used by Khinchin [9] and is useful in providing an "analytical" approach to such conditional probabilities (see, for example, [12]). More sophisticated and general measure theoretic treatments involving the definition and properties of P_0 have been given by a variety of authors (for example, [14], [15], [17], [18], [19]). In this section, we shall use a "middle of the road" approach to the definition of P_0 (based essentially on [14]) which is capable of considerable generality. Our main purpose will be to give simple proofs for formulae such as (4.1) and its generalizations to include "conditional expectations" of functions. Such results have application, for example, to the evaluation of the distributions of the times between events in terms of conditional moments [13].

We give the construction of the Palm distribution P_0 for stationary point process on the real line in the manner of [14], though, from a somewhat different viewpoint. The construction generalizes to apply to point processes on groups (see also [15]), but we consider just the real line case for simplicity relative to the later results.

Consider, then, a stationary point process (without multiple events for simplicity) with finite parameter λ on the real line. Again for simplicity, we take the sample points ω to be themselves the subsets S_ω of $T = R^1$, that is, ω is a countable set of real numbers (without finite limit points, since $\lambda < \infty$). Denote by $\mathscr{T}$ the Borel sets of $T = R^1$ and by $\mathscr{F}$ the smallest σ-field on Ω making $N(B) = N_\omega(B)$ measurable for each $B \in \mathscr{T}$. Finally, we shall again write $N(s, t)$ for $N\{(s, t]\}$, the number of events (card $\{\omega \cap (s, t]\}$) in the semiclosed interval $(s, t]$.

For any real t, and $\omega \in \Omega$, let $\omega_t \in \Omega$ denote the set of points of ω translated to the left by t; that is, if $\omega = \{t_i\}$, $\omega_t = \{t_i - t\}$. If F is any set of $\mathscr{F}$ and $\omega \in \Omega$, $\omega = \{t_i\}$, say, we define $\omega^* \in \Omega$ to consist of precisely those points $t_i \in \omega$ for which $\omega_{t_i} \in F$. In other words, to form ω^*, we "thin" ω by retaining only the points t_i such that ω_{t_i} (that is, ω translated to t_i as origin) is in F. The ω^* define a stationary point process formed from some of the events in the original point process. For example, if $F = \{\omega \colon N(0, t) \geqq 1\}$, the new process contains precisely those events t_i of the old process which are followed by a further event within a further time t (that is, no later than $t_i + t$). Write N_F for the number of events of the thinned process in the interval (0, 1), that is, card $\{\omega^* \cap (0, 1)\}$. Then the thinned process has intensity $\lambda_F = \mathscr{E}N_F$.

Now this procedure may be carried out for any $F \in \mathscr{F}$, and for fixed ω, N_F is countably additive as a function on $\mathscr{F}$. It follows at once that λ_F is a measure on $\mathscr{F}$, and hence, that

$$P_0(F) = \frac{\lambda_F}{\lambda} \tag{4.2}$$

is a probability measure on $\mathscr{F}$, $\lambda = \lambda_\Omega = \mathscr{E}N(0, 1)$. This P_0 is the desired *Palm distribution.*

To give P_0 an intuitive interpretation, one wishes to prove relations such as (4.1). Equation (4.1) is not universally true, however, as can be seen by considering a "periodic" stationary point process in which the events occur at a regular spacing h, where the distance to the first one after $t = 0$ is a uniform random variable on $(0, h)$. For this process take F to be the occurrence of at least one event in the open interval $(h - \eta, h)$. Clearly, $Pr\{F \mid N(-\delta, 0) \geqq 1\} = 1$ when $\delta < \eta$. But $N_F = 0$, and hence $P_0(F) = 0$.

We give now a class of sets for which (4.1) does hold. Specifically, we shall call a set $F \in \mathscr{F}$ *right continuous* if its characteristic function $\chi_F(\omega)$ is such that $\chi_F(\omega_t)$ is continuous to the right in t; that is $\chi_F(\omega_s) \to \chi_F(\omega_t)$ as $s \downarrow t$. Equivalently, this means that for any t, if $\omega_t \in F$ then $\omega_s \in F$ when s is sufficiently close to t on the right, and conversely.

THEOREM 4.1. *Suppose $F \in \mathscr{F}$ is a right continuous set. Then*

$$Pr\{F \mid N(-\delta, 0) \geqq 1\} \to P_0(F) \tag{4.3}$$

as $\delta \downarrow 0$.

PROOF. Let δ_m be any sequence of nonnegative numbers converging to zero as $m \to \infty$. Write r_m for the integer part $[\delta_m^{-1}]$ of δ_m^{-1}. Divide the interval (0, 1) into r_m intervals of length δ_m (with perhaps an interval of length less than δ_m left over). Write $\chi_{mi} = 1$ if $N((i-1)\delta_m, i\delta_m) \geqq 1$, $\chi_{mi} = 0$ otherwise, $i = 0, 1 \cdots r_m$. Let

$$N_m = \sum_{i=1}^{r_m} \chi_{mi}\chi_F(\omega_{i\delta_m}). \tag{4.4}$$

Then N_m denotes the number of intervals $((i-1)\delta_m, i\delta_m]$ containing an event and such that the translate $\omega_{i\delta_m}$ is in F. But by the right continuity assumption, if an event occurs at t_0, then $\omega_{t_0} \in F$ if and only if $\omega_{i\delta_m} \in F$ for that interval $((i-1)\delta_m, i\delta_m]$ containing t_0 when m is sufficiently large. Further, with probability one, when m is sufficiently large the events all lie in different intervals and there is no event in the last short interval. Hence, with probability one,

$N_m \to N_F$ as $m \to \infty$. Since $N_m \leqq N(0, 1)$ and $\mathscr{E}N(0, 1) < \infty$, it follows by dominated convergence that $\mathscr{E}N_m \to \lambda_F$ as $m \to \infty$. That is,

$$\sum_{i=1}^{r_m} Pr\{\chi_{mi} = 1, \omega_{i\delta_m} \in F\} \to \lambda_F \tag{4.5}$$

or by stationarity, $r_m Pr\{\chi_{m0} = 1, \omega \in F\} \to \lambda_F$. But

$$Pr\{\chi_{m0} = 1, \omega \in F\} = Pr\{F \mid N(-\delta_m, 0) \geqq 1\}\, Pr\{N(-\delta_m, 0) \geqq 1\}. \tag{4.6}$$

Hence, since $r_m \sim \delta_m^{-1}$ and $Pr\{N(-\delta_m, 0) \geqq 1\} \sim \lambda\delta_m$, we have

$$Pr\{F \mid N(-\delta_m, 0) \geqq 1\} \to \frac{\lambda_F}{\lambda} = P_0(F), \tag{4.7}$$

as required.

As an example, consider the set $F = \{\omega : N(0, t) \geqq r\}$, $r = 1, 2, \cdots$. This is easily seen to be right continuous, and hence the theorem applies. In this case,

$$P_0(F) = \lim_{\delta \downarrow 0} Pr\{N(0, t) \geqq r \mid N(-\delta, 0) \geqq 1\} \tag{4.8}$$

is interpreted as the distribution function for the time to the rth event after time zero, given an event occurred "at" time zero. (Note that at least r events occur in $(0, t)$ if and only if the time to the rth event after time zero does not exceed t.)

Similarly, if we take $0 < t_1 \leqq t_2 \cdots \leqq t_k$, $0 \leqq r_1 \leqq r_2 \cdots \leqq r_k$, and

$$F = \{\omega : N(0, t_1) \geqq r_1, N(0, t_2) \geqq r_2, \cdots, N(0, t_k) \geqq r_k\}, \tag{4.9}$$

then F is right continuous, leading to what could naturally be termed the joint distribution function for the time to the r_1st, r_2nd, $\cdots$, r_kth events after the origin given an event occurred at the origin.

The convergence in Theorem 4.1 does not occur for all $F \in \mathscr{F}$ in general. However, we may regard the probability space Ω as consisting of real integer valued functions increasing by unit jumps where events occur, and consider it as a subspace of D, the space of functions with discontinuities of the first kind (see [3]) where D has the "Skorohod topology." Then Theorem 4.1 may be shown to imply weak convergence of $P_\delta = P(\cdot \mid N(-\delta, 0) > 0)$ to P_0.

A slightly different definition given by Matthes ([14]) does give convergence similar to Theorem 4.1 for all $F \in \mathscr{F}$. Specifically, let $s = s(\omega)$ denote the time of the first event prior to the origin. Then instead of $P_\delta(F) = P\{\omega : \omega \in F \mid N(-\delta, 0) > 0\}$, we may consider $P_\delta^*(F) = P\{\omega : \omega_s \in F \mid N(-\delta, 0) > 0\}$. That is, the "origin is moved" slightly to the point s of $(-\delta, 0)$, where an event occurs. Then the following theorem (which is virtually identical to that of [10], Section 1(f)) holds.

THEOREM 4.2. *For each* $F \in \mathscr{F}$,

$$P_\delta^*(F) \to P_0(F) \tag{4.10}$$

as $\delta \downarrow 0$ *(hence the total variation of* $P_\delta^* - P_0$ *tends to zero* $\delta \downarrow 0$*).*

PROOF. This can be proved as in [10], Section 1. However, a very easy proof follows by a simplification of the method of Theorem 4.1. In fact, using the notation of that proof, we consider N_m^* (instead of N_m), where

$$N_m^* = \sum_{i=1}^{r_m} \chi_{mi}\chi_F(\omega_{s_{mi}}) \tag{4.11}$$

with s_{mi} denoting the position of the last event prior to (or at) $i\delta_m$. (A contribution to the sum only occurs if this event is in $((i-1)\delta_m, i\delta_m]$.) Then, with probability one, N_m^* converges to the number of events $t_i \in (0, 1)$ for which $\omega_{t_i} \in F$. It follows as before by dominated convergence and stationarity that $r_m Pr\{\chi_{m0} = 1, \omega_s \in F\} \to \lambda_F$, from which the desired result follows (for any sequence $\delta_m \downarrow 0$) using the fact that $Pr\{N(-\delta_m, 0) \geqq 1\} \sim \lambda/r_m$.

The fact that the limit in (4.10) holds for all $F \in \mathscr{F}$ is, of course, more satisfying than that in (4.1) which requires "continuity sets." However, the definition of P_δ^* is more complicated than that of P_δ and the limit in (4.1) may be more useful in practice. The difference between P_δ and P_δ^* is, of course, slight (but we feel it worthy of exploration).

Theorems 4.1 and 4.2 concerned conditional expectations of the function χ_F, given an event near the origin. One may ask whether similar results hold for other functions. To answer this in relation to Theorem 4.1, we will call a measurable function $\phi(\omega)$ *continuous to the right* if $\phi(\omega_t)$ is continuous to the right in t.

Before stating the generalization of Theorem 4.1, we give a lemma (the result of which is contained in [14]) which is useful in a number of contexts.

LEMMA 4.1. *If ϕ is a measurable function on Ω and ϕ is either nonnegative, or integrable with respect to P_0, then*

$$\lambda \int \phi dP_0 = \mathscr{E}\{\Sigma\, \phi(\omega_{t_j}) \colon t_j \in \omega \cap (0, 1)\}. \tag{4.12}$$

The statement of this lemma, when $\phi = \chi_F$, $F \in \mathscr{F}$, is just the definition of $P_0(F)$. Its truth for nonnegative measurable or P_0 integrable ϕ follows at once by the standard approximation technique.

The following result generalizes Theorem 4.2.

THEOREM 4.3. *Let ϕ be measurable (on Ω), continuous to the right and such that $|\phi(\omega_t)| < \psi(\omega)$ for all $t \in (0, 1)$, where $\mathscr{E}\{\psi N(0, 1)\} < \infty$. Then*

$$\mathscr{E}\{\phi \mid N(-\delta, 0) \geqq 1\} \to \int \phi dP_0 \tag{4.13}$$

as $\delta \downarrow 0$.

PROOF. The pattern of the proof of Theorem 4.1 applies, with ϕ written for χ_F Specifically,

$$S_m = \sum_{i=1}^{r_m} \chi_{mi}\phi(\omega_{i\delta_m}) \to \sum \{\phi(\omega_{t_j}) \colon t_j \in \omega \cap (0, 1)\} \tag{4.14}$$

with probability one. But $|S_n| \leqq \psi(\omega)\cdot N(0, 1)$ which has finite expectation, and thus, by dominated convergence and Lemma 4.1,

$$\int \phi dP_0 = \lambda^{-1} \lim_{m\to\infty} \mathscr{E}S_m = \lambda^{-1} \lim_{m\to\infty} r_m \mathscr{E}\{\chi_{m0}\phi(\omega)\} = \lim_{m\to\infty} \mathscr{E}\{\phi \,|\, \chi_{m0} = 1\} \tag{4.15}$$

since $Pr(\chi_{m0} = 1) \sim \lambda\delta_m \lambda r_m^{-1}$. This is the desired result (writing $N(-\delta_m, 0) \geqq 1$ for $\chi_{m0} = 1$).

COROLLARY 4.1. *If ϕ is a bounded, right continuous function, the result holds. For if $|\phi| \leqq K$ we may take $\psi(\omega) = K$ and $\mathscr{E}\{\psi N(0, 1)\} = K\lambda < \infty$.*

This corollary is similar to a theorem of Ryll-Nardzewski [18] (there two sided continuity of $\phi(\omega_t)$ is required and the condition $N(-\delta, 0) \geqq 1$ replaced by $N(-\delta, \delta) \geqq 1$).

COROLLARY 4.2. *Suppose $\mathscr{E}N^{k+1}(0, \tau) < \infty$ for some positive integer k, $\tau > 0$. Then*

$$\lim_{\delta\downarrow 0} \mathscr{E}\{N^k(0, \tau) \,|\, N(-\delta, 0) \geqq 1\} = \mathscr{E}_{P_0} N^k(0, \tau), \tag{4.16}$$

where $\mathscr{E}_{P_0}$ denotes expectation with respect to the Palm distribution. That is, the kth *moment of $N(0, \tau)$ with respect to the Palm distribution is simply the* kth *conditional moment (defined as a limit) given an event "at" the origin.*

The proof is immediate on noting that $\phi(\omega) = N_\omega^k(0, \tau)$ is continuous to the right, and for all $t \in [0, 1]$,

$$\phi(\omega_t) = N_\omega(t, t + \tau) \leqq N(0, 1 + \tau) = \psi(\omega), \tag{4.17}$$

where $\mathscr{E}\{\psi(\omega)N(0, 1)\} \leq \mathscr{E}N^{k+1}(0, 1 + \tau)$. This latter quantity is finite since it is easily seen by Minkowski's inequality and stationarity that $\mathscr{E}N^{k+1}(0, s) < \infty$ for all $s > 0$.

Finally, we note the corresponding generalization of Theorem 4.2. For this, the condition required above that ϕ be continuous to the right can be omitted, but the origin must "be moved" to measure from the time s of the first event prior to zero. We state this formally.

THEOREM 4.4. *Let ϕ be measurable and such that $|\phi(\omega_t)| < \psi(\omega)$ for all $t \in (0, 1)$, where $\mathscr{E}\{\psi N(0, 1)\} < \infty$. Then*

$$\mathscr{E}\{\phi(\omega_s) \,|\, N(-\delta, 0) \geqq 1\} \to \int \phi dP_0 \tag{4.18}$$

as $\delta \downarrow 0$, where $s = s(\omega)$ denotes the position of the first event prior to $t = 0$.

REFERENCES

[1] R. A. AGNEW, "Transformations of uniform and stationary point processes," Ph.D. thesis, Northwestern University, 1968.

[2] YU. K. BELYAYEV, "Elements of the general theory of random streams," Appendix to Russian edition of *Stationary and Related Stochastic Processes* by H. Cramér and M. R. Leadbetter, MIR, Moscow, 1969. (English translation under preparation for University of North Carolina Statistics Mimeograph Series, No. 703, 1970.)

[3] P. Billingsley, *Convergence of Probability Measures*, New York, Wiley, 1968.
[4] W. Fieger, "Eine für beliebige Call-Prozesse geltende Verallgemeinerung der Palmschen Formeln," *Math. Scand.*, Vol. 16 (1965), pp. 121–147.
[5] ———, "Zwei Verallgemeinerungen der Palmschen Formeln," *Transactions of the Third Prague Conference on Information Theory*, Prague,—Czechoslovak Academy of Sciences, 1964, pp. 107–122.
[6] J. R. Goldman, "Infinitely divisible point processes in R^n," *J. Math. Anal. Appl.*, Vol. 17 (1967), pp. 133–146.
[7] P. R. Halmos, *Measure Theory*, Princeton, Van Nostrand, 1950.
[8] T. E. Harris, "Random measures and motions of point processes," *Z. Wahrscheinlichkeitstheorie und verw. Gebiete*, Vol. 18 (1971), pp. 85–115.
[9] Y. A. Khinchin, *Mathematical Methods in the Theory of Queueing*, London, Griffin, 1960.
[10] D. König and K. Matthes, "Verallgemeinerungen der Erlangschen Formeln, I," *Math. Nachr.*, Vol. 26 (1963), pp. 45–56.
[11] M. R. Leadbetter, "On three basic results in the theory of stationary point processes," *Proc. Amer. Math. Soc.*, Vol. 19 (1968), pp. 115–117.
[12] ———, "On streams of events and mixtures of streams," *J. Roy. Statist. Soc. Ser. B*, Vol. 28 (1966), pp. 218–227.
[13] ———, "On certain results for stationary point processes and their application," *Bull. Inst. Internat. Statist.*, Vol. 43 (1969) pp. 309–319.
[14] K. Matthes, "Stationäre zufällige Punktfolgen I," *Jber. Deutsch. Math.-Verein.*, Vol. 66 (1963), pp. 66–79.
[15] J. Mecke, "Stationäre zufällige Masse auf lokalkompakten Abelschen Gruppen," *Z. Wahrscheinlichkeitstheorie und verw. Gebiete*, Vol. 9 (1967), pp. 36–58.
[16] R. K. Milne, "Simple proofs of some theorems on point processes," *Ann. Math. Statist.*, Vol. 42 (1971), pp. 368–372.
[17] J. Neveu, "Sur la structure des processus ponctuels stationnaires," *C. R. Acad. Sci. Paris, Ser. A*, Vol. 267 (1968), pp. 561–564.
[18] C. Ryll-Nardzewski, "Remarks on processes of calls," *Proceedings of the Fourth Berkeley Symposium on Mathematical Statistics and Probability*, Berkeley and Los Angeles, University of California Press, 1961, Vol. 2, pp. 455–466.
[19] I. M. Slivnyak, "Stationary streams of homogeneous random events," *Vestnik Harkov. Gos. Univ.*, Vol. 32 (1966), pp. 73–116. (In Russian.)
[20] F. Zitek, "On the theory of ordinary streams," *Czechoslovak Math. J.*, Vol. 8 (1958), pp. 448–458. (In Russian.)

THE DISTRIBUTION OF GENERATIONS AND OTHER ASPECTS OF THE FAMILY STRUCTURE OF BRANCHING PROCESSES

WOLFGANG J. BÜHLER
UNIVERSITY OF HEIDELBERG

1. Introduction

Throughout this paper branching processes will be viewed as models for the development of biological populations. Unless stated otherwise, it will be assumed that the population starts at time $t = 0$ with one individual in generation 0 and of age 0. Each member of the population will live for a random life time. Then he will be replaced by a random number of new individuals, his sons. These will be in generation $k + 1$ if their father was a member of the kth generation. Also we shall allow for an individual to "survive," that is, he may have himself as one offspring in generation k and start a new life. In the branching process model, it is further assumed that the lifetimes of all individuals have a common probability distribution with distribution function G, that the probability β for survival is the same for all individuals, that all individuals have the same distribution of the number of offspring given by a probability generating function h, and further, that all the random variables introduced so far are independent.

We shall follow the notations of Harris [10] who has studied many aspects of this model. Until recently the emphasis has been on studying the total population size; the possibility of an individual giving birth to offspring more than once has not usually been considered. The only exception seems to be the papers by Crump and Mode [5], [6], who consider a case somewhat more general than ours.

The questions with which this paper is concerned are about the distribution of generations in a population at a given time, the time pattern according to which generations appear and disappear, the degree of relationship between different individuals, the number of relatives of a certain degree, and so forth.

The first mention of distribution of generations with the present meaning in the literature was by Harris [10] who used the number $Z^{(k)}(t)$ of individuals in generations $0, 1, \cdots, k - 1$ alive at time t as an approximation to the total

During the author's stay at the University of California, Berkeley, this work was supported in part by Grant No. USPHS GM 10525-07 of the National Institutes of Health.

population size $Z(t)$. Thus, the probability generating function $F(\cdot, t)$ of $Z(t)$ could be approximated by that of $Z^{(k)}(t)$. Conditioning on the first event in the process, that is, on life length and number of offspring of the original ancestor, led to the recursion formula

$$(1.1)\quad F^{(k+1)}(s, t) = s[1 - G(t)] + \int_0^t h[F^{(k)}(s, t-u)]\,[\beta F^{(k)}(s, t-u) + 1 - \beta]\,dG(u)$$

with $F^{(0)}(s, t) = 1$. Formula (1.1) was given in [10] for the case $\beta = 0$. Applying the same kind of argument to the sequence $Z_k^*(t)$, defined as $Z^{(k)}(t)$ if $Z^{(k)}(t) = Z(t)$ and as ∞ otherwise, yields a sequence of probability generating functions F_k^* which also converges to F and satisfies a relation similar to (1.1), however with $F_0^*(s, t) = 0$. This sequence of probability generating functions (p.g.f.) was first introduced without a probabilistic interpretation by Levinson [16]. The asymptotic behavior of the distribution of generations after a long time and the development in time of high generations in supercritical processes (that is, in processes with $h'(1) + \beta > 1$) was studied by Martin-Löf [18] and then, independently and using different methods, by Kharlamov [13], [14], [15], Samuels [22], [23], and Bühler [4]. While Martin-Löf and Samuels considered general age dependent branching processes, Kharlamov and Bühler restricted themselves to the case of Markov branching processes (with exponential distribution of the life length) and Galton-Watson processes (with constant life length). Kharlamov allows for individuals of more than one type (admitting in [15] even an uncountable set of types). Methods similar to those of Kharlamov were used by Greig [8] in the study of chains of bacteria multiplying according to a birth and death process, where a chain is broken into two when a bacterium forming an "interior link" of the chain dies.

In all papers except [4], the survival probability β is zero. These asymptotic results will be described in Section 2. Section 3 considers the same questions in the subcritical case. While these questions themselves are not definitely answered, Theorems 3.1 and 3.2, dealing with the behavior of subcritical processes "long before extinction" may be of independent interest.

We shall then return to the supercritical case and study the family and relationship structure of the population, first extending the results given in [4] about the distribution of distant relatives, then considering the finer structure as, for example, the sizes of sibships, cousinships, and so forth. Finally, we shall reconsider and generalize a result first given by Stratton and Tucker [25] about the emergence of the first generation if the population is started by a large number N of individuals in generation zero.

2. The distribution of generations

It will be assumed throughout that the lifetime of an individual is positive with probability one, that is, that $G(0+) = 0$, and that its distribution is not concentrated on a lattice, except when we consider Galton-Watson processes. If $1 < m = h'(1) < \infty$, then there exists a unique positive α satisfying,

$$m \int_0^\infty e^{-\alpha t}\, d(G(t) = 1. \tag{2.1}$$

The subcritical case with $m < 1$ will be considered only when a (necessarily negative) solution α of (2.1) exists and when the mean $\bar{\mu}$ and variance $\bar{\sigma}^2$ of the distribution

$$\bar{F}(t) = m \int_0^t e^{-\alpha x}\, dG(x) \tag{2.2}$$

are finite. In the critical case $m = 1$, we have $\alpha = 0$ and $\bar{F} = G$. Denoting the ratio $Z_k(t)/Z(t)$ by $R_k(t)$, Samuels [23] obtained the following result.

THEOREM 2.1. *If G is not a lattice distribution, if $G(0+) = 0$, $\bar{\sigma}^2 < \infty$, $h(0) = 0$, $1 < m < \infty$, $h''(1) < \infty$, and $\beta = 0$, then*

$$\lim_{t\to\infty} \sum_{k=0}^{K_t} R_k(t) = \Phi(x) \tag{2.3}$$

in probability, whenever K_t is such that

$$\left(K_t - \frac{t}{\bar{\mu}}\right)\left(\frac{t\bar{\sigma}^2}{\bar{\mu}^3}\right)^{-1/2} \to x. \tag{2.4}$$

Here, as usual, $\Phi(t)$ denotes the distribution function of the standard normal distribution.

Kharlamov [14] proved the same result and its extension to the branching process with individuals of different types [15] for the Markov case with lifetime distribution $G(t) = 1 - \exp\{-\lambda t\}$. He did not assume $h(0) = 0$, and thus he obtained (2.3) in conditional probability given $W > 0$, where $W = \lim_{t\to\infty} Z(t) \exp\{-at\}$ with $a = \lambda(m - 1)$.

Both Kharlamov's and Samuels' proofs proceed by first showing

$$\lim_{t\to\infty} \sum_{k=0}^{K_t} \rho_k(t) = \Phi(x) \tag{2.3'}$$

with $\rho_k(t) = EZ_k(t)/EZ(t)$, where the only condition on the offspring distribution needed to establish (2.3′) is $m < \infty$. Showing then that $\Sigma_{k=0}^{K_t} Z_k(t)/E\,\Sigma_{k=0}^{K_t} Z_k(t)$ is close to $Z(t)/EZ(t)$ for large t and hence close to W, they obtain (2.3) by taking the ratio of these two expressions, cancelling W which is positive, and using (2.3′).

Using essentially the same method, Samuels establishes a local version of the above result.

THEOREM 2.2. *If the conditions of Theorem 2.1 are satisfied and if $\int x^3\, dG(x) < \infty$, then as $n \to \infty$,*

$$\bar{\sigma}\sqrt{n}\, R_n(t_n) \to \bar{\mu}\varphi(x) \tag{2.5}$$

in probability provided the sequence t_n *satisfies* $(t_n - n\bar{\mu})/\bar{\sigma}\sqrt{n} \to x$. *Again the corresponding relation*

$$\bar{\sigma}\sqrt{n}\rho_n(t_n) \to \bar{\mu}\varphi(x) \tag{2.5'}$$

requires no condition on the offspring distribution but $m < \infty$.

These theorems, loosely speaking, state that at time t the generations with generation numbers around $t/\bar{\mu}$ are the most frequent in the population or that generation n will never make up a higher proportion of the total population than around time $n\bar{\mu}$. If we denote the mean and variance of G by μ and σ^2, Samuels [23] has pointed out that for the supercritical process, while the above is true, the absolute number $Z_n(t)$ of living individuals of the nth generation achieves its maximum at a much later time, namely, around $t = n\mu$, and in fact at time $n\bar{\mu}$ is still relatively small. More precisely, she obtained the following two theorems, in which $U_n(t)$ denotes the number of generation n objects born by time t and $V_n = U_n(\infty)$ is the total size of the nth generation.

THEOREM 2.3. *If* $G(0+) = 0$ *and* $\sigma^2 < \infty$, *and if* $h(0) = 0, 1 < m < \infty$, $\beta = 0$, *and* $h''(1) < \infty$, *then as* $n \to \infty$ *and* $(t_n - n\mu)/\sigma\sqrt{n} \to x$,

$$\frac{U_n(t_n)}{V_n} \to \Phi(x) \tag{2.6}$$

in probability.

THEOREM 2.4. *If in addition to the assumptions of Theorem 2.3 we have* $\int x^3\, dG(x) < \infty$ *and if* G *is not a lattice distribution, then*

$$\frac{\sigma\sqrt{n}Z_n(t_n)}{\mu V_n} \to \varphi(x) \tag{2.7}$$

in probability.

These theorems again are established by first proving their expectation counterparts without the restrictions on h (except for $m < \infty$) and then showing that a certain sequence of differences of random variables converges to zero in quadratic mean. If we substitute $m^n W$ for V_n in the denominator of the left side of (2.6), we obtain the quantity which was shown by Martin-Löf [18] to converge to $\Phi(x)$, not only in probability but also in square mean.

The contrast between Theorems 2.1 and 2.2 on one hand and Theorems 2.3 and 2.4 on the other may be illustrated by the following corollaries, the first being a consequence of Theorem 2.3, the second following from Theorem 2.1.

COROLLARY 2.1. *Under the conditions of Theorem 2.3, the time* T_n *of the birth of a randomly chosen individual of the* nth *generations has, as* $n \to \infty$, *asymptotically a normal distribution with mean* $n\mu$ *and variance* $n\sigma^2$.

COROLLARY 2.2. *Under the conditions of Theorem 2.1, as* $t \to \infty$, *the generation* G_t *of a random individual alive at time* t *has asymptotically a normal distribution with mean* $t/\bar{\mu}$ *and variance* $t\bar{\sigma}^2/\bar{\mu}^3$.

Corollary 2.2 for the Markov case and its counterpart for the Galton-Watson process were also obtained by Bühler [4]. They will be stated separately here and proof will be indicated since it gives additional insight into the processes and the idea underlying it will again be used in later sections.

THEOREM 2.5. (i) *If in a Galton-Watson sequence with* $\beta > 0$ *and* $1 < \beta + m < \infty$, *and if* G_n *denotes the generation of an individual chosen randomly at time* n, *then*

$$\lim_{n\to\infty} P\left\{\frac{G_n - nm/(m+\beta)}{\left(nm\beta/(m+\beta)^2\right)^{1/2}} \leqq x \,\middle|\, W > 0\right\} = \Phi(x), \tag{2.8}$$

where $W = \lim_{n\to\infty} Z(n)(m+\beta)^{-n}$ *almost surely.*

(ii) *In a Markov branching process with* $G(t) = 1 - e^{-t}$, *if* $0 \leqq \beta \leqq 1$, $1 < \beta + m < \infty$ *and* $h''(1) < \infty$, *denoting by* G_t *the generation of a random individual alive at time* t, *we have*

$$\lim_{t\to\infty} P\left\{\frac{G_t - mt}{(mt)^{1/2}} \leqq x \,\middle|\, W > 0\right\} = \Phi(x), \tag{2.9}$$

where $W = \lim_{t\to\infty} Z(t) \exp\{-(\beta + m - 1)t\}$ *almost surely.*

To prove the continuous time version, one conditions first on the number $N(t)$ of "splits" that have occurred up to time t. Then given $N(t) = n$, say, we further condition on the sizes of the families produced at each split, that is, on the variables U_i, V_i, $i = 1, 2, \cdots, n$, representing the numbers of survivors (necessarily equal to zero or one) and the number of individuals newly born at the ith split. Then after choosing one of the n individuals at random following a suggestion of P. Clifford, we trace his line of ancestry back to the original ancestor present at time zero defining random variables L_i, $i = 1, 2, \cdots, n$, as follows. If the ith split produces (as a newborn individual) an ancestor of the individual considered, we let $L_i = 1$, otherwise L_i will be equal to zero. The generation G_t of our individual thus is just the sum of all the L_i. Under our conditioning the variables L_i, representing the "loss" in generation number as we go back beyond the ith split, will be independent. Let S_k denote the sum $u_1 + u_2 + \cdots + u_k + v_1 + v_2 + \cdots + v_k - k$. Then, given $(U_i, V_i) = (u_i, v_i)$, $i = 1, 2, \cdots, n$, the conditional probability of the event $L_k = 1$ is v_k/s_k and since the L_i are bounded we can apply the central limit theorem to their sum whenever the variance

$$V(n) = \sum_{k=1}^{n} \frac{v_k}{s_k}\left(1 - \frac{v_k}{s_k}\right) \tag{2.10}$$

converges to infinity. This, however, will be the case since for almost all realizations of the process V_k/S_k is of the order $1/k$ as $k \to \infty$. More precisely, invoking Toeplitz' lemma, we can replace both the conditional expectation $E(n) = \Sigma_{k=1}^{n} v_k/s_k$ and $V(n)$ by $[m/(\beta + m - 1)] \log n$. This at the same time removes the conditioning on the family sizes (U_i, V_i). The final step is unconditioning with respect to $N(t)$ using the fact, established by Athreya and Karlin [2], that $N(t)(\beta + m - 1) \exp\{-(\beta + m - 1)t\}$ converges to W almost surely.

3. The subcritical case

In the subcritical case $\beta + m < 1$, it is known that $Z(t)$ almost surely becomes zero as t becomes large. It is also well known that at least in the Galton-Watson case with $h''(1) < \infty$ and in the continuous time Markovian case the conditional distribution of $Z(t)$ given $Z(t) > 0$ has a limit as $t \to \infty$. The probability generating function of this limit will be denoted by g. Since the total population size (even after conditioning) does not go to infinity, Theorems 2.1 through 2.4 cannot possibly be true if $\beta + m < 1$ even though the assertions about ratios of expected values made in them or in connection with their proofs remain valid. It seems, however, reasonable to expect that Theorem 2.5 and possibly Corollary 2.1 should be valid in the subcritical case. If we try to imitate the proof of Theorem 2.5 in the case $\beta + m < 1$ in order to establish relation (2.10) for the process conditioned on $Z(t) \neq 0$, we might want to show that the relative frequencies of the possible values of v_k/s_k to appear up to time t have reasonable limits. A first step in this direction is to prove that in a sense we are dealing with a stationary process. While this result is relatively obvious, it seems not to have been published before. Let us first consider the Galton-Watson case, that is, let $\mathbf{Z}_n = (Z_n^1, Z_n^2, \cdots, Z_n^k)$ be a k type Galton-Watson process with basic probability generating function $\mathbf{f} = (f^1, f^2, \cdots, f^k)$,

$$(3.1) \qquad f^j(\mathbf{s}) = f^j(s_1, s_2, \cdots, s_k) = E\{s_1^{Z_1^1} s_2^{Z_1^2}, \cdots, s_k^{Z_1^k} \mid \mathbf{Z}_0 = \mathbf{e}_j\}.$$

Assume that the expectations $m_{ij} = E(Z_1^j \mid \mathbf{Z}_0 = \mathbf{e}_i)$ are finite and positive; then the matrix $\mathbf{M} = \{m_{ij}\}_{1 \leq i,j \leq k}$ has a largest positive eigenvalue ρ.

THEOREM 3.1. *If $\mathbf{Z}_0, \mathbf{Z}_1, \cdots$ is a k type Galton-Watson process with positive expectation matrix $\mathbf{M}$ and finite second moments and if $\rho < 1$, then as $n \to \infty$ and $N \to \infty$, the joint conditional distributions of $(\mathbf{Z}_{n+0}, \mathbf{Z}_{n+1}, \cdots, \mathbf{Z}_{n+r})$ given $\mathbf{Z}_0 = \mathbf{e}_i$ and $\mathbf{Z}_{N+n+r} \neq \mathbf{0}$ converge to those of a stationary process $\mathbf{X}_0, \mathbf{X}_1, \cdots$.*

The stationarity of the limit distribution, if it exists, is immediate. To prove the convergence, we first state a lemma which follows immediately from Theorem 4.3 of Jiřina [11] and its proof.

LEMMA 3.1. *Under the assumptions of Theorem 3.1, there exist constants C_i such that*

$$(3.2) \qquad \lim_{n \to \infty} \frac{f_n^i(\mathbf{s}) - 1}{c_i \rho^n} = g^i(\mathbf{s}) - 1,$$

$$(3.3) \qquad \lim_{n \to \infty} \frac{\partial^{j_1 + \cdots + j_k}}{\partial s_1^{j_1} \cdots \partial s_k^{j_k}} \frac{f_n^i(\mathbf{s})}{c_i \rho^n} = \frac{\partial^{j_1 + \cdots + j_k}}{\partial s_1^{j_1} \cdots \partial s_k^{j_k}} g^i(\mathbf{s})$$

for all $(j_1, j_2, \cdots, j_k)$. Here $g^i(\mathbf{s})$ is the limiting probability generating function of $\mathbf{Z}_n$ given $\mathbf{Z}_0 = \mathbf{e}_i$ and $\mathbf{Z}_n \neq \mathbf{0}$, and the convergence is uniform on compact subsets of the region $S = \{\mathbf{s}; |s_i| < 1, i = 1, 2, \cdots k\}$.

To avoid unnecessarily complex notation, the proof of Theorem 3.1 will be given only for the one dimensional case, $k = 1$. Let H be the conditional joint p.g.f. of $Z_n, Z_{n+1}, \cdots, Z_{n+r}, Z_{N+n+r}$ given $Z_{N+n+r} \neq 0$, then

$$\begin{aligned}(3.4)\qquad & H_{n,N}(s_0, s_1, \cdots, s_r, W) \\ &= E\left(s_0^{Z_n} s_1^{Z_{n+1}} \cdots s_r^{Z_{n+r}} W^{Z_{N+n+r}} \middle| Z_0 = 1, Z_{N+n+r} > 0\right) \\ &= \{1 - f_{N+n+r}(0)\}^{-1}\{f_n(s_0 f(s_1 f(s_2 \cdots f(s_r f_N(W)) \cdots) \\ &\qquad - f_n(s_0 f(s_1 \cdots f(s_r f_N(0)) \cdots)\}.\end{aligned}$$

Using $f_N(w) - f_N(0) \sim c_1 m^N g(w)$ and $1 - f_N(0) \sim c_1 m^N$ and applying Lemma 3.1, we obtain

$$\begin{aligned}(3.5)\qquad & \lim_{n, N \to \infty} H_{n,N}(s_0, s_1, \cdots, s_r, w) \\ &= c_1 m^{-r} g(w) s_0 s_1 \cdots s_r g'(s_0 f(s_1 f(\cdots f(s_r)) \cdots) \\ &\qquad f'(s_1 f(s_2 \cdots f(s_r)) \cdots) f'(s_2 \cdots f(s_r)) \cdots f'(s_r).\end{aligned}$$

REMARK 3.1. It is intuitively clear and confirmed by (3.5) that as $n, N \to \infty$ the variables $(Z_n, Z_{n+1}, \cdots, Z_{n+r})$ and Z_{N+n+r} become independent (conditionally given $Z_{N+n+r} > 0$).

REMARK 3.2. The joint probability distributions of $X_0 - 1, X_1 - 1, \cdots$ are the same as those of a branching process $Y_0, Y_1, \cdots$ with immigration, where Y_0 has the p.g.f. g', where the p.g.f. of the number of offspring of each individual is f, and where the number I_j of immigrants at time j has the probability generating function $f'(s)/m$.

Let us now look at Theorem 3.1 for the two type Galton-Watson process in which the two types represent the numbers of surviving and newborn individuals, respectively. Thus, the limiting process $\mathbf{X}_0, \mathbf{X}_1, \cdots$ can be written as (S_0, N_0), $(S_1, N_1), \cdots$, where obviously the number S_i of survivors at time i can be at most equal to the total number $X_{i-1} = S_{i-1} + N_{i-1}$ of individuals living at time $i - 1$. If for the process $\mathbf{X}_0, \mathbf{X}_1, \cdots$ we define the variables $L_0, L_1, \cdots$ as having the values 1 and 0 with probabilities (conditional on $\mathbf{X}_0, \mathbf{X}_1, \cdots$) N_i/X_i and S_i/X_i, respectively, then (see, for example, Doob [7], Ch. X) almost surely the sequences $(1/r)\ \Sigma_{i=1}^r N_i/X_i$ and $(1/r)\ \Sigma_{i=1}^r N_i S_i/X_i^2$ converge to the conditional expectations $M = E(N_0/X_0 | \mathscr{F})$ and $V = E(N_0 S_0/X_0^2 | \mathscr{F})$, where $\mathscr{F}$ is the σ-field of invariant events for the stationary sequence $\mathbf{X}_i$, $i = 0, 1, 2, \cdots$. Thus, we can apply the central limit theorem analogously as in the corresponding step in the proof of Theorem 2.5.

LEMMA 3.2. *Let* $\mathbf{X}_0, \mathbf{X}_1, \cdots$ *and* $L_0, L_1, \cdots$ *be defined as above and let* $G_r = L_0 + L_1 + \cdots + L_{r-1}$, *then almost surely*

$$(3.6)\qquad \lim_{r \to \infty} P\left\{\frac{G_r - rM}{(rV)^{1/2}} \leqq x \middle| \mathscr{F}\right\} = \Phi(x).$$

The asymptotic independence of remote variables (as expressed, for example, in Remark 3.1) should mean that actually, as $\mathbf{X}_0$ will be independent of $\mathscr{F}$, M

and V are nonrandom and the conditioning on $\mathscr{F}$ can be removed. Furthermore, going back to the original process $\mathbf{Z}_k = (U_k, V_k)$, $k = 0, 1, 2, \cdots$, we could let N, n, and r all be large in such a way that the first n and the last N variables do not contribute much to G_{N+n+r} and that the stretch of r variables in between would essentially behave like the stationary process $\mathbf{X}_0, \mathbf{X}_1, \cdots$.

CONJECTURE 3.1. *If $Z_0, \mathbf{Z}_1, \mathbf{Z}_2, \cdots$ is a Galton-Watson process with $\beta + m < 1$ and finite second moments, then as $I \to \infty$, the generation G_I of a random individual living at time I has asymptotically a normal distribution with expectation IM and variance IV.*

REMARK 3.3. Whereas in the supercritical case the proportion of newly born individuals almost surely converges to $m/(m + \beta)$ leading to the expressions for M and V in (2.7), this is not the case in subcritical processes where, indeed, M and V will depend on β and on the offspring distribution in a more complex way.

In the case of continuous time, we can proceed similarly. Let $\{Y_t, t \geqq 0\}$ be a subcritical Markov branching process with lifetime distribution function $1 - e^{-t}$, with h as the p.g.f. of the offspring distribution and with immigration. Assume the immigration to be compound Poisson; more precisely, let times between immigrations be independent exponential variables (having the same distribution as the lifetimes) and let the numbers of immigrants at each immigration be distributed according to $h'(s)/m$ independently of each other and of the other variables defining the process. Let finally Y_0 be distributed according to the p.g.f. $c_1 g'$, where

$$g(s) = 1 - \exp\left\{- (1 - m) \int_0^s \frac{du}{h(u) - u}\right\}, \tag{3.7}$$

and $c_1 = 1/g'(1)$.

THEOREM 3.2. *Let $\{Z(t), t \geqq 0\}$ be a Markov branching process with $G(t) = 1 - e^{-t}$ whose offspring distribution is given by $h(s)$ with $h'(1) = m < 1$ and with $\beta = 0$. Then, as t and T tend to infinity, the finite dimensional distributions of the process $\{Z(t + \tau), \tau \geqq 0\}$ conditioned on $Z(t + T) > 0$ converge to those of the process $\{Y_\tau + 1, \tau \geqq 0\}$. The joint p.g.f. of $Y_0, Y_{\tau_1}, Y_{\tau_1+\tau_2}, \cdots, Y_{\tau_1+\cdots+\tau_r}$ is given by*

$$\begin{aligned} & c_1 \exp\{- m(\tau_1 + \tau_2 + \cdots + \tau_r)\} \\ & \quad \cdot g'\big(s_0 F\big(s_1 F\big(s_2 F \cdots s_{r-1}\big(F(s_r, \tau_r), \tau_{r-1}\big), \cdots\big), \tau_1\big)\big) \\ & \quad \cdot F_t(s_r, \tau_r) F_t\big(s_{r-1} F(S_r, \tau_r), \tau_{r-1}\big) \\ & \quad \quad \cdots F_t\big(s_1 F\big(s_2 F\big(\cdots s_{r-1} F(s_r, \tau_r), \tau_{r-1}\big), \cdots\big), \tau_1\big), \end{aligned} \tag{3.8}$$

where $F(s, t)$ denotes the p.g.f. of $Z(t)$ and F_t indicates the partial derivative of F with respect to its second (time) argument.

The proof will be omitted as it basically follows that of Theorem 3.1. It can also be extended to the multiple type case.

The arguments which led to Lemma 3.2 and from there to Conjecture 3.1 have to be appropriately modified. We now define a process $\{L_\tau, \tau \geqq 0\}$ as follows. Let L_τ be constant except for those points in time at which Y_τ increases; if at $\tau = t$ the Y process jumps by the positive amount V_t, then, with probability $V_t/Y_{t+} + 1$, the process L_τ will jump by unity, otherwise it will remain constant. Thus, the increments of the process $\{L_\tau, \tau \geqq 0\}$ will be stationary. Also, given the process $\{Y_\tau, \tau \geqq 0\}$, they will be conditionally independent. An analogue of Lemma 3.1 will hold. This will allow us to argue in the same way as in the discrete time case to obtain an analogue to Conjecture 3.1.

CONJECTURE 3.2. *If $Z(t)$ is a Markov branching process with $\beta + m < 1$, then there exist constants M and V depending on the offspring distribution such that the generation of a random individual alive at time t has asymptotically, as $t \to \infty$, a normal distribution with mean Mt and variance Vt.*

4. The relationship structure, distant relatives

In this section, we shall again study supercritical branching processes in discrete or continuous time. For these we shall investigate how closely related individuals taken randomly from the population will be and how many relatives of a given large degree a random individual will have. If we take two individuals at time t, these will be in generations G_t^1 and G_t^2 and have a last common ancestor (at some time prior to t) whose generation will be denoted G_t^{12}. The number $R_t^{12} = (G_t^1 - G_t^{12}) + (G_t^2 - G_t^{12})$ will then be called the degree of relationship.

THEOREM 4.1. *Let R_n^{ij}, $1 \leqq i < j \leqq k$, be the degrees of relationship of k random individuals at time n in a Galton-Watson process with $\beta > 0$ and $1 < \beta + m < \infty$. Then asymptotically, as $n \to \infty$, given the population does not become extinct, the random variables*

$$V_n^{ij} = \left\{R_n^{ij} - \frac{2nm}{m + \beta}\right\}\left(\frac{2nm\beta}{(m + \beta)^2}\right)^{-1/2} \tag{4.1}$$

have a joint normal distribution. Asymptotically, expectations are zero, variances are unity, and the covariance of V_n^{ij} and $V_n^{\ell q}$ is $\frac{1}{2}$ if $\{i, j\}$ and $\{\ell, q\}$ have one index in common and zero otherwise.

THEOREM 4.2. *Let R_t^{ij}, $1 \leqq i < j \leqq k$, be the degrees of relationship of k individuals chosen randomly at time t in a Markov branching process satisfying the conditions of Theorem 2.5 (ii). Then, given $W > 0$, asymptotically as $t \to \infty$, the random variables*

$$V_t^{ij} = (R_t^{ij} - 2mt)(2mt)^{-1/2} \qquad 1 \leqq i < j \leqq k, \tag{4.2}$$

have the joint normal distribution with expectations zero and covariance matrix as in Theorem 4.1.

For the case $k = 2$, Theorems 4.1 and 4.2 are stated and proved in Bühler [4]. The essential step in the proof is to show that the generation G_n^A or G_t^A of the last individual who is a common ancestor of at least two of the k individuals chosen has itself a limiting distribution (the tail of which is studied in [4] for $k = 2$). The variables $G_t^1 - G_t^A, G_t^2 - G_t^A, \cdots, G_t^k - G_t^A$ are then conditionally independent given G_t^A and the R_t^{ij} are sums of two such variables plus something that can be neglected as $t \to \infty$.

We now want to choose a random individual from the population. Then, if we follow his line of ancestry back by a given number n of generations or by a given time t, how many descendents will the ancestor have and how closely will they be related to the individual chosen? To attack this question for large n or t, respectively, we shall assume that we have a supercritical process in which the age distribution is the stationary one. We shall then make use of the following lemma about renewal processes, of which part (i) is a version of the central limit theorem and part (ii) is due to Takács [26].

LEMMA 4.1. *Let $X_1, X_2, X_3, \cdots$ be nonnegative independent random variables with a common distribution function G. Let*

$$S_n = X_1 + X_2 + \cdots + X_n, \qquad N(t) = \max \{K; S_K \leqq t\}. \tag{4.3}$$

(i) *If $EX_1 = m$ and $0 < \text{Var } X_1 = \sigma^1 < \infty$, then $(S_n - nm)(n\sigma^2)^{-1/2}$ has asymptotically, as $n \to \infty$, a standard normal distribution.*

(ii) *Under the same conditions $(N(t) - t/m)(t\sigma^2/m^3)^{-1/2}$ asymptotically as $t \to \infty$, has a standard normal distribution.*

Now, as we select an individual at random, his age (under suitable conditions on G) will be a random variable X_0 with distribution function

$$A(x) = \frac{\int_0^x e^{-\alpha t}[1 - G(t)]\, dt}{\int_0^\infty e^{-\alpha t}[1 - G(t)]\, dt}. \tag{4.4}$$

At the time the individual was born he "selected his father" among the individuals present, whose ages were distributed according to A, with the risk of selecting an individual of age x proportional to the "failure rate" $g(x)/(1 - G(x))$. Thus, we are led to the following lemma.

LEMMA 4.2. *If in a supercritical branching process, $h''(1) < \infty$, and if G has a density g with $\int_0^\infty [g(t)]^p\, dt < \infty$ for some $p > 1$, then tracing the line of ancestry of a random individual, the life lengths $X_1, X_2, \cdots$ of his ancestors are independent and have the common probability distribution function*

$$B(x) = m \int_0^x e^{-\alpha t} g(t)\, dt. \tag{4.5}$$

If we choose an individual at random after a long time τ, denote his ancestor living at time $\tau - t$ his t ancestor, and let L_t be the number of generations that we "lose" when we go back to him. Two individuals alive at time τ will be called t relatives if their last common ancestor was present at time $\tau - t$. Correspondingly, we shall define the (n) ancestor of an individual by going back n generations in his line of ancestry. The (n) relatives will be those descendents of the (n) ancestor who are not also descendents of the $(n-1)$ ancestor.

THEOREM 4.3. *If $h(0) = 0$, $1 < m < \infty$, $h''(1) < \infty$, $\beta = 0$, and G has a density g with $\int_0^\infty [g(t)]^p\,dt < \infty$ for some $p > 1$ such that the distribution B of (4.5) has finite expectation $\bar{\mu}$ and variance $\bar{\sigma}^2$, and if at time zero the population consists of a (fixed or random) number of individuals whose ages are independently distributed according to stationary age distribution of the process, then asymptotically as $t \to \infty$,*

(i) *L_t has a normal distribution with mean $t/\bar{\mu}$ and variance $\sigma^2 t/\bar{\mu}^3$;*

(ii) *if D_t denotes the number of t relatives of the individual chosen, then*

$$\lim_{t\to\infty} P(D_t e^{-(m-1)t} \leqq x) = \sum_{k=1}^{\infty} \frac{kp_k}{m} F^{(k-1)}(x), \tag{4.6}$$

where F is the distribution function of $W = \lim_{t\to\infty} Z(t)e^{-(m-1)t}$ and $F^{(k)}$ is the kth convolution of F with itself, $F^{(0)}(x) = 1_{[0,\infty)}(x)$;

(iii) *the degree of relationship R_t of a random individual with a random t relative is asymptotically normally distributed with expectation $2t/\bar{\mu}$ and variance $2\bar{\sigma}^2 t/\bar{\mu}^3$.*

PROOF. Part (i) follows from Lemma 4.1 (ii) using (4.5). Part (ii) is established by conditioning on the number (k) of children of the t ancestor, where the factor kp_k/m in (4.6) will be justified in Section 5. Finally, part (iii) follows from (i) and Corollary 2.2.

Note that part (iii) is closely related to Theorem 4.2. Since the ages of individuals in the Markovian situation are immaterial, for $k = 2$, Theorem 4.2 becomes a special case of Theorem 4.3 (iii).

The study of (n) relatives is not quite as simple as that of t relatives. We shall denote by T_n the time of birth of the (n) ancestor and by R_n the degree of relationship between a random individual and a random (n) relative.

THEOREM 4.4. *Under the conditions of Theorem 4.3, we have*

(i) *asymptotically, as $n \to \infty$, $(T_n - n\bar{\mu})/\bar{\sigma}\sqrt{n}$ has a standard normal distribution;*

(ii) *if D_n is the number of (n) relatives of a random individual, then for all $x > 0$*

$$\lim_{n\to\infty} P(e^{-bn} D_n \leqq x) = \begin{cases} 0 & \text{if } b < \bar{\mu}\alpha, \\ \frac{1}{2} & \text{if } b = \bar{\mu}\alpha, \\ 1 & \text{if } b > \bar{\mu}\alpha; \end{cases} \tag{4.7}$$

(iii) *the degree of relationship R_n between a random individual and a random (n) relative is asymptotically, as $n \to \infty$, normally distributed with expectation $2n$ and variance $2n\bar{\sigma}^2$.*

PROOF. Part (i) follows from Lemma 4.2 together with Lemma 4.1 (i). To prove (4.7), we condition on T_n and then use (4.6) and part (i). As T_n becomes large

$$P\{e^{-bn}D_n \leqq x \mid T_n\} = P\{D_n \exp\{-\alpha T_n\} \leqq x \exp\{bn - \alpha T_n\} \mid T_n\} \tag{4.8}$$

is approximately equal to

$$\sum_{k=1}^{\infty} \frac{kp_k}{m} F^{(k-1)}[x \exp\{bn - \alpha T_n\}]. \tag{4.9}$$

Thus,

$$P(e^{-bn}D_n \leqq x) = EP\{e^{-bn}D_n \leqq x \mid T_n\} \tag{4.10}$$

can be approximated by

$$\frac{1}{\bar{\sigma}\sqrt{n}} \int_{-\infty}^{+\infty} \sum_{k=1}^{\infty} \frac{kp_k}{m} F^{(k-1)}[x \cdot \exp\{bn - \alpha t\}] \varphi\left(\frac{t - n\bar{\mu}}{\bar{\sigma}\sqrt{n}}\right) dt \tag{4.11}$$

which equals

$$\int_{-\infty}^{+\infty} \sum_{k=1}^{\infty} \frac{kp_k}{m} F^{(k-1)}[x \cdot \exp\{n(b - \alpha\bar{\mu}) + \alpha\bar{\sigma}\tau\sqrt{n}\}] \varphi(\tau)\, d\tau. \tag{4.12}$$

If $b - (m - 1)\bar{\mu} > 0$, the argument of $F^{(k-1)}$ will tend to $+\infty$ and therefore $F^{(k-1)}$ will tend to 1 uniformly on every finite τ interval for $k \leqq K$ for any K. Similarly, $b - (m - 1)\bar{\mu} < 0$ implies the convergence of the $F^{(k-1)}$ to 0 uniformly for $k \leqq K$ on every finite τ interval. This proves the first and last statements of (4.7). If $b = (m - 1)\bar{\mu}$, then the argument of $F^{(k-1)}$ will go to $+\infty$ or $-\infty$ according to whether $\tau > 0$ or $\tau < 0$, thus making $F^{(k-1)}$ converge to 0 on the negative halfline and to 1 on the positive halfline. With the corresponding uniformity, the whole expression converges to $\frac{1}{2}$.

Part (iii) is also proved by conditioning on T_n. We approximate for large T_n the probability

$$P\left\{\left(R_n - n - \frac{T_n}{\bar{\mu}}\right)\left(\frac{\bar{\sigma}^2 T_n}{\bar{\mu}^3}\right)^{-1/2} \leqq y \,\middle|\, T_n\right\} \tag{4.13}$$

by $\Phi(y)$ using Corollary 2.2. Therefore, we can approximate

$$\begin{aligned} &P\left\{(R_n - 2n)\left(\frac{\bar{\sigma}^2 n}{\bar{\mu}^2}\right)^{-1/2} \leqq x \,\middle|\, T_n\right\} \\ &= P\left\{\left(R_n - n - \frac{T_n}{\bar{\mu}}\right)\left(\frac{\bar{\sigma}^2 T_n}{\bar{\mu}^3}\right)^{-1/2}\right. \\ &\qquad\left. \leqq x\left(\frac{n\bar{\mu}}{T_n}\right)^{1/2} - \left(\frac{T_n}{\bar{\mu}} - n\right)\left(\frac{\bar{\sigma}^2 T_n}{\bar{\mu}^3}\right)^{-1/2} \,\middle|\, T_n\right\} \end{aligned} \tag{4.14}$$

by $\Phi(x(n\bar{\mu}/T_n)^{1/2} - (T_n/\bar{\mu} - n)(\bar{\sigma}^2 T_n/\bar{\mu}^3)^{-1/2})$. Unconditioning shows that $P\{(R_n - 2n)(\bar{\sigma}^2 n/\bar{\mu}^2)^{-1/2} \leqq x\}$ is close to

$$(4.15) \qquad \int_{-\infty}^{\infty} \Phi\left(x\left(\frac{n\bar{\mu}}{t}\right)^{1/2} - (t - n\bar{\mu})\left(\frac{\bar{\sigma}^2 t}{\bar{\mu}}\right)^{-1/2} \right. \\ \left. \cdot\, \varphi\big((t - n\bar{\mu})(n\bar{\sigma}^2)^{-1/2}\big)\right)(n\bar{\sigma}^2)^{-1/2}\, dt$$

which can be rewritten as

$$(4.16) \qquad \int_{-\infty}^{+\infty} \Phi\left\{\left(\frac{n\bar{\mu}}{n\bar{\mu} + \tau(n\bar{\sigma}^2)^{1/2}}\right)^{1/2}(x - \tau)\right\}\varphi(\tau)\, d\tau.$$

As $n \to \infty$, (4.16) converges to $\int_{-\infty}^{\infty} \Phi(x - \tau)\varphi(\tau)\, d\tau$ which is the normal distribution function with variance 2.

5. The relationship structure, close relatives

In this section, we shall study the sizes of sibships, the numbers of cousins of an individual chosen at random, and related questions. There are two methods of approaching these questions, the first makes use of the theory of multitype branching processes and yields the distribution of the number of brothers, cousins, and so forth, that a random individual will ever have; the second, which gives a more complicated method of going back to the corresponding (n) ancestor and viewing his progeny, will also enable us to determine how many close relatives are alive at the given point in time.

We shall illustrate the latter method with one example only. Suppose we want to find the joint probability distribution of the numbers S of sibs and U of uncles of a random individual. We assume that we are dealing with a supercritical process which has been developing for a long time, so that the population size is big and the distribution of ages is the stationary age distribution with distribution function

$$(5.1) \qquad A(x) = \frac{\int_0^x e^{-\alpha t}[1 - G(t)]\, dt}{\int_0^{\infty} e^{-\alpha t}[1 - G(t)]\, dt}.$$

First, condition on the age A_0 of the individual and on the life length B_0 of his father, which is distributed according to the distribution function B defined in (4.5). Given $A_0 = a$ and $B_0 = b$ our individual has s live sibs and u live uncles if his grandfather has $k \geqq u$ children out of which u survive to the age of $a + b$ and his father has $j > s$ children and s of the $j - 1$ brothers survive to the age of a. To carry our argument through, we need the conditional probability $P_k(x)$ that an individual has $k - 1$ brothers given his age is x. Thus, we select an individual at random among those of age x (or, since we are considering the continuous time case, with ages in a small interval around x). Hence, $P_k(x)$ will be proportional to the number of k sibship individuals of age x present. This is expected to be proportional to

$$p_k E\{\text{number of survivors to age } x \text{ out of a } k \text{ sibship}\} = p_k \cdot k[1 - G(x)]. \tag{5.2}$$

Therefore, $P_k(x) = kp_k/m$ for all $x \geqq 0$. Now we can simply write down the joint distribution of the numbers of uncles and sibs.

THEOREM 5.1. *In a supercritical branching process with lifetime distribution function G, with stationary age distribution according to* (5.1) *and distribution of life lengths of ancestors given by* (4.5), *the joint probability distribution of the number U of uncles and the number S of sibs of a random individual is given by*

$$\begin{aligned}P(U = u, S = s) \\ &= \int_0^\infty dA(a) \int_0^\infty dB(b) \sum_{k=s+1}^\infty \sum_{j=a+1}^\infty \frac{kp_k jp_j}{m^2\binom{k-1}{s}\binom{j-1}{u}} \\ &\quad \cdot [1 - G(a)]^s [G(a)]^{k-s-1} [1 - G(a + b)]^u [G(a + b)]^{j-u-1}.\end{aligned} \tag{5.3}$$

Using the same method, one can find the joint distribution of the ages of the individuals under consideration. As the expressions are rather complicated only a simple example will be given. If we consider the Markovian binary split process (where each individual after an exponential lifetime is replaced by two new individuals), picking a random individual given that he has a cousin, his own age A, the lifetime L of his father, and the age $\tilde{A}$ of a random live cousin have the joint probability density

$$p(a, \ell, \tilde{a}) = \frac{432}{43} \exp\{-3(a + \ell)\}(2 - \exp\{-\tilde{a}\}). \tag{5.4}$$

This density is not symmetric in a and $\tilde{a}$, since our random individual is more likely to have a brother and one cousin than no brother and two cousins.

We now turn to the consideration of multitype processes. First, we shall identify the type of an individual with the size of the sibship to which he belongs. Thus, any individual, no matter what his own type, will produce k offspring individuals all of type k, with probability p_k. The expectation matrix M then has $\mathbf{p} = (p_1, 2p_2, 3p_3, \cdots)$ in each of its rows. Therefore, $\mathbf{p}$ is also its left eigenvector with corresponding eigenvalue m. Thus, appealing to a result of Moy [20], we can find the limit distribution of sibship sizes.

LEMMA 5.1. *In a supercritical Galton-Watson process with $h''(1) < \infty$, the relative frequencies f_k of the individuals whose sibships are of size k converge in square mean to kp_k/m and the frequency of sibships of size k converges to $p_k/(1 - p_0)$.*

Since all types of individuals have the same offspring distribution, we can combine all sibship sizes higher than $K - 1$, say, into one type. The expectation matrix of the offspring then has rows

$$\left(p_1, 2p_2, \cdots, (K-1)p_{K-1}, m - \sum_{k=1}^{K-1} kp_k\right) = \mathbf{p}(K). \tag{5.5}$$

Again, $\mathbf{p}(K)$ is the left eigenvector corresponding to the eigenvalue m. For those processes with finitely many types, the convergence in square mean can be replaced by convergence almost surely (Harris [10], Theorem II, 9.2); also the continuous time Markovian case has been studied (Athreya [1]). As K is arbitrary, we have almost sure convergence for all $k \geqq 1$.

THEOREM 5.2. (i) *In a Galton-Watson process with expectation $m > 1$ and with $h''(1) < \infty$, the relative frequencies of the individuals belonging to sibships of size k converge almost surely to kp_k/m and the relative frequencies of sibships of size k converge to $p_k/(1 - p_0)$.*

(ii) *In a Markov branching process with expectation $m > 1$ and with $h''(1) < \infty$, the relative frequencies of individuals belonging to sibships whose size at birth was k converge almost surely to kp_k/m and the frequencies of the sibships of size k represented in the population by at least one live member converge almost surely to $p_k/(1 - p_0)$.*

The relationship between the frequency of sibships of size k and the frequency of individuals from such sibships is the one usually encountered when sampling individuals or sampling families in a population. That it holds in the continuous time case, where at the time of sampling families need not be complete, is essentially due to the independence of $P_k(x)$ of x.

If we want to study different types of relatives at the same time, the corresponding expectation matrix and their eigenvectors will not be quite as easy to find. However, one can manage most cases of interest. Theorem 5.3 is an example of such a case. As we shall see in its proof, it is now not possible to pool several types of individuals; therefore, we impose an additional condition probably not needed for the conclusion to hold.

THEOREM 5.3. *If in a supercritical Markov branching process, the number of offspring of an individual is less than or equal to K almost surely, then the proportion $f(n_0, n_1, n_2, \cdots, n_r)$ of individuals in the population with $n_0 - 1$ brothers, n_1 cousins, n_2 second cousins, $\cdots$, n_r rth cousins converges almost surely to*

$$p(n_0, n_1, \cdots, n_r) = \frac{n_0 p_{n_0}}{m} \sum_{i=1}^{K} \frac{ip_i}{m} P_{n_1}^{(i-1)}(1) \cdot \sum_{j=1}^{K} \frac{jp_j}{m} P_{n_2}^{(j-1)}(2) \cdots \sum_{k=1}^{K} \frac{kp_k}{m} P_{n_r}^{(k-1)}(r), \tag{5.6}$$

where $P_i^{(j)}(k)$ is the probability that j individuals of generation zero will ever have i descendants of generation k, that is, $\Sigma_{i=0}^{\infty} P_i^{(j)}(k)s^i = [h_k(s)]^j$.

PROOF. The proof shall be given only for the case $r = 1$ to simplify notation. We shall then classify an individual as being of type (i, j) when he has $i - 1$ brothers and j cousins, $i = 1, \cdots, K - 1$ and $j = 0, 1, \cdots, K(K - 1)$. Now not all rows of the expectation matrix M can be expected to be equal any more, as obviously the number of my sons' cousins will not be independent of the number of my own sibs. However, the type of my descendents will be independent of the number of my cousins and the number of sibs of my descendents will not depend on my type at all. Thus, M has a relatively simple structure and a left eigenvector can be easily found. In fact,

$$m_{(i,j)(k,\ell)} = kp_k P_\ell^{(i-1)}(1) \tag{5.7}$$

and the left eigenvector corresponding to the eigenvalue m is given by (5.6). Of course, $p(n_0, n_1, \cdots, n_r)$ is the probability for an individual to have a father with n_0 children, a grandfather with (i children, $i - 1$ of whom have a total of n_1 children), $n_0 + n_1$ grandchildren, $\cdots$, and an r ancestor with $n_0 + n_1 + \cdots + n_r$ r descendents.

6. Processes with limited rebranching

This section is concerned with a property of continuous time branching processes first discovered by Stratton and Tucker [25], subsequently put in a context which considers generations by Bühler [2], and discussed in a way closest to the present treatment by Savage and Shimi [24]. Here we consider a whole sequence of branching processes $Z_N(t)$, $t \geqq 0$, $N = 1, 2, \cdots$, with $Z_N^{(k)}(t)$ individuals in generation k, $k = 0, 1, 2, \cdots$. Assuming $Z_N(0) = Z_N^{(0)}(0) = N$, the result of [25], [2], and [24] can be stated essentially as follows. As $N \to \infty$, there will be no second generation individuals yet at time t/N: however, there will be a Poisson number of independent first generation families whose sizes will be distributed according to the probability generating function h underlying the process. Furthermore, in the limit the increments of $Z_N^{(1)}(t/N)$ become independent. We shall now extend this result and also consider the times of emergence of higher generations than the first.

THEOREM 6.1. *Let $Z_N(t)$ be a sequence of branching processes with $Z_N(0) = Z_N^{(0)}(0) = N$, with offspring probability generating function h such that $0 < m < \infty$ and with distribution of life lengths according to the distribution function G with $G(0+) = 0$ and $G'(0) = 1$. Then, as $N \to \infty$, the probability generating function $H_{k,N}(s, a_{k,N}(t))$ of the number $Z_N^{(k)}(a_{k,N}(t))$ of individuals in generation k at time $a_{k,N}(t)$ converges to*

$$\varphi_K(s, t) = \exp\left\{\frac{m^{(k-1)}t^k}{k!}\left(h(s) - 1\right)\right\}, \tag{6.1}$$

provided $a_{k,N}(t) \cdot N^{1/k} \to t$. Furthermore, under these conditions the increments of $\{Z_N^{(k)}(a_{k,N}(t)),\ t \geqq 0\}$ are asymptotically independent.

PROOF. Arguing as in Section 1, the p.g.f. $F_k(s, t)$ of the number of individuals in generation k if we start with one individual in generation zero at time 0 is shown to satisfy the relation (1.1) with $\beta = 0$ and

$$F_0(s, t) = s(1 - G(t)) + G(t). \tag{6.2}$$

From this it can be shown that

$$F_k(s, t) = 1 - \frac{m^{(k-1)}t^k(1 - h(s))}{k!} + O(t^k) \tag{6.3}$$

as $t \to 0$. Using the facts $H_{k,N}(s, t) = [F_k(s, t)]^N$ and $(1 - x/N)^N \to e^{-x}$, the first assertion is proved since $a_{k,N}(t)$ tends to zero at the correct rate. Repeating the same kind of argument for the joint distributions at different times, the independence statement can be established. Similarly, it can be shown, that for $i \neq j$, the variables $Z_N^{(i)}(a_{i,N}(t_1))$ and $Z_N^{(j)}(a_{j,N}(t_2))$ are asymptotically independent.

REMARK 6.1. Theorem 6.1 shows that, apart from a transformation of the time scale, for a large initial population, all generations emerge according to the same compound Poisson process. As for a large population, if short life lengths are possible at all, it is likely that some offspring will indeed emerge after a short time. It is not surprising that the only conditions on G are local at $t = 0$ and that we need not restrict ourselves to Markov branching processes as had been done originally.

◇ ◇ ◇ ◇ ◇

During the preparation of this manuscript I enjoyed the benefit of helpful and inspiring discussions with Professor M. L. Samuels and Mr. M. Greig.

REFERENCES

[1] K. B. ATHREYA, "Some results on multitype continuous time Markov branching processes," *Ann. Math. Statist.*, Vol. 39 (1968), pp. 347–357.

[2] K. B. ATHREYA and S. KARLIN, "Limit theorems for the split times of branching processes," *J. Math. Mech.*, Vol. 17 (1967), pp. 257–278.

[3] W. J. BÜHLER, "Slowly branching processes," *Ann. Math. Statist.*, Vol. 38 (1967), pp. 919–921.

[4] ———, "Generations and degree of relationship in supercritical Markov branching processes, *Z. Wahrscheinlichkeitstheorie und verw. Gebiete*, to appear.

[5] K. S. CRAMP and C. J. MODE, "A general age dependent branching process," *J. Math. Anal. Appl.*, Vol. 24 (1968), pp. 494–508.

[6] ———, "A general age dependent branching process II," *J. Math. Anal. Appl.*, Vol. 25 (1969), pp. 8–17.

[7] J. L. DOOB, *Stochastic Processes*, New York, Wiley, 1953.

[8] M. GREIG, "A stochastic model of the mechanism governing the distribution of lengths of chains in bacterial populations," Ph.D. thesis, Department of Statistics, University of California, Berkeley, 1972.

[9] T. E. HARRIS, "Some mathematical models for branching processes," *Proceedings of the Second Berkeley Symposium on Mathematical Statistics and Probability*, Berkeley and Los Angeles, Univeristy of California Press, 1951, pp. 305–328.

[10] ———, *The Theory of Branching Processes*, Berlin-Gottingen-Heidelberg, Springer-Verlag, 1963.

[11] M. Jiřina, "The asymptotic behavior of branching stochastic processes," *Czechoslovak Math. J.*, Vol. 7 (1957), pp. 130–153. (In Russian.) (English translation has appeared in *Selected Translations in Mathematical Statistics and Probability*, Vol. 2 (1962), pp. 87–107.)

[12] D. G. Kendall, "On super-critical branching processes with a positive chance of extinction," *Research Papers in Statistics*, Festschrift for J. Neyman (edited by F. N. David), London, Wiley, 1966, pp. 157–165.

[13] B. P. Kharlamov, "On properties of branching processes with an arbitrary set of particle types," *Teor. Verojatnost. i Primenen.*, Vol. 13 (1968), pp. 82–95. (In Russian.)

[14] ———, "On the generation numbers of particles in a branching process with overlapping generations," *Teor. Verojatnost. i Primenen.*, Vol. 14 (1969), pp. 44–50. (In Russian.)

[15] ———, "On the numbers of generations in a branching process with an arbitrary set of particle types," *Teor. Verojatnost. i Primenen.*, Vol. 14 (1969), pp. 452–467. (In Russian.)

[16] N. Levinson, "Limiting theorems for age-dependent branching processes," *Illinois J. Math.*, Vol. 4 (1960), pp. 100–118.

[17] M. Loève, *Probability Theory*, Princeton, Van Nostrand, 1963 (3rd ed.).

[18] A. Martin-Löf, "A limit theorem for the size of the *n*th generation of an age-dependent branching process," *J. Math. Anal. Appl.*, Vol. 12 (1966), pp. 273–279.

[19] Shu-Teh C. Moy, "Ergodic properties of expectation matrices of a branching process with countably many types," *J. Math. Mech.*, Vol. 16 (1967), pp. 1207–1226.

[20] ———, "Extensions of a limit theorem of Evert, Wam, and Harris on multitype branching processes with countably many types," *Ann. Math. Statist.*, Vol. 38 (1967), pp. 992–999.

[21] R. Otter, "The multiplicative process," *Ann. Math. Statist.*, Vol. 20 (1949), pp. 206–224.

[22] M. L. Samuels, "Distribution of the population among generations in an age-dependent branching process," Ph.D. thesis, Department of Statistics, University of California, Berkeley, 1969.

[23] ———, "Distribution of the branching-process population among generations," *J. Appl. Probability*, Vol. 8 (1971), pp. 655–667.

[24] I. R. Savage and I. N. Shimi, "A branching process without rebranching," *Ann. Math. Statist.*, Vol. 40 (1969), pp. 1850–1851.

[25] H. H. Stratton, Jr. and H. G. Tucker, "Limit distributions of a branching stochastic process," *Ann. Math. Statist.*, Vol. 35 (1964), pp. 557–565.

[26] L. Takács, "On a probability problem arising in the theory of counters," *Proc. Cambridge Phil. Soc.*, Vol. 52 (1956), pp. 488–498.

A METHOD FOR STUDYING THE INTEGRAL FUNCTIONALS OF STOCHASTIC PROCESSES WITH APPLICATIONS, III

PREM S. PURI
PURDUE UNIVERSITY

1. Introduction

This paper is a continuation of the results presented in two earlier papers [20], [21] and may be read as the sequel. A brief account of their results will, however, be given here in order to make this paper selfcontained. The subject under study is the distribution of the integrals of the form

$$Y(t) = \int_0^t f(X(\tau), \tau)\, d\tau, \tag{1.1}$$

where $X(t)$, $t \geqq 0$, is a continuous time parameter stochastic process defined on a probability space $(\Omega, \mathscr{A}, \mathscr{P})$, with $\mathscr{X}$ as its state space, and f is a nonnegative (measurable) function defined on $\mathscr{X} \times [0, \infty)$. Here it is assumed that the integral $Y(t)$ exists and is finite almost surely for every $t > 0$.

The integrals $Y(t)$ arise in several domains of application such as in the theory of inventories and storage (see Moran [13], Naddor [14]), and in the study of the cost of the flow stopping incident involved in the automobile traffic jams (see Gaver [9], Daley [4], and Daley and Jacbos [5]). Such integrals are also encountered in certain stochastic models suitable for the study of response time distributions arising in various live situations (see Puri [16], [18], [19]). In fact in [18], it was shown that such a distribution is equivalent to the study of an integral of the type (1.1).

In [20], the work done by several authors in the past on the integral functionals of stochastic processes was briefly surveyed. But more importantly a method was introduced for obtaining the distribution of $Y(t)$. This method is based on a "quantal response process" $Z(t)$ defined for a hypothetical animal. By definition $Z(t)$ equals one if the animal is alive at time t and is equal to zero otherwise. In particular, it is assumed that

$$P(Z(t + \Delta t = 0 | Z(t) = 1, X(t) = x) = \delta f(x, t)\Delta t + o(\Delta t), \tag{1.2}$$

This research was supported in part by research grants NONR 1100(26) and N00014-67-A-0226-0008, from the Office of Naval Research.

with $Z(0) = 1$ and δ a nonnegative constant. Here the state "zero" is an absorption state for the process $Z(t)$. It is easy to establish by using a standard argument that

$$P\big(Z(t) = 1\big) = E\left(\exp\left\{-\delta\int_0^t f\big(X(\tau), \tau\big)\,d\tau\right\}\right), \tag{1.3}$$

which in turn gives the Laplace transform (L.t.) of the integral $Y(t)$. Thus, the study of the distribution of the integral $Y(t)$ can be carried out equivalently by studying the process $Z(t)$. Note that the quantal response process $Z(t)$ does not influence the process $X(t)$ in any way, rather, as is clear from (1.2), it is influenced itself by the growth of the process $X(t)$. Again, as was pointed out in [20], f is assumed to be nonnegative without loss of generality. Finally, in [20] and [21], this method was applied to the case of Markov chains. The results obtained there are summarized in the next section for later use.

2. The case of Markov chains

Consider a time homogeneous Markov chain (M.c.) $X(t)$ with $\mathscr{X} = \{1, 2, \cdots\}$, constructively defined as follows. If $X(t_1) = i$ at some epoch t_1, the value of $X(t)$ will remain constant for an interval $t_1 \leqq t < t_1 + \tau$, whose random duration τ is exponentially distributed with density function $c_i \exp\{-c_i x\}$, $c_i \geqq 0$; the probability that $X(t_1 + \tau) = j$ is p_{ij}, where the matrix $\mathbf{p} = (p_{ij})$, is a stochastic transition matrix. By assumption, the quantities c_i and p_{ij} are independent of time. Also, we assume that $c_i < \infty$ for all i so that all the states of $\mathscr{X}$ are stable. The sample paths of the process are assumed to be right continuous. Since the process is defined constructively, it is separable. Also, it is evident from the construction that the process $\big(X(t), Z(t)\big)$ is a Markov process with state space $\mathscr{X} = \{(i, r); i = 1, 2, 3, \cdots; r = 0, 1\}$. In [20] and [21], the above method was applied to Markov chains under the assumption that f depends only on $X(t)$ and not explicitly on t, in which case, in order to specify f, we are given a sequence of numbers $f(i) = f_i$, with $0 \leqq f_i < \infty$, $i = 1, 2, \cdots$. Let

$$\begin{aligned} P_{ij}(t) &= P\big(X(t) = j \,|\, X(0) = i\big), \\ \tilde{P}_{ij}(t) &= P\big(X(t) = j, Z(t) = 1 \,|\, X(0) = i, Z(0) = 1\big); \end{aligned} \tag{2.1}$$

$$\begin{aligned} \pi_{ij}(\alpha) &= \int_0^\infty \exp\{-\alpha t\} P_{ij}(t)\,dt, \\ \tilde{\pi}_{ij}(\alpha) &= \int_0^\infty \exp\{-\alpha t\} \tilde{P}_{ij}(t)\,dt, \\ \pi(\alpha) &= \big(\pi_{ij}(\alpha)\big), \qquad \tilde{\pi}(\alpha) = \big(\tilde{\pi}_{ij}(\alpha)\big); \end{aligned} \tag{2.2}$$

$$\begin{aligned} &\mathbf{C} = (\delta_{ij} c_i), && \mathbf{I} = (\delta_{ij}), \\ &\mathbf{1} = (1, 1, 1, \cdots), && \mathbf{f} = (\delta_{ij} f_i); \end{aligned} \tag{2.3}$$

where $i, j = 1, 2, \cdots; \alpha > 0$, and δ_{ij} is the Kronecker delta.

It is known that the probabilities P_{ij} in terms of their L.t. $\pi_{ij}(\alpha)$ satisfy the backward Kolmogorov system of equations (see Feller [7])

$$(\alpha \mathbf{I} + \mathbf{C})\boldsymbol{\pi}(\alpha) = \mathbf{I} + \mathbf{C}\mathbf{p}\boldsymbol{\pi}(\alpha). \tag{2.4}$$

If the solution of (2.4) satisfies $\alpha\boldsymbol{\pi}(\alpha)\mathbf{1} = \mathbf{1}$ for $\alpha > 0$, then $\boldsymbol{\pi}$ is the unique solution of (2.4) and is also the unique solution of the forward system of equations given by

$$\boldsymbol{\pi}(\alpha)(\alpha \mathbf{I} + \mathbf{C}) = \mathbf{I} + \boldsymbol{\pi}(\alpha)\mathbf{C}\mathbf{p}. \tag{2.5}$$

Analogous to (2.4) $\tilde{\boldsymbol{\pi}}$ satisfies the backward system

$$(\alpha \mathbf{I} + \mathbf{C} + \delta\mathbf{f})\tilde{\boldsymbol{\pi}}(\alpha) = \mathbf{I} + \mathbf{C}\mathbf{p}\tilde{\boldsymbol{\pi}}(\alpha). \tag{2.6}$$

In [20], it was shown that there always exists a solution of (2.6), which is minimal among all its solutions and which also is the minimal solution of the forward system, analogue of (2.5)

$$\tilde{\boldsymbol{\pi}}(\alpha)(\alpha \mathbf{I} + \mathbf{C} + \delta\mathbf{f}) = \mathbf{I} + \tilde{\boldsymbol{\pi}}(\alpha)\mathbf{C}\mathbf{p}. \tag{2.7}$$

Let $I_j(t)$ denote the indicator function of the set $[X(t) = j]$. Since

$$\tilde{P}_{i,j}(t) = E\left(\exp\left\{- \delta \int_0^t f(X(\tau))\, d\tau\right\} I_j(t) \,\middle|\, X(0) = i\right), \tag{2.8}$$

it is evident that knowledge of the $\tilde{P}_{i,j}$ is equivalent to that of the joint distribution of $X(t)$ and $Y(t)$. With this in mind, in [20] the problem of existence and uniqueness of the solution of (2.6) and (2.7) was studied in some detail. In particular, if the chain is finite with $\mathscr{X} = \{1, 2, \cdots, N\}$, $N < \infty$, then it can be easily seen that for all $\alpha > 0$, the matrix $(\alpha\mathbf{I} + \mathbf{C} + \delta\mathbf{f} - \mathbf{C}\mathbf{p})$ has an inverse, so that from (2.6) and (2.7) we have the explicit solution for $\tilde{\boldsymbol{\pi}}(\alpha)$ as

$$\tilde{\boldsymbol{\pi}}(\alpha) = (\alpha\mathbf{I} + \mathbf{C} + \delta\mathbf{f} - \mathbf{C}\mathbf{p})^{-1}, \tag{2.9}$$

valid for $\alpha > 0$ and $\delta \geqq 0$. Let $\boldsymbol{\psi}' = (\psi_1, \psi_2, \cdots, \psi_N)$, where

$$\psi_i(\alpha) = \int_0^\infty \exp\{-\alpha t\}\, E\left(\exp\left\{- \delta \int_0^t f(X(\tau))\, d\tau\right\} \,\middle|\, X(0) = i\right) dt. \tag{2.10}$$

We then have

$$\boldsymbol{\psi}(\alpha) = \tilde{\boldsymbol{\pi}}(\alpha)\mathbf{1} = (\alpha\mathbf{I} + \mathbf{C} + \delta\mathbf{f} - \mathbf{C}\mathbf{p})^{-1}\mathbf{1}. \tag{2.11}$$

The L.t. $\psi_i(\alpha)$ is in general a rational function of α and can therefore be easily inverted to yield $E(\exp\{- \delta \int_0^t f(X(\tau))\, d\tau\} | X(0) = i)$.

Again in [20] and [21], under certain assumptions, we proved the identity

$$\boldsymbol{\pi} - \tilde{\boldsymbol{\pi}} = \delta \tilde{\boldsymbol{\pi}} \mathbf{f} \boldsymbol{\pi}, \tag{2.12}$$

which connects $\tilde{\boldsymbol{\pi}}$ and $\boldsymbol{\pi}$, allowing us to obtain the desired $\tilde{\boldsymbol{\pi}}(\alpha)$ in terms of $\boldsymbol{\pi}(\alpha)$, which may be known. The identity (2.12) is found very useful in applications particularly because of the manner in which **f** appears. In particular in [21], we used this identity to obtain the joint distribution of times spent by the Markov chain in various states of a given finite set before the process hits a given taboo set.

REMARK. A formulation alternative to the consideration of the M.c. $\{X(t), Z(t)\}$ would be to consider a modified time homogeneous M.c. $\hat{X}(t)$ with state space $\{a, 1, 2, 3, \cdots\}$ with new exponential parameters, say $\hat{c}_i$, given by

$$\hat{c}_i = (c_i + \delta f_i)(1 - \delta_{ia}), \qquad i = a, 1, 2, 3, \cdots, \tag{2.13}$$

and the new transition matrix $\hat{p}_{ij}$ given by

$$\hat{p}_{ij} = \begin{cases} c_i p_{ij}(c_i + \delta f_i)^{-1} & \text{for } i, j = 1, 2, \cdots, \\ \delta f_i (c_i + \delta f_i)^{-1} & \text{for } j = a, i = 1, 2, \cdots, \\ \delta_{aj} & \text{for } i = a, \end{cases} \tag{2.14}$$

so that the state a is an absorption state. However, since for each formulation, the relevant information concerning the distribution of $X(t)$ and $Y(t)$ is contained in the equations (2.6) and (2.7), we find no essential gain in considering this alternative formulation.

In [20], it was pointed out that in the past most of the researchers in the area touched by this paper have exploited the backward system such as (2.6) (see for instance, Gaver [9], Daley [4], Daley and Jacobs [5], and McNeil [12]). Forward equations (2.7) were not used possibly because of lack of probabilistic interpretation. The present method via the quantal response process $Z(t)$ has the advantage over the past ones in that it provides the needed probabilistic interpretation. In the present paper, we shall exploit the forward system a great deal, by applying it to the case of certain well-known processes arising in several live situations. In [20], [21] and also in the applications of the method exhibited in the present paper, we have restricted outselves mostly to Markov chains with countable state space. However, it is evident that the method is applicable to almost all types of continuous time stochastic processes. The application to certain processes such as semi-Markov processes will be dealt with elsewhere.

Finally, it may be remarked that the above method has some resemblance with the work of Kemperman [11] and also with the method of collective marks due to van Dantzig [6]; in the present case, however, the approach was motivated by the author's work on the response time distribution arising in certain biological situations (see Puri [16], [18], and [19]).

3. Birth processes

This section will be devoted to the case where the M.c. $X(t)$ is a birth process. Section 3.1 deals with the time homogeneous case, while Section 3.2 deals with linear nonhomogeneous birth processes.

3.1. *Time homogeneous birth processes.* We shall use here the notation of Section 2. Let $X(t)$ be a time homogeneous birth process with $p_{jk} = \delta_{j+1,k}$ and $X(0) = i$. Also let

$$N = \min \{j; j \geqq i, c_j = 0\}; \tag{3.1}$$

if $c_j > 0$ for all $j \geqq i$, then $N = \infty$. If $N < \infty$, the M.c. is a finite one (with $c_j > 0, j = i, i + 1, \cdots, N - 1$, and $c_N = 0$), a case which was already considered in Section 2 with an explicit answer given by (2.9). However, if $N = \infty$, we assume that $\Sigma_{j=i}^{\infty} c_j^{-1} = \infty$, so that with probability one only a finite number of jumps of the chain are allowed in any finite time interval. For the present case, the systems of equations (2.6) and (2.7) are given by

$$(\alpha + c_i + \delta f_i)\tilde{\pi}_{ik}(\alpha) - c_i\tilde{\pi}_{i+1,k}(\alpha) = \delta_{ik} \tag{3.2}$$

and

$$(\alpha + c_k + \delta f_k)\tilde{\pi}_{ik}(\alpha) - c_{k-1}\tilde{\pi}_{i,k-1}(\alpha) = \delta_{ik}, \tag{3.3}$$

respectively, with $\tilde{\pi}_{ij}(\alpha) = 0$ for $j < i$ and for $j > N$. Each of these systems can be uniquely solved for $\tilde{\pi}_{ik}(\alpha)$ recursively, yielding

$$\tilde{\pi}_{ik}(\alpha) = \begin{cases} r_i & \text{for } k = i, \\ \rho_i\rho_{i+1} \cdots \rho_{k-1}r_k & \text{for } k > i, \\ 0 & \text{for } k < i \text{ and for } k > N, \end{cases} \tag{3.4}$$

where $r_j = (\alpha + c_j + \delta f_j)^{-1}$ and $\rho_j = c_j r_j$. Because of the condition $\Sigma_{j=i}^{N} c_j^{-1} = \infty$, it is now easily seen that for $\alpha > 0$,

$$\begin{aligned} \int_0^{\infty} \exp \{-\alpha t\} E\left(\exp \left\{-\delta \int_0^t f(X(\tau))\, d\tau\right\}\middle| X(0) = i\right) dt \\ = \sum_{k=i}^{N} \tilde{\pi}_{ik}(\alpha) = \lim_{n\to N} \sum_{k=i}^{n} \tilde{\pi}_{ik}(\alpha) \\ = \frac{1}{\alpha} \lim_{n\to N} \left[1 - \rho_i\rho_{i+1} \cdots \rho_n - \delta \sum_{k=i}^{n} f_k\rho_i\rho_{i+1} \cdots \rho_{k-1}r_k\right] \\ = \frac{1}{\alpha}\left[1 - \delta \sum_{k=i}^{N} f_k\rho_i\rho_{i+1} \cdots \rho_{k-1}r_k\right], \end{aligned} \tag{3.5}$$

where if $N < \infty$, the limit of a sum as $n \to N$ is taken to be the appropriate finite sum. Here in (3.5), we have used the fact that $\alpha \Sigma_{k=i}^{N} \tilde{\pi}_{ik}(\alpha) \leqq 1$, for all $\alpha > 0$, which is known from the general theory of Markov chains. The fact that $\lim_{n\to N} (\rho_i\rho_{i+1} \cdots \rho_n) = 0$ follows from the condition $\Sigma_{j=i}^{N} c_j^{-1} = \infty$.

Expressions (3.4) can easily be inverted to yield expressions for $\tilde{p}_{ik}(t)$. For instance, if $c_j + \delta f_j$ are all distinct for $j = i, i + 1, \cdots, N$, it can be easily shown that

$$(3.6)\quad \tilde{p}_{ik}(t) = \left(\prod_{j=1}^{k-1} c_j\right) \sum_{j=i}^{k} \left[\prod_{\ell=i,\ell\neq j}^{k} (c_\ell + \delta f_\ell - c_j - \delta f_j)^{-1}\right] \exp\{-(c_j + \delta f_j)t\},$$

where by convention $(\Pi_{j=i}^{k-1} c_j) = \Pi_{\ell\neq j}^{k} (c_\ell + \delta f_\ell - c_j - \delta f_j)^{-1} = 1$ for $k = j = i$. Summing (3.6) over k, we finally obtain

$$(3.7)\quad E\left(\exp\left\{-\delta \int_0^t f(X(\tau))\, d\tau\right\}\middle| X(0) = i\right)$$
$$= \sum_{k=i}^{N} \left(\prod_{j=i}^{k-1} c_j\right) \sum_{j=i}^{k} \left[\prod_{\ell=i,\ell\neq j}^{k} (c_\ell + \delta f_\ell - c_j - \delta f_j)^{-1}\right] \exp\{-(c_j + \delta f_j)t\}.$$

Note that for the case when $N = \infty$, an interchange of the two summation signs on the right side of (3.7) is not always valid. Also, one could instead invert the L.t. given in (3.5) and obtain a different yet equivalent expression for (3.7).

Again, since $Y(t)$ is a monotone nondecreasing function of t, it almost surely converges to a random variable, say Y, as $t \to \infty$. By using a Tauberian argument it follows from (3.5) that

$$(3.8)\quad E(\exp\{-\delta Y\}|X(0) = i) = \lim_{\alpha\to 0} \alpha \sum_{k=i}^{N} \tilde{\pi}_{ik}(\alpha)$$
$$= 1 - \lim_{n\to N} \delta \sum_{k=i}^{n} f_k \rho_i^* \rho_{i+1}^* \cdots \rho_{k-1}^* r_k^*,$$

where $r_i^* = (c_i + \delta f_i)^{-1}$ and $\rho_i^* = c_i r_i^*$. Now after some manipulation with the right side of (3.8), it can be shown that

$$(3.9)\quad E(\exp\{-\delta Y\}|X(0) = i) = \lim_{n\to N} (\rho_i^* \cdots \rho_n^*) = \prod_{j=i}^{N} \left[1 + \delta\left(\frac{f_j}{c_j}\right)\right]^{-1}.$$

If $N < \infty$, this is zero (keeping in mind that $c_N = 0$) unless $f_N = 0$. Thus, if $N < \infty$ and $f_N > 0$, $Y = \infty$ a.s. Again if $N = \infty$, (3.9) is equal to zero if and only if $\Sigma_{j=i}^{\infty} (f_j/c_j) = \infty$, in which case also $Y = \infty$ a.s. Let $\Sigma_{j=i}^{\infty} (f_j/c_j) < \infty$. This means that $f_N = 0$ whenever $N < \infty$. Then (3.9) is positive for all $\delta \geqq 0$, and in particular is equal to one when $\delta = 0$, and hence Y is finite a.s. Furthermore, it is clear from (3.9) that Y has a density. In fact, if we assume that f_j/c_j, $j = i, i + 1, \cdots, N$, are all distinct, then, at least when $N < \infty$, one easily obtains the density function of Y as

$$(3.10)\quad g(y) = \sum_{j=i}^{N} \frac{c_j}{f_i} \left[\prod_{\ell=i,\ell\neq j}^{N} \left(1 - \frac{c_j f_\ell}{f_j c_\ell}\right)^{-1}\right] \exp\left\{-\frac{c_j y}{f_j}\right\}, \qquad y > 0,$$

by inverting the L.t. (3.9). As expected, it follows from (3.9) that Y can be expressed as the sum $\Sigma_{j=i}^{N} f_j \tau_j$, where $\tau_i, \tau_{i+1}, \cdots$, are independently negative exponentially distributed with parameters $c_i, c_{i+1}, \cdots$. Here $\tau_j, j = i, i+1, \cdots$, are essentially the random lengths of times that the process spends in various states.

In the following subsections, we consider certain special cases of homogeneous birth processes that arise in practice.

3.1.1. *Case of a simple epidemic.* Jerwood [10] has recently considered the case of a simple epidemic which starts with $X(0) = i$ infectives and $S(0) = N - i$ susceptibles at time $t = 0$, where $N < \infty$. If $X(t)$ denotes the number of infectives at time t, then $X(t)$ is a birth process as considered by Bailey [1] with the finite state space $(i, i+1, \cdots, N)$, N being the absorption state and

$$c_j = \beta_j(N - j), \qquad j = i, i+1, \cdots, N. \tag{3.11}$$

Jerwood [10] considers the distribution of the cost of the epidemic exhibited by

$$C_i = aW_i + bT_i, \qquad a > 0, \qquad b > 0, \tag{3.12}$$

where

$$W_i = \int_0^{T_i} X(\tau)\, d\tau, \qquad T_i = \inf\{t: X(t) = N | X(0) = i\}, \tag{3.13}$$

a is the cost per unit time per infective, and b is the cost per unit time both over the period T_i. Perhaps a more realistic situation is where the rate of the first cost varies with the number of infectives at time t, so that one would like to obtain the distribution of

$$\tilde{C}_i = aW_i' + bT_i, \qquad a > 0, \qquad b > 0, \tag{3.14}$$

where

$$W_i' = \int_0^{T_i} h(X(\tau))\, d\tau, \tag{3.15}$$

with $0 \leqq h(j) < \infty, j = i, i+1, \cdots, N-1$. Since we are concerned with the epidemic only until the first passage time T_i to state N, without loss of generality, we may take $h(N) = 0$. In order to fit this into our situation above, all we need to take is

$$f_j = \begin{cases} ah(j) + b, & j = i, i+1, \cdots, N-1, \\ 0, & \text{otherwise.} \end{cases} \tag{3.16}$$

Now since the passage to the absorption state N occurs with probability one, it is easy to see that

$$\begin{aligned} E(\exp\{-\delta\tilde{C}_i\}) &= \lim_{t\to\infty} E\left(\exp\left\{-\delta\int_0^t f(X(\tau))\, d\tau\right\}\middle| X(0) = i\right) \\ &= E(\exp\{-\delta Y\}|X(0) = i) = \prod_{j=i}^{N-1}\left[1 + \delta\left(\frac{f_j}{c_j}\right)\right]^{-1}. \end{aligned} \tag{3.17}$$

The last equality follows from (3.9). If the f_j/c_j are all distinct, then the density function of $\tilde{C}_i$ is given by

$$(3.18)\qquad g(y) = \left\{\prod_{j=1}^{N-1} \frac{c_j}{f_j}\right\} \sum_{j=1}^{N-1} \left[\prod_{\ell=i,\ell\neq j}^{N-1} \left(\frac{c_\ell}{f_\ell} - \frac{c_j}{f_j}\right)^{-1}\right] \exp\left\{-\frac{c_j y}{f_j}\right\}, \qquad y > 0,$$

the expected cost being $E(\tilde{C}_i) = \Sigma_{j=i}^{N-1}(f_j/c_j)$. Incidentally using (3.16) in (3.17), we have, with $\alpha_1 = a\delta$ and $\alpha_2 = b\delta$,

$$(3.19)\quad E(\exp\{-\alpha_1 W_i' + \alpha_2 T_i\}) = \prod_{j=i}^{N-1} \left[1 + \frac{\alpha_1 h(j) + \alpha_2}{c_j}\right]^{-1}, \alpha_1 > 0, \alpha_2 > 0,$$

which, by treating α_1 and α_2 as the dummy variables, is the joint L.t. of the random variables W_i' and T_i.

3.1.2. *Poisson process.* This process is a special case of the birth process of Section 3.1, with $c_j = \lambda > 0$ for all j. In this case, if $f_j, j = i, i + 1, \cdots$, are all distinct, then it follows from (3.6) that

$$(3.20)\quad E\left(\exp\left\{-\delta\int_0^t f(X(\tau))\,d\tau\right\} I_k(t)\middle| X(0) = i\right) = \tilde{P}_{ik}(t)$$
$$= \left(\frac{\lambda}{\delta}\right)^{k-i} \exp\{-\lambda t\} \sum_{j=i}^{k}\left[\prod_{\ell=i,\ell\neq j}^{k} (f_\ell - f_j)^{-1}\right] \exp\{-\delta f_j t\},$$
$$k = i, i + 1, \cdots.$$

A special case with $f_j = j$ has been considered elsewhere by the author (see [18]). It was shown there that for $|s| \leqq 1$ and $\delta \geqq 0$ and with $X(0) = 0$,

$$(3.21)\quad E\left(s^{X(t)} \exp\left\{-\delta\int_0^t X(\tau)\,d\tau\right\}\middle| X(0) = 0\right) = \exp\left\{-\lambda t + \frac{\lambda s}{\delta}(1 - e^{-\delta t})\right\}.$$

With $s = 1$, this can be easily inverted to yield the distribution function of $Y(t)$ given by

$$(3.22)\qquad H_t(y) = \exp\{-\lambda t\} \sum_{n=0}^{\infty} \frac{(\lambda t)^n}{n!} F_t^{*(n)}(y),$$

where $F_t^{*(n)}$ stands for the n fold convolution of F_t, the distribution function of a random variable uniformly distributed over $(0, t)$.

3.2. *Nonhomogeneous birth processes.* For the case of nonhomogeneous Markov processes, in general, one finds it more convenient to use the forward Kolmogorov system of equations than the backward. Consider a birth process $X(t)$ with $X(0) = i$, such that for $j = i, i + 1, \cdots$,

$$(3.23)\qquad \begin{aligned} P(X(t + \Delta t) = j + 1 | X(t) = j) &= c_j(t)\Delta t + o(\Delta t),\\ P(X(t + \Delta t) = j | X(t) = j) &= 1 - c_j(t)\Delta t + o(\Delta t).\end{aligned}$$

Also for the function f of the integral (1.1), let

$$(3.24)\qquad f(j, t) = f_j(t), \qquad j = i, i + 1, \cdots.$$

Here the functions $c_j(t)$ and $f_j(t)$ are assumed to be nonnegative, continuous and integrable over $(0, t)$ for every finite t. Furthermore, let the functions $c_j(t)$ be such that the process $X(t)$ has, with probability one, a finite number of jumps in any finite time interval. Then using (1.2) and (3.23), one obtains in a standard manner the forward differential equations for the probabilities $\tilde{P}_{ij}(t)$ as given by

$$\frac{d\tilde{P}_{ij}(t)}{dt} = -\big(c_j(t) + \delta f_j(t)\big)\tilde{P}_{ij}(t) + c_{j-1}(t)\tilde{P}_{i,j-1}(t), \qquad j = i, i+1, \cdots, \tag{3.25}$$

where $\tilde{P}_{ij}(t) \equiv 0$ for all $j < i$. Recursively, these equations can be solved subject to $\tilde{P}_{ij}(0) = \delta_{ij}$ to yield

$$\tilde{P}_{ii}(t) = \exp\left\{-\int_0^t \big(c_i(\tau) + \delta f_i(\tau)\big)\,d\tau\right\} \tag{3.26}$$

and

$$\tilde{P}_{ij}(t) = \int_0^t \exp\left\{-\int_\tau^t \big(c_j(u) + \delta f_j(u)\big)\,du\right\} c_{j-1}(\tau)\tilde{P}_{i,j-1}(\tau)\,d\tau, \qquad j = i+1, i+2, \cdots. \tag{3.27}$$

Unfortunately, in general, there appears to be no way of obtaining the expression for $\tilde{P}_{ij}(t)$ in a closed form. Instead, in the rest of this section, we restrict ourselves to the case of linear birth processes with

$$c_j(t) = \alpha(t) + j\nu(t), \qquad j = i, i+1, \cdots, \tag{3.28}$$

and with $f_j(t) = j\beta(t)$. Thus, here we are interested in the integral of the form

$$Y(t) = \int_0^t \beta(\tau)X(\tau)\,d\tau. \tag{3.29}$$

For this case, the equations (3.25) take the form

$$\frac{d\tilde{P}_{ij}(t)}{dt} = -[\alpha(t) + j(\delta\beta(t) + \nu(t))]\tilde{P}_{ij}(t) + [\alpha(t) + (j-1)\nu(t)]\tilde{P}_{i,j-1}(t), \tag{3.30}$$

for $j = i, i+1, \cdots$. Let

$$\tilde{G}(s, t) = \sum_{j=i}^{\infty} s^j \tilde{P}_{ij}(t), \qquad |s| \leqq 1. \tag{3.31}$$

Then multiplying both sides of (3.30) by s^j and adding over j we obtain, after some simplification, the equation

$$\tilde{G}_t - [\nu(t)s - \nu(t) - \delta\beta(t)]s\tilde{G}_s = -\alpha(t)(1-s)\tilde{G}, \tag{3.32}$$

where here and elsewhere $\tilde{G}_t$ and $\tilde{G}_s$ denote the respective partial derivatives of $\tilde{G}$. Equation (3.32) is subject to the initial condition $\tilde{G}(s, 0) = s^i$ and can be solved by standard methods yielding

$$\tilde{G}(s, t) = [\phi(s; 0, t)]^i \exp\left\{-\int_0^t \alpha(\tau)[1 - \phi(s; \tau, t)]\, d\tau\right\}, \tag{3.33}$$

where, for $0 \leqq \tau \leqq t$,

$$\begin{aligned}[\phi(s; \tau, t)]^{-1} = \frac{1}{s} \exp\left\{\int_\tau^t \big(v(u) + \delta\beta(u)\big)\, du\right\} \\ - \int_\tau^t v(u) \exp\left\{\int_\tau^u \big(v(v) + \delta\beta(v)\big)\, dv\right\} du.\end{aligned} \tag{3.34}$$

Since in the present case of (3.28), it is known that in any finite time interval, with probability one, there are only a finite number of jumps of the process $X(t)$, (3.33) with $s = 1$ yields

$$E\left(\exp\left\{-\delta \int_0^t \beta(\tau) X(\tau)\, d\tau\right\}\middle| X(0) = i\right) = \tilde{G}(1, t). \tag{3.35}$$

In the next subsection, we specialize the above results to Pólya process which arises very often in various live situations such as the theory of accident proneness (see Bates and Neyman [2]).

3.2.1. *Pólya process.* This is a special case of the linear birth process discussed above with

$$\alpha(t) = \lambda(1 + \rho\lambda t)^{-1}, \qquad v(t) = \lambda\rho(1 + \rho\lambda t)^{-1}, \qquad \lambda > 0, \qquad \rho > 0. \tag{3.36}$$

Also, we consider the special case where $\beta(t) \equiv 1$, so that we are interested in the integral $Y(t) = \int_0^t X(\tau)\, d\tau$. For this case, expression (3.33) simplifies to

$$\tilde{G}(s, t) = s^i \exp\{-i\delta t\}\left[(1 + \lambda\rho t) - \frac{\lambda\rho}{\delta}(1 - e^{-\delta t})s\right]^{-(i+\rho^{-1})}. \tag{3.37}$$

This then gives the joint distribution of $X(t)$ and $Y(t)$. In particular, from this one easily obtains, for $n = 0, 1, 2, \cdots$,

$$\begin{aligned}\tilde{P}_{i,n+i}(t) = (1 + \lambda\rho t)^{-(i+\rho^{-1})} e^{-i\delta t}\left[\frac{(i + n - 1 + \rho^{-1}) \cdots (i + \rho^{-1})}{n!}\right] \\ \cdot \left(\frac{\lambda\rho t}{1 + \lambda\rho t}\right)^n \left(\frac{1 - e^{-\delta t}}{\delta t}\right)^n.\end{aligned} \tag{3.38}$$

Again, one can easily invert (3.37) with $s = 1$ to yield the distribution function $H_t(y)$ of the integral $Y(t)$, given by

$$(3.39)\quad H_t(y) = (1 + \lambda\rho t)^{-(i+\rho^{-1})} \cdot \sum_{n=0}^{\infty} \frac{(i + n - 1 + \rho^{-1}) \cdots (i + \rho^{-1})}{n!} \left(\frac{\lambda\rho t}{1 + \lambda\rho t}\right)^n U_t * F_t^{*(n)}(y)$$

for $y > it$, and for $y \leqq it$, $H_t(y) = 0$. Here the operation $*$ denotes the convolution between two distribution functions, U_t is the distribution function corresponding to a degenerate random variable taking value as (ti), and $F_t^{*(n)}$ is as defined in Section 3.1.2. Finally, if we let $i = 0$ and $\rho \to 0$ in (3.37), we obtain the expression (3.21) for the Poisson process as expected.

4. Birth and death processes

We now consider the case of a time homogeneous birth and death process (b-d process) $X(t)$ where, in notation of Section 2, $c_k = (\lambda_k + \mu_k)$,

$$(4.1)\qquad p_{kj} = \begin{cases} \lambda_k(\lambda_k + \mu_k)^{-1} & \text{if } j = k + 1, \\ \mu_k(\lambda_k + \mu_k)^{-1} & \text{if } j = k - 1, \\ 0 & \text{elsewhere,} \end{cases}$$

and λ_k and μ_k are nonnegative constants with $\mu_0 = 0$. Let $X(0) = i$. The backward and forward equations, analogues of (2.6) and (2.7) but converted into differential equations, are given by

$$(4.2)\qquad \frac{d\tilde{P}_{ik}(t)}{dt} = -(\lambda_i + \mu_i + \delta f_i)\tilde{P}_{ik}(t) + \lambda_i\tilde{P}_{i+1,k}(t) + \mu_i\tilde{P}_{i-1,k}(t)$$

and

$$(4.3)\qquad \frac{d\tilde{P}_{ik}(t)}{dt} = -(\lambda_k + \mu_k + \delta f_k)\tilde{P}_{ik}(t) + \lambda_{k-1}\tilde{P}_{i,k-1}(t) + \mu_{k+1}\tilde{P}_{i,k+1}(t),$$

respectively. Unfortunately, we shall not consider these here in this generality. Instead, we shall consider certain special cases which arise in several practical situations. For this, we shall particularly be making use of the forward system (4.3). Since $X(0) = i$ will be kept fixed, we shall write for brevity $\tilde{P}_{ik}(t) = \tilde{P}_k(t)$.

4.1. *Linear birth and death processes with immigration.* Consider a b-d process with $\lambda_k = k\lambda + \nu$ and $\mu_k = k\mu$ for $k = 0, 1, 2, \cdots$, where λ, ν, and μ are positive constants. Such processes arise in the study of population dynamics and also with $\lambda = 0$ in the queueing theory of $M/M/\infty$ queues. Here we wish to obtain the joint distribution of $X(t)$, $\int_0^t X(\tau)\, d\tau$, and $T(t)$, the last one being the length of time during $(0, t)$ the process remains in nonzero states. In $M/M/\infty$

queue, $T(t)$ represents the time for which at least one channel remains busy during $(0, t)$; $\int_0^t X(\tau)\,d\tau$ represents the cumulative time the customers spend during $(0, t)$ while they are being served. In the study of response of host to injection of virulent bacteria, $\int_0^t X(\tau)\,d\tau$ could be regarded as a measure of the total amount of toxins produced by the live bacteria during $(0, t)$, assuming a constant toxin excretion rate per bacterium (see Puri [16], [17], and [18]).

In order to accomplish our purpose, it is sufficient to take

$$\delta f_k = \beta_1(1 - \delta_{k0}) + k\beta_2, \qquad k = 0, 1, 2, \cdots, \tag{4.4}$$

so that

$$\begin{aligned} \tilde{P}_k(t) &= E\left(\exp\left\{-\delta\int_0^t f(X(\tau))\,d\tau\right\} I_k(t)\right) \\ &= E\left(\exp\left\{-\beta_1 T(t) - \beta_2\int_0^t X(\tau)\,d\tau\right\} I_k(t)\right). \end{aligned} \tag{4.5}$$

Furthermore, for the present case, the system (4.3) takes the form

$$\frac{d\tilde{P}_k(t)}{dt} = \begin{cases} -[\nu + \beta_1 + k(\mu + \lambda + \beta_2)]\tilde{P}_k \\ \qquad + (k+1)\mu\tilde{P}_{k+1} + [\nu + (k-1)\lambda]\tilde{P}_{k-1}, & k \geqq 1, \\ -\nu\tilde{P}_0 + \mu\tilde{P}_1, & k = 0. \end{cases} \tag{4.6}$$

Let $\tilde{G}(s, t)$ be as defined in (3.31). Then from (4.6) we have

$$\tilde{G}_t - \lambda(s - r_1)(s - r_2)\tilde{G}_s = -[\nu(1 - s) + \beta_1]\tilde{G} + \beta_1\tilde{P}_0, \tag{4.7}$$

where r_1 and r_2 denote with plus and minus signs, respectively,

$$\frac{1}{2\lambda}[(\mu + \lambda + \beta_2) \pm \{(\mu + \lambda + \beta_2)^2 - 4\mu\lambda\}^{1/2}]. \tag{4.8}$$

The problem now is to solve (4.7) for $\tilde{G}$ subject to the side condition $\tilde{G}(s, 0) = s^i$. This can be accomplished by standard methods. We give here only the final answer in terms of its L.t. Let

$$\psi_1(s, t) = [h_1(s, t)]^i[h_2(s, t)]^{-\nu/\lambda}\exp\{-(\nu + \beta_1 - \nu r_2)t\} \tag{4.9}$$

and

$$\psi_2(s, t) = [h_2(s, t)]^{-\nu/\lambda}\exp\{-(\nu + \beta_1 - \nu r_2)t\}, \tag{4.10}$$

where

$$h_1(s, t) = \left\{\frac{r_2(r_1 - s) + r_1(s - r_2)\exp\{-\lambda(r_1 - r_2)t\}}{(r_1 - s) + (s - r_2)\exp\{-\lambda(r_1 - r_2)t\}}\right\} \tag{4.11}$$

and

$$h_2(s, t) = \{(s - r_2)\exp\{-\lambda(r_1 - r_2)t\} + (r_1 - s)\}(r_1 - r_2)^{-1}. \tag{4.12}$$

Also, let $\psi_1^*(s, \alpha)$, $\psi_2^*(s, \alpha)$, $\tilde{G}^*(s, \alpha)$, and $\tilde{P}_0^*(\alpha)$ be the Laplace transforms over time t of ψ_1, ψ_2, $\tilde{G}$, and $\tilde{P}_0$, respectively, defined for $\alpha > 0$. From (4.7), it is easy to show that

$$\tilde{G}(s, t) = \psi_1(s, t) + \beta_1 \int_0^t \tilde{P}_0(t - \tau)\psi_2(s, \tau)\, d\tau, \tag{4.13}$$

from which it follows that

$$\tilde{G}^*(s, \alpha) = \psi_1^*(s, \alpha) + \beta_1 \tilde{P}_0^*(\alpha)\psi_2^*(s, \alpha). \tag{4.14}$$

With $s = 0$, (4.14) yields

$$\tilde{P}_0^*(\alpha) = \psi_1^*(0, \alpha)[1 - \beta_1 \psi_2^*(0, \alpha)]^{-1}. \tag{4.15}$$

Finally, by using this in (4.14), we obtain

$$\tilde{G}^*(s, \alpha) = \psi_1^*(s, \alpha) + \beta_1 \psi_2^*(s, \alpha)\psi_1^*(0, \alpha)[1 - \beta_1 \psi_2^*(0, \alpha)]^{-1}. \tag{4.16}$$

We now consider briefly a special case with no immigration, that is, with $\nu = 0$, where we explicitly obtain

$$\tilde{P}_0(t) = \left(\frac{\mu}{\lambda}\right)^i \{\exp\{-\beta_1 t\}[J(t)]^i + \beta_1 \int_0^t \exp\{-\beta_1 \tau\}[J(\tau)]^i\, d\tau\}, \tag{4.17}$$

and

$$\tilde{G}(s, t) = \exp\{-\beta_1 t\}\left\{\beta_1 \int_0^t \exp\{\beta_1 \tau\}\tilde{P}_0(\tau)\, d\tau + [h_1(s, t)]^i\right\}, \tag{4.18}$$

with

$$J(t) = [1 - \exp\{-\lambda(r_1 - r_2)t\}][r_1 - r_2 \exp\{-\lambda(r_1 - r_2)t\}]^{-1}. \tag{4.19}$$

The case without the random variable $T(t)$ has previously been considered elsewhere by the author [15]. For the present case ($\nu = 0$), it is known that $P(X(t) \to 0 \text{ or } \infty) = 1$ and that $P(X(t) \to 0) = \min(1, \mu/\lambda)$. Also, $T(t)$ being a nondecreasing function of t tends almost surely to a random variable T as $t \to \infty$. Here T is the first passage time of the process to the state zero. Also $P(T < \infty) = \min(1, \mu/\lambda)$. Thus by using (4.17), we have

$$\begin{aligned} \lim_{t\to\infty} \tilde{G}(s, t) &= \lim_{t\to\infty} \tilde{P}_0(t) \\ &= E\left(\exp\left\{-\beta_1 T - \beta_2 \int_0^T X(\tau)\, d\tau\right\}\right) \\ &= \begin{cases} r_2^i & \text{if } \beta_1 = 0, \\ \left(\frac{\mu}{\lambda}\right)^i \beta_1 \int_0^t \exp\{-\beta_1 t\}[J(t)]^i\, dt & \text{if } \beta_1 > 0. \end{cases} \end{aligned} \tag{4.20}$$

On the other hand, in the presence of immigration ($\nu > 0$), if we wish to find the joint distribution of T and $\int_0^T X(\tau)\,d\tau$, we first modify our process (see also [21]) by taking $\lambda_0 = 0$, $\lambda_k = k\lambda + \nu$ for $k \geqq 1$, so that zero is an absorption state of the process. Here we allow immigration only as long as the process has not touched the state zero. The analogue of equation (4.7) for the modified process is then given by

$$\tilde{G}_t - \lambda(s - r_1)(s - r_2)\tilde{G}_s = -[\beta_1 + \nu(1 - s)](\tilde{G} - \tilde{P}_0). \tag{4.21}$$

The solution of (4.21) subject to $\tilde{G}(s, 0) = s^i$, $i \geqq 1$, is given by (4.13) through (4.16) with ψ_2 replaced by

$$\begin{aligned}(4.22)\quad &\hat{\psi}_2(s, t)\\ &= \frac{1}{\beta_1}\psi_2(s, t)\left[(\nu + \beta_1 - \nu r_1) + \frac{\nu(r_1 - r_2)(r_1 - s)}{(r_1 - s) + (s - r_2)\exp\{-\lambda(r_1 - r_2)t\}}\right],\end{aligned}$$

and ψ_2^* by $\hat{\psi}_2^*$. We thus have from the new (4.16), while using a Tauberian argument,

$$\begin{aligned}(4.23)\quad E\left(\exp\left\{-\beta_1 T - \beta_2\int_0^T X(\tau)\,d\tau\right\}\middle| X(0) = i\right) &= \lim_{t\to\infty}\tilde{G}(1, t) = \lim_{\alpha\to 0}\alpha\tilde{G}^*(1, \alpha)\\ &= \lim_{\alpha\to 0}\left[\alpha\psi_1^*(1, \alpha) + \beta_1\frac{\alpha\hat{\psi}_2^*(1, \alpha)\psi_1^*(0, \alpha)}{1 - \beta_1\hat{\psi}_2^*(0, \alpha)}\right].\end{aligned}$$

On the other hand, it can be easily shown that

$$(4.24)\quad \begin{aligned}\lim_{\alpha\to 0}\alpha\psi_1^*(1, \alpha) &= \lim_{t\to\infty}\psi_1(1, t) = 0,\\ \lim_{\alpha\to 0}\hat{\psi}_2^*(1, \alpha) &= \int_0^\infty \hat{\psi}_2(1, t)\,dt,\end{aligned}$$

and

$$\begin{aligned}(4.25)\quad &\lim_{\alpha\to 0}\alpha\psi_1^*(0, 1)[1 - \beta_1\hat{\psi}_2^*(0, \alpha)]^{-1}\\ &= \lim_{t\to\infty}\tilde{P}_0(t) = \tilde{P}_0(\infty)\\ &= \frac{\int_0^\infty \exp\{-(\nu + \beta_1 - \nu r_2)t\}[h_1(0, t)]^i[h_2(0, t)]^{-\nu/\lambda}\,dt}{\int_0^\infty \exp\{-(\nu + \beta_1 - \nu r_2)t\}[h_2(0, t)]^{-\nu/\lambda}\,dt}.\end{aligned}$$

Thus, we finally have from (4.23),

$$E\left(\exp\left\{-\beta_1 T - \beta_2\int_0^T X(\tau)\,d\tau\right\}\middle| X(0) = i\right) = \beta_1\tilde{P}_0(\infty)\int_0^\infty \hat{\psi}_2(1, t)\,dt. \tag{4.26}$$

4.2. *M/M/I queue.* This corresponds to the case with $\lambda_k = \lambda$ for $k \geqq 0$ and $\mu_k = \mu$ for $k \geqq 1$ with $\mu_0 = 0$. This is the case which has recently been explored by Gaver [9], Daley [4], Daley and Jacobs [5], and McNeil [12]. Most of these authors have used the backward system analogues, while we, based on our method, shall use the forward system. Also, this section will apparently have some relevance to the paper presented by Professor Gani at this Symposium. In the case of $M/M/I$ queue, $T(t)$ as defined in the previous section represents the period for which the channel remains busy during $(0, t)$ and $[\int_0^t X(\tau)\,d\tau - T(t)]$ represents the total time wasted by the customers during $(0, t)$ while standing in the queue and waiting for their turn for service. Although these random variables are of some practical importance, to the best of author's knowledge, their distributions have not been considered before. The integrals studied by Gaver, Daley, and others were restricted only to a busy period of the queue, where "zero" acts as an absorption state. We shall touch this case briefly later.

As before, we are interested in obtaining the joint distribution of $X(t)$, $\int_0^t X(\tau)\,d\tau$, and $T(t)$. The analogue of equation (4.7) for the present case is given by

$$(4.27)\qquad s\tilde{G}_s + \beta_2 s^2 \tilde{G}_s = \lambda(s - \tilde{r}_1)(s - \tilde{r}_2)\tilde{G} + [(\mu + \beta_1)s - \mu]\tilde{P}_0,$$

which is to be solved subject to $\tilde{G}(s, 0) = s^i$. Here $\tilde{r}_1$ and $\tilde{r}_2$, with positive and negative signs, respectively, are given by

$$(4.28)\qquad \frac{1}{2\lambda}[(\mu + \lambda + \beta_1) \pm \{(\mu + \lambda + \beta_1)^2 - 4\mu\lambda\}^{1/2}].$$

Unfortunately, the solution of (4.27) appears quite complex and involves Bessel functions. The author did not succeed in obtaining an explicit solution of (4.27). However, one can easily solve it when $\beta_2 = 0$, giving only the joint distribution of $X(t)$ and $T(t)$. Taking L.t. of (4.27) (with $\beta_2 = 0$) with respect to t, we obtain for $\alpha > 0$,

$$(4.29)\qquad \tilde{G}^*(s, \alpha) = [\{\mu - (\mu + \beta_1)s\}\tilde{P}_0^*(\alpha) - s^{i+1}][\lambda(s - r_1^*)(s - r_2^*)]^{-1},$$

where r_1^* and r_2^*, with positive and negative signs, respectively, are

$$(4.30)\qquad \frac{1}{2\lambda}[(\mu + \lambda + \beta_1 + \alpha) \pm \{(\mu + \lambda + \beta_1 + \alpha)^2 - 4\mu\lambda\}^{1/2}],$$

and they satisfy the relation $0 < r_2^* < 1 < r_1^*$. Since $\tilde{G}^*$ is analytic for $|s| < 1$, the first of the two expressions on the right side of (4.29) must vanish at $s = r_2^*$. This fact yields

$$(4.31)\qquad \tilde{P}_0^*(\alpha) = (r_2^*)^{i+1}[\mu - (\mu + \beta_1)r_2^*]^{-1}.$$

On substitution of this in (4.29), we obtain

$$(4.32)\qquad \tilde{G}^*(s,\alpha) = \frac{(r_2^*)^{i+1}[\mu - (\mu+\beta_1)s] - s^{i+1}[\mu - (\mu+\beta_1)r_2^*]}{\lambda(s-r_1^*)(s-r_2^*)[\mu-(\mu+\beta_1)r_2^*]}.$$

Finally, on putting $s = 1$ in (4.32) and inverting the resultant L.t. by lengthy yet standard methods (see Saaty [22]), we obtain

$$(4.33)\qquad \tilde{G}(1,t) = E\big(\exp\{-\beta_1 T(t)\}\,|\,X(0)=i\big) = \sum_{n=0}^{\infty} \tilde{P}_n(t),$$

where

$$(4.34)\ \tilde{P}_n(t) = \exp\{-(\lambda+\mu+\beta_1)t\}\Bigg\{\left(\frac{\mu}{\lambda}\right)^{(i-n)/2} B_{n-i}(\xi) + \frac{\mu+\beta_1}{\lambda}\left(\frac{\mu}{\lambda}\right)^{(i-n-1)/2} B_{n+i+1}(\xi) + \left(\frac{\mu}{\lambda}\right)^{(i-n)/2}\big(1-\lambda\mu(\mu+\beta_1)^{-2}\big)\sum_{k=2}^{\infty}[(\mu+\beta_1)(\lambda\mu)^{-1/2}]^k B_{n+i+k}(\xi)\Bigg\},$$

and $B_n(u)$ denotes the Bessel function of the first kind and $\xi = 2(\lambda\mu)^{1/2}t$.

We now consider only the busy period of the $M/M/I$ queue started with $X(0) = i$. For this we take $\lambda_k = \lambda$ for $k \geqq 1$, and $\lambda_0 = 0$, so that "zero" is an absorption state of the process $X(t)$. As before $T(t)\uparrow T$ a.s. as $t \to \infty$, where T is the length of the busy period. If $\mu \geqq \lambda$, it is known that $X(t) \to 0$ with probability one, so that $P(T < \infty) = 1$. On the other hand if $\mu < \lambda$, $P(T < \infty) = P(X(t) \to 0) = \mu/\lambda$, and $P(T = \infty) = P(X(t) \to \infty) = 1 - (\mu/\lambda)$. The analogue of (4.27) now takes the form

$$(4.35)\qquad s\tilde{G}_t + \beta_2 s^2 \tilde{G}_s = [\lambda s^2 - (\lambda+\mu+\beta_1)s + \mu](\tilde{G} - \tilde{P}_0),$$

to be solved subject to $\tilde{G}(s, 0) = s^i$. Unfortunately, the solution of this presents similar difficulties as of equation (4.27). However, once it is solved we have the desired result given by

$$(4.36)\qquad \lim_{t\to\infty} \tilde{G}(1,t) = E\left(\exp\left\{-\beta_1 T - \beta_2\int_0^T X(\tau)\,d\tau\right\}\Big|X(0)=i\right).$$

This result has been studied through other methods by Daley [4] and Daley and Jacobs [5]. Again, the equation (4.35) can be easily solved like (4.27) when $\beta_2 = 0$. However, since the distribution of the length of the busy period T is already known (see Saaty [22]), we shall not pursue this further here.

5. Illness-death processes

These processes have been extensively studied by Fix and Neyman [8] and more recently by Chiang [3]. Briefly, these are finite Markov chains with two sets of states; S_i, $i = 1, 2, \cdots, s$, are the illness states and R_θ, $\theta = 1, 2, \cdots, r$, are the death states. (In this section, symbols i, j, and k will stand for S_i, S_j, and S_k, and θ for R_θ.) Here all the death states are absorption states. Also, in terms of the notation of Section 2, for the transitions among the various states we have on adopting Chiang's notation,

$$\begin{aligned} c_i p_{ij} &= \nu_{ij}, & i \neq j,\ i, j = 1, 2, \cdots, s, \\ c_i p_{i\theta} &= \mu_{i\theta}, & i = 1, 2, \cdots, s,\ \theta = 1, 2, \cdots, r, \\ -c_i &= \nu_{ii}, & i = 1, 2, \cdots, s. \end{aligned} \tag{5.1}$$

Consider a typical person moving from one state to another according to the above M.c. until he is absorbed into one of the death states. Chiang ([3], pp. 81 and 160) has considered the lengths of this person's stay in various states within a period of length t, and has given expressions for their expected values only; while our method leads easily to their joint distribution. For this, take

$$\delta f_\ell = \begin{cases} \delta_i & \text{if } \ell \text{ is the state } S_i, \\ \sigma_\theta & \text{if } \ell \text{ is the state } R_\theta. \end{cases} \tag{5.2}$$

Analogous to Chiang's notation (see [3], p. 152), let

$$\tilde{P}_{ij}(t) = P(Z(t) = 1, X(t) = S_j | X(0) = S_i, Z(0) = 1) \tag{5.3}$$

and

$$\tilde{Q}_{i\theta}(t) = P(Z(t) = 1, X(t) = R_\theta | X(0) = S_i, Z(0) = 1), \tag{5.4}$$

where $Z(t)$ represents the "quantal response process" as defined in Section 1. Then the forward system of equations for $\tilde{P}$ are given by

$$\frac{d\tilde{P}_{ij}(t)}{dt} = -\Big(\sum_{k \neq j} \nu_{jk} + \sum_{\theta} \mu_{j\theta} + \delta_i\Big)\tilde{P}_{ij}(t) + \sum_{k \neq j} \tilde{P}_{ik}(t)\nu_{kj} \tag{5.5}$$

for $i, j = 1, 2, \cdots, s$. Similarly, one could write down the backward system. Either of these systems can be uniquely solved for $\tilde{P}_{ij}(t)$, which then can be used to obtain $\tilde{Q}_{i\theta}(t)$ by noticing that

$$\tilde{Q}_{i\theta}(t) = \sum_{j=1}^{s} \mu_{j\theta} \int_0^t \tilde{P}_{ij}(\tau) \exp\{-\sigma_\theta(t - \tau)\}\, d\tau, \qquad \theta = 1, 2, \cdots, r. \tag{5.6}$$

Here the factor $\exp\{-\sigma_\theta(t-\tau)\}$ under the integral sign denotes the probability that our hypothetical animal of the "quantal response process" $Z(t)$ does not die during (τ, t), once the process has touched the state R_θ at moment τ. Let $X(0) = S_i$, and

$$
\begin{aligned}
T_{ij}(t) &= \text{length of stay in } S_j \text{ during } (0, t), \qquad j = 1, 2, \cdots, s, \\
\tilde{T}_{i\theta}(t) &= \text{length of stay in } R_\theta \text{ during } (0, t), \qquad \theta = 1, 2, \cdots, r.
\end{aligned} \tag{5.7}
$$

Having obtained $\tilde{P}_{ij}$ and $\tilde{Q}_{i\theta}$ in the above manner, the L.t. of the joint distribution of $T_{ij}(t)$ and $\tilde{T}_{i\theta}(t)$ is then given by

$$
E\left(\exp\left\{-\sum_{j=1}^{s}\delta_j T_{ij}(t) - \sum_{\theta=1}^{r}\sigma_\theta \tilde{T}_{i\theta}(t)\right\}\right) = \sum_{j=1}^{s}\tilde{P}_{ij}(t) + \sum_{\theta=1}^{r}\tilde{Q}_{i\theta}(t), \tag{5.8}
$$

where δ_j and σ_θ act as the dummy variables for the L.t. The L.t. (5.8), in general, is a rational function of δ's and σ's and can be inverted by standard methods to yield the desired distribution. We shall illustrate the above approach through an example, where we take $s = 2$, $r = 1$, and $X(0) = S_1$. Since

$$
T_{11}(t) + T_{12}(t) + \tilde{T}_{11}(t) = t \qquad \text{a.s.}, \tag{5.9}
$$

it is sufficient to study the joint distribution of $T_{11}(t)$ and $T_{12}(t)$ only, in which case we may take $\sigma_1 = 0$. The equations (5.5) are now given by

$$
\begin{aligned}
\frac{d\tilde{P}_{11}}{dt} &= -(\nu_{12} + \mu_{11} + \delta_1)\tilde{P}_{11} + \nu_{21}\tilde{P}_{12}, \\
\frac{d\tilde{P}_{12}}{dt} &= -(\nu_{21} + \mu_{21} + \delta_2)\tilde{P}_{12} + \nu_{12}\tilde{P}_{11}.
\end{aligned} \tag{5.10}
$$

Let

$$
A = (\delta_1 + \nu_{12} + \mu_{11}), \qquad B = (\delta_2 + \nu_{21} + \mu_{21}), \tag{5.11}
$$

and a_1 and a_2, with positive and negative signs, respectively, be given by

$$
\tfrac{1}{2}[-(A + B) \pm \{(A - B)^2 + 4\nu_{12}\nu_{21}\}^{1/2}]. \tag{5.12}
$$

The solution of (5.10) is given by

$$
\begin{aligned}
\tilde{P}_{11}(t) &= [(a_1 + B)\exp\{a_1 t\} - (B + a_2)\exp\{a_2 t\}](a_1 - a_2)^{-1}, \\
\tilde{P}_{12}(t) &= \nu_{12}[\exp\{a_1 t\} - \exp\{a_2 t\}](a_1 - a_2)^{-1}.
\end{aligned} \tag{5.13}
$$

Now using (5.6) with $\sigma_\theta = 0$, we can obtain $\tilde{Q}_{11}(t)$. Finally, omitting details, we have by using (5.8),

$$
\begin{aligned}
&E(\exp\{-\delta_1 T_{11}(t) - \delta_2 T_{12}(t)\}) \\
&\quad = \tilde{P}_{11}(t) + \tilde{P}_{12}(t) + \tilde{Q}_{11}(t) \\
&\quad = \Big\{[(a_1 + B + \nu_{12} + \mu_{11}) + (\mu_{11}B + \mu_{21}\nu_{12})a_1^{-1}]\exp\{a_1 t\} \\
&\qquad - [(B + a_2 + \nu_{12} + \mu_{11}) + (\mu_{11}B + \mu_{21}\nu_{12})a_2^{-1}] \\
&\qquad \cdot\exp\{a_2 t\}\Big\}(a_1 - a_2)^{-1} + \{(\mu_{11}B + \nu_{12}\mu_{21})\}(AB - \nu_{12}\nu_{21})^{-1}.
\end{aligned} \tag{5.14}
$$

Since $T_{11}(t)$ and $T_{12}(t)$ are monotone nondecreasing functions of t, $T_{11}(t) \uparrow T_{11}$ and $T_{12}(t) \uparrow T_{12}$ almost surely as $t \to \infty$. Here the random variables T_{11} and T_{12} represent the lengths of time the person spends in S_1 and S_2, respectively, before he finally dies. Letting $t \to \infty$ and using the fact that a_1 and a_2 are negative, we obtain from (5.14)

$$(5.15)\quad E(\exp\{-\delta T_{11} - \delta_2 T_{12}\}) = \frac{\mu_{11}(\nu_{21} + \mu_{21} + \delta_2) + \nu_{12}\mu_{21}}{(\delta_1 + \nu_{12} + \mu_{11})(\delta_2 + \nu_{21} + \mu_{21}) - \nu_{12}\nu_{21}}.$$

One can easily invert this transform to give explicitly the joint distribution of T_{11} and T_{12}. However, we shall not venture into this here. Instead, we refer the reader to [21] for further details concerning the sojourn times of the type T_{11}, T_{12}, and so forth, and close with the remark that T_{11} and T_{12}, in the present case, are positively correlated. Furthermore, marginally each one of them is (negative) exponentially distributed with a positive probability mass at zero only in the case of T_{12}.

REFERENCES

[1] N. T. J. Bailey, *The Mathematical Theory of Epidemics*, New York, Hafner, 1957.

[2] G. E. Bates and J. Neyman, "Contributions to the theory of accident proneness II. True or false contagion," *Univ. California Publ. Statist.*, Vol. 1 (1952), pp. 255–276.

[3] C. L. Chiang, *Introduction to Stochastic Processes in Biostatistics*, New York, Wiley, 1968.

[4] D. J. Daley, "The total waiting time in a busy period of a stable single-server queue, I," *J. Appl. Prob.*, Vol. 6 (1969), pp. 550–564.

[5] D. J. Daley and D. R. Jacobs, Jr., "The total waiting time in a busy period of a stable single-server queue, II," *J. Appl. Prob.*, Vol. 6 (1969), pp. 565–572.

[6] D. van Dantzig, "Sur la méthode des fonctions génératrices," *Colloques Internationaux du CNRS*, No. 13 (1948), pp. 29–45.

[7] W. Feller, *An Introduction to Probability Theory and Its Applications*, Vol. 2, New York, Wiley, 1966.

[8] E. Fix and J. Neyman, "A simple stochastic model of recovery, relapse, death and loss of patients," *Human Biology*, Vol. 23 (1951), pp. 205–241.

[9] D. P. Gaver, "Highway delays resulting from flow-stopping incidents," *J. Appl. Prob.*, Vol. 6 (1969), pp. 137–153.

[10] D. Jerwood, "A note on the cost of the simple epidemic, *J. Appl. Prob.*, Vol. 7 (1970), pp. 440–443.

[11] J. H. B. Kemperman, *The Passage Problem for a Stationary Markov Chain*, Chicago, University of Chicago Press, 1961.

[12] D. R. McNeil, "Integral functionals of birth and death processes and related limiting distributions," *Ann. Math. Statist.*, Vol. 41 (1969), pp. 480–485.

[13] P. A. P. Moran, *The Theory of Storage*, London, Methuen, 1959.

[14] E. Naddor, *Inventory Systems*, New York, Wiley, 1966.

[15] P. S. Puri, "On the homogeneous birth-and-death process and its integral," *Biometrika*, Vol. 53 (1966), pp. 61–71.

[16] ———, "A class of stochastic models of response after infection in the absence of defense mechanism," *Proceedings of the Fifth Berkeley Symposium on Mathematical Statistics and Probability*, Berkeley and Los Angeles, University of California Press, 1967, Vol. 4, pp. 511–535.

[17] ———, "Some limit theorems on branching processes and certain related processes," *Sankhyā*, Vol. 31 (1969), pp. 57–74.

[18] ———, "A quantal response process associated with integrals of certain growth processes," *Proceedings of the Symposium Mathematical Aspects of Life Sciences*, held at Queen's University, to appear.

[19] ———, "Some new results in the mathematical theory of phage-reproduction," *J. Appl. Prob.*, Vol. 6 (1969), pp. 493–504.

[20] ———, "A method for studying the integral functionals of stochastic processes with applications: I. Markov chains case," *J. Appl. Probability*, Vol. 8 (1971), pp. 331–343.

[21] ———, "A method for studying the integral functionals of stochastic processes with applications: II. Sojourn time distributions for Markov chains," to appear in *Z. Wahrscheinlichkeitstheorie und verw. Gebiete*.

[22] T. L. Saaty, *Elements of Queueing Theory with Applications*, New York, McGraw Hill, 1951.

USES OF THE SOJOURN TIME SERIES FOR MARKOVIAN BIRTH PROCESS

W. A. O'N. WAUGH
UNIVERSITY OF TORONTO

1. Introduction

This paper will be concerned with the Markovian birth process, and in this section we shall establish notation and mention some properties of the process. We suppose that a sequence $\{\lambda_j : j = 1, 2, \cdots\}$ of positive constants is given. Development of the process Z_t is controlled by the conditions

$$(1.1) \qquad P\{Z_{t+\delta t} = k \mid Z_t = j\} = \begin{cases} \lambda_j \delta t + o(\delta t) & \text{when} \quad k = j + 1, \\ 1 - \lambda_j \delta t + o(\delta t) & \text{when} \quad k = j, \\ o(\delta t) & \text{when} \quad k \neq j + 1, j. \end{cases}$$

We suppose that $Z_0 = 1$. In view of well-known applications of this model, it is sometimes convenient to refer to Z_t as the population size.

Let T_n be the epoch of the nth jump in the process Z_t for $n = 1, 2, \cdots$, and write $T_0 = 0$. Let X_n be the sojourn time in state n, that is to say, $X_n = T_n - T_{n-1}$. A well-known property of the process is that the X_n are independent and that

$$(1.2) \qquad P\{X_j \leqq x\} = 1 - e^{-\lambda_j x}.$$

The mean and variance of the jth sojourn time are

$$(1.3) \qquad EX_j = \lambda_j^{-1}, \qquad \text{Var } X_j = \lambda_j^{-2},$$

respectively. In this paper, we shall make use of the random series formed by the sojourn times when centered at their means. The nth partial sum S_n of this series is given by

$$(1.4) \qquad S_n = \sum_{j=1}^{n} (X_j - EX_j) = T_n - ET_n.$$

An important property of the birth process is whether or not it is "honest," that is, whether or not

$$(1.5) \qquad \sum_{n=1}^{\infty} P\{Z_t = n\} = 1 \qquad \text{for all} \quad t \geqq 0.$$

This paper was written at the University of British Columbia, during the author's tenure of a Fellowship in the Canadian Mathematical Congress Summer Research Institute, held there in 1970. Additional support was provided by the National Research Council of Canada under NRC Grant A 5304.

The next section of the paper will provide an outline of some previous applications of the sojourn time series, including the well-known criterion for Z_t to be honest. We shall also introduce a further subdivision of the class of honest processes according to the behavior of the series (1.4).

In Section 3, we shall outline certain facts about summation formulas in a form suitable for our particular purpose. Then in the main part of the paper (Sections 4, 5, and 6) we will use the series (1.4) to study limiting behavior, for large values of the time, in one of the two classes into which we divide the honest processes. The theory will be developed in Sections 4 and 5, and some comments and numerical studies will be given in Section 6.

2. Some previous applications of the sojourn time series

The invitation to present a paper at this Symposium included the suggestion that it should contain a brief summary of results in the topic in the five years since the last Symposium. However, it seems appropriate to start earlier with its presentation in 1951 in the first edition of Feller's *Introduction to Probability Theory and Its Applications*. In that edition, the following theorem is proved.

THEOREM 2.1. *In order that* (1.5) *shall hold* (*that is, that the process shall be honest*), *it is necessary and sufficient that the series*

$$\sum_{j=1}^{\infty} \lambda_j^{-1} \tag{2.1}$$

diverge.

In accordance with the limitation to countable sample spaces which Feller adopted in his Volume 1, he did not introduce the sojourn time series, and Theorem 2.1 was proved by analytic manipulations based on the differential difference equation for $P\{Z_t = n\}$. However, in later editions (Feller [2]), the fact that $\lambda_j^{-1} = EX_j$ is noted and some heuristic remarks are added.

In [3], Theorem 2.1 reappears. It is proved as an application of Laplace transforms to the Kolmogorov differential equations but the sojourn times are also mentioned. We may quote in the present notation ([3], Chapter VIII, Section 5): "If lim $ET_n < \infty$, the distribution of T_n tends to a proper limit G. Then $G(t)$ is the probability that infinitely many jumps will occur before epoch t." Also in [3], the probability $P\{Z_t = n\}$ is obtained explicitly by noting that it is the same as $P\{T_{n-1} \leqq t,\ T_n > t\}$ and evaluating the latter as a sum of exponentials.

The sojourn time series can be made the primary tool for a proof of Theorem 2.1 and it is used thus, for example, in the text by Breiman [1]. Here a generalization to birth and death processes also occurs (see also John [6]). We will not reproduce proofs that are easily available. The idea is simple: the basic step is to use Chebyshev's inequality to show that finiteness of the series of means (2.1) implies that T_n has an a.s. finite limit.

More precise information can also be obtained from the sojourn time series. P. W. M. John [7] showed that when $\lambda_j = \lambda j^2$, the defect $1 - \Sigma P\{Z_t = n\}$ can be obtained explicitly in terms of the Jacobi theta function $\theta_4(0, e^{-\lambda t})$. The author (Waugh [8]) used the theory of Hirschman and Widder [5] about convolutions of negative exponential densities to estimate the tail of the density of T_n in the "dishonest" case and also to estimate the rate at which honest processes grow. An essential part of the latter investigation was the division of the birth processes into three classes, and we shall require this in the present paper. The classes are:

(2.2) $\bar{H}$: the dishonest or divergent birth processes, for which $\Sigma \lambda_j^{-1} < \infty$ and so *a fortiori* $\Sigma \lambda_j^{-2} < \infty$, and ΣX_j is convergent a.s., no centering being required; here we write a.s. to mean "with probability 1";

(2.3) H_c: honest processes for which $\Sigma \lambda_j^{-1} = \infty$ but $\Sigma \lambda_j^{-2} < \infty$ so that the centered sojourn time series $\Sigma (X_j - EX_j)$ converges a.s. to a random variable S;

(2.4) H_d: honest processes for which both $\Sigma \lambda_j^{-1} = \infty$ and $\Sigma \lambda_j^{-2} = \infty$.

The process with linear birth rates $\lambda_j = j\lambda$ is both a birth process belonging to H_c and a branching process. As a branching process, it is known to have associated with it a random variable W such that $Z_t/EZ_t \to W$ a.s. as $t \to \infty$. (See, for example, Harris [4].) Recently the author [9] used sojourn time considerations to show that $W = \exp\{-\lambda S - \gamma\}$, where γ is Euler's constant. The rest of the present paper is concerned with the generalization of this result to the class H_c.

3. Summation formulas

We shall require the Euler-Maclaurin summation formula and some related results. In this section, we shall state them in a form suitable for this particular problem.

Let f be a real valued function defined on $[1, \infty)$ such that

$$f(n) = \lambda_n^{-1},$$
$$f(x) > 0, \tag{3.1}$$
$$f(x) \to 0 \quad \text{as} \quad x \to \infty,$$
f is absolutely continuous.

One such function is the trapezoidal approximation given by

$$g(x) = (x - j)(\lambda_{j+1}^{-1} - \lambda_j^{-1}) + \lambda_j^{-1} \tag{3.2}$$

for $j \leqq x \leqq j + 1$, where $j = 1, 2, \cdots$. We shall be considering processes belonging to the class H_c, and in view of the conditions (2.3) on the sequence $\{\lambda_n\}$, it will be seen that g satisfies the conditions (3.1). In general, there will be an infinity of possible functions f for a given sequence $\{\lambda_n\}$. We shall make particular use of the function g, but also will find it convenient to use other functions f.

The Euler-Maclaurin summation formula is

$$\sum_{j=1}^{n} f(j) = \int_1^n f(x)\,dx + \tfrac{1}{2}[f(1) + f(n)] + \int_1^n \left(x - [x] - \tfrac{1}{2}\right) f'(x)\,dx. \tag{3.3}$$

Here a square bracket enclosing a single symbol denotes the greatest integer function. Equation (3.3) holds, of course, under wider conditions than (3.1). Let

$$F(x) = \int_1^x f(u)\,du \tag{3.4}$$

and

$$k(n) = \int_1^n \left(x - [x] - \tfrac{1}{2}\right) f'(x)\,dx + \tfrac{1}{2}[f(1) + f(n)]. \tag{3.5}$$

Then the Euler-Maclaurin formula (3.3) reads

$$\sum_{j=1}^{n} f(j) - F(n) = k(n). \tag{3.6}$$

We shall be concerned with functions f for which there is a corresponding constant k such that $k(n) \to k$ as $n \to \infty$.

EXAMPLE 3.1. If $\lambda_j = j^{-1}$ then we can take $f(x) = x^{-1}$, giving $F(x) = \log x$. It is well known that $k(n) \to \gamma$, which is Euler's constant.

EXAMPLE 3.2. For the trapezoidal function g, we have the integral

$$G(x) = \tfrac{1}{2}\lambda_1^{-1} + \sum_{j=2}^{[x]-1} \lambda_j^{-1} + \tfrac{1}{2}\lambda_{[x]}^{-1} + \tfrac{1}{2}(x - [x])\left(\lambda_{[x]}^{-1} + \lambda_{[x+1]}^{-1}\right). \tag{3.7}$$

The integral portion of the remainder term in (3.3) vanishes and (3.3) or equivalently (3.6) is just

$$\sum_{j=1}^{n} g(j) - G(n) = \tfrac{1}{2}(\lambda_1^{-1} + \lambda_n^{-1}) \to \lambda_1^{-1} \tag{3.8}$$

as $n \to \infty$.

4. A limit theorem for the class H_c

4.1. *The limit theorem.*

THEOREM 4.1. *Let Z_t be an honest birth process for which the centered sum of sojourn times is convergent, that is, $\{Z_t : t \geqq 0\} \in H_c$. Let f be a function satis-*

fying (3.1) *for which* (3.6) *converges. Let F be the integral* (3.4) *and k the corresponding constant. Then*

$$\lim_{t\to\infty} \{F(Z_t) - t\} = -S - k \qquad a.s. \tag{4.1}$$

PROOF. Since T_j is the epoch at which the population size jumps to $j + 1$, we have the equivalent events

$$\{Z_t = n\} \Leftrightarrow \{T_{n-1} \leqq t, T_n > t\}. \tag{4.2}$$

Hence,

$$\begin{aligned} Z_t &= \max \{n: T_{n-1} \leqq t\} \\ &= \max \{n: t - ET_{n-1} - S_{n-1} \geqq 0\}. \end{aligned} \tag{4.3}$$

Now $ET_{n-1} = \lambda_1^{-1} + \lambda_2^{-1} + \cdots + \lambda_{n-1}^{-1}$. Hence, for each t, there is a largest integer n for which the inequality in (4.3) holds because $t - ET_{n-1} \to -\infty$, while $S_{n-1} \to S$ which is finite a.s. Also, $Z_t \to \infty$ as $t \to \infty$. We can rewrite (4.3) as

$$Z_t = \max \{n: t - S - (S_{n-1} - S) \geqq ET_{n-1}\}. \tag{4.4}$$

Using (3.6), we have $ET_{n-1} = F(n - 1) + k(n - 1)$ which gives

$$\begin{aligned} Z_t &= \max \{n: t - S - (S_{n-1} - S) \geqq F(n-1) + k(n-1)\} \\ &= \max \{n: F(n-1) \leqq t - S - k + S - S_{n-1} - k(n-1) + k\}. \end{aligned} \tag{4.5}$$

Let $\varepsilon > 0$. Since $S_{n-1} \to S$ a.s. and $k(n - 1) \to k$, as $n \to \infty$, there is a.s. an integer n_0 such that

$$|S - S_{n-1} - k(n-1) + k| < \varepsilon \tag{4.6}$$

for all $n \geqq n_0$. Thus, for $t > T_{n_0 - 1}$, we have

$$\begin{aligned} \max \{n: F(n-1) \leqq t - S - k - \varepsilon\} &< Z_t \\ &< \max \{n: F(n-1) \leqq t - S - k + \varepsilon\} \end{aligned} \tag{4.7}$$

a.s.

Now, since f never vanishes, F is strictly monotone increasing; hence, it has a well-defined inverse function F^{-1} (also increasing) and we can write (4.7) as

$$\begin{aligned} \min \{n: n > F^{-1}(t - S - k - \varepsilon)\} &< Z_t \\ &< \min \{n: n > F^{-1}(t - S - k + \varepsilon)\}. \end{aligned} \tag{4.8}$$

Now

$$F^{-1}(t - S - k - \varepsilon) < \min \{n: n > F^{-1}(t - S - k - \varepsilon)\} \tag{4.9}$$

and the two sides of (4.9) differ by at most 1. Similarly,

$$\min \{n: n > F^{-1}(t - S - k + \varepsilon)\} \leqq F^{-1}(t - S - k + \varepsilon) + 1 \tag{4.10}$$

so that (4.8) gives

$$F^{-1}(t - S - k - \varepsilon) < Z_t < F^{-1}(t - S - k + \varepsilon) + 1. \tag{4.11}$$

Since F^{-1} is strictly monotone increasing, this gives the two inequalities

$$\begin{aligned} t - S - k - \varepsilon &< F(Z_t), \\ F(Z_t - 1) &< t - S - k + \varepsilon. \end{aligned} \tag{4.12}$$

Now

$$F(Z_t) - F(Z_t - 1) = \int_{Z_t - 1}^{Z_t} f(u)\, du \to 0 \tag{4.13}$$

a.s. as $t \to \infty$, since $f(u) \to 0$ as $u \to \infty$, while $Z_t \to \infty$ a.s. Applying this to (4.12), we can clearly obtain

$$t - S - k - 2\varepsilon < F(Z_t) < t - S - k + 2\varepsilon \tag{4.14}$$

a.s. for all sufficiently large t, and since ε was arbitrary this proves (4.1).

4.2. *Approximation based on* (4.1). *Branching processes.* The limit (4.1) can be written as an approximation valid for large t:

$$Z_t \approx F^{-1}(t - S - k) \tag{4.15}$$

so that for large values of the time, the stochastic fluctuations of the population size are accounted for by the single random variable S. Now as mentioned in Section 2, for a large class of *branching processes* it is known that there is a random variable W such that the population size Y_t suitably reduced has W as its limit

$$\frac{Y_t}{EY_t} \to W \tag{4.16}$$

a.s. as $t \to \infty$. Thus, the branching processes and the class H_c of birth processes have in common this property that their values for all sufficiently large values of time can be approximated by a function of t and of a single random variable. Later (Section 5.2, Case (i)), we shall apply our results to the Markovian binary fission process, which is both a branching process and a birth process of the class H_c, and compare (4.16) with the particular version of (4.15).

4.3. *"Stochastic lag" in the development of a population.* For many branching processes, EY_t is proportional to $e^{\rho t}$ where ρ is a constant (the "Malthusian parameter"). Thus for large t, (4.16) can be written as an approximation

$$\log Y_t \approx \rho t + \log W. \tag{4.17}$$

In biological experiments, the logarithm of the population size is sometimes plotted and observed to follow, approximately, a straight line. Here the population, perhaps supposed to descend from a single individual at some past time, has already reached a substantial size at the time of observation. If growth were actually deterministic, commencing with a single individual at time zero, the logarithm of the population size would simply be represented by a straight line through the origin. However, it will be seen that the approximate value of $\log Y_t$ in (4.17) vanishes for a time T given by

$$T = -\rho^{-1} \log W. \tag{4.18}$$

This hypothetical starting time for the population will be called the (stochastic) *lag*. It is the counterpart, for the model, of the intercept on the time axis obtained by extrapolating back an observed graph from an experiment. Such observations are frequently interpreted in the biological literature as indicative of a disturbance in the conditions of growth at the outset (while the population, for example a clone of malignant tissue cells, is too small to be observed). Thus, it is important to investigate the distribution of T.

Reference to (4.15) will show that our size dependent birth processes also exhibit a lag phenomenon, in fact, since $F(1) = 0$, T will always be given by

$$T = S + k. \tag{4.19}$$

In Section 5, we shall obtain T explicitly for the special birth processes that we introduce there.

5. Application of the limit theorem in some special cases

5.1. *General* $\{\lambda_n\}$, *trapezoidal approximation.* Substituting the function G of (3.7) in the limit (4.1), we obtain

$$\tfrac{1}{2}\lambda_1^{-1} + \sum_{j=2}^{Z_t - 1} \lambda_j^{-1} + \tfrac{1}{2}\lambda_{Z_t}^{-1} - t \to -S - \tfrac{1}{2}\lambda_1^{-1}. \tag{5.1}$$

Since $Z_t \to \infty$ and hence $\frac{1}{2}\lambda_{Z_t}^{-1} \to 0$ a.s., we can add $\frac{1}{2}\lambda_{Z_t}^{-1}$ to either side and state the resulting theorem.

THEOREM 5.1. *As t tends to infinity,*

$$\sum_{j=1}^{Z_t} \lambda_j^{-1} - t \to -S \qquad a.s. \tag{5.2}$$

PROOF. The simplicity of the limit (5.2) suggests the possibility of a proof without the use of (4.1) and this can in fact be given as follows. At any time t, T_{Z_t} is the epoch of the next jump and X_{Z_t} is the duration of the current sojourn time. Thus, we can write

$$t = T_{Z_t} - \theta X_{Z_t}, \tag{5.3}$$

where $0 \leqq \theta \leqq 1$. Thus,

$$\begin{aligned} \sum_{j=1}^{Z_t} \lambda_j^{-1} - t &= \theta X_{Z_t} - \left(T_{Z_t} - \sum_{j=1}^{Z_t} \lambda_j^{-1}\right) \\ &= \theta X_{Z_t} - S_{Z_t} \to -S \end{aligned} \tag{5.4}$$

a.s. as $t \to \infty$, since $Z_t \to \infty$ and $X_{Z_t} \to 0$ a.s.

5.2. *λ_n proportional to a power of n.* Let $\lambda_j = (j\lambda)^\alpha$ for a fixed constant $\lambda > 0$. The conditions (2.3) are satisfied provided that $\frac{1}{2} < \alpha \leqq 1$. We can take $f(x) = (\lambda x)^{-\alpha}$ for all positive x and, of course, we can obtain F in closed form. There are two cases which we shall treat separately. If $\alpha = 1$ then the birth process is also a branching process, the birth rate *per head* per unit time is independent of the population size. If $\alpha \neq 1$ this is not so.

Case (i): $\alpha = 1$. (See Example 3.1). We have $F(x) = \lambda^{-1} \log x$ and $k = \gamma/\lambda$. Hence using Theorem 4.1, we get

$$\log Z_t - \lambda t \to -\lambda S - \gamma \tag{5.5}$$

a.s. as $t \to \infty$.

To compare (5.5) with the result (4.16) from the theory of branching processes, note that $EZ_t = e^{\lambda t}$ so that (4.16) is

$$Z_t e^{-\lambda t} \to W \tag{5.6}$$

a.s. as $t \to \infty$.

Thus, (5.5) is equivalent to (5.6) with $W = \exp\{-\lambda S - \gamma\}$. As mentioned in Section 2, this special case of the limit (4.1) will be found in Waugh [9]. The limit can be stated as an approximation for large t as $Z_t \approx \exp\{\lambda t - \lambda S - \gamma\}$ and the lag is $T = S + (\gamma/\lambda)$.

Case (ii): $\frac{1}{2} < \alpha < 1$. In this case, we have

$$\begin{aligned} F(x) &= \int_1^x (\lambda u)^{-\alpha}\, du \\ &= \lambda^{-\alpha}(1-\alpha)^{-1}(x^{1-\alpha} - 1) \end{aligned} \tag{5.7}$$

and the remainder term in the Euler-Maclaurin formula (3.6) depends on α. We shall write the remainder term and its limit as $k_\alpha(n) \to k_\alpha$ as $n \to \infty$. The limit theorem (4.1) gives

$$\lambda^{-\alpha}(1-\alpha)^{-1}(Z_t^{1-\alpha} - 1) - t \to -S - k_\alpha. \tag{5.8}$$

As an approximation for large t, this can be written

$$Z_t \approx \{\lambda^\alpha(t - S - k_\alpha)(1-\alpha) + 1\}^{1/(1-\alpha)}. \tag{5.9}$$

The lag is $T = S + k_\alpha$. Since $k_\alpha \to \gamma/\lambda$ as $\alpha \to 1$, we see that this lag, and also (5.9), have the corresponding expressions for Case (i) as their limits.

Note that for a continuously growing deterministic population model given by

$$\frac{dy}{dt} = (\lambda y)^\alpha, \qquad y(0) = 1, \tag{5.10}$$

the population size is

$$y(t) = \{\lambda^\alpha t(1-\alpha) + 1\}^{1/(1-\alpha)} \tag{5.11}$$

which bears the same relation to the stochastic approximation (5.9) as the exponential $e^{\lambda t}$ does to Z_t in the branching case.

6. Comments, and numerical studies of Theorem 4.1

From the statement of Theorem 4.1 and from the examples of Section 5, it will be seen that the expression for the limit that is obtained is somewhat arbitrary, being determined by the choice of the function f. Clearly, it is only the expression obtained that is arbitrary and two approximations to Z_t stemming from different choices of f must approach one another in the limit as $t \to \infty$. Nevertheless, it has considerable implications, in particular for the stochastic lag. Recalling (4.19), that $T = S + k$, it will be seen that the constant k, which is determined by the choice of the function f, enters into the lag. For example, in Case (i) of Section 5.2 where $\lambda_j = j\lambda$ and $f(x) = (\lambda x)^{-1}$, we obtain $k = \gamma/\lambda$. The trapezoidal approximation g determined by the same sequence $\{\lambda_j\}$ gives $k = (2\lambda)^{-1}$. In any case, since $ES = 0$, we have

$$ET = k. \tag{6.1}$$

Of course, this arbitrariness of the stochastic lag corresponds to the fact that the observed lag is determined by a process of extrapolation. There is often a natural choice of fitted function, as, for example, when a biologist fits a straight line to log Y_t for a branching process. Similarly, the function (5.7) of Case (ii) and the approximation (5.9) arise naturally and, in fact, specialize to the case just mentioned, of a straight line for log Z_t, when $\alpha = 1$.

Some simulations of populations growing with birth rates as in Case (ii) of Section 5.2 were made and compared with the fitted function (5.9). These illustrate various points. The simulated populations settled down to approximately continuous growth quite quickly, say, after time $t = 6.0$ when $\alpha = 0.9$ and $t = 10.0$ when $\alpha = 0.7$. The time scale is determined by $\lambda = 1$ which gives a mean life length of 1 in the branching case.

When $\alpha < 1$, (5.9) shows that the approximation to Z_t grows as a power of t, whereas for $\alpha = 1$ growth is as $e^{\lambda t}$. Nevertheless, it will be seen that for $\alpha = 0.9$, over a range which is likely to be of interest in studying cellular colonies, growth of log Z_t is approximately linear. Thus, if growth is moderately size dependent, the error involved in treating it as a branching process will not be too serious. For $\alpha = 0.7$ the departure of log Z_t from linearity is more marked. Note that the birth rate *per head* is given by $n^{-1}\lambda_n = \lambda^\alpha n^{\alpha-1}$, which is independent of n when $\alpha = 1$ (branching case) and the greatest dependence occurs as $\alpha \to \frac{1}{2}$ when the birth rate per head is approximately proportional to $1/\sqrt{n}$.

The simulations provided a sample of values of T. Two samples of 100 each were taken in the case $\alpha = 0.7$, giving sample means of 0.4660 and 0.7222. The value of $ET = k_\alpha$ for $\alpha = 0.70$ is 0.5549. It should be noted that Var T is quite large relative to this mean, being given by

$$\text{Var } T = \lambda^{-2\alpha} \sum_{j=1}^{\infty} j^{-2\alpha} = \lambda^{-2\alpha}\zeta(2\alpha), \tag{6.2}$$

where ζ is the Riemann zeta function. For $\alpha = 0.7$ and $\lambda = 1$ we have S.D.$(T) \approx 1.75$.

The mean k_α is notably insensitive to the value of α in the range $\frac{1}{2} < \alpha \leqq 1$, being 0.5396 ($\alpha = 0.50$) and 0.5772 ($\alpha = 1.00$) and varying very nearly linearly. In view of the variance of T, an approximate mean lag of 0.55 for all degrees of size dependence in Case (ii) might be adopted without much error.

The four figures illustrate simulations as follows. Figure 1 was made with $\alpha = 1$ and is, thus, just a simulation of a branching process. Figure 2 is for $\alpha = 0.9$ and Figures 3 and 4 are both for $\alpha = 0.7$ to show two of the fitted functions which are shown as dotted lines, and which meet the time axis at different values of the lag T.

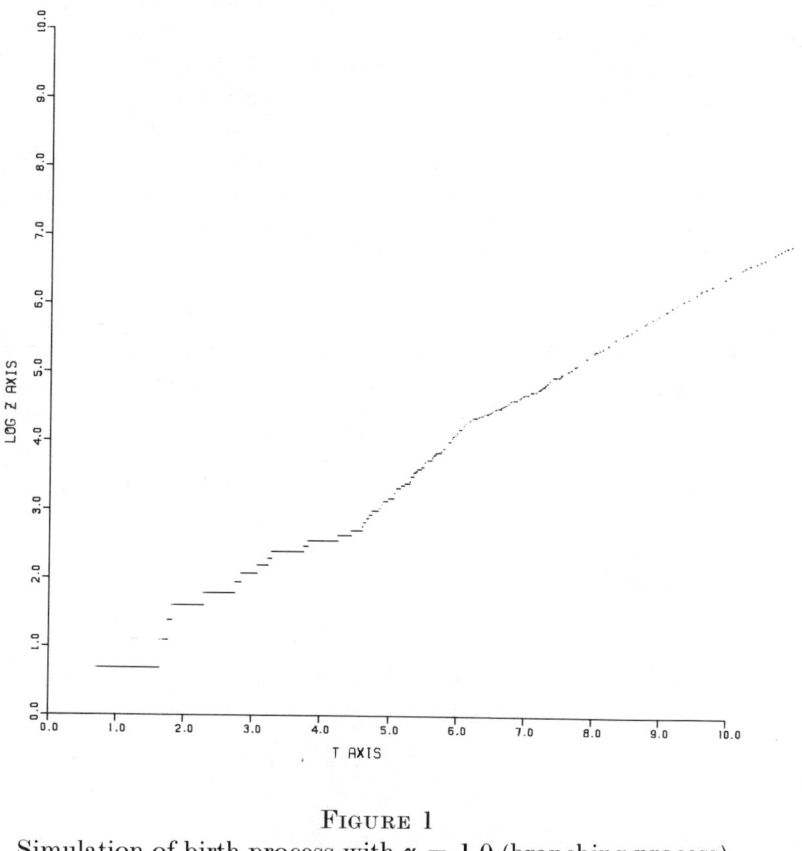

FIGURE 1
Simulation of birth process with $\alpha = 1.0$ (branching process).

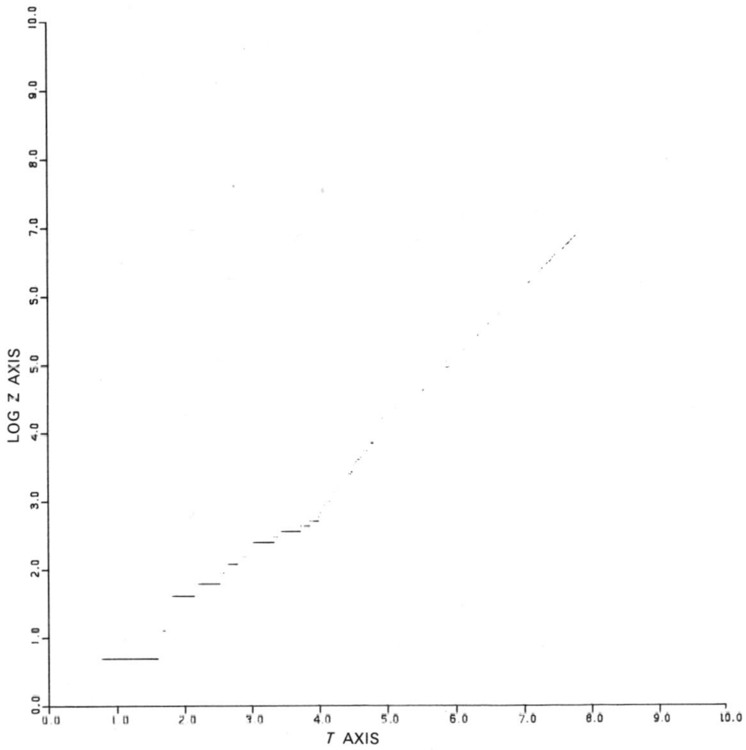

FIGURE 2
Simulation of birth process with $\alpha = 0.9$.

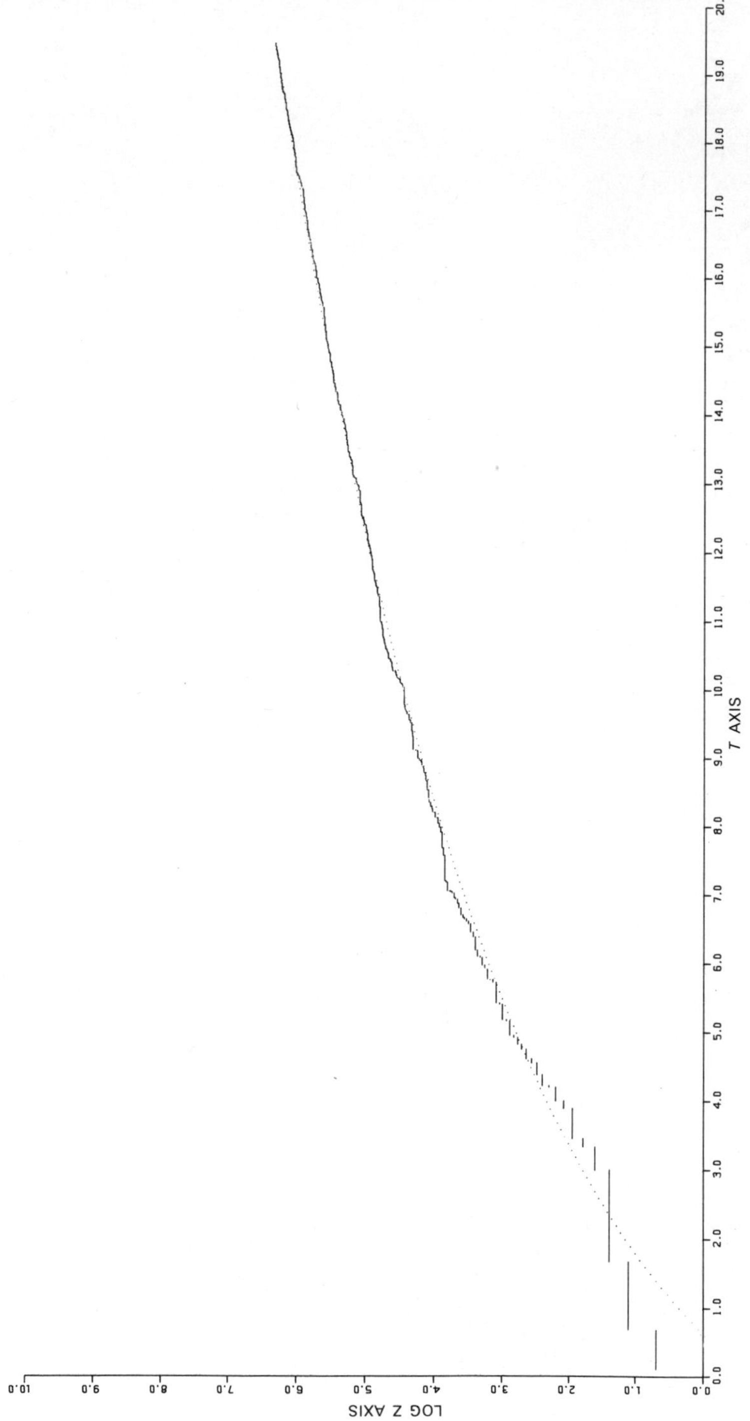

FIGURE 3
Simulation of birth process with $\alpha = 0.7$; the lag $T \approx 1.0$.

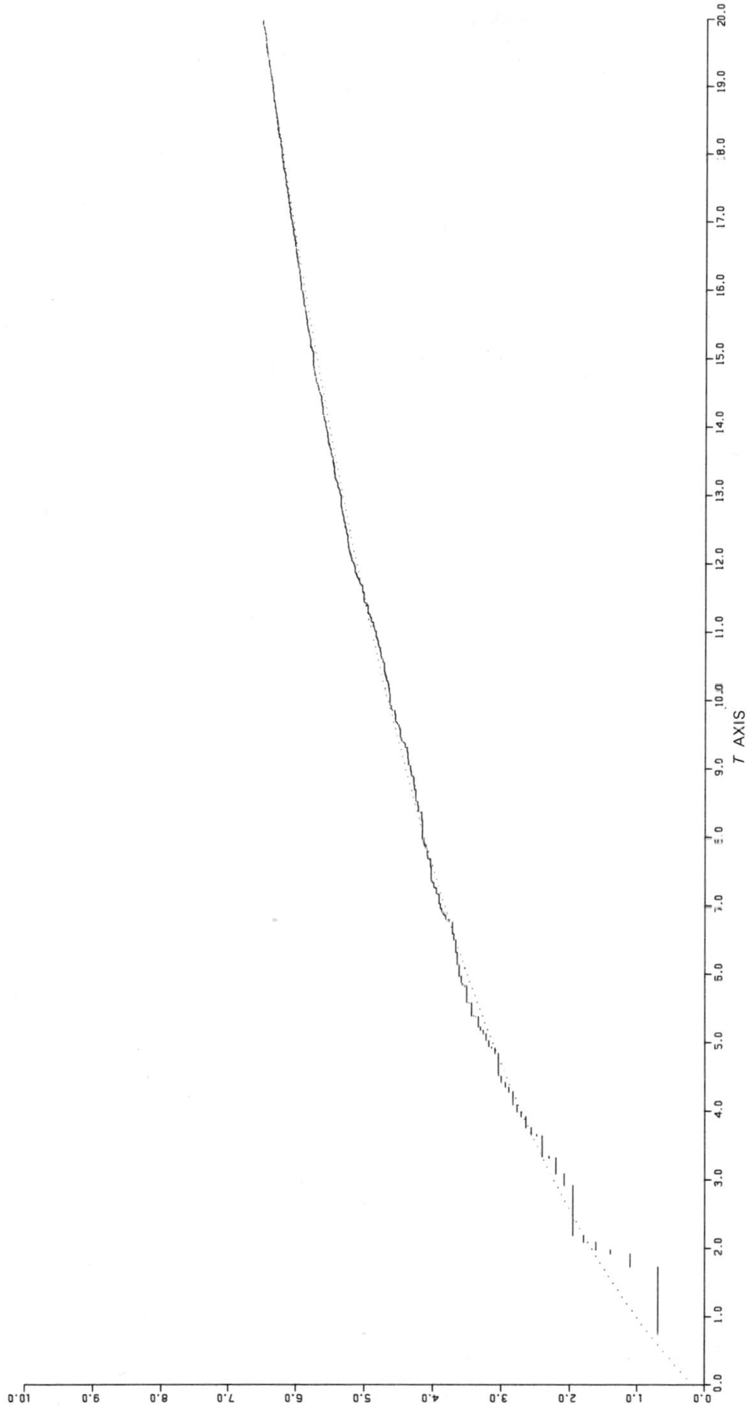

FIGURE 4

Simulation of birth process with $z = 0.7$; the lag T is negative.

REFERENCES

[1] L. BREIMAN, *Probability*, Cambridge, Addison-Wesley, 1968.
[2] W. FELLER, *An Introduction to Probability Theory and Its Applications*, Vol. 1, New York, Wiley, 1968 (3rd ed.).
[3] ———, *An Introduction to Probability Theory and Its Applications*, Vol. 2, New York, Wiley, 1966.
[4] T. E. HARRIS, *The Theory of Branching Processes*, Berlin-Heidelberg-New York, Springer-Verlag, 1963.
[5] I. I. HIRSCHMAN and D. V. WIDDER, *The Convolution Transform*, Princeton, Princeton University Press, 1955.
[6] P. W. M. JOHN, "Divergent time homogeneous birth and death processes," *Ann. Math. Statist.*, Vol. 28 (1957), pp. 514–517.
[7] ———, "A note on the quadratic birth process," *J. London Math. Soc.*, Vol. 36 (1961), pp. 159–160.
[8] W. A. O'N. WAUGH, "Growth rates for birth processes," *Abstracts of Short Communications*, International Congress of Mathematicians, Stockholm, 1962.
[9] ———, "Transformation of a birth process into a Poisson process," *J. Roy. Statist. Soc. Ser. B*, Vol. 32 (1970), pp. 418–431.

FIRST EMPTINESS PROBLEMS IN QUEUEING, STORAGE, AND TRAFFIC THEORY

J. GANI
MANCHESTER—SHEFFIELD
SCHOOL OF PROBABILITY AND STATISTICS

1. Introduction

One of the aims of the Berkeley Symposium is to encourage research workers to present a summary of results newly obtained in their fields during the previous five years. In accordance with this intention, the earlier part of the present paper will describe some interesting developments in problems of first emptiness since 1965. For simplicity, only first passage problems (to the zero state) for certain discrete time random walks on the integers $0, 1, 2, \cdots$, will be discussed. As is already known, first emptiness probabilities are of considerable importance in queueing, storage, and traffic problems. Their distributions may be interpreted

(a) in *queueing theory*, as probability distributions of the length of a busy period during which all waiting customers have been served, so that the queue is empty;

(b) in *storage theory*, as probability distributions of the times to first emptiness of a reservoir, all the stored water having been released;

(c) in *traffic theory*, as probability distributions of periods to a first gap at a "give way" intersection on a minor road, all vehicles crossing the road having passed, so that the intersection becomes empty and through traffic on the road can proceed.

A graphical representation of a random walk in discrete time of the type which arises in queueing, storage, and traffic processes is provided in Figure 1. Here $Z_0 = u$ is the initial state of the random walk at time $t = 0$; this represents the number of customers initially waiting for service in a queue, the units of water initially contained in a reservoir, or the number of vehicles initially waiting to cross a minor road, thus blocking the traffic along it.

The sequence of discrete nonnegative random variables $\{X_t\}_{t=0}^{\infty}$ constitutes the inputs into the system during the time intervals $(t, t + 1)$, $t = 0, 1, \cdots$. At the end of each time interval, there is a unit output if the random walk lies in any one of the states $1, 2, 3, \cdots$, or a zero output if it is in state zero. Inputs represent new arrivals at a queue, new water inflows into a reservoir, or new

Research supported by the Office of Naval Research and the Federal Highway Administration.

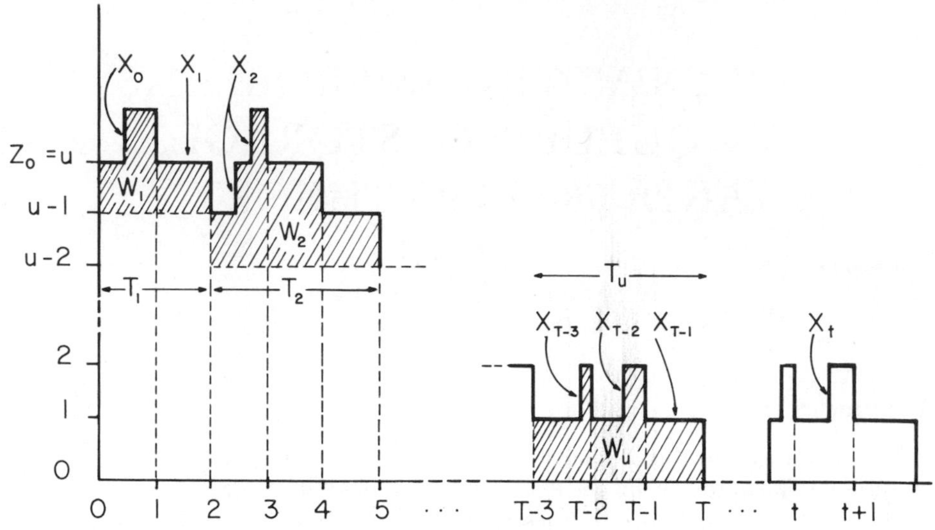

FIGURE 1
Random walk with first emptiness at $T = T(u)$.

vehicles preparing to cross a minor road in traffic; outputs denote a serviced customer in queueing, a released unit of water from a reservoir, or a vehicle which has crossed the minor road in traffic. For such processes, the random walk $\{Z_t\}_{t=0}^{\infty}$ may be characterized by the relation

$$Z_{t+1} = \{Z_t + X_t - 1\}^+, \qquad t = 0, 1, 2, \cdots, \tag{1.1}$$

where the positive index indicates the greater of $Z_t + X_t - 1$ and 0.

The time to first emptiness of the process starting from $Z_0 = u$ will be denoted by $T = T(u)$; it is clear that for $t = 0, 1, \cdots, T$,

$$Z_{t+1} = Z_t + X_t - 1, \tag{1.2}$$

with Z_t becoming zero for the first time when $t = T$. Note that in Figure 1 we have written T_j, $j = 1, \cdots, u$, for the first passage time of the random walk to state $u - j$ starting from state $u + 1 - j$, so that

$$T(u) = T_1 + T_2 + \cdots + T_u = \sum_{j=1}^{u} T_j. \tag{1.3}$$

The first part of this paper will deal mainly with recent research on the properties of $T(u)$ when the inputs $\{X_t\}$ form a Markov chain with a finite or denumerably infinite state space. We shall see that the newly derived results are similar to those previously known for independently and identically distributed (i.i.d.) inputs.

We shall also denote by W_j, $j = 1, \cdots, u$, the stochastic integral under the first passage path leading from state $u + 1 - j$ to state $u - j$ and lying above the line $u - j$ (see Figure 1). The path integral W_j corresponds to the first passage time T_j. Clearly, the stochastic integral $W(u)$ lying under the path to first emptiness starting from state u is given by

$$W(u) = \{W_1 + (u-1)T_1\} + \{W_2 + (u-2)T_2\} + \cdots + \{W_u\} \tag{1.4}$$
$$= \sum_{j=1}^{u} \{W_j + (u-j)T_j\}.$$

In the second part of the paper, we formulate some problems for stochastic path integrals of this type when the inputs $\{X_t\}$ are either i.i.d. or Markovian, and derive initial results for their probability distributions. Much remains to be done in this area; it is hoped that research workers will be encouraged to attack some of the many unsolved problems in the field.

2. Known results on first emptiness for i.i.d. inputs $\{X_t\}$

From Figure 1, it is intuitively obvious that if the inputs $\{X_t\}$ are i.i.d., the random variables T_j, $j = 1, \cdots, u$, will also be i.i.d: It follows that the distribution of $T(u)$ in (1.3) is the uth convolution of the distribution of T_i. This may be demonstrated more formally as follows. Let

$$g(\theta; u) = \sum_{T=u}^{\infty} g(T; u)\theta^T, \qquad 0 \leqq \theta \leqq 1, \tag{2.1}$$

be the probability generating function (p.g.f.) for the first emptiness time $T(u)$ of the random walk (1.1) with initial state $Z_0 = u > 1$, subject to i.i.d. inputs $\{X_t\}$. It is readily shown that

$$g(\theta; u) = \{g(\theta; 1)\}^u. \tag{2.2}$$

Assume τ to be the line of first descent of the random walk from state u to state 1; then clearly, the first emptiness probability $g(T; u)$ may be decomposed as

$$g(T; u) = \sum_{\tau=u-1}^{T-1} g(\tau; u-1)g(T-\tau; 1). \tag{2.3}$$

Forming the p.g.f. of this distribution, we obtain

$$g(\theta; u) = \sum_{T=u}^{\infty} \theta^T \sum_{\tau=u-1}^{T-1} g(\tau; u-1)g(T-\tau; 1) \tag{2.4}$$
$$= \sum_{\tau=u-1}^{\infty} g(\tau; u-1)\theta^\tau \sum_{T=\tau+1}^{\infty} g(T-\tau; 1)\theta^{T-\tau}$$
$$= g(\theta; u-1)g(\theta; 1).$$

Continuing the reduction of $g(\theta; u-1)$, we readily find result (2.2); this was originally derived in a somewhat different form by Kendall [13] in 1957.

Following (2.2), a second more interesting result, reminiscent of that holding for probabilities of first extinction in branching processes may be derived for the p.g.f. $g(\theta; 1)$. Namely, if $p(\theta) = \Sigma_{i=0}^{\infty} p_i \theta^i$ denotes the p.g.f. of any one of the random variables X_t, then $g(\theta; 1)$ satisfies the functional equation

$$g(\theta; 1) = \theta p\big(g(\theta; 1)\big) \tag{2.5}$$

subject to the condition $g(0; 1) = 0$. A proof of the equivalent result for a particular continuous time process in queueing may be found in the basic 1955 paper of Takács [24]; its derivation for the more general case was sketched by Kendall [13] in 1957. The argument in discrete time may be given very simply as follows. Let the input X_0 during the time interval (0, 1) in a random walk with initial state $Z_0 = 1$ be $i = 0, 1, 2, \cdots$. If the input is zero, first emptiness occurs at $T = 1$, but if $i \geqq 1$, the first emptiness process will continue from time $t = 1$, starting now from $Z_1 = i$. Thus, the p.g.f. $g(\theta; 1)$ will be given by the equation

$$g(\theta; 1) = p_0 \theta + \theta \sum_{i=1}^{\infty} p_i g(\theta; i); \tag{2.6}$$

but from (2.2), we see that this may be rewritten as

$$\begin{aligned} g(\theta; 1) &= p_0 \theta + \theta \sum_{i=1}^{\infty} p_i \{g(\theta; 1)\}^i \\ &= \theta p\big(g(\theta; 1)\big) \end{aligned} \tag{2.7}$$

leading to the result (2.5), where $g(0; 1) = 0$.

Precisely as in branching process theory, it is easily shown that

$$\begin{aligned} g(1-; 1) &= \zeta = 1 \quad &&\text{if} \quad E(X_t) = p'(1) \leqq 1, \\ g(1-; 1) &= \zeta < 1 \quad &&\text{if} \quad E(X_t) = p'(1) > 1. \end{aligned} \tag{2.8}$$

Takács [24] first obtained the explicit form of $g(T; 1)$ for Poisson inputs of fixed size, and Kendall [13] later derived the general formula (but in continuous time, from an integral equation) of type

$$g(T; u) = \frac{u}{T} p_{T-u}^{(T)}, \qquad u = 1, 2, \cdots, T, \tag{2.9}$$

where $p_{T-u}^{(T)} = Pr\,\{X_0 + X_1 + \cdots + X_{T-1} = T - u\}$ for arbitrary i.i.d. input distributions. Perhaps the simplest analytic method of obtaining this result is from the functional equation (2.5), using Lagrange's method of reversion of series. It has also been derived in an elementary manner by Lloyd [15] in 1963, using difference equation methods. But (2.9) is perhaps best viewed combinatorially as recording the proportion u/T of permissible paths leading to emptiness from among all those satisfying the condition $X_0 + X_1 + \cdots + X_{T-1} = T - u$.

An analysis of the restrictions on these paths is to be found in Gani [7]. Considering the case where first emptiness occurs at time T starting from $Z_0 = 1$ for simplicity, and rewriting the inputs as $Y_0 = X_{T-1}$ and $Y_1 = X_{T-2}, \cdots, Y_{T-1} = X_0$, we note that for emptiness to occur at T, it is necessary that

(2.10)

$$\begin{aligned} &Y_0 = 0, && 1 \leqq Y_{T-1} \leqq T-1,\\ &Y_0 + Y_1 \leqq 1, \quad \text{or, in slightly different terms,} && 2 \leqq Y_{T-1} + Y_{T-2} \leqq T-1,\\ &Y_0 + Y_1 + Y_2 \leqq 2, && 3 \leqq Y_{T-1} + Y_{T-2} + Y_{T-3} \leqq T-1,\\ &\vdots && \vdots\\ &Y_0 + Y_1 + \cdots + Y_{T-2} \leqq T-2, && Y_{T-1} + Y_{T-2} + \cdots + Y_1 = T-1,\\ &Y_0 + Y_1 + \cdots + Y_{T-1} = T-1, && Y_0 = 0. \end{aligned}$$

In 1963, Mott [21] showed by considering all cyclic permutations of the inputs $\{Y_j\}$ that the number of paths satisfying these conditions is precisely $1/T$ of all those for which $\Sigma_{j=0}^{T-1} Y_j = T - 1$, the probability of the latter being $p_{T-1}^{(T)}$. Thus,

$$\text{(2.11)} \qquad Pr\left\{\sum_{j=0}^{T-1} Y_j = T-1;\ \sum_{j=0}^{i} Y_j \leqq i, i = 0, 1, \cdots, T-2\right\}$$

$$= \frac{1}{T} Pr\left\{\sum_{j=0}^{T-1} Y_j = T-1\right\};$$

this can easily be generalized in a similar way to the case where $Z_0 = u$, leading to (2.9). We now outline the extension of these methods to the case of Markovian inputs $\{X_t\}$ since 1965.

3. Recent results on first emptiness for Markovian inputs $\{X_t\}$

It was Lloyd [16] in 1963 who first considered, in the context of storage theory, a random walk of the type (1.1) in which the inputs $\{X_t\}_{t=0}^{\infty}$ formed a Markov chain with a finite number of states. In a subsequent series of papers, Lloyd [17], [18], and Lloyd and Odoom [19] investigated the stationary properties of this random walk. The practical relevance of such Markovian inputs in queueing, storage, and traffic theory is obvious; a large number of arrivals for service at a queue during the time interval $(t - 1, t)$ may well discourage arrivals in the subsequent interval $(t, t + 1)$. In storage theory, there is much empirical evidence to show that annual water inflows into reservoirs are serially correlated; in traffic, advance warnings of congestion along a particular road often persuade motorists to find alternative routes to their destinations.

For convenience we shall assume, unless it is stated otherwise, that the input X_{-1} in the interval $(-1, 0)$ before the process $\{Z_t\}$ begins is zero. For inputs $\{X_t\}_{t=0}^{\infty}$ forming an irreducible Markov chain with stationary transition probabilities

$$p_{ij} = Pr\,\{X_{t+1} = j \mid X_t = i\}, \qquad i, j = 0, 1, \cdots, r, \tag{3.1}$$

it is known that the $T_j, j = 1, \cdots, u$, in Figure 1 are once again i.i.d. (see Chung [4]). It follows as before that the distribution of $T(u)$ in (1.3) will be the uth convolution of the distribution of T_1. As in the case of i.i.d. inputs, this may be proved formally as follows. Let

$$g(\theta; u, 0) = \sum_{T=u}^{\infty} g(T; u, 0)\theta^T, \qquad 0 \leqq \theta \leqq 1, \tag{3.2}$$

be the p.g.f. of the first emptiness probabilities $g(T; u, 0) = g(T; u, X_{-1} = 0)$ of $T(u)$, subject to Markovian inputs $\{X_t\}$. Again assuming τ to be the time of first descent to state 1, we may write

$$g(T; u, 0) = \sum_{\tau=u-1}^{T-1} g(\tau; u-1, 0)g(T-\tau; 1, 0). \tag{3.3}$$

Forming the p.g.f. with respect to T we find, much as in (2.2) through (2.4), that

$$\begin{aligned} g(\theta; u, 0) &= \sum_{T=u}^{\infty} \theta^T \sum_{\tau=u-1}^{T-1} g(\tau; u-1, 0)g(T-\tau; 1, 0) \\ &= g(\theta; u-1, 0)g(\theta; 1, 0) = \{g(\theta; 1, 0)\}^u. \end{aligned} \tag{3.4}$$

For Markovian inputs, Ali Khan and Gani [1] showed in 1968 that $g(\theta; 1, 0)$ satisfies a functional equation similar to (2.5), of the form

$$g(\theta; 1, 0) = \theta\lambda\big(g(\theta; 1, 0)\big) \tag{3.5}$$

subject to $g'(0; 1, 0) = p_{00}$. Here $\lambda(\theta)$ is the simple maximum eigenvalue of the positive matrix

$$\{p_{ij}\theta^j\}_{i,j=0}^r = \begin{bmatrix} p_{00} & p_{01}\theta & p_{02}\theta^2 & \cdots & p_{0r}\theta^r \\ p_{10} & p_{11}\theta & p_{12}\theta^2 & \cdots & p_{1r}\theta^r \\ \vdots & \vdots & \vdots & & \vdots \\ p_{r0} & p_{r1}\theta & p_{r2}\theta^2 & \cdots & p_{rr}\theta^r \end{bmatrix}, \qquad 0 < \theta \leqq 1, \tag{3.6}$$

such that $\lambda(1) = 1$ and $\lambda(0) = p_{00}$, where all p_{ij} may be taken positive for simplicity. This eigenvalue, though positive and strictly monotonic increasing for $\theta > 0$, is not in general a p.g.f.; it is shown in Gani [8] that when expanded in powers of θ, the generating function

$$\lambda(\theta) = \sum_{i=0}^{\infty} \frac{\lambda^{(i)}(0)\theta^i}{i!} \tag{3.7}$$

may have negative coefficients $\lambda^{(i)}(0)$ for $i \geqq 2$. When the $\{X_t\}$ are i.i.d. so that $p_{ij} = p_j$, equation (3.5) reduces to the better known functional relation (2.5). We now proceed to prove (3.5).

Following precisely the same approach as that leading to (2.6), we readily find for the random walk with Markovian inputs starting from $Z_0 = 1$ that

$$g(\theta; 1, 0) = p_{00}\theta + \theta \sum_{i=1}^{r} p_{0i} g(\theta; i, i), \tag{3.8}$$

where

$$g(\theta; i, i) = g(\theta; Z_1 = i, X_0 = i) \tag{3.9}$$

denotes the p.g.f. of first emptiness times starting from $Z_1 = i$, with prior input $X_0 = i$ instead of the usual zero. A decomposition similar to (3.4) yields

$$g(\theta; i, i) = g(\theta; 1, i)\{g(\theta; 1, 0)\}^{i-1}, \qquad i \geqq 1; \tag{3.10}$$

substituting this in (3.8), we are led directly to the relation

$$g(\theta; 1, 0) = \theta \sum_{i=0}^{r} p_{0i} g(\theta; 1, i)\{g(\theta; 1, 0)\}^{i-1}. \tag{3.11}$$

In exactly the same way, we may show that

$$g(\theta; 1, k) = \theta \sum_{i=0}^{r} p_{ki} g(\theta; 1, i)\{g(\theta; 1, 0)\}^{i-1}, \qquad k = 1, \cdots, r. \tag{3.12}$$

Hence, multiplying both (3.11) and (3.12) by $g(\theta; 1, 0)$ and setting out the results in matrix form, we obtain

$$g(\theta; 1, 0)\begin{bmatrix} g(\theta; 1, 0) \\ g(\theta; 1, 1) \\ \vdots \\ g(\theta; 1, r) \end{bmatrix} = \theta \begin{bmatrix} p_{00} & p_{01}g & \cdots & p_{0r}g^r \\ p_{10} & p_{11}g & \cdots & p_{1r}g^r \\ \vdots & \vdots & & \vdots \\ p_{r0} & p_{r1}g & \cdots & p_{rr}g^g \end{bmatrix} \begin{bmatrix} g(\theta; 1, 0) \\ g(\theta; 1, 1) \\ \vdots \\ g(\theta; 1, r) \end{bmatrix}, \tag{3.13}$$

where $g = g(\theta; 1, 0)$ is subject to the condition that $g'(0; 1, 0) = p_{00}$.

For (3.13) to hold, it is necessary that

$$|g\mathbf{I} - \theta \mathbf{p}\mathbf{G}| = 0, \tag{3.14}$$

where $\mathbf{I}$ is the unit matrix, $\mathbf{p} = \{p_{ij}\}_{i,j=0}^{r}$, and $\mathbf{G} = \text{diag}\{1, g, \cdots, g^r\}$. Hence, resolving $\mathbf{pG}$ spectrally in terms of its eigenvalues, we obtain (3.5) as required. Once again, as in (2.8),

$$\begin{aligned} g(1-; 1, 0) &= \zeta = 1 \qquad \text{if } \lambda'(1) \leqq 1, \\ g(1-; 1, 0) &= \zeta < 1 \qquad \text{if } \lambda'(1) > 1. \end{aligned} \tag{3.15}$$

It is proved in Gani [8], using Lagrange's method of reversion of series, that $g(T; u, 0)$ can be expressed in the form

$$g(T;u,0) = \frac{u}{T}\lambda^{(T)}_{T-u}, \qquad u = 1, 2, \cdots, T, \tag{3.16}$$

where $\lambda^{(T)}_{T-u}$ is the coefficient of θ^{T-u} in $\{\lambda(\theta)\}^T$. In a recent paper, Lehoczky [14] has indicated in the case of i.i.d. inputs, how this result may be given a combinatorial interpretation in terms of paths.

Result (3.5) was generalized in the summer of 1969 by Brockwell and Gani [3] to the case where the inputs $\{X_t\}$ form a Markov chain with a denumerable infinity of states. It is assumed for convenience that each $n \times n$ top left truncation of the transition probability matrix is irreducible for $n = 1, 2, 3, \cdots$. The method used is essentially that of n truncation of the relevant infinite vectors and matrices, followed by a limiting argument as $n \to \infty$. In this case, $\lambda(\theta)$ in (3.5) must be interpreted as the convergence norm of the infinite matrix

$$\{p_{ij}g^j\}_{i,j=0}^{\infty} = \begin{bmatrix} p_{00} & p_{01}g & p_{02}g^2 & \cdots \\ p_{10} & p_{11}g & p_{12}g^2 & \cdots \\ \cdots & \cdots & \cdots & \cdots \end{bmatrix}, \tag{3.17}$$

as defined by Vere-Jones [25], [26] where $g = g(\theta; 1, 0)$ remains the p.g.f. of first emptiness probabilities starting from $Z_0 = 1$ with $X_{-1} = 0$. An algorithm is obtained for the coefficients of $\lambda(\theta)$, and result (3.16) is shown to hold equally well for the case of a denumerable state space.

It may be of some interest to point out, following the analogy with extinction probabilities of the branching process mentioned in Section 2, that the present model may also be interpreted as a special type of population extinction process in discrete time. The population is now such that its progeny in consecutive time intervals is Markovian with one individual dying at the end of each interval. Whereas in a branching process each individual offspring produces its progeny independently, the present process differs in that the total progeny in one generation determines the offspring in the next.

4. New problems of stochastic integrals under first emptiness paths

The probabilistic properties of the stochastic integral under a first emptiness path have recently attracted some interest; this is the random area $W(u)$ of (1.4) enclosed under a path of the kind depicted in Figure 1. In queueing, such an area will represent the total amount of customer time (in man hours, say) lost by those waiting for service during a busy period; in storage, $W(u)$ is a measure of the total storage time capacity of a reservoir during a wet period; while for traffic it denotes the total vehicle time elapsed before an intersection is freed. Although several results are known for stochastic path integrals associated with continuous time processes, particularly of the birth and death type (see Bartlett [2], Daley [5], Daley and Jacobs [6], McNeil [20], Puri [22], [23]), few yet seem to have been obtained in the discrete time case. One of these, a result of Good's [11] in branching processes, which can also be interpreted as a stochastic path integral, has been pointed out to me by P. J. Brockwell (see also Harris [12], p. 32). In this section, we show that problems of the stochastic integral $W(u)$ effectively reduce to the study of a weighted sum of a set of constrained random variables.

Let us first examine the structure of $W(u)$ in (1.4). We have seen that for inputs $\{X_t\}$ both i.i.d. and Markovian, the passage times $T_j, j = 1, 2, \cdots, u$, are i.i.d.; the associated integrals W_j will also clearly be i.i.d., though each pair of random variables (T_j, W_j) will not be mutually independent. Thus, $W(u)$ can be considered as the sum of u independent random variables $\{W_j + (u - j)T_j\}$; if we could find the joint distribution of (T_j, W_j), the distribution of $W(u)$ would be known. In what follows, we write $W(1)$ for W_j, T for T_j, and denote by $W(1 \mid t)$ the random variable $W(1)$ conditioned on the particular value $T = t$ of the first emptiness time.

Let us assume for simplicity that during any unit time interval, inputs arrive in single units with independent uniformly distributed arrival times; then an input X_i during $(i, i + 1)$ will contribute the expected area $\frac{1}{2}X_i$. This may alternatively be assumed to be an approximation to the exact area in question. Thus for the time interval $(i, i + 1)$, the total area under the path will be $Z_i + \frac{1}{2}X_i$. It follows from (1.2), for a random walk starting from $Z_0 = 1$ and first emptying at $T = t$, that

$$
\begin{aligned}
&Z_0 = 1, \\
&Z_1 = Z_0 + X_0 - 1 = X_0, \\
(4.1)\qquad &Z_2 = Z_1 + X_1 - 1 = X_0 + X_1 - 1, \\
&\quad\vdots \\
&Z_{t-1} = Z_0 + X_0 + \cdots + X_{t-2} - (t - 1) = \sum_{i-0}^{t-2} X_i - (t - 2).
\end{aligned}
$$

Hence, we may write for $W(1 \mid t)$ the sum

$$
\begin{aligned}
(4.2)\qquad W(1 \mid t) &= \sum_{i=0}^{t-1} Z_i + \tfrac{1}{2}X_i \\
&= (t - \tfrac{1}{2})X_0 + (t - \tfrac{3}{2})X_1 + \cdots + \tfrac{3}{2}X_{t-2} - \tfrac{1}{2}t(t - 3) \\
&= \sum_{i=0}^{t-1} (i + \tfrac{1}{2})Y_i - \tfrac{1}{2}t(t - 3),
\end{aligned}
$$

where $Y_{t-i-1} = X_i$, $i = 1, \cdots, t - 1$, with $Y_0 = X_{t-1} = 0$; this sum will be subject to the usual constraints (2.10). It is clear that for a fixed value $T = t$ of the first emptiness time, $W(1 \mid t)$ is a weighted sum of constrained i.i.d. or Markovian random variables Y_i.

We can make use of (4.2) to obtain simple bounds for the moments of $W(1)$ or the joint moments of $(T, W(1))$, since, in summing the second set of inequalities in (2.10), we see that

$$\frac{1}{2}t(t-1) \leqq \sum_{i=0}^{t-1} iY_i \leqq (t-1)^2. \tag{4.3}$$

Since $\Sigma_{i=0}^{t-1} \frac{1}{2}Y_i = \frac{1}{2}(t-1)$, we finally obtain for $W(1 \mid t)$ of (4.2) the bounds

$$\frac{1}{2}(3t-1) \leqq W(1 \mid t) \leqq \frac{1}{2}(t^2+1), \tag{4.4}$$

where these represent the minimum and maximum areas contained by the stochastic paths for which $Y_{t-1} = Y_{t-2} = \cdots = Y_1 = 1$, and $Y_{t-1} = t-1$, and $Y_{t-2} = \cdots = Y_1 = 0$, respectively. We see from (4.4) that

$$E\left(\frac{1}{2^j}(3T-1)^j\right) \leqq E\big(W(1)^j\big) \leqq E\left(\frac{1}{2^j}(T^2+1)^j\right). \tag{4.5}$$

Thus, a sufficient condition for $E\big(W(1)\big)$ to be finite is that T should have a finite second moment about the origin; if T has a finite fourth moment, $W(1)$ will have a finite variance. Similarly, the joint moment of $\big(T, W(1)\big)$ lies between the bounds

$$E\big(\tfrac{1}{2}T(3T-1)\big) \leqq E\big(TW(1)\big) \leqq E\big(\tfrac{1}{2}T(T^2+1)\big). \tag{4.6}$$

Hence, for i.i.d. inputs $\{X_i\}$ with the distribution $\{p_i\}_{i=0}^{\infty}$, (4.5) reduces to

$$\sum_{t=1}^{\infty} \frac{1}{2^j t}(3t-1)^j p_{t-1}^{(t)} \leqq E\big(W(1)^j\big) \leqq \sum_{t=1}^{\infty} \frac{1}{2^j t}(t^2+1)^j p_{t-1}^{(t)}, \tag{4.7}$$

where $p_{t-1}^{(t)} = Pr\{X_0 + X_1 + \cdots + X_{t-1} = t-1\}$. For example, writing $E(X_t) = p'(1) = m < 1$ and $\text{Var}(X_t) = s^2 < \infty$ in this case, for $E\big(W(1)\big)$, we obtain the bounds

$$\frac{(2+m)}{2(1-m)} \leqq E\big(W(1)\big) \leqq \frac{1}{2(1-m)}\left\{\frac{s^2}{(1-m)^2} + \frac{2-m}{(1-m)}\right\}. \tag{4.8}$$

Similarly, for inputs $\{X_i\}$ forming a Markov chain with finite state space where $X_{-1} = 0$, we have

$$\sum_{t=1}^{\infty} \frac{1}{2^j t}(3t-1)^j \lambda_{t-1}^{(t)} \leqq E\big(W(1)^j\big) \leqq \sum_{t=1}^{\infty} \frac{1}{2^j t}(t^2+1)^j \lambda_{t-1}^{(t)}, \tag{4.9}$$

where $\lambda_{t-1}^{(t)}$ is the coefficient of θ^{t-1} in $\{\lambda(\theta)\}^t$ and $\lambda(\theta)$ is defined as the maximum eigenvalue of (3.6). While bounds such as (4.7) or (4.9) are rather wide, they may prove adequate for first approximations in practical queueing, storage, and traffic problems. We now describe an exact method for deriving the mean of $W(u)$ for i.i.d. and Markovian inputs due to Lehoczky [14].

5. Exact results for the mean $E(W(1))$

Let us first consider the case of i.i.d. inputs $\{X_i\}$; we follow Lehoczky's technique [14] of conditioning the process on the input $X_0 = i$ during the initial time interval $(0, 1)$, and so write the expectation of the path integral $W(1)$ as

$$E(W(1)) = 1 + \sum_{i=0}^{\infty} p_i \left\{\frac{i}{2} + E(W(i))\right\}. \tag{5.1}$$

Now, from (1.4),

$$E(W(i)) = E\left(\sum_{k=1}^{i} \{W_k + (i-k)T_k\}\right) = iE(W(1)) + \tfrac{1}{2}i(i-1)E(T). \tag{5.2}$$

Hence substituting this in (5.1), we finally obtain

$$E(W(1)) = 1 + \tfrac{1}{2}m + \sum_{i=0}^{\infty} ip_i E(W(1)) + \sum_{i=0}^{\infty} \tfrac{1}{2}i(i-1)p_i E(T), \tag{5.3}$$

or

$$E(W(1)) = \frac{1}{1-m} + \frac{1}{2}\frac{s^2}{(1-m)^2}, \tag{5.4}$$

where $E(X_t) = m < 1$, $\text{Var}(X_t) = s^2$ and $E(T) = 1/(1-m)$ as in Section 4. Thus for $W(u)$, we find the expectation

$$E(W(u)) = uE(W(1)) + \tfrac{1}{2}u(u-1)E(T) = \tfrac{1}{2}u\left\{\frac{u+1}{1-m} + \frac{s^2}{(1-m)^2}\right\}. \tag{5.5}$$

The technique may be applied equally well to Markovian inputs $\{X_i\}$ with finite or denumerable state space. Assume as usual that the input X_{-1} in the time interval $(-1, 0)$ is zero, and for convenience allow the number of states in the chain to be denumerably infinite. Then, once again conditioning on the input $X_0 = i$ during the interval $(0, 1)$, we obtain

$$E(W_0(1)) = 1 + \sum_{i=0}^{\infty} p_{0i}\{\tfrac{1}{2}i + E(W_i(i))\} = 1 + \tfrac{1}{2}m_0 + \sum_{i=0}^{\infty} p_{0i}E(W_i(i)), \tag{5.6}$$

where $m_0 = \Sigma_{i=0}^{\infty} ip_{0i} < 1$, and the subscripts in $W_0(1)$ and $W_i(i)$ indicate that the inputs prior to the start of the two processes are 0 and i, respectively. Now for any prior input j, considering the first descent to state $i-1$ of a process starting from state i, we obtain

$$\begin{aligned} E(W_j(i)) &= E(W_j(1)) + (i-1)E(T'_j) + E(W_0(i-1)) \\ &= E(W_j(1)) + (i-1)E(T'_j) + (i-1)E(W_0(1)) \\ &\qquad + \tfrac{1}{2}(i-1)(i-2)E(T), \end{aligned} \tag{5.7}$$

where T'_j now denotes the time to first emptiness starting from state 1 with prior input j, and we decompose $W_0(i-1)$ according to (1.4).

Thus, we may rewrite (5.6) as

$$(5.8)\qquad \begin{aligned} E(W_0(1)) & \\ &= 1 + \tfrac{1}{2}m_0 + \sum_{i=0}^{\infty} p_{0i}\{E(W_i(1)) + (i-1)E(W_0(1)) \\ &\qquad + (i-1)E(T'_j) + \tfrac{1}{2}(i-1)(i-2)E(T)\} \\ &= 1 + \tfrac{1}{2}m_0 + (m_0 - 1)E(W_0(1)) + E(T)\sum_{i-0}^{\infty} \tfrac{1}{2}(i-1)(i-2)p_{0i} \\ &\qquad + \sum_{i=0}^{\infty} p_{0i}E(W_i(1)) + \sum_{i=0}^{\infty} (i-1)p_{0i}E(T'_i). \end{aligned}$$

More generally, for $W_k(1)$ starting from state 1 with prior input k and using precisely the same arguments, we have that

$$(5.9)\qquad \begin{aligned} E(W_k(1)) = 1 + \tfrac{1}{2}m_k + (m_k - 1)E(W_0(1)) + E(T)\sum_{i=0}^{\infty} \tfrac{1}{2}(i-1)(i-2)p_{ki} \\ + \sum_{i=0}^{\infty} p_{ki}E(W_i(1)) + \sum_{i=0}^{\infty} (i-1)p_{ki}E(T'_i), \end{aligned}$$

where $m_k = \Sigma_{i=0}^{\infty}\, ip_{ki} < 1$. Thus, expressing these results in matrix form, we obtain

$$(5.10)\qquad E(\mathbf{W}(1)) = \mathbf{1} + \tfrac{1}{2}\mathbf{m} + (\mathbf{m} - \mathbf{1})E(W_0(1)) + \mathbf{p}E(\mathbf{W}(1)) + \mathbf{R},$$

where $\mathbf{W}(1)$, $\mathbf{1}$, and $\mathbf{m}$ are column vectors with kth elements $W_k(1)$, 1, and m_k, respectively, and $\mathbf{R}$ is the column vector with kth elements

$$(5.11)\qquad E(T)\sum_{i=0}^{\infty} \tfrac{1}{2}(i-1)(i-2)p_{ki} + \sum_{i=0}^{\infty} (i-1)p_{ki}E(T'_i).$$

Note that $E(T)$ and $E(T'_i)$ can be found from the p.g.f. $g(\theta; 1, 0)$ and $g(\theta; 1, i)$ of Section 3, so that $\mathbf{R}$ is assumed to be known.

If the Markov chain considered is stationary, $E(W_0(1)) = E(W(1))$ can be obtained without difficulty. Rewriting (5.10) as

$$(5.12)\qquad \{\mathbf{I} - \mathbf{p}\}E(\mathbf{W}(1)) = \mathbf{1} + \tfrac{1}{2}\mathbf{m} + (\mathbf{m} - \mathbf{1})E(W_0(1)) + \mathbf{R}$$

and premultiplying by the row vector $\boldsymbol{\pi}'$ of stationary probabilities $\{\pi_k\}$, we obtain

$$(5.13)\qquad \boldsymbol{\pi}'\{\mathbf{1} + \tfrac{1}{2}\mathbf{m}\} + \boldsymbol{\pi}'(\mathbf{m} - \mathbf{1})E(W_0(1)) + \boldsymbol{\pi}'\mathbf{R} = 0.$$

Hence,

$$(5.14)\qquad E(W_0(1)) = \frac{1 + \frac{1}{2}\sum_{i=0}^{\infty} \pi_i m_i + \sum_{i=0}^{\infty} \pi_i R_i}{1 - \sum_{i=0}^{\infty} \pi_i m_i},$$

where $\Sigma_{i=0}^{\infty} \pi_i m_i < 1$. For chains with a finite number of states, $E(\mathbf{W}(1))$ can be readily formed by inverting the matrix equation (5.12) (see Lehoczky [14]), while for chains with a denumerable state space, truncation methods will provide approximations to $W_k(1)$ for any finite k. We now proceed to discuss joint generating functions for $(T, W(1))$.

6. Joint probability generating functions for $(T, W(1))$

If results more precise than the inequalities of (4.5) or the mean values of (5.3) and (5.14) are required for the stochastic integral $W(1)$, it becomes necessary to resort to more complex methods of analysis. Equation (4.2) for $W(1 \mid t)$ suggests that an extension of the truncated polynomial technique used in [7] may provide the joint p.g.f. of $(T, W(1))$; for simplicity, we shall write W for $W(1)$ from now on.

For the first emptiness path starting from $Z_0 = 1$ and terminating at T, consider the contributions made to $T = \Sigma_{i=0}^{T-1} Y_i + 1$ and

$$W(1 \mid T) = \sum_{i=0}^{T-1} (i + \tfrac{1}{2})Y_i - \tfrac{1}{2}T(T - 3) \tag{6.1}$$

by the input $Y_0 = 0$; for inputs $\{Y_i\}$ forming an i.i.d. sequence with distribution $\{p_j\}$, starting with the input Y_0, we write the polynomial

$$G_0(\theta, \varphi) = p_0, \qquad 0 \leqq \theta, \quad \varphi \leqq 1, \tag{6.2}$$

where the zero indices of θ and φ record the contributions of $Y_0 = 0$ to T and $W(1 \mid T)$, respectively. Let us now define for the inputs $Y_1 + Y_0$ the truncated polynomial

$$G_1(\theta, \varphi) = \langle p(\theta\varphi^{3/2})G_0(\theta, \varphi)\rangle = (p_0 + p_1\theta\varphi^{3/2})p_0, \tag{6.3}$$

where $p(\theta\varphi^{3/2})$ is the p.g.f. of Y_1 with argument $\theta\varphi^{3/2}$ to record the contribution of Y_1 to T and $\frac{3}{2}Y_1$ to $W(1 \mid T)$. Here the truncation $\langle\,\rangle$ cuts off all terms in θ of degree higher than the first, since $Y_1 + Y_0 \leqq 1$.

For the remaining sums of inputs $Y_i + \cdots + Y_0$, $i = 2, \cdots, T - 1$, we define similar truncated polynomials

$$G_i(\theta, \varphi) = \langle p(\theta\varphi^{i+1/2})G_{i-1}(\theta, \varphi)\rangle, \tag{6.4}$$

where the argument $\theta\varphi^{i+1/2}$ records the contributions of Y_i to T and $(i + \frac{1}{2})Y_i$ to $W(1 \mid T)$, respectively. The truncation $\langle\,\rangle$ now cuts off all terms in θ of degree higher than i, since $Y_i + Y_{i-1} + \cdots + Y_0 \leqq i$.

It is clear that the joint probability of (T, W) will be given by the coefficient of $\theta^{T-1}\varphi^{W+T(T-3/2)}$ in $G_{T-1}(\theta, \varphi)$; thus, we may formally write the joint p.g.f. of (T, W) as

$$F(\theta, \varphi) = E(\theta^T\varphi^W) = \sum_{T=1}^{\infty} \left\{\frac{1}{2\pi i} \oint \frac{G_{T-1}(z, \varphi)}{z^T}\, dz\right\} \theta^T \varphi^{-T(T-3)/2}, \tag{6.5}$$

where $i = \sqrt{-1}$ and z is now a complex variable. Note that the integral must be taken on a suitable contour around the origin, and that $\varphi^{-T(T-3)/2}$ provides the appropriate correction to the contributions of the $\{Y_i\}$ to $W(1 \mid T)$.

When the inputs $\{Y_i\}$ form a Markov chain with transition matrix $\{p_{ij}\}$, assumed to be infinite, we may write for the input Y_0 the vector

$$\mathbf{G}_0(\theta, \varphi) = \begin{bmatrix} p_{00} \\ p_{10} \\ \vdots \end{bmatrix}, \qquad 0 \leqq \theta, \varphi \leqq 1, \tag{6.6}$$

where the zero indices of θ and φ record the contributions of $Y_0 = 0$ to T and $W(1 \mid T)$, respectively. We now define for the sums of inputs $Y_i + \cdots + Y_0$ the ith vector of truncated polynomials

$$\mathbf{G}_i(\theta, \varphi) = \left\langle \begin{bmatrix} p_{00} & p_{01}\theta\varphi^{i+1/2} & p_{02}(\theta\varphi^{i+1/2})^2 & \cdots \\ p_{10} & p_{11}\theta\varphi^{i+1/2} & p_{12}(\theta\varphi^{i+1/2})^2 & \cdots \\ \cdots & & \cdots & \cdots \end{bmatrix} \mathbf{G}_{i-1}(\theta, \varphi) \right\rangle, \tag{6.7}$$

where for $i = 1, \cdots, T - 2$, the argument $\theta\varphi^{i+1/2}$ records the contributions of Y_i to T and $(i + \frac{1}{2})Y_i$ to $W(1 \mid T)$, respectively. The truncation $\langle\,\rangle$ cuts off all terms in θ of degree higher than i, since $Y_i + Y_{i-1} + \cdots + Y_0 \leqq i$; it follows in practice that one can neglect all elements of the matrix $\{p_{kj}(\theta\varphi^{i+1/2})^j\}$ beyond those of the $(i + 1)$th column and $(i + 2)$th row. Starting with $\mathbf{G}_0(\theta, \varphi)$, this means we need only consider its first two elements $\left[\begin{smallmatrix} p_{00} \\ p_{10} \end{smallmatrix}\right]$.

Finally, since we assume once again that the input prior to the start of the process is $X_{-1} = Y_T = 0$, the truncated polynomial

$$G_{T-1}(\theta, \varphi) = \langle [p_{00}, p_{01}\theta\varphi^{T-1/2}, \cdots, p_{0T-1}(\theta\varphi^{T-1/2})^{T-1}]\mathbf{G}_{T-2}(\theta, \varphi)\rangle \tag{6.8}$$

will provide in the coefficient of $\theta^{T-1}\varphi^{W+T(T-3)/2}$ the joint probability of (T, W). Hence, as before, we may formally write the joint p.g.f. of (T, W) as

$$F(\theta, \varphi) = E(\theta^T\varphi^W) = \sum_{T=1}^{\infty} \left\{ \frac{1}{2\pi i} \oint \frac{G_{T-1}(z, \varphi)}{z^T}\, dz \right\} \theta^T \varphi^{-T(T-3)/2}, \tag{6.9}$$

where z is a complex variable and the integral is taken on a suitable contour around the origin. For a finite $(r + 1) \times (r + 1)$ matrix $\{p_{ij}\}_{i,j=0}^r$, and a finite state space, the same methods apply with appropriate modifications from $\mathbf{G}_{r-1}(\theta, \varphi)$ onwards, due to the finiteness of the transition probability matrix.

As simple illustrations of these techniques, we consider the following two random walks.

EXAMPLE 6.1. Let the $\{X_i\}$ be i.i.d. with $p(\theta) = p\theta + q$, for $0 < p < 1$ and $p + q = 1$. In this case $W = \frac{1}{2}(3T - 1)$ and the joint p.g.f. of (T, W) is

$$(6.10)\qquad F(\theta, \varphi) = \frac{q\theta\varphi}{1 - p\theta\varphi^{3/2}}, \qquad 0 \leqq \theta, \quad \varphi \leqq 1.$$

From this, we obtain, for example, that $E(W) = \frac{1}{2}(3/q - 1)$, and $\text{Var}(W) = \frac{9}{4}(1/q - 1)$.

Since, from (1.4), we know that $W(u) = \Sigma_{j=1}^{u} \{W_j + (j-1)T_j\}$, where the $V_j = W_j + (j-1)T_j$ are mutually independent, we first note that the joint p.g.f. of (T_j, V_j) is

$$(6.11)\qquad F_j(\theta, \varphi) = F(\theta\varphi^{j-1}, \varphi) = \frac{q\theta\varphi^j}{1 - p\theta\varphi^{j+1/2}}.$$

Hence, it follows that the joint p.g.f. of $W(u)$ and $T(u) = T_1 + \cdots + T_u$ is given by

$$(6.12)\qquad \prod_{j=1}^{u} F_j(\theta, \varphi) = \frac{\{q\theta\varphi^{(1/2)(u+1)}\}^u}{(1 - p\theta\varphi^{3/2})(1 - p\theta\varphi^{5/2})\cdots(1 - p\theta\varphi^{u+1/2})},$$

from which it is readily found that $E(W(u)) = (u/2q)(u + 2 - q)$ and $\text{Var}(W(u)) = (u/12q)(4u^2 + 12u + 11)(1/q - 1)$.

EXAMPLE 6.2. Let $\{X_i\}$ be a two state Markov chain with transition probability matrix $\begin{bmatrix} p_{00} & p_{01} \\ p_{10} & p_{11} \end{bmatrix}$ for $p_{ij} > 0$, $i, j = 0, 1$. With the prior input $X_{-1} = 0$, we obtain for the joint p.g.f. of (T, W) the expression

$$(6.13)\qquad F(\theta, \varphi) = \frac{\theta\varphi}{1 - p_{11}\theta\varphi^{3/2}}\{p_{00} + (p_{01}p_{10} - p_{00}p_{11})\theta\varphi^{3/2}\}.$$

Much as before, the joint p.g.f. of (T_j, V_j) is found to be

$$(6.14)\qquad F_j(\theta, \varphi) = F(\theta\varphi^{j-1}, \varphi) = \frac{\theta\varphi^j}{1 - p_{11}\theta\varphi^{j+1/2}}\{p_{00} + (p_{01}p_{10} - p_{00}p_{11})\theta\varphi^{j+1/2}\}.$$

It follows that the joint p.g.f. of $W(u)$ and $T(u)$ will take the form

$$(6.15)\qquad \prod_{j=1}^{u} F_j(\theta, \varphi) = \{\theta\varphi^{(u+1)/2}\}^u \frac{\{p_{00} + (p_{01}p_{10} - p_{00}p_{11})\theta\varphi^{3/2}\}}{1 - p_{11}\theta\varphi^{3/2}} \cdots \frac{\{p_{00} + (p_{01}p_{10} - p_{00}p_{11})\theta\varphi^{u+1/2}\}}{1 - p_{11}\theta\varphi^{u+1/2}}.$$

These results may appear somewhat slight after the complexities of the truncated polynomial technique; work is at present in progress to derive explicit results for the joint p.g.f. $F(\theta, \varphi)$ of (T, W) for input distributions such as the geometric and Poisson when the inputs $\{X_i\}$ are i.i.d.

7. Random walks imbedded in birth and death processes and an asymptotic result

Joint distribution problems for the analogous time to first emptiness $T'(u)$ and stochastic path integral $W'(u)$ have been investigated for birth and death processes in continuous time by Gani and McNeil [10]. For these, the double Laplace transform of $T'(u)$, $W'(u)$ when the birth and death parameters are, respectively, λ and μ, has been shown to be

$$\begin{aligned}(7.1)\qquad \psi_u(\alpha, \beta) &= E\big(\exp\{-\alpha T'(u) - \beta W'(u)\}\big)\\ &= \left(\frac{\mu}{\lambda}\right)^{u/2} \frac{J_{u+\nu}\big(2(\lambda\mu)^{1/2}\beta^{-1}\big)}{J_\nu\big(2(\lambda\mu)^{1/2}\beta^{-1}\big)}, \qquad \operatorname{Re} \alpha, \beta \geqq 0,\end{aligned}$$

where $\nu = (\alpha + \lambda + \mu)\beta^{-1}$. From (7.1), it is possible to find the expectation of $W'(u)$, as well as the regression of $W'(u)$ on $T'(u)$.

A similar approach is applicable to the discrete time random walk imbedded in a birth and death process. This is the random walk starting from $Z_0 = u$ with i.i.d. inputs $\{X_t\}_{t=1}^{\infty}$ of size ± 1 arriving at times $t - 0$, such that its state at times $t = 1, 2, \cdots, T(u)$ is

$$(7.2)\qquad Z_t = (Z_{t-1} + X_t),$$

with $Z_{T(u)} = 0$ for the first time. The probabilities that $X_t = +1$ and -1 are, respectively, $p = \lambda/(\lambda + \mu)$ and $q = \mu/(\lambda + \mu)$. If $F_u(\theta, \varphi) = E\big(\theta^{T(u)}\varphi^{W(u)}\big)$ is the joint p.g.f. of $\big(T(u), W(u)\big)$, it is readily seen for $u \geqq 1$, considering the input $X_1 = \pm 1$ during (0,1), that

$$(7.3)\qquad F_u(\theta, \varphi) = \theta\varphi^u\{pF_{u+1}(\theta, \varphi) + qF_{u-1}(\theta, \varphi)\},$$

where $F_0(\theta, \varphi)$ is put equal to 1 for convenience.

We can solve these difference equations by setting

$$(7.4)\qquad \frac{F_u(\theta, \varphi)}{F_{u-1}(\theta, \varphi)} = \xi_u(\theta, \varphi), \qquad u = 1, 2, \cdots;$$

whence

$$\begin{aligned}(7.5)\qquad \xi_u(\theta, \varphi) &= \frac{q\theta\varphi^u}{1 - p\theta\varphi^u\xi_{u+1}(\theta, \varphi)}\\ &= \left(\frac{q}{p}\right)^{1/2} \frac{(qp)^{1/2}\theta\varphi^u}{1 - (qp)^{1/2}\theta\varphi^u\left(\frac{p}{q}\right)^{1/2}\xi_{u+1}(\theta, \varphi)}.\end{aligned}$$

The function $\xi_u(\theta, \varphi)$ may be identified as the ratio of two Bessel functions of integer order, of the first kind, so that

$$(7.6)\qquad \xi_u(\theta, \varphi) = \left(\frac{q}{p}\right)^{1/2} \frac{J_u\big(2u(qp)^{1/2}\theta\varphi^u\big)}{J_{u-1}\big(2u(qp)^{1/2}\theta\varphi^u\big)}.$$

Hence,

$$F_u(\theta, \varphi) = \prod_{j=1}^{u} \xi_j(\theta, \varphi) = \left(\frac{q}{p}\right)^{u/2} \prod_{k=1}^{u} \frac{J_k\big(2k(qp)^{1/2}\theta\varphi^k\big)}{J_{k-1}\big(2k(qp)^{1/2}\theta\varphi^k\big)}, \tag{7.7}$$

a rather complicated explicit expression, somewhat reminiscent of (7.1).

The analogy between this random walk and the continuous time birth and death process suggests that the asymptotic normality proved for $(T'(u), W'(u))$ in [10] will extend not only to the imbedded random walk, but also to the general random walk (1.1) with regular unit output previously considered. In fact, this is the case, as is proved in detail in a recent note by Gani and Lehoczky [9]. Briefly, one begins by showing that for i.i.d. inputs contributing to the stochastic integral

$$W(u) = \sum_{j=1}^{u} W_j + (j - 1)T_j, \tag{7.8}$$

the normalized sum $u^{-3/2} \Sigma_{j=1}^{u} W_j$ tends to zero almost surely. It is then proved that, as $u \to \infty$, the normalized random variables

$$\left\{\frac{W(u) - u(u - 1)\mu}{\sigma u^{3/2}/\sqrt{3}}\right\}, \qquad \left\{\frac{T(u) - u\mu}{\sigma u^{1/2}}\right\}, \tag{7.9}$$

where $\mu = E(T_j)$ and $\sigma^2 = \text{Var}\,(T_j)$, are jointly normally distributed with correlation coefficient $\rho = \sqrt{3}/2$. For large u, this means the asymptotic regression of $W(u)$ on $T(u)$ will be known. I would conjecture that it might be possible to obtain sharper asymptotic results for the stochastic integral $W(u)$.

No account of the wide field I have tried to cover could hope to be entirely complete; despite its condensed form, I hope this brief sketch of unanswered problems in the area may encourage applied probabilists to work on them and find their solutions.

REFERENCES

[1] M. S. Ali Khan and J. Gani, "Infinite dams with inputs forming a Markov chain," *J. Appl. Prob.*, Vol. 5 (1968), pp. 72–84.

[2] M. S. Bartlett, "Equations for stochastic path integrals," *Proc. Cambridge Philos. Soc.*, Vol. 57 (1961), pp. 568–573.

[3] P. J. Brockwell and J. Gani, "A population process with Markovian progenies," *J. Math. Anal. Appl.*, Vol. 32 (1970), pp. 264–273.

[4] K. L. Chung, *Markov Chains with Stationary Transition Probabilities*, New York, Springer-Verlag, 1967.

[5] D. J. Daley, "The total waiting time in a busy period of a stable single-server queue, I," *J. Appl. Prob.*, Vol. 6 (1969), pp. 550–564.

[6] D. J. Daley and D. R. Jacobs, "The total waiting time in a busy period of a stable single-server queue, II," *J. Appl. Prob.*, Vol. 6 (1969), pp. 565–572.

[7] J. Gani, "Elementary methods for an occupancy problem of storage," *Math. Ann.*, Vol. 136 (1958), pp. 454–465.

[8] ———, "A note on the first emptiness of dams with Markovian inputs," *J. Math. Anal. Appl.*, Vol. 26 (1969), pp. 270–274.

[9] J. Gani and J. P. Lehoczky, "An asymptotic result in traffic theory," *Department of Statistics Technical Report No.* 23, Stanford University, 1970; *J. Appl. Prob.*, Vol. 8 (1971), pp. 815–820.

[10] J. Gani and D. R. McNeil, "Applications of certain birth-death and diffusion processes in traffic flow," *Department of Statistics Technical Report No.* 17, Stanford University, 1969; *Adv. Appl. Prob.*, Vol. 3 (1971), pp. 339–352.

[11] I. J. Good, "The number of individuals in a cascade process," *Proc. Cambridge Philos. Soc.*, Vol. 45 (1949), pp. 360–363.

[12] T. E. Harris, *The Theory of Branching Processes*, Berlin, Springer-Verlag, 1963.

[13] D. G. Kendall, "Some problems in the theory of dams," *J. Roy. Statist. Soc. Ser. B*, Vol. 19 (1957), pp. 207–212.

[14] J. P. Lehoczky, "A note on the first emptiness time of an infinite reservoir with inputs forming a Markov chain," *J. Appl. Prob.*, Vol. 8 (1971), pp. 276–284.

[15] E. H. Lloyd, "The epochs of emptiness of a semi-infinite discrete reservoir," *J. Roy. Statist. Soc. Ser. B*, Vol. 25 (1963), pp. 131–136.

[16] ———, "A probability theory of reservoirs with serially correlated inputs," *J. Hydrol.*, Vol. 1 (1963), pp. 99–128.

[17] ———, "Reservoirs with serially correlated inflows," *Technometrics*, Vol. 5 (1963), pp. 85–93.

[18] ———, "Stochastic reservoir theory," *Advances in Hydroscience*, Vol. 4, New York, Academic Press, 1967.

[19] E. H. Lloyd and S. Odoom, "A note on the equilibrium distribution of levels in a semi-infinite reservoir subject to Markovian inputs and unit withdrawals," *J. Appl. Prob.*, Vol. 2 (1965), pp. 215–222.

[20] D. R. McNeil, "Integral functionals of birth and death processes and related limiting distributions," *Ann. Math. Statist.*, Vol. 41 (1970), pp. 480–485.

[21] J. L. Mott, "The distribution of the time to emptiness of a discrete dam under steady demand," *J. Roy. Statist. Soc. Ser. B*, Vol. 25 (1963), pp. 137–139.

[22] P. S. Puri, "On the homogeneous birth-and-death process and its integral," *Biometrika*, Vol. 53 (1966), pp. 61–71.

[23] ———, "Some further results on the birth-and-death process and its integral," *Proc. Cambridge Philos. Soc.*, Vol. 64 (1968), pp. 141–154.

[24] L. Takács, "Investigation of waiting time problems by reduction to Markov processes," *Acta. Math. Acad. Sci. Hungar.*, Vol. 6 (1955), pp. 101–129.

[25] D. Vere-Jones, "Ergodic properties of non-negative matrices, I," *Pacific J. Math.*, Vol. 22 (1967), pp. 361–386.

[26] ———, "Ergodic properties of non-negative matrices, II," *Pacific J. Math.*, Vol. 26 (1968), pp. 601–620.

KUHN-GRÜN TYPE APPROXIMATIONS FOR POLYMER CHAIN DISTRIBUTIONS

H. E. DANIELS
UNIVERSITY OF BIRMINGHAM

1. Introduction

Polymer physics offers many problems of interest to applied probabilists because of the essentially statistical basis of many of the characteristic phenomena associated with polymers. They arise from the fact that the polymer molecules are long, more or less flexible chains which can have many configurations. In the early development of the theory it was adequate to regard polymer chains as three dimensional random walks and to use a simple Gaussian approximation to the relevant distributions. As the theory became more refined, attention had to be paid to the effect of the actual structure of the polymer chains on the distribution. Excellent accounts of the recent state of the theory are given by Volkenstein [17], Birshtein and Ptitsyn [1], and Flory [7].

Many physical properties are explainable from a knowledge of the moments, usually the second moment, of the vector length of the chain molecules, and in such cases the Gaussian approximation or some Edgeworth type expansions based on it is a suitable description of the distribution. But there are other properties, such as the elastic behavior of rubber under large strains, which require the vector length distribution to be known with comparable relative accuracy over the whole of its range, a much stronger condition which is appropriate for "large deviations." It is the latter type of property which motivates the discussion given here. The first approximation of this kind was given in a famous paper by Kuhn and Grün [14]—hence, the title of the present paper.

Apart from "stiffness" caused by the interaction of neighboring elements of the chain, we shall ignore excluded volume effects arising from the space taken up by the chain. The effect of the selfavoiding nature of the chain on the distribution of vector length is a subject of much current discussion, but the mathematical difficulties are such that only the simplest chain models can be considered. In contrast, we are concerned with distributions associated with chain models approximating to real polymer molecules.

Much of the paper is an exposition of the history and background of the subject for the benefit of applied probabilists wishing to enter the field. However, the asymptotic result sketched in Section 7 for the behavior of the distribution in the extreme tail is new and by no means fully worked out. Also, the integral equation approach is offered as a practical way of validating and generalizing the method of calculation currently in favor, which is based on the so called rotational isomeric approximation.

2. Physical background

As an example of the kind of physical problem motivating the paper we give a short account of the theory underlying the mechanical properties of rubber like substances. The kinetic theory of rubber elasticity, as originated and developed by W. Kuhn, Guth and Mark, and others (see, for example, Volkenstein [17]), assumes in its simplest form that the mechanical properties of rubber and similar polymers, like those of an ideal gas, depend only on the entropy of the system, which is a function of the distribution of configurations of the constituent chain molecules. Each molecule is made up of a long sequence of repeating units with a fair amount of freedom of independent movement between them. In the simplest version of the theory, the units are considered to be completely free to rotate in any direction relative to each other. Real chain molecules have considerable restrictions on the relative rotation imposed by fixed valence angles and steric hindrance, but they may often be conveniently regarded as "equivalent" to a smaller number of freely jointed links.

In the rubber like state, the chain molecules are connected by a relatively sparse system of cross links into a loose three dimensional network. The dimensional characteristics of such a network are determined to a large extent by the distances between the cross links, and hence by the end to end distances of the chains connecting them. The entropy of each chain is proportional to the logarithm of the total number of possible configurations for which the end separation has a specified value.

If a single random polymer chain has one end at the origin and is free to rotate about it, the distribution of the position $\mathbf{r}$ of the free end is spherically symmetrical. The probability density function of $\mathbf{r}$ is $p(r)$, where $r = |\mathbf{r}|$ and that of r is $P(r) = 4\pi r^2 p(r)$. The first of these is proportional to the required number of configurations, and the entropy S of a single chain is

$$S = \text{constant} + K \log p(r), \tag{2.1}$$

where K is Boltzmann's constant. The average tension in a single chain held at length r is then given by

$$F = -T \frac{dS}{dr}, \tag{2.2}$$

where T is the absolute temperature.

Stress-strain relationships for the whole network have been developed, with extra assumptions, by James and Guth, and many others (see Volkenstein [17]), though a completely satisfactory theory is not yet available, particularly for the treatment of large strains.

3. Freely jointed chain: the Kuhn-Grün approximation

The simplest model for a chain molecule is one of n freely jointed bonds each of length a. This is the classical problem of "random flights"; the usual

Gaussian approximation for the distribution of **r** leads to

$$S = \text{constant} - \frac{3Kr^2}{(2na^2)}, \tag{3.1}$$

and the average tension F of a single chain of length r is, from (2.2),

$$F = \frac{3KT}{na^2} r, \tag{3.2}$$

which is proportional to r. However, since the calculation of tension is based on $\log p(r)$ and not on $p(r)$, the approximation will break down as r approaches its maximum value na because the *relative* error of the Gaussian approximation to $p(r)$ cannot be uniformly bounded over the whole range of r.

Kuhn and Grün [14] were the first to provide an approximation to $\log p(r)$ which maintains its accuracy reasonably well over the whole range of r. They used a method familiar in statistical mechanics which we now outline. The following version corrects an error in the original demonstration pointed out by Jernigan and Flory [9]. No attempt is made to formulate the argument precisely.

The projection in the x direction of a completely free bond of length a is uniformly distributed over the interval $(-a, a)$. Consider a chain of n bonds where n is large. Let $-a = \xi_1 < \xi_2 < \cdots < \xi_k < \xi_{k+1} = a$ be a set of points subdividing $(-a, a)$ such that $\delta\xi_j = \xi_{j+1} - \xi_j$ are small. Let n_j of the n bonds have projections in (ξ_j, ξ_{j+1}), where $\Sigma\, n_j = n$ and n is assumed large enough for each n_j to be large. If the chain is unconstrained, the probability of the n_j is

$$P(n_1, n_2, \cdots, n_k) = \frac{n!}{n_1!\, n_2! \cdots n_k!} \prod_j \left(\frac{\delta\xi_j}{2a}\right)^{n_j}. \tag{3.3}$$

Suppose now that the total projection of the chain is constrained to have a value x. This adds the further condition $\Sigma\, n_j\xi_j = x$, and if (3.3) is summed over all values of n_j consistent with the two constraints, the result will be proportional to the probability density $f(x)$ of the projection x of the unconstrained chain. In the usual way, the summation over P is avoided by taking $P = \hat{P}$ at the most probable values $\hat{n}_j$ of n_j subject to the two constraints. Using a crude form of Stirling's approximation, one maximizes

(3.4)

$$\log P + \gamma \sum n_j + \kappa \sum n_j\xi_j$$

$$\sim \text{constant} + \sum n_j \log\left(\frac{\delta\xi_j}{2a}\right) - \sum n_j \log n_j + \sum n_j + \gamma \sum n_j + \kappa \sum n_j\xi_j,$$

and obtains

$$\hat{n}_j = \frac{1}{2a} \exp\{\gamma + \kappa\xi_j\}\, \delta\xi_j. \tag{3.5}$$

From the two constraints, we then have

$$n \sim e^{\gamma} \int_{-a}^{a} \frac{1}{2a} \exp\{\kappa\xi\}\, d\xi = \frac{1}{\kappa a} e^{\gamma} \sinh \kappa a \tag{3.6}$$

and

$$x \sim e^{\gamma} \int_{-a}^{a} \frac{1}{2a} \exp\{\kappa\xi\}\xi\, d\xi = e^{\gamma} \frac{d}{d\kappa}\left\{\frac{\sinh \kappa a}{\kappa a}\right\}. \tag{3.7}$$

The probability density $f(x)$ is approximately proportional to $\hat{P}$. Using these results, one eventually obtains

$$f(x) \sim \frac{C}{a}\left(\frac{\sinh \kappa a}{\kappa a}\right)^{n} \exp\{-\kappa x\}, \tag{3.8}$$

where κ is the unique real root of

$$\mathscr{L}(\kappa a) = \coth \kappa a - \frac{1}{\kappa a} = \frac{x}{na} = \frac{\bar{x}}{a}, \tag{3.9}$$

$\mathscr{L}(\cdot)$ is the so called Langevin function, (3.8) is usually called the Langevin approximation, C is a dimensionless normalizing constant, and $\bar{x}$ is the mean projection of a bond.

Kuhn and Grün assumed (3.8) to approximate to the probability density function (p.d.f.) $p(r)$ of r. Jernigan and Flory [9] used the relation

$$2\pi r p(r) = \left\{\frac{-df(x)}{dx}\bigg|\, x = r\right\}, \tag{3.10}$$

in the manner of Treloar [16] to obtain the more correct approximation,

$$\hat{p}(r) = \frac{A\kappa}{ra}\left(\frac{\sinh \kappa a}{\kappa a}\right)^{n} \exp\{-\kappa r\}, \tag{3.11}$$

where A is a normalizing constant and

$$\mathscr{L}(\kappa a) = \frac{r}{na} = \frac{\bar{r}}{a}. \tag{3.12}$$

The effect of using either (3.8) or (3.11) is to replace the tension-extension relation (3.2) by a formula with the reasonable property that $F \to \infty$ as $r \to na$.

4. The saddle point approximation

The form of (3.8) or (3.11) immediately suggests its connection with the saddle point approximation $\tilde{p}(r)$ to the transform of the moment generating function (or partition function) for the radial distribution. It is surprising that this fact has not been fully exploited. Dobrushin [5] mentioned that (3.8) could be derived by the essentially equivalent Cramér extension of the central limit theorem, and Kubo [11] used the saddle point method in an important study of more general chains which we refer to later. But the accuracy of the saddle point approximation is not fully appreciated. In the present case, (4.7) is more accurate than (3.11) and is applicable to quite short chains. Under conditions which are satisfied in the present application it has been shown (Daniels [4]) that the error in $\log \tilde{p}(r)$ is uniformly $O(n^{-1})$ over the whole range of r. In fact, when normalized, it is even more accurate than would be expected from this result.

For any general spherically symmetrical distribution with radial p.d.f. $P(r)$, we have

$$\Phi(\rho) = \int_0^\infty \frac{\sin r\rho}{r\rho} P(r)\, dr, \tag{4.1}$$

$$P(r) = \frac{2}{\pi} \int_0^\infty r\rho \sin r\rho \Phi(\rho)\, d\rho. \tag{4.2}$$

These are easily deduced from the polar form of the three dimensional characteristic function for $\mathbf{r}$. Since $\Phi(-\rho) = \Phi(\rho)$, (4.2) can be written as

$$P(r) = \frac{i}{\pi} \int_{\infty}^{\infty} r\rho \exp\{-ir\rho\}\, \Phi(\rho)\, d\rho. \tag{4.3}$$

On putting $\kappa = i\rho$ and $\Phi(\rho) = M(\kappa)$, (4.1) and (4.2) become

$$M(\kappa) = \int_0^\infty \frac{\sinh r\kappa}{r\kappa} P(r)\, dr, \tag{4.4}$$

and

$$P(r) = \frac{1}{\pi i} \int_{c-i\infty}^{c+i\infty} M(\kappa) \exp\{-r\kappa\}\, r\kappa\, d\kappa, \tag{4.5}$$

where c is real and $M(\kappa)$ is the spherical moment generating function in an obvious sense.

For a freely jointed chain of n units, $M(\kappa)$ has the form $\exp\{n\mu(\kappa)\}$ with $\mu(\kappa) = \log(\sinh \kappa a/\kappa a)$ and the method of steepest descents can be applied to (4.5). Choosing c to be the real root κ of

$$\mu'(\kappa) = a\mathscr{L}(\kappa a) = \bar{r} = \frac{r}{n}, \tag{4.6}$$

which is a saddle point of the integrand, we obtain as an approximation to $p(r) = P(r)/4\pi r^2$,

$$\tilde{p}(r) = \frac{\kappa \exp\{n(\mu(\kappa) - \kappa\bar{r})\}}{r(2\pi n)^{3/2}\{\mu''(\kappa)\}^{1/2}}, \tag{4.7}$$

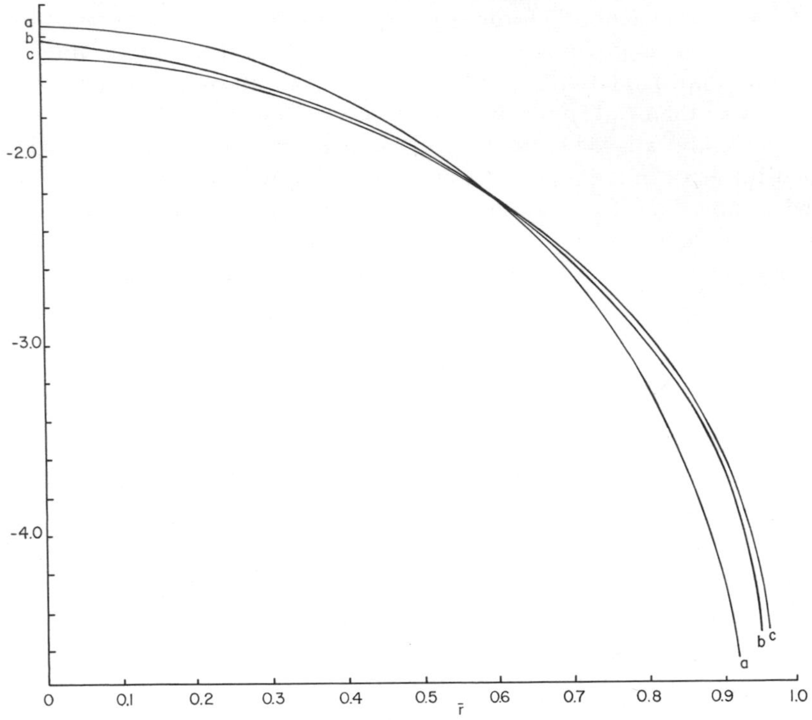

FIGURE 1
Distribution of vector length of freely jointed chain.
Comparison of approximations.
$n = 4$
a: $\log_{10} \hat{p}(r)$ (Kuhn-Grün); b: $\log_{10} p(r)$ (exact); c: $\log_{10} C\tilde{p}(r)$ (saddle point).

This differs from $\hat{p}(r)$ of (3.11) in having the extra factor $\{\mu''(\kappa)\}^{-1/2}$, where

$$\mu''(\kappa) = \operatorname{csech}^2 \kappa + \frac{1}{\kappa^2} = 1 - \bar{r}^2 - \frac{2\bar{r}}{\kappa}. \tag{4.8}$$

Its accuracy can be further improved by normalizing it to $C\tilde{p}(r)$ so that $C \int_0^\infty 4\pi r^2 \tilde{p}(r)\, dr = 1$.

Some idea of the accuracy of (4.7) can be got by comparing it with the exact values of $p(r)$,

$$p(r) = \frac{1}{2^{n+1}\pi r n(n-2)!} \sum_{s=0}^{n} (-)^s \binom{n}{s} \{(n - r - 2s)^+\}^{n-2}, \tag{4.9}$$

where $(x)^+ = \max\{x, 0\}$ (Treloar [16]). In Figure 1, $\log_{10} p(r)$ is compared with $\log_{10} \hat{p}(r)$ and $\log_{10} C\tilde{p}(r)$ for a chain of $n = 4$ units. Apart from the value at $\bar{r} = 0$, the normalized saddle point approximation is surprisingly good

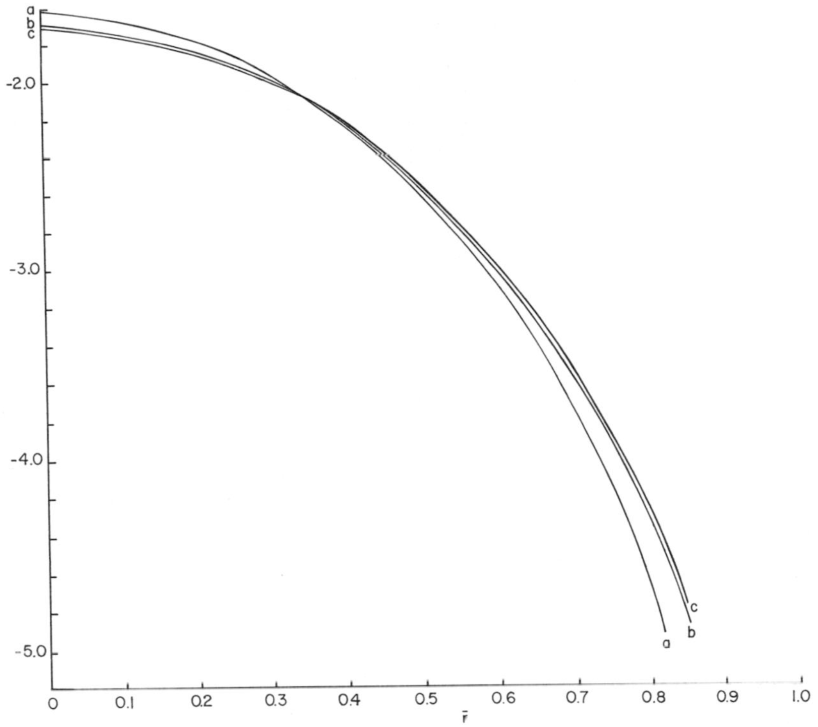

FIGURE 2
Distribution of vector length of freely jointed chain.
Comparison of approximations.
$n = 6$
a: $\log_{10} \hat{p}(r)$ (Kuhn-Grün); b: $\log_{10} p(r)$ (exact); c: $\log_{10} C\tilde{p}(r)$ (saddle point).

and has the advantage of maintaining its accuracy as $\bar{r}$ approaches its maximum, whereas $\log_{10} \hat{p}(r)$ becomes progressively worse. In the case of $n = 6$ shown in Figure 2, $\log_{10} C\tilde{p}(r)$ is practically indistinguishable from $\log_{10} p(r)$, but $\log_{10} \hat{p}(r)$ differs substantially from it, particularly as $\bar{r}$ increases.

For chains of more than about 20 units the extra factors in $\hat{p}(r)$ and $\tilde{p}(r)$ have little effect. The original Langevin approximation (3.8) (with r for x) is then quite adequate and has the merit of simplicity. The entropy takes the form

$$S = \text{constant} + nK\left\{\log\left(\frac{\sinh \kappa a}{\kappa a}\right) - \kappa\bar{r}\right\} \tag{4.10}$$

and the average force on a chain of length r is

$$F = KT\kappa = KT\mathscr{L}^{-1}\left(\frac{\bar{r}}{a}\right). \tag{4.11}$$

At large extensions, $1 - r/a$ is small and

$$\mu'(\kappa) = a\mathscr{L}(\kappa a) \sim a - \frac{1}{\kappa}. \tag{4.12}$$

Then $F \sim KT(1 - \bar{r}/a)^{-1}$ as $\bar{r}$ approaches its maximum value. We refer to this result later when discussing more general chains.

5. More realistic chain models

The simple model of a freely jointed chain of equal units is not adequate to describe the behavior of real chains except in the most general terms. The next most simple model is one where each unit is of fixed length, but free to rotate in a cone of fixed angle whose axis is the previous unit. This might be thought a suitable model for a simple molecule such as polymethylene. However, in reality the rotation is not completely free, but is restricted by the interactions between the groups of atoms making up each unit of the molecular chain. There is a nonuniform probability distribution of angular position, related by the Boltzmann formula to the nonuniform potential energy of angular position. The model assuming complete rotational freedom on the cone might be approached at high temperature.

A method of simplifying the general situation which has been extensively and fruitfully developed by Volkenstein, Flory, and their coworkers is to use the "rotational isomeric" approximation. In this approach, which originated in the work of Montroll [15] and was independently introduced by Kubo [12], the continuous distribution of angular position is replaced by a discrete set of angular states at the minima of the potential energy function, with suitable probabilities attached to them. The direction of the nth unit has a distribution governed by a Markov chain with these transition probabilities, each unit being referred to axes relative to the direction of the previous unit. (One is essentially considering a random walk on a sphere.) For a long chain, the dominant role of the principal eigenvalue is exploited.

The moments of the vector length distribution can be calculated by the methods of Flory [7], and used to approximate to the distribution by an Edgeworth expansion. But for approximations of Kuhn-Grün type the evaluation of the moment generating function itself, as described by Montroll [15], is necessary. Although essentially equivalent to Montroll's analysis, it is simpler to approach the problem directly in terms of the chain vector itself. Three similar attacks on the problem from this point of view were made independently by Kubo [11], Hermans and Ullman [8], and myself [3]. The first two specify the chain vector in terms of fixed axes, in which case in order to preserve the Markov property the vector of the final unit has also to be included in the specification of the system. My own treatment of the problem, like Montroll's, avoids this complication by the use of moving axes, and a brief account of the method now follows.

Suppose we have a chain of n bonds of fixed length a. It is convenient to choose the origin of the axes at the *end point* of the chain, and to let $\mathbf{r}$ be the vector of the *initial point*, relative to this origin, of the chain. The z axis is taken to lie along the nth bond, and the (z, x) plane contains the $(n - 1)$th unit also. If $\mathbf{r}'$ is the corresponding vector for a chain of $n - 1$ bonds, then the coordinates of $\mathbf{r}$ and $\mathbf{r}'$ are related by

$$(5.1) \qquad \begin{bmatrix} x \\ y \\ z \end{bmatrix} = \begin{bmatrix} 0 \\ 0 \\ a \end{bmatrix} + \begin{bmatrix} \cos\alpha\cos\beta & \cos\alpha\sin\beta & -\sin\alpha \\ -\sin\beta & \cos\beta & 0 \\ \sin\alpha\cos\beta & \sin\alpha\sin\beta & \cos\alpha \end{bmatrix} \begin{bmatrix} x' \\ y' \\ z' \end{bmatrix},$$

where α is the bond angle and β the rotational angle. It will be assumed that α has a fixed value and β has p.d.f. $g(\beta)$ (though both a and α could also be random variables). We have a Markov chain with successive probability distribution functions $f_n(x, y, z)$ connected by

$$(5.2) \qquad f_{n+1}(x, y, z) = \int_{-\pi}^{\pi} g(\beta)\, d\beta f_n(x', y', z').$$

The characteristic function

$$(5.3) \qquad \Phi_n(\xi, \eta, \zeta) = E \exp\{i(x\xi + y\eta + z\zeta)\}$$

satisfies

$$(5.4) \qquad \Phi_{n+1}(\xi, \eta, \zeta) = \exp\{ia\xi\}\, \Phi_n(\xi', \eta', \zeta'),$$

where

$$(5.5) \qquad [\xi, \eta, \zeta] = [\xi', \eta', \zeta'] \begin{bmatrix} \cos\alpha\cos\beta & \cos\alpha\sin\beta & -\sin\alpha \\ -\sin\beta & \cos\beta & 0 \\ \sin\alpha\cos\beta & \sin\alpha\sin\beta & \cos\alpha \end{bmatrix}.$$

The polar form of these relations is required. With $\xi = \rho \sin\psi \cos\omega$, $\eta = \rho \sin\psi \sin\omega$, $\zeta = \rho\cos\psi$, and $\phi_n(\xi, \eta, \zeta) = \Phi_n(\rho, \psi, \omega)$, (5.4) is

$$(5.6) \qquad \Phi_{n+1}(\rho, \psi, \omega) = \exp\{ia\rho\cos\psi\} \int_{-\pi}^{\pi} g(\beta)\, d\beta \Phi(\rho, \psi', \omega'),$$

where

$$(5.7) \qquad \cos\psi' = \cos\psi\cos\alpha + \sin\psi\sin\alpha\cos(\omega - \beta),$$

$$(5.8) \qquad \sin\psi'\sin\omega' = \sin\psi\sin(\omega - \beta).$$

It is preferable in the present application to work in terms of the moment generating function. Writing $\kappa = i\rho$, $M_n(\kappa, \psi, \omega) = \Phi_n(\rho, \psi, \omega)$, (5.6) becomes

$$(5.9) \qquad M_{n+1}(\kappa, \psi, \omega) = \exp\{\kappa a \cos\psi\} \int_{-\pi}^{\pi} g(\beta)\, d\beta M_n(\kappa, \psi', \omega').$$

6. Free angular rotation

In the simplest, case β is assumed to be uniformly distributed on $(-\pi, \pi)$. This represents one extreme, corresponding to very high temperatures; the other extreme is the rotational isomeric model, where β can take only discrete values with certain probabilities, an assumption which should be nearly true at low temperatures. Both may be regarded as approximations to the true situation. The first case, which is usually described (rather confusingly) as "freely rotating" has the simplifying feature that the formulae become axially symmetric, (5.9) reducing to

$$M_{n+1}(\kappa, \psi) = \exp\{\kappa a \cos \psi\} \int_{-\pi}^{\pi} \frac{1}{2\pi} M_n(\kappa, \psi')\, d\omega, \tag{6.1}$$

$$\cos \psi' = \cos \psi \cos \alpha + \sin \psi \sin \alpha \cos \omega, \tag{6.2}$$

where $\omega - \beta$ has been replaced by ω.

There is a limiting form of this model which produces the "worm like chain" of Kratky and Porod [13]. It is got by allowing the angle α and the bond length a to tend to zero in such a way that α^2/a remains finite, say $4c$, when (6.1) becomes the differential equation

$$\frac{\partial M}{\partial t} = \kappa u M + c \frac{\partial}{\partial u}\left\{(1 - u^2)\frac{\partial M}{\partial u}\right\} \tag{6.3}$$

with $t = na$ and $u = \cos \psi$. A detailed discussion of the model is given in [3] and [8] and we omit it for brevity.

The function $M(\kappa)$ to be used in (4.5) for deriving the saddle point approximation to the radial distribution is, in the present context,

$$M_n(\kappa) = \tfrac{1}{2} \int_0^{\pi} M_n(\kappa, \psi) \sin \psi\, d\psi. \tag{6.4}$$

Since (6.1) has a positive kernel, $M_n(\kappa, \psi)$, and hence $M_n(\kappa)$ will ultimately become proportional to λ_0^n where λ_0 is the unique real positive maximal eigenvalue of (6.1).

There are two ways of calculating $M_n(\kappa)$. One is to compute it directly by numerical integration from (6.1) and (6.4), starting with $M_0(\kappa, \psi) = 1$. This is quite practicable and is discussed in Section 8. The other, which was used in [3] to develop Edgeworth type expansions based on the Gaussian approximation, is to expand $M_n(\kappa, \psi)$ as a series of Legendre polynomials,

$$M_n(\kappa, \psi) = \sum_{s=0}^{\infty} M_{n,s} P_s(\cos \psi), \tag{6.5}$$

where $M_{n,0} = M_n(\kappa)$. The addition theorem for biaxial harmonics enables $P_n(\cos \psi')$ to be replaced by $P_s(\cos \psi) P_s(\cos \alpha)$ the other terms vanishing on integration, and we get

$$M_{n+1,s} = (s + \tfrac{1}{2}) \sum_{t=0}^{\infty} P_s(\cos \alpha) c_{s,t} M_{n,t}, \qquad s = 0, 1, 2, \cdots, \tag{6.6}$$

where

$$c_{s,t} = \int_{-1}^{1} \exp\{\kappa a u\}\, P_s(u) P_t(u)\, du. \tag{6.7}$$

The $c_{s,t}$ can be calculated from recurrence formulae. No generality is lost by assuming $a = 1$ which we do when convenient.

From this formulation, λ_0 is seen to be also the maximal eigenvalue of the infinite matrix $\mathbf{A}$ with elements

$$a_{s,t} = (s + \tfrac{1}{2}) P_s(\cos \alpha) c_{s,t}. \tag{6.8}$$

The simplest way of computing it was found to be the following one. If $D(\lambda) = |\mathbf{A} - \lambda \mathbf{I}|$ and $B(\lambda)$ is the principal minor of $D(\lambda)$, and if $C(\lambda) = D(\lambda)/B(\lambda)$, then A_0 is the largest real positive zero of $C(\lambda)$, and for large n,

$$M_n(\kappa) \equiv M_{0,\kappa} \sim \frac{\lambda_0^n}{C'(\lambda_0)}. \tag{6.9}$$

The saddle point approximation to $p(r)$ is given by a formula similar to (4.7), with $\mu(\kappa) = \log \lambda_0$, $\mu'(\kappa) = \bar{r}$, except that there is the additional factor $1/C'(\lambda_0)$.

The zeros of successively larger truncations of $C(\lambda)$ are computed until they settle down to a steady value λ_0, which happens quite rapidly unless κ is large. In the examples computed by this method, the next highest root was sufficiently far below λ_0 for (6.9) to be a good approximation for quite small values of n. However, when κ is large (corresponding to high extensions), convergence becomes slow and direct computation as in Section 8 is preferable.

A useful device, due to Kuhn, for comparing distributions for different chain models and different parameter values is to adjust the number n of units in each chain and the unit length a so that the chains have the same maximum length and the same mean square length. In particular, comparison with the freely jointed chain gives a reasonable definition of the number and length of "equivalent" freely jointed units. It is a well-known result that for large n the mean square length of a freely rotating chain of bond angle α is approximately

$$E(r^2) = \frac{na^2(1 + \cos \alpha)}{1 - \cos \alpha} \tag{6.10}$$

and its maximum extended length is, for even n, $r(\max) = na \cos \frac{1}{2}\alpha$, in which case the bonds have a planar zigzag configuration. For a freely jointed chain of N bonds of length b, we have $E(r^2) = Nb^2$, $r(\max) = Nb$. On equating these one finds that

$$N = n \sin^2 \tfrac{1}{2}\alpha, \qquad b = a(\sec \tfrac{1}{2}\alpha - \cos \tfrac{1}{2}\alpha). \tag{6.11}$$

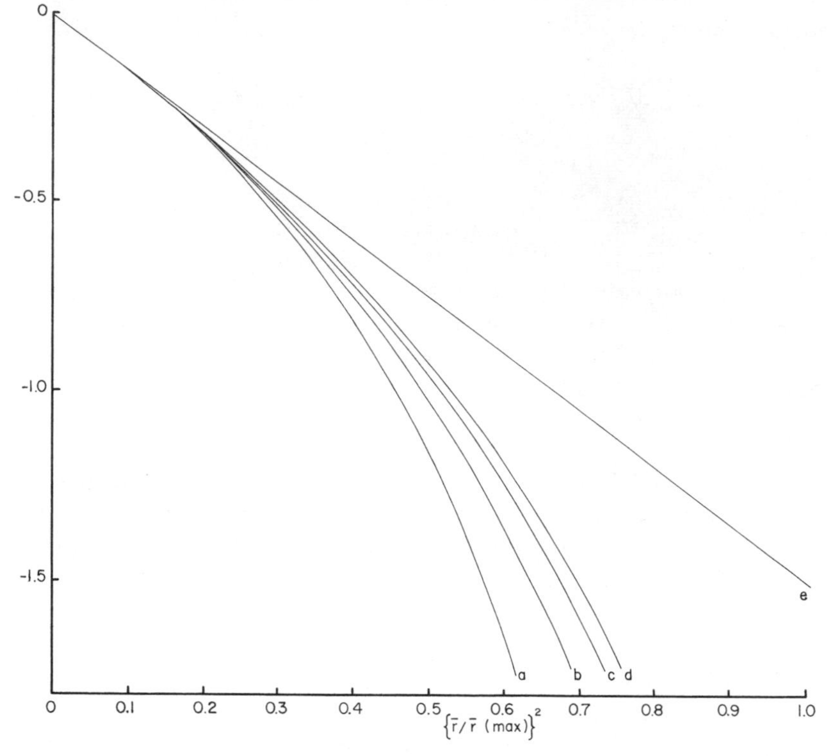

FIGURE 3

Kuhn-Grün type approximations for freely rotating chains.
$\{\mu(\kappa) - \kappa\bar{r}\} \sin^2 \frac{1}{2}\alpha$
a: $\cos\alpha = 0.9$; b: $\cos\alpha = \frac{1}{3}$; c: $\cos\alpha = 0$; d: freely jointed $\mu(\kappa) - \kappa\bar{r}$;
e: Gaussian approximation.

(In the limiting worm like chain model of (6.3), $\frac{1}{2}b$ tends to the so called "persistence length" $1/2c$ of the chain.) When n is large enough for the Kuhn-Grün approximation to be adequate the distribution is determined by $\mu(\kappa) - \kappa\bar{r}$, assuming $a = 1$, and for varying values of α the quantities to be compared are $\{\mu(\kappa) - \kappa\bar{r}\} \sin^2 \frac{1}{2}\alpha$ at the same value of $\bar{r}/\bar{r}(\max) = \bar{r} \sec \frac{1}{2}\alpha$.

The results of some calculations by my colleague R. L. Holder are shown in Figure 3. The function $\{\mu(\kappa) - \kappa\bar{r}\} \sin^2 \frac{1}{2}\alpha$ is plotted against $\{\bar{r}/\bar{r}(\max)\}^2$ for the three cases $\cos\alpha = 0.9$, $\frac{1}{3}$ and 0, together with the freely jointed (Langevin) case for comparison, and the straight line corresponding to the Gaussian approximation with which they all agree for small $\bar{r}/\bar{r}(\max)$. The values for higher $\bar{r}/\bar{r}(\max)$ were computed by the method of Section 8. It will be seen that for $\bar{r}/\bar{r}(\max)$ up to about 0.6 the distributions are not very different and the idea of "equivalent" freely jointed units is a useful one. Beyond this point the distributions begin to diverge and the idea is no longer meaningful for highly extended chains.

7. Asymptotic behavior of the distribution near maximum extension

It is of great interest to know the behavior of $\log p(r)$ in the region where r approaches its maximum value. Since it depends on the behavior of λ_0 when κ is large and convergence is slow by the methods discussed, we need to find an alternative asymptotic approach. This is a problem of some difficulty and the treatment given here is in the nature of a preliminary reconnaissance.

For freely jointed chains, we saw that when κ is large and $1 - \bar{r}/a$ is small, $\bar{r}/a \sim 1 - 1/\kappa a$. Kubo [11] stated without proof that for quite general chains a similar formula holds when κ is large, namely,

$$\frac{\bar{r}}{\bar{r}(\max)} \sim 1 - \frac{1}{\kappa a^*}, \tag{7.1}$$

where a^* is of the same order of magnitude as the unit length a and is characteristic of the chain model. I have not been able to construct a proof of this result, which can be stated in the alternative form

$$\bar{r} = a\mu'(\kappa) \sim \bar{r}(\max) - \frac{\nu}{k}, \tag{7.2}$$

where ν is a dimensionless constant, for the freely jointed chain $\nu = 1$. We now examine the limiting behavior of the integral equation (6.1) when κ is large, using heuristic arguments which strongly suggest that *for the freely rotating model* (7.2) *holds with* $\nu = \frac{3}{8}$, *whatever the bond angle* α.

When the chain is highly extended there is a considerable restriction on the possible configurations it can adopt, and indeed at its full extension it can only take the form of a planar zigzag with angles $\pm\alpha$ between successive bonds. Because of our particular choice of axes, a chain at full extension but otherwise free can rotate rigidly about one end bond lying along the z axis. With an even number $n = 2m$ of bonds, the vector $\mathbf{r}$ is uniformly distributed with fixed length $r = na \cos \frac{1}{2}\alpha$ on a cone of semiangle $\frac{1}{2}\alpha$; with an odd number $n = 2m + 1$, r is greater than $na \cos \frac{1}{2}\alpha$ and the semiangle of the cone is less than $\frac{1}{2}\alpha$, but the differences approach zero as n becomes large.

Consider a rod of length ℓ uniformly distributed on a cone of semiangle γ with axis along the z axis. It is easy to show that for this distribution,

$$\begin{aligned} M(\kappa, \psi) &= \exp\{\kappa\ell \cos \psi \cos \gamma\} I_0(\kappa\ell \sin \psi \sin \gamma) \\ &\sim \frac{\exp\{\kappa\ell \cos(\psi - \gamma)\}}{\sin \gamma (2\pi\kappa\ell)^{1/2}}, \end{aligned} \tag{7.3}$$

when κ is large. In fact, $M(\kappa, \psi)$ is appreciable only when $\psi - \gamma$ is $O(\kappa^{-1/2})$ and $\cos(\psi - \gamma)$ can be replaced by $1 - \frac{1}{2}(\psi - \gamma)^2$ in the exponent. This leads us to expect that for large values of κ corresponding to highly extended chains, $M_n(\kappa, \psi)$ will be concentrated near $\psi = \frac{1}{2}\alpha$ and negligible elsewhere.

Because chains with odd and even n behave differently, it is best to work with an equation relating chains with even n only. Assume $a = 1$ and consider

$$(7.4)\quad M_{2m+2}(\kappa, \psi) = \exp\{\kappa \cos \psi\} \int_{-\pi}^{\pi} \frac{1}{2\pi} \exp\{\kappa \cos \psi'\}\, d\omega \int_{-\pi}^{\pi} \frac{1}{2\pi} M_{2m}(\kappa, \psi'')\, d\omega',$$

where ψ, ψ', ω, and ψ', ψ'', ω' are related by (6.2). Note that

$$(7.5)\quad \cos \psi + \cos \psi' = 2 \cos \tfrac{1}{2}\alpha \cos(\psi - \tfrac{1}{2}\alpha) - \sin \alpha \sin \psi (1 - \cos \omega).$$

Analogy with (7.3) suggests that $\Phi = \psi - \frac{1}{2}\alpha$ should be $O(\kappa^{-1/2})$, so from (6.2),

$$(7.6)\quad \Phi' = -\Phi + \tfrac{1}{2}\omega^2 \sin \alpha + O(\kappa^{-1}),$$

where ω^2 must also be $O(\kappa^{-1/2})$, and hence

$$(7.7)\quad \Phi'' = \Phi + \tfrac{1}{2}(\omega'^2 - \omega^2) \sin \alpha + O(\kappa^{-1}).$$

Writing $M_n(\Phi) \equiv M_n(\kappa, \psi)$ for brevity and using (7.5), we get as an approximation to (7.4),

$$(7.8)\quad M_{2m+2}(\Phi) = \exp\{2\kappa \cos \tfrac{1}{2}\alpha - \kappa \Phi^2 \cos \tfrac{1}{2}\alpha\} \int_{-\infty}^{\infty} \frac{1}{2\pi} \exp\{-\tfrac{1}{2}\kappa \omega^2 \sin \alpha \sin \tfrac{1}{2}\alpha\}\, d\omega \cdot \int_{-\infty}^{\infty} \frac{1}{2\pi} M_{2m}(\Phi + \tfrac{1}{2}(\omega'^2 - \omega^2) \sin \alpha)\, d\omega',$$

when κ is large, the neglected terms being $O(\kappa^{-1/2})$. If we put $\sigma = \kappa \cos \frac{1}{2}\alpha$, $\Phi = \xi \sigma^{-1/2}$, $\frac{1}{2}\omega^2 \sin \alpha = \eta \sigma^{-1/2}$, $\frac{1}{2}\omega'^2 \sin \alpha = \zeta \sigma^{-1/2}$, and $M_{2m}(\Phi) \equiv Q_{2m}(\xi)$, (7.8) reduces to

$$(7.9)\quad Q_{2m+2}(\xi) = \frac{\exp\{2\sigma - \xi^2\}}{2\pi^2 \sin \alpha} \sigma^{-1/2} \int_0^{\infty} \exp\{-\eta \sigma^{1/2} \tan \tfrac{1}{2}\alpha\} \eta^{-1/2}\, d\eta \cdot \int_0^{\infty} Q_{2m}(\xi + \zeta - \eta) \zeta^{-1/2}\, d\zeta.$$

Now ω^2 was required to be $O(\kappa^{-1/2}) = O(\sigma^{-1/2})$, and hence η had to be $O(1)$. However, it will be seen from the exponential term in the integral that provided Q_{2m} is reasonably well behaved the effective range of η is in fact only $O(\sigma^{-1/2})$. We may therefore approximate again by ignoring η in $Q_{2m}(\xi + \zeta - \eta)$ and integrating out η to obtain

$$(7.10)\quad Q_{2m+2}(\xi) = \frac{\exp\{2\sigma\} \cos \frac{1}{2}\alpha}{(\pi \sin \alpha)^{3/2}} \sigma^{-3/4} \exp\{-\xi^2\} \int_0^{\infty} Q_{2m}(\xi + \zeta) \zeta^{-1/2}\, d\zeta.$$

Hence if κ is large, we can write

$$(7.11)\quad Q_{2m}(\xi) \sim \exp\{2m\kappa \cos \tfrac{1}{2}\alpha\} \kappa^{-3m/4} \frac{(\cos \frac{1}{2}\alpha)^{m/2}}{(\pi \sin \alpha)^{3m/2}} W_m(\xi),$$

where $W_m(\xi)$ is independent of κ and satisfies

$$W_{m+1}(\xi) = \exp\{-\xi^2\} \int_0^\infty W_m(\xi + \zeta)\zeta^{-1/2}\, d\zeta. \tag{7.12}$$

The conclusion from this discussion is that when κ is large enough it will occur in $M_n(\kappa, \psi)$ only within the factor $\kappa^{-3n/8} \exp\{n\kappa \cos \frac{1}{2}\alpha\}$, and hence that Kubo's result (7.2) holds with $\nu = \frac{3}{8}$ whatever the value of $\alpha > 0$. The force on a freely rotating chain at nearly maximum extension is, from (4.11),

$$F \sim \tfrac{3}{8}KT(a \cos \tfrac{1}{2}\alpha - \bar{r})^{-1}, \tag{7.13}$$

and it follows from (4.7) that in the extreme tail of the distribution $\tilde{p}(r)$ becomes proportional to $\{1 - \bar{r}/(a \cos \frac{1}{2}\alpha)\}^{3n/8-2}$.

However, apart from its heuristic nature, the discussion is not complete. A more detailed examination of the step from (7.9) to (7.10) (see Appendix) indicates that, relative to the dominant term, the term neglected at this stage is of order $\kappa^{-1/4}$. If the further approximation is to be acceptable $\kappa^{-1/4}$ must therefore be small, and this requires κ to be very large. If κ is not large enough, it appears that $\kappa(a \cos \frac{1}{2}\alpha - \bar{r})$ should lie somewhere between $\frac{3}{8}$ and $\frac{1}{2}$, depending on the value of κ. It may be that Kubo's limiting form is appropriate only for extensions so near the maximum that the assumption of a fixed bond angle is unrealistic.

There is also the fact that if both m and κ are to be large, a double limiting process is involved. One would like $W_m(\xi)$ in (7.12) to settle down for large m to the form $c^m W(\xi)$, where $c > 0$ is an eigenvalue of the equation

$$W(\xi) = c \exp\{-\xi^2\} \int_0^\infty W(\xi + \zeta)\zeta^{-1/2}\, d\zeta. \tag{7.14}$$

Unfortunately, it is known that an equation of this type has no bounded non-zero solution for any c. One can show that when $\xi > 0$,

$$W_m(\xi) = O\big((m!)^{-1/4} \exp\{-m\xi^2\}\big), \tag{7.15}$$

and when $\xi_1 > \xi_2 > 0$,

$$\frac{W_m(\xi_1)}{W_m(\xi_2)} = O\big(\exp\{m(\xi_1^2 - \xi_2^2)\}\big). \tag{7.16}$$

So on the positive axis $W_m(\xi)$ is "consumed" progressively from right to left as m increases, $W_m(0)$ dying away most slowly. Numerical iteration confirms this, and also suggests that for $\xi < 0$, $W_m(\xi)$ dies away even less rapidly (though I cannot prove it) and for large enough m, $W_m(\xi)$ will lie almost entirely on $\xi < 0$.

Nevertheless, whatever the limiting behavior of $W_m(\xi)$, our conclusions should hold for any fixed $n = 2m$ provided κ is large enough.

8. Direct computation from the integral equation

The computation of $M_n(\kappa, \psi)$ from (6.1) is a simple numerical operation most expeditiously conducted in terms of $u = \cos \psi$, so for convenience let us write $M_n(u)$ for $M_n(\kappa, \psi)$. Given that $M_n(u)$ is known for equally spaced values of u on $(-1, 1)$, $u' = \cos \psi'$ is computed from (6.2) for equally spaced values of ω and $M_n(u')$ is interpolated. Numerical integration then gives $M_{n+1}(u)$, and

$$(8.1) \qquad M_n = \frac{1}{2} \int_{-1}^{1} M_n(u)\, du$$

is evaluated at each stage. The process is started with $M_1(u) = \exp\{\kappa u\}$ (assuming $a = 1$). However, $M_n(u)$ soon becomes unmanageably large as n increases, and if κ is large, $M_n(u)$ can take very large values for all values of n. It is better to modify the procedure by calculating

$$(8.2) \qquad L_{n+1}(u) = \exp\{\kappa(u - \cos \tfrac{1}{2}\alpha)\} \int_{-1}^{1} \frac{1}{2\pi} \frac{L_n(u')}{L_n}\, d\omega,$$

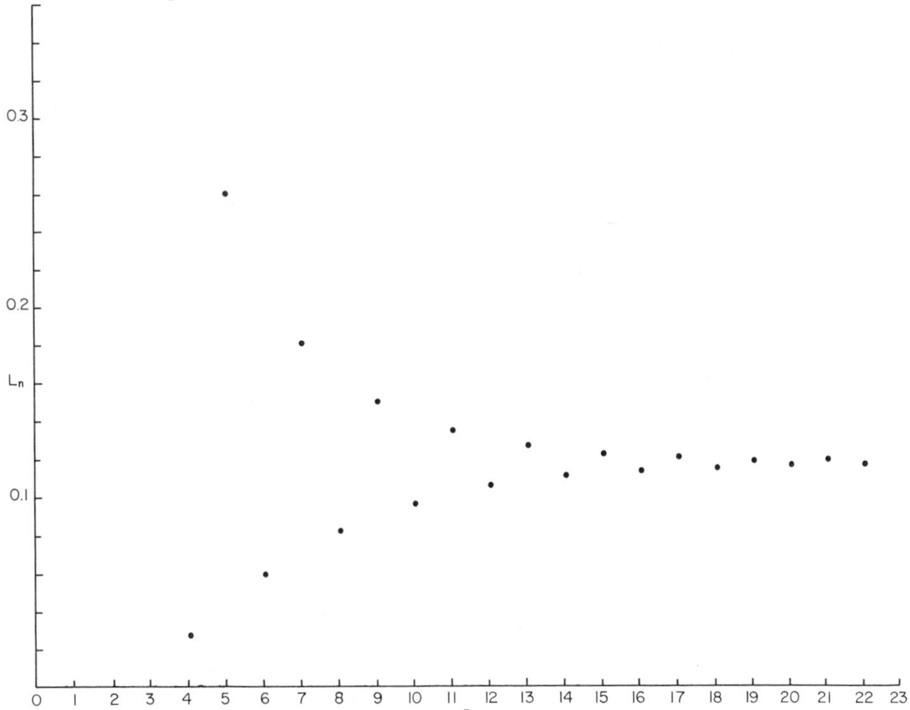

FIGURE 4

Convergence of L_n to limiting value as n increases.
$\cos \alpha_0 = \frac{1}{3}$; $\kappa = 30$; L_n against n.

where

$$L_n = \frac{1}{2}\int_{-1}^{1} L_n(u)\,du. \tag{8.3}$$

It is easily verified that

$$L_n = \frac{M_n}{M_{n-1}} \exp\{-\kappa \cos \tfrac{1}{2}\alpha\}, \tag{8.4}$$

and L_n settles down to $\exp\{\mu(\kappa) - \kappa \cos \frac{1}{2}\alpha\}$ as n increases. If n is large, this may be all that is required. The method of Section 6 produces the factor $1/C'(\lambda_0)$ in the approximation (6.6) to M_n. This factor can be recovered by the present method if $n^{-1} \Sigma_{j=1}^{n} \log L_j - \log L_n$ is computed at each stage, since it tends to $-\log C'(\lambda_0)$ for large n.

When κ is large, $L_n(u)$ becomes concentrated around $u = \cos \frac{1}{2}\alpha$ and a fine grid of values of u is needed if the numerical integration is to be accurate. Provided $C'(\lambda_0)$ is not required, it is preferable for such values of κ to start the process with an arbitrary distribution covering the expected range of values of u, to avoid using an unnecessary amount of computing time in the early stages.

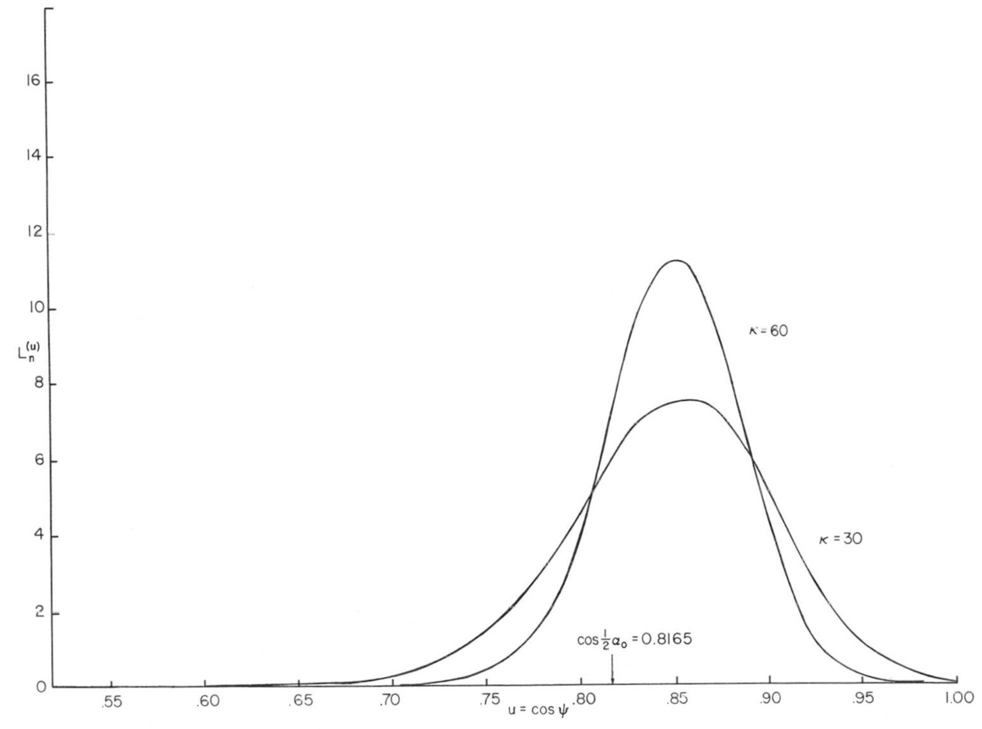

FIGURE 5

$L_n(u)$ for $\kappa = 30$ and $\kappa = 60$, $\cos \alpha_0 = \frac{1}{3}$.

L_n against $u = \cos \psi$ (n large).

The typical manner in which L_n converges to $\exp\{\mu(\kappa) - \kappa \cos \frac{1}{2}\alpha\}$ is shown in Figure 4 for $\cos \alpha = \frac{1}{3}$, $\kappa = 30$. It exhibits the oscillations to be expected for large κ because a highly extended chain behaves differently for odd and even n. The limiting forms of $L_n(u)$ for $\cos \alpha = \frac{1}{3}$ and $\kappa = 30, 60$ are also shown in Figure 5. These are interesting in view of the indication of Section 7 that when κ is very large $W_m(\xi)$ ultimately tends to lie entirely on $\xi < 0$, which implies that $L_n(u)$ should tend to lie entirely on $u > \cos \frac{1}{2}\alpha$. Clearly, $\kappa = 60$ is not large enough for this to happen, but the proportion of $L_n(u)$ for which $u > \cos \frac{1}{2}\alpha$ increases from 0.79 when $\kappa = 30$ to 0.89 when $\kappa = 60$.

9. Restricted rotation

The assumption of complete freedom of rotation is of course not true for real chains. The integral equation (5.9), or some generalization of it, can be used in a similar way when there is restricted rotation, though the abandonment of axial symmetry complicates the formulae considerably. (At the other extreme, we have the rotational isomeric model for which matrix methods are available.) For example, the approach of Section 6 can be used, but in place of (6.5), $M(\kappa, \psi, \omega)$ is expanded in a series of spherical harmonics,

$$M_n(\kappa, \psi, \omega) = \sum_{s=0}^{\infty} \sum_{\ell=-s}^{s} M_{n,s,\ell} P_s^\ell (\cos \psi) \exp\{i\ell\omega\}, \tag{9.1}$$

where $P_s^\ell(u)$ is the associated Legendre function which has an addition theorem involving Jacobi polynomials (see, for example, Edmonds [6]). The relation between the coefficients for n and $n + 1$ is

$$\begin{aligned} M_{n+1,s,\ell} &\\ &= (s + \tfrac{1}{2}) \sum_{t=0}^{\infty} \sum_{m=-t}^{t} M_{n,t,m} (\sin \tfrac{1}{2}\alpha)^{\ell-m} (\cos \tfrac{1}{2}\alpha)^{\ell+m} P_{t-m}^{(\ell-m,\ell+m)}(\cos \alpha) c_{s,t}^\ell g_\ell, \end{aligned} \tag{9.2}$$

where

$$c_{s,t}^\ell = \frac{(s-\ell)!}{(s+\ell)!} \int_{-1}^{1} \exp\{\kappa u\} P_s^\ell(u) P_t^\ell(u)\, du, \tag{9.3}$$

$$g_\ell = \int_{-\pi}^{\pi} \exp\{-im\beta\} g(\beta)\, d\beta, \tag{9.4}$$

$$P_h^{(j,k)}(x) = 2^{-h} \sum_{r=0}^{h} \binom{h+j}{r} \binom{h+k}{h-r} (x-1)^{h-r} (x+1)^r, \tag{9.5}$$

the last being the Jacobi polynomial. K. Kajiwara [10] has arrived at this formula independently in his studies of light scattering which use the characteristic function rather than the moment generating function (Burchard and Kajiwara [2]).

The required coefficient $M_{n,0,0}$ can be computed by finding λ_0 and using a formula like (6.6), but the convergence of the process is slower since the determinants are larger when truncated at a given value of s. The most practical approach seems to be to compute λ_0 directly from the integral equation (5.9) as in Section 8, though, as double integration is involved, the amount of computing time needed will be quite large. A study of the computational problems involved in both these methods is currently being carried out at Birmingham.

A theoretical investigation of the asymptotic behavior at high extensions, as in Section 7, is likely to involve new difficulties. One can imagine chain structures such that the rotational isomeric model has a different fully extended length from the real chain to which it approximates, because it rigorously excludes certain configurations. Then in cases where the rotational isomeric model is a good approximation one would expect a distinct change in the behavior of the distribution in the region beyond the maximum rotational isomeric extension.

◇ ◇ ◇ ◇ ◇

APPENDIX

The substitutions $u = \zeta + \eta$, $v = \zeta - \eta$, $u = |v| \cosh t$ reduce (7.9) to the form

$$\text{(A.1)} \qquad Q_{2m+2}(\xi) = \frac{\exp\{2\sigma - \xi^2\}}{2\pi^2 \sin\sigma}\,\sigma^{-1/2} \cdot \int_{-\infty}^{\infty} \exp\{\tfrac{1}{2}v\sigma^{1/2}\tan\tfrac{1}{2}\alpha\}\, K_0(\tfrac{1}{2}|v|\sigma^{1/2}\tan\tfrac{1}{2}\alpha) Q_{2m}(\xi + v)\, dv,$$

where

$$\text{(A.2)} \qquad K_0(x) = \int_0^{\infty} \exp\{-x\cosh t\}\, dt, \qquad x \geqq 0,$$

is a modified Bessel function of the second kind. When $x > 0$ is large, $K_o(x) \sim (\pi/2x)^{1/2} \exp\{-x\}$, and in fact $x^{1/2} \exp xK_0(x)$ increases monotonically from 0 to $(\frac{1}{2}\pi)^{1/2}$ as x increases from 0 to ∞.

Going from (7.9) to (7.10) is equivalent to ignoring the range $\sigma < 0$ in (A.1) and replacing K_0 by its asymptotic form in the range $\sigma \geqq 0$. Let us examine the effect of this operation. Consider first

$$\text{(A.3)} \qquad I_1 = \int_0^{\infty} \exp\{Nv\}\, K_0(Nv) Q_{2m}(\xi + v)\, dv,$$

where $N = \frac{1}{2}\sigma^{1/2} \tan \frac{1}{2}\alpha$ is large. Since

$$\text{(A.4)} \qquad 0 < \left(\frac{\pi}{2}\right)^{1/2} - x^{1/2} \exp x\, K_0(x) < \varepsilon$$

for all $x > X(\varepsilon)$, we have

$$\text{(A.5)} \qquad \left| I_1 - (\tfrac{1}{2}\pi)^{1/2} \int_0^\infty Q_{2m}(\xi + v)(Nv)^{-1/2}\, dv \right|$$

$$= \int_0^\infty \{(\tfrac{1}{2}\pi)^{1/2} - x^{1/2} \exp x\, K_0(x)\} Q_{2m}(\xi + v)(Nv)^{-1/2}\, dv$$

$$< (\tfrac{1}{2}\pi)^{1/2} \int_0^{X(\varepsilon)/N} Q_{2m}(\xi + v)(Nv)^{-1/2}\, dv$$

$$+ \varepsilon \int_{X(\varepsilon)/N}^\infty Q_{2m}(\xi + v)(Nv)^{-1/2}\, dv$$

$$< (2\pi)^{1/2}\{X(\varepsilon)\}^{1/2} A N^{-1} + \varepsilon \int_0^\infty Q_{2m}(\xi + v)(Nv)^{-1/2}\, dv,$$

where $Q_{2m}(\xi + v) < A$. Next consider

$$\text{(A.6)} \qquad I_2 = \int_{-\infty}^0 \exp\{Nv\}\, K_0(N|v|) Q_{2m}(\xi + v)\, dv$$

$$< (\tfrac{1}{2}\pi)^{1/2} \int_0^\infty \exp\{-2Ny\}\, (Ny)^{-1/2} Q_{2m}(\xi - y)\, dy$$

$$< (\tfrac{1}{2}\pi)^{3/2} A N^{-1}.$$

The dominant part of the integral in (A.1) is therefore of order $N^{-1/2}$ and the neglected part is of order N^{-1}. In terms of κ, this means that the dominant part approximates to the integral to within a factor $1 + O(\kappa^{-1/4})$.

◇ ◇ ◇ ◇ ◇

My principal debt is to Mr. R. L. Holder who collaborated with me extensively on the computational aspects of the work. A detailed account of our joint work will appear elsewhere. I am also grateful to Dr. D. M. G. Wishart and Professor G. E. H. Reuter for useful discussion, and to Professor M. Gordon for reviving my interest in the subject.

REFERENCES

[1] T. M. Birshtein and O. B. Ptitsyn, *Conformations of Macromolecules*, New York, Interscience, 1963.

[2] W. Burchard and K. Kajiwara, "The statistics of stiff chain molecules. I. The particle scattering factor," *Proc. Roy. Soc. Ser. A*, Vol. 316 (1970), pp. 185–199.

[3] H. E. Daniels, "The statistical theory of stiff chains," *Proc. Roy. Soc. Edinburgh, Sect. A*, Vol. 63 (1952), pp. 290–311.

[4] ———, "Saddle point approximations in statistics," *Ann. Math. Statist.*, Vol. 25 (1954), pp. 631–650.

[5] R. A. Dobrushin, Review No. 2800 in *Refer. Ž. Matematika.*, Vol. 6 (1955), p. 89, review of "Statistical physics of linear polymers," by M. V. Volkenstein and O. B. Ptitsyn, *Usephi Phys. Nauk*, Vol. 49 (1953), pp. 501–568. (In Russian.)
[6] A. R. Edmonds, *Angular Momentum in Quantum Mechanics*, Princeton, Princeton University Press, 1959.
[7] P. Flory, *Statistical Mechanics of Chain Molecules*, New York, Wiley, 1968.
[8] J. J. Hermans and R. Ullman, "The statistics of stiff chains with applications to light scattering," *Physica*, Vol. 18 (1952), pp. 951–971.
[9] R. L. Jernigan and P. J. Flory, "Distribution functions for chain molecules," *J. Chem. Phys.*, Vol. 50 (1960), pp. 4185–4200.
[10] K. Kajiwara, Private communication, 1970.
[11] R. Kubo, "Statistical theory of linear polymers. I. Intramolecular statistics," *J. Phys. Soc. Japan*, Vol. 2 (1947), pp. 47–50.
[12] ———, "Statistical theory of linear polymers, V. Paraffin-like chains," *J. Phys. Soc. Japan*, Vol. 4 (1949), pp. 319–322.
[13] O. Kratky and G. Porod, "Röntgenuntersuchung gelöster Fadenmoleküle," *Rec. Trav. Chim.*, Vol. 68 (1949), pp. 1106–1122.
[14] W. Kuhn and F. Grün, "Beziehungen zwischen elastischen Konstanten und Dehnungsdoppelbrechung hochelastischer Stoffe," *Kolloidzschr.*, Vol. 101 (1942), pp. 248–271.
[15] E. W. Montroll, "On the theory of Markoff chains," *Ann. Math. Statist.*, Vol. 18 (1947), pp. 18–36.
[16] L. R. G. Treloar, "The statistical length of long-chain molecules," *Trans. Faraday Soc.*, Vol. 42 (1946), pp. 77–82.
[17] M. V. Volkenstein, *Configurational Statistics of Polymeric Chains*, New York, Interscience, 1966.

COVERAGE OF GENERALIZED CHESS BOARDS BY RANDOMLY PLACED ROOKS

LEO KATZ
MICHIGAN STATE UNIVERSITY
and
MILTON SOBEL
UNIVERSITY OF MINNESOTA

1. Introduction

At a recent colloquium on combinatorial structures, H. Kamps and J. van Lint presented a paper [2] on the minimal number of rooks $\sigma(n, k)$ required to "cover" a generalized chessboard; the latter is represented by R_k^n, the set of n vectors (or cells) with components in the ring of integers mod k. To explain the notion of "cover" we first define the Hamming distance $d_H(\mathbf{x}, \mathbf{y})$ between two vectors ("squares" of the chessboard) as the number of components in which they differ; under the metric d_H, the board R_k^n is a metric space. The familiar chessboard is R_8^2. Then the rook domain or region *covered* by a rook at x is the unit sphere

$$B(x, 1) = \{y \in R_k^n | d_H(\mathbf{x}, \mathbf{y}) \leqq 1\}. \tag{1.1}$$

Kamps and van Lint gave the following table of $\sigma(n, k)$ which represents almost all the known results to date for the above deterministic problem.

TABLE I

KNOWN VALUES OF $\sigma(n, k)$

k \ n	3	4	5	6	7	8	...	13
2	2	4	7	12	16	2^5		
3	5	9	3^3					3^{10}
4	8	24	4^3					
5	13			5^4				
6	18	72						
7	25					7^6		

Research supported by NSF Grants GP-11021 and GP-13484.

The only general results known (see their references) are

$$\sigma(2, k) = k, \tag{1.2}$$

$$\sigma(3, k) = [(k^2 + 1)/2], \tag{1.3}$$

where $[x]$ = integer part of x, and

$$\sigma(n, k) = \frac{k^n}{1 + n(k - 1)}, \tag{1.4}$$

for $n > 3$, provided

(a) the right side of (1.4) is an integer and

(b) the integer k is the power of a prime.

For example, from (1.4), $\sigma(4, 3) = 9$ and from (1.3), we have $\sigma(3, 3) = 5$. Many values of $\sigma(n, k)$ were computed by R. Stanton [4], Stanton and J. Kalbfleisch [5], [6], and others.

We consider two stochastic versions of the rook coverage problem. Rooks are placed in cells (vectors) sequentially and independently with uniform probabilities. We consider the distribution (in particular, the expectation) of the number of rooks Y required to cover R_k^n for the first time. In the multinomial case (case M), the cells have constant probability k^{-n} and repetition of occupancy is permitted. In the hypergeometric case (case H) each successive occupancy is permitted only in one of the currently unoccupied cells, with uniform probability over these cells.

By introducing the stochastic version of the problem, we feel that the problem has been broadened in an interesting and nontrivial manner. Indeed, although the deterministic problem is trivial for $n = 2$, the corresponding stochastic problem is by no means trivial. Moreover, it is hoped that the more general approach used in the stochastic version would lead to further extensions in the deterministic version, especially in the case of higher dimensions.

2. Exact solution for the multinomial case with $n = 2$

Consider a two dimensional $k \times k$ chessboard. For case M, let Y_M denote the random number of rooks required to cover the $k \times k$ board and let y denote values of Y_M. The event "covering a row (column)" is equivalent to "occupying a row (column)."

Coverage of the board R_k^2 is characterized by occupancy either of all the rows *or* of all the columns. We also use the fact that for any given number of rooks N, the number of rows occupied is independent of the number of columns occupied. Finally, occupancy of rows (similarly for columns) is a direct consequence of of the classical Maxwell-Boltzmann statistics (see, for example, p. 59 of Feller [1]). In particular, the probability that all k rows are occupied by x randomly placed rooks is given exactly by

$$F_k(x) = \sum_{\alpha=0}^{k} (-1)^\alpha \binom{k}{\alpha} \left(1 - \frac{\alpha}{k}\right)^x, \tag{2.1}$$

and the same result holds for columns. By virtue of the independence of row and column occupancy, the cumulative distribution function (c.d.f.) $G_k(y)$ of Y_M is given by

$$G_k(y) = 1 - [1 - F_k(y)]^2. \tag{2.2}$$

The corresponding probability law $g_k(y)$ of Y_M is obtained by taking differences in equation (2.2). Expectations are then obtained from $g_k(y)$ or by summing the complement of $G_k(y)$ over $y \geqq 0$; this yields the two equivalent exact expressions

$$(2.3) \quad \begin{aligned} E\{Y_M\} &= k + \sum_{\beta=k}^{\infty} \left[\sum_{\alpha=1}^{k-1} (-1)^{\alpha} \binom{k}{\alpha} \left(1 - \frac{\alpha}{k}\right)^{\beta} \right]^2 \\ E\{Y_M\} &= k + \frac{1}{k^{2(k-1)}} \sum_{i=1}^{k-1} \sum_{j=1}^{k-1} (-1)^{i+j} \frac{\binom{k}{i}\binom{k}{j}(ij)^k}{k^2 - ij}, \end{aligned}$$

both of which are useful for computing (see Table II).

3. Exact solution of the hypergeometric case for $n = 2$

Here rooks are placed one at a time, independently, and with uniform probability in the unoccupied cells. This case requires extensive modification of the solution strategy, mainly due to the loss of independence between row occupancies and column occupancies. We employ the method of inclusion-exclusion and Fréchet sums ([1], p. 99) but the basic events have to be defined carefully.

First, we note that the k^2 vector space (chessboard) is *not* covered by y rooks if and only if at least one cell is not covered and this, in turn, holds if and only if at least one row is not occupied *and* at least one column is not occupied. The event that one particular cell is not covered, in positive terms, requires that all y rooks currently placed are in some $(k - 1) \times (k - 1)$ product subspace defined by the offending cell. Intersections of these subspaces are again product subspaces, which may be indexed by the deleted rows and columns. Thus, we define our basic events $E_{ij}^{(y)}$ as the event (row i and column j are not covered when y rooks are randomly placed). We now proceed to apply the Fréchet sum technique as follows.

In this hypergeometric setup, y rooks can be placed without repetition in $\binom{k^2}{y}$ ways. They can fall in a product subspace avoiding r specified rows and c specified columns in $\binom{(k-r)(k-c)}{y}$ ways and the probability of this event (not necessarily basic) is given by

$$\frac{\binom{(k - r)(k - c)}{y}}{\binom{k^2}{y}}, \qquad y = 0, 1, 2, \cdots. \tag{3.1}$$

Since the r rows and c columns can be specified in $\binom{k}{r}\binom{k}{c}$ ways, the Fréchet sums, for a fixed total $t = r + c, r \geqq 1, c \geqq 1$, of rows and columns not covered, are given by

$$S_t(y) = \sum_{r=1}^{t-1} \frac{\binom{k}{r}\binom{k}{t-r}\binom{(k-r)(k-t+r)}{y}}{\binom{k^2}{y}}. \tag{3.2}$$

According to the discussion above, if a cell is not covered then the sum t of the number of rows and columns not covered is at least 2 and clearly $t \leqq 2k - f(y)$ where $f(y)$ is the minimum total of rows and columns that y rooks can occupy. Hence, the probability of realization of at least one of the basic events is

$$1 - H_k(y) = \sum_{t=2}^{2k-f(y)} (-1)^t S_t(y), \tag{3.3}$$

where $H_k(y)$ is the c.d.f. of the number Y_H of rooks required for coverage in case H.

The expected value of Y_H is obtained by summing (3.3) over $y \geqq 0$. In this sum, the first k terms are all equal to 1. Since $Y_H \leqq 1 + (k-1)^2$, it follows that $1 - H_k(y) = 0$ for $y \geqq 1 + (k-1)^2$, and hence,

$$E\{Y_H\} = k + \sum_{y=k}^{(k-1)^2} (1 - H_k(y)). \tag{3.4}$$

This completes the exact solution for $E\{Y_H\}$ in case H (see Table II).

4. Asymptotic evaluations

In case M, we have from (2.1) asymptotically ($k \to \infty$)

$$F_k(x) = \sum_{\alpha=0}^{k} (-1)^\alpha \binom{k}{\alpha}\left(1 - \frac{\alpha}{k}\right)^x \sim (1 - e^{-x/k})^k \sim \exp\{-ke^{-x/k}\}. \tag{4.1}$$

Using the normalizing transformation ([1], p. 106),

$$X = k \log k + kZ, \tag{4.2}$$

we obtain for large k the limiting c.d.f. of Z (which takes on values z)

$$V_k(z) = \exp\{-e^{-z}\}, \qquad -\infty < z < \infty, \tag{4.3}$$

the (standardized) extreme value distribution.

In our application, Y_M is the smaller of two independent chance variables each having the same c.d.f. $F_k(x)$ and it follows from (4.2) that for $k \to \infty$,

$$E\{Y_M\} \sim k \log k + kE\{Z_{1:2}\} = k(C + \log k), \tag{4.4}$$

where $Z_{1:2}$ is the smaller of two independent chance variables with c.d.f. $V_k(z)$

in (4.3) and $E\{Z_{1:2}\} = C = -0.1159315$ by the table of J. Lieblein and H. Salzer [3].

In case H we no longer have independence of row and column coverage and have to resort to an "*ad hoc* method" to obtain a useful approximation which is as good as the approximation already obtained for case M. Indeed, one reason for considering the two cases together in the same paper is that we suspected that asymptotically the expectations for case M and case H would be the same to the first order approximation.

We make use of the fact that if we delete repetitions in placing Y_M rooks at random by the multinomial scheme, then the remaining observations Y_H are formally indistinguishable from a hypergeometric sample sequence. The difference $D = Y_M - Y_H$ is the redundancy in the multinomial sampling and our evaluation of $E\{Y_H\}$ arises by using

$$E\{Y_H\} = E\{Y_M\} - E\{D\}. \tag{4.5}$$

To evaluate $E\{D\}$, we first write $D = \Sigma_{i=1}^k \Sigma_{j=1}^k D_{ij}$, where D_{ij} is the redundancy due to extra rooks placed in the (i, j) cell. The total number of rooks placed in the (i, j) cell under multinomial sampling is approximately binomial with parameters Y_M and $1/k^2$. Our "*ad hoc* method" is to replace Y_M by $E\{Y_M\}$ in evaluating $E\{D_{ij}\}$; we justify this by noting that the error introduced in the last expressions of (4.7) and (4.8) below is of the order of magnitude

$$O\left(\frac{E\{Y_M\}}{k^2}\right) = O\left(\frac{C + \log k}{k}\right) \to 0 \tag{4.6}$$

as $k \to \infty$. We now obtain

$$\begin{aligned} E\{D_{ij}\} &\sim \sum_{\alpha=2}^{E\{Y_M\}} (\alpha - 1)\binom{E\{Y_M\}}{\alpha}\left(\frac{1}{k^2}\right)^{\alpha}\left(1 - \frac{1}{k^2}\right)^{E\{Y_M\} - \alpha} \\ &= \frac{1}{k^2} E\{Y_M\} - 1 + \left(1 - \frac{1}{k^2}\right)^{k(C + \log k)}. \end{aligned} \tag{4.7}$$

Using (4.4) for $E\{Y_M\}$ and expanding the last term in (4.7) gives

$$E\{D_{ij}\} \sim \frac{1}{2}\left(\frac{C + \log k}{k}\right)^2 + O\left(\frac{\log^3 k}{k^3}\right) \tag{4.8}$$

and the error term in (4.8) can also be disregarded. Thus, for the total set of k^2 cells we have from (4.8),

$$E\{D\} \sim \frac{1}{2}(C + \log k)^2 + O\left(\frac{\log^3 k}{k}\right), \tag{4.9}$$

and hence by (4.5),

$$E\{Y_H\} \sim k(C + \log k) - \frac{1}{2}(C + \log k)^2, \tag{4.10}$$

where the error, which tends to zero as $k \to \infty$, is now omitted.

TABLE II

Expected Value of the Number of Random Rooks Required to Cover the k^2 Chessboard

k	$E\{Y_M\}$	Approximation to $E\{Y_M\}$ based on (A.9)	$E\{Y_H\}$	Approximation to $E\{Y_H\}$ based on (A.10)
2	2.3333333333	1.5115	2.0000000	1.8619
3	4.1821428571	3.7886	3.5000000	3.6634
4	6,3655677654	6.1561	5.3522478	5.5832
5	8.7938685820	8.6870	7.4723892	7.7268
6	11.4171670989	11.1376	9.8091916	10.0743
7	14.2030879491	14.2070	12.3278253	12.5990
8	17.1286506847	17.1658	15.0029299	15.2784
9	20.1766249904	20.2393	17.8152024	18.0941
10	23.3335906237	23.4163	20.7494692	21.0315
11	26.5887915430	26.6878	23.7935002	24.0784
12	29.9334107812	30.0458	26.9372363	27.2250
13	33.3600877782	33.4837		30.4628
14	36.8625841610	36.9958		33.7848
15	40.4355447768	40.5770		37.1847
16	44,074322209	44.2229		40.6573
17	47.77484495	47.9297		44.1980
18	51.5335164	51.6940		47.8025
19	55.3471359	55.5125		51.4674
20	59.212836	59.3827		55.1892
21	63.12803	63.3019		58.9652
22	67.09038	67.2679		62.7925
23	71.09771	71.2786		66.6689
24	75.1481	75.3321		70.5921
25	79.2396	79.4267		74.5600
26	83.3704	83.5607		78.5710
27	87.539	87.7327		82.6231
28	91.743	91.9413		86.7150
29	95.981	96.1852		90.8450
30	100.250	100.4632		95.0119

Table II gives exact values of $E\{Y_M\}$ for $k = 2(1)30$ using (2.3) and approximate values based on (A.9). It also gives exact values of $E\{Y_H\}$ for $k = 2(1)12$ using (3.4) and approximate values based on (A.10). Roundoff errors in this table are estimated to be at most one in the last digit shown.

5. Coverage of k^n board for $n > 2$

Define a skeletal axis centered at cell C as the n mutually perpendicular lines of cells parallel to the sides of the hypercube and having the cell C in common; for $n = 3$ denote the cell by $C_{\alpha,\beta,\gamma}$, $\alpha, \beta, \gamma = 1, 2, \cdots, k$, and the corresponding skeletal axis by $C^{\alpha,\beta,\gamma}$. For any n, a cell $C_{\alpha,\beta,\gamma}$ is not covered if and only if the skeletal axis $C^{\alpha,\beta,\gamma}$ has no occupancies. Hence, we can use as our basic sets for an inclusion-exclusion argument the sets $C^{\alpha,\beta,\gamma}$, $\alpha, \beta, \gamma = 1, 2, \cdots, k$. However, the intersections of these skeletal axes are not simple and the corresponding analysis is complicated even for $n = 3$. A complete discussion of this analysis will not be considered here. Thus, the stochastic problem becomes more difficult as n increases as it does in the deterministic case of Kamps and van Lint [2]. Theodore Levy, a student of one of the authors at Michigan State University, is working on a class of such problems; the results are not yet very encouraging.

6. Use of independence in higher dimensions

It is of some interest to find a way to generalize the independence of row occupancy and column occupancy that was used above for $n = 2$. For this purpose, we define a piece that starts at a cell C in n dimensions and moves (anywhere) inside any Hamming sphere centered at C and of radius $n - 1$. For $n = 2$, this reduces to the usual rook move. For $n = 3$ and starting at cell C, the piece moves inside the horizontal plane (H plane) through C or inside the north-south plane (NS plane) through C or inside the east-west plane (EW plane) through C. Hence, one such piece covers all the cells in three mutually perpendicular slabs that contain the starting cell..

The cube R_k^3 will be covered as soon as either all L slabs or all NS slabs or all EW slabs are occupied. Hence, the same argument as for $n = 2$ (case M) gives for general n (case M) the exact solution for the c.d.f. of Y_M.

$$G_k(y) = 1 - [1 - F_k(y)]^n, \tag{6.1}$$

where $F_k(y)$ is given by (2.1). For $n = 3$, the expectation becomes

$$\begin{aligned} E\{Y_M\} &= k - \sum_{\beta=k}^{\infty} \left[\sum_{\alpha=1}^{k-1} (-1)^{\alpha} \binom{k}{\alpha} \left(1 - \frac{\alpha}{k}\right)^{\beta} \right]^3 \\ &= k - \frac{1}{k^{3(k-1)}} \sum_{\alpha=1}^{k-1} \sum_{\beta=1}^{k-1} \sum_{\gamma=1}^{k-1} (-1)^{k-\alpha-\beta-\gamma} \frac{\binom{k}{\alpha}\binom{k}{\beta}\binom{k}{\gamma}(\alpha\beta\gamma)^k}{k^3 - \alpha\beta\gamma}, \end{aligned} \tag{6.2}$$

both of which can be used for computing.

In the corresponding asymptotic ($k \to \infty$) evaluation for $n = 3$, we need the expectation of the smallest of three independent observations on the c.d.f. (4.3); this is given in [3] as -0.4036136. This analysis is easily generalized to any number of dimensions n. This type of solution became possible only after we defined a "super piece" that moved in more than one dimension. No similar analysis was found for the original definition of a rook move in the Hamming sphere of radius 1.

◇ ◇ ◇ ◇ ◇

APPENDIX

A more careful evaluation of $E\{Y_M\}$ starts as in (4.1) but, letting $x' = x - 1$ and $k' = k - 1$ we now write

$$F_k(x) = \sum_{\alpha=0}^{k} (-1)^\alpha \binom{k}{\alpha}\left(1 - \frac{\alpha}{k}\right)^x = \sum_{\alpha=0}^{k'} (-1)^\alpha \binom{k'}{\alpha}\left(1 - \frac{\alpha}{k}\right)^{x'}. \tag{A.1}$$

Letting $x'' = x - 2$, we use the approximation

$$\begin{aligned}\left(1 - \frac{\alpha}{k}\right)^{x'} &\sim \left(\exp\left\{-\frac{\alpha}{k}\right\} - \frac{1}{2}\frac{\alpha^2}{k^2}\right)^{x'} \\ &\sim \exp\left\{-\frac{\alpha x'}{k}\right\} - \frac{x'}{2}\frac{\alpha^2}{k^2}\exp\left\{-\frac{\alpha x''}{k}\right\}.\end{aligned} \tag{A.2}$$

Substituting this in (A.1) and summing gives

$$\begin{aligned}F_k(x) \sim &\left(1 - \exp\left\{-\frac{x'}{k}\right\}\right)^{k'} \\ &- \frac{x'k'(k'-1)}{2k^2}\exp\left\{-\frac{2x''}{k}\right\}\left(1 - \exp\left\{-\frac{x''}{k}\right\}\right)^{k'-2} \\ &+ \frac{x'k'}{2k^2}\exp\left\{-\frac{x''}{k}\right\}\left(1 - \exp\left\{-\frac{x''}{k}\right\}\right)^{k'-1},\end{aligned} \tag{A.3}$$

which is correct up to terms of order $\log k/k$ and $1/k$ in $F_k(x)$. These in turn yield all the terms in the final answer for $E(Y_M\}$ of order $\log k$ and 1, so that the new error will go to zero as $k \to \infty$.

Letting $X' = k \log k' + kZ$, we obtain for large k

$$F_k(x) \sim \exp\{-e^{-z}\} + \frac{(z + \log k')}{2k} e^{-z} \exp\{-e^{-z}\}(1 - e^{-z}), \tag{A.4}$$

where the leading term is the same as in (4.3) and we have dropped terms that approach zero in the final result. Using the same method as in (2.3) above, we replace the sum by an integral and obtain

$$\begin{aligned}E\{Y_M\} &\sim k + k\int_a^\infty \left[1 - \exp\{-e^{-z}\} - \frac{(z + \log k')}{2k} e^{-z}\exp\{-e^{-z}\}(1 - e^{-z})\right]^2 dz \\ &\sim k + k\int_a^\infty (1 - \exp\{-e^{-z}\})^2\, dz \\ &\quad - \int_a^\infty (1 - \exp\{-e^{-z}\})(1 - e^{-z}) \\ &\quad \cdot (z + \log k')\, e^{-z}\exp\{-e^{-z}\}\, dz,\end{aligned} \tag{A.5}$$

where $a = 1 - \log k' - 3/2k$ and we have taken a as the value of z when $x = k - \frac{1}{2}$ to get a valid approximation. Using the Euler-MacLaurin sum formula, it can be shown that this leads a correct asymptotic approximation for $k \to \infty$; we omit this proof.

Let T_1 and T_2 denote the first two terms and the last term in (A.5), respectively. We let $u = e^{-z}$ and let $b = e^{-a} = k'/\exp\{1 - 3/2k\} \sim k'/e$. Using (5.1.1), (5.1.40), and (5.1.51) of [7], we obtain

$$
\begin{aligned}
\text{(A.6)} \qquad T_1 &= k + k \int_0^b \frac{(1 - e^{-u})^2}{u}\, du \\
&= k + 2k \int_0^b \frac{(1 - e^{-u})}{u}\, du - k \int_0^{2b} \frac{(1 - e^{-u})}{u}\, du \\
&= k + 2k[E_1(b) + \log b + \gamma] - k[E_1(2b) + \log 2b + \gamma] \\
&\sim k + k(\gamma - \log 2 + \log b) \sim \tfrac{3}{2} + k(C + \log k'),
\end{aligned}
$$

where $\gamma = 0.5772156649$ is Euler's constant and $C = \gamma - \log 2$. This agrees with (4.4) in the two leading terms.

From (A.5) we obtain, for T_2,

$$
\begin{aligned}
\text{(A.7)} \qquad T_2 &\sim \int_0^b e^{-u}(1 - e^{-u})(1 - u) \log\left(\frac{u}{k'}\right) du \\
&= \left\{\left(\log \frac{u}{k'}\right)\left[ue^{-u} - \left(\frac{2u - 1}{4}\right)e^{-2u} - \frac{1}{4}\right]\right\}_0^b \\
&\quad - \left\{\frac{1}{4} e^{-2u}\right\}_0^b + \frac{1}{4}\int_0^{2b} \frac{(1 - e^{-x})}{x}\, dx \sim \frac{\gamma + 1 + \log 2k'}{4}.
\end{aligned}
$$

Combining this with T_1 in (A.6) gives

$$
\text{(A.8)} \qquad EY_M \sim k(\log k' + C) + \tfrac{1}{4}(\log 2k' + \gamma + 7),
$$

which contains all terms not approaching zero in the asymptotic expansion.

The same method used above can be extended to give the (correction) terms of order $(\log 2k')^\alpha/k$ for $\alpha = 0, 1$, and 2, and we give these without proof. The complete result, including all terms of order $1/k$ and larger, is

$$
\begin{aligned}
\text{(A.9)} \qquad E\{Y_M\} &\sim k(\log k' + C) + \frac{\log 2k' + 7 + \gamma}{4} + \frac{1}{16k}\{(\log 2k')^2 \\
&\quad + (2\gamma - 1)\log 2k' - (9 + \gamma + 2\log 2)\}.
\end{aligned}
$$

This is tabulated in Table II for comparisons with the exact answers; the error appears to be less than $\frac{1}{2}$ for all $k \geqq 2$.

An improvement is also possible in the approximation (4.10) for $E\{Y_H\}$. For large values of Y_M (for example in the neighborhood of $E\{Y_M\}$), the total number of rooks in the i, j cell under multinomial sampling is approximately binomial with parameters $Y_M - 1$ and $1/k^2$. The reason for using $Y_M - 1$ is that the last rook set down cannot be a duplicate.

If we go through the same analysis as in Section 4 for $E\{Y_H\}$ with Y_M replaced by $Y_M - 1$ (except in (4.5) and the definition of D) and use (A.9) for $E\{Y_M\}$, then we obtain instead of (4.10)

$$E\{Y_H\} \sim k(\log k' + C) - \frac{(\log k' + C)^2}{2} + \frac{\log 2k' + 7 + \gamma}{4} - \frac{(\log k' + C)}{4k}(\log 2k' + \gamma + 1). \tag{A.10}$$

This is the quantity tabulated in the last column of Table II.

◇ ◇ ◇ ◇ ◇

The authors wish to thank Mr. Theodore Levy and Ms. Elaine Frankowski for their assistance in programming Table II. Thanks are also due to Mr. Gary Simons of Stanford University for some preliminary desk computer computations and to Mr. Arthur Roth for proofreading.

REFERENCES

[1] W. Feller, *An Introduction to Probability Theory and its Applications, Volume 1*, New York, Wiley, 1968 (3rd ed.).

[2] H. J. L. Kamps and J. H. van Lint, "A covering problem," *Colloquium on Combinatorial Theory and Its Applications*, Balatonfüred, Hungary, 1969, (edited by P. Erdös, A. Rényi, and Vera T. Sós, Budapest, Janós, 1970, distributed by North Holland Publishing Co., Amsterdam; Humanities Press, New York.

[3] J. Lieblein and H. E. Salzer, "Table of the first moment of ranked extremes," *J. Res. Nat. Bur. Standards*, Vol. 59 (1957), pp. 203–206.

[4] R. G. Stanton, "Covering theorems in groups (or: How to win at football pools)," *Recent Progress in Combinatorics*, New York, Academic Press, 1969.

[5] R. G. Stanton and J. G. Kalbfleisch, "Covering problems for dichotomized matchings," *Aequationes Math.*, Vol. 1 (1968), pp. 94–103.

[6] ———, "Intersection inequalities for the covering problem," *SIAM J. Appl. Math.*, Vol. 17 (1969), pp. 1311–1316.

[7] I. A. Stegun and M. Abramowitz (editors), *Handbook of Mathematical Functions*, Appl. Math. Ser. 55, Washington, D.C., National Bureau of Standards, 1964.

PRESSURE AND HELMHOLTZ FREE ENERGY IN A DYNAMIC MODEL OF A LATTICE GAS

RICHARD HOLLEY
PRINCETON UNIVERSITY

1. Introduction

In this paper, we will study a model of an infinite volume one dimensional lattice gas. Our model differs from the usual model of a lattice gas in that the configuration of particles is a stochastic process. That is, the particles in the system will move around, and we will be studying properties of the system which are related to the motion. The particular interaction which governs the behavior of each particle will be introduced in Section 2. This interaction was discovered by F. Spitzer [4].

In Section 2, we will define the Helmholtz free energy in the usual way and prove that at constant temperature the Helmholtz free energy does not increase with time. In thermodynamics, this is usually derived as a consequence of the second law of thermodynamics. In Section 3, we will use the results obtained in Section 2 to prove that all shift invariant equilibrium states are limiting Gibbs distributions. Finally, in Section 4, we use the intuitive description of the interaction of the particles to motivate a definition for the pressure of a state. The usual definition of pressure used in statistical mechanics is only given for limiting Gibbs distributions, and the two definitions do not agree there. However, we will show that when they are both defined, they are both strictly increasing functions of the particle density at constant temperature. In the case of the usual definition this is well known.

In order to keep the notation as simple as possible, we will only consider one model in this paper. This model can clearly be generalized in several ways. Many of these generalizations can be found in [4]. The techniques in this paper are adequate to handle some, though by no means all, of these generalizations.

2. Helmholtz free energy

2.1. *Intuitive description.* We begin by giving an intuitive description of the stochastic process. For a careful proof that this process really exists the reader is referred to [1].

This work was prepared while the author was a Miller Fellow in the Statistics Department, University of California, Berkeley.

Let Z represent the integers, and give $\{0, 1\}$ the discrete topology. We will take $E = \{0, 1\}^Z$ with the product topology for the state space. If $\eta \in E$, η will be interpreted as a configuration of particles on the integers with a particle at x if and only if $\eta(x) = 1$.

Let V be a real valued function on the nonnegative integers such that for some positive integer L, $V(n) = 0$ if $n > L$. We define a pair potential $U(x, y)$ by the formula $U(x, y) = V(|x - y|)$. We think of $U(x, y)$ as the potential energy due to particles at x and y, and attribute half of this energy to each of the particles. If the system is in configuration η and $\eta(x) = 1$, then the particle at x has energy equal to

$$\tfrac{1}{2} \sum_y U(x, y)\eta(y), \tag{2.1}$$

$\frac{1}{2}U(x, x)$ represents the chemical potential of the particle at x.

Now let β be a positive constant which will represent the reciprocal of the temperature. Then if the system is in configuration η at time t, the particle at x attempts to make a jump during the time interval $(t, t + \Delta t)$ with probability

$$\exp\{\beta \sum_y U(x, y)\eta(y)\}\Delta t + o(\Delta t). \tag{2.2}$$

When a jump is attempted, the particle tries to move to the right one or to the left one, each with probability $\frac{1}{2}$. The direction of the attempted jump is independent of the time of the jump and the position of the particle. If the site where it is trying to go is unoccupied, it goes there. Otherwise, it remains where it is and starts over. It turns out that although the motion of each individual particle is not Markovian, the stochastic process of the configuration η_t is Markovian; we will denote the probability that the system goes from configuration η to a configuration in a Borel set $A \subseteqq E$ in time t by $P^\eta(\eta_t \in A)$.

The probability measures on the Borel subsets of E will be called states. If μ_0 is any initial state, we will define μ_t to be the state which assigns measure

$$\mu_t(A) = \int P^\eta(\eta_t \in A)\mu_0(d\eta) \tag{2.3}$$

to the set A.

A measure μ_0 will be called an equilibrium state, if $\mu_0 = \mu_t$ for all $t \geqq 0$. In [1], a set C_0 of equilibrium states for this Markov process is given. We will describe below a set C which certainly contains C_0 and at first glance looks as though it may be strictly larger than the closed convex hull of C_0. It is probably true that C is equal to the closed convex hull of C_0, although we have been unable to prove this. A proof that all of the measures in C are equilibrium states can be accomplished by a slight modification of the proof in [1].

2.2. *The set C.* Let N be an integer greater than L and let Y be a subset of the integers contained in $[-N, N]\backslash[-N + L, N - L]$. Define

$$(2.4) \quad S(N, n, Y) = \left\{ \eta \in E \,\middle|\, \sum_{-N+L}^{N-L} \eta(x) = n, \text{ if } N - L < |y| \leqq N \text{ then } \eta(y) = 1 \text{ if and only if } y \in Y, \text{ and if } |z| > N \text{ then } \eta(z) = 0 \right\}.$$

Now let $\nu_{N,n,Y}$ be the probability measure on $S(N, n, Y)$ given by the formula

$$(2.5) \quad \nu_{N,n,Y}(\{\eta\}) = \psi(N, n, Y) \exp\left\{ -\tfrac{1}{2}\beta \sum_{a=-N}^{N} \sum_{b=-N}^{N} U(a, b)\eta(a)\eta(b) \right\}.$$

Here $\psi(N, n, Y)$ is the normalizing constant.

We may think of $\nu_{N,n,Y}$ as a probability measure on E which gives zero measure to the complement of $S(N, n, Y)$. Now let ν_N be any convex combination of the $\nu_{N,n,Y}$, where n is allowed to vary between 0 and $2N - 2L + 1$ and Y varies over all subsets of $[-N, N]\backslash[-N + L, N - L]$.

If we do this for each N, we get a sequence of probability measures on a compact metric space. The set C consists of all the possible weak limit points of sequences obtained in this way.

We still need a little more notation. Let Λ be a finite subset of Z and let μ be a probability measure on E. Then μ^Λ will denote the probability measure on the subsets of Λ defined by

$$(2.6) \quad \mu^\Lambda(X) = \mu(\{\eta \mid \text{ if } x \in \Lambda, \text{ then } \eta(x) = 1 \text{ if and only if } x \in X\}).$$

If $\Lambda = [-N, N]$, we will use μ^N instead of $\mu^{[-N,N]}$.

Let us recall the definition of Helmholtz free energy in thermodynamics. If U represents the internal energy of a system in a certain state, and S and T are, respectively, the entropy and temperature of that state, then the Helmholtz free energy of that state is defined to be $U - ST$. For an infinite volume system, such as the one with which we are dealing, both the internal energy and the entropy may be infinite. In that case, this definition does not make any sense; however, we can define the Helmholtz free energy per site.

Let μ be a state on E. We define its Helmholtz free energy per site, $A(\mu)$, as follows

$$(2.7) \quad A(\mu) = \limsup_{N\to\infty} \frac{1}{2N + 1}\left(\sum_{X \subseteq [-N,N]} U(X)\mu^N(X) + T \sum_{X \subseteq [-N,N]} \mu^N(X) \log \mu^N(X) \right).$$

Here $U(X) = \frac{1}{2} \sum_{x,y\in X} U(x, y)$. The first sum in the definition of $A(\mu)$ represents the internal energy of the state μ between $-N$ and N, while the second sum represents the negative of the entropy of μ between $-N$ and N.

Throughout this paper 0 log (0) is understood to be zero. Some of our proofs require special attention to the case when 0 log (0) appears in an expression. The modifications necessary then will be left to the reader.

An essential tool which we will need is the infinitesimal generator of the Markov process η_t. It is proved in [1] that if f is a continuous function on E which depends on only finitely many coordinates, then the infinitesimal generator Ω operating on f is given by

$$(2.8)\qquad \Omega f(\eta) = \sum_{x,y\in Z, |x-y|=1} \eta(x)[1 - \eta(y)]c(x, \eta)[f(\eta_{x,y}) - f(\eta)],$$

where $c(x, \eta) = \frac{1}{2}\exp\{\beta \sum_{w\in Z}\eta(w)U(x, w)\}$, and

$$(2.9)\qquad \eta_{x,y}(w) = \begin{cases} \eta(w) & \text{if } w \neq x, w \neq y, \\ 0 & \text{if } w = x, \\ 1 & \text{if } w = y. \end{cases}$$

The use we will make of this is given in the following lemma.

LEMMA 2.1. *Let* $W(X\cup Y, a) = \frac{1}{2}\exp\{\beta \sum_{c\in X\cup Y} U(a, c\}$ *and*

$$(2.10)\qquad D(N, X, Y) = \{(a, b)|a \in X\cup Y, b \notin X \cup Y, |a - b| = 1, \textit{and } |a| \leqq N \textit{ or } |b| \leqq N\}.$$

Then

$$(2.11)\qquad \frac{d}{dt}\mu_t^N(X) = \sum_{Y}\sum_{(a,b)\in D(N,X,Y)} W(X\cup Y\cup b\backslash a, b)\mu_t^{N+L}(X\cup Y\cup b\backslash a) - \sum_{Y}\sum_{(a,b)\in D(N,X,Y)} W(X\cup Y, a)\mu_t^{N+L}(X\cup Y).$$

In (2.10) *the summation over* Y *is over all subsets* Y *of* $[-N - L, -N - 1]\cup [N + 1, N + L]$.

Throughout this paper if X and Y are subsets of the integers, and a and b are integers, we will write $X\cup Y\cup b\backslash a$ instead of $(X\cup Y\cup \{b\})\backslash\{a\}$.

The proof of Lemma 2.1 is simply an application of (2.8) and will be left to the reader. Recall that we are assuming that $U(a, b) = 0$ if $|a - b| > L$.

THEOREM 2.1. *With* $A(\cdot)$ *and* μ_t *as defined above,* $A(\mu_t)$ *is a nonincreasing function of* t.

PROOF. Recalling that $\beta = 1/T$, we may revise (2.7) slightly to obtain

$$(2.12)\qquad A(\mu_t) = \limsup_{N\to\infty}\frac{1}{\beta}\frac{1}{2N+1}\sum_{X\subseteqq[-N,N]}\mu_t^N(X)[\log \mu_t^N(X) + \beta U(X)] = \limsup_{N\to\infty}\frac{1}{\beta}\frac{1}{2N+1}\sum_{X\subseteqq[-N,N]}\mu_t^N(X)\log\frac{\mu_t^N(X)}{P(X)},$$

where $P(X) = \exp\{-\beta U(X)\}$. To finish the proof, it will be sufficient to show that

$$(2.13)\qquad \frac{d}{dt}\sum_{X\subseteqq[-N,N]}\mu_t^N(X)\log\frac{\mu_t^N(X)}{P(X)}$$

is bounded above by some constant which is independent of μ_t and N. Now

$$\sum_{X \subseteq [-N,N]} \frac{d}{dt} \mu_t^N(X) = 0, \tag{2.14}$$

and therefore, interchanging the summation and differentiation in (2.13) yields

$$\sum_{X \subseteq [-N,N]} \left\{\frac{d}{dt} \mu_t^N(X)\right\} \log \frac{\mu_t^N(X)}{P(X)}. \tag{2.15}$$

Note that we are sure that the expression in (2.13) exists for all t only if we permit the derivative to take the value minus infinity. If for some X, $\mu_t^N(X) = 0$ and $(d/dt)\ \mu_t^N(X) > 0$, then both (2.13) and (2.15) are minus infinity. If for some X, $\mu_t^N(X) = 0$ and $(d/dt)\ \mu_t^N(X) = 0$, then by first using (2.8) to prove that $\mu_t^N(X)$ has two continuous derivatives, it can be seen that $(d/dt)\big(\mu_t^N(X) \log \mu_t^N(X)\big) = 0$. Thus, using our convention about 0 log (0), we can then write

$$\frac{d}{dt}\big(\mu_t^N(X) \log \mu_t^N(X)\big) = \left(\frac{d}{dt} \mu_t^N(X)\right) \log \mu_t^N(X), \tag{2.16}$$

and it is still true that (2.15) equals (2.13).

We will omit the subscript t from the notation during the rest of the proof. Substituting (2.10) into (2.15), we have

$$\sum_X \left[\sum_Y \sum_{(a,b)} W(X \cup Y \cup b\backslash a, b)\mu^{N+L}(X \cup Y \cup b\backslash a) - \sum_Y \sum_{(a,b)} W(X \cup Y, a)\mu^{N+L}(X \cup Y)\right] \log \frac{\mu^N(X)}{P(X)}, \tag{2.17}$$

where the summations are on $(a, b) \in D(N, X, Y,)$. Now set

$$D_1(N, X, Y) = D(N, X, Y) \cap \{(a, b) \,|\, |a| \leq N - L \text{ and } |b| \leq N - L\} \tag{2.18}$$

and $D_2(N, X, Y) = D(N, X, Y)\backslash D_1(N, X, Y)$. Note that $D_1(N, X, Y)$ does not depend on Y; hence, we will write it $D_1(N, X)$. Expression (2.17) can be broken into two terms, one with D_1 replacing D and the other with D_2 replacing D. We first consider the expression resulting when D is replaced by D_2:

$$\begin{aligned} &\sum_X \left[\sum_Y \sum_{(a,b)} W(X \cup Y \cup b\backslash a, b)\mu^{N+L}(X \cup Y \cup b\backslash a) - \sum_Y \sum_{(a,b)} W(X \cup Y, a)\mu^{N+L}(X \cup Y)\right] \log \frac{\mu^N(X)}{P(X)} \\ &= \sum_X \sum_Y \sum_{(a,b)} W(X \cup Y \cup b\backslash a, b)\mu^{N+L}(X \cup Y \cup b\backslash a) \left[\log \frac{\mu^N(X)}{P(X)} - \log \frac{\mu^N(X \cup b\backslash a)}{P(X \cup b\backslash a)}\right], \end{aligned} \tag{2.19}$$

where the summations are on $(a, b) \in D_2(N, X, Y)$. One of a or b may not be in

$[-N, N]$. In that case, we must understand $\mu^N(X \cup b\backslash a)$ to be $\mu^N(X \cup b)$ if $a \notin [-N, N]$ or to be $\mu^N(X\backslash a)$ if $b \notin [-N, N]$. Similarly, for $P(X \cup b\backslash a)$.

If there exists X, Y, and $(a, b) \in D_2(N, X, Y,)$ such that $\mu^N(X) = 0$ and $\mu^{N+L}(X \cup Y \cup b\backslash a) > 0$, then (2.19) is $-\infty$. Therefore, since we are trying to show that (2.19) is bounded above, we may assume that if $\mu^{N+L}(X \cup Y \cup b\backslash a) > 0$, then $\mu^N(X) > 0$.

Now

$$(2.20) \quad \mu^{N+L}(X \cup Y \cup b\backslash a)\left[\log \frac{\mu^N(X)}{P(X)} - \log \frac{\mu^N(X \cup b\backslash a)}{P(X \cup b\backslash a)}\right]$$

$$= \mu^{N+L}(X \cup Y \cup b\backslash a)\Bigg[\log \frac{\mu^N(X)}{\mu^N(X \cup b\backslash a)} + \tfrac{1}{2}\beta \sum_{m,n \in X} U(m, n) - \tfrac{1}{2}\beta \sum_{m,n \in X \cup b\backslash a} U(m, n)\Bigg]$$

$$\leqq - \frac{\mu^{N+L}(X \cup Y \cup b\backslash a)}{\mu^N(X)} \log \left(\frac{\mu^N(X \cup b\backslash a)}{\mu^N(X)}\right) \mu^N(X) + \mu^{N+L}(X \cup Y \cup b\backslash a)K,$$

where $K = 2\beta \sum_{k=-L}^{L} |U(0, k)|$.

Since $\mu^{N+L}(X \cup Y \cup b\backslash a) \leqq \mu^N(X \cup b\backslash a)$ and $-x \log x \leqq e^{-1}$, the right side of (2.20) is bounded above by $e^{-1}\mu^N(X) + K\mu^{N+L}(X \cup Y \cup b\backslash a)$.

From the definition of $W(X \cup Y, a)$, we see that $W(X \cup Y, a) \leqq \frac{1}{2} \exp \frac{1}{2}K$.

Substituting this bound and the bound for (2.20) into (2.19), we see that (2.19) is bounded above by

$$(2.21) \quad \tfrac{1}{2} \exp \{\tfrac{1}{2}K\} \sum_X \sum_Y \sum_{(a,b) \in D_2(N,X,Y)} [\exp\{-1\}\mu^N(X) + K\mu^{N+L}(X \cup Y \cup b\backslash a)]$$

$$\leqq \tfrac{1}{2} \exp \{\tfrac{1}{2}K - 1\}2^{2L}4L + \tfrac{1}{2}K \exp \{\tfrac{1}{2}K\}4L.$$

In the future, we will denote the right side of (2.21) by K_1.

We now return to (2.17) and consider the expression obtained if $D(N, X, Y)$ is replaced by $D_1(N, X)$. The first thing to note is that

$$(2.22) \quad \sum_X \Bigg[\sum_Y \sum_{(a,b) \in D_1(N,X)} W(X \cup Y \cup b\backslash a, b)\mu^{N+L}(X \cup Y \cup b\backslash a) - \sum_Y \sum_{(a,b) \in D_1(N,X)} W(X \cup Y, a)\mu^{N+L}(X \cup Y)\Bigg] \log \frac{\mu^N(X)}{P(X)}$$

$$= \sum_X \Bigg[\sum_{(a,b) \in D_1(N,X)} W(X \cup b\backslash a, b)\mu^N(X \cup b\backslash a) - \sum_{(a,b) \in D_1(N,X)} W(X, a)\mu^N(X)\Bigg] \log \frac{\mu^N(X)}{P(X)}.$$

For if $(a, b) \in D_1(N, X)$, then neither $W(X \cup Y, a)$ nor $W(X \cup Y \cup b\backslash a, b)$ depend on Y. Hence, we may first perform the summation on Y.

LEMMA 2.2. *Let D_1, W, and P be as above. For $z \geqq 0$ set $F(z) = z - z\log(z) - 1$. Then*

$$\sum_X \Big[\sum_{(a,b)\in D_1(N,X)} W(X\cup b\backslash a, b)\mu^N(X\cup b\backslash a) - \sum_{(a,b)\in D_1(N,X)} W(X,a)\mu^N(X)\Big] \log\frac{\mu^N(X)}{P(X)} \tag{2.23}$$

$$= \sum_X \sum_{D_1(N,X)} F\left(\frac{P(X)}{\mu^N(X)}\frac{\mu^N(X\cup b\backslash a)}{P(X\cup b\backslash a)}\right) W(X\cup b\backslash a, b)\frac{P(X\cup b\backslash a)}{P(X)}\mu^N(X).$$

REMARK 2.1. For (2.23) to be correct, we must make the following convention. For $a, b, c > 0$ we understand $F\left(\frac{a}{0}\frac{b}{c}\right)\cdot 0$ to be minus infinity and $F\left(\frac{a}{0}\frac{0}{c}\right)\cdot 0$ to be zero.

Let us assume Lemma 2.2 for the moment. One easily checks that $F(z) \leqq 0$ for all $z \geqq 0$, and thus the expression appearing in (2.23) is nonpositive. Since (2.17) is the sum of (2.22) and (2.19), we see from the above results that (2.17) is bounded above by K_1, which is independent of N and μ.

The proof will be complete as soon as we prove Lemma 2.2.

PROOF OF LEMMA 2.2. Let each $X \subseteqq [-N, N]$ be represented as $X_0 \cup X_1$, where $X_0 \subseteqq [-N, N]\backslash[-N+L, N-L]$ and $X_1 \subseteqq [-N+L, N-L]$. Then we can rewrite the left side of (2.23) to get

$$\sum_{X_0}\Bigg[\sum_{X_1}\sum_{(a,b)\in D_1(X_1)} W(X_0\cup X_1\cup b\backslash a, b)\mu^N(X_0\cup X_1\cup b\backslash a)\log\frac{\mu^N(X_0\cup X_1)}{P(X_0\cup X_1)} - \sum_{X_1}\sum_{(a,b)\in D_1(X_1)} W(X_0\cup X_1, a)\mu^N(X_0\cup X_1)\log\frac{\mu^N(X_0\cup X_1)}{P(X_0\cup X_1)}\Bigg]. \tag{2.24}$$

We write $D_1(X_1)$ instead of $D_1(N, X_0\cup X_1)$, since $D_1(N, X_0\cup X_1)$ depends only on X_1 and N is fixed throughout the proof.

We next notice that for fixed X_0 there is a kernel $\mathfrak{U}_{X_0}(\cdot,\cdot)$ defined on the subsets of $[-N+L, N-L]$ by the formula

$$\mathfrak{U}_{X_0}(A,B) = \begin{cases} W(X_0\cup A, a) & \text{if } B = A\cup b\backslash a \text{ for some } a\in A,\ b\notin A \text{ with } |a-b| = 1, \\ -\sum_{(a,b)\in D_1(A)} W(X_0\cup A, a) & \text{if } B = A, \\ 0 & \text{otherwise,}\end{cases} \tag{2.25}$$

and that

$$\sum_A P(X_0\cup A)\mathfrak{U}_{X_0}(A,B) = 0 \tag{2.26}$$

for all X_0 and all B.

This last assertion is an easy computation which will be left to the reader.

Now (2.24) may be rewritten to yield

$$\sum_{X_0}\left[\sum_A\sum_B \mathfrak{U}_{X_0}(A, B)\mu^N(X_0\cup A)\log\frac{\mu^N(X_0\cup B)}{P(X_0\cup B)}\right]. \tag{2.27}$$

Since $\Sigma_B\, \mathfrak{U}_{X_0}(A, B) = 0$, we have $\Sigma_A\, \Sigma_B\, \mu^N(X_0\cup A)\mathfrak{U}_{X_0}(A, B) = 0$ and

$$\sum_A\sum_B \mu^N(X_0\cup A)\log\left(\frac{\mu^N(X_0\cup A)}{P(X_0\cup A)}\right)\mathfrak{U}_{X_0}(A, B) = 0. \tag{2.28}$$

Using these observations and (2.26), we see, after some simplification, that (2.27) is equal to

$$\begin{aligned}(2.29)\quad &\sum_{X_0}\left[\sum_B\sum_A F\left(\frac{P(X_0\cup B)}{\mu^N(X_0\cup B)}\,\frac{\mu^N(X_0\cup A)}{P(X_0\cup A)}\right)\mathfrak{U}_{X_0}(A, B)\frac{P(X_0\cup A)}{P(X_0\cup B)}\,\mu^N(X_0\cup B)\right]\\ &= \sum_{X_0}\sum_B\sum_{A\neq B} F\left(\frac{P(X_0\cup B)}{\mu^N(X_0\cup B)}\,\frac{\mu^N(X_0\cup A)}{P(X_0\cup A)}\right)\mathfrak{U}_{X_0}(A, B)\frac{P(X_0\cup A)}{P(X_0\cup B)}\,\mu^N(X_0\cup B).\end{aligned}$$

We may delete the terms where $A = B$ because $F(1) = 0$. Now by substituting the formula for $\mathfrak{U}_{X_0}(A, B)$ into the right side of (2.29), the lemma is proved.

REMARK 2.2. Since $F \leqq 0$, and for $z \neq 1$, $F(z) < 0$, it is clear from Lemma 2.2 that if the expression in (2.23) is equal to zero and if A and B are two subsets of $[-N + L, N - L]$ with the same number of elements, then for all X_0 contained in $[-N, -N + L - 1]\cup[N - L + 1, N]$,

$$\frac{\mu^N(X_0\cup A)}{P(X_0\cup A)} = \frac{\mu^N(X_0\cup B)}{P(X_0\cup B)}. \tag{2.30}$$

Indeed, there is a finite sequence $A = B_0, B_1, \cdots, B_n = B$ such that $B_{i+1} = B_i\cup b_i\backslash a_i$ for some $a_i\in B_i$, $b_i\notin B_i$ with $|a_i - b_i| = 1$; and it is immediate that if the expression in (2.23) is equal to zero, then

$$\frac{\mu^N(X_0\cup B_i)}{P(X_0\cup B_i)} = \frac{\mu^N(X_0\cup B_{i+1})}{P(X_0\cup B_{i+1})}. \tag{2.31}$$

3. Shift invariant states

If $X \subseteqq Z$, we will set $X + a = \{x + a | x\in X\}$. A state μ is shift invariant if

$$\mu^\Lambda(X) = \mu^{\Lambda+a}(X + a) \tag{3.1}$$

for all finite subsets $\Lambda \subset Z$, all $X \subseteqq \Lambda$, and all $a\in Z$.

Let $\mathscr{M}$ be the space of all states on E and give $\mathscr{M}$ the weak topology. Denote by $\mathscr{M}_1$ the closed subspace of all shift invariant states.

We will need the following facts about shift invariant states.

PROPOSITION 3.1. *If μ_0 is shift invariant, then μ_t is also shift invariant for all $t \geqq 0$.*

PROPOSITION 3.2. *In the definition of $A(\mu)$ (see (2.7)), the limit supremum is actually the limit. Thus, if μ is shift invariant,*

$$A(\mu) = \lim_{N\to\infty} \frac{1}{\beta}\frac{1}{2N+1}\left[\sum_{X\subseteqq[-N,N]} \mu^N(X)\log\frac{\mu^N(X)}{P(X)}\right]. \tag{3.2}$$

A proof of (1) based on the techniques in [1] is routine and is left to the reader. A proof of (2) can be found in [2], Section 7.2.

If m is large enough so that $2^m - 1 \geqq L$, let

$$H_m(\mu) = \sum_X \sum_{(a,b)} F\left(\frac{P(X)}{\mu^{2^m-1}(X)}\frac{\mu^{2^m-1}(X\cup b\backslash a)}{P(X\cup b\backslash a)}\right) \cdot W(X\cup b\backslash a, b)\frac{P(X\cup b\backslash a)}{P(X)}\mu^{2^m-1}(X), \tag{3.3}$$

where $X \subseteqq [-2^m+1, 2^m-1]$ and $(a,b) \in D_1(2^m-1, X)$. Using the convention in the Remark 2.1, it is easily seen that $H_m(\cdot)$ is an upper semicontinuous function on $\mathscr{M}$.

We will also need the following fact, which can be easily proved from the results in [1].

PROPOSITION 3.3. *The map $(\mu_0, t) \to \mu_t$ is continuous in the product topology on $\mathscr{M} \times [0, \infty)$.*

LEMMA 3.1. *Let $\mu \in \mathscr{M}_1$. Then $H_m(\mu) \leqq 2H_{m-1}(\mu)$.*

PROOF. Let

$$A_1(X) = \{(a,b) \in D_1(2^m-1, X) \mid a \leqq -L-1 \text{ and } b \leqq -L-1\} \tag{3.4}$$

and

$$A_2(X) = \{(a,b) \in D_1(2^m-1, X) \mid a \geqq L+1 \text{ and } b \geqq L+1\}. \tag{3.5}$$

Then, since $F \leqq 0$, $H_m(\mu)$ is less than or equal to the sum of the two expressions obtained if D_1 in (3.3) is replaced by A_1 and subsequntly by A_2.

Now each $X \subseteqq [-2^m+1, 2^m-1]$ can be written as $X = V_1 \cup V_2$, where $V_1 \subseteqq [-2^m+1, -1]$ and $V_2 \subseteqq [0, 2^m-1]$. We then notice that $A_1(V_1\cup V_2)$ depends only on V_1, and hence may be written $A_1(V_1)$. In fact,

$$A_1(V_1) = D_1(2^{m-1}-1, V_1+2^{m-1}) - (2^{m-1}, 2^{m-1}). \tag{3.6}$$

Similarly, if $(a,b) \in A_1(V_1)$, then $W(V_1\cup V_2, a)$ depends only on V_1, and in fact,

$$W(V_1\cup V_2, a) = W(V_1+2^{m-1}, a+2^{m-1}). \tag{3.7}$$

And finally, if $(a,b) \in A_1(V_1)$, then it is easy to check that

$$\frac{P(V_1\cup V_2\cup b\backslash a)}{P(V_1\cup V_2)} = \frac{P(V_1\cup b\backslash a)}{P(V_1)}. \tag{3.8}$$

To simplify the notation, we will write $D_1(V_1)$ instead of $D_1(2^{m-1}, V_1+2^{m-1}) - (2^{m-1}, 2^{m-1})$, $W(V_1, a)$ instead of $W(V_1+2^{m-1}, a+2^{m-1})$, and $\mu'(X)$ instead of $\mu^{2^m - 1}(X)$.

$$(3.9)\quad \sum_X \sum_{(a,b)\in A_1(X)} F\left(\frac{P(X)}{\mu'(X)} \frac{\mu'(X\cup b\backslash a)}{P(X\cup b\backslash a)}\right) W(X\cup b\backslash a, b) \frac{P(X\cup b\backslash a)}{P(X)} \mu'(X)$$

$$= \sum_{V_1}\sum_{V_2} \sum_{(a,b)\in D_1(V_1)} F\left(\frac{P(V_1)}{\mu'(V_1\cup V_2)} \frac{\mu'(V_1\cup V_2\cup b\backslash a)}{P(V_1\cup b\backslash a)}\right) W(V_1\cup b\backslash a, b)$$

$$\cdot \frac{P(V_1\cup b\backslash a)}{P(V_1)} \frac{\mu'(V_1\cup V_2)}{\mu^{[-2^m+1,-1]}(V_1)} \mu^{[-2^m+1,-1]}(V_1)$$

$$\leqq \sum_{V_1} \sum_{(a,b)\in D_1(V_1)} F\left(\frac{P(V_1)}{\mu^{[-2^m+1,-1]}(V_1)} \frac{\mu^{[-2^m+1,-1]}(V_1\cup b\backslash a)}{P(V_1\cup b\backslash a)}\right)$$

$$\cdot W(V_1\cup b\backslash a, b) \frac{P(V_1\cup b\backslash a)}{P(V_1)} \mu^{[-2^m+1,-1]}(V_1).$$

The last inequality follows from Jensen's inequality, since F is concave. In the above argument, we have assumed that all $\mu^{[-2^m+1,-1]}(V_1) > 0$. If this is not the case, then the inequality in (3.9) follows directly from the convention given in Remark 2.1.

Since μ and P are both shift invariant, the right side of (3.9) is equal to $H_{m-1}(\mu)$. Similarly, if D_1 in (3.3) is replaced by A_2, the result is less than or equal to $H_{m-1}(\mu)$, and the lemma is proved.

LEMMA 3.2. *The function*

$$(3.10)\quad H(\mu) = \frac{1}{\beta} \lim_{m\to\infty} \frac{1}{2^{m+1}-1} H_m(\mu)$$

exists on $\mathscr{M}_1$ (it is possibly minus infinity) and is upper semicontinuous there. Moreover, if $\mu_0 \in \mathscr{M}_1$, then $A(\mu_t) - A(\mu_0) \leqq \int_0^t H(\mu_s)\,ds$.

PROOF. Let $G(m) = \Pi_{j=m}^{\infty} [(2^{j+2} - 2)/(2^{j+2} - 1)]$. Then by Lemma 3.1, if $\mu \in \mathscr{M}_1$,

$$(3.11)\quad G(m+1)\frac{1}{2^{m+2}-1} H_{m+1}(\mu) \leqq G(m+1)\frac{2^{m+2}-2}{2^{m+2}-1}\frac{1}{2^{m+1}-1} H_m(\mu)$$

$$= G(m)\frac{1}{2^{m+1}-1} H_m(\mu).$$

Therefore, $G(m)(1/2^{m+1} - 1)\, H_m(\mu)$ is a decreasing sequence of upper semicontinuous functions on $\mathscr{M}_1$. Hence, the limit exists and is upper semicontinuous on $\mathscr{M}_1$. Since $G(m)$ goes to one as m goes to infinity, this limit when divided by β is equal to $H(\mu)$.

In the proof of Theorem 2.1, we showed that

$$(3.12)\quad \sum_{X\subseteqq[-N,N]} \mu_t^N(X) \log \frac{\mu_t^N(X)}{P(X)} - K_1 t$$

is a nonincreasing function of t. Therefore an application of Lebesgue's theorem and Fatou's lemma yields

$$\text{(3.13)} \quad \sum_{X \subseteqq [-N,N]} \mu_t^N(X) \log \frac{\mu_t^N(X)}{P(X)} - \sum_{X \subseteqq [-N,N]} \mu_0^N(X) \log \frac{\mu_0^N(X)}{P(X)} - K_1 t$$

$$\leqq \int_0^t \left[\frac{d}{ds} \left(\sum_{X \subseteqq [-N,N]} \mu_s^N(X) \log \frac{\mu_s^N(X)}{P(X)} \right) - K_1 \right] ds.$$

Now if $\mu_0 \in \mathscr{M}_1$, we may use (3.1) and (3.2) together with this inequality to get

$$\text{(3.14)} \quad A(\mu_t) - A(\mu_0)$$

$$\leqq \lim_{m \to \infty} \frac{1}{\beta(2^{m+1} - 1)} \int_0^t \frac{d}{ds} \left[\sum_X \mu_s^{2^m - 1}(X) \log \frac{\mu_s^{2^m - 1}(X)}{P(X)} \right] ds$$

$$\leqq \lim_{m \to \infty} \frac{1}{\beta(2^{m+1} - 1)} \int_0^t [H_m(\mu_s) + K_1]\, ds.$$

In the middle expression, the summation extends to all $X \subseteqq [-2^m + 1, 2^m - 1]$. The last inequality follows exactly as in the proof of Theorem 2.1.

Now since $0 < G(m) < 1$ and $H_m(\mu) \leqq 0$, $H_m(\mu) \leqq G(m)H_m(\mu)$. Therefore, using this inequality and monotone convergence, we have

$$\text{(3.15)} \quad A(\mu_t) - A(\mu_0) \leqq \lim_{m \to \infty} \int_0^t \frac{1}{\beta(2^{m+1} - 1)} [G(m)H_m(\mu_s) + K_1]\, ds$$

$$= \int_0^t H(\mu_s)\, ds,$$

and the proof is complete.

LEMMA 3.3. *Let $\mu \in \mathscr{M}_1$. Then if $\mu \notin C$, $H(\mu) < 0$.*

PROOF. We may think of μ^{2m-1} as a measure on $\bigcup_Y \bigcup_n S(2^m - 1, n, Y)$ (here $S(N, n, Y)$ is as in Section 2.2), and it is easily seen that μ is the weak limit of the μ^{2m-1}. From Remark 2.2, it is clear that if $H_m(\mu) = 0$, then μ^{2m-1} is equal to one of the ν_{2m-1} used to describe the set C. Thus, if $H(\mu) = 0$, then all $H_m(\mu)$ are zero, and therefore $\mu \in C$.

THEOREM 3.1. *Let $\mu_0 \in \mathscr{M}_1$ and suppose that $t_n \to \infty$ and that μ_{t_n} converges weakly to μ. Then $\mu \in C$.*

PROOF. Since $\mathscr{M}_1$ is weakly closed and each $\mu_{t_n} \subset \mathscr{M}_1$, $\mu \in \mathscr{M}_1$. Suppose that $\mu \notin C$. Then by Lemma 3.3, $H(\mu) < 0$. Therefore, there is a $\delta > 0$ such that if

$$\text{(3.16)} \quad G_\mu = \{\nu \in \mathscr{M}_1 \mid H(\nu) < -\delta\},$$

then $\mu \in G_\mu$. Since H is upper semicontinuous, G_μ is open. Therefore, there is an open subset $\hat{G}_\mu$ of $\mathscr{M}_1 \times [0, \infty)$ containing $(\mu, 0)$ and such that if $(\nu_0, s) \in \hat{G}_\mu$ then $\nu_s \in G_\mu$ (see (3.4)). Since $(\mu, 0) \in \hat{G}_\mu$, there is an open set $\bar{G}_\mu \subset \mathscr{M}_1$ and an $\varepsilon > 0$ such that $\mu \in \bar{G}_\mu$ and $\bar{G}_\mu \times [0, \varepsilon) \subseteqq \hat{G}_\mu$. Thus, if $\nu_0 \in \bar{G}_\mu$ and $0 \leqq s < \varepsilon$, then $H(\nu_s) < -\delta$.

Since μ_{t_n} converges weakly to μ, $\mu_{t_n} \in \bar{G}_\mu$ for all sufficiently large n. Thus, by Lemma 3.2, for all sufficiently large n, $A(\mu_{t_n+\varepsilon}) - A(\mu_{t_n}) \leqq -\delta\varepsilon$.

This together with Theorem 2.1 implies that $\lim_{t\to\infty} A(\mu_t) = -\infty$. But it is easily seen from the definition of $A(\nu)$ that $\inf_{\nu\in\mathcal{M}} A(\nu) > -\infty$. This is a contradiction and completes the proof.

COROLLARY 3.1. *Let G be a weakly open subset of $\mathcal{M}$ containing C, and let $\mu_0 \in \mathcal{M}_1$. Then for all sufficiently large t, $\mu_t \in G$.*

PROOF. This follows immediately from the compactness of $\mathcal{M}$ and Theorem 3.1.

COROLLARY 3.2. *All shift invariant equilibrium states are elements of C.*

REMARK 3.1. It is clear from the proof of Lemma 3.1 that if the state at time t is shift invariant, but not an equilibrium state, then the Helmholtz free energy at all future times is strictly less than it is at time t.

4. Pressure

We first modify the Markov process η_t introduced in Section 1 in order to motivate our definition of pressure. We need a "wall" from which the particles "rebound," so we let Z^- be the negative integers and take $D = \{0, 1\}^{Z^-}$ as the state space. The intuitive description is the same as before except that no particle is allowed to jump from minus one to zero. A proof that such a process exists can be given by imitating the one in [1]. The attempted jumps from minus one to zero are to be thought of as collisions with the wall, and we will take the pressure of a state to be twice the expected number of collisions with the wall per unit time. Thus, if μ is a state, its pressure is given by the formula

$$p(\mu) = \int \eta(-1) \exp\left\{\sum_{k=-\infty}^{-1} U(-1, k)\eta(k)\right\} \mu(d\eta). \tag{4.1}$$

Pressure is usually defined only for equilibrium states, so perhaps this should be thought of as instantaneous pressure.

The temperature will no longer play a role in this section; thus, we have absorbed the β into the potential U. However, it will be important to display the chemical potential explicitly and so we will require $U(x, x) = 0$ for all x.

Now let γ be any real number and set $S_N = \{\eta \in D \mid \eta(X) = 0 \text{ if } x < -N\}$. We define a probability measure $\nu_{N,\gamma}$ with support S_N by the formula

$$\nu_{N,\gamma}(A) = \sum_{\eta\in A\cap S_N} \frac{1}{\theta(N, \gamma)} \exp\left\{\gamma \sum_{x=-N}^{-1} \eta(x) - \tfrac{1}{2} \sum_x \sum_y \eta(x)\eta(y)U(x, y)\right\}. \tag{4.2}$$

Here $\theta(N, \gamma)$ is the normalizing constant. It can be proved (see [3], footnote 7) that there is a probability measure ν_γ on D which is the weak limit of the $\nu_{N,\gamma}$. Just as in the case where the state space is E instead of D, it can be shown that the ν_γ are equilibrium states for the Markov process η_t. Then ν_γ are moreover the only states for which the pressure is usually defined.

The usual definition of pressure is given in terms of the normalizing constants $\theta(N, \gamma)$, and we will make use of the following result (see [2], Section 5.6).

LEMMA 4.1. *There are constants* $\lambda(\gamma) > 1$ *and* $c(\gamma) > 0$ *such that*

$$\lim_{N\to\infty} \lambda^{-N}(\gamma)\theta(N, \gamma) = c(\gamma). \tag{4.3}$$

The usual definition of pressure associated with the state ν_γ is taken to be $P(\gamma) = \log \lambda(\gamma)$ (see [2], Section 3.4).

THEOREM 4.1. *The functions* $P(\gamma)$ *and* $p(\nu_\gamma)$ *are both strictly increasing in* γ.

PROOF. A proof that $P(\gamma)$ is a strictly increasing analytic function of γ can be found in [2], Section 5.6; therefore, we will restrict our attention to $p(\nu_\gamma)$.

Since $\eta(-1)\exp\{\Sigma_{x=-\infty}^{-1} U(-1, x)\eta(x)\}$ is a continuous function on D, we have

$$\begin{aligned}\int \eta(-1)\exp\left\{\sum_{x=-\infty}^{-1} U(-1, x)\eta(x)\right\}\nu_\gamma(d\eta) &= \lim_{N\to\infty}\int \eta(-1)\exp\left\{\sum_{x=-\infty}^{-1} U(-1, x)\eta(x)\right\}\nu_{N,\gamma}(d\eta)\\ &= \lim_{N\to\infty}\frac{1}{\theta(N, \gamma)}\sum_{\eta\in S'_N}\left[\exp\left\{\sum_{x=-\infty}^{-1} U(-1, x)\eta(x)\right\}\right.\\ &\qquad\left.\exp\left\{\gamma\sum_{x=-N}^{-1}\eta(x) - \tfrac{1}{2}\sum_x\sum_y \eta(x)\eta(y)U(x, y)\right\}\right].\end{aligned} \tag{4.4}$$

Here we have set $S'_N = \{\eta \in S_N \mid \eta(-1) = 1\}$. Some elementary manipulations reduce this to

$$\lim_{N\to\infty}\frac{1}{\theta(N, \gamma)}\exp\{\gamma\}\sum_{\eta\in S''_N}\exp\left\{\gamma\sum_{x=-N}^{-2}\eta(x) - \tfrac{1}{2}\sum_x\sum_y \eta(x)\eta(y)U(x, y)\right\}. \tag{4.5}$$

In our last expression, $S''_N = S_N \backslash S'_N$. Now

$$\sum_{\eta\in S''_N}\exp\left\{\gamma\sum_{x=-N}^{-2}\eta(x) - \tfrac{1}{2}\sum_x\sum_y \eta(x)\eta(y)U(x, y)\right\} = \theta(N - 1, \gamma), \tag{4.6}$$

and thus

$$\begin{aligned}p(\nu_\gamma) &= \lim_{N\to\infty}\exp\{\gamma\}\frac{1}{\theta(N, \gamma)}\theta(N - 1, \gamma)\\ &= \exp\{\gamma\}\frac{1}{\lambda(\gamma)} = \exp\{\gamma - P(\gamma)\}.\end{aligned} \tag{4.7}$$

We have now reduced the problem to proving that $\gamma - P(\gamma)$ is a strictly increasing function of γ. To do this we clearly need more information about P, and this can be found in [2], Section 3.4. The crucial fact is that there is a function $f(\rho)$ defined on the interval $[0, 1)$ such that

$$P(\gamma) = \sup_{0\leq\rho<1}(\rho\gamma - f(\rho)). \tag{4.8}$$

Therefore, $\gamma - P(\gamma) = \inf_{0 \leqq \rho < 1} [(1 - \rho)\gamma + f(\rho)]$, which is clearly a nondecreasing function of γ. As mentioned in the first line of the proof, $P(\gamma)$ is analytic; and thus if $\gamma - P(\gamma)$ is not strictly increasing, then it is constant. But this cannot be since $P(\gamma) > 0$ for all γ. Hence, $p(\nu_\gamma)$ is also a strictly increasing function of γ.

We conclude this section by giving a physical interpretation of Theorem 4.1. The parameter γ determines what the density $\rho(\gamma)$ of the state ν_γ will be; and in the case which we are considering, it is known that $\rho(\gamma)$ is a continuous strictly increasing function of γ. Thus, Theorem 4.1 tells us that at constant temperature the pressure is an increasing function of the density. Moreover, the pressure is a strictly increasing function of the density, and this is interpreted to mean that there is no change of phase for the model with which we are dealing.

The pressure $p(\nu)$ is defined for states other than the ν_γ, and Theorem 4.1 says nothing about how the pressure varies with the density for the other states. However, if ν is not an equilibrium state, one would not expect any nice relationship between the pressure and the density. Because of the results in Section 3, we feel that it is highly unlikely that there are any equilibrium states besides the ν_γ and convex combinations of the ν_γ. We are unfortunately unable to prove this.

REFERENCES

[1] R. Holley, "A class of interactions in an infinite particle system," *Advances in Math.*, Vol. 5 (1970), pp. 291–309.

[2] R. Ruelle, *Statistical Mechanics*, New York, Benjamin, 1969.

[3] ———, "Statistical mechanics of a one-dimensional lattice gas," *Comm. Math. Phys.*, Vol. 9 (1968), pp. 267–278.

[4] F. Spitzer, "Interaction of Markov processes," *Advances in Math.*, Vol. 5 (1970), pp. 246–290.

THE RATE OF SPATIAL PROPAGATION OF SIMPLE EPIDEMICS

DENIS MOLLISON
KING'S COLLEGE RESEARCH CENTRE, CAMBRIDGE

1. Introduction

The work described here concentrates on one aspect of the development of epidemics, namely, spatial propagation, ignoring such features as variable density of population, the gradual introduction of fresh susceptibles, and, for the most part, the removal of infected cases. (For an introduction to more sophisticated models for epidemics see Bailey [1].)

The basic feature of the mathematical models considered here is that the rate of infection of susceptibles is assumed to be proportional to the product of the number of susceptibles with the number of infectious individuals. This follows immediately from the assumption that the infectious influence of an infectious individual on a susceptible is independent of the state of other members of the population. Thus, if there are X susceptibles and Y infectious individuals living at an isolated point—the significance of "at a point" is that they should live so close together as to affect each other equally—then $\dot{X}$, the rate of change of X with time, is proportional to (minus) XY.

If we wish to study the spatial propagation of infection for such an epidemic model, we must allow for the dependence of this infectious influence on the distance between the individuals concerned, so that the rate of infection of susceptibles at a point s at time t, namely, $-\dot{X}(s, t)$, is proportional to the product $X\bar{Y}$ of the number of susceptibles at s with an average value $\bar{Y}$ of the numbers of infectious individuals at all points, weighted according to their distances from s. This weighting function may be taken to be a probability distribution function V; then $\bar{Y}$ is the convolution of Y with dV, that is, $\int_{\text{space}} Y(s - r)\, dV(r)$.

The introduction of such a weighted average $\bar{Y}$ to our equations causes considerable difficulties in their analysis which have not, to the best of my knowledge, been tackled hitherto (Neyman and Scott [12] make allowance for such a dependence on distance as is considered here, but their approach otherwise differs widely). I have, accordingly, concentrated on the most simple type of epidemic model which incorporates this feature, namely, a *simple epidemic* in which there are only two types of individual, *susceptible* and *infected*; infected and infectious individuals are taken to be the same. For the most part, too, I have restricted attention to a deterministic model.

The work described here was carried out during the tenure of an S.R.C. grant in the Department of Pure Mathematics and Mathematical Statistics at Cambridge University, and (more recently) of a Research Centre Fellowship at King's College.

Previous authors have made considerable progress on the manner of spatial propagation of epidemics (Kendall [7]) and of a dominant gene (Kolmogorov, Petrovsky, and Piscounov [9]) through a linearly distributed population by using a local (diffusion) approximation for the effects of cross infection at a distance; they used $Y + k(\partial^2 Y/\partial s^2)$ instead of $\bar{Y}$. They discovered that propagation as a travelling wave is possible at and above a certain critical velocity. I have shown [11] that under a negative exponential weighting function $\left(v(s) = \frac{1}{2}\exp\{-|s|\}\right)$ waves are possible exactly as for the local approximation, the critical velocity being only slightly higher.

In the first part of this paper, it is shown that the negative exponential weighting function is a borderline case: while for less spread weighting functions we can find waveform upper bounds (Theorem 2(i)), outbreaks of epidemics which do not satisfy the condition

$$(1.1) \qquad \int_{\mathscr{R}} e^{ks}\, dV(s) < \infty \qquad \text{for some } k > 0$$

progress at arbitrarily high rate as $t \to \infty$, unbounded by any wave (Theorem 2(ii) and Lemma 4)).

It appears then that (1.1) is a necessary condition for diffusion approximations to be any guide to the behavior of epidemics. This is quite a restrictive condition, which suggests that the faith of previous authors in diffusion approximations is unjustified. What is more, the importance of (1.1) appears to depend merely on the basic, roughly linear, dependence of $\dot{Y}$ on $\bar{Y}$ for small values of Y, so that there is every reason to suppose that the same qualitative results will hold for other models for geographical spread, such as those of Kolmogorov, Petrovsky, and Piscounov [9], and Fisher [5] for the spread of an advantageous gene (which are discussed in (v) of Section 2.1), of Marris [10] (see Chapter 4, especially pp. 149–175) for the spread of consumer demand, and of Zeldovitch [16] for flame propagation. Also there appears to be no difficulty in extending the (qualitative) results of Section 2 to two (or more) dimensions (see (iv) of Section 2.1, and Kendall [7]).

Section 3 is devoted to the (more realistic) discrete stochastic analogue of the continuous deterministic model of Section 2. This is, not surprisingly, more resistant to analysis, and while the theoretical framework (Section 3.2) is considerably more elegant than that for the deterministic model (Section 2.2), the results section (also 3.2) is noticeably thinner than Section 2.3. Under rather stronger conditions on the initial situation than those of Theorem 2 and Lemma 4, we find (Theorem 3) that in the stochastic case we can replace (1.1) by *the variance of* V *is finite* as the condition for propagation at a finite rate.

The last two subsections of Section 3 are devoted to simulations of the stochastic model, which have been carried out on TITAN, the computer at the Cambridge University Mathematical Laboratory. For several weighting functions which are less spread out than the negative exponential function, and for the negative exponential function itself (or rather its discrete equivalent), progress is observed as might be expected at a steady rate, with a front which, averaged over a period of time, is wavelike. The simulated outbreaks with V of "just infinite" variance progress in wilder and wilder *great leaps forward*, as again might be expected from Theorem 3.

The interesting case here seems to be the intermediate one. In this case V is of finite variance but not negative exponentially bounded, and epidemics appear to progress in a mixture of steady progress and great leaps forward which would not be forecast by local approximation equations. If, for example, one could show that the distributions for light windborne objects such as some kinds of germs and plant seeds are of this type, one might throw new light on quite a number of problems of geographical spread. It would explain, for instance, why outbreaks of epidemics, or mutant species, sometimes appear to have several origins (see, for example, Chamberlain [2]; Davies, Lewis, and Randall [3]; Norris and Harper [13], and Tinline [14] on the spread of foot and mouth disease).

2. The velocity of simple deterministic epidemics

2.1. *Introduction.* The result of Mollison ([11], Theorem 4.1) that waves of all velocities above a certain minimal velocity are possible for a simple epidemic under negative exponential weighting, is a full answer to a rather specialized question. In this paper, we try to answer a vaguer but more general question: under what conditions does an outbreak of a simple epidemic propagate at a finite rate?

In Section 2.2, we first consider various possible definitions for the rate of propagation of an epidemic, including the *mean velocity* (2.8) and *velocity at level* α (2.9). Second, having defined a *simple epidemic* by its differential equation, we define as *pseudoepidemics* a class of differential equations, and prove a result (Theorem 1) which will allow us profitably to compare outbreaks of a simple epidemic with "outbreaks" of pseudoepidemics (for both epidemics and pseudoepidemics we define an *outbreak* as a particular solution of the relevant differential equation). The advantage of this is that we shall be able to choose pseudoepidemics with differential equations which are easier to handle than that for a simple epidemic (mainly in that we can evade the convolution $\bar{y}$).

Section 2.3 is devoted to the connection between "the mean velocity is eventually finite" (that is, "$\limsup_{s\to\infty}$ (mean velocity) $< \infty$") and two conditions, one on the epidemic (1.1) and one on the initial situation of the particular outbreak (2.18). Roughly speaking, each condition says that the relevant function ($dV(s)$ and $y(s, 0)$, respectively) should tail off at least exponentially fast as $s \to \infty$. The results of Section 2.3 nearly add up to:

the mean velocity is eventually finite if and only if both (1.1) *and* (2.18) *hold.*

The exceptions to this are listed under Corollary 2 (the two conditions are necessary, but one alone might be sufficient provided it holds sufficiently strongly).

The analysis of Sections 2.2 and 2.3 may appear more comprehensible when regarded as an extension of results which hold for the linear equation, $\dot{y} = \bar{y}$, which approximates the epidemic equation $(\dot{y} = \bar{y}(1 - y))$ for small values of y. Thus, Theorem 2 (i) and Lemma 5 are based on the following theorem on waveform solutions of $\dot{y} = \bar{y}$.

We look for waveform solutions to $\dot{y}(s, t) = \bar{y}(s, t)$, s and $t \in \mathscr{R}$, that is, solutions for which $\dot{y} = -cy'$, where c is a constant (the velocity of the wave). We may then consider y as a function of just one variable, s, say.

Suppose that $f_v(k) \equiv \int_{\mathscr{R}} ke^{ks}(1 - V(s))\,ds$ *converges for* $0 < k < k^*$. *Let*

$$y_{\varepsilon,k}(s) \equiv \begin{cases} y(s)e^{ks} & \text{for } s \leqq 0, \\ y(s)e^{\varepsilon s} & \text{for } s \geqq 0. \end{cases} \tag{2.1}$$

Then the only waveforms $y(s)$ *for which* $y_{\varepsilon,k}(s) \in L^2(\mathscr{R})$, *some* ε, k *with* $0 < \varepsilon < k < k^*$, *are of the form* $\Sigma_\nu \Sigma_{p=1}^{q} C_{\nu,p} s^{p-1} e^{k_\nu s}$, *where the* $C_{\nu,p}$ *are constants, and* k_ν *runs through the solutions of* $f_v(k) = ck$ *for which* $0 <$ *real part of* $k < k^*$, *and* q *is the multiplicity of the root* k_ν.

The proof proceeds as follows. We have $y'(s) = (1/c)\int_{\mathscr{R}} y(s - u)\,dV(u)$, whence by integration, with the boundary conditions $y(-\infty) = \infty$, $y(\infty) = 0$, suitable to a wave of positive velocity,

$$y(s) = -\frac{1}{c}\int_{\mathscr{R}} y(s - u)\big(V(u) - 1\big)\,du. \tag{2.2}$$

Let $0 < \varepsilon' < \varepsilon$, $k < k' < k^*$; then

$$f_{\varepsilon',k'}(u) \equiv \begin{cases} \dfrac{1}{c}(1 - V(u))e^{\varepsilon' u} & \text{for } u \leqq 0, \\[2ex] \dfrac{1}{c}(1 - V(u))e^{k' u} & \text{for } u \geqq 0, \end{cases} \tag{2.3}$$

is in $L^1(\mathscr{R})$, and the overlapping of the intervals $(-\infty, \varepsilon)$, (ε', k'), (k, ∞), ensures that the Fourier transforms $\hat{y}_-$, $(1/c)\,[1 - V(u)]\hat{}$ and $\hat{y}_+$ have overlapping regions of regularity so that we can apply Wiener–Hopf technique to obtain the theorem as stated (see Titchmarsh [15], Theorem 146, p. 305, whose proof adapts almost word for word to our problem).

In Theorem 2 (i), we shall show that such waveforms of $\dot{y} = \bar{y}$ can be used as upper bounds for the propagation of epidemic outbreaks, and in Lemma 5 we shall consider more exactly the existence of roots k_ν for varying values of c (and varying V). The other main results of Section 2.3 are similarly based on results for $\dot{y} = \bar{y}$, and the reader may find their proofs more easy to understand if he at first ignores the terms corresponding to the $(1 - y)$ factor which occur in these proofs (for example, the $(1 - \alpha)$ factor in the proof of Theorem 2 (ii)).

To conclude this introduction, we mention some lines of research which will not be written up here, as they have either not been taken very far or have proved unprofitable.

(i) Pursuing the lines of Section 2.3, one would like to know more about the eventual behavior of outbreaks of epidemics for which both (1.1) and (2.18) hold. In Theorem 2 (i) we define a *critical velocity* c_v for each such epidemic. It is tempting to conjecture that under fairly weak conditions on the initial situation of the outbreak $\left(y(s, 0) \leqq ae^{-k_v s}\right.$ and perhaps $y(s, 0) \geqq$ some lower bound$\left.\right)$ the mean velocity tends to c_v. In particular one might conjecture that it tends to a waveform, which brings us to (ii).

(ii) Do waveforms exist? Clearly, not unless (1.1) holds. Theorem 2 (ii) shows this. Several approaches look possible.

(1) A specific differential equation approach as in Section 2. (Kolmogorov, Petrovsky, and Piscounov [9] use such an approach in their work on a diffusion model for genetic spread.)

(2) We might also mention here their approach to the analogue of (i), which is to take the simple initial conditions

$$(2.4) \qquad y(s, o) = \begin{cases} 1 & \text{if } s \leqq 0, \\ 0 & \text{otherwise}, \end{cases}$$

and show that $y(s, t)$ tends to the waveform of minimal velocity. We might, for instance, be able to show that the slope of $y(s, t)$ for fixed y, namely, $y'(y)(t)$, tends down to a limit. The difficult step appears to be proving that $y'(y)$ cannot increase (of course this may not be true!). It would then be easy to prove that this limit must be a waveform of velocity $\leqq c_v$ (applying Lemma 5 (i)), and easy to extend to a class of initial conditions on y, certainly to those of Theorem 2 (ii). If it turned out to be unnecessary to assume the existence of a waveform, this would furnish another approach to (ii).

(3) An elegant alternative approach is to consider waveforms as fixed points of the continuous function $T_c(f) \equiv 1 - \exp\{-\bar{f}^c\}$, where $\bar{f}^c$ denotes convolution with $V(ct)$ rather than $V(s)$; and try to apply Tychonov's theorem that a continuous function from a compact convex subset of a topological vector space to itself has a fixed point ([4], pp. 456–459). If $c \geqq c_v$, the set $\{f : f_0 \leqq f \leqq \min(e^{ckt}, 1)\}$ will do, provided only that f_0 is monotone nondecreasing, positive, and $\leqq T_c(f_0)$; but such an f_0 has so far eluded discovery.

I have given more space here to (ii) because it appears more tractable than (i). The latter is, however, surely the more important question, and it may be that a direct approach, possibly finding a sequence of lower bounds for the outbreak with initial conditions (2.4), might provide an adequate answer which sidestepped the "existence of waves" question.

(iii) It is easy to extend the qualitative results of Section 2.3 to two (or more) dimensions, at least if V is radially symmetric. For lower bounds for rate of propagation, we can consider a strip of constant width, for upper bounds a strip of infinite width. We then produce pseudoepidemics for comparison which are essentially one dimensional epidemics multiplied by an appropriate constant. Then the qualitative parts of the results of Section 2.3 (for example, those referring to whether the mean velocity is eventually finite) will apply, suitably adjusted in their statement, to epidemics in the plane. Clearly, they will also hold for any asymmetric V which can be sandwiched between two symmetric distributions.

(iv) Apart from the problems raised in (1) and (2) above, the matter of epidemics with removal is clearly the next problem deserving attention in our line of research into continuous deterministic models for epidemics. Provided that the differential equations defining the removal rate are sufficiently regular to allow comparison theorems such as Theorem 1, it seems clear that the progress of an outbreak of an epidemic with removal will be bounded above by that of a simple epidemic with similar initial conditions, as it seems implausible that removals should speed up an outbreak.

Since Theorem 2 (ii) relies in its proof mainly on events in the forward tail of the outbreak, where the proportion of susceptibles ≈ 1, there seems hope of extending this result also.

Rather than continuing with problems (i), (ii), (iii), and (iv), it has seemed to me more profitable to raise one's eyes from the problems of continuous deterministic models for epidemics, and to look at stochastic models, more resistant to analysis but more realistic; Section 3 is, accordingly, devoted to these.

(v) Before abandoning deterministic models, it seems worth mentioning the problem of genetic spread. Diffusion models have been considered in [9] and [5] (the former paper considers the case of dominance among genes, the latter only a particular case of partial dominance, which is in fact covered by the general type of equation analyzed in [9]). If we replace the diffusion approximation of Kolmogorov [9] by the exact convolution equation, we find that the analysis of Section 2.3 applies at least qualitatively, Thus, Kolmogorov, Petrovsky, and Piscounov are wrong in stating that it is a sufficient condition for propagation at a finite rate that the first three moments of V should converge ([9], p. 4). This merely ensures that their *equation* is a good approximation to the convolution equation.

A more serious error is their failure to point out an assumption that the Hardy–Weinberg law holds for variably interacting populations with varying proportions of the different genotypes (which it does not). Nevertheless, whether it is a reasonable approximation I am unsure; certainly the corrected equations are horrible. In this uncertain situation, I have preferred to omit work on the genetic problem from this paper.

2.2. *Analytic preliminaries.* We consider simple epidemics among a population of uniform density σ on the line $\mathscr{R}$. If $y(s, t)$ denotes the *proportion* infected at s at time t, the basic equation of propagation is $\dot{y} = \alpha\sigma\bar{y}(1 - y)$, where $\bar{y}(s, t)$ denotes the weighted average $\int_{\mathscr{R}} y(s - u, t)\, dV(u)$, and V denotes some probability distribution function. We may, without loss of generality, take the constant $\alpha = 1/\sigma$, so that our basic equation becomes

$$\dot{y} = \bar{y}(1 - y). \tag{2.5}$$

The first half of this section will be concerned with defining "the rate of propagation" of a simple epidemic. Let us first define the epidemic itself more precisely.

DEFINITION 1. *Speaking mathematically, we define a* simple epidemic *as a function which determines how infectability depends on the relative positions of each possible (infectious, susceptible) pair.*

In the present case this is just the distribution function V. Then, given a population P distributed with measurable density over some metric space, we can set up the equation of propagation analogous to (2.5).

DEFINITION 2. *We define an* outbreak, *of a simple epidemic E among a population P, as a solution of the epidemic equation for a particular initial condition, specifying the numbers $y(s, t_0)$ infected at each point s at time t_0.*

When, as here, the equation of propagation satisfies conditions ensuring uniqueness for its solutions (here a Lipschitz condition), an outbreak will be completely determined by its initial condition; the contrast between this and the more realistic state of affairs attending a *stochastic* outbreak (Definition 5) should be noted.

Without loss of generality, we may take $t_0 = 0$. Also, we need only consider propagation in the direction of increasing s (on $\mathscr{R}$); to apply our results to propagation in the other direction, one need only transfer attention from $y(s, t_0)$ to $y(-s, t_0)$, and from $V(s)$ to $1 - V(-s)$.

We must now define our criteria for saying that an outbreak propagates at finite (or, respectively, infinite) rate. For the whole outbreak the best measure of *rate of propagation* would seem to be

$$c(t) \equiv \int_{\mathscr{R}} \dot{y}(s, t)\, ds. \tag{2.6}$$

NOTE. We choose $\int \dot{y}$ rather than $(\partial/\partial t)(\int y)$ because $\int y$ can diverge and yet $\int \dot{y}$ converge, but not vice versa, since $\int \dot{y} = \int \bar{y}(1 - y) \leqq \int \bar{y} = \int y$. Of course, when both converge, $(\partial/\partial t)(\int y) = \int \dot{y}$.

For the rate of propagation *in the direction of increasing* s then, we want to take an integral of $\dot{y}$ with respect to some measure which tends to ordinary Lebesgue measure as $s \to \infty$, and to the zero measure as $s \to -\infty$. Since we are only interested here in the behavior of outbreaks as $t \to \infty$, it matters little which we take, so we may as well take the simplest, which gives

$$c^+(t) = \int_{s_0}^{\infty} \dot{y}\, ds. \tag{2.7}$$

Further, since we have as yet no distinguished point on the space axis, this s_0 has only spurious generality; we may as well take $s_0 = 0$.

Now consider the type of result we might hope to prove regarding the eventual rate of propagation of outbreaks of simple epidemics. Suppose we have conditions (1) *on the values of* $y(s, 0)$ and (2) *on the type of weighting function*. Then we *might* have that "$c^+(t) \to$ *some infinite value as* $t \to \infty$ *if* (1) *and* (2) *hold; otherwise* $c^+(t) \to \infty$". In practice (Section 2.3), the only part of this ideal which we shall attain will be a realization of condition (2); as regards (1), we shall have to be content with mutually exclusive, but not exhaustive, conditions. Also, we shall not be able to prove that $c^+(t)$ has a limit when our conditions hold, or even that it has a finite (upper) bound. In this section, I shall only deal with the inadequacies of Section 2.3 as regards the velocity c^+, and prepare some of the apparatus with which we shall investigate c^+.

From the type of proof I provide, it is not possible to tell about the short term behavior of c^+; instead we shall deal with

$$\bar{c}^+(t) \equiv \frac{1}{t}\left(\int_0^t c^+(\tau)\,d\tau\right), \qquad t > 0, \tag{2.8}$$

which I shall call the *mean velocity* (*to time* t). The velocity $\bar{c}^+(t)$ is not an easy object to analyze directly, so we introduce one more type of velocity $c_\alpha(t)$, $0 < \alpha < 1$, and prove that if $c_\alpha(t) \to \infty$, so does $\bar{c}^+(t)$ (under rather trivial conditions).

We define the *velocity at level* α,

$$c_\alpha(t) \equiv \frac{1}{t}\big(\sup\,(s : y(u, t) \geqq \alpha \text{ for } 0 \leqq u \leqq s)\big). \tag{2.9}$$

(This is, of course, again a definition for propagation in the direction of increasing s.) We have the following connection between c_α and $\bar{c}^+$.

LEMMA 1. *Let* $y(s, t)$ *be an outbreak for which* ess lim $\sup_{s\to\infty} y(s, 0) < 1$. *Then, for* α *such that* ess lim sup $y(s, 0) < \alpha < 1$,

$$c_\alpha(t) \to \infty \textit{ as } t \to \infty \Rightarrow \bar{c}^+(t) \to \infty \textit{ as } t \to \infty. \tag{2.10}$$

NOTE. The ess lim sup of $f(s)$ as $s \to \infty$ is defined as $\inf\{h : \mu\big(f^{-1}((h, \infty))\cap(s, \infty)\big) = 0 \text{ for some } s\}$; compare lim sup $f(s)$, $= \inf\{h : \big(f^{-1}((h, \infty))\big)\cap(s, \infty) = \varnothing \text{ for some } s\}$.

PROOF. Choose β with ess lim sup $y(s, 0) < \beta < \alpha$. Then there exists s^* such that $y(s, 0) < \beta$ for $s > s^*$, except on a set of measure zero. Let $s(t)$ denote $\sup\big(s : y(u, t) \geqq \alpha \text{ for } 0 \leqq u \leqq s\big)$. Then

$$\begin{aligned} t\bar{c}^+(t) &\equiv \int_0^\infty \big(y(s, t) - y(s, 0)\big)\,ds \geqq \int_{s^*}^{s(t)} \big(y(s, t) - y(s, 0)\big)\,ds \\ &\geqq \int_{s^*}^{s(t)} (\alpha - \beta)\,ds = (\alpha - \beta)\big(s(t) - s^*\big) = (\alpha - \beta)\big(tc_\alpha(t) - s^*\big). \end{aligned} \tag{2.11}$$

Therefore, $\bar{c}^+(t) \geqq (\alpha - \beta)c_\alpha(t) - (\alpha - \beta)s^*/t$; whence it follows that if $c_\alpha(t) \to \infty$ as $t \to \infty$, so does $\bar{c}^+(t)$. *Q.E.D.*

Our last preliminary comment on velocities is directed to the converse problem: circumstances under which $\bar{c}^+(t)$ is bounded.

LEMMA 2. *Suppose* $y(s, t)$ *is an outbreak bounded by a travelling wave of velocity* c, *that is, a function* $z(s, t)$ *such that* $y(s, t) \leqq z(s, t)$ *and* $z(s, t) =$

$z(s - ct, 0)$, *for all* $t \geqq 0$ *and all* s. *And suppose that* $Z, \equiv \int_0^\infty z(s, 0)\, ds < \infty$. *Then* $\limsup \bar{c}^+(t) \leqq c$.

PROOF. We have

$$(2.12) \qquad t\bar{c}^+(t) \leqq \int_0^\infty y(s, t)\, ds \leqq \int_0^\infty z(s, t)\, ds \leqq \int_0^\infty z(s, 0)\, ds + ct.$$

Therefore, $\bar{c}^+(t) \leqq c + Z/t$. Since the latter term $\to 0$ as $t \to \infty$, our result follows.

We now turn to a more basic problem. Consider two outbreaks y_1 and y_2 of the same epidemic, for which it is given that $y_1(s, 0) \leqq y_2(s, 0)$ for all s. We should expect (intuitively) that $y_1(s, t) \leqq y_2(s, t)$ for all $t \geqq 0$. Rather than proving this directly, we shall derive it as a corollary of a more general result (Theorem 1) which allows us to compare solutions of two different equations, which need not be epidemic equations.

DEFINITION 3. *We define a* pseudoepidemic *as an autonomous differential equation*, $\dot{y} = f(y)$, *where* f *is a function from the space of measurable functions on* $\mathscr{R}$ *bounded by* 0 *and some constant* $b > 0$, *to the space of bounded positive measurable functions on* $\mathscr{R}$, *such that:*

(i) *there exists* $K < \infty$ *such that* $\sup_s |f(g(s))| \leqq K$ *and* $\sup_s |f(g_1)(s) - f(g_2)(s)| \leqq K \sup_s |g_1(s) - g_2(s)|$ *for all* g, g_i *in the domain of* f, *except (possibly) at* s *for which either* $g_i = b$;

(ii) $g \geqq 0 \Rightarrow f(g) \geqq 0$; *and* $f(g(s_0)) = 0$ *if* $g(s_0) = b$.

NOTE. This definition of pseudoepidemic suffices for this section; it is not, however, clear that it is the right analogue of the more elegant definition given in the stochastic case (Definition 4), so it should be regarded as provisional.

As in the epidemic case, we define an *outbreak* as a particular solution of the pseudoepidemic. We can apply the well-known fixed point theorem on the existence and uniqueness of solutions to differential equations (see, for example, [8], pp. 46–47), to obtain Lemma 3.

LEMMA 3. *Given a pseudoepidemic* $\dot{y} = f(y)$, *and an initial condition* $y(s, 0)$ *(specified for all* s, *and bounded):*

(i) *the outbreak* $y(s, t)$ *is uniquely defined for all* s, *and all* $t \geqq 0$;

(ii) *there exists a (fairly) concrete representation of* $y(s, t)$, *as follows: there exists* $\tau > 0$ *such that if for* $0 \leqq t^* \leqq t \leqq t^* + \tau$ *we define*

$$(2.13) \qquad \psi_*(g)(s, t) \equiv y(s, t^*) + \int_{t^*}^{\tau} f(g(s, \theta))\, d\theta,$$

for g *a measurable bounded function defined for all* s *for* $t \in [t^*, \tau]$, *then*

$$(2.14) \qquad y(s, t) = \lim_{n \to \infty} \psi_*^n(x)(s, t),$$

where $x(s, t) \equiv y(s, t^*)$.

PROOF. (i) To apply the fixed point theorem cited above, we must strengthen it slightly, to make it applicable to a space of functions which is more complex in two ways, but, if we refer to the fixed point theorem for a complete metric space itself ([8], p. 43), we shall find that only routine adaptation of the proof given on pp. 46–47 there is necessary. More specifically, (a) we are dealing here with functions of s and t, rather than just of t; however, with the sup norm (taken over s) replacing the modulus, the mathematics is formally identical; (b) for the outbreak to be properly defined we require that $y(s, t)$ should be $\geqq 0$: the condition "$g \geqq 0 \Rightarrow f(g) \geqq 0$" ensures this, provided that we are considering increasing t as we are here. (*Since we cannot guarantee that* $y(s, t)$ *remains* $\geqq 0$ *as we decrease* t, *we cannot in general extend the outbreak to times* <0.) This completes part (i) of the lemma.

(ii) The proof of this part consists of footnotes to the fixed point theorem on solutions to differential equations. The proof of the metric space theorem ([8], p. 43) involves showing that, for any g in the appropriate metric space $\psi^n(g)$ tends to the fixed point of ψ (the solution of the differential equation) as $n \to \infty$; clearly, x is in this space. The only remaining point is that τ may be chosen independent of t^*, because we have a bounding constant K independent of g, g_i in the conditions of Definition 3 for a pseudoepidemic. *Q.E.D.*

We shall say that one pseudoepidemic E_1 is *dominated* by another E_2, written as $E_1 \ll E_2$, if and only if there exists $K_{12}(<\infty)$ such that

$$f_1(g_1)(s) - f_2(g_2)(s) \leqq K_{12}(g_2(s) - g_1(s)), \tag{2.15}$$

for each s, whenever $g_1 \leqq g_2$. (*Note that any pseudoepidemic is dominated by itself.*) We shall say that an outbreak O_1 of E_1 is dominated by O_2 of E_2 if and only if $y_1(s, t) \leqq y_2(s, t)$ for all s and t for which the outbreaks are both defined.

THEOREM 1. *Let* O_i, $i = 1, 2$, *be outbreaks of pseudoepidemics* E_i. *If* $E_1 \ll E_2$ *and* $y_1(s, 0) \leqq y_2(s, 0)$, *then* $O_1 \ll O_2$ (*regarding* O_i *as being defined only for* $t \geqq 0$).

PROOF. Suppose first that $y_1(s, 0) \leqq y_2(s, 0)$. Consider $z(s, t)$, defined by $(y_2(s, t) - y_1(s, t)) \exp\{K_{12}t\}$. Then

$$\begin{aligned} \dot{z}(s, t) = {} & (f_2(y_2)(s, t) - f_1(y_1)(s, t)) \exp\{K_{12}t\} \\ & + K_{12}(y_2(s, t) - y_1(s, t)) \exp\{K_{12}t\}, \end{aligned} \tag{2.16}$$

and

$$f_2(y_2) - f_1(y_1) \geqq -K_{12}(y_2 - y_1), \tag{2.17}$$

since $E_1 \ll E_2$, so $\dot{z} \geqq 0$ (provided $z \geqq 0$); $z(s, 0) > 0$, so $z(s, t) > 0$ for all $t \geqq 0$.

Second, suppose $y_1(s, 0) = y_2(s, 0)$ for at least one s. In this case, define $y_{2,n}(s, 0) \equiv y_2(s, 0)(1 + 1/n)$ for $n > 0$. By the above $y_1(s, t < y_{2,n}(s, t)$ for all $t \geqq 0$, all n. Now for each fixed t, $y_{2,n} \to y_2(s, t)$ as $n \to \infty$ (since $y_{2,n}(s, t) - y_2(s, t) \leqq (y_{2,n}(s, 0) - y_2(s, 0)) \exp\{K_2 t\}$, where K_2 is the Lipschitz constant for E_2): whence $y_1(s, t) \leqq y_2(s, t)$ for all $\geqq 0$. *Q.E.D.*

2.3. *Conditions for finite propagation.* Here we establish the importance of the condition given in (1.1), $\int_{\mathscr{R}} e^{ks}\, dV(s) < \infty$ for some $k > 0$, in determining the nature of the propagation (in the direction of increasing s) of simple epidemics on the line $\mathscr{R}$. Over a wide range of initial conditions, (1.1) turns out to be a necessary and sufficient condition for propagation at a finite rate ($\limsup_{t\to\infty} \bar{c}^+(t) < \infty$ being taken as criterion), though there are initial conditions under which $\bar{c}^+ \to \infty$ almost regardless of V (see Corollary 1).

We shall start with a result (Theorem 2) which contains the two essential ideas connected with the influence of (1.1); to keep the proofs as clear as possible, we leave refinements of both halves of this result to a series of corollaries and lemmas which follow this theorem.

THEOREM 2 (i). *If E is a simple epidemic for which* (1.1) *holds, and if O is an outbreak of E for which the initial condition*

$$y(s, 0) \leqq ae^{-bs}, \tag{2.18}$$

for some $a, b > 0$, holds, then $\limsup_{t\to\infty} \bar{c}^+(t) < \infty$.

(ii) *If O is an outbreak of a simple epidemic E, for which there exist δ, β such that $y(s, 0) \geqq \delta > 0$ for $s \leqq 0$, and* $\operatorname{ess\,lim\,sup}_{s\to\infty} y(s, 0) \leqq \beta < 1$, *and if* $\limsup_{t\to\infty} \bar{c}^+(t) < \infty$, *then* (1.1) *holds.*

PROOF. (i) Let $f_v(k) \equiv \int_{\mathscr{R}} e^{ks}\, dV(s)$; from (1.1) we have that $f_v(k)$ converges for some $k > 0$. The convergence or otherwise of $f_v(k)$ for $k \geqq 0$ depends on that of $\int_0^\infty e^{ks}\, dV(s)$, since then $\int_{-\infty}^0 e^{ks}\, dV(s) \leqq \int_{-\infty}^0 dV(s) \leqq 1$. The integral $\int_0^\infty e^{ks}\, dV(s)$ is monotone with k, so the set of positive k for which $f_v(k)$ converges is an interval, $[0, k^*)$ or $[0, k^*]$. Choose k such that $k \leqq b$ and $0 < k < k^*$; let $c \equiv f_v(k)/k$. Consider $g(s) \equiv a'e^{-ks}$. Then

$$\begin{aligned}\bar{g}(s) &\equiv \int_{\mathscr{R}} g(s-u)\, dV(u) \\ &= g(s) \int_{\mathscr{R}} e^{ku}\, dV(u) = f_v(k) g(s).\end{aligned} \tag{2.19}$$

(We are motivated here by a desire to produce a solution to the "linearized" equation $\dot{y} = \bar{y}$.) Now suppose we turn $g(s)$ into a travelling wave of velocity c by multiplying by the factor e^{kct}. We shall then have a function, $h(s, t)$ say, for which $\dot{h} = ckh = f_v(k)h = \bar{h} \geqq \bar{h}(1-h)$. Thus, at each moment the rate of increase of h is greater than it would be if h were an outbreak of a simple epidemic (h being a solution of the linearized equation). Also,

$$h(s, 0) = a'e^{-ks} \geqq \min(ae^{-bs}, 1) \geqq y(s, 0) \tag{2.20}$$

for all s if we choose a' appropriately.

To turn these observations to advantage, we must adapt h to become an outbreak of a pseudoepidemic which dominates E. Let E^* be the equation

$$\dot{y} = \begin{cases} \max\left(f_v(k)y, \bar{y}\right) & \text{for } 0 \leqq y < 1, \\ 0 & \text{for } y = 1. \end{cases} \tag{2.21}$$

Then it is easy to see that E^* is a pseudoepidemic as defined in Definition 3, and that $E \ll E^*$ (we may take the relevant constant $K_{12} = 0$, see (2.15)).

Let $y^*(s, t) \equiv \min\left(h(s, t), 1\right)$. This satisfies E^* (note that for $y^* \leqq 1$, $f_v(k)y^* = f_v(k)h = \bar{h} \geqq \bar{y}^*$, so that $f_v(k)y^* \geqq \bar{y}^*$ for all s and t), so it is an outbreak, O^* say, of a pseudoepidemic which dominates E. Also (by our choice of a', k) $y^*(s, 0) \geqq y(s, 0)$ for all s.

Hence, $O \ll O^*$ (Theorem 1). Thus, O is dominated by a travelling wave of velocity c. Since $\int_0^\infty y^*(s, 0)\, ds = a'/k < \infty$, we may apply Lemma 2 to conclude that $\limsup_{t\to\infty} \bar{c}^+(t) \leqq c < \infty$ as desired.

(ii) If $V(0) = 1$, $\int_{\mathscr{R}} e^{ks}\, dV(s) \leqq \int_{-\infty}^0 dV(s) = 1$ for all positive k; so (1.1) holds. So we assume $V(0) \neq 1$ henceforth. Here also we apply Theorem 1, this time comparing O with outbreaks of pseudoepidemics dominated by E. We divide the growth of O into two stages. Choose α such that $\beta < \alpha < 1$.

(a) For $0 \leqq t \leqq 1$ we compare O with an outbreak O_* of the pseudoepidemic E_* defined for $0 \leqq y_*(s, t) \leqq \alpha$ by

$$f_*\big(y_*(s, t)\big) = \begin{cases} \bar{y}_*(s, 0)(1 - \alpha) & \text{if } y_* \leqq \alpha \text{ and } s > 0, \\ 0 & \text{if } y_* = \alpha \text{ or } s \leqq 0. \end{cases} \tag{2.22}$$

Then $f_*\big(y_*(s, t)\big) \leqq \bar{y}_*(s, t)\big(1 - y_*(s, t)\big)$ for all s, t, whence $E_* \ll E$ (by Theorem 1). Define O_* by

$$y_*(s, 0) = \begin{cases} \delta & \text{for } s \leqq 0, \\ 0 & \text{for } s > 0. \end{cases} \tag{2.23}$$

Then $y_*(s, 0) \leqq y(s, 0)$, so (by Theorem 1) $y(s, 1) \geqq y_*(s, 1)$. Therefore,

$$y(s, 1) \geqq \begin{cases} \delta & \text{for } s \leqq 0, \\ \big(\delta \bar{V}(s)\big)(1 - \alpha) & \text{for } s > 0, \end{cases} \tag{2.24}$$

where $\bar{V}(s)$ denotes $\int_s^\infty dV(s)$, $= 1 - V(s)$.

(b) Define E_u for $0 \leqq y_u \leqq \alpha$ by

$$f_u(y_u) = \begin{cases} (1 - \alpha) \min\big(\bar{V}(0) y_u, \bar{y}_u\big) & \text{if } s \leqq u \text{ and } y_u < \alpha, \\ 0 & \text{if } s > u \text{ or } y_u = \alpha. \end{cases} \tag{2.25}$$

This pseudoepidemic is also easily seen to be dominated by E. Define an outbreak O_u of E_u (for $t \geqq 1$) by

$$y_u(s, 1) = \begin{cases} \delta(1 - \alpha)\bar{V}(u) & \text{if } s \leqq u, \\ 0 & \text{if } s > u. \end{cases} \tag{2.26}$$

Since $y_u(s, 1) \leqq y(s, 1)$, $O_u \ll O$. Now we may write

$$y_u(s, t) = \begin{cases} \delta(1 - \alpha)\bar{V}(u) \exp\{(1 - \alpha)\bar{V}(0)(t - 1)\} & \text{if } s \leqq u, \\ 0 & \text{if } s > u. \end{cases} \tag{2.27}$$

(Clearly, this solves the equation E_u, with initial condition $y_u(s, 1)$, and, therefore, it is the unique solution for $y_u(s, t)$.)

Restricting attention to u sufficiently large that $y_u(s, 1) < \alpha$, let t_u be the time at which $y_u(s, t) = \alpha$ for $s \leqq u$. Since $O_u \ll O$, we have that $y(s, t_u) \geqq \alpha$ for $s \leqq u$; whence, $u \leqq t_u c_\alpha(t_u)$ (see (2.9)).

Now since $\limsup_{t\to\infty} \bar{c}^+(t)$ is finite, so is $\limsup_{t\to\infty} c_\alpha(t)$ (by Lemma 1); let the latter $\equiv c$. Then for all $\varepsilon > 0$, we can choose t_* such that $c_\alpha(t) < c + \varepsilon$ for $t \geqq t_*$. Choose u_* such that $t_u \geqq t_*$ for $u \geqq u_*$. Then for such u, $u \leqq t_u(c + \varepsilon)$. Inserting this inequality in the equation which defines t_u ((2.27) with $y_u = \alpha$), we have

$$\alpha \geqq \delta(1 - \alpha)\bar{V}(u) \exp\left\{(1 - \alpha)\bar{V}(0)\left(\frac{u}{c + \varepsilon} - 1\right)\right\}. \tag{2.28}$$

Hence,

$$\bar{V}(u) \leqq k_1 \exp\{-k_\varepsilon u\} \text{ for all } u \geqq u_*, \tag{2.29}$$

where $k_1 \equiv \alpha \exp\{(1 - \alpha)\bar{V}(0)\}/\delta(1 - \alpha)$, $k_\varepsilon \equiv (1 - \alpha)\bar{V}(0)/(c + \varepsilon)$. Therefore, $\int_{u*}^\infty e^{ks}\, dV(s)$ converges for $k < k_\varepsilon$ and thus for $k < (1 - \alpha)\bar{V}(0)/c$ (*since* ε *may be taken arbitrarily small*); whence, $\int_{\mathscr{R}} e^{ks}\, dV(s)$ converges for $0 \leqq k < (1 - \alpha)\bar{V}(0)/c$. Thus, condition (1.1) holds, as was to be proved. *Q.E.D.*

COROLLARY 1. *Provided that* $V(0) \neq 1$, *if* $\limsup_{t\to\infty} \bar{c}^+(t) < \infty$, *then* (2.18) *holds.*

PROOF. This follows the line of reasoning of the second stage of the proof of Theorem 2 (ii), replacing $\bar{V}(u)$ by $y(u, 0)$.

Thus, we start (at time 0) with the outbreak of E_u defined by

$$y_u(s, 0) = \begin{cases} y(u, 0) & \text{if } s \leqq u, \\ 0 & \text{if } s > u, \end{cases} \tag{2.30}$$

and deduce that $y(u, t) \leqq y(u, 0) \exp\{(1 - \alpha)\bar{V}(0)t\}$ (compare (2.27)), whence we may derive $y(u, 0) \leq k_1 \exp\{-k_\varepsilon u\}$ for $u \geq u_*$, where now $k_1 \equiv \alpha$, $k_\varepsilon = (1 - \alpha)\bar{V}(0)/(c + \varepsilon)$ (compare (2.29)).

Hence, $y(s, 0) \leqq ae^{-bs}$ for all s, where a and b (both > 0) must be chosen such that (1) $b < (1 - \alpha)\bar{V}(0)/c$, and (2) $a \geqq \alpha$, and $y(s, 0) \leqq ae^{-bs}$ for $s < u_*$ (note that u_* depends on ε, and hence on the choice of b). At worst $a = \exp\{bu_*\}$ will suffice. *Q.E.D.*

The conditions on $y(s, 0)$ required for Theorem 2 (ii) seem excessive, as they exclude any outbreak where the initial set of infected locations is bounded. Ideally, we should like to replace them by the absolutely minimal condition that the numbers initially infected should be nonzero, that is, that $\int_{\mathscr{R}} y(s, 0)\, ds > 0$. It turns out that we can do this provided that V has a density v which is monotone decreasing with s for s positive (we are, as usual, thinking just of propagation in the direction of increasing s); in fact, it will be clear from the proof that it would suffice for v to be greater than some scalar multiple of such a density. Thus, we offer the following alternative to Theorem 2 (ii).

LEMMA 4. *Suppose* V *has a density* v, *monotone decreasing for positive* s, *and that* $V(0) \neq 0$ *or* 1. *Suppose* O *is a nontrivial outbreak, one for which* $\int_{\mathscr{R}} y(s, 0)\, ds > 0$. *Then if* $\limsup_{y\to\infty} \bar{c}^+(t) < \infty$, (1.1) *must hold.*

PROOF. Again we follow the lines of the proof of Theorem 2 (ii).

(i) Without loss of generality, we may assume that $\int_0^\delta y(s, 0) > 0$, some finite δ. Considering E_* for $0 \leqq t \leqq 1$, we obtain

$$y(s, 1) \geqq y_*(s, 1) = (1 - \alpha) \int_0^\delta y(u, 0) v(s - u)\, du. \tag{2.31}$$

(ii) In place of $y(s, 1) \geqq y_*(u, 1)$ for $s \leqq u$, we have (from the condition on v) that $y(s, 1) \geqq y_*(u, 1)$ for $0 \leqq s \leqq u$. So we define E_u^*, which replaces E_u, by

$$f_u^*(y_u^*) = \begin{cases} (1 - \alpha) \min\left(k_u y_u^*, y_u^*\right) & \text{for } 0 \leqq y_u^* < \alpha, \\ 0 & \text{for } y_u^* = \alpha, \end{cases} \tag{2.32}$$

where $k_u \equiv \inf_{0 \leqq s \leqq u} \left(V(s) - V(s - u)\right)$ which is monotone increasing with u and >0 for sufficiently large u, since $V(0) \neq 0$ or 1.

Define O_u^* by

$$y_u^*(s, 1) = \begin{cases} y_*(u, 1) & \text{for } 0 \leqq s \leqq u, \\ 0 & \text{elsewhere.} \end{cases} \tag{2.33}$$

Then $f_u^*(y_u^*) = (1 - \alpha) k_u y_u^*$ since $\bar{y}_u^* \geqq k_u y_u^*$ everywhere, and so, for $t \geqq 1$,

$$y_u^*(s, t) = \begin{cases} y_*(u, 1) \exp\{(1 - \alpha) k_u (t - 1)\} & \text{for } 0 \leqq s \leqq u, \\ 0 & \text{elsewhere,} \end{cases} \tag{2.34}$$

whence $y(u, t) \geqq (1 - \alpha)\left(\int_0^\delta y(s, 0) v(u - s)\, ds\right) \exp\{(1 - \alpha) k_u (t - 1)\}$ (compare (2.27)).

From this point we follow the same line exactly as for Theorem 2 (ii), to deduce that

$$\int_0^\delta y(s, 0) v(u - s)\, ds \leqq k_1 \exp\{-k_\varepsilon u\} \qquad \text{for } u \geqq u_*, \tag{2.35}$$

where $k_1 \equiv \alpha/(1 - \alpha) \exp\{(1 - \alpha) k_{u_*}\}$, $k_\varepsilon \equiv (1 - \alpha) k_{u_*}/(c + \varepsilon)$ (compare (2.29)). Therefore

$$\int_{u_0}^\infty \int_0^\delta y(s, 0) v(u - s)\, ds\, du \leqq \frac{k_1 \exp\{-k_\varepsilon u_0\}}{k_\varepsilon} \tag{2.36}$$

for $u_0 \geqq u_*$. Interchanging the order of integration on the left side, we have $\int_0^\delta y(s, 0)\, \bar{V}(u_0 - s)\, ds$, which is $\geqq \bar{V}(u_0) \int_0^\delta y(s, 0)\, ds$.

Thus, $\bar{V}(u) \leqq k_* c^{-k_\varepsilon u}$ for $u \geqq u_*$, where $k_* = k_1\left(k_\varepsilon \int_0^\delta y(s, 0)\, ds\right)^{-1}$, whence (1.1) holds (for $0 \leqq k < k_\varepsilon$). *Q.E.D.*

With Theorem 2, Corollary 1, and Lemma 4, we have nearly established that the lim sup of the mean velocity is finite for a simple epidemic if and only if (1.1) and (2.18) both hold. It may be convenient to list the exceptions to this.

COROLLARY 2 (Corollary to Theorem 2, Corollary 1, and Lemma 4). *For a simple epidemic on the line* $\mathscr{R}$, $\limsup_{t \to \infty} \bar{c}^+(t) < \infty$ *if and only if* (1.1) *and* (2.18) *hold, except possibly when* (i) $\int_{\mathscr{R}} y(s, 0)\, ds = 0$, *or* $V(0) = 0$ or 1, *or* (ii) *there is no* $\delta > 0$ *and no* s_* *such that* $y(s, 0) \geqq \delta$ *or* $s \leqq s_*$; and V *does not have a density greater than some scalar multiple of a density which is monotone decreasing for positive s.*

Note that inspection of the proof of Corollary 2 reveals that we could replace "positive" by "sufficiently large".

Lastly, we refine Theorem 2 (i). We there established that, if (1.1) and (2.18) hold, the outbreak is bounded by a wave of finite velocity c, where we may take $c = f_v(k)/k$ for any value of k for which $f_v(k)$ converges and $k \leqq b$. It is of interest to know how low we can choose c.

LEMMA 5. (i) *If* $V(0) \neq 1$, $f_v(k)/k$ *has a minimal value* $c_v > 0$. *There exists* $k_v > 0$ *such that if* $y(s, 0) \leqq a \exp\{-k_v s\}$ *for some* $a > 0$, *the outbreak is bounded by a wave of velocity* c_v. *If we only have that* $y(s, 0) \leqq ae^{-bs}$, $0 < b < k_v$, *the best we can do is some* $c_v(b) > c_v$.

(ii) *If* $V(0) = 1$ *and if* $y(s, 0) \leqq ae^{-bs}$ *for arbitrarily large b, the outbreak is bounded by waves of arbitrarily low velocity. If we only have that* $y(s, 0) \leqq ae^{-bs}$, $0 < b < \infty$, *the best we can do is some* $c_v(b) > 0$.

(iii) *If V is symmetric,* $c_v/\sqrt{w_2} \geqq c_* = \sinh k_*$, *where* k_* *is the positive root of* $\sinh k = \cosh k/k$, *and* w_2 *is the variance of* V; $c_* \approx 1.509$.

(iv) *For the negative exponential distribution* (*density* $\frac{1}{2}\beta e^{-|\beta|s}$) c_v *coincides with the* minimal velocity *found in* [11]; *thus,* $c_v/\sqrt{w_2} = 3\sqrt{3}/2\sqrt{2} \approx 1.834$ (note how close to c_* this is).

PROOF. (i, a) If $V(0) \neq 1$, $f_v(k)$ together with all its derivatives ($f_v^n(k) = \int_{\mathscr{R}} s^n e^{ks}\, dV(s)$), tends to ∞ as $k \to \infty$. Hence, $f_v(k)/k \to \infty$ both as $k \downarrow 0$ and as $k \to \infty$; it is > 0 on $(0, \infty)$, so it has a minimum value, c_v say, which it attains because it is a continuous function of k. The set $\{k : f_v(k)/k = c_v\}$ is a closed nonnull set bounded below by 0, and, therefore, has a least member, k_v say. It is then immediate from the proof of Theorem 2 (i) that if $y(s, 0) \leqq ae^{-k_v s}$, the outbreak is bounded by a wave of velocity c_v; and that, with the line of argument of that theorem, we can do no better.

(i, b) If it is only given that $y(s, 0) \leqq ae^{-bs}$ for some $b < k_v$, we cannot even do so well. In this case what we want is $c_v(b) \equiv \min\{f_v(k)/k : k \in (0, b]\}$. Since $f_v(k)/k$ is continuous, >0, and $\to\infty$ as $k \downarrow 0$, this minimum is >0 and is attained, at $k_v(b)$ say, so we can find a wave of velocity $c_v(b)$ which bounds the outbreak.

(ii) If $V(0) = 1, f_v(k) = \int_{-\infty}^{0} e^{ks}\, dV(s) \leqq 1$ for all $k \geqq 0$. Therefore, $f_v(k)/k \to 0$ as $k \to \infty$. Thus, if $y(s, 0) \leqq ae^{-bs}$ where b can be taken arbitrarily large (for example, if the initial set of infected is bounded on the right), we can take c arbitrarily small. Otherwise (i, b) applies and we have some minimum value $c_v(b)\,(>0)$ for c.

(iii) We use the expansion $f_v(k) = \Sigma_0^\infty w_n k^n/n!$, where $w_n \equiv \int_{\mathscr{R}} s^n\, dV(s)$, valid for any k for which both $f_v(k)$ and $f_v(-k)$ converge. If V is symmetric, $f_v(-k) = f_v(k)$, and $w_n = 0$ for all odd n, so $f_v(k) = \Sigma_0^\infty w_{2n} k^{2n}/(2n)!$, for all k for which $f_v(k)$ converges, and hence, for all k.

Contracting V by a factor $\sqrt{w_2}$, that is, putting $V_{(1)}(s) = V(s\sqrt{w_2})$, is equivalent to multiplying the time scale by $\sqrt{w_2}$; thus, $c_v = \sqrt{w_2}c_{v_{(1)}}$, and it suffices to consider V whose variance $(w_2) = 1$.

We next prove that the sequence (w_{2n}) is convex, that is, that $w_{2n+2} \geqq \frac{1}{2}(w_{2n} + w_{2n+4})$; since $w_0 = w_2 = 1$, it will follow that (w_{2n}) is monotone increasing with n. Rewrite w_{2n} as

$$2\int_0^1 \left(s^{2m}\,dV(s) + s^{-2m}\,dV\left(\frac{1}{s}\right)\right). \tag{2.37}$$

For all s, $(s^2 + s^{-2}) \geqq 2$, so

$$\begin{aligned} w_{2n+2} &\leqq \int_0^1 (s^2 + s^{-2})\left(s^{2n+2}\,dV(s) + s^{-(2n+2)}\,dV\left(\frac{1}{s}\right)\right) \\ &= \tfrac{1}{2}(w_{2n} + w_{2n+4}). \end{aligned} \tag{2.38}$$

Thus, (w_{2n}) is monotone increasing with n, and thus, $w_{2n} \geqq 1$ for all n. Therefore, $f_v(k) \geqq \Sigma_0^\infty k^n/n!$ for all symmetric V (of variance 1). Thus, $f_*(k) \equiv \Sigma_0^\infty k^n/n!$ is the minimal $f_v(k)$ (among such V); it corresponds to the distribution concentrated on ± 1. Then $f_*(k)/k \leqq f_v(k)/k$, so the minimal value for c_v will be that of $f_*(k)/k$, c_* say.

The velocity c_* may be found by solving $f_*'(k_*) = f_*(k_*)/k_*(=c_*)$, k_* positive. Since $f_*(k) = \cosh k$, this equation is $\sinh k_* = \cosh k_*/k_*$. The solution of this is approximately $c_* = 1.509$. Thus, for all symmetric V, $c_v \geqq c_*\sqrt{w_2} \approx 1.509\sqrt{w_2}$.

(iv) For the negative exponential distribution with density $\frac{1}{2}\beta e^{-\beta|s|}$, $w_{2n} = (2n)!/\beta^{2n}$. Thus, if $w_2 = 1$,

$$f_v(k) = \sum_0^\infty \frac{k^{2n}}{2^n} = \frac{1}{1 - \frac{1}{2}k^2}. \tag{2.39}$$

We find c_v by solving $f_v' = f_v/k$:

$$\frac{k_v}{(1 - \frac{1}{2}k_v^2)^2} = \frac{1}{k_v(1 - \frac{1}{2}k_v^2)}, \tag{2.40}$$

whence, $k_v = (\frac{1}{3})^{1/2}$, $c_v = (\frac{3}{2})^{3/2} \approx 1.834$. *Q.E.D.*

Lemma 5 (iv) agrees with [11], where it is shown that, when V is negative exponential, waveforms exist for exactly those velocities for which $\dot{y} = \bar{y}$ (the linear approximation to (2.5) for small y) has waveform solutions, that is, for $c = c_v(=3\sqrt{3}/2\sqrt{\beta}$ for the negative exponential). This encourages the conjecture that waves of (only) velocities $\geqq c_v$ exist for simple epidemics for which condition (1.1) holds or, further, that a large class of outbreaks of such epidemics tend to waves of velocity c_v as $t \to \infty$. We have as yet nothing substantial towards proving such results.

Indeed, in the case where (1.1) holds, we have found no lower bound for the rate of propagation. We can make good this deficiency, in some degree, provided that $\bar{V}(s)e^{ks}$ diverges as $s \to \infty$, for some $k > 0$.

Suppose, as for Theorem 2 (ii), that there exists δ such that $y(s, 0) \geqq \delta > 0$ for $s \leqq 0$; for simplicity we assume here that ess lim $\sup_{s \to \infty} y(s, 0) = 0$.

COROLLARY 3 (Corollary to Theorem 2 (ii)). *Under the above conditions* $\liminf_{t \to \infty} c_\alpha(t) \geqq (1 - \alpha)\bar{V}(0)/k$.

PROOF. The analysis of the proof of Theorem 2 (ii) yields

$$\bar{V}(u) \leqq k_1 \exp\left\{-\frac{(1-\alpha)\bar{V}(0)u}{c_\alpha(t)}\right\} \tag{2.41}$$

for $u \geqq u_*$, and this contradicts "$\bar{V}(s)e^{ks}$ diverges" unless $c_\alpha(t) \geqq (1 - \alpha)\bar{V}(0)/k$. *Q.E.D.*

NOTE. If we divide the growth of our outbreak into three stages rather than the two used in the proof of Theorem 2 (ii), we can do slightly better. Let stage (c) consist of growth from level ε to level α; for this $\dot{y} \geqq (1 - \alpha)\bar{V}(0)y$, so this stage only takes time $\leqq (1/(1 - \alpha)\bar{V}(0)) \log(\alpha/\varepsilon)$—a constant. Thus, we can obtain $c_\alpha(t) \geqq (1 - \varepsilon)\bar{V}(0)/k$, for all ε, α with $0 < \varepsilon \leqq \alpha < 1$. Letting $\varepsilon \to 0$ we have $c_\alpha(t) \geqq \bar{V}(0)/k$. Applying Lemma 1, we have that $\bar{c}^+(t) \geqq \bar{V}(0)/k$ (for sufficiently large t), since we may take the constants α and β of that lemma as near to 1 and 0, respectively, as we like.

Thus, for instance, for the negative exponential distribution of variance 1, for which we may take any $k < \sqrt{2}$, we have the lower bound $1/(2\sqrt{2})$ (≈ 0.354) for lim inf $\bar{c}^+(t)$.

3. Simple stochastic epidemics

3.1. *Introduction.* The study of spatial propagation in a simple deterministic epidemic has, I trust, now been carried far enough for it to seem necessary to offer some indication of how far the results obtained apply to more realistic models. Consistent with the emphasis I have laid on comparison of exact convolution equations with their diffusion approximations, I shall again begin with a simple epidemic among a population of uniform density on an infinite line. The difference here will be that the population will consist of discrete individuals, with a constant number σ living at each integer point of $\mathscr{R}$, and that infection will be a stochastic rather than deterministic process.

Not being a statistician, it was my original intention merely to simulate such a model on a computer, to obtain some idea of how closely its behavior was related to what one might expect from a deterministic model. However, it turned out to be unexpectedly easy to obtain necessary and sufficient conditions for propagation at a finite rate (Theorem 3), corresponding to those obtained for the deterministic case (Theorem 2). The derivation of this result is preceded by the precise setting up of a stochastic model. It is followed by the presentation of a program for computer simulation of the stochastic model (Section 3.3), and by an account of some results obtained from it (Section 3.4); the program can, suitably modified, be used to simulate general epidemics (where allowance is made for removed cases) as well; indeed, the modified program for this is in some ways simpler than the original program.

I shall end this introductory section with an account of the details of the stochastic model sufficient for the reader who wishes to go straight on to the sections dealing with computer simulation.

As in the deterministic case we make the following assumption.

ASSUMPTION 1. *The infectious influence of one individual on another depends solely on their states (infectious or susceptible) and on their spatial separation.*

Thus, for each (infectious, susceptible) pair we assume that cross infection is a Poisson process of frequency $\alpha(n)$, where n is the separation of the pair. Of course, we must regard this process as being operative only between the times when the former and the latter, respectively, become infected; it will never be operative at all if they become infected in the other order.

Looking at the role of the infectious individual, we introduce a notion of *germs*; we may think of each individual as emitting germs in a Poisson process of frequency $\sigma \Sigma_{n=-\infty}^{\infty} \alpha(n)$; let $\alpha \equiv \Sigma_{-\infty}^{\infty} \alpha(n)$. If $\alpha = \infty$, a single infectious individual would almost surely infect an infinite number of susceptibles in any nonzero time interval, so that the epidemic would proceed at what can only be described as an infinite rate. Accordingly, I shall ignore this case. Let $v(n) \equiv \alpha(n)/\alpha$; $v(n)$ is a probability density on the integers, corresponding to the weighting function $dV(s)$ of the deterministic model.

These germs are only active if the emitting individual is infectious, and they travel instantaneously to a random individual at relative position N, where N is a random variable with the distribution $v(n)$; naturally, a germ only causes a new infection if the victim chosen is susceptible.

The convenience of this way of looking at the model is that it is easy to see how to program a computer to simulate an epidemic, using a random number routine to produce and distribute germs (see Section 3.3). It also provides a concept to hold onto as we float through the abstract sample spaces of the next section.

In one respect, however, this notion of germs runs counter to intuition: the rate at which an individual emits germs increases linearly with the population density σ. A realistic situation to which our model corresponds would be one where each individual emits a large number of germs, whose probability of causing an infection at their terminal location is proportional to the number of susceptibles there; then the "successful germs" would correspond to what I have called germs.

We should also note that, as in the deterministic case, changing the rate of emission of germs by a scale factor is equivalent to speeding events up by that factor, so that the differences in our model resulting from different assumptions about the dependence of the rate of emission on α may be easily calculated. In contrast, because of the discrete nature of the stochastic model, the dependence of the rate of development of the epidemic upon σ is nonlinear; and the linear dependence on the first moment of v of the deterministic case has no relevance here, as we cannot vary a discrete distribution in the way we can a discontinuous one. Thus, the proportionality to σ/β of the deterministic rate of events can have no correlate here. However, it is still true that scale changes to α, σ, and any reasonable approximate substitute for β, will not affect the necessary and sufficient condition for finite propagation (see Theorem 3).

3.2. *Pseudoepidemics and finite propagation.* If P is a countable population, we may define a class of models on P, of which simple stochastic epidemics will form a subclass.

DEFINITION 4. *We define a* pseudoepidemic among P, E *say, as consisting of:*

(i) *a product space* $(\Omega, A, \mu) \equiv \Pi\, (\Omega_{pq}, A_{pq}, \mu_{pq})$, *where the product is taken over all ordered pairs* (p, q) *of distinct members of* P, *and each probability triple* $(\Omega_{pq}, A_{pq}, \mu_{pq})$ *represents a Poisson process of frequency* $\alpha(p, q)$; *we impose the conditions* $\Sigma_{p\in P}\, \alpha(p, q) < \infty$ *and* $\Sigma_{q\in P}\, \alpha(p, q) < \infty$;

(ii) almost sure *rules (that is, ones which work except on a subset of measure* 0) *for deducing, given* ω *in* Ω *and the set* Q *of individuals infected at time* τ, *the set* $Q(t)$ *of individuals infected at time* t, *for all* $t \geqq \tau$.

DEFINITION 5. *An* outbreak *of* E *is then defined as a triple* (ω, Q, τ), *where* $\omega \in \Omega$, $Q \subseteqq P$, *and* $\tau \in \mathscr{R}$.

We give as examples the three pseudoepidemics with which we shall be concerned in Theorem 3. In each case, P is a homogeneous population of density σ on the integers. Let $n(p)$ denote the integer at which the individual p lives. In each case the product space ((i) above) will be the same; it is the infection rules ((ii)) that we shall vary.

We shall, in the first instance, assume that $\alpha(p, q)$ depends solely on the spatial separation of p and q (Assumption 1), that is, that $\alpha(p, q) = \alpha v(n(q) - n(p))$, where v is a probability density on the integers. We shall not, in general, demand that v be symmetric, which would represent a *directionally unbiased* pseudoepidemic.

EXAMPLE A. The infection rules for simple stochastic epidemics are given in terms of the notion of a chain of infection. A *chain of infection* is a strictly increasing sequence $\{t_{q_i,q_{i+1}} : 0 \leqq i < n\}$ for which $t_{q_i,q_{i+1}} \in \omega_{q_i,q_{i+1}}$ for each i, which means that $t_{q_i,q_{i+1}}$ is a time at which a germ passes from q_i to q_{i+1}. A chain of infection from p to r between times t_1 and t_2 is one for which $q_0 = p$, $q_n = r$, and $t_1 < t_{q_0,q_1} < \cdots < t_{q_{n-1},q_n} \leqq t_2$. Then, given (ω, Q, τ), let $Q(t) \equiv \{r :$ for some $p \in Q$ there exists a chain of infection from p to r between times τ and $t\}$.

EXAMPLE B. For a *cliff edge pseudoepidemic*, we relax the infection rules somewhat. We define a *chain of *infection* as a sequence $\{t_{q_i,q_i+1}\}$ for which for each i *either* $t_{q_i,q_{i+1}} > t_{q_{i-1},q_i}$ and $t_{q_i,q_{i+1}} \in \omega_{q_i,q_{i+1}}$ (as for the simple epidemic) *or* $t_{q_i,q_{i+1}} \geqq t_{q_{i-1},q_i}$ and $n(q_{i+1}) \leqq n(q_i)$. Then the set of infected at time t, $Q^*(t)$ is defined as $\{r :$ for some $p \in Q$ there exists a chain of *infection from p to r between times τ and $t\}$.

The effect of the alternative infection rule is to ensure that as soon as an individual p becomes infected, so do all the remaining susceptibles q with $n(q) \leqq n(p)$. Thus, the set $Q^*(t) = \{p : n(p) \leqq m\}$, for some m depending on t, so that a diagram showing the numbers infected at each integer always has a "cliff edge" shape.

EXAMPLE C. For a *noninfectious pseudoepidemic*, we have infection rules more restrictive than in Example A. We define a *chain of* $_*$*infection* as a chain of infection for which $q_i \in Q$ for $i \leqq n - 1$, and define $Q_*(t)$ similarly to Q^* of Example B.

In this pseudoepidemic, the only effectively infectious individuals are those in Q, that is, those initially infected.

Even without the conditions we have imposed on the partial sums of $\{\alpha(p, q)\}$, all three pseudoepidemics are well defined, and satisfy the following two conditions which we might reasonably demand.

CONDITION 1. *The set* $Q(t)$ *is nondecreasing with both* Q *and* t.

CONDITION 2. *If* $t \geqq \theta \geqq \tau$, $Q(\theta)(t) = Q(t)$, *where* $Q(\theta)(t)$ *denotes the set of infected at time t for the outbreak* $(\omega, Q(\theta), \theta)$.

First, if $Q_1(t_1) \subseteqq Q_2(t_1)$, it is immediate from the definition of chains of infection that a chain of infection from a member of $Q_1(t_1)$ to r between times t_1 and t_2 is also a chain of infection from a member of $Q_2(t_1)$ to r between times t_1 and t_3 for any $t_3 \geqq t_2$; so $Q_1(t_2) \subseteqq Q_2(t_3)$; whence Condition 1. For Condition 2 note that if c_1 is a chain from p to q between τ and θ and c_2 a chain from q to r between θ and t, then the concatenation $c_1 \circ c_2$ is a chain from p to r between τ and t; while conversely, if c_3 is a chain from p to r between τ and t, we can split it into $c_1 \circ c_2$ by taking $c_1 \equiv c_3 \cap (t_{q_i, q_{i+1}} \leqq \theta)$, $c_2 \equiv c_3 \cap (t_{q_i, q_{i+1}} > \theta)$. Note that in the case of the noninfectious pseudoepidemic, the time τ was distinguished by the further property that only members of the set Q were infectious; thus, for Condition 2 to hold in this case, we must cheat by retaining Q as the set of infectious individuals, rather than taking $Q(\theta)$ (as we should if we were honest), for the outbreak $(\omega, Q(\theta), \theta)$. *Q.E.D.*

The restrictions on $\{\alpha(p, q)\}$ yield two further conditions.

CONDITION 3. *Define* $t_q \equiv \inf\{t : q \in Q(t)\}$: *the time at which* q *becomes infected. If we ensure that the probability of q receiving an infinite number of germs in a finite time interval is zero, by demanding that* $\Sigma_{p \in Q} \alpha(p, q) < \infty$, *we can deduce that there exists, almost surely, a chain of infection from p to q between times* τ *and* t_q, *for some* $p \in Q$; *of course* t_{q_{n-1}, q_n} *must* $= t_q$ *for such a chain.*

CONDITION 4. *Similarly, we can ensure that the probability of an individual* emitting *an infinite number of germs in a finite time interval is zero, by demanding that* $\Sigma_{q \in Q} \alpha(p, q) < \infty$; *whence, if* Q *is finite, so (almost surely) is* $Q(t)$ *for all finite* t $(\geqq \tau)$.

(These conditions are equivalent for pseudoepidemics among a homogeneous population on the integers, with $\alpha(p, q) = \alpha v(n(q) - n(p))$, such as we are considering, since all of the partial sums, of both kinds, equal $\alpha\sigma$.)

Consider an outbreak in which Q includes no individuals p with $n(p) > 0$; we interest ourselves in its progress among the positive integers.

DEFINITION 6. *Let us define the* front *of the pseudoepidemic at time t as the set of integers* >0 *at which there exist both infected and susceptible individuals.*

Two possible measures of the progress of the pseudoepidemic are the least and greatest integers in the front; it seems reasonable to suppose that the latter,

$F(t) \equiv \sup \{n(p): p \in Q(t)\}$, is the more interesting. A third possibility, intermediate between these two, is the *mean front* $M(t)$, defined as $(1/\sigma)|Q(t) \cap \{n(p) > 0\}|$; this is more obviously a good measure of the progress of the pseudoepidemic ($M(t)/t$ corresponds to the *mean velocity* defined in the deterministic case—see (2.8)) but is less susceptible of analysis.

For a full analysis of how simple stochastic epidemics progress, we should also need to consider statistics of the size and distribution of the front, to mention but one deficiency of the present investigation, in which we shall merely consider $F(t)$.

Consider outbreaks (ω, Q, τ) for fixed Q and τ. If the expectation of $F(t)$ is differentiable with respect to t, let $e(t) \equiv (d/dt)[\mathbf{E}(F(t))]$, the expected velocity of the front. If $\mathbf{E}(F(t))$ is not differentiable, we take $e(t) \leqq k$ to mean $[\mathbf{E}(F(t + dt) - F(t))]/dt \leqq k$ for sufficiently small dt.

THEOREM 3. *Consider the outbreaks* $\{(\omega, Q, \tau): \omega \in \Omega\}$ *of a simple stochastic epidemic*. If Q *satisfies the conditions*

(i) $F(\tau)$ *is finite, and*

(ii) *there exists* k *such that for each* $M \leqq k$ *there exists* $q \in Q$ *with* $n(q) = m$, then $e(t)$ *is finite if and only if* $\Sigma_{s=1}^{\infty} s^2 v(s)$ *is finite*.

More precisely,

(i′) *if* $\Sigma_1^{\infty} s^2 v(s)$ *is finite*, $e(t) \leq \alpha\sigma^2 \Sigma_{s=1}^{\infty} \frac{1}{2}s(s + 1)v(s)$ (*for* all $t \geqq \tau$)*;*

(ii′) *if* $\Sigma_1^{\infty} s^2 v(s)$ *is infinite, so is the expectation of* $F(t)$ *for all* $t > \tau$.

REMARKS. If v is, for example, directionally unbiased, we may replace $\Sigma_1^{\infty} s^2 v(s)$ by the variance of v.

We prove a stronger result, that condition (i) implies conclusion (i′) and condition (ii) implies conclusion (ii′). These two independent parts have been stated as one theorem partly for convenience, and partly because of an interest in outbreaks with pretensions to moving as a waveform; it seems reasonable to suppose that such an outbreak would satisfy both conditions, whatever definition of "moving approximately as a waveform" we adopt for the stochastic case.

Another theoretical approach might lay its chief emphasis on outbreaks for which Q is finite—clearly a case of practical importance; such a Q of course satisfies Condition (i), but not (ii).

PROOF. Conditions (i) and (ii) may be restated as $Q \subseteq Q_1$ and $Q \supseteq Q_2$, respectively, where $Q_1 \equiv \{q: n(q) \leqq F(\tau)\}$, and Q_2 contains for each $m \leqq k$ exactly one individual with $n(q) = m$, and no individuals with $n(q) > k$; without loss of generality, $k = 0$.

(i) We compare the epidemic outbreak (ω, Q, τ) with the outbreak (ω, Q_1, τ) of the cliff edge pseudoepidemic defined earlier (Example B). From the definitions, any chain of infection is a chain of *infection, whence, *for each* ω and all $t \geqq \tau$, $Q_1(t) \subseteq Q^*(t)$ and $Q(t) \subseteq Q_1(t)$ (Condition 1), so $Q(t) \subseteq Q^*(t)$. Hence $F^*(t)$, $\equiv \sup \{n(p): p \in Q_1^*(t)\}$, $\geqq F(t)$, again for each ω and all $t \geqq \tau$.

Now the epidemic and this pseudoepidemic share the same product space with positive measure μ (see the paragraph following Definition 5). Hence,

$$\mathbf{E}(F(t)) = \int_\Omega F(t)\, d\mu(\omega) \leqq \int_\Omega F_1^*(t)\, d\mu(\omega) = \mathbf{E}(F_1^*(t)). \tag{3.1}$$

Here we might pause to point out that exactly similar arguments will apply, with the inequalities reversed, to a comparison of the outbreaks (ω, Q, τ) with the outbreaks (ω, Q_2, τ) of the noninfectious pseudoepidemic (Example C), so in that case we shall have

$$\mathbf{E}(F(t)) \geqq \mathbf{E}(F_{2*}(t)) \qquad \text{for all } t \geqq \tau. \tag{3.2}$$

Returning to the proof of (i), we investigate $\mathbf{E}(F_1^*(t))$. The cliff edge pseudoepidemic advances in jumps, with $F_1^*(t)$ increasing from m to $m + s$, say. Before the jump $Q_1^*(t) = \{q : n(q) \leqq m\}$. Infections ahead of m by a distance s occur as a Poisson process of frequency $\Sigma_s^\infty \alpha\sigma^2 v(u)$, since each of the σ individuals p with $n(p) = m + s - u$ has cross infection frequency $\alpha v(u)$ with each of the σ individuals q with $n(q) = s$. Thus, jumps of $F_1^*(t)$ take place in a Poisson process of frequency $\alpha\sigma^2 \Sigma_1^\infty \Sigma_s^\infty v(u)$ (convergent because $\Sigma_1^\infty uv(u) \leqq \Sigma_1^\infty u^2 v(u)$, which is convergent by (i)); and *independently of the time interval between jumps*, jumps have the distribution $\Sigma_s^\infty v(u)/\Sigma_1^\infty \Sigma_s^\infty v(u)$, where s takes values $\geqq 1$. Therefore, the expected increase in $F_1^*(t)$ in any time interval $(t_1, t_2]$ is

$$(t_2 - t_1)\alpha\sigma^2 \sum_1^\infty s \sum_s^\infty v(u) = \alpha\sigma^2(t_2 - t_1) \sum_1^\infty \tfrac{1}{2}s(s + 1)v(s). \tag{3.3}$$

It is finite because $\Sigma_1^\infty s^2 v(s)$ is finite.

Now $F(\tau) = F_1^*(\tau)$ and $\mathbf{E}(F(t)) \leqq \mathbf{E}(F_1^*(t))$ for all $t \geqq \tau$; whence,

$$\mathbf{E}(F(t) - F(\tau)) \leqq \alpha\sigma^2(t - \tau) \sum_1^\infty \tfrac{1}{2}s(s + 1)v(s). \tag{3.4}$$

The behavior of simple epidemics is homogeneous with time (Condition 2), so the same applies with τ replaced by any t_0 with $\tau \leqq t_0 \leqq t$. Therefore,

$$\frac{\mathbf{E}(F(t)) - \mathbf{E}(F(t_0))}{t - t_0} \leqq \alpha\sigma^2 \sum_1^\infty \tfrac{1}{2}s(s + 1)v(s) \tag{3.5}$$

if $\tau \leqq t_0 < t$, which we may paraphrase as

$$e(t) \leqq \alpha\sigma^2 \sum_1^\infty \tfrac{1}{2}s(s + 1)v(s). \tag{3.6}$$

(ii) As remarked above (3.2), comparison of (ω, Q, τ) with the outbreaks (ω, Q_2, τ) of the noninfectious pseudoepidemic yields

$$\mathbf{E}(F(t)) \geqq \mathbf{E}(F_{2*}(t)) \qquad \text{for all } t \geqq \tau. \tag{3.7}$$

We prove that $\mathbf{E}(F_{2*}(t))$ is infinite if $\Sigma_1^\infty s^2 v(s)$ is infinite by showing that the expected location of the first infection with $n(q) > 0$ is infinite.

The individuals with $n(q) = s > 0$ are exposed to a Poisson process of germs of frequency $\alpha\sigma \sum_s^\infty v(u)$. This is the same as in (i) except for the loss of a factor σ, so we may skip several stages in the argument to arrive at

$$\mathbf{E}(F_{2*}(t)) \geqq \left(1 - \exp\{-k_1(t - \tau)\}\right)\left(\alpha\sigma \sum_1^\infty \tfrac{1}{2}s(s+1)v(s)\right), \tag{3.8}$$

where $k_1 = \alpha\sigma \sum_1^\infty \sum_s^\infty v(u)$, so that $(1 - \exp\{-k_1(t - \tau)\})$ is the probability of having at least one infection to the right of 0 in $(\tau, t]$; k_1 may diverge, but this will not distress us as all we require is that it be nonzero. Now we are assuming that $\sum_1^\infty s^2 v(s)$ is infinite, so $\sum_1^\infty \frac{1}{2}s(s+1)\,v(s)$ diverges. Therefore, $\mathbf{E}(F_{2*}(t))$ diverges; whence, $\mathbf{E}(F(t))$ also diverges (for all $t > \tau$). This concludes the proof of Theorem 3. *Q.E.D.*

The immediate theoretical problem which offers itself beyond this theorem is whether some similar result to conclusion (ii′) holds when condition (ii) is dropped; this problem is analogous to that satisfactorily answered in Lemma 4. The other interesting question that I can see concerns the behavior of epidemics which satisfy the condition that $\sum_1^\infty s^2 v(s)$ be finite, but not the corresponding condition for the deterministic model considered in earlier chapters (Condition (1.1): $\sum_{-\infty}^\infty p^n v(n)$ converges for some $p > 1$). The computer simulations of such an epidemic (3.10) suggest that this is an interesting problem, whose theoretical analysis (if possible) should prove rewarding.

3.3. *An epidemic simulating program.* Our starting point here is the germ model first described in Section 3.1.

First, consider just one infected individual q, at location $n(q)$. Individual q produces germs in a Poisson process of frequency $\alpha\sigma$, which are distributed to locations $n(q) + s$ according to the probability density $v(s)$; a germ causes a new infection with probability $X(n(q) + s)/\sigma$, where $X(k)$ denotes the number of susceptibles at k.

Simulating a Poisson process or a choice between two events of given probability (such as whether or not a germ causes a new infection) is trivial, given a computer subroutine which produces (pseudo-)random numbers between 0 and 1. So we shall be able to simulate the infections caused by a single individual, provided only that we can also use this subroutine to simulate the probability density $v(s)$, as we shall if v can be characterized in a finite manner acceptable to the computer.

Next, suppose we have finitely many infected individuals, say m of them. Since they emit germs as *independent* Poisson processes (each of frequency $\alpha\sigma$), the cumulative effect is of a germ emittive Poisson process of frequency $m\alpha\sigma$, with conditional probabilities $1/m$ of each particular individual being responsible for a particular germ, independent of the past history of the outbreak. Thus, it is hardly more difficult to simulate the infections caused by finitely many individuals than those caused by one.

Clearly, we cannot allow for an infinite set of infected individuals if they have to be dealt with separately; for a start, we could not store their locations.

However, what we can and shall do is to simulate simple epidemics in which all individuals to the left of some location ℓ are infected. Since there are then no susceptibles to the left of ℓ, we need only simulate those germs which terminate to the right of, or at, ℓ. Summing over all locations to the left of ℓ, we see that such germs terminate at $\ell - 1 + s$ in a Poisson process of frequency $\alpha\sigma^2 \Sigma_s^\infty V(u)$. Thus, we can allow for such an infinite set of infected individuals, provided that two conditions are satisfied:

(i) the overall frequency of such germs, say $(*m)\alpha\sigma, = \alpha\sigma^2 \Sigma_1^\infty \Sigma_1^\infty v(u) = \alpha\sigma^2 \Sigma_1^\infty sv(s)$, is finite (and calculable);

(ii) the conditional densities $(\Sigma_s^\infty v(u))/\Sigma_1^\infty sv(s)$ can be simulated.

Germs from the left of ℓ and other germs are emitted in *independent* Poisson processes, so no difficulty arises in their simultaneous simulation; we just simulate a Poisson process of frequency $\alpha\sigma(*m + m)$ and assign each germ to (being from) the left of ℓ with probability $*m/(*m + m)$. If all the individuals at $\ell + 1$ become infected, we can replace ℓ by $\ell + 1$, and it is convenient to do so.

This completes the theoretical background to the simulation of simple stochastic epidemics among a population of uniform density inhabiting the integers. The program I have used to implement the simulation is best explained by a flow diagram (see Figure 1). The complete program for the computer differs from this mainly in possessing output sections and error catching devices, and in being rearranged in an illogical order to facilitate alteration. A glossary for Figure 1 is as follows:

$*t$ is used for time (instead of t);

ℓ and k are, respectively, the least and greatest integers in the front (see Definition 6); thus, $k \equiv F(t)$;

$.si$ is used for σ;

$Y[Q]$ is the number of infected at $Q, = .si - X[Q]$.

The computer only has finite store, and so cannot deal with more than a certain number of separate locations, here taken $= 3001$. By reusing storage for locations to the left of ℓ, it is possible to keep going indefinitely, as long as $k - \ell \leqq 3000$, but this sophistication has been regarded as unnecessary for the present program.

Lastly, we deal with the adaptation of our program to deal with the simplest type of epidemic with removal, where each infected individual is liable to removal with probability $\beta\, dt$ in time dt, independent of the behavior of others. This just gives us one more Poisson process to combine with those we already have for germs.

Now in an epidemic with removal we have no need to deal with infinite numbers of infected individuals such as we had in the simple epidemic case; for example, a waveform will contain only finitely many infected individuals, in contrast to that case. This is perhaps just as well, as the complications of removals would now prevent us from using the notion of "germs from the left of ℓ" even if we wished to.

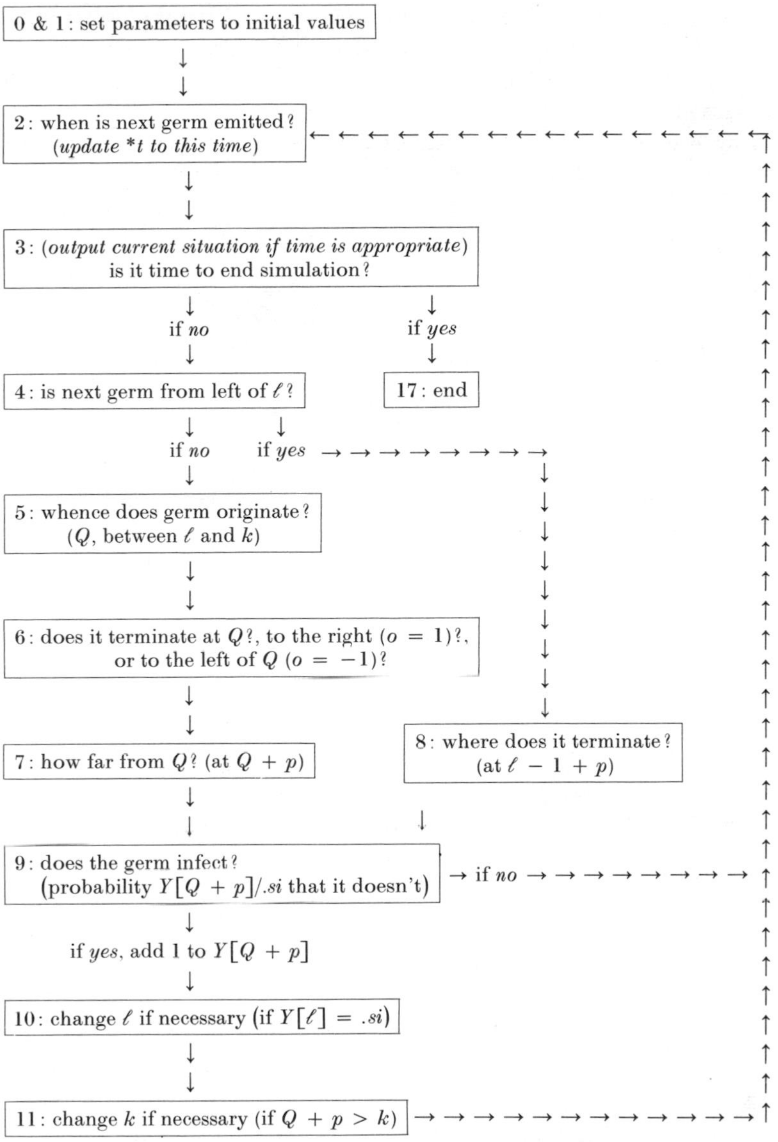

FIGURE 1

Flow diagram for simulation of simple epidemic.

Let us refer to both germ emissions and removals as *happenings*. When there are m infected individuals, to obtain the next happening we simulate a Poisson process of frequency $(\alpha\sigma + \beta)m$, and choose the next happening to be a removal with probability $\beta/(\alpha\sigma + \beta)$, a germ emission otherwise.

Again we explain the program for implementing this by a flow diagram (see Figure 2). The terminology differs from that of Figure 1 as follows:

k is used for $\sup_{t \leqq {}^*t} F(t)$ (because of removal, $F(t)$ is no longer necessarily monotone with time);

ℓ now becomes the *leftmost infected location*, its former definition being appropriate only to simple epidemics.

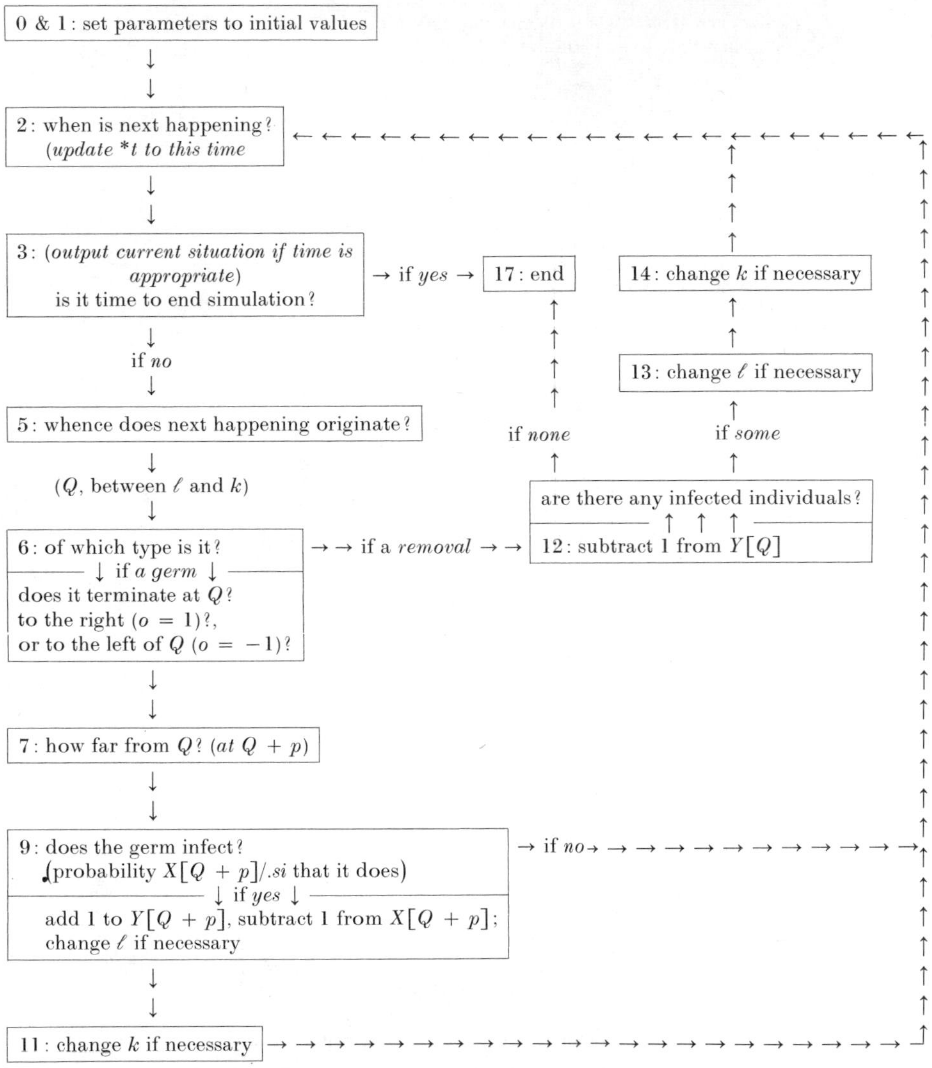

FIGURE 2

Flow diagram for simulation of epidemic with removal.

3.4. *Results of simulations.* It has been my aim here, not to provide statistical tests of specific hypotheses about the behavior of epidemic outbreaks, but to take a quick look at a varied selection of epidemics to see whether they conform to the expectations aroused by the theoretical work of this paper, and to look for phenomena which may suggest further lines of research (this approach has already yielded Theorem 3, which was provoked by the results of the first few simulations). Despite the lack of statistical tests, we refer to the results of our simulations as though, for each epidemic considered, they covered all facets of its behavior. We may justify this approach, apart from its convenience for descriptive purposes, by noting that, with the exception of E_3 (see (3.11)), outbreaks of all the epidemics simulated return with considerable frequency to roughly the same state, where the front of the outbreak is of small extent; thus, while great deviations from the types of progress observed *may* be possible, they may well be of great rarity, and thus, would be more suitably investigated theoretically than experimentally.

The first question that springs to mind is whether outbreaks of simple epidemics with expected finite velocity travel in a regular, approximately wave-like manner. So we consider first two epidemics where v is of finite variance:

(3.9) $$E_1,\ \text{defined by } v_1(s) = \left(\tfrac{1}{2}\right)^{|s|}/3$$

(*geometric* distribution of variance 4), which satisfies (1.1); and

(3.10) $$E_2,\ \text{defined by } v_2(s) = (72/5)\left(\prod_{u=1}^{4}\left(|s| + u\right)\right)^{-1}$$

(roughly inverse fourth power, of variance 4), which does not satisfy (1.1). Inspection of graphs of their progress (Figures 3 and 4) shows that O_1 advances regularly with approximately constant velocity, while O_2 does so only intermittently, being interrupted by *great leaps forward.*

The average velocity of O_1 is approximately $1.21\sqrt{w_2}$ (w_2, the variance, $= 4$), rather less than the minimal velocity found in Theorem 4.1 of [11] ($c_0 \approx 1.834\sqrt{w_2}$) for its deterministic equivalent. A graph of velocity for outbreaks of E_1 with varied population densities σ (O_1 was with $\sigma = 10$) (Figures 5 and 6 show the mean waveforms for these values of σ) is consistent with the conjecture that the (average) velocity for E_1 tends upwards to a value near c_0 as $\sigma \to \infty$. We cannot expect such a result for E_2, though its velocity does not appear to increase at the rate of the upper bound ($=1.2\sigma$) guaranteed by Theorem 3 (see Figure 7).

To return to E_1, at any one time the shape of the front is naturally subject to relatively large stochastic variations; but if we average over 20 epochs evenly spread over 100 time units, using the mean front $M(t)$ (see discussion following Definition 6) as origin, we obtain pretty regular sigmoid curves which we may call *mean waveforms* (Figure 8). An obvious question is whether we can provoke E_1 into travelling at any velocity other than the average velocity found for O_1. So, for instance, we may choose an outbreak O_1' for which initially the front is

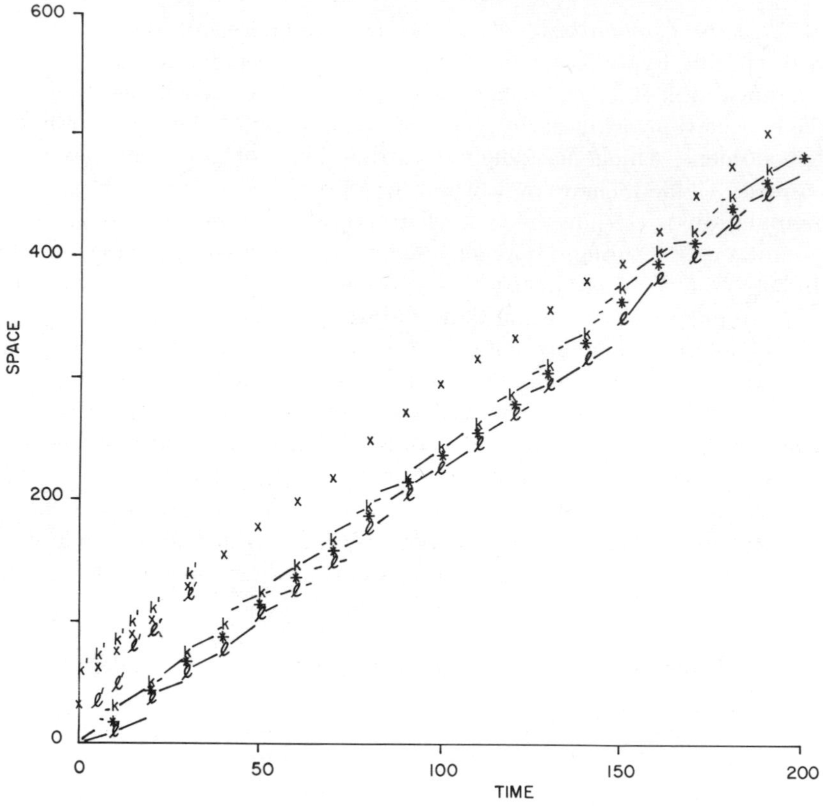

FIGURE 3
Simple epidemic O_1 with $\sigma = 10$, $v(s) = \frac{1}{2}^{|s|}/3$.
The mean front $M(t)$ is indicated by $*$; ℓ and k are the two ends of the front (sketched in at shorter time intervals to show all sizeable discontinuities).
The x, ℓ', and k' are for O_1' (see Figure 9).

of the same sigmoid form as the mean waveform of O_1 but of three times the extent, which should give roughly three times the velocity. The effect of this initial "fast waveform" turns out to extend rarely beyond its initial nose, and the behavior of the two outbreaks subsequently appears as identical as one could hope (Figure 9). We conclude that E_1 appears to have just one mean waveform, of velocity closely comparable to the minimal velocity suggested by deterministic analysis.

Before leaving E_1, we might mention the dependence of this velocity on w_2. In the deterministic case, the dependence of velocities on $\sqrt{w_2}$ is exactly linear; here it is not, because the negative exponential has been replaced by a *discrete* density, but the average velocity does not seem to deviate much from linearity (Figure 10). Two other examples of epidemics satisfying (1.1) have been

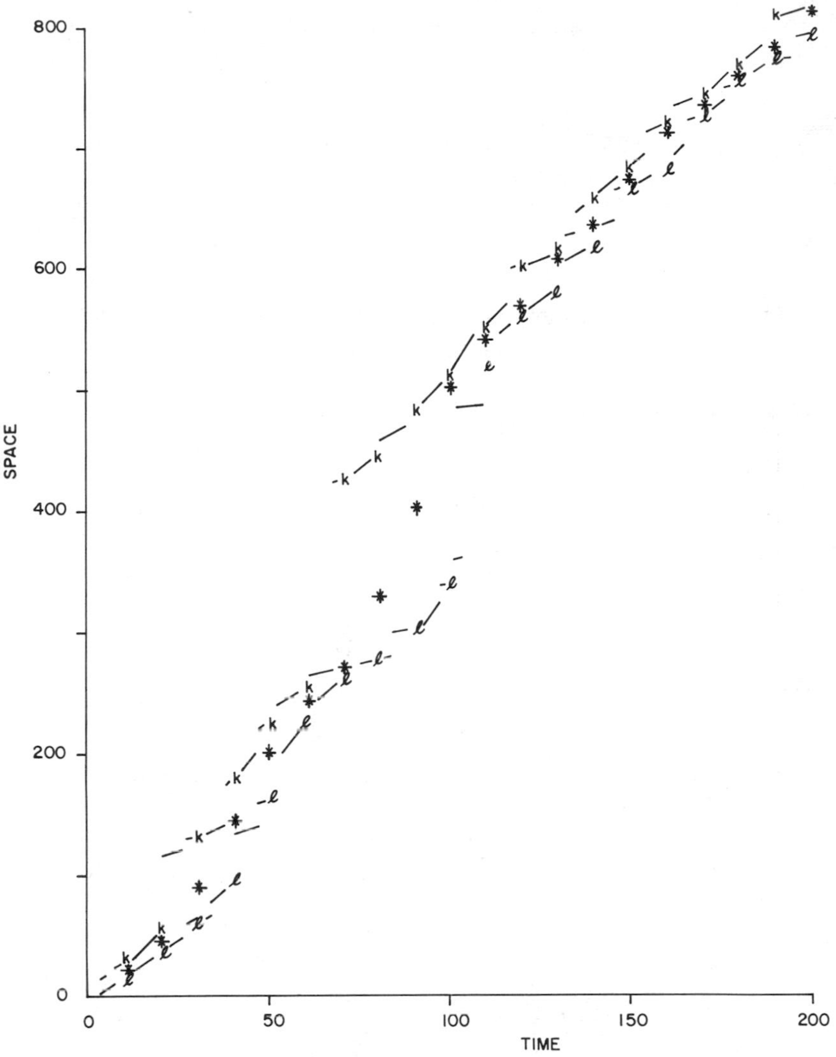

FIGURE 4
Simple epidemic O_2 with $\sigma = 10$, $v(s) = (72/5)\left(\Pi_{u=1}^{4}(|s| + u)\right)^{-1}$.
The mean front $M(t)$ is indicated by $*$; ℓ and k are the two ends of the front (sketched in at shorter time intervals to show all sizeable discontinuities).

simulated, the uniform density on $[-3, 3]$ (E_4, say) and the density concentrated on ± 1 (E_*). Outbreaks of both of these progress much like O_1 (*not* appreciably more regularly), but with slightly lower velocities, O_* having the least (roughly $0.87\sqrt{w_2}$ when $\sigma = 10$; this is in roughly the same ratio to that

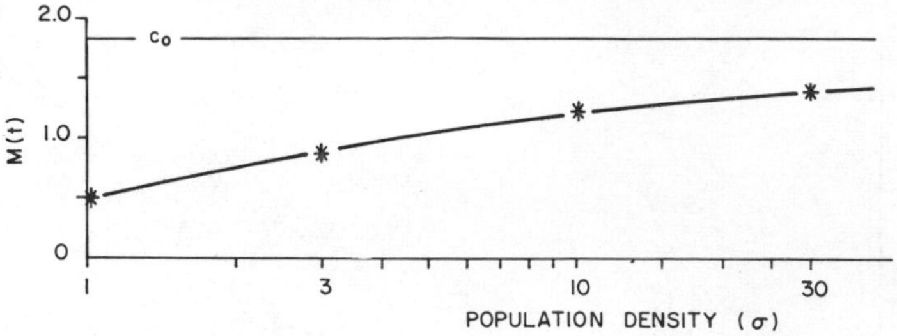

FIGURE 5
Velocity of simple epidemic as a function of population density (average of $M(t)$ against σ), plotted on semilog grid.

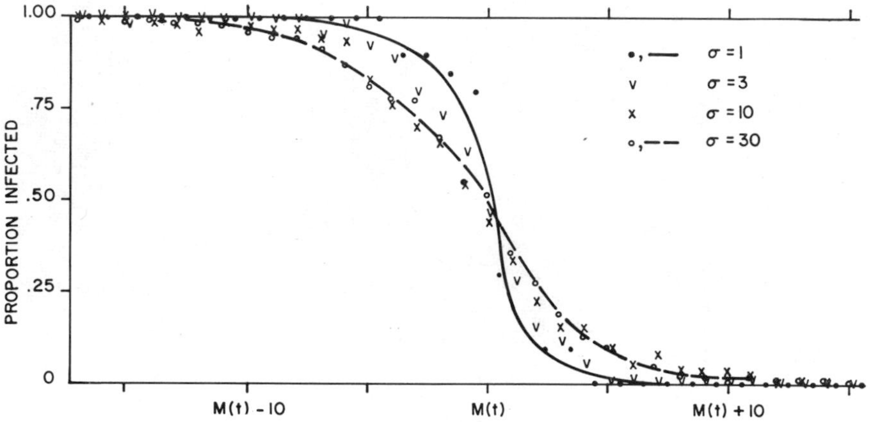

FIGURE 6
Mean waveforms of simple epidemic for varying σ.

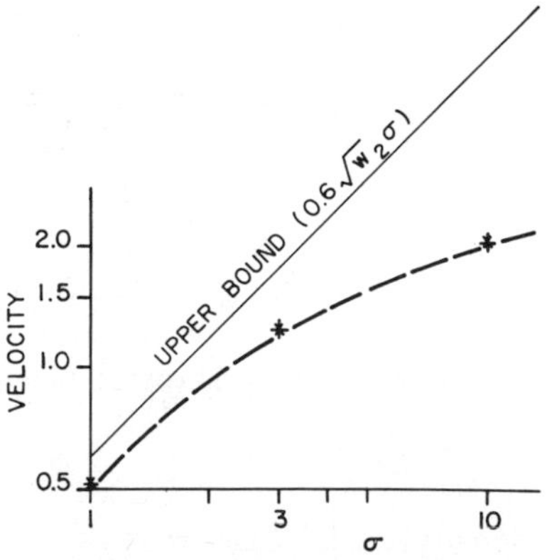

FIGURE 7
Velocity of simple epidemic as a function of σ;
$v(s) = (72/5)\left(\Pi_{u=1}^{4}\left(|s| + u\right)\right)^{-1}$ (plotted on log log grid).

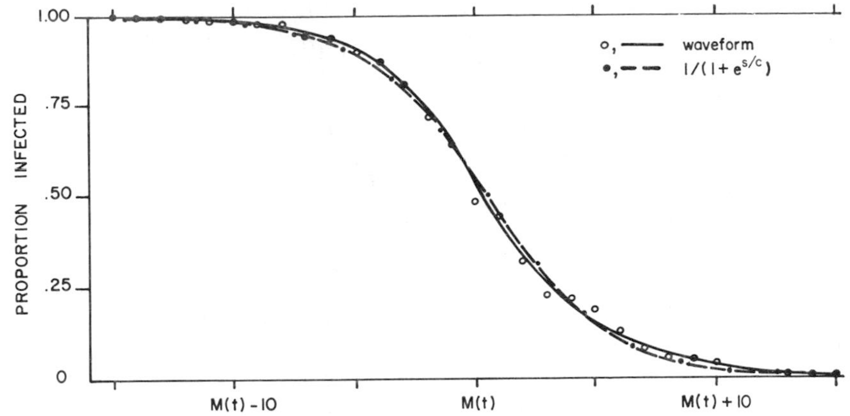

FIGURE 8
Mean waveform for simple epidemic with $\sigma = 30$, $v(s) = \frac{1}{2}^{|s|}/3$. Shown for comparison, $1/(1 + e^{s/c})$, the waveform of $\dot{y} = y(1 - y)$ of same velocity.

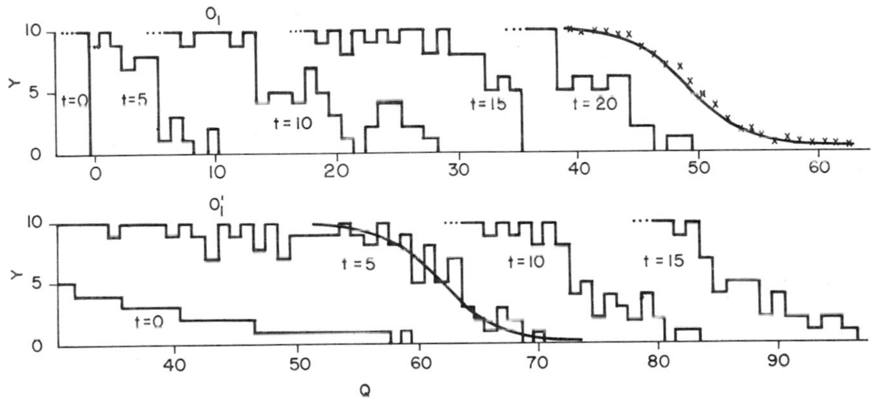

FIGURE 9
Attempt to provoke the simple epidemic E to velocities higher than usual. (Plotted on semilog grid. See Figure 3 also).

of O_1, as that of the corresponding critical velocities c_v for the deterministic case—see Lemma 5 (iii) and (iv)).

The great leaps forward of O_2 are in each case initiated by a single infection far ahead of the front of the outbreak. Between these leaps the outbreak seems each time to settle to roughly the same velocity, and for a period such as $(100, 200]$ of the outbreak O_2 (Figure 4) we can evaluate a "mean waveform" which is not much more irregular, or indeed faster, than that for O_1 (Figure 11, crosses) (for a period covering a large leap such as $(0, 100]$ for O_2, such a procedure is nearly meaningless (Figure 11, dots)). This is, in a way, more

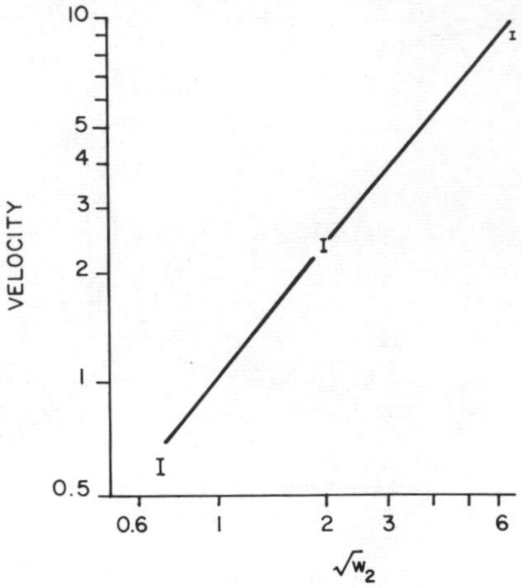

FIGURE 10
Velocity of geometrically weighted simple epidemic for varying w_2.
(Plotted on log log grid.)
Symbol I indicates range of values of velocity.

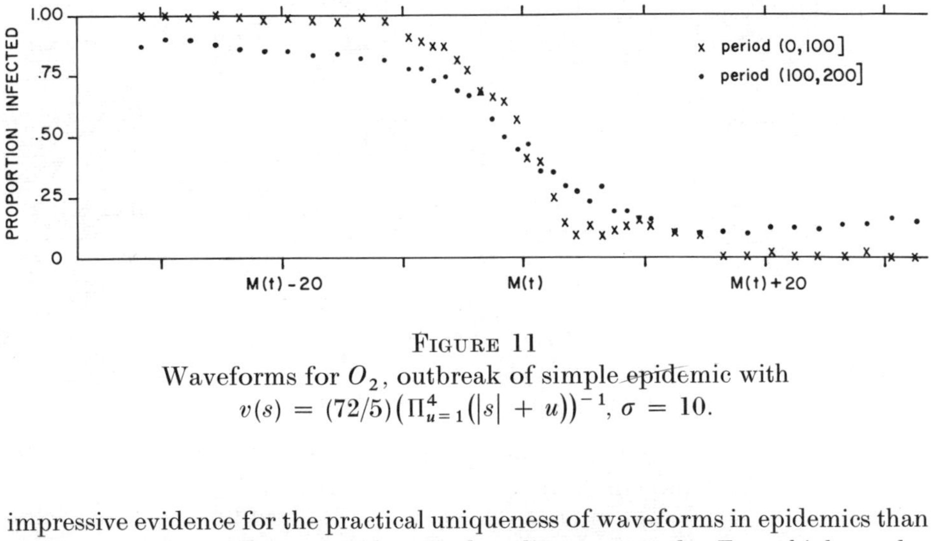

FIGURE 11
Waveforms for O_2, outbreak of simple epidemic with
$v(s) = (72/5)\left(\Pi_{u=1}^{4}(|s| + u)\right)^{-1}$, $\sigma = 10$.

impressive evidence for the practical uniqueness of waveforms in epidemics than that provided by the attempt described earlier to provoke E_1 to higher velocities, since in E_2 the outbreak is continually provoking itself in the direction of speeding up by the leaps it takes.

Lastly among simple epidemics we consider E_3, defined by

$$v_3(s) \equiv 3\left(\prod_{u=1}^{3} (|s| + u)\right)^{-1}. \tag{3.11}$$

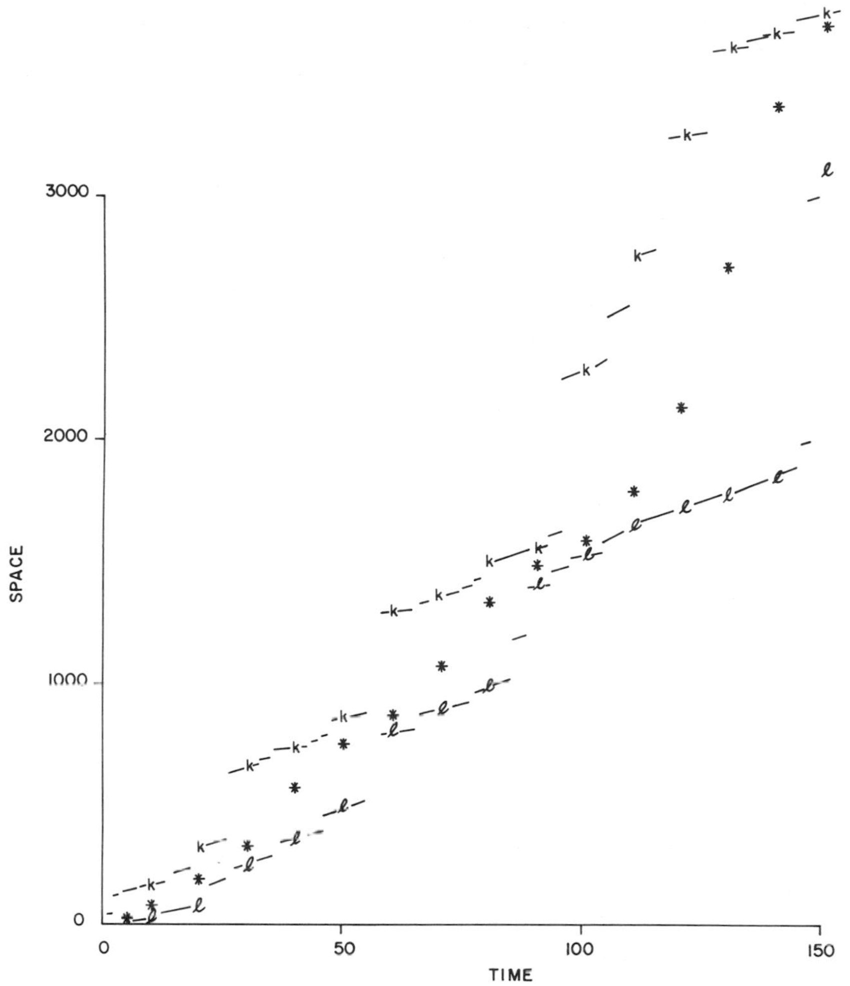

FIGURE 12
Simple epidemic O_3 with $\sigma = 10$, $v(s) = 3\left(\Pi_{u-1}^{4}\left(|s| + u\right)\right)^{-1}$.
The mean front $M(t)$ is indicated by $*$; ℓ and k are the two ends of the front (sketched in at shorter time intervals to show all sizeable discontinuities).
For details of times [60–80], see Figure 13.
(Note different scales from Figures 3 and 4.)

The density v_3 is of infinite variance (though "only just": $\Sigma_{-\infty}^{\infty} s^{2-\varepsilon}v(s)$ converges for arbitrarily small ε), and as we might expect from comparison with O_2, O_3 progresses in wilder and wilder leaps forward (Figure 12), and shows no sign of ever settling to a steady velocity—not surprisingly as its expected velocity is infinite. The output for O_3 for $t = 60, 65, 70, 75$, and 80 is given in Figure 13, to illustrate the process of catching up on a great leap forward.

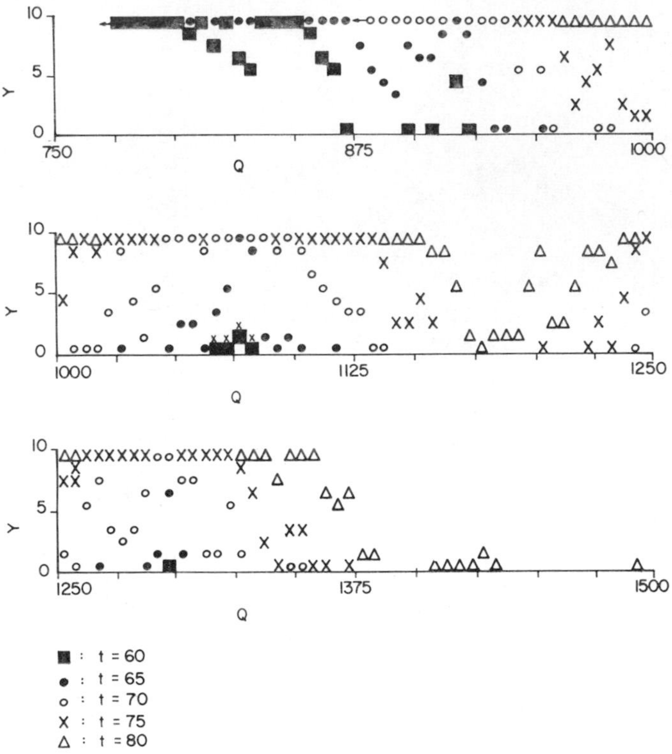

FIGURE 13
Details of O_3 for times 60–80
(to illustrate how an outbreak catches up with itself after a great leap forward; in this case the front has jumped from 867 to 1298 in two leaps which have taken place just before this diagram begins).
Locations are lumped together in sets of 5;
averages of $Y[Q]$ are rounded up to the nearest integer.

As a check that it is indeed the tails of the densities v_2 and v_3 that cause the irregular behavior of their respective epidemics, simulations have been run for the deficient densities v_1^*, v_2^*, and v_3^*, which are obtained by setting $v_i^*(s) = 0$ for $|s| > 10$ and reallocating the missing weight to $s = \pm 10$. Outbreaks of all these three progress in a manner not noticeably different from that of O_1; O_1^* is, in fact, the fastest of the three (their mean waveforms are shown in Figure 14).

Simulations of epidemics with removal are not yet at a stage where it is worth reporting on the results at any length (for a start, the possibility of extinction—for any waveform eventual extinction is a certainty—means that larger numbers of simulations are required). The only lead so far is that it appears that removal reduces the velocity (during any period when the

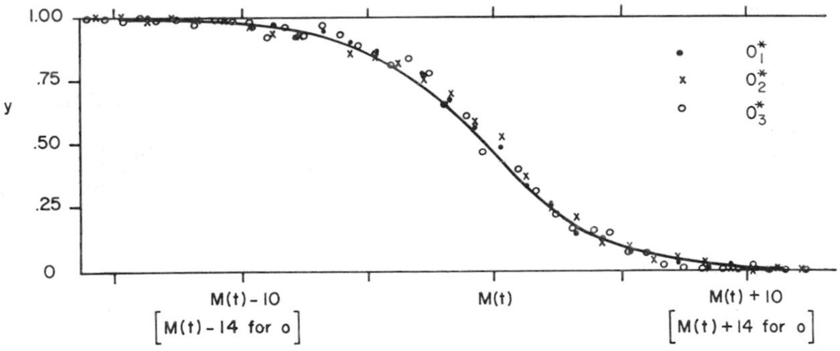

FIGURE 14
Mean waveforms for O_1^*, O_2^*, and O_3^*.
The scale for O_3^* is different because it is of noticeably larger variance: $\sqrt{w_2} \approx 2.81$, as opposed to ≈ 1.99 for each of the other epidemics. Contraction by a factor $\approx 2.81/1.99$ is, thus, needed to make its waveform comparable.

epidemic travels approximately as a wave) more than is suggested by the deterministic analysis of Kendall [7]; for example, for E_1 with removal rate $(b) = 0.5$, the velocity is reduced to about one fifth that for the simple epidemic, instead of to the proportion $(1 - b)^{1/2} \approx 0.7$ suggested by theory.

The most unexpected and interesting discovery of these simulations is the behavior of E_2. It would be interesting to know whether this mixture of steady progress and great leaps forward is typical of all epidemics with densities which are of finite variance, but do not satisfy (1.1). Further simulations would of course help here, but I think the next step should be a return to theory; the first problem there will be to characterize the two types of progress with sufficient precision.

REFERENCES

[1] N. T. J. BAILEY, *The Mathematical Theory of Epidemics*, London, Griffin, 1957.

[2] A. C. CHAMBERLAIN, "Deposition and uptake of windborne particles," *Nature*, Vol. 225 (1970), pp. 99–100.

[3] W. K. DAVIES, G. B. LEWIS, and H. A. RANDALL, "Some distributional features of the Foot and Mouth epidemic," *Nature*, Vol. 219 (1968), pp. 121–125.

[4] N. DUNFORD and J. T. SCHWARTZ, *Linear Operators, Part 1: The General Theory*, New York, Interscience, 1958.

[5] R. A. FISHER, "The wave of advance of advantageous genes," *Ann. Eugen.*, Vol. 7 (1937), pp. 355–369.

[6] D. G. KENDALL, Discussion of "Measles periodicity and community size," by M. S. Bartlett, *J. Roy. Statist. Soc. Ser. A* (1957), pp. 64–67.

[7] ———, "Mathematical models of the spread of infection," *Mathematics and Computer Science in Biology and Medicine*, London, H.M. Stationery Office, 1965, pp. 213–225.

[8] A. N. Kolmogorov and S. V. Fomin, *Elements of the Theory of Functions and Functional Analysis*, Vol. 1, Rochester, Graylock, 1957.

[9] A. N. Kolmogorov, I. G. Petrovsky, and N. S. Piscounov, "Étude de l'équation de la diffusion avec croissance de la quantité de matière et son application à une problème biologique," *Bull. de l'Univ. d'état à Moscou (ser. intern.)*, Sec. A (1937), pp. 1–25.

[10] R. L. Marris, *The Economic Theory of "Managerial" Capitalism*, London, Macmillan, 1970.

[11] D. Mollison, "Possible velocities for a simple epidemic," *Advances in Appl. Probability*, Vol. 4 (1972).

[12] J. Neyman and E. L. Scott, "A stochastic model of epidemics," *Stochastic Models in Medicine and Biology*, Madison, University of Wisconsin Press, 1964, pp. 45–83.

[13] K. P. Norris and G. J. Harper, "Windborne dispersal of Foot and Mouth virus," *Nature*, Vol. 225 (1970), pp. 98–99.

[14] R. Tinline, "Lee wave hypothesis for the initial pattern of spread during the 1967–68 Foot and Mouth epizootic," *Nature*, Vol. 227 (1970), pp. 860–862.

[15] E. C. Titchmarsh, *The Theory of Fourier Integrals*, Oxford, Clarendon Press, 1937.

[16] Y. B. Zeldovitch, *Theory of Flame Propagation*, Washington DC, NACA T.M. 1282, 1951.

ASYMPTOTIC DISTRIBUTION OF EIGENVALUES OF RANDOM MATRICES

W. H. OLSON[1]
and
V. R. R. UPPULURI[2]
OAK RIDGE NATIONAL LABORATORY

1. Introduction

The impetus for this paper comes mainly from work done in recent years by a number of physicists on a statistical theory of spectra. The book by M. L. Mehta [10] and the collection of reprints edited by C. E. Porter [14] are excellent references for this work. The discussion in Section 1.1 is an attempt to present a rationale for such investigations. Our interpretation of linear operators as used in quantum mechanics is based largely on the book by T. F. Jordan [8].

1.1. *Statistical theory of spectra.* In quantum mechanics knowledge of the value of measurable quantities of a system is expressed in terms of probabilities. A state of the system specifies these probabilities. Measurable quantities are represented by self-adjoint linear operators on a separable Hilbert space. The only possible values of the measurable quantities are those in the spectrum of the self-adjoint operator which represents the measurable quantity.

Experience indicates that energy is represented by the Hamiltonian operator. We are interested in the point spectrum of the Hamiltonian, which is its set of eigenvalues. The eigenvalues E of the Hamiltonian operator H, which are real since H is self-adjoint, are those values of energy for which some state of the system specifies a probability of one that the energy is exactly equal to E [8]. This is expressed in the Schrödinger time independent equation,

$$H\psi = E\psi, \tag{1.1.1}$$

where ψ is an eigenvector associated with E.

In ordinary statistical mechanics, renunciation of exact knowledge of the state of a system is made and only properties of averages are considered. An exact knowledge of the laws governing the system is assumed known; it is the impossibility in practice of observing the state of the system in all its detail that leads to the consideration of properties of averages.

[1] Research partially supported by the Oak Ridge Associated Universities and the National Institute of Environmental Health Sciences through grant numbers 5 T01 ES 00033-03, 04 and 05.

[2] Research sponsored by the U.S. Atomic Energy Commission under contract with the Union Carbide Corporation.

An analogous situation exists with respect to the Hamiltonian operator. It is possible to choose an orthonormal basis for the separable Hilbert space in such a way that the matrix representation of the Hamiltonian with respect to this basis is in a form with blocks (finite dimensional square matrices) along the diagonal and zeros elsewhere (see [10]). Each block corresponds uniquely to each set of values of a certain set of parameters. These parameters are variables which may be used to describe certain aspects of the system, whatever state it may be in. We are interested in the eigenvalues of the very large blocks. There are two difficulties. First, we do not know the Hamiltonian and, second, even if we did, it would be far too complicated to attempt to solve it. These difficulties lead to a renunciation of an exact knowledge of the system itself, that is, of the Hamiltonian. The basic statistical hypothesis is this: the statistical behavior of energy levels in a simple sequence (a simple sequence is one whose levels all have the same set of values of the parameters mentioned above) is identical with the behavior of the eigenvalues of a random matrix. It is desirable, due to our ignorance of the system, that the statistical properties of the eigenvalues be independent of as many of the properties of the distributions of the elements of the matrices as possible. At best the elements of these matrices are random variables whose distributions are restricted only by the general symmetry properties we might impose on the ensemble of operators.

1.2. *Outline of contents.* There are three basic parts. Section 3 contains the combinatorial arguments which are essential for the proofs of the theorems in the second part, Sections 4, 5, and 6. These sections all deal with the asymptotic distribution of the empirical distribution function of the eigenvalues of a symmetric random matrix from the points of view of weakening the conditions placed on the distribution of the elements of the matrix and of strengthening the mode of convergence of the empirical distribution functions. The last part, Section 7, discusses results of the same type, that is, asymptotic distributions of the empirical distribution function of the eigenvalues of random matrices, for the Gaussian orthogonal ensemble [10], a Toeplitz ensemble [5], and a Wishart ensemble.

2. Notation, definitions and preliminaries

2.1. *Random matrices.* Let $(\Omega, \mathscr{F}, P)$ denote a probability space, that is, Ω is a nonempty abstract set, $\mathscr{F}$ is a σ-algebra of subsets of Ω, and P is a probability measure on $\mathscr{F}$; and let (R_n, β_n) be the measurable space where R_n is n dimensional Euclidean space and β_n is the Borel σ-algebra of subsets of R_n.

A mapping $X : \Omega \rightsquigarrow R_n$ is called a *random vector* if $\{\omega \in \Omega : X(\omega) \in B\} \in \mathscr{F}$ for all $B \in \beta_n$. When $n = 1$, X is called a *random variable.*

A mapping $A : \Omega \times R_n \rightsquigarrow R_n$ is called a *random operator* if $A(\omega)[x]$ is for every $x \in R_n$ a random vector. A random operator A is said to be linear if $A(\omega)[\alpha x_1 + \beta x_2] = \alpha A(\omega)[x_1] + \beta A(\omega)[x_2]$ for every $\omega \in \Omega$, x_1, $x_2 \in R_n$, and $\alpha, \beta \in R_1$.

A linear random operator defined by the $n \times n$ matrix

$$A = (a_{ij})_{i,j=1}^{n}, \tag{2.1.1}$$

where the a_{ij} are random variables is called a *random matrix*. Thus, a random matrix is a linear random operator on $\Omega \times R_n$ to R_n.

Throughout the paper all random quantities will be assumed to be defined on some fixed probability space $(\Omega, \mathscr{F}, P)$.

2.2. *Continuity of ordered eigenvalues.* It is established in this section that ordered eigenvalues of symmetric random matrices are indeed random variables.

The following lemma is needed. Denote by $\lambda_i(A)$, $i = 1, \cdots, n$, the eigenvalues of any $(n \times n)$ matrix A.

LEMMA 2.2.1. *Let A be an $(n \times n)$ matrix and suppose $\varepsilon > 0$ is given. Then there exists a $\delta > 0$ such that for any matrix $D = (d_{ij})_{i,j=1}^{n}$ such that $\Sigma_{i,j=1}^{n} |d_{ij}| < \delta$, and there exists a permutation σ of $\{1, 2, \cdots, n\}$ for which*

$$|\lambda_i(A) - \lambda_{\sigma(i)}(A + D)| < \varepsilon, \qquad i = 1, \cdots, n. \tag{2.2.1}$$

A proof of this lemma may be found in A. M. Ostrowski [13].

Denote by $\lambda_1(A) \leqq \lambda_2(A) \leqq \cdots \leqq \lambda_n(A)$ the ordered eigenvalues of any $(n \times n)$ Hermitian matrix A.

COROLLARY 2.2.1. *The ordered eigenvalues $\lambda_1, \lambda_2, \cdots, \lambda_n$ are continuous functions of the elements of the Hermitian matrix $A = (a_{ij})$.*

PROOF. The proof is by contradiction. Let $A = (a_{ij})$ be given. By the above lemma it is known that for a given $\varepsilon > 0$ there exists a $\delta > 0$ such that for any $A' = (a'_{ij})$ such that $\Sigma_{i,j=1}^{n} |a_{ij} - a'_{ij}| < \delta$ one has for a suitable permutation σ of $\{1, 2, \cdots, n\}$,

$$|\lambda_i(A) - \lambda_{\sigma(i)}(A')| < \varepsilon \tag{2.2.2}$$

for $i = 1, \cdots, n$. Assume $|\lambda_i(A) - \lambda_i(A')| > \varepsilon$; to be definite assume $\lambda_i(A') > \lambda_i(A)$. Then $\lambda_1(A) \leqq \cdots \leqq \lambda_i(A) < \lambda_i(A') \leqq \cdots \leqq \lambda_n(A')$. With each $\lambda_j(A)$, $j = 1, \cdots, i$, is associated $\lambda_{\sigma(j)}(A')$ such that $|\lambda_j(A) - \lambda_{\sigma(j)}(A')| < \varepsilon$. But only $\lambda_1(A'), \cdots, \lambda_{i-1}(A')$ are available for this purpose since $\lambda_i(A') - \lambda_i(A) > \varepsilon$, and hence $\lambda_i(A') - \lambda_j(A) > \varepsilon$, $j = 1, \cdots, i$. Thus, one must conclude $|\lambda_i(A) - \lambda_i(A')| < \varepsilon$, $i = 1, \cdots, n$. This completes the proof.

Let $A = (a_{ij})_{i,j=1}^{n}$ be a random matrix such that $a_{ij} = a_{ji}$ a.s. (referred to as symmetric random matrix), and denote by $\lambda_1(A) \leqq \lambda_2(A) \leqq \cdots \leqq \lambda_n(A)$ its ordered eigenvalues. Then by the above corollary the ordered eigenvalues $\lambda_i(A)$ are random variables since they are continuous functions of random variables.

2.3. *Modes of convergence.* Three types of convergence of a sequence of random variables are considered in this paper: convergence in law, convergence in probability, and convergence almost surely. Let X be a random variable and let $\{X_n\}_{n=1}^{\infty}$ be an infinite sequence of random variables; let F_n and F denote the distribution functions of X_n and X, respectively.

The sequence $\{X_n\}_{n=1}^{\infty}$ is said to converge in law to X as $n \to \infty$, written $X_n \xrightarrow{\mathscr{L}} X$ as $n \to \infty$, if $F_n(x) \to F(x)$ as $n \to \infty$ at all points of continuity of F.

The sequence $\{X_n\}_{n=1}^{\infty}$ is said to converge in probability to X as $n \to \infty$, written $X_n \xrightarrow{P} X$ as $n \to \infty$, if for any given $\varepsilon > 0$,

$$P(|X_n - X| > \varepsilon) \to 0 \qquad \text{as } n \to \infty. \tag{2.3.1}$$

The sequence $\{X_n\}_{n=1}^{\infty}$ is said to converge a.s. to X as $n \to \infty$, written $X_n \to X$ a.s. as $n \to \infty$, if

$$P(\{\omega : \lim_{n \to \infty} X_n(\omega) = X(\omega)\}) = 1. \tag{2.3.2}$$

The following implication structure exists among these modes of convergence: $X_n \to X$ a.s. as $n \to \infty$ implies $X_n \xrightarrow{P} X$ as $n \to \infty$ implies $X_n \xrightarrow{\mathscr{L}} X$ as $n \to \infty$.

2.4. *Empirical distribution function.* Let $\{X_1, X_2, \cdots, X_n\}$ be a set of random variables. For any $B \in \mathscr{F}$, let I_B denote the indicator function of B,

$$I_B(\omega) = \begin{cases} 1 & \text{if } \omega \in B, \\ 0 & \text{if } \omega \notin B. \end{cases} \tag{2.4.1}$$

The empirical distribution function of $\{X_1, X_2, \cdots, X_n\}$ is a mapping $F_n : R_1 \times \Omega \rightsquigarrow [0, 1]$ defined by

$$F_n(x)(\omega) = \frac{1}{n} \sum_{i=1}^{n} I_{[X_i \in (-\infty, x)]}(\omega). \tag{2.4.2}$$

Let $A_n = (a_{ij})_{i,j=1}^n$ be a random Hermitian matrix. Let $\lambda_1(A_n) \leqq \lambda_2(A_n) \leqq \cdots \leqq \lambda_n(A_n)$ denote the a.s. real ordered random eigenvalues of A_n. Denote by W_n the empirical distribution function of $\{\lambda_1(A_n), \lambda_2(A_n), \cdots, \lambda_n(A_n)\}$. The basic question examined in this paper is (for any $x \in R_1$): how does $W_n(x)$ behave as $n \to \infty$?

2.5. *Some lemmas.* In this section are listed some lemmas which will be used below.

Given a random variable X and a sequence of random variables $\{X_n\}_{n=1}^{\infty}$, let F_n and F denote the distribution functions of X_n and X, respectively. Furthermore, let

$$\alpha_k = \int_{R_1} x^k \, dF(x), \qquad \alpha_{k,n} = \int_{R_1} x^k \, dF_n(x) \tag{2.5.1}$$

define the kth moment of the distribution functions F and F_n, respectively, if they exist.

LEMMA 2.5.1. *If, for all $k \geqq k_0$ arbitrary but fixed, the sequence $\alpha_{k,n} \to \alpha_k$ finite, then these sequences converge for every value of k, and if the sequence $\{\alpha_k\}_{k=1}^{\infty}$ uniquely determines F, then $F_n(x) \to F(x)$ as $n \to \infty$ at all points of continuity of F.*

A proof of this lemma may be found in M. Loève [9].

For any infinite sequence of sets, $\{A_n\}_{n=1}^{\infty}$, $A_n \in \mathscr{F}$, define

$$\limsup_{n \to \infty} A_n = \bigcap_{m=1}^{\infty} \bigcup_{n=m}^{\infty} A_n. \tag{2.5.2}$$

LEMMA 2.5.2. *The sequence* $X_n \to 0$ a.s. *as* $n \to \infty$ *if and only if for all* $\varepsilon > 0$

$$P\left(\limsup_{n\to\infty} \{\omega : |X_n(\omega)| > \varepsilon\}\right) = 0. \tag{2.5.3}$$

A proof of this lemma may be found in K. L. Chung [3].

LEMMA 2.5.3 (Borel-Cantelli). *If* $\Sigma_{n=1}^{\infty} P(A_n) < \infty$, *then* $P(\limsup_{n\to\infty} A_n) = 0$.

A proof of this may be found in Loève [9].

Let W_n be the empirical distribution function of $\{X_1, X_2, \cdots, X_n\}$. Let W be a distribution function which is uniquely determined by its sequence of moments, $\{\alpha_k\}_{k=1}^{\infty}$. Let us define

$$\begin{aligned} M_{k,n}(\omega) &= \int_R x^k \, dW_n(x)(\omega) \\ &= \frac{1}{n} \sum_{i=1}^{n} X_i^k(\omega). \end{aligned} \tag{2.5.4}$$

LEMMA 2.5.4. *If* $M_{k,n} \xrightarrow{P} \alpha_k$ *as* $n \to \infty$ *for all* $k = 1, 2, \cdots$, *then* $W_n(x) \xrightarrow{P} W(x)$ *as* $n \to \infty$, *at all points of continuity of* W.

PROOF. The following result is used to establish the lemma: $X_n \xrightarrow{P} X$ as $n \to \infty$ if and only if every subsequence $\{X_{n_i}\}$ contains a subsequence which converges a.s. to X. Let $\{n_i\}$ be any subsequence of the positive integers. Then

$$\int_R x^k \, dW_{n_i}(x) \xrightarrow{P} \int_R x^k \, dW(x) \tag{2.5.5}$$

for all $k = 1, 2, \cdots$. By the diagonal procedure, it is possible to select a subsequence $\{n_i'\}$ of $\{n_i\}$ such that

$$\int_R x^k \, dW_{n_i'}(x) \to \int_R x^k \, dW(x) \qquad \text{a.s.} \tag{2.5.6}$$

for all $k = 1, 2, \cdots$. Then by Lemma 2.5.1,

$$W_{n_i'}(x) \to W(x) \qquad \text{a.s.} \tag{2.5.7}$$

The above quoted result then gives $W_n(x) \xrightarrow{P} W(x)$ as $n \to \infty$ at all points of continuity of W. This completes the proof.

3. Combinatorial arguments

The following combinatorial lemmas are of central importance in the proofs of the limit theorems to follow. They are slight extensions of results given by E. P. Wigner [17].

Denote by $A_{k,n}$, $k > 1$, $n \geqq 1$, the class of all finite sequences $f: \{1, 2, \cdots, k+1\} \to \{1, 2, \cdots, n\}$. Any ordered pair of positive integers, (i, j), will be called a *step*. The step (j, i) will be called the reverse step of (i, j). With each $f \in A_{k,n}$ is associated a sequence $g_f : \{1, 2, \cdots, k\} \to \{\text{all steps}\}$ defined as follows: $g_f(v) = (f(v), f(v+1))$, $1 \leqq v \leqq k$. The sequence g_f will be called the sequence of steps associated with f. The cardinality of any set A will be denoted by $\#A$. Let

$$D_f = \{f(i), 2 \leqq i \leqq k + 1 : f(i) \notin \{f(1), \cdots, f(i - 1)\}\}, \tag{3.1}$$

and let $d_f = \#D_f + 1$. By definition, f has b *different members* if and only if $d_f = b$. Let $\#(i, j)_f$ denote

$$\#\{g_f(\nu) = (f(\nu), f(\nu + 1)), 1 \leqq \nu \leqq k : f(\nu) = i, f(\nu + 1) = j\}. \tag{3.2}$$

For $1 \leqq \nu \leqq k$, $g_f(\nu) = (f(\nu), f(\nu + 1))$ is called a *free step* if and only if $f(\nu + 1) \notin \{f(1), \cdots, f(\nu)\}$ and a repetitive step if and only if $f(\nu + 1) \in \{f(1), \cdots, f(\nu)\}$. Let

$$\begin{aligned} F_f &= \{g_f(\nu), 1 \leqq \nu \leqq k : g_f(\nu) \text{ is free}\}, \\ R_f &= \{g_f(\nu), 1 \leqq \nu \leqq k : g_f(\nu) \text{ is repetitive}\}. \end{aligned} \tag{3.3}$$

It is immediate that

$$\begin{aligned} \#F_f + \#R_f &= k, \\ \#F_f &= \#\{g_f(\nu), 1 \leqq \nu \leqq k : g_f(\nu) \text{ is free}\} \\ &= \#\{f(\nu + 1), 1 \leqq \nu \leqq k : f(\nu + 1) \notin \{f(1), \cdots, f(\nu)\}\} \\ &= \#\{F(\nu), 2 \leqq \nu \leqq k + 1 : f(\nu) \notin \{f(1), \cdots, f(\nu - 1)\}\}, \end{aligned} \tag{3.4}$$

and

$$\#F_f = \#D_f = d_f - 1. \tag{3.5}$$

LEMMA 3.1. *Let $f \in A_{k,n}$ be such that if $(i, j) \in \{g_f(1), g_f(2), \cdots, g_f(k)\}$, then $\#(i, j)_f + \#(j, i)_f \geqq 2$. Then $d_f \leqq [\frac{1}{2}k] + 1$.*

PROOF. Let $f \in A_{k,n}$ satisfy the conditions of the lemma. If $f(i) \in \{f(\nu), 2 \leqq \nu \leqq k + 1 : f(\nu) \notin \{f(1), \cdots, f(\nu - 1)\}\}$, then $(f(i - 1), f(i))$ is a free step. The condition of the lemma implies at least one step among $g_f(i), \cdots, g_f(k)$ must equal $(f(i - 1), f(i))$ or $(f(i), f(i - 1))$ (no step among $g_f(1), \cdots, g_f(i - 2)$ equals $(f(i - 1), f(i))$ or $(f(i), f(i - 1))$ since $f(i) \notin \{f(i), \cdots, f(i - 1)\}$). Any such occurrence, say $(f(\ell - 1), f(\ell))$, must be repetitive since $f(\ell) \in \{f(1), \cdots, f(\ell - 1)\}$. Hence, with each free step is associated a repetitive step which is equal to the free step or its reverse. This implies $\#F_f \leqq \#R_f$, since all free steps are different. This, with (3.4), implies $\#F_f \leqq [\frac{1}{2}k]$. Hence, by (3.5), $d_f - 1 = \#F_f \leqq [\frac{1}{2}k]$ or $d_f \leqq [\frac{1}{2}k] + 1$. This completes the proof of Lemma 3.1.

LEMMA 3.2. *Let $f \in A_{k,n}$ be such that:*

(i) *if $(i, j) \in \{g_f(1), g_f(2), \cdots, g_f(k)\}$, then $\#(i, j)_f + \#(j, i)_f \geqq 2$;*

(ii) *$f(\ell) = f(\ell + 1)$ for some ℓ, $1 \leqq \ell \leqq k$.*

Then $d_f \leqq [\frac{1}{2}k]$.

PROOF. If f is constant one is through. Assume f is not constant. If $f(\ell) = f(\ell + 1)$ for some ℓ, $1 \leqq \ell \leqq k$, a new sequence of steps may be formed from $g_f(1), g_f(2), \cdots, g_f(k)$ by omitting all those steps equal to $(f(\ell), f(\ell + 1))$

(there will be two or more such steps, by condition (i)). The sequence of steps thus formed is associated with a sequence $h: \{1, 2, \cdots, i\} \to \{1, 2, \cdots, n\}$, $2 \leqq i \leqq k - 1$, (a lower bound of 2 since f is not constant) which satisfies condition (i) and which is such that $d_h = d_f$. Lemma 3.1 then gives $d_f = d_h \leqq [\frac{1}{2}(i - 1)] + 1 \leq [\frac{1}{2}(k - 2)] + 1 = [\frac{1}{2}k]$. This completes the proof of Lemma 3.2.

Lemma 3.3. *Let k be even, say $k = 2v$. Let $f \in A_{2v,n}$ be such that:*
(i) *if $(i, j) \in \{g_f(1), g_f(2), \cdots, g_f(2v)\}$, then $\#(i, j)_f + \#(j, i)_f \geqq 2$;*
(ii) $f(1) = f(2v + 1)$;
(iii) $d_f = v + 1$.
If $(i, j) \in \{g_f(1), g_f(2), \cdots, g_f(2v)\}$, then $\#(i, j)_f = 1$, $\#(j, i)_f = 1$.

Proof. Let $f \in A_{2v,n}$ satisfy conditions (i), (ii), and (iii) of the lemma. (For $n \geqq v + 1$ such an f is easily constructed. For example, let $f(1) = 1$, $f(2) = 2, f(v) = v, f(v + 1) = v + 1, f(v + 2) = v, \cdots, f(2v) = 2, f(2v + 1) = 1$.) Lemma 3.2 shows $d_f \leqq v + 1$. By Lemma 3.3 one must have $f(v) \neq f(v + 1)$, $1 \leqq \ell \leqq 2v$. Equation (3.5) holds:

$$\#F_f = \#D_f = d_f - 1 = v. \tag{3.6}$$

Consider the first step $g_f(1) = (f(1), f(2))$. If $f(1) \notin \{f(3), \cdots, f(2v - 1)\}$, then by condition (i) the last step must be the reverse of the first since $f(1) \notin \{f(2), \cdots, f(2v)\}$. On the other hand, if $f(1) \in \{f(3), \cdots, f(2v - 1)\}$ the following argument applies. Let ℓ, $3 \leqq \ell \leqq 2v - 1$ be the least integer such that $f(\ell) = f(1)$. Assume $f(\ell - 1) \neq f(2)$. Condition (i) implies the repetitive step $g_f(\ell - 1) = (f(\ell - 1), f(1))$ must be matched by at least one further occurrence among $g_f(\ell), \cdots, g_f(2v)$ of a step equal to $(f(\ell - 1), f(1))$ or $(f(1), f(\ell - 1))$, these occurrences being repetitive steps, since no free step equals $(f(\ell - 1), f(1))$ or $(f(1), f(\ell - 1))$; which is so because: (1) the first step does not since $f(\ell - 1) \neq f(2)$; (2) no step among $g_f(2), \cdots, g_f(\ell - 2)$ involves an $f(1)$; and (3) any further free step among $g_f(\ell), \cdots, g_f(2v)$, say $(f(i - 1), f(i))$, could not have $f(i) = f(1)$ or $f(i) = f(\ell - 1)$ because in either case $f(i) \in \{f(1), \cdots, f(i - 1)\}$. For each free step there is an occurrence in the sequence of steps of a repetitive step equal to the free step itself or its reverse, by condition (i). Since there are v different free steps, one must have at least $2v$ steps in the sequence equaling these or their reverses. This is apart from the 2 or more repetitive steps equaling $(f(\ell - 1), f(1))$ or $(f(1), f(\ell - 1))$, since no free step equals either. Altogether one would need at least $2v + 2$ steps; but only $2v$ are available. Hence, one must have $f(\ell - 1) = f(2)$. Thus, the reverse of the first step occurs.

Now define a sequence $h: \{1, 2, \cdots, 2v + 1\} \to \{1, 2, \cdots, n\}$ as follows:

$$\begin{aligned} &h(1) = f(2), \qquad h(2) = f(3), \cdots, \qquad h(i) = f(i + 1), \cdots, \\ &h(2v) = f(2v + 1) = f(1), \qquad h(2v + 1) = f(2). \end{aligned} \tag{3.7}$$

Associated with h is the sequence of steps $g_h(1) = (f(2), f(3))$, $g_h(2) = (f(3), f(4)), \cdots, g_h(2v - 1) = (f(2v), f(1)), g_h(2v) = (f(1), f(2))$. It is immediate that h satisfies conditions (i), (ii), and (iii) of the lemma. The above argument shows that the reverse of $g_h(1) = (f(2), f(3))$ occurs among $g_h(2) = (f(3), f(4)), \cdots, g_h(2v) = (f(1), f(2))$. Continuing in the same manner, one concludes if $(i, j) \in \{g_f(1), g_f(2), \cdots, g_f(2v)\}$, then $(j, i) \in \{g_f(1), g_f(2), \cdots, g_f(2v)\}$. Since there are v different free steps and $2v$ steps altogether, one must have $\#(i, j)_f = 1$, $\#(j, i)_f = 1$ for each $(i, j) \in \{g_f(1), g_f(2), \cdots, g_f(2v)\}$. This completes the proof of Lemma 3.3.

LEMMA 3.4. *Let k be odd, say $k = 2v + 1$. Let $f \in A_{2v+1,n}$ be such that:*

(i) *if $(i, j) \in \{g_f(1), g_f(2), \cdots, g_f(2v + 1)\}$, then $\#(i, j)_f + \#(j, i)_f \geqq 2$;*

(ii) $f(1) = f(2v + 2)$.

Then $d_f \leqq v = [\frac{1}{2}k]$.

PROOF. Let $f \in A_{2v+1,n}$ satisfy conditions (i) and (ii) of the lemma. Lemma 3.1 shows $d_f \leqq v + 1$. Assume $d_f = v + 1$. Then by Lemma 3.2, $f(\ell) \neq f(\ell + 1)$, $1 \leqq \ell \leqq 2v + 1$. There are $\#F_f = \#D_f = d_f - 1 = v$ different free steps. For each free step there is a repetitive step equal to the free step itself or its reverse. This occupies $2v$ of the $2v + 1$ steps associated with f. By condition (i) the remaining step must equal one of the free steps or its reverse. In other words, $\#(i, j)_f + \#(j, i)_f = 2$ for all $(i, j) \in \{g_f(1), \cdots, g_f(2v + 1)\}$ except one, say (k, ℓ), for which $\#(k, \ell)_f + \#(\ell, k)_f = 3$.

All possibilities are now considered. First consider the case $\#(k, \ell)_f = 3$. (The case $\#(\ell, k)_f = 3$ is the same.) With f is associated a sequence of steps $g_f(1), \cdots, g_f(r), \cdots, g_f(s), \cdots, g_f(t), \cdots, g_f(2v + 1)$, where $g_f(r) = g_f(s) = g_f(t) = (k, \ell)$ and $s - r \geqq 2$, $t - s \geqq 2$. Let $g_f^*(i)$ denote the reverse of $g_f(i)$. From the sequence of steps $g_f(1), g_f(2), \cdots, g_f(2v + 1)$ form a sequence of steps associated with a sequence $h: \{1, 2, \cdots, 2v - 1\} \to \{1, 2, \cdots, n\}$ in the following manner:

(3.8)
$$\begin{aligned}
g_h(1) &= g_f(t + 1) = (\ell, f(t + 2)) = (h(1), h(2))\\
g_h(2) &= g_f(t + 2) = (f(t + 2), f(t + 3)) = (h(2), h(3))\\
&\vdots\\
g_h((2v + 1) - t) &= g_f(2v + 1) = (f(2v + 1), f(2v + 2))\\
&= (h((2v + 1) - t), h((2v + 1) - (t - 1)))\\
g_h((2v + 1) - (t - 1)) &= g_f(1) = (f(1), f(2))\\
&= (h((2v + 1) - (t - 1)), h((2v + 1) - (t - 2)))\\
&\vdots\\
g_h((2v + 1) - (t - r + 1)) &= g_f(r - 1) = (f(r - 1), k)\\
&= (h((2v + 1) - (t - r + 1)), h((2v + 1) - (t - r)))\\
g_h((2v + 1) - (t - r)) &= g_f^*(s - 1) = (k, f(s - 1))\\
&= (h((2v + 1) - (t - r)), h((2v + 1) - (t - r - 1)))\\
g_h((2v + 1) - (t - r - 1)) &= g_f^*(s - 2) = (f(s - 1), f(s - 2))\\
&= (h((2v + 1) - (t - r - 1)), h((2v + 1) - (t - r - 2)))
\end{aligned}$$

$$\vdots$$

$$\begin{aligned} g_h\big((2v+1)-(t-s+2)\big) &= g_f^*(r+1) = \big(f(r+2), \ell\big) \\ &= \big(h\big((2v+1)-(t-s+2)\big), h\big((2v+1)-(t-s+1)\big)\big) \\ g_h\big((2v+1)-(t-s+1)\big) &= g_f(s+1) = \big(\ell, f(s+2)\big) \\ &= \big(h\big((2v+1)-(t-s+1)\big), h\big((2v+1)-(t-s)\big)\big) \end{aligned}$$

$$\vdots$$

$$\begin{aligned} g_h\big((2v+1)-4\big) &= g_f(t-2) = \big(f(t-2), f(t-1)\big) \\ &= \big(h\big((2v+1)-4\big), h\big((2v+1)-3\big)\big) \\ g_h\big((2v+1)-3\big) &= g_f(t-1) = \big(f(t-1), k\big) \\ &= \big(h\big((2v+1)-3\big), h\big((2v+1)-2\big)\big). \end{aligned}$$

The steps associated with h are $g_f(1), \cdots, g_f(r-1), g_f^*(r+1), \cdots, g_f^*(s-1), g_f(s+1), \cdots, g_f(t-1), g_f(t+1), \cdots, g_f(2v+1)$, in other words, the same as those associated with f except all steps equaling (k, ℓ) have been dropped and some of the steps associated with f have been reversed. It is easily seen that if $(i, j) \in \{g_h(1), \cdots, g_h(2v-2)\}$, then $\#(i, j)_h + \#(j, i)_h = 2$, and $d_h = d_f = v+1$. But Lemma 3.1 shows $d_h \leqq v$. One must conclude, by contradiction, that $d_f \leqq v$.

The other possibility is $\#(k, \ell)_f = 2$, $\#(\ell, k)_f = 1$ (or, what is the same thing, $\#(k, \ell)_f = 1$, $\#(\ell, k)_f = 2$) for which an argument similar to the above may be given. The details are not given here. This completes the proof of Lemma 3.4.

WIGNER'S COMBINATORIAL THEOREM. *Let $B_{2v,n}$ be the set of all $f \in A_{2v,n}$ such that:*

(i) *if $(i, j) \in \{g_f(1), g_f(2), \cdots, g_f(2v)\}$, then $\#(i, j)_f + \#(j, i)_f \geqq 2$;*
(ii) *$f(1) = f(2v+1)$;*
(iii) *$d_f = v+1$.*

Then

$$\#B_{2v,n} = \frac{(2v)!}{v!(v+1)!} n^{v+1} + o(n^{v+1}). \tag{3.9}$$

PROOF. Let $f \in B_{2v,n}$. By Lemma 3.2, $f(\ell) \neq f(\ell+1)$, $1 \leqq \ell \leqq 2v$. By Lemma 3.3, if $(i, j) \in \{g_f(1), \cdots, g_f(2v)\}$, then $\#(i, j)_f = 1$, $\#(j, i)_f = 1$. A sequence $t: \{1, 2, \cdots, m\} \to \{\text{integers}\}$ is called a *type sequence* if and only if $t(\ell) \geqq 0$, $1 \leqq \ell \leqq m$, $t(1) = 1$, $t(m) = 0$, and $t(\ell+1) - t(\ell) = \pm 1$, $1 \leqq \ell \leqq m-1$. For each $f \in B_{2v,n}$ define the type sequence $t_f: \{1, 2, \cdots, 2v\} \to \{\text{integers}\}$ as follows:

$$\begin{aligned} t_f(\ell) = {} & \#\{g_f(i), 1 < i < \ell : g_f(i) \text{ is free}\} \\ & - \#\{g_f(i), 1 < i < \ell : g_f(i) \text{ is repetitive}\}. \end{aligned} \tag{3.10}$$

For a given type sequence $t: \{1, 2, \cdots, 2v\} \to \{\text{integers}\}$ one has

$$\#\{f \in B_{2v,n} : t_f(\ell) = t(\ell), 1 \leqq \ell \leqq 2v\} = n(n-1) \cdots (n-v). \tag{3.11}$$

This is so because: (1) there are n choices for $f(1)$; (2) for $2 \leqq i \leqq 2\nu$, if $t(i) - t(i-1) = 1$, then $(f(i-1), f(i))$ is a free step and $f(i)$ may be any number which has not been used yet; and (3) for $2 \leqq i \leqq 2\nu$, if $t(i) - t(i-1) = -1$, then $(f(i-1), f(i))$ is a repetitive step and must be the reverse of the step which originally led to $f(i-1)$ (Lemma 3.3), and hence $f(i)$ is completely determined. Let S_ν denote the number of type sequences with domain $\{1, 2, \cdots, 2\nu\}$. Then

$$\#B_{2\nu,n} = S_\nu n(n-1)\cdots(n-\nu) = S_\nu n^{\nu+1} + o(n^{\nu+1}). \tag{3.12}$$

To find S_ν, one argues as follows. The number of type sequences t such that $t(i) > 0$, $1 \leqq i \leqq 2\nu - 1$, $t(2\nu) = 0$, that is, no 0 before the last value, will be denoted by S'_ν. From such sequences, one can obtain a type sequence with domain $\{1, 2, \cdots, 2\nu - 2\}$ by omitting $t(1)$, $t(2\nu)$ and subtracting 1 from each $t(i)$, $2 \leqq i \leqq 2\nu - 1$. Hence,

$$S'_\nu = S_{\nu-1}, \qquad S'_1 = S_0 = 1. \tag{3.13}$$

Given a type sequence with domain $\{1, 2, \cdots, 2\nu\}$, let $2k$ be the smallest integer such that $t(2k) = 0$ for the first time. Then $t_1 : \{1, 2, \cdots, 2k\} \to \{\text{integers}\}$ forms a 0 free type sequence while $t_2 : \{2k+1, \cdots, 2\nu\} \to \{\text{integers}\}$ forms an arbitrary type sequence. Hence,

$$S_\nu = \sum_{k=1}^{\nu} S'_k S_{\nu-k} = \sum_{k=1}^{\nu} S_{k-1} S_{\nu-k}, \qquad \nu = 1, 2, \cdots. \tag{3.14}$$

These recursive equations permit the successive calculation of the S_ν. Formally, one can obtain a closed formula for them by writing $t(x) = \Sigma_{\nu=0}^{\infty} t_\nu x^\nu$. The recursive formula (3.14) then gives

$$t(x) = 1 + x t^2(x). \tag{3.15}$$

The 1 on the right side is necessary because (3.14) is not valid for $\nu = 0$. It follows that

$$t(x) = \frac{1 \pm (4 - x)^{1/2}}{2x}. \tag{3.16}$$

Actually, the lower sign has to be taken. It gives

$$S_\nu = \tfrac{1}{2}\binom{\frac{1}{2}}{\nu+1}(-4)^{\nu+1} = \frac{(2\nu)!}{\nu!(\nu+1)!}, \qquad \nu = 1, 2, \cdots. \tag{3.17}$$

And finally,

$$\#B_{2\nu,n} = \frac{(2\nu)!}{\nu!(\nu+1)!} n^{\nu+1} + o(n^{\nu+1}). \tag{3.18}$$

This completes the proof of the theorem.

Let $C_{2k,n}$ denote the set of all $f \in A_{2k,n}$ such that:

(i) $f(1) = f(2k+1)$;

(ii) $f(i) \neq f(i+1)$, $1 \leqq i \leqq 2k$;

(iii) if $(i, j) \in \{g_f(1), g_f(2), \cdots, g_f(2\nu)\}$, then $\#(i, j)_f + \#(j, i)_f$ is even.

Let $C^j_{2k,n}$ denote the set of all $f \in C_{2k,n}$ such that $d_f = j$. By Lemma 3.2, if $f \in C_{2k,n}$, then $d_f \leqq k + 1$. Thus, $C_{2k,n} = \bigcup_{j=1}^{k+1} C^j_{2k,n}$ and

$$\#C_{2k,n} = \sum_{j=1}^{k+1} \#C^j_{2k,n}. \tag{3.19}$$

LEMMA 3.5. *The sets just defined satisfy the relation* $\#C^j_{2k,n} = \binom{n}{j}(\#C^j_{2k,j})$.

PROOF. The relation $\sim$ determined by $f \sim f^*$ if and only if $f \in C^j_{2k,n}$, $f^* \in C^j_{2k,n}$ and $\{f(1), f(2), \cdots, f(2k+1)\} = \{f^*(1), f^*(2), \cdots, f^*(2k+1)\}$ is an equivalence relation. The set $C^j_{2k,n}$ is split into $\binom{n}{j}$ equivalence classes by $\sim$, each containing $\#C^j_{2k,j}$ members. Hence, $\#C^j_{2k,n} = \binom{n}{j}(\#C^j_{2k,j})$. This completes the proof of the lemma.

Using Lemma 3.5, one has

$$\#C_{2k,n} = \sum_{j=1}^{k+1} \binom{n}{j} (\#C^j_{2k,j}). \tag{3.20}$$

Thus, $\#C_{2k,n}$ is determined for all n by $\#C^1_{2k,1}$, $\#C^2_{2k,2}$, $\cdots$, $\#C^{k+1}_{2k,k+1}$. An unsuccessful attempt to determine these numbers in a closed form was made. In an attempt to solve the problem, the enumerations found in Table I were made on a computer. It will be pointed out in the next section in what context these numbers may be of interest.

TABLE I

ENUMERATIONS

The numbers in the body of the table are $(1/n)(\#C^j_{3k,n}) = (1/n)(\#C^j_{2k,j})$.

k, n \ j	2	3	4	5	6	7
3, 2	1					
3, 3	2	20				
3, 4	3	60	30			
3, 5	4	120	120			
3, 6	5	200	300			
4, 2	1					
4, 3	2	84				
4, 4	3	252	390			
4, 5	4	504	1,560	336		
4, 6	5	840	3,900	1,680		
5, 2	1					
5, 3	2	340				
5, 4	3	1,020	3,840			
5, 5	4	2,040	15,360	8,544		
5, 6	5	3,400	38,400	42,720	5,040	
6, 2	1					
6, 3	2	1,364				
6, 4	3	4,092	34,980			
6, 5	4	8,184	139,920	153,600		
6, 6	5	13,640	349,800	768,00	214,080	
6, 7	6	20,460	699,600	2,304,000	1,284,480	95,040

4. Random sign ensemble and Wigner's conjecture

4.1. *Wigner's* 1955 *paper*. In 1955, Wigner [16] proved the result discussed below.

Let $A_n = (a_{ij})_{i,j=1}^n$ be a random matrix such that:

(i) $a_{ij} = a_{ji}$ a.s.;

(ii) $\{a_{ij}, i \leqq j\}$ is independent;

(iii) $P(a_{ij} = \sigma) = \frac{1}{2}, i \neq j, P(a_{ij} = -\sigma) = \frac{1}{2}, i \neq j, P(a_{ii} = 0) = 1.$

Let B_n denote the normalized matrix

$$B_n = \frac{1}{2\sigma\sqrt{n}} A_n. \tag{4.1.1}$$

Denote by $\lambda_1(B_n) \leqq \lambda_2(B_n) \leqq \cdots \leqq \lambda_n(B_n)$ the ordered random eigenvalues of B_n and by $W_n(x)$ the empirical distribution function of $\{\lambda_1(B_n), \lambda_2(B_n), \cdots, \lambda_n(B_n)\}$, that is,

$$W_n(x)(\omega) = \frac{1}{n} \sum_{i=1}^{n} I_{[\lambda_i(B_n) \in (-\infty, x)]}(\omega). \tag{4.1.2}$$

Then one has the following theorem.

THEOREM 4.1.1 (Wigner [16]). *For all* $x \in R_1$, $\lim_{n \to \infty} E(W_n(x)) = W(x)$, *where* W *is the absolutely continuous distribution function with semicircle density*

$$w(x) = \begin{cases} \frac{2}{\Pi}(1 - x^2)^{1/2}, & |x| \leqq 1, \\ 0, & |x| > 1. \end{cases} \tag{4.1.3}$$

PROOF. The distribution function W is uniquely determined by its moment sequence since

$$\begin{aligned} \sum_{k=0}^{\infty} \frac{\gamma_k}{k!}(it)^k &= \sum_{k=0}^{\infty} \frac{(-1)^k}{k!(k+1)!}\left(\frac{t}{2}\right)^{2k} = \frac{2}{t} J_1(t) \\ &= \frac{2}{\Pi} \int_{-1}^{1} e^{itx}(1 - x^2)^{1/2}\, dx, \end{aligned} \tag{4.1.4}$$

where J_1 denotes the Bessel function of order 1 of the first kind and

$$\gamma_k = \int x^k\, dW(x) = \begin{cases} 0 & \text{for } k \text{ odd}, \\ \dfrac{k!}{2^k(\frac{1}{2}k)!(\frac{1}{2}k + 1)!} & \text{for } k \text{ even}. \end{cases} \tag{4.1.5}$$

It is immediate that $EW_n(x)$ is a distribution function in x. Thus, if it can be established that

$$\int x^k\, dEW_n(x) \to \gamma_k \text{ as } n \to \infty \tag{4.1.6}$$

for all $k = 1, 2, \cdots$, then Lemma 2.5.1 will yield the desired result.

Consider the set T of all ordered $\frac{1}{2}(n-1)n$ tuples of the numbers $+\sigma$ and $-\sigma$. For each $(i_{12}, \cdots, i_{1n}, i_{23}, \cdots, i_{2n}, \cdots, i_{n-1,n}) \in T$, define $d_{(i_{12},\cdots,i_{n-1,n})} = \{\omega \in \Omega : a_{11}(\omega) = 0,\ a_{12}(\omega) = i_{12}, \cdots, a_{1n}(\omega) = i_{1n},\ a_{22}(\omega) = 0,\ a_{23}(\omega) = i_{23}, \cdots, a_{2n}(\omega) = i_{2n}, \cdots, a_{n-1,n}, a_{nn}(\omega) = 0\}$. Then using assumptions (i), (ii), and (iii), we have

$$P(D_{(i_{12},\cdots,i_{n-1,n})}) = \frac{1}{h}, \tag{4.1.7}$$

for all points in T, where $h = 2^{(n-1)n/2}$ One has $W_n(x) = N_n(x)/n$, where $N_n(x)$ equals the number of eigenvalues of B_n less than x. On each D_i, $i \in T$, $W_n(x)$ is constant; denote these values by $W_n(x)_i = N_n(x)_i/n$, $i \in T$. Then, since $\Omega = (\bigcup_{i\in T} D_i) \cup N$, where $(\bigcup_{i\in T} D_i) \cap N = \varphi$ and $P(N) = 0$,

$$\begin{aligned} EW_n(x) &= \int_{\bigcup_{i\in T} D_i} W_n(x)\, dP \\ &= \sum_{i\in T} \int_{D_i} W_n(x)\, dP \\ &= \sum_{i\in T} \frac{N_n(x)_i}{nh}. \end{aligned} \tag{4.1.8}$$

This shows that $EW_n(x)$ is a discrete distribution function with jumps of length $1/nh$ or multiples thereof at the eigenvalues of the h possible (that is, occurrence with positive probability) values of B_n. Each $i \in T$ represents one of the possible h values of B_n on Ω; denote these values by $B_n(i)$, $i \in T$. Let $e_1, e_2, \cdots, e_{nh}$ be the set of all eigenvalues of all possible values of B_n. Then we have

$$\begin{aligned} \int x^k\, dEW_n(x) &= \frac{1}{nh} \sum_{j=1}^{nh} e_j^k \\ &= \frac{1}{nh} \sum_{i\in T} \operatorname{tr}(B_n^k(i)) \\ &= \frac{1}{h(2\sigma)^k n^{1+k/2}} \sum_{i\in T} \sum_{j_1=1}^{n} \cdots \sum_{j_k=1}^{n} \prod_{\ell=1}^{k} a_{j_\ell j_{\ell+1}}(i), \end{aligned} \tag{4.1.9}$$

where $j_{k+1} = j_1$ and $a_{jk}(i)$ equals the value a_{jk} assumes on D_i. Interchanging the order of summation and denoting by $A_{k,n}$ the class of all sequences f: $\{1, 2, \cdots, k+1\} \to \{1, 2, \cdots, n\}$, one has

$$\int x^k\, dEW_n(x) = \frac{1}{(2\sigma)^k n^{1+k/2}} \sum_{f\in B_{k,n}} \frac{1}{h} \sum_{i\in T} \prod_{\ell=1}^{k} a_{f(\ell)f(\ell+1)}(i) \tag{4.1.10}$$

where $B_{k,n} = \{f \in A_{k,n} : f(1) = f(k+1)\}$, or

$$(4.1.11)\quad \int x^k\, dEW_n(x) = \frac{1}{(2\sigma)^k n^{1+k/2}} \sum_{f\in B_{k,n}} \sum_{i\in T} \int_{D_i} \prod_{\ell=1}^{k} a_{f(\ell)f(\ell+1)}\, dP$$
$$= \frac{1}{(2\sigma)^k n^{1+k/2}} \sum_{f\in B_{k,n}} E \prod_{\ell=1}^{k} a_{f(\ell)f(\ell+1)}.$$

Since $a_{ii} = 0$ a.s. this becomes

$$(4.1.12)\quad \int x^k\, dEW_n(x) = \frac{1}{(2\sigma)^k n^{1+k/2}} \sum_{f\in C_{k,n}} E \prod_{\ell=1}^{k} a_{f(\ell)f(\ell+1)},$$

where

$$(4.1.13)\quad C_{k,n} = \{f\in B_{k,n} : f(\ell) \neq f(\ell+1),\ \ell = 1, \cdots, k\}.$$

Two cases are now considered, k odd and k even. Note that all the random variables a_{ij} are symmetric about 0, so that all odd moments vanish. Let $k = 2v + 1$. For $f \in C_{2v+1,n}$ one has $\prod_{\ell=1}^{2v+1} a_{f(\ell)f(\ell+1)} = a^m_{f(i)f(i+1)} \prod a_{jk}$ a.s., for some i, where m is odd and the product $\prod a_{jk}$ involves no $a_{f(i)f(i+1)}$ or $a_{f(i+1)f(i)}$. Then, by independence and symmetry,

$$(4.1.14)\quad E \prod_{\ell=1}^{2v+1} a_{f(\ell)f(\ell+1)} = E\, a^m_{f(i)f(i+1)}\, E \prod a_{jk} = 0.$$

Thus,

$$(4.1.15)\quad \int_R x^{2v+1}\, dEW_n(x) = 0 = \gamma_{2v+1}.$$

Now assume $k = 2v$. One need only consider those $f \in C_{2v,n}$ for which if $(i, j) \in \{g_f(1), g_f(2), \cdots, g_f(2v)\}$, then $\#(i, j)_f + \#(j, i)_f$ is even, for otherwise the argument of the odd case applies and the term vanishes. Thus,

$$(4.1.16)\quad \int_R x^{2v}\, dEW_n(x) = \frac{1}{(2\sigma)^{2v} n^{v+1}} \sum_{f\in D_{2v,n}} E \prod_{\ell=1}^{2v} a_{f(\ell)f(\ell+1)},$$

where $D_{2v,n} = \{f \in C_{2v,n} :$ if $(i, j) \in \{g_f(1), g_f(2), \cdots, g_f(2v)\}$, then $\#(i, j)_f + \#(j, i)_f$ is even$\}$. For $f \in D_{2v,n}$ one has, by Lemma 3.3

$$(4.1.17)\quad E \prod_{\ell=1}^{2v} a_{f(\ell)f(\ell+1)} = \sigma^{2v}.$$

Thus,

$$(4.1.18)\quad \int_R x^{2v}\, dEW_n(x) = \frac{\# D_{2v,n}}{2^{2v} n^{v+1}}.$$

By Lemma 3.1, $f \in D_{2v,n}$ is such that $d_f \leqq v + 1$. Thus,

$$(4.1.19)\quad \# D_{2v,n} = \sum_{j=1}^{v+1} \# D^j_{2v,n},$$

where $D^j_{2\nu,n} = \{f \in D_{2\nu,n} : d_f = j\}$. As with Lemma 3.5, one has

(4.1.20) $$\# D^j_{2\nu,n} = \binom{n}{j} (\# D^j_{2\nu,j}).$$

Let

(4.1.21) $$f_1(\nu, n) = \frac{\# D^{\nu+1}_{2\nu,n}}{2^{2\nu} n^{\nu+1}},$$

(4.1.22) $$f_2(\nu, n) = \frac{\sum_{j=1}^{\nu} \binom{n}{j} (\# D^j_{2\nu,j})}{2^{2\nu} n^{\nu+1}}.$$

Then

(4.1.23) $$\int_R x^{2\nu} \, dE W_n(x) = f_1(\nu, n) + f_2(\nu, n).$$

Since $\binom{n}{j}/n^{\nu+1} \to 0$ as $n \to \infty$ for $j = 1, \cdots, \nu$, one has $f_2(\nu, n) \to 0$ as n $\to \infty$. By Wigner's combinatorial theorem

(4.1.24) $$\# D^{\nu+1}_{2\nu,n} = \frac{(2\nu)!}{\nu!(\nu+1)!} n^{\nu+1} + o(n^{\nu+1}).$$

Thus,

(4.1.25) $$f_1(\nu, n) = \frac{(2\nu)!}{2^{2\nu}\nu!(\nu+1)!} + \frac{o(n^{\nu+1})}{2^{2\nu} n^{\nu+1}}$$

and

(4.1.26) $$f_1(\nu, n) \to \frac{(2\nu)!}{2^{2\nu}\nu!(\nu+1)!}$$

as $n \to \infty$. Thus,

(4.1.27) $$\int_R x^{2\nu} \, dE W_n(x) \to \frac{(2\nu)!}{2^{2\nu}\nu!(\nu+1)!} = \gamma_{2\nu}$$

as $n \to \infty$. This completes the proof of Theorem 4.1.1.

Note the equation,

(4.1.28) $$\int_R x^{2\nu} \, dE W_n(x) = \frac{\# D_{2\nu,n}}{2^{2\nu} n^{\nu+1}},$$

derived during the course of the proof. A knowledge of $\# D_{2\nu,n}$ would give the sequence of moments of the distribution function $EW_n(x)$. As mentioned at the end of Section 3, these numbers are determined for all n by a knowledge of only $\# D^1_{2\nu,1}, \# D^2_{2\nu,2}, \cdots, \# D^{\nu+1}_{2\nu,\nu+1}$.

4.2. *Wigner's* 1958 *conjecture.* In 1958 Wigner [18] conjectured the following result. Let $A_n = (a_{ij})_{i,j=1}^n$ be a random matrix such that:

(i) $a_{ij} = a_{ji}$ a.s.;

(ii) $\{a_{ij}, i \leqq j\}$ is independent;

(iii) the distribution function of each a_{ij} is absolutely continuous with density p_{ij};

(iv) each a_{ij} is symmetric;

(v) $E\, a_{ij}^2 = \sigma^2$ for all $1 \leqq i, j \leqq n$;

(vi) $E\, |a_{ij}|^k < C_k$ for all $1 \leqq i, j \leqq n$, where C_k is independent of n.

Let $B_n = (2\sigma\sqrt{n})^{-1}A_n$ and denote by $W_n(x)$ the empirical distribution function of $\{\lambda_1(B_n), \lambda_2(B_n), \cdots, \lambda_n(B_n)\}$ where $\lambda_1(B_n) \leqq \lambda_2(B_n) \leqq \cdots \leqq \lambda_n(B_n)$ are the ordered random eigenvalues of B_n. Under the above conditions one has:

WIGNER'S CONJECTURE. *For all* $x \in R_1$, $EW_n(x) \to W(x)$ *as* $n \to \infty$, *where* $W(x)$ *is the absolutely continuous distribution function with semicircle density*

$$w(x) = \begin{cases} \frac{2}{\pi}(1 - x^2)^{1/2}, & |x| \leqq 1, \\ 0, & |x| > 1. \end{cases} \tag{4.2.1}$$

In the next section, we shall discuss work of U. Grenander [7] who sketched a proof of convergence in probability of the empirical distribution functions to the semicircle law. We shall also discuss the work of L. Arnold [2] in this connection.

5. The results of Grenander and Arnold

5.1. *Convergence in probability.* Grenander [7] sketches a proof leading to the result given in this section.

Let $A_n = (a_{ij})_{i,j=1}^n$ be a random matrix such that:

(i) $a_{ij} = a_{ji}$ a.s.;

(ii) $\{a_{ij}, i \leqq j\}$ is independent;

(iii) a_{ij} is symmetric;

(iv) $E\, a_{ij}^2 = \sigma^2$;

(v) $|E\, a_{ij}^k| \leqq C_k$, $k = 1, 2, \cdots$, where C_k is independent of n.

Let $B_n = (2\sigma\sqrt{n})^{-1}A_n$ and denote by $W_n(x)$ the empirical distribution function of $\{\lambda_1(B_n), \lambda_2(B_n), \cdots, \lambda_n(B_n)\}$, where $\lambda_1(B_n) \leqq \lambda_2(B_n) \leqq \cdots \leqq \lambda_n(B_n)$ are the ordered random eigenvalues of B_n.

THEOREM 5.1.1 (Grenander [7]). *For all* $x \in R_1$, $W_n(x) \xrightarrow{P} W(x)$ *as* $n \to \infty$, *where* W *is the absolutely continuous distribution function with semicircle density*

$$w(x) = \begin{cases} \frac{2}{\pi}(1 - x^2)^{1/2}, & |x| \leqq 1, \\ 0, & |x| > 1. \end{cases} \tag{5.1.1}$$

PROOF. Let

$$M_{k,n} = \int_{R_1} \lambda^k \, dW_n(x). \tag{5.1.2}$$

By Lemma 2.5.4, it will be sufficient to prove $M_{k,n} \to \gamma_k = \int x^k \, dW(x)$ as $n \to \infty$ for all $k = 1, 2, \cdots$. This will be achieved in two steps. First it will be shown that $EM_{k,n} \to \gamma_k$ as $n \to \infty$ as for all $k = 1, 2, \cdots$. Then it will be shown that $E(M_{k,n} - EM_{k,n})^2 \to 0$ as $n \to \infty$ for all $k = 1, 2, \cdots$. Chebyshev's inequality,

$$P(|M_{k,n} - EM_{k,n}| > \varepsilon) \leqq \frac{E(M_{k,n} - EM_{k,n})^2}{\varepsilon^2}, \tag{5.1.3}$$

then gives $M_{k,n} - EM_{k,n} \xrightarrow{P} 0$ as $n \to \infty$ for all $k = 1, 2, \cdots$. This and $EM_{k,n} \to \gamma_k$ imply $M_{k,n} \xrightarrow{P} \gamma_k$ as $n \to \infty$ for all $k = 1, 2, \cdots$.

It is now shown that $EM_{k,n} \to \gamma_k$ as $n \to \infty$. Let $A_{k,n}$ denote the class of all sequences $f: \{1, 2, \cdots, k+1\} \to \{1, 2, \cdots, n\}$. Then

$$\begin{aligned} EM_{k,n} &= E \int \lambda^k \, dW_n(x) = E \frac{1}{n} \sum_{i=1}^{n} \lambda_i(B_n) \\ &= E \frac{1}{n} \operatorname{tr} B_n^k \\ &= E \frac{1}{(2\sigma)^k n^{1+k/2}} \sum_{f \in B_{k,n}} \prod_{\ell=1}^{k} a_{f(\ell)f(\ell+1)} \\ &= \frac{1}{(2\sigma)^k n^{1+k/2}} \sum_{f \in B_{k,n}} E \prod_{\ell=1}^{k} a_{f(\ell)f(\ell+1)}, \end{aligned} \tag{5.1.4}$$

where $B_{k,n} = \{f \in A_{k,n} : f(1) = f(k+1)\}$. As in Theorem 4.1.1, two cases are considered, k odd and k even. For exactly the reasons given in the proof of Theorem 4.1.1 one concludes immediately that for $k = 2\nu + 1$

$$EM_{2\nu+1,n} = 0 = \gamma_{2\nu+1}. \tag{5.1.5}$$

Now let $k = 2\nu$. If $f \in B_{2\nu,n}$ is such that there exists an $(f(i), f(i+1))$ such that $\#(f(i), f(i+1))_f + \#(f(i+1), f(i))_f = 1$, then by independence and symmetry,

$$E \prod_{\ell=1}^{2\nu} a_{f(\ell)f(\ell+1)} = E a_{f(i)f(i+1)} E \prod_{\ell \neq i, \ell=1}^{2\nu} a_{f(\ell)f(\ell+1)} = 0. \tag{5.1.6}$$

Thus,

$$EM_{2\nu,n} = \frac{1}{(2\sigma)^{2\nu} n^{\nu+1}} \sum_{f \in C_{2\nu,n}} E \prod_{\ell=1}^{2\nu} a_{f(\ell)f(\ell+1)}, \tag{5.1.7}$$

where $C_{2\nu,n} = \{f \in B_{2\nu,n} : \text{if } (i,j) \in \{g_f(1), g_f(2), \cdots, g_f(2\nu)\}, \text{ then } \#(i,j)_f + \#(j,i)_f \geqq 2\}$. By Lemma 3.1, $d_f \leqq \nu + 1$ for all $f \in C_{2\nu,n}$. Let

$$\begin{aligned} f_1(\nu, n) &= \frac{1}{(2\sigma)^{2\nu} n^{\nu+1}} \sum_{f \in C_{2\nu,n}^{\nu+1}} E \prod_{\ell=1}^{2\nu} a_{f(\ell)f(\ell+1)}, \\ f_2(\nu, n) &= \frac{1}{(2\sigma)^{2\nu} n^{\nu+1}} \sum_{f \in \bigcup_{j=1}^{\nu} C_{2\nu,n}^{j}} E \prod_{\ell=1}^{2\nu} a_{f(\ell)f(\ell+1)}, \end{aligned} \tag{5.1.8}$$

where $C^j_{2\nu,n} = \{f \in C_{2\nu,n} : d_f = j\}$. Then

$$EM_{2\nu,n} = f_1(\nu, n) + f_2(\nu, n). \tag{5.1.9}$$

By assumption (v), one has

$$|f_2(\nu, n)| \leqq \frac{D_{2\nu}}{(2\sigma)^{2\nu} n^{\nu+1}} \sum_{j=1}^{\nu} \# C^j_{2\nu,n} \tag{5.1.10}$$

for some constant $D_{2\nu} < \infty$, or

$$|f_2(\nu, n)| \leqq \frac{D_{2\nu} \sum_{j=1}^{\nu} \binom{n}{j} (\# C^j_{2\nu,j})}{(2\sigma)^{2\nu} n^{\nu+1}}. \tag{5.1.11}$$

Since $\binom{n}{j}/n^{\nu+1} \to 0$ as $n \to \infty$ for $j = 1, \cdots, \nu$, one has $f_2(\nu, n) \to 0$ as $n \to \infty$. Let $f \in C^{\nu+1}_{2\nu,n}$. Then, using Lemma 3.3, one has

$$E \prod_{\ell=1}^{2\nu} a_{f(\ell)f(\ell+1)} = \sigma^{2\nu}. \tag{5.1.12}$$

Hence,

$$f_1(\nu, n) = \frac{\# C^{\nu+1}_{2\nu,n}}{2^{2\nu} n^{\nu+1}}. \tag{5.1.13}$$

By Wigner's combinatorial theorem,

$$f_1(\nu, n) = \frac{(2\nu)!}{2^{2\nu} \nu! (\nu + 1)!} + \frac{o(n^{\nu+1})}{2^{2\nu} n^{\nu+1}}. \tag{5.1.14}$$

Hence,

$$f_1(\nu, n) \to \frac{(2\nu)!}{2^{2\nu} \nu! (\nu + 1)!} = \gamma_{2\nu} \tag{5.1.15}$$

as $n \to \infty$. Thus, $EM_{k,n} \to \gamma_k$ as $n \to \infty$ for all $k = 1, 2, \cdots$.

It will now be shown that $E(M_{k,n} - EM_{k,n})^2 \to 0$ as $n \to \infty$ for all $k = 1, 2, \cdots$. For $f \in A_{2k+1,n}$, let

$$\begin{aligned} E(f) = {} & E \prod_{i=1}^{k} a_{f(i)f(i+1)} \prod_{j=k+2}^{2k+1} a_{f(j)f(j+1)} \\ & - E \prod_{i=1}^{k} a_{f(i)f(i+1)} E \prod_{j=k+2}^{2k+1} a_{f(j)f(j+1)}. \end{aligned} \tag{5.1.16}$$

One has after some manipulation

$$E(M_{k,n} - EM_{k,n})^2 = \frac{1}{(2\sigma)^{2k} n^{k+2}} \sum_{f \in B_{2k+1,n}} [E(f)], \tag{5.1.17}$$

where $B_{2k+1,n} = \{f \in A_{2k+1,n}$: (i) $f(1) = f(k + 1)$; (ii) $f(k + 2) = f(2k + 2)$;

(iii) if $(i, j) \in \{g_f(1), \cdots, g_f(k), g_f(k+2), \cdots, g_f(2k+1)\}$, then $\#(i,j)_f + \#(j,i)_f \geqq 2\}$. Condition (iii) follows from the independence and symmetry conditions. It allows one to conclude, by arguments exactly as those of the proof of Lemma 3.1, that $d_f \leqq k+1$ for all $f \in B_{2k+1,n}$. By assumption (v),

$$|E(f)| \leqq \delta_k < \infty, \tag{5.1.18}$$

where δ_k is independent of n. Thus,

$$E(M_{k,n} - EM_{k,n})^2 \leqq \frac{\delta_k(\# B_{2k+1,n})}{(2\sigma)^{2k} n^{k+2}}. \tag{5.1.19}$$

Now

$$\begin{aligned} \# B_{2k+1,n} &= \sum_{j=1}^{k+1} \# B^j_{2k+1,n} \\ &= \sum_{j=1}^{k+1} \binom{n}{j} (\# B_{2k+1,j}), \end{aligned} \tag{5.1.20}$$

where $B^j_{2k+1,n} = \{f \in B_{2k+1,n} : d_f = j\}$. Thus,

$$E(M_{k,n} - EM_{k,n})^2 \leqq \frac{\delta_k \sum_{j=1}^{k+1} \binom{n}{j} (\# B^j_{2k+1,n})}{(2\sigma)^{2k} n^{k+2}}. \tag{5.1.21}$$

Since $\binom{n}{j}/n^{k+2} \to 0$ as $n \to \infty$ for $j = 1, 2, \cdots, k+1$, one has

$$E(M_{k,n} - EM_{k,n})^2 \to 0 \tag{5.1.22}$$

as $n \to \infty$. This completes the proof of Theorem 5.1.1.

5.2. *Convergence almost surely.* Arnold [2] sketches a proof leading to the result given in this section.

Let $A_n = (a_{ij})^n_{i,j=1}$ be a random matrix such that:

(i) $a_{ij} = a_{ji}$ a.s.;

(ii) $\{a_{ij}, i \leqq j\}$ is independent;

(iii) the a_{ij}, $i \neq j$ are identically distributed with distribution function F, and the a_{ii} are identically distributed with distribution function G;

(iv) $Ea_{ij} = \int x\, dF = 0$, $i \neq j$;

(v) $Ea^2_{ij} = \int x^2\, dF = \sigma^2$, $i \neq j$;

(vi) (a) $Ea^2_{ii} = \int x^2\, dG < \infty$, $Ea^4_{ij} = \int x^4\, dF < \infty$;

(b) $Ea^4_{ii} = \int x^4\, dG < \infty$, $Ea^6_{ij} = \int x^6\, dF < \infty$.

Let $B_n = (2\sigma\sqrt{n})^{-1} A_n$ and denote by $W_n(x)$ the empirical distribution function of $\{\lambda_1(B_n), \lambda_2(B_n), \cdots, \lambda_n(B_n)\}$, where $\lambda_1(B_n) \leqq \cdots \leqq \lambda_n(B_n)$ are the ordered random eigenvalues of B_n. Arnold [2] then gives the following theorem.

THEOREM 5.2.1. *Under conditions* (i) *to* (v) *and* (vi) (a), $W_n(x) \xrightarrow{P} W(x)$ *as* $n \to \infty$ *for all* $x \in R$, *and under conditions* (i) *to* (v) *and* (vi) (*b*), $W_n(x) \to W(x)$ *a.s. as* $n \to \infty$ *for all* $x \in R_1$, *where* W *is an absolutely continuous distribution function with semicircle density*

$$w(x) = \begin{cases} \frac{2}{\pi}(1 - x^2)^{1/2}, & |x| \leqq 1, \\ 0, & |x| > 1. \end{cases} \tag{5.2.1}$$

6. On Wigner's 1958 conjecture

6.1. *A limit theorem.* We shall prove in this section the following theorem. Let $A_n = (a_{ij})_{i,j=1}^n$ be a random matrix such that:
(i) $a_{ij} = a_{ji}$ a.s.;
(ii) $\{a_{ij}, i \leqq j\}$ is independent;
(iii) $Ea_{ij} = 0, 1 \leqq i, j \leqq n$;
(iv) $Ea_{ij}^2 = \sigma^2, 1 \leqq i \neq j \leqq n$;
(v) $E|a_{ij}|^k \leqq M_k, 1 \leqq i, j \leqq n$, where M_k is not dependent on n.

Let $B_n = (2\sigma\sqrt{n})^{-1}A_n$ and denote by $W_n(x)$ the empirical distribution function of $\{\lambda_1(B_n), \lambda_2(B_n), \cdots, \lambda_n(B_n)\}$, where $\lambda_1(B_n) \leqq \lambda_2(B_n) \leqq \cdots \leqq \lambda_n(B_n)$ are the ordered random eigenvalues of B_n.

THEOREM 6.1.1. *For all $x \in R_1$, $W_n(x) \to W(x)$ a.s. as $n \to \infty$, where W is an absolutely continuous distribution function with semicircle density*

$$w(x) = \begin{cases} \frac{2}{\pi}(1 - x^2)^{1/2}, & |x| \leqq 1, \\ 0, & |x| > 1. \end{cases} \tag{6.1.1}$$

PROOF. It is to be proved that

$$P(\lim_{n\to\infty} W_n(x) = W(x)) = 1. \tag{6.1.2}$$

Let

$$M_{k,n} = \int_{-\infty}^{\infty} x^k \, dW_n(x) \tag{6.1.3}$$

and

$$\gamma_k = \int_{-\infty}^{\infty} x^k \, dW(x) = \begin{cases} 0, & \text{odd } k, \\ \dfrac{k!}{2^k(\frac{1}{2}k)!(\frac{1}{2}k + 1)!}, & \text{even } k. \end{cases} \tag{6.1.4}$$

By Lemma 2.5.1, it will be sufficient to prove

$$P(\lim_{n\to\infty} M_{k,n} = \gamma_k, k \geqq 1) = 1. \tag{6.1.5}$$

This will be true if

(6.1.6) $$P\left(\lim_{n\to\infty} M_{k,n} = \gamma_k\right) = 1$$

for all $k > 1$. By the triangle inequality,

(6.1.7) $$|M_{k,n} - \gamma_k| \leqq |M_{k,n} - EM_{k,n}| + |EM_{k,n} - \gamma_k|,$$

and it will be sufficient to prove:

(6.1.8)
$$\text{(a)}\ \lim_{n\to\infty} EM_{k,n} = \gamma_k, \qquad k \geqq 1;$$
$$\text{(b)}\ P\left(\lim_{n\to\infty} (M_{k,n} - EM_{k,n}) = 0\right) = 1, \qquad k \geqq 1.$$

For (b), it will be sufficient to prove

(6.1.9) $$\sum_{n=1}^{\infty} E(M_{k,n} - EM_{k,n})^2 < \infty, \qquad k \geqq 1.$$

This is seen as follows. The statement

(6.1.10) $$P\left(\lim_{n\to\infty} (M_{k,n} - EM_{k,n}) = 0\right) = 1$$

is equivalent to

(6.1.11) $$P\left(\limsup_{n\to\infty} \{\omega : |M_{k,n}(\omega) - EM_{k,n}| > \varepsilon\right) = 0$$

for every $\varepsilon > 0$, by Lemma 2.5.2. Let $A_n = \{\omega : |M_{k,n}(\omega) - EM_{k,n}| > \varepsilon\}$. It is to be shown that $P(\limsup_{n\to\infty} A_n) = 0$. Chebyshev's inequality gives

(6.1.12) $$P(|M_{k,n} - EM_{k,n}| > \varepsilon) \leqq \frac{E(M_{k,n} - EM_{k,n})^2}{\varepsilon^2}.$$

This and $\Sigma_{n=1}^{\infty} E(M_{k,n} - EM_{k,n})^2 < \infty$ implies $\Sigma_{n=1}^{\infty} P(A_n) < \infty$. Then Lemma 2.5.3 (Borel–Cantelli) gives $P(\limsup_{n\to\infty} A_n) = 0$. Altogether, then, it will be sufficient to prove:

(6.1.13)
$$\text{(a)}\ \lim_{n\to\infty} EM_{k,n} = \gamma_k, \qquad k \geqq 1;$$
$$\text{(b)}\ \sum_{n=1}^{\infty} E(M_{k,n} - EM_{k,n})^2 < \infty, \qquad k \geqq 1.$$

The proof of (6.1.13) (a) follows.

Letting $A_{k,n}$ denote the class of all sequence $f: \{1, 2, \cdots, k+1\} \to \{1, 2, \cdots, n\}$, one has

$$EM_{k,n} = E\int \lambda^k \, dW_n(x) = E\frac{1}{n}\sum_{i=1}^{n} \lambda_i^k(B_n) \tag{6.1.14}$$

$$= E\frac{1}{n}\text{tr } B_n^k$$

$$= E\frac{1}{(2\sigma)^k n^{1+k/2}} \sum_{f\in B_{k,n}} \prod_{\ell=1}^{k} a_{f(\ell)f(\ell+1)}$$

$$= \frac{1}{(2\sigma)^k n^{1+k/2}} \sum_{f\in B_{k,n}} E\prod_{\ell=1}^{k} a_{f(\ell)f(\ell+1)},$$

where $B_{k,n} = \{f \in A_{k,n} : f(1) = f(k+1)\}$. Let $f \in B_{k,n}$ be such that there exists $(f(i), f(i+1)) \in \{g_f(1), g_f(2), \cdots, g_f(k)\}$ such that $\#(f(i), f(i+1))_f + \#(f(i+1), f(i))_f = 1$. Then, by the independence and zero mean assumptions,

$$E\prod_{\ell=1}^{k} a_{f(\ell)f(\ell+1)} = Ea_{f(i)f(i+1)}E\prod_{\ell\neq i,\ell=1}^{k} a_{f(\ell)f(\ell+1)} = 0. \tag{6.1.15}$$

Thus, one has

$$EM_{k,n} = \frac{1}{(2\sigma)^k n^{1+k/2}} \sum_{f\in C_{k,n}} E\prod_{\ell=1}^{k} a_{f(\ell)f(\ell+1)}, \tag{6.1.16}$$

where $C_{k,n} = \{f \in B_{k,n}:$ if $(i,j) \in \{g_f(1), g_f(2), \cdots, g_f(k)\}$, then $\#(i,j)_f + \#(i,j)_f \geqq 2\}$. Two cases are now considered, k odd and k even. Let $k = 2\nu + 1$. By Lemma 3.1, $d_f \leqq \nu + 1$ for all $f \in C_{2\nu+1,n}$. Thus,

$$EM_{2\nu+1,n} = \frac{1}{(2\sigma)^{2\nu+1} n^{\nu+3/2}} \sum_{j=1}^{\nu+1}\sum_{f} E\prod_{\ell=1}^{2\nu+1} a_{f(\ell)f(\ell+1)}, \tag{6.1.17}$$

where the summation is taken over $f \in C_{2\nu+1,n}^j = \{f \in C_{2\nu+1,n} : d_f = j\}$. By assumption (v),

$$\left|E\prod_{\ell=1}^{2\nu+1} a_{f(\ell)f(\ell+1)}\right| \leqq D_\nu < \infty, \tag{6.1.18}$$

for some constant D_ν. Hence,

$$|EM_{2\nu+1,n}| \leqq \frac{D_\nu \sum_{j=1}^{\nu+1} \# C_{2\nu+1,n}^j}{(2\sigma)^{2\nu+1} n^{(\nu+1)+1/2}} \tag{6.1.19}$$

$$= \frac{D_\nu \sum_{j=1}^{\nu+1} \binom{n}{j} (\# C_{2\nu+1,j}^j)}{(2\sigma)^{2\nu+1} n^{(\nu+1)+1/2}}.$$

Since $\binom{n}{j}/n^{(\nu+1)+1/2} \to 0$ as $n \to \infty$ for $j = 1, 2, \cdots, \nu + 1$, one has $EM_{2\nu+1,n} \to 0$ as $n \to \infty$. Now consider $k = 2\nu$. By Lemma 3.1, $d_f \leqq \nu + 1$ for all $f \in C_{2\nu,n}$. Let

(6.1.20)
$$f_1(\nu, n) = \frac{1}{(2\sigma)^{2\nu} n^{\nu+1}} \sum_{f \in C_{2\nu,n}^{\nu+1}} E \prod_{\ell=1}^{2\nu} a_{f(\ell)f(\ell+1)},$$
$$f_2(\nu, n) = \frac{1}{(2\sigma)^{2\nu} n^{\nu+1}} \sum_{f \in \bigcup_{j=1}^{\nu} C_{2\nu,n}^j} E \prod_{\ell=1}^{2\nu} a_{f(\ell)f(\ell+1)},$$

where $C_{2\nu,n}^j = \{f \in C_{2\nu,n} : d_f = j\}$. Note that $EM_{2\nu,n} = f_1(\nu, n) + f_2(\nu, n)$. By assumption (v),

(6.1.21)
$$\left| E \prod_{\ell=1}^{2\nu} a_{f(\ell)f(\ell+1)} \right| \leqq \delta_\nu < \infty,$$

for some constant δ_ν. Thus,

(6.1.22)
$$|f_2(\nu, n)| \leqq \frac{\delta_\nu \sum_{j=1}^{\nu} \# C_{2\nu,n}^j}{(2\sigma)^{2\nu} n^{\nu+1}}$$
$$= \frac{\delta_\nu \sum_{j=1}^{\nu} \binom{n}{j} (\# C_{2\nu,j}^j)}{(2\sigma)^{2\nu} n^{\nu+1}}.$$

Since $\binom{n}{j}/n^{\nu+1} \to 0$ as $n \to \infty$ for $j = 1, \cdots, \nu$, one has $f_2(\nu, n) \to 0$ as $n \to \infty$. By Lemma 3.3,

(6.1.23)
$$E \prod_{\ell=1}^{2\nu} a_{f(\ell)f(\ell+1)} = \sigma^{2\nu}$$

for all $f \in C_{2\nu,n}^{\nu+1}$. Thus,

(6.1.24)
$$f_1(\nu, n) = \frac{\sigma^{2\nu} (\# C_{2\nu,n}^{\nu+1})}{(2\sigma)^{2\nu} n^{\nu+1}}.$$

By Wigner's combinatorial theorem,

(6.1.25)
$$f_1(\nu, n) = \frac{(2\nu)!}{2^{2\nu} \nu! (\nu + 1)!} + \frac{o(n^{\nu+1})}{(2\sigma)^{2\nu} n^{\nu+1}}.$$

Thus,

(6.1.26)
$$f_1(\nu, n) \to \frac{(2\nu)!}{2^{2\nu} \nu! (\nu + 1)!} = \gamma_{2\nu}$$

as $n \to \infty$. Altogether

(6.1.27)
$$EM_{2\nu,n} \to \frac{(2\nu)!}{2^{2\nu} \nu! (\nu + 1)!} = \gamma_{2\nu}$$

as $n \to \infty$. This completes the proof of (i). The proof of (ii) follows.

Consider $E(M_{k,n} - EM_{k,n})^2$. For $f \in A_{2k+1,n}$ let $E(f)$ denote

$$(6.1.28) \qquad E \prod_{i=1}^{k} a_{f(i)f(i+1)} \prod_{j=k+2}^{2k+1} a_{f(j)f(j+1)} - E \prod_{i=1}^{k} a_{f(i)f(i+1)} E \prod_{j=k+2}^{2k+1} a_{f(j)f(j+1)}.$$

One has

$$(6.1.29) \qquad E(M_{k,n} - EM_{k,n})^2 = \frac{1}{(2\sigma)^{2k} n^{k+2}} \sum_{f \in B_{2k+1,n}} E(f),$$

where $B_{2k+1,n} = \{f \in A_{2k+1,n}$: (i) $f(1) = f(k+1)$; (ii) $f(k+2) = f(2k+2)$; (iii) $\{g_f(1), g_f(2), \cdots, g_f(k)\} \cap \{g_f(k+2), \cdots, g_f(2k+1), g_f^*(k+2), \cdots, g_f^*(2k+1)\} \neq \varnothing$, where $g_f^*(\ell)$ denotes the reverse of $g_f(\ell)$; (iv) $E(f) = 0\}$. Reasons for conditions (i) and (ii) are obvious. If condition (iii) is not met by f, the term in the summation corresponding to f will be zero, by the independence assumption. Condition (iv) is trivial. It will now be shown that if $f \in B_{2k+1,n}$ then $d_f \leqq k$. Using condition (iii), suppose, for the sake of definiteness, that $g_f(s) = g_f(t)$ for some s, $1 \leqq s \leqq k$, and some t, $k + 2 \leqq t \leqq 2k + 1$. (The only other case to consider is when $g_f(s) = g_f^*(t)$ for some s, $1 \leqq s \leqq k$, and some t, $k + 2 \leqq t \leqq 2k + 1$, for which the following argument also applies.) Define a new sequence $h \in A_{2k+1,n}$ as follows: $h(1) = f(s)$, $h(2) = f(s+1)$, $\cdots, h(k-s+1) = f(k), h(k-s+2) = f(1), h(k-s+3) = f(2), \cdots, h(k) = f(s-1)$, $h(k+1) = f(s)$, $h(k+2) = f(t+1)$, $h(k+3) = f(t+2), \cdots$, $h(2k-t+2) = f(2k+1), h(2k-t+4) = f(k+3), \cdots, h(2k+1) = f(t-1)$, $h(2k+2) = f(t)$. It is immediate that $d_h = d_f$. The sequence of steps associated with h is

(6.1.30)

$$\begin{aligned} g_h(1) &= (f(s), f(s+1)), \cdots, g_h(k-s+1) = (f(k), f(1)), \\ g_h(k-s+2) &= (f(1), f(2)), \cdots, g_f(k) = (f(s-1), f(s)), \\ g_h(k+1) &= (f(s), f(t+1)) = (f(s), f(s+1)), \cdots, \\ g_h(2k-t+2) &= (f(2k+1), f(k+2)), \\ g_h(2k-t+3) &= (f(k+2), f(k+3)), \cdots, g_h(2k+1) \\ &= (f(t-1), f(t)). \end{aligned}$$

It is true that:

(i) $h(1) = h(2k+2)$;

(ii) if $(i, j) \in \{g_h(1), g_h(2), \cdots, g_h(2k+1)\}$, then $\#(j, i)_f \geqq 2$.

Assertion (i) is immediate. To see (ii) one proceeds as follows. If (i, j) equals $g_h(k+1) = (f(t), f(t+1))$ or $g_h^*(k+1) = (f(t+1), f(t))$, then $\#(i, j)_h + \#(j, i)_h \geqq 2$ since $g_h(1) = g_h(k+1)$. On the other hand, if (i, j) equals any other step among $g_h(1), \cdots, g_h(k), g_h(k+2), \cdots, g_h(2k+1)$, and $\#(i, j)_h + \#(j, i)_h = 1$, then the independence assumption implies

$$(6.1.31) \quad E\left(\prod_{i=1}^{k} a_{f(i)f(i+1)} \prod_{j=k+2}^{2k+1} a_{f(j)f(j+1)}\right) - E\left(\prod_{i=1}^{k} a_{f(i)f(i+1)} \mathop{E}_{j=k+2}^{2k+1} a_{f(j)f(j+1)}\right)$$

$$= E\left(\prod_{i=1}^{k} a_{h(i)h(i+1)} \prod_{j=k+2}^{2k+1} a_{h(j)h(j+1)}\right)$$

$$- E\left(\prod_{i=1}^{k} a_{h(i)h(i+1)} \underset{j=k+2}{\overset{2k+1}{E}} a_{h(j)h(j+1)}\right) = 0,$$

contrary to the assumption that for $f \in B_{2k+1,n}$ this term is nonzero. Hence, h must satisfy condition (ii) above. Lemma 3.4 then applies, giving $d_f = d_h \leqq k$. Now consider

$$E(M_{k,n} - EM_{k,n})^2 = \frac{1}{(2\sigma)^{2k} n^{k+2}} \sum_{j=1}^{k} \sum_{f \in B^j_{2k+1,n}} E(f), \tag{6.1.32}$$

where $B^j_{2k+1,n} = \{f \in B_{2k+1,n} : d_f = j\}$. By assumption (v), one has $|E(f)| \leqq G_\nu < \infty$ for some constant G_ν. Thus,

$$|E(M_{k,n} - EM_{k,n})^2| \leqq \frac{G_\nu \sum_{j=1}^{k} (\# B^j_{2k+1,n})}{(2\sigma)^{2k} n^{k+2}} \tag{6.1.33}$$

$$= \frac{G_\nu \sum_{j=1}^{k} \binom{n}{j} (\# B^j_{2k+1,j})}{(2\sigma)^{2k} n^{k+2}}.$$

Since

$$\sum_{n=1}^{\infty} \frac{\binom{n}{j}}{n^{k+2}} < \infty \tag{6.1.34}$$

for $j = 1, 2, \cdots, k$, one has, by the comparison test for series

$$\sum_{n=1}^{\infty} |E(M_{k,n} - EM_{k,n})^2| < \infty, \tag{6.1.35}$$

which implies

$$\sum_{n=1}^{\infty} E(M_{k,n} - EM_{k,n})^2 < \infty \tag{6.1.36}$$

which was to be proved. This completes the proof of Theorem 6.1.1.

6.2. *Comments.* A little reflection will reveal that the assumption of zero means for the diagonal elements is not necessary. For, in proving $EM_{k,n} \to \gamma_k$ as $n \to \infty$, it was established that the only sequences of interest were those $f \in A_{k,n}$ for which (i) $f(1) = f(k+1)$ and (ii) if $(i, j) \in \{g_f(1), \cdots, g_f(k)\}$, then $\#(i,j)_f + (j,i)_f \geqq 2$. Condition (ii) alone implies $d_f \leqq [\frac{k}{2}] + 1$. If one assumes, however, that (iia) if $(i, j) \in \{g_f(1), \cdots, g_f(k)\}$, *where* $i \neq j$, then $\#(i,j)_f + \#(j,i)_f \geqq 2$, then conditions (i) and (iia) together imply $d_f \leqq [\frac{1}{2}k] + 1$. For odd k, say $k = 2\nu + 1$, one has $EM_{2\nu+1,n} \to 0 = \gamma_{2\nu+1}$ as $n \to \infty$ exactly as before. For even k, say $k = 2\nu$, the only sequences of interest are those f such that $d_f = v + 1$. If $d_f = v + 1$, then $f(i) = f(i + 1)$ for some i is not possible, since under conditions (i) and (iia) arguments similar to those of the proof of Lemma 3.2 would give $d_f \leqq v$. Thus, Wigner's combinatorial theorem holds under conditions (i), (iia), and (iii) $d_f = v + 1$; application of this theorem then gives $EM_{2v,n} \to \gamma_{2v}$ as $n \to \infty$ exactly as before. Note that use of the property of zero expectation of diagonal elements has been eliminated by substitution of condition (iia) for condition (ii). Similar arguments also hold for the proof that $\Sigma_{n=1}^{\infty} E(M_{k,n} - EM_{k,n})^2 < \infty$.

That the off diagonal elements all have second moments equal to σ^2 is not necessary. An examination of the proof shows that it is sufficient to assume that the ratio of the number of elements of the matrix having the same second moment to the total number of elements of the matrix approach 1 as the dimension becomes arbitrarily large.

It should be noted that Wigner's conjecture of 1958 is a special case of Theorem 6.1.1. Wigner's conjecture is not a special case of the theorem indicated by Arnold, for Arnold assumes the diagonal random variables are identically distributed and the off diagonal random variables are identically distributed. Arnold does drop the requirements, given by Wigner, of symmetric random variables and the existence of higher order moments. Theorem 6.1.1 is not only more general than the result conjectured by Wigner in the sense that it deals with almost sure convergence, but it also drops Wigner's requirement of symmetric random variables.

7. Related results

7.1. *The Gaussian orthogonal ensemble.* In quantum mechanics, under certain symmetry conditions, energy is represented by a real symmetric matrix X. If for a first observer energy is represented by X, then for a second observer with a rotated coordinate system energy is represented by OXO', where O is the orthogonal matrix relating the axes of the observers. Descriptions based on X and OXO' are completely equivalent physically. Thus, if a statistical hypothesis is made on X, then it is natural to make the same statistical hypothesis on OXO'. The following makes this precise and characterizes the possible statistical hypotheses.

Let $\{x_{ij}\}_{i \leqq j}$, $i, j = 1, 2, \cdots, n$ be an independent set of random variables on a probability space $(\Omega, \mathscr{F}, P)$. Let

$$X = \begin{bmatrix} x_{11} & x_{12} & \cdots & x_{1n} \\ x_{21} & x_{22} & \cdots & x_{2n} \\ \vdots & \vdots & \ddots & \\ x_{n1} & x_{n2} & & x_{nn} \end{bmatrix}, \tag{7.1.1}$$

where $x_{ij} = x_{ji}$ a.s., and let

$$Y_0 = (y_{ij}^0) = OXO', \tag{7.1.2}$$

where O is any orthogonal matrix. Let $x = (x_{11}, \cdots, x_{1n}, x_{22}, \cdots, x_{2n}, \cdots, x_{nn})$ and $y_0 = (y_{11}^0, \cdots, y_{1n}^0, y_{22}^0, \cdots, y_{2n}^0, \cdots, y_{nn}^0)$. Let β_n denote the Borel σ-algebra of subsets of the n dimensional Euclidean space R_n. We do not consider the case where $x = 0$ a.s.

This theorem seems to have been first proved in this context under more restrictive conditions than those given here by C. E. Porter and N. Rosenzweig [15].

THEOREM 7.1.1. *For all $B \in \beta_{n(n+1)/2}$ and all orthogonal O,*

$$P(a \in B) = P(y_0 \in B) \tag{7.1.3}$$

if and only if x_{ii} is normal with mean μ and variance $2a^2$ and x_{ij}, $i < j$, is normal with O and variance a^2, for some constants μ and $a^2 > 0$.

A proof of this may be found in Olson and Uppuluri [12].

If one assumes that X is a random matrix such that: (i) X is symmetric; (ii) the set of diagonal and superdiagonal elements of X form an independent set of random variables; and (iii) the distribution of X is invariant under orthogonal similarity transforms, then Theorem 7.1.1 allows one to say that the elements of X are normally distributed as indicated in the theorem. The physicists call this model the Gaussian orthogonal ensemble.

For the particular Gaussian orthogonal ensemble $X_{ii} \sim n(0, 1)$ and $X_{ij} \sim n(0, \frac{1}{2})$ the probability density function of the $n \times n$ symmetric random matrix $X = (X_{ij})$ is given by

$$\text{const} \exp \{-\tfrac{1}{2} \operatorname{tr} X^2\}. \tag{7.1.4}$$

By using standard methods of multivariate analysis, one can show that the probability density function of the eigenvalues $\varepsilon_1, \varepsilon_2, \cdots, \varepsilon_n$ of X is given by

$$\frac{1}{2^{n/2} n! \prod_{j=1}^{n} \Gamma(\frac{1}{2}j)} \exp\left\{-\tfrac{1}{2} \sum_{i=1}^{n} \varepsilon_i^2\right\} \prod_{i<j} |\varepsilon_i - \varepsilon_j|. \tag{7.1.5}$$

We note from this explicit form of the density function that the eigenvalues $\varepsilon_1, \varepsilon_2, \cdots, \varepsilon_n$ in this case are exchangeable (for definition of exchangeability see [6]). M. L. Mehta and M. Gaudin [11] exploit this property by using the technique of integration over alternate variables (see N. G. de Bruijn [4]) to obtain the density function of a single eigenvalue (for the case $n = 2m$) as

$$\sigma_{2m}(\varepsilon) = \sum_{i=0}^{2m-1} \varphi_i^2(\varepsilon) + \sqrt{m}\varphi_{2m-1}(\varepsilon) \int_0^{\varepsilon} \varphi_{2m}(y)\, dy, \tag{7.1.6}$$

where

$$\varphi_j(\varepsilon) = (2^j j! \sqrt{\pi})^{-1/2} e^{\varepsilon^2/2} \left(-\frac{d}{d\varepsilon}\right)^j e^{-\varepsilon^2}. \tag{7.1.7}$$

Then it is claimed by Mehta and Gaudin [11] that $\sigma_{2m}(x)$ is asymptotically equal to $\sigma(x)$, where

$$\sigma(x) = \begin{cases} \frac{1}{\pi}(4m - x^2)^{1/2}, & |x| < (4m)^{1/2}, \\ 0, & \text{otherwise.} \end{cases} \tag{7.1.8}$$

Indications of why this holds are also outlined in an appendix to Mehta's book [10]. For a different approach to the convergence to the semicircle law for a Gaussian orthogonal ensemble one may refer to Wigner [17].

For a normalized Gaussian orthogonal ensemble, Theorem 6.1.1 gives the semicircle law as the almost sure limit of the empirical distribution function of the eigenvalues of the normalized random matrix $X/\sqrt{2n}$. This, however, does not imply the convergence of the corresponding probability density functions mentioned above.

7.2. *A random Toeplitz ensemble.* It is of interest to know whether there exist random ensembles whose empirical distribution functions of their eigenvalues converge to limiting distributions other than Wigner's semicircle distribution. Such an ensemble was recently discussed by V. M. Dubner [5]. He considered the random Toeplitz ensemble described below.

Let $\{Z_k, k = 0, \pm 1, \cdots, \pm 2m\}$ be a set of complex valued random variables such that:

(i) $Z_k = \bar{Z}_{-k}$;

(ii) $Z_k = x_k + iy_k$, where $\{x_k, y_k, k = 0, 1, \cdots, 2m\}$ is an independent set of random variables each of which has a Gaussian distribution with mean 0 and variance σ^2 (except, $y_0 = 0$).

Let $A_{2m+1} = (a_{ij})_{i,j=1}^{2m+1}$, be a random matrix such that:

(i) $a_{ij} = \bar{a}_{ji}$

(ii) For $i < j$, $a_{ij} = Z_{j-i}$, $0 \leqq j - i \leqq [\frac{1}{2}(2m + 1)]$, $a_{ij} = \bar{Z}_{(2m+1)-(j-i)}$, for $[\frac{1}{2}(2m + 1)] + 1 \leqq j - i \leqq 2m$.

For instance, when $m = 2$, we have the 5×5 random matrix

$$\begin{bmatrix} Z_0 & Z_1 & Z_2 & \bar{Z}_2 & \bar{Z}_1 \\ \bar{Z}_1 & Z_0 & Z_1 & Z_2 & \bar{Z}_2 \\ \bar{Z}_2 & \bar{Z}_1 & Z_0 & Z_1 & Z_2 \\ Z_2 & \bar{Z}_2 & \bar{Z}_1 & Z_0 & Z_1 \\ Z_1 & Z_2 & \bar{Z}_2 & \bar{Z}_1 & Z_0 \end{bmatrix} \tag{7.2.1}$$

For this random Toeplitz ensemble, Dubner [5] has indicated that the asymptotic distribution of the sequence of empirical distribution functions of the set of eigenvalues is Gaussian.

7.3. *A Wishart ensemble.* In general, statisticians are interested in the distribution of the eigenvalues of a sample variance-covariance type matrix, in contrast to the physicists' interest in the distribution of the eigenvalues of a random matrix of the most general type. Recently, C. Stein considered the limiting distribution of the expected value of the empirical distribution function of the eigenvalues of random matrices of the variance-covariance type (Stein's result appears in Technical Report No. 42, December 2, 1969, Department of Statistics, Stanford University).

Stein's result may be stated as follows. Let $X = (X_{ij})$ be a $p \times n$ random matrix such that:

(i) $\{X_{ij}, 1 \leqq i \leqq p, 1 \leqq j \leqq n\}$ is an independent set of random variables;
(ii) $EX_{ij} = 0$;
(iii) $EX_{ij}^2 = 1$;
(iv) $E|X_{ij}|^k \leqq C_k < \infty$ for $k = 1, 2, \cdots$.

Let $B = (1/n)XX'$ and denote by $\lambda_1 \leqq \lambda_2 \leqq \cdots \leqq \lambda_p$ the ordered eigenvalues of B. Denote the empirical distribution function of $\lambda_1, \lambda_2, \cdots, \lambda_n$ by $W_{p,n}(x)$ so that

$$W_{p,n}(x) = \frac{1}{p}(\#\lambda_i < x). \tag{7.3.1}$$

THEOREM 7.3.1. *Let F_β be the absolutely continuous distribution function with density*

$$f_\beta(x) = \begin{cases} \dfrac{\beta}{2\pi x}[(x-a)(b-x)]^{1/2} & a \leqq x \leqq b, \\ 0, & \textit{elsewhere}, \end{cases} \tag{7.3.2}$$

where $a = [1 - \beta^{-1/2}]^2$ and $b = [1 + \beta^{-1/2}]^2$, then

$$EW_{p,n}(x) \to F_\beta(x) \tag{7.3.3}$$

as $p \to \infty$, $n \to \infty$ in such a way that $n/p \to \beta > 1$.

It is interesting to note that when $\beta = 1$ there is a relation between this result and Wigner's semicircle distribution. If X is a random variable with a semicircle distribution, then $Y = 4X^2$ has the probability density function

$$g(y) = \begin{cases} \dfrac{1}{2\pi}(4-y)^{1/2}y^{-1/2}, & 0 \leqq y \leqq 4, \\ 0, & \text{elsewhere}. \end{cases} \tag{7.3.4}$$

REFERENCES

[1] T. W. ANDERSON, *An Introduction to Multivariate Statistical Analysis*, New York, Wiley, 1958.
[2] L. ARNOLD, "On the asymptotic distribution of the eigenvalues of random matrices," *J. Math. Anal. Appl.*, Vol. 20 (1967), pp. 262–268.
[3] K. L. CHUNG, *A Course in Probability Theory*, New York, Harcourt, Brace and World, 1968.
[4] N. G. DE BRUIJN, "On some multiple integrals involving determinants," *J. Indian Math. Soc.*, Vol. 19 (1955), pp. 133–151.
[5] V. M. DUBNER, "The distribution of the eigenvalues of random cyclic matrices," *Theor. Probability Appl.*, Vol. 14 (1969), pp. 342–344.

[6] William Feller, *An Introduction to Probability Theory and its Applications*, Vol. 2, New York, Wiley, 1966.

[7] U. Grenander, *Probabilities on Algebraic Structures*, New York, Wiley, 1963.

[8] T. F. Jordan, *Linear Operators for Quantum Mechanics*, New York, Wiley, 1969.

[9] M. Loève, *Probability Theory*, Princeton, Van Nostrand, 1963 (3rd ed.).

[10] M. L. Mehta, *Random Matrices*, New York, Academic Press, 1967.

[11] M. L. Mehta and M. Gaudin, "Addendum and Erratum," *Nuclear Phys.*, Vol. 22 (1961), p. 340.

[12] W. H. Olson and V. R. R. Uppuluri, "Characterization of the distribution of a random matrix by rotational invariance," *Sankhyā Ser. A*, Vol. 32 (1970), pp. 325–328.

[13] A. M. Ostrowski, *Solution of Equations and Systems of Equations*, New York, Academic Press, 1960.

[14] C. E. Porter, *Statistical Theories of Spectra: Fluctuations*, New York, Academic Press, 1965.

[15] C. E. Porter and N. Rosenzweig, "Statistical properties of atomic and nuclear spectra," *Ann. Acad. Sci. Fennicae A VI*, Vol. 44 (1960). (Reprinted in [14], pp. 235–299.)

[16] E. P. Wigner, "Characteristic vectors of bordered matrices with infinite dimensions," *Ann. of Math.*, Vol. 62 (1955), pp. 548–564.

[17] ———, "Statistical properties of real symmetric matrices with many dimensions," *Proceedings of the Canadian Mathematical Congress*, Toronto, University of Toronto Press, 1957, pp. 174–184.

[18] ———, "On the distribution of the roots of certain symmetric matrices," *Ann. of Math.*, Vol. 67 (1958), pp. 325–326.

A PRIORI BOUNDS FOR THE RICCATI EQUATION

R. S. BUCY
UNIVERSITY OF SOUTHERN CALIFORNIA

1. Introduction and general results

Since the appearance of [6] and [7], the theory of linear filtering has experienced a renaissance. This theory, although evidently well known to statisticians in terms of "least squares estimates," has found many applications in the early sixties, largely because of the realization and synthesis methods provided in [6] and [7]. For an indication of some of the aerospace applications to guidance of spacecraft, the interested reader may find detailed information in [3]. Although the theory of linear filtering has changed little from that given in [7] for the continuous time problem, the practical realization of the so-called "correlated noise problem" as treated *mathematically* in [6], has recently found a solution in [4]. The full solution of this discrete time filtering problem and its meaning is described in detail in [5]. For readers desirous of a survey of recent results in linear and nonlinear filtering, it is available in [5], while more detailed information can be found in [3].

In this paper, our interest will center on the discrete matrix Riccati equation with emphasis on the study of the asymptotic behavior of its covariance matrix solution. A major tool in this study will be the Duffin parallel resistance of two nonnegative definite matrices A and B denoted by $A:B$. This operation is described in detail in [1] and provides for us a link between the Riccati equation and the classical continued fraction theory described in [9] and [10].

We have undertaken to study the discrete Riccati equation from the point of view of continued fractions because this technique provides considerable generality in that the nonsingular theory becomes a rather special case (see [3], Chapter 5) and much deeper results are obtained for singular problems; also the methods are striking generalizations of classical continued fraction methods.

We will be concerned with the cone C of $d \times d$ real entry symmetric nonnegative definite matrices. The cone C induces a natural partial ordering as for $A \in C$ and $B \in C$, $A \geqq B$ when and only when $A - B \in C$. The object of study will be the map of C;

This research was supported in part by the United States Air Force, Office of Aerospace Research, Applied Mathematics Division, under Grant AF–AF OSR–1244–67.

$$\tau_n(A) = \phi_n H'_{n-1}(H_{n-1}AH'_{n-1} : R_{n-1})H_{n-1}\phi'_n + C_n, \tag{1.1}$$

where with $M_{\ell,m}(R)$ the set of $\ell \times m$ real entry matrices.

The matrices A and ϕ_n belong to $M_{d,d}(R)$, $H_{n-1} \in M_{s,d}(R)$, $R_{n-1} \in M_{s,s}(R)$, and R_{n-1} is positive definite. The matrix $C_n \in M_{d,d}(R)$ with $s \geqq d$ and $H'_{n-1}H_{n-1} = I_d$, the $d \times d$ identity. We will restrict C_n and A to be members of the cone C and will center our attention on iterates of the mapping τ. We will consider the case where $s \geqq d$ and provide motivation for (1.1) in terms of the filtering problem in a later section. We recall that $(A : B) = A(A + B)^{\#}B$ (see [1]), where $^{\#}$ denotes the Moore–Penrose pseudoinverse determined by the axioms:

(a) $AA^{\#}A = A$,
(b) $A^{\#}AA^{\#} = A^{\#}$,
(c) $(AA^{\#})' = AA^{\#}$,
(d) $(A^{\#}A)' = A^{\#}A$,

with $'$ denoting transpose.

The following series of lemmas will prove useful in the sequel.

LEMMA 1.1. *For* $A_i \in M_{\ell,k}(R)$,

$$A_i\left[\sum_{j=1}^{n} A'_jA_j\right]\left[\sum_{j=1}^{n} A'_jA_j\right]^{\#} = A_i \tag{1.2}$$

for $1 \leqq i \leqq n$.

PROOF. See [8].

LEMMA 1.2. *For A and $B \in C$ and $A \geqq B$, $\tau_n(A) \geqq \tau_n(B)$, and in particular $\tau_n(C) \subseteqq C$.*

PROOF. Denoting by $\|\mathbf{x}\|_A^2$ the quadratic form induced by $A \in C$, it is quite easily seen that

$$\|\mathbf{x}\|^2_{\tau_n(A)} = \min_{\mathbf{r} \in R^s} \{\|\phi'_n\mathbf{x} + H'_{n-1}\mathbf{r}\|_A^2 + \|\mathbf{r}\|^2_{R_{n-1}} + \|\mathbf{x}\|^2_{C_n}\}, \tag{1.3}$$

so that

$$\tau_n(A) = \phi_n(A - AH'_{n-1}(H_{n-1}AH'_{n-1} + R_{n-1})^{\#}H_{n-1}A)\phi'_n + C_n. \tag{1.4}$$

In particular, if $A \geqq B$, then

$$\|\mathbf{x}\|^2_{\tau_n(A)} = \min_{\mathbf{r} \in R^s} \{\|\phi'_n\mathbf{x} + H'_{n-1}\mathbf{r}\|_B^2 + \|\mathbf{r}\|^2_{R_{n-1}} + \|\mathbf{x}\|^2_{C_n} + \|\phi'_n\mathbf{x} + H'_m\mathbf{r}\|^2_{A-B}\}; \tag{1.5}$$

the result follows. Of course, if $S_{n-1}(A) = A - AH'_{n-1}(H_{n-1}AH'_{n-1} + R_{n-1})^{\#}H_{n-1}A$, then

$$H_{n-1}S_{n-1}(A)H'_{n-1} = H_{n-1}AH'_{n-1}(H_{n-1}AH'_{n-1} + R_{n-1})^{\#}R_{n-1} = (H_{n-1}AH_{n-1}){:}R_{n-1} \tag{1.6}$$

in view of Lemma 1.1, or

$$S_{n-1}(A) = H'_{n-1}(R_{n-1} - R_{n-1}(H_{n-1}AH'_{n-1} + R_{n-1})^{\#}R_{n-1})H_{n-1} \tag{1.7}$$

since by our assumption, $H'_{n-1}H_{n-1} = I_d$.

We will be concerned with the iterated composition mapping $\tau_1(\tau_2(\cdots(\tau_n(A))\cdots)$. It is clear that $\tau_1\,\tau_2\cdots\tau_n(0)$ is monotone increasing in C, while $\tau_1\cdots\tau_n(\infty)$ is monotone decreasing in C. We remark that Riesz has shown bounded monotone sequences in C converge, that is, if $A_n \geqq A_{n-1}$ and $A_n \leqq \alpha I_d$ then $\lim_{n\to\infty} A_n$ exists.

For a direct generalization of the scalar continued fraction theory see Wall [10].

THEOREM 1.1. *Let* $t_i(A) = D_1 + A$ *if* $i = 1$, *and*

$$(1.8)\qquad t_i(A) = -\phi_{i-1}H'_{i-2}R_{i-2}(H_{i-2}(A + D_i)H'_{i-2} + R_{i-2})^{\#}R_{i-2}H_{i-2}\phi'_{i-1},$$

otherwise then

$$(1.9)\qquad \begin{aligned}&\tau_1\tau_2\cdots\tau_n(A)\\ &= t_1\cdots t_n(-\phi_nH'_{n-1}R_{n-1}(H_{n-1}AH'_{n-1} + R_{n-1})^{\#}R_{n-1}H_{n-1}\phi'_n),\end{aligned}$$

where $D_i = C_i + \phi_iH'_{i-1}R_{i-1}H_{i-1}\phi'_i$.

PROOF. An induction proof will be used to establish (1.9). For $n = 1$, it follows that,

$$(1.10)\qquad \begin{aligned}\tau_1(A) &= D_1 - \phi_1H'_0R_0(H_0AH'_0 + R_0)^{\#}R_0H_0\phi'_1\\ &= t_1(-\phi_1H'_0R_0(H_0AH'_0 + R_0)^{\#}R_0H_0\phi'_1).\end{aligned}$$

Now assume (1.9) is true for all $n \leq k$, then

$$(1.11)\qquad \begin{aligned}&\tau_1\cdots\tau_{k+1}(A)\\ &= \tau_1\cdots\tau_k(\tau_{k+1}(A))\\ &= t_1\cdots t_k(-\phi_kH'_{k-1}R_{k-1}(H_{k-1}\tau_{k+1}(A)H'_{k-1} + R_{k-1})^{\#}R_{k-1}H_{k-1}\phi'_k),\end{aligned}$$

in view of the induction hypothesis. Now by the definition of t_{k+1}, (1.11) is equal to $t_1\cdots t_{k+1}(\tau_{k+1}(A) - D_{k+1})$; however,

$$(1.12)\qquad \tau_{k+1}(A) - D_{k+1} = -\phi_{k+1}H'_kR_k(H_kAH'_k + R_k)^{\#}R_kH_k\phi'_{k+1},$$

in view of the proof of Lemma 1.2, so that the assertion follows.

COROLLARY 1.1. *For* $A \in C$,

$$(1.13)\qquad \begin{aligned}\tau_1\cdots\tau_n(0) &= t_1\cdots t_n(-\phi_nH'_{n-1}R_{n-1}H_{n-1}\phi'_n)\\ &\leqq \tau_1\cdots\tau_n(A) \leqq t_1\cdots t_n(0) = \tau_1\cdots\tau_n(\infty).\end{aligned}$$

PROOF. The proof is immediate.

COROLLARY 1.2. *Let* $t_i^*(A) = D_i + A$. *The limits* $V_i = \lim_{n\to\infty} t_i^*t_{i+1}\cdots t_n(0)$ *and* $L_i = \lim_{n\to\infty} t_i^*t_{i+1}\cdots t_n(-\phi_nH'_{n-1}R_{n-1}H_{n-1}\phi'_n)$ *exist and* $L_i \leqq V_i$.

PROOF. The map $t_i^*t_{i+1}\cdots t_n(\infty) = \lim_{A\uparrow\infty}\tau_i\cdots\tau_n(A)$ is monotone non-increasing in n and bounded above by D_1 and below by 0 and hence converges. A similar argument demonstrates the existence of the other limit.

REMARK 1.1. Notice that V_i and L_i are equilibrium solutions in that from the relation

$$(1.14)\qquad \tau_i\tau_{i+1}\cdots\tau_n(A) = \tau_i(\tau_{i+1}\cdots\tau_n(A))$$

and as τ_i preserves order in C, it follows that $V_i = \tau_i(V_{i+1})$ and $L_i = \tau_i(L_{i+1})$ when $L_i \in \dot{C}$, $\dot{C}$ the interior of C. Further, when τ_i is autonomous, L_i and V_i are constant.

COROLLARY 1.3. Suppose $[\Pi_{j=i}^{n-1} \rho_{j,j+1}^n]\|D_n - C_n\|$ tends to zero as $n \to \infty$, where

$$\rho_{i,i+1}^n = \|\phi_i H'_{i-1} R_{i-1}[(H_{i-1} L_{i+1,n} H_{i-1} + R_{i-1})^{\#}]^2 R_{i-1} H_{i-1} \phi'_i\| \tag{1.15}$$

and $L_{i,n} = \tau_i \cdots \tau_n(0)$, then

$$L_i = \lim_{n\to\infty} \tau_i \cdots \tau_n(0) = \lim_{n\to\infty} \tau_i \cdots \tau_n(A) = \lim_{n\to\infty} \tau_i \cdots \tau_n(\infty) = V_i \tag{1.16}$$

for all $A \in C$.

PROOF. Let $E_{i,n} = \tau_i \cdots \tau_n(\infty) - \tau_i \cdots \tau_n(0)$; then with $E_{n,n} = D_n - C_n$ and with $L_{i,n} = \tau_i \cdots \tau_n(0)$

$$\begin{aligned} E_{i,n} = \phi_i H'_{i-1}\{(H_{i-1}[E_{i+1,n} + L_{i+1,n}]H'_{i-1} : R_{i-1}) \\ - (H_{i-1} L_{i+1,n} H'_{i-1} : R_{i-1})\} H_{i-1} \phi'_i \end{aligned} \tag{1.17}$$

or in view of Lemma 27 in [1]

$$E_{i,n} = \phi_i H'_{i-1} R_{i-1} C_{i,n}^{\#}(E_{i+1,n} : C_{i,n}) C_{i,n}^{\#} R_{i-1} H_{i-1} \phi'_i, \tag{1.18}$$

where $C_{i,n} = H_{i-1} L_{i+1,n} H'_{i-1} + R_{i-1}$. Hence, if $\varepsilon_{i,n} = \|E_{i,n}\|$, then $\varepsilon_{i,n} \leqq \rho_{i,i+1}^n \varepsilon_{i+1,n}$, where

$$\rho_{i,i+1}^n = \|\phi_i H'_{i-1} R_{i-1}[(H_{i-1} L_{i+1,n} H'_{i-1} + R_{i-1})^{\#}]^2 R_{i-1} H_{i-1} \phi'_i\| \tag{1.19}$$

so that

$$\varepsilon_{i,n} \leqq \left[\prod_{j=i}^{n-1} \rho_{j,j+1}^n\right] \|D_n - C_n\| \tag{1.20}$$

by Corollary 1.1.

EXAMPLE 1.1. Consider the scalar Riccati equation

$$p_n = \alpha_n^2 p_{n-1} - \frac{\alpha_n^2 p_{n-1}^2}{r_{n-1} + p_{n-1}} + q_n, \tag{1.21}$$

where $q_n \geqq 0$ and $r_n \geqq 0$. The associated continued fraction is

$$b_0 + \frac{a_1|}{|b_1} + \frac{a_2|}{|b_2} \cdots, \tag{1.22}$$

where

$$\begin{aligned} b_0 &= q_0 + \alpha_0^2 r_{-1}, \\ b_i &= q_i + \alpha_i^2 r_{i-1} + r_{i-2}, \\ a_i &= \alpha_{i-1}^2 r_{i-2}^2. \end{aligned} \tag{1.23}$$

In view of Corollary 1.3, it suffices that

$$\left[\prod_{j=1}^{n-1} \frac{\alpha_j^2 r_{j-1}^2}{(q_{j+1} + r_{j-1})^2}\right] \alpha_n^2 r_{n-1} \tag{1.24}$$

tends to zero as $n \to \infty$. Notice this result is far stronger than those of [2]. In the autonomous case the associated continued fraction is

$$y = \beta_0^* + \frac{\alpha^*|}{|\beta^*} + \frac{\alpha^*|}{|\beta^*} + \cdots, \tag{1.25}$$

where $\beta_0^* = \beta^* - r = q + \alpha^2 r$ and $\alpha^* = -\alpha^2 r^2$ and it is well known that the continued fraction converges to $y = x - r$ with x the positive root of $x^2 - \beta^* x - \alpha^* = 0$. For detailed information see [10], problem 1.4, page 23. Explicitly, y is given by

$$y = \tfrac{1}{2}(q + (\alpha^2 + 1)r + [(q + (\alpha^2 + 1)r)^2 - 4\alpha^2 r]^{1/2}) - r. \tag{1.26}$$

Further, if $\alpha = e^{\Delta f}$, $r = r^*/\Delta$, and $q = q^*\Delta$, it is easily verified that as $\Delta \to 0$

$$y(\Delta) \downarrow fr^* + (f^2(r^*)^2 + r^*q^*)^{1/2}, \tag{1.27}$$

the equilibrium solution of the analogous continuous time Riccati equation (see Example 1, p. 98, [7]).

COROLLARY 1.4. *For* $n \geqq 1$, $C_1 \leqq \tau_1 \cdots \tau_n(A) \leqq D_1$.

THEOREM 1.2. *If* $S_\alpha^+ = \{A \in C \mid \tau_n(A) \geqq A$ *for all* $n \geqq \alpha\}$ *and* $S_\alpha^- = \{A \in C \mid \tau_n(A) \leqq A$ *for all* $n \geqq \alpha\}$, *then* $\tau_\alpha \cdots \tau_n(A)$ *is monotone nondecreasing for* $A \in S^+$ *and monotone nonincreasing for* $A \in S^-$. *Further, if there exist real* γ_α *and* β_α *such that* $D_n \leqq \gamma_\alpha I$, $C_n \geqq \beta_\alpha I$, *for all* $n \geqq \alpha$, *then*

$$\begin{aligned} S_\alpha^+ &\supseteq \{A \in C \mid A \leqq \beta_\alpha I\}, \\ S_\alpha^- &\supseteq \{A \in C \mid A \geqq \gamma_\alpha I\}. \end{aligned} \tag{1.28}$$

PROOF. Since τ_n preserves the ordering of C, if $\tau_n(A) \geqq A$, then $\tau_{n-1}\tau_n(A) \geqq \tau_{n-1}(A)$, and so forth, so that $\tau_\alpha \cdots \tau_n(A) \geqq \tau_\alpha \cdots \tau_{n-1}(A)$ for $A \in S_\alpha^+$. The other assertions are obvious.

REMARK 1.2. In view of the double argument convexity of $(A:B)$, S_α^+ is convex (see [1], Theorem 24).

We will now specialize further and assume that ϕ_i, C_i, and R_i are invertible for all i; in view of Corollary 1.4, it suffices to consider $\tau_n(A)$ for $A \in C$ the interior of C.

Now

$$\begin{aligned} \tau_n(A) &= \phi_n(A^{-1} + O_n^{-1})^{-1}\phi_n' + C_n \\ &= \phi_n(A : O_n)\phi_n' + C_n \end{aligned} \tag{1.29}$$

with $O_n = (H_{n-1}' R_{n-1}^{-1} H_{n-1})^{-1}$. This relation follows since

$$\begin{aligned} &(H_{n-1}AH_{n-1}' : R_{n-1}) \\ &\quad = H_{n-1}AH_{n-1}'(H_{n-1}AH_{n-1}' + R_{n-1})^{-1}R_{n-1} \\ &\quad = H_{n-1}AH_{n-1} - H_{n-1}AH_{n-1}'(H_{n-1}AH_{n-1}' + R_{n-1})^{-1}(H_{n-1}AH_{n-1}'), \end{aligned} \tag{1.30}$$

so that

$$
\begin{aligned}
(1.31)\qquad & H'_{n-1}(H_{n-1}AH'_{n-1} : R_{n-1})H_{n-1} \\
&= A - AH'_{n-1}(H_{n-1}AH'_{n-1} + R_{n-1})^{-1}H_{n-1}A \\
&= (A^{-1} + O_n^{-1})^{-1} = (A : O_n)
\end{aligned}
$$

by the Schur lemma (see [3], Theorem (8.6)). Hence, (1.1) demonstrates (1.29) in view of (1.30). Again, it is of interest to study $\tau_1 \cdots \tau_n(A)$, however, now for $A \geqq C_0 > 0$.

In order to simplify this study, we introduce the following transformation

$$
(1.32)\qquad \begin{aligned}
\sigma_i^*(A) &= +A + S_i, \\
\sigma_i(A) &= -\phi_{i-1}O_{i-1}(A + S_i + O_{i-1})^{-1}O_{i-1}\phi'_{i-1}
\end{aligned}
$$

with $S_i = C_i + \phi_i O_i \phi'_i$.

It will be convenient to rewrite (1.29) as

$$
(1.33)\qquad \tau_n(A) = S_n - \phi_n O_n(O_n + A)^{-1}O_n\phi_n.
$$

THEOREM 1.3. *For* $A \in C$,

$$
(1.34)\qquad \sigma_i^*\sigma_{i+1} \cdots \sigma_n\big(-\phi_n O_n(O_n + A)^{-1}O_n\phi'_n\big) = \tau_i \cdots \tau_n(A).
$$

PROOF. We prove this by induction. For $n = i$,

$$
(1.35)\qquad \sigma_i^*\big(-\phi_i O_i(O_i + A)^{-1}O_i\phi'_i\big) = S_i - \phi_i O_i(O_i + A)^{-1}O_i\phi'_i = \tau_i(A)
$$

in view of (1.33). Suppose (1.34) is valid for all $n \leqq k$, then

$$
(1.36)\qquad \tau_i \cdots \tau_{k+1}(A) = \tau_i \cdots \tau_k\big(\tau_{k+1}(A)\big),
$$

which equals by the induction hypothesis

$$
\begin{aligned}
(1.37)\qquad & \sigma_i^* \cdots \sigma_k\big(-\phi_k O_k(O_k + \tau_{k+1}(A))^{-1}O_k\phi'_k\big) \\
&= \sigma_i^*\sigma_{i+1} \cdots \sigma_{k+1}\big(\tau_{k+1}(A) - S_{k+1}\big) \\
&= \sigma_i^*\sigma_{i+1} \cdots \sigma_{k+1}\big(-\phi_{k+1}O_{k+1}(O_{k+1} + A)^{-1}O_{k+1}\phi'_{k+1}\big),
\end{aligned}
$$

demonstrating the assertion for invertible A. Since (1.33) holds for all $A \in C$, the theorem is valid.

COROLLARY 1.5. *The maps* σ *and* τ *satisfy the relations*

$$
\begin{aligned}
(1.38)\qquad & \sigma_1^*\sigma_2 \cdots \sigma_n\big(-\phi_n O_n(O_n + C_{n+1})^{-1}O_n\phi'_n\big) \\
&= \tau_1 \cdots \tau_n(C_{n+1}) = \tau_1 \cdots \tau_{n+1}(0) \\
&\leqq \tau_1 \cdots \tau_{n+1}(A) \leqq \tau_1 \cdots \tau_{n+1}(\infty) = \tau_1 \cdots \tau_n(S_{n+1}) \\
&= \sigma^*\sigma_2 \cdots \sigma_n\big(-\phi_n O_n(O_n + S_{n+1})^{-1}O_n\phi'_n\big).
\end{aligned}
$$

PROOF. Since $\tau_1 \cdots \tau_{n+1}(0) \leqq \tau_1 \cdots \tau_{n+1}(A) \leqq \tau_1 \cdots \tau_{n+1}(\infty)$, the result follows from the theorem and the obvious relations

$$
(1.39)\qquad \begin{aligned}
\tau_1 \cdots \tau_{n+1}(0) &= \tau_1 \cdots \tau_n(C_{n+1}), \\
\tau_1 \cdots \tau_{n+1}(\infty) &= \tau_1 \cdots \tau_n(S_{n+1}).
\end{aligned}
$$

Now for $n \geqq 1$, it is clear that

$$\sigma_1^* \sigma_2 \cdots \sigma_{n+1}(\infty) = \sigma_1^* \sigma_2 \cdots \sigma_n(0) \equiv A_n B_n^{-1}, \tag{1.40}$$

so that we will suppose that

$$\sigma_1^* \sigma_2 \cdots \sigma_{n+1}(\Gamma) = (A_{n+1} + A_n K_n \Gamma)(B_{n+1} + B_n K_n \Gamma)^{-1}. \tag{1.41}$$

In fact, if the following relations are satisfied

$$\begin{aligned} K_n &= \phi_n'^{-1} O_n^{-1}, \\ A_{n+1} &= A_n K_n (S_{n+1} + O_n) - A_{n-1} K_{n-1} \phi_n O_n, \\ B_{n+1} &= B_n K_n (S_{n+1} + O_n) - B_{n-1} K_{n-1} \phi_n O_n, \end{aligned} \tag{1.42}$$

with $A_0 = K_0^{-1}$, $A_1 = S_1$, $B_0 = O$, $B_1 = I$, the relationship (1.41) can be proved by induction. The only point here to be verified is that $B_{n+1} + B_n K_n \Gamma$ is invertible, and this follows by a simple disconjugacy argument analogous to that given in [3], Theorem (5.1).

REMARK 1.3. Equation (1.41) is the analogy of the symplectic system

$$\begin{bmatrix} X_n \\ Y_n \end{bmatrix} = \begin{bmatrix} \phi_n'^{-1} & \phi_n'^{-1} O_n^{-1} \\ C_n \phi_n'^{-1} & \phi_n - C_n \phi_n'^{-1} O_n^{-1} \end{bmatrix} \begin{bmatrix} X_{n-1} \\ Y_{n-1} \end{bmatrix} \tag{1.43}$$

associated with the mapping $\tau_k(\Gamma)$ given by (1.29) as $\tau_n(\Gamma) = Y_n X_n^{-1}$ with $X_{n-1} = I$ and $Y_{n-1} = \Gamma$.

Another point of interest is that of investigation of the equilibrium solutions of (1.33). Previously, we have shown the existence of two equilibrium solutions L_i and V_i. Let E_i denote general equilibrium solution, then if $\Sigma_i = E_i + O_{i-1}$ it follows from (1.33) that

$$\Sigma_i = S_i + O_{i-1} - \phi_i O_i (\Sigma_{i+1})^{-1} O_i \phi_i' \tag{1.44}$$

and in the autonomous case

$$\Sigma = S + O - \phi O (\Sigma)^{-1} O \phi'. \tag{1.45}$$

These equations generalize the quadratic fixed points of scalar continued fractions.

EXAMPLE 1.2. If $\Phi = I$ in (1.45), then

$$[O^{-1/2} \Sigma O^{-1/2}] = O^{-1/2} C O^{-1/2} + 2I - [O^{-1/2} \Sigma O^{-1/2}]^{-1} \tag{1.46}$$

and

$$\begin{aligned} O^{-1/2} \bar{P} O^{-1/2} &= -I + T\{1 + \tfrac{1}{2}\mu_i + [\mu_i + (\tfrac{1}{2}\mu_i)^2]^{1/2}\}T' \\ &= T\{\tfrac{1}{2}\mu_i + [\mu_i + (\tfrac{1}{2}\mu_i)^2]^{1/2}\}T, \end{aligned} \tag{1.47}$$

where $T' O^{-1/2} C O^{-1/2} T = \{\mu_i\}$, the notation $\{\lambda_i\}$ denotes the diagonal matrix

$$\begin{bmatrix} \lambda_1 & & 0 \\ & \ddots & \\ 0 & & \lambda_n \end{bmatrix}. \tag{1.48}$$

The equilibrium solution of $\tau(A)$ is unique and equals $\bar{P}$ with $\bar{P} = L = V$, if O and C are positive definite.

Explicitly, in terms of the unique positive semidefinite square root, we obtain

$$(1.49) \qquad \bar{P} = \tfrac{1}{2}C + O^{1/2}(\tfrac{1}{4}O^{-1/2}CO^{-1}CO^{-1/2} + O^{-1/2}CO^{-1/2})^{1/2}O^{1/2}.$$

The continuous version of (1.49) when $C = A\Delta$ and $O = B(1/\Delta)$ and Δ tends to zero is

$$(1.49') \qquad \bar{P} = B^{1/2}(B^{-1/2}AB^{-1/2})^{1/2}B^{1/2}.$$

This equilibrium solution of $\dot{P} = -PB^{-1}P + A$ has been obtained by Reid, Bellman, and others and has lead to the comment that in this particular case the Riccati equation is a continuous "square rooter."

We rewrite (1.29) in the more convenient form for L_i and V_i as

$$(1.50) \qquad L_i = \phi_i^* L_{i+1}\phi_i^{*\prime} + \phi_i(L_{i+1}: O_i)O_i^{-1}(L_{i+1}: O_i)\phi_i' + C_i$$

with $\phi_i^{*\prime} = (L_{i+1} + O_i)^{-1}O_i\phi_i'$ and

$$(1.51) \qquad V_i = \phi_i^{**}V_{i+1}\phi_i^{\prime **} + \phi_i(V_{i+1}: O_i)O_i^{-1}(V_{i+1}: O_i)\phi_i' + C_i$$

with $\phi_i^{**\prime} = (V_{i+1} + O_i)^{-1}O_i\phi_i'$.

It also follows easily that $E_i = V_i - L_i$ satisfies

$$(1.52) \qquad E_i = \phi_i^* E_{i+1}\phi_i^{**\prime}$$

(see [1], Lemma 27). Now the following theorem establishes the asymptotic theory for iterates of τ_i.

THEOREM 1.4. *Suppose ϕ_i, R_i and, O_i are invertible and further that there exist α and β real positive numbers such that*

(i) $0 < \alpha I \leqq C_i$,
(ii) $V_i \leqq \beta I$,

and that $\|\phi_i - I\| \leqq C_1$, *then* $V_i = L_i = \bar{P}_i$ *and*

$$(1.53) \qquad \lim_{n\to\infty} \tau_i \cdots \tau_n(A) = \bar{P}_i,$$

the convergence being exponential.

Further (1.44) *has a unique supnorm bound solution with* $\Sigma_i \geqq O_{i-1}$ *given by* $\Sigma_i = \bar{P}_i + O_{i+1}$ *and* $\bar{P}_i \in C$.

PROOF. From (1.50) and (1.51) and our assumptions, it is easily checked that the quadratic forms $\|\mathbf{x}\|_{L_i}^2$ and $\|\mathbf{x}\|_{V_i}^2$ are Liapounov functions for $\mathbf{x}_{(i+1)} = \phi_i^{*\prime}\mathbf{x}_i$ and $\mathbf{y}_{(i+1)} = \phi_i^{**\prime}\mathbf{y}_i$, respectively. In particular, this implies

$$(1.54) \qquad \|\phi_i^* \cdots \phi_n^*\| \leqq c_* \exp\{-\gamma_*(n - i)\}$$

for $n \geqq i$ for some c_* and γ_* positive and analogous relations for $\phi_i^{**\prime}$ hold with parameters c_{**} and γ_{**}. But since $V_i - L_i$ satisfies (1.52), it follows that

$$E_i = \phi_i^* \cdots \phi_k^* E_{k+1} \phi_k^{**\prime} \cdots \phi_i^{**\prime}, \tag{1.55}$$

and hence,

$$\|E_i\|^2 \leqq c_* c_{**} \exp\{-(\gamma_* + \gamma_{**})(k - i)\} \|E_{k+1}\|^2. \tag{1.56}$$

However, as k is arbitrary and $\|E_{k+1}\|^2$ is uniformly bounded, E_i must be zero. The validity of equation (1.53) follows as $L_{i,n} \leqq \tau_i \cdots \tau_n(A) \leqq V_{i,n}$, since $L_{i,n}$ and $V_{i,n}$ converge to $\bar{P}_i = V_i = L_i$, as $E_i = 0$. Now suppose (1.44) has two solutions Σ_i and T_i, then $x_i = \Sigma_i - O_{i-1}$ and $y_i = T_i - O_{i-1}$ are equilibrium solutions of (1.29) both supnorm bounded and members of C. Now

$$\begin{aligned} \tau_i \cdots \tau_n(x_{n+1}) &= x_i, \\ \tau_i \cdots \tau_n(y_{n+1}) &= y_i, \end{aligned} \tag{1.57}$$

and hence, as x_{n+1} and y_{n+1} are members of C,

$$\begin{aligned} L_{i,n} \leqq x_i \leqq V_{i,n}, \\ L_{i,n} \leqq y_i \leqq V_{i,n}, \end{aligned} \tag{1.58}$$

but the last equation implies $x_i = y_i = \bar{P}_i$.

REMARK 1.4. Theorem 1.4 is essentially unchanged when R_i is singular and provides stability and a unique equilibrium solution for (1.1) when L_i is uniformly positive definite and V_i is uniformly bounded in C.

REMARK 1.5. Notice that conditions (i) and (ii) of the theorem depend in general on the existence of uniform *a priori* bounds, which we developed previously.

2. Applications to the theory of filtering

In order to apply the theory of iterates of the mapping $\tau_n(A)$ of the last section to problems in linear filtering theory, we must study the mapping

$$\Delta_n(A) = \phi_n A \phi_n' - K_n(H_{n-1} A H_{n-1}' + R_{n-1})^{\#} K_n' + G_n G_n' \tag{2.1}$$

with $K_n = \phi_n A H_{n-1}' + G_n L_{n-1}'$, $R_{n-1} \geqq L_{n-1} L_{n-1}'$ for $A \in C$. This mapping is similar to τ_n except that $H \in M_{s,d}(R)$ and $G \in M_{d,r}(R)$ with r and s for many problems *less* than d. Note also the more general nonlinear term which arises from the observation noise being correlated with the signal process. Iterates of Δ determine the error covariance matrix of the optimal filter (see [3], Chapters 4 and 9). The following lemma relates $\Delta_n(A)$ to the simpler Riccati mapping T_n.

LEMMA 2.1. *For* $A \in C$, $\Delta_n(A) = T_n(A)$, *where*

$$T_n(A) = \psi_n S_n(A) \psi_n' + G_n(I - L_{n-1}' R_{n-1}^{\#} L_{n-1}) G_n' \tag{2.2}$$

and

$$\begin{aligned} S_n(A) &= A - A H_{n-1}'(H_{n-1} A H_{n-1}' + R_{n-1})^{\#} H_{n-1} A, \\ \psi_n &= \phi_n - G_n L_{n-1}' R_{n-1}^{\#} H_{n-1}. \end{aligned} \tag{2.3}$$

PROOF. From the definition, $\Delta_n(A)$ can be written as

$$\begin{aligned}(2.4)\qquad \Delta_n(A) = \phi_n S_n(A)\phi'_n - G_n L'_{n-1}(H_{n-1}AH_{n-1} + R_{n-1})^{\#} H_{n-1}A\phi'_n \\ + G_n L'_{n-1}(H_{n-1}AH_{n-1} + R_{n-1})^{\#} L_{n-1}G'_n \\ - \phi_n A H_{n-1}(H_{n-1}AH'_{n-1} + R_{n-1})^{\#} L_{n-1}G'_n + G_n G'_n.\end{aligned}$$

But by the definition of $S_n(A)$ and Lemma 1.1, it follows that

$$\begin{aligned}(2.5)\qquad & S_n(A)H'_{n-1}R^{\#}_{n-1}L_{n-1}G'_n \\ & = AH_{n-1}(H_{n-1}AH'_{n-1} + R_{n-1})^{\#} R_{n-1}R^{\#}_{n-1}L_{n-1}G'_n \\ & = AH_{n-1}(H_{n-1}AH'_{n-1} + R_{n-1})^{\#} L_{n-1}G'_n,\end{aligned}$$

and hence,

$$\begin{aligned}(2.6)\qquad & G_n L'_{n-1}R^{\#}_{n-1}H_{n-1}S_n(A)H'_{n-1}R^{\#}_{n-1}L_{n-1}G'_n \\ & = G_n L'_{n-1}R^{\#}_{n-1}L_{n-1}G'_n - G_n L'_{n-1}(H_{n-1}AH_{n-1} + R_{n-1})^{\#} L_{n-1}G'_n.\end{aligned}$$

In view of the above equalities and (2.4), the lemma follows.

REMARK 2.1. Lemma 2.1 is the discrete time *generalization* of continuous time equivalence of Riccati equations (see [3], especially page 90).

REMARK 2.2. Notice that if $R_{n-1} > 0$, then Q controllability and R observability of (2.1) hold when and only when (2.2) is Q^* controllable and R^* observable. In fact, in the general case, it seems appropriate to call (2.1) controllable and observable when these conditions hold for (2.2).

In order to overcome difficulty that H has in general less sensors than the state dimension, we process the observations in blocks of k corresponding to k sequential time observations. In other words, if $\pi(n, \Gamma, n_0)$ represents solution of the Riccati equation, we find the recursion for $\pi(n_0 + kv, \Gamma, n_0)$ in terms of $\pi(n_0 + k(v-1), \Gamma, n_0)$. This recursion equation is of the general form of (2.1) with $H_n \in M_{ks,d}(R)$, and hence for $sk \geqq d$ using Lemma 2.1, the results of Section 1 are applicable.

As an example, we consider the mapping

$$(2.7)\qquad T(A) = \phi_*\{A - AH'_*\{H_*AH'_* + R_*\}^{\#} H_*A\}\phi'_* + G_*G'_*.$$

Then $T^k(A) = \phi^k\{A - (AH' + \phi^{-k}GL')(HAH' + R + LL')^{\#}(HA + LG'\phi'^{-k})\phi'^k + GG'$, where

$$(2.8)\qquad L = \begin{Bmatrix} 0 & 0 & \cdots & & & \\ u_0 & 0 & \cdots & & & \\ u_1 & u_0 & 0 & \cdots & & \\ u_2 & u_1 & u_0 & 0 & \cdots & \\ \cdot & \cdot & \cdot & \cdot & \cdot & \cdot \\ u_{k-2} & u_{k-3} & \cdots & u_0 & 0 & \cdots \end{Bmatrix}, \qquad u_i = H_*\phi^i G_*,$$

$$(2.9)\qquad R = \begin{bmatrix} R_* & 0 & 0 & 0 \\ 0 & R_* & 0 & \cdots \\ 0 & 0 & R_* & 0 \\ \cdot & \cdot & \cdot & \cdot \\ \cdot & \cdot & \cdot & 0R_* \end{bmatrix},$$

$$(2.10)\qquad H = \begin{bmatrix} H_* \\ H_*\phi_* \\ \vdots \\ H_*\phi_*^{k-1} \end{bmatrix}, \qquad G = [\phi_*^{k-1}G_*, \cdots, G_*].$$

If ϕ_* is invertible, define M via the equation $H\phi_*^{-k}G = M + L$, using properties of the pseudoinverse

$$(2.11)\qquad \|\mathbf{x}\|^2_{T^k(A)} = \min_{\mathbf{y}\in R^s} \{\|\psi'\mathbf{x} + H'\mathbf{y}\|^2_A + \|\mathbf{y}\|^2_{R+LL'} + \|G'x\|^2_{I-L'(R+LL')^{\#}L}\}$$

for all $x \in R^d$.

An interesting problem is that of characterizing $J_\ell = \{\underline{x} \in R^d \mid T^j(A)\underline{x} = T^\ell(0)\underline{x}$, all $A \in C$, all $j \geqq \ell\}$. This problem has been solved for H_* a vector and $R_* = 0$ (see [4]). In [5], the general solution has been given for $R = 0$. From (2.11) the following theorem determines J_k in general.

THEOREM 2.1. *With matrices given by* (2.8), (2.9), *and* (2.10),

$$(2.12)\qquad \begin{aligned} J_k &= \{\mathbf{x} \in R^d \mid T^k(A)\mathbf{x} = T^k(0)\mathbf{x}, \ \textit{for all} \ \ A \in C\} \\ &= \{\mathbf{x} \in R^d \mid \psi'\mathbf{x} = H'\ell, \ell \in R^s, \|\ell\|^2_{(R+LL')} = 0\}. \end{aligned}$$

PROOF. It is clear that $\{\mathbf{x} \in R^d \mid T^k(A)\mathbf{x} = T^k(0)\mathbf{x}$, for all $A \in C\} \supseteqq J_k$. Since for $A = T^\ell(B)$ and x such that $T^k(0)\mathbf{x} = T^k(A)\mathbf{x}$, it follows that $T^k(0)\mathbf{x} = T^{k+\ell}(B)\mathbf{x}$ for arbitrary $B \in C$, so that first set equality is valid. The second set equality follows from (2.11) by considering $T^k(0)$ and $T^k(A)$ for $A \in \dot{C}$.

REMARK 2.3. Notice that the *invertibility* of ϕ is unnecessary for the validity of Theorem 2.1.

The general technique of enlarging the sensor by block processing is valid in the time dependent case and leads to a structure analogous to (2.8), (2.9), and (2.10). Because of this the *a priori* bounds of Section 1 as well as the asymptotic results apply in general with the only restriction being that there exists a k such that rank $[H'_{n_0}, \cdots \phi'(n_0, n_0 + k)H'_{n_0+k}] = d$ for all n_0.

3. Conclusions

We have shown that the theory of the Riccati equation which arises in the discrete time linear filtering problem can be easily obtained by considering the temporal evolution of k fold iterates. A generalized theory of continued fractions in semidefinite matrices has been given, which provides best possible upper and lower *a priori* bounds for the Riccati equation solutions. It would seem that the upper and lower approximates would provide interesting ways to compute suboptimal filters in environments where the prior variance is unknown.

In a future paper, we will study the analogous continuous time situation.

REFERENCES

[1] W. N. Anderson and R. J. Duffin, "Series and parallel addition of matrices," *J. Math. Anal. Appl.*, Vol. 26 (1969), pp. 576–593.

[2] R. S. Bucy, "Global theory of the Riccati equations," *J. Comput. Systems Sci.*, Vol. 1 (1967), pp. 349–361.

[3] R. S. Bucy and P. D. Joseph, *Filtering for Stochastic Processes with Application to Guidance*, New York, Interscience, 1968.

[4] R. S. Bucy, D. Rappaport, and L. M. Silverman, "Correlated noise filtering and invariant directions for the Riccati equation," *IEEE Trans. Automatic Control*, Vol. AC–15 (1970), pp. 535–540.

[5] R. S. Bucy, "Linear and non-linear filtering," *Proc. IEEE*, Vol. 58 (1970), pp. 854–864.

[6] R. E. Kalman, "A new approach to linear filtering and prediction problems," *ASME J. Basic Eng.*, Vol. 82 (1960), pp. 35–45.

[7] R. E. Kalman and R. S. Bucy, "New results in linear filtering and prediction theory," *ASME J. Basic Eng.*, Vol. 83 (1961), pp. 95–108.

[8] R. E. Kalman and T. S. Englar, "A user's manual for A.S.P.-*C*," NASA CR–475, Ames Research Center, Moffet Field, California, 1965.

[9] O. Perron, *Die Lehre von den Kettenbruecken*, Stuttgart, Teubner, 1954.

[10] H. S. Wall, *Analytic Theory of Continued Fractions*, New York, Chelsea, 1967.

LOSE A DOLLAR OR DOUBLE YOUR FORTUNE

THOMAS S. FERGUSON
UNIVERSITY OF CALIFORNIA, LOS ANGELES

1. Summary and introduction

A gambler with initial fortune x (a positive integer number of dollars) plays a sequence of identical games in which he loses one dollar with probability π, $0 < \pi < 1$, and doubles his fortune with probability $1 - \pi$. Playing continues until, if ever, the gambler is ruined (his fortune drops to zero). Let q_x denote the probability that the gambler starting with initial fortune x will eventually be ruined. Then, q_x satisfies the difference equation

$$q_x = \pi q_{x-1} + (1 - \pi)q_{2x}, \qquad x = 1, 2, \cdots, \tag{1.1}$$

with boundary conditions $q_0 = 1$ and $\lim_{x\to\infty} q_x = 0$. In Section 2, a solution to this difference equation subject to these boundary conditions is explicitly exhibited, and it is shown that there is only one such solution. In Section 3, the equation is extended to allow arbitrary noninteger values for the fortune, and again a solution is found. Section 4 contains several other extensions.

Equation (1.1) arises in connection with the following more general gambling problem described in [2]. A gambler is confronted with a sequence of games affording him even money bets at varying probabilities of success, $p_1, p_2, \cdots$, chosen independently from a distribution function F known to the gambler. The probability of winning the jth game p_j is told to the gambler after he plays game $j - 1$ and before he plays game j. The gambler must decide how much to bet in the jth game as a function of the past history, his present fortune, and the win probability p_j. He may bet any amount not exceeding his present fortune; however, he must bet at least one dollar on each game (called Model 2 in [2]). The problem of the gambler is to choose a betting system (a sequence of functions $b_1, b_2, \cdots$, where b_j is the amount bet in game j) that minimizes the probability of eventual ruin. Theorems relating to the dynamic programming solution of this problem may be found in Truelove [4].

Let q_x denote the infimum, over all betting systems, of the probability of ruin given the initial fortune x. It was shown in [2] that q_x tends to zero exponentially as x tends to infinity, and it was conjectured that, for some $0 < r < 1$ and $c > 0$,

$$q_x r^x \to c \quad \text{as} \quad x \to \infty. \tag{1.2}$$

The preparation of this paper was supported in part by NSF Grant No. GP-8049.

Under this conjecture, the asymptotic (for large x) form of the optimal betting system was found to be

$$b(p) = \max\left\{1, \frac{\log(1-p) - \log p}{\log r^2}\right\}, \tag{1.3}$$

where r is the unique root between zero and one of the equation

$$\int_0^1 \left[pr^{b(p)} + (1-p)r^{-b(p)}\right] dF(p) = 1. \tag{1.4}$$

In [1], Breiman proved conjecture (1.2) to be valid under the condition that F give no mass to some neighborhood of 1, that is, that there exists an $\varepsilon > 0$ such that $F(1-\varepsilon) = 1$.

There remains the question of the validity of the conjecture (1.2), or more generally the validity of the asymptotic form of the optimal betting function (1.3), when F gives mass to all neighborhoods of 1.

In this paper, the simplest such F is considered, namely, the F that gives mass π to 0 and mass $1-\pi$ to 1, where $0 < \pi < 1$. Under this F, the form of the optimal betting function for all x is obvious: bet the minimum value 1 when the probability of win p is 0 and bet the maximum value x when p is 1. If p is 0, you lose a dollar; if p is 1, you double your fortune.

This leads to the recurrence equation (1.1) for q_x, with the boundary condition $q_0 = 1$. As noted before, it is known that q_x tends to zero exponentially. We add as a boundary condition the weaker assertion $\lim_{x\to 0} q_x = 0$, that together with $q_0 = 1$ is sufficient to insure the unicity of the solution to (1.1), for integer values of x. However, it is of interest, in connection with the gambling system problem from which equation (1.1) originally arose, to investigate (1.1) for noninteger values of x. This investigation leads to a useful solution for rational x. If the initial boundary conditions on q_x for $0 \leqq x < 1$ are continuous in x, then approximations may be found for q_x for x irrational. Furthermore, it is observed that the conjecture (1.2) is not valid for the boundary condition $q_x = 1$ for $0 \leqq x < 1$.

2. A solution of difference equation (1.1)

We present in Theorem 1, the unique solution to the difference equation (1.1) subject to the boundary conditions $q_0 = 1$ and $\lim_{x\to\infty} q_x = 0$, as a series in $\pi^{2^j x}\, j = 0, 1, 2, \cdots$, with coefficients depending on π.

The systems of equations (1.1), even with q_0 fixed equal to one, is vastly underdetermined. In fact, the values of q_x for x odd, $x = 1, 3, 5, \cdots$, may be chosen arbitrarily and the system (1.1) will serve only to determine the values of q_x for x even. However, the addition of the single boundary condition $\lim_{x\to\infty} q_x = 0$ determines all the q_x uniquely as follows.

THEOREM 1. *Let $0 < \pi < 1$. There exists a unique solution to the difference equation $q_x = \pi q_{x-1} + (1-\pi)q_{2x}$, $x = 1, 2, 3, \cdots$, subject to the boundary conditions*

$$q_0 = 1 \tag{2.1}$$

and

$$\lim_{x\to\infty} q_x = 0. \tag{2.2}$$

That solution is

$$q_x = \alpha \sum_{j=0}^{\infty} c_j \pi^{2^j x}, \qquad x = 0, 1, 2, 3, \cdots, \tag{2.3}$$

where the c_j are defined inductively by

$$c_0 = 1, \qquad c_j = \frac{1-\pi}{1-\pi^{1-2^j}} c_{j-1}, \tag{2.4}$$

and where

$$\alpha = \left(\sum_{j=0}^{\infty} c_j\right)^{-1}. \tag{2.5}$$

PROOF. The following proof of unicity is taken from Theorem 1 of MacQueen and Redheffer [3].

If q_x and q'_x both satisfy (1.1), (2.1), and (2.2), then the difference $d_x = q_x - q'_x$ satisfies (1.1), $d_0 = 0$, and (2.2). Suppose that $q_x \neq q'_x$ for some x, and assume, without loss of generality, that $q_x > q'_x$ for some x. Let $x_0 \geqq 1$ be the largest value of x at which d_x achieves its maximum value, assumed to be positive. Then $q_{x_0-1} \leq q_{x_0}$, and $q_{2x_0} < q_{x_0}$. This contradicts (1.1), thus proving unicity.

To see that (2.3) is a solution, note first from (2.4) that $|c_j| \leqq |c_{j-1}|$ so that (2.3) converges absolutely for $0 < \pi < 1$. Thus, (2.2) is obviously satisfied and (2.1) is satisfied from (2.5). Using (2.3), we compute the right side of (1.1):

$$\begin{aligned} &\pi q_{x-1} + (1-\pi)q_{2x} \\ &= \pi\alpha \sum_{j=0}^{\infty} c_j \pi^{2^j(x-1)} + (1-\pi)\alpha \sum_{j=1}^{\infty} c_j \pi^{2^j 2x} \\ &= \alpha \sum_{j=0}^{\infty} c_j(\pi^{1-2^j})\pi^{2^j x} + \alpha \sum_{j=1}^{\infty} c_{j-1}(1-\pi)\pi^{2^j x} \\ &= \alpha c_0 \pi^x + \alpha \sum_{j=1}^{\infty} \left(c_j \pi^{1-2^j} + c_{j-1}(1-\pi)\right)\pi^{2^j x}. \end{aligned} \tag{2.6}$$

From the recurrence relation (2.4), this is equal to q_x, thus showing that (1.1) is satisfied and completing the proof.

The solution (2.3) is quite suitable for computational purposes, provided π is not too close to one. Table I, for which I am greatly indebted to David Cantor, gives an indication of the behavior of q_x for moderate values of π. As π tends to one, $q_{z/(1-\pi)}$ tends to $H(z) = \alpha' \Sigma_0^\infty c_j' \exp\{-2^j z\}$, where $c_0' = 1$, $c_j' = \Pi_1^j (1 - 2^i)^{-1}$ and $\alpha' = (\Sigma_0^\infty c_j')^{-1}$. In Section 4.5, it is seen that $1 - H(z)$ is the cumulative distribution function of $Z = \Sigma_0^\infty 2^{-j} Y_j$, where $Y_0, Y_1, Y_2, \cdots$ are independent identically distributed exponential variables.

A simple asymptotic approximation to q_x as π tends to zero is $q_x \sim \pi^x(1 + \pi)$, $x = 1, 2, \cdots$.

TABLE I

TABLE OF q_x

	.5	.6	.7	.8	.9
1	.7044	.8436	.9421	.9895	.9997
2	.4087	.6089	.8071	.9477	.9974
3	.2188	.4043	.6446	.8740	.9901
4	.1131	.2569	.4921	.7802	.9760
5	.0574	.1593	.3649	.6788	.9541
10	.0018	.0129	.0681	.2734	.7577
15	.0001	.0010	.0116	.0952	.5248
20		.0001	.0020	.0318	.3380
25			.0003	.0105	.2096
30				.0035	.1272
35				.0012	.0763
40				.0004	.0455
45				.0001	.0270
50					.0160

3. Fractional fortunes

The problem of the preceding section may be extended to the case in which the initial fortune x may be any nonnegative number. The difference equation to be considered is thus

$$q_x = \pi q_{x-1} + (1 - \pi) q_{2x}, \qquad x \geqq 1. \tag{3.1}$$

If the gambler is declared ruined when his fortune falls below one, then the boundary condition (2.1) is replaced by $q_x = 1$ if $0 \leqq x < 1$. We consider in this section an arbitrary bounded q_x for $0 \leqq x < 1$.

We first note that if q_x satisfies (3.1) and (2.2), then q_x is bounded for $x \geqq 0$. Since $\lim_{x\to\infty} q_x = 0$, there exists an integer N such that $|q_x| < 1$ if $x \geqq N$. From (3.1),

$$|q_{x-1}| = \pi^{-1}|q_x - (1 - \pi) q_{2x}| \leqq \pi^{-1}|q_x| + \pi^{-1}(1 - \pi)|q_{2x}|, \tag{3.2}$$

so that $|q_x| < \pi^{-1}(2 - \pi)$ for $x \geqq N - 1$. By induction therefore, $|q_x| < (\pi^{-1}(2 - \pi))^N$ for $x \geqq 0$.

We next note that probabilistic considerations alone allow us to infer the existence of a solution to (3.1) and (2.2) subject to an arbitrary initial condition on q_x for $0 \leqq x < 1$, provided $0 \leqq q(x) \leqq 1$. Then, since the product of a solution to (3.1) and (2.2) by a scalar, and the sum of two solutions to (3.1) and (2.2) are also solutions to (3.1) and (2.2), we may achieve any bounded initial condition on q_x for $0 \leqq x < 1$.

Finally, we note that an extension of the argument of the previous section allows us to conclude that there exists a unique solution to (3.1) and (2.2) subject to an arbitrary bounded initial condition on q_x for $0 \leqq x < 1$. For if $\hat{q}_x$ represents the difference of any two solutions, then $\hat{q}_x$ satisfies (3.1) and (2.2) and the boundary condition $\hat{q}_x = 0$ for $0 \leqq x < 1$. We must show that $\hat{q}_x$ vanishes for all $x \geqq 1$. Since $\hat{q}_x$ must be bounded, let $\beta = \sup_{x \geqq 0} \hat{q}_x$, suppose, without loss of generality, that $\beta > 0$, and let x_0 be the largest number for which there exists a sequence $x_n \to x_0$ and $\hat{q}_{x_n} \to \beta$. Then equation (3.1) shows that $\hat{q}_{2x_n} \to \beta$, contradicting the assertion that x_0 was the largest number with this property, and completing the proof of unicity.

The following theorem presents a class of solutions to (3.1) and (2.2), but it is not known whether an arbitrary bounded initial condition on q_x, $0 \leqq x < 1$, can be achieved as one of them.

THEOREM 2. *Let $\alpha(x)$ be a bounded periodic function of period one, and let, for $0 < \pi < 1$,*

$$q_x = \sum_{j=0}^{\infty} c_j \alpha(2^j x) \pi^{2^j x}, \qquad x \geqq 0, \tag{3.3}$$

where $c_0 = 1$ and $c_j = (1 - \pi)(1 - \pi^{1-2^j})^{-1} c_{j-1}$ for $j = 1, 2, \cdots$. Then q_x satisfies (3.1) and (2.2).

The proof, being similar to the corresponding part of Theorem 1, is omitted. The hypothesis that α be bounded is equivalent to (2.2).

This theorem may be used to find q_x for all x with dyadic fractional part for an arbitrary boundary condition on q_x for $0 \leqq x < 1$. Such a solution depends only on q_x, $0 \leqq x < 1$, through the dyadic rational values of x. To accomplish this, it is sufficient to find $\alpha(\frac{1}{2})$, $\alpha(\frac{1}{4})$, $\alpha(\frac{3}{4})$, $\cdots$. As in Theorem 1, equation (3.3) with $x = 0$ enables us to find $\alpha(0)$. For $x = \frac{1}{2}$, equation (3.3) becomes

$$q_{1/2} = c_0 \alpha(\tfrac{1}{2}) \pi^{1/2} + \sum_{j=1}^{\infty} c_j \alpha(0) \pi^{2^{j-1}} \tag{3.4}$$

from which $\alpha(\frac{1}{2})$ may be found. For $x = \frac{1}{4}$, equation (3.3) becomes

$$q_{1/4} = c_0 \alpha(\tfrac{1}{4}) \pi^{1/4} + c_1 \alpha(\tfrac{1}{2}) \pi^{1/2} + \sum_{j=2}^{\infty} c_j \alpha(0) \pi^{2^{j-2}}, \tag{3.5}$$

from which $\alpha(\frac{1}{4})$ may be found. It is clear that $\alpha(x)$ may be found for all dyadic rational x by this method.

In addition, one can find $\alpha(x)$, and thus $q(x)$ for rational x directly. We illustrate for $x = \frac{1}{3}$. Evaluating (3.3) at $x = \frac{1}{3}$ and $x = \frac{2}{3}$ yields

$$q_{1/3} = \alpha(\tfrac{1}{3}) \sum_{j \text{ even}} c_j \pi^{2^j/3} + \alpha(\tfrac{2}{3}) \sum_{j \text{ odd}} c_j \pi^{2^j/3} \tag{3.6}$$
$$q_{2/3} = \alpha(\tfrac{1}{3}) \sum_{j \text{ odd}} c_j \pi^{2^{j+1}/3} + \alpha(\tfrac{2}{3}) \sum_{j \text{ even}} c_j \pi^{2^{j+1}/3}$$

from which one can solve for $\alpha(\frac{1}{3})$ and $\alpha(\frac{2}{3})$ provided the determinant of this system of equations does not vanish. In this case, it is clear that the determinant is not zero, since in each row the summation over j even is larger than the absolute value of the summation over j odd.

The general problem of finding q_x for x irrational seems more difficult. If the boundary condition on q_x for $0 \leqq x < 1$ specifies a bounded continuous function of x, in particular, if it is assumed that $q_x = 1$ for $0 \leqq x < 1$, then q_x for $x \geqq 1$ has discontinuities only at the dyadic rationals, and hence can be approximated as closely as desired by q_x for x dyadic rational.

To see that if the boundary condition specifies a bounded continuous function of x, then q_x is continuous at all x for which x is not dyadic rational, we may argue as follows. Let q_x be a solution to (3.1) and (2.2), suppose that q_x is continuous for $0 \leqq x < 1$, and let $q_x^+ = \limsup_{t \to x} q_t$ and $q_x^- = \liminf_{t \to x} q_t$ for $x > 0$. Then, for $x > 0$,

$$q_x^+ \leqq \pi q_{x-1}^+ + (1 - \pi) q_{2x}^+, \tag{3.7}$$
$$q_x^- \geqq \pi q_{x-1}^- + (1 - \pi) q_{2x}^-.$$

Hence, letting $d_x = q_x^+ - q_x^-$, we see that

$$0 \leqq d_x \leqq \pi d_{x-1} + (1 - \pi)\, d_{2x}, \qquad x > 0, \tag{3.8}$$

$$d_x = 0, \qquad 0 < x < 1, \tag{3.9}$$

$$\lim_{x \to \infty} d_x = 0. \tag{3.10}$$

If we consider this equation on the set of all positive x not dyadic rational, then the argument used to prove the unicity of (3.1) may be used here again to show $d_x = 0$ for all x not dyadic rational even though there is an inequality in (3.8) (as in MacQueen and Redheffer [3]) instead of equality. Hence, q_x is continuous at all x not dyadic rational.

Finally, we note that the conjecture (1.2) is false in the present case. In fact

$$\frac{q_x}{\pi^x} = \alpha(x) + \sum_{j=1}^{\infty} \alpha(2^j x) c_j \pi^{2^j x - x} \sim \alpha(x) \tag{3.11}$$

which is not constant in x, but periodic of period one. However, the trivially optimal betting system is still of the (asymptotic) form (1.3).

4. Other extensions

The method of expanding q_x as a series in π^{2jx} appears to be quite useful. We mention several other related difference equations that can be solved by the use of such an expansion.

4.1. Given $\pi_{-1} > 0$, $\pi_i \geqq 0$ for $i = 0, \cdots, n$, $\pi_{n+1} > 0$, and $\Sigma_{-1}^{n+1} \pi_i = 1$, find q_x to satisfy

$$q_x = \sum_{i=-1}^{n} \pi_i q_{x+i} + \pi_{n+1} q_{2x}, \qquad x = 1, 2, \cdots, \tag{4.1}$$

subject to

$$q_0 = 1, \qquad \lim_{x \to \infty} q_x = 0. \tag{4.2}$$

The unicity of the solution follows as before. We attempt a series for q_x of the form

$$q_x = \sum_{j=0}^{\infty} c_j r^{2jx}, \qquad x = 0, 1, 2, \cdots. \tag{4.3}$$

Formal substitution of this series into (4.1) yields

$$\sum_{j=0}^{\infty} c_j r^{2jx} = \sum_{j=0}^{\infty} c_j \left(\sum_{i=-1}^{n} \pi_i r^{2ji} \right) r^{2jx} + \pi_{n+1} \sum_{j=1}^{\infty} c_{j-1} r^{2jx}. \tag{4.4}$$

We equate the coefficients of r^{2jx}. For $j = 0$, we obtain

$$c_0 = c_0 \sum_{i=-1}^{n} \pi_i r^i. \tag{4.5}$$

This equation in r has a unique root between zero and one, since the right side is convex in r, tends to $+\infty$ as r tends to zero, and to $c_0 \Sigma_{-1}^{n} \pi_i < c_0$ as r tends to one. For arbitrary $j \geq 1$, we obtain

$$c_j = c_j \sum_{i=-1}^{n} \pi_i r^{2ji} + c_{j-1} \pi_{n+1}. \tag{4.6}$$

Thus, the solution of (4.1) subject to (4.2) is (4.3), where r satisfies (4.5) and the c_j are defined recursively by (4.6), and where c_0 is chosen to satisfy $q_0 = 1$.

4.2. The model called Model I in [2] leads in the present context to the equation

$$q_x = \pi q_{x-1} + (1 - \pi) q_{2x-1}, \qquad x = 1, 2, \cdots, \tag{4.7}$$

subject to $q_0 = 1$ and $\lim_{x \to \infty} q_x = 0$. Since no new ideas are involved, we omit the details.

4.3. Analogous methods can be used to treat difference equations involving more terms of the form q_{x-n} and correspondingly more boundary conditions.

As an example, consider the equation ($\pi_1 > 0, \pi_2 > 0, \pi_1 + \pi_2 < 1$)

$$q_x = \pi_1 q_{x-1} + \pi_2 q_{x-2} + (1 - \pi_1 - \pi_2) q_{2x}, \qquad x = 1, 2, \cdots, \tag{4.8}$$

subject to specified values of q_0 *and* q_{-1} and subject to $\lim_{x \to \infty} q_x = 0$. The unicity of the solution follows as before. Again we try a solution of the form (4.2). Formal substitution into (4.8) yields

$$\sum_{j=0}^{\infty} c_j r^{2^j x} = \sum_{j=0}^{\infty} c_j (\pi_1 r^{-2^j} + \pi_2 r^{-2^{j+1}}) r^{2^j x} + (1 - \pi_1 - \pi_2) \sum_{j=1}^{\infty} c_{j-1} r^{2^j x}. \tag{4.9}$$

Again, we equate coefficients of $r^{2^j x}$. For $j = 0$, we obtain $1 = \pi_1 r^{-1} + \pi_2 r^{-2}$. There are two roots to this equation,

$$r_1 = \tfrac{1}{2}[\pi_1 + (\pi_1^2 + 4\pi_2)^{1/2}], \qquad r_2 = \tfrac{1}{2}[\pi_1 - (\pi_1^2 + 4\pi_2)^{1/2}]. \tag{4.10}$$

For $j > 0$, we obtain for each r_i a recursion formula for the c_j, denoted by $c_{j,i}$:

$$c_{j,i} = c_{j,i}(\pi_1 r_i^{-2^j} + \pi_2 r_i^{-2^{j+1}}) + c_{j-1,i}(1 - \pi_1 - \pi_2) \tag{4.11}$$

for $j = 1, 2, \cdots$ and $i = 1, 2$. Hence, the general solution of (4.8) subject to $\lim_{x \to \infty} q_x = 0$ is

$$q_x = \sum_{j=0}^{\infty} c_{j,1} r_1^{2^j x} + \sum_{j=0}^{\infty} c_{j,2} r_2^{2^j x}. \tag{4.12}$$

The values of $c_{0,1}$ and $c_{0,2}$ may be chosen to obtain specified values of q_0 and q_{-1}.

4.4. Finally, we note that in expanding q_x in terms of $r^{2^j x}$, the "2" appears because we are concerned with the doubling of the fortune. To solve an equation of the type $q_x = \pi q_{x-1} + (1 - \pi) q_{3x}$ an expansion as a series in $\pi^{3^j x}$ works. The details involve no new ideas. However, it might be worthwhile to give an explicit solution to the equation

$$q_x = \pi_1 q_{x-1} + \pi_2 q_{2x} + \pi_3 q_{3x}, \qquad x = 1, 2, \cdots, \tag{4.13}$$

where $\pi_1 > 0$, $\pi_2 > 0$, $\pi_3 > 0$, and $\pi_1 + \pi_2 + \pi_3 = 1$ subject to the condition $\lim_{x \to \infty} q_x = 0$. This equation requires an expansion of the form

$$q_x = \sum_{i=0}^{\infty} \sum_{j=0}^{\infty} c_{i,j} \pi_1^{2^i 3^j x}, \qquad x = 0, 1, 2, \cdots. \tag{4.14}$$

By substituting this expansion into (4.13) and equating coefficients of $\pi_1^{2^i 3^j}$, we find the following recurrence relations for the $c_{i,j}$,

$$c_{i,j} = \frac{c_{i-1,j}\pi_2 + c_{i,j-1}\pi_3}{1 - \pi_1^{1-2^i3^j}}, \qquad i = 1, 2, \cdots, \quad j = 1, 2, \cdots,$$

(4.15)
$$c_{i,0} = \frac{c_{i-1,0}\pi_2}{1 - \pi_1^{1-2^i}}, \qquad i = 1, 2, \cdots,$$

$$c_{0,j} = \frac{c_{0,j-1}\pi_3}{1 - \pi_1^{1-3^j}}, \qquad j = 1, 2, \cdots.$$

The value of $c_{0,0}$ may be chosen to obtain a specified value of q_0.

4.5. Let $Y_0, Y_1, Y_2, \cdots$ be independent identically distributed random variables with exponential densities

(4.16)
$$f(y) = \begin{cases} e^{-y}, & y \geqq 0, \\ 0, & y < 0. \end{cases}$$

Let $0 < \beta < 1$, and let $Z = \Sigma_0^\infty \beta^j Y_j$. We find the distribution function of Z, $G(z) = P(Z \leqq z)$, as follows. Note that Z may be written as $Z = Y_0 + \beta Z_1$, where $Z_1 = Y_1 + \beta Y_2 + \cdots$ has the same distribution as Z, and is independent of Y_0. Therefore, for $z > 0$,

(4.17)
$$G(z) = \int_0^z e^{-y} G\left(\frac{z - y}{\beta}\right) dy = \beta \int_0^{z/\beta} e^{-z+\beta x} G(x)\, dx.$$

Multiplying both sides by e^z and differentiating with respect to z, yields

(4.18)
$$G(z) + G'(z) = G\left(\frac{z}{\beta}\right).$$

We solve this equation subject to the boundary conditions

(4.19)
$$G(0) = 0, \qquad \lim_{z \to \infty} G(z) = 1,$$

and subject to the existence of one continuous derivative of G. (In the present case G, being the convolution of an infinite number of absolutely continuous random variables, must be C^∞.) The unicity of the solution to (4.18) subject to (4.19) and the continuity of G' follows as in the previous problems. It is then a straightforward matter to check that the expansion

(4.20)
$$G(z) = 1 - \alpha \sum_{j=0}^{\infty} c_j \exp\{-\beta^{-j} z\},$$

where

$$c_0 = 1, \qquad c_j = \prod_{i=1}^{j} (1 - \beta^{-i})^{-1}, \qquad \alpha = \left(\sum_{j=0}^{\infty} c_j \right)^{-1}, \tag{4.21}$$

satisfies (4.18) and (4.19), and thus represents the distribution function of Z.

◇ ◇ ◇ ◇ ◇

The author gratefully acknowledges useful discussions on this problem with David Cantor and James B. MacQueen.

REFERENCES

[1] L. BREIMAN, "On random walks with an absorbing barrier and gambling systems," Western Management Science Institute, University of California, Los Angeles, Working paper No. 71, 1965.

[2] T. S. FERGUSON, "Betting the systems which minimize the probability of ruin," *J. Soc. Indust. Appl. Math.*, Vol. 13 (1965), pp. 795–818.

[3] J. MACQUEEN and R. M. REDHEFFER, "Some applications of monotone operators in Markov processes," *Ann. Math. Statist.*, Vol. 36 (1965), pp. 1421–1425.

[4] A. J. TRUELOVE, "Betting systems in favorable games," *Ann. Math. Statist.*, Vol. 41 (1970), pp. 551–566.

NECESSARY CONDITIONS FOR DISCRETE PARAMETER STOCHASTIC OPTIMIZATION PROBLEMS

HAROLD J. KUSHNER
BROWN UNIVERSITY

1. Introduction

Consider the following formal optimization problem. Let $\{\xi_i\}$ denote a sequence of random vectors, and define the sequence (1.1) of n dimensional vectors $\{X_i, i = 0, \cdots, k\}$, $X_i = \{X_i^1, \cdots, X_i^n\}$, where k is a fixed integer and u_i is a control, which is an element of an abstract set $\tilde{U}_i$:

$$X_{i+1} = X_i + f_i(X_i, u_i, \xi_i). \tag{1.1}$$

The object is to find the $\{u_i\}$ which minimizes

$$\begin{aligned} EX_k^0 &\equiv \sum_{i=0}^{k-1} f_i^0(X_i, u_i, \xi_i), \\ X_{i+1}^0 &= X_i^0 + f_i^0(X_i, u_i, \xi_i), \qquad X_i^0 \text{ fixed}, \end{aligned} \tag{1.2}$$

subject to certain constraints. Sometimes it is convenient to augment the vector X_i by adding X_i^0, the "cost" component. Then, we write ${}^+\underline{X}_i = (X_i^0, X_i)$, $\underline{f}_i = (f_i, f_i^0)$ and

$$\underline{X}_{i+1} = \underline{X}_i + \underline{f}_i(X_i, u_i, \xi_i). \tag{1.1$'$}$$

The constraints are

$$r_0(X_0) \equiv E\tilde{r}_0(X_0) = 0, \qquad q_0(\underline{X}_0) \equiv E\tilde{q}_0(\underline{X}_0, E\underline{X}_0) \leqq 0, \tag{1.3}$$

$$\begin{aligned} q_i(X_k) &\equiv E\tilde{q}_i(X_i, EX_i) \leqq 0, \qquad i = 1, \cdots, k, \\ r_k(X_k) &\equiv E\tilde{r}_k(X_k, EX_k) = 0, \end{aligned} \tag{1.4}$$

where $\tilde{r}_0$, $\tilde{q}_0$, $\tilde{r}_k$, and $\tilde{q}_i$ are vector valued functions. The q_0 is allowed to depend on X_0^0 in order to fix or limit X_0^0 in some way. That is, some component of $\tilde{q}_0(\underline{X}_0)$ may be $\tilde{q}_0^0(\underline{X}_0) = -X_0^0 \leqq 0$.

This research was supported in part by the National Science Foundation under Grant No. GK 2788, in part by the National Aeronautics and Space Administration under Grant No. NGL 40-002-015, and in part by the Air Force Office of Scientific Research under Grant No. AF-AFOSR 67-0693A.

The constraints $E\tilde{q}_i(X_i, EX_i) \leqq 0$ of (1.4) can be used to model or approximate a variety of constraints. For example, we can approximate the constraint $X_n \in A$ with probability 1 by letting q_n be the expectation of a suitably smooth approximation to the indicator of A. The constraint $P\{X_n \notin A$, some $n = 1, \cdots, k\} \leqq \varepsilon$ can be modelled letting $\tilde{g}(\cdot)$ denote a suitably smooth approximation to the indicator of A and admitting the constraint $g(X_1, \cdots, X_k) = E \max_{k \geqq n \geqq 1} \tilde{g}(X_n) \geqq 1 - \varepsilon$. Note that g may have a "convex differential," although not necessarily a linear differential. See the comment after Theorem 3.1.

Necessary conditions for optimality in the form of Kuhn-Tucker conditions or Lagrange multiplier rules are well developed for very general deterministic discrete and continuous parameter problems [4], [11], and much of the recent work depends heavily on abstractions of the well-known geometric methods of nonlinear programming. In this paper, we apply some of the recent developments in abstract programming to obtain necessary conditions for (local) optimality for several discrete parameter optimization problems. The results are only typical of the possibilities and do not exhaust them. Hopefully, the results will suggest useful computational procedures, although our investigations along these lines are only beginning.

In [8] and [9], the author derived some necessary conditions for optimality for a class of continuous parameter stochastic problems, and in [10] for a discrete problem. The results in [8] and [9] are true "maximum principles" or "minimum principles" in the sense used in control theory, while the result in [5] is a necessary condition for a stationary point. Subsequent work was reported in [1], [2], [3], [5], [12], [13]. The development in [3], for an essentially linear problem (f_i linear) with a convex cost, and where the u_i are real numbers, seems to be the only work in which programming ideas are explicitly used. However, the programming approach gives better results with reasonable effort. Indeed, by properly identifying quantities in the abstract work [11] with quantities in the stochastic problems, we obtain and extend most previous discrete parameter results. Continuous parameter results will be reported elsewhere.

Section 2 cites the basic results from [11], which will be heavily used in the sequel. Sections 3 to 5 deal with the discrete parameter problem. In Section 2, the u_i are measurable with respect to given σ-algebras $\tilde{\mathscr{B}}_i$; in Section 3, the u_i are allowed to depend explicitly on the states, X_i, and so forth; and in Section 5 a maximum principle is derived, analogous to the deterministic discrete parameter maximum principle [4].

2. Mathematical background

This section describes a somewhat weakened version of a result of Neustadt [11], on an abstract variational problem which underlies the development of the sequel. Let $\mathscr{T}$ be a Banach space which contains the sets B and Q. The structures introduced next are abstract counterparts of these used in nonlinear programming in Euclidean space. The terminology is slightly changed from that of [11].

DEFINITION 2.1. *Let Z be a convex cone with vertex $\{0\}$ in $\mathscr{T}$. If ρ is an arbitrary ray of Z, let there be a cone Z_ρ with a nonempty interior and vertex $\{0\}$ and ρ internal to Z_ρ, and also a neighborhood N_ρ of $\{0\}$, such that $Z_\rho \cap N_\rho \subset B$. Then Z is an internal cone to B at $\{0\}$.*

DEFINITION 2.2. *Let P^ν denote the set $\{\beta: \beta_i \geqq 0, \Sigma_1^\nu \beta_i \leqq 1\}$. Let K be a convex set in $\mathscr{T}$ which contains $\{0\}$ and some point other than $\{0\}$. Let $w_1, \cdots, w_\nu$ be in K and let N be an arbitrary neighborhood of $\{0\}$. Let there exist an $\varepsilon_0 > 0$ (depending on ν, $w_1, \cdots, w_\nu$, and N) so that, for each ε in $(0, \varepsilon_0]$, there is a continuous map $\zeta_\varepsilon(\beta)$ from P^ν to $\mathscr{T}$ with the property*

$$\zeta_\varepsilon(\beta) \subset \left\{\varepsilon\left(\sum_{i=1}^{\nu} \beta_i w_i + N\right)\right\} \cap Q. \tag{2.1}$$

Then K is a first order convex approximation to Q.

2.1. *A basic optimization problem.* Let $\mathscr{T}$ contain the set Q'. Find the element $\hat{w}$ in Q' which minimizes $\varphi_0(w)$ subject to the constraints $\varphi_i(w) = 0$, $i = 1, \cdots, m$, $\varphi_{-i}(w) \leqq 0$, $i = 1, \cdots, t$. We say that $\hat{w}$ is a *local solution* to the optimization problem (or, more loosely, the *optimal solution*) if, for some neighborhood N of $\{0\}$, $\varphi_0(w) \geqq \varphi_0(\hat{w})$ for all w in $\hat{w} + N$ which satisfy the constraints. Let $\hat{w}$ denote the optimal solution. The constraints φ_{-i} for which $\hat{\varphi}_{-i} \equiv \varphi_{-i}(\hat{w}) = 0$ for $i = 1, \cdots, t$ are called the *active constraints*. Define the set of indices $J = \{i: \varphi_{-i}(\hat{w}) = 0, i > 0\} \cup \{0\}$.

2.2. *The basic necessary condition for optimality.* First we collect some assumptions.

ASSUMPTION 2.1. *The $\varphi_i(w)$, $i \geqq 1$, are continuous at $\hat{w}$ and have Fréchet derivatives ℓ_i at $\hat{w}$, and $\ell_1, \cdots, \ell_m$ are continuous and linearly independent.*

Thus, $[\varphi_i(\hat{w} + \varepsilon w) - \varphi_i(\hat{w})]/\varepsilon - \ell_i(w) \to 0$ uniformly for w in any bounded neighborhood of $\mathscr{T}$.

ASSUMPTION 2.2. *There is a neighborhood N of $\{0\}$ in $\mathscr{T}$ so that, for all* inactive *constraints, we still have $\varphi_{-i}(\hat{w} + w) < 0$ for $w \in N$.*

ASSUMPTION 2.3. *Let the active constraints and also φ_0 be continuous at $\hat{w}$. Let*

$$\frac{\varphi_{-i}(\hat{w} + \varepsilon w) - \varphi_{-i}(\hat{w})}{\varepsilon} \to c_i(w) \tag{2.2}$$

for all w in $\mathscr{T}$, and uniformly for w in any bounded neighborhood of $\{0\}$, where $c_i(w)$ is a continuous and convex functional. There is some w and $j \in J$ for which $c_j(w) > 0$. There is a w for which $c_j(w) < 0$ for all $j \in J$.

A case of particular importance is where the $c_i(w)$ are linear. Then we substitute the stronger Assumption 2.3′.

ASSUMPTION 2.3′. *Let the* active *constraints and also φ_0 be continuous at $\hat{w}$ and have Fréchet derivatives c_i at $\hat{w}$ (corresponding to φ_{-i}) which are continuous, and suppose that there is a $w \in \mathscr{T}$ for which $c_i(w) < 0$ for all $i \in J$.*

We now have a particular case of Neustadt [11], Theorem 4.2. The *local solution* here is called a *totally regular local solution* in [11].

THEOREM 2.1. *Let Assumptions 2.1 to 2.3 hold. Let $\hat{w}$ be a local solution to the optimization problem. Then there exist $\alpha_1, \cdots, \alpha_m, \alpha_0, \alpha_{-1}, \cdots, \alpha_{-t}$ not all zero with $\alpha_{-1} \leqq 0$ for $i \geqq 0$, so that*

$$\sum_{i=1}^{m} \alpha_i \ell_i(w) + \sum_{i \in J} \alpha_{-i} c_i(w) \leqq 0 \tag{2.3}$$

for all w in $\bar{K}$, where K is a first order convex approximation to $Q' - \hat{w} \equiv Q$, and $\bar{K}$ is the closure of K in $\mathcal{T}$.

OBSERVATION. Let $\varphi_i(\cdot) = 0, i > 0$. If there is a $w \in K$ for which $c_j(w) < 0$ for all active j, then $\alpha_0 < 0$, and we can set $\alpha_0 = -1$.

Define

$$\begin{aligned} B &= \{w: \varphi_{-i}(\hat{w} + w) < \varphi_{-i}(\hat{w}), i \in J\} \cup \{0\}, \\ \pi &= \{w: \ell_i(\hat{w} + w) = 0, i = 1, \cdots, m\}. \end{aligned} \tag{2.4}$$

Then Theorem 2.1 is essentially a consequence of the result (see [11]) that the intersection of π and any internal cone to B can be separated from $K \cap \pi$ by a continuous linear functional.

3. The stochastic variational formula when the controls are measurable over fixed σ-algebras

In the first part of this section, a stochastic optimization problem will be treated in a fairly general way. We introduce only those assumptions which are required to apply Theorem 2.1. Then, more specific conditions which guarantee some of these assumptions are introduced.

3.1. *A stochastic optimization problem. Definitions and assumptions.* Let $\xi_0, \cdots, \xi_i, \cdots$ be a sequence of random variables, where $\xi_0, \cdots, \xi_i$ are measurable on the σ-algebra $\mathcal{B}(\xi_0, \cdots, \xi_i)$, and define the random sequence $\{\underline{X}_i\}$ by (1.1′). The measures on the $\mathcal{B}(\xi_0, \cdots, \xi_i)$ do not depend on the selected control sequence; the ξ_i are of the nature of "exogenous inputs." We seek the $\underline{X}_0, \cdots, \underline{X}_k$, $u_0, \cdots, u_{k-1}$ which minimizes (1.2) subject to the constraints (1.3) and (1.4).

3.2. *The admissible controls.* For a vector Y with components Y^i write $|Y| = \Sigma_i |Y^i|$ and $\|Y\|_q = \Sigma_i E^{1/q}|Y^i|^q$. Denote $L_q(\mathcal{B})$ the Banach space of $\mathcal{B}$ measurable random functions Y with norm $\|Y\|_q$. Let $\underline{L}_q(\mathcal{B})$ be the Banach space of $n + 1$ dimensional vectors $\underline{X}_i = (X_i^0, X_i)$ with norm $\|\underline{X}_i\|_q \equiv E|X_i^0| + \|X_i\|_q$. For a random matrix $M = \{M_{ij}\}$, define $\|M\|_q = \Sigma_{i,j} \|M_{ij}\|_q$. Suppose that $\{\tilde{\mathcal{B}}_i\}$ and $\mathcal{B}_0$ are a sequence of given σ-algebras, and U_i a sequence of convex sets. The $\tilde{\mathcal{B}}_i$, $\mathcal{B}_0$ and the measures on them do not depend on the chosen controls. In this section the admissible control set, denoted by $\tilde{U}_i$, are the random variables in $L_{p'}(\tilde{\mathcal{B}}_i)$ which take values in U_i for given $p' \geqq 1$. Then the $\underline{X}_i$ are measurable over $\mathcal{B}_i$, where $\mathcal{B}_i \equiv \mathcal{B}_{i-1} \cup \tilde{\mathcal{B}}_{i-1} \cup \mathcal{B}(\xi_{i-1})$ and $\underline{X}_0$ is a random variable measurable over the given σ-algebra $\mathcal{B}_0$. The set of admissible controls covers at least the three cases:

(i) the u_i depend explicitly on some function of the $\xi_0, \cdots, \xi_{i-1}$;

(ii) the u_i depend explicitly on noise corrupted observations of the $\xi_0, \cdots, \xi_{i-1}$, where the corrupting noise does not depend on the selected control sequence;

(iii) a randomized version of (i) and (ii).

It is well known from linear programming on Markov chains that a randomized control may give a smaller cost in a constrained stochastic optimization problem, than a nonrandomized control. Our controls can be randomized by a suitable choice of $\tilde{\mathscr{B}}_i$. Let $\tilde{v}_0, v_0, \cdots, v_k$ denote a sequence of independent random variables, which are also independent of the $\{\xi_i\}$ sequence and each of which has, say, a uniform distribution on $[0, 1]$. (We suppose that the underlying probability space is big enough to carry these random variables.) Suppose that the data field $\hat{\mathscr{B}}_i \subset \mathscr{B}(\xi_0, \cdots, \xi_{i-1})$ is available to the controller at time i. (That is, $\hat{\mathscr{B}}_i$ measures the information upon which the control depends.) Randomization is achieved by letting $\tilde{\mathscr{B}}_i = \hat{\mathscr{B}}_i \cup \mathscr{B}(v_i)$ and $\mathscr{B}_0 = \mathscr{B}(\tilde{v}_0)$. To determine the actual control value $u_i(\omega)$, we need to draw a value of v_i at random.

3.3. *Assumptions and notation.* Notation will frequently be abused by using the same term for a function and for its values. Let $u_i \in \tilde{U}_i$. Let IC_i denote the *pointwise internal cone* to $\tilde{U}_i - \hat{u}_i$ at $\{0\}$; that is, IC_i is a convex cone of random variables in $L_{p'}(\tilde{\mathscr{B}}_i)$ with the property that, if $\delta u_i^s \in IC_i$, for $s = 1, \cdots, v$, then

$$\hat{u}_i + \varepsilon \sum_{s=1}^{v} \beta_s \delta u_i^s \in U_i \text{ for all } \omega \text{ for } \beta_s \geqq 0, \sum_s \beta_s \leqq 1 \text{ and } 0 \leqq \varepsilon \leqq \varepsilon_0, \tag{3.1}$$

where $\varepsilon_0 > 0$ may depend on the δu_i^s. Also, $\delta u_i^s \in L_{p'}(\tilde{\mathscr{B}}_i)$.

Let $\delta u^s = (\delta u_0^s, \cdots, \delta u_{k-1}^s) \in IC_u \equiv IC_0 \times \cdots \times IC_{k-1}$. Write

$$\begin{aligned} \delta u_i(\beta) &\equiv \sum_{s=1}^{v} \beta_s \delta u_i^s, \qquad \delta u(\beta) \equiv \sum_{s=1}^{v} \beta_s \delta u^s, \\ \delta \underline{X}_{i+1}^s &= \delta \underline{X}_i^s + \underline{f}_{i,x} \cdot \delta X_i^s + \underline{f}_{i,u} \cdot \delta u_i^s, \\ \delta \underline{X}_i(\beta) &= \sum_s \beta_s \delta \underline{X}_i^s. \end{aligned} \tag{3.2}$$

We have

$$\underline{X}_{i+1}(\beta) = \underline{X}_i(\beta) + \underline{f}_i(X_i(\beta), \hat{u}_i + \varepsilon \delta u_i(\beta), \xi_i). \tag{3.3}$$

Let $r_{0,x}$ denote the matrix $\partial r_0(x)/\partial x$ and $\hat{r}_{0,x}$ denote $r_{0,x}$ evaluated at $\hat{x}_0$. Let $\tilde{q}_{i,e}$ denote $\partial \tilde{q}_i(x, e)/\partial e$, $i > 0$, the derivatives with respect to the second vector argument of $\tilde{q}_i(\cdot, \cdot)$. We also use $\hat{q}_{i,x} = \tilde{q}_{i,x}(\hat{X}_i, E\hat{X}_i)$ and $q_{0,x} = \partial q_0(\underline{x})/\partial \underline{x}$. Also

$$\underline{f}_{i,x} = \frac{\partial f(x, u, \xi)}{\partial x}, \qquad \underline{f}_{i,x} = \frac{\partial \underline{f}(x, u, \xi)}{\partial x}, \tag{3.4}$$

and $\hat{q}_i = q_i(\hat{X}_i)$.

Fix $\delta u_i^s \in IC_i$ for all $i = 0, \cdots, k-1$ and $s = 1, \cdots, \ell$.

ASSUMPTION 3.1. *Assume* $u_i \in \tilde{U}_i$, *and for any sequence* $u_i \in \tilde{U}_i$, *and any* $\underline{X}_0$ *satisfying the constraints, assume that the* $\underline{X}_i$ *given by* (1.1′) *are in* $\underline{L}_p(\mathscr{B}_i)$ *for given* $p \geqq 1$ *and* $i = 0, \cdots, k$. *The* $\delta\underline{X}_i$ *given by* (3.14) *are in* $\underline{L}_p(\mathscr{B}_i)$ *for any* $\delta u_i^s \in IC_i$.

ASSUMPTION 3.2. *The* IC_i *contain at least one point other than the origin.*

ASSUMPTION 3.3. *For* $\varepsilon_0 \geqq \varepsilon > 0$, *where* $\varepsilon_0 > 0$ *depends on the* δu_i^s, *suppose that the* $X_i(\beta)$ *given by* (3.3) *are continuous in* β *in* $L_p(\mathscr{B}_i)$, *and that*

$$\|X_i(\beta) - \hat{X}_i - \varepsilon\delta X_i(\beta)\|_p = o(\varepsilon) \tag{3.5}$$

uniformly in $\beta = (\beta_1, \cdots, \beta_m)$, *for* $\beta_s \geqq 0$, $\Sigma_s \beta_s = 1$.

ASSUMPTION 3.4. *For a real number* K_1,

$$\begin{aligned} E|q_i(X_i)| &\leqq K_1(1 + E|X_i|^p), \qquad i = 1, \cdots, k, \\ E|r_i(X_i)| &\leqq K_1(1 + E|X_i|^p). \end{aligned} \tag{3.6}$$

ASSUMPTION 3.5. *Let* $\tilde{q}_{i,x}$, $\tilde{q}_{i,e}$, $\tilde{r}_{i,x}$, *and* $\tilde{r}_{i,e}$ *exist and be continuous, and* $\|\hat{q}_{i,e}\|_1 < \infty$, $\|\hat{q}_{i,x}\|_{p/(p-1)} < \infty$. *Let* N_i *denote an arbitrary bounded neighborhood of* $\{0\}$ in $\mathscr{T}$. *Then all the following tend to zero as* $\varepsilon \to 0$, *uniformly for* $\underline{v}_i$ *in* N_i (*and also for* $\tilde{r}_{i,x}$, $\tilde{r}_{i,e}$ *replacing* $\tilde{q}_{i,x}$, *and* $\tilde{q}_{i,e}$, *respectively*),

$$\begin{aligned} &\|\tilde{q}_{i,e}(\hat{X}_i + \varepsilon\underline{v}_i, E\hat{X}_i + \varepsilon E v_i) - \tilde{q}_{i,e}(\hat{X}_i, E\hat{X}_i)\|_1, \\ &\|\tilde{q}_{i,x}(\hat{X}_i + \varepsilon\underline{v}_i, E\hat{X}_i + \varepsilon E v_i) - \tilde{q}_{i,x}(\hat{X}_i, E\hat{X}_i)\|_{p/(p-1)}. \end{aligned} \tag{3.7}$$

ASSUMPTION 3.6. *Define the linear maps* $\hat{R}_0$, $\hat{R}_k$ (*from* $y_0 \in L_p(\mathscr{B}_0)$ *and* $y_k \in L_p(\mathscr{B}_k)$ *to the appropriate Euclidean space*), *and suppose that the components are linearly independent for each* i. *Then*

$$\hat{R}_i \cdot y_i \equiv E[\hat{r}_{i,x} \cdot y_i + \hat{r}_{i,e} E y_i]. \tag{3.8}$$

ASSUMPTION 3.7. *For the inactive constraints* q_i^j, *suppose that there is a neighborhood* N_i *of the origin in* $L_p(\mathscr{B}_i)$ *for* $i > 0$ *and in* $\underline{L}_p(\mathscr{B}_0)$ *for* $i = 0$, *for which* $q_i^j(\hat{X}_i + y_i) < 0$, $q_0^j(\underline{\hat{X}}_0 + \underline{y}_0) < 0$, *for* $y_i \in N_i$, $i > 0$, $\underline{y}_0 \in N_0$. *Suppose that there is an* X_i *in* $L_p(\mathscr{B}_i)$, $i > 0$, *and* $\underline{X}_0 \in \underline{L}_p(\mathscr{B}_0)$ *so that*

$$\begin{aligned} E[\tilde{q}_{i,x}^j \cdot X_i + \tilde{q}_{i,e} E X_i] &< 0 \qquad \text{for all active } q_i^j, \\ E[\tilde{q}_{0,\underline{x}}^j \cdot \underline{X}_0 + \tilde{q}_{0,\underline{e}} E\underline{X}_0] &< 0 \qquad \text{for all active } q_0^j. \end{aligned} \tag{3.9}$$

ASSUMPTION 3.8. *Assume that* $f_{i,x}^0$, $f_{i,u}^0$ *are continuous in* x *and* u *and* $\|\hat{f}_{i,x}^0\|_{p/(p-1)} < \infty$ *and* $\|\hat{f}_{i,u}^0\|_{p'/(p'-1)} < \infty$. *For a real* K_1,

$$|f_i^0(X_i, u_i, \xi_i)| \leqq K_1(1 + |X_i|^p + |u_i|^{p'}) \tag{3.10}$$

and

$$\begin{aligned} &\|f_{i,x}^0(\hat{X}_i + \varepsilon v_i, \hat{u}_i + \varepsilon\delta u_i(\beta)) - \hat{f}_{i,x}^0\|_{p/(p-1)} \to 0, \\ &\|f_{i,u}^0(\hat{X}_i + \varepsilon v_i, \hat{u}_i + \varepsilon\delta u_i(\beta)) - \hat{f}_{i,u}^0\|_{p'/(p'-1)} \to 0, \end{aligned} \tag{3.11}$$

as $\varepsilon \to 0$, *uniformly for* v_i *in* N_i *and in* β, *for* $i = 0, \cdots, k - 1$.

3.4. *Identification with the definition in Section* 2. Define $\mathscr{T}$ to be the space in which $\underline{X}_0, \cdots, \underline{X}_k$ lie, namely, $\mathscr{T} = \underline{L}_p(\mathscr{B}_0) \times \cdots \times \underline{L}_p(\mathscr{B}_k)$, and let Q' denote the set of all sequences in $\mathscr{T}$ which are solution to (1.1′) for the class of allowed controls and initial conditions.

Assumption 3.8 implies that (3.5) can be replaced by

$$\|\underline{X}_i(\beta) - \hat{\underline{X}}_i - \varepsilon\delta\underline{X}_i(\beta)\|_p = o(\varepsilon), \tag{3.12}$$

since, by (3.5), we can show that

$$\begin{aligned} E\big|f_i^0(X_i(\beta), \hat{u}_i + \varepsilon\delta u_i(\beta), \xi_i) - f_i^0(\hat{X}_i, \hat{u}_i, \xi_i) - \varepsilon\hat{f}_{i,x}^0\cdot\delta X_i(\beta) - \varepsilon\hat{f}_{i,x}^0\cdot\delta u_i(\beta)\big| \\ \leqq \varepsilon E\big|f_{i,x}^0\big(\hat{X}_i + \theta_{\varepsilon,\beta}\big(X_i(\beta) - \hat{X}_i\big), \hat{u}_i + \varepsilon\theta_{\varepsilon,\beta}\delta u_i(\beta), \xi_i\big) - \hat{f}_{i,x}^0\big|\cdot\big|\delta X_i(\beta)\big| \\ + \varepsilon\big|f_{i,u}^0\big(\hat{X}_i + \theta_{\varepsilon,\beta}\big(X_i(\beta) - \hat{X}_i\big), \hat{u}_i + \varepsilon\theta_{\varepsilon,\beta}\delta u_i(\beta), \xi_i\big) - \hat{f}_{i,u}^0\big|\cdot\big|\delta u_i(\beta)\big|, \end{aligned} \tag{3.13}$$

where $\theta_{\varepsilon,\beta}$ is a random variable in $[0, 1]$, and we can complete the assertion by using Hölder's inequality. Then it is straightforward to verify that the set $K \in \mathscr{T}$ (given by (3.2) or (3.14)) of all vectors $\delta\underline{X}_0, \cdots, \delta\underline{X}_k$ corresponding to $\delta u_i \in IC_i$, $\delta\underline{X}_0 \in \mathscr{B}_0$, is a first order convex approximation to $Q \equiv Q' - \{\hat{\underline{X}}_0, \cdots, \hat{\underline{X}}_k\} \subset \mathscr{T}$. One can write

$$\begin{aligned} \delta\underline{X}_{i+1} &= \delta\underline{X}_i + \hat{\underline{f}}_{i,\underline{x}}\delta\underline{X}_i + \hat{\underline{f}}_{i,u}\delta u_i, \\ \delta\underline{X}_i &\equiv \sum_{j=1}^{i} F(j, i)\hat{\underline{f}}_{j-1,u}\delta u_{j-1} + F(0, i)\delta\underline{X}_0, \\ F(j, i) &= (I + \hat{\underline{f}}_{i-1,x}) \cdots (I + \hat{\underline{f}}_{j,x}), \qquad j < i, \\ F(i, i) &= I. \end{aligned} \tag{3.14}$$

$$\hat{\underline{f}}_{i,\underline{x}} = \left[\begin{array}{c|c} 0 & \hat{f}_{i,x^1}^0, \cdots, \hat{f}_{i,x^n}^0 \\ \hline 0 & \\ \vdots & \hat{f}_{i,x} \\ 0 & \end{array}\right]. \tag{3.15}$$

Identify the components of r_0 and r_k with $\varphi_1, \cdots$, and φ_{-i}, $i > 0$, with the components of the q_i, $i \geqq 0$. Also $\varphi_0 \equiv EX_k^0$. The $\hat{R}_i$ of Assumption 3.6 is the Fréchet derivative of the vector valued map $r_i(X_i)$. The following $\hat{Q}_i$, $i \geqq 0$,

$$\begin{aligned} \hat{Q}_i\cdot y_i &\equiv E[\hat{q}_{i,x}\cdot y_i + \hat{q}_{i,e}Ey_i]. \\ \hat{Q}_0\cdot \underline{y}_0 &\equiv E[\hat{q}_{i,\underline{x}}\cdot \underline{y}_0 + \hat{q}_{0,\underline{e}}E\underline{y}_0] \end{aligned} \tag{3.16}$$

are the Fréchet derivatives of the vector valued maps q_i at $\hat{X}_i$. Thus, Assumption 2.1 is implied by Assumptions 3.4, 3.5, and 3.6. Assumption 3.7 implies Assumptions 2.2, and 2.3′ is implied by 3.1 and 3.4 through 3.8.

That $\hat{Q}_i$ is a Fréchet derivative can be seen from the following brief calculation. Let N_i denote an arbitrary bounded neighborhood of $\{0\}$ in $L_p(\mathscr{B}_i)$. There are random variables $\theta \in [0, 1]$ (depending on ε, v_i) so that, for $i > 0$,

$$\begin{aligned} e \equiv \varepsilon^{-1} |E\tilde{q}_i(X_i + \varepsilon v_i, EX_i + \varepsilon Ev_i) - E\tilde{q}_i(X_i, EX_i) \\ - \varepsilon E\tilde{q}_{i,x}(X_i, EX_i) \cdot v_i - \varepsilon E\tilde{q}_{i,e}(X_i, EX_i) EV_i| \\ \leqq |E[\tilde{q}_{i,x}(X_i + \varepsilon\theta v_i, EX_i + \varepsilon\theta Ev_i) - \tilde{q}_{i,x}(X_i, EX_i)]v_i \\ + E[\tilde{q}_{i,e}(X_i + \varepsilon\theta v_i, EX_i + \varepsilon\theta v_i) - \tilde{q}_{i,e}(X_i, EX_i)]Ev_i|. \end{aligned} \tag{3.17}$$

By using Assumption 3.5 and Hölder's inequality, we can show that $e \to 0$ as $\varepsilon \to 0$ uniformly in v_i, completing the calculation.

Note that, for the Fréchet derivatives of the equality constraints to be linearly independent, it is enough to consider $r_0(X)$ and $r_k(X_k)$ separately, since r_0 does not depend on X_k and r_k does not depend on X_0.

Theorem 3.1 is the main result of this section. Let P' denote the $(n + 1)$ row vector $(1, 0, \cdots, 0)$. The prime on P' denotes transpose. While r_0, r_k, q_i, $i > 0$, do not actually depend on the X_i^0, it is convenient to write (3.19) and subsequent formulas as though they did. Thus, we write $\underline{r}_k(\underline{X}_k, E\underline{X}_k)$ for $r_k(X_k, EX_k)$ and $\tilde{\underline{r}}_{k,\underline{x}}(\underline{X}_k, E\underline{X}_k)$ for

$$\begin{bmatrix} 0 & \cdot & \\ 0 & \vdots & r_{k,x}(X_k, EX_k) \\ 0 & & \end{bmatrix}, \qquad \text{and so forth.} \tag{3.18}$$

THEOREM 3.1. *Let Assumptions* 3.1 *through* 3.8 *hold. There exists a scalar* $p^0 \leqq 0$, *and there exist vectors* α_0, α_k, *and* $\psi_i \leqq 0$, $i = 0, \cdots, k$, *not all zero, such that*

$$\begin{aligned} p^0 E\delta\underline{X}_k^0 + E\alpha_0'[\hat{\underline{r}}_{0,\underline{x}} + (E\hat{\underline{r}}_{0,\underline{e}})]\delta\underline{X}_0 + E\alpha_k'[\hat{\underline{r}}_{k,\underline{x}} + (E\hat{\underline{r}}_{k,\underline{e}})]\delta\underline{X}_k \\ + E \sum_{i=0}^{k} \psi_i'[\hat{\underline{q}}_{i,\underline{x}} + (E\hat{\underline{q}}_{i,\underline{e}})]\delta\underline{X}_i \leqq 0 \end{aligned} \tag{3.19}$$

for $\delta\underline{X}_0, \cdots, \delta\underline{X}_k \in \bar{K}$, *where* $\psi_i'\hat{q}_i = 0$. *Define the vectors* $\underline{p}_k, \cdots, \underline{p}_0$:

$$\begin{aligned} \underline{p}_k = p^0 P + [\hat{\underline{r}}_{k,\underline{x}}' + (E\hat{\underline{r}}_{k,\underline{e}}')]\alpha_k + [\hat{\underline{q}}_{k,\underline{x}}' + (E\hat{\underline{q}}_{k,\underline{e}}')]\psi_k, \\ \underline{p}_{i-1} = (I + \hat{\underline{f}}_{i-1,\underline{x}}')\underline{p}_i + [\hat{\underline{q}}_{i-1,\underline{x}}' + (E\hat{\underline{q}}_{i-1,\underline{e}}')]\psi_{i-1} \\ + [\hat{\underline{r}}_{i-1,\underline{x}}' + (E\hat{\underline{r}}_{i-1,\underline{e}}')]\alpha_{i-1}, \qquad k \geqq i \geqq 1. \end{aligned} \tag{3.20}$$

Then

$$E[\underline{p}_i'\hat{\underline{f}}_{i-1,u} \mid \bar{\mathscr{B}}_{i-1}]\delta u_{i-1} \leqq 0 \tag{3.21}$$

for all $\delta u_{i-1} \in \overline{IC}_{i-1}$ *and*

$$E[\underline{p}_0 \mid \mathscr{B}_0] = 0. \tag{3.22}$$

PROOF. Equation (3.19) follows from Theorem 2.1 and the discussion preceding Theorem 3.1. Equations (3.21) and (3.22) are specializations of (3.19), as follows. Let $\delta \underline{X}_0 = 0$, $\delta u_j = 0, j \neq i - 1$. Then $\delta \underline{X}_j = F(i, j)\underline{f}_{i-1,u}\delta u_{i-1}$, and (3.19) yields

$$E\{p^0 P' F(i, k) + \alpha_k'[\hat{\underline{r}}_{k,\underline{x}} + (E\hat{\underline{r}}_{k,\underline{e}})]F(i, k) + \sum_{j=1}^{k} \psi_j'[\hat{\underline{q}}_{j,\underline{x}} + (E\hat{\underline{q}}_{j,\underline{e}})]F(i, j)\}\hat{\underline{f}}_{i-1,u}\cdot \delta u_{i-1} \leqq 0. \tag{3.23}$$

The bracketed term in (3.23) is $\underline{p}_i'$. The closure of the first order convex approximation given by (3.2) and (3.3) is merely the set of solutions $(\delta \underline{X}_0, \cdots, \delta \underline{X}_k)$ of (3.2) and (3.3) which can be obtained by using $\{\delta u_i\}$ in the closure in $L_{p'}(\tilde{\mathscr{B}}_i)$ of $\{IC_i\}$. Thus,

$$E[\underline{p}_i'\hat{\underline{f}}_{i-1,u}\delta u_{i-1}] \leqq 0 \tag{3.24}$$

for all $\delta u_{i-1} \in \overline{IC}_{i-1}$. Let $B \in \tilde{\mathscr{B}}_{i-1}$ and suppose that (χ_B is the characteristic function of B)

$$E\chi_B \underline{p}_i'\hat{\underline{f}}_{i-1,u}\delta u_{i-1} > 0. \tag{3.25}$$

Then $\delta \tilde{u}_{i-1} \equiv \chi_B \delta u_{i-1} \in \overline{IC}_{i-1}$ and we have $E\underline{p}_i'\hat{\underline{f}}_{i-1,u}\cdot \delta \tilde{u}_{i-1} > 0$, which contradicts (3.24). Thus, (3.21) holds.

Next, let $\delta u_i = 0$, $i = 0, \cdots, k - 1$. Then substituting $\delta \underline{X}_i = F(0, i)\delta \underline{X}_0$ into (3.19) yields

$$E\underline{p}_0'\delta \underline{X}_0 \leqq 0 \tag{3.26}$$

for all δX_0 in $\underline{L}_p(\mathscr{B}_0)$. Using the argument which proved (3.21) and the fact that $-\delta \underline{X}_0 \in \underline{L}_p(\mathscr{B}_0)$ if $\delta \underline{X}_0$ is, gives (3.22). *Q.E.D.*

3.5. *Remark on generalizations.* The spaces $\underline{L}_p(\mathscr{B}_i)$ can easily be replaced by less restrictive spaces where, for example, each of the components X_i^j has its own integrability property, (that is, $X_i^j \in L_{p_{ji}}(\mathscr{B}_{ji})$). Assumption 2.3 requires only that the $c_i(x)$ be smooth and convex, whereas the "derivatives" Q_i of the $q_0, \cdots, q_k, EX_k^0$, were linear operators. The "convex" derivatives of Assumption 2.3 arise, for example, where, the cost to be minimized, or the state space constraints take the form $E \max_i \|X_i - t_i\|$, and Theorem 3.1 can be extended to include constraints or costs of these forms. Constraints of the type $P\{X_n \in A\} > 1 - \varepsilon$ can conceivably be inserted into the definition of Q', but we do not know how to find a first order convex approximation to such a constrained Q'.

For illustrative purposes, we verify Assumption 3.3 under a specific set of conditions on the f_i.

THEOREM 3.2. *Let* $u_i \in \tilde{U}_i$ *with* $p' \geqq p \geqq 1$, *and* $\tilde{\mathscr{B}}_i \subset \mathscr{B}(\xi_0, \cdots, \xi_{i-1}) \cup \mathscr{B}(v_i)$, *where the independent sequence* $\{v_i\}$ *is independent of the independent sequence of matrices* $\{\xi_i\}$ *and*

$$X_{i+1} = X_i + f_i(X_i, u_i, \xi_i) = g_i(X_i, u_i) + \xi_i h_i(X_i, u_i). \tag{3.27}$$

The moments satisfy $E|\xi_i|^q < \infty$ *for all* $q = 1, 2, \cdots$. *Let* g_i *and* h_i *be continuous with bounded and continuous derivatives in* X_i, u_i. *Then Assumption* 3.3 *holds.*

PROOF. From the following estimate, for some real K,

$$|X_{i+1}| \leqq |X_i| + K(|X_i| + |u_i| + 1) + K(|X_i| + |u_i| + 1)|\xi_i|), \tag{3.28}$$

we can deduce that all moments of $|X_i|$ exist up to order p', and similarly for the moments of the δX_i given by $\delta X_{i+1} = \delta X_i + \hat{f}_{i,x}\delta X_i + \hat{f}_{i,u}\delta u_i$, or for the moments of $\delta X_i(\beta)$.

Fix $\varepsilon > 0$ and write

$$X_{i+1}(\beta) = X_i(\beta) + g_i[X_i(\beta), \hat{u}_i + \varepsilon\delta u_i(\beta)] + \xi_i h_i[X_i(\beta), \hat{u}_i + \varepsilon\delta u_i(\beta)]. \tag{3.29}$$

From the relation, for some real K,

$$\begin{aligned} |X_{i+1}(\beta) - X_{i+1}(\tilde{\beta})| \leqq & K|X_i(\beta) - X_i(\tilde{\beta})|(1 + |\xi_i|) \\ & + \varepsilon K|\delta u_i(\beta) - \delta u_i(\tilde{\beta})|(1 + |\xi_i|), \end{aligned} \tag{3.30}$$

and the relations $|\delta u_i(\beta) - \delta u_i(\tilde{\beta})| \to 0$ in $L_{p'}(\mathscr{B}_i)$ as $\tilde{\beta} \to \beta$, we conclude that $X_i(\beta)$ is a continuous $L_p(\mathscr{B}_{i-1})$ valued function of β, for any $\varepsilon > 0$. Next, define the sequence $Y_i = X_i(\beta) - \hat{X}_i$,

$$\begin{aligned} Y_{i+1} = Y_i + [g_i(\hat{X}_i + Y_i, \hat{u}_i + \varepsilon\delta u_i(\beta)) + \xi_i h_i(\hat{X}_i + Y_i, \hat{u}_i + \varepsilon\delta u_i(\beta))] \\ - [g_i(\hat{X}_i, \hat{u}_i) + \xi_i h_i(\hat{X}_i, \hat{u}_i)]. \end{aligned} \tag{3.31}$$

From (3.31), we can easily show that $E^{1/p}|Y_i|^p = O(\varepsilon)$, uniformly in β. Next, $Z_i \equiv Y_i - \varepsilon\delta X_i$ satisfies, for random $\theta_i \in [0, 1]$, which may depend on ε and β,

$$\begin{aligned} Z_0 &= 0 \\ Z_{i+1} &= Z_i + [\hat{g}_{i,x} + \xi_i \hat{h}_{i,x}]Z_i \\ &\quad + [g_{i,x}(\hat{X}_i + \theta_i Y_i, \hat{u}_i + \varepsilon\theta_i\delta u_i(\beta)) - \hat{g}_{i,x}]Y_i \\ &\quad + \xi_i[h_{i,x}(\hat{X}_i + \theta_i Y_i, \hat{u}_i + \varepsilon\theta_i\delta u_i(\beta)) - \hat{h}_{i,x}]Y_i \\ &\quad + \varepsilon[g_{i,u}(\hat{X}_i + \theta_i Y_i, \hat{u}_i + \varepsilon\theta_i\delta u_i(\beta)) - \hat{g}_{i,u}]\delta u_i(\beta) \\ &\quad + \varepsilon\xi_i[h_{i,u}(\hat{X}_i + \theta_i Y_i, \hat{u}_i + \varepsilon\theta_i\delta u_i(\beta)) - \hat{h}_{i,u}]\delta u_i(\beta). \end{aligned} \tag{3.32}$$

This expression together with $E^{1/p}|Y_i|^p = O(\varepsilon)$, implies that $E^{1/p}|Z_i|^p = O(\varepsilon)$. The proof is straightforward and only the following observation is needed.

$$\left(\frac{Y_i}{\varepsilon}\right)^p [g_{i,x}(\hat{X}_i + \theta_i Y_i, \hat{u}_i + \varepsilon\theta_i\delta u_i(\beta)) - \hat{g}_{i,x}]^p \tag{3.33}$$

is uniformly integrable with parameters ε and β, and goes to zero as $\varepsilon \to 0$ with probability 1. Thus, the expectation of the term goes to zero as $\varepsilon \to 0$, uniformly in β. *Q.E.D.*

4. The multiplier rule when the control depends explicitly on the state

In Section 3, the controls u_i were measurable over the fixed σ-algebras $\tilde{\mathscr{B}}_i$, and did not depend explicitly on the state. If we allow the controls u_i to depend on the X_i, then some condition must be imposed on the u_i which guarantees that replacing $u_i(X_i)$ by $u_i(X_i + \delta X_i) + \varepsilon\delta u_i(X_i + \delta X_i)$ in (1.1) (where $X_{i+1} + \delta X_{i+1} = X_i + \delta X_i + f(X_i + \delta X_i, u_i + \varepsilon\delta u_i, \xi_i)$) alters the paths only the order of ε. In Section 3, $u_i(X_i + \delta X_i) = u_i(X_i)$. Thus, some smoothness on the u_i is required. In Theorem 4.1, we assume the form (3.27).

For simplicity of notation, it is assumed that u_i depends explicitly on X_i, and is not randomized. Subsequently, several extensions are stated.

4.1. *Assumptions and notation.* Let $p = p'$ and let $\mathscr{T}$ be as in Section 3, where $\mathscr{B}_i = \mathscr{B}(\xi_0, \cdots, \xi_{i-1})$ and $\mathscr{B}_0$ is the trivial σ-algebra.

ASSUMPTION 4.1. *Let U_i be a convex set, and let $\tilde{U}_i$ denote the convex set of controls which can be used at time i. We have $u_i \in \tilde{U}_i$ if $u_{i,x}$ is bounded and continuous, and $u_i(x) \in U_i$ for each x.*

Again, let $\hat{X}_0, \cdots, \hat{X}_k$, $\hat{u}_0(\hat{X}_0) = \hat{u}_0, \cdots, \hat{u}_{k-1}(\hat{X}_{k-1}) = \hat{u}_{k-1}$ denote the optimal solution. Assume that IC_i, the internal cone to $\tilde{U}_i - \hat{u}_i$ at $\{0\}$ exists and contains some point other than $\{0\}$. Then, for any $\delta u_i^s \in IC_i$, $\delta u_{i,x}^s$ is bounded and continuous and $\hat{u}_i(x) + \varepsilon \sum_{s=1}^{v} \beta_s \delta u_i^s(x) \in U_i$ for sufficiently small ε, for all x and $\beta = (\beta_1, \cdots, \beta_v) \in P^v$.

ASSUMPTION 4.2. *Assume that $h_{i,x}, g_{i,x}, h_{i,u}, g_{i,u}$ are bounded and are continuous in their arguments. The $\{\xi_i\}$ are mutually independent, and all of their moments exist.*

ASSUMPTION 4.3. *Assume that $f_{i,x}^0, f_{i,u}^0$ are continuous in their variables and, for some real $K < \infty$,*

$$\begin{aligned} &|f_i^0(x, u)| \leqq K(1 + |x|^p + |u|^p), \\ &|f_{i,x}^0(x, u)| + |f_{i,u}^0(x, u)| \leqq K(1 + |x|^{p-1} + |u|^{p-1}). \end{aligned} \tag{4.1}$$

Define $\delta \underline{X}_0(\beta) = \Sigma_s \beta_s \delta \underline{X}_0^s$,

$$\begin{aligned} &\delta u_i(\beta, X_i) = \sum_s \beta_s \delta u_i^s(X_i), \qquad \delta u_i^s(x) \in IC_i \\ &\delta \underline{X}_{i+1} = \delta \underline{X}_i + [\hat{\underline{f}}_{i,\underline{x}} + \hat{\underline{f}}_{i,u} \cdot \hat{u}_{i,\underline{x}}]\delta \underline{X}_i + \hat{\underline{f}}_{i,u} \cdot \delta \hat{u}_i, \end{aligned} \tag{4.2}$$

where we write $\delta \hat{u}_i$ for $\delta u_i(\hat{X}_i)$ and also $\delta \underline{X}_i(\beta)$ for $\delta \underline{X}_i$ if $\delta \hat{u}_i$ takes the form $\delta u_i(\beta, \hat{X}_i)$. With

$$\begin{aligned} &F_u(j, i) \equiv (I + \hat{\underline{f}}_{i-1,\underline{x}} + \hat{\underline{f}}_{i-1,u}\hat{u}_{i-1,\underline{x}}) \cdots (I + \hat{\underline{f}}_{j,\underline{x}} + \hat{\underline{f}}_{j,u}\hat{u}_{j,\underline{x}}), \quad j \leqq i, \\ &F_u(i, i) = I, \end{aligned} \tag{4.3}$$

we have

$$\delta \underline{X}_{i+1} = F_u(i, i + 1)\delta \underline{X}_i + \hat{\underline{f}}_{i,u} \cdot \delta \hat{u}_i \tag{4.4}$$

and

$$\delta \underline{X}_i = \sum_{j=1}^{i} F_u(j, i)\underline{\hat{f}}_{j-1,u}\delta\hat{u}_{j-1} + F_u(0, i)\delta\underline{X}_0. \tag{4.5}$$

We will use the notation $\underline{\hat{f}}_i = \underline{f}(\hat{X}_i, \hat{u}_i(\hat{X}_i))$, and so forth. If arguments of a function are other than $\hat{X}_i$, $\hat{u}_i(\hat{X}_i)$, or $\underline{\hat{X}}_i$, they will be explicitly inserted.

THEOREM 4.1. *Let Assumptions* 4.1, 4.2, 4.3, and 3.4, 3.5, 3.6, 3.7 *hold. Define* $\underline{p}_k$ *by* (3.20) *and* $\underline{p}_i$, $i < k$, *by*

$$\underline{p}_{i-1} = (I + \underline{\hat{f}}'_{i-1,\underline{x}} + \underline{\hat{u}}'_{i-1,\underline{x}}\underline{\hat{f}}'_{i-1,u})\underline{p}_i + [\underline{q}'_{i-1,\underline{x}} + (E\underline{\hat{q}}'_{i-1,\underline{e}})]\psi_{i-1} + [\underline{\hat{r}}'_{i-1,\underline{x}} + (E\underline{\hat{r}}_{i-1,\underline{e}})]\alpha_{i-1}. \tag{4.6}$$

Then (4.7) *and* (4.8), *the analogs of* (3.21) *and* (3.22), *hold, for all* $\delta\hat{u}_{i-1} \in \overline{IC}_{i-1}$,

$$E[\underline{p}_0|\mathscr{B}_0] = E\underline{p}_0 = 0, \tag{4.7}$$

$$E[\underline{p}'_i\underline{\hat{f}}'_{i-1,u}|\hat{X}_{i-1}]\delta\hat{u}_{i-1} \leqq 0. \tag{4.8}$$

PROOF. First we verify that Assumption 3.3 holds. By Assumption 4.2,

$$|X_{i+1}| \leqq K(1 + |\xi_i|)(|X_i| + |u_i(X_i)|) \tag{4.9}$$

and, since $|u_i(x)| \leqq K(1 + |x|)$, all moments of X_i exist; similarly, so do all moments of δX_i, where δX_i is given by (4.2) for $\delta u_i \in IC_i$ and δX_0 is an arbitrary n vector.

Next, fix both $\varepsilon > 0$ and the δu_i^s, and write

$$X_{i+1}(\beta) = X_i(\beta) + f_i[X_i(\beta), \hat{u}_i(X_i(\beta)) + \varepsilon\delta u_i(\beta, X_i(\beta))]. \tag{4.10}$$

Using the Lipschitz conditions on f_i, namely,

$$|f_i(a, b, \xi) - f_i(\tilde{a}, \tilde{b}, \xi)| \leqq K(1 + |\xi|)(|a - \tilde{a}| + |b - \tilde{b}|), \tag{4.11}$$

and the bounds ($|\beta - \tilde{\beta}| = \Sigma_s|\beta_s - \tilde{\beta}_s|$),

$$\begin{aligned} |\delta X_0(\beta) - \delta X_0(\tilde{\beta})| &\leqq K|\beta - \tilde{\beta}|, \\ |\delta u_i(\beta, x) - \delta u_i(\tilde{\beta}, \tilde{x})| &\leqq \sum_s |\beta_s\delta u_i^s(x) - \tilde{\beta}_s\delta u_i^s(x)| \\ &\leqq \sum_s \{|\beta_s - \tilde{\beta}_s|\cdot|\delta u_i^s(x)| + |\delta u_i^s(x) - \delta u_i^s(\tilde{x})|\tilde{\beta}_s\}, \\ |\delta u_i^s(x) - \delta u_i^s(\tilde{x})| &\leqq K|x - \tilde{x}|, \end{aligned} \tag{4.12}$$

we have that $\|X_i(\beta) - X_i(\tilde{\beta})\|_p \to 0$ as $|\beta - \tilde{\beta}| \to 0$ for any $p \geqq 1$, and any $\varepsilon > 0$. Thus, the $X_i(\beta)$ given by (4.10) are continuous in β in the $L_p(\mathscr{B}_i)$ sense.

Write

$$Y_{i+1} = Y_i + f_i[\hat{X}_i + Y_i, \hat{u}_i(\hat{X}_i + Y_i) + \varepsilon\delta u_i(\beta, \hat{X}_i + Y_i)] - \hat{f}_i \tag{4.13}$$

(see (3.31)). Again, using the bounds on $f_{i,x}, f_{i,u}$ and $\hat{u}_{i,x}$, (for example, $|f_{i,x}(x, u)| \leqq K(1 + |\xi_i|)(|x| + |u|)$) and the bound on $\hat{u}_{i,x}$ and $\delta u_{i,x}$, it is straightforward to show that $\|Y_i\|_p = O(\varepsilon)$ for any $p \geqq 1$.

Next, defining $Z_i = Y_i - \varepsilon\delta X_i(\beta)$, as in Theorem 3.2, we can show that $\|Z_i\|_p = o(\varepsilon)$ uniformly in β. Thus, Assumption 3.3 holds.

Next, we show that $X_i^0(\beta)$ is continuous in β in the $L_1(\mathscr{B}_i)$ sense for any $\varepsilon > 0$. This follows from (4.14) by an application of Assumption 4.3, Hölder's inequality, the Lipschitz conditions on $\hat{u}(x)$ and $\delta u(\beta, x)$, and the continuity of $X_i(\beta)$ in β in the $L_p(\mathscr{B}_i)$ sense. We have

$$
\begin{aligned}
(4.14)\qquad & f_i^0[X_i(\beta), \hat{u}_i(X_i(\beta)) + \varepsilon\delta u_i(\beta, X_i(\beta))] \\
&\qquad - f_i^0[\tilde{X}_i(\tilde{\beta}), \hat{u}_i(\tilde{X}_i(\tilde{\beta})) + \varepsilon\delta u_i(\beta, \tilde{X}_i(\beta))] \\
&= f_{i,x}^0(\alpha_1, \alpha_2)(\tilde{X}_i(\tilde{\beta}) - X_i(\beta)) \\
&\quad + f_{i,u}^0(\alpha_1, \alpha_2)[\hat{u}_i(\tilde{X}_i(\tilde{\beta})) - \hat{u}_i(X_i(\beta)) \\
&\qquad + \varepsilon\, \delta u_i(\tilde{\beta}, \tilde{X}_i(\tilde{\beta})) - \varepsilon\delta u_i(\beta, X_i(\beta))],
\end{aligned}
$$

where, for some random θ_i with values in $[0, 1]$,

$$
\begin{aligned}
& \alpha_1 = X_i(\beta) + \theta_i(\tilde{X}_i(\tilde{\beta}) - X_i(\beta)), \\
(4.15)\qquad & \alpha_2 = \hat{u}_i(X_i(\beta)) + \varepsilon\delta u_i(\beta, X_i(\beta)) \\
&\qquad + \theta_i[\hat{u}_i(\tilde{X}_i(\tilde{\beta})) - \hat{u}_i(X_i(\beta)) + \varepsilon\delta u_i(\tilde{\beta}, \tilde{X}_i(\tilde{\beta})) - \varepsilon\delta u_i(\beta, X_i(\beta))].
\end{aligned}
$$

We will not complete the details (which are quite straightforward), but it can be shown that $\|Z_i^0 - Y_i^0\|_1 = o(\varepsilon)$. Thus, the set $\{\delta \underline{X}_0, \cdots, \delta \underline{X}_k\}$ given by (4.2) is a first order convex approximation K to $Q' - \{\underline{X}_0, \cdots, \underline{X}_k\}$.

Now, (3.19) holds for $(\delta \underline{X}_0, \cdots, \delta \underline{X}_k)$ in $\bar{K}$, the closure of K in $\mathscr{T}$. By specializing (3.19), we get (4.7) and $Ep_i' f_{i-1,u}' \delta\hat{u}_{i-1} \leqq 0$ for $\delta\hat{u}_{i-1} \in IC_{i-1}$. But $\bar{K}$ contains those $(\delta \underline{X}_0, \cdots, \delta \underline{X}_k)$ which can be obtained by using the $\delta u_i(\cdot)$ in the $L_p(\mathscr{B}_i)$ closure $\overline{IC}_i$ of IC_i and $\overline{IC}_i$ contains pointwise limits of uniformly bounded sequences in $\overline{IC}_i$. Thus, if $\chi_A(\cdot)$ is the characteristic function of an n dimensional Borel set A and $\delta u_i(\cdot) \in IC_i$, then $\chi_A(\cdot)\delta u_i(\cdot) \equiv \delta\tilde{u}_i(\cdot) \in \overline{IC}_i$. Equation (4.8) is obtained by combining the last statement together with the argument which led from (3.24) to (3.21). *Q.E.D.*

4.2. *Extensions.* Let $y_i(\cdot)$ be a continuous vector valued function with uniformly bounded and continuous derivatives. Let u_i depend on $y_i(X_i)$, rather than on X_i directly. Then Theorem 4.1 remains true if the $\hat{u}_{i,\underline{x}}$ term in (4.6) is replaced by $\hat{u}_{i,y} \cdot \hat{y}_{i,\underline{x}}$, the conditioning in (4.8) is on $y_i(\hat{X}_i)$, and the $\delta u_i(\cdot)$ are functions of $y_i(X_i)$.

If the control has the form $u_i[y_i(X_i, X_{i-1}, \cdots, X_0)]$, it is still possible to derive a multiplier result, but the expressions are considerably more complicated, since $\delta \underline{X}_i$ may depend explicitly on $\delta \underline{X}_{i-1}, \cdots, \delta \underline{X}_0$.

The controls and initial condition can be randomized in the following way. Let $\tilde{v}_0, v_0, \cdots, v_{k-1}$ be independent random variables with values in $[0, 1]$ and which are independent of the $\{\xi_i\}$ sequence. Let $\mathscr{B}_0 = \mathscr{B}(\tilde{v}_0)$. In addition to the conditions in Theorem 4.1, let u_i depend on X_i and v_i. Suppose that $u_i(x, v_i)$ is differentiable in x and measurable in both variables, and that $u_{i,x}(x, v)$ is bounded and continuous, uniformly in v in $[0, 1]$. Also $u_i(x, v) \in U_i$, a convex set. Then Theorem 4.1 remains true if the conditioning on $\hat{X}_{i-1}$ in (4.8) is replaced by conditioning on $\hat{X}_{i-1}$ and v_{i-1}.

5. A stochastic maximum principle

For the *continuous time deterministic problem*, where $\dot{\underline{x}} = \underline{f}(x, u)$ and $\underline{p}_t$ denotes the adjoint vector, relation (3.24) is $p_t' \underline{f}(\hat{x}_t, u_t) \leqq p_t' \underline{f}(\hat{x}_t, \hat{u}_t)$ for all $u_t \in U_t$ or, equivalently, $\hat{u}_t$ is the u which maximizes $p_t' f(\hat{x}_t, u)$. Under a convexity condition, Halkin [6] and Holtzman [7] have proved a similar relation for the discrete time deterministic case. The stochastic analogy of this result is straightforward to derive, and we closely follow the treatment in Canon, Cullum, and Polak ([4], pp. 84–93).

For the sake of concreteness, we treat essentially the analog of Theorem 3.1, with a more specific form of Assumption 3.3, although generalizations are possible.

DEFINITION 5.1. *With the* $\tilde{U}_i$ *defined in Section* 3, *and system* (1.1′) *with constraints* (1.3), (1.4), *the control problem is directionally convex if, for each* $0 \leqq \lambda \leqq 1$ *and* u_i', u_i'' *in* $\tilde{U}_i$, *there is a* $u_i(\lambda) \in \tilde{U}_i$ *so that, with probability* 1, *for each* $X_i \in L_p(\mathscr{B}_i)$,

$$
\begin{aligned}
\lambda f_i(X_i, u_i', \xi_i) + (1 - \lambda) f_i(X_i, u_i'', \xi_i) &= f_i(X_i, u_i(\lambda), \xi_i), \\
\lambda f_i^0(X_i, u_i') + (1 - \lambda) f_i^0(X_i, u_i'') &\geqq f_i^0(X_i, u_i(\lambda)).
\end{aligned}
\tag{5.1}
$$

EXAMPLE 5.1. A common and important example of a directionally convex problem is

$$
\begin{aligned}
f_i(x, u, \xi) &= g_i(x, \xi) + k_i(x, \xi) u, \\
f_i^0(x, u) &= g_i^0(x) + u' Q u,
\end{aligned}
\tag{5.2}
$$

where Q is nonnegative definite. Then $u_i(\lambda) = \lambda u_i' + (1 - \lambda) u_i''$.

5.1. *A comment on Theorem* 2.1. Using the notation of Section 2, let B_i denote the set $\{w : \varphi_i(\hat{w} + w) < \varphi_i(\hat{w})\} \cup \{0\}$, and let Z_i denote a nonempty internal cone to B_i. Define

$$
Z' = \Big[\bigcap_{i>0} \{w : \ell_i(w) = 0\}\Big] \bigcap_{i \in J} Z_i.
\tag{5.3}
$$

and assume that it contains a point other than $\{0\}$. Theorem 2.1 is a consequence of the fact that, if $\hat{w}$ is optimal, then Z' and K (a first order convex approximation to $Q = Q' - \hat{w}$) can be separated by a continuous linear functional. (See Theorems 2.1 and 4.2 in [11].) Indeed, the proofs of Theorems 2.1 and 4.2 in [11] imply that if Theorem 2.1 does not hold at a given $\hat{w}$, (namely, if there is a ray which is internal to both K and Z'), then for any neighborhood N of $\{0\}$ in $\mathscr{T}$, there is a $\tilde{w} \in Q' \cap \{N + \hat{w}\}$ which satisfies the constraints for which $\varphi_0(\tilde{w}) < \varphi_0(\hat{w})$. Thus, if Theorem 2.1 does not hold at $\hat{w}$, then $\hat{w}$ is *not* an optimal solution.

5.2. *A transformation of the control problem.* The stochastic optimization problem of Section 3 is equivalent to the following problem. Find the $\underline{X}_i, \underline{v}_i$ satisfying $\underline{v}_i \in \underline{f}_i(X_i, \tilde{U}_i, \xi_i)$ and $\underline{X}_{i+1} = \underline{X}_i + \underline{v}_i$, for which $r_0(X_0) = r_k(X_k) = 0$, $q_0(\underline{X}_0) \leq 0$, $q_i(X_i) \leq 0$, $i > 0$, and for which $E\ \Sigma_{i=0}^{k-1} v_i^0$ is a minimum. Denote the optimizing variables by $\hat{\underline{X}}_0, \cdots, \hat{\underline{X}}_k, \hat{\underline{v}}_0, \cdots, \hat{\underline{v}}_{k-1}$.

Since the variables to be chosen are now $\underline{X}_0, \cdots, \underline{X}_k, \underline{v}_0, \cdots, \underline{v}_{k-1}$, with both $\underline{X}_i$ and $\underline{v}_i$ in $\underline{L}_p(\mathscr{B}_i)$, redefine $\mathscr{T}$ to be

$$\mathscr{T} = \underline{L}_p(\mathscr{B}_0) \times \cdots \times \underline{L}_p(\mathscr{B}_k) \times \underline{L}_p(\mathscr{B}_1) \times \cdots \times \underline{L}_p(\mathscr{B}_{k-1}). \tag{5.4}$$

Let the problem be directionally convex, and define

$$\tilde{Q}' = \{\underline{X}_0, \cdots, \underline{X}_k, \underline{v}_0, \cdots, \underline{v}_{k-1} : v_i \in \text{co}\, \underline{f}_i(X_i, \tilde{U}_i, \xi_i), \underline{X}_{i+1} = \underline{X}_i + \underline{v}_i\}, \tag{5.5}$$

where co S is the convex hull of the set S. Namely, co $\underline{f}_i(X_i, \tilde{U}_i, \xi_i)$ is the convex hull of the set of random variables $\{\underline{f}_i(X_i, u_i, \xi_i), u_i \in \tilde{U}_i\}$. Let $\tilde{K}$ denote a first order convex approximation to

$$\tilde{Q}' - \{\hat{\underline{X}}_0, \cdots, \hat{\underline{X}}_k, \hat{\underline{v}}_0, \cdots, \hat{\underline{v}}_{k-1}\} = \tilde{Q}' - \hat{w} \equiv \tilde{Q}. \tag{5.6}$$

Suppose that the inequality in Theorem 2.1 does not hold for some suitable set of constants where $\tilde{K}$ replaces K (using the identification of terms and boundedness and continuity conditions in Section 3). Then the comment of the last subsection implies that there is a ray which is internal to both Z' and $\tilde{K}$, a neighborhood N of $\hat{w}$, and a $\tilde{w} = \{\tilde{\underline{X}}_0, \cdots, \tilde{\underline{X}}_k, \tilde{\underline{v}}_0, \cdots, \tilde{\underline{v}}_{k-1}\} \in \tilde{Q} \cap \{N + \hat{w}\}$ for which the constraints hold and

$$\begin{aligned} \varphi_0(\tilde{w}) &= E \sum_{i=0}^{k-1} \tilde{v}_i^0 < E \sum_{i=0}^{k-1} \hat{v}_i^0 = \varphi_0(\hat{w}), \\ \tilde{\underline{X}}_{i+1} &= \tilde{\underline{X}}_i + \tilde{\underline{v}}_i. \end{aligned} \tag{5.7}$$

There are $u_i^s \in \tilde{U}_i$, $\lambda_i^s \geq 0$, and $\Sigma_s \lambda_i^s = 1$ so that

$$\begin{aligned} \tilde{v}_i^0 &= \sum_s \lambda_i^s f_i^0(\tilde{X}_i, u_i^s), \\ \tilde{v}_i &= \sum_s \lambda_i^s f_i(\tilde{X}_i, u_i^s, \xi_i). \end{aligned} \tag{5.8}$$

By directional convexity, there is a $\tilde{u}_i \in \tilde{U}_i$ for which

$$\begin{aligned} \tilde{v}_i &= f_i(\tilde{X}_i, \tilde{u}_i, \xi_i), \\ \tilde{v}_i^0 &\leq f_i^0(\tilde{X}_i, \tilde{u}_i). \end{aligned} \tag{5.9}$$

Thus, by combining (5.8) and (5.9), one gets $\tilde{X}_{i+1} = \tilde{X}_i + f_i(\tilde{X}_i, \tilde{u}_i, \xi_i)$ and

$$E \sum_{i=0}^{k-1} f_i^0(\tilde{X}_i, \tilde{u}_i) < E \sum_{i=0}^{k-1} f_i^0(\hat{X}_i, \hat{u}_i), \tag{5.10}$$

which contradicts the optimality of $\{\hat{\underline{X}}_i, \hat{u}_i\}$. Thus, the inequality in Theorem 2.1 holds for $\tilde{K}$ replacing K. Also, (3.19) holds for all $\delta\underline{X}_i$ for which $\{\delta\underline{X}_0, \cdots, \delta\underline{X}_k, \delta\underline{v}_0, \cdots, \delta\underline{v}_{k-1}\} \in \tilde{K}$.

Define the set $\tilde{K} \in \mathscr{T}$:

$$\begin{aligned}\tilde{K} = \{\delta\underline{X}_0, \cdots, \delta\underline{X}_k, \delta\underline{v}_0, \cdots, \delta\underline{v}_{k-1}\colon \delta\underline{X}_{i+1} = \delta\underline{X}_i + \delta\underline{v}_i, \quad \text{such that}\\ \lambda[\delta\underline{v}_i - \hat{\underline{f}}_{i,\underline{x}}\cdot\delta\underline{X}_i] \in \operatorname{co} \underline{f}_i(\hat{X}_i, \tilde{U}_i, \xi_i) - \hat{\underline{v}}_i, \delta\underline{X}_0 \in L_p(\mathscr{B}_0)\}\end{aligned} \tag{5.11}$$

for sufficiently small λ. Theorem 5.1 gives conditions under which $\tilde{K}$ is a first order convex approximation to $\tilde{Q}$.

Let

$$\lambda[\delta\underline{v}_i^s - \hat{\underline{f}}_{i,\underline{x}}\cdot\delta\underline{X}_i^s] \in \operatorname{co} \underline{f}_i(\hat{X}_i, \tilde{U}_i, \xi_i) - \hat{\underline{v}}_i \tag{5.12}$$

for $s = 1, \cdots, v$, and all sufficiently small λ. The elements ($\lambda_i^\ell \geqq 0$, $\Sigma_\ell \lambda_i^\ell = 1$)

$$\begin{aligned}\delta\underline{X}_{i+1}^s &= \delta\underline{X}_i^s + \hat{\underline{f}}_{i,\underline{x}}\cdot\delta\underline{X}_i^s + [\sum_\ell \lambda_i^\ell \underline{f}_i(\hat{X}_i, u_i^{\ell,s}, \xi_i) - \hat{\underline{v}}_i]\\ &\equiv \delta\underline{X}_i^s + \delta\underline{v}_i^s,\end{aligned} \tag{5.13}$$

and $\delta\underline{v}_i^s$ and their convex combinations for $\beta_s \geqq 0$, $\Sigma_s \beta_s = 1$, namely,

$$\begin{aligned}\delta\underline{X}_{i+1}(\beta) &= \delta\underline{X}_i(\beta) + \hat{\underline{f}}_{i,\underline{x}}\cdot\delta\underline{X}_i(\beta) + \sum_s \beta_s[\sum_\ell \lambda_i^\ell \underline{f}_i(\hat{X}_i, u_i^{\ell,s}, \xi_i) - \hat{\underline{v}}_i]\\ &\equiv \delta\underline{X}_i(\beta) + \delta\underline{v}_i(\beta),\end{aligned} \tag{5.14}$$

and $\delta\underline{v}_i(\beta) = \Sigma_s \beta_s \delta\underline{v}_i^s$ are in $\tilde{K}$. We may write

$$\begin{aligned}\delta\underline{X}_{i+1}(\beta) &= [I + \hat{\underline{f}}_{i,\underline{x}}]\delta\underline{X}_i(\beta) + \delta\underline{W}_i(\beta),\\ \delta\underline{W}_i(\beta) &= \sum_s \beta_s[\sum_\ell \lambda_i^\ell \underline{f}_i(\hat{X}_i, u_i^{\ell,s}, \xi_i) - \hat{\underline{v}}_i],\\ \delta\hat{X}_i(\beta) &= \sum_{j=1}^{i} F(j, i)\delta\underline{W}_{j=1}(\beta) + F(0, i)\delta\underline{X}_0(\beta).\end{aligned} \tag{5.15}$$

THEOREM 5.1. *Let Assumptions* 3.4 *through* 3.7 *hold and assume that the control problem is directionally convex. Also make the following assumptions:*

(i) $\tilde{U}_i$ *is the convex set of functions in* $L_p(\tilde{\mathscr{B}}_i)$ *with values in the convex set* U_i; IC_i *contains some point other than zero;*

(ii) *the* $\{\xi_i\}$ *are mutually independent and all of their moments are finite;*

(iii) $|f_i(x, u, \xi)| \leqq K(1 + |\xi|)(1 + |u| + |x|)$ *and* $|f_i^0(x, u)| \leqq K(1 + |u|^{p'} + |x|^p)$ *for a real* K;

(iv) $|f_i(x, u, \xi) - f_i(\tilde{x}, u, \xi)| \leqq K(1 + |\xi|)(|x - \tilde{x}|)$ *and* $f_i^0(X_i, u_i)$ *is continuous in* X_i *in the* $\|\cdot\|_p$ *norm for any* u_i *in* $L_{p'}(\tilde{\mathscr{B}}_i)$;

(v) *$f_{i,x}(x, u)$ is uniformly bounded and is continuous in x for each vector u and $f^0_{i,x}(x, u)$ is continuous in x in the $\|\cdot\|_{p/(p-1)}$ norm for each fixed u in $L_{p'}(\tilde{\mathscr{B}}_i)$.*

Then, for p_k, p_i given by (3.20), equation (3.22) holds and (3.21) is replaced by the maximum principle.

$$E[\underline{p}'_{i+1}\underline{f}_i(\hat{X}_i, u_i, \xi_i) \,|\, \tilde{\mathscr{B}}_i] \leqq E[\underline{p}'_{i+1}\underline{f}_i(\hat{X}_i, \hat{u}_i, \xi_i) \,|\, \tilde{\mathscr{B}}_i] \tag{5.16}$$

with probability 1 for any u_i in $\tilde{U}_i$.

PROOF. Suppose that $\tilde{K}$ is a first order convex approximation to $\tilde{Q}$. By the discussion prior to the theorem, equation (3.19) must hold for all $\delta\underline{X}_i$ of the form (5.15). Setting $u_i^{\ell,s} = 0$ and $\delta\underline{X}_0 \neq 0$, we get (3.22) as in Theorem 3.1. Equation (5.12) follows by letting $u_j^{\ell,s} = \hat{u}_j, j \neq i, \delta\underline{X}_0 = 0$ and $u_i^{\ell,s} = u_i \neq \hat{u}_i$, substituting (5.15) into (3.19), and using the definitions of $\hat{\underline{v}}_i$ and $\underline{p}_j$. We have only to show that $\tilde{K}$ is a first order convex approximation to $\tilde{Q}$.

Clearly, $\tilde{K}$ is a convex cone, with typical elements $\{\delta\underline{X}_0^s, \cdots, \delta\underline{X}_k^s, \delta\underline{v}_0^s, \cdots, \delta\underline{v}_{k-1}^s\}$, and their convex combinations $\{\delta\underline{X}_0(\beta), \cdots, \delta\underline{X}_k(\beta), \delta\underline{v}_0(\beta), \cdots, \delta\underline{v}_{k-1}(\beta)\}$ are given by (5.14). Consider the mapping $\{\underline{X}_0(\beta), \cdots, \underline{X}_k(\beta), \underline{v}_0(\beta), \cdots, \underline{v}_{k-1}(\beta)\}$ from P^v to $\mathscr{T}$, for the fixed sequence of controls $\{u_i^{\ell,s}\}$:

$$\begin{aligned} \underline{X}_{i+1}(\beta) &= \underline{X}_i(\beta) + \underline{v}_i(\beta), \\ \underline{v}_i(\beta) &= \underline{f}_i(X_i(\beta), \hat{u}_i, \xi_i) \\ &\quad + \varepsilon \sum_s \beta_s[\sum_\ell \lambda_i^\ell \underline{f}_i(X_i(\beta), u_i^{\ell,s}, \xi_i) \quad \underline{f}_i(X_i(\beta), \hat{u}_i, \xi_i)] \\ \underline{X}_0(\beta) &= \hat{\underline{X}}_0 + \varepsilon\delta\underline{X}_0(\beta), \end{aligned} \tag{5.17}$$

where $\lambda_i^\ell \geqq 0$ and $\Sigma_\ell \lambda_i^\ell = 1$. Under (iv) of the theorem, the maps $\underline{X}_i(\beta)$ to $\underline{L}_p(\mathscr{B}_i)$ and $\underline{v}_i(\beta)$ to $\underline{L}_p(\mathscr{B}_{i+1})$ are continuous functions of β, for $\beta \in P^v$, and any $1 > \varepsilon > 0$. Thus, the composite map (taking $\{\underline{X}_0(\beta), \cdots, \underline{X}_k(\beta), \underline{v}_0(\beta), \cdots, \underline{v}_{k-1}(\beta)\}$ into $\mathscr{T}$) is a continuous $\mathscr{T}$ valued function of β.

Using (v) it can be shown that

$$\begin{aligned} \hat{X}_i(\beta) &= \hat{X}_i + \varepsilon\delta\hat{X}_i(\beta) + O_{1,i} \\ \underline{v}_i(\beta) &= \hat{\underline{v}}_i + \varepsilon\delta\underline{v}_i(\beta) + O_{2,i} \end{aligned} \tag{5.18}$$

where $O_{1,i}$ and $O_{2,i}$ are of the order of $o(\varepsilon)$ in $\underline{L}_p(\mathscr{B}_i)$ and $\underline{L}_p(\mathscr{B}_{i+1})$, respectively. Then, $\tilde{K}$ is indeed a first order convex approximation. The details of the last two steps involve straightforward expansions and estimates, as in Theorems 3.1, 3.2, and 4.1, and are omitted. They are probabilistic versions of the cited result ([4], pp. 84–93). *Q.E.D.*

The definition of a directionally convex problem holds if the control u_i depends on a function of the state X_i. Under directional convexity and the conditions of Theorem 4.1, Theorem 4.1 holds with equation (4.8) replaced by

$$E[\underline{p}'_{i+1}\underline{f}_i(\hat{X}_i, u_i, \xi_i) \,|\, \hat{X}_i] \leqq E[\underline{p}'_{i+1}\underline{f}_i(\hat{X}_i, \hat{u}_i, \xi_i) \,|\, \hat{X}_i]. \tag{5.19}$$

6. A relation with dynamic programming

For simplicity of presentation, this section will be largely formal. Suppose that the problem is directionally convex, and there are no constraints r_i and q_i. Let u_i depend on X_i and define the (dynamic programming) costs

$$\begin{aligned} V_i(\underline{x}) &= \inf_{u_i, \cdots, u_{k-1}} E[X_k^0 | \underline{X}_i = \underline{x}] = E[\hat{X}_k^0 | \underline{X}_i = \underline{x}], \\ \tilde{V}_i(x) &= V_i(\underline{x}) - x^0. \end{aligned} \tag{6.1}$$

Define

$$W_i(\hat{X}_i; \xi_i, \cdots, \xi_{k-1}) = \hat{X}_k^0 - \hat{X}_i^0 = \sum_{j=i}^{k-1} f_i^0(\hat{X}_i, \hat{u}_i). \tag{6.2}$$

Then drop some arguments for notational simplicity and write

$$\text{grad } W_i = W_{i,x} = \text{grad } W_i(X_i; \xi_i, \cdots, \xi_{k-1}) \tag{6.3}$$

evaluated at $x = \hat{X}_i$; similarly, for $V_{i,\underline{x}}$. Then grad $W_k = W_{k,x} = 0$ and

$$W_{i,x} = (I + \hat{f}'_{i,x} + \hat{f}'_{i,u}\hat{u}_{i,x})W_{i+1,x} + \hat{f}^0_{i,x}. \tag{6.4}$$

Thus,

$$W_{i,x} = -p_i, \qquad (W_{i,x}, 1) = -\underline{p}_i, \tag{6.5}$$

$$\tilde{V}_i(x) = E[W_i | \hat{X}_i = x], \tag{6.6}$$

and

$$\tilde{V}_{i,x}(x) = E(I + \hat{f}'_{i,x})\tilde{V}_{i+1,x} + E\hat{f}^0_{i,x}. \tag{6.7}$$

We must have $p^0 < 0$, since there are no constraints r_i, q_i, and not all the p^0, α_i, ψ_i can be zero. Thus, we set $p^0 = -1$.

By the principle of optimality, $EV_{i+1}(\underline{x} + \underline{f}_i(x, \hat{u}_i, \xi_i)) \leqq EV_{i+1}(\underline{x} + \underline{f}_i(x, u_\varepsilon, \xi_i))$, where u_ε is the control which, for given $u_i \neq \hat{u}_i$, satisfies

$$\begin{aligned} (1 - \varepsilon)f_i(x, \hat{u}_i, \xi_i) + \varepsilon f_i(x, u_i, \xi_i) &= f_i(x, u_\varepsilon, \xi_i), \\ (1 - \varepsilon)f_i^0(x, \hat{u}_i) + \varepsilon f_i^0(x, u_i) &\geqq f_i^0(x, u_\varepsilon). \end{aligned} \tag{6.8}$$

Noting that $V_{i+1}(\tilde{\underline{x}}) \leqq V_{i+1}(\underline{x})$ if $\tilde{x} = x$, and $\tilde{x}^0 \leqq x^0$, we get

$$\begin{aligned} EV_{i+1}(\underline{x} + \underline{f}_i(x, \hat{u}_i, \xi_i)) &\leqq EV_{i+1}(\underline{x} + \underline{f}_i(x, u_\varepsilon, \xi_i)) \\ &\leqq EV_{i+1}(\underline{x} + (1 - \varepsilon)f_i(x, \hat{u}, \xi_i) + \varepsilon f_i(x, u_i, \xi_i)). \end{aligned} \tag{6.9}$$

Thus,

$$0 \leqq EV'_{i+1,\underline{x}}(\underline{x} + \underline{f}_i(x, \hat{u}_i, \xi_i))[\underline{f}_i(x, u_i, \xi_i) - \underline{f}_i(x, \hat{u}_i, \xi_i)], \tag{6.10}$$

where $V_{i+1,\underline{x}} = \text{grad } V_{i+1}(\underline{x})$, evaluated at $\underline{x} + \underline{f}_i(x, \hat{u}_i, \xi_i)$. With the identification (6.5) and $\tilde{V}_{i+1,x}(\hat{X}_{i+1}) = E[W_{i+1,x} | \hat{X}_{i+1}]$, we get precisely the maximum principle

$$E[\underline{p}'_{i+1}(\underline{f}_i(\hat{X}_i, \hat{u}_i, \xi_i) - \underline{f}_i(\hat{X}_i, u_i, \xi_i)) | \hat{X}_i] \geqq 0. \tag{6.11}$$

REFERENCES

[1] R. F. Baum, "Optimal control systems with stochastic boundary conditions," Report 69–23, Department of Industrial Engineering, University of Michigan, Ann Arbor, 1969.

[2] F. Brodeau, "Contribution à l'étude du controle optimal stochastique," Ph.D. thesis, University of Grenoble, 1968.

[3] G. F. Bryant and D. Q. Mayne, "A minimum principle for a class of discrete time stochastic systems," *IEEE Trans. Automatic Control*, Vol. 14 (1969), pp. 401–403.

[4] M. D. Canon, C. D. Cullum, Jr., and E. Polak, *Theory of Optimal Control and Mathematical Programming*, New York, McGraw-Hill, 1970.

[5] W. Fleming, "Stochastic Lagrange multipliers," *Mathematical Theory of Control* (edited by A. V. Balakrishnan and L. W. Neustadt), New York, Academic Press, 1967.

[6] H. Halkin, "A maximum principle of the Pontriagin type for systems described by nonlinear difference equations," *SIAM J. Control*, Vol. 4 (1966), pp. 90–111.

[7] J. M. Holtzman, "On the maximum principle for nonlinear discrete time systems," *IEEE Trans. Automatic Control*, Vol. 4 (1966), pp. 528–547.

[8] H. J. Kushner, "On the stochastic maximum principle: fixed time of control," *J. Math. Anal. Appl.*, Vol. 11 (1965), pp. 78–92.

[9] ———, "On the stochastic maximum principle with average constraints," *J. Math. Anal. Appl.*, Vol. 12 (1965), pp. 13–26.

[10] H. J. Kushner and F. Schweppe, "A maximum principle for stochastic control systems," *J. Math. Anal. Appl.*, Vol. 8 (1964), pp. 287–302.

[11] L. W. Neustadt, "An abstract variational theory with applications to a broad class of optimization problems," *SIAM J. Control*, Vol. 4 (1966), pp. 505–527.

[12] D. D. Sworder, "On the control of stochastic systems," *Internat. J. Control*, Vol. 6 (1967), 179–188.

[13] ———, "On the stochastic maximum principle," *J. Math. Anal. Appl.*, Vol. 24 (1968), pp. 627–640.

DIFFERENTIAL GAMES

P. P. VARAIYA
UNIVERSITY OF CALIFORNIA, BERKELEY

1. Introduction

In this paper, we shall give a brief account of the main ideas in the theory of differential games. Our presentation will be limited to two player, zero sum, differential games, except for a few words regarding the situations where there are more than two players or where the game is not zero sum.

To begin with, recall the game formulation when there are two players, I and II. The *actions* (*strategies*) available to I are represented by a set $A = \{\alpha\}$, whereas those available to II are described by the set $B = \{\beta\}$. There is specified a *payoff* function $P : A \times B \rightsquigarrow R$, and I chooses α in A to maximize P while II chooses β in B to minimize P. In general, we know that the order in which the choices are made is essential, and we can only assert that

$$\sup_A \inf_B P(\alpha, \beta) \leqq \inf_B \sup_A P(\alpha, \beta). \tag{1.1}$$

When equality holds in (1.1), we say that the game (P, A, B) has a *saddle value*, and this common number is called the (saddle) value of the game. If there exist α^* in A and β^* in B such that,

$$P(\alpha, \beta^*) \leqq P(\alpha^*, \beta^*) \leqq P(\alpha^*, \beta), \qquad \alpha \in A, \beta \in B \tag{1.2}$$

then we say that (α^*, β^*) constitute a *saddle point* for the game. It is easy to see that in this case the game has a saddle value, and it is equal to $P(\alpha^*, \beta^*)$. Let us call games of this kind *matrix games*, since P has an obvious matrix representation when A and B are finite sets. The theory of matrix games started with the result of von Neumann, and since then many generalizations have appeared. A typical result is the following.

THEOREM 1.1. *Suppose A and B are convex, compact topological spaces. Suppose for fixed β in B, $P(\cdot, \beta)$ is concave, and upper semicontinuous, over A, and for fixed α in A, $P(\alpha, \cdot)$ is convex, and lower semicontinuous, over B. Then, the game (P, A, B) has a saddle point.*

Although the concave-convex assumption on P can be weakened slightly [14], it appears that in general this assumption (or an equivalent hypothesis) is essential [12]. This is in apparent striking contrast with the situation in differential games.

Research sponsored by the National Aeronautics and Space Administration under Grant NGL-05-003-016 (Sup 8).

We can think of a differential game as matrix games played continuously over some time interval, say $[0, 1]$. At each time t, the *states* (positions) of I and II are represented by n dimensional vectors $x(t)$ and $y(t)$, respectively. Having observed the past history $\{(\tau, y(\tau))|0 \leqq \tau \leqq t\}$ of his opponent, player I chooses an action (*control*) $u(t)$ from a specified set U, to guide his state according to the differential equation

$$\dot{x}(t) = f(t, x(t), u(t)), x(0) = x_0. \tag{1.3}$$

Similarly, II chooses a control $v(t)$ from a set V, based upon the past history of I, to steer his own state according to the differential equation

$$\dot{y}(t) = g(t, y(t), v(t)), y(0) = y_0. \tag{1.4}$$

At the end of the time interval $[0, 1]$, we obtain the continuous *trajectories* $x : [0, 1] \rightsquigarrow R^n$, $y : [0, 1] \rightsquigarrow R^n$ of the two players, respectively, and II gives to I the amount $P(x, y)$ where $P : C \times C \rightsquigarrow R$ is a specified *payoff* function. Here, C is the space of all continuous functions from $[0, 1]$ into R^n.

Difficult technical problems arise when we try to specify precisely the set of strategies available to each player. Since the controls $u(t)$ and $v(t)$ will in general be functionals of the past histories of y and x, respectively,

$$u(t) = F(t, y_{[0,t]}), \qquad v(t) = G(t, x_{[0,t]}), \tag{1.5}$$

when we insert these back into the differential equations, we are likely to lose the results of existence and uniqueness of solutions even if f and g are "nice" functions. Furthermore, elementary examples [10] show that there is no natural way to limit the arbitrariness of the functionals F and G. In the next section, we shall give a natural extension of the notion of solution which avoids these difficulties.

After defining strategies accurately, we shall discuss the questions of existence of saddle values and saddle points. We will indicate why saddle values and saddle points exist for a large class of differential games. In Section 3, we shall consider the synthesis problem: how do we find the saddle value and a saddle point when we know they exist? We shall also present a stochastic version of a differential game which sheds some light on the synthesis problem, and also exhibits an intriguing connection with Theorem 1.1. In the final section, we shall discuss generalizations.

2. Game formulation and existence results

Consider the differential equations:

$$\dot{x}(t) = f(t, x(t), u(t)), \qquad x(0) = x_0, \tag{2.1}$$

$$\dot{y}(t) = g(t, y(t), v(t)), \qquad y(0), y_0. \tag{2.2}$$

We assume that U and V are compact subsets of R^m and $f : [0, 1] \times R^n \times U \to R^n$ is measurable in the first, Lipschitz in the second, and continuous in the third variable. Furthermore, $|f(t, x, u)|$ grows at most linearly in $|x|$, uniformly in t, u. Similar assumptions are placed on g. Finally, we assume that the sets $f(t, x, U)$ and $g(t, y, V)$ are convex.

By an admissible control for player *I* (*II*), we mean any measurable function $u : [0, 1] \to U(v : [0, 1] \to V)$. For any admissible control u, there is defined a unique continuous *trajectory* $x : [0, 1] \to R^n$ satisfying (2.1) for almost all t in $[0, 1]$. Let X be the set of trajectories obtained from all the admissible controls of *I*. Similarly, let Y be the set of trajectories obtained from all the admissible controls v of player *II*. We consider X and Y as subsets of the Banach space C of all continuous functions $z : [0, 1] \to R^n$ under the norm, $\|z\| = \max \{|z(t)| \mid t \in [0, 1]\}$. The following result is well known [13].

THEOREM 2.1. *The sets X and Y are compact subsets of C.*

At each time t, player *I* chooses $u(t) \in U$ based on the past history of the trajectory y of *II*. Since the result of his actions determines a trajectory x in X, we can define a strategy for *I* as a map from Y into X, taking care to insure that only the past history is used, that is, the maps should be causal. This motivates the following definition.

DEFINITION 2.1. *An (admissible)* strategy *for I is any map $\alpha : Y \to X$ such that if y, y' in Y, satisfy $y(\tau) = y'(\tau)$, $0 \leqq \tau \leqq t$, then $\alpha(y)(\tau) = \alpha(y')(\tau)$, $0 \leqq \tau \leqq t$.*

Let A be the set of all strategies of *I*. Similarly, we define the set $B = \{\beta\}$ of strategies of *II*.

The notion of solution of an ordinary differential equation is generalized as follows.

DEFINITION 2.2. *Let $\alpha \in A$ and $\beta \in B$. A pair of trajectories (x, y) of $X \times Y$ is said to be an outcome of (α, β) if there exist sequences $\{x_n\} \subset X$, and $\{y_n\} \subset Y$ such that*

$$\lim_{n\to\infty} x_n = \lim_{n\to\infty} \alpha(y_n) = x \tag{2.3}$$

and

$$\lim_{n\to\infty} y_n = \lim_{n\to\infty} \beta(x_n) = y. \tag{2.4}$$

Let $0(\alpha, \beta)$ be the set of all outcomes of (α, β).

PROPOSITION 2.1 (Varaiya and Lin [16]). *For all (α, β), $0(\alpha, \beta)$ is a nonempty, closed subset of $X \times Y$.*

We remark that the most natural definition of an outcome of (α, β) should be a pair (x, y) such that $\alpha(y) = x$, $\beta(x) = y$. Indeed, if α, β are continuous, we get the same definition. The generalization consists in considering the closures of the graphs of α and β instead of α and β.

The payoff function is any map $P : C \times C \to R$.

DEFINITION 2.3. *Let π be the set valued map defined on $A \times B$ by*

$$\pi(\alpha, \beta) = \{P(x, y) | (x, y) \in 0(\alpha, \beta)\}. \tag{2.5}$$

DEFINITION 2.4. *The game (A, B, P) has a saddle value if*

$$\sup_{\alpha \in A} \inf_{\beta \in B} [\inf \pi(\alpha, \beta)] = \inf_{\beta \in B} \sup_{\alpha \in A} [\sup \pi(\alpha, \beta)] \tag{2.6}$$

and this number is called the value of the game. The game has a saddlepoint (α^*, β^*) if

$$\pi(\alpha, \beta^*) \leqq \pi(\alpha^*, \beta^*) \leqq \pi(\alpha^*, \beta) \tag{2.7}$$

for all $\alpha \in A$, $\beta \in B$.

Here we adopt the convention that $\pi(\alpha_1, \beta_1) \leqq \pi(\alpha_2, \beta_2)$ if $p_1 \leqq p_2$ for $p_i \in \pi(\alpha_i, \beta_i)$, $i = 1, 2$.

A basic result of the theory of differential games is the following theorem [16].

THEOREM 2.2. *If the differential equations* (2.1) *and* (2.2) *satisfy the assumptions stated earlier, and if $P : C \times C \rightsquigarrow R$ is continuous, then the game (A, B, P) has a saddle point.*

We shall sketch the basic idea involved in showing existence of the saddle value for the game (A, B, P). The idea is due to Fleming [3]. We approximate the continuous time game $G = (A, B, P)$ by a sequence of discrete time games $G^\delta = (A^\delta, B_\delta, P)$ and $G^\delta = (A_\delta, B^\delta, P)$ in such a way that in the game G^δ, the information pattern is biased in favor of player I whereas the situation is reversed in the game G_δ. The bias in information vanishes as δ approaches 0. More precisely, consider the following definitions.

DEFINITION 2.5. *Let δ be any number of the form 2^{-k}, $k = 0, 1, \cdots$. Let A_δ (respectively, A^δ) be the set of all functions α_δ(respectively, α^δ) : $Y \rightsquigarrow X$ such that if y, y' in Y satisfy*

$$y(\tau) = y'(\tau), \qquad 0 \leqq \tau \leqq i\delta, \tag{2.8}$$

then

$$\alpha_\delta(y)(\tau) = \alpha_\delta(y)(\tau), \qquad 0 \leqq \tau \leqq (i + 1)\delta, i = 1, \cdots, \frac{1}{\delta} - 1, \tag{2.9}$$

{respectively, $\alpha^\delta(y)(\tau) = \alpha^\delta(y)(\tau)$, $0 \leqq \tau \leqq i\delta$, $i = 1, \cdots, 1/\delta$}.

Similarly, we define the sets B_δ and B^δ. Note that,

$$\begin{aligned} A_{\delta_2} &\subset A_{\delta_1} \subset A \subset A^{\delta_1} \subset A^{\delta_2}, \\ B_{\delta_2} &\subset B_{\delta_1} \subset B \subset B^{\delta_1} \subset B^{\delta_2} \end{aligned} \tag{2.10}$$

for $\delta_1 \leqq \delta_2$.

The game G^δ is played as follows. Player I chooses $\alpha^\delta \in A_\delta$ and II chooses $\beta_\delta \in B_\delta$. The outcome is a unique pair (x, y) in $X \times Y$, which we write $o(\alpha^\delta, \beta_\delta)$, such that $\alpha^\delta(y) = x$ and $\beta_\delta(x) = y$. We define

$$V^\delta = \inf_{\beta_\delta \in B_\delta} \sup_{\alpha^\delta \in A^\delta} P(o(\alpha^\delta, \beta_\delta)). \tag{2.11}$$

Dually, we define the game $G_\delta = (A_\delta, B^\delta, P)$ and the number

$$V_\delta = \sup_{\alpha_\delta \in A_\delta} \inf_{\beta^\delta \in B^\delta} P(o(\alpha_\delta, B^\delta)). \tag{2.12}$$

It is useful and elementary to observe that we also have

$$V^\delta = \sup_{A^\delta} \inf_{B_\delta} P(o(\alpha^\delta, \beta_\delta)) \tag{2.13}$$

and

$$V_\delta = \inf_{B^\delta} \sup_{A_\delta} P(o(\alpha_\delta, \beta^\delta)). \tag{2.14}$$

Also from (2.10), we can immediately conclude that $V_{\delta_2} \leqq V_{\delta_1} \leqq V^{\delta_1} \leqq V^{\delta_2}$ when $\delta_1 \leqq \delta_2$, so that we can define

$$\underline{V} = \lim_{\delta \to 0} V_\delta, \qquad \overline{V} = \lim_{\delta \to 0} V^\delta. \tag{2.15}$$

We can now see that the original game $G = (A, B, P)$ has a saddle value if $\underline{V} = \overline{V}$.

The crucial observation at this point is to note that when the dynamics are modeled by (2.1) and (2.2), the advantage due to the information bias in favor of player I in the game G^δ, and to II in G_δ, disappears as δ approaches 0. Specifically, we can prove the following proposition [16].

PROPOSITION 2.2. *For every $\varepsilon > 0$, there is a $\delta > 0$ and a map $\pi_\delta : X \rightsquigarrow X$ such that*

$$\|\pi_\delta(x) - x\| \leqq \varepsilon \qquad \textit{for all } x \in X, \tag{2.16}$$

and if x and x' in X satisfy

$$x(\tau) = x'(\tau) \qquad \textit{for } 0 \leqq \tau \leqq t, \tag{2.17}$$

then, $\pi_\delta(x)(\tau) = \pi_\delta(x')(\tau)$ for $0 \leqq \tau \leqq t + \delta$.

As a corollary we get the following crucial result.

COROLLARY 2.1. *If $\alpha^\delta \in A^\delta$, and $\beta^\delta \in B^\delta$, then $(\pi_\delta \circ \alpha^\delta) \in A_\delta$ and $(\beta^\delta \circ \pi_\delta) \in B_\delta$. Furthermore,*

$$\|(\pi_\delta \circ \alpha^\delta)(y) - \alpha^\delta(y)\| \leqq \varepsilon, \qquad y \in Y, \alpha^\delta \in A^\delta \tag{2.18}$$

whenever π_δ satisfies (2.16).

Now let $\eta > 0$ be arbitrary and let $\hat{\varepsilon} > 0$ be such that

$$|P(x, y) - P(x', y')| < \tfrac{1}{2}\eta, \tag{2.19}$$

whenever

$$\begin{aligned} |x - x'| &\leqq \hat{\varepsilon}, \\ |y - y'| &\leqq \hat{\varepsilon}, \end{aligned} \tag{2.20}$$

for $x, x' \in X$, $y, y' \in Y$. Next let δ be small enough so that (2.16) is satisfied with $\varepsilon = \hat{\varepsilon}$. Let $\underline{\alpha}^\delta \in A^\delta$ be such that (see (2.13))

$$P(o(\underline{\alpha}^\delta, \beta_\delta)) \geqq V^\delta - \tfrac{1}{2}\eta \qquad \text{for all } \beta_\delta \in B^\delta \tag{2.21}$$

and define

$$\underline{\alpha}_\delta = \pi_\delta \circ \underline{\alpha}^\delta. \tag{2.22}$$

Finally, let β^δ in B^δ be arbitrary and let $(x, y) = o(\underline{\alpha}_\delta, B^\delta)$, that is,

$$\begin{aligned} x &= \underline{\alpha}_\delta(y), \\ y &= \beta^\delta(x). \end{aligned} \tag{2.23}$$

But, if we let $\underline{\beta}_\delta = \beta^\delta \circ \pi_\delta$ and $\underline{x} = \underline{\alpha}^\delta(y)$, we see from (2.33) that $(\underline{x}, y) = o(\underline{\alpha}^\delta, \underline{\beta}_\delta)$ so that $P(\underline{x}, y) \geqq V^\delta - \frac{1}{2}\eta$ by (2.21). However, $|x - x'| \leqq \hat{\varepsilon}$ by (2.18), so that we get

$$P\big(o(\underline{\alpha}_\delta, \beta^\delta)\big) = P(x, y) \geqq P(\underline{x}, y) - \tfrac{1}{2}\eta \geqq V^\delta - \eta \tag{2.24}$$

from (2.19). Since β^δ is arbitrary, we can take the infinum over β^δ on the left and conclude that $V_\delta \geqq V^\delta - \eta$. This proves that the two limits in (2.15) coincide and the game (A, B, P) has a saddle value. It is fairly straightforward to show from this fact that the game also has a saddle point. See [16] for details.

Subsequently, in a series of papers [5], [6], [7], Friedman extended Theorem 2.2 to the case where the dynamics are not separated but are mixed as in

$$\dot{z} = h\big(t, z(t), u(t), v(t)\big), \qquad x(0) = z_0. \tag{2.25}$$

The payoff function P is assumed to be of the form

$$P(u, v) = \int_0^1 h_0\big(t, z(t), u(t), v(t)\big)\, dt. \tag{2.26}$$

If we let U be the set of all admissible controls of I and V be the set of all admissible controls of II, then for this case strategies of I are defined as causal maps $\alpha : V \rightsquigarrow U$, and strategies for II are causal maps $\beta : U \rightsquigarrow V$, just as in Definition 2.1. Using approximations, similar to the games G^δ, G_δ, Friedman shows that a saddle value exists *provided* that the functions h, h_0 in (2.25) and (2.26) separate in the form

$$h(t, z, u, v) = h^1(t, z, u) + h^2(t, z, v), \tag{2.27}$$

$$h_0(t, z, u, v) = h_0^1(t, z, u) + h_0^2(t, z, v). \tag{2.28}$$

Such a separation is crucial in showing that the advantage to players in the games G^δ, G_δ vanishes as δ approaches 0. This point is easily demonstrated by considering the following example. Take $\dot{z}(t) = \big(u(t) - v(t)\big)^2$, $z(0) = 0$, $0 \leqq t \leqq 1$, and $P(u, v) = z(1)$. We require that $u(t) \in [0, 1]$, $v(t) \in [0, 1]$. Now suppose player II chooses a strategy first. Player I then chooses his strategy according to,

$$u(t) = \begin{cases} 1 & \text{if } 0 \leqq v(t) \leqq \frac{1}{2}, \\ 0 & \text{if } \frac{1}{2} < v(t) \leqq 1. \end{cases} \tag{2.29}$$

Clearly then, independent of the choice of II, $P \geqq \frac{1}{4}$. On the other hand, if I chooses a strategy first and II follows according to the formula $v(t) \equiv u(t)$, evidently $P = 0$. Hence a saddle value cannot exist.

3. The synthesis problem

Let us consider games with dynamics of the form (2.25) and payoff in the form (2.26). We assume that (2.27) and (2.28) hold. Furthermore, h and h_0 are required to be continuous, $u(t) \in U$, $v(t) \in V$, where U and V are compact. Instead of starting in a fixed initial condition z_0 at time 0, let us consider games starting in different initial states z at times t in $[0, 1]$, the game being defined over $[t, 1]$. Let $\pi(z, t)$ be the value of the game starting in the initial condition (z, t). It can be shown [5] that π is continuously differentiable and satisfies the partial differential equation

$$0 = \frac{\partial \pi(z, t)}{\partial t} + \min_{v \in V} \max_{u \in U} \left[\left\langle \frac{\partial v}{\partial z}(z, t), h(t, z, u, v) \right\rangle + h_0(t, z, u, v) \right]. \tag{3.1}$$

Of course, we have the boundary data

$$\pi(z, 1) \equiv 0. \tag{3.2}$$

Note that because of (2.26) and (2.27) the min and max in (3.1) can be interchanged. Equations (3.1) and (3.2) have been solved formally in numerous cases by Isaacs [10]. Indeed, he was the first to discover this equation and it is called Isaacs' equation. In the case of one player, it reduces to the Hamilton–Jacobi equation of the calculus of variation.

In a significant paper, Berkovitz [1] considers solving (3.1) and obtaining a saddle point as a pair of "feedback" strategies

$$\begin{aligned} u(t) &= u(t, z(t)), \\ v(t) &= v(t, z(t)), \end{aligned} \tag{3.3}$$

by constructing a field of trajectories.

In a different direction, Fleming [4] studies the parabolic equation obtained from (3.1) by adding the term

$$\varepsilon \sum_i \frac{\partial^2 \pi}{\partial z_i^2}(z, t), \qquad \varepsilon > 0, \tag{3.4}$$

to the right side. If we denote by π^ε the solution to this equation, then Fleming shows that π^ε converges to the value π as ε approaches 0, uniformly on compact sets. Furthermore, he shows that π^ε can be regarded as the value of the stochastic game with dynamics

$$dz(t) = h(t, z, u, v)\, dt + (2\varepsilon)^{1/2}\, dB(t), \tag{3.5}$$

and with payoff

$$P(u, v) = E \int_0^1 h_0(t, z, u, v)\, dt. \tag{3.6}$$

In (3.5), $B(t)$ is a standard n dimensional Brownian motion, and in (3.6), E denotes expectation.

This brings us to the final part of this section. Consider the stochastic game defined by the differential equation

$$dx(t) = f\big(t, x(t), u(t), v(t)\big)\, dt + dB(t), \qquad x(0) = x_0, 0 \leqq t \leqq 1, \tag{3.7}$$

where f is of the form

$$f(t, x, u, v) = \begin{pmatrix} f_1(t, x, u) \\ f_2(t, x, v) \end{pmatrix}. \tag{3.8}$$

In (3.7), $B(t)$ is an n dimensional Brownian motion process. We shall define the solution of (3.7) in such a way that $x(t)$ has continuous sample paths, so that we will suppose that the sample paths of x belong to C, the Banach space of continuous functions from $[0, 1]$ into R^n. For each $t \in [0, 1]$, let $\mathscr{C}_t$ be the σ-field generated by all subsets of the form

$$\{z \mid z \in C, z(\tau) \in A\}, \tag{3.9}$$

where $0 \leqq \tau \leqq t$, and A is a Borel subset of R^n.

Suppose that u and v are m dimensional and U and V are compact subsets of R^m.

DEFINITION 3.1. *A strategy for player I is any function* $\alpha : [0, 1] \times C \rightsquigarrow U$ *such that:*

(i) α *is measurable with respect to the product* σ*-algebra* $\mathscr{L} \otimes \mathscr{C}_1$ *on* $[0, 1] \times C$ *(here* $\mathscr{L}$ *is the set of Lebesgue measurable subsets of* $[0, 1]$*);*

(ii) *for each fixed t in* $[0, 1]$, $\alpha(t, \cdot)$ *is measurable with respect to* $\mathscr{C}_t$.

Let A denote the set of all strategies of player I. Similarly, we define B as the set of all strategies of player II consisting of all jointly measurable, causal maps $\beta : [0, 1] \times C \rightsquigarrow V$.

We impose the following conditions on f:

(i) $f(t, x, u, v)$ is measurable in (t, x, u, v), continuous in (u, v) for fixed (t, x);

(ii) $f(t, x, U, v)$ and $f(t, x, u, V)$ are convex;

(iii) there is an increasing function $f_0 : R \rightsquigarrow R$ such that $|f(t, x, u, v)| \leqq f_0(|x|)$ for all $x \in R^n$, $u \in U$, $v \in V$, $t \in [0, 1]$.

The following fundamental existence result is due to Girsanov [8].

THEOREM 3.1. *For each* $\alpha \in A$, $\beta \in B$, *there exists a solution* $x(t)$ *of* (3.7) *with sample paths in C such that the measure* $\mu_{(\alpha, \beta)}$ *induced by* x *on* $(C, \mathscr{C}_1)$ *is mutually absolutely continuous with respect to the Wiener measure* μ *on* $(C, \mathscr{C}_1)$ *and the density* $\eta_{(\alpha,\beta)} = d\mu_{(\alpha,\beta)}/d\mu$ *is given by*

$$\eta_{(\alpha,\beta)} = \exp\left\{\int_0^1 \langle f(t, B(t), \alpha(t, B), \beta(t, B)), dB(t)\rangle - \tfrac{1}{2}\int_0^1 |f(t, B(t), \alpha(t, B), \beta(t, B))|^2\, dt\right\}. \tag{3.10}$$

(In the above formula the first integral should be interpreted as a stochastic integral.)

When the function f has the form (3.8), we see that (3.10) becomes

$$\eta_{(\alpha,\beta)}(x) = \eta_\alpha(x)\eta_\beta(x), \tag{3.11}$$

where

$$\eta_\alpha = \exp\left\{\int_0^1 \langle f_1(t, B(t). \alpha(t, B)), dB_1(t)\rangle - \tfrac{1}{2}\int_0^1 |f_1(t, B(t), \alpha(t, B))|^2\, dt\right\} \tag{3.12}$$

and

$$\eta_\beta = \exp\left\{\int_0^1 \langle f_2(t, B(t), \beta(t, B)), dB_2(t)\rangle - \tfrac{1}{2}\int_0^1 |f_2(t, B(t), \beta(t, B))|^2\, dt\right\}. \tag{3.13}$$

Let $P : C \twoheadrightarrow R$ be any bounded function measurable with respect to $\mathscr{C}_1$. For any $\alpha \in A$, $\beta \in B$, we define the payoff to I as $E_{(\alpha,\beta)}(P)$, where $E_{(\alpha,\beta)}(P)$ is the expectation of P with respect to the measure $\mu_{(\alpha,\beta)}$ induced on C by (α, β) via x. In view of (3.11), we see that

$$E_{(\alpha,\beta)}(P) = \int_C P(x)\eta_\alpha(x)\eta_\beta(x)\, d\mu(x). \tag{3.14}$$

From (3.14), we see that the choice of α or β affects the payoff only through the densities η_α, η_β. Let

$$\begin{aligned} D_I &= \{\eta_\alpha \mid \alpha \subset A\}, \\ D_{II} &= \{\eta_\beta \mid \beta \in B\}. \end{aligned} \tag{3.15}$$

The following result is a useful characterization.

THEOREM 3.2 (Duncan and Varaiya [2]). *The sets* D_I *and* D_{II} *are strongly closed, convex subsets of* $L_1(C, \mathscr{C}, \mu)$.

As a corollary of Theorems 1.1 and 3.2, we have the following existence result.

THEOREM 3.3. *There exists a saddle point for the above stochastic game.*

PROOF. The sets D_I and D_{II} are weakly compact. The payoff (3.14) is linear and continuous in η_α for fixed η_β and linear and continuous in η_β for fixed η_α. By Theorem 1.1, there exists a saddle point.

4. Generalizations and comments

In the results presented in the previous sections, the game started at a fixed initial state and ended at a fixed final time. It is important in many cases to have a variable end time. Usually, this situation is formulated by requiring that the termination of the game occur when the state of the players enters a specific target set. A specific game of this kind—the pursuit-evasion game—has been considered in [16]. Generalizations have been studied in [6] and [7].

Essentially, the only class of games for which explicit solutions are available is the case where the dynamics in (2.25) are linear and the integrand of the payoff function (2.26) is quadratic. Numerous special cases have been solved and the literature on differential games is growing rapidly. An exhaustive bibliography classifying the literature, up to October 1969, appears in [11].

As is well known, unlike the situation in zero sum games where saddle point is the natural solution concept, in the case of more than two players or where the payoff is not zero sum, there are many solution concepts. Roughly speaking, these concepts separate into two classes, cooperative and noncooperative, but the distinction becomes less clear in dealing with a dynamic situation. Although most results in general (as opposed to two person, zero sum) differential games are concerned with noncooperative solutions, there are the beginnings of a theory (or theories) in the richer area of cooperative solutions. Reference [15] exhibits the wealth of possible solution concepts in the general case, and [9] is an attempt to place differential games within the context of decision theory.

REFERENCES

[1] L. D. Berkovitz, "Necessary conditions for optimal strategies in a class of differential games and control problems," *SIAM J. Control*, Vol. 5 (1967), pp. 1–25.

[2] T. E. Duncan and P. Varaiya, "On the solutions of a stochastic control system," *SIAM J. Control*, Vol. 9 (1971), pp. 354–371.

[3] W. H. Fleming, "The convergence problem for differential games," *J. Math. Anal. Appl.*, Vol. 3 (1961), pp. 102–116.

[4] ———, "The Cauchy problem for degenerate parabolic equations," *J. Math. Mech.*, Vol. 13 (1964), pp. 987–1008.

[5] A. Friedman, "On the definition of differential games and the existence of value and of saddle points," *J. Differential Equations*, Vol. 7 (1970), pp. 69–91.

[6] ———, "Existence of value and of saddle points for differential games of pursuit and evasion," *J. Differential Equations*, Vol. 7 (1970), pp. 92–110.

[7] ———, "Existence of value and of saddle points for differential games of survival," *J. Differential Equations*, Vol. 7 (1970), pp. 111–125.

[8] I. V. Girsanov, "On transforming a class of stochastic processes by absolutely continuous substitutions of measures," *Theor. Probability Appl.*, Vol. 5 (1960), pp. 285–301.

[9] Y. C. Ho, "Differential games, dynamic optimization, and generalized control theory," *J. Optimization Theory Appl.*, Vol. 6 (1970), pp. 179–209.

[10] R. Isaacs, *Differential Games*, New York, Wiley, 1965.

[11] *Proceedings of the First International Conference on Theory and Application of Differential Games*, Amherst, University of Massachusetts Press, 1969.

[12] R. T. ROCKAFELLAR, "Saddle-points and convex analysis," *Differential Games and Related Topics* (edited by H. W. Kuhn and G. P. Szegő), Amsterdam, North-Holland, 1971.

[13] E. ROXIN, "The existence of optimal controls," *Michigan Math. J.*, Vol. 9 (1962), pp. 109–119.

[14] M. SION, "On general minimax theorems," *Pacific J. Math.*, Vol. 8 (1958), pp. 171–176.

[15] A. W. STARR and Y. C. HO, "Further properties of nonzero-sum differential games," *J. Optimization Theory Appl.*, Vol. 3 (1969), pp. 184–206.

[16] P. VARAIYA and J. LIN, "Existence of saddle-points in differential games," *SIAM J. Control*, Vol. 7 (1969), pp. 141–157.

EPSILON ENTROPY OF PROBABILITY DISTRIBUTIONS

EDWARD C. POSNER and EUGENE R. RODEMICH
JET PROPULSION LABORATORY
CALIFORNIA INSTITUTE OF TECHNOLOGY

1. Introduction

This paper summarizes recent work on the theory of epsilon entropy for probability distributions on complete separable metric spaces. The theory was conceived [3] in order to have a framework for discussing the quality of data storage and transmission systems.

The concept of data source was defined in [4] as a probabilistic metric space: a complete separable metric space together with a probability distribution under which open sets are measurable, so that the Borel sets are measurable. An ε partition of such a space is a partition by measurable ε sets, which, depending on context, can be sets of diameter at most ε or sets of radius at most $\frac{1}{2}\varepsilon$, that is, sets contained in spheres of radius $\frac{1}{2}\varepsilon$. The entropy $H(U)$ of a partition U is the Shannon entropy of the probability of the distribution consisting of the measures of the sets of the partition. The (one shot) epsilon entropy of X with distribution μ, $H_{\varepsilon;\mu}(X)$, is defined by

$$H_{\varepsilon;\mu}(X) = \inf_U \{H(U);\ U \text{ an } \varepsilon \text{ partition}\} \tag{1.1}$$

and, except for roundoff in the entropy function, a term less than 1, $H_{\varepsilon;\mu}(X)$ is the minimum expected number of bits necessary to describe X to within ε when storage is not allowed. The inf in (1.1) was shown to be a min in [4].

For X a compact metric space, Kolmogorov's epsilon entropy $H_\varepsilon(X)$ is defined as

$$H_\varepsilon(X) = \min_U \{\log \text{card}\,(U);\ U \text{ an } \varepsilon \text{ partition}\} \tag{1.2}$$

and, except for roundoff in the logarithm, is the minimum number of bits necessary to describe X to within ε when words of fixed length are used.

Suppose one does experiments from X independently and then attempts storage or transmission. That is, take a cartesian product $X^{(n)}$ of X, with product measure $\mu^{(n)}$ and supremum metric. Thus, ε sets in the product are the subsets

This paper presents the results of one phase of research carried out under Contract NAS 7-100, sponsored by the National Aeronautics and Space Administration.

of products of ε sets. This is the notion that insures that knowledge of outcomes to within ε in $X^{(n)}$ forces knowledge to within ε in each of the factors. Then the following limit can be proved to exist:

$$I_{\varepsilon;\mu}(X) = \lim_{n\to\infty} \frac{1}{n} H_{\varepsilon;\mu^{(n)}}(X^{(n)}), \tag{1.3}$$

and is called the absolute epsilon entropy of X. It represents the minimum expected number of bits per sample needed to describe a sequence of independent outcomes of X when arbitrary storage between experiments can be used. Similarly, define the absolute epsilon entropy of the compact metric space X as

$$I_{\varepsilon}(X) = \lim_{n\to\infty} \frac{1}{n} H_{\varepsilon}(X^{(n)}), \tag{1.4}$$

with the same definition for the metric on $X^{(n)}$.

2. Relations with channel coding

It was shown in [3] that

$$I_{\varepsilon;\mu}(X) = \inf_{\rho} \{I(\rho);\, \rho \in R_{\varepsilon}(X)\}, \tag{2.1}$$

where $R_{\varepsilon}(X)$, in the case of the radius definition, is the class of probability distributions on $X \times X$ which are supported within $\frac{1}{2}\varepsilon$ of the diagonal, with a more complicated definition for the diameter case. Here $I(\rho)$ stands for mutual information. We do not know if the inf need be attained. However, (2.1), coupled with the continuity of $I_{\varepsilon;\mu}(X)$ from above in ε, proved in [3], allows us to prove a strong channel coding theorem and its converse [3]:

> "If K is a memoryless channel with capacity Γ less than $I_{\varepsilon;\mu}(X)$ (of finite capacity if $I_{\varepsilon;\mu}(X)$ is infinite), then it is not possible to transmit outcomes of X over K such that, with probability approaching 1, an arbitrarily large fraction of a long sequence of outcomes are known to within an error not much more than ε. But if Γ is greater than $I_{\varepsilon;\mu}(X)$ (assuming $I_{\varepsilon;\mu}(X)$ is finite), then it is possible to transmit outcomes of X over the channel such that, with probability approaching 1, all the outcomes are known to within ε."

3. A useful inequality

An important and useful inequality relating $H_{\varepsilon;\mu}$ and $I_{\varepsilon;\mu}$ was proved in [3]:

$$H_{\varepsilon;\mu}(X) \leqq I_{\varepsilon;\mu}(X) + \log^{+} I_{\varepsilon;\mu}(X) + C, \tag{3.1}$$

where C is a universal constant. In other words, it doesn't help much to store independent experiments if the entropy is large. This result, coupled with (2.1), makes it easy to obtain asymptotic bounds on $H_{\varepsilon;\mu}(X)$, which we shall do in subsequent sections.

4. Relation between $I_{\varepsilon;\mu}$ and I_ε

Here is a surprising relation between $I_{\varepsilon;\mu}(X)$ and $I_\varepsilon(X)$ for X compact [1]:

$$I_\varepsilon(X) = \max_\mu I_{\varepsilon;\mu}(X) \tag{4.1}$$

for all but countably many ε, a condition which we now know cannot be removed. What this result means is that Nature can choose a μ on X so bad that nothing can be saved by using variable length coding. The proof uses von Neumann's minimax theorem to prove, as an intermediate step, that

$$\max_\mu I_{\varepsilon;\mu}(X) = \log \frac{1}{v(\varepsilon)}, \tag{4.2}$$

where $v(\varepsilon)$ is the value of the zero sum two person game, which, in the radius case, has as its space of pure strategies the points of X, with payoff 0 or 1 to the first player; the payoff is 1 if and only if the points chosen by the two players have distance at most $\frac{1}{2}\varepsilon$.

5. Finite dimensional spaces

The differential entropy of a density function p an Euclidean n space is defined, when it exists, as

$$H(p) = \int p \log \frac{1}{p}\, dm(x), \tag{5.1}$$

where $m(x)$ is Lebesgue measure. The relation between differential entropy and epsilon entropy was considered in [2]. The metric can be any norm $\|\cdots\|_S$ on n space, where S, of Lebesgue measure v_1, is a compact symmetric convex set in E^n, and is the unit sphere under $\|\cdots\|_S$. Let p be a sufficiently nice density function on E^n, so that E^n is a probabilistic metric space with probability μ given by $\mu(A) = \int_A p\, dm$. Then as $\varepsilon \to 0$,

$$H_{\varepsilon;\mu}(E^n) = n \log \frac{2}{\varepsilon} + \log \frac{1}{v_1} + H(p) + C(S) + o(1) \tag{5.2}$$

for a constant $C(S)$ called the entropic packing constant. Furthermore, $C(S)$ is between 0 and 1, and is 0 if and only if translates of S fill E^n. Moreover, $C(S)$ as a function of S is continuous in the Hausdorff metric, in which the distance between two compact sets is the maximum over the two sets of the distance of a point in one of the two sets from the other set. A somewhat analogous result holds for H_ε of the unit n cube, but the analogous $C'(S)$, the deterministic packing constant, is not bounded by 1 but rather can be as large as $(1 - o(1)) \log n$ as $n \to \infty$.

If a Borel probability μ on E^n with Euclidean distance has mean 0 and a second moment $\sigma^2 = E\|x\|^2$, $H_{\varepsilon;\mu}(E^n)$ is finite, even though (5.2) does not hold [4]; in fact, for small ε,

$$H_{\varepsilon;\mu}(E^n) \leqq n \log \frac{1}{\varepsilon} + \frac{n}{2} \log n + n \log \sigma + 1 + \log (2 + \sqrt{\pi}). \tag{5.3}$$

The normal distribution comes under either of those two cases. When $n = 1$, the unique minimizing partition is known [6]. It is the partition by consecutive intervals of length ε such that the mean is in the center of one of the intervals. The proof is hard; a simpler one would be nice. For $n > 1$, minimizing partitions are not known, even for the independent normal of equal variances.

6. Epsilon entropy in $L_2[0, 1]$

The bound of (5.3) depends on n, and, in fact, examples can be given that show this dependence can actually occur. It is not surprising then, that if $L_2[0, 1]$ is made into a probabilistic metric space by the measure induced by a mean continuous separable stochastic process on the unit interval, then the epsilon entropy can be infinite, even though the expectation of $\|x\|^2$ is always finite for such a process. In fact, [4] proved that a given convergent sequence $\{\lambda_n\}$ of nonnegative numbers written in nonincreasing order is the set of eigenvalues of some mean continuous stochastic process on the unit interval of infinite epsilon entropy for some $\varepsilon > 0$ if and only if

$$\sum n\lambda_n = \infty. \tag{6.1}$$

Conversely, if $\Sigma\, n\lambda_n = \infty$, there is a process with infinite epsilon entropy for every $\varepsilon > 0$. Thus, a slightly stronger condition than finite second moment is necessary to insure finite epsilon entropy in the infinite dimensional case.

7. Product entropy of Gaussian processes

In this section, we shall consider the definition of *product entropy* if $X = L_2[0, 1]$, but only for mean continuous Gaussian processes. We shall defer the case of the epsilon entropy of Gaussian processes until Section 9. Product entropy $J_\varepsilon(X)$ is defined as the minimum entropy over all *product* epsilon partitions of $L_2[0, 1]$. A product epsilon partition is a (countable) epsilon partition of all of X except a set of probability zero by sets which are hyperrectangles (with respect to eigenfunction coordinates) of diagonal epsilon. Thus,

$$J_\varepsilon(X) \geqq H_\varepsilon(X). \tag{7.1}$$

Surprisingly, $J_\varepsilon(X)$ is infinite for one ε if and only if it is infinite for all ε, and is finite if and only if the "entropy of the eigenvalues" $\Sigma\, \lambda_n \log 1/\lambda_n$ is finite, where the eigenvalues are written in nonincreasing order [5]. We have no good explanation of why the entropy of the eigenvalues occurs as the condition for finite epsilon entropy. Furthermore, $\Sigma\, \lambda_n \log 1/\lambda_n$ is necessary and sufficient in order that there be a product epsilon partition, and is also the condition that there be a hyperrectangle of positive probability and finite diameter. Incidentally, $J_\varepsilon(X)$ depends only on the eigenvalues of the Gaussian process, as do $H_{\varepsilon;\mu}(X)$ and $I_{\varepsilon;\mu}(X)$, where μ is the measure on $L_2[0, 1]$ induced by the mean continuous Gaussian process. In [6], product entropy is estimated rather precisely in terms of the eigenvalues and the optimum product partitions found by a variational argument. By remarks of Section 5, the optimal partitions are products of centered partitions on each coordinate axis.

The interpretation of product entropy is as follows. One wishes to transmit sample functions of the process so that one knows outcomes to within ε in the L_2 norm, but only wishes to consider methods which involve correlating the sample function with the eigenfunction and then sending a quantized version of these correlations. Since the diagonal of the product sets has diameter ε, the method guarantees knowledge of the sample function to within ε. Unfortunately, as we shall see, this method is not very good compared to the optimal compression schemes which are not restricted to product partitions but can use arbitrary ε partitions.

In [5], conditions are given which guarantee either

$$J_\varepsilon(X) = O(H_{\varepsilon;\mu}(X)) \tag{7.2}$$

or

$$J_\varepsilon(X) \sim H_{\varepsilon;\mu}(X) \tag{7.3}$$

as $\varepsilon \to 0$ for a mean continuous Gaussian process. The condition for (7.2) is that the sum of the eigenvalues beyond the nth (in nonincreasing order) be $O(n\lambda_n)$. For (7.3), the condition is that the sum be $o(n\lambda_n)$. In the first case, in fact,

$$J_\varepsilon(X) = O(L_\varepsilon(X)) \tag{7.4}$$

and in the second

$$J_\varepsilon(X) \sim L_\varepsilon(X), \tag{7.5}$$

where $L_\varepsilon(X)$ is a general lower bound for the epsilon entropy of a mean continuous Gaussian process to be discussed later.

For a stationary band limited Gaussian process on the unit interval with continuous spectral density,

$$\lambda_n \sim n^{-1}(Cn)^{-2n}, \tag{7.6}$$

C constant, as $n \to \infty$. Thus, the $o(n\lambda_n)$ condition is satisfied and $L_\varepsilon(X)$ can be evaluated. The final result is

$$J_\varepsilon(X) \sim \frac{\frac{1}{2}\left(\log \frac{1}{\varepsilon}\right)^2}{\log \log \frac{1}{\varepsilon}} \tag{7.7}$$

as $\varepsilon \to 0$. Now

$$J_\varepsilon(X) \sim n \log \frac{1}{\varepsilon} \tag{7.8}$$

as $\varepsilon \to 0$ if the process has only finitely many nonzero eigenvalues, n in number. So in the case at hand where there are infinitely many nonzero eigenvalues, the growth of $J_\varepsilon(X)$ had to be faster than any constant times $\log(1/\varepsilon)$. The rate of growth given by (7.8) is not much faster, however. This is an expression of the fact that the sample functions from such a process are entire functions, hence not very random, and they should be easy to approximate.

8. Entropy in $C[0, 1]$

The case of $C[0, 1]$ is much more difficult than $L_2[0, 1]$, partly because it is hard to determine whether a given process has continuous paths. However, in [7] it is proved that if the mean continuous separable stochastic process on $[0, 1]$ satisfies,

$$\begin{aligned} E(x(0))^2 &\leqq A, \\ E(x(s) - x(t))^2 &\leqq A|s - t|^a, \qquad s, t \in [0, 1], \end{aligned} \tag{8.1}$$

for some $A \geqq 0$, $1 < a \leqq 2$, then the paths are continuous with probability 1, a known result, and, if μ is the measure induced by the process on $C[0, 1]$, then

$$H_{\varepsilon;\mu} \leqq C(a) A^{1/a} \varepsilon^{-2/a} \tag{8.2}$$

for $\varepsilon < \sqrt{A}$, where $C(a)$ depends only on a. The proof was achieved by constructing an ε partition of $C[0, 1]$ using uniformly spread points on $[0, 1]$ and facts about the modulus of continuity of the process that are forced by the given conditions on the covariance function.

Conditions for finite entropy in function spaces other than $L_2[0, 1]$ and $C[0, 1]$ have not been looked for or found, even in the case of Gaussian processes. In that case, a bound like (8.2) is found for $C[0, 1]$, which is valid under weaker conditions on the covariance function of the process. Multidimensional time processes have not been considered at all.

9. Bounds for L_2 Gaussian processes

Various lower bounds have been found for L_2 Gaussian processes. The first is the bound L_ε of [6]. This bound is defined as follows. Let $\{\lambda_n\}$ in nonincreasing order be the eigenvalues of the process. Define $b = b(\varepsilon)$ for $\varepsilon > 0$ by

$$\begin{aligned} \sum \frac{\lambda_n}{1 + b(\varepsilon)\lambda_n} &= \varepsilon^2, \qquad \varepsilon^2 < \sum \lambda_n, \\ b(\varepsilon) &= 0, \qquad \varepsilon^2 \geqq \sum \lambda_n. \end{aligned} \tag{9.1}$$

Then

$$L_\varepsilon(X) = \frac{1}{2}\sum \log\left[1 + \lambda_n b(\varepsilon)\right] \tag{9.2}$$

is a lower bound for $H_{\varepsilon;\mu}(X)$. It was derived from the obvious inequality

$$H_{\varepsilon;\mu}(X) \geqq E \log \frac{1}{\mu[S_\varepsilon(x)]}, \tag{9.3}$$

and so is quite weak. (The term $S_\varepsilon(x)$ is the ball of center x and radius ε.) A stronger bound $M_\varepsilon(X)$, never any worse, is

$$M_\varepsilon(X) = L_{\varepsilon/2}(X) - \frac{1}{8}\varepsilon^2 b\left(\frac{1}{2}\varepsilon\right). \tag{9.4}$$

This bound is derived by bounding the probability density drop in a Gaussian distribution under translation, and using the fact, proved in [6], that the sphere of radius $\frac{1}{2}\varepsilon$ about the origin in L_2 under a Gaussian distribution is at least as probable as any set of diameter ε. The $M_\varepsilon(X)$ bound is the best asymptotic lower bound we have for arbitrary L_2 Gaussian processes, but, for special ones, the next section gives an improvement.

The $L_\varepsilon(X)$ lower bound was introduced chiefly because it is also proved in [5] that

$$H_\varepsilon(X) \lesssim L_{m\varepsilon}(X) \qquad \text{for any } m < \frac{1}{2}. \tag{9.5}$$

This difficult proof uses products of partitions of finite dimensional eigensubspaces of the process where the dimension increases without bound. In each finite dimensional subspace, "shell" partitions are used, partitions which are composed of partitions between regions of concentric spheres of properly varying radii. The deterministic epsilon entropy of the n sphere in Euclidean space needed to be estimated in the proof. This proves the finiteness of $H_\varepsilon(X)$ for X a mean continuous Gaussian process on $[0, 1]$.

From results of Section 3, we know $H_\varepsilon(X) \lesssim I_\varepsilon(X)$. The measure ρ on $X \times X$, which is defined by choosing a point of the first factor according to μ and then assigning probability in $S_{\varepsilon/2}(x)$ according to μ, has, as in [1], mutual information

$$I(\rho) \leqq E\left(\log \frac{1}{\mu[S_{\varepsilon/2}(x)]}\right). \tag{9.6}$$

Thus, for any probabilistic metric space,

$$H_{\varepsilon;x}(X) \lesssim E\left(\log \frac{1}{\mu[S_{\varepsilon/2}(x)]}\right). \tag{9.7}$$

This, coupled with (9.3), gives a pair of bounds valid in general, but not readily computable.

10. Examples of L_2 entropy bounds

The bounds of the previous section lead to some interesting asymptotic expressions which yield quick answers when the eigenvalues of the process are known asymptotically. For example, a Gaussian process arising as the solution of a linear differential equation with constant coefficients driven by white Gaussian noise can be handled. For such processes, [8] shows that

$$\lambda_n \sim An^{-p}, \qquad p > 1, A > 0, \tag{10.1}$$

and shows how to find A and p from the equation with simple calculations. The results of the previous section then yield

$$M_\varepsilon(X) \leqq H_\varepsilon(X) \lesssim L_{\varepsilon/2}(X), \tag{10.2}$$

$$M_\varepsilon(X) \sim (p-1)\left(\frac{\pi}{p \sin \pi/p}\right)^{p/(p-1)} 2^{2/(p-1)-1}\left(\frac{A}{\varepsilon^2}\right)^{1/(p-1)}, \tag{10.3}$$

$$L_\varepsilon(X) \sim \frac{p}{2}\left(\frac{\pi}{p \sin \pi/p}\right)^{p/(p-1)} \left.\frac{A}{\varepsilon^2}\right)^{1/(p-1)} \tag{10.4}$$

For example, if X is the Wiener process

$$E[X(s)X(t)] = \min(s, t), \qquad s, t \in [0, 1], \tag{10.5}$$

then

$$\lambda_n = \frac{1}{\pi^2(n - 1/2)^2}, \qquad n \geqq 1, \tag{10.6}$$

and so $A = 1/\pi^2, p = 2$, and

$$\frac{1}{2\varepsilon^2} \lesssim H_\varepsilon(X) \lesssim \frac{1}{\varepsilon^2}. \tag{10.7}$$

A better lower bound valid in general for L_2 is given by

$$\begin{aligned} H_\varepsilon(X) &\geqq M_\varepsilon(X) + \frac{1}{2}\sum E\left\{\frac{\lambda_n x_n^2 q^2(x)}{[\varepsilon + \lambda_n q(x)]^2}\right\} \\ &= N_\varepsilon(X), \end{aligned} \tag{10.8}$$

say, where $q(x) = 0$, $\|x\| \leqq \varepsilon$, and

$$\sum \frac{x_n^2}{[\varepsilon + \lambda_n q(x)]^2} = 1, \qquad \|x\| > \varepsilon. \tag{10.9}$$

Here x_n denotes the component of x along the nth eigenfunction. Equation (10.8) is a refined version of (9.4), and is useful only when the eigenvalues decrease slowly enough so that $q(x)$ is almost deterministic. The condition turns out to be satisfied if (10.1) is, and allows $N_\varepsilon(X)$ to be found asymptotically as

$$(10.10)\quad N_\varepsilon(X) \sim (p-1)\left(\frac{\pi}{p\sin \pi/p}\right)^{p/(p-1)}\left[2^{2/(p-1)} + \frac{1}{2}p^{-p/(p-1)}\right]\left(\frac{A}{\varepsilon^2}\right)^{1/(p-1)}$$

Thus, for X the Wiener process, (10.7) can be improved to

$$(10.11)\qquad \frac{17}{32\varepsilon^2} \lesssim H_\varepsilon(X) \lesssim \frac{1}{\varepsilon^2}.$$

The result given by (10.11) is all we know about the entropy of the Wiener process. Our lower bounds just are not good enough to prove our conjecture

$$(10.12)\qquad H_\varepsilon(X) \sim \frac{1}{\varepsilon^2}$$

for X the Wiener process.

Notice, however, that for the Wiener process on $C[0, 1]$, where one might expect the entropy to be much larger because of the more stringent covering requirement, Section 8 yields

$$(10.13)\qquad H_\varepsilon(X \text{ in } C[0, 1]) = O(H_\varepsilon(X \text{ in } L_2[0, 1])).$$

In fact, (10.13) holds for any Gaussian process satisfying the eigenvalue condition (10.1). With this surprising result we close the paper.

REFERENCES

[1] R. J. McEliece and E. C. Posner, "Hide and seek, data storage, and entropy," *Ann. Math. Statist.*, Vol. 2 (1971), pp. 1706–1716.

[2] E. C. Posner and E. R. Rodemich, "Differential entropy and tiling," *J. Statist. Phys.*, Vol. 1 (1969), pp. 57–69.

[3] ———, "Epsilon entropy and data compression," *Ann. Math. Statist.*, Vol. 42 (1971), pp. 2079–2125.

[4] E. C. Posner, E. R. Rodemich, and H. Rumsey, Jr., "Epsilon entropy of stochastic processes," *Ann. Math. Statist.*, Vol. 38 (1967), pp. 1000–1020.

[5] ———, "Product entropy of Gaussian distributions," *Ann. Math. Statist.*, Vol. 40 (1969), pp. 870–904.

[6] ———, "Epsilon entropy of Gaussian distributions," *Ann. Math. Statist.*, Vol. 40 (1969), pp. 1272–1296.

[7] E. R. Rodemich and E. C. Posner, "Epsilon entropy of stochastic processes with continuous paths," in preparation.

[8] H. Widom, "Asymptotic behavior of eigenvalues of certain integral operators," *Arch. Rational Mech. Anal.*, Vol. 17 (1967), pp. 215–229.

AUTHOR REFERENCE INDEX

This index includes authors from Volumes I, II, III only. A reference index for Volumes IV, V, and VI is contained in those Volumes.